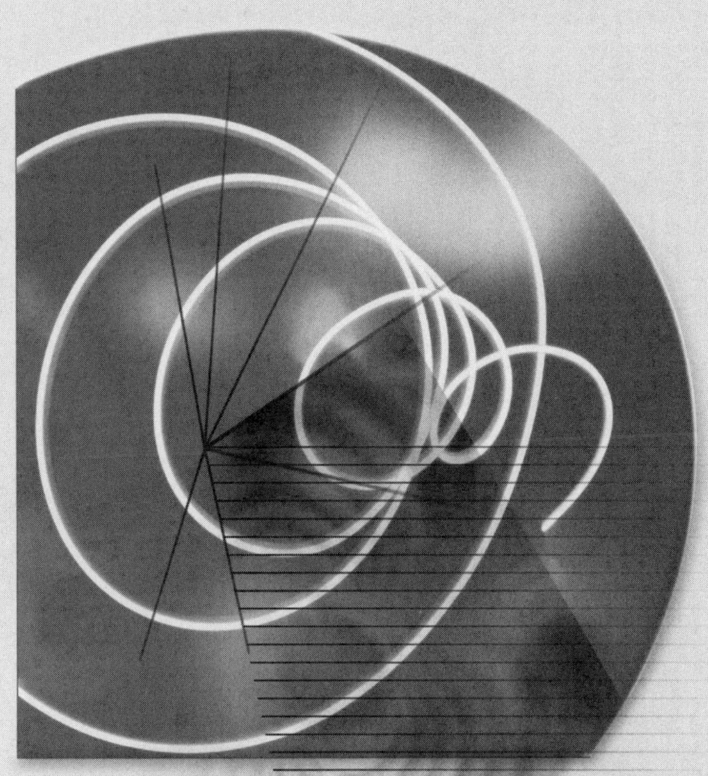

AutoCAD® 2002 and the Fundamentals of Mechanical Drafting

AutoCAD® 2002 and the Fundamentals of Mechanical Drafting

ALAN JEFFERIS
TEREASA JEFFERIS
MICHAEL JONES

autodesk® press

Australia • Canada • Mexico • Singapore • Spain • United Kingdom • United States

autodesk Press

Autocad® 2002 and the Fundamentals of Mechanical Drafing
by Alan Jefferis, Tereasa Jefferis and Michael Jones

Autodesk Press Staff

Business Unit Director:
Alar Elken

Executive Editor:
Sandy Clark

Acquisitions Editor:
James DeVoe

Development Editor:
John Fisher

Editorial Assistant:
Jasmine Hartman

Executive Marketing Manager:
Mary Johnson

Marketing Coordinator:
Karen Smith

Executive Production Manager:
Mary Ellen Black

Production Manager:
Larry Main

Production Editor:
Stacy Masucci

Art/Design Coordinator:
Mary Beth Vought

Printed in Canada

1 2 3 4 5 XXX 05 04 03 02 01

For more information contact Delmar,
3 Columbia Circle, PO Box 15015,
Albany, NY 12212-5015.

Or find us on the World Wide Web at http://www.delmar.com

Library of Congress Cataloging-in-Publication Data:

ISBN 0-7668-1515-3

CONTENTS

PREFACE

INTRODUCTION TO DRAWING COMMANDS xxii
PREREQUISITES xxiii
ONLINE COMPANION xxiii
ABOUT THE AUTHORS xxiii
ACKNOWLEDGMENTS xxiv
TO THE STUDENT xxv

SECTION 1 DRAWING CONSTRUCTION

CHAPTER 1 EXPLORING AUTOCAD TOOLS AND DISPLAYS

CONTROLLING THE WORK ENVIRONMENT 4
- Equipment 4
- Work Habits 5

AN INTRODUCTION TO CONTROLLING AUTOCAD 5
- Making Your Selection 5
- Starting AutoCAD 8
- Exploring the Startup Options of AutoCAD 9

EXPLORING THE DRAWING AREA TOOLS AND DISPLAYS 11
- Drawing Area 11
- Title Bar 14
- Menu Bar 15
- Command Line 16
- Status Bar 17
- Scroll Bars 18
- The User Coordinate System Icon 18
- Cursors 19

COMMAND SELECTION METHODS 21
- Toolbars 22
- Menus 26
- Keyboard Entry 29
- Dialog Boxes 30
- Shortcut Menus 36
- Function Keys 37
- Control Keys 38

GETTING HELP 39
- Using Active Assistance 39
- Using the Help Menu 41

Index 43
AutoCAD Learning Assistance 44
EXPLORING THE SAVE COMMAND 44
Ending a Drawing Session 46
WHAT IT ALL MEANS 46
CHAPTER 1 QUIZ 46

CHAPTER 2 CREATING DRAWING AIDS

ENTERING THE DRAWING ENVIRONMENT 50
Open an Existing Drawing 51
CREATING NEW DRAWINGS 55
Drawing Setup Using a Wizard 55
Drawing Setup Using a Template 62
Start from Scratch 67
STARTING A NEW DRAWING FROM WITHIN AUTOCAD 67
Displaying Multiple Windows 68
Working with Partial Files 70
Working with Files without AutoCAD 72
CONTROLLING DRAWING PARAMETERS FROM WITHIN A DRAWING 72
Setting Units 73
Setting Limits 73
Setting Grids 75
Setting Snap 78
CHAPTER 2 QUIZ 80

CHAPTER 3 DRAWING AND CONTROLLING LINES

LINES 82
Entering the Line Command 82
Command Line Input 83
Menu Entry 84
LOCATING LINES USING COORDINATE ENTRY METHODS 84
Cartesian Coordinates 86
Relative Coordinates 88
Polar Coordinates 89
CONTROLLING LINES 91
Line Continuation 91
Line Closure 91
Controlling Lines with the Status Bar Tools 95
AN INTRODUCTION TO EDITING LINES 98
Erase 99
Oops 101
Undo 101
CHAPTER 3 EXERCISES 102
CHAPTER 3 QUIZ 110

CHAPTER 4 WORKING WITH DRAWING FILES

CREATING AND SAVING DRAWING TEMPLATES 114
Using an Existing Template 114
Creating Your Own Template 119
SAVING USER PREFERENCES 120
Creating and Saving a Profile 124
SAVING DRAWINGS 126
The Save Command 127
The Saveas Command 129
Qsave 130
Automatic Saving 131
Saving a Template 131
Leaving AutoCAD 132
MANAGING FILES AND FOLDERS 133
Files 133
Folders 133
Office Procedure 134
Storage Problems 134
CHAPTER 4 EXERCISES 137
CHAPTER 4 QUIZ 138

CHAPTER 5 DRAWING ORGANIZATION

AN INTRODUCTION TO LINETYPES 140
Linetypes Standards 140
AN INTRODUCTION TO LINETYPES 152
Linetype Characteristics 152
Loading Linetypes into a Drawing 153
Making a Linetype Current 154
Altering the Linetype Scale 156
AN INTRODUCTION TO LINEWEIGHT 157
Exploring Lineweights 158
Assigning Lineweights 158
Displaying Lineweights 160
AN INTRODUCTION TO COLOR 161
Color Numbers and Names 162
Setting Object Color 162
AN INTRODUCTION TO LAYERS 164
The Layer Command 166
Creating a New Layer 166
Making a Layer Current 168
Altering Layer Qualities 169
Controlling the Visibility of Layers 170
AN INTRODUCTION TO PRINTING 172
Exploring the Plotting Process 172

Layout and Page Setup Names 172
Paper Size and Paper Units 173
Plot Area 173
Drawing Orientation 174
Plot Scale 175
Previewing 177
Producing the Print 178
CHAPTER 5 EXERCISES 180
CHAPTER 5 QUIZ 186

CHAPTER 6 MULTIVIEW DEVELOPMENT
COMMON TYPES OF DRAWINGS 189
THE THEORY OF ORTHOGRAPHIC PROJECTION 190
The Glass Cube 191
Placement of Drawings 194
SELECTING THE REQUIRED VIEWS 197
Selecting the Front View 198
CREATING VIEWS 201
Using Offset 201
Using Trim to Complete an Object 205
Using Fillet to Adjust Line Length 206
Projection of Objects 207
Line Precedence 210
CONSTRUCTING AUXILIARY VIEWS 210
Using Snap to Create an Auxiliary View 213
Projecting Auxiliary Views 215
Placement of Auxiliary Views 216
FREEHAND SKETCHING 219
Sketching Tools 219
Sketching Techniques 220
Keeping Proportions in Sketches 226
Multiview Sketches 228
Isometric Sketches 232
CHAPTER 6 EXERCISES 236
CHAPTER 6 QUIZ 253

CHAPTER 7 DRAWING GEOMETRIC SHAPES
CIRCLES 255
Center and Radius 258
Center and Diameter 259
Two-Point Circles 259
Three-Point Circles 261
Tangent, Tangent, and Radius 261

Tangent, Tangent, Tangent 262
Circle@ 263
DRAWING ARCS 264
Three-Point Arcs 266
Start, Center, End Arcs 267
Start, Center, Angle Arcs 269
Start, Center, Length Arcs 269
Start, End, Angle Arcs 270
Start, End, Direction Arcs 271
Start, End, Radius Arcs 272
Center, Start, End Arcs 273
Center, Start, Angle Arcs 275
Center, Start, Length of Chord Arcs 275
Continue 276
DRAWING DONUTS 279
USING THE ELLIPSE COMMAND 281
DRAWING POLYGONS 285
DRAWING RECTANGLES 289
Chamfer 289
Fillet 291
Width 292
USING THE SPLINE COMMAND 293
Drawing Closed Splines 294
Altering the Fit tolerance 296
Altering the Tangent Points 297
Converting a Spline-Fitted Polyline 298
Editing Splines 298
Altering Splines using Splinedit 298
CHAPTER 7 EXERCISES 302
CHAPTER 7 QUIZ 310

CHAPTER 8 CONTROLLING DRAWING ACCURACY

CONTROLLING THE APERTURE BOX 311
Size 313
Altering the Aperture Color 314
CONTROLLING DRAWING SETTINGS 314
AutoSnap Settings 315
AutoTrack Settings 317
OBJECT SNAP 318
OBJECT SNAP MODES 320
Endpoint 320
Midpoint 321
Intersection 323
Extension 324

Apparent Intersection 326
Center 327
Node 328
Quadrant 329
Insert 330
Perpendicular 330
Tangent 331
Nearest 332
Parallel 333
Quick 334
From 335
Tracking 336
CONTINUOUS OSNAP 336
CHAPTER 8 EXERCISES 339
CHAPTER 8 QUIZ 344

CHAPTER 9 DRAWING DISPLAY OPTIONS

ZOOM 345
Realtime Command 347
Zoom Window 349
Transparent Zoom 350
Zoom All 352
Zoom Center 352
Zoom Extents 353
Zoom Previous 353
Zoom Scale 353
Zoom In 354
Zoom Out 354
Zoom Dynamic 355
Pan 356
Pan Realtime 356
View 359
Exploring the Named Views Tab 359
AERIAL VIEW 363
Accessing the Aerial View Window 363
Aerial View Window Menus 364
Using the Aerial View Window 365
REFRESHING THE SCREEN 366
Redraw 367
Regen 367
Regenall 368
Regenauto 368
View Resolution 369
CHAPTER 9 EXERCISES 371
CHAPTER 9 QUIZ 372

SECTION 2 ENHANCING DRAWINGS

CHAPTER 10 BASIC METHODS OF SELECTING AND MODIFYING DRAWING OBJECTS

SELECTING OBJECTS FOR EDITING 375
Selecting Objects Using a Pick Box 376
Selecting with a Window 377
Auto Selection Mode 378
Selecting with a Crossing Window 379
Selecting Objects with the WPolygon Option 379
Selecting Objects with the CPolygon Option 381
Selecting Objects Using a Fence 381
Selecting All Drawing Objects 383
Selecting the Last Object 383
Adding Objects to the Selection Set 383
Removing Objects from the Selection Set 384
Multiple Selection Set 384
DEFINING THE SELECTION MODE 384
Noun/Verb Selection 385
Using Shift to Add to Selection 385
Press and Drag Selection 386
Implied Windowing 386
Object Grouping 386
Associative Hatch 386
MODIFYING A DRAWING USING ERASE 386
MODIFYING A DRAWING BY ADDING OBJECTS 389
Copy 389
Mirror 394
Offset 400
Array 403
CHAPTER 10 EXERCISES 414
CHAPTER 10 QUIZ 419

CHAPTER 11 MODIFYING THE POSITION AND SIZE OF DRAWING OBJECTS

ALTERING THE POSITION OR SIZE OF OBJECTS 422
Move 422
Rotate 425
Align 428
Scale 430
Stretch 434
Lengthen 438

MODIFYING OBJECTS BY ALTERING LINES 443
Trim 444
Extend 448
Break 451
Editing with the Chamfer Command 454
Editing with the Fillet Command 459
SELECTING OBJECTS WITH GRIPS 462
Using Grips 462
Grip Options 464
Changing Grip Options 466
SELECTING OBJECTS FOR MULTIPLE EDITS 467
NAMING OBJECT GROUPS 468
Creating and Naming a Group 468
Identifying a Group 470
Creating a New Group 471
Changing a Group 471
CHAPTER 11 EXERCISES 473
CHAPTER 11 QUIZ 487

CHAPTER 12 SUPPLEMENTAL DRAWING COMMANDS

MARKING AND DIVIDING SPACE 489
Placing Points 489
Exploring the Divide Command 491
Exploring the Measure Command 493
DRAWING LINES OF VARIED WIDTH AND LENGTH 494
Creating Lines Using the Ray Command 495
Creating Lines Using the Xline Command 496
Creating Lines with the Sketch Command 499
THE U COMMAND 504
UNDO 504
Auto 504
Control 505
Begin and End 505
Back 505
Mark 505
Number 506
REDO 506
CHAPTER 12 EXERCISES 506
CHAPTER 12 QUIZ 510

CHAPTER 13 CONTROLLING LINES AND LAYERS

CONTROLLING LINETYPES 511
Changing the Line Segment Size 511

Changing Individual Line Segment Sizes ... 513
NAMING LAYERS ... 514
Layer Names ... 514
CONTROLLING LAYERS ... 514
Assigning Color by Layer ... 514
Assigning Linetypes to Layers ... 516
Setting the Current Layer ... 519
Setting Layers ON/OFF ... 519
Thawing and Freezing Layers ... 520
Renaming Layers ... 522
Deleting Layers ... 524
ALTERING LINETYPE, COLOR, AND LAYERS ... 524
Layer Control Menu ... 524
Making an Object's Layer Current ... 525
Altering Layers using Layer Previous ... 526
Altering the Display with Match Properties ... 526
Changing Object Properties ... 527
Changing Objects with Qselect ... 527
CHAPTER 13 EXERCISES ... 532
CHAPTER 13 QUIZ ... 536

CHAPTER 14 INQUIRY COMMANDS

DETERMINING DISTANCE ... 538
DETERMINING AREA ... 538
First Corner Point ... 538
Object ... 540
Add and Subtract ... 540
MASS PROPERTIES ... 543
LISTING INFORMATION ABOUT AN OBJECT ... 545
OBTAINING INFORMATION ABOUT A DRAWING ... 548
IDENTIFYING A POINT ... 548
TRACKING DRAWING TIME ... 549
Time Display Values ... 549
Prompt Options ... 550
STATUS ... 550
PURGE ... 552
USING THE AUTOCAD CALCULATOR ... 555
Performing Mathematical Calculations ... 555
Using the Calculator to Create or Edit Points ... 557
CHAPTER 14 EXERCISES ... 558
CHAPTER 14 QUIZ ... 564

SECTION 3 PLACING TEXT AND DIMENSIONING

CHAPTER 15 PLACING TEXT ON A DRAWING

GENERAL TEXT CONSIDERATIONS 567
- Text Similarities 568
- Text Placement 569
- Lettering Height and Scale Factor 574

USING THE TEXT COMMAND 576
- Start Point 576
- Height 577
- Rotation Angle 577

PLACING TEXT 577
- Justifying Text 579
- Style Option 583

WORKING WITH STYLE 584
- Naming a Style 584
- Lettering Fonts 586
- Defining Style Properties 588
- Special Characters 591

EDITING EXISTING TEXT 591
- Text and the Mirror Command 594

QUICK TEXT 594

CREATING TEXT WITH MTEXT 596
- Controlling Text Characters 597
- Adjusting Text Properties 602
- Line Spacing 604
- Using Find/Replace 605

EDITING MULTILINE TEXT 606
- Using Ddedit 606
- Editing with Properties 607

USING THE SPELL CHECKER 609
- Changing the Dictionary 610

FINDING AND REPLACING TEXT 612

INCORPORATING TEXT INTO TEMPLATE DRAWINGS 613

CHAPTER 15 EXERCISES 613

CHAPTER 15 QUIZ 615

CHAPTER 16 INTRODUCTION TO DIMENSIONS

DIMENSIONING STANDARDS AND GOALS 618
- Dimensioning Goals 619

DIMENSIONING COMPONENTS 619
- Elements of a Dimension 619
- Dimensioning Text Placement 626
- Methods of Placing Dimensions 626

ACCESSING DIMENSIONING COMMANDS 631
DIMENSIONING OPTIONS 632
Linear Dimensions 633
Aligned Dimensions 639
Ordinate Dimensions 641
Radius Dimensions 642
Diameter Dimensions 643
Creating a Center Mark 646
Angular Dimensions 646
Baseline 649
Continuous Dimensions 651
QUICK DIMENSIONS 653
Quick Dimension Placement Options 656
LEADER LINES 659
Using Quick Leaders 659
Adjusting Qleader Settings 663
CHAPTER 16 EXERCISES 668
CHAPTER 16 QUIZ 670

CHAPTER 17 PLACING DIMENSIONS ON DRAWINGS
CONTROLLING DIMENSIONS VARIABLES WITH STYLES 671
Introduction to Dimension Styles 672
CONTROLLING DIMENSION VARIABLES 672
Dimension Style Manager 672
CREATING A NEW STYLE 675
Lines and Arrows Tab 676
Text Tab 680
Fit Tab 686
Primary Units Tab 690
Alternate Units Tab 694
Tolerances Tab 694
COMBINING VARIABLES TO CREATE A STYLE 695
Creating Dimensioning Styles 696
Modifying a Style 699
Overriding Existing Styles 699
Comparing Styles 699
Renaming or Deleting Styles 700
Applying Styles 701
EDITING DIMENSION PLACEMENT 701
Stretching Dimensions 702
Editing Objects and Dimensions with the Modify Menu 704
EDITING DIMENSION TEXT 705
Editing Dimension Text Using Properties 705
Editing Dimension Text Using Dimedit 706

Editing Dimension Text Using Dimtedit 708
Updating Dimensions 708
WHAT IT ALL MEANS 709
CHAPTER 17 EXERCISES 710
CHAPTER 17 QUIZ 711

SECTION 4 GROUPING AND PRESENTING INFORMATION

CHAPTER 18 REPRESENTING MACHINED FEATURES AND FINISHES

REPRESENTING MACHINED FEATURES 716
Drills, Reams, Cores, and Bores 716
Representing Circular Features 716
Representing Rectangular Features 725
Projection of Machined Features 728
SURFACE FINISHES 730
Surface Waviness and Lay 731
Specifying Surface Finish 732
CHAPTER 18 EXERCISES 734
CHAPTER 18 QUIZ 739

CHAPTER 19 SECTIONING AND HATCHING

CREATING SECTION VIEWS 741
Placing the Cutting Plane 742
Placing Section Lines 746
Types of Sections 749
HATCHING METHODS 754
USING THE QUICK TAB 755
Pattern Type 756
Selecting a Pattern 756
Pattern Angle 759
Pattern Scale 759
Selecting Objects for Hatching 760
Using the Preview Option 764
Removing Islands 766
View Selection 767
Inherit Properties 767
Composition 768
Controlling Pattern Placement 768
DEFINING BOUNDARY HATCH USING THE ADVANCED TAB 769
Island Detection Style 769
Object Type 770

Boundary Set 771
Island Detection Method 772
EDITING HATCH PATTERNS 772
CHAPTER 19 EXERCISES 773
CHAPTER 19 QUIZ 782

CHAPTER 20 CREATING BLOCKS AND WBLOCKS

BLOCKS 783
Benefits of Blocks 784
Creating a Block 786
Specifying the Base Point 791
Description 791
Preparing Blocks for Insertion 792
Inserting Blocks 796
WORKING WITH BLOCKS 797
Minsert—Making Multiple Copies 797
Making an Entire Drawing a Block 798
Nesting Blocks 798
Using Blocks in Multiple Drawings 799
Controlling Blocks with DesignCenter 799
Using Hyperlinks 800
EDITING BLOCKS 803
Editing Block Descriptions 803
Redefining a Block 803
Exploding Blocks 804
WBLOCKS 805
Creating Wblocks 806
Inserting Wblocks 808
Editing Wblocks 808
CHAPTER 20 EXERCISES 809
CHAPTER 20 QUIZ 812

CHAPTER 21 WORKING WITH MULTIPLE DRAWINGS

THE MULTIPLE DOCUMENT ENVIRONMENT 813
General Considerations for Multiple Document Environments 813
Using the Cut, Copy, and Paste Commands 816
Dragging Objects 819
Matching Properties 820
EXPLORING THE AUTOCAD DESIGNCENTER 821
Features of DesignCenter 822
Examining the DesignCenter Toolbar 823
Opening Drawings Using the DesignCenter 827
Adding Content to Drawings Using the DesignCenter 828
Accessing Favorite Contents 830

CHAPTER 21 EXERCISES 831
CHAPTER 21 QUIZ 831

CHAPTER 22 WORKING WITH LAYOUTS AND VIEWPORTS

COMPARING MODEL AND PAPER SPACE 834
An Overview of Paper Space 834
WORKING WITH LAYOUTS 835
Creating New Layouts 836
Altering Layouts 840
WORKING WITH VIEWPORTS 843
Displaying Multiple Tiled Viewports 843
WORKING WITH FLOATING VIEWPORTS 848
Altering Viewport Properties 849
Viewport Scaling 849
Creating Multiple Floating Viewports 850
Creating a Single Floating Viewport 850
Controlling the Display in Floating Viewports 857
CREATING VIEWPORTS WITH IRREGULAR SHAPES 860
Creating Polygonal Viewports 860
Using Vpclip to Adjust the Viewport 860
Converting Existing Objects to a Viewport 861
CHAPTER 22 EXERCISES 863
CHAPTER 22 QUIZ 864

CHAPTER 23 CONTROLLING OUTPUT

AN OVERVIEW OF PLOTTING 865
Reviewing the Plotting Process 867
Exploring the Plot Dialog Box 867
Exploring the Plot Device Tab 868
Exploring the Plot Settings Tab 871
WORKING WITH PLOT STYLES 877
Exploring Plot Style Modes 878
Creating a Plot Style Table 879
Attaching a Plot Style Table to a Layout 882
Defining a Plot Style 884
Editing Plot Styles 887
CHAPTER 23 EXERCISE 888
CHAPTER 23 QUIZ 888

CHAPTER 24 AUTOCAD AND THE INTERNET

INTERNET ACCESS 890
Working with Uniform Resource Locators 890

OPENING AND SAVING INTERNET FILES 891
Changing the Default Web Site 892
DRAWINGS ON THE INTERNET 893
Opening Drawings from the Internet 893
Inserting a Block from the Internet 896
Accessing Other Files on the Internet 898
Saving the Drawing to the Internet 898
USING HYPERLINKS WITH AUTOCAD 899
Hyperlinks inside AutoCAD 900
Hyperlinks outside AutoCAD 907
THE DRAWING WEB FORMAT (DWF) 907
Other DWF Viewing Options 908
Viewing DWG Files 908
Creating a DWF File 909
VIEWING DWF FILES 911
DWF Plug-in Commands 912
PUBLISHING DRAWINGS TO THE WEB 916
Publish to the Web 916
eTransmit 921
MEET NOW 922
CHAPTER 24 QUIZ 923

Preface

AutoCAD® 2002 and the Fundamentals of Mechanical Drafting is a practical, comprehensive text that is easy to use and understand. The contents are designed to teach the skills needed to master the 2D drawing commands used on most mechanical-related drawings. The content can be used as presented to gain an understanding of the use of AutoCAD. This text can also be used with older releases of AutoCAD, but it is best suited to AutoCAD 2000 and newer releases. All concepts are presented using mechanical examples so that each command can be correlated to a specific skill that will be required of engineers and drafters. Commands and drawing skills are presented in a sequence that is easy to understand and follow with each chapter building on the skills presented in the previous chapter. Other specific features of this text include the following:

- How national standards will control drawings is explained.
- Methods for creating free-hand sketches are described.
- Orthographic drafting concepts are introduced.
- Professional drawings by CAD drafters are used as illustrations.
- Chapter exercises review chapter contents in addition to reinforcing previous material.
- Each drawing chapter contains drawing problems that will reinforce the chapter concepts.

Individual chapters will provide knowledge to master the following drawing skills:

- Drawing lines and geometric shapes
- Producing orthographic projections and sketches
- Using varied linetype and width
- Editing basic geometric shapes
- Combining skills to produce drawings such as orthographic projection, auxiliary views, and sections
- Drawing, notation, and dimensioning of common machined features and fastening methods
- Editing an entire drawing to produce other similar drawings

- Understanding dimensioning and tolerancing, including geometric tolerancing methods and techniques
- Placing text within a drawing
- Creating varied text and dimensioning styles
- Creating symbol libraries to increase drawing speed and efficiency

This text provides a user-friendly environment for mastering computer drafting skills. In addition, the authors have written an *Instructor's Guide* that contains supplemental drawing exercises. For more information, instructors can contact Delmar Publishers.

INTRODUCTION TO DRAWING COMMANDS

AutoCAD 2002 is a powerful program that contains hundreds of commands to meet a variety of drawing needs. In this text, only the 2D command options will be explored. Several methods are available to execute each drawing command. Data entry can be made by using toolbars, menus, shortcut menus, and the keyboard. The selection method will depend on individual preference. Each command will be presented initially through the method that provides the fastest access. For many commands, this will be the use of the keyboard, with available options used to supplement the basic procedure. With the keyboard presented as the primary option, the entire command sequence can be presented, displaying the many options that are available within a specific command.

Commands, options, or values that must be typed at the keyboard are listed with **BOLDFACE** capital letters. Once the command is entered, pressing the Enter or Return key is required and will be represented by the symbol ENTER throughout the text. Command sequences will be presented in the following manner:

Command: **L** ENTER *(Or click the Line button on the Draw toolbar.)*

LINE

Specify first point: **4,8** ENTER

Specify next point or [Undo]: **8,8** ENTER

Specify next point or [Undo]: ENTER

When input is required to complete a command, and the selected point can be any point, it will be presented using italics in the following manner:

Command: **L** ENTER *(Or click the Line button on the Draw toolbar.)*

LINE

Specify first point: *(Select a point.)*

Specify next point or [Undo]: *(Select a point.)*

Specify next point or [Undo]: ENTER

Command:

PREREQUISITES

Although this text can serve as an excellent reference for experienced professionals, no knowledge of mechanical drafting is required to learn the command structure of AutoCAD.

ONLINE COMPANION

The Online Companion™ is your link to AutoCAD on the Internet. Updates are posted monthly, including a command of the month, tutorials, and FAQs. You can find the Online Companion at

http://www.autodeskpress.com/resources/olcs/index.html

A separate Online Companion for *AutoCAD® 2002 and the Fundamentals of Mechanical Drafting* contains additional content provided in five web chapters entitled: Polylines; Dimension Tolerances and Geometric Tolerancing; Fastening Methods; Adding Attributes to Enhance Blocks; and Combining Drawings Using Xref.

ABOUT THE AUTHORS

Michael Jones is the CAD Systems Administrator and a drafting instructor at Clackamas Community College, which is an Autodesk Premier Training Center in Oregon City, Oregon. Michael has introduced new users and experienced professionals to the commands and options of AutoCAD for over 19 years. His training center experience is enhanced by 21 years of professional drafting and CAD system management experience in private industry. While working in the industry of machine design and manufacturing, Michael has worked in a variety of drafting and design areas. He has worked on designs including: oil rigging equipment, marine crane equipment, food processing machinery, and heavy equipment for the timber industry. He has also played a fundamental part in converting several companies from manual drafting facilities to CAD technology.

Tereasa Jefferis is a graduate of Clackamas Community College and the Mechanical Systems Institute. She has been a drafter for five years and has experience with commercial plumbing drawings and HVAC systems. She currently works as an independent contractor with several mechanical engineer firms.

Alan Jefferis has been a drafting instructor at Clackamas Community College for 23 years, including 6 years of introducing students to the basics of mechanical drafting. His professional experience includes 8 years of drafting experience with engineers, and 19 years as a professional designer. He is co-author of *Architectural Drafting and Design, Print Reading for Architectural and Construction Technology, Commercial Drafting and Detailing, AutoCAD for Architecture Release 12, AutoCAD for Architecture Release 13, AutoCAD for Architecture Release 14, AutoCAD for Architecture Release 2000*, and *AutoCAD For Architecture 2002*. All are available through Delmar Publishers.

ACKNOWLEDGMENTS

Thanks to Michael J. Durrer, Vice President of Mt. Hood Community College, for his encouragement and for providing an opportunity to teach mechanical drafting. In addition to the knowledge of mechanical drafting shared by Mr. Durrer, the quality of this text has been greatly enhanced by the donations of many professional offices and individuals within private agencies. Special thanks to the following:

Paul Bissell

Michael J. Easley

Jill Heilman

Floyd D. Miller

Cathy Schlesser

Hanh Tran

All of these people have made substantial contributions to the success of this text. We're also especially thankful for the insight, work, and efforts of John Fisher and Sandy Clark of Delmar Publishers, who had the vision to make this book a reality. This project was enhanced by the technical knowledge and contributions of Terrel Broiles, David Byrnes, Vince Potenza, Lee Seroka, and Gail Taylor. No one has been more helpful and supportive than Janice Ann. Thanks for the help and support. But most importantly, thanks to 'he who is able to keep you from falling, and to present you before his glorious presence without fault and with great joy.'

TO THE STUDENT

AutoCAD® 2002 and the Fundamentals of Mechanical Drafting is designed for students in mechanical drafting and engineering programs. The contents are presented in an order that has been tested within the classroom. Information is based on common drafting practices from the offices of engineers and designers. To make the best use of the information within the text, students should follow these few helpful hints.

Read the Text.

I know it sounds so simple, but if you'll read a chapter prior to sitting at a computer, you'll find your drawing time will be much more productive. Read through the chapter and make note of things that you're not quite sure of. With the reading complete, now try to mentally work through the commands that you've been reading about.

Study the Examples.

One of the many features of this text are the many illustrations and examples of how to complete each command sequence. Carefully study and compare the command sequence with the illustration of how to complete the command.

Practice at a Workstation.

Once you've read the chapter and studied the illustrations, complete the command sequences at a computer workstation. Practice each command several times and don't be afraid to explore all of the options within a command. Because each chapter is based on the previous chapters, don't hesitate to reread past material. Above all else, don't be afraid to experiment. It's very common for new computer users to be filled with a fear of breaking something. If you save your drawings often, there is very little that can go wrong by experimenting. Relax and have fun!

SECTION 1

DRAWING CONSTRUCTION

Exploring AutoCAD Tools and Displays

This chapter will introduce methods for controlling the work environment, starting AutoCAD, and the tools and displays of AutoCAD. These include

- Drawing area
- Title bar
- Menu bar
- Command line
- Status bar
- Scroll bars
- User coordinate icons
- Cursors

Command selection methods to be introduced include

- Toolbars
- Menus
- Keyboard entry
- Dialog boxes
- Shortcut menus
- Function keys
- Control keys

Commands to be introduced include

- **Line**
- **Arc**

- **Cursorsize**
- **Erase**
- **Filedia**
- **Assist**
- **Help**
- **Save**

You're about to start exploring the most popular computer-aided drafting program on the market. If you're like many new AutoCAD users, it will only take a few minutes of exploring to realize that AutoCAD offers lots of options, and the task of learning the program might seem a bit overwhelming. Relax! It's really not that bad. CAD drafting, and AutoCAD in particular, tends to become addictive. AutoCAD may seem frustrating for the first few weeks, but I think you'll love it.

An understanding of the tools you'll be using will help to eliminate the natural frustration that comes with learning. This chapter will introduce key components of the hardware and the software that you will be using to create and control your drawings.

CONTROLLING THE WORK ENVIRONMENT

Although becoming a proficient AutoCAD user will offer many exciting opportunities, spending long hours at a workstation also has its pitfalls. You might start to notice problems with your eyes, back and wrist. Fortunately, most physical problems arising from computer use can be minimized or eliminated by proper workstation configuration, good work habits, and good posture.

EQUIPMENT

You might have little control over your workstation at school, but if you're setting up your own workstation, it should meet the following requirements:

- Your chair should be adjusted to a height that allows your feet to be supported on an inclined footrest or to rest on the floor. The chair should also provide a tilt adjustment so that your back is firmly supported as you work. It should also have an armrest to provide support for your wrists. The height of the chair should allow your wrist to remain level as you type.
- The monitor should be placed so that the top of the monitor is level with your line of eyesight and between 18 and 30 inches from your eyes. Lighting should be placed so that it will not produce a glare on the display screen.
- Use a desk with a keyboard drawer that allows your wrists to remain level.
- Use an ergonomic keyboard, a foam wrist support or forearm supports.

WORK HABITS

You can have a state-of-the-art workstation, but if you don't maintain good work habits, your body will wear out from fatigue. To keep your productivity high and maintain your sanity, you'll need to control your work habits and posture.

- Avoid looking at a monitor for long periods of time. To ease eyestrain, several times an hour look away from the monitor and force your eyes to focus on an object several feet away from your workstation.
- Avoid resting your hand on the mouse when you're not actually using it. When you are using the mouse, avoid resting your hand on the desk. Get in the habit of keeping your wrist level, or adjust the armrest of your chair to provide support.
- Take time to stretch your legs, neck, shoulders, back, and wrist several times each hour. In addition to reducing tension throughout your body, a stretch provides an excellent opportunity to get your eyes off the monitor.
- Most importantly, take time to walk away from your workstation at least once an hour. It's easy to get caught up with a deadline and convince yourself that you just don't have time for a break, but short breaks will help keep you productive and help to eliminate drawing errors.

AN INTRODUCTION TO CONTROLLING AUTOCAD

A workstation consists of neat stuff, but these gadgets can do nothing without software. Software consists of programs such as AutoCAD that are loaded into the computer to tell it what to do. The platform coordinates the computer and the software so that each can function. The operating systems that coordinate the software programs in your computer include Microsoft Windows XP, Microsoft Windows 2000 Professional, Windows Millennium Edition, and Windows 98. If you've recently purchased a personal computer (PC), it will come loaded with Windows XP, ME or 98 as the controlling platform. When a computer is booted using Windows 2000 or Windows 98, a screen similar to Figure 1.1 will be displayed.

MAKING YOUR SELECTION

No matter which Windows platform you use to run AutoCAD, they each have common features that must be understood to control AutoCAD. Before AutoCAD can be efficiently used, it's important that you understand the basics of Windows—including buttons, clicking, double-clicking, and dragging—before you tackle AutoCAD.

Icon Buttons

Pictures called icons are displayed to identify what program or command in a program is to be activated. Windows and AutoCAD use an abundance of icons as buttons that

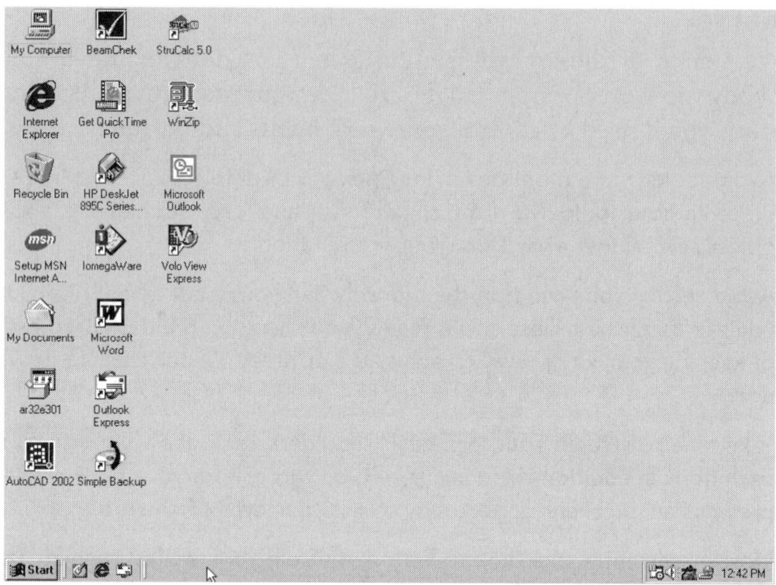

Figure 1.1 *Although the display will vary greatly depending on what programs are loaded on the machine, the program icons are displayed as Windows is entered. Icon buttons are used throughout Windows and AutoCAD to represent specific programs or components within a program.*

cause actions when they are selected with the left mouse button. The icon buttons allow greater productivity and are used in place of written commands. Icon buttons are also used within AutoCAD to identify specific portions of the programs to be executed. Icons that represent programs on your workstation desktop are shown in Figure 1.1. Figure 1.2 shows the buttons on the **Standard** toolbar of AutoCAD. Later portions of this chapter will discuss the use of each of the buttons.

Clicking and Double-Clicking

A mouse is the pointing device used to input commands with Windows applications. On a two-button mouse, the left button is primarily used for the select button. The

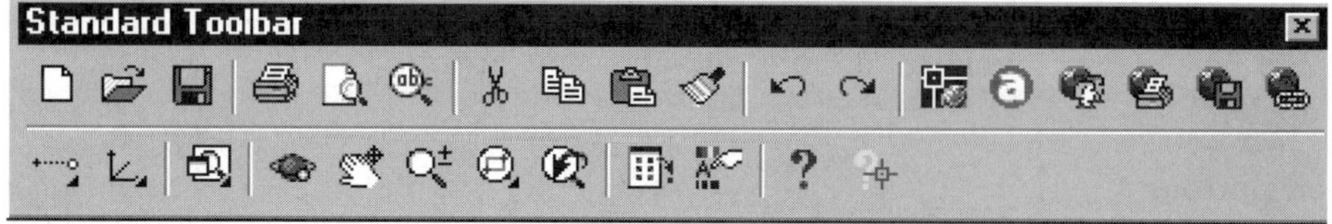

Figure 1.2 *Buttons of the AutoCAD* **Standard** *toolbar can be used to perform specific functions.*

right button is used to terminate a command or to select a shortcut menu to alter a command. Later chapters will introduce additional functions of each button. The mouse can be used to select commands from menus that are displayed beside, above, or inside the drawing area, or to select starting and ending points while in a command. Most often, you will be asked in this text to select a command from a menu. This is the process of going to a menu, placing the cursor on the desired option, and pressing the left button once as you would a telephone button. A single click of a mouse button is referred to as a *click* or a *pick*.

Menu items are typically selected with a single click. If a menu is displayed with six options, selecting one of the options with a single click will highlight that menu option. Figure 1.3 shows the **Draw** menu with the **Line** command highlighted. Clicking once again will start the command sequence.

Clicking a mouse button twice in rapid succession is referred to as double-clicking. If a double click is entered too slowly, it will be interpreted as two single clicks. A double click is used to select a program, folder, or file for an immediate use.

Double-clicking the **AutoCAD** icon button of the original Windows display shown in Figure 1.1 will remove that display and start the program. If you find yourself unsure about when to click or double-click, start with a single click. If the desired results are not achieved with a single click, use a double click.

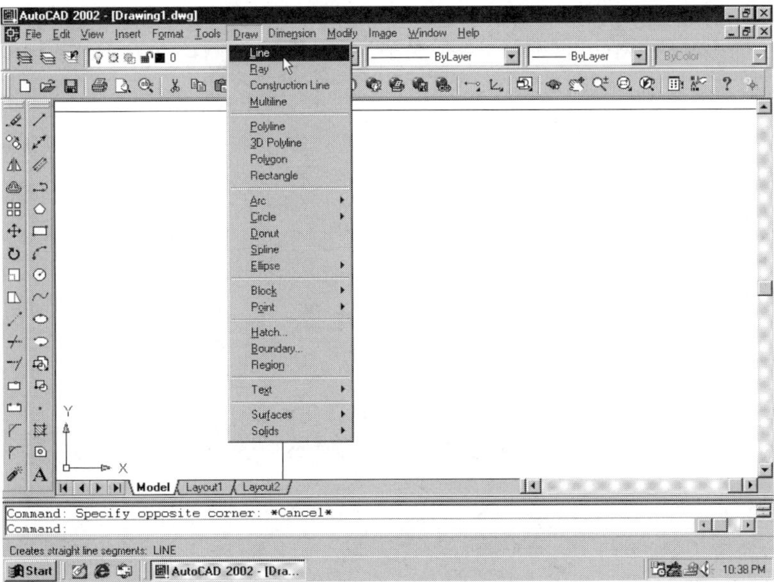

Figure 1.3 *Selecting* **Draw** *on the menu bar displays the* **Draw** *menu. As the cursor is moved to Line command, the command is highlighted for selection.*

Most commands and options are executed by clicking the left mouse button to make selections in Windows and AutoCAD. The right mouse button can produce several responses depending on what you're doing. Clicking a blank portion of the screen will typically reveal a hidden menu. The menu will be altered depending if you are in the middle of a command sequence, or if the right button is pressed before entering a command. Variations of the right button, or shortcut menus, will be discussed in later chapters as each command is discussed. Clicking with the left button will remove the menu.

STARTING AUTOCAD

The easiest method of starting AutoCAD is to double-click the **AutoCAD** icon button from the desktop. See Figure 1.1. You can also start AutoCAD by double-clicking AutoCAD 2002 in the AutoCAD 2002 folder of the Program Files folder in Windows Explorer, or the **Run** menu will open AutoCAD. Each of these methods can be accessed by selecting the **Start** button. Double-clicking the **AutoCAD** button will produce the AutoCAD 2002 banner, followed by the AutoCAD drawing screen shown in Figure 1.4. Three areas are displayed on the screen as AutoCAD is opened: the Active Assistant, the TODAY window, and the drawing area.

The Active Assistant provides automatic or on-demand context sensitive help. Depending on the settings of your computer, it might not be displayed as AutoCAD is entered. It will be further explored later in this chapter.

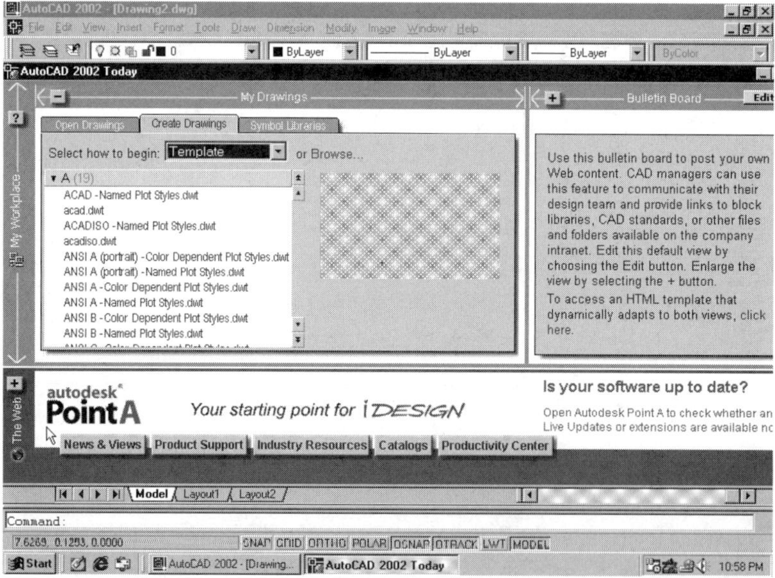

Figure 1.4 *Double-clicking the* **AutoCAD** *button will produce the default AutoCAD startup window. In the default setting, the drawing window, the TODAY window, and the Active Assistance window are displayed.*

The TODAY window can be used to access drawings or the Autodesk web site. Key portions of this window include the following:

- Access to Autodesk Point A provides personalized content that is designed to help you accomplish your daily work with AutoCAD. Point A allows you to access the Internet while working in AutoCAD, giving you the ability to access industry news, links, and resources, utilize tips and techniques, new feeds, live updates, and content posted by your CAD manager.
- The Bulletin Board allows your CAD manager to post messages to the drawing team via the company's Intranet. Project information, new software patches, or drawings components to complete your project can be posted on the bulletin board.
- The tools for accessing drawings will be discussed in the next portion of this chapter. The benefits and the tools for connecting to Autodesk Point A will be discussed in Chapter 24.

The drawing area is used to create and display drawings. Each portion of the display will be the subject of the balance of this chapter.

The exact settings that are displayed as AutoCAD is opened will vary depending on how the program in your workstation has been configured. Settings can be altered using the General Options portion of the **System** tab of the **Options** dialog box—this process will be explored later in this chapter. Settings for how AutoCAD will appear when opened include the following:

- **Show TODAY startup window**—This option can be seen in Figure 1.4.
- **Show the traditional startup window**—This option can be seen in Figure 1.5.
- **Do not show a startup dialog**—Figure 1.6 shows this display option.

Most users will prefer one of the first two options. If you're using the TODAY startup window, select the **Create Drawings** tab. Notice that **Template** is now displayed in the **Select how to begin** list box. Selecting the arrow on the right side of the list box provides the options of **Template**, **Start from Scratch**, and **Wizards**. These same options, as well as **Open an Existing Drawing**, are available if the traditional startup window is used to access AutoCAD.

EXPLORING THE STARTUP OPTIONS OF AUTOCAD

The drawing access options of AutoCAD include the following:

- **Open a Drawing**—Opens an existing drawing. This option provides a list of the four most recently opened AutoCAD drawings.
- **Start from Scratch**—Starts a new drawing with the expectation that you will provide all required measurement and drawing parameters.

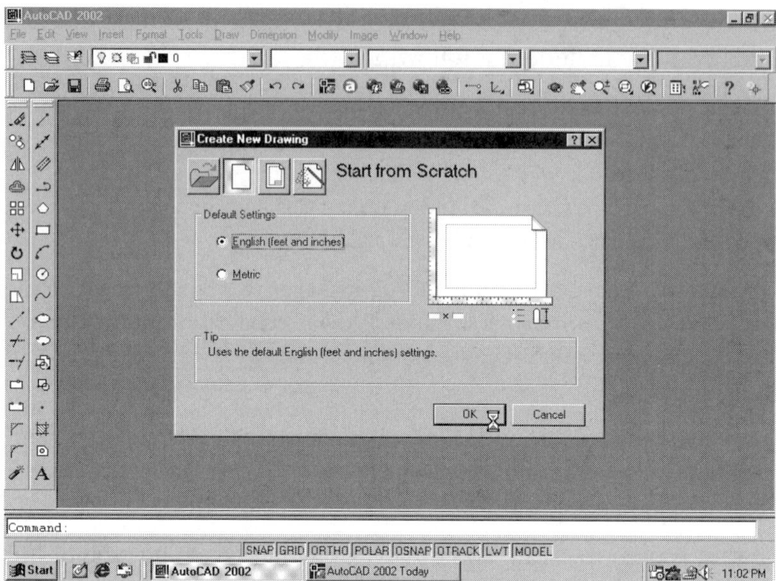

Figure 1.5 *The result of selecting the traditional startup window.*

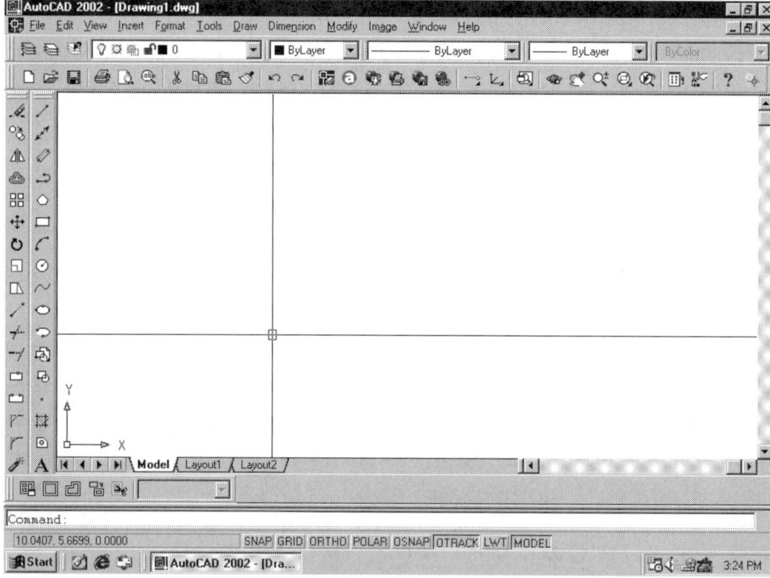

Figure 1.6 *The result of selecting* **Do not show a startup dialog**.

- **Use a Template**—Opens an existing drawing template for the creation of a new drawing. This option allows you to begin a drawing session using an existing template, which contains preset measurement and drawing parameters.

- **Use a Wizard**—Creates a new drawing with drawing parameters set by the use of either the **Quick Setup** wizard or the **Advanced Setup** wizard.

Chapter 2 will introduce each of these tools for controlling the parameters of a drawing session. For now, click the **Close** button (the X in the upper right corner) to remove the TODAY window. The drawing area shown in Figure 1.6 is now displayed. If your machine displays the traditional startup display, press ESC, or click the **OK** button or the **Close** button to remove the **Startup** dialog box. This will allow access to the default version of AutoCAD, displaying a drawing area of approximately 12" × 9" with measuring units set to four-place decimal inches. AutoCAD will provide either a light gray or black drawing screen, depending on how your machine has been configured. The exact size of the drawing area will also vary, depending on how the program is configured. Later chapters will explore how these settings can be set to match your preferences.

The drawing area contains several important tools for controlling AutoCAD. To use AutoCAD effectively, you need an understanding of each of the methods of interacting with the software. This includes knowledge of the drawing area displays, toolbars, types of menus available, common components of a dialog box, function keys, and control keys. Don't let all of the different menus, boxes, and buttons confuse you. Each will become quite easy to use if you are willing to explore the program.

Tip: Push a few keys and explore some options. As long as you keep liquid away from the keyboard, there is very little you can do to hurt the hardware, software, or yourself.

EXPLORING THE DRAWING AREA TOOLS AND DISPLAYS

AutoCAD offers the user many tools and displays to enter information into the program, and to determine the status of the current drawing file. These tools include the drawing area, title bar, menu bar, command line, status bar, scroll bars, user coordinate icon, and cursors.

DRAWING AREA

Central to each of the AutoCAD tools is the drawing area where the material that you enter is displayed. The size of the drawing area will vary depending on how many toolbars or dialog boxes are displayed. In the default setting of AutoCAD, it is the white area, surrounded by gray areas. Methods of altering the colors of each portion of the display will be introduced in later chapters.

Multiple Windows

One of the features that made Microsoft Windows so popular is that many programs can be opened at the same time. One of the features of AutoCAD is the ability to open several drawing projects in one AutoCAD session. Figure 1.7 shows an example of a

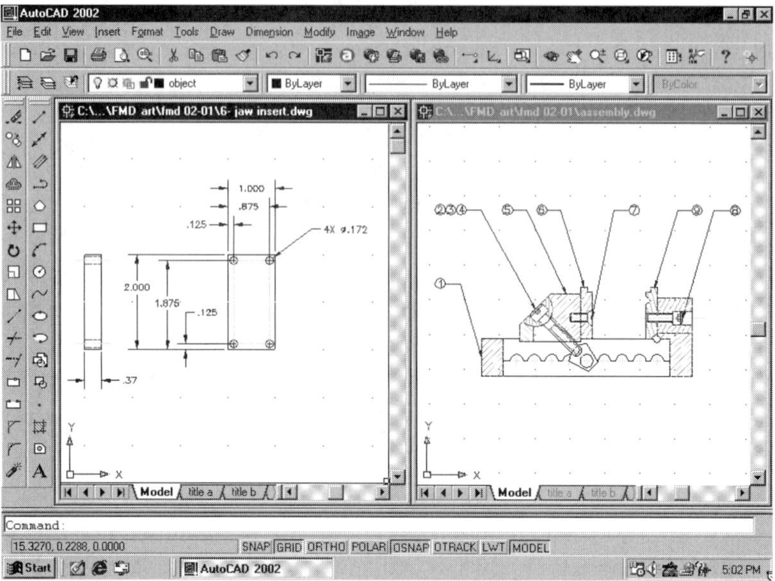

Figure 1.7 *AutoCAD allows multiple drawings to be opened. The name of each drawing is listed in the title bar of each window. The active drawing (JAW INSERT) is displayed in a blue title bar, and the inactive drawing (ASSEMBLY) has a gray title bar. Courtesy Cathy Schlesser.*

drawing session with two drawings displayed. AutoCAD refers to the side-by-side arrangement as vertical tiles. As with other windows, the arrangement and size of the drawing window can be altered. The size of a window is changed by using the **Minimize/Maximize** buttons or by activating the window, and then dragging the edge to the desired size. Selecting **Window** from the menu bar will provide options for altering the arrangement of each window. Figure 1.8 shows the drawings arranged in a horizontal display. The arrangement of windows can be altered at any point during the drawing session. Later chapters explore the process for opening and working with multiple files.

Model and Paper Space Layouts

At the bottom of the drawing area are the Model and Layout tabs. Drawings are created in model space at full scale. If a valve is to be 6" wide, the lines that represent the edges of the valve are drawn 6" apart. If you're drawing an airplane wing that is 100' long, the drawing area can be adjusted so that the entire wing can be seen. Figure 1.9 shows a vice assembly displayed in the drawing area. Paper space is used to plot drawings that have been reduced to a scale such as 1/4"=1'–0". With a 100' long wing, the wing will be 25" long. Paper space allows a 100' long wing to fit on a 22" × 34" sheet of paper. Figure 1.10 shows the part from Figure 1.9 in paper space. This is how the drawing will appear when plotted.

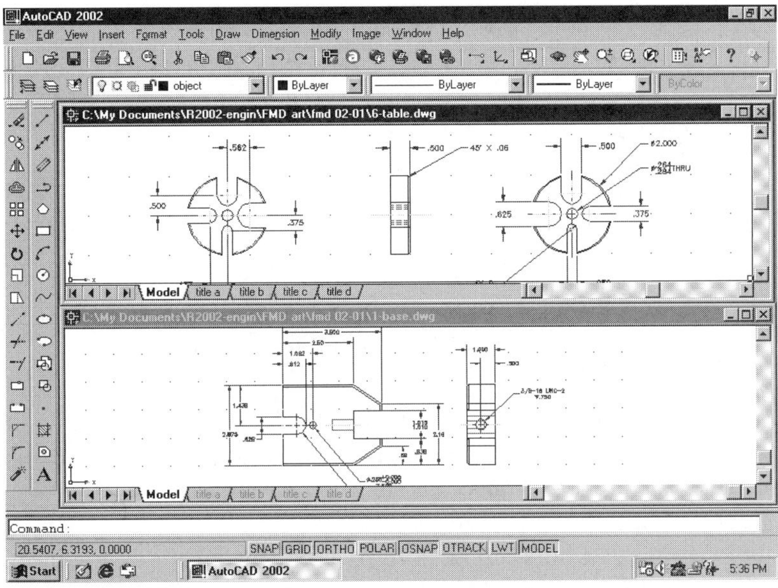

Figure 1.8 *AutoCAD drawings can be arranged in horizontal window arrangements. Courtesy Cathy Schlesser.*

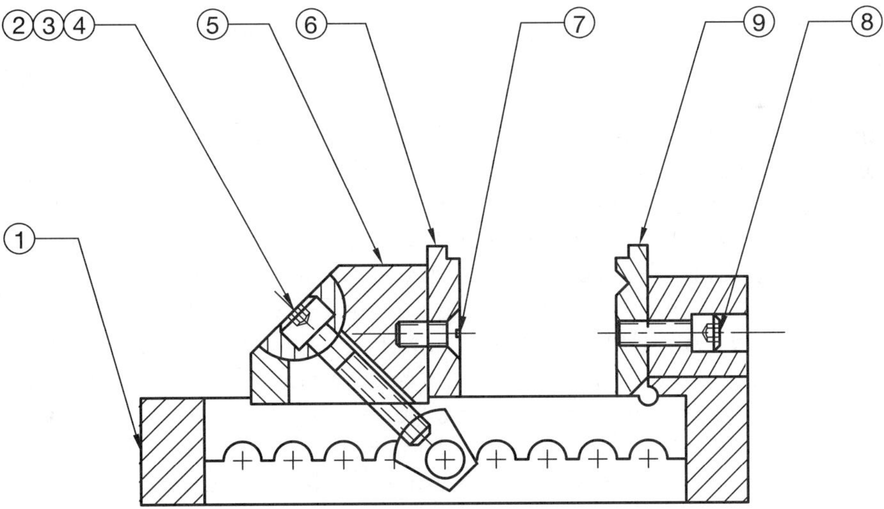

Figure 1.9 *Drawings are created using model space at full scale. Enlarging the drawing area will show the entire drawing. Courtesy Cathy Schlesser.*

AutoCAD allows model and multiple paper space layouts to be displayed in a single drawing. When multiple sheets are combined in a single drawing, the number of individual drawings needed to produce and easily generate different views of a project can

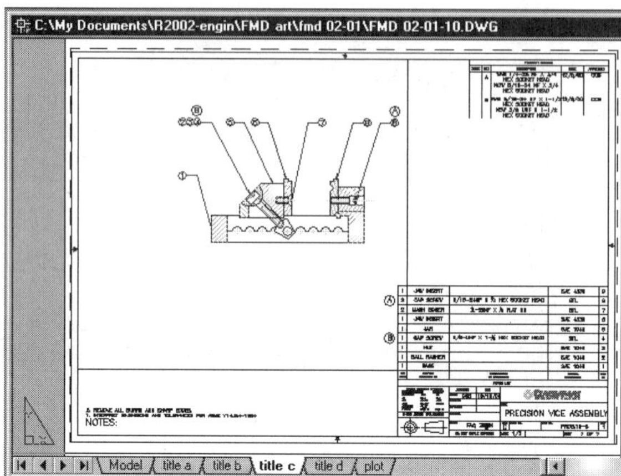

Figure 1.10 *The drawing from Figure 1.9 is shown in paper space. Paper space scales a drawing to fit within the limits of a sheet of paper. Courtesy Cathy Schlesser.*

be consolidated. Multiple layouts can be used to aid in the layout of an object, and in the plotting of an object. Uses for model and paper space layouts will be further discussed through the balance of this text.

TITLE BAR

A title bar is located across the top of every window. It displays the icon and name of the current program as well as the name of the current file. The title bar serves as a handle to drag the window to a different location on the screen. Notice that in Figure 1.7, two drawings are opened and displayed in two windows within the AutoCAD program. The AutoCAD title bar is at the top of the drawing screen, and each drawing window has its own title bar. If multiple drawings are open, the title bar of the active window is highlighted with a blue bar, while inactive windows are displayed in another color. The name of each drawing session is listed in the title bar. Notice in Figure 1.7 that the active drawing is named JAW INSERT and the inactive drawing is named ASSEMBLY. At the far right end of each title bar are three buttons that can be used to control the size of the current window. These buttons are the **Minimize**, **Maximize**, **Restore**, and **Close** file buttons. The **Minimize** and **Close** buttons will always be displayed, but only the **Maximize** or the **Restore** button is displayed, depending on the current window setting. You'll notice that sometimes there are two rows of buttons. Although they function in a similar manner, the buttons at the end of the title bar control the display of the current program window, and the buttons at the end of the menu bar control the display of the current file.

The Minimize Button

Selecting the **Minimize** button (at A in Figure 1.11) with the select (left) button removes the current program or drawing from the display screen and places the program icon in the taskbar at the bottom of the drawing area. To restore the window, use a single click of the select button to select the program icon in the taskbar. The taskbar will be discussed later in this chapter.

Tip: It's important to remember that if you select the **Minimize** button and remove the file from the screen, the file still exists–it's just been placed out of your view. It hasn't been saved or destroyed, just moved for convenience so you can do something else.

The Maximize Button

Selecting a window corner or border of a window with the cursor allows the size of the window display to be altered. When the size of a window has been reduced, selecting the **Maximize** button (at B in Figure 1.11) restores the current window to its largest size. Other windows that might be open are still open; they've just been covered by the current window. Notice that as soon as the window is maximized, the **Maximize** button is changed to the Restore button.

The Restore Button

The **Restore** button, shown at C in Figure 1.11, is displayed only when a window is maximized. Selecting the **Restore** button allows the window to be reduced to its previous size.

The Close Button

Selecting the **Close** button (at D in Figure 1.11) closes the current program. When the **Close** button is selected while in a program such as AutoCAD, a box will be displayed with options to save or not to save the current file, or cancel the **Close** command.

MENU BAR

The menu bar is located below the title bar and is used to display menus specific to the open program. The menu bar for AutoCAD can be seen in Figure 1.8. Each word of the menu bar represents a separate menu with options related to the key word.

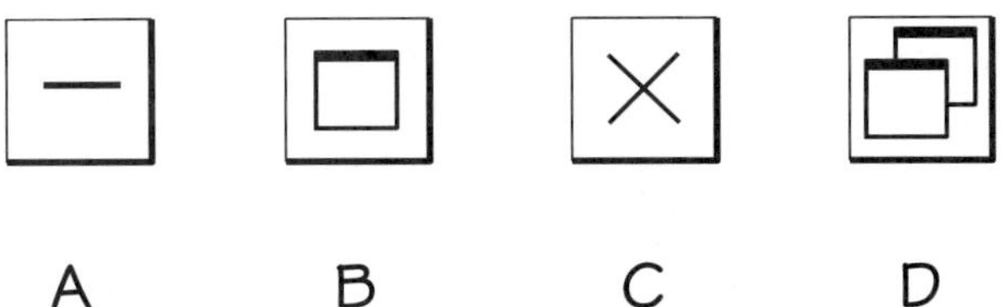

Figure 1.11 *Four buttons for controlling windows are located on the right end of the title bar. Selecting the* **Minimize** *button (A) will remove a window or drawing from the screen and place the title in the taskbar. Selecting the* **Maximize** *button (B) will increase the window to its largest possible size. The* **Restore** *button (C) is displayed only when a window is maximized. Selecting the* **Restore** *button will return the window to its previous size. Selecting the* **Close** *button (D) will close a window and remove the program button from the taskbar.*

Moving the cursor to one of these words and clicking will display the corresponding menu. Figure 1.12 shows the **Draw** menu for AutoCAD. Some commands such as **Arc** have a solid black triangle on the right side. The triangle indicates that moving the mouse to this command will produce another menu of options related to that command. Moving the mouse to the **Arc** listing will produce the menu shown in Figure 1.13. Select the **Format** menu with a left click of the mouse. Notice that the **Plot Style** listing appears different from other listings. This option is inactive and is not currently available for selection. Inactive options only become active as other options are activated.

COMMAND LINE

The command line displays the prompt **Command:** and allows you to enter your request into AutoCAD using the keyboard. Command entry methods will be discussed later in this chapter. As you explore AutoCAD, it's important to remember that the command line displays prompts for completing commands. Even if you enter commands by using a toolbar or menu, AutoCAD will display prompts and a command history in the command window. The command line also shows the results of selecting a button. If the **Line** command is selected by clicking the **Line** button,

_line Specify first point

is displayed at the command line.

Tip: Users new to AutoCAD often fail to read the Command prompt and miss AutoCAD's attempt to communicate with them.

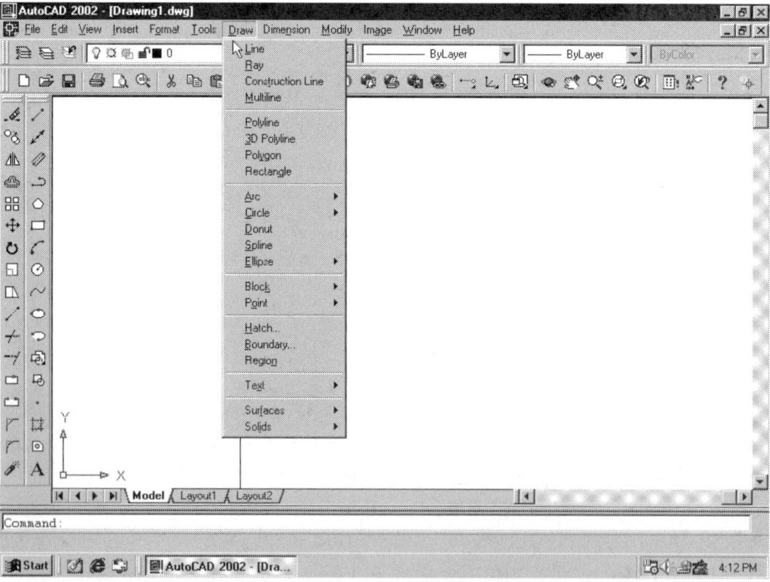

Figure 1.12 *The* **Draw** *menu of AutoCAD.*

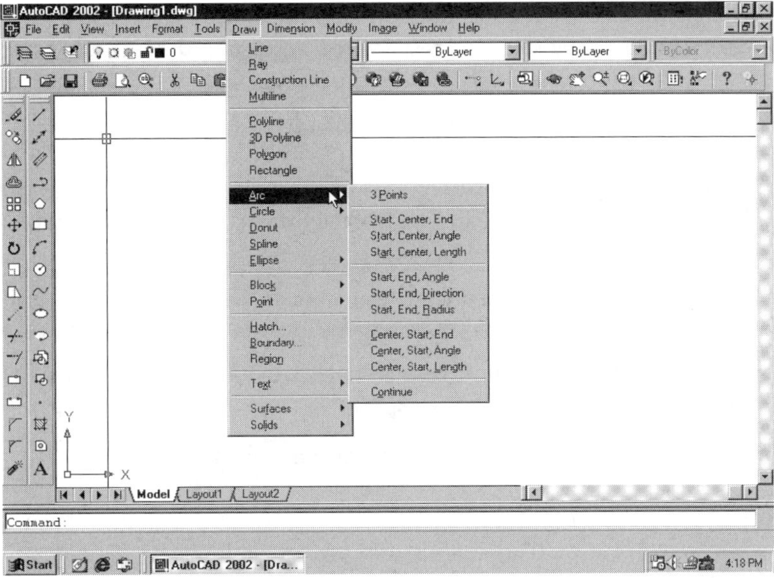

Figure 1.13 *Listings in a menu followed by a triangle will produce a flyout menu. The* **Arc** *fly-out menu is displayed when Arc is highlighted.*

STATUS BAR

The status bar is located below the command line. Key elements of the status bar can be seen in Figure 1.14 and include the coordinate display and toggle buttons that can be used to control a drawing aid. The drawing aids in the status bar affect the placement or display of your drawings. Options are in the ON position when the button appears to be pressed in. Buttons include the following:

SNAP—Toggles an invisible grid ON/OFF for placing lines at precise intervals. Snap increments can be set using the **Snap and Grid** tab of the **Drafting Settings** dialog box. Display the dialog box by selecting **Drafting Settings** from the **Tools** menu.

GRID—Toggles a visible grid ON/OFF for placing modular objects. Grid increments can be set using the **Snap and Grid** tab of the **Drafting Settings** dialog box.

ORTHO—Toggles the ability to draw lines that are vertical and horizontal ON) or at any angle (OFF).

POLAR—Toggles the polar tracking feature that will describe the position of the cursor in distance and angle relative to another point. By default, the tracking angle is set to a 90° increment angle. The angle can be altered using the **Polar Tracking** tab of the **Drawing Settings** dialog box.

OSNAP—Toggles on the ability to select drawing objects such as endpoints, intersections, or eleven other options. AutoCAD automatically snaps to the endpoints of existing lines as the cursor is moved toward those lines. Other modes can be set by selecting the desired button on the **Object Snap** toolbar, or by selecting the desired mode from the **Drafting Settings** dialog box. Modes can also be selected by holding down SHIFT and the right mouse button at the same time.

OTRACK—Toggles the tracking feature that allows specific angles to be tracked along a projected path based on an object endpoint to increase drafting precision.

LWT—Toggles the display of lineweight. When ON, assigned lineweights are displayed. When OFF, all lines are shown using the same lineweight.

MODEL/PAPER—Toggles between model and Paper space. In model space, click MODEL to change to paper space.

Each button will be explained in detail in subsequent chapters.

SCROLL BARS

The scroll bars consist of a vertical bar located on the right side of the drawing area and a horizontal bar located below the drawing area. The vertical bar consists of an up arrow at the top of the bar, a down arrow at the bottom of the bar, and a slide box that changes its position to reflect the current location's position. Scroll bars are used to alter the material viewed in the drawing area. They are also used in dialog boxes that are displayed throughout AutoCAD to alter program settings. For the drawing area, the scroll bars are one of many ways to alter the display in the drawing area. There are better, faster ways than scroll bars to alter the displayed view. In a dialog box, scroll bars offer a quick method of moving through a long list of options. Dialog boxes will be discussed later in this chapter.

THE USER COORDINATE SYSTEM ICON

The User Coordinate System icon describes the current coordinate system and orientation of the drawing area. AutoCAD drawings are imposed over an invisible coordinate system based on X, Y, and Z coordinates. AutoCAD uses both a fixed world coordinate system (WCS) and a movable user coordinate system (UCS). Each method will be explained in later chapters. In its standard setting, as you draw in AutoCAD, you're working in model space. Model space can best be thought of as drawing at full scale. When a drawing is to be plotted, it is plotted using paper space. Paper space is space created so that one or several drawings at one or more scales can be plotted on one sheet of paper. Model and paper space will be discussed in detail in later chapters. Figure 1.15 shows the model and paper space USC icons.

Tip: As you explore AutoCAD, the important thing to remember is that the icon is a reminder of what mode you are currently using to create your drawing. Unless your instructor tells you otherwise, work in model space as you create a drawing, and work in paper space to plot a drawing.

Figure 1.14 *The status bar contains the coordinate display and eight buttons for controlling drawing aids.*

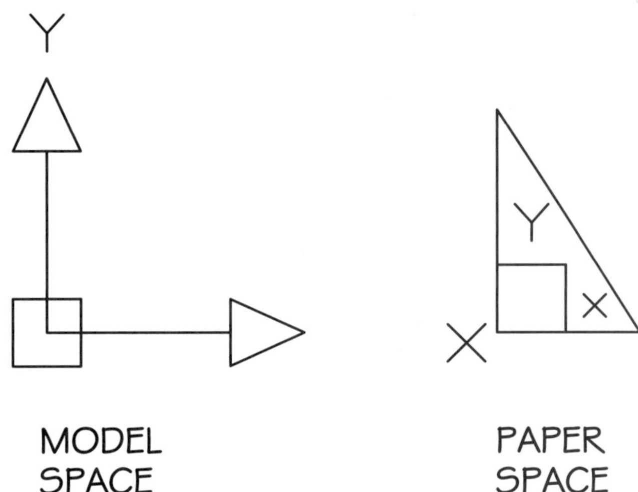

Figure 1.15 *The user coordinate system (UCS) icons for model and paper space.*

CURSORS

The term *cursor* refers to the pointing device that is placed on the screen to indicate the current mouse position. As seen in Figure 1.13, the cursor for the AutoCAD drawing area is crosshairs. As the mouse is moved, the location of the crosshairs is moved. The cursor provides a method of entering the location for drawing objects. When commands such as **Line** are started, the cursor is the primary method of selecting the beginning and ending points for a line.

Altering the Crosshairs Size

Depending on the complexity of the drawing being completed, some operators like to adjust the size of the cursor. Crosshairs that extend across the screen will aid in projecting objects from one view to another. Projecting heights from one surface to another is much easier with enlarged crosshairs. The size of the cursor can be altered using either of the following methods:

Using a dialog box:

1. Select **Options** from the **Tools** menu. This will display the **Options** dialog box.

2. Select the **Display** tab. Use the slider bar in the **Crosshairs size** edit box to alter the cursor size.
3. When adjusted, click the **OK** button to return to the drawing area.

From the command line:

- Type: **CURSORSIZE** (ENTER) at the Command prompt. This will produce the prompt:

  Command: CURSORSIZE ENTER

  Enter new value for CURSORSIZE <5>:

No matter how the command is entered, the default value is displayed as 5. The 5 represents the percentage of the screen the cursor will occupy. Select and hold the slider bar with the left mouse button and move it to the right to increase the size of the cursor. If you're working at the command line, type the desired value and press ENTER. Changing the value to 10 will double the size of the crosshairs. Entering a value of 2.5 will decrease the size of the cursor to half the original size. Entering a value of 100 will cause the lines forming the crosshairs to cross the entire screen.

Alternate Cursors

Just as with other Windows programs, the cursor changes shapes depending on where it is located in AutoCAD. When the cursor moves outside the drawing area, the crosshairs turn into an arrow. It can also appear as an hourglass, pointing finger, flashing vertical line, double arrow, or four-sided arrow. Each cursor can be seen in Figure 1.16.

- The cursor changes to an hourglass (at B in Figure 1.16) each time AutoCAD or Windows is processing information. While the hourglass is present, the current program is unable to process additional information. Once the command

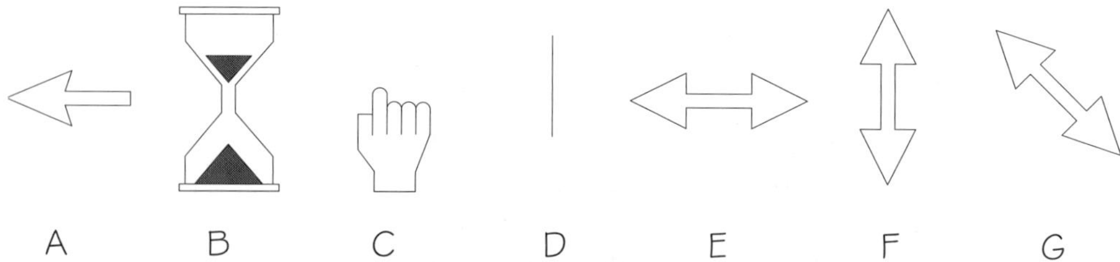

Figure 1.16 *The cursors of a Windows platform include (A) the arrow—used to make selections; (B) the hourglass— indicating a command is being processed; (C) the pointing finger—for making a selection; (D) the flashing bar—for entering text; (E) the horizontal double arrow—for altering a window size in the left/right direction; (F) the vertical double arrow—for altering a window in the top/ bottom direction; and (G) the diagonal double arrow—for altering a window in two directions in one movement.*

is executed, the hourglass reverts to the previous cursor, and new commands can be processed.

- The pointing finger (at C in Figure 1.16) is the politically correct symbol used to indicate that there is additional information available for a specified command or option. Clicking the mouse will display additional information related to the designated topic.
- The arrow cursor changes to a flashing bar cursor (at D in Figure 1.16) when the cursor is moved into a portion of the screen where text input is required. Any text that is entered at the keyboard will be entered just to the left of the cursor bar.
- As the cursor is placed on the border of a window, it will change to a double arrow (at E, F, and G in Figure 1.16). The direction the arrow points will depend on the placement of the cursor. If the select button of the mouse is selected and held, the size of the window can be enlarged or decreased depending on the movement of the mouse. Using the arrow to move a side border will alter the horizontal size. Selecting the bottom or top border allows the vertical size to be altered. If the pointer is placed in the corner of a window and the select button is held, the size of the window can be altered in both directions in one movement. The size of a window can be altered throughout a drawing session.

COMMAND SELECTION METHODS

AutoCAD contains more than 250 commands related to drawing or editing a drawing. Five options are available to enter commands including selecting buttons from toolbars, menus, keyboard entry, dialog boxes and shortcut menus. Consideration will also be given to using Function and Control keys to alter drawing settings or alter commands. Each method has its advantages. Throughout the balance of the text, commands will be presented in the order that tends to be the fastest method of entry. For most commands the fastest entry method is by selecting a button from a toolbar or by keyboard entry. Remember that the key word is fastest, not best. As you're learning each command, take the time to explore each option. You may decide the keyboard is the fastest method of command entry.

As you begin to work with AutoCAD menus, you will see common terms used repeatedly. You will need to understand these so you can progress through AutoCAD. Common terms include the following:

- **Click**—To use a pointing device to make a selection from a toolbar, menu, or in a dialog box.
- **Command**—An instruction to be carried out by the computer, such as **Line** or **Arc**.

- **Default**—A value that will remain constant until a new value is entered. Default values are shown on the command line in brackets: < 3 >.
- **Option**—A portion of a command that requires a selection.
- **Pick**—To use a pointing device to make a selection.
- **Enter**—ENTER is used to denote pressing the appropriate key (key names vary, depending on the keyboard manufacturer) after typing a request at the command line.
- **Escape**—ESC is used to exit a command.
- **Select**—To make a choice from a menu of commands or options.
- **Specify**—To enter a value, such as a coordinate on the command line.

TOOLBARS

A toolbar contains buttons that represent commands. In its default display, the **Standard** and **Object Properties** toolbars are displayed above the drawing area just below the menu bar, and the **Modify** and **Draw** toolbars are displayed on the left side of the drawing screen. Key features of these four toolbars include the following:

- The **Standard** toolbar contains the names of frequently used buttons. In addition to common methods for controlling AutoCAD, the toolbar contains standard Microsoft buttons such as **Open**, **Save**, **Print**, **Cut**, **Copy**, and **Paste**.
- The **Object Properties** toolbar can be used to set the properties of objects such as color, lineweight, linetype, and the layer an object is displayed on.
- The **Draw** toolbar contains buttons to represent each of the major drawing commands, including **Line**, **Polygon**, **Rectangle**, **Arc**, **Circle**, **Ellipse**, and **Point**.
- The **Modify** toolbar contains buttons to represent commands that can be used to edit drawing objects such as **Erase**, **Copy**, **Mirror**, **Offset**, **Array**, **Move**, and **Rotate**.

Because of their ease of use, and availability, for most AutoCAD users the toolbars will become the primary method of selecting commands. AutoCAD offers 24 different toolbars that can be displayed to meet various drawing needs. Toolbars can be displayed in two ways:

- Select **Toolbars** from the **View** menu.
- Type **TOOLBAR** (ENTER) at the command line.

Each method will produce the toolbar list shown in Figure 1.17. A similar toolbar list can be displayed by moving the cursor to any open toolbar and right-clicking. Selecting **Dimension** in the menu will produce the **Dimension** toolbar shown in Figure

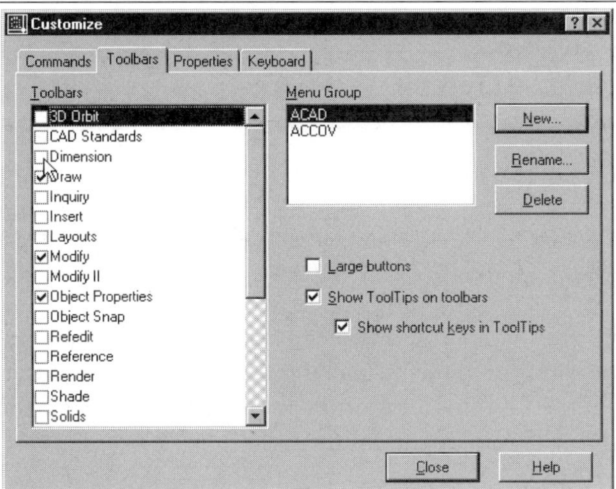

Figure 1.17 *Additional toolbars can be displayed by selecting* **Toolbars** *from the View menu.*

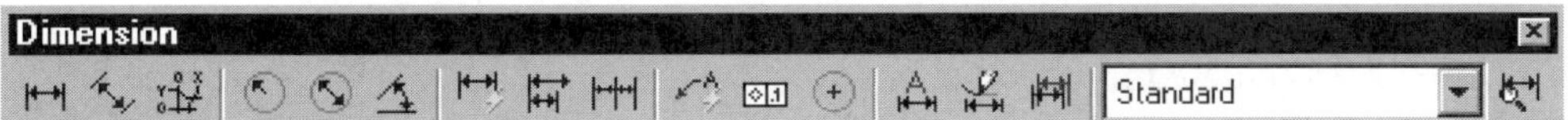

Figure 1.18 *The* **Dimension** *toolbar is displayed by making the selection started in Figure 1.17.*

1.18. Each of the toolbars listed in Figure 1.17 will be discussed as specific commands are introduced throughout the text. As you get started using AutoCAD, it is more important that you feel comfortable using the toolbars that are currently displayed.

To remove the **Customize** dialog box, click the **Close** button at the bottom of the dialog box or click the **Close** button in the upper right corner. The **Dimension** toolbar can also be removed from the screen by clicking the **Close** button in the toolbox title bar.

As the cursor is moved from the drawing screen to a toolbar, it changes from crosshairs to the arrow cursor. The cursor can now be used to select the desired button. You'll also notice as the cursor passes over, or rests on a button in the toolbar, that the button takes on a 3D appearance. Move the cursor to the button in the left corner beside the drawing window. As the cursor is placed above the button, the title of the button will be displayed below the cursor, similar to the display in Figure 1.19. The title is referred to as a tooltip. Tooltips provide a written display of the results that will occur if the button is selected. In this case, the cursor is over the **Line** button. Select the **Line** button with a single-click to start the **Line** command. This will display the following prompt at the command line:

```
Command:_line Specify first point:
```

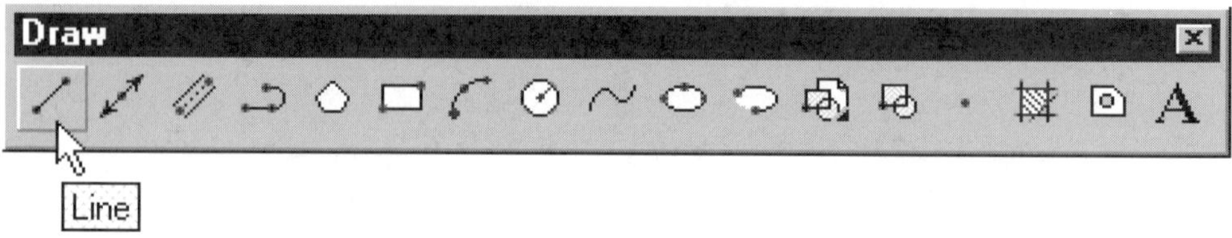

Figure 1.19 *Resting the cursor over a button provides a written description of the button's function called a tooltip. Notice that below the command line a brief description is also displayed.*

The program is now waiting for you to select a starting point for a line. Move the cursor and select an endpoint for a line by clicking the select button of the mouse. As the endpoint is selected, the Command prompt is changed to request the next point of the line. The prompt will display the following:

Specify next point or [undo]:

AutoCAD is waiting for you to select an endpoint for a line. Move the cursor to a new location and pick a point. The prompt will continue to request additional points after each new point is entered until you choose to stop drawing lines. As you draw one line, the command will continue to allow you to add to the existing line. To terminate the **Line** command, press SPACEBAR or ENTER, or click the right mouse button, which will produce the menu shown in Figure 1.20. Selecting the **Enter** or **Cancel** option will end the **Line** command. The other options of this menu will be explored in Chapter 3 as the **Line** command is explored further.

To start another line in a different location, press ENTER, which will automatically restart the last command sequence. Repeating this process will allow an unlimited number of line segments to be drawn without having to select the **Line** button for each line. Take time to enter AutoCAD and draw several different continuous line segments. Once you've drawn several lines, end the command, and then reenter the **Line** command. Now draw several individual line segments. This process can be used with each of the buttons on the **Draw** toolbar.

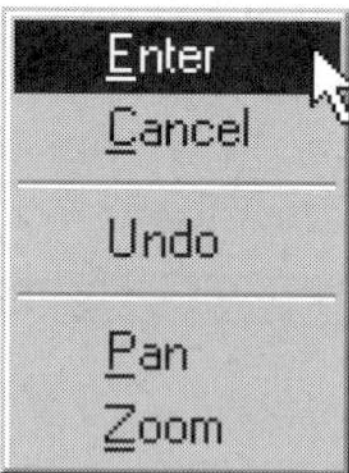

Figure 1.20 *Right-clicking while in the middle of the* **Line** *command will produce a command-mode menu. Selecting* **Enter** *or* **Cancel** *will terminate the* **Line** *command at the last point that was specified.*

Although the command will be discussed in detail in later chapters, take time to explore the **Arc**, **Circle**, and **Rectangle** commands on the **Draw** toolbar. As you're exploring, you might also want to investigate the **Erase** command on the **Modify** toolbar. Selecting the **Erase** button will allow you to select lines and remove them from the drawing screen.

As you're exploring, notice that several of the buttons on the **Standard** toolbar have a small triangle in the corner of the button. Clicking and holding down the cursor on a button with a small black triangle in the lower right corner will produce a flyout menu, which is a submenu of the selected menu. Figure 1.21 shows the flyout menu for the **Zoom Window** button.

Altering Toolbars

Toolbars are used most often to enter commands because they are extremely flexible. In addition to being able to display multiple toolbars on the screen, you can move, dock, float, resize, and alter the contents of toolbars to allow for more convenient usage.

Moving Toolbars

A toolbar can be moved to any convenient area of the drawing screen. Placing the cursor in the title bar of the menu and then selecting and holding the title bar moves the toolbar. Pressing the select button and moving the mouse alters the location of the toolbar.

Floating Toolbars

A toolbar is referred to as a floating toolbar once it is moved from the border of the drawing display. Once a toolbar is floating, it can be resized or docked, or the contents altered. When the **Dimension** toolbar is activated (see Figure 1.18), it is placed in the middle of the drawing screen, making it a floating tool bar. Selecting the title bar and holding the select button allows the toolbar to be moved to any desired location. Move the cursor to the top border of the **Draw** toolbar. Press and hold the select button and move the toolbar into the drawing area.

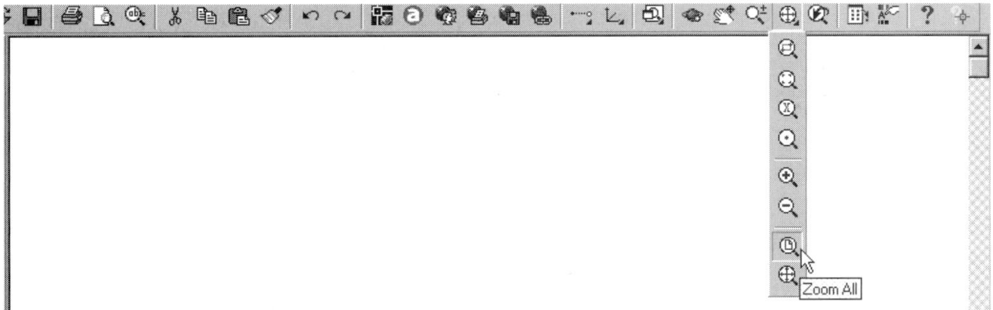

Figure 1.21 *Clicking and holding a button that contains a triangle will produce a flyout menu.*

Resizing Toolbars
If the toolbar interferes with the drawing display, the size of a toolbar can be altered. Placing the cursor anywhere on the border of the toolbar and dragging it in the direction you want the bar to be resized alters the shape of a toolbar.

Docking Toolbars
A toolbar is docked by selecting and holding the title bar and then moving the toolbar to a position above, below, or on either side of the drawing screen. An alternative to docking a toolbar is to totally remove it. A toolbar can be removed from the screen while it is floating by selecting the **Close** button in the upper right corner. To restore the window, select the check box of the desired toolbar from **Toolbars** in the **View** menu.

Refloating Docked Toolbars
A docked toolbar can be refloated by placing the cursor anywhere on the border of the toolbar and then dragging and dropping the box in its desired position on the drawing screen.

Altering Toolbar Content
It might seem unimportant now, but one of the nice features of AutoCAD is that it allows you to modify existing toolbars or to create a new one with your favorite buttons in it. You can change the size or location of toolbar buttons. You can also add or delete buttons from a toolbar. Use the following method to modify a toolbar:

1. Select **Toolbars** from the **View** menu, and then activate the desired toolbar.
2. Change the size of the buttons by activating the **Large button** option. Keep in mind this will reduce the size of the drawing area.
3. While the **Customize** dialog box is still displayed, select the **Commands** tab. This will display the different categories of commands featured on toolbars.
4. Select the category that you would like displayed in the image box.
5. Change the location of the buttons within the toolbar by selecting the desired button and dragging it to the desired location in the toolbar.
6. Add a button to a toolbar by choosing the desired button and then dragging it to the desired location in the toolbar.
7. Remove a button from a toolbar by choosing the desired button and then dragging it into the drawing area.

MENUS

As the cursor is moved into the menu bar, a key method of interacting with AutoCAD is presented. The menu bar displays the names of eleven menus used to control AutoCAD. As with other Windows menus, selecting a specific name will produce a display of that menu. Selecting a menu displays a complete listing of related commands. Move the cursor to the **Draw** menu and select **Draw** to display the menu shown in Figure 1.22. If you accidentally select a wrong menu, move to the desired menu. Menus automatically open and close as the mouse is moved over the menu item.

Once in the **Draw** menu, you'll notice that three different types of commands are listed:

- Typical commands where the command is selected and executed
- Commands with a cascading menu of related commands or options
- Commands that produce a dialog box

Selecting one of the first eight options from the **Draw** menu removes the menu and returns the crosshairs to the screen so that the selected command can be executed. Select Line from the menu. This will remove the menu from the screen and start the Line command. Notice the command line has now changed, to display

_line Specify first point

AutoCAD is waiting for you to select a point. Draw a few lines. You'll notice that the command works the same as when you entered the command using the **Line** button on the **Draw** toolbar.

Exit the **Line** command by pressing ENTER and return to the **Draw** menu. Select **Arc** from the **Draw** menu to display the menu shown in Figure 1.23. Select the **3Points**

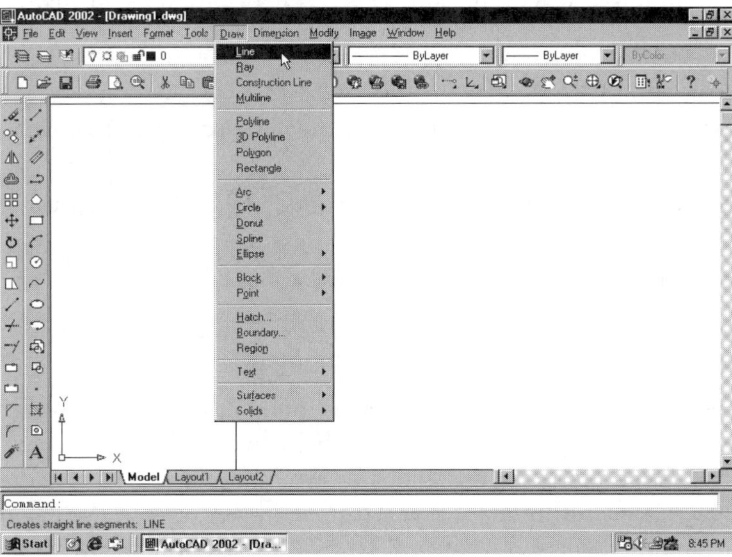

Figure 1.22 *Moving the cursor to the menu bar and selecting* **Draw** *will display the* **Draw** *menu.*

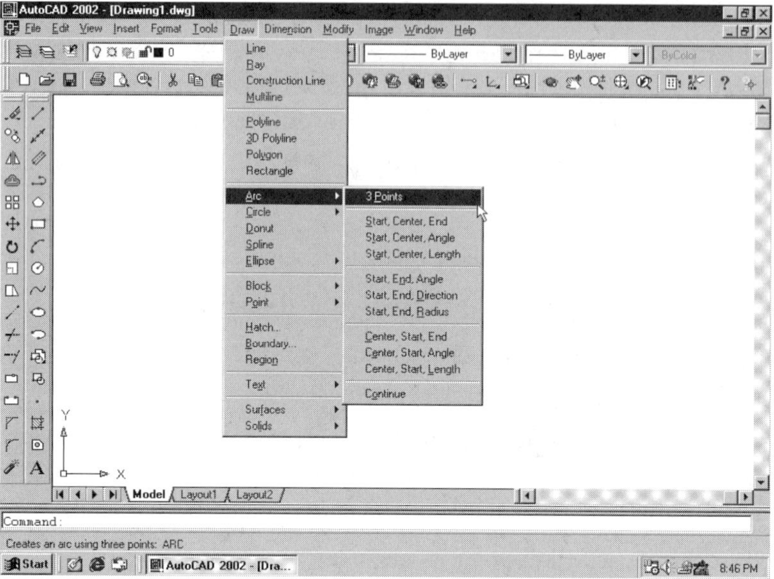

Figure 1.23 *The submenu produced when the* **Arc** *option is selected.*

option. This will remove the menu, return the crosshairs, and allow the selected command option to be executed.

Once the **Arc** option is selected, the command sequence to draw a three-point arc is as follows:

> Command: *(Select arc from the Draw menu.)*
>
> _arc Specify start point of arc or [Center]: *(Select a start point for the arc with the mouse.)*
>
> Specify second point of arc or [Center/End]: *(Select a mid point for the arc.)*
>
> Specify end point of arc: *(Select the end point of the arc.)*

As the third point of the arc is selected, the arc will be displayed in the drawing area and the Command prompt awaits your next command.

Open the **Draw** menu. A third type of display is found in this menu. Notice that two of the options are followed by an ellipsis (...). Selecting an option such as **Hatch...** with a single click will produce the dialog box shown in Figure 1.24. Using dialog boxes will be introduced later in this chapter and explained throughout the balance of the text as specific commands are introduced. To exit the **Boundary Hatch** dialog box, move the arrow to the **Cancel** button, and press the select button to return to the drawing screen.

KEYBOARD ENTRY

The command line allows you to enter your request into the AutoCAD using the keyboard. AutoCAD will provide prompts on the command line to complete the command. Even with all of the buttons and toolbars provided by Windows and AutoCAD, the keyboard is the fastest and easiest method of executing many commands. This is true even if you're a two-finger typist. Throughout this text, commands will be discussed by providing examples of toolbar button and keyboard entry. To enter a command by keyboard, type the name of the command followed by pressing ENTER. To draw a line, type the word **LINE** (ENTER) at the Command prompt. The text will automatically be entered at the Command prompt. For most commands, an alias can be used to reduce typing. For the **Line** command, an alias of L can be used to start the command. Once the command alias is typed, pressing ENTER will activate the specified command. Notice that once ENTER is pressed, the command line display will be altered. The command sequence is as follows:

Command: **L** ENTER

LINE Specify first point:

Use the cursor to select a starting point for the line just as you did when the command was started from the toolbar or menu.

Note: AutoCAD is not case sensitive. You can use all caps, all lower case or a combination.

Expanding the Command Line

By default, the command window will display three lines of text. Many AutoCAD users prefer to have more lines of text displayed so that an entire command sequence can be seen at one time. Additional lines of command text can be viewed by select-

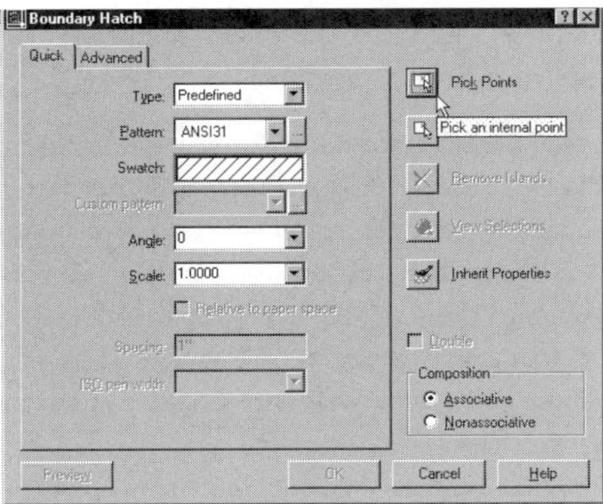

Figure 1.24 *Menu options followed by an ellipsis (…) will produce a dialog box. Dialog boxes allow the functions performed by certain commands to be adjusted.*

ing the top edge of the window and dragging it to a new location. Selecting **Options** from the **Tools** menu can also alter the size of the command window. This will produce the **Options** dialog box that is displayed in Figure 1.25. Select the **Display** tab—this will produce six groupings for controlling the screen display. (See Figure 1.26). Notice that the box in the upper left corner is titled **Window Elements** and allows the size of the command line to be altered. Other options for controlling the display will be discussed in future chapters. To change the command line from three to four lines use the following sequence:

1. Move the cursor to the text box for **Text lines in the command line window**.
2. Highlight the 3 by placing the cursor beside the 3, and then moving the cursor over the 3 while pressing the select button. This will place the 3 in a blue box, indicating that the 3 can be altered.
3. Enter the desired number using the keyboard, in this case 4.
4. Select the **OK** button at the bottom of the dialog box. This will close the **Option** dialog box and return the drawing area. The command line will now display four lines of text.

As the size of the window is increased, a scroll bar is added to the window to allow viewing of previous commands. Up to 100 lines of text may be displayed, but a display of over 35 lines will totally fill the drawing area. The scroll bar on the right edge of the display can be used to show additional lines of text. In addition to displaying information about the current command sequence, AutoCAD can display a list of commands used throughout the drawing. This can be done by using the F2 function key. (The function keys are located across the top of the keyboard. The use of each function key will be discussed later in this chapter.) Pressing F2 will produce a display of the AutoCAD Text window that contains the drawing history. The window can be used to display all of the commands used to create the current drawing file. You can remove the display by pressing F2 again. Like other windows, the drawing history display can be moved to a new location or resized.

Note: You've just used **Options** from the **Tools** menu to change the size of the command window. One of the most useful tools for AutoCAD users who share a workstation is to be able to have the settings that control the work environment saved for future work sessions. The **Options** dialog box can be used to set and save user profiles for future drawing sessions. Although these options will not be introduced until Chapter 4, it is important to remember your favorite settings can be saved as you're exploring AutoCAD's options in this chapter. If you can't wait until Chapter 4, see the **Help** menu.

DIALOG BOXES

Although not a menu of commands, dialog boxes provide a convenient way to adjust certain command parameters. You encountered your first dialog box when AutoCAD

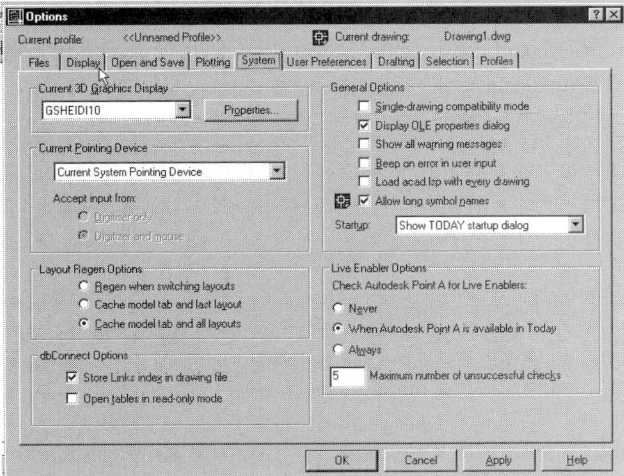

Figure 1.25 *Selecting* **Options** *from the* **Tools** *menu will display the* **Options** *dialog box.*

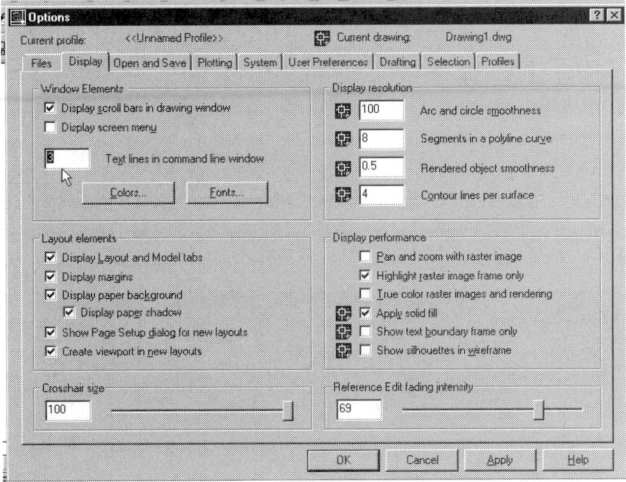

Figure 1.26 *Select the* **Display** *tab of the* **Options** *dialog box to adjust the command window.*

opened and the TODAY or **Startup** dialog box filled the screen. You've also opened the **Customize** and **Options** dialog boxes. A dialog box provides options for controlling commands, options, or portions of the drawing program. Figure 1-27 is a display of the dialog box for controlling layers. Access this dialog box by selecting the **Layers** button on the **Object Properties** toolbar, or by selecting the **Layer** from the **Format** menu. Dialog boxes can also be opened from the command line by typing the name of the desired dialog box to be displayed. The **Layer Properties Manager** dialog box is opened by typing **LAYER** (ENTER). Don't worry about what a layer is for now. Layers will be explained in later chapters. For now, it is important to understand the features found in dialog boxes.

WARNING: AutoCAD provides system variables to control the display of some dialog boxes. If dialog boxes are not displayed, type **FILEDIA** (ENTER) at the command line. The default value is 1. Options include

0 — does not display dialog boxes.

1— displays file dialog boxes.

Other system variables will be introduced as the commands they affect are introduced.

When a dialog box is selected, the crosshairs of the drawing area changes to an arrow. The arrow can be used to select an option by placing the arrow in the box or button by the desired option and pressing the select button. More specific information regarding the actual use of each dialog box will be discussed when each command is introduced. Common components of a dialog box are seen in Figure 1.28; they often include **OK** and **Cancel** buttons, scroll bars, buttons, radio buttons, check boxes, edit boxes, image tiles, and alerts.

OK/Cancel

As you display a dialog box, you will notice that the cursor no longer responds to the drawing area. To return to the drawing and remove the dialog box, select either the

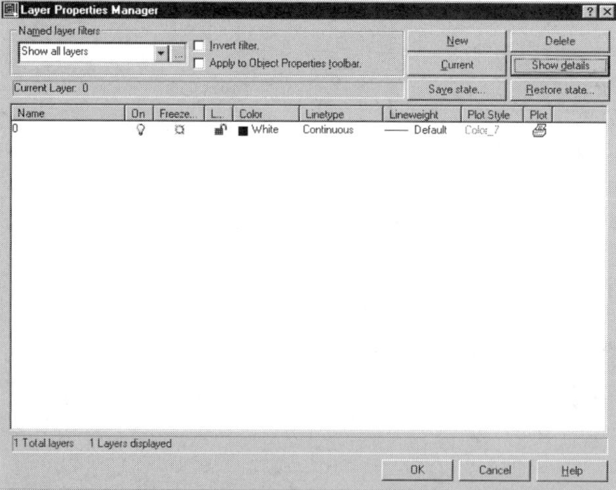

Figure 1.27 *A dialog box provides options for controlling commands, options, or portions of the drawing program. Display the* **Layer Properties Manager** *by selecting the* **Layers** *button on the* **Object Properties** *toolbar, or by selecting* **Layer** *from the* **Format** *menu.*

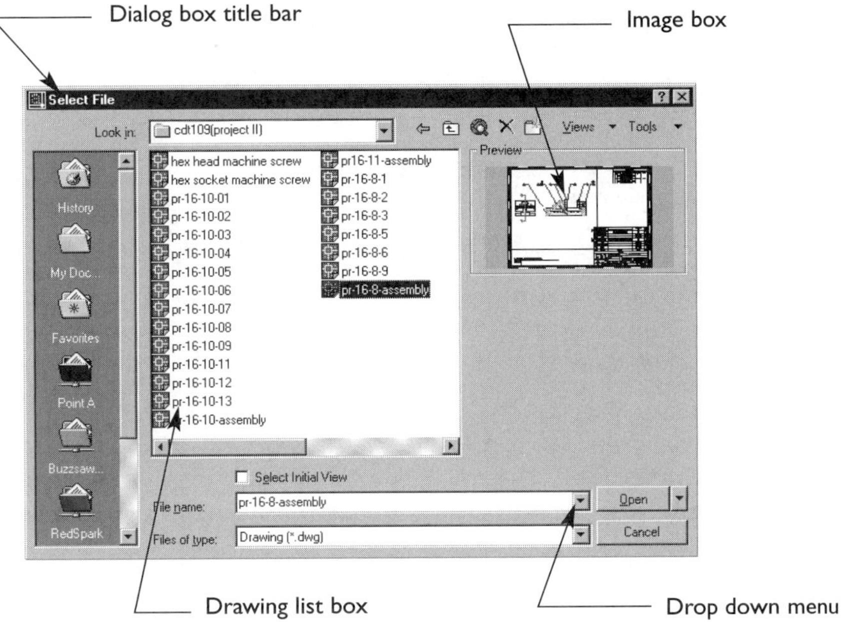

Figure 1.28 *Common components of a dialog box.*

OK or other similar action button, or the **Cancel** button. If you don't want to save the changes in a dialog box, close the box using the **Cancel** button.

- The **OK** option updates the dialog settings and removes the box from the drawing area.
- The **Cancel** option removes the box from the screen, but it will also delete any changes you may have made. Pressing ESC has the same effect as selecting **Cancel**.

Buttons

Several buttons often are included in a dialog box in addition to **OK** and **Cancel**. Examples are seen in Figure 1.29. The **Plot** dialog box can be displayed by clicking the **Plot** button on the **Standard** toolbar or by typing PLOT (ENTER). Selecting one of these buttons will cause the selected option to be performed.

- Buttons with a wide border are default options.
- A button with an option followed by ... will display another dialog box that must be addressed before proceeding.
- A button with an option followed by < indicates that an action needs to be performed in the graphics area before proceeding. This might include selecting

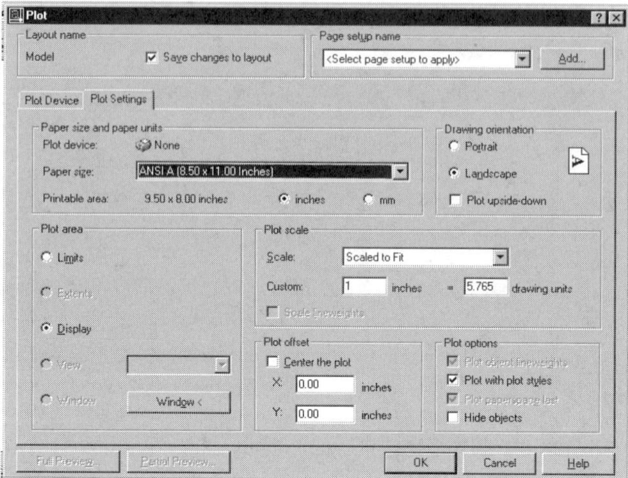

Figure 1.29 *Check boxes and radio buttons in a dialog box serve as toggle switches.*

an object or point in the drawing field. When an action button is selected, the dialog box is removed temporarily to allow for selecting an object from the drawing. Selecting the **Windows<** button provides prompts for selecting the area to be plotted.

Radio Buttons

Radio buttons are buttons that list mutually exclusive options. Examples can be seen in Figure 1.29. **Limits**, **Extents**, **Display**, **View**, and **Window** are radio buttons found in the **Plot** dialog box. Each will present a different area to be plotted, but only one of the four options can be used per plot. Notice in Figure 1.29 that the **Extents**, **View**, and **Window** options are not displayed with black text. Gray options are inactive options and can't be selected until another option has been completed. Select one of the active or black options by moving the cursor to the desired button and clicking the select button.

Check Boxes

Check boxes serve as toggle switches. Open the **Drafting Settings** dialog box by selecting **Drafting Settings** from the **Tools** menu. Once it is open, select the **Object Snap** tab to produce the display shown in Figure 1.30. In the **Drafting Settings** dialog box, each of the modes is represented by a check box. Each mode is independent of the other listings. Selecting the **Endpoint** mode will allow the cursor to jump to the nearest endpoint of existing lines as the cursor is moved. Each of these options, as well as many others, will be presented throughout the text. For now, remember that if a box has an "x" or check in it, that option is enabled. If the box is empty, the option is disabled.

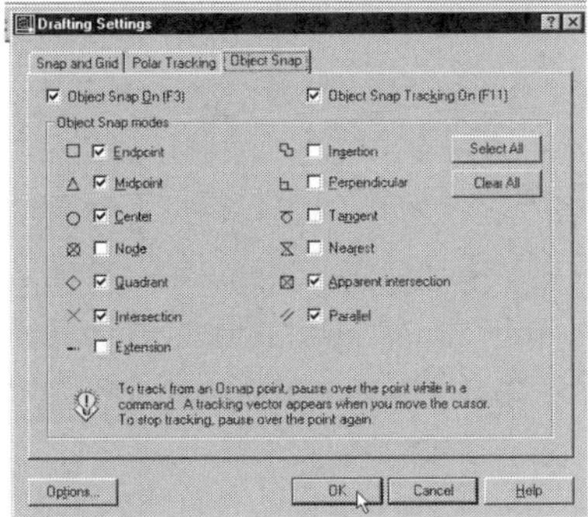

Figure 1.30 Object snap *check boxes are displayed in the* **Object Snap** *tab of the* **Drafting Settings** *dialog box.*

Edit Boxes

An edit box is a text entry box that allows information to be entered by keyboard. Figure 1.28 shows an example of an edit box. The **Select File** dialog box is displayed by selecting **Open** from the **File** menu. When the edit box is selected, it contains a vertical line (a cursor) that flashes slowly. The flashing cursor means AutoCAD is waiting to be told the name of the file. Information can be typed into the box, in this case the **File name** edit box. If errors are made in typing, the cursor can be moved using the left and right arrow keys. Activate the text in the box by pressing ENTER.

Edit Keys

Several keys can be used to aid input in a dialog box. These include text keys, arrow keys, and control keys. These functions include the following:

Text—Symbols from the keyboard are inserted at the cursor. As a letter is inserted, the cursor and existing text move to the right. When the cursor reaches the right side limit, the cursor remains and text is moved to the left.

LEFT ARROW—Moves the cursor to the left without changing the existing text.

RIGHT ARROW—Moves the cursor to the right without changing the existing text.

DEL—Deletes the character at the cursor and moves all remaining text to the left. The cursor will remain stationary.

BACKSPACE—Deletes the character to the left of the cursor while the cursor and all remaining text is moved to the left.

Image Boxes

An image box is a portion of a dialog box that displays an image of the object or pattern that is available. Figure 1.31 shows the image box used to show the pattern that will be placed if this hatching option is selected. This box can be found by selecting **Hatch** from the **Draw** menu.

Alerts

An alert is a message that is displayed in the lower left corner of a dialog box, or in a separate box that is superimposed over the dialog box. Alerts warn the user that they have made a mistake by displaying a message such as "Invalid filename." Another common type of alert is a warning that a file is about to be changed, for example:

> The specified file already exists.
>
> Do you want to replace it?
>
> Yes   No   Cancel

SHORTCUT MENUS

Similar to most Microsoft Windows programs, AutoCAD contains menus that provide rapid access to options for tasks that are underway. Click the right mouse button to display these shortcut menus. As the name implies, the contents of the menu will change depending on when the menu is selected. If you're in the middle of a drawing command and right-click, the menu will be different than if you right-click while trying to edit a drawing. Five different shortcut menus are available while in AutoCAD. Each can be customized. The default settings of these menus are as follows:

- The default menu is displayed when a right click is performed with the cursor in the drawing area prior to starting a command sequence.
- The command-mode menu appears when a right-click is performed in the middle of a command sequence. Options specific to the command will be displayed.

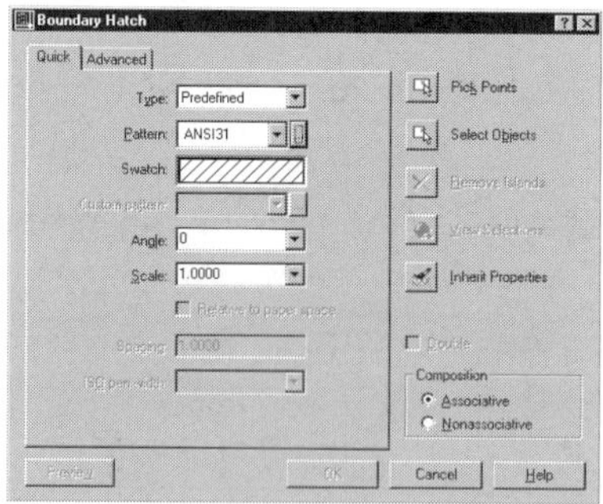

Figure 1.31 *An image box is a portion of a dialog box that displays an image of the selected item.*

- The Edit-mode menu is displayed with a right click if objects have been selected, but no command is in progress.
- The Dialog-mode menus appear if a right-click is performed while in a dialog box.
- Other menus will be displayed when right-clicking based on the cursor location.
- Right-clicking with the cursor over a toolbar produces a list of all toolbars.
- Right-clicking while over the command line, layout tabs or status bar will each produce different context-specific menus.

FUNCTION KEYS

Twelve function keys are located across the top of the keyboard or to the left of the keyboard. The actual layout depends on the keyboard manufacturer. Ten function keys are used by AutoCAD to provide access to commands. Each is used as a toggle from ON to OFF and will be explained further in Chapter 5. Other keys can be programmed to perform specific commands. The following function keys are used by AutoCAD:

F1—Help. Selecting this key clears half of the screen and displays the Help window. Using the Help options of AutoCAD will be discussed later in this chapter. Select the **Cancel** button or the **Close** box to remove the display from the screen.

F2—Flip Screen. Pressing F2 in the middle of a drawing will remove the drawing from the screen and display a written list of the commands that were used to generate the drawing. Press F2 a second time and the text window will be removed.

F3—Object Snap. Snap allows the cursor to be moved to specified points on drawing objects such as an endpoint or midpoint of a line. Later chapters will explore how to set the object snap settings. Press F3 to toggle Object Snap ON/OFF.

F4—Tablet mode. This function key serves as a toggle if a digitizing tablet is used. Some professional offices use a digitizer tablet menu to control command options specific to their company. A digitizing tablet can be used as an alternative to a mouse for selecting objects from screen menus and controlling the crosshairs. If you're working at a station with no tablet, you'll see the following prompt:

CALibrate the TABLET before turning it ON <off>

If you're at a machine with no tablet, ignore the prompt and continue exploring.

F5—Isoplane crosshairs mode. This is used when creating isometric drawings. This button toggles between the top, left, right, and bottom planes of an isometric square.

F6—Coordinate display. The coordinate display can be seen on the status line in Figure 1.8. This function key will toggle the coordinated display ON or OFF.

F7—Grid display. In later chapters you will learn how to create a visible grid to aid in drawing layout. Once you have selected the size of the grid to be displayed, display or remove the grid by pressing F7. Grids can also be toggled ON/OFF by selecting the GRID button on the status bar.

F8—ORTHO. ORTHO can be used to control the movement of the cursor. When it is ON, only vertical and horizontal lines can be drawn. When OFF, lines at any angle may be drawn. These options will be explored in Chapter 5. By switching ORTHO OFF, angular features can be drawn.

F9—Snap mode. This key activates the ability to move the crosshairs within the drawing area at specific intervals. Snap will be explained in Chapter 5. Snap can also be toggled ON/OFF by selecting the SNAP button on the status bar.

F10—Polar Tracking. This key toggles polar tracking ON/OFF.

Note: Remember that function keys can be used at any time throughout the life of a drawing including in the middle of a command. Although ORTHO in the ON setting will make rectangular features much easier to draw, it can be quickly toggled OFF to allow angled lines to be drawn, and then reset to the ON position to finish other perpendicular features.

CONTROL KEYS

Like many other software programs, AutoCAD uses control keys to perform common tasks. These jobs are achieved when the control key is pressed in conjunction with a specified letter key. Many of these tasks duplicate the jobs performed by the function keys. The control keys are located in the lower corners of the keyboard and may be labeled CTRL. The control key functions are as follows:

CTRL+A—Selects all objects for editing.

CTRL+B—Snap mode toggle ON/OFF.

CTRL+C—Copyclip command copies selection to the clipboard.

CTRL+D—Coordinate display of status line toggle ON/OFF.

CTRL+E—Crosshairs in isoplane position toggle left/top/right.

CTRL+F—OSNAP mode toggle ON/OFF.

CTRL+G—Grid toggle ON/OFF.

CTRL+H—Toggles the **Pickstyle** value between 0 and 1.

CTRL+J—Same as using ENTER.

CTRL+K—**Hyperlink** command.

CTRL+L—ORTHO mode toggle ON/OFF.

CTRL+M—Same as using ENTER.

CTRL+N—Executes the **New** command.

CTRL+O—**Open** command, used to open a new file.

CTRL+P—Displays the **Plot** dialog box.

CTRL+R—Toggles between viewports.

CTRL+S—**Save** command.

CTRL+T—Tablet mode toggle ON/OFF.

CTRL+U—Toggles Polar mode ON/OFF.

CTRL+V—**Pasteclip** command, used to paste clipboard locations to a specified location.

CTRL+W—Toggles Object Snap Tracking ON/OFF.

CTRL+X—**Cutclip** cuts selected objects to the clipboard.

CTRL+Y—**Redo** command.

CTRL+Z—**Undo** command.

CTRL+1—Toggles the display of the Properties window.

CTRL+2—Toggles AutoCAD DesignCenter.

CTRL+6—Toggles dbConnect Manager.

GETTING HELP

One of the best features of AutoCAD is that it offers free advice with no ridicule. You're going to be exposed to hundreds of commands, options, and menus, and sometimes it's easy to feel lost or overwhelmed. If you can avoid panic in the first 30 seconds you'll probably do fine. Your chances for success are even better if you remember that there has been an Active Assistance window as well as a **Help** button in every dialog box you've looked at.

USING ACTIVE ASSISTANCE

The Active Assistance window provides command related on-demand help. Depending on how your program is configured, the Active Assistance window will be displayed. Options for display include the following:

- **All commands**
- **New and enhanced commands**
- **Dialogs only**
- **On Demand**

The Active Assistance window can be relocated, enlarged, or reduced just as any other window.

If you've never used AutoCAD, the **All commands** option will be helpful. In this setting, each time a command is started, the window displays information related to that command. Type L (ENTER) to start the **Line** command. As ENTER was pressed, the display changed to display the information shown in Figure 1.32. Typing **C** (ENTER) will change the display to show information about the **Circle** command.

As you gain experience with AutoCAD, you might find the window distracting. Display options for Active Assistance can be altered by moving the mouse anywhere in the assistance window and right-clicking. Selecting **Settings** from the menu will produce the display shown in Figure 1.33. Two important options alter the display:

- The **New and enhanced commands** radio button will display the Assistance box only as new or altered AutoCAD commands are started. This will allow you to draw lines without assistance, but receive help when a command new to this release of AutoCAD is entered.
- The **Show on Start** check box, if cleared, will make the default option inactive, and the window will not be displayed the next time an AutoCAD session is started.

If you decide you want to alter the Assistance display, type **ASSIST** (ENTER) at the Command prompt. This will display the Active Assistance window shown in Figure 1.34. Changes can be made by right-clicking in the window and selecting Settings.

Three other options that you should investigate include **Dialogs only**, **Hover Help** and **On Demand**.

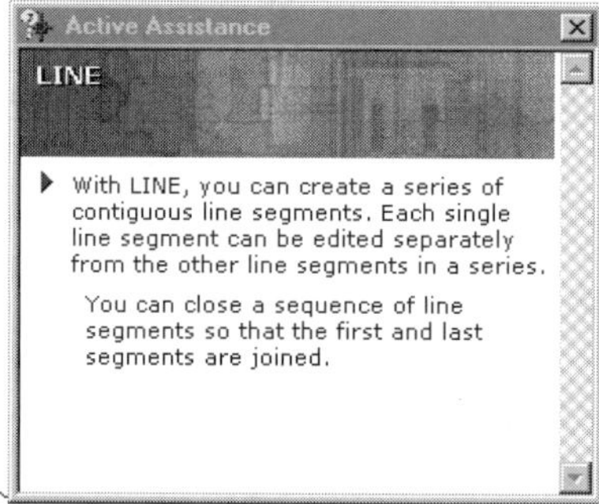

Figure 1.32 *The Active Assistance window provides command related on-demand help.*

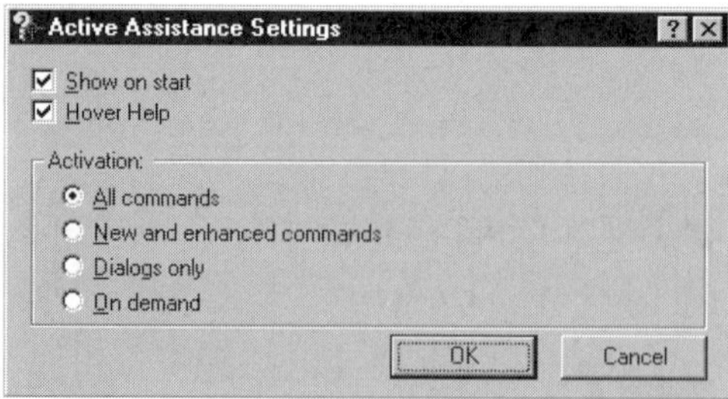

Figure 1.33 *The Active Assistance window automatically displays information about the active command.*

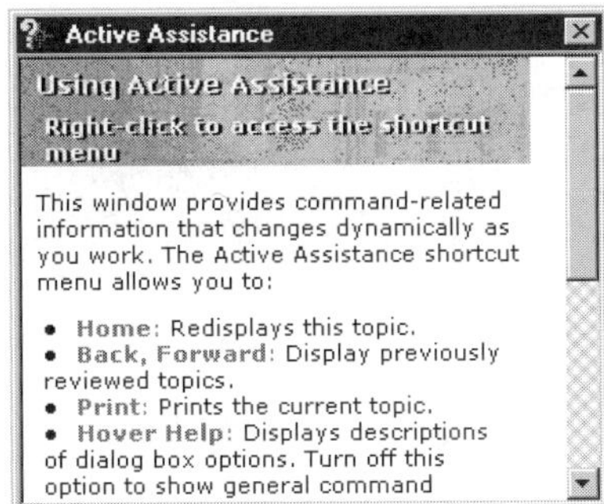

Figure 1.34 *The Active Assistance dialog box can be used to determine when and how the Assistance box will be displayed.*

Dialogs only—When this radio button is selected, the assistance window is displayed only when a dialog box is displayed.

Hover Help—This check box controls how assistance will be displayed when a dialog box is open. With the box active (the check mark), whenever the cursor is over an option in a dialog box, the Active Assistance window displays information relative to that option.

On demand—With this radio button selected, the Active Assistance window is displayed only when the Active Assistance button is clicked. Clicking the **Active Assistance** button on the **Standard** toolbar or typing **ASSIST** (ENTER) at the command line will produce the Active Assistance window.

USING THE HELP MENU

Additional help is available for all areas of AutoCAD by selecting the **Help** command. Access the **Help** command by pressing F1, by selecting **Help** from the menu bar, by clicking the **?** button on the **Standard** toolbar, or by typing ? (ENTER) or HELP (ENTER) at the Command prompt. Each method will produce the AutoCAD Help: User Documentation dialog box shown in Figure 1.35. The **Help** command is a transparent command, one that can be completed in the middle of another command sequence. Start a transparent command by typing an apostrophe before entering the command. To get help in the middle of the **Line** command, make the following entry:

Command: **L** ENTER *(Or click the Line button on the Draw Toolbar.)*

LINE Specify first point: *(Select desired starting point.)*

Specify next point or [undo]: **'HELP** ENTER

The F1 key can also be used in the middle of another command. When the **Help** command is accessed in the middle of a command, the **Help** command automatically provides command-related help. Asking for help in the middle of the **Line** command will produce the Help display shown in Figure 1.35, specific to the command.

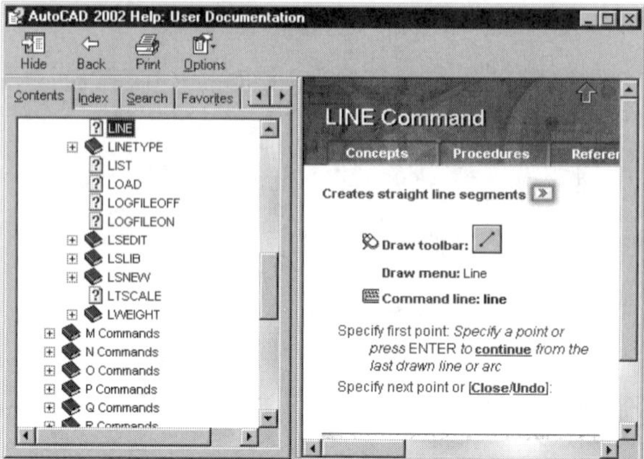

Figure 1.35 *Typing* **HELP** *(*ENTER*) at the command line, clicking the* **Help** *button on the* **Standard** *toolbar, or pressing F1 will produce the Help window.*

The left portion of the display is used for finding information in the Help window. The tabs at the top of the left screen provide the methods for using the Help window and the right side of the window is where the help text will be displayed. In the top left corner of the Help window is the **Hide** button. Selecting this button will hide the left half of the Help window. Once the desired material has been displayed, the window can be reduced. Selecting **Show** will restore the left portion of the Help window, allowing additional use of the Help window. Selecting the **Back** button will restore the previous help screen. The Help window offers help in the areas of **Contents**, **Index**, **Search**, **Favorites**, and **Ask Me**.

Contents

The **Contents** folder of the Help window can be used to obtain information about the Help program, or individual portions of the AutoCAD program. Notice that 10 listings are displayed in the contents menu, and each listing is preceded by a closed book. The **Contents** folder contains a How to Guide and a Users Guide to provide instruction on using each aspect of the program. The Command Reference portion of the folder provides information on each command used by AutoCAD. As you scroll through the listing of **Contents**, each listing will be highlighted. Clicking a highlighted menu item will provide a brief description of the selected item on the right side of the window. Double-clicking a menu item will provide a submenu of the contents. Use the following sequence to receive information about the **Line** command:

1. Display the Help menu by pressing F1 or by any one of the other methods introduced in this chapter.
2. Double-click the Command Reference listing; it provides a list of commands presented in alphabetical order. To find information on the **Line** command, select the L Commands listing from the menu display.

3. Select LINE from the listing of L Commands. This will produce the display shown in Figure 1.35, and provide a brief description of the **Line** command.

To obtain a detailed description of the command, click the **>>** (**Click for more** button). Figure 1.36 shows the LINE command description. Clicking the **<<** (**Click to hide** button) will return to the original menu. Click the **Close** button in the Help title bar to close the Help window and return to the drawing screen.

INDEX

Selecting the **Index** tab will produce a menu that is like the index found at the end of a book. Look up a key term and the book index will list several pages where information can be found. With the Help index, you don't even have to know the word you're looking for. If you enter the first few letters of a command or option, the Help window will display a list of commands and sources of information about that command. Entering **line** in the edit box will produce the display shown in Figure 1.37. This display box can be used to select information about the command. Double-clicking LINE command will produce a listing of related options. Double-clicking LINE [command Reference:ARC] will produce the help display for the **Line** command. You can also display the text by highlighting the desired option and pressing the **Display** button.

Search

The **Search** option of Help produces a search of the Help based on words entered in the search box. Entering the word **LINE** in the search box restricts the search to the LINE command, as opposed to other line options. The search field can be restricted with the three options at the bottom of the display. These include **Search previous results**, **Match similar words**, and **Search titles only**.

Favorites

The **Favorites** tab allows you to assemble a list of favorite topics for quick access. Place a help topic selected with the **Content**, **Index** or **Search** tab on the Favorites list by

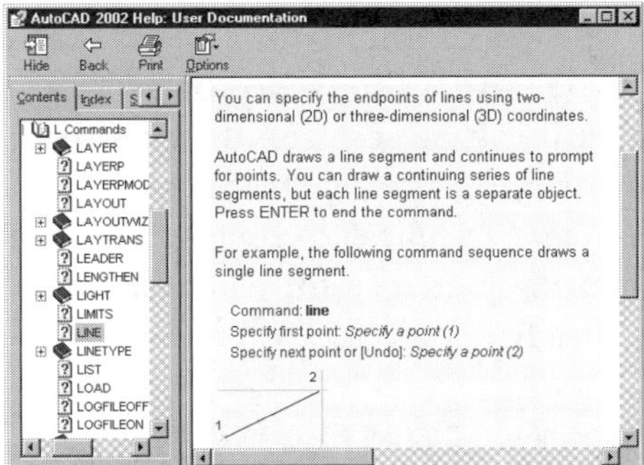

Figure 1.36 *Clicking the* **>>** *button from the LINE display produces a detailed description of the command.*

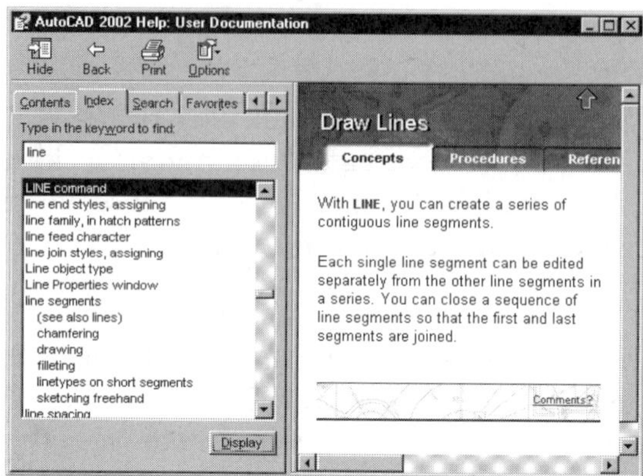

Figure 1.37 *Selecting the* **Index** *tab will produce a menu similar to an index at the end of a book. As commands are typed in the* **Keyword** *edit box, the display is altered to reflect the entry.*

picking the **Add** button at the bottom of the page. Remove topics from the list by highlighting the subject to be deleted and clicking the **Remove** button.

Ask Me

The **Ask Me** tab allows you to enter a question or a word in the **Topics** box. Once it is entered, select the search parameters from the **List of components to search** edit box. Enter **draw a line** in the top edit box, and select the Users Guide in the **Components to search** edit box. The display will be altered to list common solutions to the specified problem. For this search, Drawing Linear Objects is the first recommended resource. Double-clicking this option produces a display of drawing options related to **Draw**.

AUTOCAD LEARNING ASSISTANCE

In addition to the Help menu, help can be obtained through the AutoCAD Learning Assistance tutorial, and the AutoCAD Support Assistant. AutoCAD provides additional help through a tutorial-based learning tool named the AutoCAD 2000/2000i Learning Assistance (ALA). The ALA is provided on a CD separate from AutoCAD and must be installed before it can be accessed through the Help menu The ALA includes lessons for learning many of the new features of AutoCAD. The AutoCAD Support Assistant (ASA) contains technical documentation, solutions to common support issues and trouble shooting guidance for common AutoCAD error messages. AutoCAD users can periodically update the ASA through Web access.

EXPLORING THE SAVE COMMAND

Throughout this chapter you've explored several methods of controlling commands. The **Line** command was used as methods of using toolbars, menus, and keyboard entry

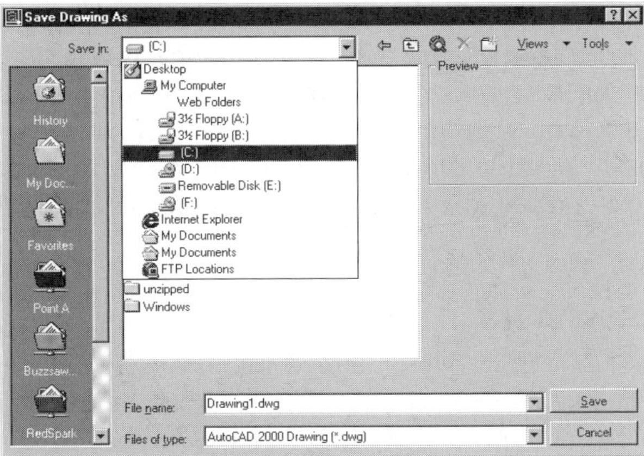

Figure 1.38 *Display the* **Save Drawing As** *dialog box by selecting* **Save** *from the* **File** *menu.*

for command entry were explored. The lines you've created thus far will be destroyed if you end your drawing session without saving them. Chapter 4 will explore how and where files can be saved. This chapter will introduce the procedure for saving a drawing file.

If you have not already done so, enter the drawing area and draw several lines. Because this is a chapter for exploring, use the **Draw** toolbar to draw a circle and a rectangle. Have you now created a work of art that must be saved for future generations to enjoy? Probably not, but it will create a file that can be used in Chapter 2 as you explore methods for opening an existing drawing. Use the following procedure to save to the hard drive:

1. Select **Save** from the **File** menu. This will display the **Save Drawings As** dialog box shown in Figure 1.38.
2. Select the arrow in the **Save in** box. This display is similar to display in Windows Explorer and allows the storage destination for the file to be selected. Select the C: drive for storage on a hard drive or A: for storage on a diskette. Only save to a diskette at the end of a drawing session.
3. The name of Drawing1 is currently displayed. To alter the name, click in the edit box, move the cursor so that Drawing 1 is highlighted (in a blue box.), and type the desired name, in this case, **EXE1-1**.
4. Select the type of file to be used if the drawing is to be saved to be compatible with older releases of AutoCAD.
5. Click the **Save** button. The dialog box will be removed from the screen, the drawing will be saved, and the drawing area will be restored.

ENDING A DRAWING SESSION

You can continue to add objects to this drawing for as long as you so desire. When you want to end the drawing session, save the drawing again using by selecting the Save button to update the file on your hard drive. Chapter 4 will introduce other options for saving drawings. To exit the drawing session, select **Exit** from the **File** menu or select the **Close** button on the title bar. For now, continue in AutoCAD and explore the Help window.

WHAT IT ALL MEANS

By now, you might be feeling overwhelmed by all of the bells and whistles of AutoCAD. In this chapter you've been introduced to several methods of entering a command, and explored a multitude of buttons, boxes, and controls for altering the commands and options. Each has been introduced with the **Line** command. By the time you explore each of these controls in relations to the hundreds of commands, options, and buttons, your brain may be on overload. The good news is that you don't have to memorize every command, option, and control method. You only have to remember your favorites and the ones you use on a regular basis. Use the Active Assistance window, the Help window, and this book to help keep track of options that you don't use on a regular basis. Another key point of this chapter is that there are usually several ways to do one thing. Generally, one method is as good as another, but don't limit yourself to the first method you find. The key to becoming a successful AutoCAD operator is to explore. The following quiz will require you to do just that. Many of the questions will deal with information not introduced in this chapter. This is your first real opportunity to explore AutoCAD.

CHAPTER 1 QUIZ

1. List three sources for obtaining help with AutoCAD.
2. Select the **My Computer** button from the startup display. Select the **Printer** button and determine which printers or plotters are configured to the current workstation.
3. Explain the difference between a click and a double click.
4. List the four screen areas displayed after entering AutoCAD.
5. Is a click or double click required to start AutoCAD using the **Programs** menu?
6. Describe key functions of the left and right mouse buttons.
7. List three different methods that can be used for selecting commands and tell how they are accessed.
8. You've displayed a dialog box with 40 listings. Describe two methods that can be used to move through the list to the end.
9. List three areas for obtaining help with AutoCAD.

10. Explain the procedure to float the **Draw** toolbar.
11. Explain the following terms:

    Default

    Pick

    Command
12. List in order the first eight commands on the **Draw** menu.
13. You have just finished drawing a line and would like to draw an arc. Explain how to get the arc options using the menu.
14. What is the last **Arc** option listed in the menu?
15. List the options in the **File** menu.
16. You've displayed the **Insert** menu, but you wanted the **View** menu. How can you get **View** to be displayed?
17. List the tabs that are seen on the status line.
18. Is **Ellipse**, on the **Draw** menu, a command or an option? How can you be sure?
19. A button showing a paintbrush is on the **Standard** toolbar. What is the button, and what does it do?
20. List the four toolbars shown as AutoCAD is loaded.
21. What key will execute the following request?

    Display grid

    ORTHO

    Snap mode
22. Someone has asked you what the last three commands were that you used on your drawing. How could you get a record if you don't remember?
23. List five items that can be found in a dialog box.
24. Search through each menu and find the location of these items:

    Erase

    Tablet

    Toolbars
25. What letter key would be needed when using the control key (CTRL) to complete the following options?

    Grid toggle

    Save

    Backspace

    Coordinate display
26. List and describe the five shortcut menus of AutoCAD.

27. How many sides will be drawn if the default value is selected for a polygon?
28. The **Draw** toolbar has been removed from the screen. Explain the procedure to display the **Draw** toolbar.
29. How many different types of arcs are listed in the **Draw** menu?
30. You've discovered, after terminating the command, that you actually wanted to draw one more line. What is the quickest way to reenter the **Line** command?

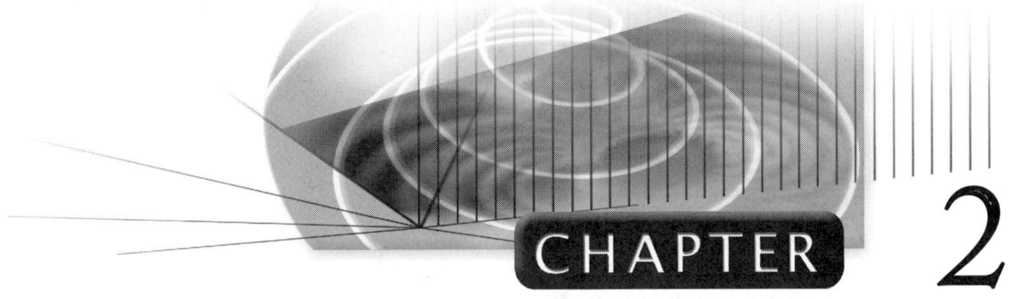

CHAPTER 2

Creating Drawing Aids

This chapter will introduce methods for

- Entering the Drawing Environment, including the TODAY and traditional Startup window
- Starting a new drawing from within AutoCAD, including Wizards, Templates, and Start From Scratch
- Controlling drawing parameters from within a drawing

Volo View Express will also be briefly examined.

Commands to be introduced include

- **New**
- **Open**
- **Partialopen**
- **Partiaload**
- **Units**
- **Limits**
- **Zoom**
- **Grid**
- **Snap**
- **Pan**

In Chapter 1 you opened AutoCAD and were exposed to the TODAY window and the traditional **Startup** dialog box. To quickly start drawing, you closed the TODAY and Active Assistance windows and went directly to the drawing screen. It is the equivalent to taking a piece of vellum with no other tools and starting to draw. This system worked well in kindergarten but is not suitable for the precision required on mechanical drawings. This chapter will introduce methods of starting a drawing session as well as explore methods for controlling the drawing environment.

ENTERING THE DRAWING ENVIRONMENT

The TODAY window and the traditional **Startup** dialog box offer four alternatives to help set up a drawing file. Each can be seen in Figure 2.1. The decision as to which to use will depend on how you usually manage a drawing session and on personal preference. Both the TODAY window and the **Startup** dialog box will get you to the same point of a drawing session. AutoCAD offers two initial options for opening a drawing session using the TODAY window. These include the **Open Drawings** and **Create Drawings** tabs. If the **Create Drawings** tab is selected, options for starting a drawing session are given for **Template**, **Start from Scratch**, and **Wizards**. If the traditional **Startup** dialog box is used, options include **Open Drawing**, **Start from Scratch**, **Use a Template**, and **Use a Wizard** to organize drawings. Each startup method has advantages for drawing setup.

Open Drawing, (**Open Drawings**, tab)—Opens a drawing file created in a previous drawing session and allows editing or additions to be made to the existing drawing.

Wizards—Starts a drawing session when you know the scale or the paper size that will be needed to plot the project. This will allow you to use the setup assistance from AutoCAD. This is an excellent option to use for the design stage of a drawing. Drawing wizards offer helpful prompts to guide in drawing setup but do not provide title blocks. Later chapters will help you provide a title block to drawings created without a template. **Wizards** offers the **Quick Setup** and **Advanced Setup** options for determining how the new drawing environment will be established. Each option will be explored in the next section of this chapter.

Template—Allows predetermined drawing parameters to be used to control the drawing environment. This is an excellent design tool to use when you have planned the drawing requirements and can choose the space required to display the needed views. As you gain an understanding of the options that are available, you can create your own templates. These templates can include the company logo or similar items that can be created and saved for use on each drawing. Once you've created and stored your own template, the drawing parameters will be preset to meet your standards each time you start a new drawing. Chapter 4 will help you create your own template.

Start from Scratch—Starts a drawing with minimal drawing controls. This option is the equivalent of grabbing a blank sheet of paper with no title block. It is best suited for users who like to plan as they go. Use **Start from Scratch** for sketching, line drawings that will not be drawn to scale, or for

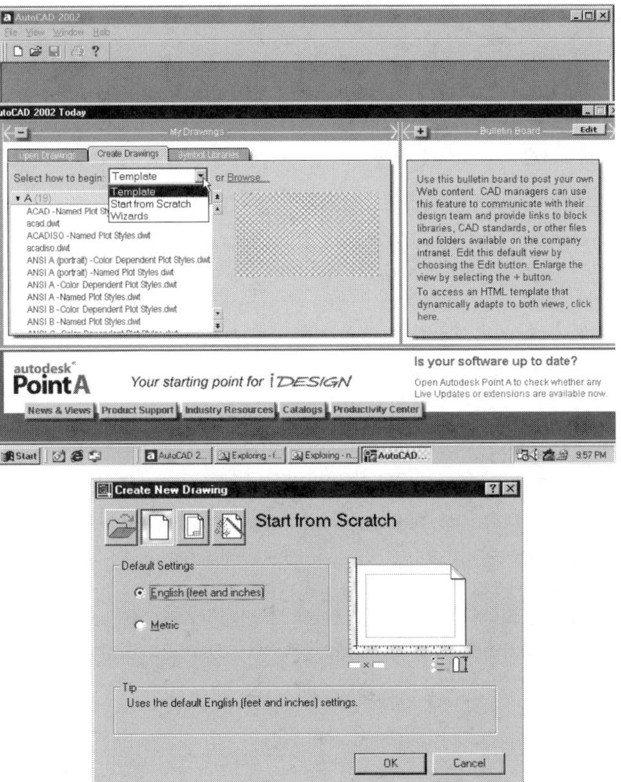

Figure 2.1 *The TODAY window or the traditional* **Startup** *dialog box will be displayed as AutoCAD is entered to provide help in setting drawing parameters.*

drawings that do not require a high level of accuracy. This option also works well when you want to establish the drawing's parameters as they are needed. Although this option provides some basic drawing controls, you will need to add many other controls and variables to increase your drawing efficiency.

Note: If the **Startup** dialog box is not displayed as you enter AutoCAD, someone has hidden it from view. When it is hidden, you will automatically enter AutoCAD using the method that was last used to access the drawing area. You display the **Startup** dialog box or hide it from view by selecting **Options** from the **Tools** menu, and then selecting the **System** tab of the **Options** dialog box. Use the **Show Startup Dialog** button in the **General Options** box to control the display of the startup menu. This can be a setting that you adjust and save as you create your own profile in Chapter 4.

OPEN AN EXISTING DRAWING

Much of your time in an office will be spent adding to or editing existing drawings. The **Open Drawing** option of the TODAY window is slightly different from the traditional method. In the default setting shown in Figure 2.2a, the names of drawings created in the last week are displayed. Moving the cursor over a file name will display the contents of the file in the image display box and list the full file name and the date and time of the last drawing session. The display of drawings to be opened can be altered using the **Select how to begin** list box. Display options include the following:

> **Most Recently Used**—Provides a list of the drawings most recently used. Because you've only worked on drawing EXE 1-1, your list is restricted to this single listing. As you work on more drawings, this list will be expanded to include up to nine drawing names. More names can be displayed by using the slide bar on the right side of the listing.
>
> **History (by Date)**—Displays a list of drawings for a specified date. The list is based on the most recent week. As the size of the list of drawings increases, the list is expanded to reflect drawings created in the last four weeks.
>
> **History (by Filename)**—Displays a list of drawings based on the drawing file name and the number of drawings in that file. A listing of numbers and letters is displayed representing the starting number or letter of current file names. Clicking on a number or letter will display the names of files that have recently been used. Double-clicking a letter or number will close the listing.

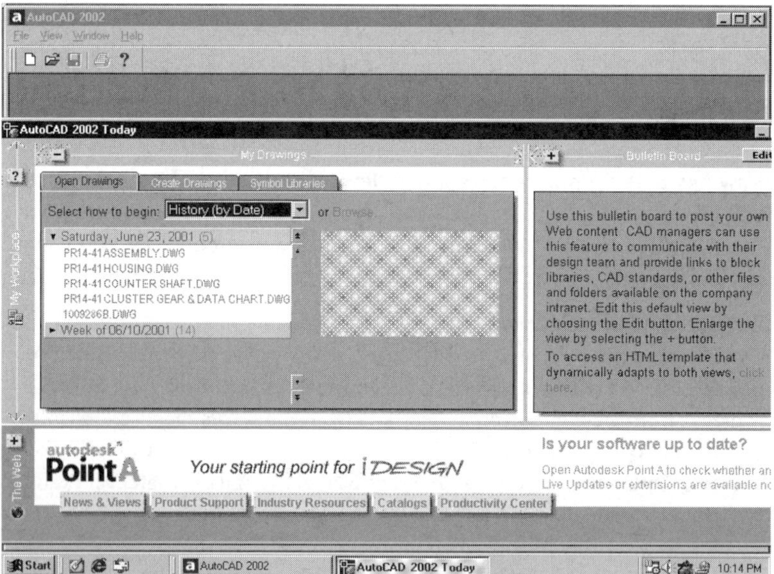

Figure 2.2a *Selecting* **Open Drawings** *in the TODAY window starts a search of the most recently used drawings. The search for existing drawings can also be defined by the file name or location.*

History (by Location)—Displays a list of drawings based on the drawing storage location. Letters are displayed to represent each of the drives and folders that contain drawings that have recently been used.

Browse—Displays a list of drawings similar to Windows Explorer using the location of the last drawing that was opened. Altering the location to be examined in the **Look in** list box will allow other folders and drives to be examined.

Selecting a file name with a single click will automatically open the file, placing the drawing in the drawing area.

If the traditional **Startup** dialog box is used, this option will display a startup menu similar to Figure 2.2b. The **Select a File** list box will display a list of the last four, and up to nine, drawings that have been opened. In your dialog box, select the file created in Chapter 1, titled EXE1-1. Clicking the **OK** button will remove the dialog box and open the drawing that was created in Chapter 1. As you develop more drawings, the drawing file to be opened might not be displayed in the **Select a File** list box. Select the **Browse** button to find an existing drawing file that is not listed in the **Startup** dialog box. Selecting **Browse** will produce a **Select File** dialog box similar to Figure 2.3. The display will vary depending on the number of drives and the file contents.

The **Select File** dialog box can be used to list and select a drawing file from any of your storage locations. By default, files will be selected from the folder that was last

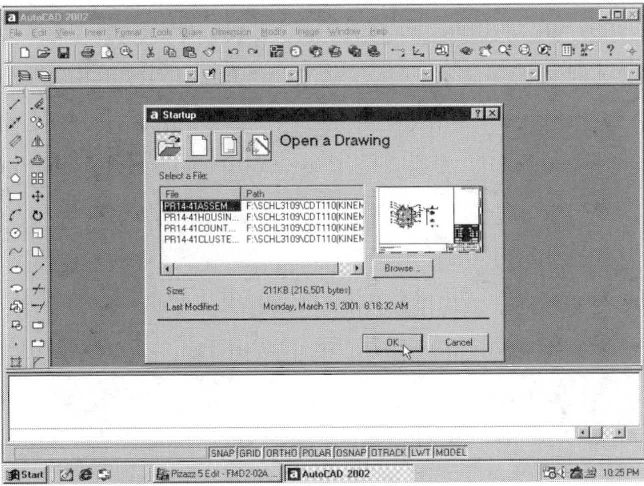

Figure 2.2b *Selecting the* **Open an Existing File** *button in the traditional Startup dialog box will display this dialog box. The last four drawings that you have worked on will be displayed in the* **Select a File** *edit box. Click the* **Browse** *button to search for files not listed. To open an existing drawing, select the name of the desired drawings and click the* **OK** *button.*

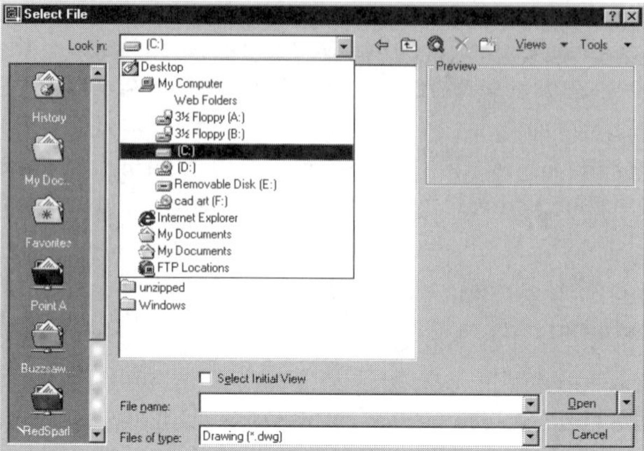

Figure 2.3 *The* **Look in** *list is used in a manner similar to using Windows Explorer.*

accessed. Use the following steps to open an existing file that is not listed in the **Select a File** listing:

1. Select the arrow beside the **Look in** box to select the desired drive or folder that contains the drawing file to be opened. As the arrow is selected, a drop-down list is displayed, reflecting the contents of Windows Explorer. (See Figure 2.4.)
2. Move to the desired drive just as you would using Windows Explorer and select the file to be opened.

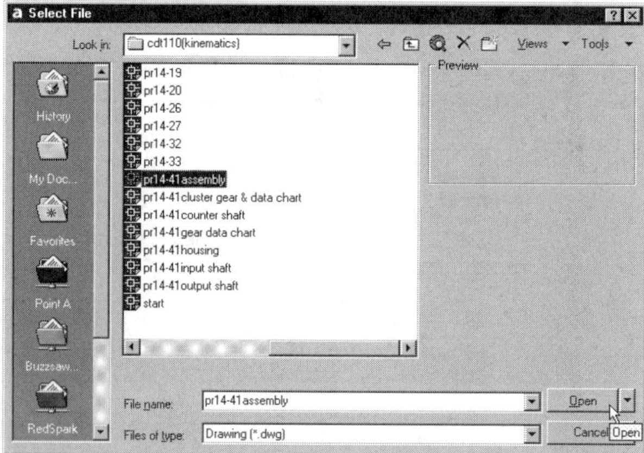

Figure 2.4 *Selecting the* **Browse** *button will produce a listing of available drives. Once in the desired drive, select the desired file to be opened. An image of the selected file will be displayed in the* **Preview** *box. If this is the file you want to open, select the* **Open** *button. If this is not the file you expected, select another file name. Once you've found the desired file, select the* **Open** *button.*

- Click the desired drive to be searched to display a list of the drawing folders contained in the selected drive.
- Double-click a folder to display a list of files contained in that folder. Highlighting a file with a single click will display an image of the file in the **Preview** box.

3. If this is the file you want to open, click the **Open** button to remove the dialog box and open the selected drawing on the drawing area, similar to Figure 2.5. If this is not the file you expected, highlight another filename. Once you've found the desired file, select the **Open** button to bring the file into the drawing area. You can also open a drawing by double-clicking the file name.

Note: Remember AutoCAD allows multiple drawings to be opened. An existing drawing can also be opened in the middle of another drawing session by selecting the **Open** button on the **Standard** toolbar, by selecting **Open** from the **File** menu, or by typing **OPEN** (ENTER) at the Command prompt. Each method will produce the dialog box shown in Figure 2.3.

CREATING NEW DRAWINGS

The **Create Drawings** tab of the TODAY window and the remaining three buttons of the traditional startup window offer three major options for creating a new drawing. These options include **Wizards**, **Start from Scratch**, and **Template**.

DRAWING SETUP USING A WIZARD

Using a wizard is an excellent method for starting a new drawing session if you know the size of paper the project will be plotted on. AutoCAD provides the **Quick**

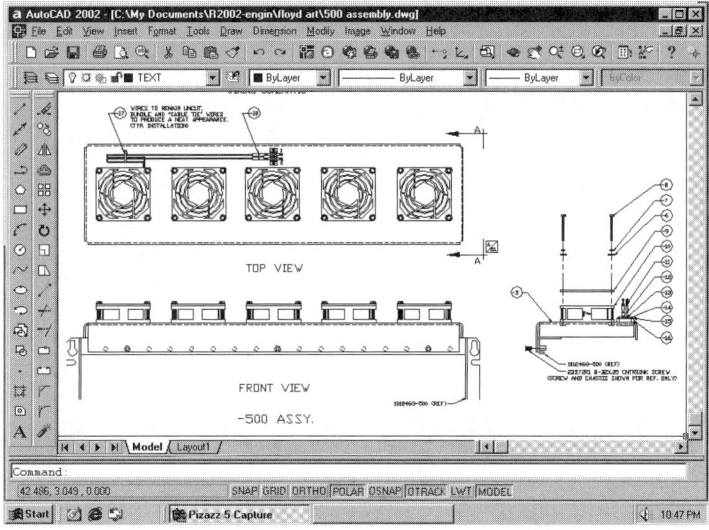

Figure 2.5 *The result of selecting the file in Figure 2.4. Courtesy Floyd D. Miller.*

Startup wizard and the **Advanced Setup** wizard. The method chosen will depend on how much help you would like. Choosing **Quick Startup** will assist you in setting the drawing units and the drawing area. The **Advanced Setup** wizard will provide six start-up prompts.

Quick Setup

The **Quick Setup** option produces a drawing area in model space. This option works well for drawing, but it requires some adjustment for plotting. Selecting **Quick Setup**, as shown in Figure 2.6, allows the selection of the drawing units and the drawing area.

Setting the Drawing Units

One of the unique features of AutoCAD is that as you work in model space you will be drawing at full scale. The drawing will not be scaled until you change to paper space to plot your drawing. If you want to draw an airplane wing that is 50' long, you will draw it full size, and then choose a scale as you plot, using AutoCAD's paper space option. (This will be covered in later chapters.) Units allow you to set the measuring method to be used to establish the drawing. In its default settings, your computer is set to measure lines in four-place decimal inch units. You can confirm this by looking at the coordinate display in the lower left corner of the status bar. To describe the current location of the crosshairs, two four-place decimal numbers are used. The default setting of AutoCAD is to provide measurements in four-place decimals. Mechanical drawings units are governed by the American Society of Mechanical Engineers (ASME). ASME Y14.5M–1994 is the current standard for controlling drawing

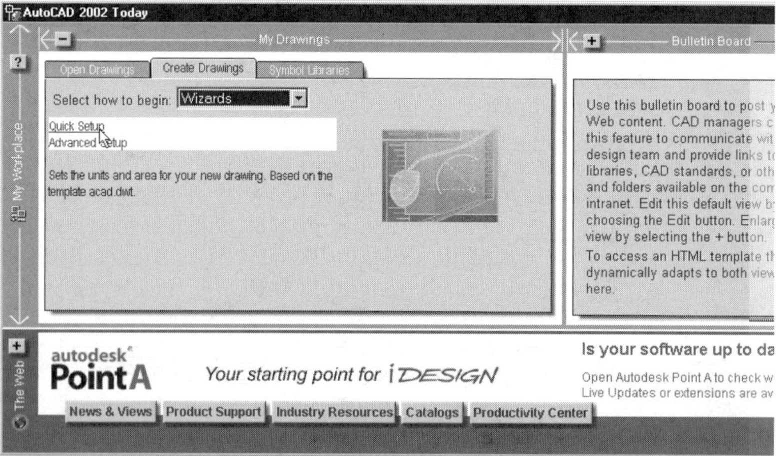

Figure 2.6 *Selecting* **Wizards** *from the* **Create Drawings** *tab of the TODAY window or selecting the* **Use a Wizard** *button in the traditional* **Startup** *dialog box provides two choices for drawing setup.*

units and recommends two-place decimals be used. Values greater than two places should only be used when the use of the part requires greater accuracy. Options for unit settings and their display include the following:

> **Decimal – 15.0000**—This option can be used to create drawings in decimal or metric units. Both options can be found on mechanical drawings, depending on client needs. Decimal dimensions should be expressed as two-place decimals unless the project requires a higher degree of accuracy. Future chapters will introduce other ASME dimensioning standards.
>
> **Engineering – 1'–3.5000"**—These units are best suited to civil drawings such as site plans, topography drawings, bridge and dam construction or other drawings dealing with large land areas. Engineering units are not appropriate for mechanical drawings.
>
> **Architectural – 1'–3 1/2"**—Architectural units express distance in feet and inches and are used for drawing structures for architectural and structural clients. Architectural units are not appropriate for mechanical drawings.
>
> **Fractional – 15 1/2**—This option is not used on mechanical drawings. This option can be used on drawings that express measurements in fractions such as cabinet drawings but are not appropriate for mechanical drawings.
>
> **Scientific – 1.5500E+01**—This option is not used on mechanical drawings but is suitable for drawings used in chemical engineering. This unit of measurement is suitable when very large or very small measurements must be expressed on a drawing. Numbers are expressed using a base number and a multiplier. The E+01 represents a base number multiplied by 10 to the first power.

The units can be set by placing the cursor in the desired setting and pressing the select button. This will place a black dot in the button, indicating the selection is activated. For now, leave **Units** as four-place decimal. Methods of altering the accuracy will be introduced later in this chapter.

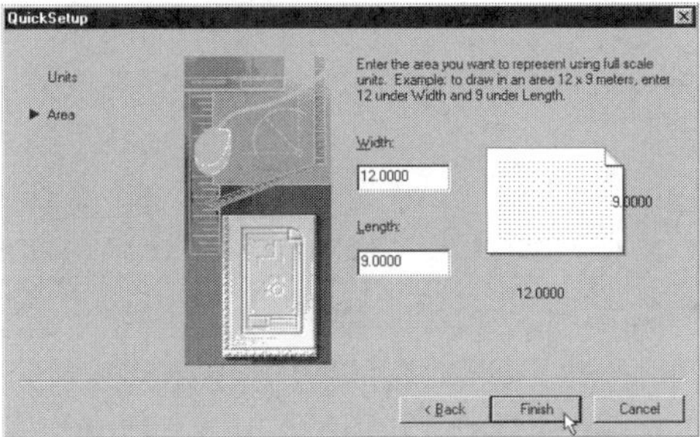

Figure 2.7 *Most drawings can be drawn using Decimal or Engineering units. Setting the area defines how much space is displayed on the screen. Set the area slightly larger than what you think you will need to complete the drawing.*

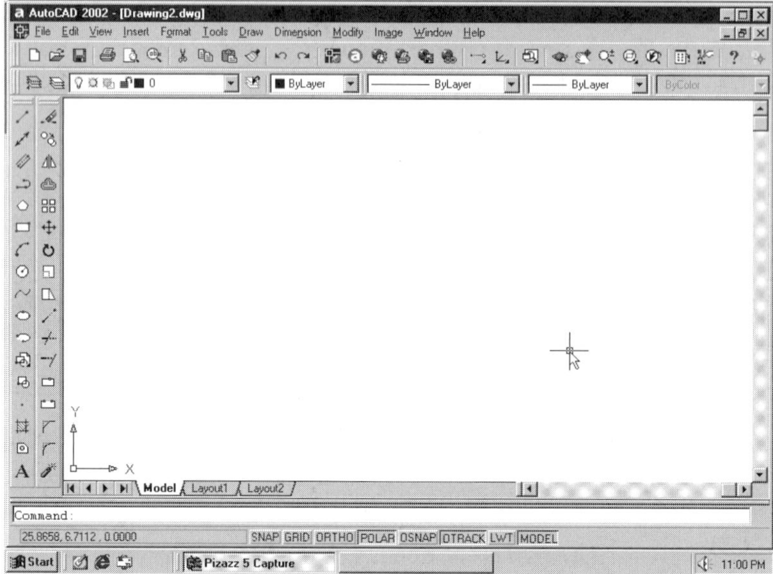

Figure 2.8 *Accepting the default value of 12" x 9" and selecting the* **Finish** *button will display the drawing screen. Display the grid in the drawing area by selecting the GRID button on the status bar.*

Setting the Drawing Area

Select the **Next>** button to continue the drawing setup. This will produce the display shown in Figure 2.7, showing a default value of 12" × 9". The drawing area should be based on what you expect to draw. For now, accept the default value and select the **Finish** button. This will display the AutoCAD drawing screen shown in Figure 2.8.

Advanced Setup

Selecting this option will display the dialog box shown in Figure 2.9. The **Advanced Setup** wizard will help you set six controls related to the units of measurement, angle of measurement, direction of angle, and area. Select the **Next>** button to start the wizard.

Units

The **Units** control folder shown in Figure 2.9 is displayed as the **Advanced Setup** is opened. The default value is decimal. Choose the value that best meets the need of the drawing to be started using the methods described using the **Quick Setup** Wizard.

Unit Precision

The **Precision** edit box is located at the bottom of the **Units** dialog box. The precision is now set at four-place decimal units. To change the setting, select the arrow on the right side of the list box. This displays the **Precision** list shown in Figure 2.10. Selecting the down arrow will scroll through the list, allowing the degree of accura-

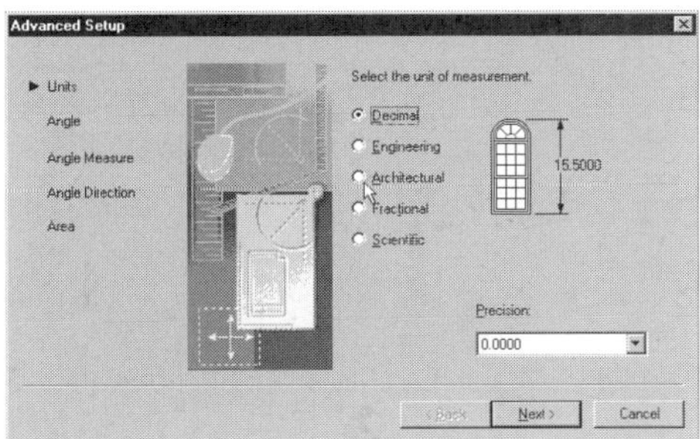

Figure 2.9 *Selecting the* **Advanced Setup** *wizard option will aid in setting six drawing aids. Notice that unlike the* **Quick Setup** *wizard, this method allows the precision of the units to be measured.*

cy to be selected. Settings range from whole units to eight decimal places. For now, use the up arrow to scroll through the menu and select the 0.00 setting by highlighting it with the cursor. This will return you to the **Units** box and allow you to continue the drawing setup. Notice, as the unit precision was changed from 0.0000 to 0.00, the display of units was also changed. The units will now display 15.50. If you're satisfied with the units, select the **Next>** button to continue the drawing setup. The units can be changed at any point during the life of the drawing. Clicking the **Cancel** button will exit the wizard, discard any changes to the unit settings, and place the drawing area on the screen.

Note: You might consider 1.0 and 1.00000 as equal. In the manufacturing world, adding zeros to the right side of the decimal adds to the precision required to manufacture the part. As the precision is increased, so is the cost. Unless specified by your supervisor, two place decimals are adequate for most parts.

Angles

Once the **Units** are selected, the **Angle** display for controlling angles shown in Figure 2.11 is displayed. Most mechanical drawings can be completed easily using decimal angles. Options for measuring angles include the following:

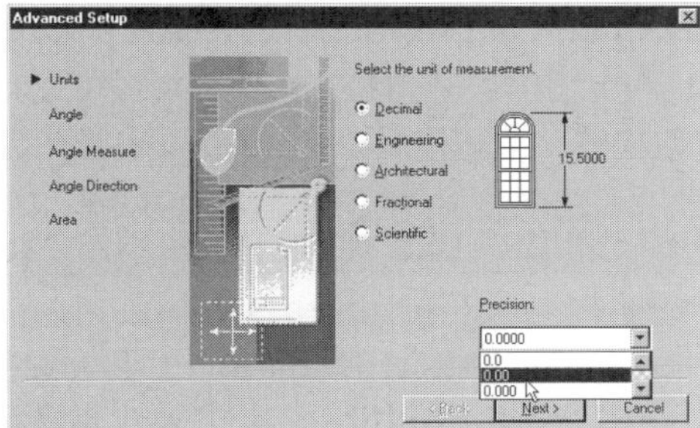

Figure 2.10 *Selecting the* **Precision** *arrow button displays the* **Precision** *list. Settings range from whole units to eight decimal places. Selecting 0.0 will measure units in whole inches.*

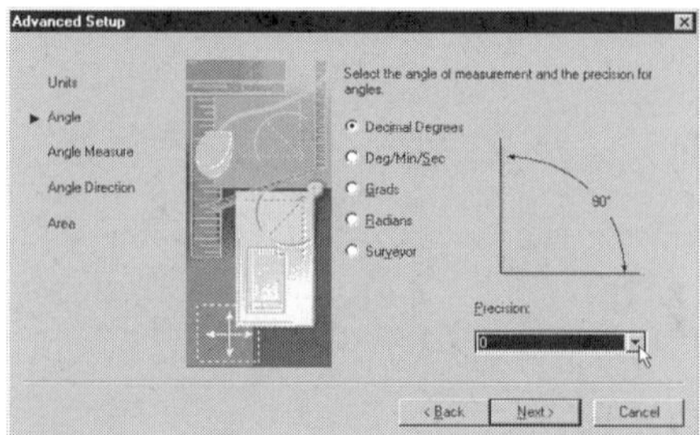

Figure 2.11 *The* **Advanced Setup** *menu allows the units for measuring angles and the precision to be selected.*

The Decimal Degrees – 90.00—This option is best suited for mechanical drawings. It will display angles in degrees and decimal parts of a degree.

Deg/Min/Sec – 90d00'00"—This option is best suited for land measurements, such as site plans. This option will display angles in degrees, minutes, and seconds.

Grads – 100.00g—Grads is the abbreviation for a gradient. One hundred grads equal one-quarter of a circle. This option of angle measurement is rarely used with mechanical drawings.

Radians – 1.57r—A radian is an angular unit measurement where 2p = 360° and a 90° is p/2 radians. This option of angle measurement should not be used with mechanical drawings.

Surveyor – N0d00'00"E—This option presents surveyor's units, which are used to represent land, street, or sewer layouts, with angles displayed as bearings displayed as <N/S> <angle> <E/W>. The angle is based on north or south and will always be less than 90°. This option of angle measurement should not be used with mechanical drawings.

The method of angle measurement can be set throughout the life of the drawing with the methods that will presented at the end of this chapter. The method of angle measurement and precision is set using similar methods that were used to set the drawing units.

For now, accept the default option of Decimal Degrees. The precision for measuring angles can be set using the same method used to set the unit precision. Selecting the down arrow will provide options for zero to eight decimal places, with the default of 0. Normally, two-place accuracy is sufficient for mechanical drawings. Accept the default by clicking the **Next>** button to proceed to the next page of the wizard.

Angle Measurement

The **Advanced Setup** wizard allows the starting point of angle measurement to be selected. The default option is east. Options are shown in Figure 2.12. Standard AutoCAD layout of angles places 0° at the three o'clock position and then moves counterclockwise. Figure 2.13 shows a drawing with angular surfaces.

You can specify that angle measurement start in any location. To change the starting point of angles to north, select the **North** radio button. If you select north as the new starting direction, 0° will now be placed at north, 90° will be west, 180° will be south, and east will be at 270°. An angle other than one of the four compass points can be selected by choosing the **Other** button, and then entering the desired starting angle in the edit box. Selecting **Other** and entering a starting angle of 45° will place 0° halfway between north and east. When the desired starting point has been selected, clicking **Next>** will allow the wizard to continue.

Angle Direction

Figure 2.14 shows the **Advanced Setup** wizard for selecting the directions that angles will be measured. The default setting for angle measurement is counterclockwise. Selecting the **Clockwise** button will change the direction of angle measurement. For now, accept the default. Remember that the defaults are there because they fit the needs of most users. The direction, like each of the other variables, can be altered at any time throughout the life of the drawing. Once the direction is selected, click the **Next>** button to proceed.

Area

The final setting of the **Advanced Setup** wizard allows the drawing area to be specified using the display shown in Figure 2.15. In the default setting, the drawing will be 12" × 9". Enter the values for the width and length of the area of the intended drawing to be completed in the appropriate edit box. Enter the desired value, and click the **Finish** button. Clicking the **Finish** button closes the wizard, establishes each of the parameters that were entered, and returns you to the drawing area.

Note: New CAD users are often fearful of making a wrong decision while using a wizard. With each setting it was stated that the setting could be adjusted throughout the life of the drawing. Don't make light of that fact. There are no wrong settings, only inconvenient settings. If the current setting is troublesome, change it, don't fight it.

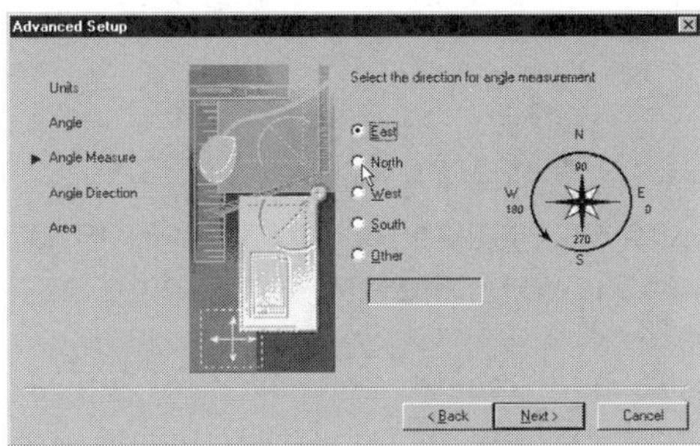

Figure 2.12 *The default starting point for measuring angles is east. You can select one of the other three quadrants, or any point in between, by choosing* **Other** *and providing an angle.*

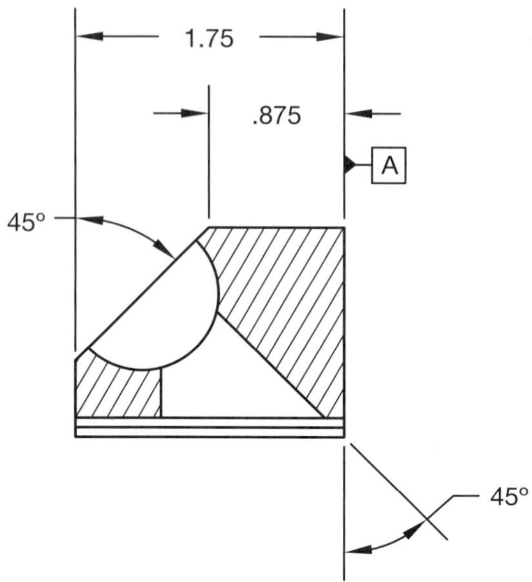

Figure 2.13 *Angular features are a common feature of mechanical drawings. Courtesy Jill Heilman.*

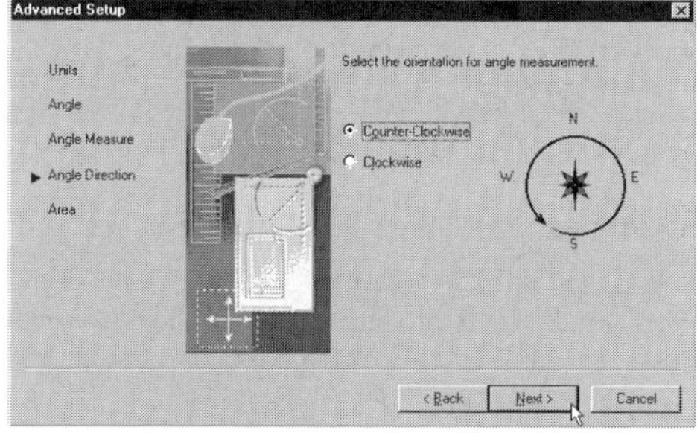

Figure 2.14 *The* **Advanced Setup** *wizard allows the direction for angle measurement to be entered. For most drawings, the default setting of counterclockwise works well.*

DRAWING SETUP USING A TEMPLATE

Drawing templates with predetermined drawing values can be used to start a drawing. Each template contains values for drawing units, limits, and other values that can be altered throughout the life of a drawing. A template supplied by AutoCAD can be customized to meet specific requirements and saved for future use. Select **Template** from the **Create Drawing** tab of TODAY to use a template to start a drawing session. Selecting **Template** will produce the display shown in Figure 2.16. The templates that start with A are listed. Scroll through the list using the arrows on the side. Double-clicking the A will display a listing of the letters representing the names of various templates contained by AutoCAD. Letters and their contents include the following:

A—ACAD templates

D—Din templates

G—Generic templates

I—ISO templates

J—JIS templates

M—Metric layout templates

Beside each category designation is the number of templates in parentheses for that category. Click on the category to display a listing of templates in that category. Moving the mouse over a listing displays an image of the selected template. Clicking the **Browse** button will display the **Select File** dialog box (see Figure 2.17). This box contains a listing of the templates contained in the default AutoCAD Drawing Template Folder. This dialog box can be used to search for templates stored in other folders, other drives, other computers on the network, or on the Internet.

Select the **Template** tab from the startup menu to display the menu shown in Figure 2.18. Accept the default setting of ACAD-Named Plot Styles.dwt and clicking the **OK** button will produce a display similar to Figure 2.8, with units measured in four-place decimal inches. Scroll through the list of templates and select the ANSI C-Color Dependent Plot Styles.dwt and click the **OK** button. This will produce a drawing display as shown in Figure 2.19.

Selecting a Template

The template options are listed in groupings indicating the drafting standard used, the size of the title block in the preset layout, and the plot style. Listed drafting standards include ACAD, ANSI, Architectural, Din (based on the German Institute of Standardization), Generic, ISO (International Organization for Standardization), and JIS (Japanese Industry Standard). ANSI, and generic templates provided with

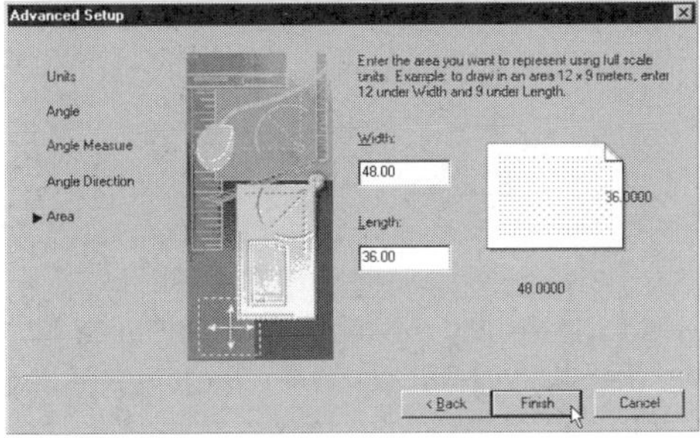

Figure 2.15 *Entering drawing units of 48,36 will allow most parts to be started.*

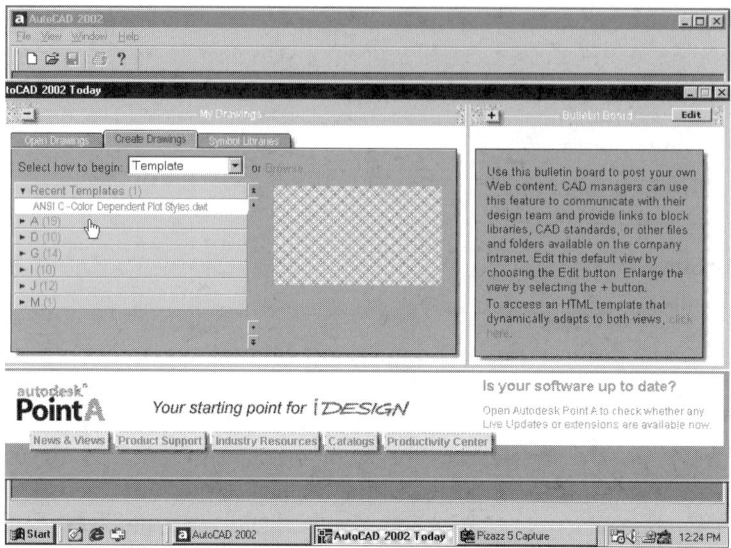

Figure 2.16 *Selecting* **Template** *from the* **Create Drawing** *tab of TODAY will produce the display that starts with A. Scroll through the list using the arrows on the side. Double-clicking the* **A** *will display a listing.*

AutoCAD are based on decimal units of measurement. ACAD templates are based on decimal measurements, but do not contain a title block. Din, ISO and JIS templates are based on metric measurements. Architectural templates are based on feet and inch measurements.

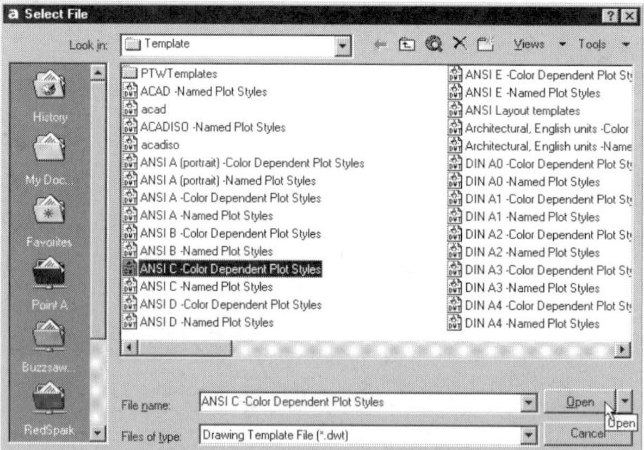

Figure 2.17 *Selecting* **Browse** *will display the* **Select File** *dialog box containing a listing of the templates contained in the default AutoCAD Drawing Template folder. This dialog box can be used to search for templates stored in other folders, other drives, other computers on the network, or on the Internet.*

Choosing a Sheet Size

The sheet size that is referenced in the template description is the size of paper that the drawing will eventually be plotted on. Although plotting will not be discussed until Chapters 22 and 23, the size of paper to be used should be considered as the template is selected. Decimal inch paper sizes for mechanical projects are governed by ASME/ANSI Y14.1-1995. Standard flat paper sizes include

- **A**—8.5 × 11 (horizontal format)
- **A**—11 × 8.5 (vertical format)
- **B**—11 × 17
- **C**—17 × 22
- **D**—22 × 34
- **E**—34 × 44
- **F**—28 × 40

Roll sizes of paper are also listed by ASME/ANSI. The letter and vertical heights of standard sizes include G (11), H (28), J (34), and K (40). Notice that only templates for ANSI sizes A through E are provided in AutoCAD.

Metric paper sizes for mechanical projects are governed by ASME/ANSI Y14.1M-1995. Basic flat paper sizes in millimeters include

- **A0**—841 × 1189
- **A1**—594 × 841
- **A2**—420 × 594

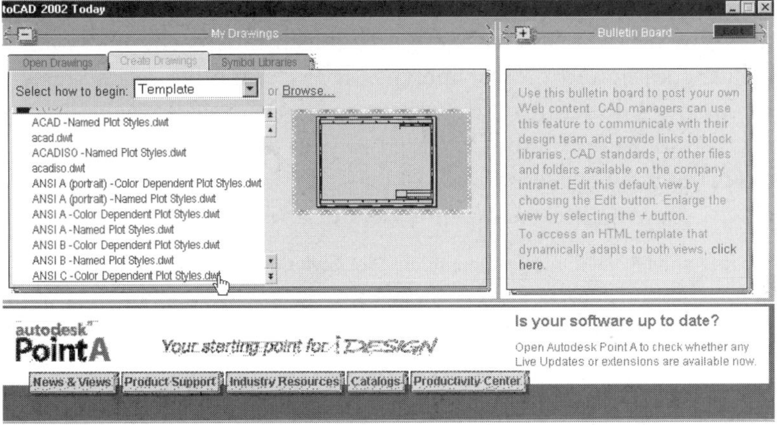

Figure 2.18 *Selecting the* **Use a Template** *button allows a drawing with preset drawing parameters to be used. Selecting* **Browse** *allows user-created templates to be accessed. Scroll through the list of templates and highlight the ANSI C-Color Dependent Plot Styles.dwt.*

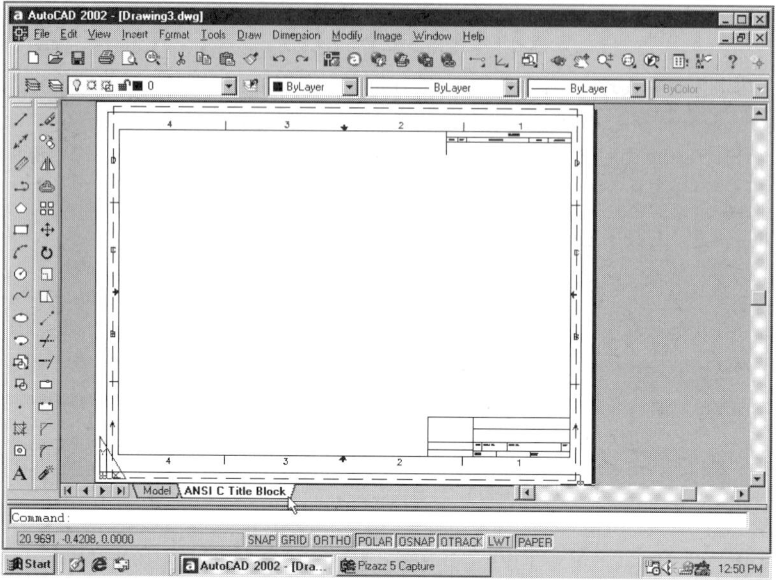

Figure 2.19 *The ANSI C-Color Dependent plot styles.dwt template is suitable for basic mechanical drawing projects. Chapter 4 will introduce options for creating and saving your own title block.*

A3—297 × 420

A4—210 × 297

Metric sizes listed by ASME/ANSI also include elongated and extra-elongated. In addition to using the templates supplied with AutoCAD, you can develop your own template to meet specific needs of a project, your class or your employer. In addition to each of the drawing controls that were introduced in this chapter, your template could also include common text fonts, linetypes, layers, and other related features. Each of these components will be introduced throughout the text. As each is introduced, features that you find useful should be added to your template. Methods for creating and saving templates will be covered in Chapter 4. For now, if you choose to use a template to enter the drawing area, accept the default and click OK.

Note: As the ANSI C-Color Dependent Plot Styles.dwt template is entered, you enter the drawing session in paper space. To create a drawing, select the Model tab at the bottom of the drawing area. This will remove the title block and produce a blank drawing screen. Now draw a few lines.

Select the ANSI C Title Block tab and the lines that you drew will be displayed on the template. Click the PAPER button on the status bar (not the tab at the bottom of the drawing area) to display the drawing in the template in model space. If your drawing does not fit into the template, select the **Pan Realtime** button (the hand) from the Standard toolbar. Pan allows the drawing to be moved to the desired location. After you select **Pan**, a hand will be displayed in the drawing area. Press the select button, and move the cursor in the direction you would like to move the drawing.

When you're satisfied with the location, press ENTER to exit the **Pan** command. If the drawing is too big to be seen in the template, click the **Zoom** button (the magnifying lens) on the **Standard** toolbar. After you select **Zoom**, a magnifying lens will be displayed in the drawing area. Press the select button and move the cursor down to reduce the size of the object. When you're satisfied with the location, press ENTER to exit the **Zoom** command. The full power of these commands will be introduced later.

START FROM SCRATCH

Selecting **Start from Scratch** from the TODAY window or the **Start from Scratch** button from the **Startup** dialog box allows a drawing to be started with a minimum of predetermined values. Setup is allowed in English or Metric units. English units are displayed in four-place decimal units, with drawing limits of approximately 12" × 9". Rather than being prompted by a wizard or restricted by a template, with **Start from Scratch** you can set drawing controls as needed using a dialog box or from the Command prompt. Selecting English will produce a display similar to Figure 2.8.

STARTING A NEW DRAWING FROM WITHIN AUTOCAD

In addition to the four methods you've just explored, AutoCAD allows a drawing to be started from inside an AutoCAD drawing. Use one of the methods introduced earlier in this chapter to start a new drawing. Once the drawing is started, draw a triangle so that the screen resembles Figure 2.20a. A new drawing can be started without harming the current drawing. To open another drawing window, select the **New** button at the far left end of the **Standard** toolbar, or select **New** from the **File** menu. You can also type NEW (ENTER) at the command prompt. Each method will hide the existing drawing file and begin the process of starting a new file. Start the new drawing using any one of the four startup methods you prefer. Your original drawing will be removed from the drawing area, and a new drawing screen will be displayed.

Use the **Line** command to draw a rectangle so that your drawing area resembles Figure 2.20b. Click the **Minimize** button, and the original drawing will be displayed. Click the **Minimize** button for this drawing, and the display will resemble Figure 2.20c. The drawing tabs of both the open drawings are displayed. Several new drawings can be started, or you can open several existing files in one AutoCAD session. Although this might not seem important now, several of AutoCAD's features will make this very important: AutoCAD allows you to transfer objects, settings, and other types of information from one drawing to another. AutoCAD allows drawing attributes to be transferred between drawings using methods similar to those used with Windows Explorer. Drawing features can be moved between a drawing stored on your hard drive, a drawing on a network, or a drawing located across the Internet. These features will be addressed throughout the balance of the text. For now, remember that two or more drawings can be worked on in the same drawing session.

DISPLAYING MULTIPLE WINDOWS

As with other Windows programs, AutoCAD provides choices on how open windows and drawings will be displayed. Individual program or drawing windows within AutoCAD can be arranged in horizontal, vertical, or cascading tiles. You can display multiple layouts by selecting **Window** from the menu bar, as shown in Figure 2.21. The lower portion of the menu will display a listing of the current open drawing layouts. Selecting the name of the desired view will remove the current drawing and display the selected drawing. Choose a display for multiple windows by selecting one of the options from the top portion of the menu. A cascading or tiled display is often used when two or more windows are to be displayed.

Cascading Windows

Selecting **Cascade** from the **Window** menu displays the drawings similar to those shown in Figure 2.22a. Cascading drawings are stacked over each other in descending order. Drawings are displayed in the order that they are selected. To display one of the lower drawings, click the title bar of the desired drawing to move it to the top of the stack. Clicking the title bar and holding the select button down allows the window to be dragged anywhere on the screen.

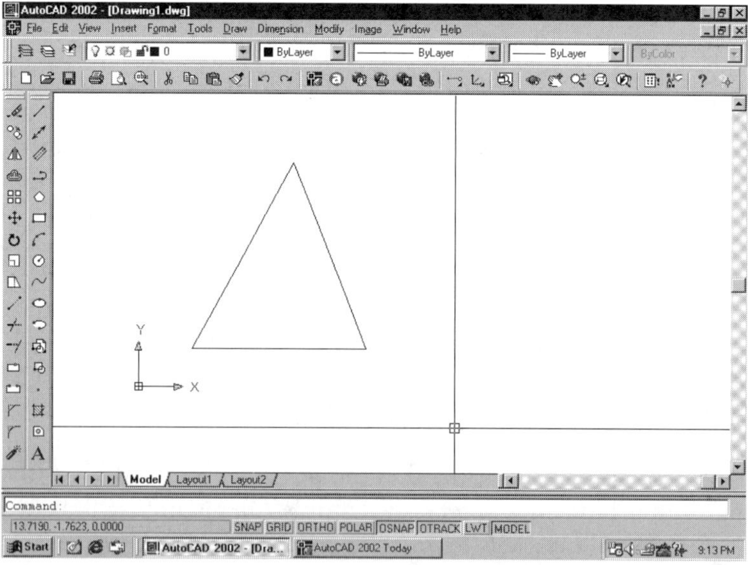

Figure 2.20a *AutoCAD allows a drawing to be started from inside an active AutoCAD drawing. Use one of the methods introduced earlier in this chapter to start a new drawing. Once the drawing is started, draw a triangle.*

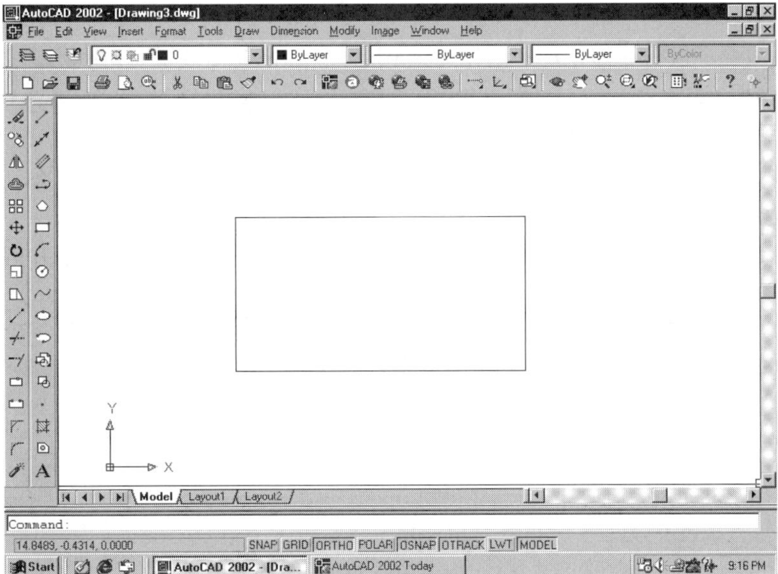

Figure 2.20b *A new drawing can be started without harming the current drawing. Another drawing window can be opened by selecting the* **New** *button on the* **Standard** *toolbar, by selecting* **New** *from the* **File** *menu, or by typing* **NEW** (ENTER) *at the command prompt. Each method will remove the existing drawing from view and begin the process of starting a new file.*

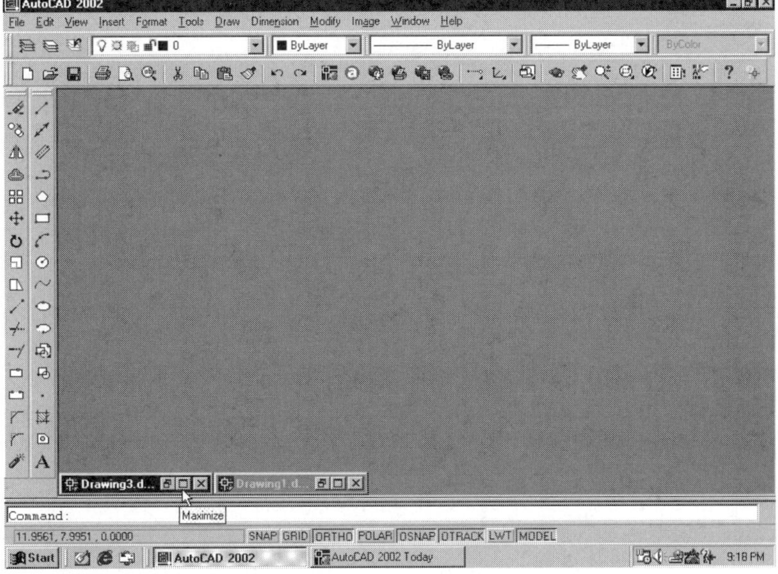

Figure 2.20c *Click the* **Minimize** *button, and the current display will be minimized to display the drawing tabs for all current drawings.*

Tile Displays

The **Tile** options allow drawings to be arranged in horizontal (Figure 2.22b) or vertical displays (Figure 2.22c). Tiled windows permit each window to be viewed at the same time. Just as with other programs, if you're unhappy with the size of a window, alter the size by selecting and dragging the edge to the desired size.

WORKING WITH PARTIAL FILES

AutoCAD offers two additional methods of working with drawing files. Although each method is beyond your current needs as you explore AutoCAD, you should at least be aware of the **Partialopen** and **Partiaload** commands. Both commands can be used when you work with large files to allow only a specific portion of the drawing to be accessed. The projects that you work on throughout this text will typically be too small to benefit from **Partialopen**. On extremely detailed drawings, **Partialopen** will allow specific information to be viewed in the drawing area. If you're working on a complex project, the drawing can be divided into zones, and then each zone can be saved by what AutoCAD refers to as a view. **Partialopen** allows you to open a specific zone of the drawing and display some of the information, while hiding other information. For more information about these commands, see the **Help** menu.

Note: You now have several methods that can be used to start a drawing session. Just as important as how you start a drawing session, is where the file that you want to open is located. When you're working with existing files, always use files stored on your hard drive to open the drawing session. AutoCAD needs work space as you work on a drawing session. If the drawing and the work space exceed the storage space on the disk, you might

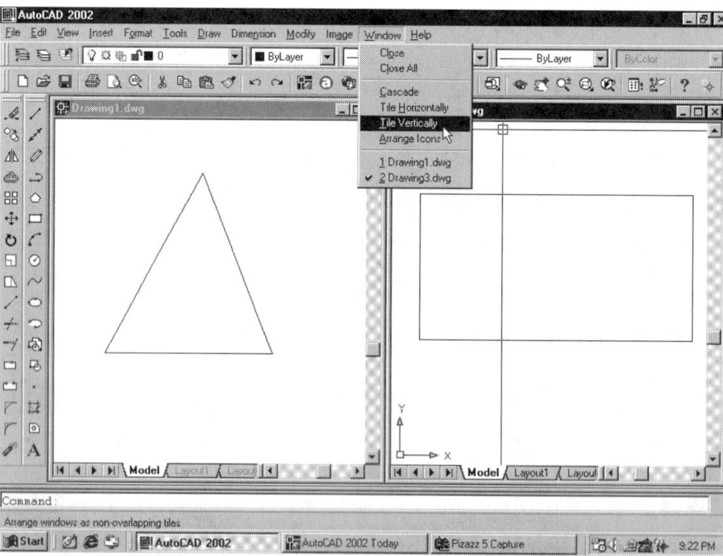

Figure 2.21 *Multiple layouts can be displayed from the* **Window** *menu.*

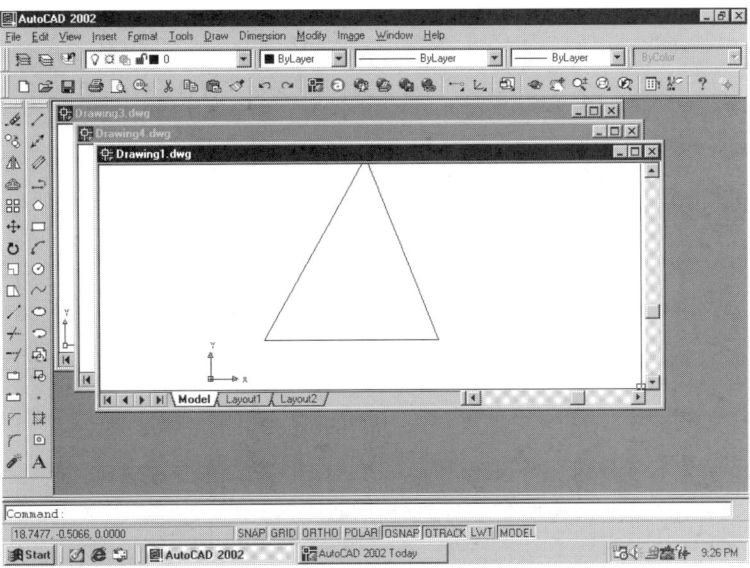

Figure 2.22a *Selecting* **Cascade** *from the* **Window** *menu displays drawing windows that are stacked over each other in descending order.*

encounter unwanted results. If the drawing is stored on a network, you'll typically have no shortage of work space, but your hard drive is generally faster. If the most current file is saved on a diskette or in a network folder, copy the file to the hard drive prior to opening the file.

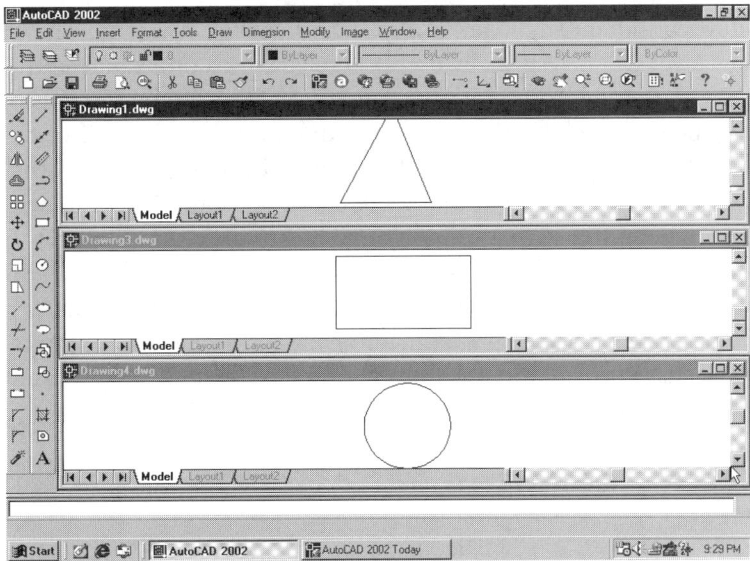

Figure 2.22b *The* **Tile Horizontal** *option displays drawing windows in horizontal rows.*

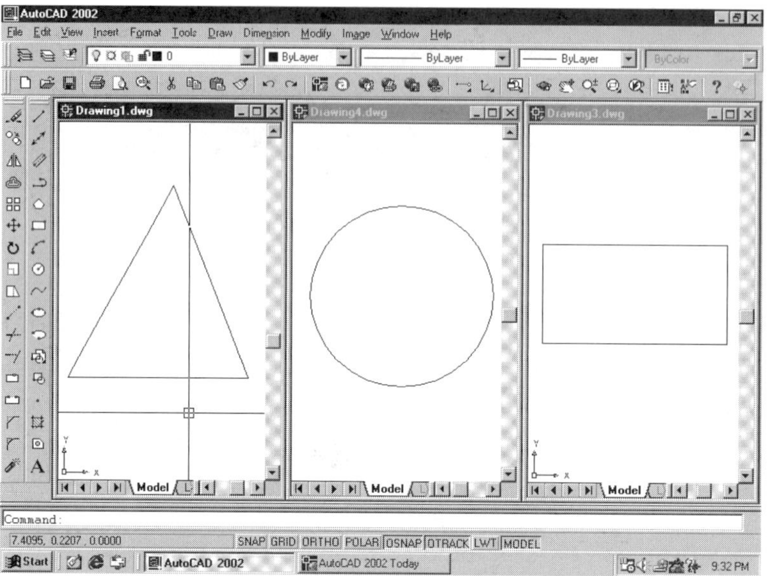

Figure 2.22c *The* **Tile Vertical** *option displays drawing windows in vertical columns*

WORKING WITH FILES WITHOUT AUTOCAD

When Volo View Express is loaded, Autodesk now provides the option of being able to view AutoCAD drawings from the office network or from a Web site without having to use AutoCAD. Volo View Express is supplied with AutoCAD and can be used to view, review, and print AutoCAD drawings without loading AutoCAD. This feature will allow consultants to review AutoCAD drawings away from the work environment. The viewer will be especially useful in reviewing a project using the **Markup** without opening AutoCAD. To review a drawing, a client can be given a copy of the base drawing to be reviewed. Posting a drawing on a secure Web site (see chapter 24) will allow access by multiple consultants. With one of the options from the **Tools** or **Markup** tab of Volo View Express, text or a sketch can be added to the drawings. Once the review is complete, the suggested edits can be saved to a file written in redline markup language format (.rml) that is linked to the original file. The original drawing remains unaltered, giving you the final say over which suggested changes get added to the drawing base. Volo View Express can also be used to view drawings created using Inventor, and Mechanical Desktop 4.0 if object enablers are downloaded from the Web site of the desired application.

CONTROLLING DRAWING PARAMETERS FROM WITHIN A DRAWING

No matter what method is used to begin a drawing, the drawing parameters can be altered at any time. Parameters can be altered through dialog boxes, menus, or the keyboard.

SETTING UNITS

The **Drawing Units** dialog box (see Figure 2.23) can be used to change the drawing units. The **Drawing Units** dialog box can be displayed at any point of a drawing session using one of the following methods:

- Select **Units** from the **Format** menu.
- Type **UN** (ENTER) at the Command prompt.

The drawing units are set by selecting the desired style from the **Type** list. Selecting the arrow on the right side of the **Precision** edit box will display the drop-down list. Select the desired precision by choosing the desired value from the Precision list. The dialog box can also be used to select the method for measuring angles and for setting the angle precision. Angle precision accuracy is set using the same methods as for changing the unit precision. Selecting the **Direction** button will display the **Direction Control** dialog box (see Figure 2.24). This dialog box allows the angle starting point and the angle direction to be set, by selecting the appropriate radio button or highlighting the desired accuracy. Once it is selected, the accuracy will be displayed in the edit box and the menu will be closed.

SETTING LIMITS

Limits are two-dimensional points that represent the lower left and upper right boundaries of the drawing area. Changing the limits alters the size of the drawing area. Before you change the limits, draw a square inside the drawing area, similar to Figure 2.25a. This will help you to visualize the change that occurs when limits are set.

To change the size of the drawing limits, select **Drawing Limits** from the **Format** menu or type LIMITS (ENTER) at the command line. The drawing limits can also be altered in the middle of another command if you type LIMITS (ENTER) at the Command prompt. Each method of entering the command will produce the display:

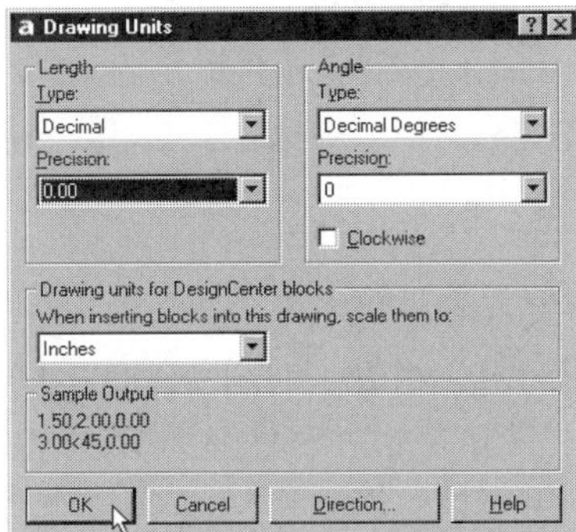

Figure 2.23 *Drawing units can be altered at any time using the* **Drawing Units** *dialog box. Display the dialog box by selecting* **Units** *from the* **Format** *menu or by typing* **UN** *(ENTER) at the Command prompt.*

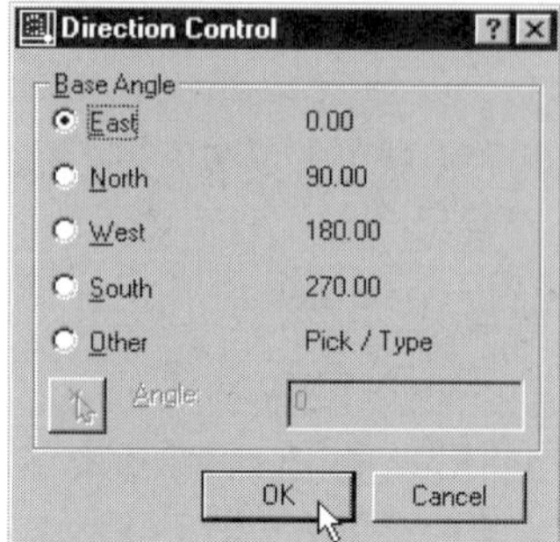

Figure 2.24 *Selecting the* **Direction** *button in the* **Drawing Units** *dialog box allows the orientation for angle direction to be altered.*

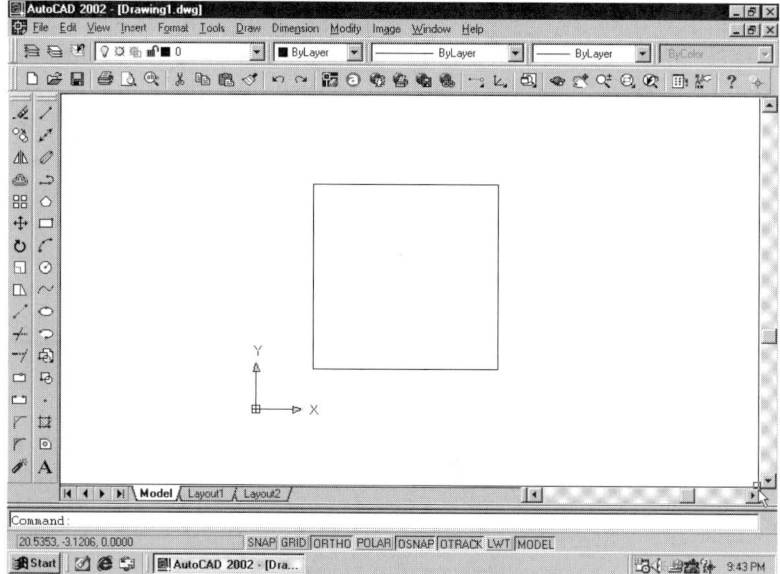

Figure 2.25a *With the drawing limits set at the default units of 12,9, an object might appear quite large.*

Reset Model space limits:

Specify lower left corner or [ON/OFF] <0.0000,0.0000>:

Acceptable responses are ON, OFF, and ENTER. ON will keep the current values (0.0000,0.0000) and activate limit checking. Limit checking will not allow objects to be drawn that are outside the limits. Think of limit checking as a border on your drawing paper to keep your project on the paper. OFF will disable limit checking but retain the limit values for future use of limit check.

Pressing ENTER retains the current location of the lower left drawing limit of (0.0000,0.0000) This will be the most typical response. Pressing ENTER produces the prompt:

Specify upper right corner (12.0000,9,0000):

The default value is 12.0000,9.0000, with the first number representing the horizontal dimension and the second number representing the vertical dimension. This is your opportunity to change the size of the drawing area to fit your current project. Type **60,48** (ENTER), and the Command prompt will return with no apparent change. Type **Z** (ENTER) at the keyboard, then **A** (ENTER) (the A represents all). Now the effect of your new limits can be seen in Figure 2.25b. Remember that the box is still the same size but it appears smaller because the area to be viewed has increased.

SETTING GRIDS

Another helpful tool for drawing layout is a grid. This will provide the equivalent of the nonrepro blue grid that is preprinted on drafting vellum. The **Grid** command will produce a visible grid of dots at any desired spacing. The grid size can be adjusted throughout the life of the drawing, depending on the size of the object being designed. The grid also can be displayed or removed to aid in visual clarity, but it will not be produced when the drawing is printed or plotted, even if it is displayed on the screen. Figure 2.26 shows an example of a drawing with the grid display.

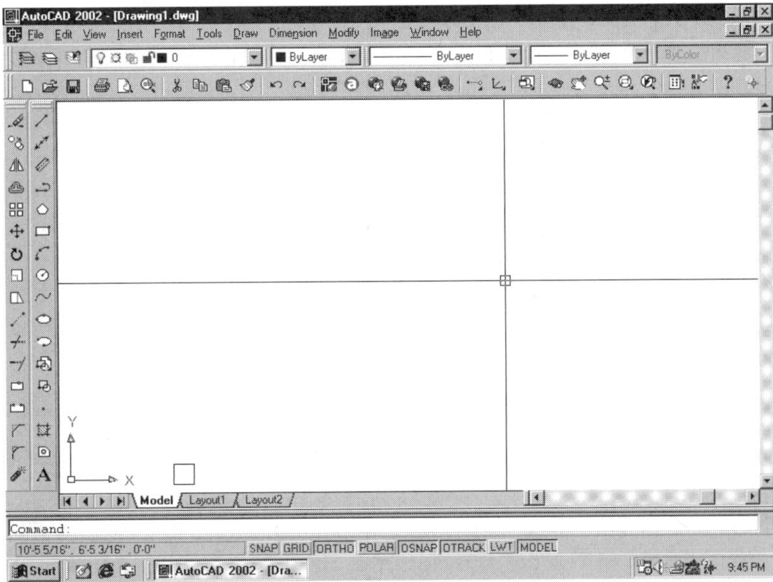

Figure 2.25b *By changing the location of the upper right corner and pressing* ENTER, *you can allow viewing of a larger area. In this example, the limits are set to 60,48.*

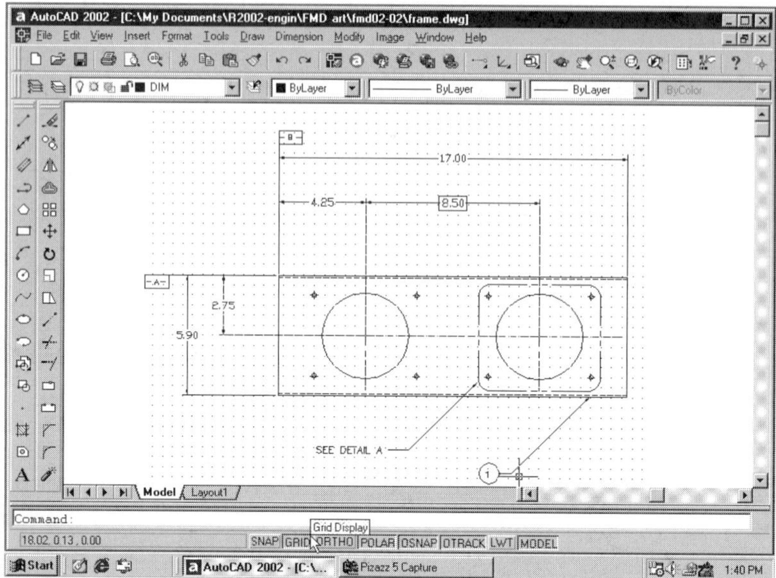

Figure 2.26 *Drawing limits of 24,18 with a grid value of 0.50". Courtesy Floyd D. Miller.*

The grid can be set by dialog box or from the command line. The current status is displayed on the status bar. Once the values have been set, the grid can be toggled ON/OFF by selecting the GRID button on the status bar. The grid can also be toggled ON/OFF in the middle of another command by pressing F7 or by pressing CTRL+G.

Assigning Grid Values Using a Dialog Box

Grid values can be set using the **Drafting Settings** dialog box, accessed by selecting **Drafting Settings** from the **Tools** menu. The dialog box, shown in Figure 2.27, can also be displayed by moving the cursor to the GRID or SNAP button on the status bar, right-clicking, and then choosing **Settings** from the shortcut menu, or by typing RM (ENTER) at the Command prompt. Once the **Drafting Settings** dialog box is displayed, grids are set using the following procedure:

1. Select the **Snap and Grid** tab.
2. Enter the desired value for the X (left-to-right) grid spacing.
3. Enter the desired value for the Y (top-to-bottom) grid spacing. If you want the Y value to match the X value, click the select button. Pressing ENTER will also match the X and Y values and close the dialog box. To enter a different Y value, enter the desired value.
4. Click the **OK** button.

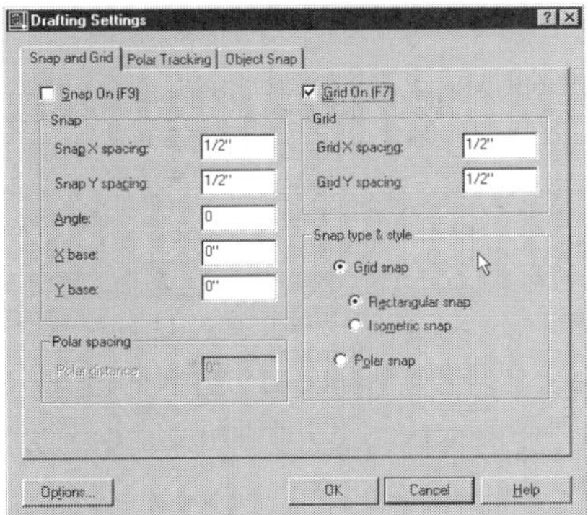

Figure 2.27 *Values for the grid can be altered on the* **Snap and Grid** *tab of the* **Drafting Settings** *dialog box by right-clicking the GRID button on the status bar and then selecting the* **Settings** *option. Grids can also be altered by selecting* **Drawing Aids** *from the* **Tools** *menu or by typing* **RM** *(*ENTER*) at the Command prompt and then selecting the* **Snap and Grid** *tab.*

The dialog box also offers methods of altering the angle used to display the grid. Type the desired angle for the grid in the **Angle** edit box. Figure 2.28 shows an example of a drawing with a grid rotated to 45°. Selecting the **Isometric Snap** setting will establish a grid suitable for isometric drawings. Isometric drawings will be discussed later in this text.

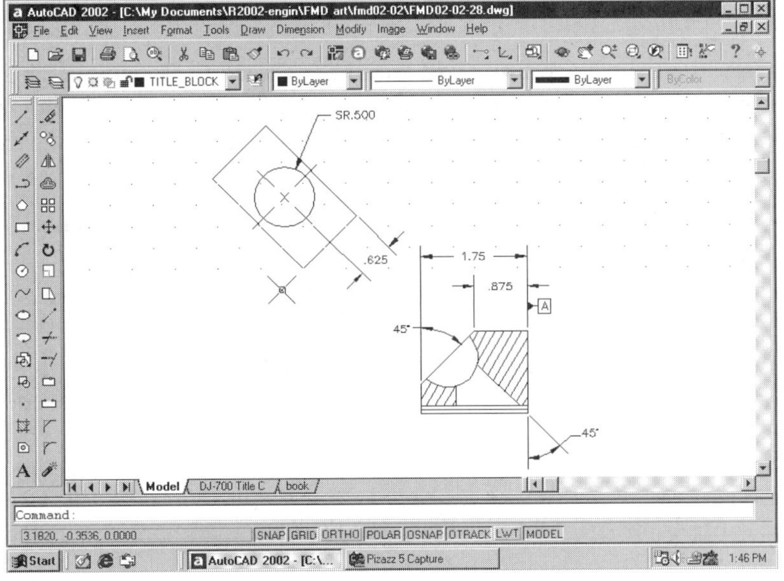

Figure 2.28 *The grid can be rotated to ease layout of an irregular shape. Courtesy Jill Heilman.*

Setting Grids at the Command Line

The grid values can be established, altered, and displayed from the Command prompt. The controls for the grid are accessed at the Command prompt by typing **GRID** (ENTER). The command line will now read

Command: **GRID** ENTER

Specify grid spacing (X) or [ON/OFF/Snap/Aspect] <0.50>:

Type **1** (ENTER). This will return the command prompt and display the grid field with dots at 1" spacing on the drawing screen. Grid should be set in units common to the object to be drawn.

SETTING SNAP

When working on a modular component, it is helpful to be able to control the accuracy of lines entered with the mouse. The **Snap** command sets up an imaginary rectangular grid that controls placement of the drawing cursor in precise intervals. The distance between cursor movements is defined as snap resolution. Snap intervals can be set at any interval and can be changed throughout the life of the drawing. A change in the snap grid affects only the points entered after the change in resolution. Points entered prior to the change in resolution might no longer line up with the new snap location. The size of the grid setting and the size of the object to be drawn will affect the snap value. If the grid is set at 1", a snap value of .50" or .25" will be helpful in producing accuracy. If a detail of a small part is drawn, a snap of 1/2" might be used to accurately move through the drawing. If a 50' long wing is to be drawn, a 12" or 24" snap will be more helpful than the .50" accuracy used for a detail.

Snap can be set using the **Snap and Grid** tab of the **Drafting Settings** dialog box. It can also be set by right-clicking the SNAP button on the status bar and selecting **Settings** from the shortcut menu. Snap is adjusted using the same methods that were used to adjust the grid.

Values for the X and Y spacing can be entered in the appropriate edit box. Once the values have been set, snap can be toggled ON/OFF by double-clicking the SNAP button on the status bar. Snap can also be toggled ON/OFF in the middle of another command by using F9 or by pressing CTRL+B. Entering an angle in the **Snap Angle** edit box controls the display angle for snap and grid. Entering a value in the X or Y **Base** edit boxes allows the base point for the snap pattern to be established.

Setting Snap at the Command Line

The snap values can be established, altered, and displayed from the Command prompt. The controls for snap can be accessed at the Command prompt by typing SN (ENTER). This will produce the prompt:

Specify snap spacing or [ON/OFF/Aspect/Rotate/Style/Type] <0.50>:

Entering a numeric value will set the spacing of the grid for entry points. Type **.75** (ENTER). The command line returns and allows for drawing. Type **L**(ENTER) and draw a line. You'll notice that the crosshairs jump across the screen in .75" intervals rather than moving smoothly as they had before. In addition to ON/OFF, settings include Aspect, Rotate, Style and Type.

Aspect
Some projects are set up where different values for the horizontal (X) and vertical (Y) directions are used. To do this, type **SN** (ENTER). Type **A** (ENTER) for Aspect, and the prompt will display the additional listings:

Specify horizontal spacing <0.75>: **50** ENTER

Specify vertical spacing <0.75>: **1** ENTER

Command:

Rotate
Because not all projects are rectangular, such as the project in Figure 2.29, it might be helpful to rotate the snap grid to match the drawing. Type **SN** (ENTER) and then type **R** (ENTER) for Rotate at the prompt. This will produce the additional display:

Specify snap spacing or [ON/OFF/Aspect/Rotate/Style/Type]
<A>: **R** ENTER

Specify base point <0.00, 0.00>:

AutoCAD is asking for the point of the drawing where you would like the snap grid to be rotated. To select a location, move the crosshairs to the desired location in the drawing and press the select button. This will now provide the prompt:

Rotation angle <0>:

A number between –90 and 90 can be entered. Using a negative angle will rotate the snap grid counterclockwise. Using a positive angle will rotate the snap grid clockwise. Type a value and press ENTER. The snap grid will be rotated around the base point just selected.

Style
The format of the snap grid is adjusted by using the style option. Type **S** (ENTER) for Style at the snap prompt, and the prompt will display

Specify snap spacing or [ON/OFF/Aspect/Rotate/Style/Type]
<A>: **S** ENTER

Enter Snap grid style [Standard/Isometric] <S>:

For orthographic drawings the style will not need to be set. For isometric drawings type **I** (ENTER).

Type

This setting toggles between polar and grid snap. Polar snaps align with polar angles relative to the Polar tracking point. Grid snaps along the grid. Type **T** (ENTER) for Type at the snap prompt, and the prompt will display

Specify snap type [Polar/Grid]<Grid>:

Note: Coordinating the grid and snap will provide great accuracy and speed. If the grid value is set at zero, the grid spacing automatically adjusts to the snap resolution. To specify the grid spacing as a multiple of the snap resolution, type **X** (ENTER) after the grid value. For instance, if the snap value is .5", typing **6X** (ENTER) at the command prompt produces a grid of 3".

CHAPTER 2 QUIZ

1. What command will set an invisible grid to aid layout?
2. List the five options of the units.
3. What is the default unit for angle measurement?
4. What is the default for angle direction?
5. List three methods to toggle between **Grid** ON/OFF.
6. List three methods to toggle between **Snap** ON/OFF.
7. What is the main difference between grid and snap?
8. You've just entered AutoCAD and don't want to establish any drawing parameters. How can you get to the drawing screen?
9. When would **Start from Scratch** be a useful method of entering a drawing?
10. What does **Area** control?
11. What is the default method of entering AutoCAD?
12. How can you use the **Advanced Setup** wizard to start a drawing?
13. What command is used to access the **Drafting Settings** dialog box from the Command prompt?
14. What is the range of precision for measuring angles?
15. What is the effect of typing **UNITS** (ENTER) at the Command prompt?

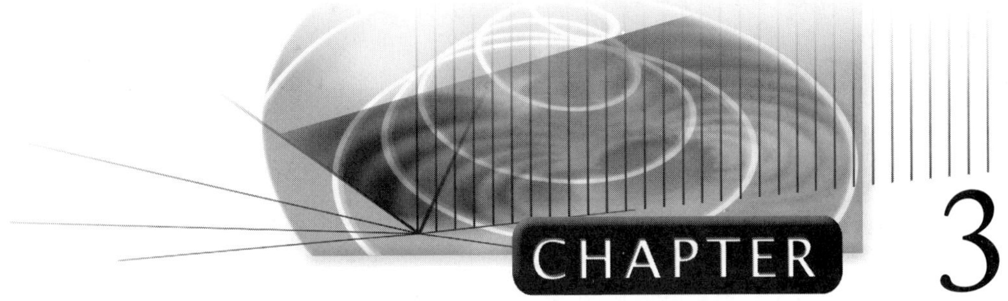

3

Drawing and Controlling Lines

This chapter will introduce methods for

- Line command, including toolbars, keyboard, and the **Draw** menu
- Controlling the placement of lines, including absolute coordinates, relative coordinates, and polar coordinates
- Controlling lines with the tools of the status bar and for editing lines

Commands and options to be introduced include

- **Line**
- **Close**
- SNAP
- GRID
- ORTHO
- POLAR
- OSNAP
- **Erase**
- **Oops**
- **Undo**
- **Redo**

Now that you're able to set up drawing units and limits, basic drawing components can be explored. The main components of the drawings you'll be working with will consist of lines. Chapter 1 introduced the **Line** command using the **Draw** toolbar, the keyboard, and the **Draw** menu. This chapter introduces methods for making effective use of the **Line** command, as well as methods for locating and altering lines.

LINES

The basic components of every drawing are lines, which are used to represent the edge and surface changes of a part. Figure 3.1 shows the lines required to represent a portion of a mechanical part. Although many of these lines are represented with different line patterns or thickness, each can be drawn using the **Line** command. Lines are used to represent different types of information within a drawing. Establishing and controlling varied linetypes will be introduced in later chapters.

ENTERING THE LINE COMMAND

The **Line** command is accessed by toolbar, keyboard, or menu. Each method was introduced in Chapter 1. To start the command, select the **Line** button on the **Draw** toolbar. This will produce the prompt:

Command: (Select the Line button.)

_line Specify first point:

Move the cursor to where you would like to start a line and click the select button. The "first point" will become the beginning point of a line segment. As the first point is selected, the prompt displays

Specify next point or [Undo]:

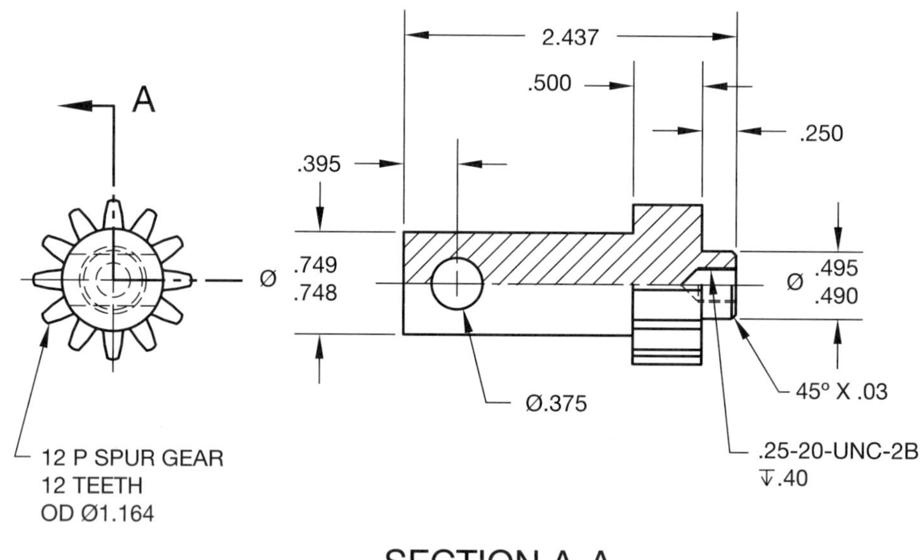

Figure 3.1 *Lines of varied widths and patterns are used on most mechanical drawings. Courtesy Hanh Tran.*

Move the cursor to the desired location and click the select button to specify the endpoint. As the cursor is moved, a rubber-band line is displayed from the first point to the crosshairs location. This can be seen in Figure 3.2. The rubber-band line will help you to visualize the placement of the line to be drawn and to experiment with its placement. The rubber-band line will remain as the crosshairs are moved until the next point is selected or the Line command is terminated. To terminate the **Line** command, press ENTER, or SPACEBAR, or click the right mouse button.

- Pressing ENTER or SPACEBAR will draw the indicated line and then exit the command.
- Right-clicking and selecting ENTER will produce the same results as pressing the SPACEBAR.
- Right-clicking and selecting **Cancel** will draw the indicated lines and terminate the command.

COMMAND LINE INPUT

A second method of entering commands is by keyboard. The **Line** command is entered by typing L (ENTER) at the Command prompt. The sequence for entering the Line command is as follows:

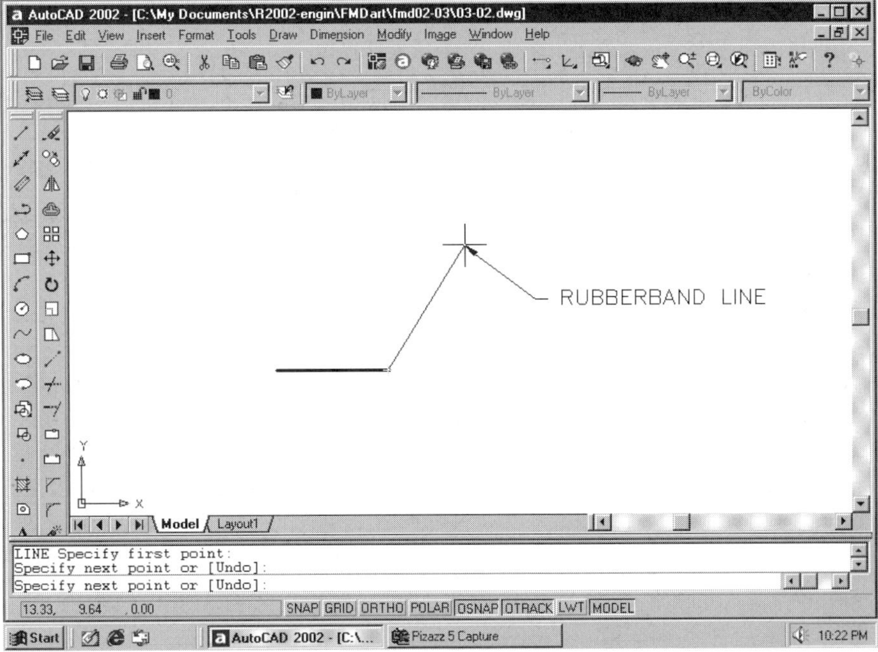

Figure 3.2 *When a "next point" is provided, a rubber-band line extends from the "first point" to the present crosshairs location. This line indicates the line that will be drawn if the current crosshairs location is used.*

Command: **L** ENTER

LINE specify first point: *(Specify an end point.)*

Specify next point or [Undo]: *(Specify an end point.)*

The **Line** command functions exactly the same no matter how it is entered.

MENU ENTRY

A third method for entering a command is to select **Line** from a menu. When you select **Line** from the **Draw** menu, the menu will disappear and the Command prompt will show

Command: _line Specify first point:

Notice that four other options for lines are presented in the **Draw** menu—**Ray**, **Construction Line**, **Multilines**, and **Polylines** commands. A fifth method of drawing lines is **Sketch**. The **Sketch** command allows freehand drawing methods to be reproduced. Each of these commands will be introduced in later chapters.

No matter which method is used to start the command, a request for the "first" and "next" points is displayed at the command line, and can be entered with the mouse, or by keyboard. Specific coordinates can be entered by keyboard rather than by selecting the points with the mouse. Methods of selecting line and point by keyboard will be discussed later in this chapter. As you start to experiment with drawing lines, you should draw objects combining all entry methods.

Note: You've entered the **Line** command using three different methods. As you start to experiment with drawing lines, you should draw objects combining all entry methods. Each method works well to initially start the command. The easiest way to repeat a command is to press ENTER or SPACEBAR; both will repeat the previous command.

LOCATING LINES USING COORDINATE ENTRY METHODS

As you start to complete drawings in this chapter, it will be important to understand the user coordinate system (UCS) of AutoCAD. The UCS is based on a Cartesian coordinate system, dividing space into four quadrants. Figure 3.3 shows the division of space based on the Cartesian coordinates. Points in these quadrants are located in turn by using their points in a horizontal (X) and vertical (Y) direction along the plane. Points are always specified by the X coordinate followed by the Y coordinate.

Figure 3.4 shows the relationship of the drawing screen to the Cartesian coordinates. As you begin to draw, AutoCAD will ask you to specify a "first point." As you move to the "next point," the coordinate display at the left end of the status line will reflect the crosshairs movement. To activate coordinate display, click the coordinate display, or press F6 or CTRL +D.

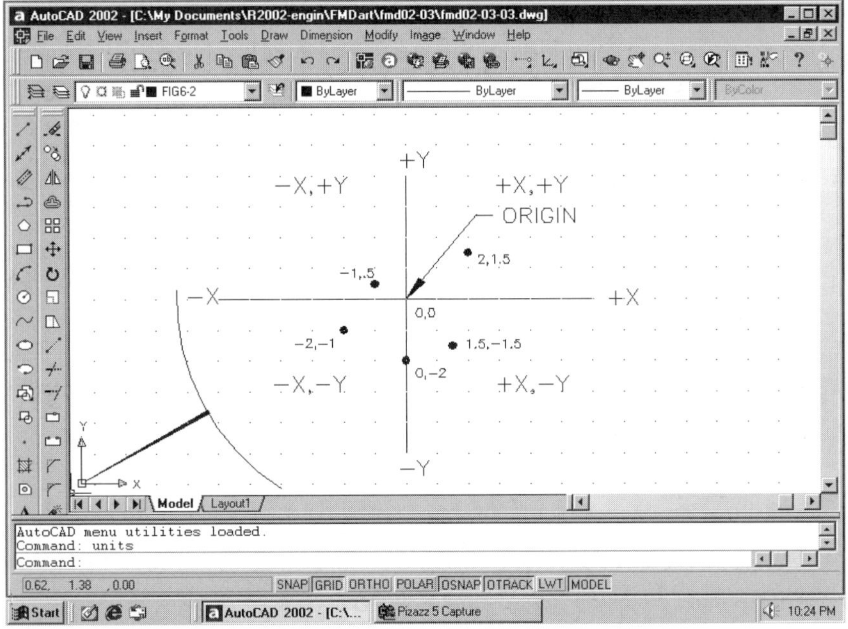

Figure 3.3 *The division of space using the Cartesian coordinate system. The intersection of the X and Y coordinates is assumed to be 0,0, which is the lower left corner of the screen. The grid has been moved to the center of the screen to represent negative X and Y locations.*

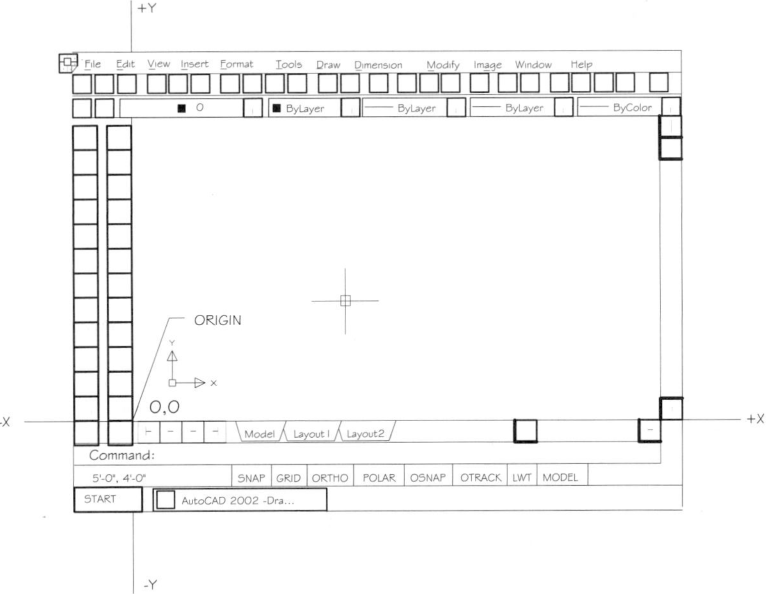

Figure 3.4 *The relationship of the drawing screen to Cartesian coordinates with the Limits set at the lower left = 0,0, and the upper right values set as two positive values.*

Because mechanical drawings deal with very specific locations, a means of controlling the point placement is desirable. This can be done by using Cartesian, relative, or polar coordinates for point entry methods.

CARTESIAN COORDINATES

The Cartesian coordinate system locates all points from an origin assumed to be 0,0. The 0,0 origin point is assumed to be the same 0,0 point that the limits, snap, and grid are based on, although it can be changed. The axes for the X and Y coordinates intersect at 0,0. Each point on the screen is located by a numeric coordinate based on the original axis. Objects are drawn by entering the coordinates of X and Y separated by a comma. In the example in Figure 3.5 a line is drawn between 2,2 and 2,4. The next "to point" is 4,4, based on the cursor location and the current coordinate display. To complete the drawing in Figure 3.6 use the following command sequence:

Command: **L** ENTER *(Or click the Line button on the Draw toolbar.)*

LINE

Specify first point: **4,1.5** ENTER

Specify next point or [Undo]: **6,1.5** ENTER

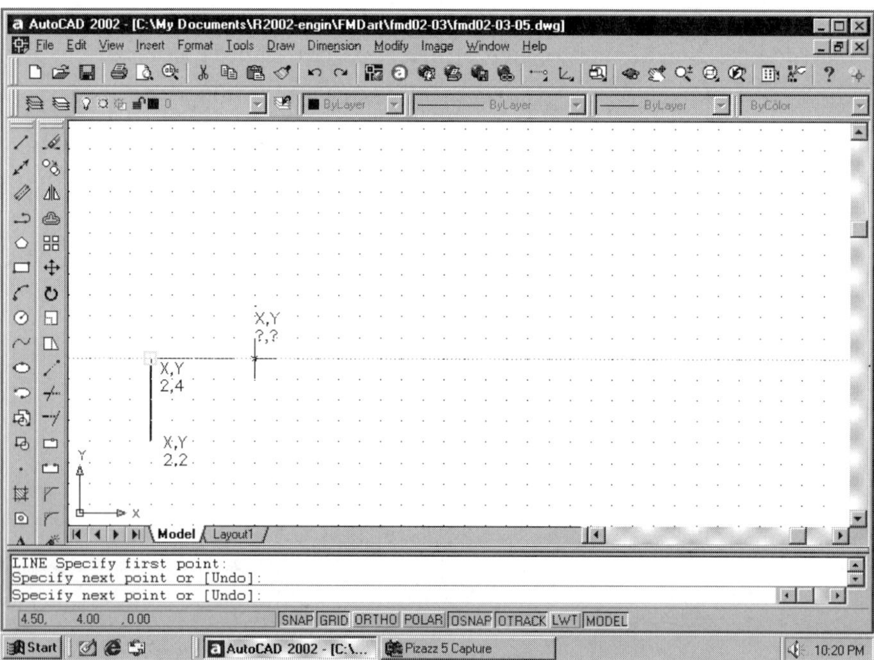

Figure 3.5 *With absolute coordinates, and the 0,0 location as the lower left corner of the drawing screen, the line drawn starts at 2" over (+X) and 2" up (+Y) and extends to a location 2" over (+X) and 4" up (+Y).*

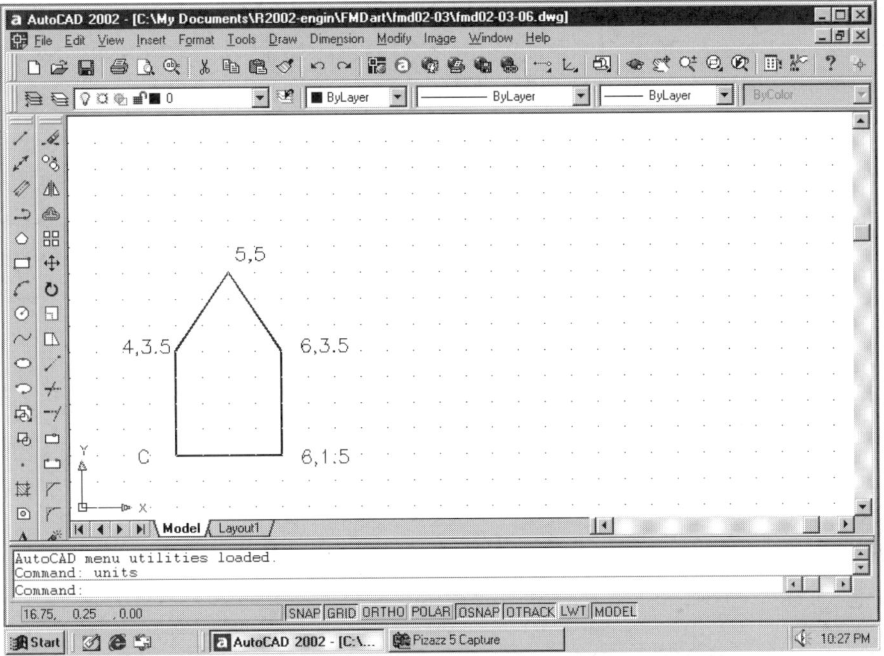

Figure 3.6 *A simple shape and the absolute coordinates required to describe it.*

Specify next point or [Undo]: **6, 3.5** ENTER

Specify next point or [Close/Undo]: **5,5** ENTER

Specify next point or [Close/Undo]**: 4,3.5** ENTER

Specify next point or [Close/Undo]: **C** ENTER

Command:

Notice that C and ENTER were typed at the last command, rather than numeric coordinates. C represents close. C (ENTER) will end the **Line** command sequence at the point where the line sequence was begun. With Object Snap activated (click the OSNAP button on the status bar), moving the cursor to the desired endpoint will also allow easy closure.

Note: An alternative to using the C option to close the object is to use the Object Snap (OSNAP) option of AutoCAD. As the cursor is moved near the end of the original line a yellow box is placed around the end of the original line. The color of the box can be altered in the **Options** dialog box by selecting the **AutoSnap Settings** area of the **Drafting** tab. The box indicates that the nearest endpoint has been selected to become the 'next point'. Object Snap will be discussed in detail in Chapter 8.

Because Cartesian coordinates are based on an origin of 0,0, they are often not convenient for most mechanical projects. Many CAD users find relative coordinates more suited to a mechanical drafting project.

RELATIVE COORDINATES

Relative coordinates locate "to points" based on the last point instead of the 0,0 origin. Relative coordinates are entered as @ X,Y. In addition to preceding the coordinate value with the @ symbol, with relative coordinates, you need to be mindful of positive and negative values.

Figure 3.7 shows an object drawn using relative coordinates. Positive coordinates will move the crosshairs to the right or above the origin point. Negative coordinates will move the crosshairs down or to the left of the last entry point. The object in Figure 3.7 was drawn using the following command sequence:

Command: **L** ENTER *(Or click the Line button on the Draw toolbar.)*

LINE

Specify first point: **4,1.5** ENTER

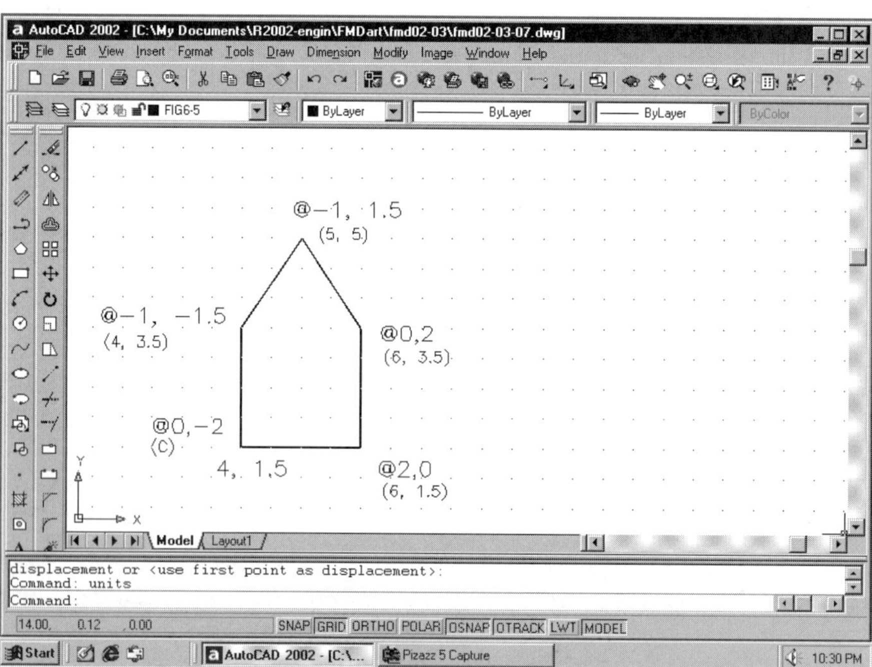

Figure 3.7 *Relative coordinates describe the location of the next point based on the last point. Relative coordinates must be preceded by the @ symbol. The value @2,0 would extend a line 2" to the right with no vertical rise. Relative coordinates are a common entry choice for many projects.*

Specify next point or [Undo]: **@2,0** ENTER

Specify next point or [Undo]: **@0,2** ENTER

Specify next point or [Close/Undo]: **@–1,1.5** ENTER

Specify next point or [Close/Undo]: **@–1,–1.5** ENTER

Specify next point or [Close/Undo]: **@–1,–1.5** ENTER

Specify next point or [Close/Undo]: ENTER

Command:

The first point was entered using Cartesian coordinates, with the following coordinate points entered by relative coordinates. The final coordinates could have been omitted and the object closed using Object Snap or by typing **C** (ENTER).

POLAR COORDINATES

Polar coordinates use the last entry point as the origin and combine a distance and an angle. This is useful on drawings with angled surfaces. Polar coordinates are entered as

@3<27.5 ENTER

This would produce a line 3" long at a 27.5° angle, as shown in Figure 3.8. To draw the opposite side of this angle would require typing **@3<332.5** (ENTER). Remember, polar coordinates are based on the current origin of the crosshairs. In Figure 3.9 the location of line 2 at 332.5° is determined by subtracting 27.5° from 360°.

To complete the triangle started in Figure 3.9, use the Object Snap feature of the cursor or type **C** (ENTER). Figure 3.10 shows an object drawn using polar coordinates. The required command sequence would be as follows:

Command: **L** ENTER *(Or click the Line button on the Draw toolbar.)*

LINE

Specify first point: **4,1.5** ENTER

Specify next point or [Undo]: **@2<0** ENTER

Specify next point or [Undo]: **@2<90** ENTER

Specify next point or [Close/Undo]: **@1.8<124** ENTER

Specify next point or [Close/Undo]: **@1.8<236** ENTER

Specify next point or [Close/Undo]: *(With OSNAP ON, move the cursor to the origin point and press the pick button or type C* ENTER.*)*

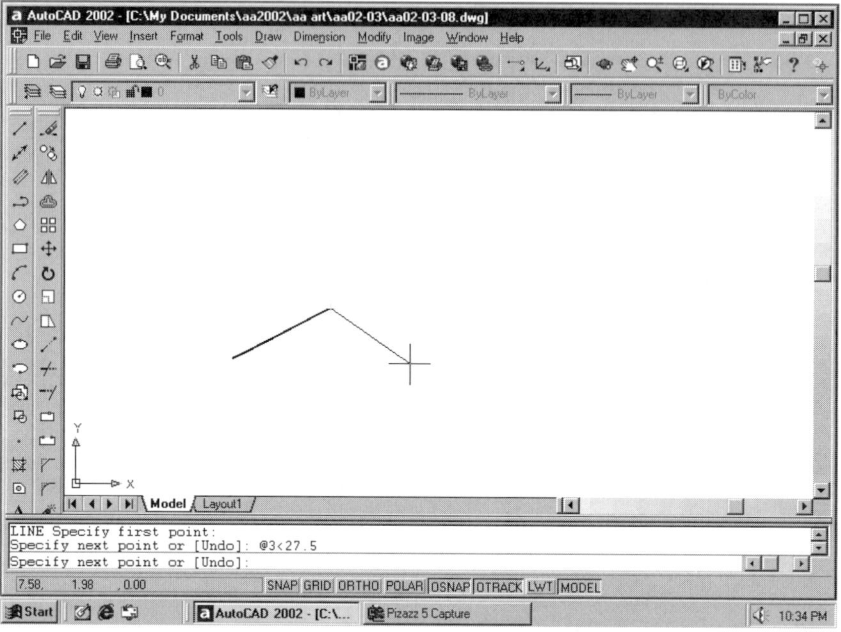

Figure 3.8 *Polar coordinates describe a point relative to the last position using a length and an angle. The @ symbol precedes the length, and the < symbol precedes the angle specification in degrees. Polar coordinates are useful in drawing angled surfaces.*

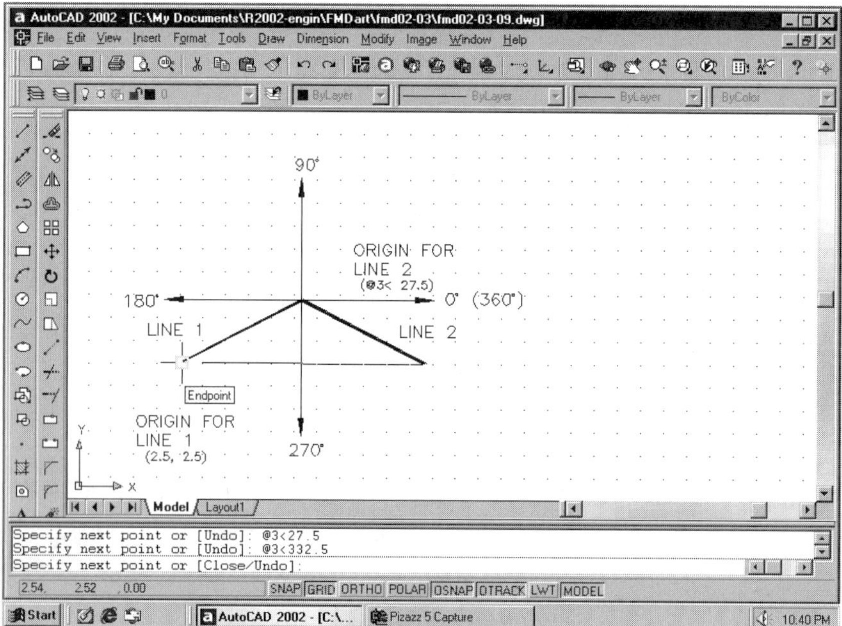

Figure 3.9 *Specifying angles using polar coordinates.*

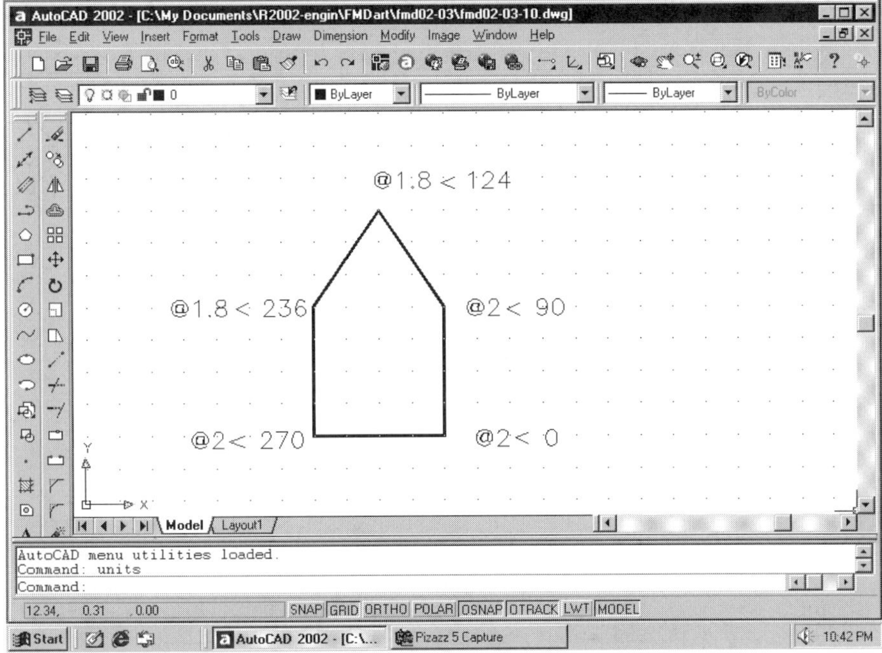

Figure 3.10 *The use of polar coordinates to establish a simple shape.*

CONTROLLING LINES

In addition to the aids for placing lines accurately, AutoCAD offers help for placing lines easily. Line continuation, closure, and the tools of the status bar can each be used to control the placement of lines.

LINE CONTINUATION

Once the **Line** command has been terminated, it can be resumed from the end of the last line segment that was drawn. This can be done even if other commands were used. The horizontal line segment drawn in Figure 3.11 was drawn prior to the circle. The circle was drawn by clicking the **Circle** button on the **Draw** toolbar. It could also be drawn by typing C (ENTER) at the Command prompt and selecting a center or radius point. This command will be further explored in the Chapter 7. If ENTER is typed for the "first point," the new line will begin at the end of the last line segment.

LINE CLOSURE

If a sequence of lines will form a closed polygon, the last line can be drawn without providing endpoints or coordinates for the termination point. You've already used the Object Snap Endpoint mode to select an object. Rather than struggling to pick the exact end point of the first line, typing C (ENTER) at the "next point:" will close the

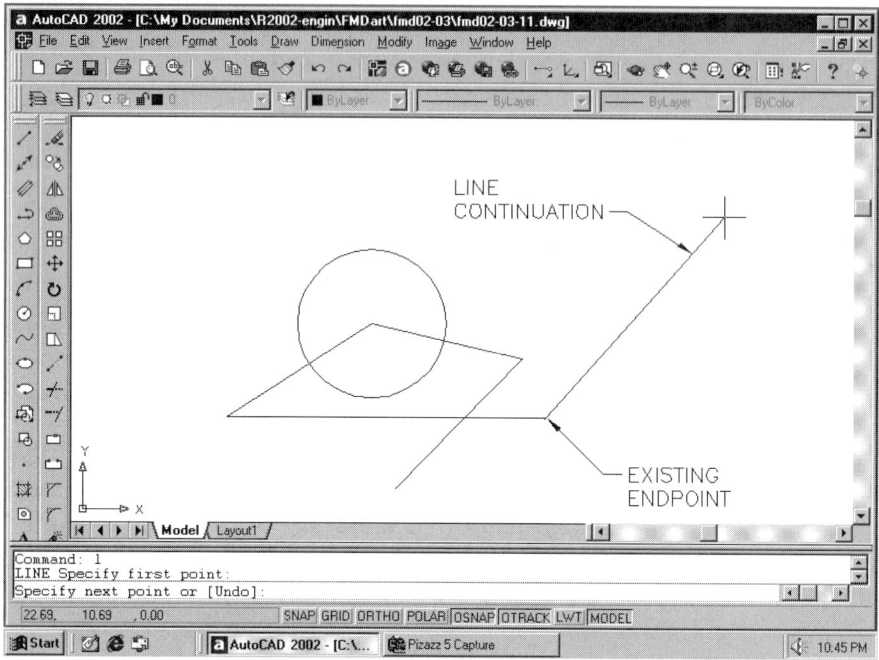

Figure 3.11 *A line can be continued from the end of the last line drawn by pressing* ENTER *for the "first point."*

polygon precisely. This can be seen in Figure 3.12 and in the following command sequence:

Command: **L** ENTER *(Or click the Line button on the Draw toolbar.)*

LINE

Specify first point: **3,3** ENTER

Specify next point or [Undo]: **@6,0** ENTER

Specify next point or [Undo]: **@0,2** ENTER

Specify next point or [Close/Undo]: **@–6,0** ENTER

Specify next point or Close/[Undo]: **C** ENTER

Command:

Typically, the polygons you will draw for projects will be much more complex. It's not uncommon to have to interrupt the drawing of the shape to execute another command, such as Erase. Once the polygon has been interrupted, C (ENTER) cannot be used. This can be seen in Figure 3.13 and in the following command sequence:

Command: **L** ENTER (Or click the Line button on the Draw toolbar.)

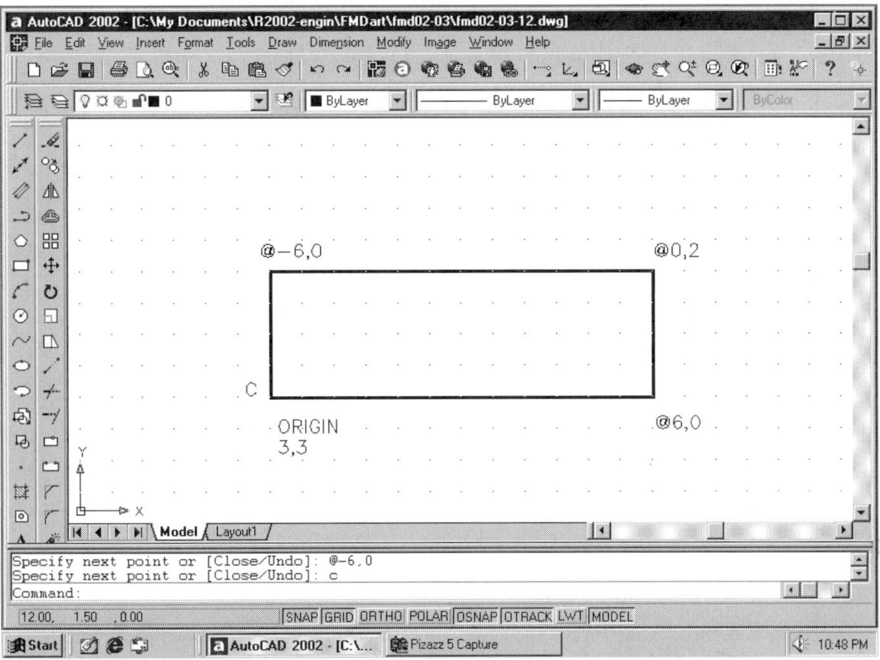

Figure 3.12 *Typing* **C** (ENTER) *for the last "to point" will close the polygon.*

LINE

Specify first point: **4,2** ENTER

Specify next point or [Undo]: **@2,0** ENTER

Specify next point or [Undo]: **@0,2** ENTER

Specify next point or [Close/Undo]: **@2,2** ENTER

Specify next point or [Close/Undo]: **@–2,2** ENTER

Specify next point or [Close/Undo]: **@–2,–2** ENTER *(Wrong coordinate entered and a line drawn.)*

Specify next point or [Close/Undo]: ENTER

Command: **E** ENTER *(Or click the Erase button on the Modify toolbar.)*

ERASE

Select objects: ENTER

Command: **L** ENTER *(Or click the Line button on the Draw toolbar.)*

Specify next point or [Undo]: ENTER *(This uses the endpoint of the last line drawn as the new endpoint.)*

Specify next point or [Undo]: **@–2,0** ENTER

Specify next point or [Undo]: **@–2,–2** ENTER

Specify next point or [Close/Undo]: **@0,–2** ENTER

Specify next point or [Close/Undo]: **@2,0** ENTER

To point: **C** ENTER

Command:

Although the intended close is at the origin point of 4,2, the polygon will close at the start of the second sequence of lines that were drawn. Chapter 8 will offer other drawing tools to select exact points of a drawing entity as alternatives to C (ENTER).

In Figure 3.13 an error was made and the **Line** command was exited to correct the mistake. A better way to correct the mistake would have been to use the **Undo** command when the "next point: @–2,–2 (ENTER)" was entered. This would have returned the origin point to the end of the last segment drawn (–2,–2) and the polygon could have been continued, as shown in Figure 3.14.

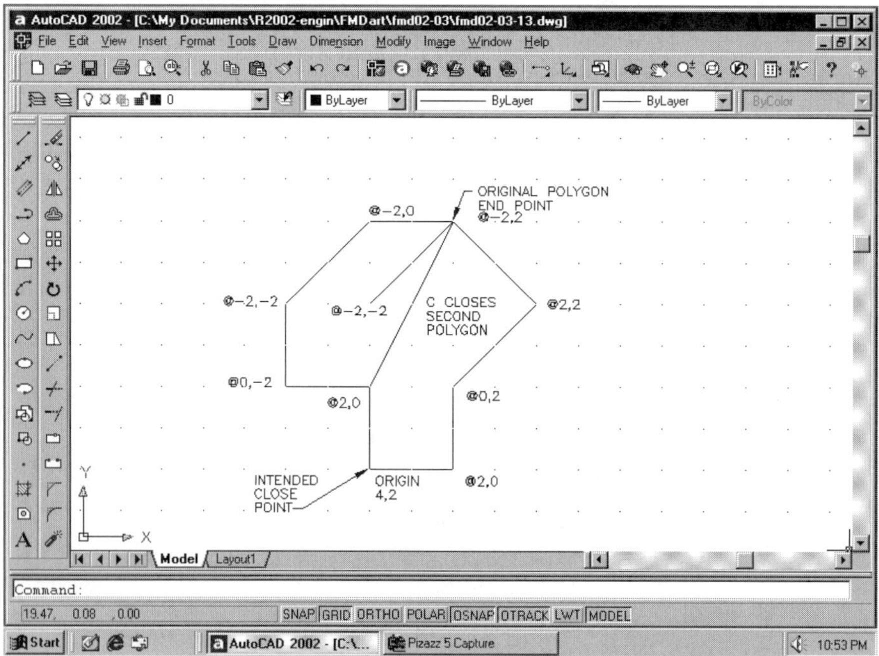

Figure 3.13 *If the drawing of the polygon has been interrupted,* **C** *(ENTER) will terminate the line at the original endpoint rather than at the close of the total polygon. The desired close point can be easily selected with OSNAP activated by moving the cursor near the desired point. As the cursor nears the endpoint, a box will be placed around the crosshairs, indicating an endpoint will be selected.*

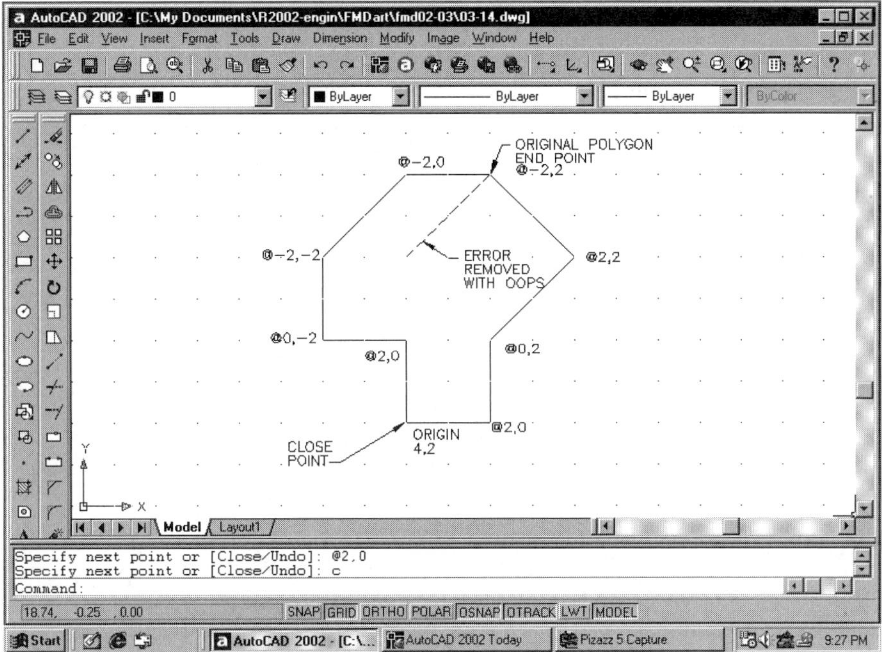

Figure 3.14 *The* **Oops** *command should be used to remove mistakes rather than exiting the* **Line** *command and using the* **Erase** *command.*

CONTROLLING LINES WITH THE STATUS BAR TOOLS

The status bar was introduced in Chapter 2. This discussion will help you make practical application of these tools. Snap, grid, and polar tracking can each be set using the **Drafting Settings** dialog box. Access the dialog box by selecting **Drafting Settings** from the **Tools** menu, and then selecting the desired tab.

Snap

Snap was introduced in the last chapter. Although the snap grid is invisible, it can be a valuable tool for placing lines and other drawing elements. Snap should be adjusted to a modular size of the object being drawn. When working on a modular part, a snap setting of 1" would be helpful for the initial layout of the basic shapes. A setting of .25" allows lines to be easily placed as the plan is being fine-tuned. Snap will be most helpful if you adjust the value throughout the life of the drawing to meet specific needs. Keep in mind that when snap is on, you might have a difficult time drawing angled lines. Snap can be toggled ON/OFF using the SNAP button on the status bar at any point of a command sequence.

Grid

The visible grid introduced in Chapter 2 should be used to plan and lay out a drawing. Setting the grid value as a multiple of the snap setting will greatly aid in placing drawing objects. The grid serves as a visual reference as well as an aid in placing lines. A grid of 2" and a snap of 1" will greatly aid in the layout of simple multi-view projects. The 2" spacing of the grid will provide a reference to keep track of object sizes, and the snap will provide a useful grid for placing lines. Grid can be toggled ON/OFF using the GRID button of the status bar at any point of a command sequence.

ORTHO

Most of the beginning projects that you will draw will be composed of perpendicular lines. Type L (ENTER) at the Command prompt and draw a few lines. You'll notice that as a line is drawn from the "first point" to the "next point" a rubber-band line is extended to the crosshairs and moves as the crosshairs move. The ORTHO command will allow you to place the end of the rubber-band line in an exact horizontal or vertical position based on the current snap or grid pattern. If the snap and grid are rotated, ORTHO will respond by producing perpendicular lines based on the new grid. This will be necessary when auxiliary drawings are drawn. See Chapter 6. Toggle ORTHO ON/OFF by clicking the ORTHO button on the status bar.

Polar Tracking

Earlier in this chapter you were introduced to using polar coordinates to place a line. Activating polar tracking by clicking the POLAR button on the status bar will produce a display at specified intervals to aid in placing lines. The display will show the distance and the angle of the line in progress. Figure 3.15 shows an example of a polar display for placing a line. The display can be a helpful alternative to entering coordinates at the keyboard. Settings can be adjusted on the **Polar Tracking** tab of the **Drafting Settings** dialog box (choose **Drafting Settings** from the **Tools** menu). The **Polar Tracking** tab can be seen in Figure 3.16. Key elements of the dialog box include the following:

Increment angle—This edit box allows the desired angle to be used with the tracker to be set. Default angles include 5, 10, 15, 18, 22.5, 30, 45, and 90 degrees. Selecting 10 will display a tracking indicator as the cursor is moved in 10° increments.

Additional angles—Click the **New** button to see this dialog box, which allows up to 10 additional angles to be added to the default list. If you're drawing an object with an angle different than the default angles, selecting an angle such as 27.5° would aid in the layout of the project. The setting could be entered as follows:

1. Click the **New** button.

2. Enter 27.5 in the edit box.
3. Click the **OK** button.

Settings can be removed from the default list by highlighting the setting to be removed and then clicking the **Delete** button.

Object Snap Tracking Settings—This option selects options for object track settings. Options include orthogonal and polar angle tracking settings.

> **Track orthogonally only**—With this option active, the object tracker displays only horizontal or vertical object snap tracking paths for acquired object snap points when Object Snap Tracking is on.
>
> **Track using all polar angles**—Allows the cursor to track along any polar angle tracking path when Object Snap Tracking is on.

Polar Angle Measurements—Sets the method to be used to measure polar angles. Options include Absolute and Relative.

> **Absolute**—Displays polar tracking angles based on the current UCS.
>
> **Relative to last segment**—Displays polar tracking angles on the most recent object.

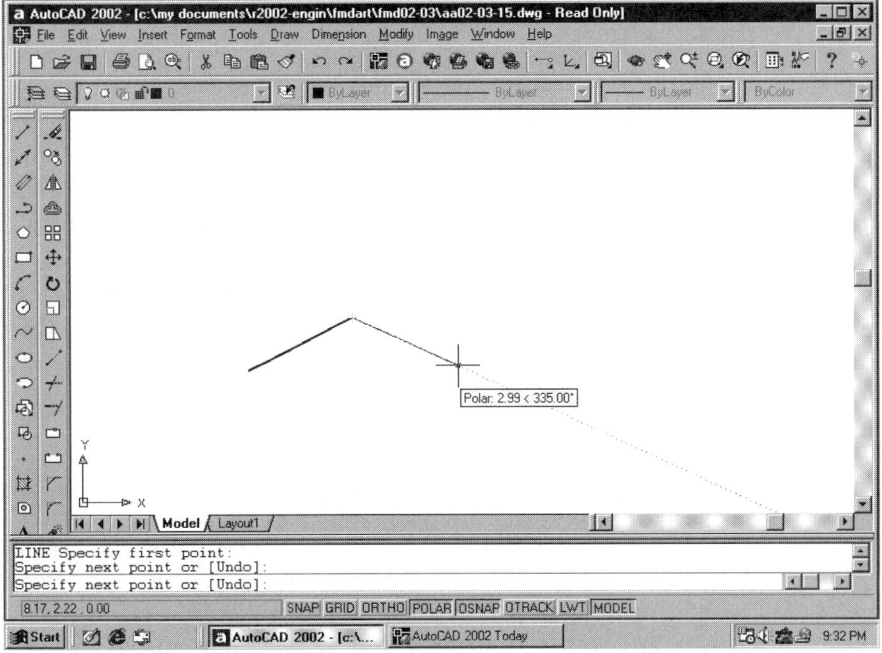

Figure 3.15 *Polar display for tracking a line. The display can be a helpful alternative to entering coordinates at the keyboard. Settings can be adjusted on the* **Polar Tracking** *tab of the* **Drafting Settings** *dialog box.*

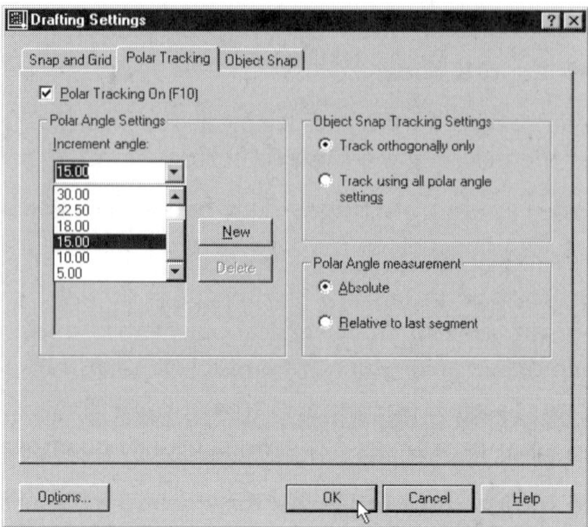

Figure 3.16 *The* **Polar Tracking** *tab of the* **Drafting Settings** *dialog box can be used to control the display used with polar tracking.*

Options—Clicking this button displays the **Options** dialog box. This dialog box can be used to adjust the color and size of the object snap marker. The dialog box also provides a convenient method of altering the size of the aperture box.

Object Snap

Activating Object Snap provides one of the easiest and fastest means of accurately drawing in AutoCAD. With this option ON (click OSNAP on the status bar), a marker will be placed on the crosshairs as it is moved near an endpoint, midpoint, or intersection of lines. With the marker displayed, the "next point" of a line will be placed exactly at the endpoint. Figure 3.17 shows an example of the display as the cursor is moved near the endpoint of an existing line. Rarely will you want to turn this feature off. Chapter 8 will explore methods of altering the default settings of Object Snap. If you want to explore on your own, select **Drafting Settings** from the **Tools** menu and select the **Object Snap** tab to view the options. Use the **Help** menu if you just can't wait until Chapter 8.

AN INTRODUCTION TO EDITING LINES

This section will introduce three methods of removing drawing errors from a drawing file. These methods include the **Erase**, **Oops**, and **Undo** commands. The **Redraw** command will also be introduced as a method of cleaning the drawing display. Chapter 9 will present methods for using these commands more effectively.

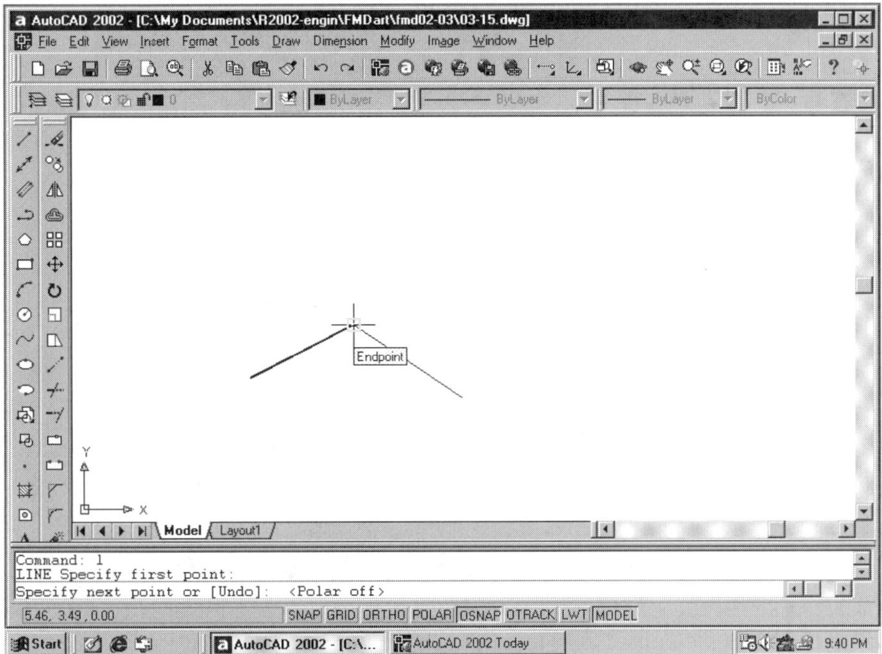

Figure 3.17 *Object Snap is an excellent method for increasing drawing accuracy. With the OSNAP button on the status bar active, a square will be added to the crosshairs as it moves near the endpoint of a line. The "next point" will automatically snap to the endpoint of the selected line.*

ERASE

The **Erase** command will allow you to remove unwanted drawing objects. To remove objects using the **Erase** command, select the **Erase** button on the **Modify** toolbar, type E (ENTER) or select **Erase** from the **Modify** menu. This will display the prompt:

> Command: **E** ENTER *(Or click the Erase button on the Modify toolbar.)*
>
> Select objects:

As the prompt is displayed, the crosshairs turn into a pick box. The pick box can be used to select objects to be removed. Place the box on the object to be removed and left-click. In Figure 3.18 the circle has been selected for removal. Once selected, the object will now appear highlighted. An unlimited number of objects to be erased can be selected. The command line will continue to show "Select objects:" until ENTER is pressed. Pressing ENTER will remove the selected lines and restore the Command prompt, as seen in Figure 3.19.

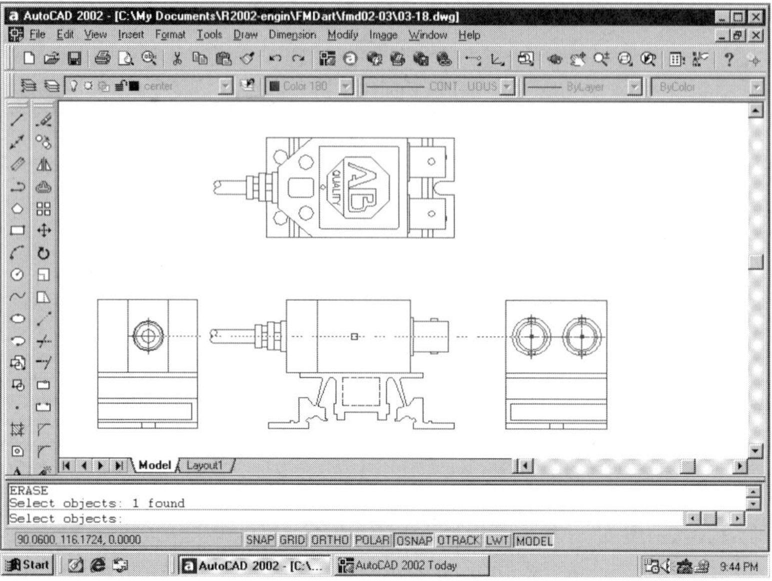

Figure 3.18 *As the* **Erase** *command is entered, the crosshairs will be replaced by a pick box. Placing the pick box on the desired item to be erased and selecting it, will select that object for removal. Notice that the centerline has been selected for removal. Once an item has been selected for removal from the drawing, it will be displayed as short dashed lines. Courtesy Floyd D. Miller.*

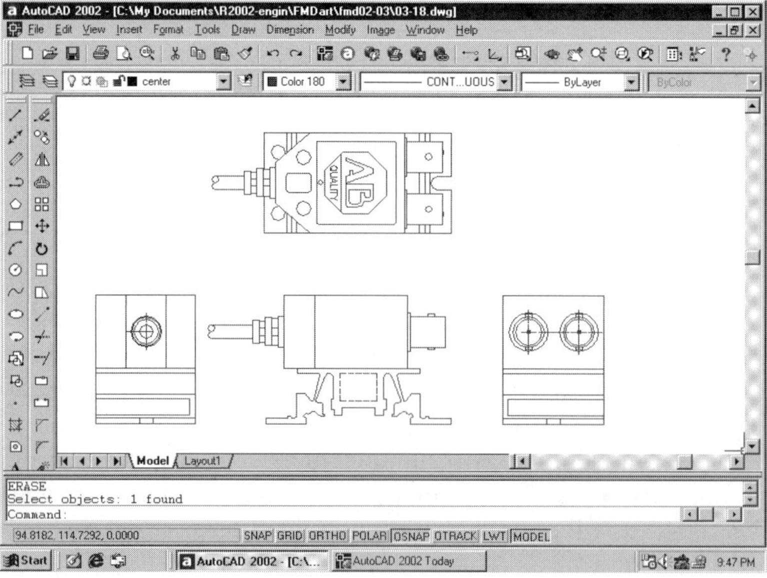

Figure 3.19 *Pressing* ENTER *or right-clicking will remove the selected object from the drawing screen. Courtesy Floyd D. Miller.*

Erase allows for the last drawing object to be erased without having to select the object. This can be done by selecting the **Erase** button or typing E (ENTER) and then L (ENTER). The Command prompt is as follows:

Command: **E** ENTER *(Or click the Erase button on the Modify toolbar.)*

ERASE

Select object: **L** ENTER

I found:

Select objects: ENTER

Command:

As the command line is restored, the last object drawn is eliminated. If you would like to remove the objects drawn with the last drawing sequence, type the following:

Command: **E** ENTER *(Or click the Erase button on the Modify toolbar.)*

ERASE

Select objects: **L** ENTER

Select objects: *(Select object.)*

I found

Select objects *(Continue to select objects or press enter to end sequence.)* ENTER

Command:

Selecting objects to be erased can be continued indefinitely until the entire drawing is removed, but this process is not the most efficient way to remove multiple objects. Later chapters will provide in-depth coverage of **Erase** and other editing options, as well as more efficient methods of selecting objects to be edited.

OOPS

Occasionally an object will be erased by mistake. The **Oops** command will restore the object removed by a previous **Erase** command. Chapter 11 will provide options for modifying drawing objects.

UNDO

An alternative to **Erase** or **Oops** is the **Undo** command. **Undo** will remove the command that was just executed. This command can be used as many times as you wish, and it will continue to move you backward through the drawing sequence one step at a time. Select **Undo** on the **Standard** toolbar, from the **Edit** menu, or by keyboard. When the toolbar is used to enter the command, simply select the button when you wish to remove the command that has just been executed. If the button is selected in the middle of a command sequence or after the command is ended, the entire line sequence will be removed. The **Undo** command can be entered by keyboard in the middle of the **Line** command using the following sequence:

Specify next point or [Close/Undo]: **@–2,–2** ENTER

Specify next point or [Close/Undo]: **U** ENTER

Specify next point or [Close/Undo]: **@–2,0** ENTER

If you decide that you really do want the line after you have used the **Undo** command, the **Redo** command can be used to correct your change of mind. **Redo** will reverse the effects of **Undo**. Perform a **Redo** by selecting the **Redo** button on the **Standard** toolbar or by typing REDO (ENTER) at the Command prompt.

CHAPTER 3 EXERCISES

Set limits, units, snap, and grid to best suit each project. Save each exercise to a diskette and to a folder on the hard drive or network, as per your instructor's directions.

1. Draw the following outline using the coordinates that are provided.

   Point of beginning: 1,1

point b	@4.2426<45
point c	@3.1623<342
point d	@2.5<0
point e	@4.0311<120
point f	@4.2720<159
point g	@3.9051<230

   C (ENTER)

   Save your drawing with a file name of E-3-1.

2. Draw the following outline using the coordinates that are provided.

   Point of beginning: 2,1

point b	@2.5<10
point c	@3.5<52
point d	@2.5<310
point e	@1,1
point f	@1,–1
point g	@4<90
point h	@–6,1

point i @3.5<231

C (ENTER)

Save your drawing with a file name of E-3-2.

3. Draw the following outline using the coordinates that are provided. Save the drawing as file name E-3-3.

   Point of beginning: 9,9

   point b @–5,–2.5

   point c @0,–1.5

   point d @4<230

   point e @–1.5,–2

   point f @4.5,1.5

   point g @1.5<90

   point h @2<335

   point i @2.5<120

   point j @5.5<330

   point k @1,1

   point l @–2.5,1

   point m @–.5,–.5

   point n @3<151

   point o @3.8125,0

   point a **C** (ENTER)

4. Use the following coordinates to lay out the required shape. Save as file name E-3-4.

   Starting point: 7.75, 7.95

   @1.91<127

   @–2,0

   @0,–2

   @–1.50, 1.75

   @–.96,0

   @1.62<137°

   @.46<227°

   @1.22<311°

@–.92,0

@1.62,–.5

@-1.11,-4.13

@3.5,0

@1.62,1.62

@1.75<308

@3.32<80.73

5. Use the dimensions to draw the object below. Save the drawing with the file name E-3-5.

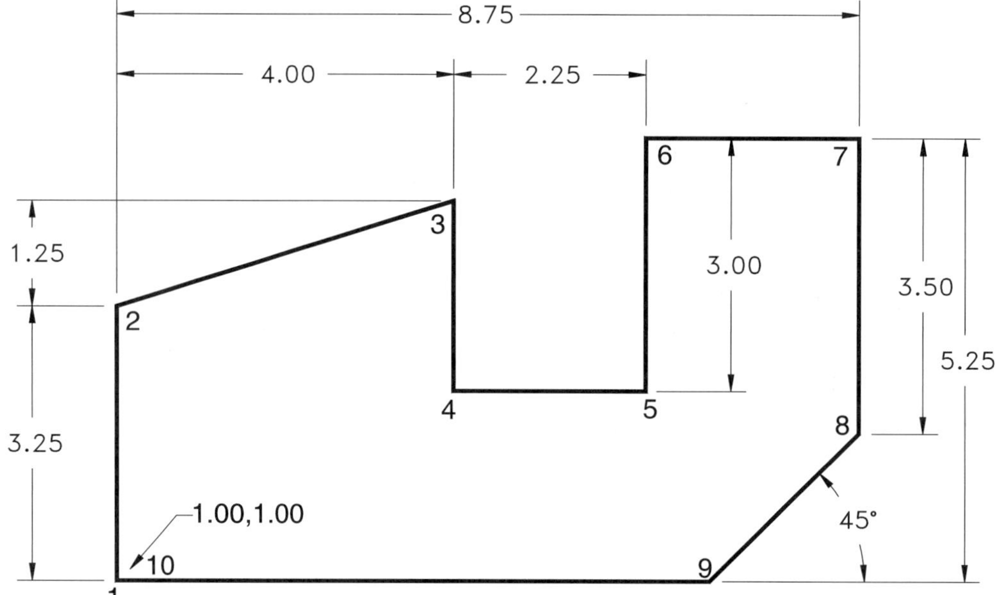

6. Use the attached drawing to draw the object below. Save the drawing as E-3-6.

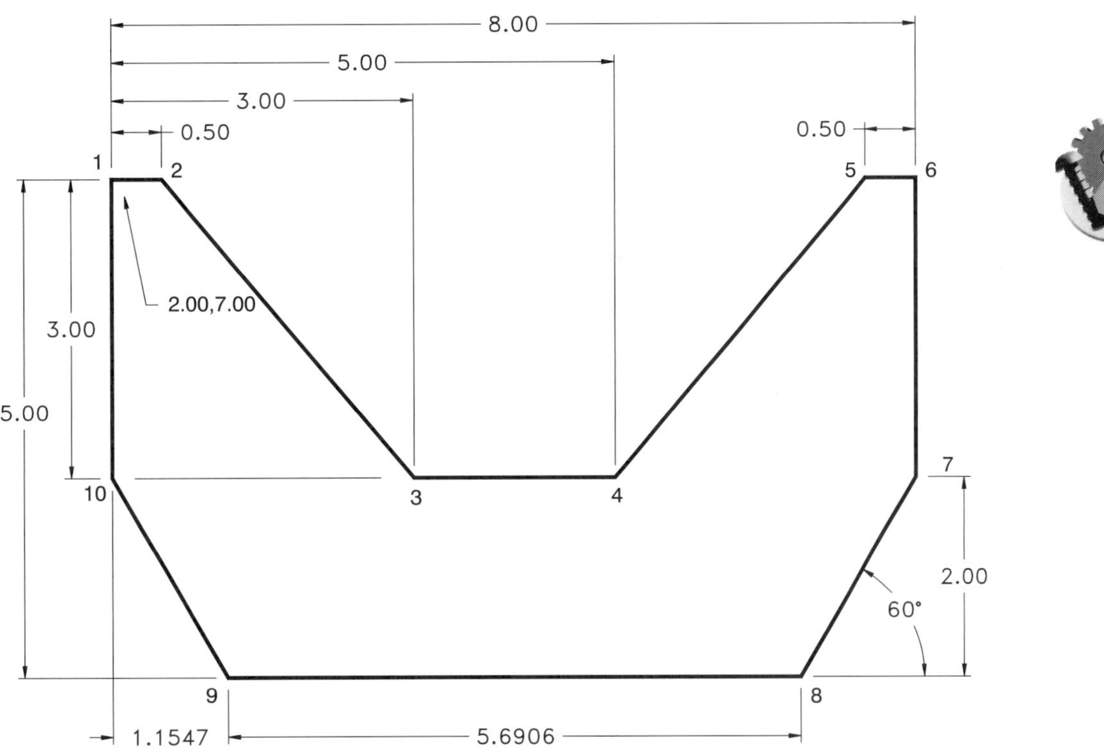

7. Use the attached drawing as a guide to draw the object below. Save the drawing as E-3-7.

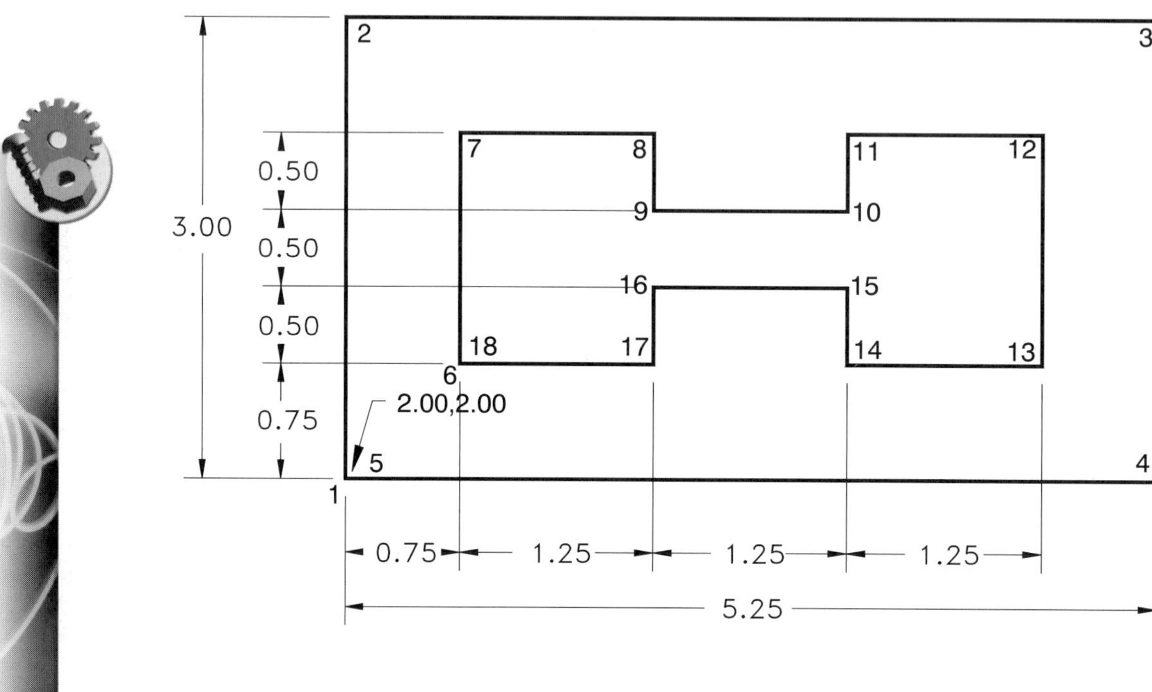

8. Use the attached drawing as a guide to draw the object below. Save the drawing as file name E-3-8.

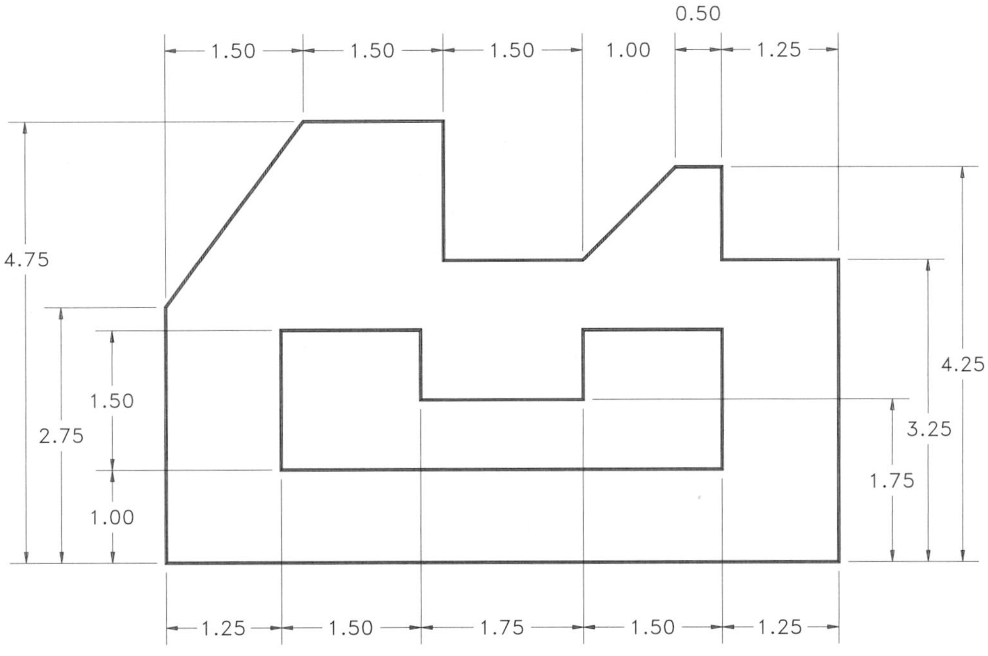

9. Use the attached drawing as a guide to complete the stencil. Save the drawing as file name E-3-9.

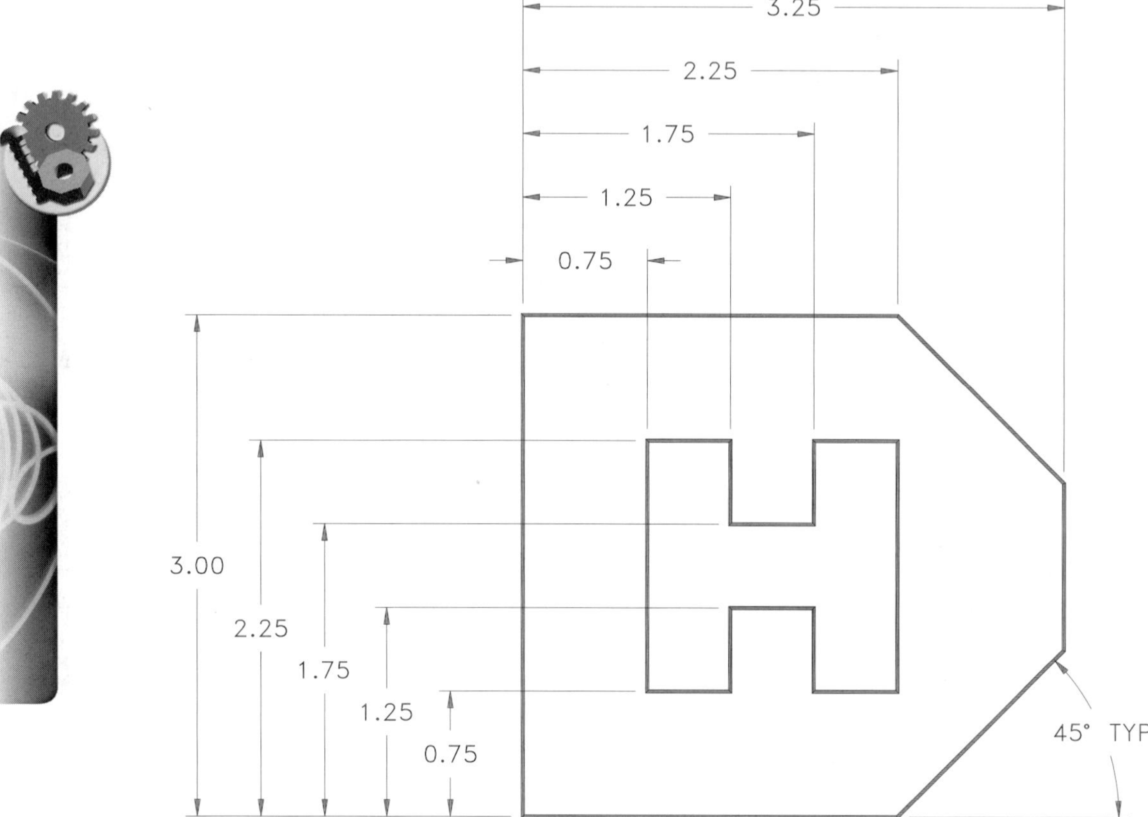

10. Use the attached drawing as a guide to complete the drawing. Save the drawing as file name E-3-10.

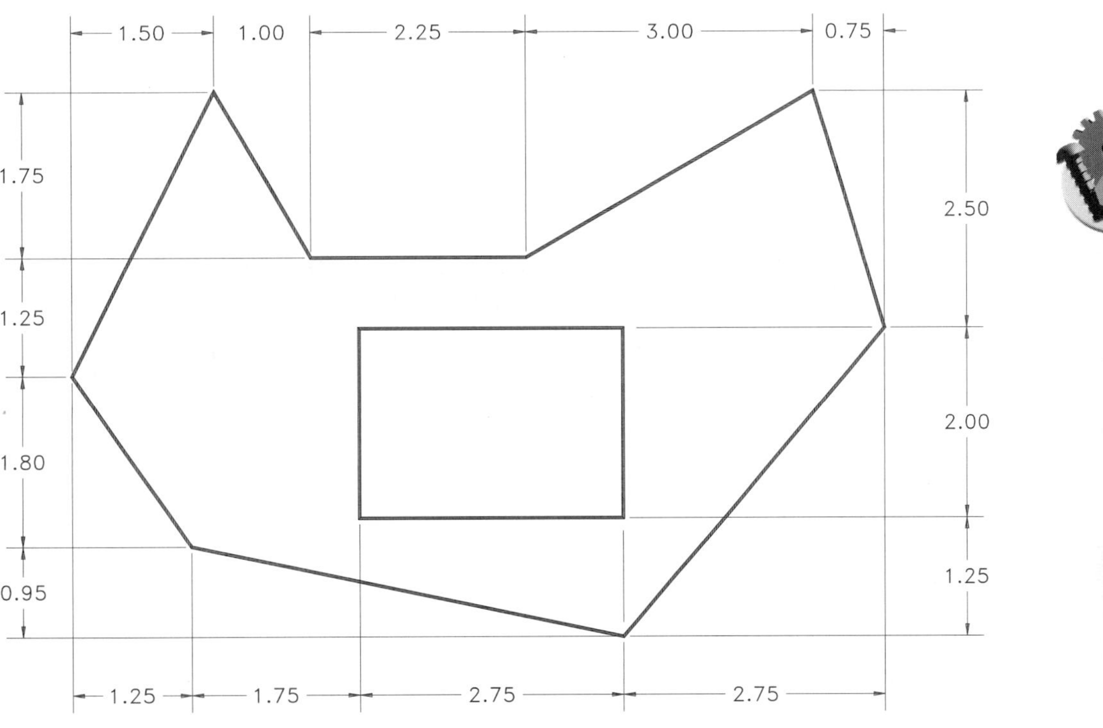

CHAPTER 3 QUIZ

1. Use the drawing below to provide absolute coordinates. Grid=.5", Snap=.25"

   a. 2.5,2
   b. ________
   c. ________
   d. ________
   e. ________
   f. ________
   g. ________
   h. ________
   i. ________
   j. ________
   k. ________
   l. ________
   a. ________

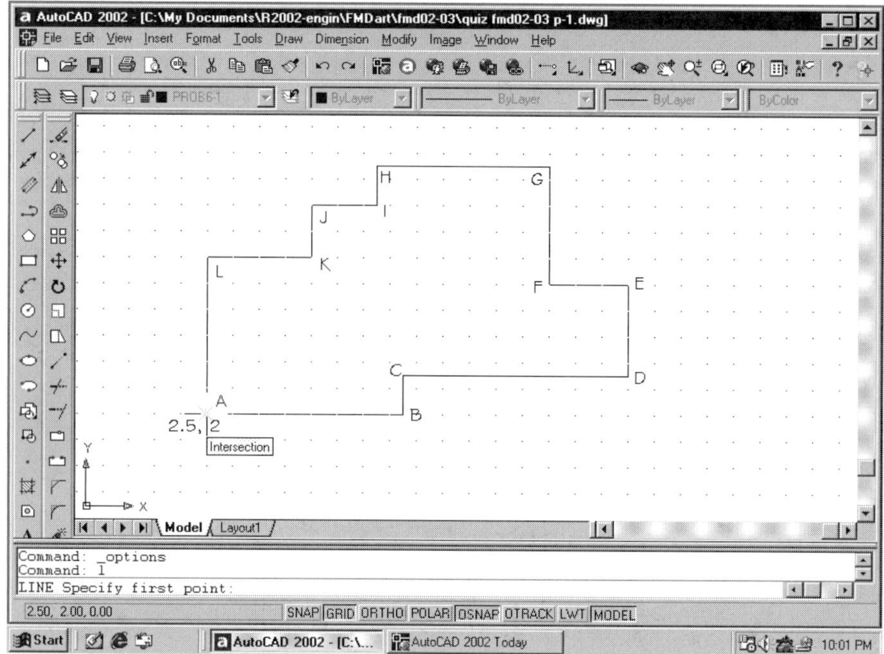

2. Use the drawing from Quiz problem 1 to provide relative coordinates.

   a. 2.5,2
   b. ________
   c. ________
   d. ________
   e. ________
   f. ________
   g. ________
   h. ________
   i. ________
   j. ________
   k. ________
   l. ________
   a. ________

3. Use the drawing from Quiz problem 1 to provide polar coordinates.

   a. 2.5,2
   b. ________
   c. ________
   d. ________
   e. ________
   f. ________
   g. ________
   h. ________
   i. ________
   j. ________
   k. ________
   l. ________
   a. ________

4. Use the drawing below to provide relative coordinates.

   a. 1.5,1.5
   b. ________
   c. ________
   d. ________
   e. ________
   f. ________
   g. ________
   h. ________
   i. ________
   j. ________
   a. ________

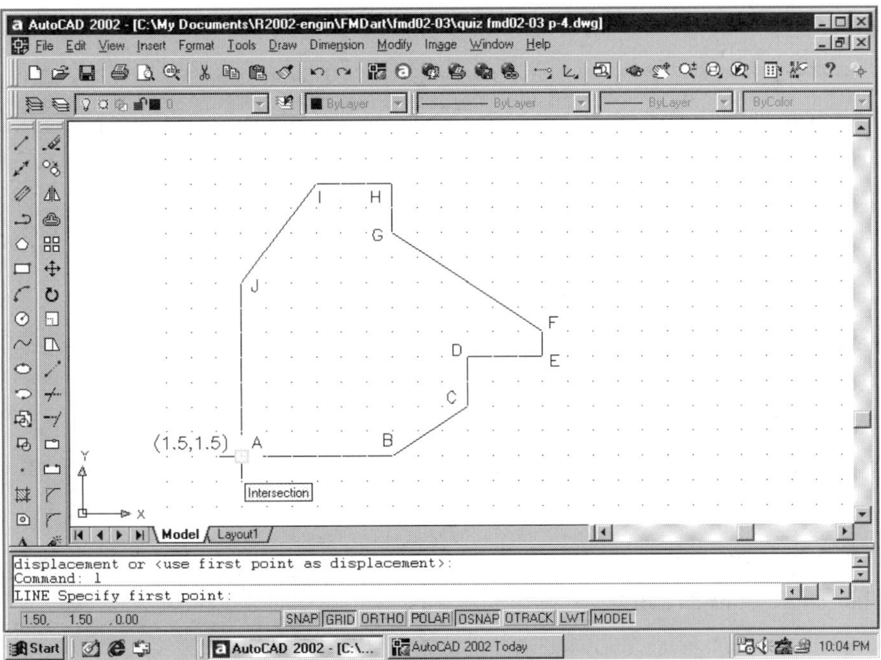

5. Use the drawing from Quiz problem 4 to provide absolute coordinates.

a. 1.5,1.5	e. ________	i. ________
b. ________	f. ________	j. ________
c. ________	g. ________	a. ________
d. ________	h. ________	

6. Use the drawing from Quiz problem 4 to provide polar coordinates. Use relative coordinates where insufficient information is available to determine polar entries.

a. 1.5,1.5	e. ________	i. ________
b. ________	f. ________	j. ________
c. ________	g. ________	a. ________
d. ________	h. ________	

7. What entry method uses the X,Y format to enter coordinates?
8. What entry method uses the @X<# format to enter coordinates?
9. What entry method uses the @X,Y format to enter coordinates?
10. Can a true vertical line be drawn with ORTHO OFF?
11. Can an angled line be drawn with snap and grid both on?
12. Describe the line that is formed between 2,0 and 4,0.
13. Describe the line that is formed between 2,0 and 4,2.
14. List two methods for activating the coordinate display.
15. What settings must be changed to use Cartesian, relative, and polar coordinates on the same drawing?
16. How can the **Line** command be accessed from the menu?
17. List four methods for accessing the **Line** command.
18. What steps are required to erase a line?
19. If a line has been erased by mistake, how can it be restored?
20. What effect does **Erase** have on the cursor?
21. Describe the difference between **Oops** and **Undo**.
22. What effect will typing **C** (ENTER) have on a line segment?
23. How does the snap setting affect the use of the **Line** command?
24. What type of coordinate entry is best suited for drawing E-3-5?
25. Use the **Help** command and list four methods of accessing the **Erase** command.

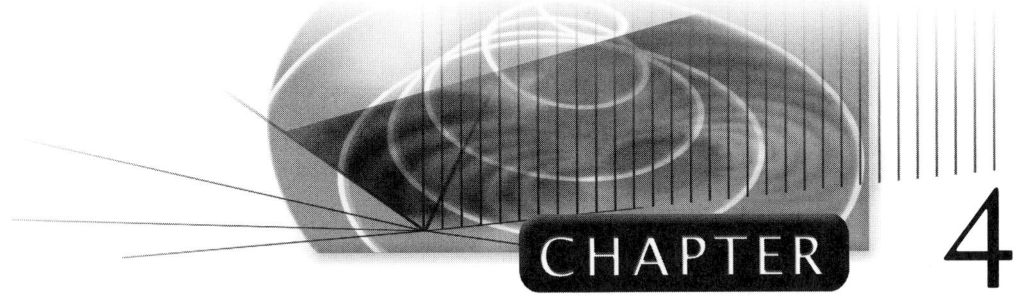

4

Working with Drawing Files

This chapter will introduce methods for

- Creating and saving drawing templates
- Saving user preferences
- Saving drawing files
- Managing files and folders

Commands and major options to be explored in this chapter include

- Paper space
- Model space
- Viewports
- **Units**
- **Limits**
- **Pan**
- **Options**
- **Save**
- **Saveas**
- **Qsave**
- **Savetime**
- **Quit**
- **Exit**
- **Close**

- **Closeall**
- **Audit**
- **Recover**

In Chapter 2 you explored methods of entering the AutoCAD drawing area when a new drawing file is created with the wizard or a template. In Chapter 3 you started to create great works of art. This chapter will introduce you to methods for creating a template drawing, explore methods to save drawing properties for future use, expand your methods for saving and managing drawing files, and explore methods to merge existing drawings.

CREATING AND SAVING DRAWING TEMPLATES

In Chapter 2 you were introduced to the templates that AutoCAD provides to help make each new drawing that you start. The template can be created and saved using the TODAY window or the **Startup** dialog box, and then used each time a drawing requiring those parameters is needed. Several useful drawing parameters were introduced in Chapter 2. Other useful values such as linetype, lettering size and styles, and dimensioning styles and format will be introduced in chapters that follow. Necessary values can be added to the drawing template for future use. Once created and saved, they can be opened to form the base for a new drawing file and save time.

USING AN EXISTING TEMPLATE

AutoCAD provides several templates to ease drawing setup. The default template is the ACAD-named plot styles. The names of recently used templates will be displayed above the listing as various templates are used. Scroll through the A listing and you'll find several different ANSI templates. These templates will provide a drawing screen showing a sheet of paper with a border that is suitable for most mechanical drawings. You'll also notice that in addition to variation by paper size, the templates are either "named plot styles" or "color dependent." Plot styles alter the way a plotted drawing will appear. Each object in a drawing has a plot style property. These styles can be assigned by color or by named groups. With a change in the plot style, the color, linetype, and lineweight of an object can be altered on the print. This will allow objects within the drawing to be highlighted. The benefits of plotting using assigned styles will be addressed in detail in Chapter 23 as plotting is explored. For now, select the **Template** tab in the TODAY window or use the traditional **Startup** dialog box and select the **Use a Template** option. Select the A menu, and then select the ANSI C-color dependent plot style template. This will produce the template shown in Figure 4.1. Although the title implies that this template is for mechanical drawings, it should not be used without some basic changes. Three areas of the template should be examined before you complete a drawing. These areas are paper space, model space, and viewport.

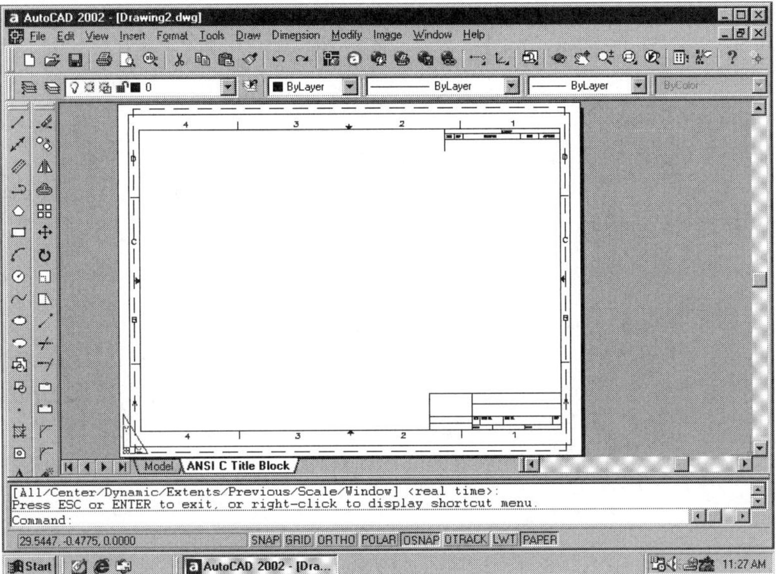

Figure 4.1 *The ANSI-C Color Dependent Plot Style template can be used for a base layout that will meet the needs for most mechanical projects. Display the template by selecting the* **Use a Template** *option from the* **Startup** *dialog box and selecting the desired paper size option.*

Paper Space

As the template is displayed, you're working in paper space (notice the paper space icon in the lower left corner) with the ANSI C-Title Block tab active. The paper is 17" × 22" with a drawing area of approximately 20.0" × 14.5". In effect, you're now working on a blank sheet of paper. Move the cursor and you'll notice that it can be moved to any point in the display area. Paper space will allow you to start a line in the drawing area and finish the line at any point outside the drawing border.

Note: Paper space will be a great aid for plotting the drawing, but it should not be used to create a drawing.

Model Space

Model space is the area of the template where the drawing should be completed. Model space allows you to work in "real" space. Select the Model tab below the drawing area to switch from paper to model space. This will hide the template and display a blank screen with the model space icon in the lower left corner of the display. With model space active, you'll have a drawing area of approximately 22" × 12.5". The size will vary depending on the number of command lines displayed, and which toolbars you have displayed. By changing the size of the space, you can display the 100" long part on the screen in its entirety. The size of the drawing space is altered using the Limits command. Before you explore new areas, enter the ANSI-C-Color Dependent Plot Styles template, select the Model tab, and use the **Line** command to draw a box that is 15" × 12". This box will be useful in visualizing how model space and the viewport work together. Notice that the drawing can be seen, but the template title block is not displayed. You're still working in the template, but in model space, you're only displaying the space between the borders.

Viewports

The viewport is a 'hole' in paper space that allows the drawing in model space to be viewed. A viewport allows you to place a 100" long part on a 22" long sheet of paper. Figure 4.2 shows a representation of how the viewport and the template and model space work together. Switch back to paper space by selecting the ANSI-C Title Block tab. Select the PAPER button on the status bar. The PAPER button will now change to read MODEL, and the drawing will change to display the drawing shown in Figure 4.3.

The template appears similar to Figure 4.1, but it has changed to reflect model space inside the template borders. The bold black line represents the edge of the viewport.

Move the crosshairs to the border of the template. As soon as the crosshairs touch the border, it changes to an arrow. You're not allowed to draw outside the viewport. You're now working with approximately a 12" × 9" drawing area or viewport. Nice looking border, but not many projects will fit into the drawing area.

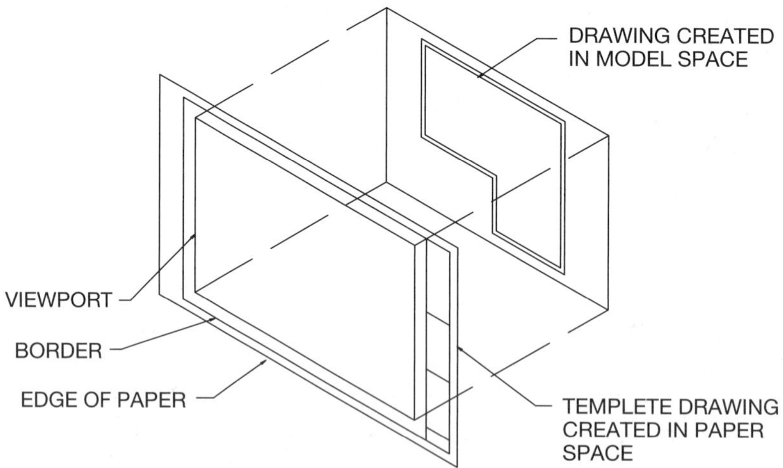

Figure 4.2 *A viewport is provided in each drawing template created in paper space. Additional layouts can be created to allow for different plotting displays.*

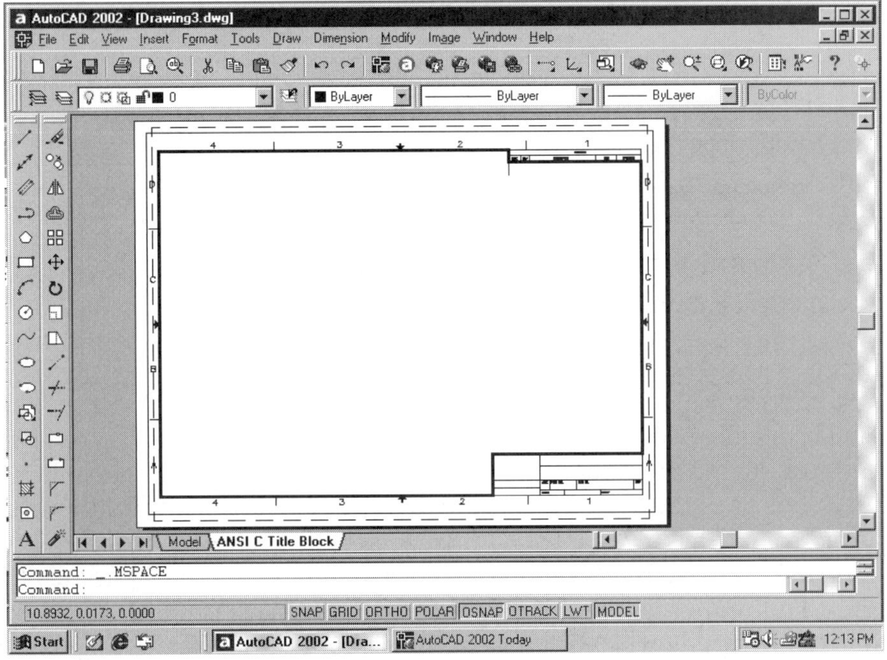

Figure 4.3 *Selecting the PAPER button on the status bar will toggle the display to a viewport with model space.*

Note: Even though the viewport display is in model space, your drawing should not be created or edited here. This area should only be used to adjust the relationship of the completed drawing (created in model space using the model tab) to the border.

Adjusting the Viewport Size

The size of the viewport must be adjusted based on the desired scale to be used when the drawing is plotted. This chapter will help you adjust the size of the viewport so that a scale of 1=1 or full scale can be used. Chapter 22 will explain how to adjust the viewport for varied scales.

If you haven't already done so, open the ANSI-C Color Dependent Plot Styles template and select the Paper tab. Use the methods described in Chapter 2 to set the limits as 20" × 14.5". This size was determined by using the **Line** command and the coordinate display to determine the distance between the borders. Use the following procedures to update the template to mechanical standards:

- Type **UNITS** (ENTER) at the Command prompt. Use the **Drawing Units** dialog box to adjust the units to the desired precision.
- Type **LIMITS** (ENTER) at the Command prompt. Use the following sequence to alter the drawing limits:

Command: **LIMITS** ENTER

Reset Model space limits:

Specify lower left corner [ON/OFF]:<0.00,0.00">: ENTER

Specify upper right corner <12.00,9.00>: **20,14.5** ENTER

Command: **Z** ENTER *(Or select the Zoom button on the Object Properties toolbar.)*

Specify corner of window, enter a scale factor (nX or nXP), or

All/Center/Dynamic/Extents/Previous/Scale/Window/

<Realtime>: **A** ENTER

Command:

You should be familiar with the effects of using the **Limits** command. As you typed Z (ENTER), you started the **Zoom** command. Think of zoom as a telephoto lens on a camera. Zoom alters the area of a drawing to be displayed. (The **Zoom** command will be explored in detail in Chapter 9.) If you enter the **Zoom** command and type A (ENTER), you've selected the All option. You've asked AutoCAD to display all of the display area. This process will alter the size of the drawing area from 12 × 9 to 20 × 14.5. Now you have a drawing area that is suitable for drawing. Later chapters will show you how to add other elements to the template to ease drawing setup.

Note: It's important to remember that not all of this drawing area should be used. ANSI guidelines recommend that a 0.5" space be provided on each side of the drawing area and the border. In addition, the drawing should not be allowed to extend into the area above or beside the title block.

Comparing Model Space, Paper Space, and the Template

Now that you've explored each aspect of a template, you can combine those features to create a useful drawing tool for future drawings. As you've explored each aspect of a template, a box has been drawn in model space, and the limits that have been set are for the viewport. Select the Model tab to restore the view of the 15" × 12" box. The box that was drawn earlier as you examined model space can be seen, but the template title block is not displayed. You're still working in the template, but in the model space, you're displaying only the space between the borders.

Selecting the ANSI tab displays the template with the title block and border, but all of the box can't be seen. Click the PAPER button on the status bar, and the viewport is displayed. See Figure 4.4a. The viewport can now be used to adjust the location of the drawing to the border using the **Pan** command. Select the **Pan** button (the hand) on the **Standard** toolbar. As the command is activated, a hand button will be placed in the drawing area. Pressing and holding the select button allows the drawing to be relocated in the viewport. When the drawing is in the desired location, press

ENTER to end the **Pan** command. Chapter 9 will explore the **Pan** command in detail. Figure 4.4b shows the completed process.

Note: As a new AutoCAD operator, you might find this a bit overwhelming. If nothing else, keep in mind that a template is a great place to store drawing tools. It's like the junk drawer we all have in a desk. All the good stuff goes in the drawer for future use. A template is a drawing base where you will begin to store the settings that you use over and over. No one should be setting the precision of units or the drawing limits each time a drawing is started.

The second thing to remember is that you want to work in model space to create objects and paper space to plot them. Each of the preceding chapters has reminded you of this key point. It's important.

CREATING YOUR OWN TEMPLATE

Any drawing can be saved and used as a template. An existing AutoCAD template can be modified and saved with a new name or a template drawing can be created using the methods used to create a drawing file. Open a drawing and establish the drawing units, limits, snap, and grid values. Special consideration will need to be given to the drawing limits as you establish a drawing template. In Chapter 2, as the limits were established, common paper sizes were introduced. The size of paper that the completed project will be printed or plotted on must be considered as the drawing template is

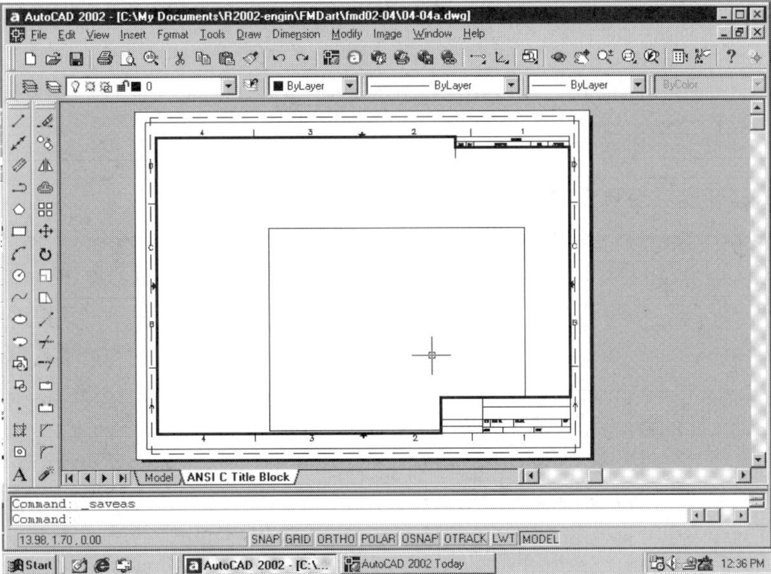

Figure 4.4a *Clicking the ANSI C Title Block tab or the Model button will switch from model space to paper space with the box (in model space) partially displayed in the template. Selecting the PAPER button in the status bar allows the display to be altered.*

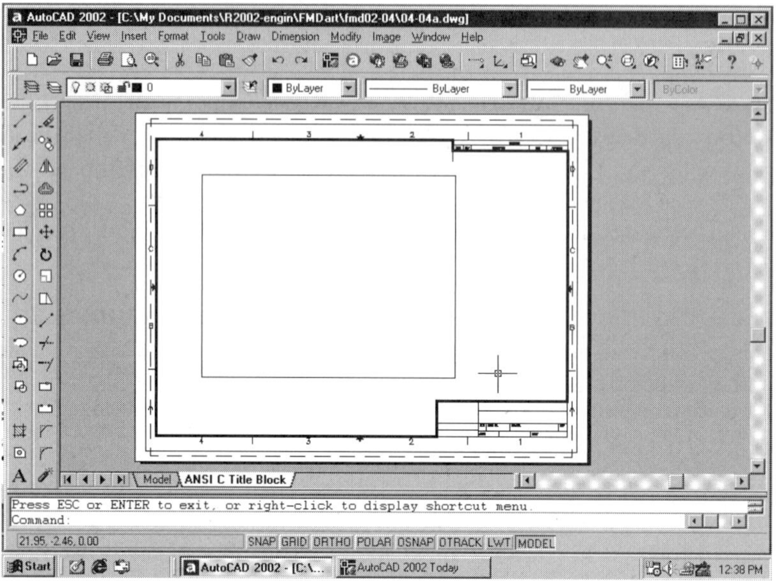

Figure 4.4b *The location of objects drawn in model space can be adjusted relative to the template using the* **Pan** *or* **Zoom** *command. This drawing is now ready for plotting.*

established. Although the drawing limits of model space can be altered at any time throughout the life of the drawing, establishing limits of the viewport that match standard paper sizes will aid in the use of the template. Common limits for A, B, C, and D material are shown in Figure 4.5. Once the desired settings are established, the template can be saved and recalled for future use. Methods of saving a template will be introduced later in this chapter.

SAVING USER PREFERENCES

A drawing template works well for saving drawing settings that must remain constant for each member of the drawing team. There are many features of the drawing environment that do not affect the final project, but that affect the productivity of each team member.

Drawing profiles allow each user to adjust and save the drawing settings to meet personal preferences. Drawing profiles can be especially helpful if you're working at a computer that you share with several other people. If you like a black background and you share your computer with someone who likes a white background, AutoCAD allows these differences to be stored so that they can be changed quickly. Items that might be included in a profile include the **Startup** dialog box display, crosshairs size, display resolution of objects, intervals between automatic saves, text display format, screen colors, and plotter or printer information. Using a profile, you'll be able to adjust, save, and retrieve drawing settings the next time you use the workstation.

SHEET SIZES AND SETTINGS

MECHANICAL SCALE FACTORS					METRIC SCALE FACTORS				
SCALE	TEXT HEIGHT		SCALE FACTOR	LTSCALE FACTOR	SCALE	TEXT HEIGHT		SCALE FACTOR	LTSCALE FACTOR
	.125"	.25				.125"	.25		
.125=1	1.00	2.00	8	4	1 = 50	6.25	12.5	1270	25
25=1	.50	1.00	4	2	1 = 20	2.5	5	508	10
.50=1	.25	.50	2	1	1 = 10	1.25	2.5	254	5
.75=1	.166	.33	1.33	.67	1 = 5	.625	1.25	127	2
2=1	.0625	.125	.50	.25	1 = 2	.25	.5	50.8	1

MECHANICAL SHEET SIZES				METRIC SHEET SIZES			
PAPER SIZE IN INCHES	APPROXIMATE DRAWING AREA	SCALE	APPROXIMATE DRAWING LIMITS	PAPER SIZE IN MM	APPROXIMATE DRAWING AREA	SCALE	APPROXIMATE DRAWING LIMITS
ANSI A 11 X 8.5	9 X 7	.125=1	72" X 56"	A4 297 X 210	277 X 190	1 = 50	13850 X 9500
		25=1	36 X 28"			1 = 20	5540 X 3800
		.50=1	18" X 14"			1 = 10	2770 X 1900
		.75=1	12" X 9.3"			1 = 5	1385 X 950
		2=1	4.5" X 3.5"			1 = 2	554 X 380
ANSI B 17 X 11	15 X 10	.125=1	120" X 80"	A 3 420 X 297	400 X 277	1 = 50	20000 X 13850
		25=1	60 X 40"			1 = 20	8000 X 5540
		.50=1	30" X 20"			1 = 10	4000 X 2770
		.75=1	20" X 13"			1 = 5	2000 X 1385
		2=1	7.5" X 50"			1 = 2	780 X 554
ANSI C 22 X 17	20 X 15	.125=1	160" X 120"	A 2 594 X 420	577 X 400	1 = 50	28700 X 20000
		25=1	80 X 60"			1 = 20	11480 X 8000
		.50=1	40" X 30"			1 = 10	5740 X 4000
		.75=1	26.7" X 20"			1 = 5	2870 X 2000
		2=1	10" X 7.5"			1 = 2	1148 X 800
ANSI D 22 X 17	32 X 20	.125=1	256" X 160"	A 1 841 X 594	801 X 554	1 = 50	41050 X 28700
		25=1	128 X 80"			1 = 20	16420 X 11480
		.50=1	64" X 40"			1 = 10	8210 X 5740
		.75=1	42.6" X 26.6"			1 = 5	4105 X 2870
		2=1	16" X 10"			1 = 2	1642 X 1148

Figure 4.5 *The scale that a drawing will be plotted at must be considered as a template is constructed. The drawing limits will be affected by the scale to be used and the paper size.*

To set user profiles, select **Options** from the **Tools** menu. This will display the **Options** dialog box, with nine tabs to choose from. The tab that was used last is displayed by default. Each of these tabs will be explored in detail as commands are introduced throughout the text. The goal of this chapter is to introduce the **Options** dialog box and to help you save a few of your favorite settings. As new features are introduced, they can be added to your profile. Highlights of the **Options** tabs include the following:

Files—This tab is used to control the directories where AutoCAD will search for driver, menu, and other files. At this point of your AutoCAD career, only a few of the options need to be considered. Both points will be discussed later in this chapter.

Display—This tab allows the display of AutoCAD to be customized. Several of the options of this tab have been discussed in previous chapters. When the values are set and saved in a profile, they will automatically be reset when the profile is selected as the active profile. Take time to evaluate each portion of this box. If in doubt, click the **Help** button at the bottom of the tab for an explanation. Items you should adjust include the following:

Text Lines in the Command Line Window—Specifies the number of lines displayed in the docked window (two or three are usually adequate).

Colors—Allows the color of each aspect of the display screen to be altered. Selecting this option will display the **Color Options** dialog box and allow you to set the color of each element in the drawing area.

Crosshairs size—Allows the size of the crosshairs to be altered by entering a number that represents the percentage of the screen to be filled. (100 allows easy projection from one drawing view to another).

Display Resolution—The settings in this box affect the quality of the display of objects in the monitor. Do you want to display a circle as a circle, or as a series of straight line segments that appear to make a circle? The effects of this area will be discussed in Chapter 7 (100 is adequate for most drawings).

Open and Save—This tab will affect how and where drawing files are saved. This tab will be covered in detail later in this chapter.

SaveAs—This portion of the dialog box can be used to determine the format used when saving your drawings. If you work at multiple workstations, this is your opportunity to save files to older drawing formats.

Automatic Save—This option controls the time interval between automatic saves. This option is one of the greatest safety measures you can have to protect your drawing files (15 minutes is adequate for most drawings).

Create Backup Copy with Each Save—This option toggles the creation of backup files. Typically, having a backup copy is a good safety feature. If your machine is short of storage space, deactivating this option will save your file over the existing copy of the file. If you decide to delete backup files, make sure you make multiple copies of your files on the hard drive, multiple diskettes, or a Zip drive.

Plotting—This tab controls all of the options that affect plotting. Components of this tab will be explored in the next chapter and discussed in detail in Chapter 23.

System—This tab controls how the system will operate. Take time to examine the **General Options** box. These settings control general options that relate to drawing operation. Key elements to consider:

Single Drawing Compatibility Mode—One of the features of AutoCAD is that you can truly work on two drawings at once. This only holds true if you have this setting in the inactive mode.

Show Startup Dialog—This setting will control the display of the **Startup** dialog box. Most new users prefer to have the dialog box displayed until they have developed their own work habits. Keep this setting activated in your profile.

Beep on Error in User Input—In the active setting, AutoCAD will sound an alarm (a small beep) when an invalid response is given. If you're in the middle of the Line command, and AutoCAD is expecting a "to point" to be provided, the alarm will sound if you try and use the **Circle** command. As a new user, keep this activated. When it is deactivated, many new users get frustrated when the desired results are not achieved. The alarm often serves to bring you back to reality, reminding you to look at the Command prompt so that the desired information can be provided.

User Preferences—This tab affects options that optimize your performance in AutoCAD. One feature that experienced users of AutoCAD might want to explore is the Right-click Customization box. Some experienced users are so accustomed to using the right-click button in place of ENTER that they might be willing to forsake the new shortcut menus. If you're a new user, leave the menus active. Other options will be addressed in later chapters.

Drafting—Most of the options in this box have been introduced in previous chapters, and will be covered in detail in later chapters. Options that you should address as you set your own profile include the following:

AutoSnap Marker Color—This control can be used to alter the color of the snap box.

AutoSnap Marker Size—This control can be used to alter the size of the snap box. As the size of the box is increased, it gets easier to see. As the drawing complexity increases, a large snap box can lead to inaccurate point selection.

Aperture Size—This slide bar can be used to control the size of the aperture box. The aperture box is the AutoCAD selection tool when a command is started.

Selection—These features control how AutoCAD will select drawing objects. This box will be explored in detail in Chapter 10. A feature that you might want to set in your initial profile is the pick box size.

Profiles—This tab controls the creation and use of profiles. The **Profiles** tab can be seen in Figure 4.6. The next section of this chapter will help you explore this tab and create your first profile. As you gain experience and discover new settings that you would like to include in the profile, you can update the profile using this tab.

CREATING AND SAVING A PROFILE

To create a user profile, select **Options** from the **Tools** menu. Select the **Profiles** tab to display the dialog box shown in Figure 4.6. Go through each of the nine tabs and set the options to meet your preferences. Once you have selected the desired settings, follow this procedure:

1. Select the **Add to List** option. This will display the **Add Profile** dialog box shown in Figure 4.7.
2. Enter a profile name in the **Profile name** edit box. Use a name that will quickly describe the user or drawing type.

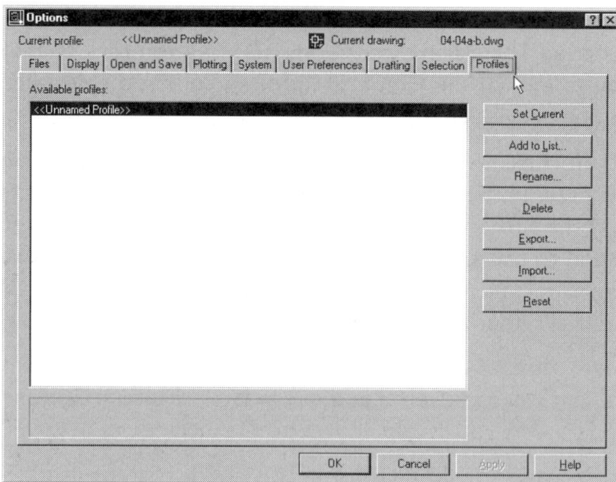

Figure 4.6 *The* **Profiles** *tab controls the creation and use of profiles. As you gain experience and discover new settings that you would like to include in the profile, the profile can be updated using this tab.*

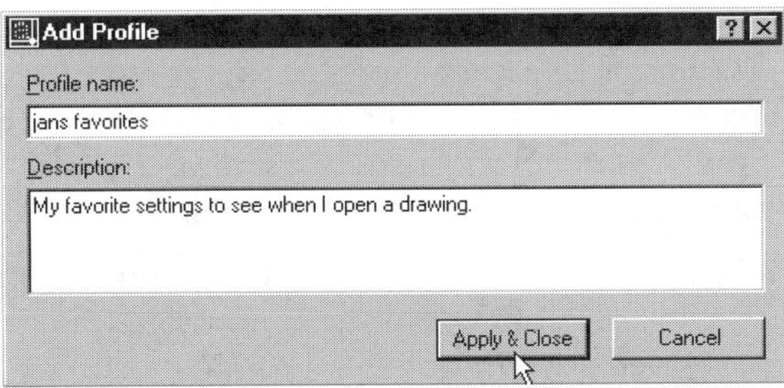

Figure 4.7 *Selecting the* **Add to List** *option will display the* **Add Profile** *dialog box. The box is used to name and describe the profile to be created.*

3. Provide a short description in the **Description** edit box that will quickly describe the contents.
4. Click the **Apply & Close** button to accept the settings. When the button is clicked, the dialog box will be closed and the profile name you provided will be added to the **Available profiles** list.

Once more than one profile has been created, you can quickly select the desired profile from the **Profiles** tab of the **Options** dialog box by highlighting the listing of the desired profile and then clicking the **Set Current** button. Other buttons in this menu that should be considered include the following:

Rename—This option allows an existing profile to be renamed. Clicking this button will produce the **Change Profile** dialog box shown in Figure 4.8. Alter the name as desired and then click the **Apply & Close** button to alter the profile name and close the dialog box.

Delete—This option allows an existing profile to be removed from the **Available Profiles** list. Highlight the profile to be deleted and click the **Delete** button. This will produce a warning to verify that you really want to delete the profile. Selecting the **No** button will terminate the delete process. Selecting the **Yes** button will delete the selected profile.

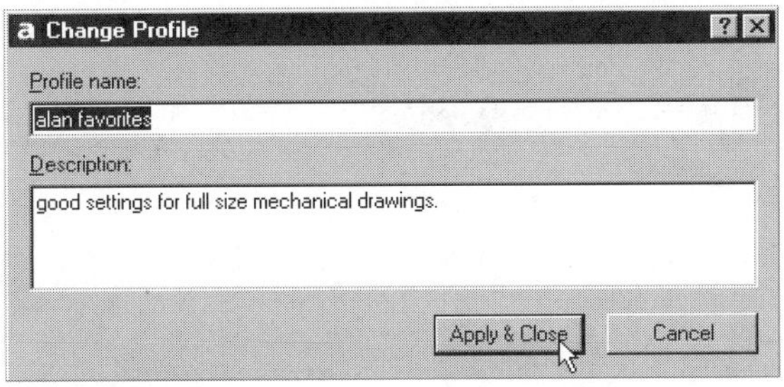

Figure 4.8 *Selecting the* **Rename** *button will produce the* **Change Profile** *dialog box. The box is used to alter the name of existing profiles.*

Export—Selecting this option displays the **Export Profile** dialog box. (See Figure 4.9.) This dialog box can be used to export a profile as a file with an extension of .ARG. The profile can be saved and then shared with another computer. If you're working on a network, save your profile to your secure folder. Backup copies of profiles should also be saved to a diskette.

Import—Selecting this option will display the **Import Profile** dialog box. Similar to the **Export Profile** dialog box, the display can be used to import a profile that was saved using **Export**.

Reset—Selecting this option will reset the computer to the AutoCAD default settings.

SAVING DRAWINGS

Chapter 1 introduced the **Save** command, and Chapter 3 gave you practice using it. This chapter will help you expand the use of the command. Once you have created your drawing, it will need to be saved. This means saving your drawings in a folder on the hard drive, on a diskette, or in your own folder on a network. Check with your instructor to see if saving on the hard disk is allowed. If so, set up a folder so that all of your drawings can easily be accessed. If not, select the letter of the appropriate drive so files will be saved to the diskette. Avoid saving a file to a diskette, except when the drawing session will be ended. A file should be saved to the hard drive throughout the drawing session to save time and avoid a possible problem caused by a full diskette.

You might be saving on the hard drive, tape drives, Zip drives, as well as on diskettes. Because working on the hard drive is so much faster, you will be saving to the hard drive during the drawing session, and saving on the diskette and the hard drive at the

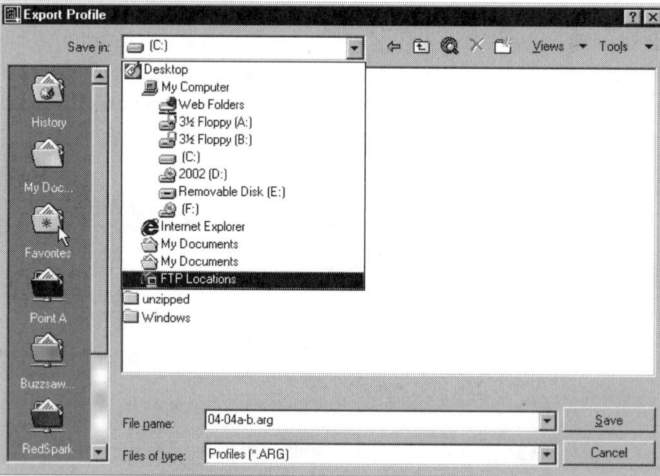

Figure 4.9 *The* **Export Profile** *dialog box can be used to export a profile as an .ARG file. The profile can be saved and then shared with another computer.*

end of the drawing session. Many professionals save on the hard drive and make two diskette backups, which are each stored in different locations. The open drawing file should be saved to the hard drive approximately every 15 minutes. When the drawing session is done, the file should again be saved to update the existing file on the hard drive. The file should also be saved on another source, such as a diskette. This is especially true for students who must transport their drawing files between home and school.

No matter where you save your drawing, the most important thing to remember is to save often. Saving every 15 to 20 minutes will help you avoid losing your work if the power supply to your computer is interrupted. As you start saving your drawings, you will have four options: **Save**, **Saveas**, **Qsave**, and **Export**. The first three options will be discussed in this chapter. Review the **Help** menu for methods of exporting drawing files.

THE SAVE COMMAND

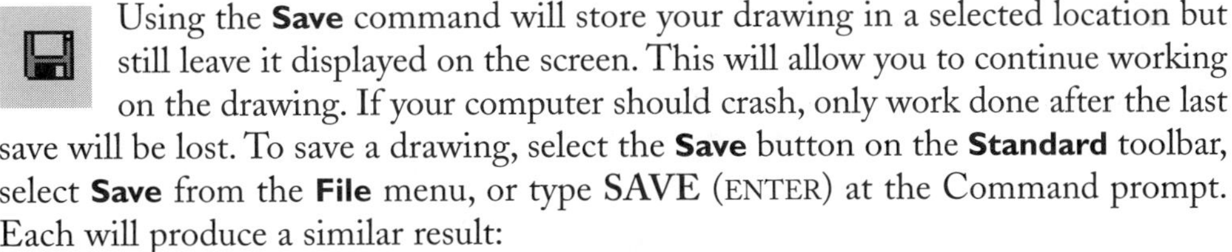

Using the **Save** command will store your drawing in a selected location but still leave it displayed on the screen. This will allow you to continue working on the drawing. If your computer should crash, only work done after the last save will be lost. To save a drawing, select the **Save** button on the **Standard** toolbar, select **Save** from the **File** menu, or type SAVE (ENTER) at the Command prompt. Each will produce a similar result:

- The first time you save the drawing, the **Save Drawing As** dialog box shown in Figure 4.10a is displayed.
- The next time the drawing is saved, and all subsequent times, the save will occur automatically, with the name that was assigned during the initial save.

During the initial save, selecting the **Save in** edit box arrow will display the contents of Windows Explorer. When the destination drive has been selected, the contents of the specified drive will be listed. Enter the desired file name in the **File name** edit box. The top five icons down the left side of the window can be used to provide quick access to common folders on your computer. Buzzsaw, RedSpark, and FTP, icons farther down the left side, allow quick access to the Internet for saving files, posting information, and conducting meetings. Use the **Save as Type** edit box to control the drawing type to be used. Figure 4.10b shows a list of alternative drawing types that can be used. Once the type has been selected, click the **Save** button. The .DWG extension is added to the drawing file and it is saved to the designated drive. This will be the desired method of saving for most AutoCAD users.

File Names

As you save drawings, you might misname a drawing. If you haven't pressed ENTER, just use BACKSPACE to remove the unwanted name. If the name is entered, the file can

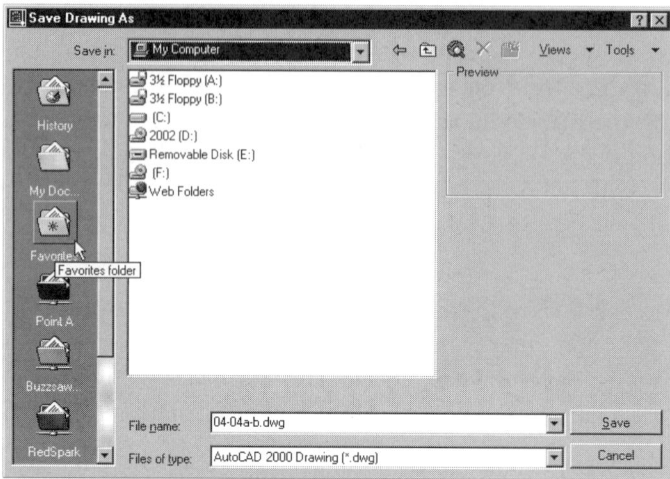

Figure 4.10a *Display the* **Save Drawing As** *dialog box by selecting the* **Save** *button on the* **Standard** *toolbar, selecting* **Save** *or* **Save As** *from the* **File** *menu, or by typing* **SAVE** *(ENTER) or* **SAVEAS** *(ENTER) at the Command prompt. The dialog box is displayed only the initial time the drawing is saved with the* **Save** *command. The second time the drawing is saved, the save will occur automatically using the name that was assigned during the initial save.*

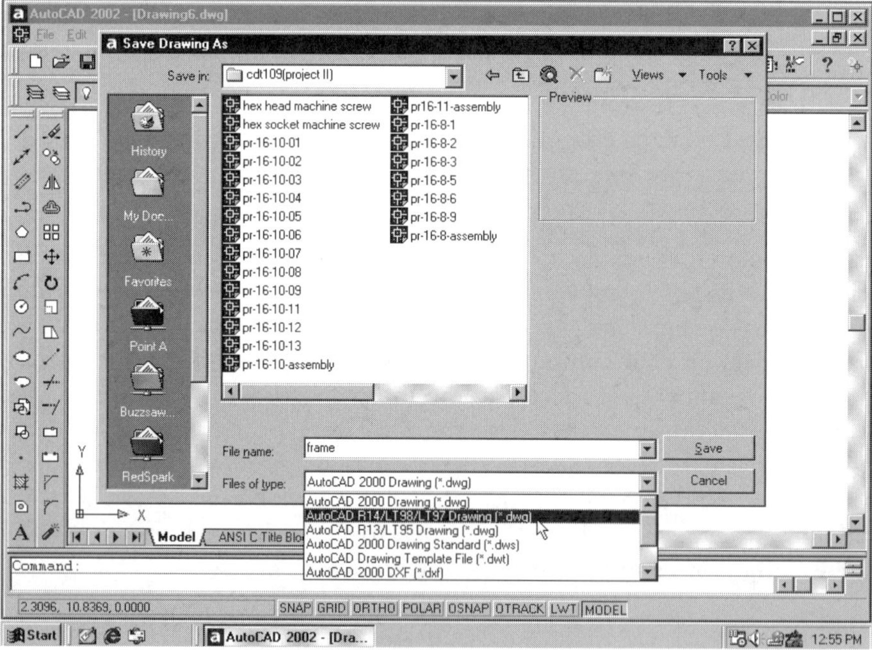

Figure 4.10b *The* **Save as type** *edit box allows drawings to be saved and be compatible with older versions of AutoCAD.*

be renamed using Windows Explorer. Occasionally you will assign a drawing file a name that already has been used. In this case you will see a display similar to the one in Figure 4.11. If you proceed, you will destroy the existing file. The name will remain in the folder and your present drawing will be saved using the old file name, but the existing drawing will vanish. This is a fine way to update an old drawing, but be sure you can live without the old drawing.

Backup Files

Earlier you discovered that when AutoCAD saves a file it adds an extension of .DWG or .BAK. The .DWG stands for a drawing file. The .BAK is the extension for a backup file. These are created automatically by AutoCAD each time you end a drawing. When a drawing is terminated using **Save**, the existing .DWG file becomes the .BAK file and a new .DWG file is created. The .BAK file might not contain all of your recent changes, but it should be only 15 minutes out of date if you've been saving at proper intervals. What a motivator to keep saving your drawings! Backup files are also great if your disk has been damaged. The .BAK file can be renamed to become a .DWG file using Windows Explorer and then reloaded. If the .BAK file cannot be converted to a .DWG file, or if it was scrambled, this will serve as one last reminder to save your files in more than one location. If you have a limited storage space, or if you save to multiple locations, you might not want to create backup files. If you decide you really don't want .BAK files created, you can deactivate the option by clearing the **Create backup copy with each save** check box on the **Open and Save** tab of the **Options** dialog box

THE SAVEAS COMMAND

The **Saveas** command is similar to the **Save** command.

- With **Save**, the **Save Drawing As** dialog box is displayed only the first time the drawing is saved.
- With **Saveas**, the dialog box is displayed each time the **Saveas** command is accessed.

Saveas allows the name of the file, the file destination, and the file type to be altered. **Saveas** can be accessed only from the **File** menu or from the Command prompt. Select **Save As** from the **File** menu, or type SAVEAS (ENTER) at the Command prompt. This

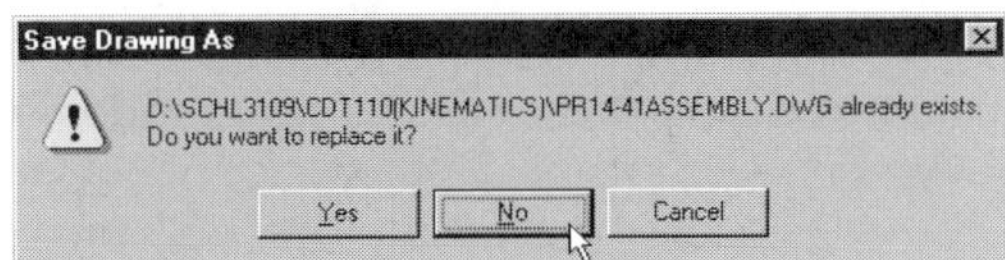

Figure 4.11 *Attempting to save a drawing using a name that has been assigned to another file will result in this alert message. If the* **Yes** *option is accepted, the drawing will be saved using the selected name, and the existing drawing will be overwritten.*

will produce the **Save Drawing As** dialog box, seen in Figure 4.10a. The default name in the **File name** edit box will be the name you selected the last time the drawing was saved. If you like that name, click the **Save** button. If you want to change the name of the drawing file, enter the new name in the **File name** edit box.

Be sure that the correct drive destination is current. Note that the destination is listed in the **Save in** edit box. To change the current drive, use the arrow beside the **Save in** edit box to display the desired drive. Clicking the **Save** button will save the drawing to the specified drive. When a drawing is saved, it will be placed in the current drive. AutoCAD will assign the extension .DWG following the file name to help identify the contents of your directory.

Another benefit of **Saveas** is that it allows the file type to be altered. Examine Figure 4.10b again: it shows seven other alternatives for saving a file. These alternatives allow drawings created in AutoCAD 2002 to be saved so they will be compatible with different versions of AutoCAD. These options can be especially useful for students whose home machines are not current with those at school or for engineering firms with different versions of the software. To save a drawing created in AutoCAD 2002 to be used on a machine running R14, select the AutoCAD R14/LT98 Drawing (*.DWG) option. When a drawing is saved using a version other than the default, AutoCAD remembers the file format that was used, and uses the specified format in subsequent **Save**, **Qsave**, or **Autosave** operations that do not call for a dialog box.

Note: You can set the type of drawing format to be used for saving using the **Open and Save** tab of the **Options** dialog box and then save the setting as part of your profile.

Open a drawing session, draw a few lines, and then save the file with a name of JUNK. As you assign a name, you do not have to type the .DWG at the end of the drawing file because it will be added automatically by AutoCAD. Once the drawing is saved, add a few more lines to the drawing and save the drawing again. If you save the drawing using the Save command, the **Save Drawing As** dialog box will not be displayed; the current file of JUNK.DWG is automatically updated. If you save the drawing using the **Saveas** command, the **Save Drawing As** dialog box will be displayed, allowing a new drawing name to be assigned to the drawing. Draw a few more lines and save the file using the **Saveas** command. When the **Save Drawing As** dialog box is displayed, change the drawing name from JUNK to TRASH, using the **File name** edit box. You will now have files named JUNK and TRASH. When you're done with the drawing session, be sure to delete these files using Windows Explorer.

QSAVE

Qsave, or quick save, will probably be your most frequently used method of saving a drawing. Selecting the **Save** button on the **Standard** toolbar or typing QSAVE (ENTER) at the Command prompt will save the current drawing using its given name without displaying the **Save Drawing As** dialog box. If **Qsave** is used on an unnamed drawing file, the **Save Drawing As** dialog box of Figure 4.10a will be displayed.

Note: Save, Saveas, and **Qsave** will each save a file for future use. Use **Qsave** for saving a drawing that you want to continue working on. It's the fastest method of saving because you only need to select the **Save** button on the **Standard** toolbar. **Qsave** saves named drawings without requesting a file name. Use **Saveas** when you want to save a drawing by giving it a new name or to save the drawing to a new destination. **Saveas** renames the current drawing and saves it to a file name. **Save** does the same thing as **Qsave** but is slightly slower to access because you need to select both the **File** menu and **Save**. **Save** saves the current drawing to a file name.

AUTOMATIC SAVING

By default, the existing drawing will be saved every 120 minutes. You can alter the time between automatic saves by using the **Open and Save** tab of the **Options** dialog box (select **Options** from the **Tools** menu). Use the following procedure to alter the interval between saves:

1. Display the **Options** dialog box.
2. Select the **Open and Save** tab.
3. Highlight the current value in the **Minutes between Saves** edit box.
4. Enter the desired value.
5. Click the **OK** button.

The save time can also be adjusted through the **Savetime** command. Type SAVETIME (ENTER) at the Command prompt. This will produce the prompt:

Command: SAVETIME ENTER

Enter new value for SAVETIME <120>: **15** ENTER

AutoCAD allows a value between 1 and 120 minutes to be entered. The value entered will start the timer when a change is made to the current drawing file. The commands **Save, Saveas,** and **Qsave** will reset the timer. With a value of 20, your drawing file will be automatically saved every 20 minutes with a name of AUTO.SV$. If this file needs to be accessed, Windows Explorer can be used to rename the file from a .SV$ file to a .DWG file. You can disable **Savetime** by entering a value of 0, but this could prove to be a huge mistake if your computer ever crashes.

SAVING A TEMPLATE

Earlier in this chapter you learned how to create a template. Once each of the values for the template is established, the drawing can be saved, to be reused whenever suitable values are required. Save the template using **Saveas** with the Drawing Template File (*.DWT) option selected. (See Figure 4.10b.) Enter the name of the drawing just as you would do with any other drawing. A name that describes the scale or paper size such as CSIZE will be helpful for sorting templates. Once the name has been entered, click the **Save** button. Provide a brief description of the template, such as "template for multi views." This description is displayed whenever you select this template from the **Create New Drawing** dialog box. The final step to save the template is to select the **OK** button. The template can now be reused when a new drawing is opened.

LEAVING AUTOCAD

Once a drawing is saved, you might want to leave the drawing and start another project. AutoCAD allows you to exit the program using the **Close** button or by using **Quit**, **Exit**, **Close**, and **Closeall** commands.

The Quit and Exit Commands

The **Quit** and **Exit** commands will exit the drawing and close AutoCAD. Execute the Quit command by selecting the **Close** button on the AutoCAD title bar or by typing QUIT (ENTER) at the Command prompt. If you attempt to leave a drawing with unsaved changes, you will see the alert box shown in Figure 4.12. If you click the **Yes** button, AutoCAD will save the changes to the default file. Clicking **No** will exit the drawing, close AutoCAD, and return you to other opened programs—any changes made since the last save will be destroyed. **Cancel** will terminate the **Quit** command and return you to the drawing. **Exit** functions in the same manner as **Quit**. It must be entered from the keyboard by typing EXIT (ENTER).

The Close Command

The **Close** command can be used to close the current drawings but remain in AutoCAD. The command must be entered from the Command prompt by typing CLOSE (ENTER) at the prompt. If there have been no changes, the drawing will be closed automatically but AutoCAD will remain open. If changes have been made since the drawing was last saved, you will be prompted to save or discard the changes.

The Closeall Command

If you are working in multiple drawings, the **Closeall** command can be used to close all open drawings. The command must be entered from the Command prompt by typing CLOSEALL (ENTER) at the prompt. If there have been no changes, the draw-

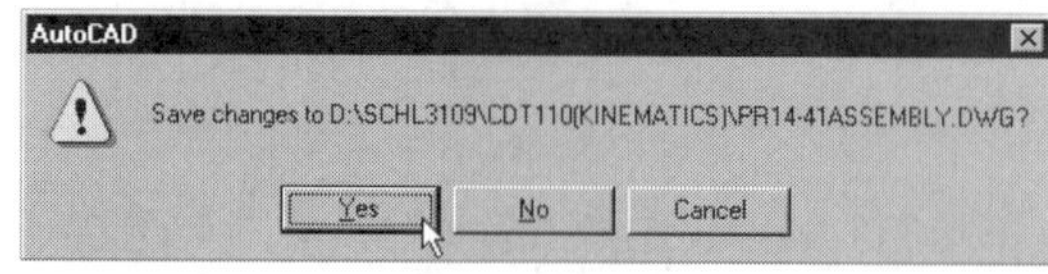

Figure 4.12 *An alert message will be displayed if you attempt to open a new drawing without saving the existing drawing.*

ings will be closed automatically but AutoCAD will remain open. If changes have been made since each drawing was last saved, you will be prompted to save or discard the changes.

MANAGING FILES AND FOLDERS

No office could function if the drafters took their finished drawings and threw them into a room. If a drawing were needed, some poor drafter would have to wade in and hunt through thousands of sheets of paper to find the specific project. Engineering offices typically use well-organized filing systems to keep track of each project for quick, easy retrieval.

FILES

No matter what method is used to save a drawing, information that is to be kept on your disk must be stored in some orderly manner to allow quick, easy retrieval. Disks are sorted in files, folders, and paths. Your drawings will be stored as files and will need names to distinguish them from other files. File names for AutoCAD can contain up to 256 characters. Other guidelines for naming files include the following:

- Layer names can contain letters, digits, spaces, and special characters such as $ (dollars), - (hyphen), and _ (underscore).
- Special characters used by Microsoft, Windows, or AutoCAD cannot be used. Do not use the following symbols:

*	(asterisk)	< >	(less-than and greater-than symbols)
‘	(back quote)	?	(question mark)
:	(colon)	“ ”	(quotation marks)
,	(comma)	;	(semicolon)
=	(equal sign)	|	(vertical bar)
/ \	(forward and back slashes)		

FOLDERS

Files should be organized in folders to provide efficient storage. Folders are the equivalent of dividers in a file cabinet drawer. By default, your drawings will be stored in the Acad2002 folder. Information about files such as file size, time of creation, and the time of last file revision is automatically kept as files are stored. Folders can be created using Windows Explorer to organize information based on client names, class requirements, or any other similar criteria. Folder naming should follow the same guidelines used to name drawing files. Names should be used that will quickly define the contents such as DRAWINGS, MAPS, REPORTS, or CLIENTS. Folders can be placed inside other folders to better organize information. Subfolders can be created and named using the same guidelines used to create a folder. Figure 4.13 shows an example of how folders are set up within folders created by Windows Explorer in a hierarchical method. This use of paths in saving material will help in retrieval.

OFFICE PROCEDURE

Many small offices keep work for each client in separate folders or disks and label the disk by the client name. Drawings are then saved by contents such as valve casing or a part number. Some offices assign a combination of numbers and letters to name each project. Numbers are usually assigned to represent the year the project is started, as well as a job number with letters representing the client or the type of drawing. For example, 0253VICE would represent the fifty-third project started in 2002. Later chapters will introduce External Reference as a method of saving drawing files, and methods of posting drawings on the Internet.

STORAGE PROBLEMS

One of the great thrills for a new CAD operator is seeing the message, "DISK FULL." It's not a serious problem, but it is discouraging. As a drawing is started, AutoCAD examines the disk to find a suitable storage space. Because AutoCAD creates space for several files, it will need to perform expected drawing commands and there might not be enough space to create the intended drawing. In the middle of a command you might find that a DISK FULL error is displayed. Your drawing will be terminated, but all work up to that point will be saved. You will typically encounter this problem only if you save to a diskette in the middle of a drawing session.

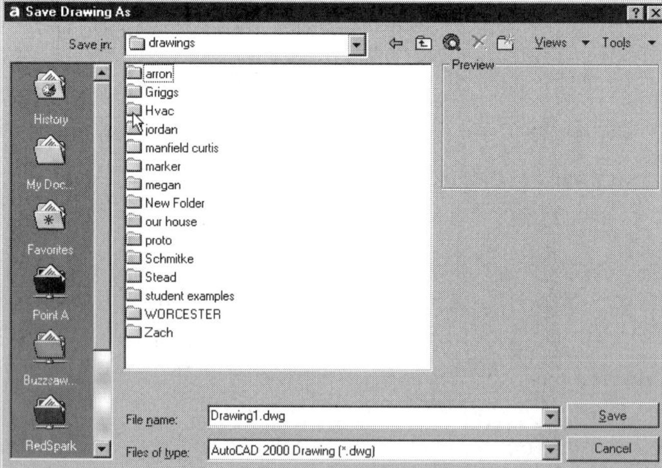

Figure 4.13 *Folders should be created to contain similar drawing files to ease retrieval. Notice, in the C drive, that all drawings are stored in the DRAWING folder. The DRAWING folder contains subfolders named for clients to ease retrieval.*

Occasionally a full disk will cause FATAL ERROR to flash across the screen just before your machine destroys two hours worth of drawing. Sure, you know that you should save every 15 minutes, but sometimes you've got to learn the hard way. The unfortunate aspect of a fatal error is that it might disrupt or destroy other files on your diskette. You'll find that you only have FATAL ERROR when you're 99 percent done with a project.

Note: One Last Reminder: As you are working on a drawing file, save the drawing to the hard drive. Only save to a diskette when you are done with the drawing session and are closing the file. Working from a diskette will slow the drawing session and can lead to storage problems.

Diskette-Handling Procedures

Diskettes should serve only to back up projects that have been saved on the hard disk or network. Because diskettes can be damaged easily, extreme care must be taken with them. The biggest destroyers of information on a diskette are dust, smoke, magnets, and bending.

Dust and smoke interfere with the head in the disk drive as it attempts to interpret the diskette. Magnetic fields from such common items as a stereo speaker or a telephone can scramble the information on a disk into an unusable form. In addition to destroying information, a bend can hinder retrieval of a diskette from the drive unit. Disk labels that are not affixed well to the disk also can cause the disk to jam in the drive.

Guidelines for Safe Diskette Handling

- ALWAYS keep the disk labels in the outer corners of the disk.
- ALWAYS label diskettes clearly, indicating the contents.
- ALWAYS store diskettes in a protective carrying case and in a cool, dry location.
- ALWAYS store diskettes between 50° and 125° Fahrenheit (10° to 52° centigrade).
- NEVER place diskettes on top of a monitor.

Protecting Disks

Diskettes can be protected so that no one can accidentally enter information or change the original contents. There is a notch in the top right corner of the disk when you look at it from the front. Turn the diskette over and slide the black tab to its upper position. This will allow you to see through the notch and make the diskette write protected.

Correcting File Errors

No matter how carefully you follow the guidelines for handling a disk, they will occasionally become corrupted. AutoCAD provides the **Audit** and **Recover** commands that attempt to make corrupted drawing files useable. Each is located in **Drawing Utilities** on the **File** menu. The **Purge** command can be used to clean unneeded information from the drawing base and will be introduced in Chapter 14.

Examining a Drawing File with Audit

The **Audit** command can be used to examine the current drawing file for errors and to correct any errors that exist. These errors are not spelling mistakes or drawing errors, but errors that occasionally are made by the software. This command will detect errors, generate extensive descriptions of the problems, and recommend actions to correct the errors. Use Audit by selecting **Audit** from **Drawing Utilities** on the **File** menu or by typing AUDIT (ENTER) at the Command prompt. The command sequence is as follows:

Command: **AUDIT** ENTER

Fix any errors detected? [Yes/No] <N>

Responding with an N (ENTER) will produce a report of the file and list the errors but will not repair the errors. The final listing of the report resembles

Total errors found 3 fixed 0

Responding Y (ENTER) at the prompt will produce a report of the file and list the errors. The listing resembles

19 Blocks audited		
Pass 1	1793	objects audited
Pass 2	1793	objects audited
Pass 3	3800	objects audited
Total errors found 2		fixed 2

With the **Auditctl** system variable set to 1, an Audit report file will be created using the file name with .ADT as the extension. This file will be placed in the directory containing the original drawing file and will list all of the functions required to fix file errors.

Recovering a Drawing

If errors are discovered that cannot be repaired by **Audit**, the **Recover** command can be used to try and salvage the file. Access **Recover** by selecting **Recover** from **Drawing Utilities** on the **File** menu or by typing RECOVER (ENTER) at the Command prompt. AutoCAD will automatically attempt to repair damaged drawings as they are opened. If the drawing can't be opened, a message will be displayed that the drawing must be recovered. Once the file to be recovered has been selected,

AutoCAD will begin a series of diagnostic procedures. When the process is complete, a listing of problems and the corrective measures taken by the program will be displayed. Generally, the recovery process will open a damaged drawing. If the **Recover** command fails, it's time to get out the backup file. This is one last opportunity to remind you that no matter how fast and good computers have become, they occasionally fail, and generally only at times of great stress such as the end of a school term or the due date of a big project. Backup files can greatly reduce the stress when you have a file that can't be recovered.

CHAPTER 4 EXERCISES

1. Start a new drawing. You will be using this exercise as a base for a drawing of a small subdivision. The drawing will be drawn at a scale of 1"=30' and plotted on D-size paper. North will be at the top of the paper (twelve o'clock). Set all applicable defaults to two-place decimals.

   What unit setting should be used for this drawing? ____________

   Set the limits at the proper value. What option number will be required?

   ____

   Many of the property lines are described with angles measured in degrees, minutes, and seconds. Set the angle measurement system in the appropriate system. What default number was selected? ____________

2. Start a new drawing. Draw a square, a rectangle, and a triangle. Save the drawing using the name E-4-2.
3. Open drawing EXER4-3 and add another circle and a rectangle. Save as E-4-3.
4. Edit drawing E-4-3 by adding any figure that you would like. Save file as E-4-4.
5. Get a directory listing of your diskette; list the following:

   The contents

   The time your first exercise was finished

   The date of your last exercise

   The number of bytes used

   The number of bytes free

6. Open the ANSI-C color dependent plot style.DWT template.

   What is the current snap resolution? X _____ Y _____

   What is the current grid setting? X _____ Y _____

   What layer will new lines be added to? ____________

   You will be using this exercise to set up a drawing for a stencil. Two-place decimal angles are desired.

Set the units at the proper value. What option number will be required? ____________ What was the default? ____________

Give the current system for measuring angles. ____________

Will the current setting allow for two-place decimals? ____________

In which direction will angles be drawn? ____________

What are the current drawing limits? ____________

What limits will be needed to achieve the listed parameters of this exercise?

____

What is the current snap value? ____________

Set the snap value to 0.50".

What is the current grid default value? ____________

Set the grid to 1.00".

CHAPTER 4 QUIZ

1. What is the default value for **Savetime**, and is this value adequate?
2. What command will store the drawing on the hard disk but leave the drawing on the screen?
3. How many characters can be used for a file using AutoCAD 2002?
4. What command will store the drawing on a disk and terminate the drawing?
5. List three different methods for accessing the **Save** command.
6. You have a drawing file labeled CASING. You draw a detail and start to save it as CASING. What will AutoCAD respond and how will your drawings be affected?
7. Give six forms of information listed when the dialog box is used to save a drawing.
8. What two extensions are assigned to a drawing file?
9. What does **Quit** do differently from **Save**?
10. What is a fatal error?
11. Give two reasons for backing up a drawing in multiple locations.
12. List two reasons why you might receive an alert message while saving a drawing.
13. You've opened an existing drawing named BASECLAMP.DWG, made some changes, and now want to save the drawing while keeping BASECLAMP.DWG intact. How can this be done?
14. You've saved a file named LOST.DWG somewhere on your hard drive, but you're not sure where. How can you find the file?
15. Explain the difference between **Save** and **Saveas**.

5

Drawing Organization

This chapter will introduce

- Common linetypes used on construction drawings
- Methods of loading varied linetypes into a drawing
- Methods of controlling lineweight
- Methods of assigning color to drawing objects
- The use of layers for controlling lines, lineweight and color
- Making a check print

Commands to be introduced include

- **Linetype**
- **Properties**
- **Lineweight**
- **Color**
- **Layer**
- **Plot**

Now that you're able to set up drawing units and limits, basic drawing components can be explored. The main component of the drawings you will be working with will be lines. This chapter will introduce you to various linetypes used on mechanical drawings and to methods for assigning and controlling varied linetypes. As linetypes are added to a drawing, a method of managing these linetypes must also be considered. Color can be assigned to lines to help distinguish types of building components. Most professional AutoCAD users sort linetypes by the use of layers. The **Linetype**, **Color**, and **Layer** commands will be introduced in this chapter, with a detailed discussion presented in later chapters. In addition to exploring the methods of controlling lines, this chapter will explore the process of making a non-scaled check print. Plotting a drawing to scale will be explored in later chapters.

AN INTRODUCTION TO LINETYPES

Varied linetypes are used in every type of drafting and consulting firm involved in the manufacturing process. Lines are used to describe the shape of an object, changes in surface height, hidden surfaces or objects, centerline locations, the location of where a part has been broken open to aid viewing, and varied positions the part can be moved to. This chapter will introduce you to common linetypes used to make mechanical drawings. It will explain how to load linetypes to be used in a drawing session and how to alter and control linetypes that are part of a drawing.

LINETYPES STANDARDS

Guidelines established by the American National Standards Institute (ANSI) and the American Society of Mechanical Engineers (ASME) are used throughout the drawing and design industry to ensure drawing uniformity. The width and pattern used to create each line type is specified by document ANSI Y14.2M, Line Conventions and Lettering. Figure 5.1 shows the line types established by ANSI. Recommended line

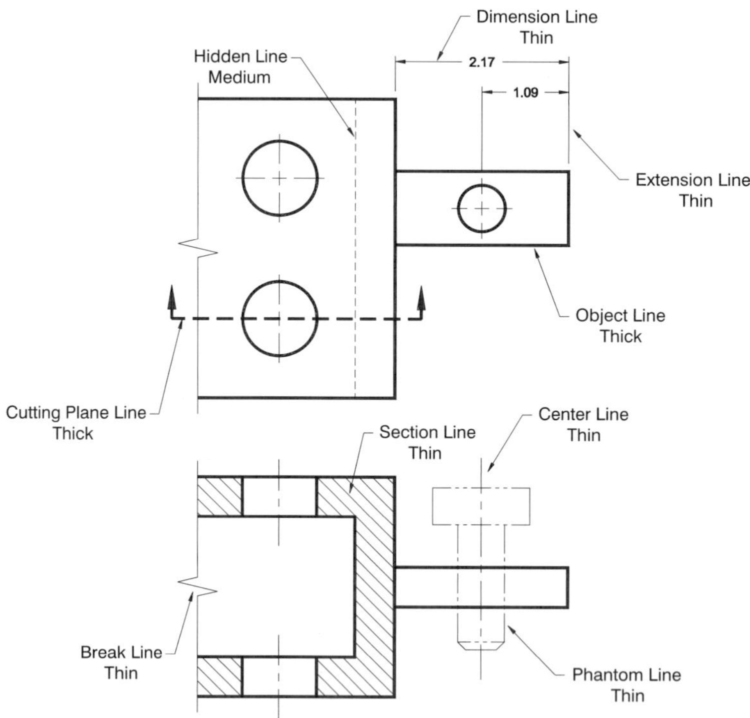

Figure 5.1 *Common linetypes used on mechanical drawings.*

width for thick lines is 0.6 mm and 0.3 mm wide lines for thin lines. Methods of altering line width will be introduced in this chapter and later in Chapter 13 and in Chapter 23 as the drawing is plotted. Common line types found on drawings include object, hidden, center, viewing plane, cutting plane, section, break, phantom lines, chain, extension, dimension and leader lines. Specific uses for each line type will be discussed throughout the text.

Object Lines

Object lines are thick, continuous lines used to describe the shape of an object. They can also be used to show changes in the surface of an object. Figure 5.2 shows both uses of object lines. Each of the lines that form the outer shape is an object line. Line 1 not only defines the shape of surface A and B, but also indicates that the surfaces are not level with each other. With only this view of the object, it cannot be determined which surface is closer to the print reader, only that there is a difference in the heights of the two surfaces. Multiview drawings will be introduced in the next chapter. Using the multiview drawings in Figure 5.3, it can be seen that surface C separates surface A from surface B.

Hidden Lines

Dashed or *hidden lines* are thin lines used to represent a surface that is hidden from view. ANSI specifies that the lines should be .125" (3 mm) long with .06" (1.5 mm) space between lines. The size of the line can be controlled using the **Ltscale** command. AutoCAD will automatically control the spacing, based on the line length. Other guidelines for hidden lines include the following:

- When a hidden line ends at an object line, the two lines should touch. See Figure 5.4 at A.

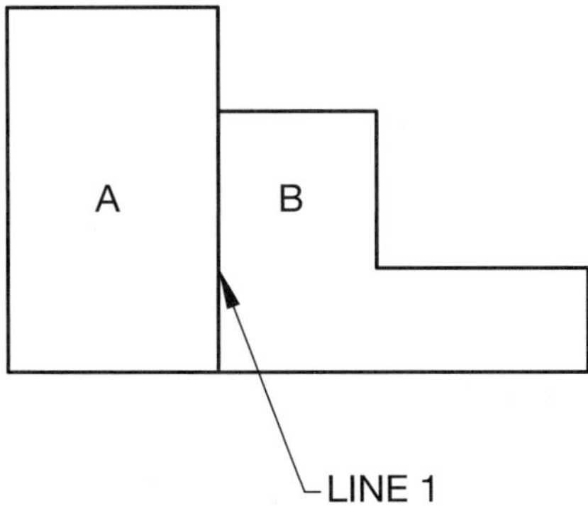

Figure 5.2 *Object lines are continuous lines used to describe the shape of an object or to show changes in the surface of an object.*

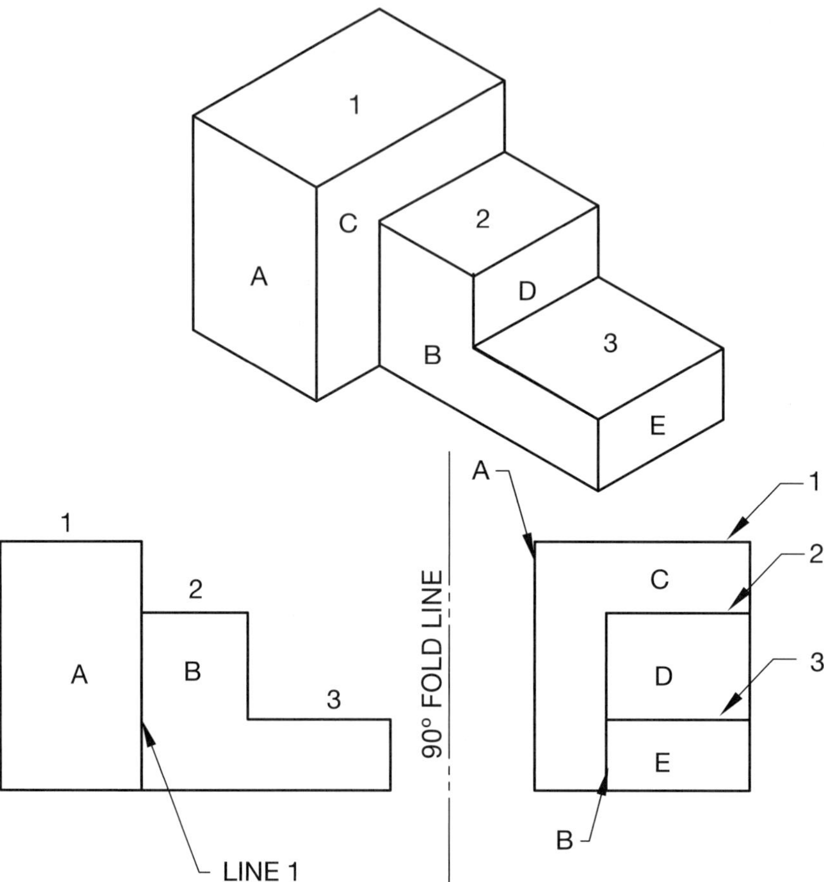

Figure 5.3 *Using multiview drawings allows the print reader to determine changes in surfaces.*

- When a hidden line is parallel to an object line at an intersection, the lines should not touch. See Figure 5.4 at B.
- When two or more hidden surfaces intersect, the hidden lines used to represent these surfaces should touch. See Figure 5.4 at C.
- When two or more hidden lines cross each other, but the surfaces they represent do not touch each other, the hidden lines should not touch each other. See Figure 5.4 at D. This function is controlled by AutoCAD. AutoCAD may have the lines touching depending on the length of line to be used. You may need to draw three specific lines to avoid an intersection if your school or company follows ANSI to the letter of the law.
- When a hidden curved surface intersects an object line, the two should not touch. See Figure 5.4 at E.

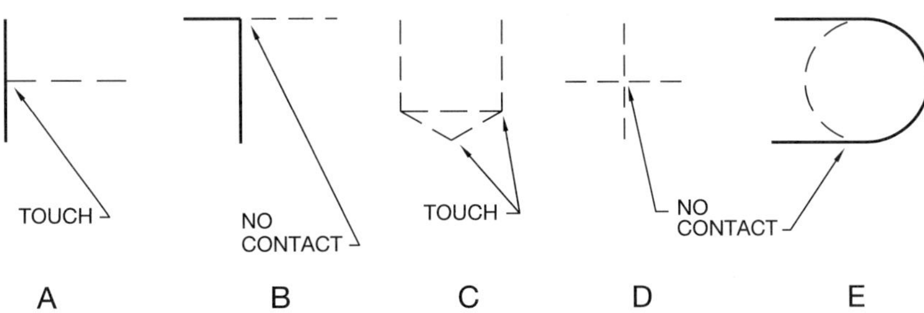

Figure 5.4 *Dashed or hidden lines are thin lines used to represent a surface or object that is hidden from view. Their relationship to other lines will determine how an intersection is represented.*

Centerlines

A *centerline* is composed of thin lines that create a long-short-long pattern. Centerlines are used to locate the center axis of circular features such as drilled holes or cylinders. Figure 5.5 shows the recommended ANSI sizes for lines and spaces. ANSI specifications for placing centerlines include the following:

- The short line should be .125" long, the space should be .06", and the long segment must be between .75" and 1.50" long. Although the length of the long segment can vary from drawing to drawing, all long segments in a drawing should remain constant. Figure 5.6 shows common requirements.
- Pairs of perpendicular lines should be used to indicate a center point.
- The short lines of the centerline should intersect at the center point of the object being described.
- The centerline pattern should always end with the long line segment.
- The long line segment should extend .125" past the object being described.
- The centers for holes should be located along a bolt circle (Figure 5.7 at A) or from the center of a circular feature (Figure 5.7 at B). The method to be used depends on how the holes will be machined. Both options will be discussed later in this chapter and in Chapter 18.

Viewing Plane Lines

A *viewing plane line* is used to indicate the location of special views that may need to be drawn to clarify a part that is not clearly shown in other views or if a view is drawn at a different scale than the other views. Chapter 6 will introduce the six views of orthographic projection and auxiliary views that might need to be created to clearly show an inclined surface. If the auxiliary view is placed in line with the plane that it is projected from, no viewing plane line is required. If the auxiliary view is moved to a position so that it is no longer in line with the surface that it is projected from,

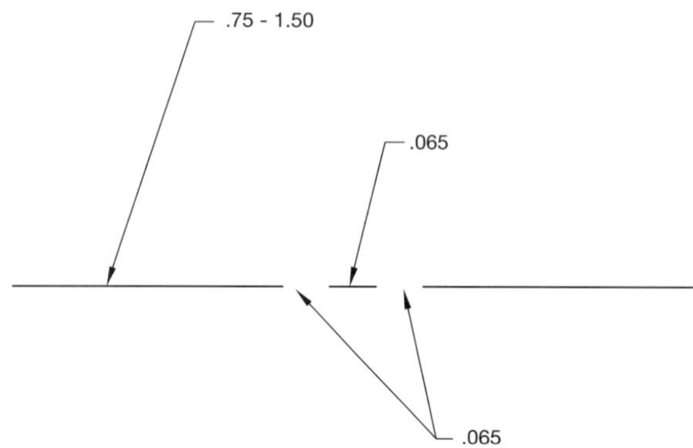

Figure 5.5 *A centerline is composed of thin lines that create a long-short-long pattern. Centerlines are used to locate the center axis of circular features such as drilled holes, bolts, and shafts.*

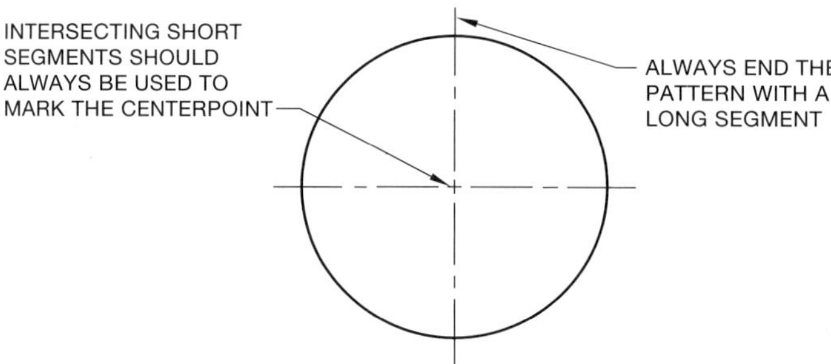

Figure 5.6 *When the size of the circle permits, a center mark should be used to indicate the center of the circle. Centerlines should extend .125" past the object they describe.*

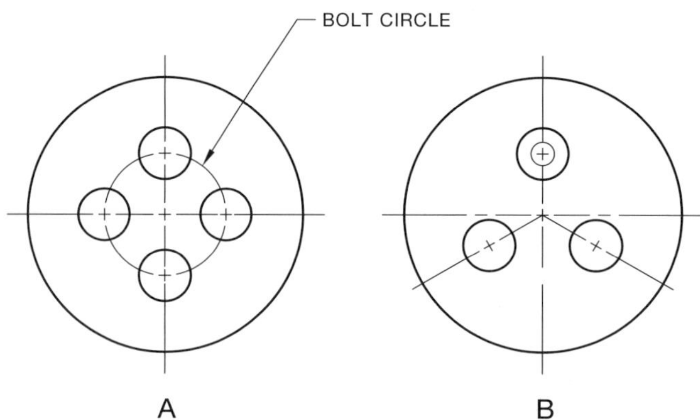

Figure 5.7 *Centerlines can be used to describe the location of circular features on a bolt circle, or their angular location from a center point.*

a viewing plane must be provided to show how the view relates to the balance of the object. Figure 5.8 shows the use of viewing plane lines. ANSI specifications for a viewing plane line include the following:

- Use a long, short, short, long pattern.
- The long line segment should be between .75" and 1.5" long. The short lines should be .125" long, and the space between lines should be .06".
- Use thick lines (thicker than object lines).
- Terminate the line with arrows to indicate the viewing direction.
- Place a letter using .25" high text to indicate the view at each end of the line.

Cutting Plane Lines

To see interior features of a part, the part is often cut or sectioned. A *cutting plane line* is used to indicate the location of the section and also the direction of sight. Figure 5.9 shows an example of the use of a cutting plane. Chapter 19 will provide additional examples of how a cutting plane can be used. ANSI specifications for cutting plane lines include the following:

- Use a long, short, short, long pattern.
- The long line segment should be between .75" and 1.50" long. The short lines should be .125" long, and the space between lines should be .06".
- Use thick lines (thicker than object lines).

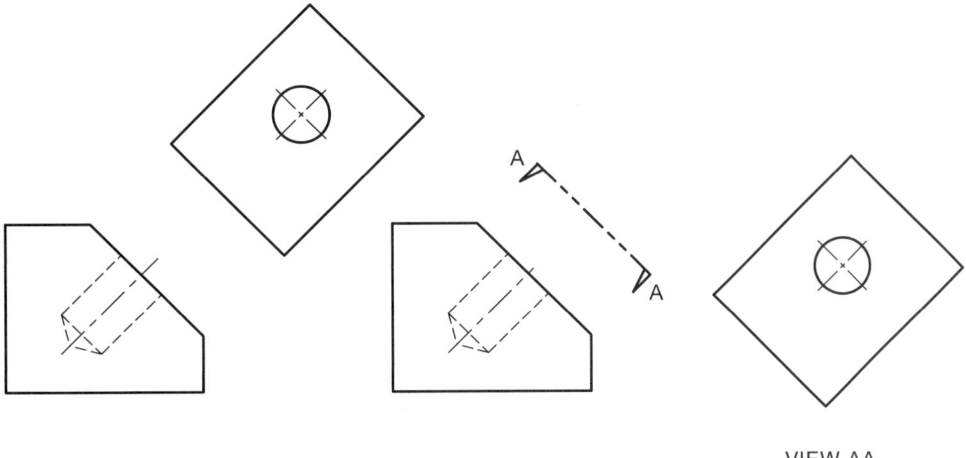

VIEW AA

Figure 5.8 *A viewing plane line is used to indicate the true projection plane when an auxiliary view has been moved to an alternate position.*

- Terminate the line with arrows to indicate the viewing direction.
- Place a letter using .25" high text to indicate the view at each end of the line.
- Cutting plane lines can be eliminated if the location of the section is clearly understood by the location of the section.
- When a cutting plane line passes through a circular feature, the cutting plane line is shown in place of the centerline.

Section Lines

When an object has been sectioned to reveal its cross section, a pattern is placed to indicate the portion of the part that has been sectioned. In their simplest form, *section lines* are thin parallel lines used to indicate what portion of the object has been cut by the cutting plane and are placed using the Bhatch command. Figure 5.9 shows a sectioned object and the placement of section lines to indicate the portion of the object that has been affected by the cutting plane. Chapter 19 will discuss representing sectioned material. ANSI guidelines for placement of section lines include the following:

- Section lines should be drawn at a 45° angle but should not be parallel or perpendicular to any part of the object that has been sectioned.
- If an angle other than 45° is used, it should be between 15° and 75°.

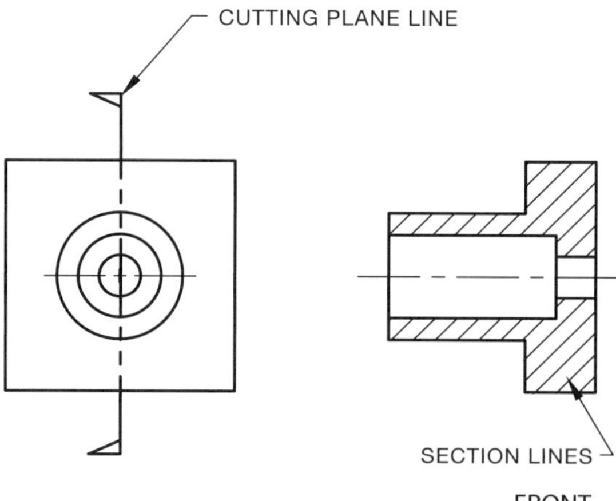

Figure 5.9 *A cutting plane line is used to indicate the location of a section and the direction of sight when viewing the section. The line is thicker than object lines and must terminate with arrows to indicate the viewing direction. If more than one cutting plane is required for a drawing, each arrow must be marked with a letter. Section lines are thin parallel lines drawn at a 45° angle to indicate what portion of the object the cutting plane has cut.*

- If two adjacent parts have been sectioned, section lines should be placed at opposing angles. See Figure 5.10.
- Although most materials have their own special pattern, lines placed at 45° are the industry standard for representing most materials. Chapter 19 will introduce the patterns used to represent various materials.
- Although spacing between section lines can be altered depending on the size of the part that is sectioned, typically a spacing of .125" should be provided between section lines.
- An object line should never be placed through sectioned material.

Break Lines

Break lines are used to remove unimportant portions of an object from a drawing so that the drawing will fit into a specific space. If you consider a standard wood pencil, between the eraser and the point, the pencil is the same. If you had to draw a 10" pencil at full scale in a 5" space, break lines can be used to remove 5" of the pencil between the point and the eraser. Three common break lines used on mechanical drawings include the short, long, and cylindrical break lines.

- A short break line is a thick jagged line placed where material has been removed. See Figure 5.11 at A.
- A long break line is a thin line with a zigzag shape or inverted S shape inserted into the line at uniform intervals. See Figure 5.11 at B.

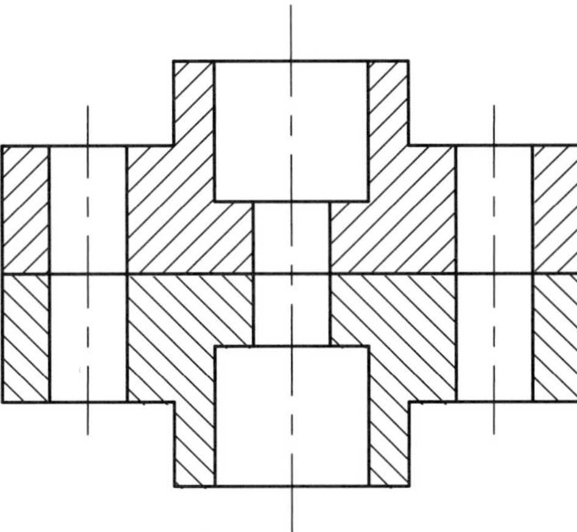

Figure 5.10 *If two adjacent parts have been sectioned, section lines should be placed at opposing angles to help clearly define each part.*

- A cylindrical break line is a thick line resembling a backward S. See Figure 5.11 at C. Notice that the line is altered for solid and hollow cylindrical shapes. See Figure 5.11 at D.

Phantom Lines

Phantom lines are thin lines in a long-short-short-long pattern that are used to show motion, or to show an alternative position of a moving part. See Figure 5.12.

- The long line segment should be between .75" and 1.50" long, but remain a constant length within each drawing.
- The short lines should be .125" long.
- The space between lines should be .06".

Chain Lines

If a portion of a surface is to receive a special treatment, a *chain line* can be used to indicate the limits of the treatment. Chain lines are thick lines in a long, short, long pattern. Figure 5.13 shows the use of a chain line to show the surface finish of a portion of a cylinder. Chapter 18 will explain surface finishes.

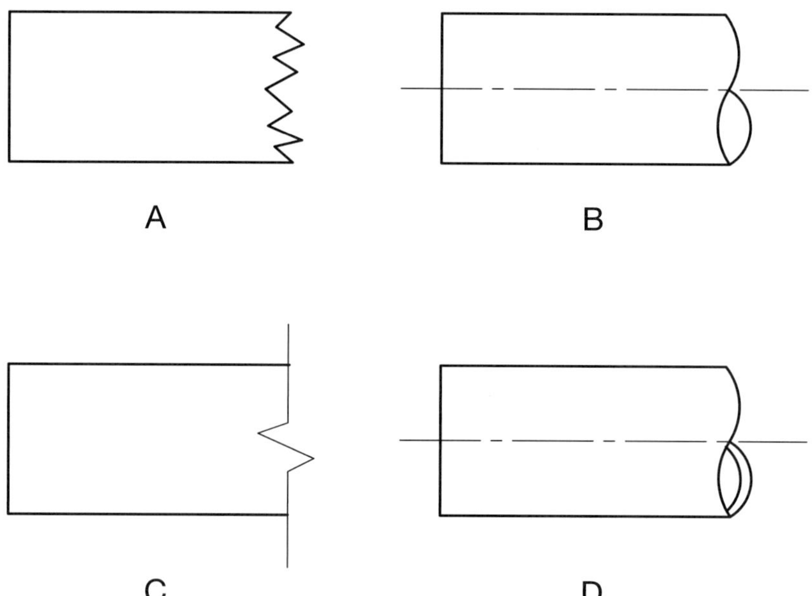

Figure 5.11 *Common break lines of mechanical drawings: a short break line (A) is a thick jagged line placed where material has been removed; a long break line (B) is a thin line with a zigzag shape at uniform intervals; a cylindrical break line (C) is a thin line resembling a backward S; a cylindrical break line for hollow objects (D) is a pair of thin lines resembling a backward S.*

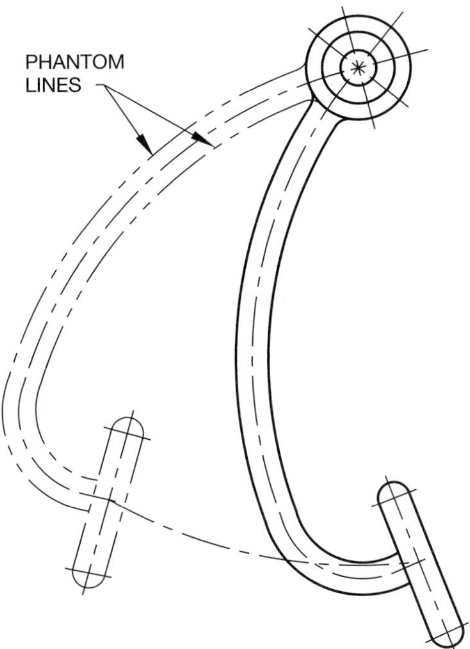

Figure 5.12 *Phantom lines are used to indicate motion or alternative positions of a moving part.*

- The long line segment should be between .75" and 1.50" long, but remain a constant length within each drawing.
- The short lines should be .125" long.
- The space between lines should be .06".

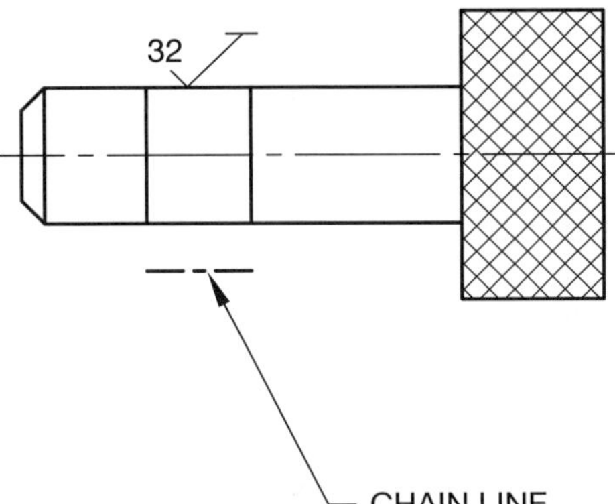

Figure 5.13 *Chain lines are used to indicate where a special treatment will be applied to a surface.*

Extension Lines

Extension lines are thin lines used to relate dimension text to a specific surface of a part that are placed using the **Dimension** commands. See Figure 5.14. Extension lines are created using the commands on the **Dimension** toolbar.

- Provide a space between .06" and .125" between the surface of the part being described and the start of the extension line.
- Extend the extension line .125" past the dimension line.
- An extension line can cross other extension lines, object lines, centerlines, hidden lines and any other line **with one exception**:
    - An extension line can never cross a dimension line. Never!

Chapters 16 and 17 will introduce other guidelines for placing extension lines using the various commands for placing dimensions.

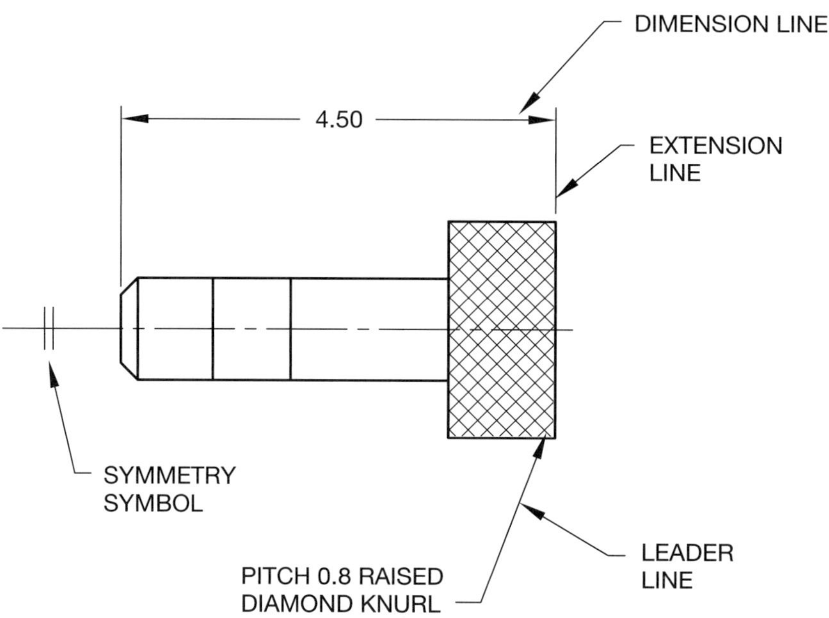

Figure 5.14 *Four types of lines are used to aid in the placement of text and dimensions. The dimension lines are thin lines used to relate dimension text to a specific part of an object. Extension lines extend from the object to define the length of the dimension. Symmetry lines are place on a centerline to indicate that each half of a part is exactly the same. Leader lines are thin lines used to relate a dimension or note to a specific portion of a drawing.*

Dimension Lines

Dimension lines are thin lines used to relate dimension text to a specific part of an object (see Figure 5.14). These lines are created using the commands on the **Dimension** toolbar. Chapters 16 and 17 will provide specific uses of dimension lines and dimension text.

- Dimension lines extend from one extension line to another.
- Dimension lines require a line terminator where the line intersects the extension line. The terminator is usually a solid arrowhead.
- Dimension text is typically .125" high text.
- Dimension text should be placed centered from end to end in the dimension line. Text might need to be placed off center to aid clarity when several lines of text are near each other.
- Dimension text should be centered so that it is half above and half below the dimension line.
- The space between the dimension text and the dimension line should be equal to half the height of the dimension text.

Leader Lines

Leader lines are thin lines used to relate a dimension or note to a specific portion of a drawing. Selecting the **Leader** button on the **Dimension** toolbar creates leader lines. See Figure 5.14 and Chapters 16 and 17.

- A solid arrowhead should be used at the object end of the leader line.
- A horizontal line approximately .125" long should be used at the note end of the leader line.
- Leader lines can be placed at any angle, but angles between 15° and 75° are most typical.
- Vertical and horizontal leader lines should never be used.
- The arrow end of the leader line should touch the edge of the part it describes. When used to describe a circle, the arrow should end at the edge of the circle, but if extended, the line would pass through the center of the circle.
- The leader line should extend far enough away from the object it is describing so that the attached note is a minimum of .75" from the outer surface of the object.

AN INTRODUCTION TO LINETYPES

You've been introduced to the varied types of lines that are required to produce engineering drawings. AutoCAD contains many of the linetypes required by ANSI for mechanical drawings in the Standard linetype library. Access these linetypes by selecting **Linetype** from the **Format** menu or by typing LINETYPE (ENTER) or LT (ENTER) at the Command prompt. Each method will produce the **Linetype Manager** dialog box shown in Figure 5.15. The Linetype library is displayed by selecting the **Load** button. Figure 5.16 shows the linetype display.

LINETYPE CHARACTERISTICS

The **Linetype** command will allow you to vary the linetype pattern for each object you draw or for the entire layer. Most offices assign a specific linetype to a specific layer. As you start to assign linetypes to drawing objects, you need to remember the following two key elements regarding linetypes:

Effects

Objects that will be affected by a change of linetype include line, circle, arcs, polygons, donuts, ellipse, hatch, and 2D polylines. Any other drawing object will be displayed

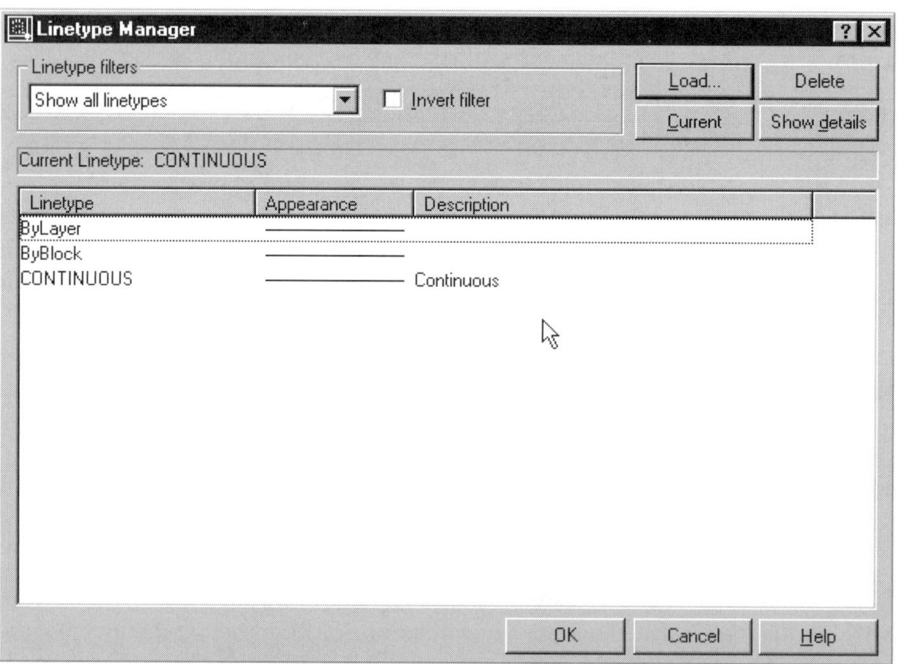

Figure 5.15 *The* **Linetype Manager** *dialog box can be used to change and load linetypes. Access it by selecting* **Other** *from the* **Linetype Control** *menu on the* **Object Properties** *toolbar, or by selecting* **Linetype** *from the* **Format** *menu.*

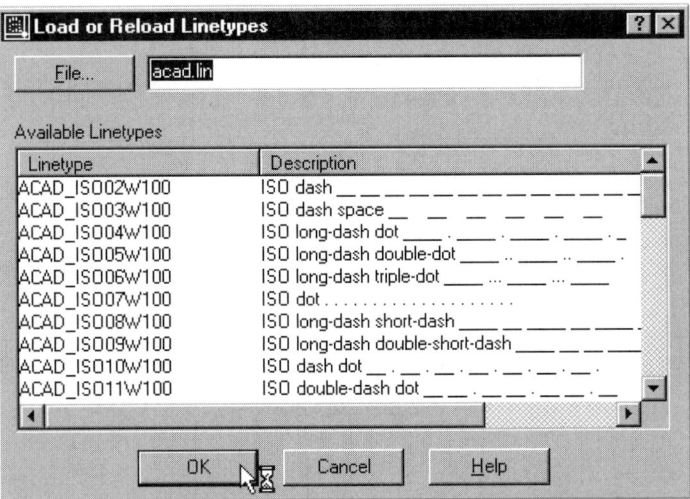

Figure 5.16 *Selecting the* **Load** *button will display the* **Load or Reload Linetypes** *dialog box with a list of AutoCAD linetypes. Use the scroll bar to see the balance of the list.*

as a continuous line. That will be good news as you start to add dimensions or text to a drawing and you're working on a layer drawn with centerlines or another line pattern as the current linetype.

Loading

A second concept to keep in mind as you work with linetypes is that you can't use a linetype unless it has been loaded into your drawing. The linetypes will be loaded automatically as a function of the **Linetype Manager** dialog box when the **Linetype** command is used.

LOADING LINETYPES INTO A DRAWING

As you display the **Linetype Manager** dialog box, you'll notice that three listings are displayed: ByLayer, ByBlock, and Continuous. Additional linetypes are loaded by using the **Load** button in the **Linetype Manager** dialog box. Use the following procedure to load the center, dashed, hidden, and phantom linetypes.

1. Display the **Linetype Manager** dialog box by selecting **Other** from the **Linetype Control** box on the **Object Properties** toolbar, by selecting **Linetype** from the **Format** menu, or by typing **LT** (ENTER) at the Command prompt.
2. Click the **Load** button. This will display the **Load or Reload Linetypes** dialog box shown in Figure 5.16.
3. Select the name, appearance, or description of the line to be loaded into the drawing to highlight the indicated linetype. To load more than one linetype,

press and hold CTRL while selecting the names of the desired linetypes with the select button of the mouse.

4. When you're through selecting the linetypes to be loaded into a drawing, click the **OK** button.

The dialog box will be removed, and the selected linetypes will be displayed in the **Linetype** window of the **Linetype Manager** dialog box (see Figure 5.17). The lines are also shown in the **Linetype Control** box on the **Object Properties** toolbar when the display arrow is selected. Figure 5.18 shows the linetypes currently loaded into the drawing base.

MAKING A LINETYPE CURRENT

Remember that all you've done so far is to load linetypes into the drawing. To use the linetype, you must select one of the loaded linetypes as the current linetype. The following process will allow you to alter the lines that are used to create a drawing. Later in this chapter you'll be introduced to methods of assigning linetypes to a specific layer. Most professionals find linetypes easier to control when they are associated to a specific layer rather than scattered throughout one layer.

By default, you've been drawing with a continuous black line. Use the following procedure to alter the current linetype:

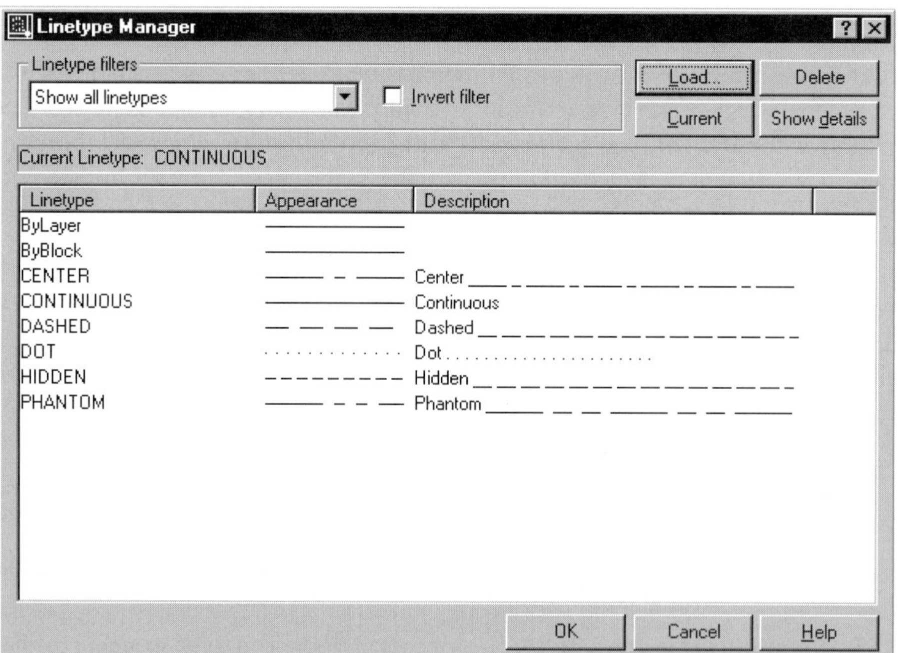

Figure 5.17 *Once a linetype has been selected to load, it will be displayed in the Linetype list.*

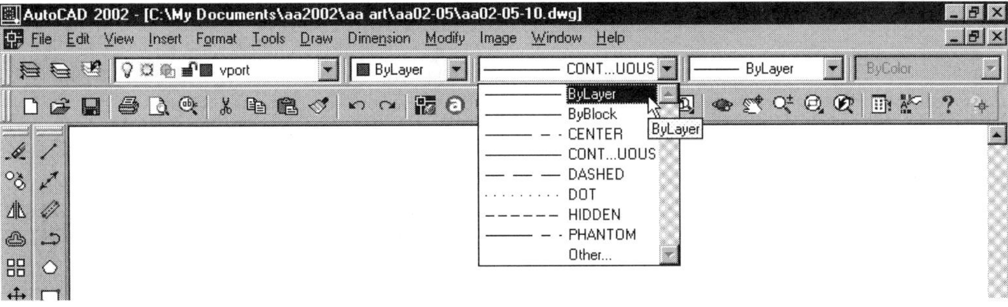

Figure 5.18 *The linetypes currently loaded into the drawing base can be found in the* **Linetype Control** *box on the* **Object Properties** *toolbar.*

1. Select the linetype that will become the current linetype in the **Linetype Manager** dialog box. Select Center for now.
2. Click the **Current** button.
3. Click the **OK** button.

This process will make the selected linetype the current linetype to be used, close the dialog box, and return you to the drawing screen. Notice the name and pattern of the selected line will now be displayed in the **Linetype Control** box on the **Object Properties** toolbar. See Figure 5.19. The linetypes that are loaded into a drawing can also be displayed by selecting the **Properties** button on the **Standard** toolbar, by selecting **Properties** from the Modify menu or by typing PROPERTIES (ENTER) at the Command prompt. Each method will produce the **Properties** dialog box. Selecting the current linetype will display a list arrow similar to Figure 5.20. Selecting the arrow will display a listing of the current linetypes similar to Figure 5.21.

Note: New users of AutoCAD often expect the **Linetype** command to work miracles. Once a linetype has been made the current linetype, it will affect only drawing objects that are added after the linetype is changed. Objects drawn prior to the selection will not be affected. Objects drawn prior to the selection can be transformed to a different linetype using the **Match Properties** command, which will be discussed in Chapter 13.

Figure 5.19 *The current linetype is displayed on the* **Object Properties** *toolbar. The* **Linetype Control** *box can be used to alter the linetypes that will be drawn.*

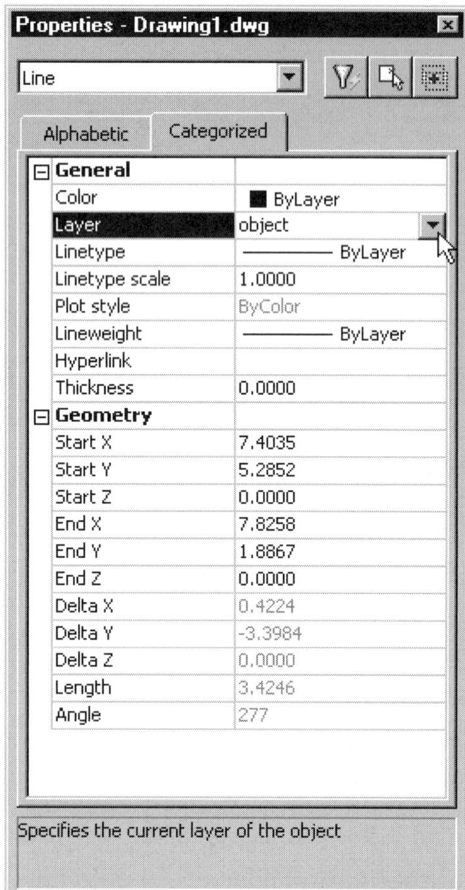

Figure 5.20 *The linetypes that are loaded into a drawing can be displayed by selecting the* **Properties** *button on the* **Standard** *toolbar or by selecting* **Properties** *from the* **Modify** *menu.*

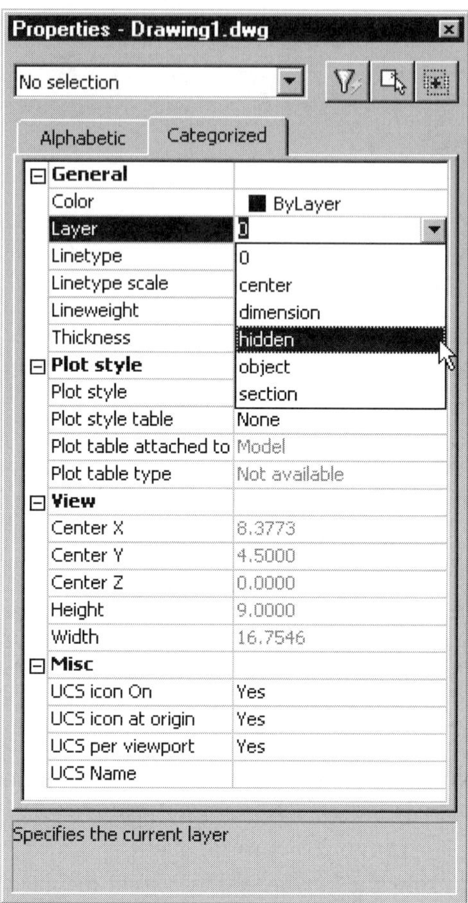

Figure 5.21 *Selecting the current linetype will display a menu arrow similar to Figure 5.20. Selecting the arrow will display a listing of the current linetypes.*

ALTERING THE LINETYPE SCALE

Start the **Line** command and draw several lines. The new linetype should now be reflected on the drawing screen. If the line still appears as a continuous line, check the **Object Properties** toolbar to verify that you did select the current button. If the name and linetype show the desired linetype, most likely the linetype does not appear altered because you've altered the drawing limits from the original 12" × 9" screen. Use the following steps to alter the linetype scale:

1. Redisplay the **Linetype Manager** dialog box and click the **Show Details** button. This will produce a dialog box similar to Figure 5.22.

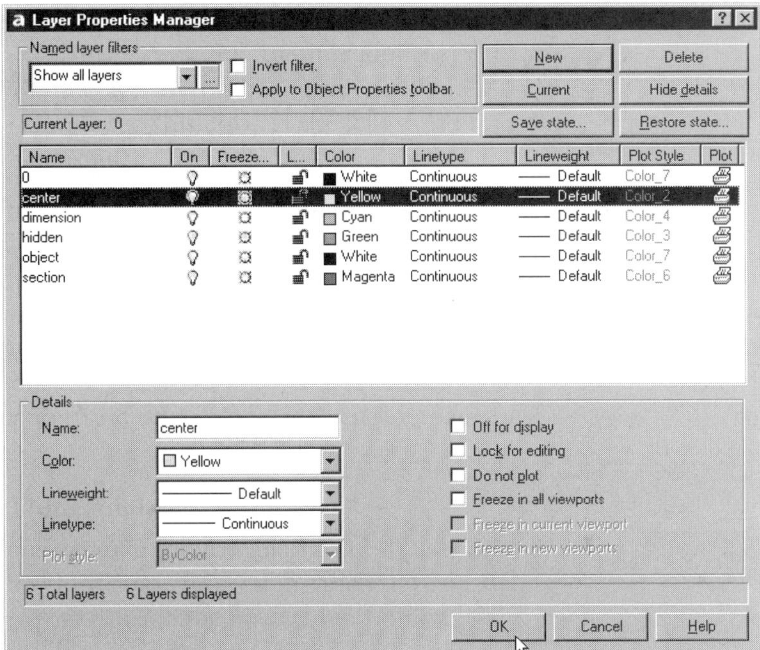

Figure 5.22 *Selecting the* **Show Details** *button in the* **Linetype Manager** *dialog box will display controls for linetype scaling. If you've altered the limits of the drawing screen from the 12" × 9", a line drawn with a centerline may appear to be drawn as continuous.*

2. Alter the **Global scale factor** if you're working with a scale other than 1=1. If you're working on a template set for plotting at a scale of 1=2, change the **Global scale factor** to .25.
3. Click the **OK** button.

As you return to the drawing screen, lines should now reflect the selected pattern, as shown in Figure 5.23.

Note: Although it will be meaningless now, open your drawing template and type **PSLTSCALE** (ENTER) at the Command prompt. Change the default value from 1 to 0. This will ensure proper display of linetypes as drawings are plotted in the future. This process will need to be repeated for each layout that is created with a drawing.

AN INTRODUCTION TO LINEWEIGHT

You've been introduced to common linetypes and line widths earlier in this chapter. Varied line widths will help you and the print reader keep track of various components of a complex drawing. As you assign lineweight to drawings, remember that you're using thick and thin lines to add contrast, not represent thickness of an object. Web

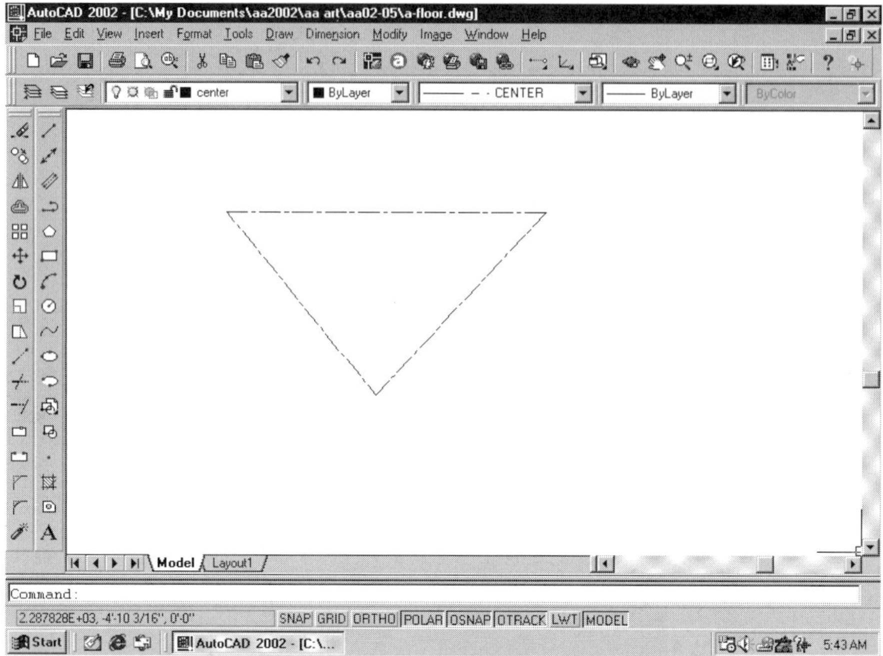

Figure 5.23 *The result of setting the* **Global scale factor***. If you're working on a drawing that will be plotted at a scale of 1/2 = 1, the scale factor is 2. Once the linetype scale is altered, the object will now be displayed using centerlines.*

Chapter 1 will introduce an alternative method of using a thickness to accurately represent objects.

EXPLORING LINEWEIGHTS

Lineweights can be assigned to a layer by selecting **Lineweights** from the **Format** menu, or by typing LW (ENTER) or LWEIGHT (ENTER) at the Command prompt. Each method will produce the **Lineweight Settings** dialog box. Once you've changed some basic settings, lineweights can be altered using the **Lineweight Control** box on the **Object Properties** toolbar. For now, display the **Lineweight Settings** dialog box by selecting **Lineweights** from the **Format** menu. This will produce a display similar to Figure 5.24. Common lineweights and their width in millimeters and inches, point size, and pen size can be seen in Figure 5.25.

ASSIGNING LINEWEIGHTS

The **Lineweight Settings** dialog box can be used to set the units for listing line width, the default lineweight, the display scale, and as a display of the current settings. As you display the dialog box, notice that the **Display Lineweight** check box is not selected. In its current state, you can draw using varied lineweights, but the width will

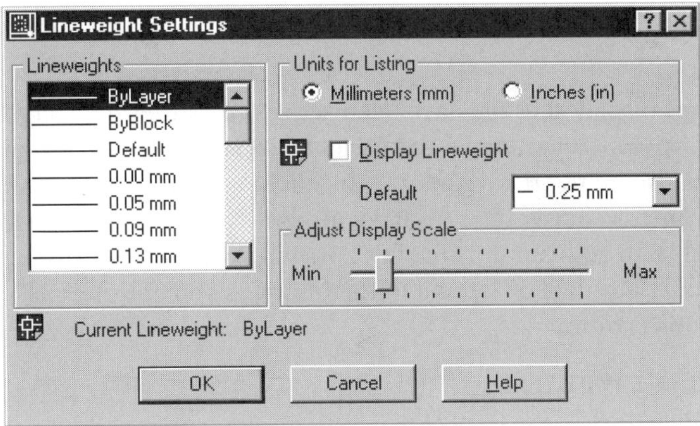

Figure 5.24 *The* **Lineweight Settings** *dialog box can be used to control drawing lineweights. Access the dialog box by selecting* **Lineweights** *from the* **Format** *menu or by typing* **LWEIGHT** *(ENTER) at the Command prompt.*

LINEWEIGHT	MILLIMETERS	INCHES	POINTS	PEN SIZE
	.25	.010	3/4	000
	.30	.012		00
	.35	.014	1	0
	.50	.020		1
	.53	.021	1-1/2	
	.60	.024		2
	.80	.031		3
	1.00	.039		3-1/2
	1.20	.047		4

Figure 5.25 *Common lineweights and their width in metric and inches, point size and pen size.*

be reflected only on a print or plot and not in the drawing display. For most users, this should be the first adjustment you make. Select the **Display Lineweight** check box to make it active, and varied lineweights will be displayed on the screen. Most users are also more comfortable working in inches rather than metric units. Make this change by choosing the **Inches** radio button. With this change, you'll notice that the lineweights are now displayed in inches. To assign a lineweight, scroll down the list of lineweights, find the desired lineweight, and click the **OK** button. As you use the

Line command, any line added to the drawing will now reflect the current linetype and lineweight.

Once the unit and display settings have been set, you can alter lineweights by selecting the **Lineweight Control** arrow on the **Object Properties** toolbar (see Figure 5.26). Lineweights can also be toggled ON/OFF by clicking LWT, the lineweight display control button on the status bar. A third method to control lineweight display, already discussed, is to activate the **Display Lineweight** check box in the **Lineweight Settings** dialog box. This will be helpful in the future, as you enlarge areas of the drawing using the **Zoom** command.

DISPLAYING LINEWEIGHTS

In addition to activating **Display Lineweight** in the **Lineweight Settings** dialog box, you also must consider the difference in lineweight display between model and paper space. In model space, lineweights are displayed in relation to pixels. Lineweights in paper space are displayed in the exact plotting width. For work in model space, lineweights should be selected that are proportional to their pixel value and not their real world width. Chapter 13 will explore the display scale for lineweights. For now, use the following guide for determining lineweight.

mm	In.	Description	Use
0.35	.014	thin	Extension, dimension, leader lines
0.60	.024	thick	Object lines
0.80	.031	very thick	Section and cutting plane lines
1.20	.047	very, very thick	Borders

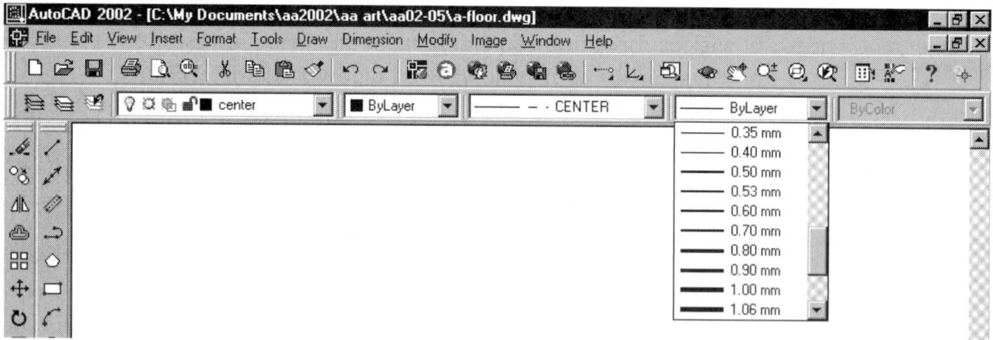

Figure 5.26 *Once the unit and display settings have been set, lineweights can also be altered by selecting* **Lineweight Control** *on the* **Object Properties** *toolbar.*

AN INTRODUCTION TO COLOR

Up to this point, all of your drawings have been done with black lines on a white background, or white lines on a black background. The choice of screen background can be altered by the following procedure:

1. Select **Options** from the **Tools** menu. This will display the **Options** dialog box.
2. Select the **Display** tab in the **Options** dialog box.
3. Click the **Colors** button from the **Window Elements** portion of the **Display** tab. This will produce the **Color Options** dialog box shown in Figure 5.27.
4. The box allows the color of each of the drawing window displays to be altered. To change the color of the drawing area, select the desired color tile.
5. Click the **Apply & Close** button to close the **Color Options** dialog box.
6. Click the **OK** button in the **Options** dialog box. The dialog box will close and the screen will display the selected background color.

AutoCAD allows the color of each 2D drawing object to be altered using one of 255 colors. A name or an AutoCAD Color Index (ACI) number ranging from 1 through 255 can identify the color for objects. Rendered 3D drawings use the same 255 colors but can display millions of colors by altering the tint. The actual number of colors that are displayed will depend on third party software and how the computer is configured.

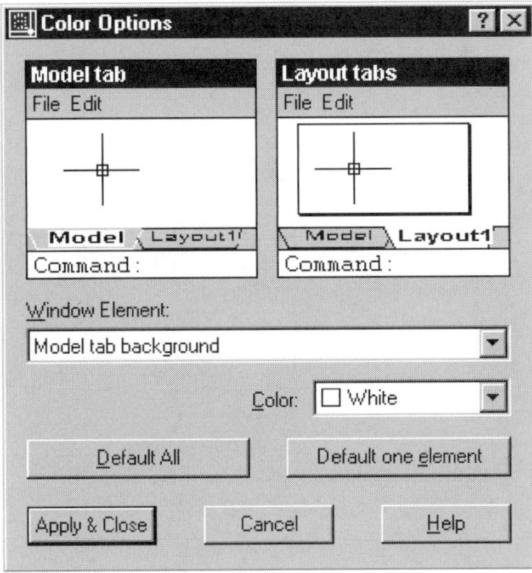

Figure 5.27 *The Colors button on the* **Display** *tab will produce the* **Color Options** *dialog box.*

COLOR NUMBERS AND NAMES

Each drawing object or layer can be assigned a color. The seven basic colors used by AutoCAD are listed in Figure 5.28. AutoCAD identifies colors using numbers from 1 through 255, with the selection based on the color displayed in the color palette. Multiple objects or layers can have the same color number. Objects of the same color can be assigned to a specific pen during the plotting process. For instance, all red objects can be assigned to a thick pen, and all blue objects can be assigned to a thin pen to produce good line contrast on the plotted drawing. Plot styles can be used to control plotting and colors, and they will be explored later.

SETTING OBJECT COLOR

The **Color** command can be used to assign colors to an entire layer or to individual objects on a layer. Setting colors for layers is the default method and will be explained as you proceed through layer options later in this chapter. The easiest method for assigning color is to select the **Color** button in the middle of the **Object Properties** toolbar. You can select the color tile, the color name, or the down arrow. Each method will display the list shown in Figure 5.29. Selecting one of the listed colors will close the list and return the drawing display. Now, if a line is drawn, it will be drawn using the selected color. Using the current settings, the line would be a red line drawn with a center linetype. Select the **Color** button again to redisplay the color list. Notice the last option is **Other**. Selecting this option will display a menu similar to Figure 5.30. The **Select Color** dialog box can also be displayed by selecting **Color** from the **Format** menu or by typing COLOR (ENTER) at the Command prompt. The

COLOR NUMBER	COLOR NAME
1	RED
2	YELLOW
3	GREEN
4	CYAN
5	BLUE
6	MAGENTA
7	BLACK

Figure 5.28 *The eight major colors of AutoCAD. The default color is 7. Depending on the background color, 7 will produce either a black or white line.*

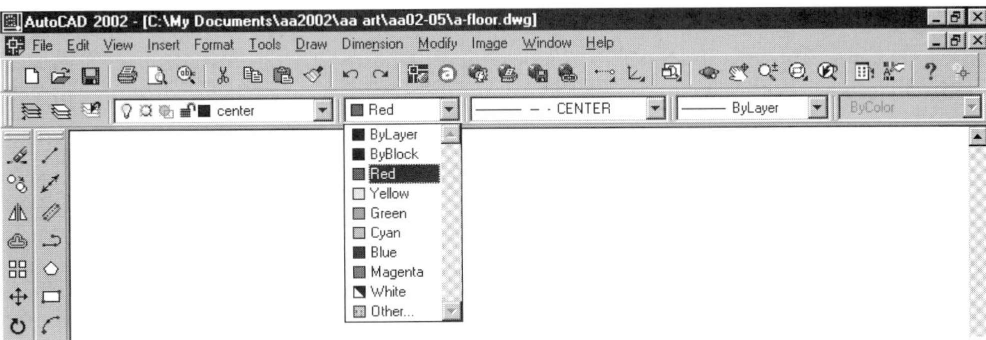

Figure 5.29 *Selecting the* **Color Control** *arrow on the* **Object Properties** *toolbar will display the color list and allow the color of the current layer to be changed.*

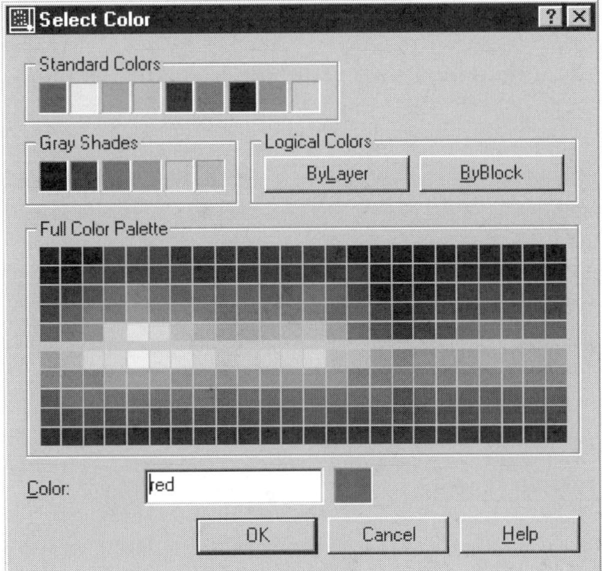

Figure 5.30 *Selecting* **Other** *from the* **Color list** *displays the color palette and allow a wide variety of color tints to be selected for drawing objects.*

Select Color dialog box allows shades of the standard colors to be selected. Notice the current color is listed as red. Selecting any of the color tiles will alter the current color display tile and list the color by number. Clicking the **OK** button will close the display and make that the current color. Using the **Line** command now will create lines using the selected color.

Notice that the **Select Color** dialog box contains buttons for **ByLayer** and **ByBlock**. The **ByBlock** option will be discussed in Chapter 20. If **ByLayer** is accepted as the default, the color will be assigned once each layer has been defined. Most offices set colors using the **ByLayer** option rather than assigning colors to individual objects. Although there are no current ANSI specifications for setting colors, unless your instructor tells you otherwise, the following guidelines can be used for drawing projects in this text.

Linetype	Color
Center	yellow
Cutting Plane	blue
Dimension components	cyan
Hidden	green
Object	black
Section	magenta

Drawing components should be drawn using the following colors:

Item	Color
Borders	blue
Drawing notes	cyan
Title block (new text)	cyan
Title block (text headings)	blue
Title block tolerance charts	blue
Parts lists	blue
Revision list	blue
View and section names	cyan

AN INTRODUCTION TO LAYERS

The use of layers in a CAD drawing is similar to overlay drafting procedures of manual design. Overlay drawing will have a base drawing, such as the valve assembly seen in Figure 5.31, on one sheet of Mylar®, and have overlay sheets of notes and dimensions on separate sheets of Mylar, as seen in Figure 5.32. Depending on who needs what, different sheets can be attached to the base and various prints made. Layers allow all the flexibility with none of the alignment problems of overlay drafting procedures. Until now you have been drawing on the 0 layer. AutoCAD will allow you to create different layers, assign each layer a different name, linetype, lineweight, and color, and allow some layers to be made invisible.

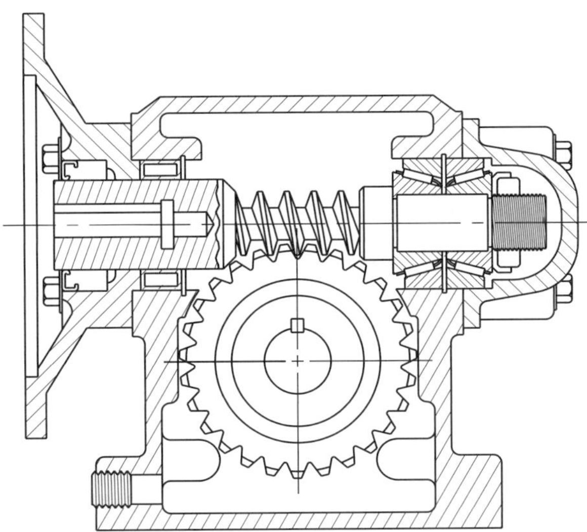

Figure 5.31 *The use of layers is similar to placing information on several sheets of paper in overlay drafting. This drawing shows the basic drawing with no supplemental information provided. Courtesy Michael J. Easley.*

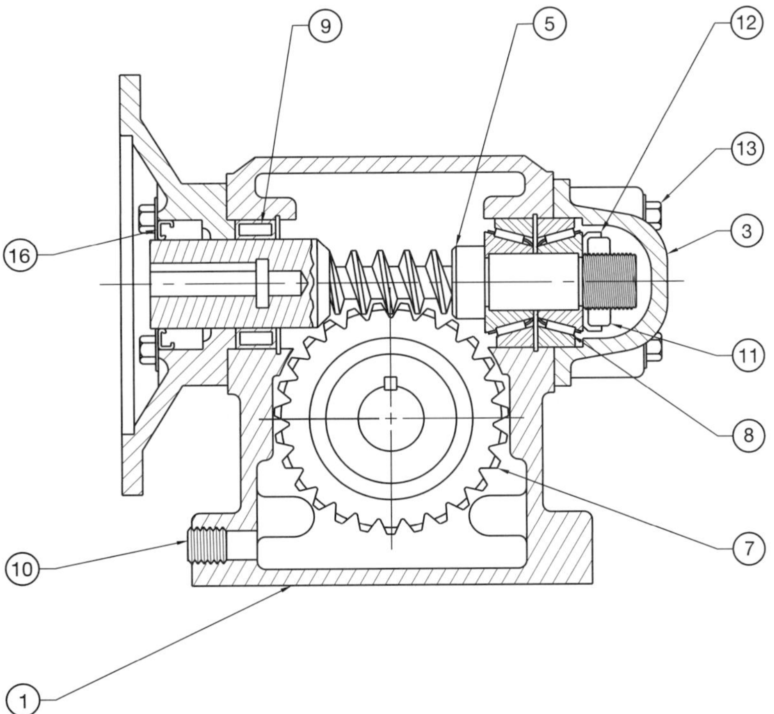

Figure 5.32 *Supplemental information added to the base drawing. Courtesy Michael J. Easley.*

THE LAYER COMMAND

Start the **Layer** command by selecting the **Layers** button on the **Object Properties** toolbar, by selecting **Layer** from the **Format** menu, or by typing LA (ENTER) at the Command prompt. Each will display the **Layer Properties Manager** dialog box shown in Figure 5.33.

CREATING A NEW LAYER

Once the **Layer Properties Manager** dialog box is displayed, press ENTER or click the **New** button. Each method will add the highlighted name of Layer 1 in the **Name** edit box (see Figure 5.34a). With Layer 1 highlighted, a new layer name can be created by keyboard. Click the select button to accept the new layer name. If you would like to add multiple layers, press ENTER, which adds the previously changed name to the drawings and adds a new layer titled Layer1. In Figure 5.34b, Layer1 is assigned the name of COOLSTUFF. Guidelines for naming layers will be discussed in Chapter 13. For now, keep the names short, simple, and descriptive, using names that will identify the contents. Once the name has been supplied, click the **OK** button to save the layer and return to the drawing area. Just as with **Linetype**, all you've done is to create a new layer. The current layer is still the default layer of 0.

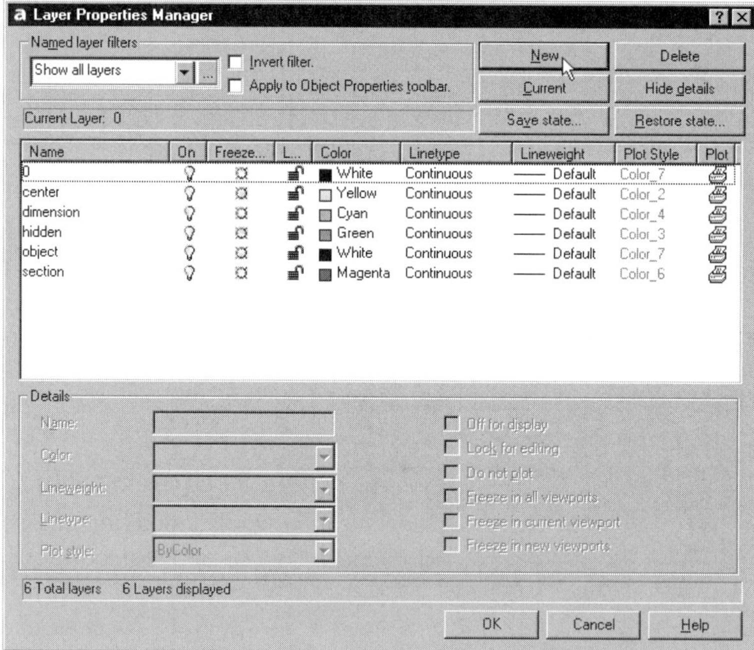

Figure 5.33 *Clicking the* **Layer** *button on the* **Object Properties** *toolbar or typing* **LA** *(ENTER) will display the* **Layer Properties Manager** *dialog box. The box can be used to create and control the properties of each drawing layer.*

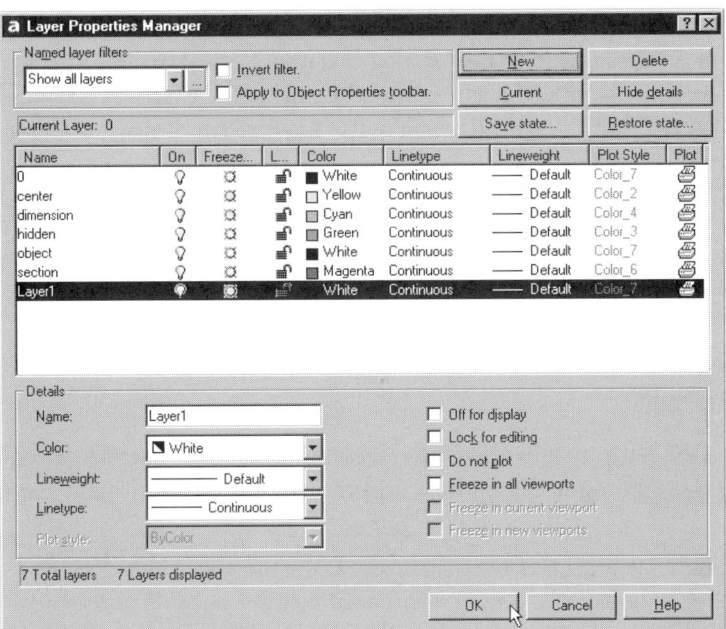

Figure 5.34a *Pressing* ENTER *or clicking the* **New** *button will display a new layer titled Layer1.*

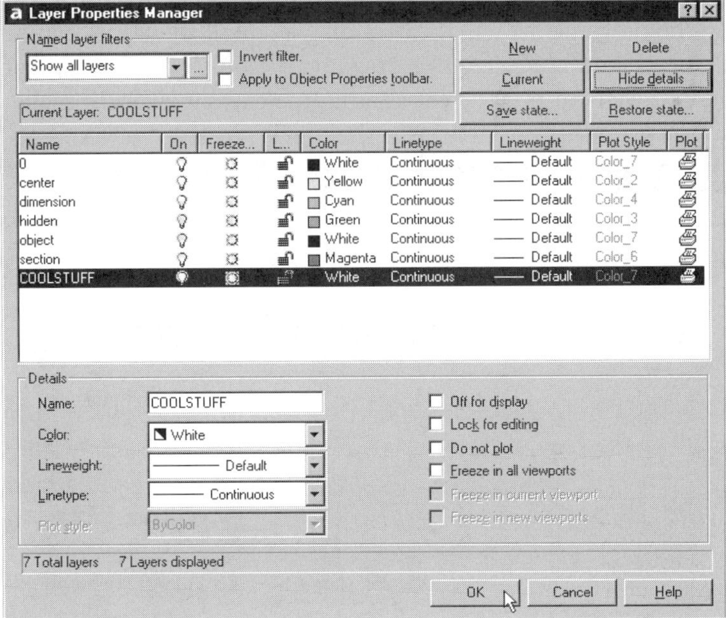

Figure 5.34b *The selected layer name in Figure 6.34a can be altered to reflect the desired layer name. Guidelines for naming layers are the same as those used to name a drawing file.*

MAKING A LAYER CURRENT

Close the **Layer Properties Manager** dialog box and reenter the drawing area and add a few lines. The lines will be the same linetype and color that was drawn before you created the COOLSTUFF layer. Redisplay the **Layer Properties Manager** by selecting the **Layers** button on the **Object Properties** toolbar. The dialog box displays 0 and COOLSTUFF as the only layers in the drawing. The box surrounds the 0 layer, indicating that it is the current layer. Use the following steps to make COOLSTUFF the current layer:

1. Select COOLSTUFF.
2. Click the **Current** button.
3. Click **OK**.

Clicking the **OK** button will make the selected layer the current layer, close the **Layer Properties Manager** dialog box, and return you to the drawing area.

Note: You can also make a layer current by displaying the **Layer Properties Manager** dialog box and double-clicking the name of the layer to be made current.

Notice that just to the right of the **Layers** button on the **Object Properties** toolbar is an edit bar that describes the current layer status. (See Figure 5.35a.) You'll see icons that describe the current status of the layers, with COOLSTUFF listed as the current layer. Clicking the down arrow will create a display of all the layers contained in the drawing, similar to Figure 5.35b. In this drawing, the only other option is the 0 layer, but this is a very convenient method for altering the current layer. To make a layer the

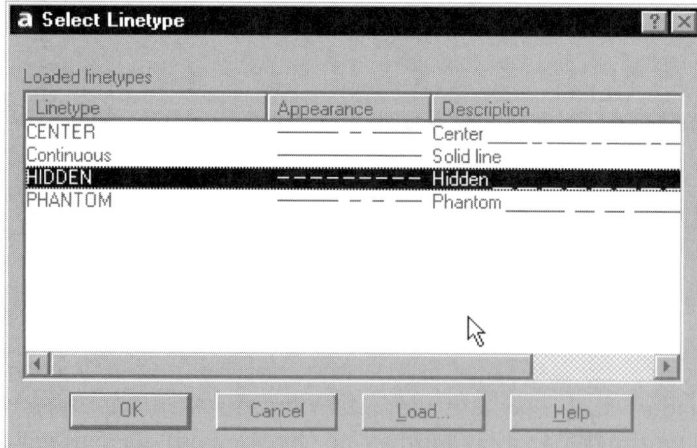

Figure 5.35a *The* **Layer** *box on the* **Object Properties** *toolbar describes the properties of the current layer.*

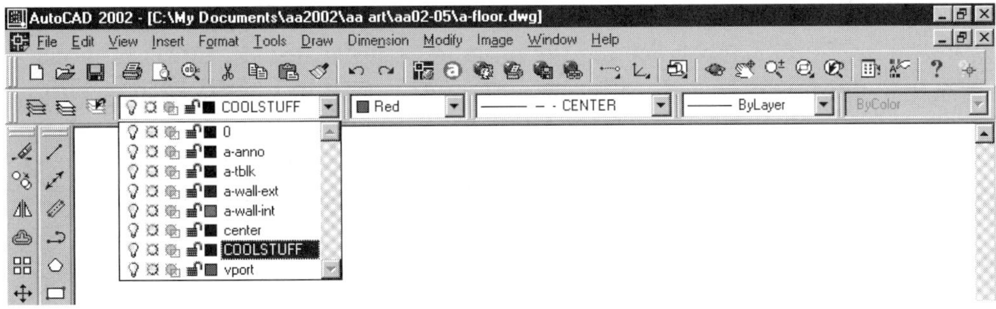

Figure 5.35b *Selecting the arrow displays the status of each drawing layer.*

current layer, select the down arrow, and then select the name of the layer you want as the current layer.

ALTERING LAYER QUALITIES

Redisplay the **Layer Properties Manager** dialog box, select the COOLSTUFF layer and then click the **Current** button. COOLSTUFF has the same color, linetype, and lineweight qualities as the 0 layer. Any object created on the COOLSTUFF layer will appear the same as it would if it were drawn on the 0 layer. Layers become a powerful drawing tool when varied linetypes, lineweights, and colors are assigned to specific layers. Creating a layer with thin, blue, hidden lines can be a great way of easily separating hidden features from visible objects that are drawn with thick, continuous red lines.

Linetype qualities can be assigned to all objects created, when this layer is the current layer, by selecting the current Linetype display. Use the following steps to alter the linetype of the COOLSTUFF layer:

1. Display the **Layer Properties Manager** dialog box.
2. Highlight the layer that you wish to alter.
3. With the layer highlighted, select the current linetype for that layer. This will produce the **Select Linetype** dialog box shown in Figure 5.36a. Only linetypes that have been loaded into the drawing will be displayed.
4. Select the desired linetype to be used.
5. Click the **OK** button to close the **Select Linetype** dialog box.
6. Click **OK** to close the **Layer Properties Manager** dialog box and return to the drawing area.

Selecting one of the linetypes will use that linetype for all drawing objects created using that layer. The lineweight and color can be altered using the same procedure.

Selecting the color display for COOLSTUFF will produce the **Select Color** dialog box shown in Figure 5.36b. Selecting the lineweight display for COOLSTUFF will display the **Lineweight** dialog box similar to Figure 5.36c. Once the desired qualities of a layer have been assigned, click the **Current** button if you want to use that layer as you return to the drawing editor. Clicking the **OK** button will close the **Layer Properties Manager** dialog box and return you to the drawing screen.

Note: Assigning colors, linetypes, and lineweights will affect all objects drawn on the layer that is altered, even objects drawn prior to making the changes.

CONTROLLING THE VISIBILITY OF LAYERS

Eight icons can be displayed by each layer name. These buttons represent various button controls for the use of each layer. The number of buttons displayed will vary depending on how much space is assigned for each button. The lightbulb button is a toggle for the ON/OFF setting of a layer. For the drawing shown in Figure 5.31, the Text layer was toggled OFF, and for Figure 5.32, the Text layer was toggled ON. Clicking the lightbulb will change the bulb from yellow to gray. The yellow bulb represents ON, the gray OFF

The second button, a sun, can also be used to control the visibility of a layer—it is a toggle between Thaw/Freeze. When the sun is displayed, the contents of that layer will be displayed. Selecting the sun button will change it to a snowflake, which represents a frozen layer. The contents of a frozen layer will not be displayed. The ben-

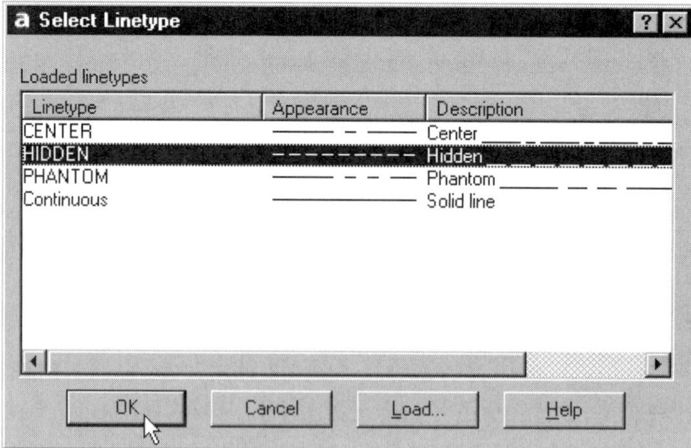

Figure 5.36a *Selecting the current linetype for the layer will produce the* **Select Linetype** *dialog box. Only linetypes that have been loaded into the drawing will be displayed. You can load additional linetypes into the drawing by selecting the* **Load** *button and following the guidelines presented early in this chapter.*

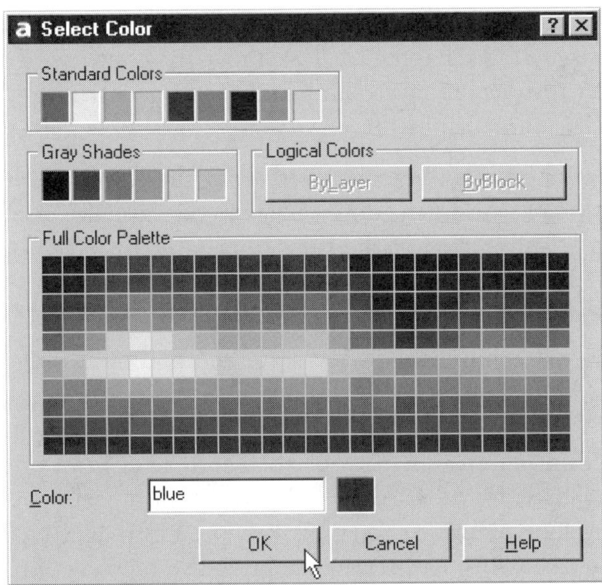

Figure 5.36b *The color of a layer can be adjusted by selecting the current color display for that line. The* **Select Color** *dialog box allows the color to be altered.*

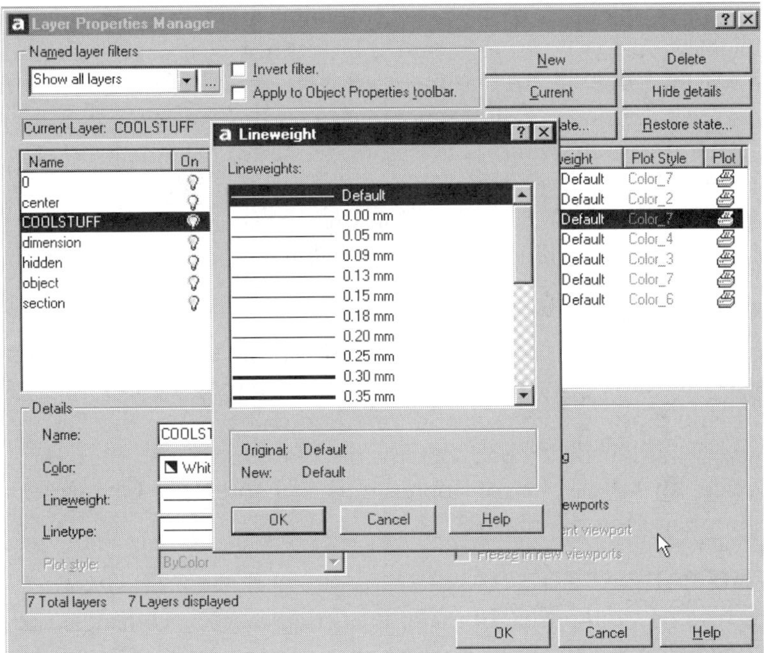

Figure 5.36c *The lineweight of a layer can be adjusted by selecting the current lineweight display for that line. The* **Lineweight** *dialog box allows the line width to be altered.*

efits of using the ON/OFF and Thaw/Freeze options, as well as each of the other layer controls, will be discussed in Chapter 13. To remove the contents of a layer from the screen display, select the desired layer name and then select the lightbulb or the sun button and then select the **OK** button.

The visibility of layers can also be controlled from the **Layer Control** box on the **Object Properties** toolbar. Select the down arrow to display a list of current layers and select the lightbulb or sun button of the layer to be altered.

AN INTRODUCTION TO PRINTING

Once you've created a drawing, you'll need some method of passing the contents of the file on to those who will put it to use. This might include sending the drawing electronically over the Internet to another firm or sending the drawing by diskettes. For many, it also means making a hard copy such as a paper print or plot. No matter how technology advances, for most AutoCAD operators, a paper copy is still a common method for checking drawings. You've already been introduced to the layout templates as you explored drawing setup. Layouts provide a preview of how the drawing will appear prior to plotting. Each layout displays a border and the drawing as it will appear at the current lineweight and scale factor. Although it will be a bit overwhelming now, AutoCAD will allow you to establish multiple plotting layouts of the same drawing project. The same drawing can be plotted at different scales or configured for different plotting devices by using layouts. The settings to repeat the plotting process can be saved with the drawing, providing a valuable saving of time. The purpose of this chapter is not to explore all of the plotting functions, but to help you produce a quick check print of your drawing. This chapter will introduce the process of making a paper print using the system printer configured using your Windows platform. Later chapters will provide additional information on choosing additional printers, configuring a plotter, and controlling the many variables associated with plotting.

EXPLORING THE PLOTTING PROCESS

Start the print process by selecting the **Plot** button on the **Standard** toolbar, by selecting **Plot** from the **File** menu, or by typing PLOT (ENTER) at the Command prompt. Each selection method will produce the **Plot** dialog box shown in Figure 5.37. The dialog box contains the **Plot Device** and **Plot Settings** tabs. The **Plot Device** tab will not be considered in this chapter. See Chapter 23 for further discussion of altering the plotter.

LAYOUT AND PAGE SETUP NAMES

The **Plot Settings** tab contains eight major areas that must be considered to produce a print. The **Layout name** and **Page setup name** portions of the **Plot** dialog box are used to specify layout and plotting device settings. The name of the current layout (Model, Layout1, or Layout2) is displayed on the left. Options for listing and choos-

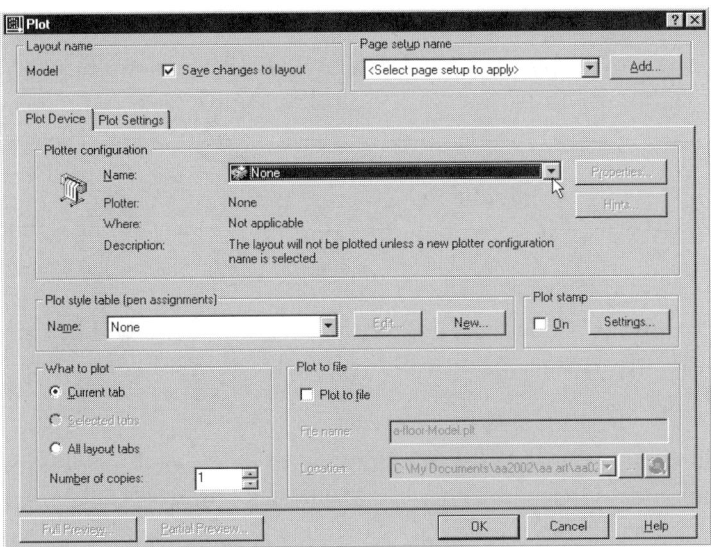

Figure 5.37 *Selecting the* **Plot** *button on the* **Standard** *toolbar, selecting* **Plot** *from the* **File** *menu, or typing* **PLOT** *(*ENTER*) at the command line will display the* **Plot** *dialog box. The dialog box can be used to control each aspect of creating a plot.*

ing page setups for plotting are listed on the right. Each option will be explored as plotting is explored in later chapters. Chapter 22 will introduce methods of creating new page setups.

PAPER SIZE AND PAPER UNITS

The **Paper size and paper units** portion of the **Plot Settings** tab displays the current printing device, the paper size the plot will be made on, and the size of the printable work area. Selecting **Options** from the **Tools** menu and selecting the **Plotting** tab in the **Options** dialog box can also control printer and plotter selections. For the current discussion, the default printer will be used. The box also displays the plotting units in inches and millimeters.

PLOT AREA

The **Plot** area provides options for describing what part of the current drawing will be plotted. Because often the entire drawing does not need to be plotted, five options are given to determine what portion of the drawing will be plotted.

Limits

Using this option will plot all of the drawing objects that lie within the current drawing limits, defined during the drawing setup.

Extents

Choosing this option will plot the current drawing with the extent of the drawing objects as the maximum limits that will be displayed in the plot.

Display

The default setting is to plot only the material that is currently displayed on the screen. If you have zoomed into a drawing, only that portion of the drawing will be plotted.

View

This option is currently not active. Methods of creating views will be introduced in Chapter 9 and the use of this option will be introduced in Chapter 22.

Window

This option will allow a specific area of a drawing to be defined for plotting. Initially the **Window** radio box is inactive. When the **Window<** button is clicked, the **Plot** dialog box will be removed from the screen, allowing you to select what is to be plotted. The command line will show the prompt:

> Command:_plot:
>
> *(Select the Window< button.)*
>
> Specify window for printing
>
> Specify first corner:

AutoCAD is waiting for you to define a window or box that will surround what you would like to plot. To create the plotting window, move the cursor and select a point. Once the first corner of the window is selected, AutoCAD will provide a prompt to specify the second corner. Move the cursor to the opposite corner. As the cursor is moved, a box (the window) will be formed. Figure 5.38 shows the process for defining the window. Anything that lies within the window will be plotted. Once the window is defined, the **Plot** dialog box will be redisplayed, allowing you to continue defining the plot to be created. The **Window** radio button will be activated, allowing other options to be set. Each option will be discussed in later chapters.

DRAWING ORIENTATION

The **Drawing orientation** portion allows the drawing orientation to be specified. The current orientation of the box shows that the drawing will be printed in the normal orientation for mechanical drawing, with the long paper edge horizontal. The drawing will be read looking from the top to bottom, and the title block will be along the right side of the print. This setting is referred to as Landscape. When the print is rotated so that the long edge of the paper is vertical, the print is referred to as a Portrait print. Adjust the orientation of the drawing to the paper by selecting the appropriate radio button. The **Plot upside-down** option plots the drawing in reverse of the

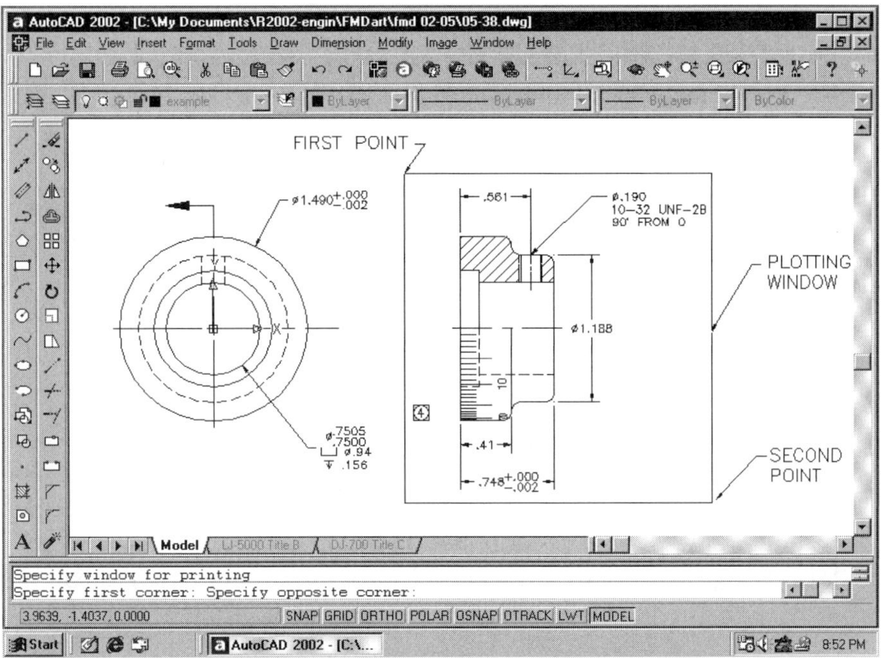

Figure 5.38 *When objects are selected to be plotted using the* **Window** *option, the size of the window can be set by selecting the window corners with the cursor. Selecting the Window< button will remove the menu and return the drawing to the screen. Once the corners of the select window are located, the* **Plot** *dialog box will be displayed again so that other print options can be set. Courtesy Paul Bissell.*

selected position from the 0,0 origin. If **Landscape** is the selected position, and **Plot upside-down** is active, the top right corner of the drawing will be placed in the 0,0 position rather than the lower left corner of the drawing.

PLOT SCALE

The **Plot scale** portion controls settings for the scale of the plot. Its options include **Scale** and **Custom**.

Selecting a Scale

With the default setting, **Scaled to Fit**, in the **Scale** edit box, when a plot is made, it will be reduced in size to fit on the current printer or plotter paper. Unless you'll be plotting a small detail, most mechanical drawings can't be printed at full size on a printer using A or B-size material. You can, however, print an entire component on A-size material if the scale is reduced. The **Scaled to Fit** option reduces the selected material to be printed to an appropriate size to fill the current paper. You can print portions of a drawing to scale by selecting the desired scale and using the window selection method. To print a drawing to a scale, select the arrow beside the edit box to reveal the scale menu. (See Figure 5.39.) Select the desired scale, and the list will close to allow you to continue setting the print parameters.

Custom Scale

The Custom edit box allows you to print a drawing to a scale that you create.

Note: Keep in mind that the goal of this chapter is to help you make a quick, non-scaled, check print so that you can evaluate your work. Stick with **Scaled to Fit** or one of the common mechanical or metric scales until plotting is discussed in detail.

Plot Offset

The **Plot offset** portion of the **Plot** dialog box can be used to specify an offset of the plotting area from the lower left corner of the paper. The default setting for plotting origin is X=0, Y=0. Printers typically specify the origin from the upper right corner. Specifying various locations for the origin can be a useful technique for altering the drawing location on the paper. With the **X** or **Y** origin box selected, new values can be entered. Entering an X value of 12 and a Y value as 10 would place the new origin 12 units to the right and 10 units above the original plotting origin.

Plot Options

The **Plot options** portion of the **Plot** dialog box can be used to specify options for lineweights, plot styles, and the current plot style table. By default, the **Plot with**

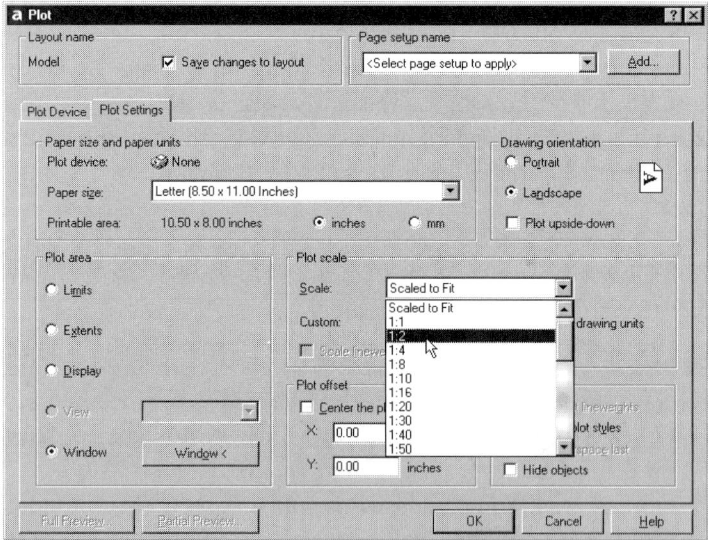

Figure 5.39 *To plot a drawing to a scale, select the arrow beside the edit box to reveal the scale list. Select the desired scale. Once the scale is selected, the list will close to allow you to continue setting the plot parameters.*

lineweights and the **Plot with plot styles** options are active. Each plot option will be discussed in later chapters.

PREVIEWING

Selecting a preview allows the outcome of the current plot settings to be viewed prior to making the print. Two options are available for previewing the print: **Partial Preview** and **Full Preview**.

Partial Preview

Clicking the **Partial Preview** button will produce a display similar to the one seen in Figure 5.40. A white box represents the paper size and the limits of the effective printable area are specified on the screen by a blue box. A warning also will be given of any problems that might be encountered during the plot. Common warnings displayed in the warning box include the following:

Effective area too small to display

Origin forced effective area off display

Plotting area exceeds plotter maximum

A small triangular Rotation Icon is displayed in the lower left corner of the paper image to represent a landscape rotation. As the rotation is set to **Portrait**, the icon is moved to the lower left corner of the paper image.

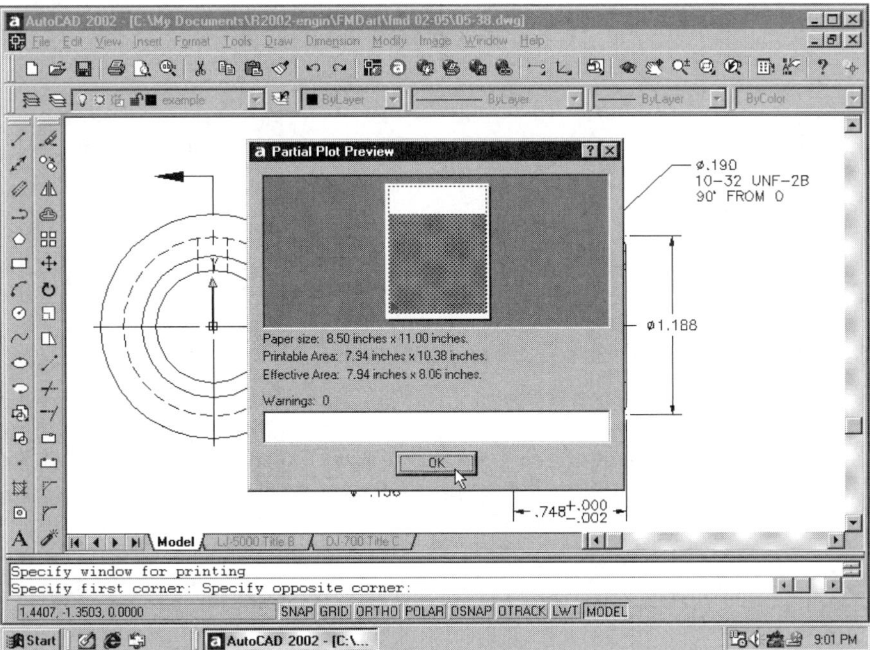

Figure 5.40 *A partial preview of a drawing to be printed.*

Full Preview

Clicking the **Full Preview** button will produce a display of the drawing as it will appear when printed or plotted. Figure 5.41 shows an example of a full preview. An outline of the paper size is drawn on the screen, which displays the portion of the drawing to be printed. The cursor is changed to show a magnifying glass with a + and – sign beside it. This represents the **Zoom** command and allows the image for viewing to be enlarged or reduced. It will not affect the plotted image. Press and hold the select button to alter the size of the image. Moving the mouse toward the top of the screen enlarges the image, and moving the mouse toward the bottom of the screen reduces the size of the image. Right-clicking will display the shortcut menu shown in Figure 5.42, allowing other options to be selected. Selecting the **Pan** option allows the image to be shifted, but the magnification of the image is not altered. Each of the options will be further discussed in Chapter 22. To continue with the plot process, select the **Exit** option to remove the menu.

PRODUCING THE PRINT

Seven areas of the **Plot** dialog box have been examined. If you change your mind on the need for a print, select the **Cancel** button; the dialog box will be terminated, and the drawing will be redisplayed. If you are satisfied with the plotting parameters that

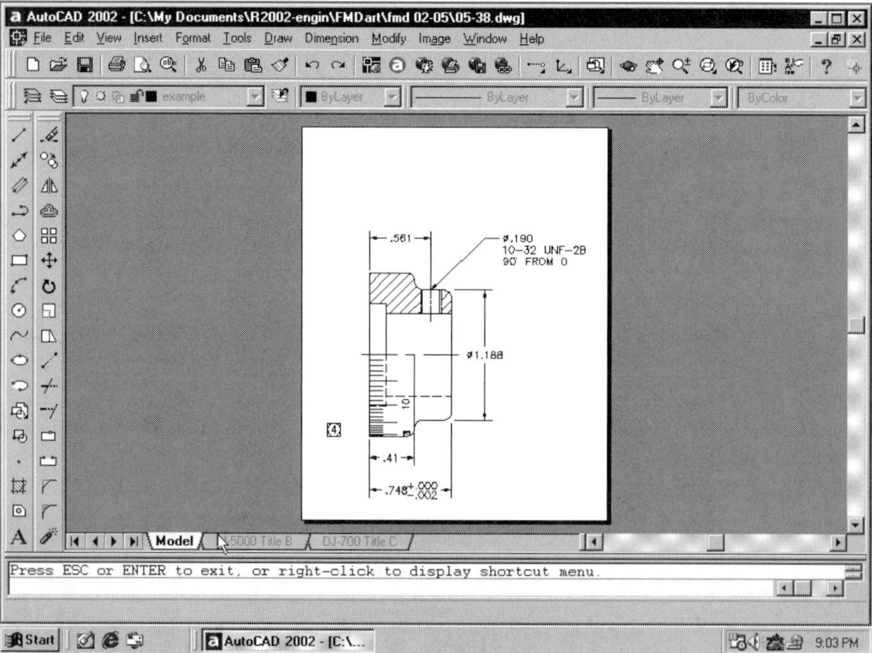

Figure 5.41 *A full preview will show the limits of the paper and the drawing to be plotted. Courtesy Paul Bissell.*

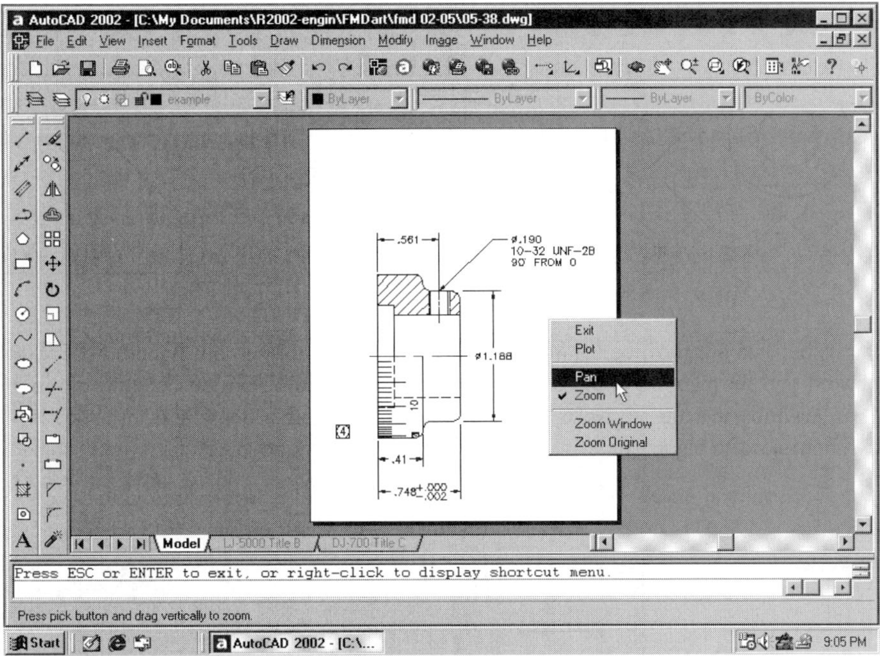

Figure 5.42 *Right-clicking while using a full preview will display this menu. The options can be used to control the display to be printed. Courtesy Michael J. Easley.*

have been established, click the **OK** button. The dialog box will be removed from the screen and the print will be produced. Figure 5.43 shows the result of the selected plot settings.

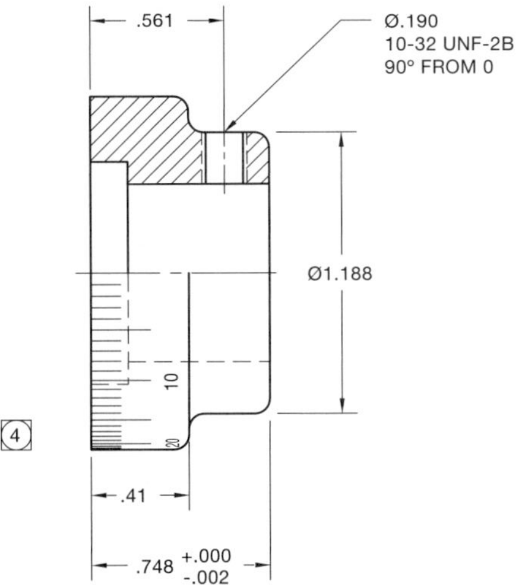

Figure 5.43a *The result of the preview shown in Figures 6.41 and 6.42. Courtesy Michael J. Easley.*

CHAPTER 5 EXERCISES

Use an appropriate ANSI template to complete the following drawings. Set limits, units, snap, and grid to best suit each project. Assume a grid size of .25 unless noted. Use appropriate linetypes and lineweights that best suit each project. Dimensions, and text do not need to be placed at this time. Save the following exercises in a safe location.

1. Use the attached sketch and create the gasket. Use a grid size of .50 to determine the drawing sizes. Save the drawing as E-5-1.

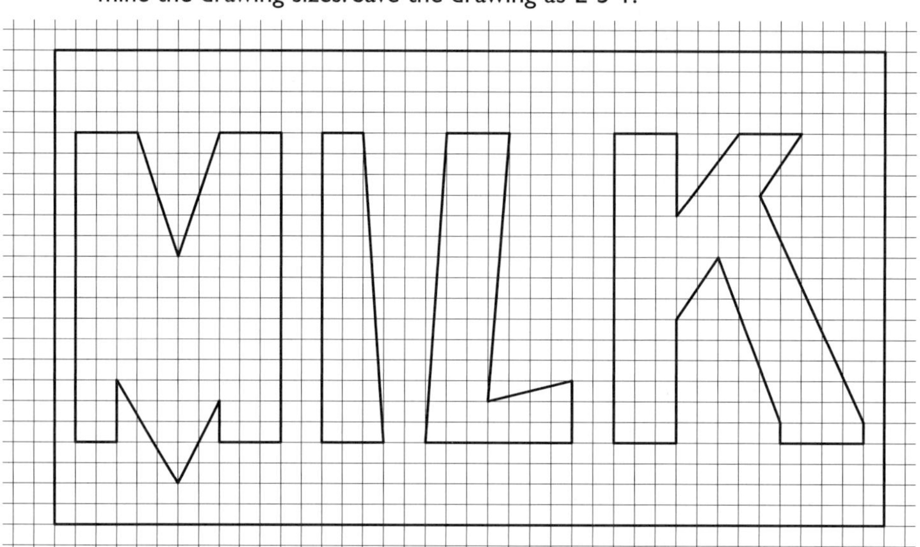

2. Use the attached sketch and create the bracket. Save the drawing as E-5-2.

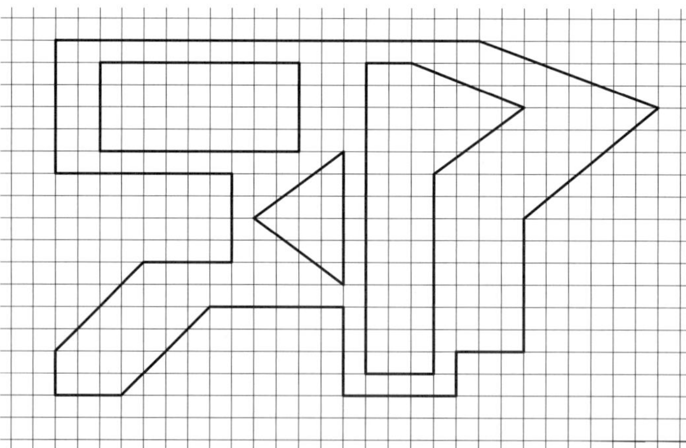

3. Use the attached sketch and create the guide plate. Save the drawing as E-5-3.

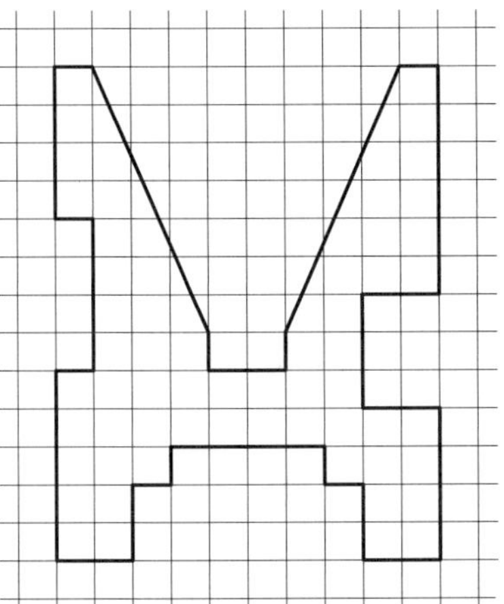

4. Use the attached sketch to create the guide block. Save the drawing as E-5-4.

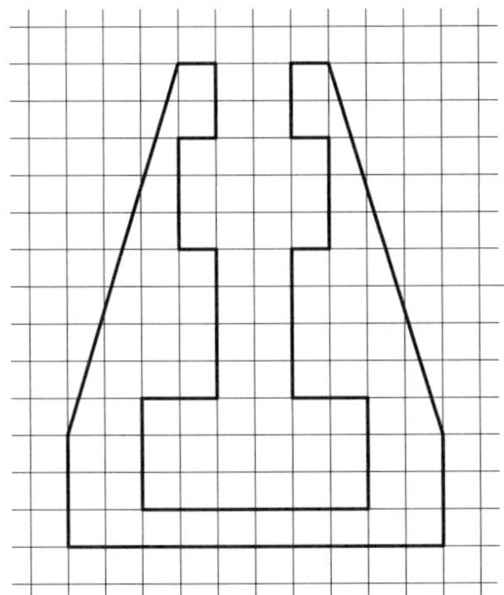

5. Use the attached sketch to create the saw guide. Save the drawing as E-5-5.

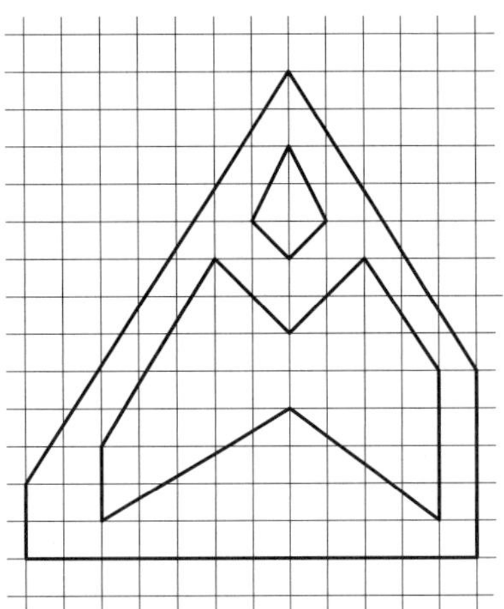

6. Use the attached sketch to create the gasket. Save the drawing as E-5-6.

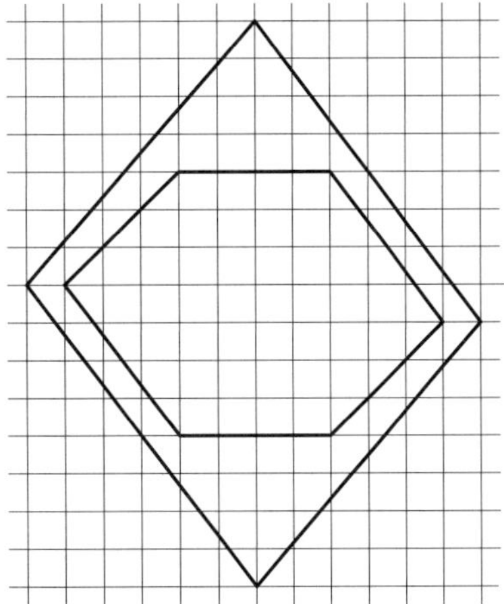

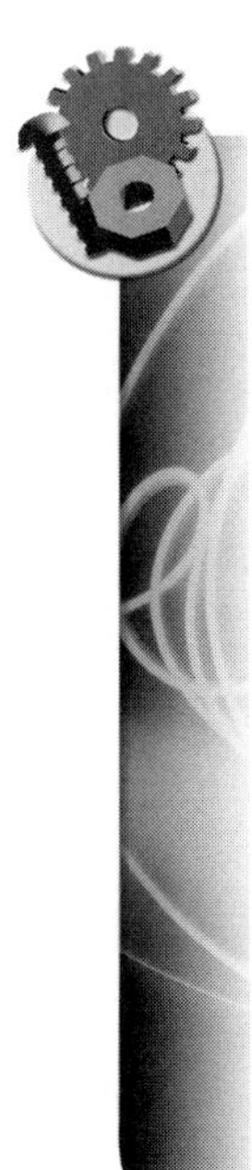

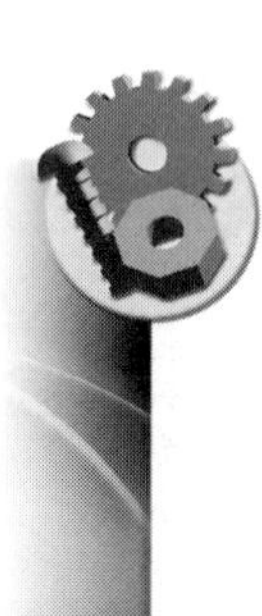

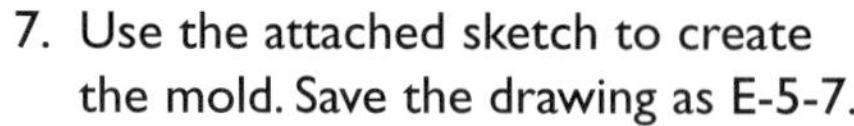

7. Use the attached sketch to create the mold. Save the drawing as E-5-7.

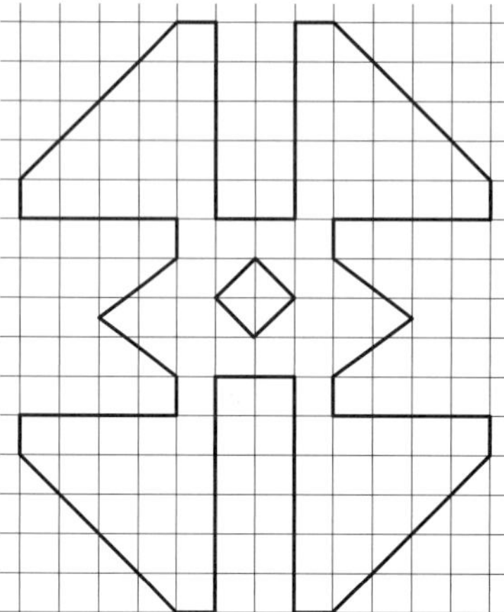

8. Use the attached sketch to create the warp drive shield modulator. Save the drawing as E-5-8.

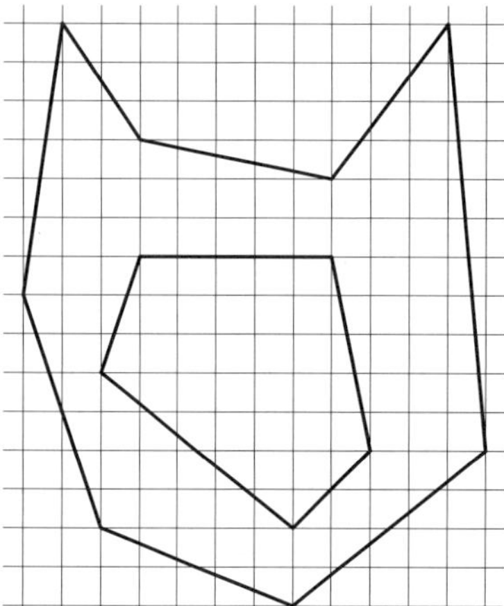

9. Use the attached sketch to create the support base. Save the drawing as E-5-9. Make a print of your drawing using the **Scaled to Fit** option.

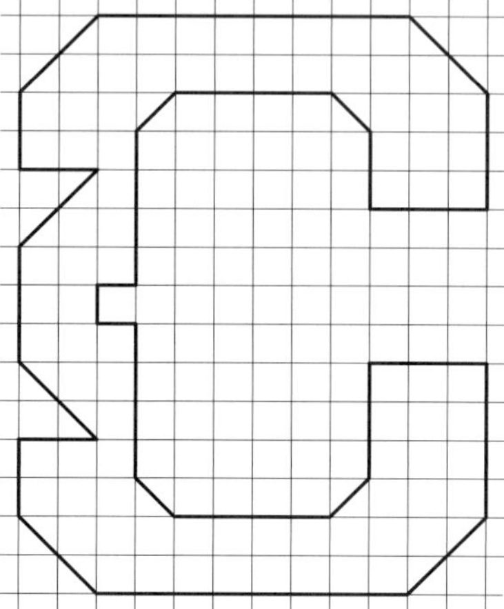

10. Use the attached sketch to create the V-block. Save the drawing as E-5-10. Make a print of your drawing using a scale of 1=1.

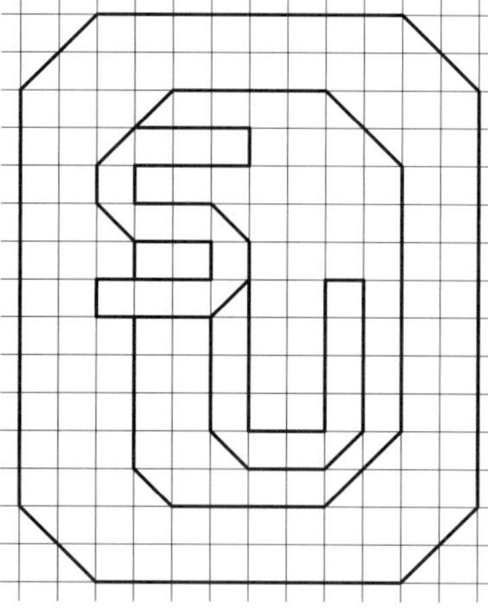

Exercises 11 through 15: Use the attached sketch and the dimensions and angles specified by the engineer to complete the attached drawings. Save each drawing with an appropriate name.

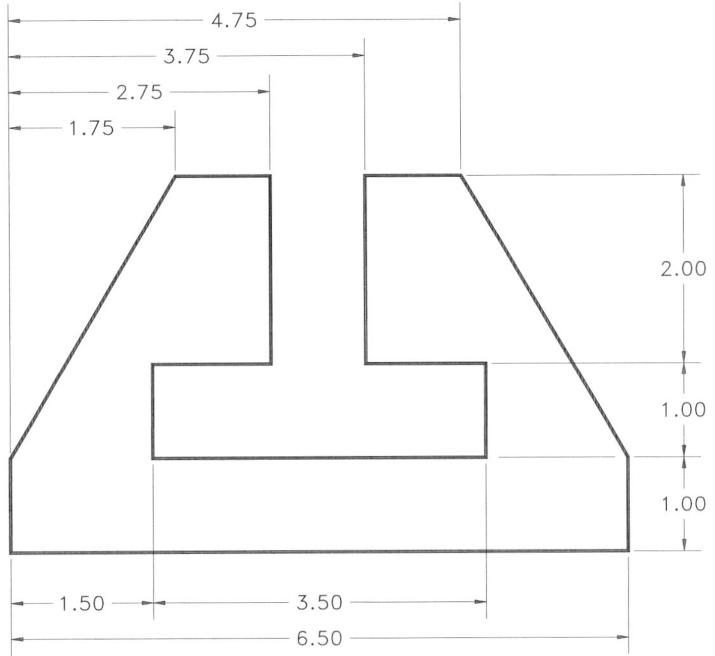

Problem 5-11

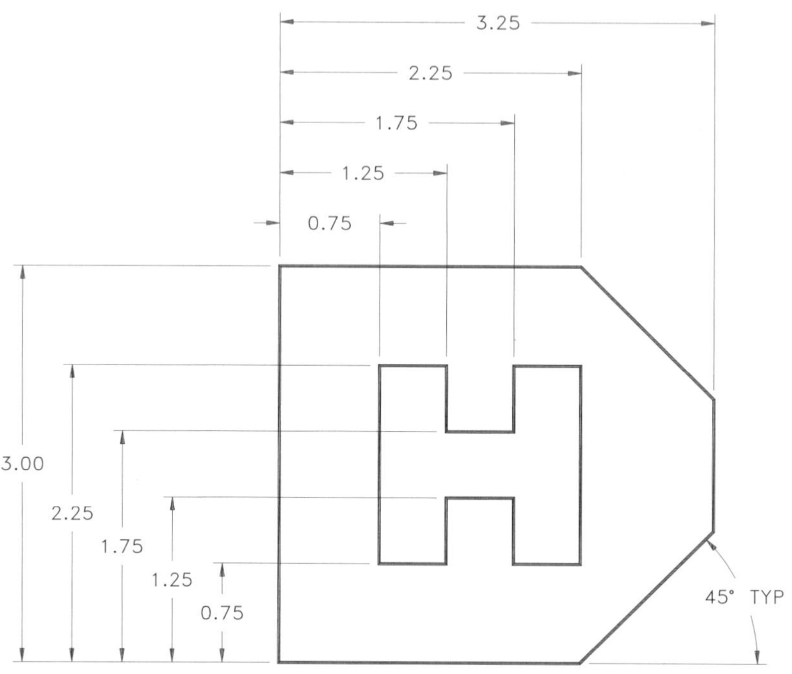

Problem 5-12

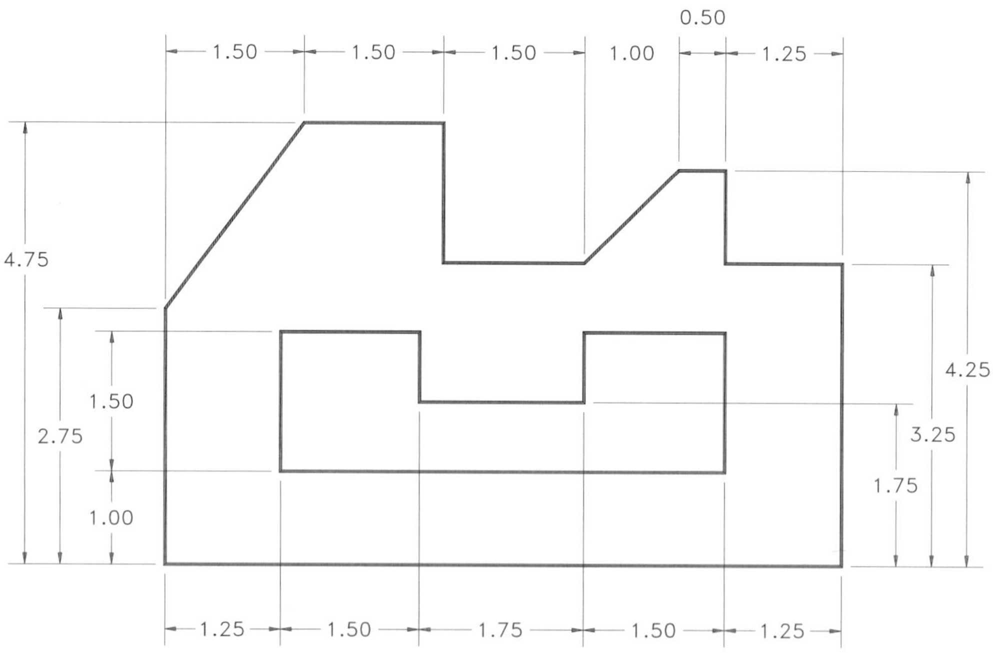

Problem 5-13

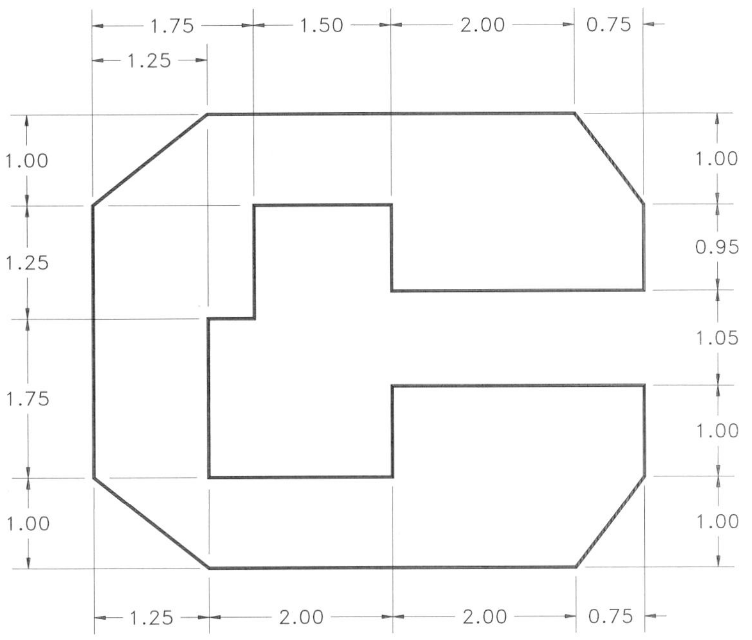

Problem 5-14

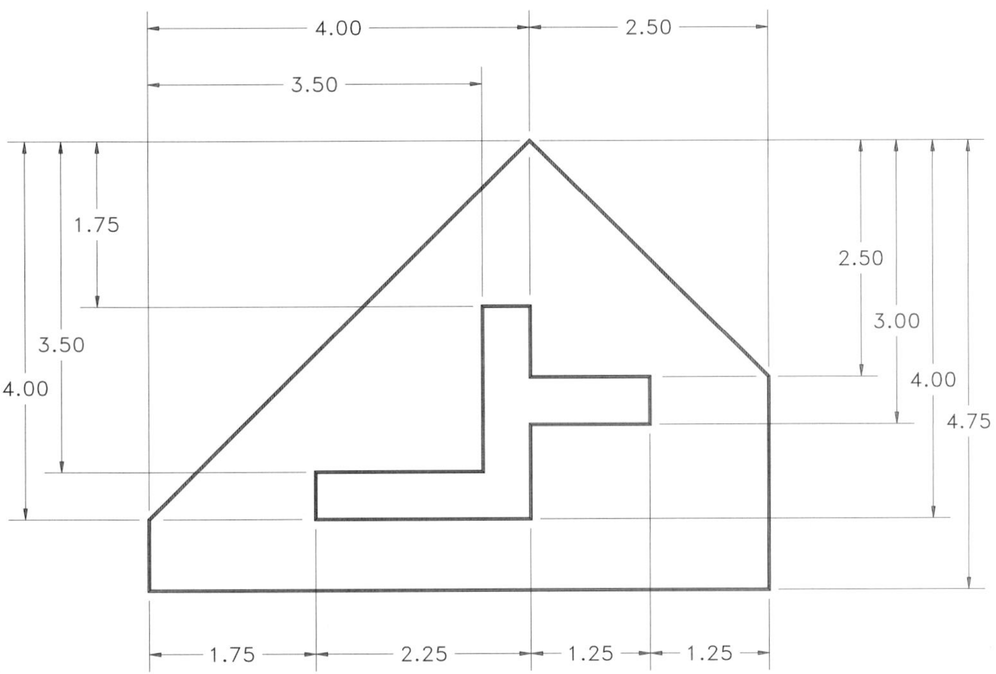

Problem 5-15

CHAPTER 5 QUIZ

1. List the methods for accessing varied linetypes.
2. List two possible uses for an object line.
3. Describe the length and thickness of each line used in a centerline pattern.
4. Use the drawing below and identify the following types of lines.

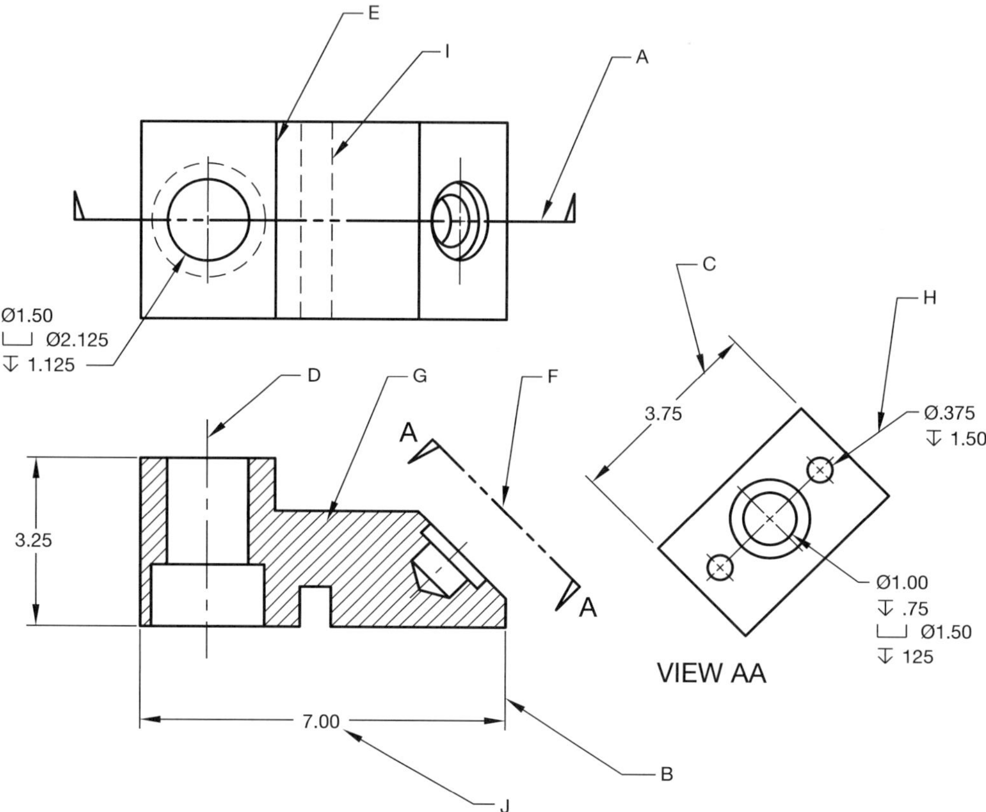

5. Describe the length and thickness of each line used in a hidden line pattern.
6. Describe the use of a viewing plane and cutting plane lines.
7. A special view of a surface needs to be created, but space does not allow it to be placed in its normal location. What kind of line should be used to describe where the view originated?
8. A note must be placed to describe the surface of a part. What kind of line is used to reference the text to the drawing? In addition to this line, what else is needed to complete the correct placement?

9. Sketch how long and short break lines should be drawn.
10. A hole is drilled in the top of an object but the hole does not go all of the way through the part. What kind of line would be used to represent the hole when looking at the bottom of the part?
11. Describe the differences between a section line and a section cutting plane line.
12. A portion of a shaft needs to be removed so that it will fit in a drawing. Name and describe the line used where the portion of the object has been removed.
13. Describe the options for a plot preview.
14. What command will allow the line pattern to be altered?
15. Explain the difference between plotting using the **Display** and **Window** options.
16. List four options for controlling the visibility of a layer.
17. You've just loaded five linetypes, and have entered the **Line** command. Only continuous lines are being drawn. List two possible problems.
18. What is the difference between assigning a color to an object or a layer?
19. List four options for entering the **Layer** command.
20. What is the current layer name, and how can a new layer be named?
21. What are the limitations to assigning layer names?
22. A friend needs help to load five linetypes into a drawing. How can this be done quickly?
23. Centerline has just been made the current linetype. You return to the drawing and nothing is different? What's happened and why did it happen?
24. You've used the **Layer Properties Manager** dialog box to make centerline the current linetype. You return to the drawing. What's happened and why did it happen?
25. Use the **Help** menu and explore how the global scale factor will affect linetypes.
26. You've used **Lineweight Control** on the **Object Properties** toolbar to alter the lineweight. You return to the drawing area and draw several lines, but no change has been made to the lineweight. Why is there no change, and what must be done to see varied lineweights?
27. List the suggested lineweight for thick _____, moderate _____, and thin lines _____.
28. You want to draw one red line. How can it be done?
29. You want all of your lines to be red. What is the best way to do this?
30. You want to plot a drawing at 1/2 = 1. How can this be done?
31. List the line types that ANSI requires to be thick and those it requires to be thin.
32. List and describe the type of line that should be used to show the movement of a brake pedal.

33. A length of 6.00" must be placed to describe the length of a part. What kind of line is used to reference the text to the drawing? In addition to this line, what else is needed to complete the correct placement?
34. The surface of a shaft is to have a very smooth texture. What type of line is used to describe the limits of the texture?
35. A 1" × 1" × 2" deep square feature will be cut into the surface of a part. Would a centerline be appropriate to describe the shape? Explain your answer.

CHAPTER 6

Multiview Development

Up to this point, the drawings that you've done have been of gaskets, patterns, or other flat objects. Parts that are to be machined are typically more complex and require drawings that show each surface of the part. These drawings are referred to as multiview projections. This chapter will introduce

- Common types of drawings used in mechanical drafting
- The theory of orthographic projection
- Selection of the views to be created for a multiview layout
- Methods of constructing auxiliary views
- Using freehand sketching to improve your computer drawings

Commands to be introduced include

- **Offset**
- **Trim**
- **Fillet**
- **Stretch**
- **Snap**

COMMON TYPES OF DRAWINGS

The two major types of drawings used to machine and manufacture parts are orthographic projections and isometric drawings. Isometric drawings appear to be three-dimensional drawings by showing three surfaces of an object in one view. Figure 6.1 shows an isometric drawing that represents the front, top, and right side of an object. A total of ten different surfaces can be seen in this one view, giving the impression that the object is three dimensional. Methods of creating isometric sketches will be introduced later in this chapter. Isometric drawings are primarily used in sales brochures and non-production drawings. Isometric drawings will be used throughout the balance of the text to introduce many of the orthographic drawing projects, but they will not be considered as a drawing option in this text.

Orthographic projections show only one side of an object. Several different planes can be represented in one view, but only the portion of the object seen from a specific side is represented. Figure 6.2 shows the front view of the object shown in Figure 6.1. Notice that only surfaces 1 and 2 are seen as planes, and that surfaces 3 through 14 are seen as lines that represent the edge of the plane. The balance of this chapter will introduce methods for creating orthographic projections.

THE THEORY OF ORTHOGRAPHIC PROJECTION

Orthographic projection is a drawing method that projects the features of an object onto an imaginary plane. In theory, an object is placed in a glass cube similar to Figure

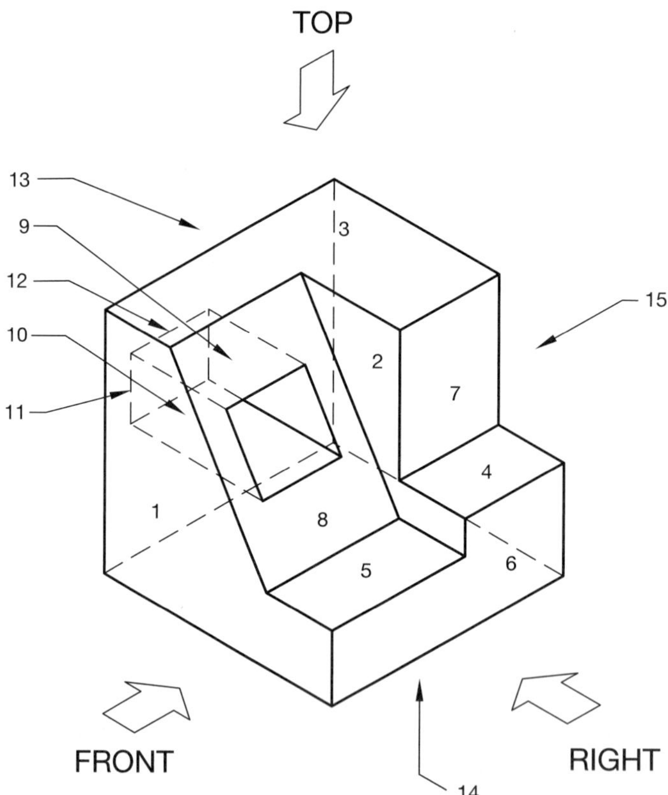

Figure 6.1 *An isometric drawing gives the impression that the object is three dimensional.*

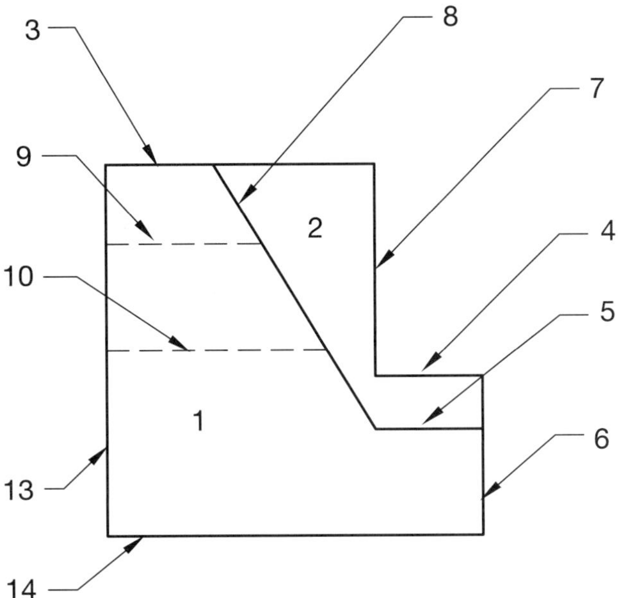

Figure 6.2 *Orthographic projections show only one side of an object. Several different planes can be represented in one view, but only the portion of the object seen from a specific side is represented. This orthographic projection shows the front view of the object shown in Figure 6.1.*

6.3, and each surface of the object is projected to a parallel surface of the glass cube. Any portion of the object that is not parallel to the surface of the cube will not be seen in true size and shape. Methods of viewing non-parallel surfaces will be discussed later in this chapter as auxiliary views are explored. Once each surface of the object is projected to the surface of the glass cube, the cube is unfolded to provide six views of the object. The six views are the front, top, right side, back, left side, and bottom. Figure 6.4a shows the unfolding of the glass cube. Figure 6.4b shows the box completely unfolded.

THE GLASS CUBE

The key to completing an orthographic projection is the placement of the object in the glass cube so that the sides of the object are parallel to the sides of the cube. When parallel surfaces of the object are projected to the cube, the 3D object will be reduced to a 2D projection. Figure 6.5 shows the right side view of the object from Figure 6.4a projected to a surface of the glass cube. Only surfaces of the object that are parallel to a surface of the box will be projected in true size and shape. Non-parallel surfaces will be projected to the cube, but will appear distorted on the surface of the cube. The planes A, C, and D are parallel to a surface of the cube and will be represented in their

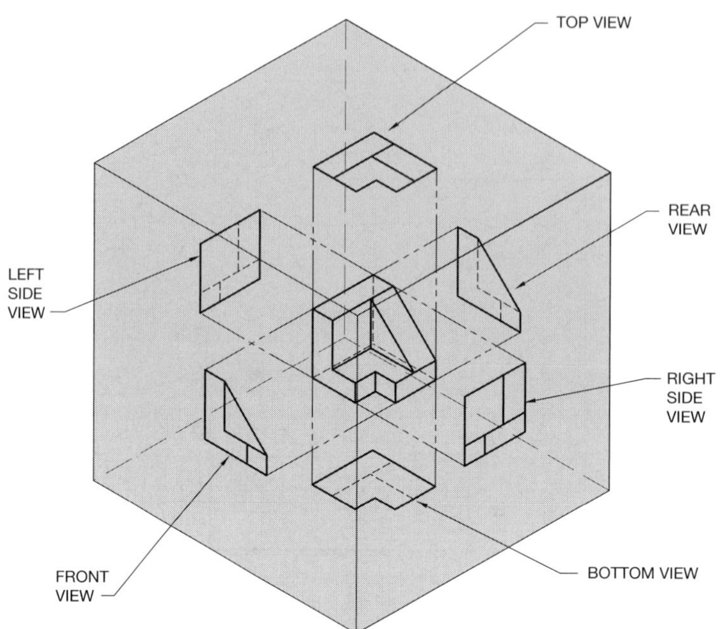

Figure 6.3 *Orthographic projection in theory places an object in a glass cube, and each surface of the object is projected to a parallel surface of the glass cube.*

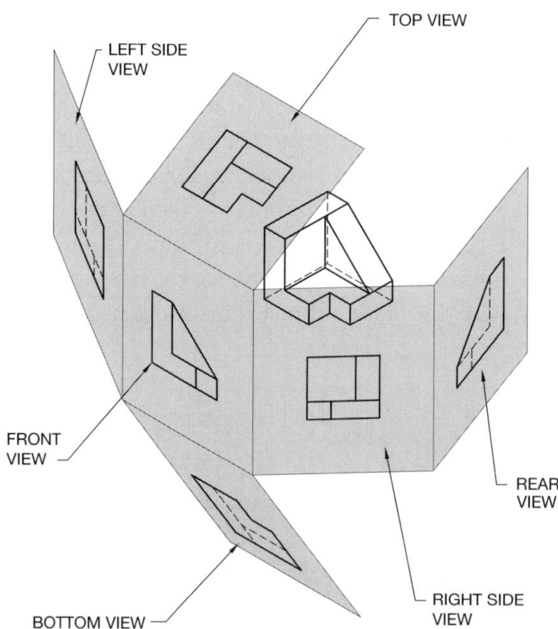

Figure 6.4a *Once each surface of the object is projected to the surface of the glass cube, the cube is unfolded to provide six views of the object.*

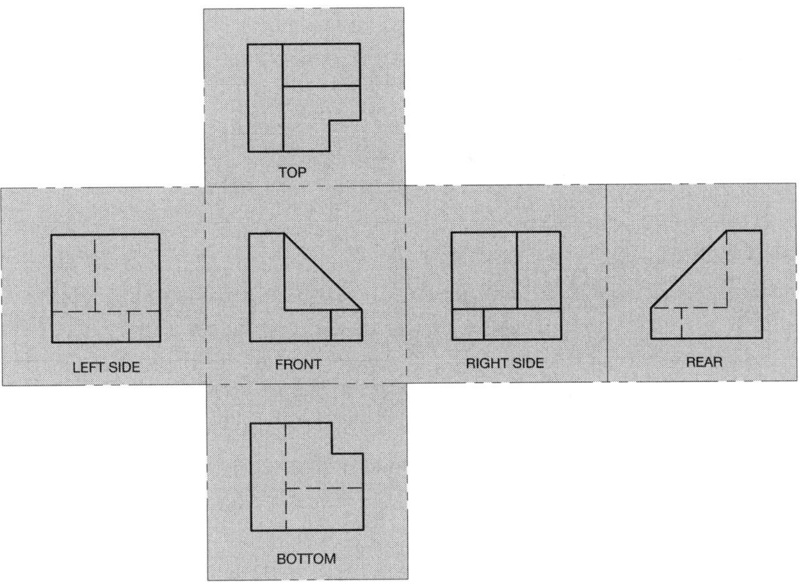

Figure 6.4b *The front, top, right side, back, left side, and bottom views are unfolded so that they align with each of the other views.*

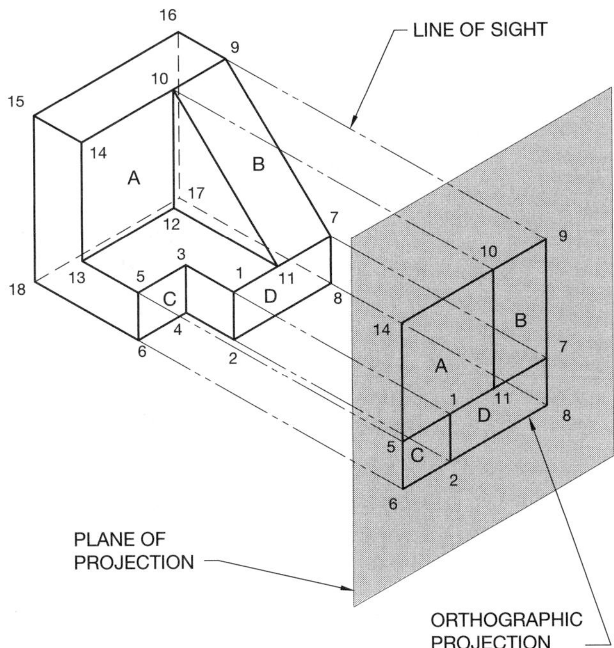

Figure 6.5 *The right side view of the object from Figure 6.4a projected to a surface of the glass cube. Only surfaces of the object that are parallel to a surface of the box will be projected in true size and shape.*

true size and shape. Plane B is also projected to the plane, but because it is not parallel to the cube, it is distorted. Given only one projection, it is impossible to determine the true size and shape of plane B. By combining information of a variety of projections, the print reader can accurately determine the true size and shape of an object. In the front view of Figure 6.4b, you can clearly see that surface B is inclined, but you can no longer tell how wide the surface is.

Unfolding the Cube

In theory, the object in Figure 6.4a was projected to the surface of the glass cube and then the cube was unfolded to allow viewing of each plane of the object. The edges of the cube are referred to as fold lines. To accurately view an object that has been projected to the cube, the cube is unfolded along the fold lines and viewed as seen in Figure 6.4b. As the cube is unfolded to lie flat, a 90° bend is placed between each drawing view. In is important to remember as you start drawing orthographic projections that the bend is always 90° between views. If any other angle is used, the projection will be distorted.

Comparing Surfaces

Once projected, surfaces on the object can be found on each of the six planes of the glass cube. In two views, the surface will be represented as a plane, and in the other views the surface will be represented as a line. Figure 6.6 shows the projections for an object with no inclined surfaces. Find surface 1,2,3,4 in the isometric view. Notice in the top view, only edge 1-3 is visible. The plane is represented by plane 1,2,6,5 in the front view. Because line 3-4 in the isometric drawing is behind line 5-6 in the front view, line 3-4 can't be seen in the front view. Plane 1,2,3,4 is represented by line 1-3 in the top view, line 1-2 in the right side view, line 3-4 in the left view and by line 2-4 in the bottom view. Plane 1,2,3 4 in the isometric view is represented in the rear view by plane 3,4,8,7. In the rear view, line 1-2 is hidden behind line 7-8, and a hidden line represents line 3-4.

PLACEMENT OF DRAWINGS

The second major consideration of orthographic projection is the placement of the views once the cube is unfolded. Notice that in Figure 6.6, the top view is placed directly above the front view and the right side view is placed on the right side of the front view. For easy interpretation of information, engineering drawings must be placed in the proper position to each other. Proper placement includes the relationship of two views to each other by position and relationship.

Location of Views by Position

Figure 6.7 shows the proper relationship for front, side, and top views by position. The top view is placed directly above the front view and the side view is aligned with the front view. If you were drawing this object, once the front view is created, lines

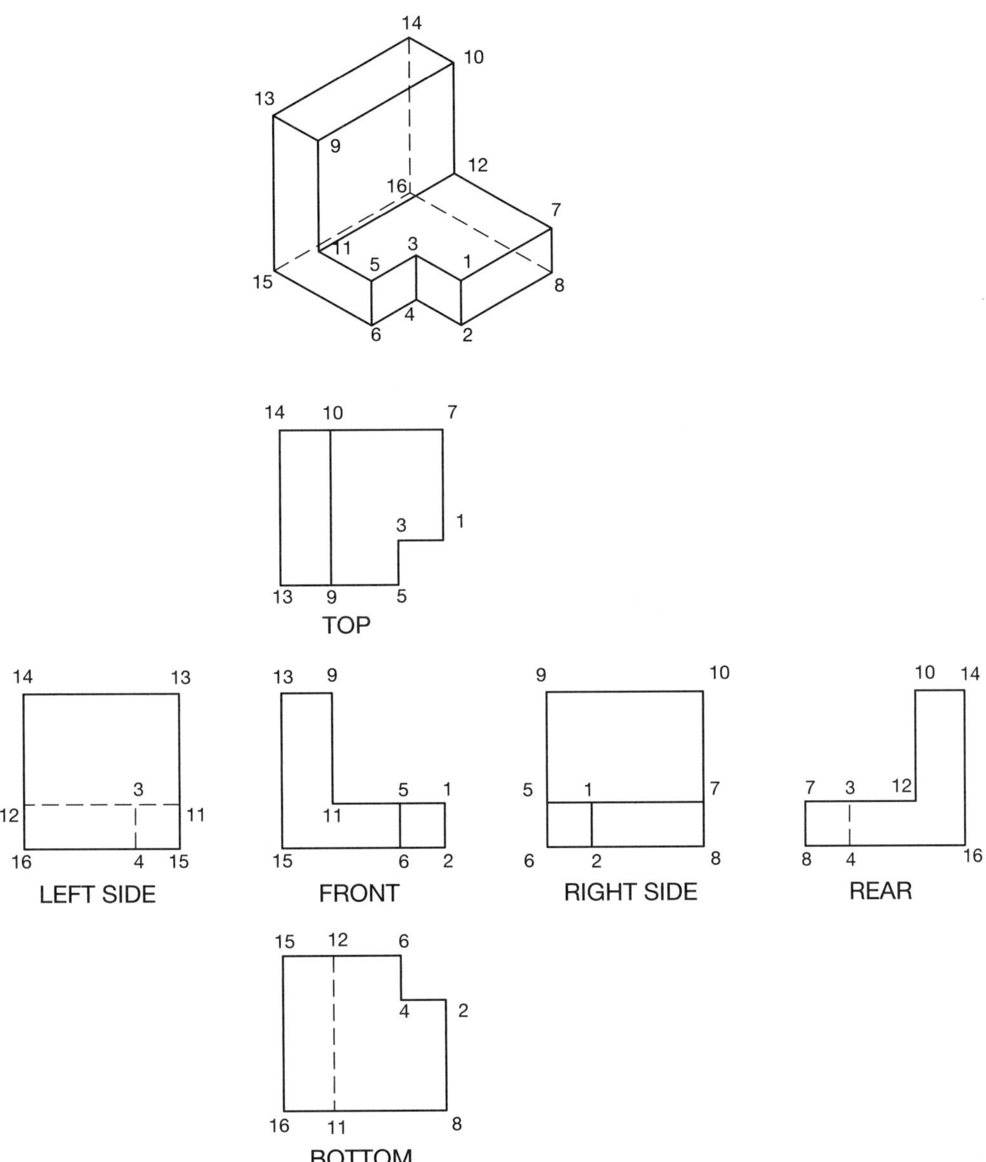

Figure 6.6 *The projections for an object with no inclined surfaces.*

could be extended from points 1, 5, 9, and 13 to establish the widths required for the top view. Direct alignment of views will minimize measurements that must be made and eliminate errors.

Note: It might seem obvious, but the width of the object in one view must be equal in all other views. The part cannot change size as you move from one view to another. Proper placement of views will allow projection of sizes to eliminate errors. If widths are different in your drawings, you know for sure that you've made a mistake.

Selecting Points with OSNAP

A common error when projecting objects is to allow the object to change size from one view to another. In Figure 6.8, line 9-10 in the top view was placed by measur-

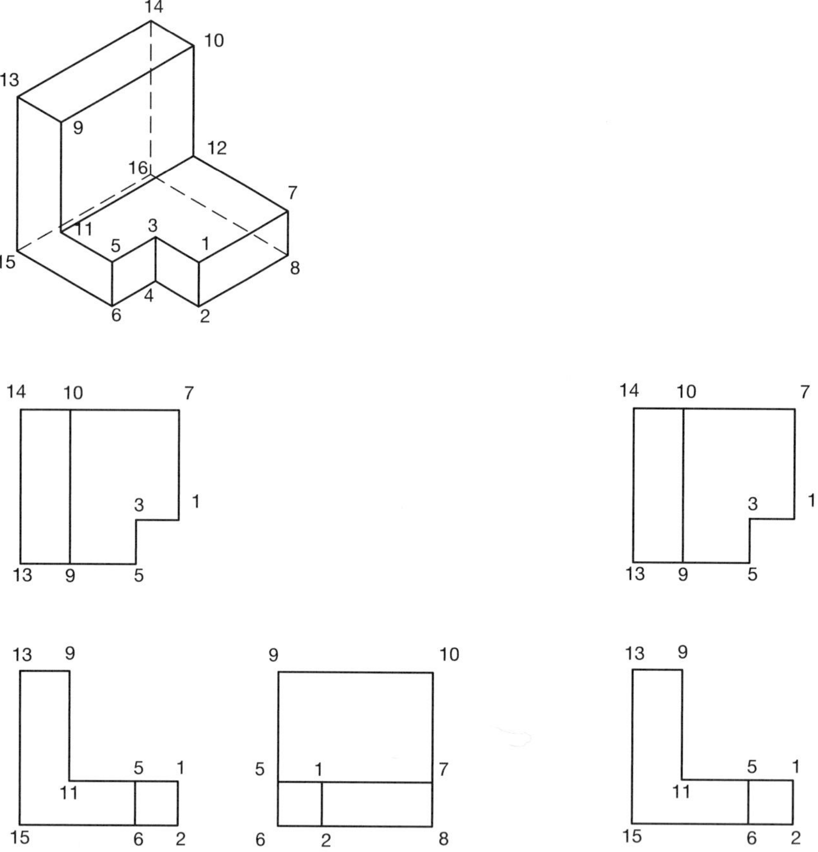

Figure 6.7 *The proper relationship of views places the top view directly above the front view and the side view is aligned with the front view. Direct alignment of views minimizes the measurements that must be made and eliminates the chance of measurement errors.*

Figure 6.8 *A common error when projecting objects is to allow the object to change size from one view to another. Projecting size from one object to another rather than measuring each object in each view could have eliminated this error. If the length is projected to the top view, the length will remain constant.*

ing a distance of .50" from line 13-14. Unfortunately, the CAD operator used a distance of .75" instead of .50" and introduced an error into the drawing. Projecting size from one object to another rather than measuring each object in each view could have eliminated this error. If the length is projected to the top view, the length will remain constant.

If you attempt to draw a line from point 9 in the front view, you may or may not place the crosshairs exactly on the point as you select the "First point" for the projection line. The exact point can be easily selected by clicking the OSNAP button on the status bar. This will activate the object snaps that were introduced in Chapter 3. Now if you select point 9 in the front view, the starting point for the line will automatically be selected from the intersection of line 9-11 and 9-13. Chapter 8 will explain other OSNAP options and other methods to ensure that exact placement of "First" or "Next" points.

First and Third Angle Projection

Another consideration for view placement deals with the theory of drawing placement using first or third angle projection. Each of these methods deals with the projection of the object once inside the glass cube. Third angle projection is used in the United States. If you imagine you favorite cereal box unfolded, third angle projection will fold the box so that all of the pictures and nutritional facts are displayed on the outside of the box. First angle projection is used throughout the balance of the world, and folds the box so that all of the images are on the inside. First angle projection places the object in front of a plane, and then projects back to the surface. Third angle projection places the object behind the glass plane and projects to the surface. Figure 6.9 shows how first angle projection would affect the same object. Notice in third angle projection (Figure 6.6), that the right side view is on the right side of the front view, and the bottom view is on the bottom side of the front view. With first angle projection, the right side view is placed on the left side of the front view, and the top view is placed on the bottom side of the front view. ANSI standards dictate that one of the symbols shown in Figure 6.10 be placed in the title block of mechanical drawings to describe the method of projection. The balance of this text will only consider third angle projection.

SELECTING THE REQUIRED VIEWS

Whatever the method of projection, the selection of the views to be used is critical if the CAD operator is to communicate clearly. The selection of the front view and the number of views to be represented are the two critical elements that must be considered in determining what views will need to be drawn to represent the part to be manufactured.

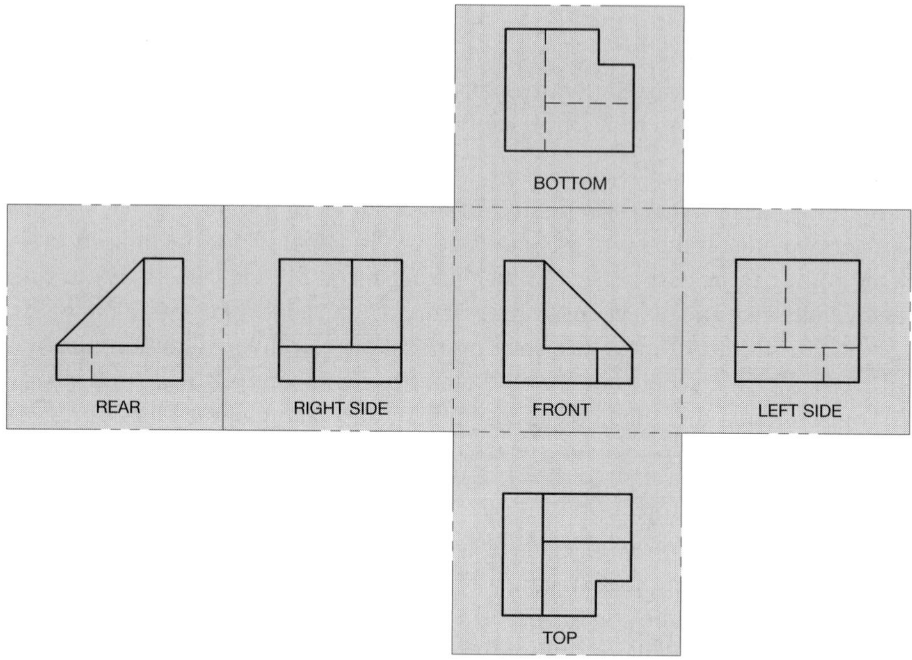

Figure 6.9 *First angle projection places the right side view on the left side of the front view, and the top view is placed on the bottom side of the front view.*

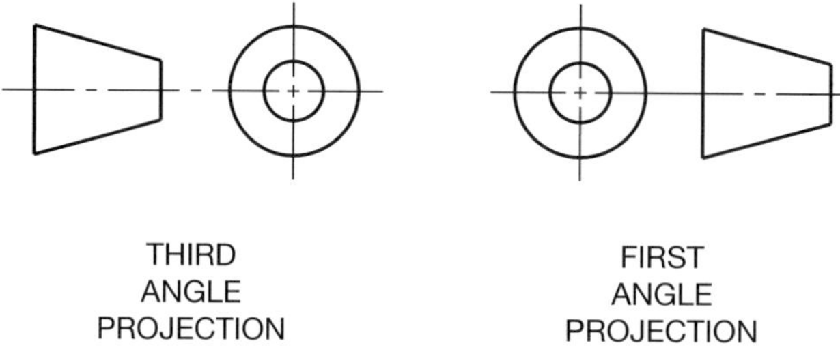

Figure 6.10 *ANSI standards dictate that one of the symbols be placed in the title block to describe the method of projection.*

SELECTING THE FRONT VIEW

Because the location for all required views is based on the front view, selection of the front view is crucial in proper drawing planning. The front view should be based on the following six guidelines. The front view should

- Represent the most natural position of use.
- Best describe the overall contour of the component.

- Show the longest side.
- Have the fewest hidden features.
- Be in the most stable position.
- Provide the best clarity in locating information.

Each of these guidelines can be applied to the component shown in Figure 6.11. The longest sides of the object can be seen in Figure 6.12. View F can be eliminated because of the hidden lines and its length. View E has no hidden line, but is shorter than the other four sides. View C and D show the long edge, and some of the contour, but each contains many hidden lines. Views A and B have no hidden features, and each shows a different aspect of the contour. If the engineer sketched the part in its natural position of use, view B would make an excellent front view because it shows

- The longest side (views A, C and D also meet this requirement).
- Least amount of hidden lines (view A also requires no hidden lines to represent a surface).

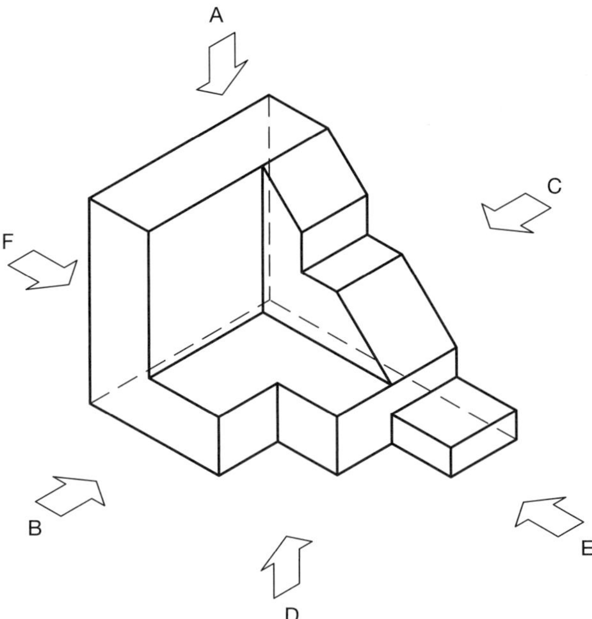

Figure 6.11 *The location for all required views is based on the selection of the front view. The front view should represent the most natural position of use, best describe the contour of the component, show the longest side, have the fewest hidden features, represent the object in the most stable position, and provide the best clarity in locating information.*

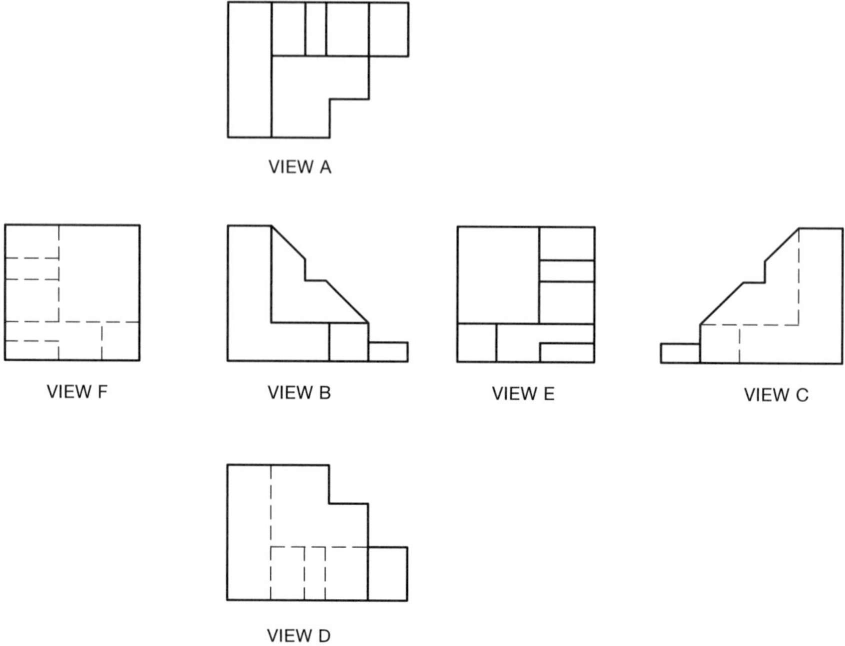

Figure 6.12 *Selecting the front view. View F can be eliminated because of the hidden lines and its length. View E has no hidden line, but is shorter than the other four sides. View C and D show the long edge, and some of the contour, but each contains many hidden lines. Views A and B have no hidden features, and each shows a different aspect of the contour. If the engineer sketched the part in its natural position of use, view B would make an excellent front view.*

- Represents the contour (view B shows the inclined surface and the notch in that surface better than A and D).
- Shows the part in natural position of use (view C also meets this requirement).
- Shows the part in its most stable position (with surface D resting on a level surface, the part would be very difficult to turn over).

The final guideline will be discussed in Section 3 as text and dimensions are added to drawings.

Determining the Number of Views to Be Drawn

Just because there are six possible views of each object that can be drawn does not mean the CAD operator must draw each view. Remembering the axiom "time is money"—it is crucial to select only the views needed to represent each feature accurately. If you were to describe a basketball, only one view is needed to describe the diameter. As seen in Figure 6.13, if the proper linetypes and symbols are used, a shaft can also be rep-

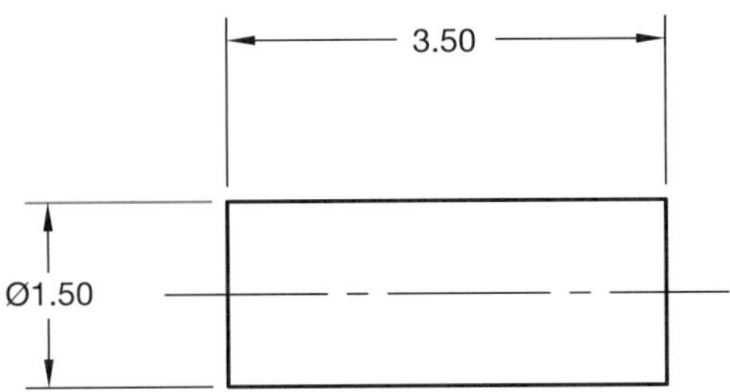

Figure 6.13 *Just because there are six possible views of each object does not mean that each must be drawn. Select only the views needed to represent each feature accurately. If the proper linetypes and symbols are used, a shaft can be represented using only one view.*

resented using only one view. The centerline and the diameter each tell the print reader that this is a cylinder and not a rectangle. Some of the complex gaskets drawn in Chapter 5 can be represented in one view if the thickness is listed in the title block. Most machined features will take a minimum of two views. The actual number of views to be drawn will be determined by the number of views required to place each of the notes and the dimensions needed to fully describe the part. Using the part shown in Figure 6.11, it can be seen in Figure 6.14 that all of the dimensions needed to machine the part can be placed in two views. Because the right side view needs no dimensions, it should be omitted from the final drawing.

CREATING VIEWS

As the top view of Figure 6.7 was created, lines were projected from points 1, 5, 9 and 13 of the front view to establish the lengths of the top view. A line can be drawn anywhere above the front view to represent plane 13-5. The placement of each view can be altered throughout the life of the drawing based on the information needing to be placed. (Space requirements for plotted drawings will be discussed later.) Once plane 13-5 has been drawn, the outline of the top of the object can be completed by drawing plane 5-3, 3-1, 1-7, 7-12, etc. until the outline of the entire object has been completed. An easier method is to draw a line, and then use the Offset, Trim, and Fillet commands to complete the object. Each of these commands will be introduced to aid in drawing construction, and then explained in detail in subsequent chapters.

USING OFFSET

The Offset command allows a line, circle, arc, or polygon to be copied and relocated parallel to, and at a specific distance from, the original object. The uses of this command are almost endless. Given a horizontal and vertical line, an entire rectangular object can be created by offsetting lines and then cleaned up using other editing commands presented in later chapters. Offset allows lines to be drawn in the exact location without giving thought to their exact length. Get the location right without

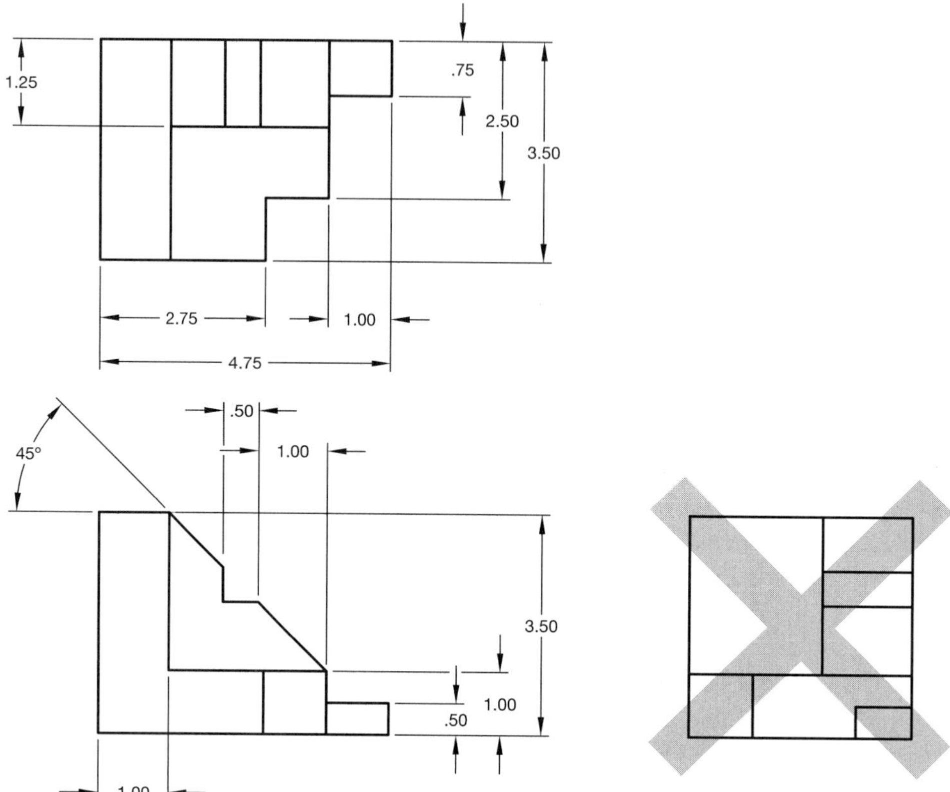

Figure 6.14 *The actual number of views to be drawn will be determined by the number of views required to place each of the notes and dimensions needed to fully describe the part. With the part shown in Figure 6.11, all of the dimensions needed to machine the part can be placed in two views. Because the right side view needs no dimensions, it should be omitted from the final drawing.*

worrying that the line might be too long, and then edit the length with **Fillet**, **Stretch**, or **Trim**.

The **Offset** command can be used to create the front view or any subsequent views. For our example, we'll use it to create the top view. Once the front view has been completed, draw the projection lines points 13, 9, 5, and 1. Now draw a line anywhere above the front view to represent line 5-13. Because the part is 2" wide, the **Offset** command can be used to create a line that is 2" away from the line drawn to represent plane 7-14.

Start the **Offset** command by clicking the **Offset** button on the **Modify** toolbar, by typing O (ENTER) at the Command prompt, or by selecting **Offset** from the **Modify** menu. Each will produce the prompt:

Specify offset distance or [Through]<1.00>:**2** ENTER

Figure 6.15 shows how the **Offset** process works. For this example, enter the desired distance, in this case 2". Now you'll be asked to select the object to offset. Select the line drawn to represent plane 13-5. The final step to offset the line is to select the location for the new line. Select the top side of line 13-5, and a new line of equal length will be offset 2" above the existing line. The process can be seen in Figure 6.16. The complete command sequence for the **Offset** command is as follows:

Command: **O** ENTER *(Or click the Offset button on the Modify toolbar.)*

OFFSET

Specify offset distance or [Through] <1.00>**2** ENTER

Select object to offset or <exit>: *(Select line to be offset.)*

Specify point on side to offset: *(Select a point in the direction to be offset.)*

Select object to offset or <exit>: ENTER

Command:

Repeat the process to place lines to draw plane 3-1 so that the drawing looks like Figure 6.17.

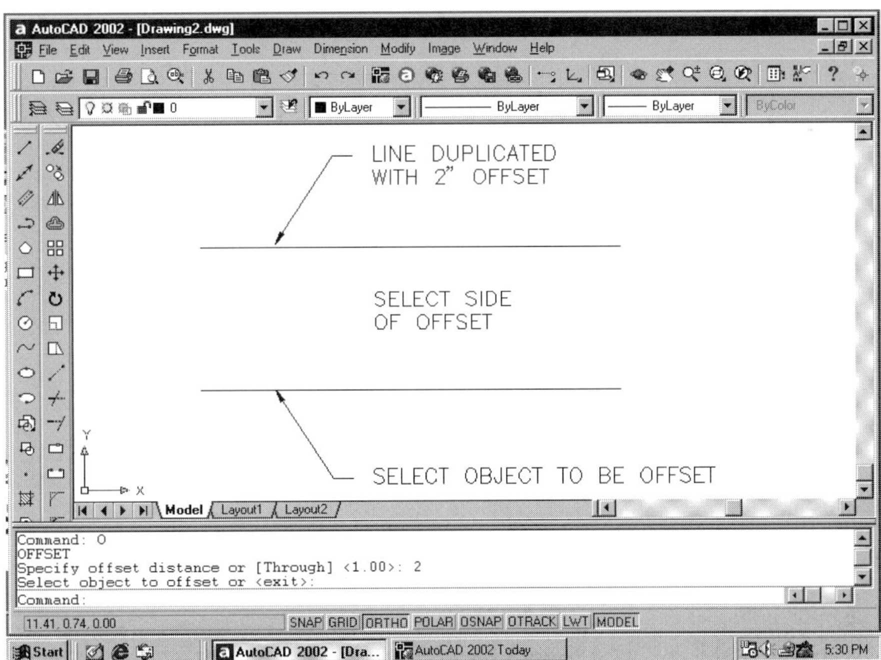

Figure 6.15 *The* **Offset** *command allows a line, circle, arc, or polygon to be copied and relocated parallel to, and at a specific distance from, the original object.*

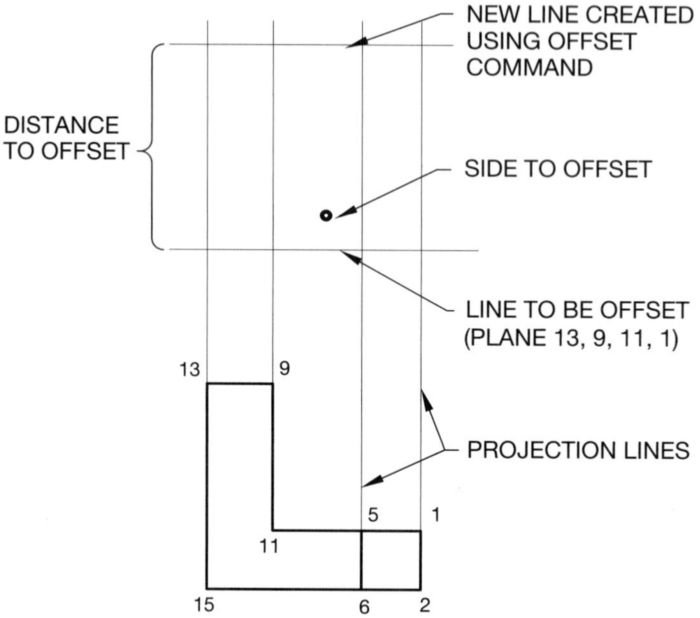

Figure 6.16 *Using the* **Offset** *command to lay out the required width of an object.*

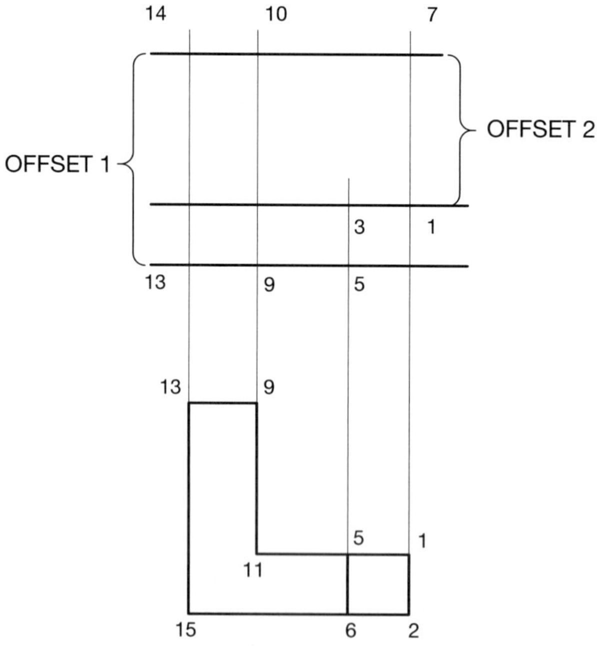

Figure 6.17 *The* **Offset** *command is repeated to locate an interior surface.*

USING TRIM TO COMPLETE AN OBJECT

Because the lines in the top view were drawn with no consideration of length, they must now be trimmed to the desired length. This can be done using either the **Trim** or **Fillet** command. The left end of the drawing will be adjusted using **Trim** and the right end of the drawing will be altered using the **Fillet** command. Each is effective, and they can be used interchangeably. Both will be discussed in detail in Chapter 11.

The **Trim** command can be used to remove portions of lines that lie between planes, or cutting edges. **Trim** can be used to clean the corners of intersecting lines or to modify the size of objects. The command allows you to edit lines, arcs, or circles using a cutting edge. Any portion of the selected object extending past the designated cutting edge will be removed.

In our example, we want to shorten the line representing the left end of the part so that it is 2" long. Start the command by selecting **Trim** on the **Modify** toolbar, by typing TR (ENTER) at the Command prompt, or by selecting **Trim** from the **Modify** menu. The command process to trim the line is as follows:

Command: **TR** ENTER *(Or click Trim on the Modify toolbar.)*

TRIM

Current settings: Projection=UCS, Edge=None

Select cutting edges . . .

Select objects: *(Select the desired cutting edge, line 14-7.)*

1 found

Select objects: *(Select the desired cutting edge, line 13-5.)*

Select objects: 1 found, 2 total

Select objects: ENTER

Select object to trim or shift-select to extend or
[Project/Edge/Undo]: *(Select the line to be trimmed.)*

Select object to trim or shift-select to extend or
[Project/Edge/Undo]: *(Continue to select lines to be trimmed.)*

Select object to trim or shift-select to extend or
[Project/Edge/Undo]: ENTER

Command:

Once the cutting edges are defined, the command will continue until ENTER is pressed. Any object that extends past the cutting edge can be trimmed. The process can be seen in Figure 6.18. Notice that an object can be selected to be a cutting edge, and it can also be an object to be trimmed. Repeat the process as needed to trim lines until the desired shape is achieved. Figure 6.19 shows the left end of the top view completed using the **Trim** command.

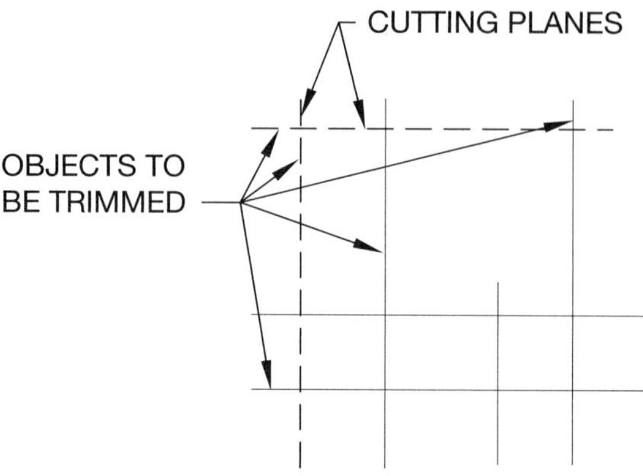

Figure 6.18 *Once the cutting edges are defined, the* **Trim** *command will continue until* ENTER *is pressed. Any object that extends past the cutting edge can be trimmed. Notice that an object can be selected to be a cutting edge and it can also be an object to be trimmed.*

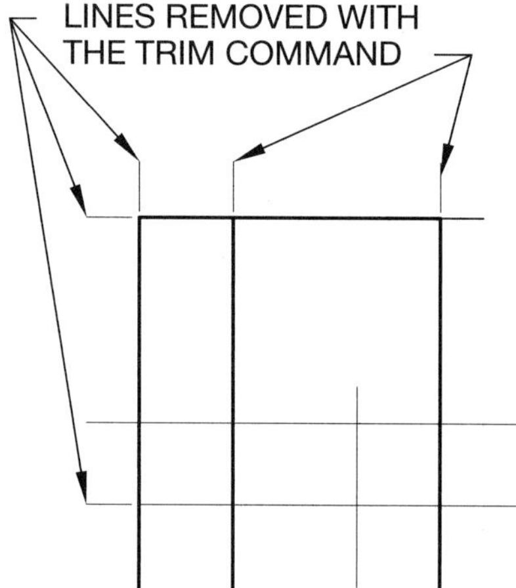

Figure 6.19 *The left end of the top view completed using the* **Trim** *command.*

USING FILLET TO ADJUST LINE LENGTH

The **Fillet** command connects two lines to form a corner or allows two lines, arcs, or circles to be fitted together smoothly in rounded corners. The command can be used to control the length of lines, even if they don't intersect, by using a radius of zero. The command requires two steps the first time it is used: The first step defines the size of the radius, and the second step allows selecting two lines to receive the fillet. Once the size of the radius is entered, it remains the default until changed. For this example, the **Fillet** command will be used to square two lines using a radius of 0. Start the command by selecting the **Fillet** button on the **Modify** toolbar, by typing F (ENTER) at the Command prompt, or by selecting **Fillet** from the **Modify** menu. The command sequence to fillet lines 14-7 and 7-1 is as follows:

Command: **F** ENTER *(Or click Fillet on the Modify toolbar.)*

FILLET

Current settings: Mode = TRIM, Radius = 0.50

Select First object or [Polyline/Radius/Trim]: **R** ENTER

Specify fillet radius <0.50>: **0** ENTER

Select First object or [Polyline/Radius/Trim]: *(Select first line to be filleted, plane 14-7.)*

Select second object: *(Select second line to be filleted, plane 7-1.)*

Command:

The result of the **Fillet** command can be seen in Figures 6.20a and 6.20b. The completed top and front views started in Figure 6.7 can be seen in Figure 6.21. The object was completed using **Line**, OSNAP, **Offset**, **Trim**, and **Fillet**.

PROJECTION OF OBJECTS

Orthographic projections offer one other aid in drawing layout. As the first view is constructed, all dimensions must come from a sketch and be entered into the drawing. As the second view is constructed, all of the sizes for length can be projected from the front view, and only the widths need to be measured. Once two views have been created, all sizes for the three of the four remaining views can be projected, and no measurements are required. If the rear side is needed, it must be constructed using the same methods used to create the top view.

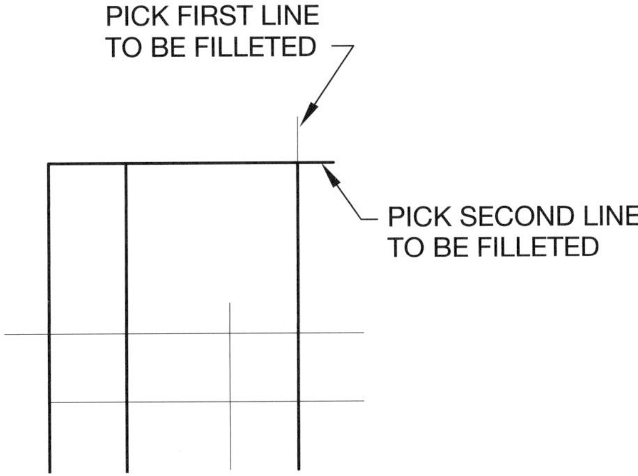

Figure 6.20a *Using the* **Fillet** *command to square a corner.*

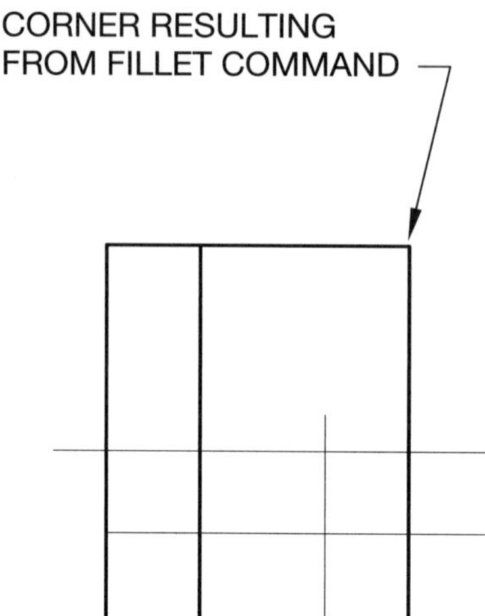

Figure 6.20b *Using the* **Fillet** *command with a 0 radius provides square corners.*

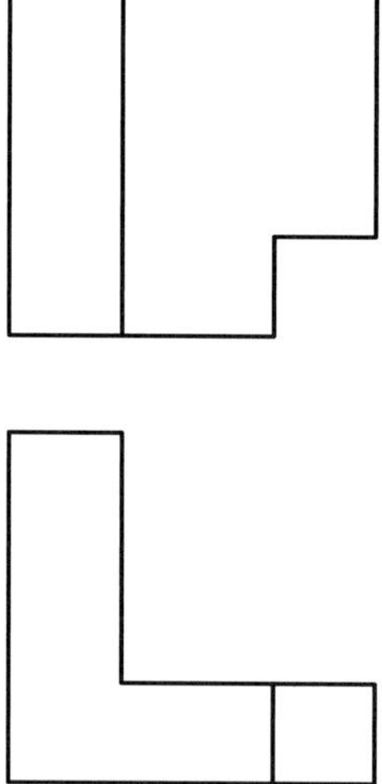

Figure 6.21 *The top and front views started in Figure 6.7 were completed using* **Line**, *OSNAP,* **Offset**, **Trim**, *and* **Fillet**.

Using the front view, all heights can be projected to help form the right side view. Now draw a line to represent the left edge of the right side view. Remember that the line can be placed anywhere on the right side of the front view, and that the completed view can be moved later. Figure 6.22 show the process that will be used to project the side view. Extend a line from plane 5-13 until it crosses the line used to represent the left

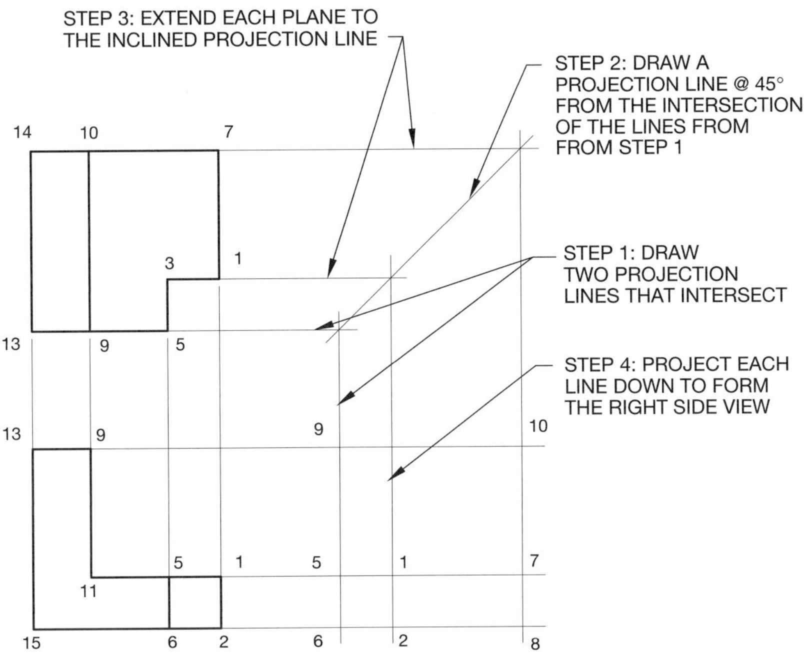

Figure 6.22 *Using orthographic projection to project the side view. Extend a line from plane 5-13 until it crosses the line used to represent the left side of the right view (planes 5-6 and 9-11). Draw a line at 45° that extends through the intersection of the two lines. Now project all widths in the top view to this inclined line.*

side of the right view (planes 5-6 and 9-11). Draw a line at 45° that extends through the intersection of the two lines. Now project all widths in the top view to this inclined line so that your drawing resembles Figure 6.22.

What might seem like of a maze of crossing lines is the base of the right side view. Remembering that lines in one view must be of equal length in another view will help in locating features in the right side view. Line 1-7 in the top view must be equal to the length of line 1-7 in the right side view. Line 1-7 in the right side view must be the same distance above line 2-8 because point 1 is above point 2 in the front view. Projecting each line will ensure proper location and length. Also remember that if you projected a point such as point 1 in the top view, in the right side view you will see the length of that point, in this case line 1-2. It's similar to looking at the end of a pencil and seeing the point. When you view it from the end, you can't tell how long the pencil is. Now rotate the pencil 90° and you see the profile of the pencil, but you can no longer tell if the pencil is an octagon or round. Figure 6.23 shows the completed right side view.

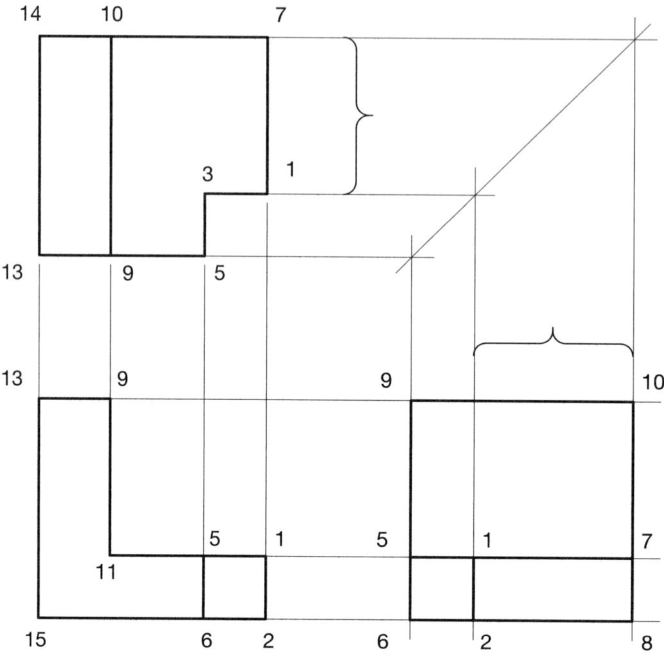

Figure 6.23 *The completed right side view.*

LINE PRECEDENCE

As multiview drawings are created, surfaces represented by lines will often be located on the same plane. As a CAD operator, your job is to communicate clearly. To maintain drawing clarity, you will often need to decide which line of several to draw, and which to delete. Use the follow guidelines to maintain drawing clarity:

- Object lines take precedence over hidden lines and centerlines.
- Hidden lines take precedence over centerlines.
- Cutting plane lines take precedence over centerlines.

Figure 6.24 shows how these guidelines are applied.

CONSTRUCTING AUXILIARY VIEWS

Orthographic projection works well for drawing objects when all of the surfaces are parallel to the glass box. A surface that is not parallel to one of the surfaces of the glass cube will appear distorted in the orthographic projections where it appears as a plane. It will not be distorted in views where the inclined surface is represented as a line. Figure 6.25 shows an object with an inclined surface. The true length of the inclined plane is represented in the front view by line 6-7. The inclined surface is rep-

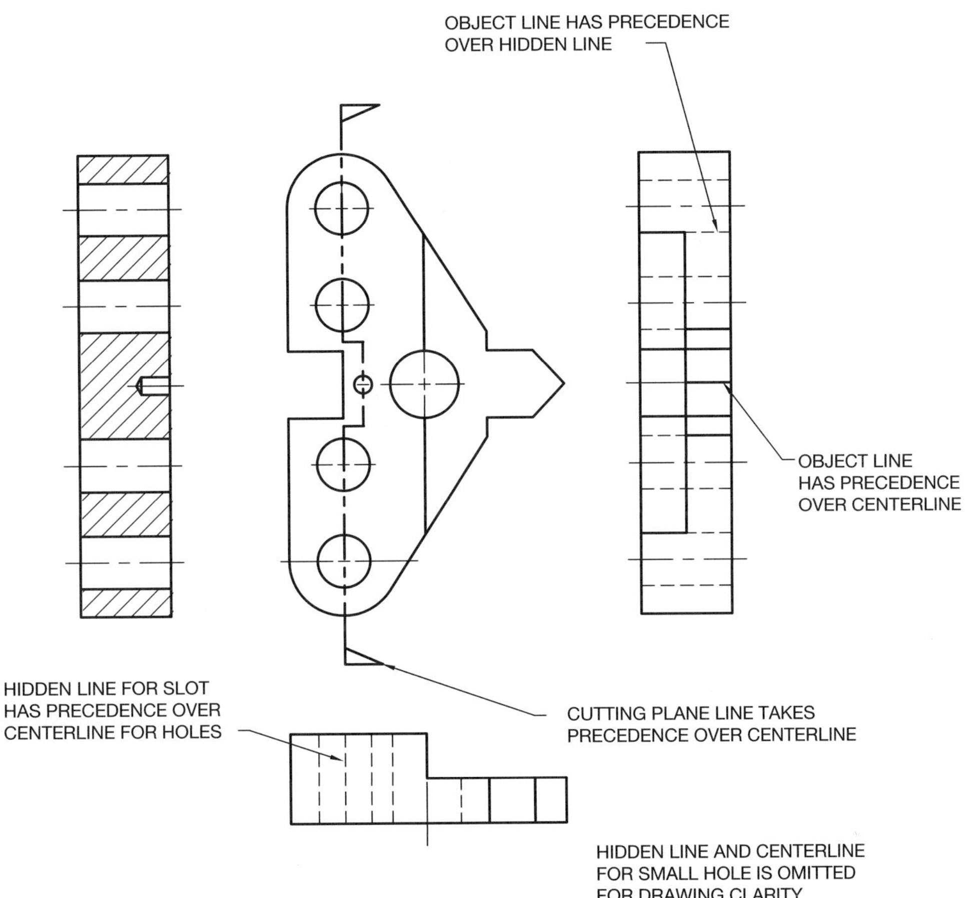

Figure 6.24 *To maintain drawing clarity, you will often need to decide which line of several to draw, and which to delete. Object lines take precedence over hidden lines and centerlines, hidden lines take precedence over centerlines, and cutting plane lines take precedence over centerlines.*

resented in the top and side views by a plane that will appear foreshortened because of the incline. To see the inclined surface in its true shape will require an auxiliary view. An auxiliary view must be projected at 90° from where the inclined surface shows as a line. Figure 6.26 shows the true shape of the plane represented by line 6-7 in the front view. This auxiliary view was created by projecting lengths from the front view (the view where the inclined plane shows as a line) and the widths were determined from the side view and drawn using the **Offset** command. There is no way to project the widths of features to the auxiliary view. The width of features can only be placed by measuring. The projection of the auxiliary view can be easily done by using the Rotate option of the **Snap** command.

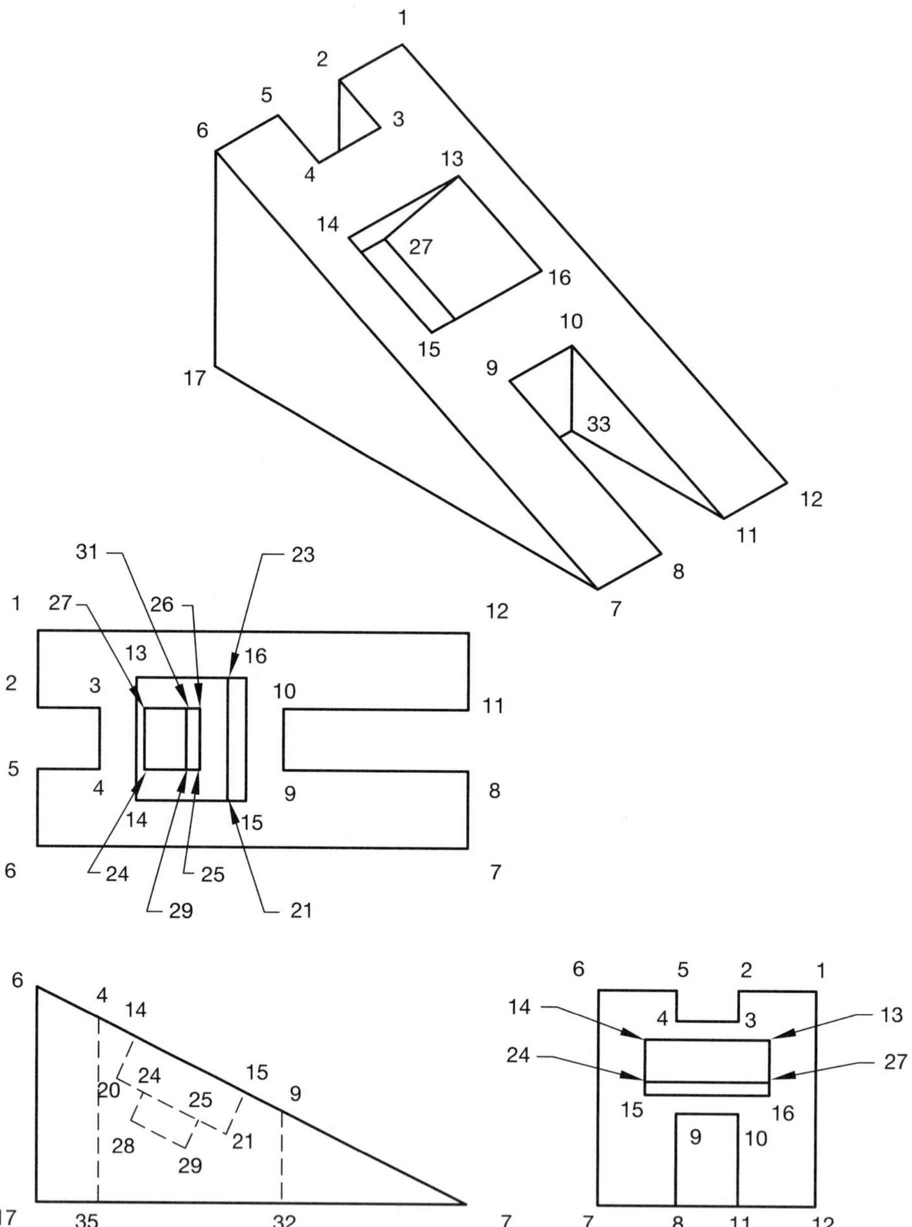

Figure 6.25 *A surface that is not parallel to one of the surfaces of the glass cube will appear distorted in the orthographic projections where it appears as a plane. It will only be seen in true size where the inclined surface is represented as a line (line 6-7 in the front view).*

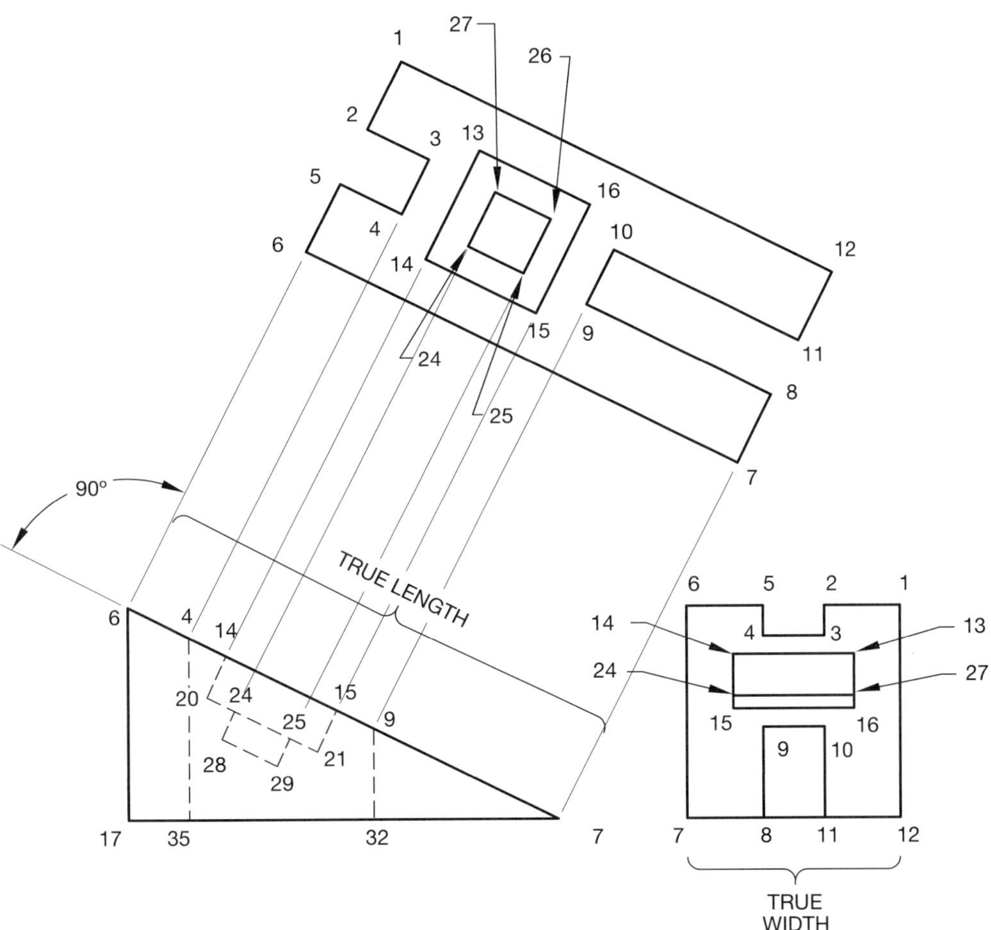

Figure 6.26 *The true shape of the plane represented by line 6-7 in the front view. The auxiliary view was created by projecting lengths from the front view, the view where the inclined plane shows as a line, and the widths were determined from the side view and drawn using the Offset command.*

USING SNAP TO CREATE AN AUXILIARY VIEW

Up to this point the grid and snap pattern has been displayed in horizontal and vertical patterns. This method works well for objects that are parallel to a surface of the glass cube. For an auxiliary view to be drawn easily, the grid should be rotated to lie parallel to the surface that will be used to project the auxiliary view. The **Snap** command allows the grid to be rotated to any angle. You can alter the grid by selecting **Drafting Settings** from the **Tools** menu, and then selecting the **Snap and Grid** tab. The desired angle can then be entered in the **Angle** edit box. The angle can also be set using the following command sequence:

Command :**SN** ENTER

Specify snap spacing or [ON/OFF/Aspect/Rotate/Style]
<0.50>: **R** ENTER

Specify base point <0.00,.000>: *(Select one end of the inclined line that will be the projection plane. Be sure the Intersection option of OSNAP is on.)*

Specify rotation angle <0>: *(Enter the angle if known or select the other end of the projection plane.)*

Command:

The grid, snap, and cursor will now be parallel to the projection plane. The grid can be rotated back to its normal position by repeating the command and entering an angle of 0. If the grid is not displayed, click the GRID button on the status bar. Figure 6.27 shows the effects of rotating the grid using the **Snap** command.

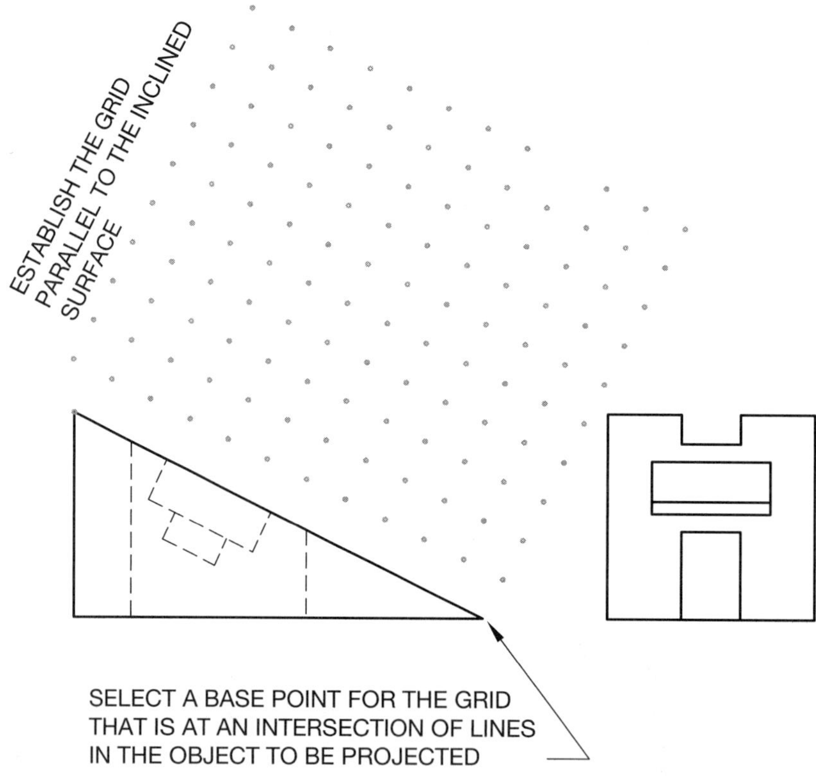

Figure 6.27 *An auxiliary view can easily be drawn by rotating the grid to lie parallel to the surface that will be used to project the auxiliary view. Snap can be altered by selecting the* **Snap and Grid** *tab of the* **Drafting Settings** *dialog box (select* **Drafting Settings** *from the* **Tools** *menu.*

PROJECTING AUXILIARY VIEWS

With the grid established, project each feature of the object from the inclined surface. Figure 6.28 shows the initial steps in projecting an auxiliary view. Once the lengths are projected, draw a line to represent the left edge of the surface. This new line can be placed at any convenient distance from the surface to be projected and moved later. Use the same process that was used earlier to draw the front view. Use **Offset**, **Trim**, and **Fillet** to complete the drawings. The completed auxiliary view can be seen in Figure 6.29. Typically auxiliary views are drawn only when a machined feature is located on the inclined surface. Machined features will be introduced in Chapter 18.

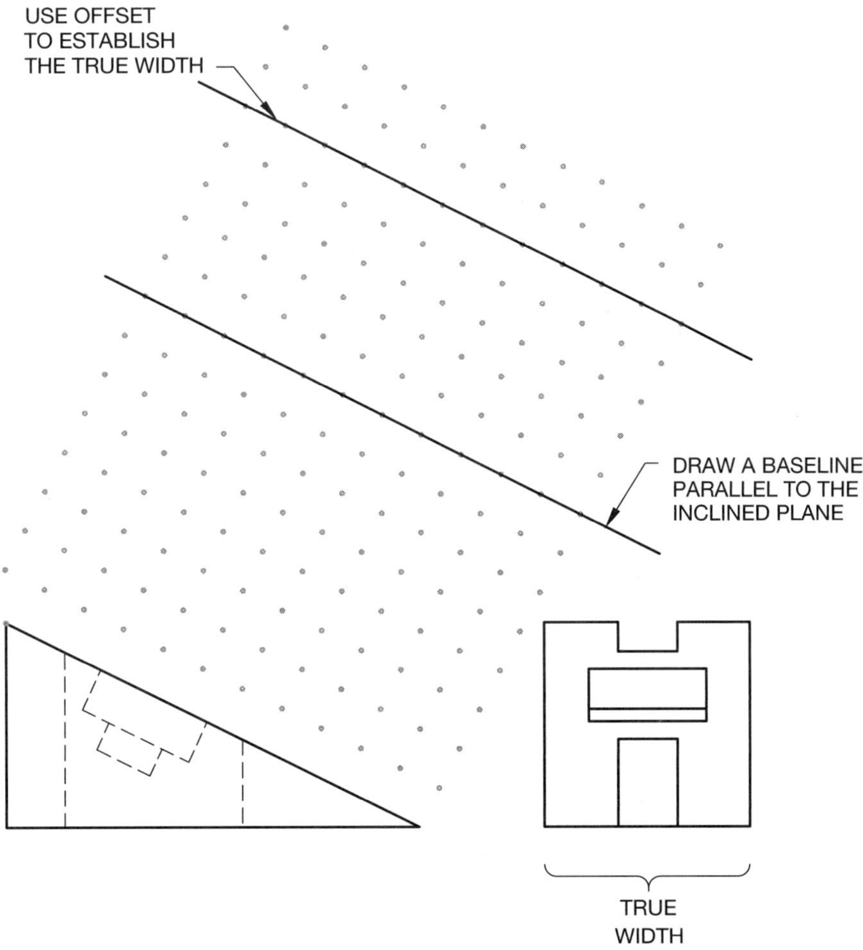

Figure 6.28 *The initial step of projecting an auxiliary view is to project each feature of the object from the inclined surface.*

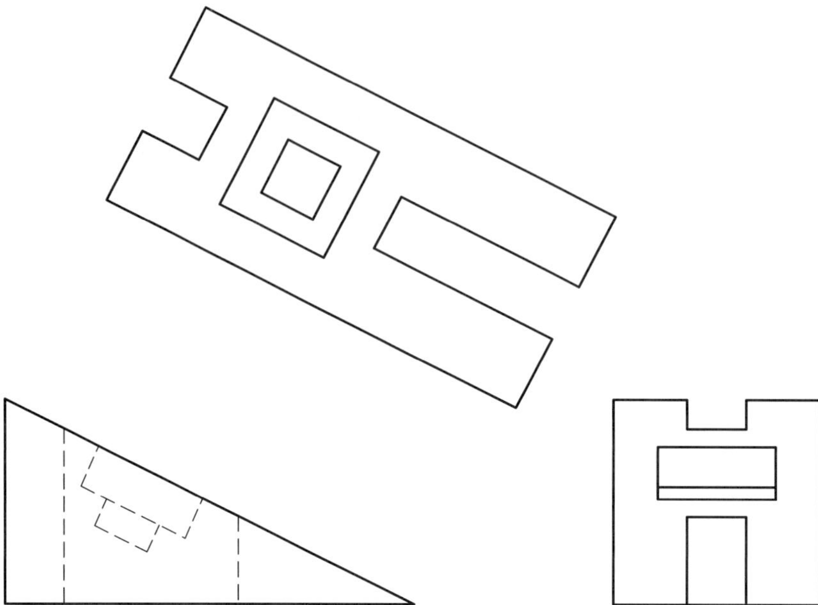

Figure 6.29 *Use* **Offset, Trim,** *and* **Fillet** *to complete the auxiliary view.*

One of the unique features of an auxiliary view is that you often see more than one surface at a time. It is important to remember that if the surface is not parallel to a surface of the glass cube, the surface is distorted. Figure 6.30 shows an auxiliary view with several distorted surfaces surrounding the true surface. The auxiliary view is projected to an auxiliary plane, not one of the six original sides of the cube. Any surface in the auxiliary view not parallel to the auxiliary plane will be distorted. Figure 6.31 shows only the surfaces that are parallel to the auxiliary plane. Distorted surfaces should be omitted by stopping the drawing at an edge of the plane that represents the true shape or at the centerline of round features. An alternative to breaking the object at an edge is to show a portion of the distorted surface for a source of reference. A break line is used so that the entire edge does not need to be drawn, as seen in the drawing in Figure 6.32.

PLACEMENT OF AUXILIARY VIEWS

Each of the auxiliary views presented up to this point has been projected from the drawing where the inclined surface appears as a line. Although the auxiliary view must always be projected from the inclined surface, the actual view can be moved once it is fully projected to aid in page layout and provide additional space for notes and dimensions. Although notes and dimension will not be discussed until Section 3, it is important to understand placement of auxiliary views. Figure 6.33 shows an example of a drawing with an auxiliary view that has been moved from its projected loca-

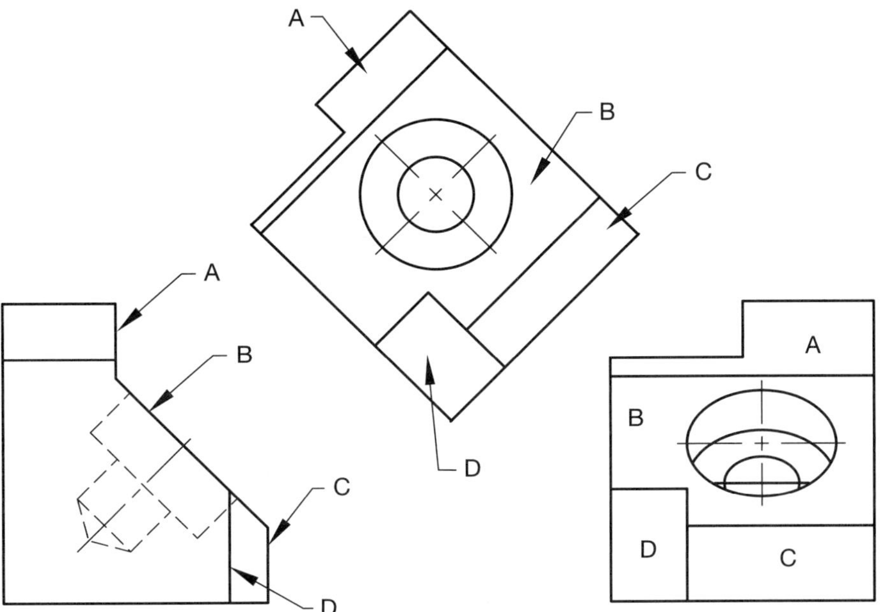

Figure 6.30 *One of the unique features of an auxiliary view is that you often see more than one surface at a time. It is important to remember that if the surface is not parallel to a surface of the glass cube, the surface is distorted.*

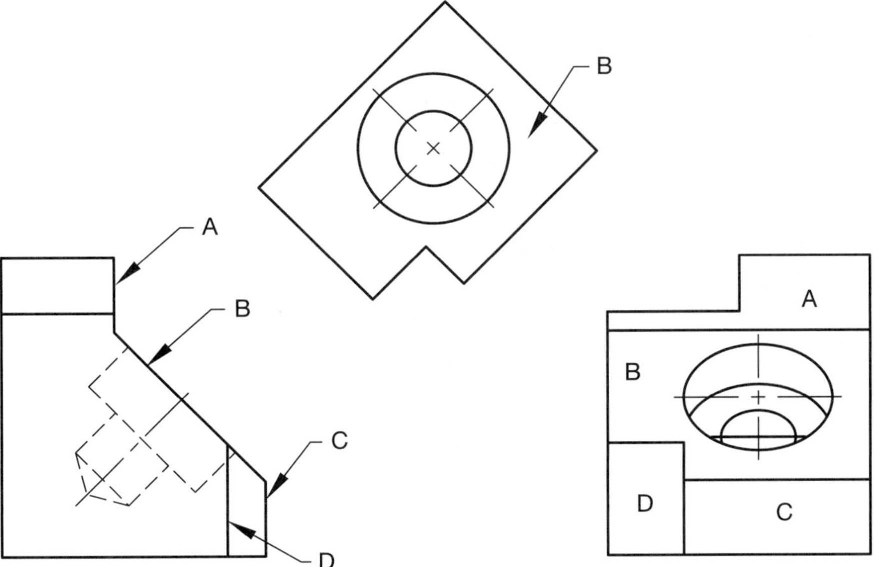

Figure 6.31 *Only the surfaces that are parallel to the auxiliary plane should be represented.*

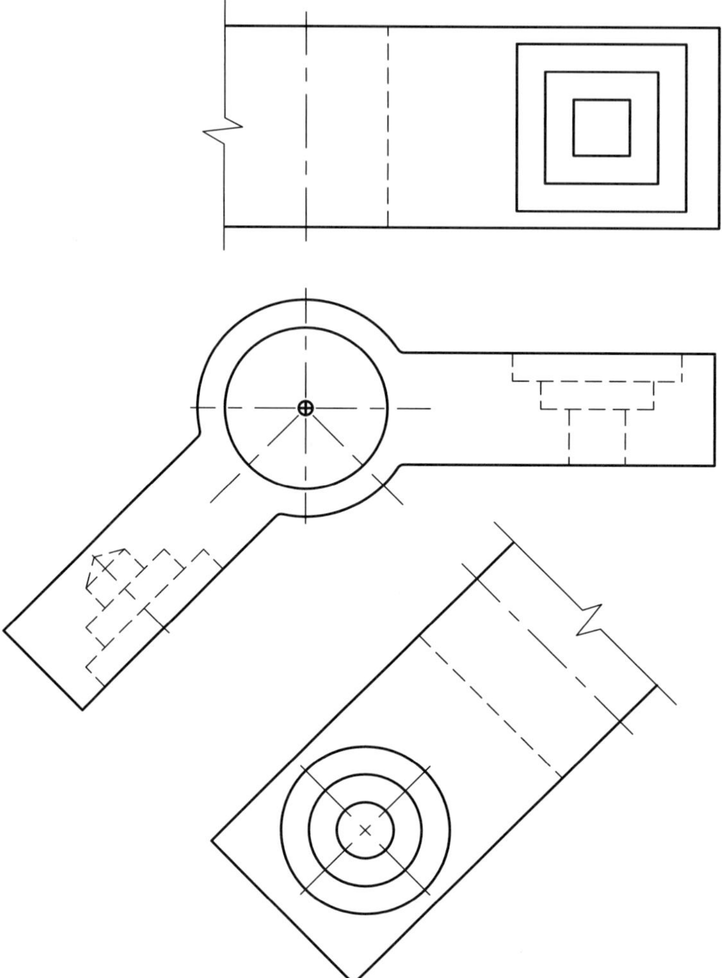

Figure 6.32 *Distorted surfaces in an auxiliary view should be omitted by stopping the drawing at an edge of the plane that represents the true shape or at the centerline of round features. An alternative to breaking the object at an edge is to show a portion of the distorted surface for a source of reference.*

tion. ASME standards allow the auxiliary drawing to be moved to any convenient location in the drawing area as long as the angle of projection is not altered. If the position is altered, a viewing plane must be added to the drawing to show the origin of the projected view. A .25" high letter is placed at each end of the viewing plane to indicate the name of the new view. If only one new view has been created, the letters can be omitted from the viewing plane.

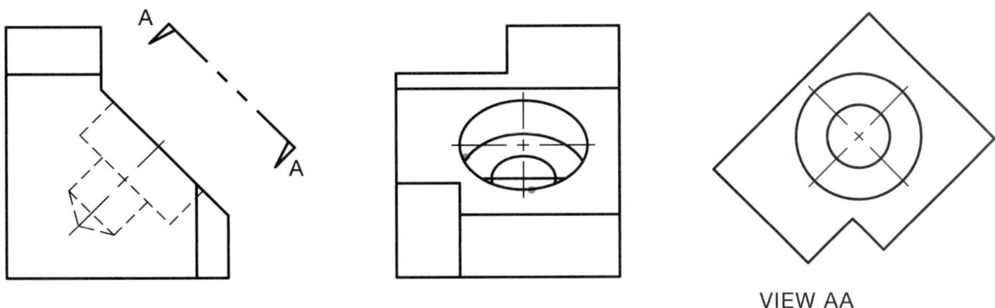

Figure 6.33 *Although the auxiliary view must always be projected from the inclined surface, the actual drawing can be moved once it is fully projected, to aid in page layout and provide additional space for notes and dimensions. ASME standards allow the auxiliary drawing to be moved to any convenient location in the drawing area as long as the angle of projection is not altered.*

FREEHAND SKETCHING

Students often assume that because they are learning to draw using a computer, mastering a skill such as sketching will not be necessary. Nothing could be further from the truth. The computer is just a drawing tool. Sketching is a drawing skill, and a tool for organizing your thoughts for drawing layout. A sketch can be a valuable tool for communicating ideas between the engineer and the CAD operator during the initial stages of design. As you progress through the creation of a multiview drawing, a sketch will be a valuable tool in helping to apply the six criteria in selecting the front view and for determining which views can be eliminated from the formal drawing. As you progress through the advanced chapters in this text, a sketch will be an invaluable tool for planning where notation and dimensions should be placed. Even with the use of palm-sized computers, sketching is invaluable in creating drawings in the field for existing parts. Your ability to master sketching common geometric shapes in proportion is a tool just as valuable as the CAD commands that have been introduced throughout this text.

SKETCHING TOOLS

Sketches can be created using a variety of materials and tools, including AutoCAD. The AutoCAD **Sketch** command will be introduced in Chapter 12. The most common tools for freehand sketching are paper, pencil, and eraser. Sketches for designs worth millions of dollars have been created on a used lunch sack or a paper napkin. When lunch sacks aren't available, paper with a slight matte finish is easier to use than paper with a glossy finish. Graph paper works well for sketching because of its texture, and the obvious aid for creating lines using a uniform grid. Pencils with a soft lead such as a common number 2 lead, or mechanical pencils with F, H, or HB leads are excellent for sketching. If a common pencil is used, the lead should be worn to a smooth, rounded edge before you start a sketch—a sharp lead point can tear the drawing paper. Forming the lead into a wedge and then drawing with the edge of the wedge allows thin lines to be made. Using the full width of the wedge will create wide lines. The pressure applied to the pencil will control lightness and darkness of the line. For correcting mistakes, a soft eraser will work best. An eraser can be either at the end of the pencil, rectangular, or stick. If you work at keeping construction lines very light, erasing can be kept to a minimum.

SKETCHING TECHNIQUES

The following guidelines will aid in the development of sketches:

- Hold the pencil firmly, but not so tightly that tension is created. If your fingers are sore after a few minutes of sketching, you're choking the pencil rather than holding it.
- Grip the pencil approximately 1" to 1.5" up from the pointed end.
- Draw horizontal lines from left to right using short, light, overlapping strokes.
- Draw vertical lines from top to bottom using short, light, overlapping strokes.
- Create square boxes to frame a circle, and then make short, overlapping strokes.
- Notice that short, light, overlapping strokes are a key to good sketching.

Sketching Straight Features

Straight features can be made easily if graph paper is being used. If a straight line is to be made on plain paper, place two points to represent the ends of the line, and then connect the dots using a series of short, light, connected segments. Imperfections in the line can be erased and corrected. As you sketch, keep your eye on the point you are moving toward. If you're right-handed, this will typically mean pulling the pencil from top to bottom for vertical lines, and from left to right for horizontal lines. Once the complete line has been successfully placed, it can be retraced in one movement to get a dark line. The process can be seen in Figure 6.34.

Using one continuous motion to connect the dots is an alternative method for sketching short lines. As the length of line is increased, care must be taken to avoid placing a curve in the line. Short, straight lines between 2" and 3" in length can be made in one continuous motion of the wrist. Slightly longer lines require the movement of the elbow. Lines longer than about 12" will require movement of the entire arm. Experiment with each method to see which method seems most comfortable. For most users, placing short segments will provide the best change of success.

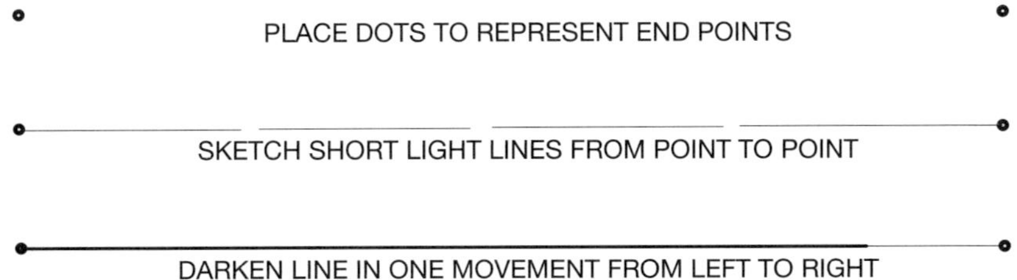

Figure 6.34 *To sketch a straight line, place two points to represent the ends of the line, and then connect the dots using a series of short, light, connected segments. Imperfections in the line can be erased and corrected.*

Parallel Lines

If graph paper is not available, placing a series of dots parallel to a line will help to form the second line. To draw two parallel lines, draw a line in the desired position, and then place dots parallel to the original line. The distance from the original line to the string of dots can be kept uniform by measuring with the end of a pencil to another point on the pencil marked by a finger, or by marking the space between two dots with a second piece of paper. Once the original distance is determined, slide the makeshift ruler along the original line and mark a new location. The process can be repeated several more times until a series of dots is parallel to the original line. Short, light lines can then be sketched between the set of dots. The process can be seen in Figure 6.35.

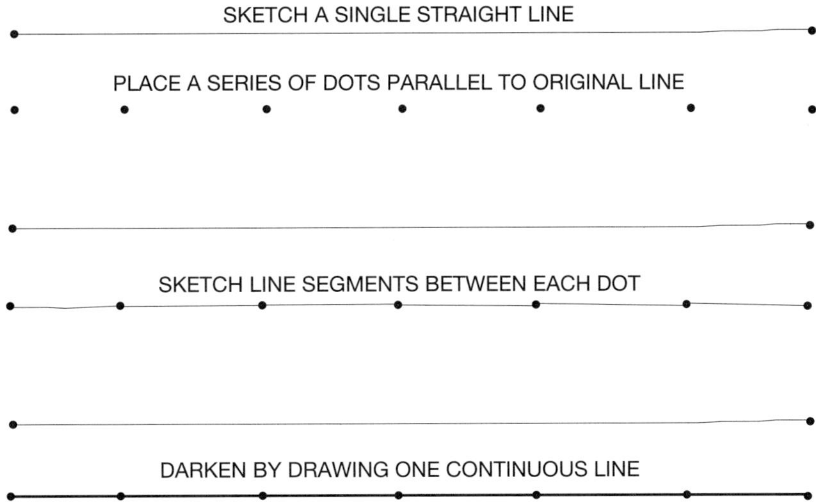

Figure 6.35 *To draw two parallel lines, draw a line in the desired position, and then place dots parallel to the original line.*

Angular Lines

Whatever the angle to be drawn, lightly drawing a 90° angle is an excellent way to start the process of sketching angled lines. Be sure to keep each leg of the perpendicular lines of equal length. Next draw a line between end points of the perpendicular lines similar to Figure 6.36. By placing a dot at the middle of the inclined line and then drawing a line from the intersection to this dot, you create a 45° angle. By dividing the original inclined line into thirds, you can draw lines at 30° and 60°. By dividing the inclined line into thirds, and then dividing each of these spaces in half, you can determine 15° increments. Angles smaller than 15° can be determined by breaking the original 90° angle into 15° increments, then breaking a 15° segment into 5° increments, and then breaking this space into 1° units. Figure 6.37 shows an example of a line sketched at approximately 42°. Once of the angle has been determined, the easiest method of drawing the inclined line is to rotate the paper so that the line to be drawn will be horizontal or vertical.

Sketching Circular Features

Common circular features that occur throughout mechanical drawings include arcs, circles, ellipses, and cylinders.

Sketching Arcs

An arc can be completed by drawing a box equal to the size of the radius, and then forming the arc around the center point. As the process is started, the closer the original box is to square, the better the resulting arc. Be sure to use marks on a second sheet

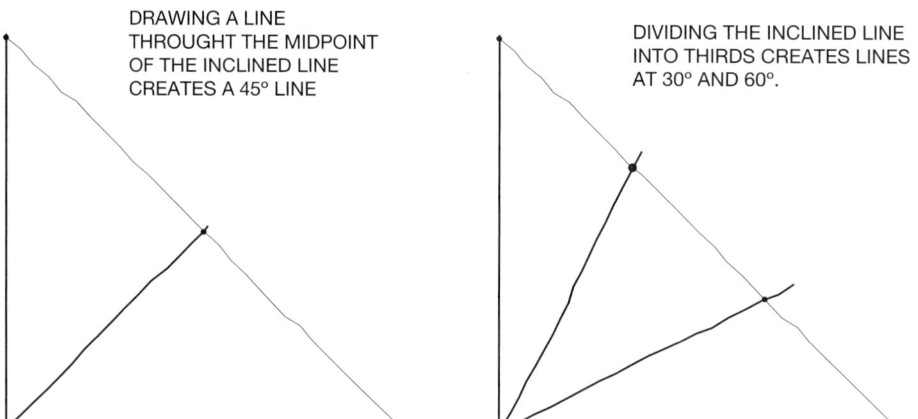

Figure 6.36 *Whatever the angle to be drawn, lightly draw a 90° angle to start the process of sketching angled lines. Next draw a line between end points of the perpendicular lines. Placing a dot at the middle of the inclined line and then drawing a line from the intersection to this dot will create a 45° angle.*

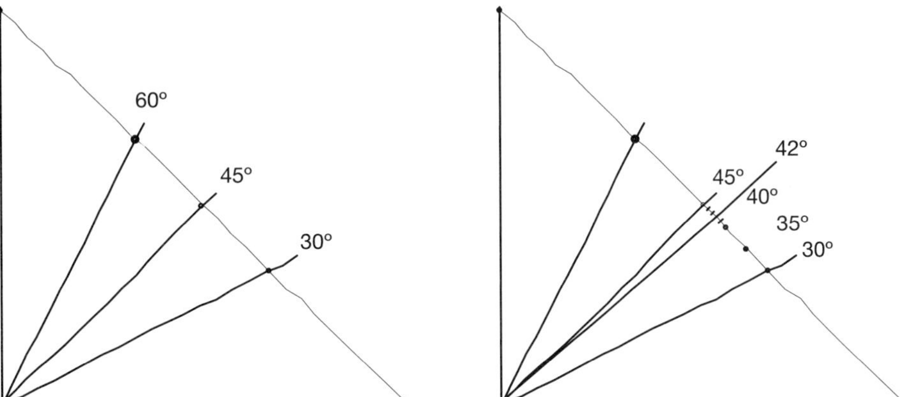

Figure 6.37 *By dividing the original inclined line into thirds, you can draw lines at 30° and 60°. By dividing the inclined line into thirds and then dividing each of these spaces in half, you can determine 15° increments. Angles smaller than 15° can be determined by breaking the original 90° angle into 15° increments, then breaking a 15° segment into 5° increments, and then breaking this space into 1° units.*

of paper to ensure equal lengths for the square. An arc can be completed using the following steps:

1. Draw a square to represent the desired radius.
2. Draw a diagonal line from one corner of the square to mark the center point of the arc.
3. Using a second sheet of paper as a ruler, mark the radius on the diagonal line.
4. Lightly sketch an arc from one corner of the box to the centerline.
5. Complete the layout by sketching the remaining portion of the arc.
6. Darken the rough draft by completing the arc in one movement.

The process can be seen in Figure 6.38

Sketching Circles

Circular features can be completed by drawing a box, drawing its center points, and then forming the circle around these center points. As the process is started, the closer the original box is to square, the better the resulting circle. Be sure to use marks on a second sheet of paper to ensure equal lengths for the square. A circle can be completed using the following steps:

1. Draw a square to represent the desired diameter.
2. Mark the center of each edge of the square.
3. Draw diagonal lines from each corner of the square to mark the center point of the circle.

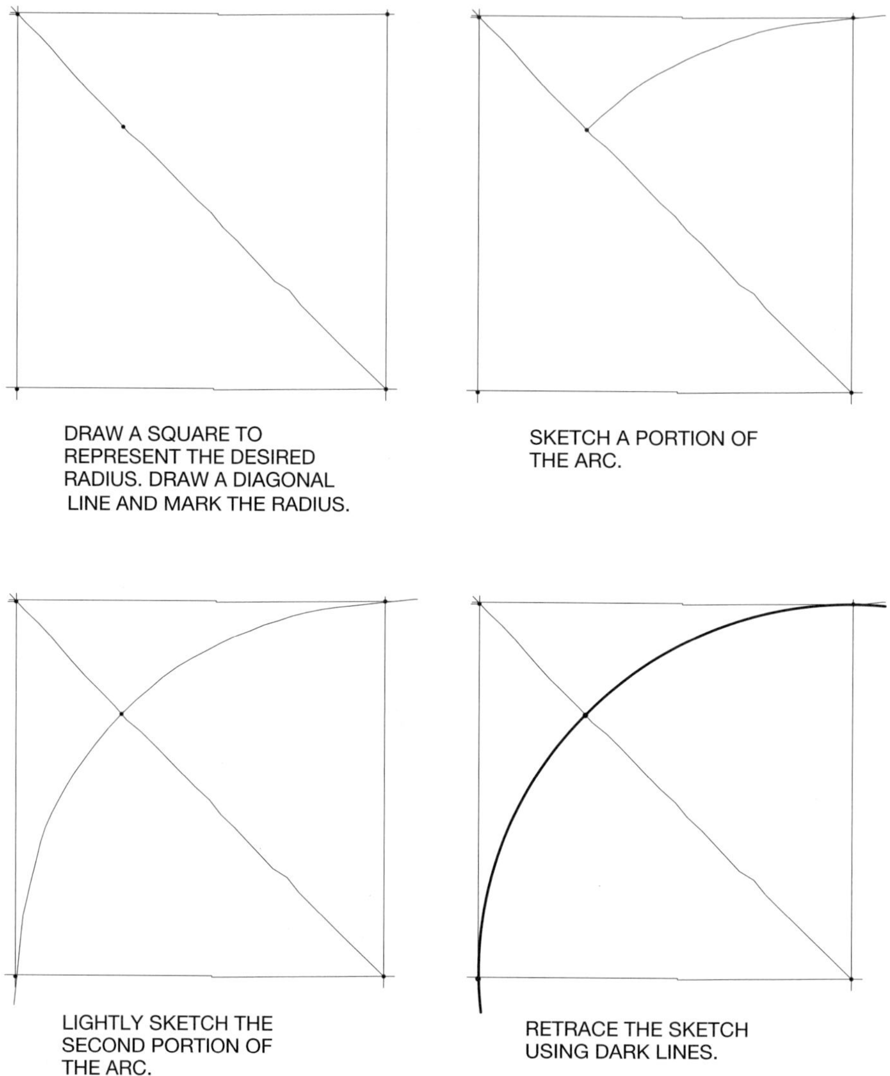

Figure 6.38 *An arc can be completed by drawing a box equal to the size of the radius and then forming the arc around the center point.*

4. Using a second sheet of paper as a paper ruler, mark a distance on the diagonal line equal the distance from the center point to an edge.
5. Lightly sketch an arc from one of the centerlines to a point on the nearest diagonal line.
6. Complete the layout by sketching the seven remaining arcs.

7. Darken the rough draft by completing one-half of the circle in one movement, and then darken the other side.

The process can be seen in Figure 6.39.

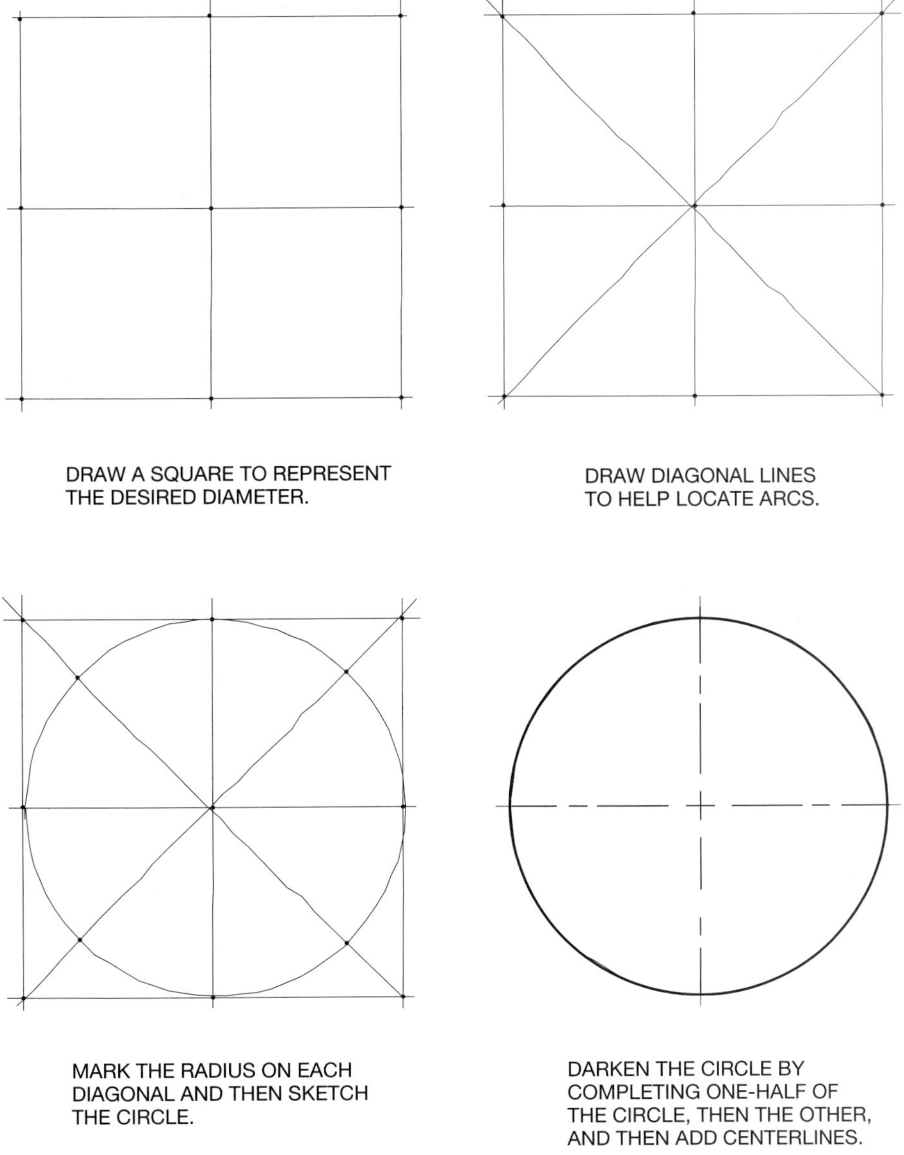

Figure 6.39 *Circular features can be completed by drawing a box, drawings its center points, and then forming the circle around these center points.*

Sketching Ellipses

An ellipse is completed using steps similar to those used to draw a circle. Start by drawing a rectangle that represents the lengths of the major and minor diameters. An ellipse can be completed using the following steps:

1. Draw a rectangle that represents the lengths of the major and minor diameters.
2. Find the center point of the rectangle by drawing diagonal lines through the rectangle.
3. Define the center points of the rectangle.
4. Draw an arc to represent one quadrant of the ellipse.
5. Measure the point where the arc passes through the diagonal, and then mark this distance on the remaining diagonal lines.
6. Sketch arcs in the remaining grids.
7. Darken the ellipse using similar methods that were used to darken a circle.

The process can be seen in Figure 6.40.

Sketching Cylinders

Sketching the rectangle that will form the outline of the cylinder will help ensure the accuracy of the sketch. Follow the steps to draw an ellipse to create the top of the cylinder. Repeat the process for drawing the ellipse at the other end of the rectangle. The portion of the bottom ellipse that lies beyond the edges of the rectangle can be erased, or left if it is sketched lightly. The lines forming the cylinder can be darkened using methods that were used to darken arcs, ellipses, and straight lines. The process can be seen in Figure 6.41.

KEEPING PROPORTIONS IN SKETCHES

Even though sketches are not drawn to scale, if they are to be a useful production tool, it is important that they be proportionately accurate. It has been mentioned that a pencil or a piece of paper can be used as measuring tools, but using these tools is time consuming. With practice, accurate measurements can often be done by eye. As you sketch, remember that the lines that make up the object are related to each other by size and direction. As you add a line to the sketch, visually compare the new line to previous lines. Taking time to verify that lengths and direction are kept in proportion will enable you to clearly communicate through your sketch. The following guidelines will aid in keeping sketches accurate:

- Before starting to sketch, mentally divide the part into its major components.
- Start the sketch by drawing the largest component, and work to the smallest details.

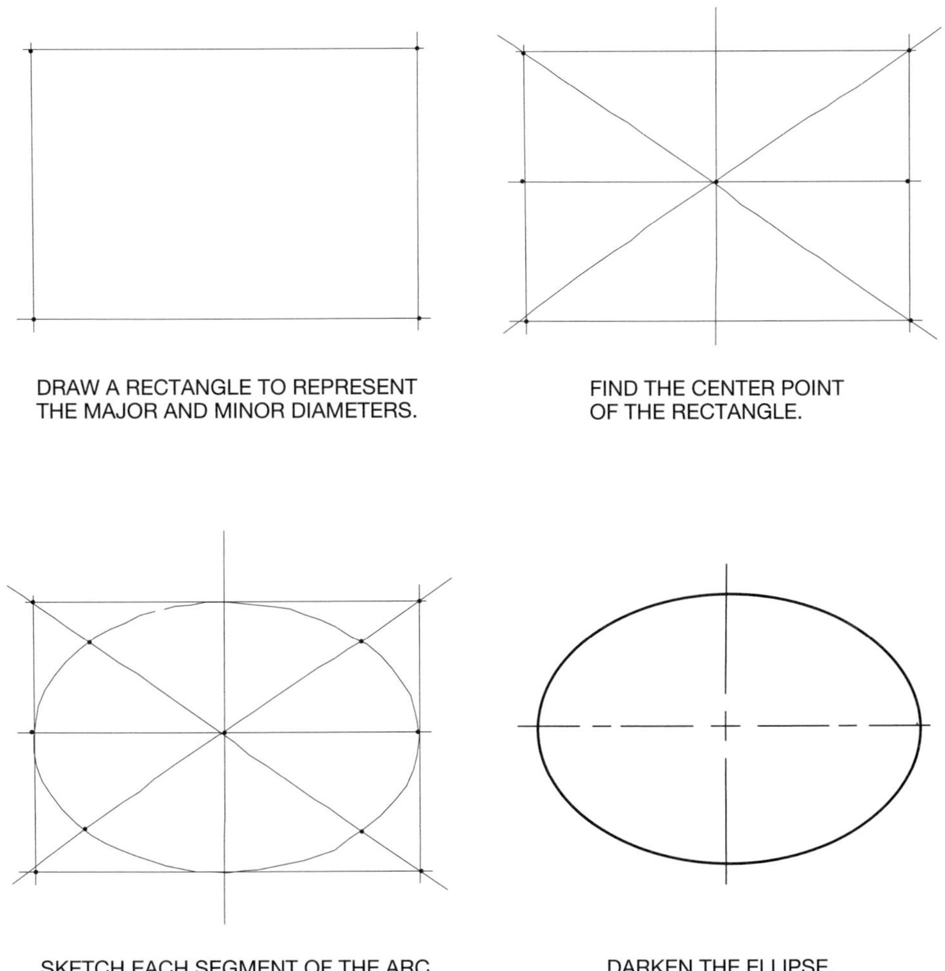

Figure 6.40 *An ellipse is completed using steps similar to those used to draw a circle. Start by drawing a rectangle that represents the lengths of the major and minor diameters.*

- Use squares and rectangles to form a base for irregular shapes.
- The first line drawn determines the size of other lines. Keep all portions of the object in proportion.

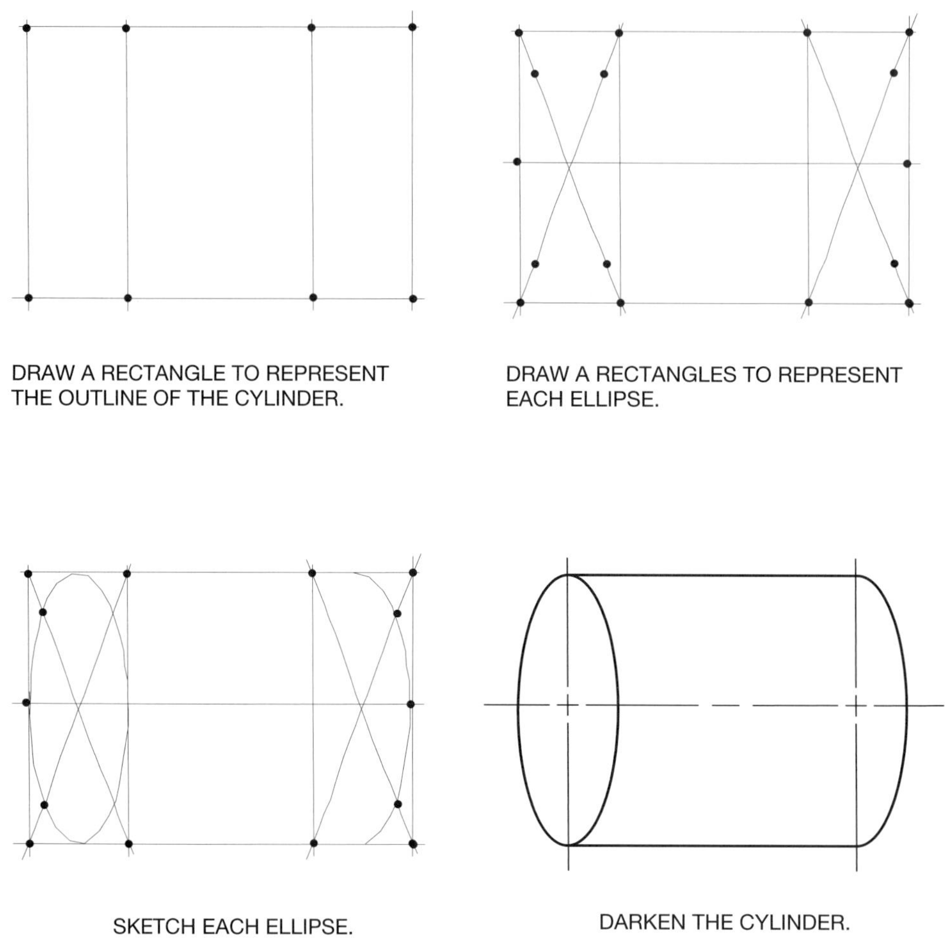

Figure 6.41 *Sketch the rectangle that will form the outline of the cylinder to ensure the accuracy of the sketch. Follow the steps to draw an ellipse to create the top of the cylinder. Repeat the process for drawing the ellipse at the other end of the rectangle.*

MULTIVIEW SKETCHES

Orthographic sketching involves drawing one or more of the six views of the object. The views are developed using the same methods and line types that were described in the creation of multiview drawings. Figure 6.42 shows an object that will be used to develop multiview sketches. Once the views to be created have been determined, multiview sketches can be developed using the following steps:

1. Sketch proportional rectangles for each of the views to be created. Typically this will require boxes for the top, front, and a side view. See Figure 6.43.

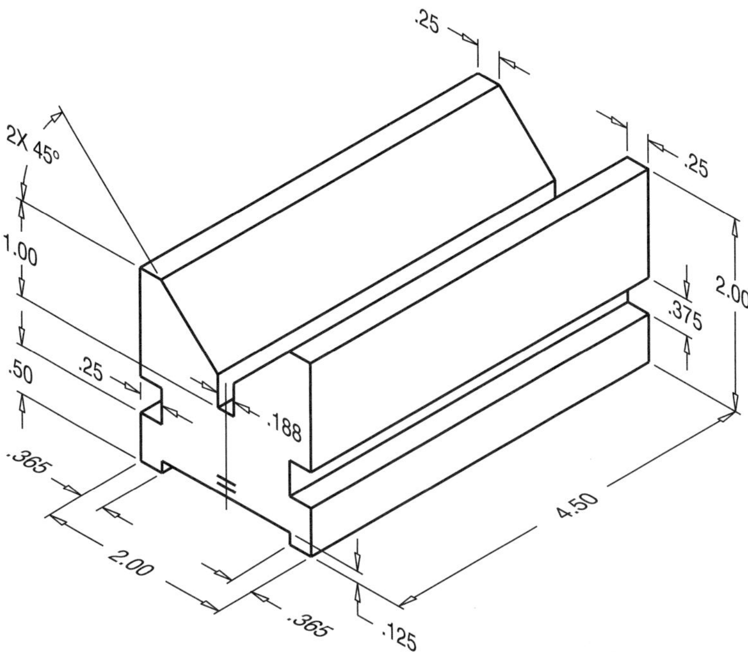

Figure 6.42 *Use this object to develop multiview sketches.*

2. Use a 45° line projecting from the corner of the front view to help keep the width of the top and side views the same. See Figure 6.43.
3. Divide the rectangle for each view into smaller squares and rectangles to represent features that will be represented in each view. See Figure 6.44.
4. Lightly sketch circles and irregular shaped features. See Figure 6.45.
5. Darken each feature using appropriate line types and line weights.

Figure 6.46 shows the completed sketches of the top, front, and side view for the object in Figure 6.42.

Figure 6.43 *Sketch proportional rectangles for each of the views to be created. Typically this will require boxes for the top, front, and a side view. Use a 45° line projecting from the corner of the front view to help keep the width of the top and side views the same.*

Figure 6.44 *Divide the rectangle for each view into smaller squares and rectangles to represent features that will be represented in each view.*

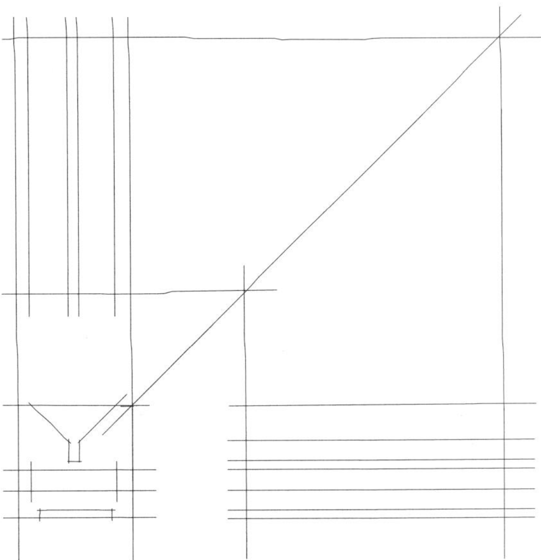

Figure 6.45 *Lightly sketch the interior features.*

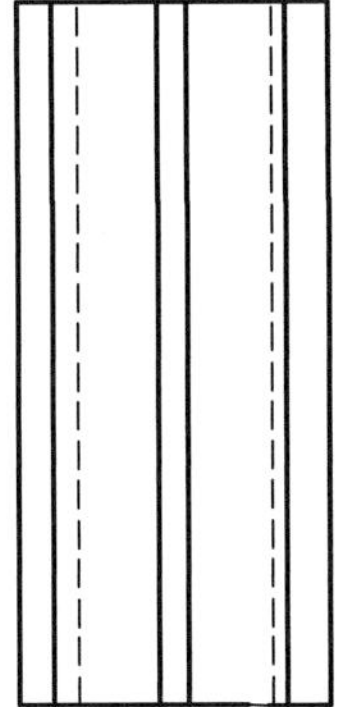

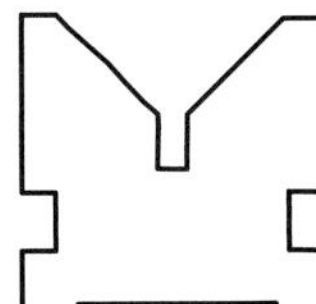

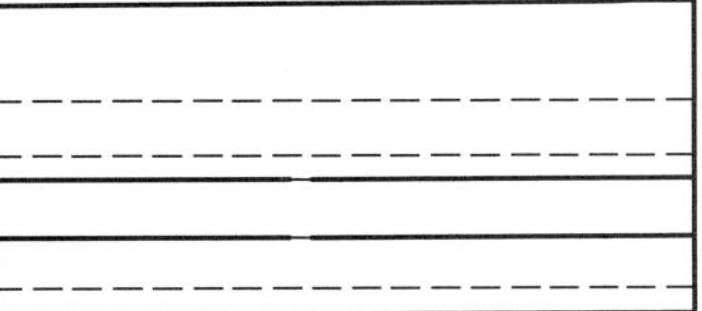

Figure 6.46 *The completed sketches of the top, front, and side views using appropriate line-types and lineweights.*

ISOMETRIC SKETCHES

Isometric sketches are shown throughout this chapter. Isometric sketches are three-dimensional representations that display the objects similar to the way they appear to the eye. Although not used for production drawings, isometric drawings are an ideal tool to help in planning multiview drawings. An isometric sketch can be used to visualize how each feature of an object will affect other features of the part and to visualize what should be represented in each view of a drawing.

Forming an Isometric Drawing

The base of an object is rotated to be at a 30° angle to the horizontal plane to form an isometric drawing or sketch. All horizontal planes are represented by a 30° angle, and vertical lines represent all vertical planes. Figure 6.47 shows the formation of an isometric cube. A cube should be created to start each isometric sketch. The cube can be created by the following steps:

1. Sketch a horizontal reference line.
2. Sketch a vertical reference line from the mid-point of the horizontal line.
3. Sketch an inclined line at 30° from horizontal on each side of the vertical reference line. Sketch lines to represent the top of the cube that are parallel to the two lines just drawn.
4. Sketch the remaining two lines to close the cube.

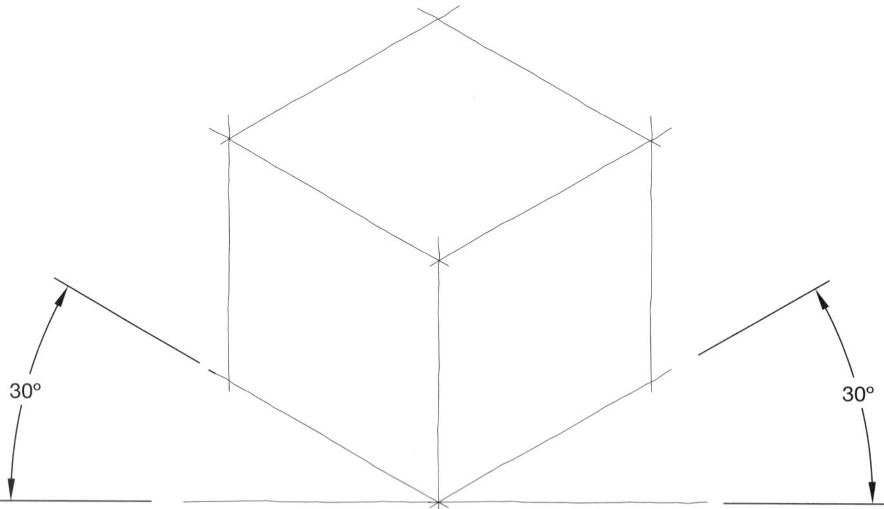

Figure 6.47 *The base of an object is rotated to be at a 30° angle to the horizontal plane to form an isometric drawing or sketch. All horizontal planes are represented by a 30° angle, and vertical lines represent all vertical planes.*

Once the cube is completed, an isometric sketch can be completed. The object shown in Figure 6.14 can be completed using the following steps:

1. Divide the cube into smaller cubes to represent features that must be represented. See Figure 6.48.
2. Darken each feature using appropriate line types and line weights. Figure 6.49 shows a completed isometric sketch.

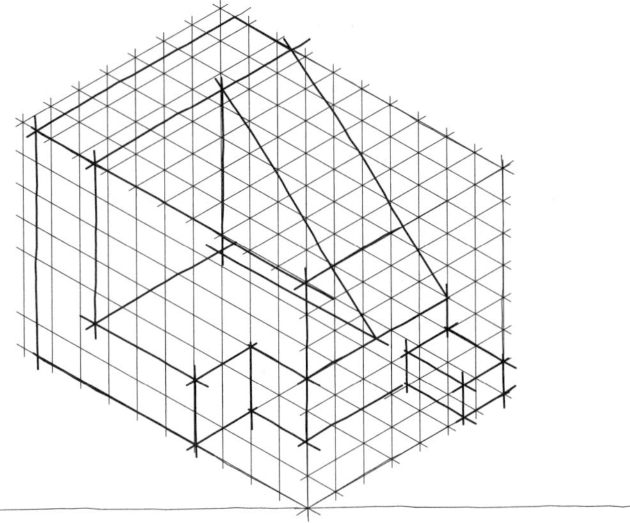

Figure 6.48 *A cube should be created to start each isometric sketch. Next divide the cube into smaller cubes to show features that must be represented.*

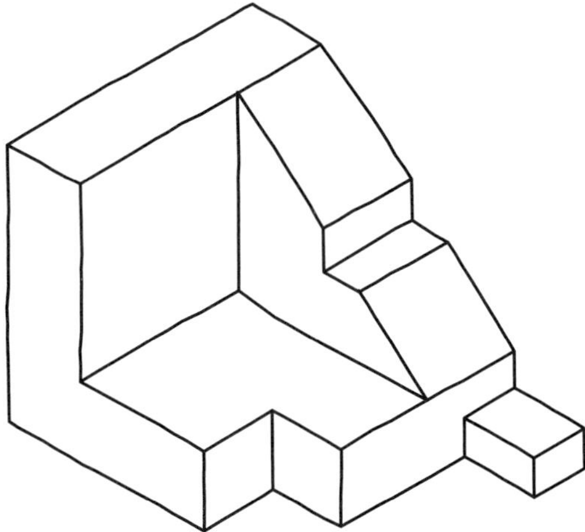

Figure 6.49 *The sketch is completed by darkening each feature using appropriate linetypes and lineweights.*

Representing Angles in Isometric Sketches

Angled features of an object cannot be accurately measured in isometric sketches, but they can be accurately represented. When a surface is not parallel to one of the surfaces of the glass cube, it can't be measured directly. Each end of the inclined plane must be located and then connected by an object line. Figure 6.49 shows an inclined surface in a multiview drawing.

Isometric Circles

A circle in an isometric sketch will appear as an ellipse. The position of the ellipse will vary depending on which surface of the cube the circle is located. Figure 6.50 shows the three common orientations for isometric circles. Each circle can be sketched by dividing the circle into four smaller arcs, and drawing these arcs first. Figure 6.51 shows how the three planes can be divided to locate the centerpoint for each of the four arcs that will form the ellipse. Figure 6.52 shows how the arcs can be created using the following steps:

1. Draw the cube that will contain the circle.
2. Draw lines from the corner of the cube to the midpoint of the opposite lines.
3. Repeat the last step using the opposite corner.

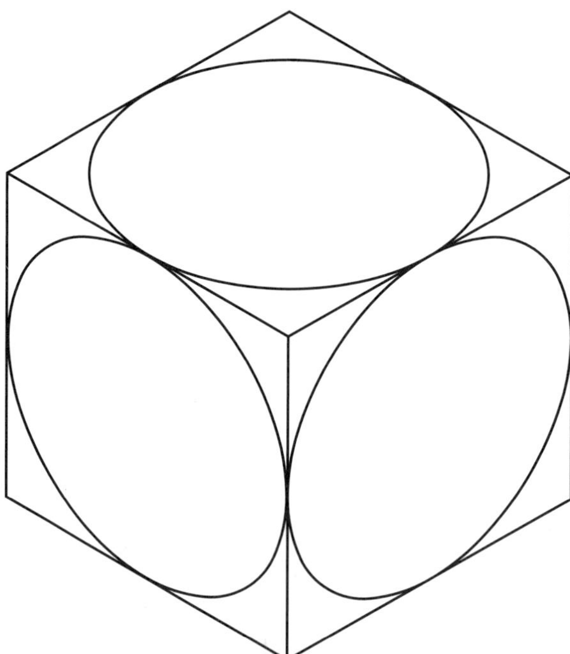

Figure 6.50 *Circles in an isometric sketch will appear as an ellipse. The position of the ellipse will vary depending on which surface of the cube the circle is located.*

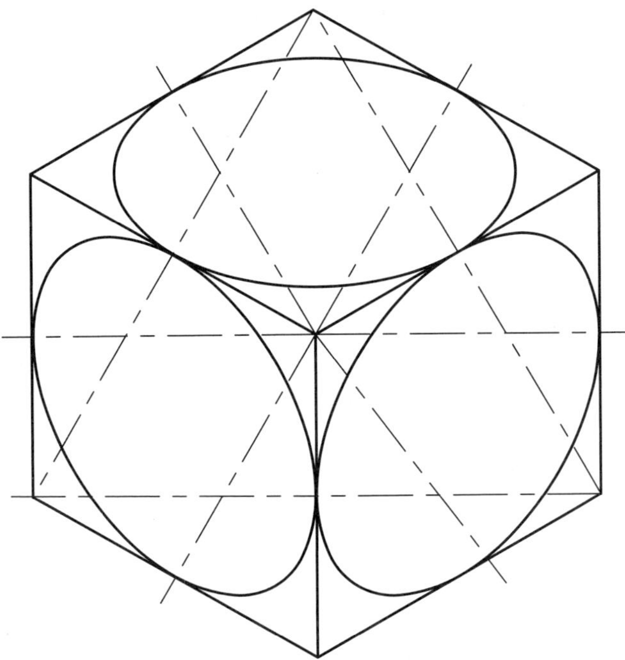

Figure 6.51 *The three planes can be divided to locate the center point for each of the four arcs that will form each ellipse.*

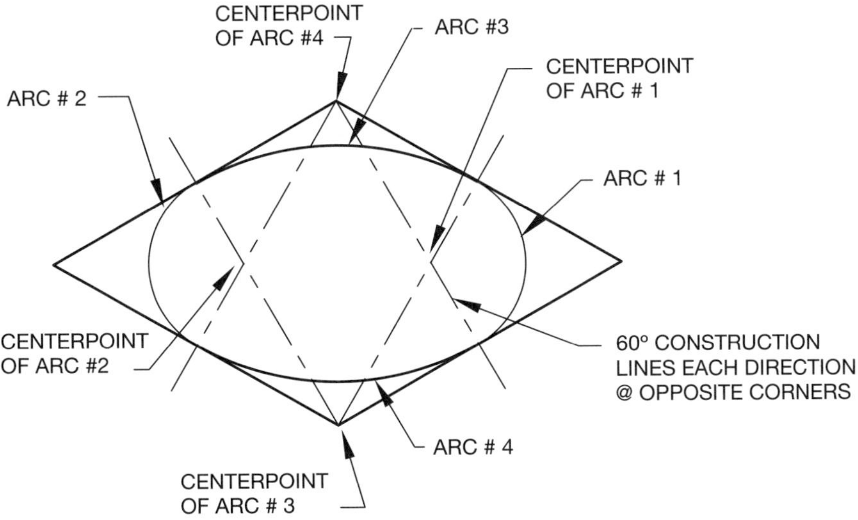

Figure 6.52 *The steps for sketching an isometric circle.*

4. Use points 1 and 2 to form the centers for the large arcs.
5. Use points 3 and 4 to draw the smaller arcs that form the end of the ellipse.
6. Notice that the minor diameter of the ellipse is always lined up on the centerline that is parallel to one of the edges of the cube.

CHAPTER 6 EXERCISES

Part I—Missing Lines

Exercises 1 through 5:

Complete the attached drawings by redrawing the following objects on grid paper. Use the isometric drawing to provide the missing lines and planes in the orthographic views. Use proper linetypes. Do not draw the isometric drawings. Use the completed sketch as a guide to complete the drawing using AutoCAD. Save the drawing using an appropriate file name that reflects the exercise number.

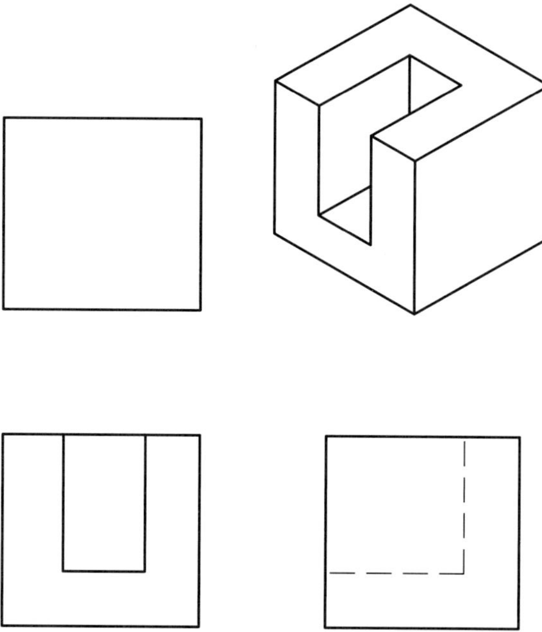

Problem 6-1

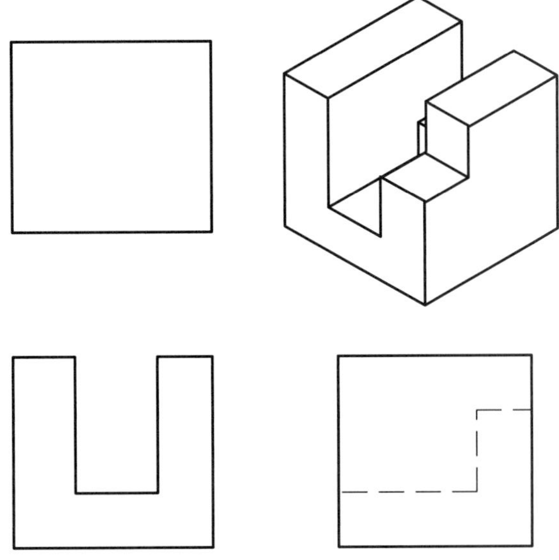

Problem 6-2

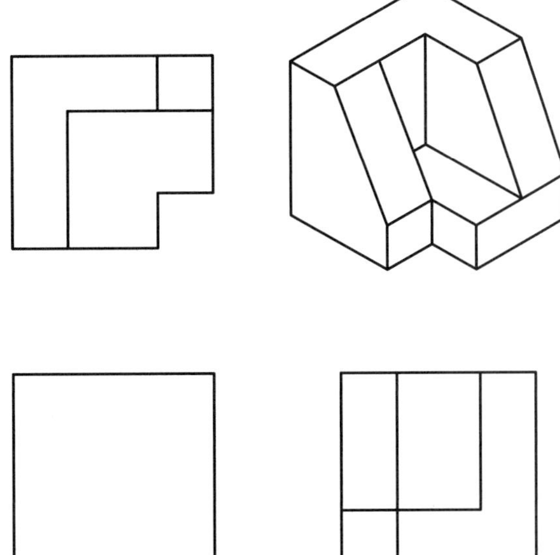

Problem 6-3

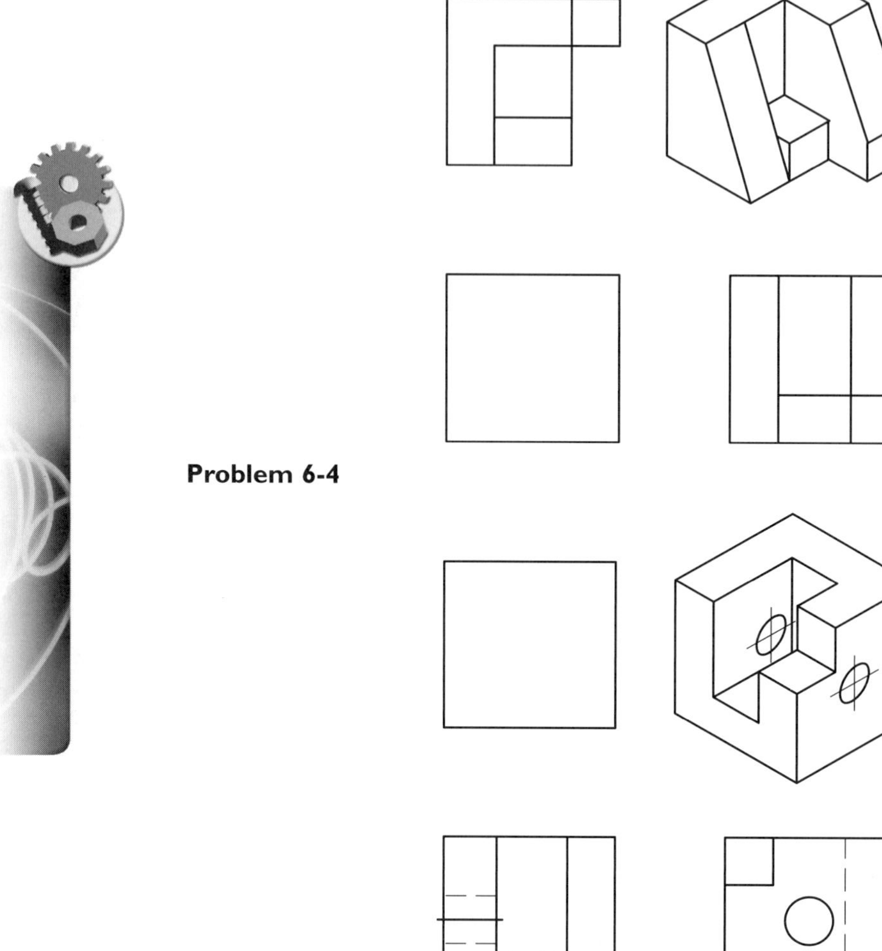

Problem 6-4

Problem 6-5

Part II—Missing Views
Exercises 6 through 10:

Complete the attached multiview drawings by sketching the following objects. Use the isometric drawing to provide the missing orthographic views. Assume each grid equals 0.25 units. Use proper linetypes. When the sketch is complete, complete the multiview drawing using AutoCAD. Hand in your sketch attached to each drawing assignment. Do not draw the isometric drawings. Save the drawing using the proper exercise number. Use the following steps to complete the orthographic projections.

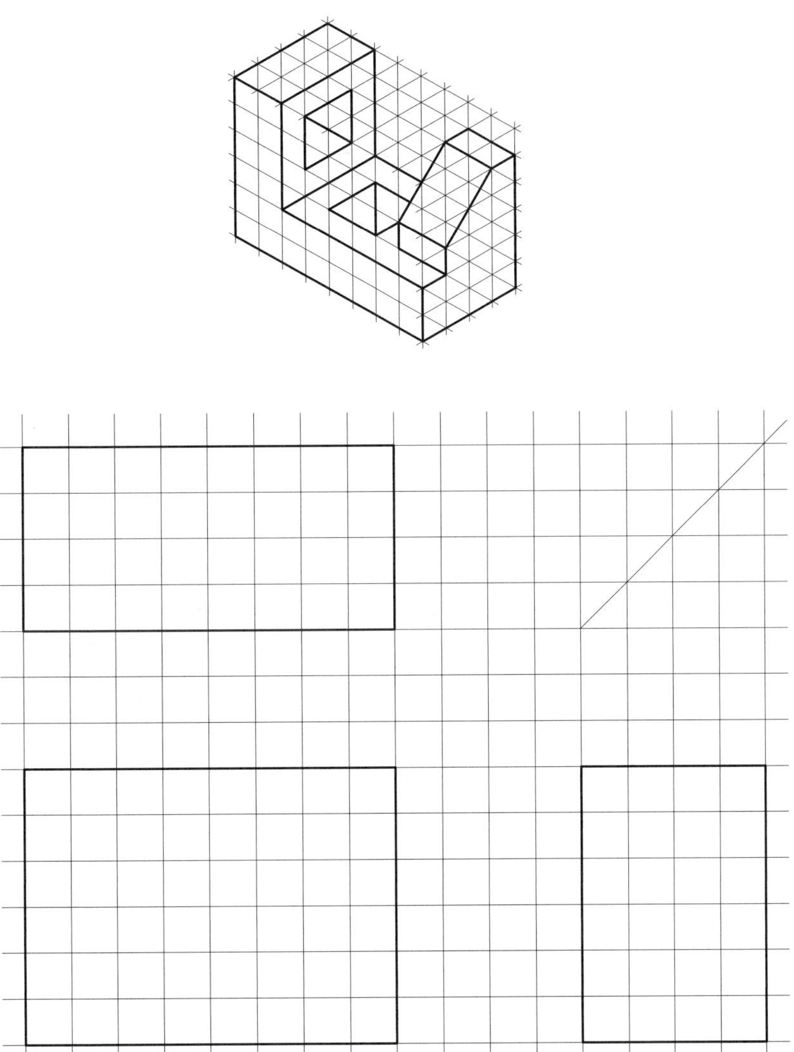

Step 1. *Begin each problem by creating the outline of the required views.*

STEP 1

Begin each problem by creating the outline of the required views. Do not attempt to include interior details at this time. For this example, create the top, front, and side views. By counting the grids in the isometric drawing, you can determine that the front view should be 8 units wide and 6 units high. The right side view should be 4 units wide and 6 units high. The size of the top view can now be determined by projecting sizes from the front and right views. The top view will be 8 units long and 4 units wide.

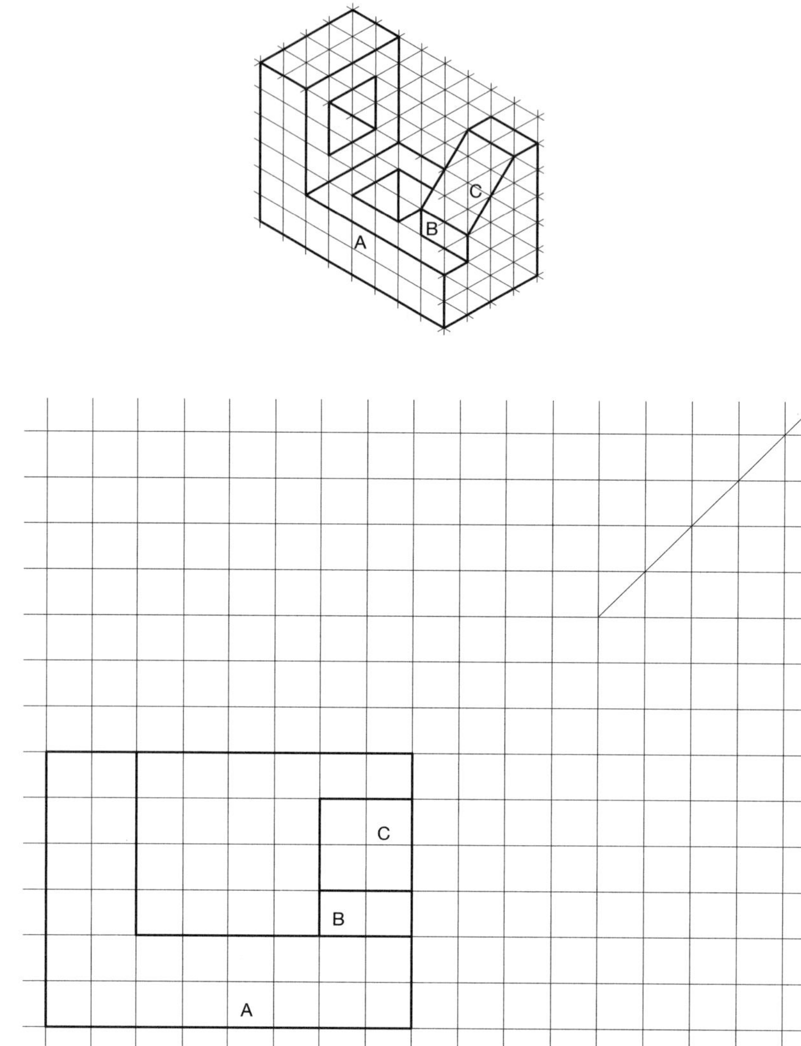

Step 2. *With the overall shape complete, draw the outline of all visible objects in the front view.*

STEP 2

With the overall shape complete, draw the outline of all visible objects in the front view. Draw the surfaces (A) that are closest to the viewing plane first. Now draw the next closest plane (B), followed by plane C. Do not attempt to include hidden features at this time.

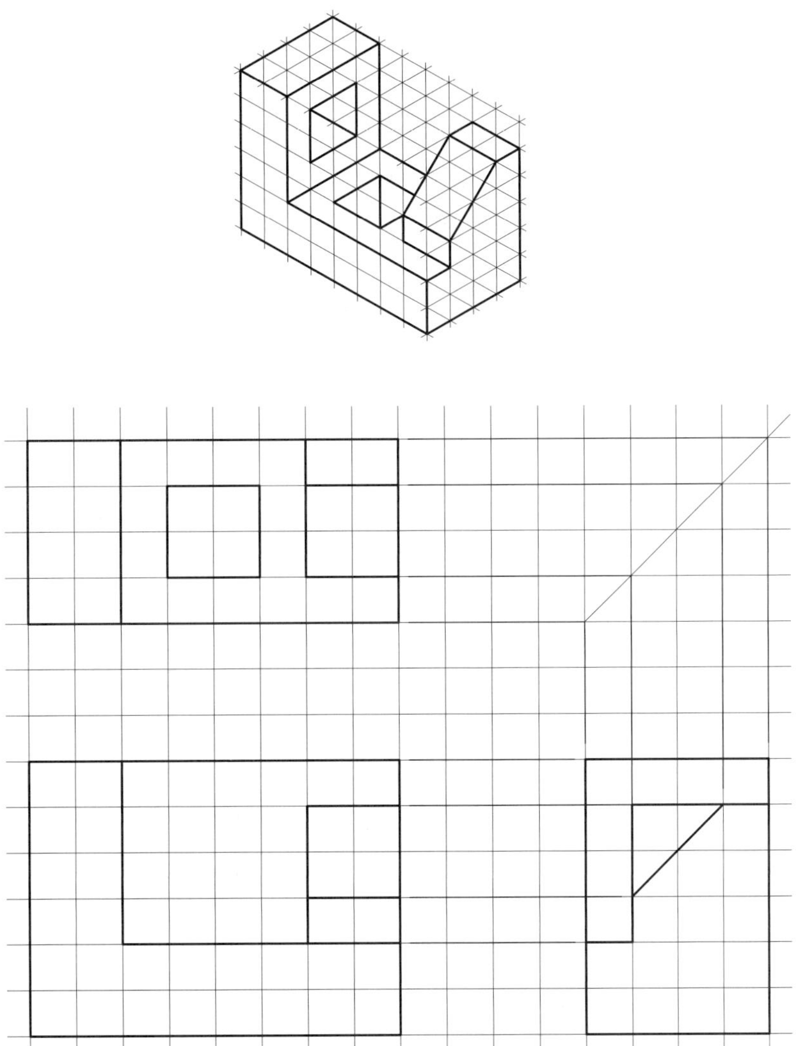

Step 3. *Project each of the shapes in the front view to the right side and top views.*

STEP 3

Project each of the shapes in the front view to the right side and top views. All of the visible surfaces should now be represented.

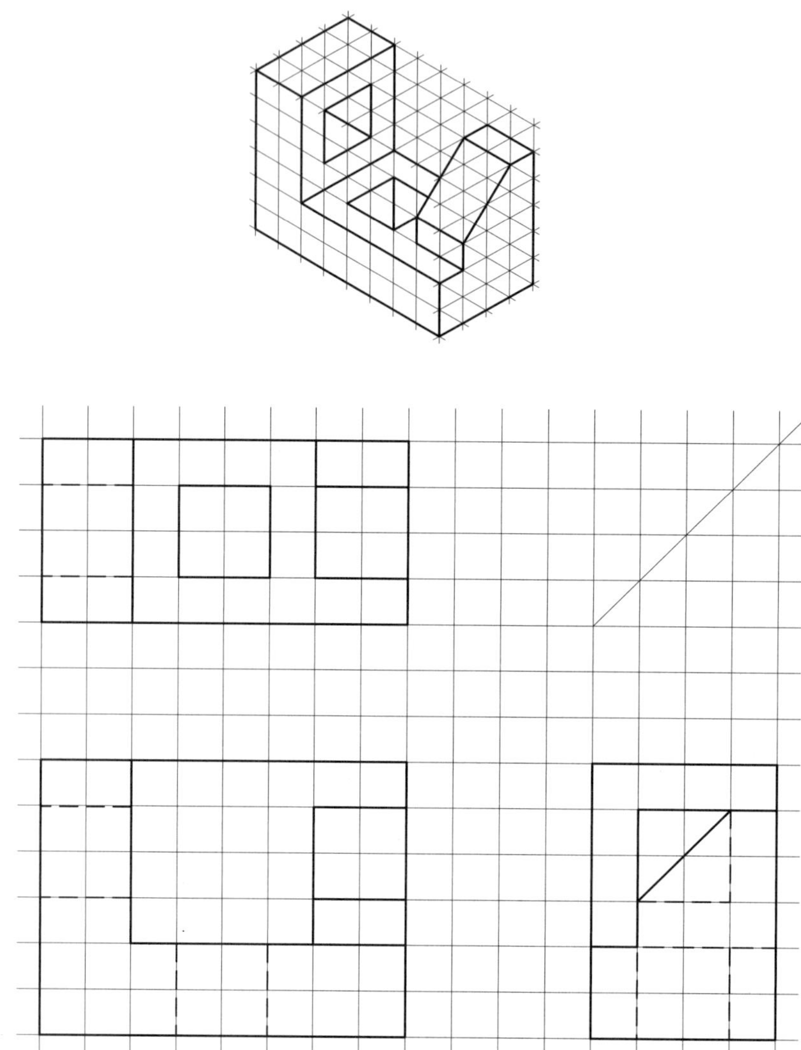

Step 4. *Start with the front view and draw all hidden features. Now project these hidden features to the other views.*

STEP 4

Start with the front view and draw all hidden features. Now project these hidden features to the top and right side views.

Part III—Multi-view Drawings
Exercises 11 through 25:

Use the attached isometric drawing and select the view that will represent the best front view. Sketch a multiview drawing using a minimum of three views. Use suitable linetypes to complete the drawings. Use your sketch as a guide to complete the multiview drawing using AutoCAD. Save the drawing using the appropriate exercise number. Assume each grid equals 0.25 units. Do not draw the isometric drawings or the grids. Do not try to place the dimensions on the drawing at this time. Hand in your sketch attached to each drawing assignment.

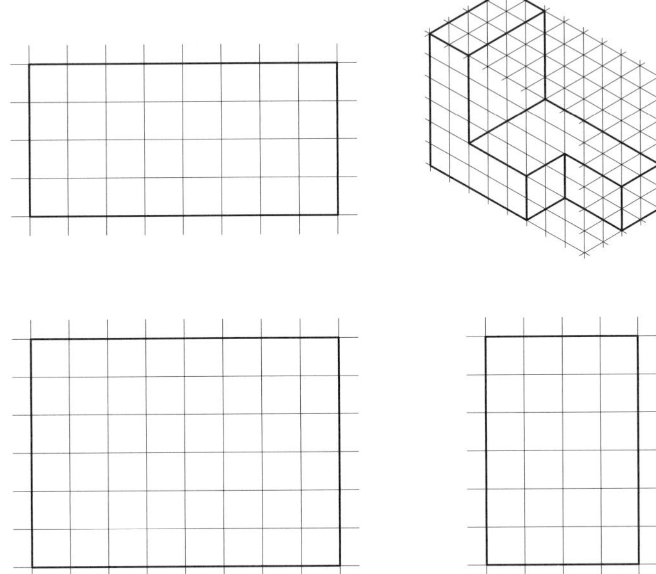

Problem 6-6

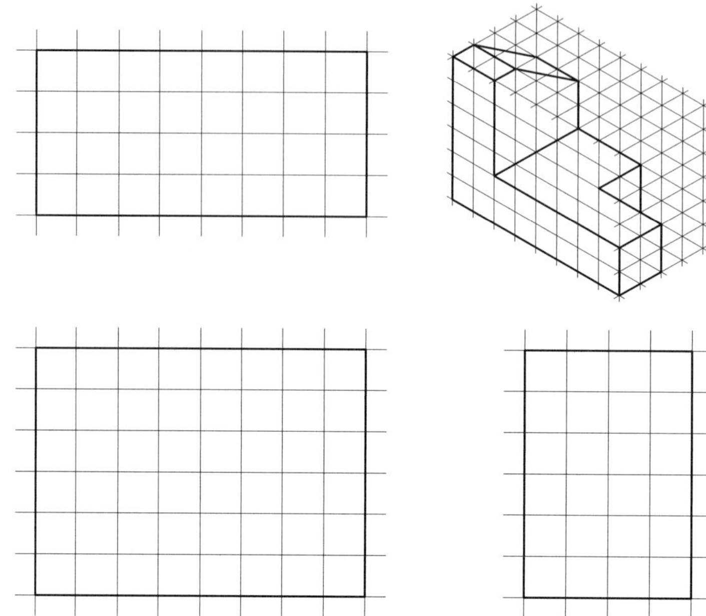

Problem 6-7

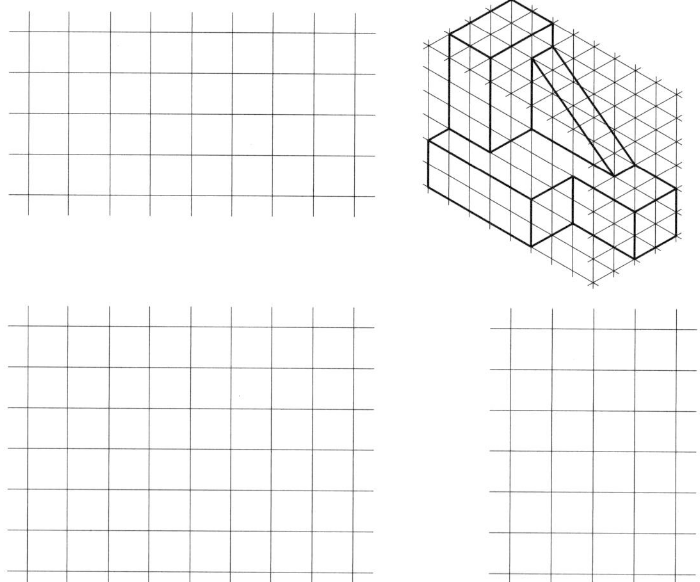

Problem 6-8

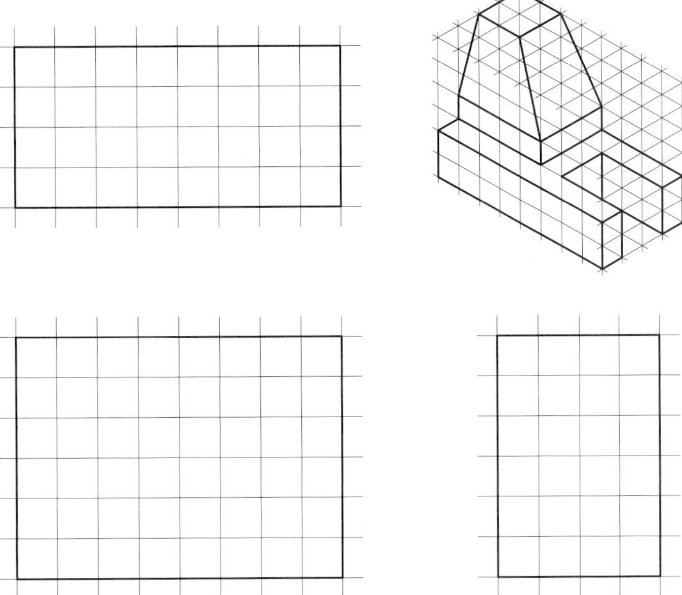

Problem 6-9

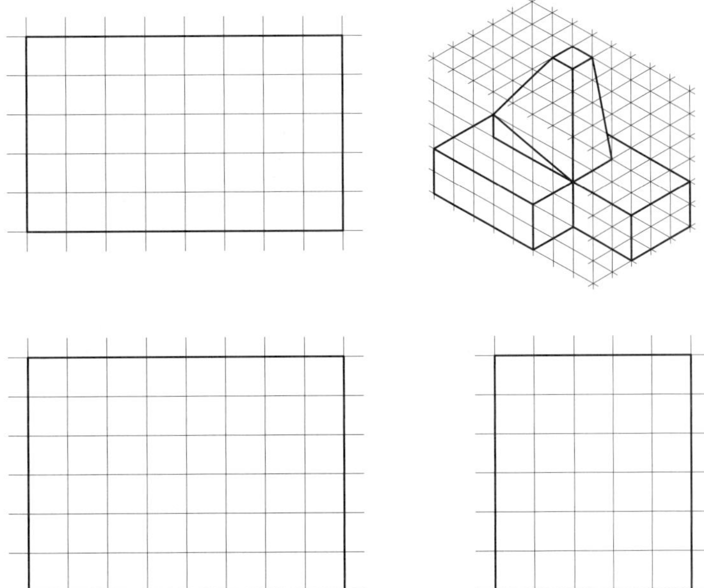

Problem 6-10

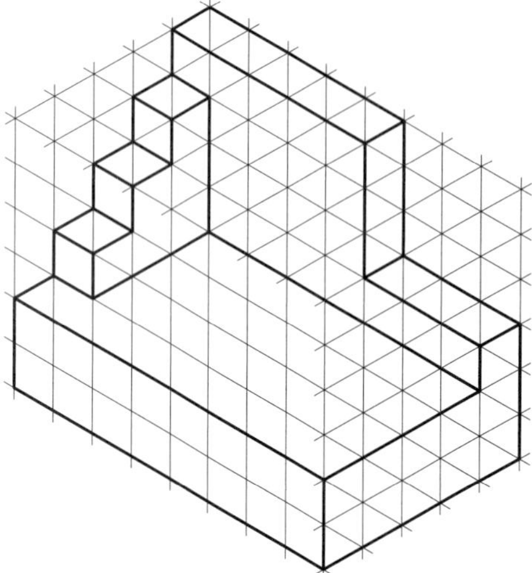

Problem 6-11

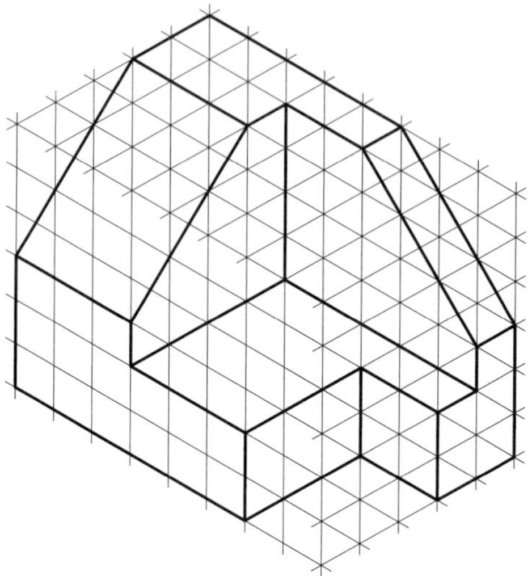

Problem 6-12

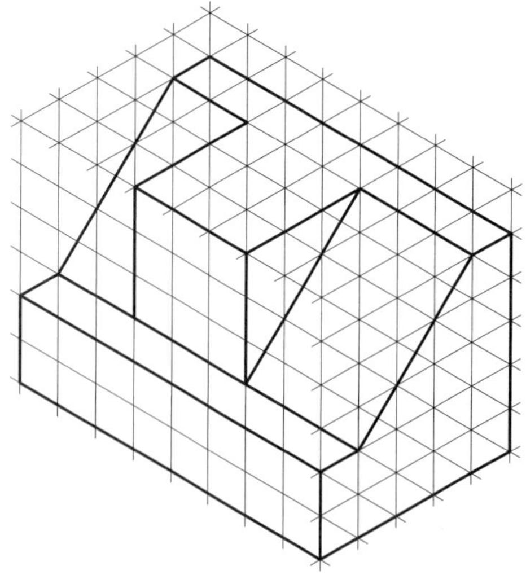

Problem 6-13

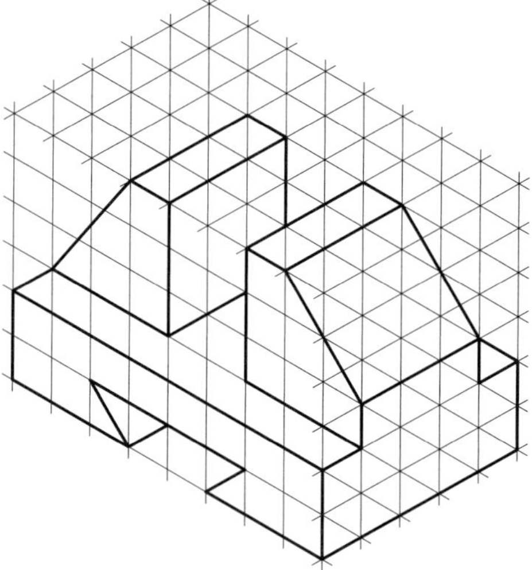

Problem 6-14

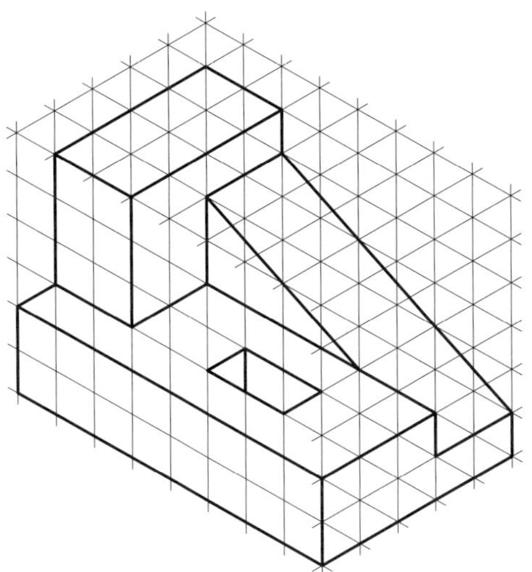

Problem 6-15

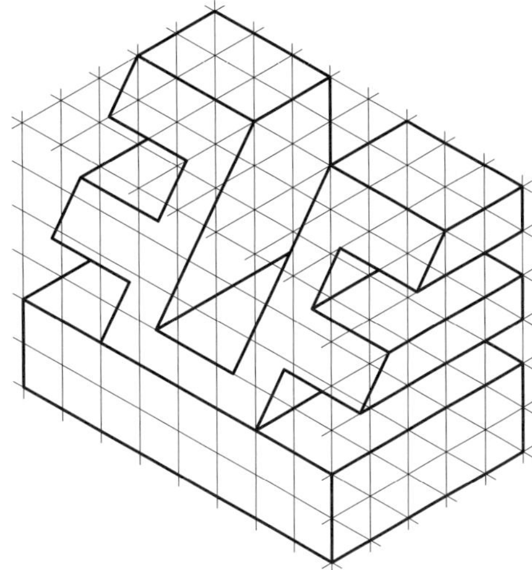

Problem 6-16

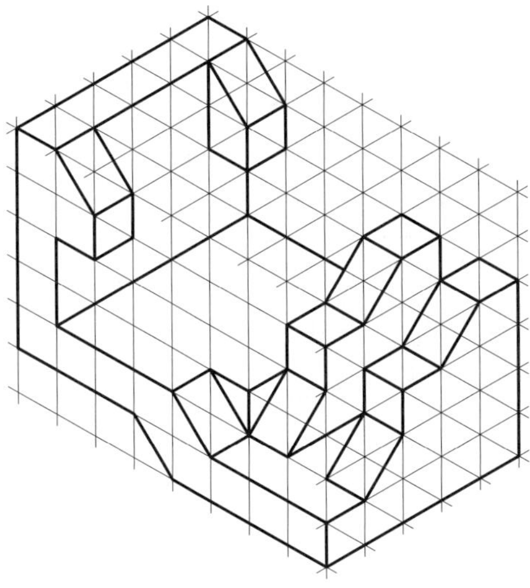

Problem 6-17

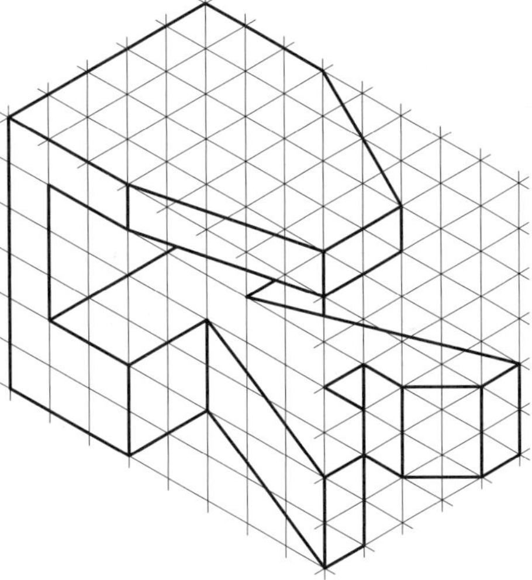

Problem 6-18

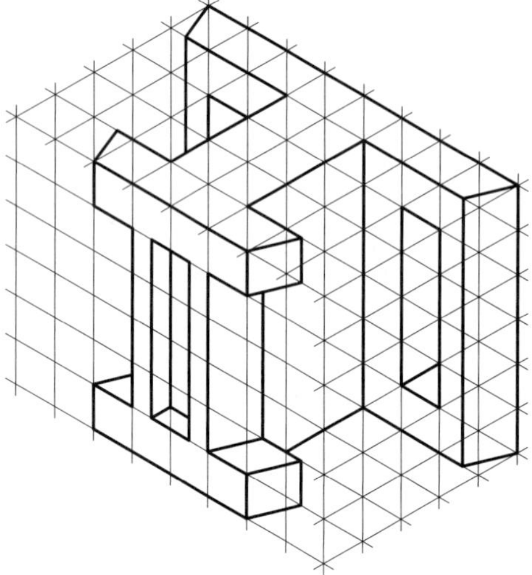

Problem 6-19

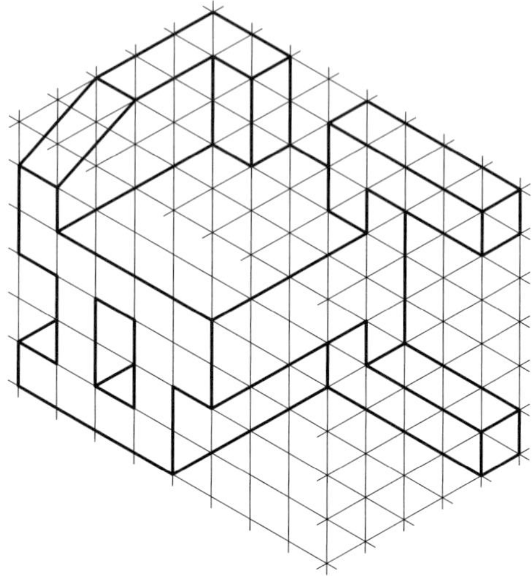

Problem 6-20

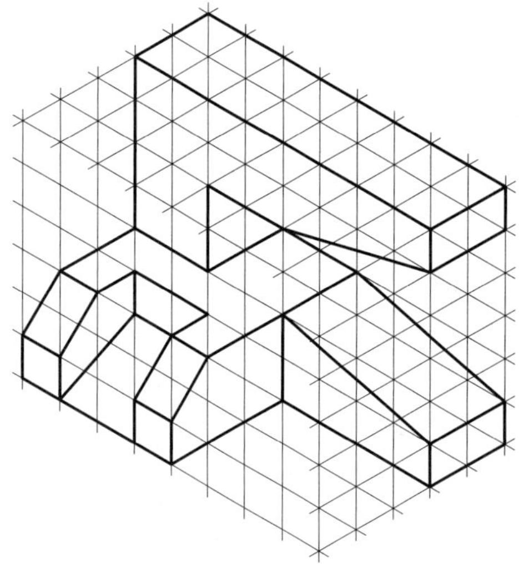

Problem 6-21

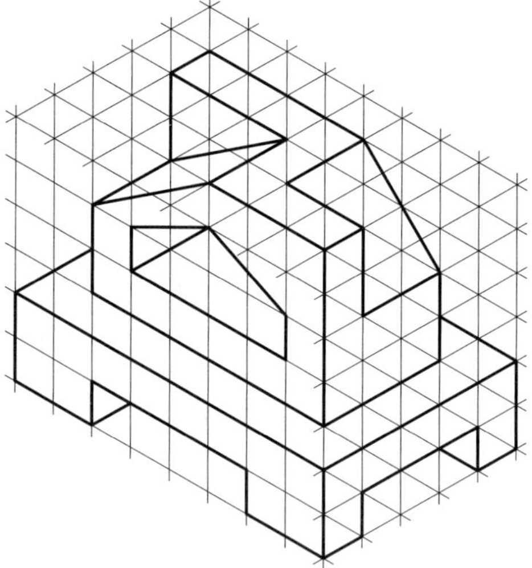

Problem 6-22

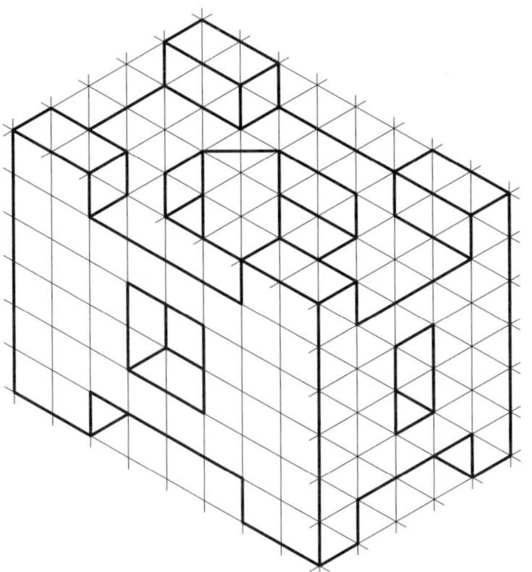

Problem 6-23

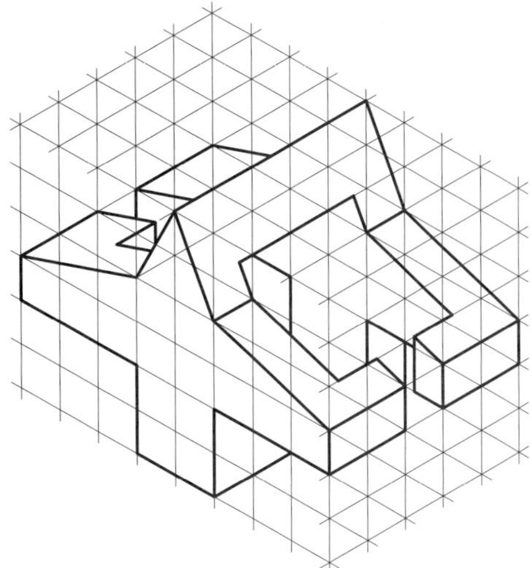

Problem 6-24

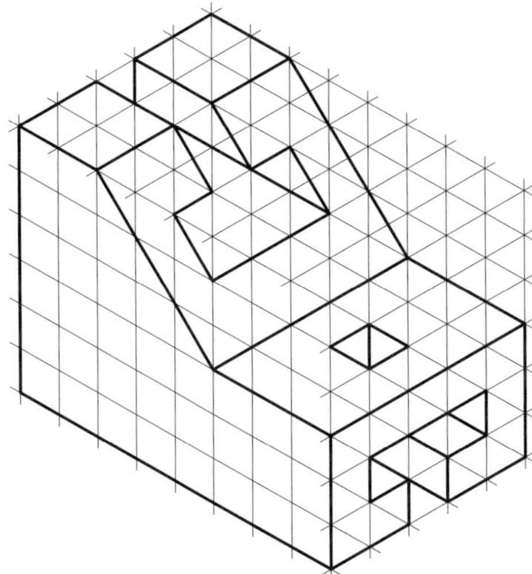

Problem 6-25

Part IV—Auxiliary Views
Exercises 26 through 30:

Use the isometric drawings and suitable linetypes to complete the required views. Select the view that will represent the best front view and sketch the required views, including an auxiliary view. Use suitable linetypes to complete the drawings. Use your sketch as guide to complete the multiview drawing and auxiliary view using AutoCAD. Save the drawing using the appropriate exercise number. Hand in your sketch attached to each drawing assignment. Assume each grid equals 0.25 units. Do not draw the isometric drawings or the grids. Do not try to place the dimensions on the drawing at this time.

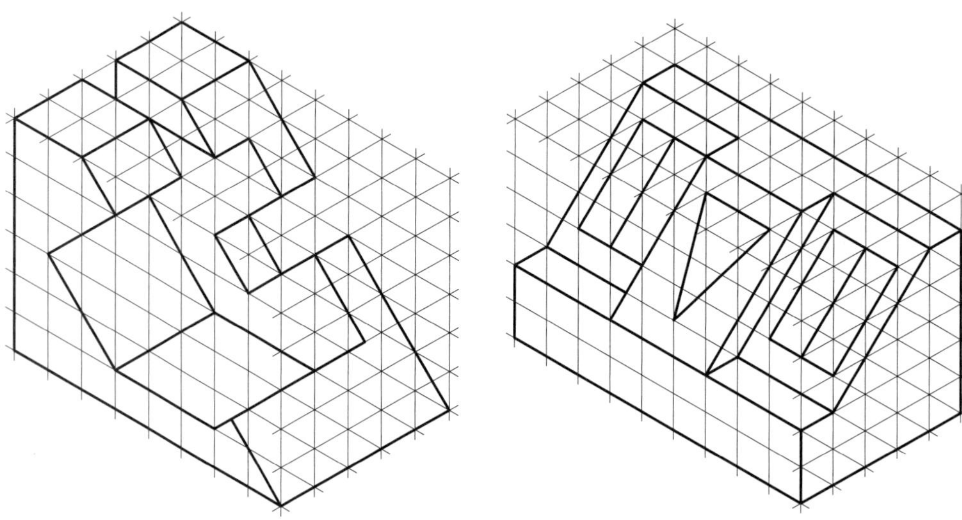

Problem 6-26 **Problem 6-27**

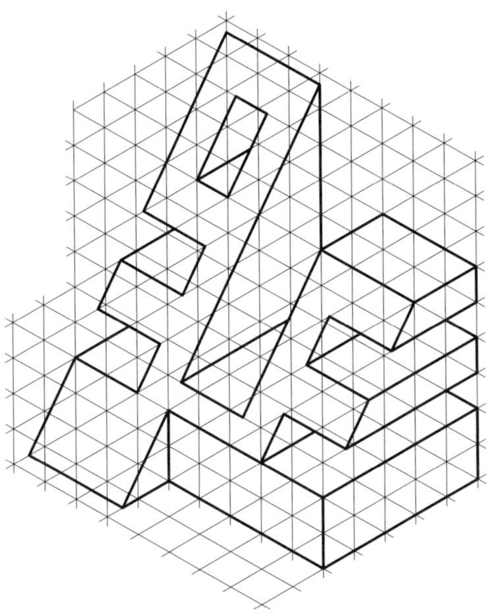

Problem 6-28

Problem 6-29

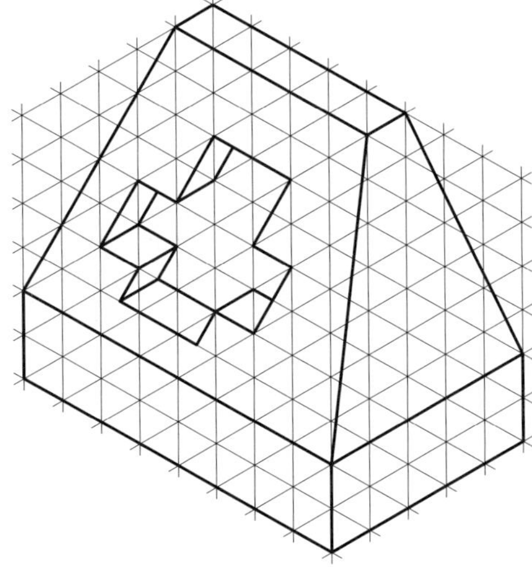

Problem 6-30

Part V—Isometric Sketches

Exercises 31 through 35:

Use the multiview drawings and suitable linetypes to complete the isometric sketch. Assume each grid equals 0.25 units. Do not draw the isometric drawings using AutoCAD at this time. Save the drawing using the proper exercise number.

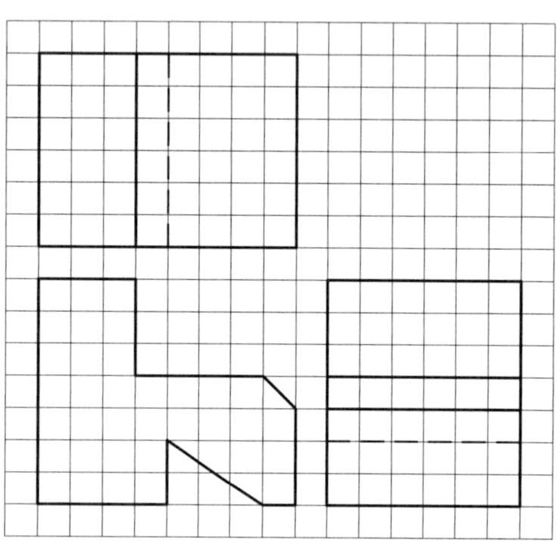

Problem 6-31

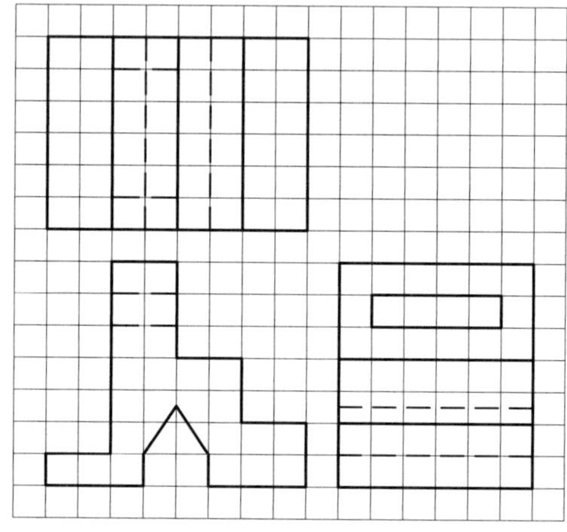

Problem 6-32

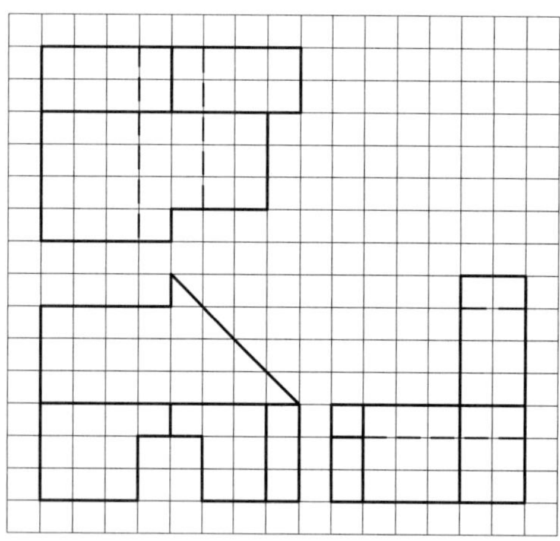

Problem 6-33

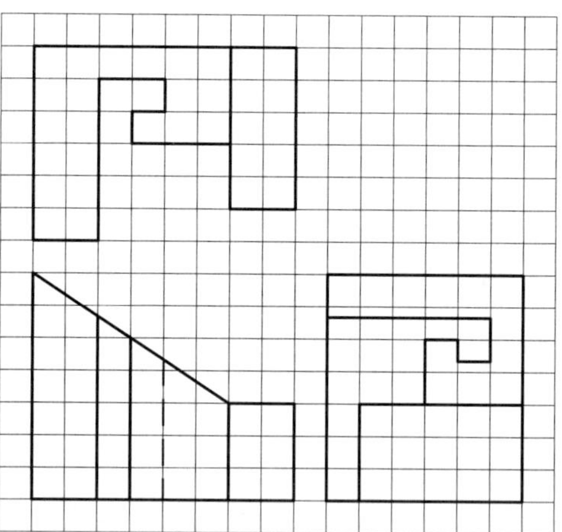

Problem 6-34

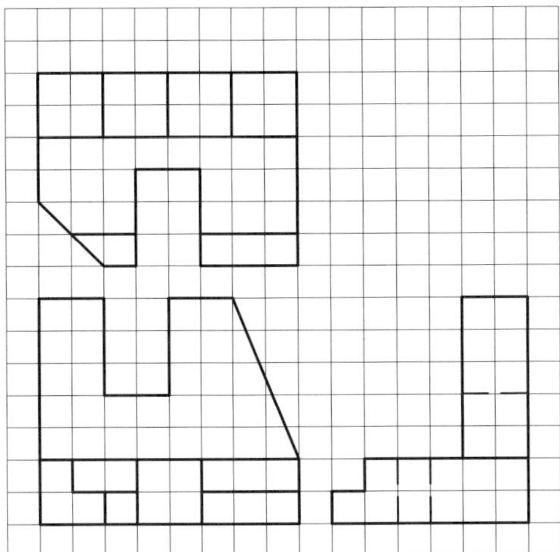

Problem 6-35

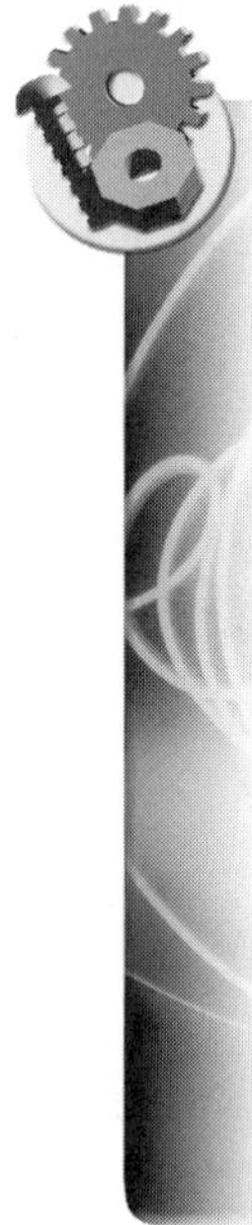

CHAPTER 6 QUIZ

1. What is the key to visualizing the projection of objects for multiview projections?
2. You're looking at a drawing and the views appear backwards. Other than bad layout, how would you explain this?
3. If a plane appears as a line in the front view, how would it appear in the top view? In the right side view?
4. Explain the difference between first and third angle projection.
5. The artist in you wants to place the top, front, and side views of a drawing so that they are neatly aligned in a side-by-side arrangement. What are the problems with this layout?
6. A surface is at an incline to the glass cube. How will appear in an orthographic projection?
7. How can OSNAP aid in drawing construction?
8. Why should a sketch be used when the drawing will be created using AutoCAD?
9. You're trying to pick a front view for a drawing layout. The shortest side shows the best contour, and the longest side would be filled with hidden surfaces. Which view would you use? Explain your answer.
10. What will be the deciding factor in choosing the number of views required to represent a part?
11. List the three common tools of sketching and the guidelines for selecting each.

12. How can the **Offset** command be used in drawing construction?
13. Explain the difference between **Fillet** and **Trim**.
14. An object contains an inclined surface that must be represented at 37° from vertical. Explain how to represent the surface in a sketch.
15. You've drawn the top and front views. How can you draw a side view without measuring?
16. When should an auxiliary view be created?
17. Describe how to create parallel lines when sketching.
18. Describe two principles to projecting auxiliary views.
19. How can **Snap** be used to create an auxiliary view?
20. How many surfaces should be projected to the auxiliary view?
21. What angle forms the base of an isometric drawing or sketch?
22. Summarize the process for deciding where to put an auxiliary view.
23. How is the true location of an auxiliary view indicated?
24. What is the difference between an orthographic projection and an isometric drawing?
25. Describe the differences in what will be seen when a point, a line, or a plane is rotated in orthographic projections.

Drawing Geometric Shapes

Lines and points may be the base of all drawings, but it's difficult to imagine a large engineering project without curved features. This chapter will introduce drawing methods for creating common geometric shapes, including circles, arcs, ellipses, polygons, and rectangles.

Commands to be introduced include

- **Dragmode**
- **Circle**
- **Arc**
- **Donut**
- **Fill**
- **Regen**
- **Ellipse**
- **Polygon**
- **Rectangle**

CIRCLES

Circles are found in a variety of projects throughout the engineering world. Circular features will be prevalent throughout your drawings that are drilled, reamed, or bored. Other round features that might be specified include a tapered hole, counterbore, spotface, countersink, counterdrill, threads or a boss. Each of these terms represents machined features that will be explored in Chapter 18. Figure 7.1 uses circles to represent drilled features. Start the **Circle** command by selecting the **Circle** button on the **Draw** toolbar, by selecting **Circle** from the **Draw** menu (as seen in Figure 7.2), or by keyboard. To draw a circle, select the **Circle** button on the **Draw**

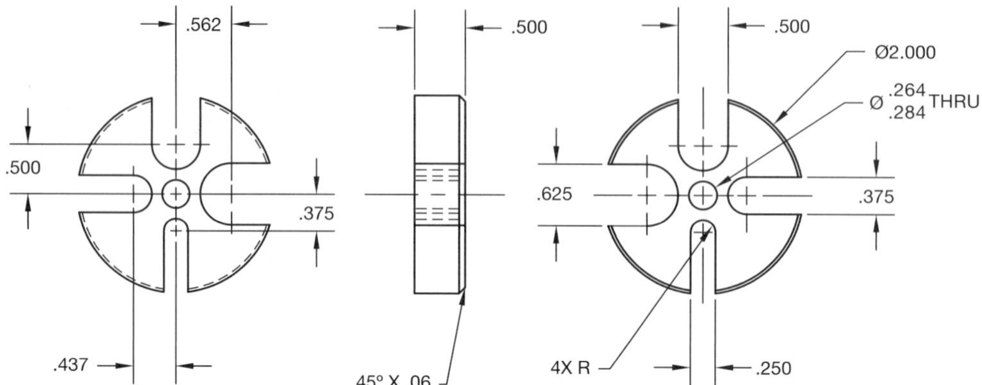

Figure 7.1 *Circles are used throughout mechanical drawings to represents holes, shafts, and many other circular features. Courtesy Cathy Schlesser.*

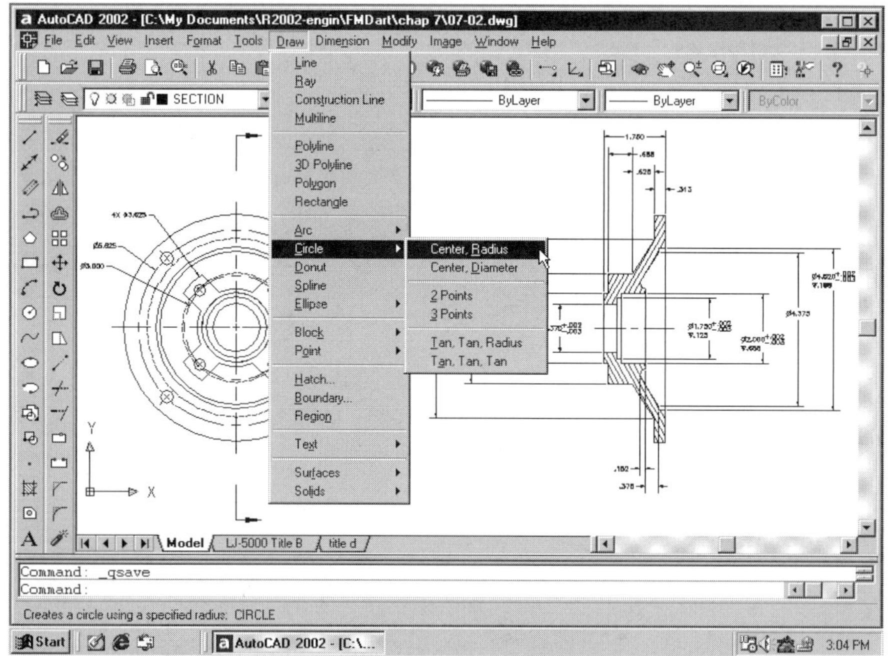

Figure 7.2 *To draw a circle, select the* **Circle** *button on the* **Draw** *toolbar, select* **Circle** *from the* **Draw** *menu, or type* **C** *(ENTER) at the Command prompt. Courtesy Michael J. Easley.*

toolbar, or type C (ENTER) at the Command prompt. Entering the **Circle** command on the toolbar or by keyboard will display the prompt:

Command: **C** ENTER *(Or click the Circle button on the Draw toolbar.)*

CIRCLE Specify center point for circle or [3P/2P/Ttr (tan tan radius)]:

AutoCAD is waiting for you to select a center point for the origin point of the circle. Once the center point is selected, you're prompted for the edge of a circle. The process can be seen in Figure 7.3.

Circles can be drawn by entering seven different combinations of information about the circle. Six of the circles are listed in the **Circle** submenu shown in Figure 7.2. The seventh, the **Donut** command on the **Draw** menu, produces a circle with varied line width and will be discussed later in this chapter.

Notice that once the center point is selected, a circle will appear on the screen, but the size will be altered as the mouse is moved. AutoCAD refers to this as drag or rubber band. Moving the mouse and changing the circle size is called dragging the circle into position. This drag image is typically helpful in determining the final locations of lines and other geometric shapes. If you find it distracting, type **DRAGMODE** (ENTER) at the Command prompt and select OFF to disable it.

Note: As you draw circular features, you might want to review the Display tab of the Options dialog box. The **Arc and Circle Smoothness** setting can be used to control the smoothness of the display of arcs, circles, and ellipses. The default value is 100, with a range of 1 through 20,000. A higher number will produce a smoother display but will require

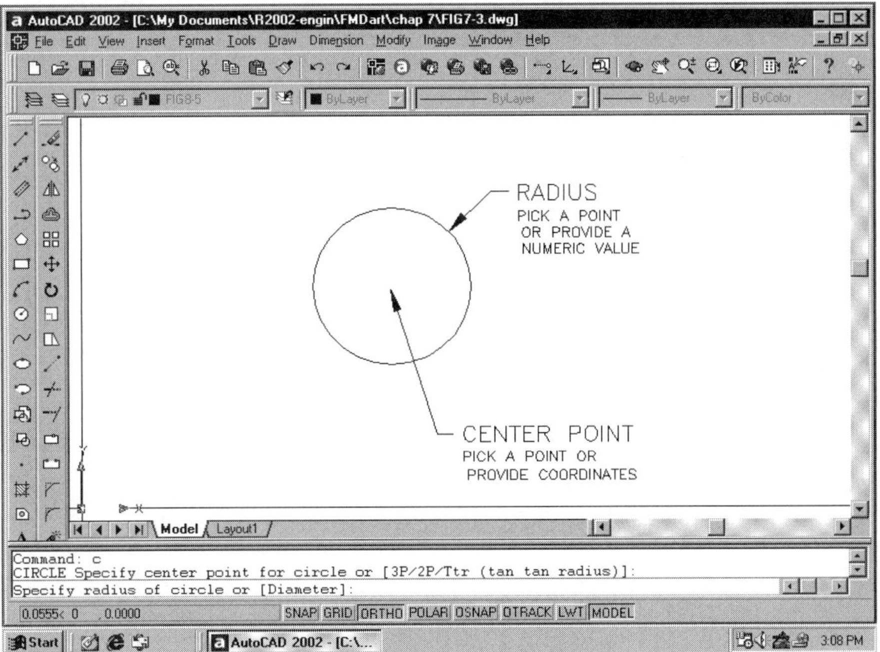

Figure 7.3 *Control the size of a circle by entering a value for the radius or the diameter.*

longer regeneration times on larger drawings. A lower number will display circular features with a series of straight lines. The true shape of objects will be displayed on plots, regardless of the resolution.

CENTER AND RADIUS

The default method of drawing a circle is to enter the center point and a radius. As the **Circle** command is started, enter the center point by typing coordinates or by selecting with the pointing device. In Figure 7.4 three circles have been drawn, and a fourth has been dragged into position. Notice that the rubber-band line extends from the chosen center point to the radius indicated by the crosshairs. Move the crosshairs to change the size of the circle. Type a numeric value and press ENTER to draw a circle with the desired radius. To draw a 4" diameter circle, the sequence is as follows:

Command: **C** ENTER *(Or click the Circle button on the Draw toolbar.)*

CIRCLE Specify center point for circle or [3P/2P/Ttr (tan tan radius)] *(Select center point.)*

Specify radius of circle or [Diameter]: **2** ENTER

Command:

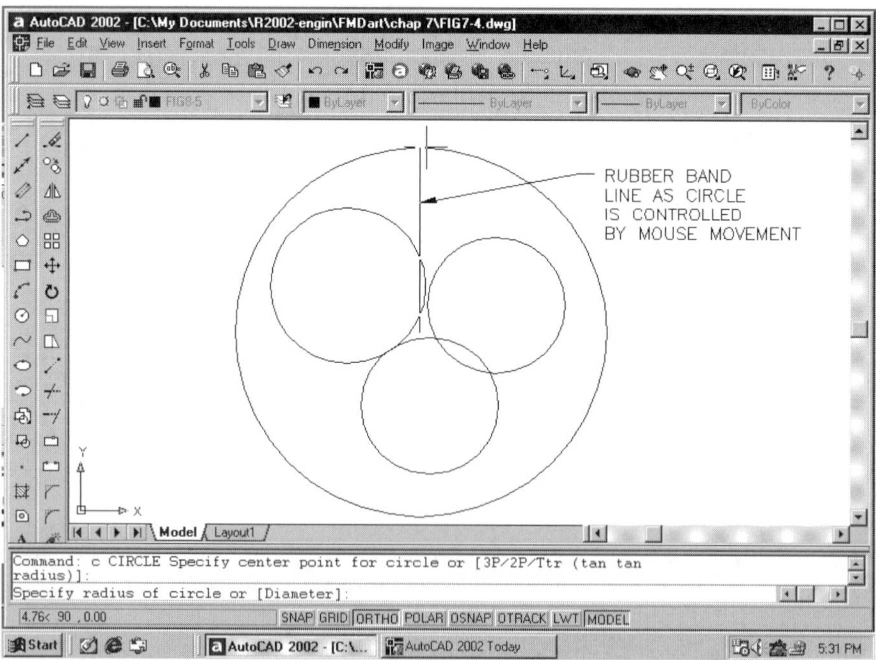

Figure 7.4 *Rather than entering a specific size, use the rubber-band line to see the effect on the circle to be drawn.*

Note: To draw another circle, continue the command by pressing ENTER or SHIFT, or right-clicking and then selecting **Repeat CIRCLE**.

CENTER AND DIAMETER

Sizes of circular features such as bolts, drilled features, and cylindrical features are typically given as a diameter. At the CIRCLE prompt, type D (ENTER) to switch the prompt from radius to diameter. When entered by toolbar or keyboard, the command sequence is as follows:

Command: **C** ENTER *(Or click the Circle button on the Draw toolbar.)*

CIRCLE Specify center point for circle or [3P/2P/Ttr (tan tan radius)] *(Select center point.)*

Specify radius of circle or [Diameter] <2.00>: **D** ENTER

Specify diameter of circle:<4.00> *(Enter a diameter value or move the crosshairs to the desired location.)*

Command:

The command sequence is similar when **Center, Diameter** is selected from the **Circle** submenu. AutoCAD simplifies the process with the menu so that the Diameter option is automatic.

Once the circle center is selected, the circle will be dragged across the screen but will not line up with the crosshairs. Figure 7.5 shows an example of a circle being created using the diameter. Notice that the length of the line extending from the center point is the diameter of the circle to be drawn. The line will disappear once the diameter is selected. The diameter entered will become the default setting for the next circle drawn.

TWO-POINT CIRCLES

Drawing a two-point circle can be useful if the diameter is known, but the center point is not. Draw the two-point circle by selecting a point and then selecting a point on the opposite side of the circle, as seen in Figure 7.6. The points can be selected with the crosshairs, or coordinates can be entered. The command sequence is as follows:

Command: **C** ENTER *(Or click the Circle button on the Draw toolbar.)*

CIRCLE Specify center point for circle or [3P/2P/Ttr (tan tan radius)]: **2P** ENTER

Specify first end point of circle's diameter: *(Enter coordinates or select a point.)*

Specify second end point of circle's diameter: *(Enter coordinates or select a point.)*

Command:

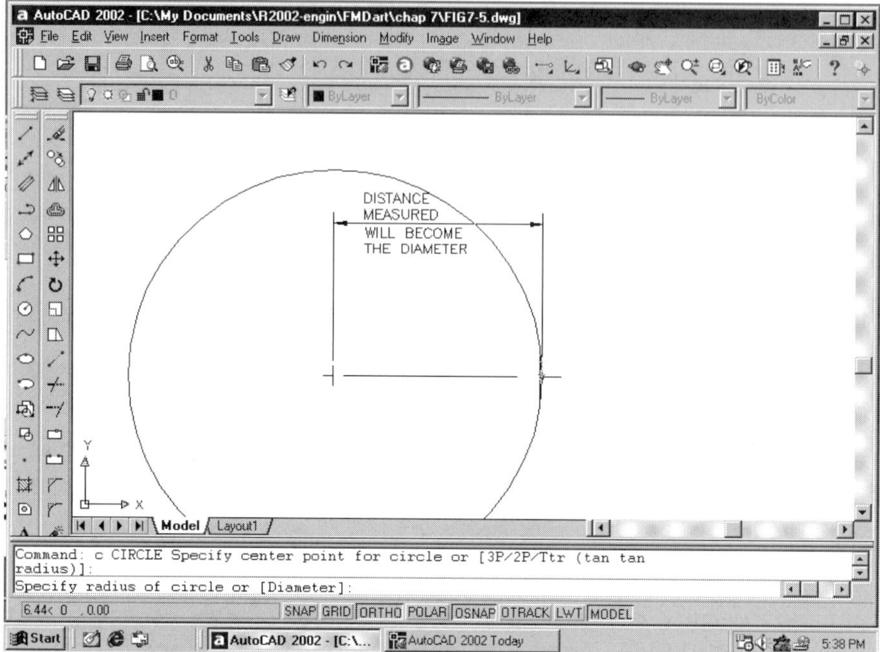

Figure 7.5 *With the Diameter option, the distance that the cursor is moved from the center point will become the diameter of the circle.*

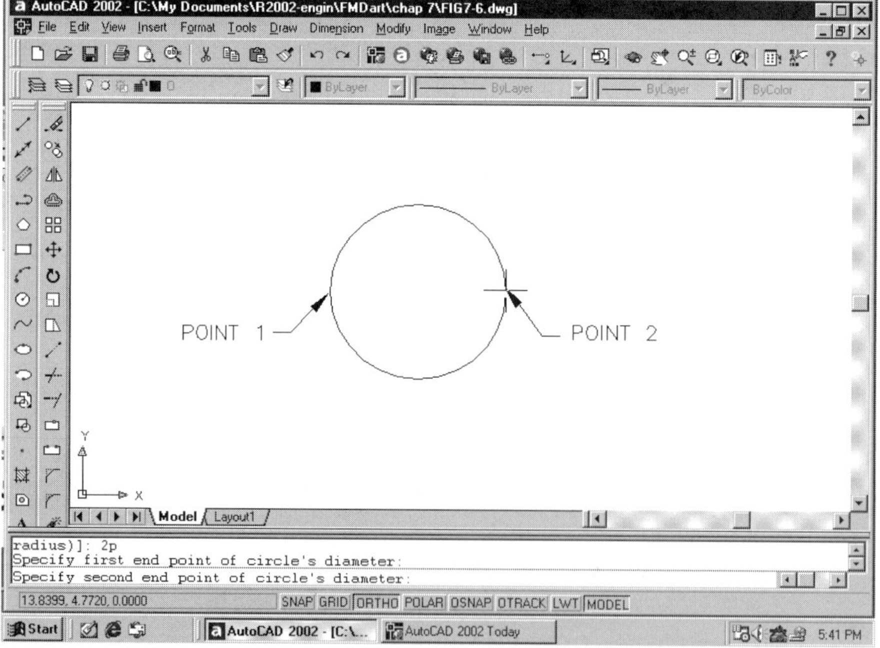

Figure 7.6 *A circle can be drawn by specifying the endpoints of a line that would be the diameter.*

THREE-POINT CIRCLES

If certain points of a circle are known, a circle can be drawn by entering any three points. Figure 7.7 shows an example of a three-point circle. The points can be selected with the crosshairs, or coordinates can be entered. The process for drawing a three-point circle is as follows:

Command: **C** ENTER *(Or click the Circle button on the Draw toolbar.)*

CIRCLE Specify center point for circle or [3P/2P/Ttr (tan tan radius)]: **3P** ENTER

Specify first point on circle: *(Enter coordinates or select a point.)*

Specify second point on circle: *(Enter coordinates or select a point.)*

Specify third point on circle: *(Enter coordinates or select a point.)*

Command:

TANGENT, TANGENT, AND RADIUS

An object is tangent if it intersects another object at one and only one point. A circle can be drawn tangent to existing features by selecting two features that the circle will be tangent to and then specifying a radius. This can be seen in Figure 7.8. The command sequence is as follows:

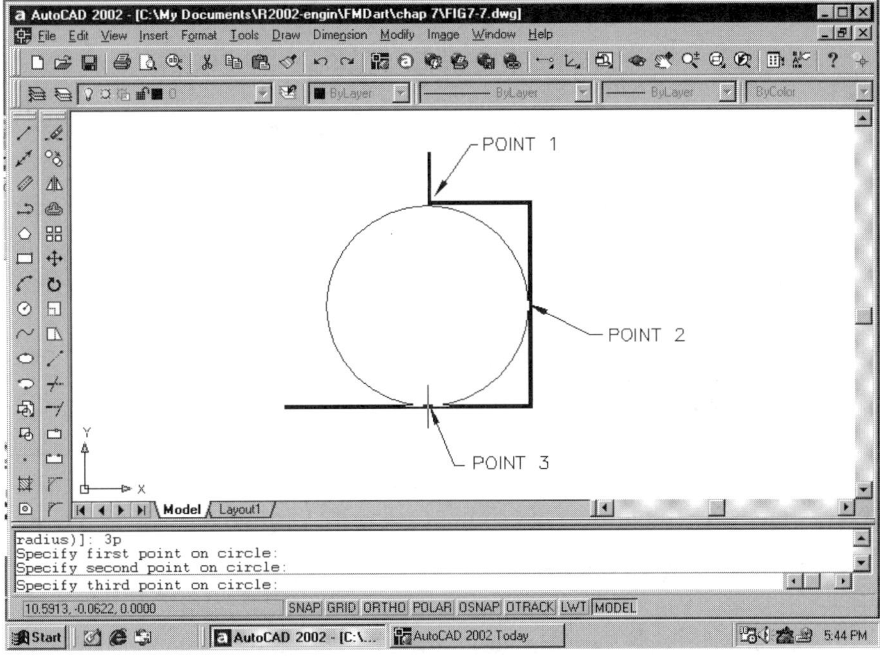

Figure 7.7 *The size of a circle can be determined by specifying three tangent points.*

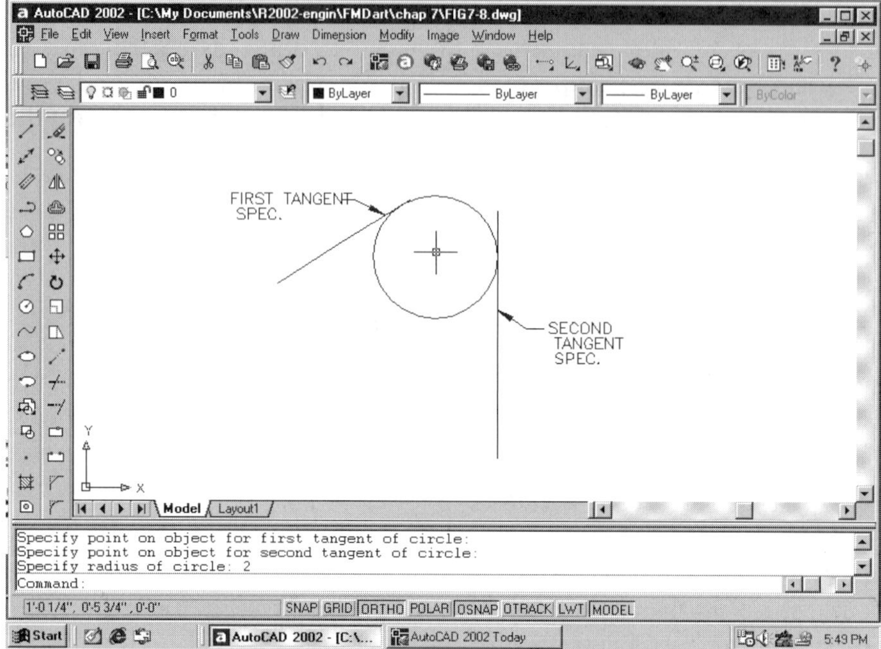

Figure 7.8 *A circle can be drawn by specifying two tangent surfaces and a radius.*

Command: **C** ENTER *(Or click the Circle button on the Draw toolbar.)*

CIRCLE Specify center point for circle or [3P/2P/Ttr (tan tan radius)]: **TTR** ENTER

Specify point on object for first tangent of circle: *(Select a line or circle.)*

Specify point on object for second tangent of circle: *(Select a line or circle.)*

Specify radius of circle: <2.00> *(Enter a numeric value or select a point.)* ENTER

Specify second point: *(Enter a numeric value or select a point.)* ENTER

Command:

In Figure 7.9 each line was selected as a tangent point and a radius was entered. Because the radius is smaller than the distance between the lines, the first line does not touch the circle. If the line were extended, it would be tangent to the circle.

TANGENT, TANGENT, TANGENT

This option works well for drawing a circle that is tangent to three objects. This option can be accessed only through **Tan, Tan, Tan** on the **Circle** submenu of the **Draw** menu. The command sequence is as follows:

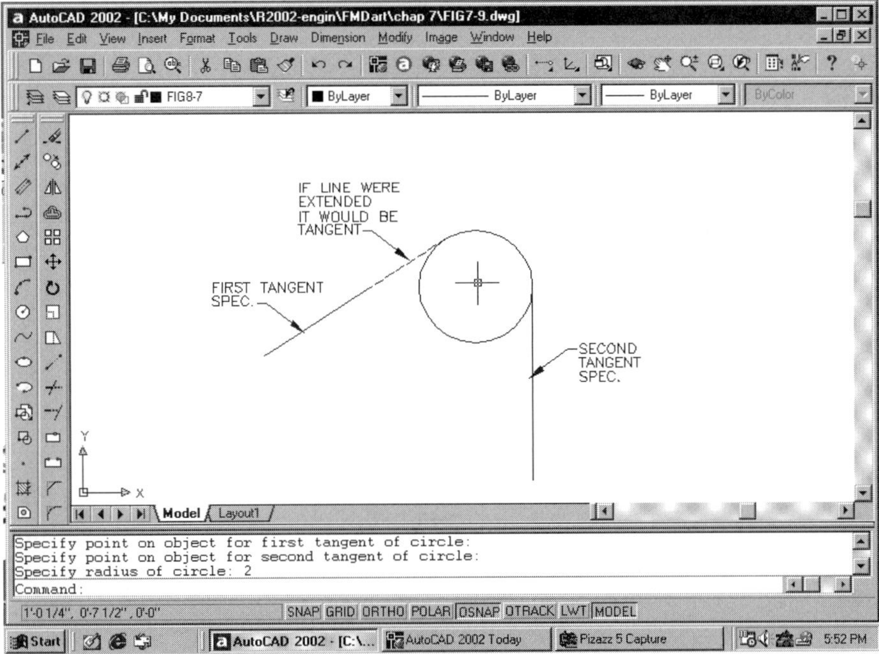

Figure 7.9 *If a radius is selected that is not large enough for the specified circle to touch each line, a circle is drawn tangent to one line and tangent to where the other line would be if it were extended.*

Command: *(Select Tan, Tan, Tan from Circle on the Draw menu.)*

_circle Specify center point for circle or [3P/2P/Ttr (tan tan radius)]: _

_3p Specify first point on circle: _tan to *(Select the first object the circle is to be tangent to.)*

Specify second point on circle: _tan to *(Select the second object the circle is to be tangent to.)*

Specify third point on circle: _tan to *(Select the third object the circle is to be tangent to.)*

Command:

Figure 7.10 shows the development of a circle tangent to three surfaces.

CIRCLE@

This option allows a circle to be drawn using the last endpoint of the last line drawn for the center point of a new circle. The option is best used for drawing concentric circles associated with pipes and circular columns. The command sequence is as follows:

Command: **C** ENTER *(Or click the Circle button on the Draw toolbar.)*

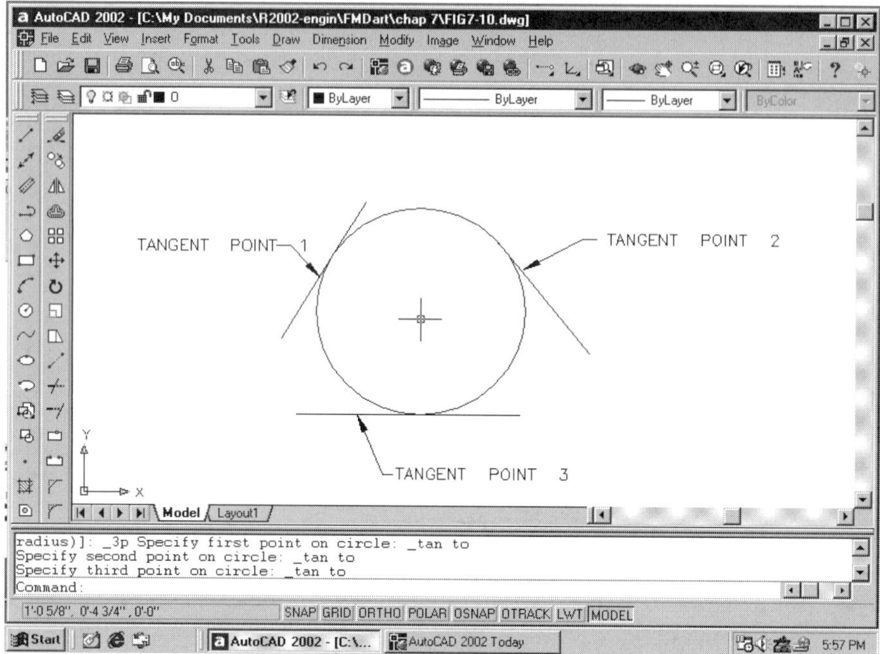

Figure 7.10 *Tan,Tan,Tan can be used to draw circles that are tangent to three lines.*

CIRCLE 3P/2P/Ttr/<Centerpoint>: @ ENTER

Diameter/<radius> <1.00> *(Specify desired diameter.)*

Command:

DRAWING ARCS

Arcs are used where a full circle is not required. Curved components are found throughout the object seen in Figure 7.11. Start the command by selecting the **Arc** button on the **Draw** toolbar or by selecting **Arc** from the **Draw** menu. You can also start the command by typing A (ENTER) at the Command prompt. The menu can be seen in Figure 7.12. Eleven different options are available for drawing arcs, depending on what information is available or on personal preference. Selecting **Arc** on the toolbar will produce the prompt:

Command: _arc Specify start point of arc or [Center]:

The default setting of the toolbar and keyboard is the three-point arc. With the exception of the three-point arc, you can specify each arc drawing method by typing a letter and pressing ENTER or SPACEBAR when prompted for a point of the arc. Letters and the function of the arc they represent are as follows:

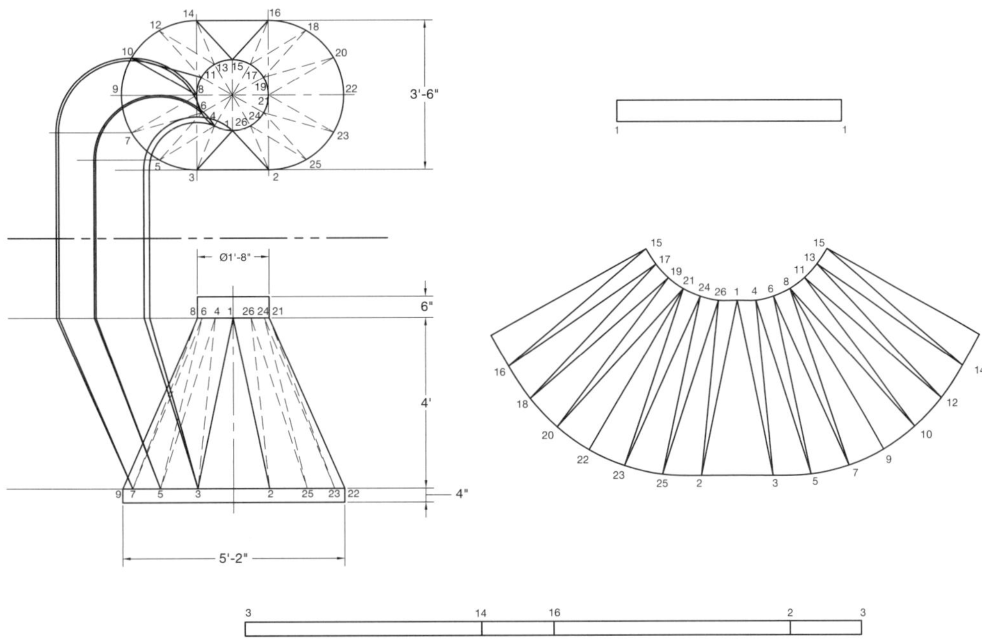

Figure 7. 11 *Arcs are found throughout many manufactured parts. Courtesy Jill Heilman.*

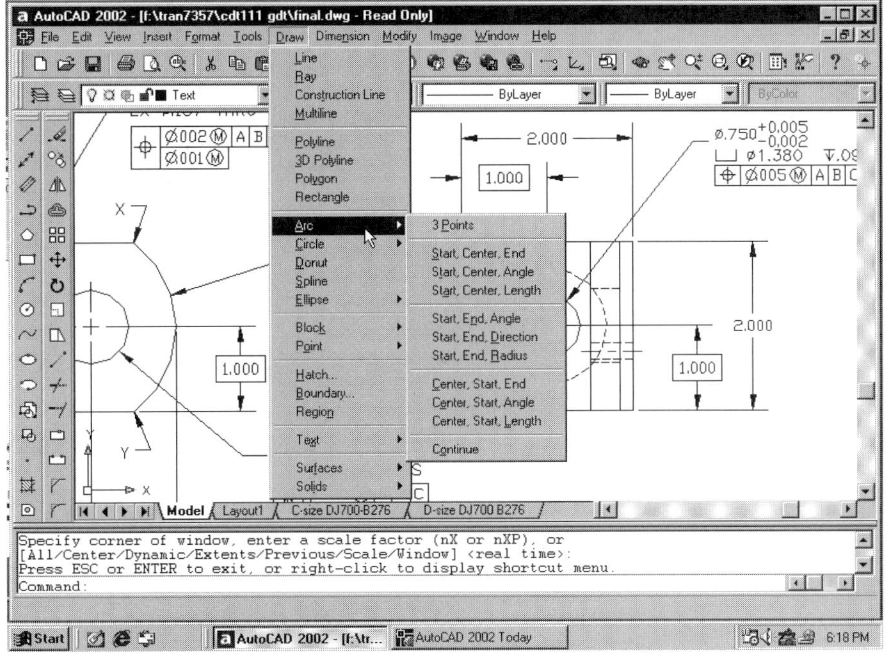

Figure 7.12 *The* **Arc** *menu. Courtesy Hanh Tran.*

A: included Angle

C: Center

D: starting Direction

E: Endpoint

L: Length of chord

R: Radius

Some of the terms used to describe an arc can be seen in Figure 7.13 Notice that "C" describes the center of the circle that the arc is part of, not the center of the arc. When the **Arc** command is started from the menu and the method is specified, prompts are automatically provided to match the option selected.

THREE-POINT ARCS

Drawing a three-point arc is similar to drawing a 3P circle and can be seen in Figure 7.14. The first and third points are the arc endpoints; the second point is any point on the arc. The command sequence for a three-point arc is as follows:

Command: **A** ENTER *(Or click the Arc button on the Draw toolbar.)*

ARC Specify start point of arc or [Center]: *(Select any point or enter coordinate.)*

Specify second point of arc or [Center/ENd]: *(Select second point.)*

Specify end point of arc: *(Select the arc's endpoint or enter coordinates.)*

Command:

Notice that once the second point is selected, the arc will drag across the screen with the crosshairs. The arc will continue to change size until the endpoint is selected.

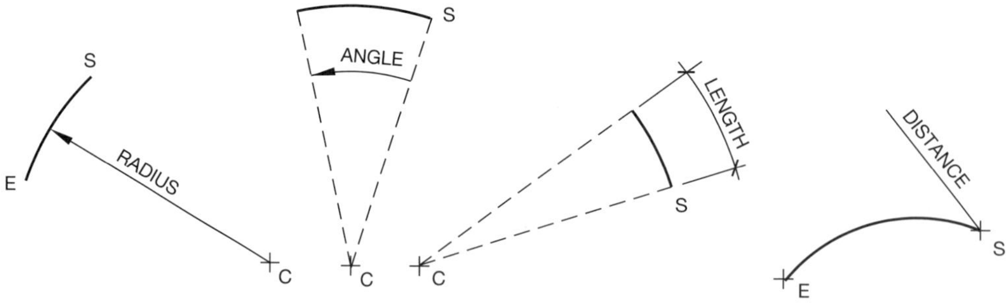

Figure 7.13 *Terms used to describe arcs are Start (S), Center (C), End (E), Angle, Length, Radius, and Distance.*

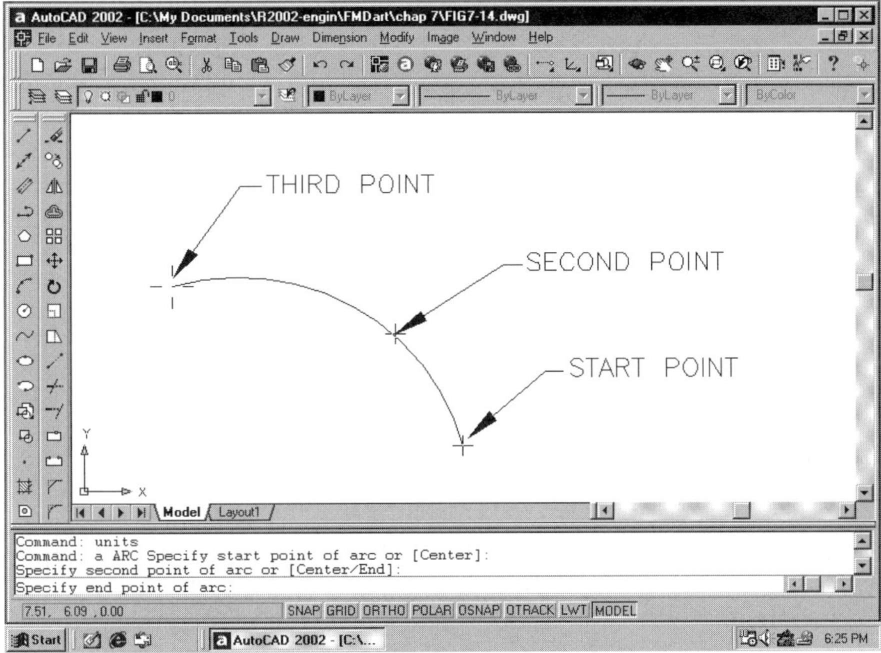

Figure 7.14 *Drawing a three-point arc is similar to drawing a three-point circle.*

START, CENTER, END ARCS

When the ends and center of an arc are known, the Start, Center, End arc option can be used. The command sequence is as follows:

> Command: **A** ENTER *(Or click the Arc button on the Draw toolbar.)*
>
> ARC Specify start point of arc or [Center]: *(Select any point or enter coordinate.)*
>
> Specify second point of arc or [Center/End]: **C** ENTER
>
> Specify center point of arc: *(Select second point.)*
>
> Specify end point of arc [Angle/chord Length]: *(Select the arc's endpoint or enter coordinates.)*
>
> Command:

As the last point is entered, the arc will be drawn and the Command prompt will be returned. The process can be seen in Figure 7.16. Notice that as the last point is entered, the arc cannot end at the point entered. The endpoint that was specified is used only to determine the angle at which the arc will end. The actual radius was determined by the start and center points.

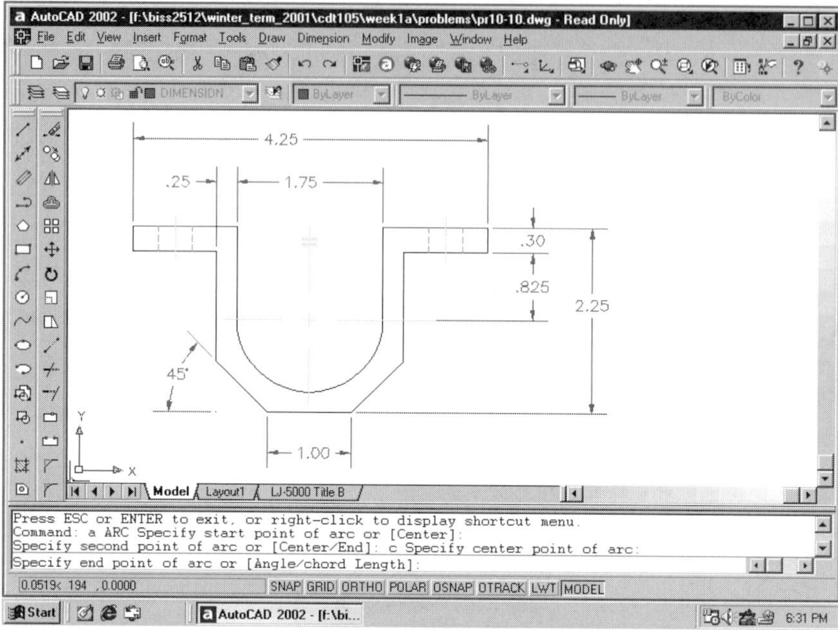

Figure 7.15 *An arc drawn using the Start, Center, End option. Courtesy Paul Bissell.*

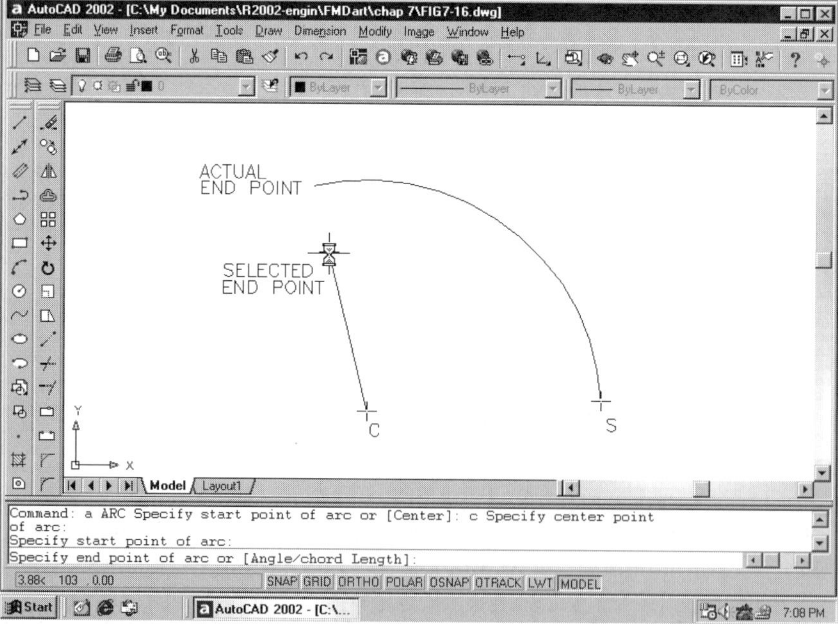

Figure 7.16 *The selected endpoint is used to determine the angle of the specified circle. The actual size of the circle is determined by the distance from the start to the center point.*

START, CENTER, ANGLE ARCS

It often is necessary to draw an arc using lengths or the included angle. If a positive value is used for the angle, the angle will be measured in a counterclockwise direction. If a negative value is used, the angle will be measured in a clockwise direction. The effects of positive and negative values can be seen in Figure 7.17. The command sequence for a Start, Center, Angle arc is as follows:

Command: **A** ENTER *(Or click the Arc button on the Draw toolbar.)*

ARC Specify start point of arc or [Center]: *(Select any point or enter coordinates.)*

Specify second point of arc or [Center/End]: **C** ENTER

Specify center point of arc: *(Select center point of arc.)*

Specify end point of arc [Angle/chord Length]: **A** ENTER

Specify included angle: **30** ENTER

Command:

When the desired angle is entered, the arc will be drawn and the command line will return.

START, CENTER, LENGTH ARCS

A chord is the straight line connecting the endpoints of an arc. These arcs are drawn by specifying the start and center points and then supplying the chord length. A positive or negative value must be entered with the length. Figure 7.18 shows the difference between a chord of +3" or –3". The command sequence for a Start, Center, Length arc is as follows:

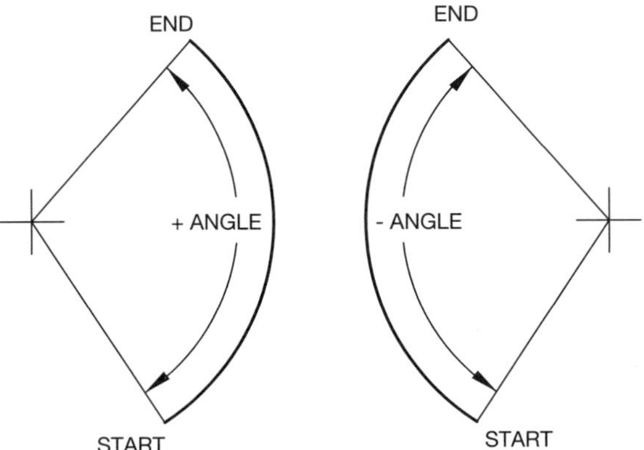

Figure 7.17 *The direction of layout is controlled by entering a positive (counterclockwise) or negative (clockwise) angle when drawing arcs with the Start, Center, Angle arc option.*

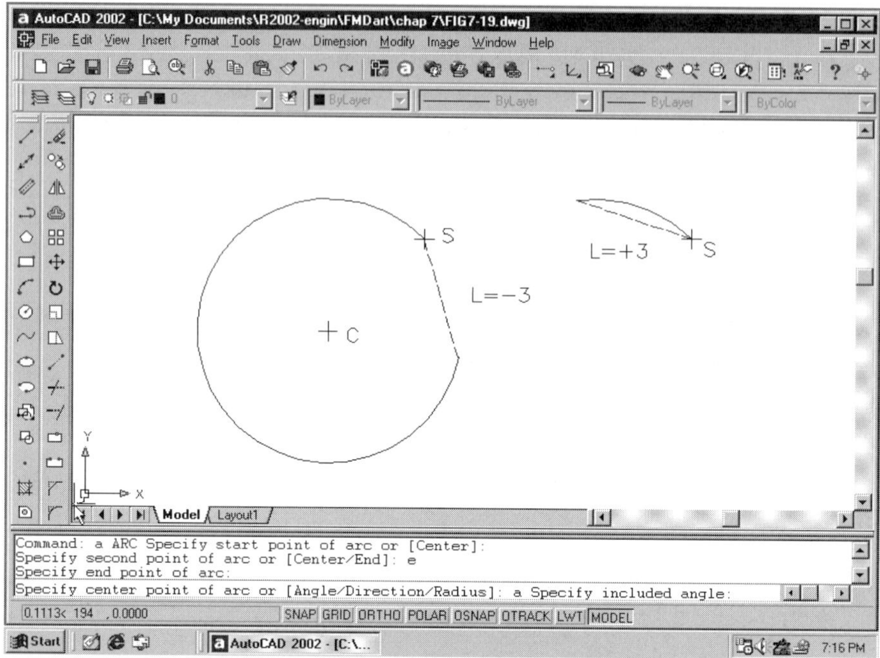

Figure 7.18 *Entering a positive or negative chord length determines which portion of the arc will be drawn using the Start, Center, Length arc option.*

Command: **A** ENTER *(Or click the Arc button on the Draw toolbar.)*

ARC Specify start point of arc or [Center]: *(Select start point or enter coordinates.)*

Specify second point of arc or [Center/ENd]: **C** ENTER

Specify center point of arc: *(Select any point or enter coordinates.)*

Specify end point of arc [Angle/chord Length]: **L** ENTER

Specify length of chord: **3** ENTER

Command:

When a length is specified or a point is selected, the arc will be drawn and the command line will return.

START, END, ANGLE ARCS

An arc drawn using Start, End, and included Angle is drawn counterclockwise from the start point to the endpoint if a positive angle value is used. If a negative angle is used, the angle will be drawn clockwise. The results of coordinate entry can be seen in Figure 7.19. The command sequence is as follows:

Command: **A** ENTER *(Or click the Arc button on the Draw toolbar.)*

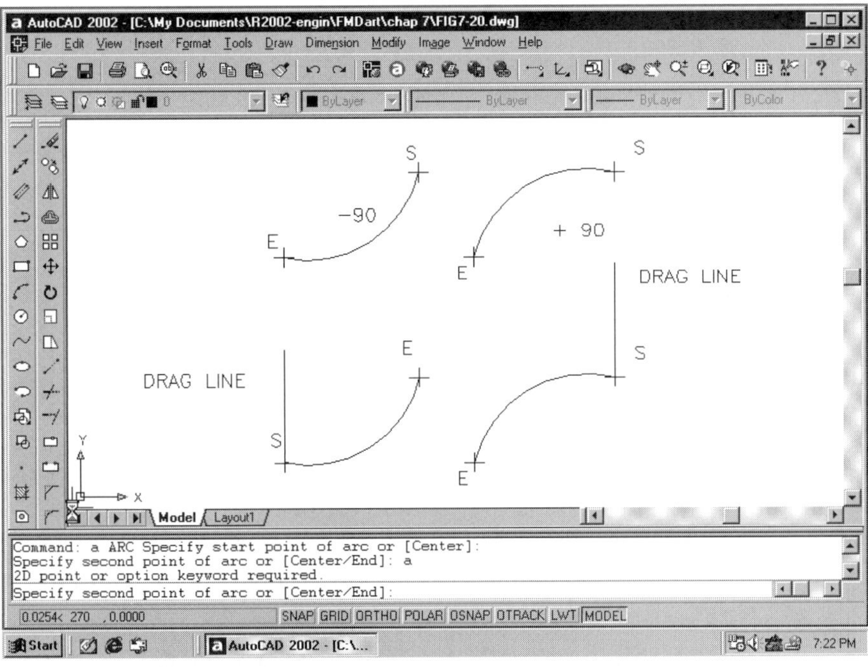

Figure 7.19 *Entering a positive or negative angle determines which direction an arc will be drawn using the Start, End, Angle arc option.*

ARC Specify start point of arc or [Center]: *(Select start point or enter coordinates.)*

Specify second point of arc or [Center/End]: **E** ENTER

Specify end point of arc: *(Select the desired endpoint or enter coordinates.)*

Specify center point of arc [Angle/chord Length]: **A** ENTER

Specify Included angle: *(Select desired angle.)*

Command:

When a length is specified or a point is selected, the arc will be drawn and the command line will return.

START, END, DIRECTION ARCS

The Start, End, Direction option allows the direction of the arc to be selected. Major or minor arcs can be drawn in any orientation to the start point. Effects can be seen in Figure 7.20. The command sequence is as follows:

Command: **A** ENTER *(Or click the Arc button on the Draw toolbar.)*

ARC Specify start point of arc or [Center]: *(Select start point or enter coordinates.)*

Specify second point of arc or [Center/End]: **E** ENTER

Specify end point of arc: *(Select the desired endpoint or enter coordinates.)*

Specify center point of arc or [Angle/Direction/Radius]: **D** ENTER

Specify tangent direction for start of arc: *(Select desired direction.)*

Command:

When a length is specified or a point is selected, the arc will be drawn and the command line will return.

START, END, RADIUS ARCS

Start, End, Radius arcs can only be drawn counterclockwise. Because the variables can produce different options, the arc is drawn counterclockwise from the start point. Entering a positive or negative value for the radius will determine if the major or minor arc will be drawn. The effects of the value can be seen in Figure 7.21. The command sequence is as follows:

Command: **A** ENTER *(Or click the Arc button on the Draw toolbar.)*

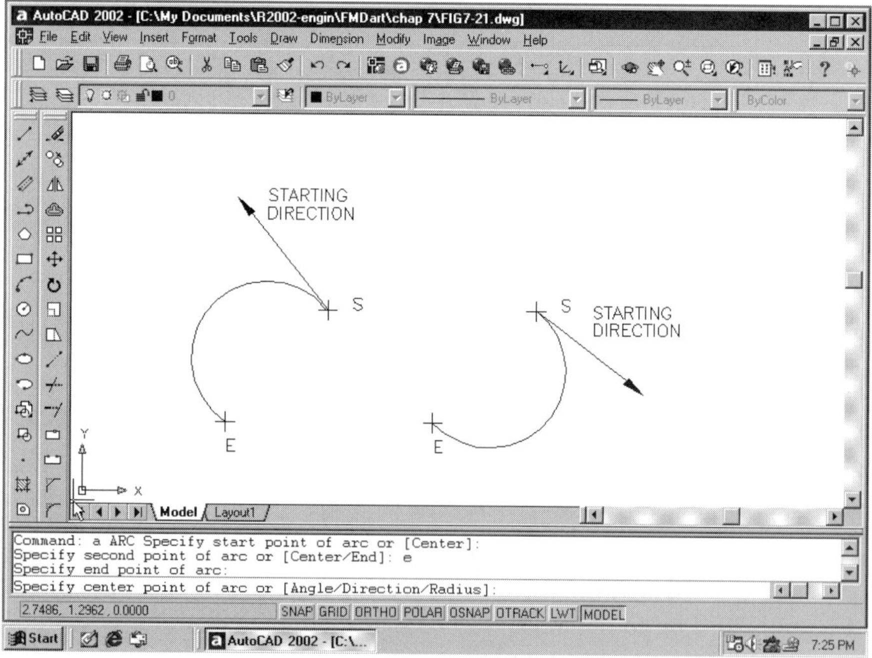

Figure 7.20 *The Start, End, Direction arc option allows the direction of the arc to be selected.*

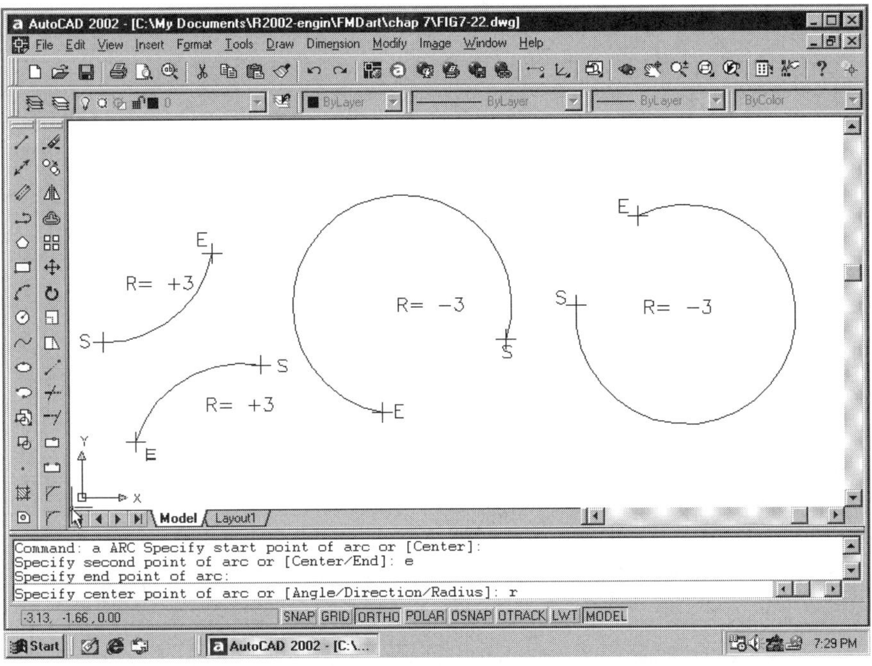

Figure 7.21 *Entering a positive radius determines which portion of the arc will be drawn using the Start, End, Radius arc option. The direction will always be counterclockwise.*

> ARC Specify start point of arc or [Center]: *(Select start point or enter coordinates.)*
>
> Specify second point of arc or [Center/End]: **E** ENTER
>
> Specify end point of arc: *(Select the desired endpoint or enter coordinates.)*
>
> Specify center point of arc or [Angle/Direction/Radius]: **R** ENTER
>
> Specify radius of arc: *(Select desired radius or enter a point.)*
>
> Command:

When a length is specified or a point is selected, the arc will be drawn and the command line will return (see Figure 7.22).

CENTER, START, END ARCS

The Center, Start, End option is similar to the Start, Center, End option. The effects of each command can be seen in Figure 7.23. The command sequence is as follows:

> Command: **A** ENTER *(Or click the Arc button on the Draw toolbar.)*
>
> ARC Specify start point of arc or [Center]: **C** ENTER

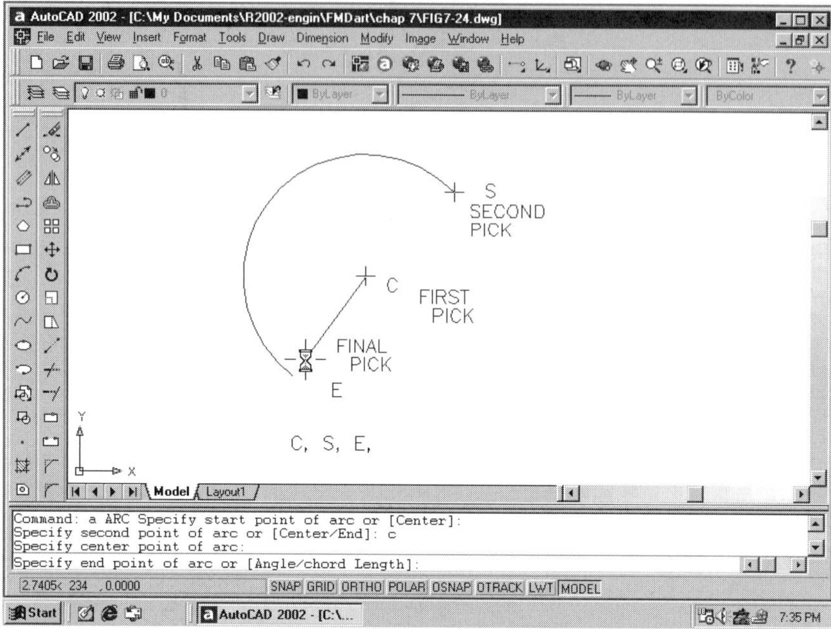

Figure 7.22 *The Center, Start, End arc option is similar to the Start, Center, End option.*

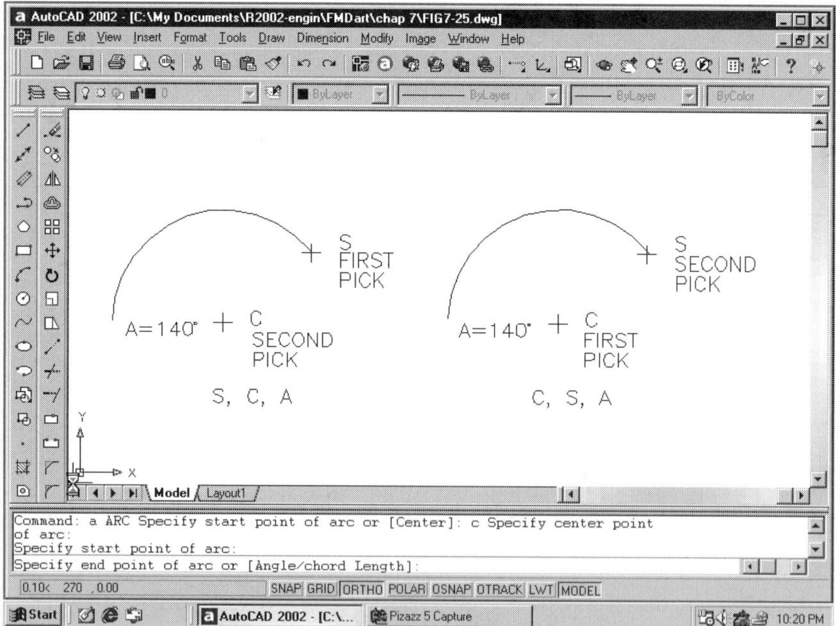

Figure 7.23 *The Center, Start, Angle arc option produces results similar to the Start, Center, Angle option.*

Specify center point of arc: *(Select center point or enter coordinates.)*

Specify start point of arc: *(Select the start point or enter coordinates.)*

Specify end point of arc or [Angle/chord Length]: *(Select the desired endpoint or enter coordinates.)*

Command:

When a length is specified or a point is selected, the arc will be drawn and the command line will return.

CENTER, START, ANGLE ARCS

The Center, Start, Angle option is similar to the Start, Center, Angle option. The effects of each command can be seen in Figure 7.23. The command sequence for this arc is as follows:

Command: **A** ENTER *(Or click the Arc button on the Draw toolbar.)*

ARC Specify start point of arc or [Center]: **C** ENTER

Specify center point of arc: *(Select center point or enter coordinates.)*

Specify start point of arc: *(Select the start point or enter coordinates.)*

Specify end point of arc or [Angle/chord Length]: **A** ENTER

Specify included angle: *(Select the desired angle.)*

Command:

When a length is specified or a point is selected, the arc will be drawn and the command line will return.

CENTER, START, LENGTH OF CHORD ARCS

The Center, Start, Length of chord option is similar to the Start, Center, Length option. The effects of each command can be seen in Figure 7.24. The command sequence is as follows:

Command: **A** ENTER *(Or click the Arc button on the Draw toolbar.)*

ARC Specify start point of arc or [Center]: **C** ENTER

Specify center point of arc: *(Select center point or enter coordinates.)*

Specify start point of arc: *(Select the start point or enter coordinates.)*

Specify end point of arc or [Angle/chord Length]: **L** ENTER

Specify length of chord: *(Select desired length.)*

Command:

When a length is specified or a point is selected, the arc will be drawn and the command line will return.

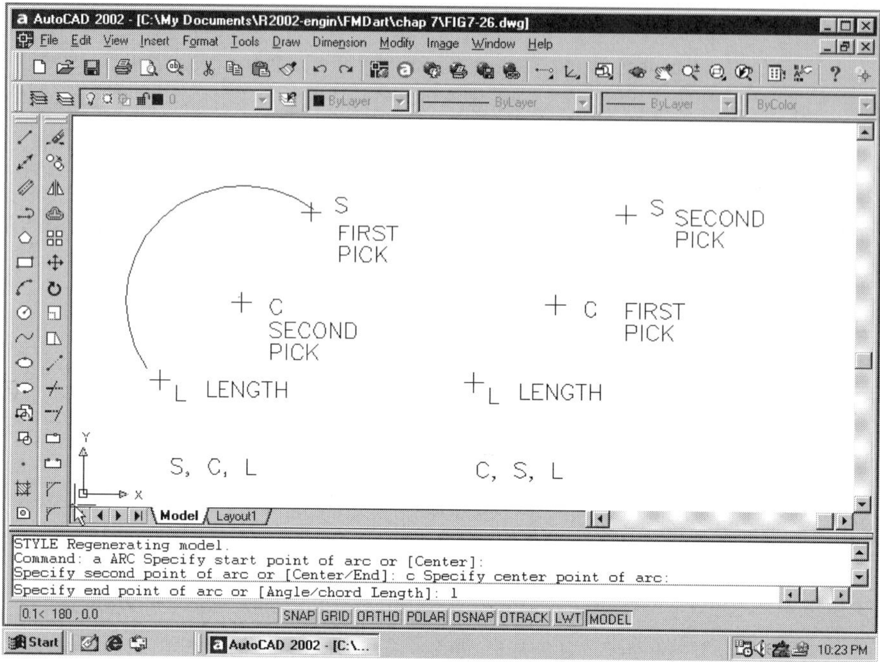

Figure 7.24 *The Center, Start, Length arc option produces results similar to the Start, Center, Length arc option.*

CONTINUE

An arc can be continued from the previous arc. The Arc command can be continued by pressing ENTER or SPACE, or by selecting **Repeat ARC** from the shortcut menu instead of selecting the first arc entry point. When continued from a previous arc, the current arc's start point and direction are taken from the endpoint and ending direction of the last arc. Figure 7.25 shows an example of four continuous arcs. The command sequence is as follows:

ARC 1

Command: **A** ENTER *(Or click the Arc button on the Draw toolbar.)*

ARC Specify start point of arc or [Center/End]: *(Select any point or enter coordinates.)*

Specify second point of arc or [Center/End]: *(Select second point.)*

Specify end point of arc: *(Select the arc's endpoint or enter coordinates.)*

Command:

ARC 2

Command: ENTER *(To resume Arc command.)*

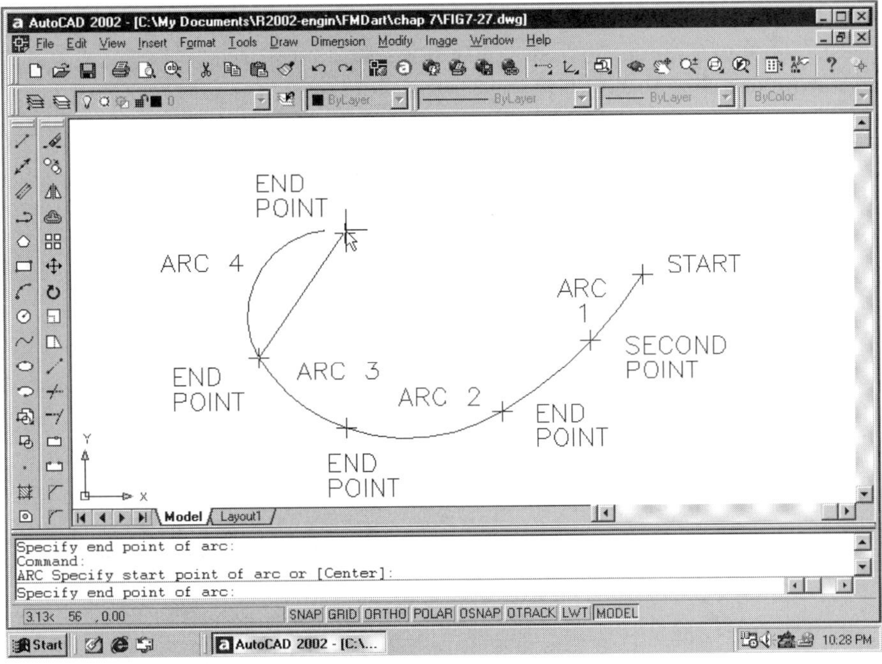

Figure 7.25 *The Continue option of the* **Arc** *command allows for multiple arcs to be drawn.*

ARC Specify start point of arc or [Center/End]: ENTER *(To resume arc from last endpoint.)*

Specify end point of arc: *(Select the arc's endpoint or enter coordinates.)*

Command:

ARC 3

Command: ENTER *(To resume Arc command.)*

ARC Specify start point of arc or [Center/End]: ENTER *(To resume arc from last endpoint.)*

Specify end point of arc: *(Select the arc's endpoint or enter coordinates.)*

Command:

ARC 4

Command: ENTER *(To resume Arc command.)*

ARC Specify start point of arc or [Center/End]: ENTER *(To resume arc from last endpoint.)*

Specify end point of arc: *(Select the arc's endpoint or enter coordinates.)*

Command:

The Continue option also can be used to extend an arc tangent to the endpoint of a line. Figure 7.26 shows an example of continuous arcs drawn from a straight line. The command sequence is as follows:

LINE 1

Command: **L** ENTER *(Or click the Line button on the Draw toolbar.)*

From point: *(Select starting point or enter coordinates.)*

Specify next point or [Undo]: *(Select point or coordinates.)*

Specify next point or [Undo]: ENTER

Command:

ARC 1

Command: **A** ENTER *(Or click the Arc button on the Draw toolbar.)*

ARC Specify start point of arc or [Center]: ENTER *(This will start the arc command at the line endpoint.)*

Specify endpoint of arc: *(Select the arc's endpoint or enter coordinates.)*

Command:

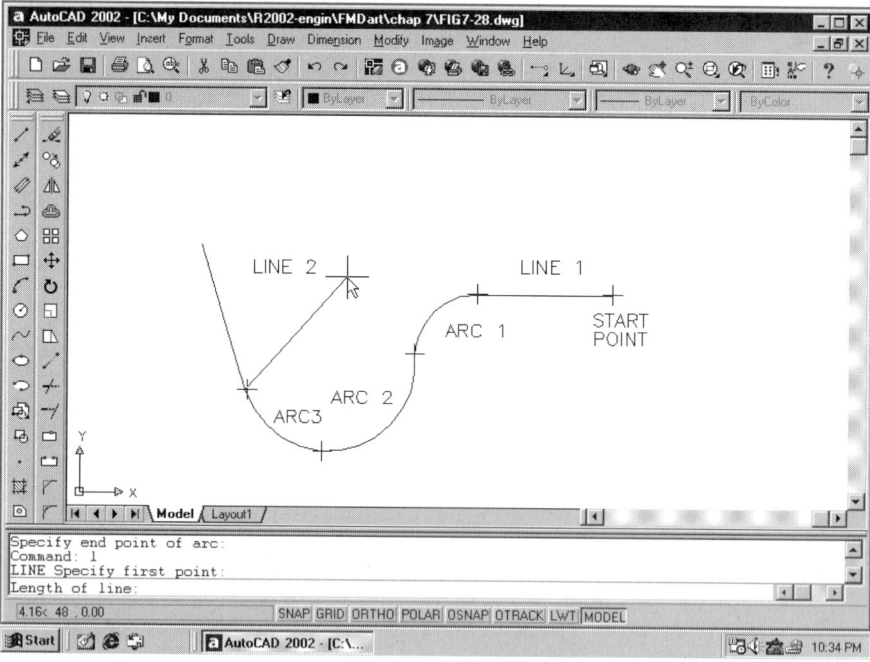

Figure 7.26 *Continue can be used to extend an arc tangent to the last line segment.*

ARC 2

Command: ENTER *(This will resume the Arc command.)*

ARC Specify start point of arc or [Center]: ENTER *(This will start the arc at the arc endpoint.)*

ENTER *(This will start at the end of the last line drawn.)*

Specify endpoint of arc: *(Select the arc's endpoint or enter coordinates.)*

Command:

ARC 3

Command: ENTER *(This will resume the Arc command.)*

ARC Specify start point of arc or [Center]: ENTER *(This will continue the arc from the endpoint of the last arc.)*

Specify endpoint of arc: *(Select the arc's endpoint or enter coordinates.)*

Command:

LINE 2

Command: *(Select the Line button or type)* **L** ENTER

LINE Specify first point: ENTER

Specify next point or [Undo]: *(Select point or coordinates.)*

Specify next point or [Undo]: ENTER *(To end drawing sequence.)*

Command:

Notice in Figure 7.26 that line 2 was drawn tangent to arc 3 because the Continue option was used. To place a line that continues from an arc but is not tangent to the arc will require the arc endpoint to be selected as the line "next point," using the methods described in Chapter 9.

DRAWING DONUTS

A donut in AutoCAD is a filled circle or ring. Start the command by typing **DO** (ENTER) at the Command prompt, or by selecting **Donut** from the **Draw** menu. Each can be seen in Figure 7.27. Although not on a toolbar, the **Donut** button can also be added to any toolbar if you use it frequently. Donuts are used to show connections on electrical layout drawings. The command sequence to create a donut is as follows:

Command: DO ENTER

DONUT

Specify inside diameter of donut <0.50>**0** ENTER

Specify outside diameter of donut <1.00>**.75** ENTER

Specify center of donut or <exit>: *(Select the desired location for the donut.)*

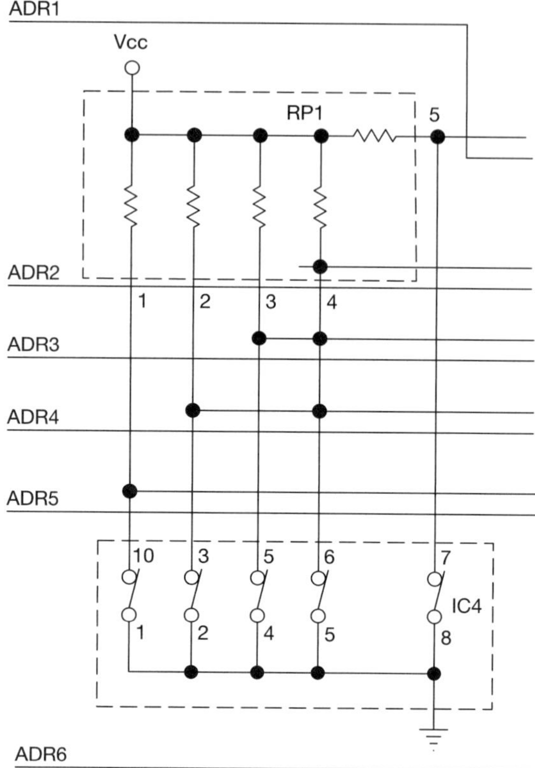

Figure 7.27 *Donuts are used to show connections on electrical drawings. They can be drawn by selection from the* **Draw** *menu or by typing* **DO** (ENTER) *at the Command prompt. Courtesy Jill Heilman.*

Specify center of donut or <exit>: *(The command will continue indefinitely; press* ENTER *to end the command.)*

Command:

Three different options are available for drawing donuts: a ring, a filled circle that is solid black, or a circle filled with lines. Each can be seen in Figure 7.28. The fill material is determined by the status of the **Fill** command. Typing FILL (ENTER) at the Command prompt will allow for a choice to be made for Donut and several other options. If OFF is entered, filled objects will resemble the bottom circle in Figure 7.28. With **Fill** ON, filled objects will resemble the middle circle in Figure 7.28.

Note: Fill will affect several of the commands you will soon be using. In addition to donuts, **Fill** will affect the appearance of polylines, arrows created with **Leader**, and the display of hatch patterns. In older, slower versions of AutoCAD, many users set **Fill** to OFF to a gain speed when working with large drawings. Unless you're trying to reproduce hollow objects, leave **Fill** in the ON setting.

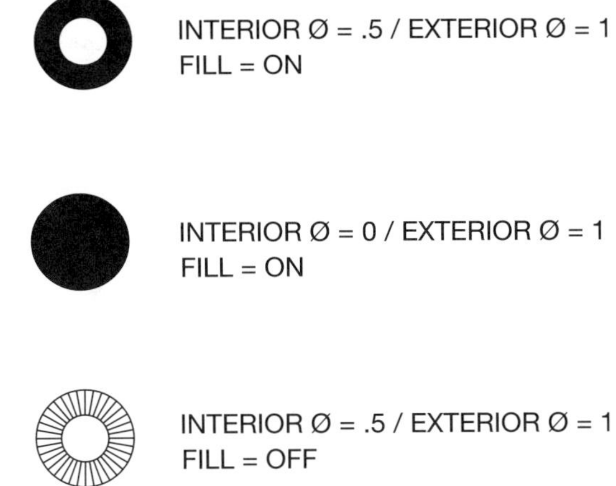

Figure 7.28 *Three options are available for the display of a donut.*

Another thing to remember about **Fill** is that the effect of changing the setting is usually not seen immediately. To see the effect of **Fill**, enter the command, alter the setting, and press ENTER. No change shows until you type REGEN (ENTER) at the Command prompt. Now the effect of **Fill** can be seen.

USING THE ELLIPSE COMMAND

Ellipses are also used to show round objects that are not being seen in true orthographic projection, such as a round feature that is not perpendicular to the viewing plane of the drawing. To start the **Ellipse** command, select the **Ellipse** button on the **Draw** toolbar, select **Ellipse** from the **Draw** menu, or type EL (ENTER) at the Command prompt. Menu options for **Ellipse** can be seen in Figure 7.29. Figure 7.30 shows the major and minor axes that will be referred to in laying out an ellipse. The command sequence is as follows:

Command: **EL** ENTER *(Or click the Ellipse button on the Draw toolbar.)*

ELLIPSE

Specify axis endpoint of ellipse or [Arc/Center]: *(Select a point or enter coordinates for the major axis endpoint.)*

Specify other endpoint of axis: *(Select a point or enter coordinates for the opposite major axis endpoint.)*

Specify distance to other axis or [Rotation]: *(Select a point or enter coordinates for the minor axis endpoint.)*

Command:

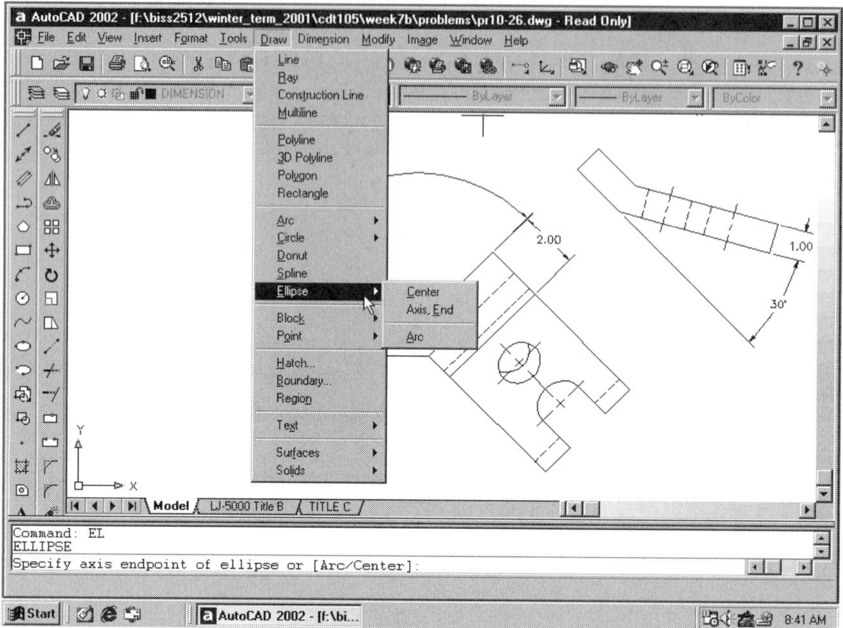

Figure 7.29 *The menu options for Ellipse. Courtesy Paul Bissell.*

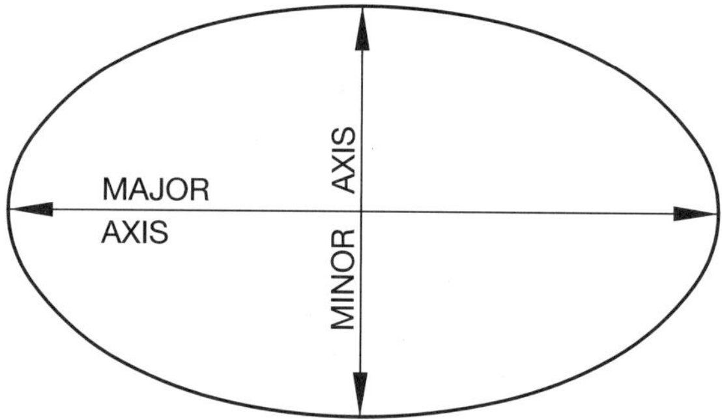

Figure 7.30 *An ellipse is defined by the major and minor axis.*

An ellipse is now shown on the screen between the two selected points. The process can be seen in Figure 7.31. AutoCAD is waiting for you to enter the axis or rotation. If the angle is not known, the crosshairs can be moved until the ellipse represents the desired shape.

If you know the ellipse represents a drilled feature that is at a 60° angle to the viewing plane, type R (ENTER). The command sequence is as follows:

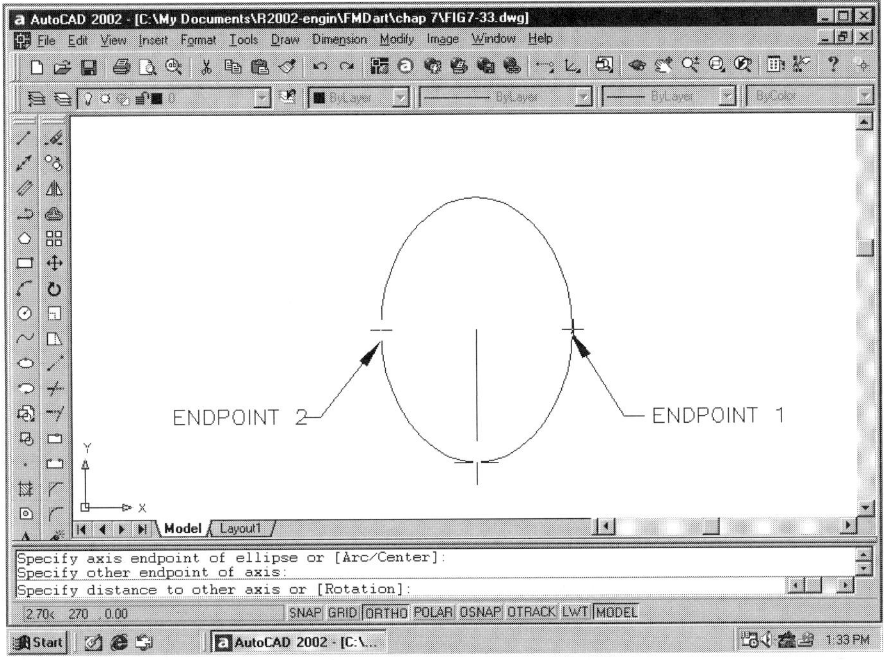

Figure 7.31 *Once the two endpoints are specified, the ellipse will change as the cursor is moved.*

Command: **EL** ENTER *(Or click the Ellipse button on the Draw toolbar.)*

ELLIPSE

Specify axis endpoint of ellipse or [Arc/Center]: *(Select a point or enter coordinates for the major axis endpoint.)*

Specify other endpoint of axis: *(Select a point or enter coordinates for the opposite major axis endpoint.)*

Specify distance to other axis or [Rotation]: **R** ENTER

Specify rotation around major axis: *(Enter desired angle of rotation.)* **60** ENTER

Command:

The command sequence will produce the ellipse shown in Figure 7.32. Figure 7.33 shows the effect of rotating a cylinder at 10° increments. An ellipse with a 90° rotation cannot be drawn using the **Ellipse** command because it would produce a single straight line.

An ellipse also can be drawn by locating the center point and then specifying an endpoint for each axis. This process can be seen in Figure 7.34. The command sequence is as follows:

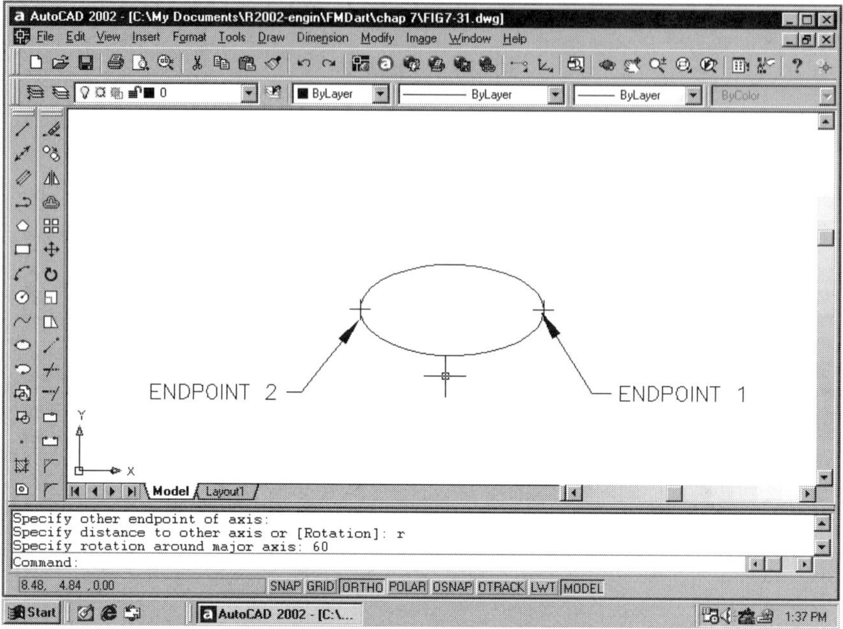

Figure 7.32 *Specifying the angle of rotation of the viewing plane will produce an ellipse. Using a 60° rotation produces the ellipse shown.*

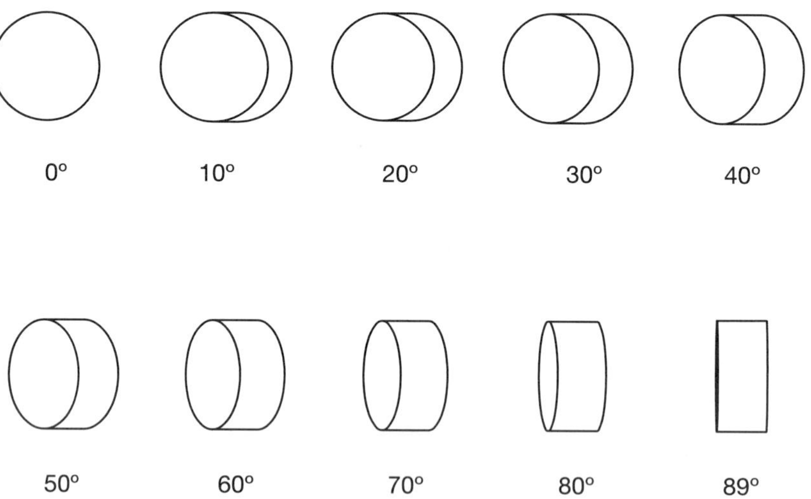

Figure 7.33 *The effects of rotating a circular object in 10° increments.*

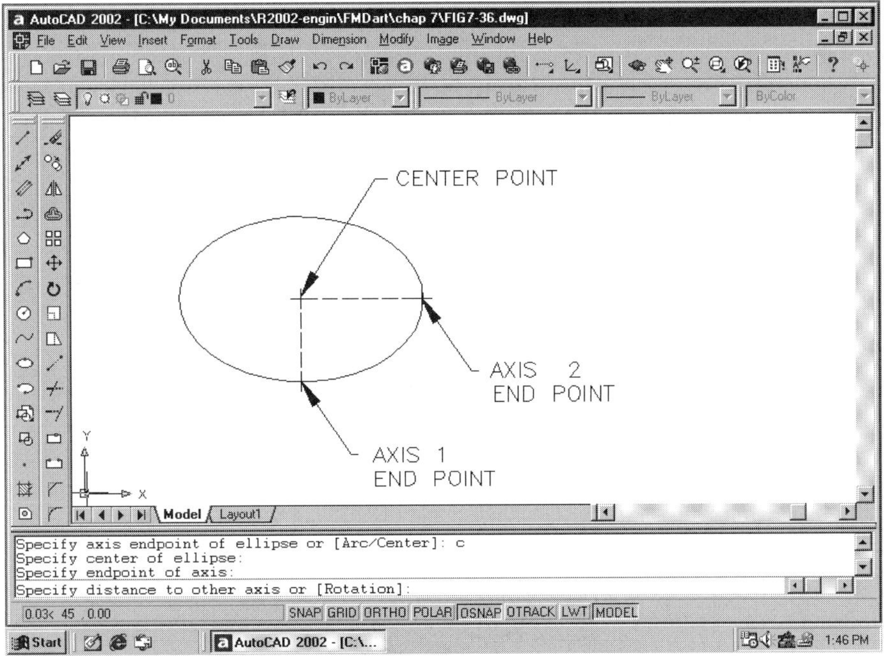

Figure 7.34 *An ellipse can be drawn by locating the center point followed by an endpoint for the major and minor axis.*

Command: **EL** ENTER *(Or click the Ellipse button on the Draw toolbar.)*

ELLIPSE

Specify axis endpoint of ellipse or [Arc/Center]: **C** ENTER

Specify center of ellipse: *(Select a point or enter coordinates for the major axis center point.)*

Specify endpoint of axis: *(Select a point or enter coordinates for the endpoint of major axis.)*

Specify distance to other axis or [Rotation]: *(Select a point or enter coordinates for the minor axis endpoint.)*

Command:

When the second axis endpoint is entered, the Command prompt is returned and the ellipse is drawn on the screen.

DRAWING POLYGONS

A polygon is any geometric shape lying on one plane, bound by three or more straight lines. Triangles, squares, pentagons, hexagons, and octagons are used as symbols and material shapes in many drawings. AutoCAD draws polygons

with equal length lines and angles having between 3 and 1,024 sides. To draw a polygon, you will be asked to choose the number of sides and whether the object is to be inscribed in or circumscribed around a circle. When the polygon is inscribed in a circle, the circle diameter provides the distance across the polygon points. When the polygon is circumscribed around a circle, the circle diameter provides the distance across the flat surfaces of the polygon. The effects of a hexagon inscribed in and circumscribed around a 2.5" diameter circle can be seen in Figure 7.35.

Start the **Polygon** command by selecting the **Polygon** button on the **Draw** toolbar, by selecting **Polygon** from the **Draw** menu, as shown in Figure 7.36, or by typing POL (ENTER) at the Command prompt. The command sequence is as follows:

Command: **POL** ENTER *(Or click the Polygon button on the Draw toolbar.)*

POLYGON

Enter number of sides <4>: *(Select the desired number of sides.)* **8** ENTER

Specify center of polygon or [Edge]: *(Select a point or enter coordinates for the center point.)*

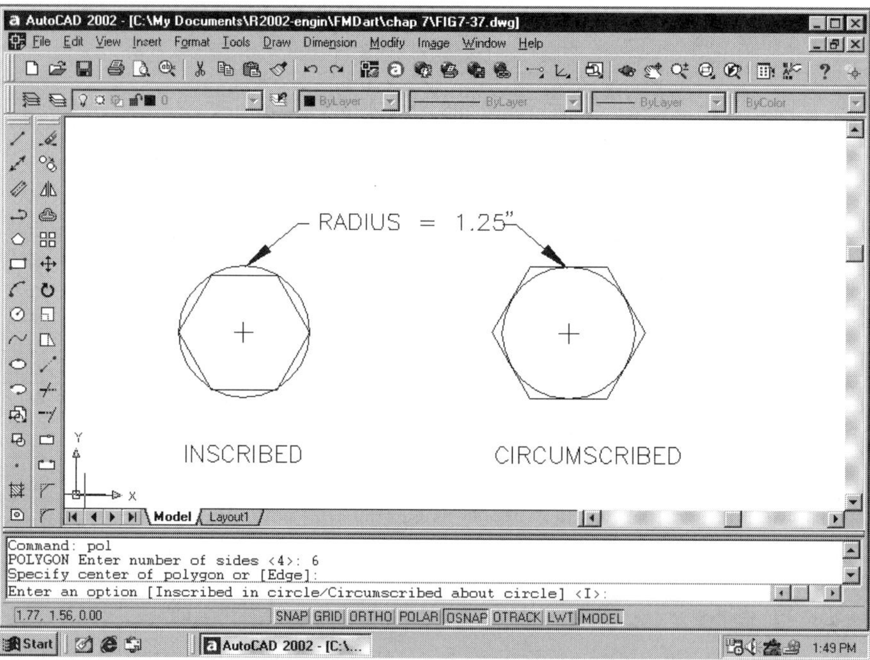

Figure 7.35 *Polygons are drawn based on a circle. Once the number of sides has been specified, you will be asked to specify whether the polygon will be placed inside or outside the circle.*

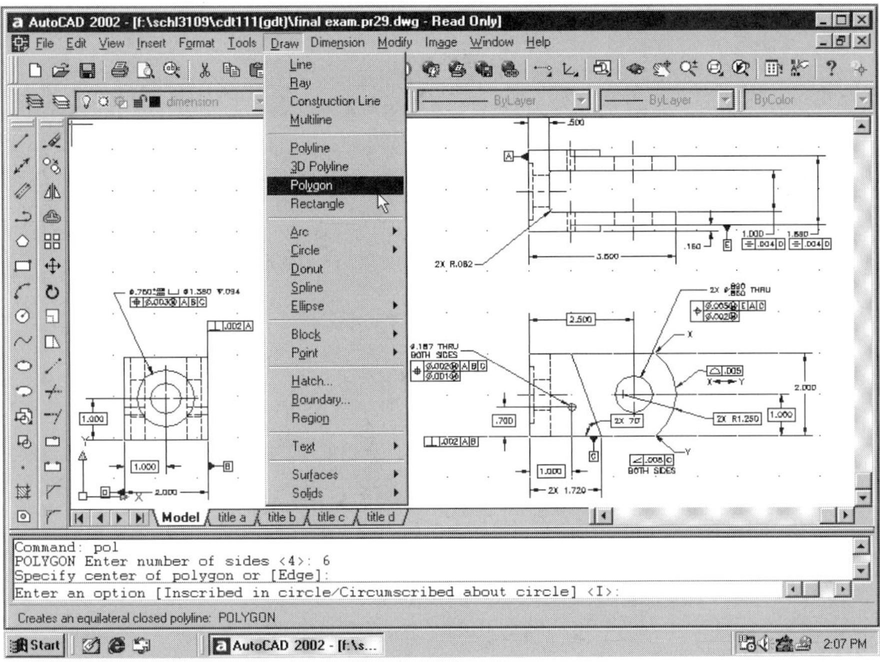

Figure 7.36 *The Polygon command on the Draw menu. The command is useful for drawing bolt heads. Courtesy Cathy Schlesser.*

> Enter an option [Inscribed in circle/Circumscribed about circle] <I>:
> *(Accept I or select C depending on your need and press* ENTER*)*
>
> Specify radius of circle: *(Enter desired radius.)* **2.75** ENTER
>
> Command:

The result of the sequence can be seen in Figure 7.37. The setting of ORTHO will affect how the polygon will be placed. When ORTHO is ON, the edges of the polygon will be parallel or perpendicular to the drawing screen. When ORTHO is OFF, the edges of the polygon will rotate based on the movement of the mouse. To activate ORTHO, click the ORTHO button on the status bar or press F8.

A polygon also can be drawn based on the location of one edge, rather than the location of the center point. Once the number of sides has been provided, you will be asked for an edge or center point. When E (ENTER) is typed at the prompt, a polygon similar to the one in Figure 7.38 can be drawn. The command sequence is as follows:

> Command: **POL** ENTER *(Or click the Polygon button on the Draw toolbar.)*
>
> POLYGON
>
> Enter number of sides <8>: **6** ENTER

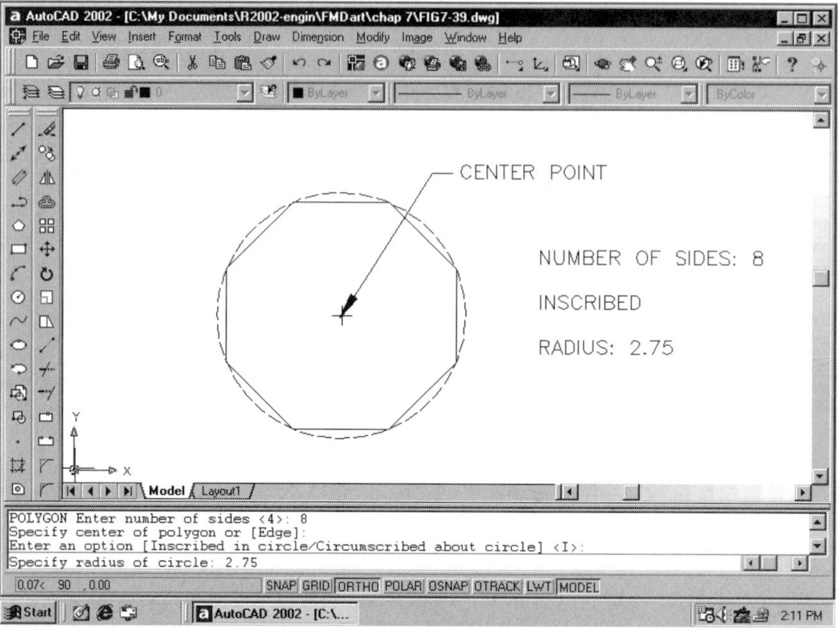

Figure 7.37 *The result of drawing an inscribed octagon with a 2.75" radius.*

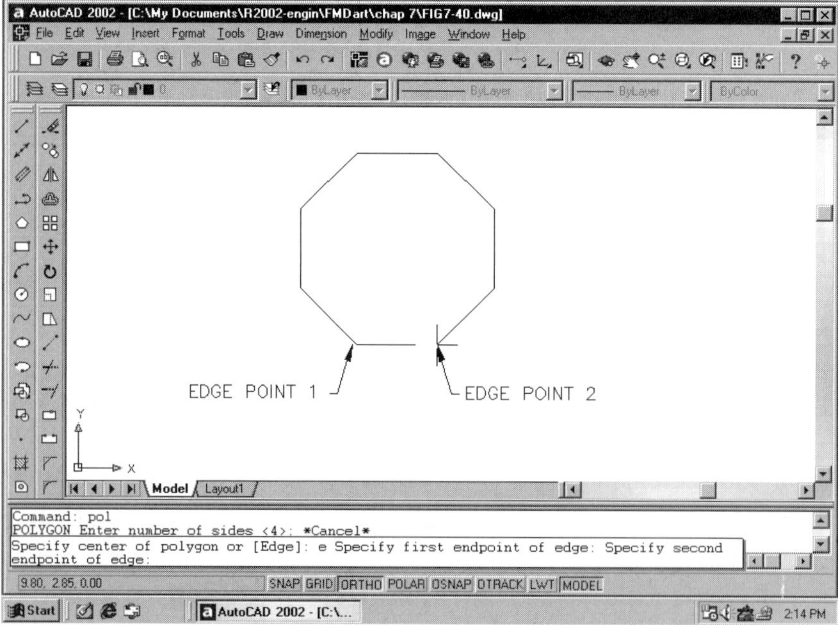

Figure 7.38 *A polygon also can be drawn by specifying the points that define one edge.*

Specify center of polygon or [Edge]: **E** ENTER

Specify first endpoint of edge: *(Select a point or enter coordinates for the first endpoint.)*

Specify second endpoint of edge: *(Select a point or enter coordinates for the second endpoint.)*

Command:

As the first point is entered, a polygon will be displayed that will vary in size as the mouse is moved to the second point. As the second point is entered, the Command prompt is returned and the polygon is drawn. The setting of ORTHO will affect the placement of a polygon using the Edge option.

DRAWING RECTANGLES

A rectangle is drawn by selecting the **Rectangle** button on the **Draw** toolbar, by selecting **Rectangle** from the **Draw** menu, or by typing REC (ENTER) at the Command prompt. (A square can be drawn using the **Polygon** command.) The command sequence for drawing a rectangle is as follows:

Command: **REC** ENTER *(Or click the Rectangle button on the Draw toolbar.)*

RECTANGLE

Specify first corner point [Chamfer/Elevation/Fillet/Thickness/Width]: *(Select a corner of the rectangle with the mouse.)*

Specify other corner point or [Dimensions]: *(Select the opposite corner of the rectangle.)*

Command:

As the cursor is moved between the first and second corner, a rectangle will be displayed on the screen allowing the exact size to be determined. The sequence can be seen in Figure 7.39. Selecting either the Chamfer or Fillet option allows the corners of the rectangle to be altered. Selecting the Width option allows the width of the line to be altered. The other options affect only a rectangle drawn in 3D and will not be discussed in this chapter.

CHAMFER

A chamfer is an angled or mitered corner. Drawing a rectangle with the Chamfer option active will place a chamfer on each corner of the rectangle. Once the Chamfer option is selected, prompts will be given for the first and second chamfer distances. If the distances are equal, the result will be as shown in Figure 7.40. Providing different values will produce a rectangle similar to the example on the right in Figure 7.40. The command sequence to draw a rectangle with a chamfer is as follows:

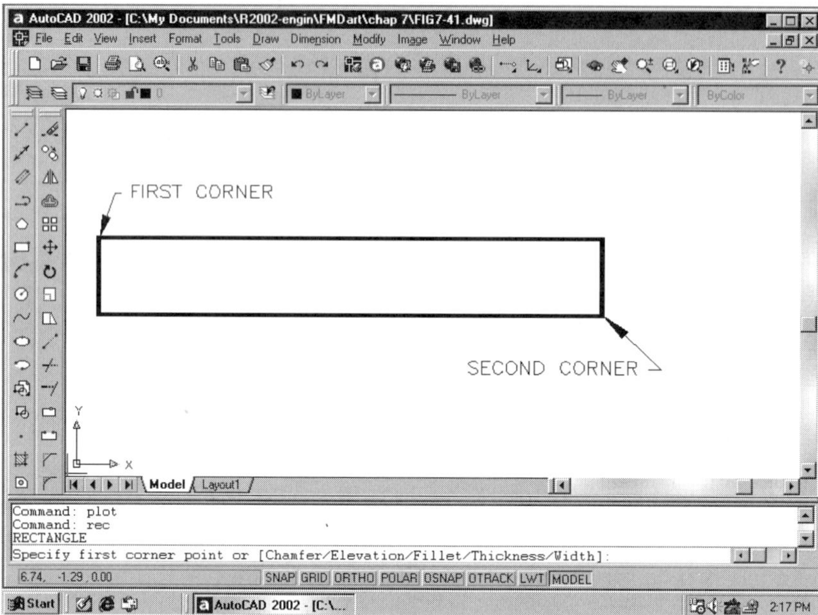

Figure 7.39 *Rectangular shapes can be drawn as one object using the Rectangle command on the* **Draw** *toolbar or the* **Draw** *menu, or by typing* **REC** *(*ENTER*) at the command line.*

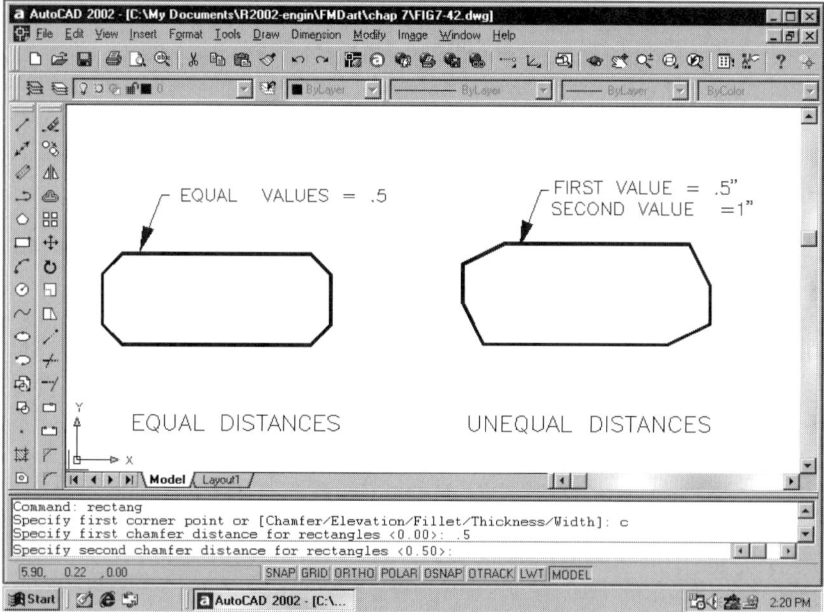

Figure 7.40 *The Chamfer option of the* **Rectangle** *command can be used to provide a chamfer at each corner of a rectangle. The rectangle on the left shows equal values for the chamfers while the rectangle on the right shows unequal values.*

Command: **REC** ENTER *(Or click the Rectangle button on the Draw toolbar.)*

RECTANGLE

Specify first corner point or [Chamfer/Elevation/Fillet/Thickness/Width]: **C** ENTER

Select first chamfer distance for rectangles <0.00>: *(Enter the desired size.)* **.5** ENTER

Second chamfer distance for rectangles <0.50>: ENTER

Specify first corner point or [Chamfer/Elevation/Fillet/Thickness/Width]: *(Select a corner of the rectangle.)*

Specify other corner point or [Dimensions]: *(Select the opposite corner of the rectangle or enter a dimension to define the length.)*

Command:

FILLET

A fillet is a rounded corner. Drawing a rectangle with the Fillet option active will place a fillet on each corner of the rectangle. Once the Fillet option is selected, a prompt will be given to provide a radius for the fillet. Figure 7.41 shows an example of a rectangle with filleted corners. The command sequence to draw a rectangle with a fillet is as follows:

Command: **REC** ENTER *(Or click the Rectangle button on the Draw toolbar.)*

RECTANGLE

Specify first corner point or [Chamfer/Elevation/Fillet/Thickness/Width]: **F** ENTER

Select fillet radius for rectangles <0.50>: *(Enter the desired size.)* ENTER

Specify first corner point or [Chamfer/Elevation/Fillet/Thickness/Width]: *(Specify a corner of the rectangle.)*

Specify other corner point or [Dimensions]: *(Select the opposite corner of the rectangle.)*

Command:

Notice that the default radius value is .50". The value that was provided for the chamfer will become the default value for all future chamfers and fillets until a new value is provided.

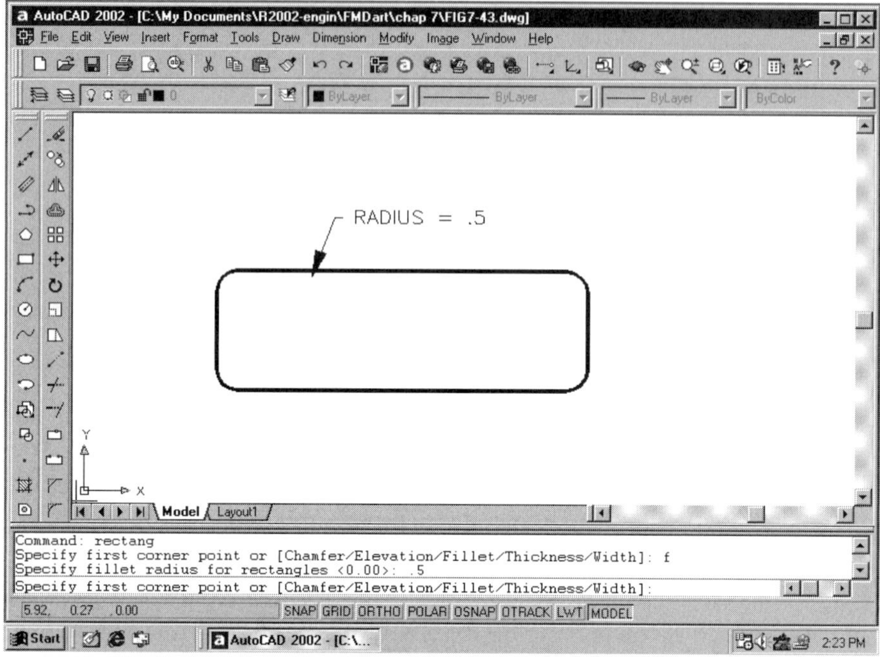

Figure 7.41 *The Fillet option of the* **Rectangle** *command can be used to provide a fillet at each corner of a rectangle. The value used for the radius will remain constant for future use until it is altered.*

WIDTH

Selecting the Width option allows the line thickness to be altered. Because the Fillet option was just used, setting the thickness for a rectangle will result in a rectangle with .50" radius corners to be drawn with a thick line. To draw a rectangle with square corners, enter the **Rectangle** command and alter the value for Fillet to 0, and then set the thickness, or vice versa. The command sequence is as follows:

Command: **REC** ENTERr *(Or click the Rectangle button on the Draw toolbar.)*

RECTANGLE

Specify first corner point or [Chamfer/Elevation/Fillet/Thickness/Width]: **W** ENTER

Specify line width for rectangles <0.00>: **.25** ENTER

Specify first corner point or [Chamfer/Elevation/Fillet/Thickness/Width]: **F** ENTER

Select fillet radius for rectangles <0.00>: **0** ENTER *(The desired size entered.)*

Specify first corner point or [Chamfer/Elevation/Fillet/Thickness/Width]: *(Specify a corner of the rectangle.)*

Specify other corner point or [Dimensions]: *(Select the opposite corner of the rectangle.)*

Command:

The result of this command sequence can be seen in Figure 7.42. The width will remain the current width for all future rectangles until the value is altered. Methods of providing thickness to other drawing features will be discussed as polylines are introduced.

USING THE SPLINE COMMAND

The Spline option of the **Pline** command converts straight-line segments to a curved line. The curve is an approximation of a true spline and is not accurate. The **Spline** command can be used to create a smooth curve given a series of points. AutoCAD creates a curve referred to as a NURB (non-uniform rational B-spline curve). Splines are created by providing a series of points that lie on the desired path of the curve. These curves are especially helpful in automotive and naval design, and in the creation of land drawings. Access the command by selecting the **Spline** button on the **Draw**

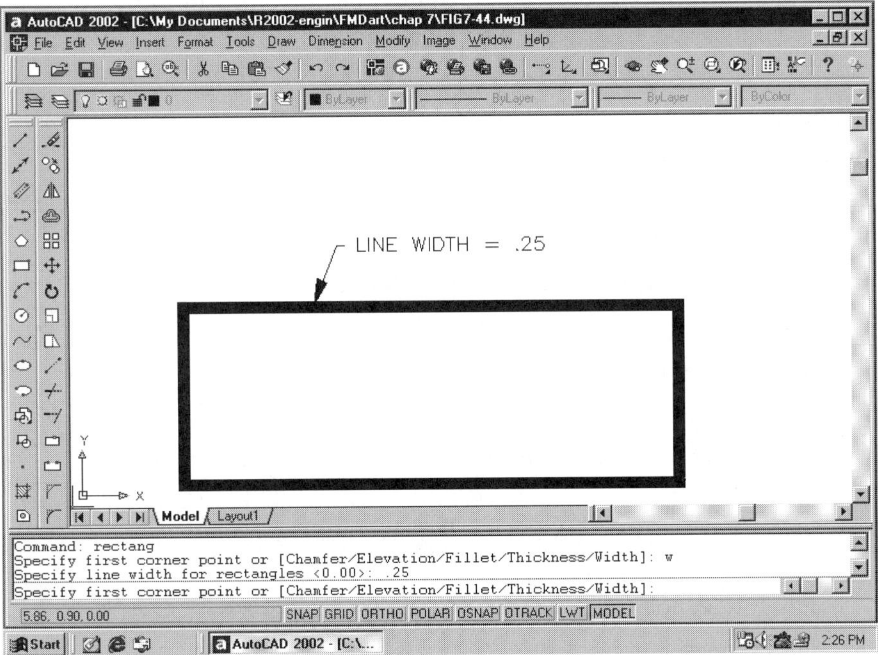

Figure 7.42 *The Width option can be used to assign a line width to the* **Rectangle** *command.*

toolbar, by typing SPL (ENTER) at the Command prompt, or by selecting **Spline** from the **Draw** menu. The command sequence to construct a spline is as follows:

Command: **SPL** ENTER *(Or click the Spline button on the Draw toolbar.)*

SPLINE

Specify first point or [Object]: *(Select point.)*

Specify next point: (Select point.)

Specify next point or [Close/Fit tolerance] <start tangent>: *(Select point.)*

Specify next point or [Close/Fit tolerance] <start tangent>: *(Select point.)*

Specify next point or [Close/Fit tolerance] <start tangent>: ENTER

Specify start tangent: ENTER

Specify end tangent: ENTER

Command:

The location for each of the points can be specified by selecting the point or by providing coordinate entry points. Once each of the points of the curve has been specified, pressing ENTER allows the start tangency and end tangency to be specified. Specifying the start and end tangents changes the direction in which the spline curve is started and stopped. Pressing ENTER at each tangent prompt will accept the AutoCAD default values. Figure 7.43 shows the result of drawing a spline curve using the default values.

DRAWING CLOSED SPLINES

The Close option of **Spline** allows a closed spline similar to Figure 7.44 to be drawn. The command sequence is as follows:

Command: **SPL** ENTER *(Or click the Spline button.)*

SPLINE

Specify first point or [Object]: *(Select point.)*

Specify next point: *(Select point.)*

Specify next point or [close/Fit tolerance] <start tangent>: *(Select point.)*

Specify next point or [close/Fit tolerance] <start tangent>: *(Select point.)*

Specify next point or [close/Fit tolerance] <start tangent>: **C** ENTER

Specify tangent: ENTER

Command:

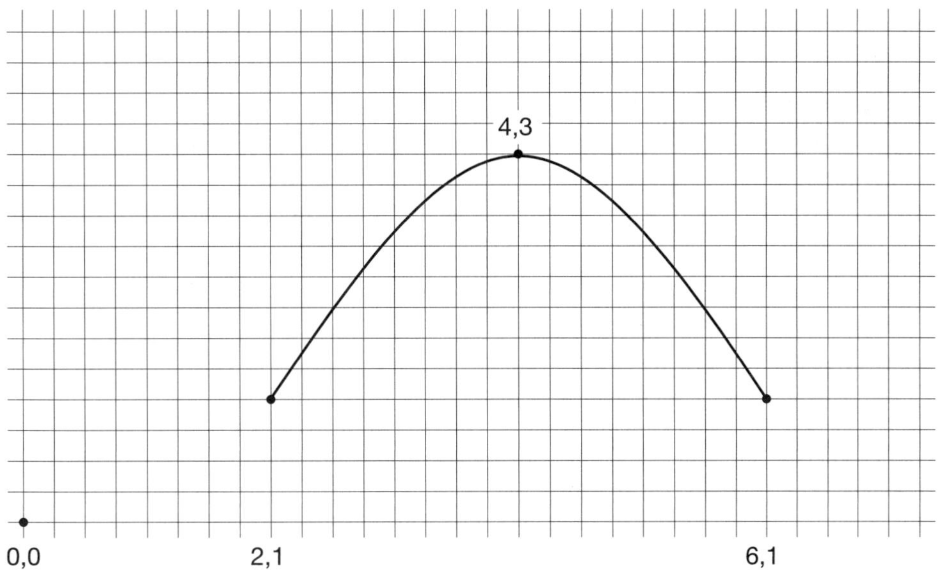

Figure 7.43 *A curve created using the* **Spline** *command will pass through each of the control points when each of the default values are used.*

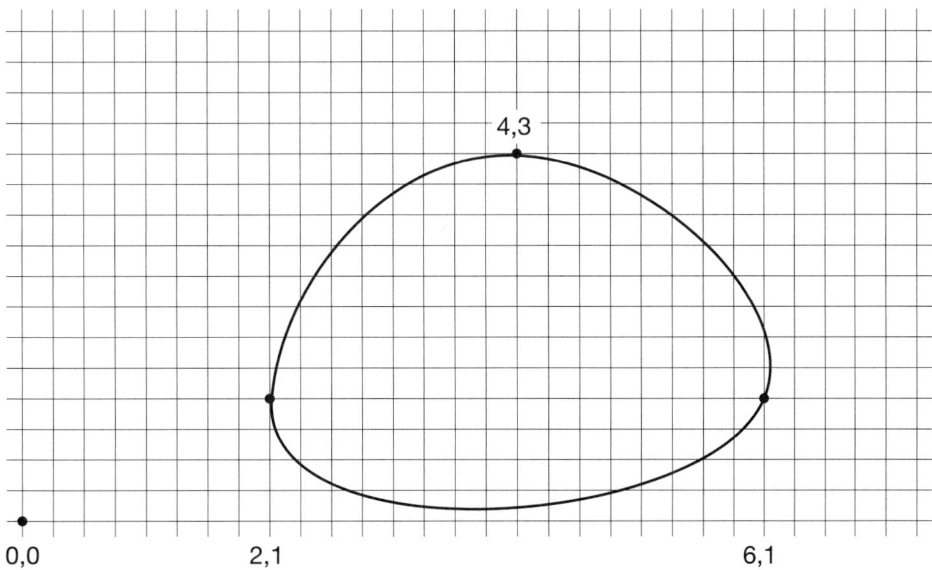

Figure 7.44 *A spline can be closed using the Close option of the* **Spline** *command.*

Typing C (ENTER) for the final point will close the spline. Pressing ENTER for the tangent point accepts the default value.

ALTERING THE FIT TOLERANCE

The spline fit tolerance determines how closely the spline curve fits the specified spline points. With the default tolerance of 0, the spline will pass through the points. As the tolerance is altered, the shape of the spline curve will be altered. Figure 7.45 shows the spline from Figure 7.43 drawn with a tolerance of .125. The Fit tolerance option can be selected at any of the prompts throughout the selection process. The command sequence would resemble the following:

> Command: **SPL** ENTER *(Or click the Spline button.)*
>
> SPLINE
>
> Specify first point or [Object]: *(Select point.)*
>
> Specify next point: *(Select point.)*
>
> Specify next point or [Close/Fit tolerance] <start tangent>: **F** ENTER
>
> Specify fit tolerance <0.00>: **.0125** ENTER
>
> Specify next point or [Close/Fit tolerance] <start tangent>: *(Select point.)*

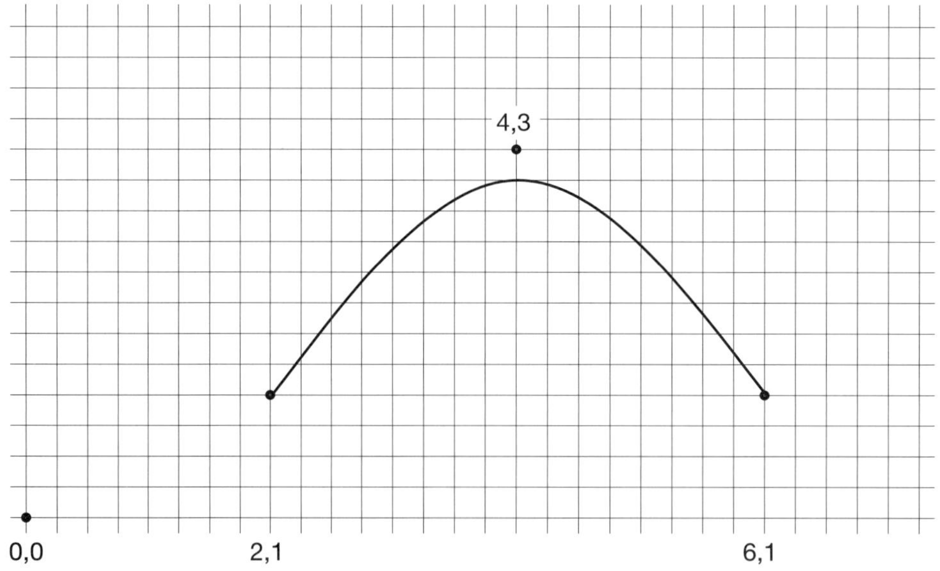

Figure 7.45 *The Fit tolerance determines how closely the spline curve fits the specified control points. In this example, a tolerance of 0.125 was used.*

Specify next point or [Close/Fit tolerance] <start tangent>: *(Select point.)*

Specify next point or [Close/Fit tolerance] <start tangent>: ENTER

Specify start tangent: ENTER

Specify end tangent: ENTER

Command:

ALTERING THE TANGENT POINTS

Overriding the default start and end tangent directions allows the shape of the spline to be altered. Rather than accepting the default values, you can specify tangent directions by moving the mouse in the desired direction. Figure 7.46 shows an example of three splines created using the same coordinates with different tangent points. The command sequence to alter tangent end points is as follows:

Command: **SPL** ENTER *(Or click the Spline button.)*

SPLINE

Specify first point or [Object]: *(Select point.)*

Specify next point: *(Select point.)*

Specify next point or [close/Fit tolerance] <start tangent>: *(Select point.)*

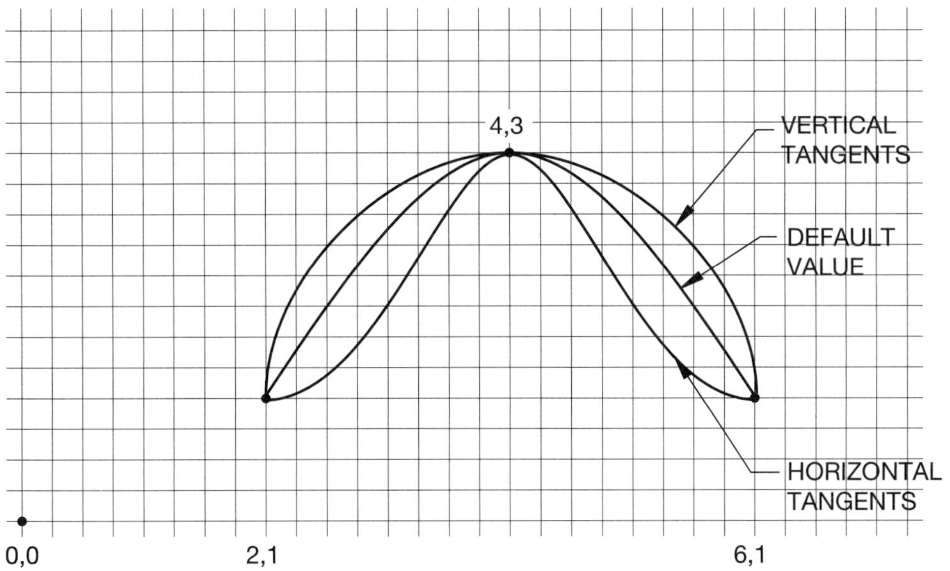

Figure 7.46 *The shape of the spline will vary greatly depending on the tangent points. Each of the three splines uses the same control points, but with different tangent settings.*

Specify next point or [close/Fit tolerance] <start tangent>: ENTER

Specify start tangent: *(Move cursor in desired tangent direction.)* ENTER

Specify end tangent: *(Move cursor in desired tangent direction.)* ENTER

Command:

CONVERTING A SPLINE-FITTED POLYLINE

A spline created using the **Pline** command can be converted to a true spline using the Object option of the **Spline** command. The sequence to convert a spline-fitted polyline object to a spline is as follows:

Command: SPL ENTER *(Or click the Spline button.)*

SPLINE

Specify first point or [Object]: **O** ENTER

Select objects to convert to splines...

Select objects: *(Select the spline-fitted polyline.)*

Select objects: ENTER

Command:

EDITING SPLINES

Splines can be edited using either grips or the **Splinedit** command. Figure 7.47 shows the grips for a spline. Highlighting a spline will display a grip at each end, peak and valley. Activating a grip will allow the hot grip of the spline to be stretched or moved. The **Splinedit** command allows a spline to be edited using several different options.

ALTERING SPLINES USING SPLINEDIT

The **Splinedit** command also places grips on a spline, allowing the object to be edited using five different options. Start the command by selecting the **Edit Spline** button on the **Modify II** toolbar, by typing SPE (ENTER) at the command prompt, or by selecting **Spline** from the **Modify** menu. Each will produce the following display:

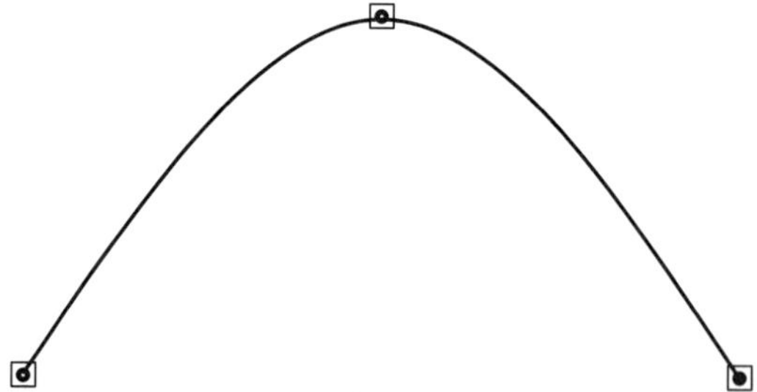

Figure 7.47 *Selecting a spline will display grips at each control point.*

Command: **SPE** ENTER *(Or click Edit Spline button on the Modify II toolbar.)*

Select spline: *(Select spline to be edited.)*

Enter an option [Fit Data/Close/Move vertex/Refine/rEverse/Undo]:

As a spline is selected, grips will be displayed similar to Figure 7.48. One of seven options can now be applied to the spline. These options include Fit data, Close, Open, Move vertex, Refine, Reverse, and Undo. The Open option is not currently displayed because an open spline has been selected.

Fit Data

The Fit data option of **Splinedit** can be used to edit spline control points called fit points. The fit points are displayed as grips. Modes for altering fit points include Add, Close, Delete. Move, Purge, Tangents, toLerance, and eXit. The command sequence to edit a spline is as follows:

Command: **SPE** ENTER

Select spline: *(Select spline to be edited.)*

Enter an option [Fit Data/Close/Move vertex/Refine/rEverse/Undo]: **F** ENTER

Enter a Fit data option

[Add/Close/Delete/Move/Purge/Tangents/toLerance/eXit]<exit>:

Add

The Add mode allows new fit points to be added between two existing fit points. As the first fit point is selected, it and the next fit point will be highlighted. The new control point will be placed between the two selected points. The Undo option can be used to remove the last point selected. The command sequence is as follows:

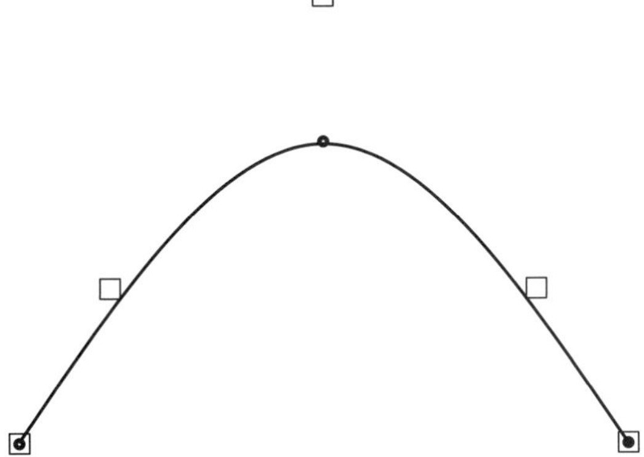

Figure 7.48 *Using* **Splinedit**, *grips are displayed at each of the fit points.*

Enter an option [Fit Data/Close/Move vertex/Refine/rEverse/Undo]:
F ENTER

Enter a Fit data option

[Add/Close/Delete/Move/Purge/Tangents/toLerance/eXit]<exit>:
A ENTER

Specify control point <exit>: *(Specify a control point or press* ENTER*)*

Specify new control point:

The new control point can be added by selecting a point or by entering coordinates. If the new point lies on the spline, a new control point is added between the two existing control points. If the selected point does not lie on the spline, the control point will be added and the spline will be altered to pass through the new control point. Figure 7.49 shows the results of adding a new fit point.

Close/Open
The mode that is displayed will depend on the selected spline. If the spline is open, the Close mode will be displayed, and vice versa. The modes allow closed splines to be opened and open splines to be closed.

Delete
The Delete mode removes an existing control point and refits the spline between the remaining fit points.

Move
The Move mode allows an existing fit point to be relocated. As the Move mode is selected, the spline start point is highlighted. This point can now be moved to the

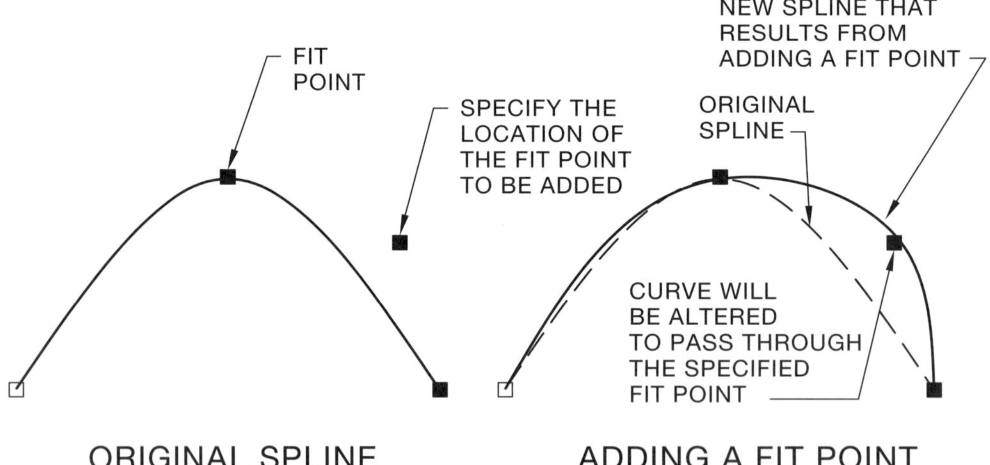

Figure 7.49 *Fit points can be added to alter the shape of the spline using the Add mode of the* **Splinedit Fit Data** *option.*

desired location. Alternatives for moving the hot point are displayed at the Command prompt and include the following:

> **Specify new location**—In the default mode, the selected control point can be moved to the desired location.
>
> **Next (N)**—Changes the highlighted change point to the next change point on the spline. Each time ENTER is pressed, a new change point is selected.
>
> **Previous (P)**—Changes the highlighted change point to the previous change point on the spline. Each time ENTER is pressed a new change point is selected.
>
> **eXit (X)**—Selecting eXit returns the Fit Data prompt.

Purge

The Purge mode removes the data for the selected change point to be removed from the drawing database. This mode can be helpful on very complex drawings because the drawing size is reduced. Once data for a selected spline has been purged, the Fit Data option will not be displayed.

Tangents

The Tangents mode allows the start and end tangents of a spline to be altered.

Tolerance

The toLerance mode allows the tolerance value to be redefined.

Exit

This mode returns to the main **Splinedit** prompt.

Close/Open

The Splinedit option that is displayed will depend on the selected spline. If the spline is open, the Close option will be displayed, and vice versa. The options allow a closed spline to be opened and an open spline to be closed.

Move Vertex

The Move vertex option allows the control points of a spline to be altered. As the option is selected, options identical to the Move mode of Fit data are displayed.

Refine

The Refine option allows the spline to be redefined. Control points can be added to reshape the spline. Modes include Add control point, Elevate order, and Weight.

Add Control Point

The Add control point mode allows the number of control points for a portion of the spline to be altered. Points can be specified by selecting a point with the mouse or by providing coordinates.

Elevate Order

Elevating the order of the spline will display more control points on the spline. As more control points are added, greater control over the location of the spline is obtained. As the name implies, control points can be added to the spline, with a maximum value of 26. The value can only be adjusted upward using this mode of the Refine option.

Weight

The Weight mode allows the pull, or weight of a specific spline control point to be adjusted. In effect, it allows the tolerance of one control point to be accented above the others. The default value of 1.00 can be raised or lowered to alter the effect of each control point. The higher the weight value, the closer the spline will be pulled to the control point. Specifying the weight is done in the same manner as setting the **Splinedit Move** vertex option. The selected spline is highlighted and the control point is highlighted by using the Next and Previous modes. Select Point is used to select the control point to be altered. Exit returns to the original prompt.

Reverse

The rEverse option alters the order of the spline control points. The original start point becomes the endpoint and the original endpoint becomes the start point.

Undo

As with the Undo option of other commands, this option will undo the previous alterations made to a spline.

CHAPTER 7 EXERCISES

Save the following exercises using the indicated file names.

1. Draw a 3" diameter circle, a circle with a 1" radius, and a 1" diameter circle that is tangent to two 1" long perpendicular lines. Make the smallest circle green with continuous lines. Make the largest circle red with hidden lines. Give one of the lines a width of .065". Save the drawing as E-7-1.
2. Draw two parallel lines that are 1" long and 1" apart, and provide an arc at each end of the lines to form a slotted hole. Draw these objects on a layer named SLOTS, using dashed blue lines, with a thin lineweight. Save the drawing as E-7-2.
3. Draw a 4" long line. Use the continuous option to place a 1" diameter circle at each end of the line. Save the drawing as E-7-3.
4. Draw a box formed by three 1" long lines. Draw an arc to close the polygon. Save as drawing E-7-4.
5. Draw an arc with a start point that is 2" to the right of the endpoint with an included angle of 232°. Save as drawing E-7-5.

6. Draw an arc with an endpoint that is @2,2 from the center point with an included angle of 45°. Save as drawing E-7-6.
7. Draw an ellipse with a 3" long axis at a 42° rotation. Save as drawing E-7-7.
8. Draw an ellipse with a center point that is 2.57" from the end of the major axis and 1" from the minor axis. Draw a four-sided polygon using the center of the ellipse as the center of the polygon. Make two edges of the square touch the ellipse. Save as drawing E-7-8.
9. Draw the following polygons:

3 sides	inscribed	2" diameter
3 sides	circumscribed	2" diameter
4 sides	inscribed	.5" radius
4 sides	circumscribed	.5" radius
6 sides	inscribed	.75" radius
6 sides	circumscribed	.75" radius
8 sides	inscribed	1.25" radius
8 sides	circumscribed	1.25" radius

Save the drawing as E-7-9.

10. Draw a donut with an inside diameter of .25", outside diameter of .75", and Fill set at ON. Draw another donut with an inside diameter of .0" and an outside diameter of .375" with Fill set at OFF. Draw a donut with an inside diameter of 0 and an outside diameter of .5" with Fill set to ON. Save the drawing as E-7-10.
11. Draw a rectangle with filleted corners using a 2" radius. Save the drawing as E-7-11.
12. Draw a rectangle with chamfered corners using a 1" distance for both settings. Save the drawing as E-7-12.
13. Draw a rectangle with chamfered corners using a 1.5" distance for the first distance and a 2.5" value for the second setting. Assign a line width of .065 and save the drawing as E-7-13.
14. Start a new drawing and draw at least three connected combinations of lines and arcs. Save the drawing as E-7-14.
15. Draw two perpendicular lines that are 2" long and form a corner. Draw a 3" diameter circle that is tangent to each of the lines. Draw three lines that are not connected and are neither parallel nor perpendicular to each other. Draw a circle that is tangent to all three line segments. Save this drawing as E-7-15.

Exercises 16 through 20: Use the attached drawings to complete each problem. Create separate layers using different colors for each required linetype. Assume that each grid equals .25" unless noted.

16. Use the **Circle**, **Arc**, **Fillet**, and **Trim** commands to complete the arc guide. Use centerlines where required. Each grid equals .25". Save the drawing as E-7-16.

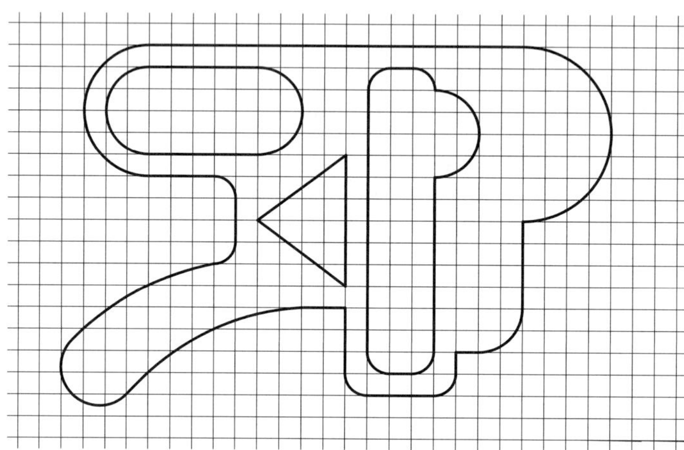

Problem 7-16

17. Use the **Circle**, **Arc**, **Fillet**, and **Trim** commands to complete the V-guide. Use centerlines where required. Each grid equals .125". Save the drawing as E-7-17.

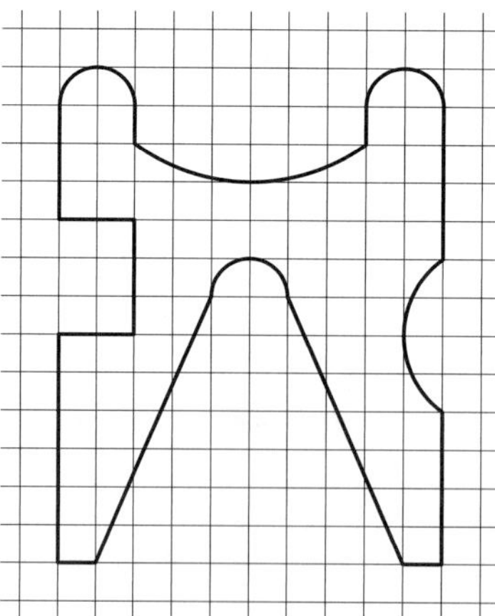

Problem 7-17

18. Draw the mold using the attached drawing. Save the drawing as E-7-18.

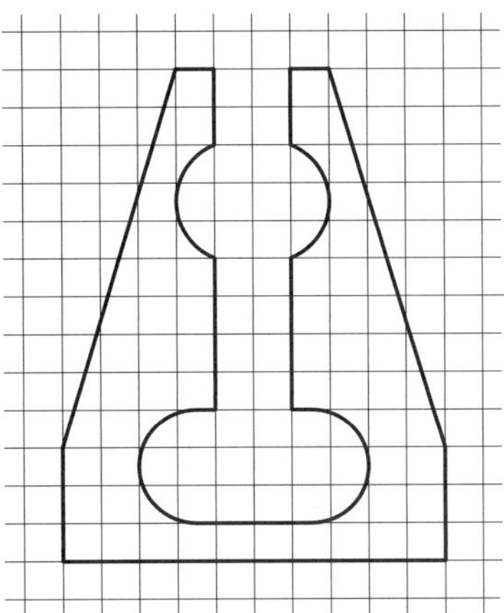

Problem 7-18

19. Draw the M-template using the attached drawing. Save the drawing as E-7-19.

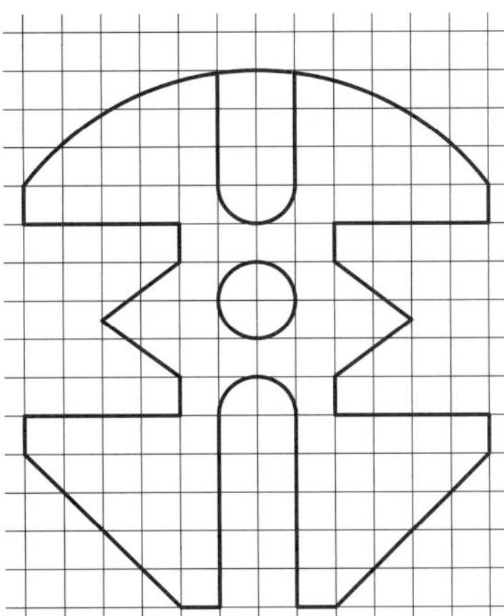

Problem 7-19

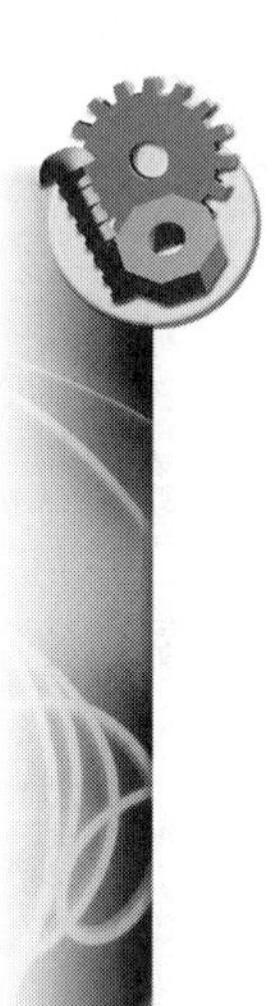

20. Draw the U-template using the attached drawing. Rotate the drawing 90° so that the opening is at the top. Save the drawing as E-7-20.

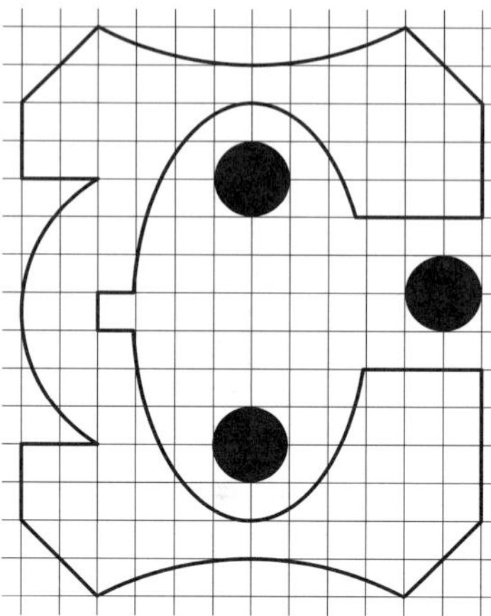

Problem 7-20

Exercises 21 through 25: Use the attached isometric drawings to complete a minimum of three views for each problem. Use proper linetype and lineweight to represent each feature. Assume that each grid equals .25". Save each problem using the corresponding problem number.

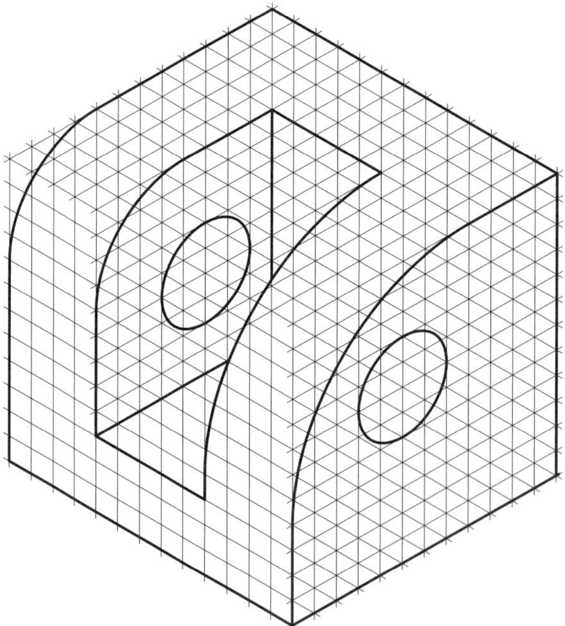

Problem 7-21

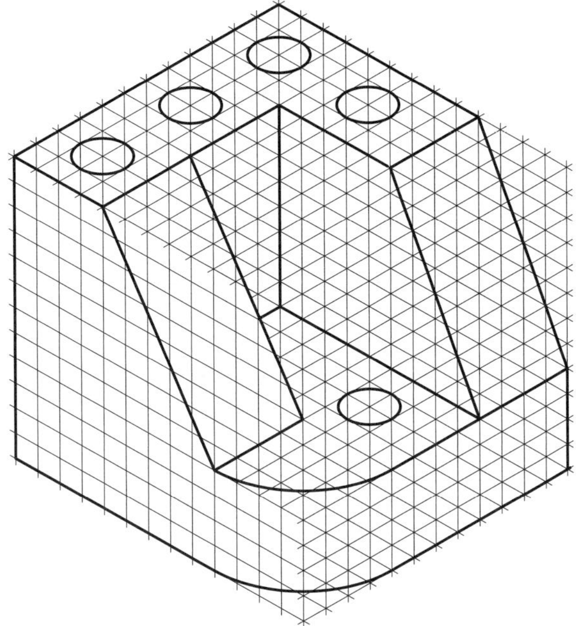

Problem 7-22

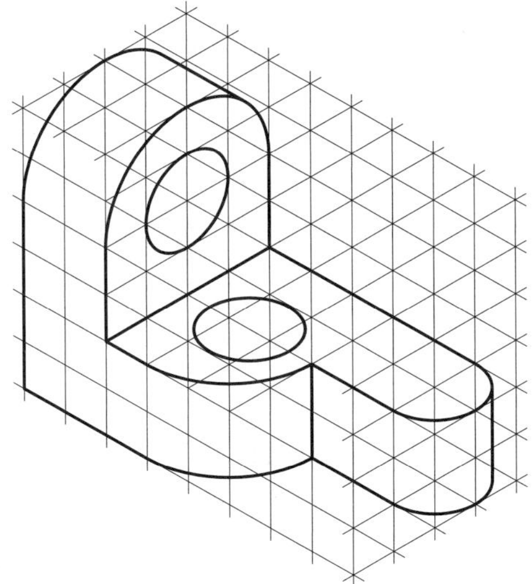

Problem 7-23

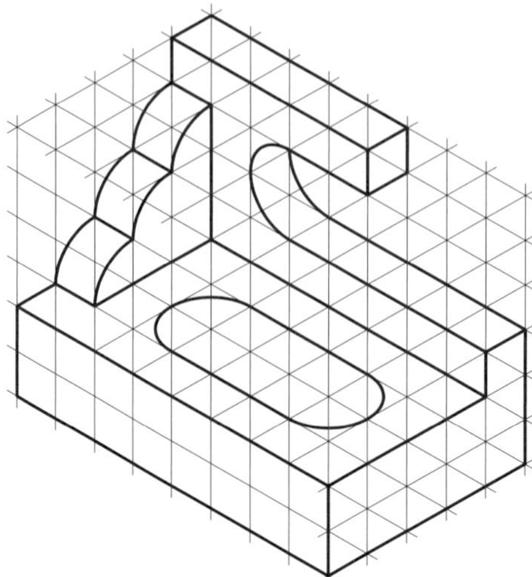

Problem 7-24

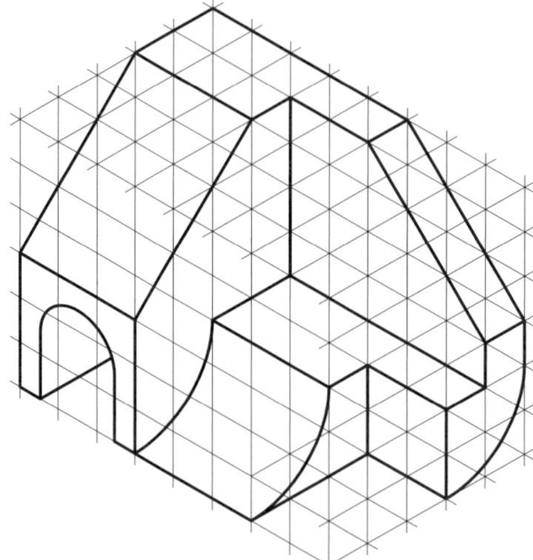

Problem 7-25

26. Open the drawing that was created in exercise 6.8. The part currently contains a 1.5" × .5" projection on the right side of the base. Change each of the corners of this slot to be .125"R. Change each end of the inclined surface to provide a .25"R tangent to each surface. Save the drawing as E-7-26.
27. Open the drawing that was created in exercise 6.11. The part currently has several square steps. Redraw these steps as tangent curves with a radius of .125". Redraw the rectangular slot in the base to have .50" radius ends. Save the drawings as E-7-27.

28. Open the drawing that was created in exercise 6.13. Add a .50" diameter circle in the center of the .75" × 1.00" flat surface in the front view. Add three holes with a diameter of .375" equally spaced to the front of the base. Show all of these holes extending through the part. Save this drawing as E-7-28

29. Open the drawing that was created in exercise 6.14. Add a .375" diameter hole centered in both of the upper inclined surfaces. Draw these holes so that they are perpendicular to each inclined plane and extend through the part. Draw an auxiliary view to clearly show the placement of one of the holes. Save this drawing as E-7-29.

30. Open the drawing that was created in exercise 6.31. Add a .375" diameter hole that extends through the entire part. Center the hole in the .75" × .75" area in the upper left end of the front view. Add two .25" diameter holes to the right side view and place the center of the holes up .25" from the bottom of the part. Place the center the holes so that they are .50" from each edge. Extend each hole through the part. Save this drawing as E-7-30.

31. Start a new drawing and draw a spline using the following coordinates: 3,3; 6,6; 9,3 and the default start and end point tangents. Repeat the process and draw a closed spline with the default tangents. Save the drawing as W-E-1-6.

32. Start a new drawing and draw a spline using the following coordinates: 3,3; 6,6; 9,3 and use vertical tangents for the start and end points. Repeat the process using horizontal tangents for the end points. Save the drawing as W-E-1-7.

33. Draw a spline with at least four control points using the default values for the tangents. Make five additional copies of the original and make the following alterations:

    a) Edit one copy so that it is a closed spline.

    b) Edit one copy by adding a control point.

    c) Edit one copy by deleting a control point.

    d) Edit one copy by altering the tolerance of each control point.

    e) Edit the final copy by moving the location on one of the control points.

    Save the drawing as W-E-1-9.

CHAPTER 7 QUIZ

1. What is the maximum number of sides that can be drawn for a polygon?
2. When the **Arc** command is used, what do the following letters represent?

   St　　Ce　　D

   Len　　E　　R

   Ang
3. What is the default method for drawing arcs from the toolbar?
4. What is the command sequence to draw three continuous arcs?
5. What command determines whether the interior of a donut will be black or hatched with lines?
6. What is the difference between inscribed and circumscribed, and what are the effects of each on a polygon?
7. What four components can be used to construct an ellipse?
8. When would TTR be used for drawing a circle?
9. What are the options available for drawing circles?
10. What command controls the rubber-band effect of geometric shapes?
11. What options must be entered before a donut can be drawn?
12. What is the largest angle rotation that can be used with an ellipse?
13. What shape would a polygon with 500 sides appear as?
14. How does ORTHO affect the drawing of a polygon?
15. Use the **Help** menu and determine the use of Elevation as it relates to a rectangle.
16. What is a NURB?
17. You've drawn a spline and are not happy with the results in one portion of the spline. List options for altering the shape.
18. How can spline data be removed from the drawing database?
19. Other than altering the tolerance, how can you control how close the spline comes to a control point?
20. What are the benefits of Fit tolerance?

CHAPTER 8

Controlling Drawing Accuracy

This chapter will introduce methods of controlling drawing accuracy by

- Controlling the aperture box
- Controlling drawing settings using the **Drafting Settings** dialog box
- Effectively using Object Snap options

Commands and options introduced in this chapter include

- Aperture
- **Dsettings**
- OSNAP

Up to this point, you've been creating drawing objects with little regard to drawing accuracy. In previous chapters, you were introduced to using the tools on the status bar to make the crosshairs move at specific intervals. This included the default settings of Object Snap. As you begin to combine geometric shapes into drawings, it is often helpful to join the shapes at exact locations. Joining objects by eye is difficult and should rarely be done. Objects that appear to be tangent might overlap when the **Zoom** command is used (see Figure 8.1). Perfect intersections can be achieved by using the Object Snap modes that were introduced in Chapter 3.

CONTROLLING THE APERTURE BOX

The normal cursor for the drawing area is the crosshairs surrounded by a box. If you enter a drawing command such as **Line**, the pick box is removed from the crosshairs as you select the First and Next points. As OSNAP is activated, an aperture box is added to the crosshairs, as seen in Figure 8.2. The marker is displayed as the cursor is moved near one of the active Object Snap points such as an endpoint or midpoint. The shape of the marker will vary depending on the mode being used. Default settings were introduced in Chapter 6, and various Object Snap modes will be introduced

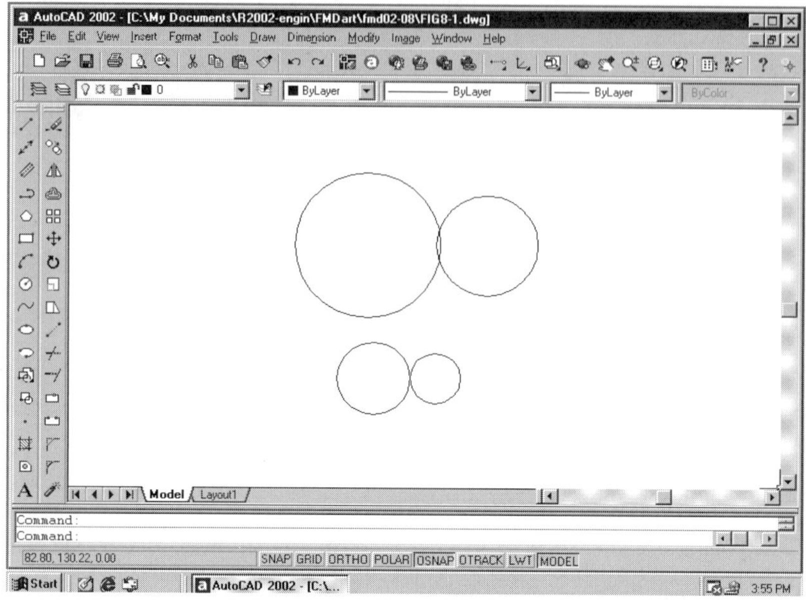

Figure 8.1 *Objects such as the lower circles that are placed by eye often appear to be perfectly placed. As the drawing is magnified using the* **Zoom** *command, circles that appear to be tangent are crossing.*

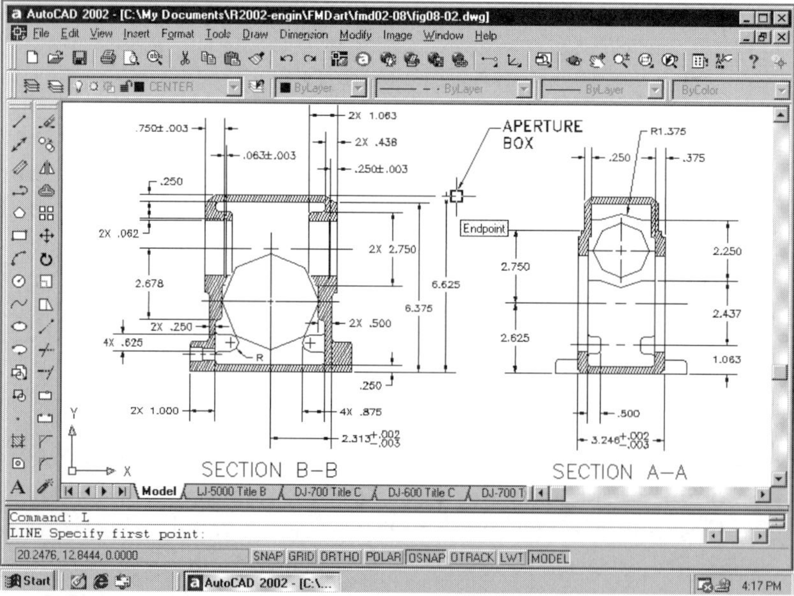

Figure 8.2 *As Object Snap is activated, an aperture box is added to the crosshairs. The box is used for selecting existing drawing objects to be used as a selection point for a new line, circle, arc, or ellipse. Courtesy Michael J. Easley.*

later in this chapter. Any object that lies within the target or aperture box is subject for the drawing or editing command being used. Although objects can be added or deleted from within the aperture box, that is time consuming. To obtain higher accuracy in object selection, you can adjust the size of the aperture box.

Note: The aperture box will not be displayed unless the **Display AutoSnap aperture box** check box is activated. Even if it is not displayed, an invisible aperture box will still function as you draw. The aperture box is displayed on the **Drafting** tab of the **Options** dialog box. To display the dialog box, select **Options** from the **Tools** menu.

SIZE

The size of the aperture box can be adjusted using the **Drafting** tab of the **Options** dialog box or by keyboard. Access the **Options** dialog box by selecting **Options** from the **Tools** menu and select the **Drafting** tab. This will produce the dialog box shown in Figure 8.3. Select the slide bar in the **Aperture Size** area and hold the select button down. Dragging the bar to the left will decrease the size. Moving the bar to the right will increase the size. The size will be indicated in the **Aperture Size** window display.

To adjust the size of the aperture box at the Command prompt, type

Command: **APERTURE** ENTER

Object snap target height (1-50 pixels) <10>:

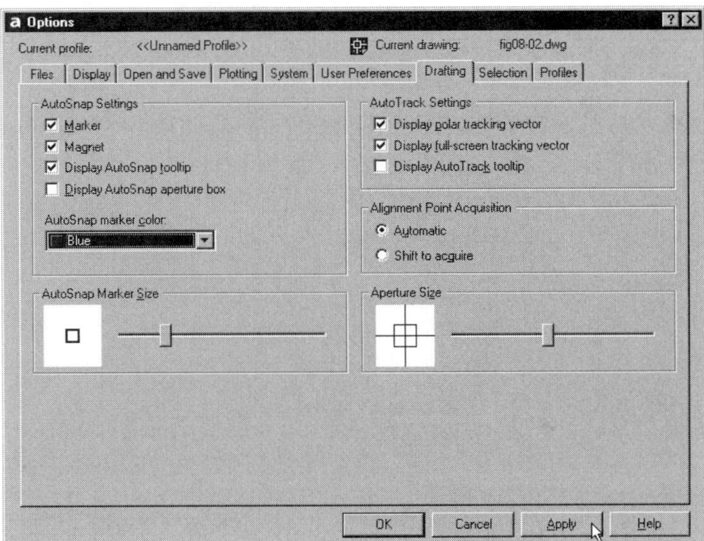

Figure 8.3 *The Options dialog box is accessed by selecting* **Options** *from the* **Tools** *menu. The dialog box can be used to adjust the aperture box, the marker size and color, AutoSnap settings, and AutoTrack settings.*

A numeric value can now be entered. The larger the number of pixels selected, the larger the box. Figure 8.4 shows the effect of changing the value. The value entered will now be the default for all future Object Snap commands until a new value is entered. Changing the default for the Object Snap aperture box will not affect the size of the select box used for **Erase**.

ALTERING THE APERTURE COLOR

In addition to adjusting the size of the box, you can adjust the aperture color. Select **Options** from the **Tools** menu and click the **Display** tab of the **Options** dialog box. Select the **Color** button. This will display the **Color Options** dialog box similar to Figure 8.5. Select the list arrow beside the **Window Element** display. Selecting the **Model Tab Pointer** option will display a color menu. Once the desired color has been selected, clicking the **Apply & Close** button will change the crosshairs and box to reflect the selected color.

CONTROLLING DRAWING SETTINGS

To be used effectively, Object Snap settings need to be understood and put to use. Major areas to be considered include the **AutoSnap** settings and **AutoTrack Settings**. Each can be set using the **Options** dialog box shown in Figure 8.3.

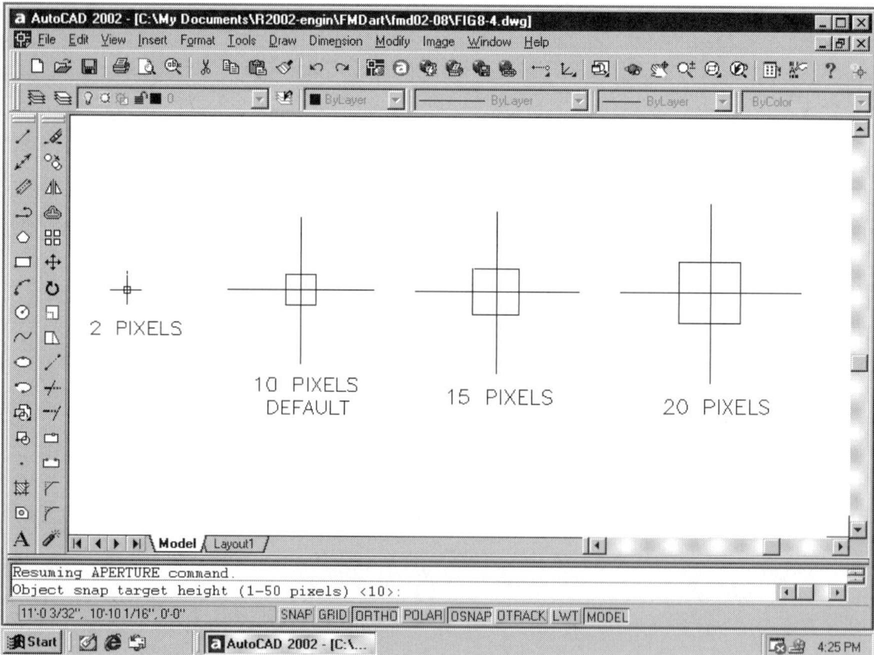

Figure 8.4 *As the value for the height of the aperture box is altered, the height is changed.*

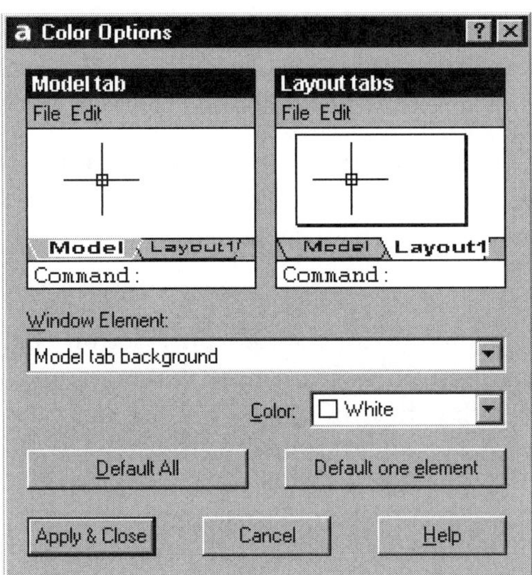

Figure 8.5 *The* **Color Options** *dialog box can be used to adjust the marker color. Access the dialog box by choosing* **Options** *from the* **Tools** *menu, selecting the* **Display** *tab, and clicking the* **Color** *button.*

AUTOSNAP SETTINGS

The AutoSnap settings control the settings for each Object Snap mode. Settings include the marker, magnet, and tooltip.

Marker

As an Object Snap command is used, the marker for the current Object Snap mode will be displayed by the aperture box. AutoCAD uses a different geometric shape to represent each Object Snap mode. Specific markers will be introduced later in this chapter. With the **Marker** check box activated, each time the cursor is moved to a point controlled by Object Snap, the appropriate Object Snap marker will be displayed. For example, when you use the **Line** command, each time the cursor is moved toward the end of a line, the Endpoint marker is added to the cursor to indicate that the endpoint will be selected. If you deactivate this setting, the marker box will not be displayed, but the current Object Snaps will be unaffected. AutoCAD has it ON by default for a reason. It is usually very helpful for new users to be reminded of the current settings.

Altering the Marker Size

The size of the Object Snap marker can be adjusted using the **Drafting** tab of the **Options** dialog box. Select the slider bar in the **AutoSnap Marker Size** area and hold the select button down. Dragging the bar to the left will decrease the marker size.

Moving the bar to the right will increase the marker size. The size will be indicated in the **AutoSnap Marker Size** window display.

Altering the Marker Color

Selecting the arrow in the **AutoSnap marker color** edit box will display a menu of the seven basic drawing colors. Selecting a color name will automatically change the marker to that color and close the color menu.

Magnet

The **Magnet** check box is a toggle for the ON/OFF setting. By default, AutoCAD magnet is ON. As you've used a drawing command, the cursor often moved toward the end of a line. That is the effect of having the magnet ON. The magnet automatically moves the cursor to lock it onto the nearest Object Snap point as the cursor is moved. Keeping the magnet on will increase your drawing speed.

Tooltip

The **Display AutoSnap tooltip** check box is a toggle for the ON/OFF setting. It is ON By default. Each time the cursor is moved near an active Object Snap mode, the Object Snap mode to be used will be displayed in a tooltip below the cursor. Figure 8.6 shows the tooltip for the Endpoint Object Snap mode. On simple drawings, the

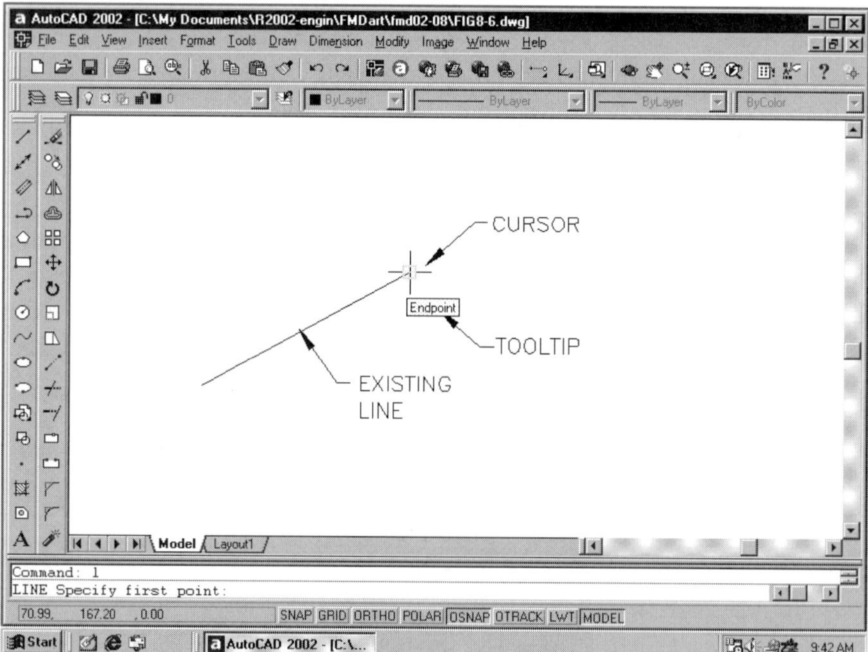

Figure 8.6 *Each time the cursor is moved near an active Object Snap mode, a tooltip below the cursor shows the Object Snap mode to be used.*

tooltip might not seem important. On complex drawings with several Object Snap modes active, the tooltip can be helpful in determining what Object Snap mode will be used. Once you recognize each of the Object Snap markers, you might find the tooltips unnecessary.

AUTOTRACK SETTINGS

The AutoTrack settings control the settings that can enhance each Object Snap mode. Settings include the polar tracking vector, full-screen tracking vector, and AutoTrack tooltip displays. Each option is an ON/OFF toggle switch. By default, each is set to ON.

Polar Tracking Vector

If Polar Tracking is ON, the **Display polar tracking vector** check box, when activated, sets Polar Tracking so that lines can be drawn along angles relative to the drawing "first" or "next" points. Angle settings must be 90° divisors, such as 15°, 30°, and 45°. Tracking increments are set on the **Polar Tracking** tab of the **Drafting Settings** dialog box (see Figure 8.7). You can access the dialog box by selecting **Drafting Settings** from the **Tools** menu or by typing OS (ENTER) or DSETTINGS (ENTER) at the Command prompt. With this option ON, tracking angles will be displayed similar to Figure 8.8. The polar tracking display can be toggled ON/OFF using the POLAR button on the status bar.

Full-Screen Tracking Vector

The **Display full-screen tracking vector** check box controls the display of tracking vectors. Tracking vectors are construction lines that can be used as a base, enabling

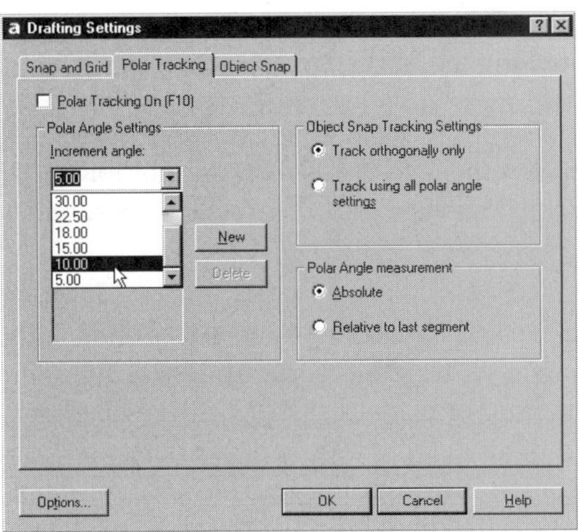

Figure 8.7 *Tracking increments can be set using the* **Increment angle** *edit box on the* **Polar Tracking** *tab of the* **Drafting Settings** *dialog box.*

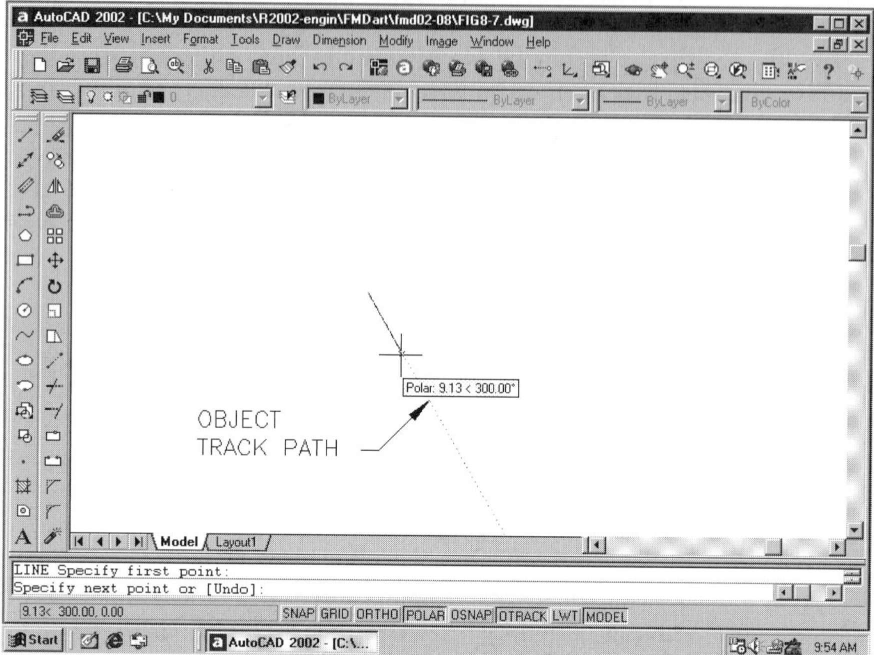

Figure 8.8 *If the Polar Tracking option is ON, tracking angles will displayed below the cursor.*

objects to be drawn at a specific angle (see Figure 8.8). They can also be used to draw objects at a specific relationship to other objects. In their default setting, tracking vectors extend to the edge of the drawing display area. Alignment paths can be shortened or eliminated. When this check box is not selected, the path is displayed only from the Object Snap point to the cursor. Controls for setting Object Snap Tracking are found on the **Polar Tracking** tab of the **Drafting Settings** dialog box.

AutoTrack Tooltip

The AutoTrack tooltip is similar to the AutoSnap tooltip. The **Display AutoTrack tooltip** check box controls the display of the tracking coordinates.

OBJECT SNAP

Object Snap modes can be used to select an exact start or endpoint for connecting lines, rather than hoping you're close. Object snaps allow a line or other object to be snapped to a specific point. For instance, Object Snap will allow a line to be started exactly at the midpoint of an existing line, with absolutely no guessing. Object Snap has two modes of operation: single-point override and continuous. The continuous mode allows modes to be set and used throughout the drawing session. The override mode allows you to select an Object Snap mode to be used once in the middle of a command.

By default, Endpoint, Midpoint, Center, Intersection, and Extension are set on the ON position. The desired override Object Snap can be activated by selecting its button on the **Object Snap** toolbar, by selecting its button from the **Temporary Tracking** flyout menu of the **Object Properties** toolbar, by pressing SHIFT and right-clicking, or by keyboard. Display the toolbar by selecting **Object Snap** from **Toolbars** on the **View** menu. Figure 8.9 shows a listing of each Object Snap mode found on the **Object Snap** toolbar and the marker that will be displayed with each mode. An Object Snap mode can be used in the middle of a command by entering three letters to represent the mode for the "next" point of a command sequence. For instance, to draw a line that is perpendicular to an existing line would require the following sequence:

Command: **L** ENTER *(Or click the Line button.)*

LINE

Specify first point: *(Select any point for the starting point of the new line).*

Specify next point or [Undo]: **PERP** ENTER *(Or click the Perpendicular button.)*

to: *(Select an endpoint for the starting point of the new line).*

Specify next point or [Undo]:

Continuous Object Snap allows the desired Object Snap mode to be selected for continuous use. Continuous Object Snap options are set on the **Object Snap** tab of the **Drafting Settings** dialog box. (Choose **Drafting Settings** from the **Tools** menu or type OS (ENTER) at the Command prompt to access the dialog box.) Continuous Object

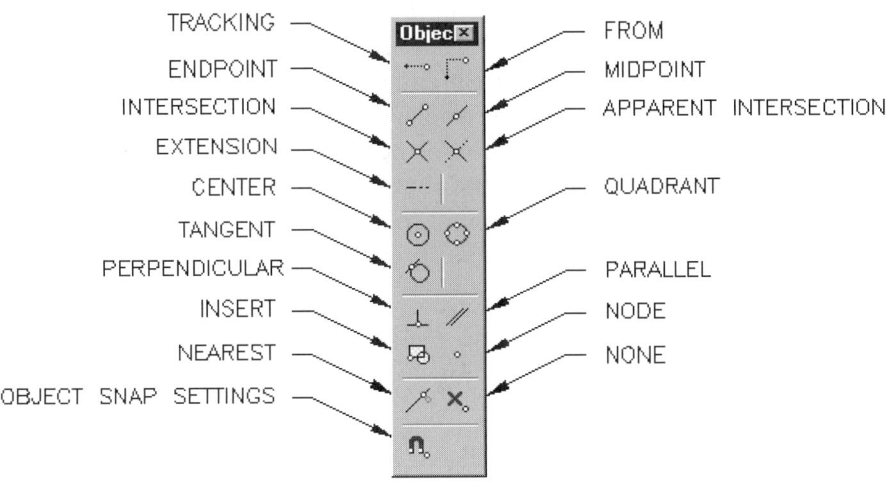

Figure 8.9 *The* **Object Snap** *toolbar and the marker box for each mode.*

Snap will be discussed later in this chapter. No matter how they are accessed, using the Object Snap commands will save time and greatly improve your accuracy. Once OSNAP is activated, the marker box is used to select objects to snap to. Move the crosshairs so that the desired object lies within the marker box, and then click the select button. If more than one object lies within the marker box, the closest object will be selected.

OBJECT SNAP MODES

Object Snap tools include ENDpoint, MIDpoint, INTersection, EXTension, APParent Intersection, CENter, NODe, QUAdrant, INSertion, PERpendicular, TANgent, NEArest, PARallel, QUIck, and NONe. These modes can be used for drawing lines or arcs that connect at certain points to existing lines, circles, arcs, or ellipses. Object Snap modes also can be used any time a point needs to be specified, and for **Copy**, **Move**, and **Insert** commands, which will be discussed in later chapters. Notice in the modes list that the first three letters of some of the modes are written in capital letters. Typing these letters at the prompt for start point or endpoint will activate the desired Object Snap mode.

ENDPOINT

The Endpoint mode allows an arc, line, or circle to be drawn to or from the endpoint of a previous line or arc by snapping to the desired endpoint. Start the process by selecting the **Endpoint** button on the **Object Snap** toolbar when you are prompted for a "first point." To add a line that extends between the endpoints of two existing lines, the command sequence would be as follows:

Command: **L** ENTER *(Or click the Line button.)*

LINE Specify first point: *(Select the Endpoint button.)*

Now move the cursor near the end of the line that you would like the new line to connect to. The cursor does not have to touch the end of the line, but it must be between the midpoint and endpoint so the correct end of the line is selected. As the cursor touches a line, the square marker box is displayed around the nearest endpoint. (See Figure 8.10.) Click the select button, and the new line segment will be connected to the selected endpoint. Select the **Endpoint** button again at the prompt for the "next point," to connect the new line to the end of the second line. The new line will automatically end at the selected end of the existing line segment.

The endpoint can also be selected at the keyboard, by typing END (ENTER) as the "first point" of the desired command. Move the marker box and select the desired line. To draw a line from the endpoint of an existing line to the endpoint of a second line, use the following command sequence:

Command: **L** ENTER *(Or click the Line button.)*

LINE Specify first point: **END** ENTER *(Or click the Endpoint button.)*

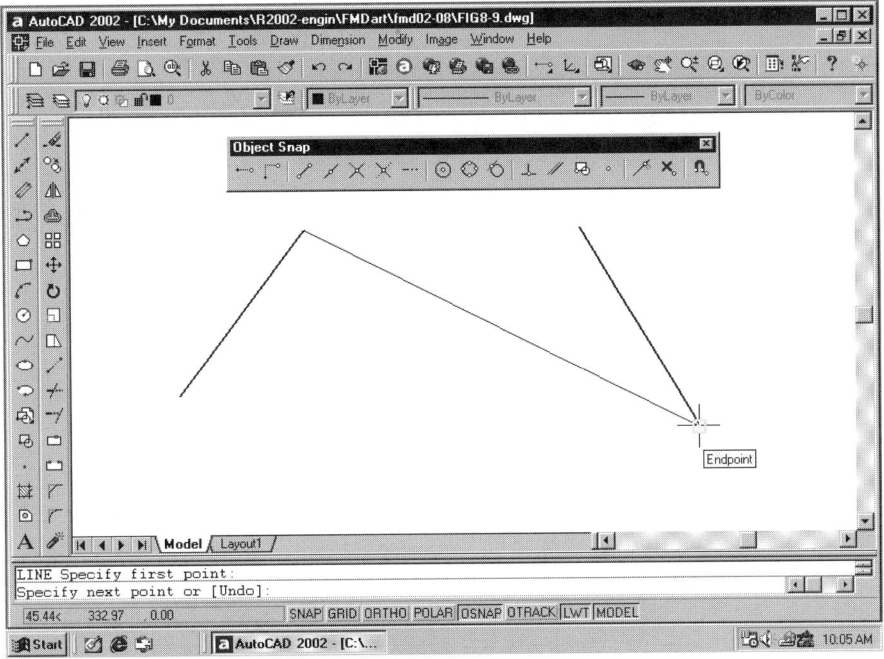

Figure 8.10 *When the Endpoint mode of Object Snap is selected, a marker is placed on the endpoint of the selected line to indicate the selected "first" or "next" point.*

_end of (Select desired line.)

Specify next point or [undo]: **END** ENTER *(Or click the Endpoint button.)*

of *(Select desired line.)*

Specify next point or [undo]: ENTER

Command:

The command sequence can be seen in Figure 8.11. If you're thinking, "This is extra work," you're right. By default, AutoCAD automatically draws lines from endpoint to endpoint. However, the process will work wonders as different Object Snap modes are entered.

MIDPOINT

Midpoint allows the selection of a line or arc at its midpoint. The **Midpoint** button is selected in response to the "first point" prompt. To add a line that extends between the midpoints of two existing lines, the command sequence would be as follows:

Command: **L** ENTER *(Or click the Line button.)*

LINE

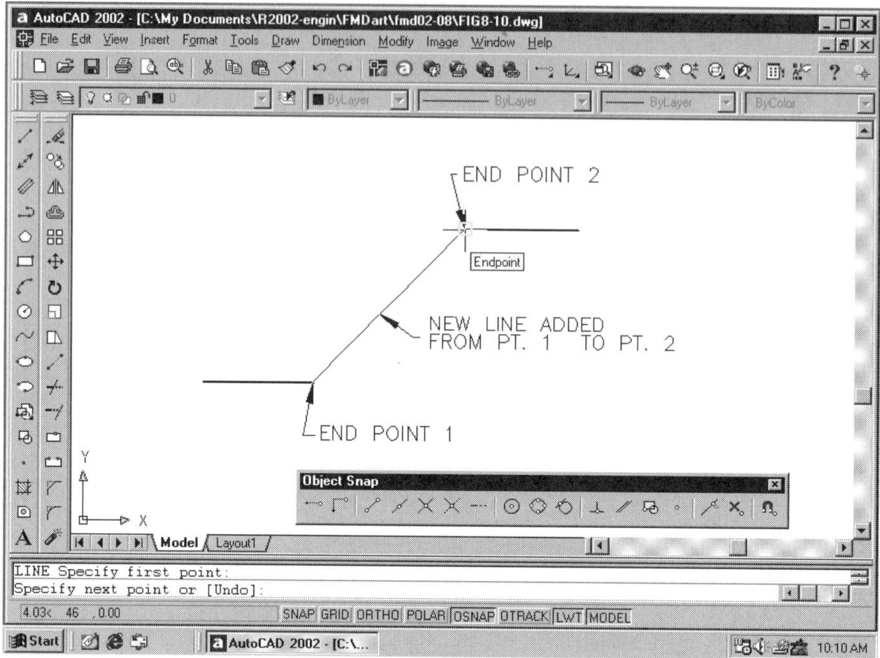

Figure 8.11 *To draw a line from the endpoint of one line to the endpoint of another, click the* **Endpoint** *button or type* **END** (ENTER) *when the "first point "and the "next point" prompts of the* **Line** *command are displayed.*

Specify first point: *(Select the Midpoint button on the Object Snap toolbar.)*

As the cursor touches a line, the marker box is displayed around the nearest midpoint. Click the select button, and the new line segment will be connected to the selected midpoint of the existing line. Select the **Midpoint** button as the "next point" is requested at the prompt and the new line will automatically end at the selected midpoint of the existing line segment.

The midpoint for an arc or line can also be selected by typing **MID** (ENTER) as the "first point" or "next point" prompt is displayed. To draw a line from the midpoint of a line to the midpoint of a second line, use the following command sequence:

Command: **L** ENTER *(Or click the Line button.)*

LINE

Specify first point: **MID** ENTER *(Or click the Midpoint button.)*

_mid of (Select desired line.)

Specify next point or [Undo]: **MID** ENTER *(Or click the Midpoint button.)*

of *(Select desired line.)*

Specify next point or [Undo]: ENTER

Command:

The command sequence can be seen in Figure 8.12.

INTERSECTION

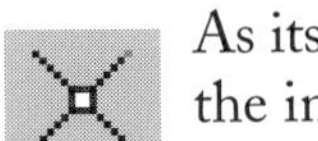

As its name implies, Intersection allows a line or arc to be extended to or from the intersection of any lines, arcs, or circles. The command sequence is similar to other Object Snap command entries and can be seen in Figure 8.13. By default, this option is ON. To draw a line from the intersection of two lines, move the cursor so the desired intersection lies within the target box. This will cause the cursor to snap to the intersection, allowing the next line to be drawn based on the selected intersection.

The Intersection mode can also be used by selecting the **Intersection** button on the **Object Snap** toolbar or by entering INT (ENTER) as a "first point" or "next point"; the desired intersection can be snapped to. The command sequence is as follows:

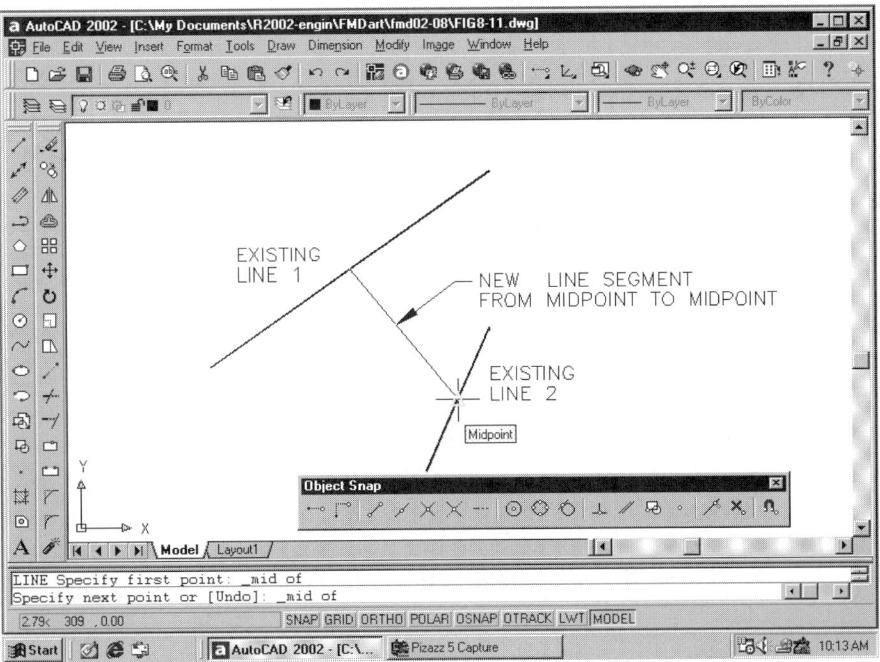

Figure 8.12 *The Midpoint mode can be used to draw a line from the midpoint of one line to the midpoint of another by selecting the* **Midpoint** *button or by typing* **MID** *(*ENTER*) for the "first" and "next" point prompts of the* **Line** *command.*

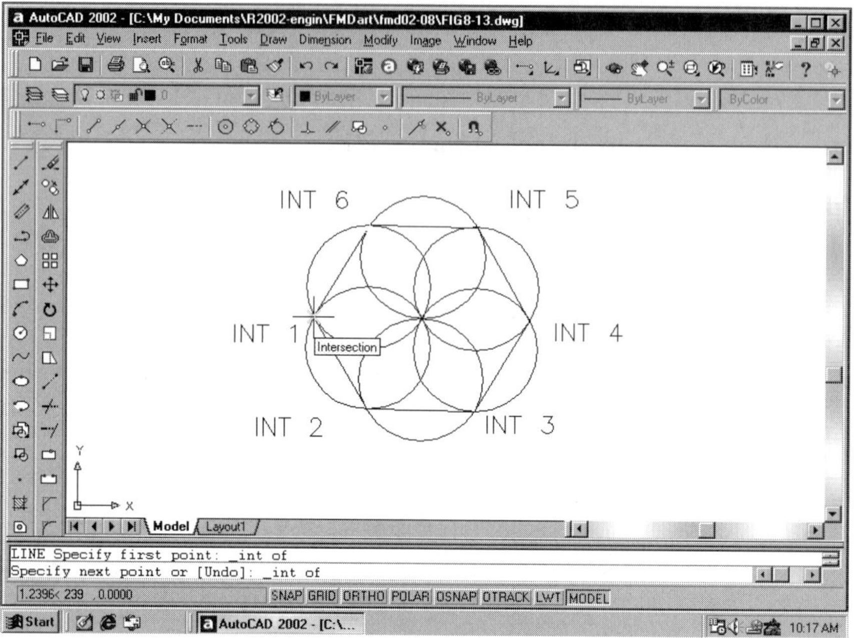

Figure 8.13 *The Intersection mode of Object Snap is used to draw a line from the intersection of each circle.*

Command: **L** ENTER *(Or click the Line button.)*

LINE

Specify first point: **INT** ENTER *(Or click the Intersection button.)*

_int of *(Select desired to point.)*

Specify next point or [Undo]: *(Select desired to point.)*

Specify next point or [Undo]: ENTER

Command:

EXTENSION

The Extension mode of Object Snap will use a point that lies along an extension path of a line as a First, Next, or Center point for an object. By default, this option is ON. This mode has two methods of operation. The first method will extend a line from an existing line. (See Figure 8.14.) This method can be used to lengthen a line. To extend a line from the end of an existing line, use the following procedure:

Command: **L** ENTER *(Or click the Line button.)*

LINE

Specify first point: **EXT** ENTER *(Or click the Extension button.)*

_ext of *(Select desired end point.)*

Specify next point or [Undo]: *(Select desired to point. Be sure OTRACK is ON.)*

Specify next point or [Undo]: ENTER

Command:

The second method will use a point that lies on the path as the line is extended from the existing line. This method will work well to place a point lying on a projected path. (See Figure 8.15.) Use the following procedure to select a point that lies on an extension of an existing line:

Command: **L** ENTER *(Or click the Line button.)*

LINE

Specify first point: **EXT** ENTER *(Or click the Extension button.)*

_ext of *(Move the cursor over the endpoint of the existing line but do not pick the endpoint. Extension will automatically select the indicated endpoint and project a path. Select a point that lies on the projected path.)*

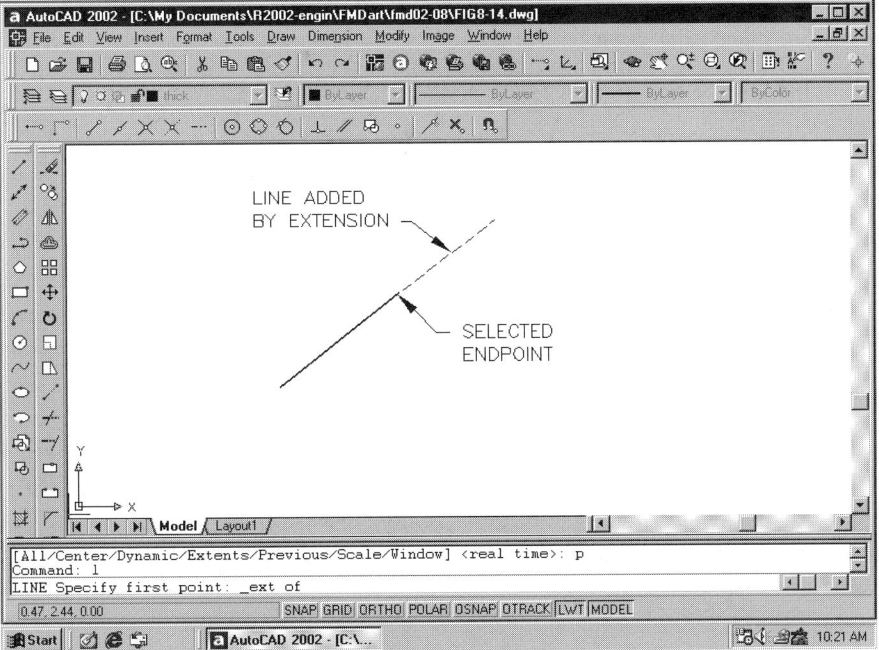

Figure 8.14 *The Extension mode of Object Snap will use a point that lies along an extension path of a line as a "next" point. When the endpoint of an existing line is selected for the extension point, the existing line will be lengthened along the projected path to the "next" point.*

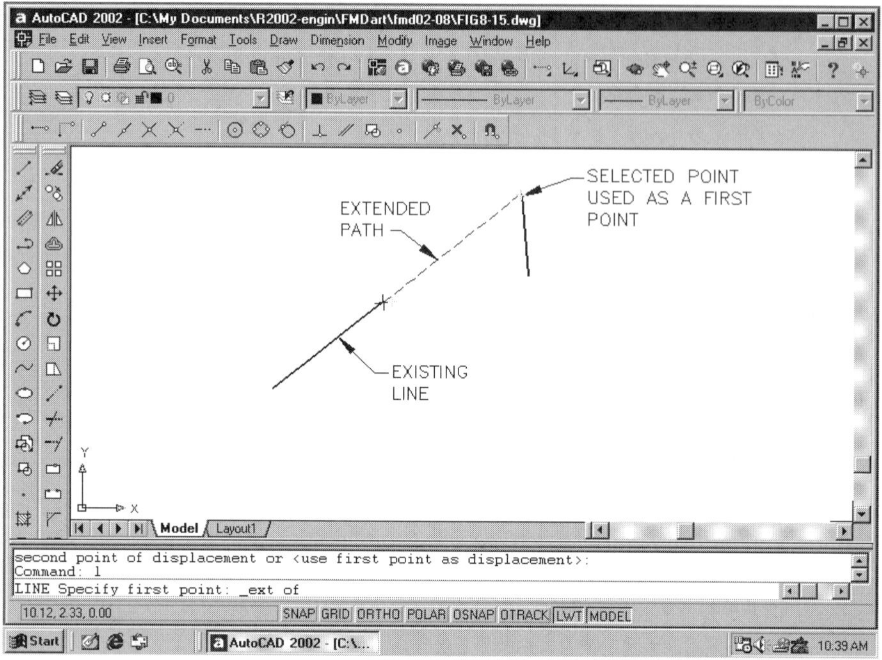

Figure 8.15 *Passing the EXT marker over the endpoint, but not selecting it as an endpoint, will use that point as a projection point for the extension path. Extension will use a point that lies on the path as the line is extended from the existing line.*

Specify next point or [Undo]: *(Select desired to point.)*

Specify next point or [Undo]: ENTER

Command:

APPARENT INTERSECTION

The Apparent Intersection mode finds the apparent intersection, or where two lines would intersect if one or both were extended. To find the apparent intersection, select the **Apparent Intersect** button or use the keyboard to type **APP** (ENTER) at the "first" or "next" prompt. Because the line will be generated from where two lines will intersect, two "first" points will need to be selected. The prompts will read " _appint of" and "and" to remind you to select two lines. The new line will be placed where the selected lines would meet. From the Command prompt, use the following command sequence:

Command: **L** ENTER *(Or click the Line button.)*

LINE

Specify first point: **APP** ENTER *(Or click the Apparent Intersect button.)*

_appint of *(Select first line to be projected.)*

and *(Select second line to be projected.)*

Specify next point or [Undo]: *(Select desired to point.)*

Specify next point or [Undo]: ENTER

Command:

Pressing ENTER will extend the line. The command sequence can be seen in Figure 8.16. Chapter 11 will present two alternatives to this option.

CENTER

By default, the Center mode is ON, allowing an object to be drawn based on the center of a circle. A line or arc often needs to be drawn from the center of an existing circle, or several circles might need to be drawn around one center point, as seen in Figure 8.17. This drawing could be easily drawn using the Center mode of Object Snap. To draw a line from the center of a circle, enter the **Line** command, and move the cursor over the edge of the circle. This will highlight the center of the circle. Move the cursor to the center point, and the cursor will snap to the center, allowing the center point to be used as the first point of the line to be drawn.

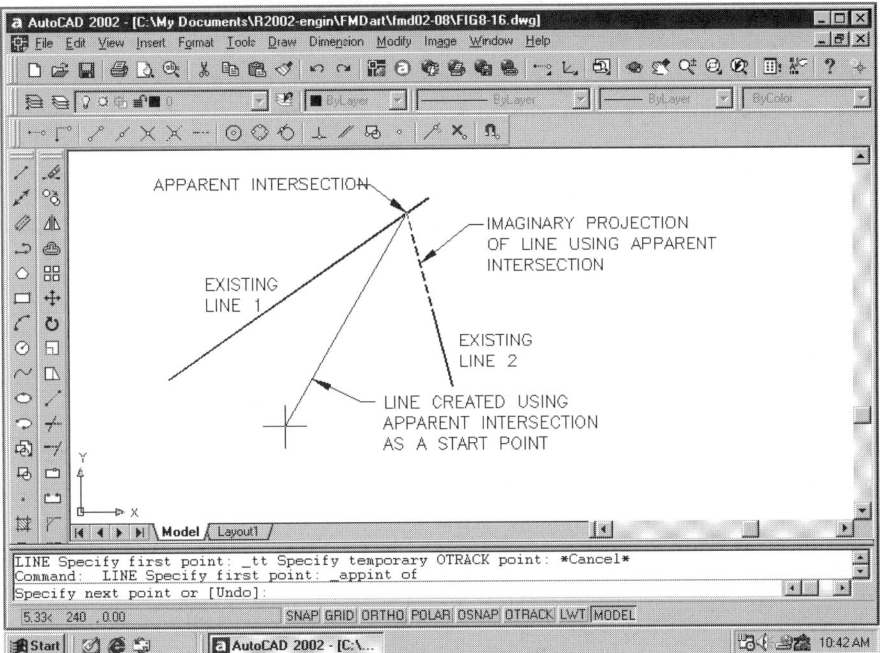

Figure 8.16 *The Apparent Intersection mode of Object Snap can be used to extend a line between two lines that would intersect. The line was started by using the Endpoint mode to select the end of line 2. Then the* **Apparent Intersect** *button was selected or* **APP** (ENTER) *was typed for the "to" point. The marker box was used to select line 1 as the line to intersect. The endpoint of the original line must be selected to display the new line segment.*

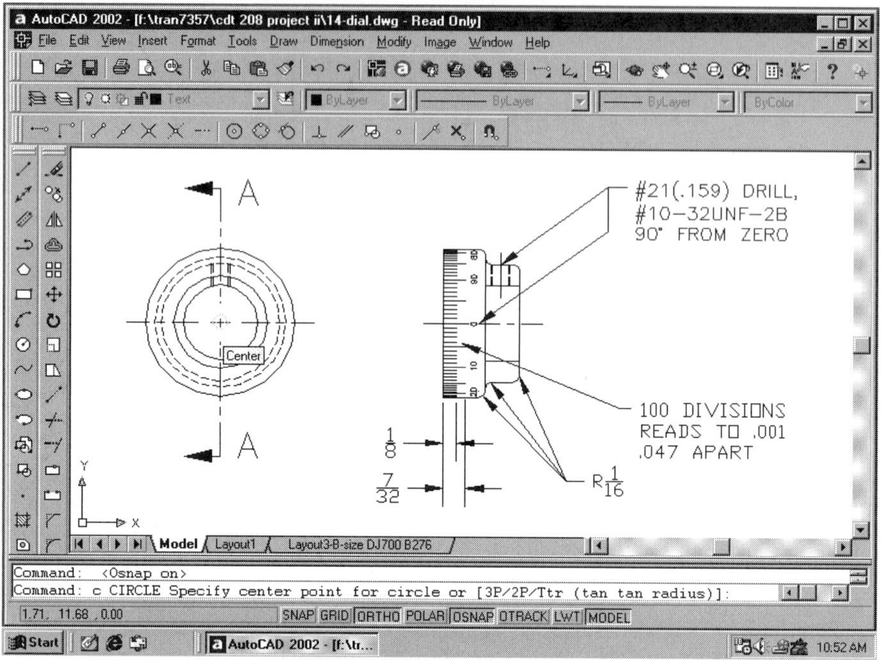

Figure 8.17 *The Center mode of Object Snap can be used whenever a drawing object needs to be placed based on the center of an existing circle. Courtesy Hanh Tran.*

With the option OFF, a line or arc can be extended from the center point of a circle by selecting the **Center** button on the **Object Snap** toolbar. The Center mode can also be selected by keyboard using the following command sequence:

Command: **L** ENTER *(Or click the Line button.)*

LINE

Specify first point: **CEN** ENTER *(Or click the Center button.)*

_cen of: *(Select a circle.)*

Specify next point or [Undo]: *(Select the desired endpoint.)*

Specify next point or [Undo]: ENTER

Command:

Touch any part of the circle, and a rubber-band line now extends from the exact center of the circle until a "next" point is selected. This process can be seen in Figure 8.18.

NODE

The Node mode can be used to snap to a point—lines or arcs can be extended from or to predetermined points. Methods of drawing a point will be introduced in Chapter 12. This could be used, for instance, to lay out points on a

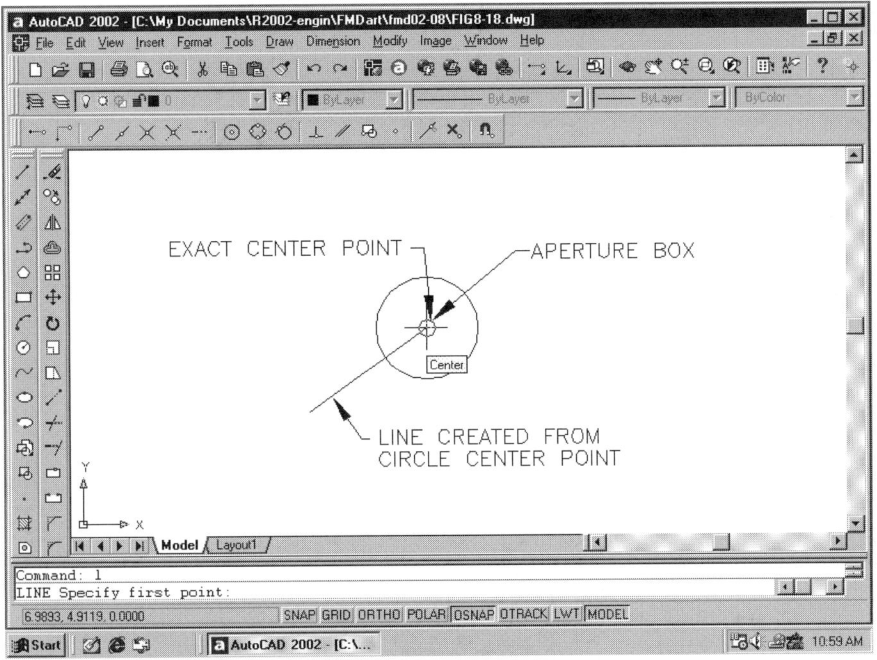

Figure 8.18 *To extend a line from the center of a circle, enter the* **Line** *command. When prompted for a "first" point, select the* **Center** *button or type* **CEN** *(*ENTER*) at the Command prompt. This will produce the center marker that can be used to select the desired circle. When you place the marker anywhere on the circle and select the circle, AutoCAD will automatically use the center point of that circle as the "first" point for a line.*

pattern layout. To activate this mode, select the Node button on the **Object Snap** toolbar or type NOD (ENTER) at either the "first " or "next" point.

QUADRANT

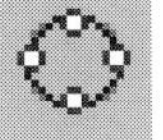

The Quadrant mode will allow a line or arc to be snapped to the 0°, 90°, 180°, and 270° positions of a circle or arc. Start this mode by selecting the **Quadrant** button on the toolbar or by typing QUA (ENTER) as the "first" or "next" point. Place the aperture box on the circle or arc near the quadrant point, and the new object will automatically be joined at the quadrant point, as seen in Figure 8.19. To draw a line from the quadrant point of an existing circle, use the following command sequence:

Command: **L** ENTER *(Or click the Line button.)*

LINE

Specify first point: **QUA** ENTER *(Or click the Quadrant button.)*

_qua of *(Select desired circle.)*

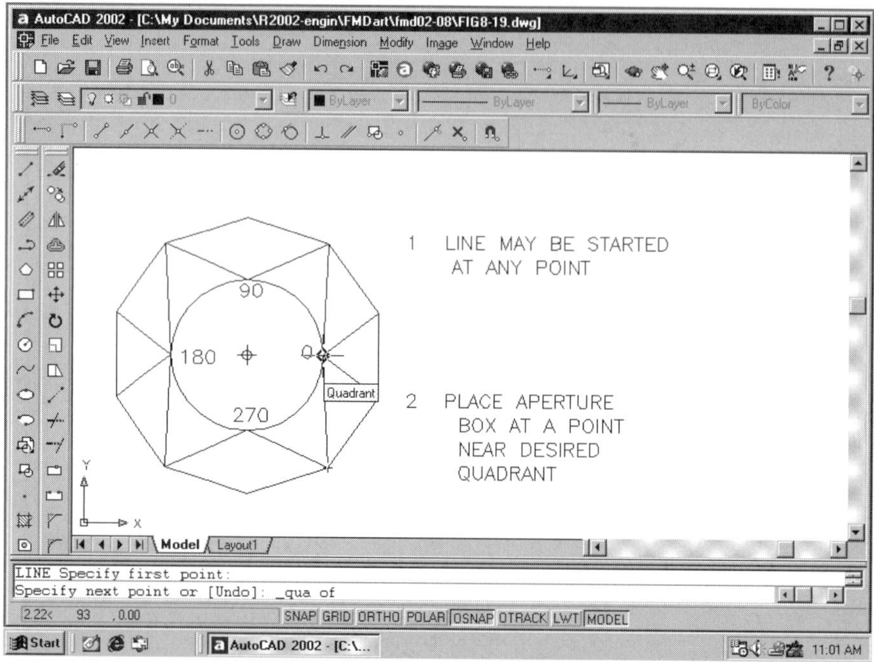

Figure 8.19 *The Quadrant mode of Object Snap can be used to attach a line or arc to the quadrants of an existing circle.*

Specify next point or [Undo]: *(Select desired to point.)*

Specify next point or [Undo]: ENTER

Command:

INSERT

The Insert mode will snap to the insertion point of a shape, text, attribute, or block. This might seem like meaningless stuff now, but combined with the information in later chapters, it will allow information to be inserted at an exact point.

PERPENDICULAR

Previously introduced Object Snap modes could be used to bring a line or an arc toward or away from an object. The Perpendicular mode is used to draw a line that is perpendicular to an existing line. The process can be seen in Figure 8.20. To draw a line from any point perpendicular to an existing line, select the **Perpendicular** button when prompted for a "first" or "next" point, or type PER (ENTER) at the Command prompt and use the following command sequence:

Command: **L** ENTER *(Or click the Line button.)*

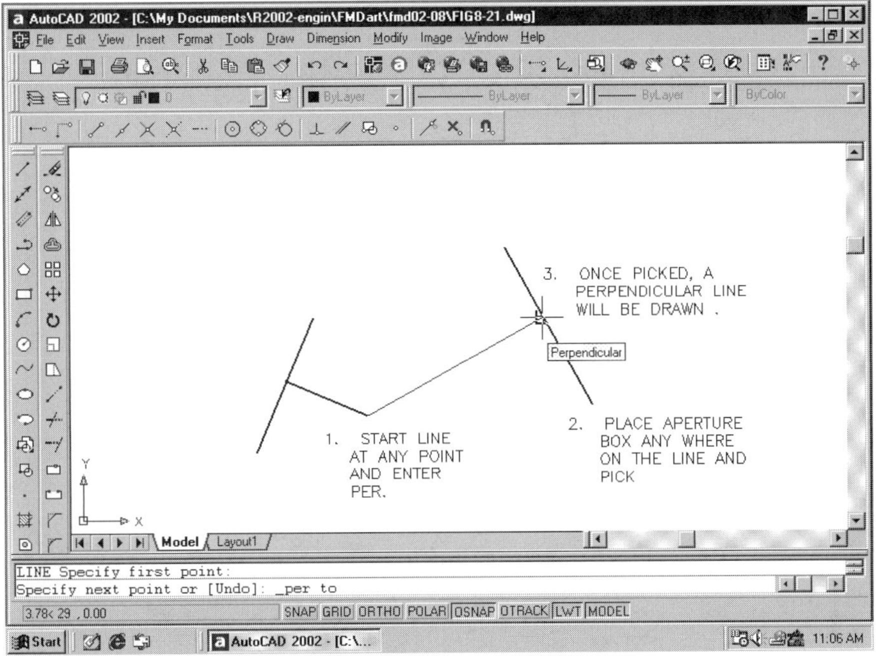

Figure 8.20 *The Perpendicular mode can be used to project a new line so that it will be at a 90° angle to an existing line.*

LINE

Specify first point: *(Select desired point.)*

Specify next point or [Undo] **PER** ENTER *(Or click the Perpendicular button.)*

_per *(Select desired line.)*

Specify next point or [Undo]: ENTER

Command:

TANGENT

The Tangent mode will allow a line to be snapped tangent to a circle or an arc. Start this mode by selecting the **Tangent** button on the **Object Snap** toolbar or by typing TAN (ENTER) at the "first" or "next" point prompt. Figure 8.21 shows an example of how Tangent can be used. To draw a line tangent to an existing circle, use the following command sequence:

Command: **L** ENTER *(Or click the Line button.)*

LINE

Specify first point: *(Select desired point.)*

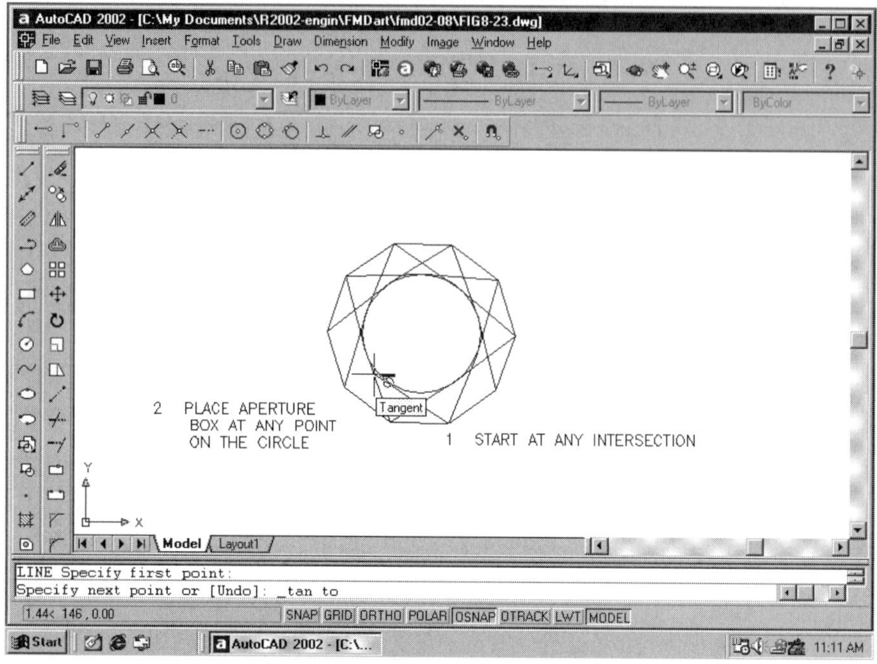

Figure 8.21 *Tangent is started by selecting the* **Tangent** *button or by typing* **TAN** *(*ENTER*) for the "first" or "next" point of the new line or arc.*

Specify next point or [Undo]: **TAN** ENTER *(Or click the Tangent button.)*

_tan to *(Select desired arc or circle.)*

Specify next point or [Close/Undo]: ENTER

Command:

NEAREST

The Nearest mode can be used to snap to the circle, line, or arc that is nearest to the crosshairs. To activate this mode, select the **Nearest** button on the **Object Snap** toolbar or type NEA (ENTER) at either the "first" or "next" point. To draw a line from any point of an existing line, use the following command sequence:

Command: **L** ENTER *(Or click the Line button.)*

LINE

Specify first point: **NEA** ENTER *(Or click the Nearest button.)*

_nea of *(Select desired line.)*

Specify next point or [Undo]: *(Select desired to point.)*

Specify next point or [Undo]: ENTER

Command:

The command sequence can be seen in Figure 8.22.

PARALLEL

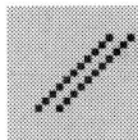

The Parallel mode will draw a line parallel to an existing line. Unlike other modes, Parallel is specified after the "first" point is selected. Chapter 10 will explore an alternative method of placing parallel lines using the **Offset** command.

Command: **L** ENTER *(Or click the Line button.)*

LINE

Specify first point: *(Select desired to point.)*

Specify next point or [Undo]:**PAR** ENTER *(Or click the Parallel button.)*

_par to *(Move the cursor over, but do not select the endpoint of the desired line. As the aperture box moves over the endpoint, the point is marked and cursor snaps to a parallel path.)*

Specify next point or [Undo]: ENTER

Command:

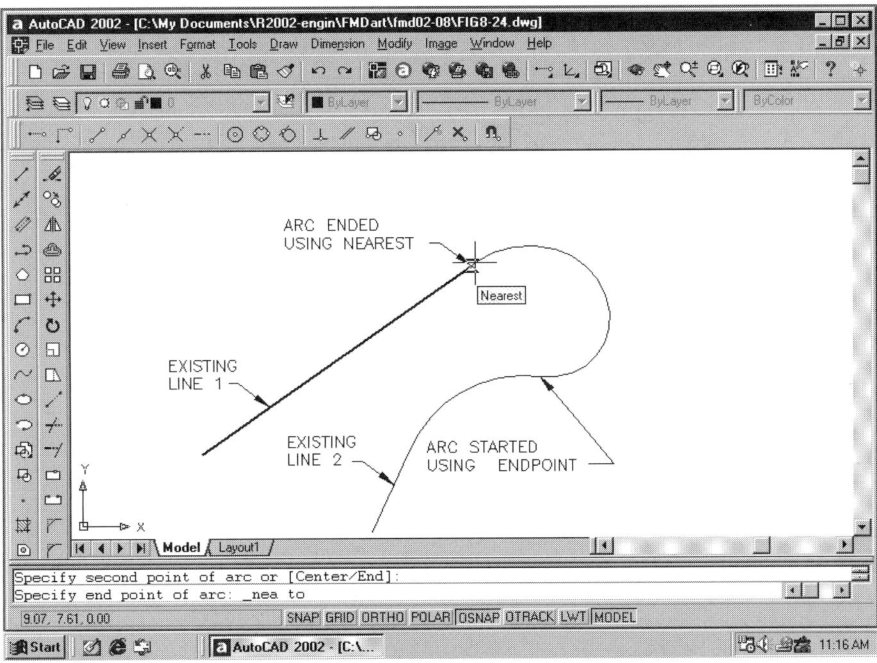

Figure 8.22 *The Nearest mode can be used to select the nearest drawing object for a "first" or "next" point when lines, arcs, circles, and other geometric shapes are placed.*

The command sequence can be seen in Figure 8.23.

QUICK

Although it might seem that objects are selected very quickly, AutoCAD must search each object of your drawing to find the designated objects of the aperture box. This process can be simplified by using Quick. Quick is used in conjunction with another Object Snap mode and allows for AutoCAD to stop searching as soon as it finds one object with at least one point of the specified type. Object Snap searches for all objects crossing the target and selects the closest potential snap points. If two or more objects that match the Object Snap request are in the aperture box, Quick mode will choose the first one it sees. On a complex drawing, this might not always give the desired results. The command sequence to use Quick and draw a line from one line endpoint to the midpoint of another line is as follows:

> Command: **LINE** ENTER *(Or click the Line button.)*
>
> LINE
>
> Specify first point: **QUI,END** ENTER

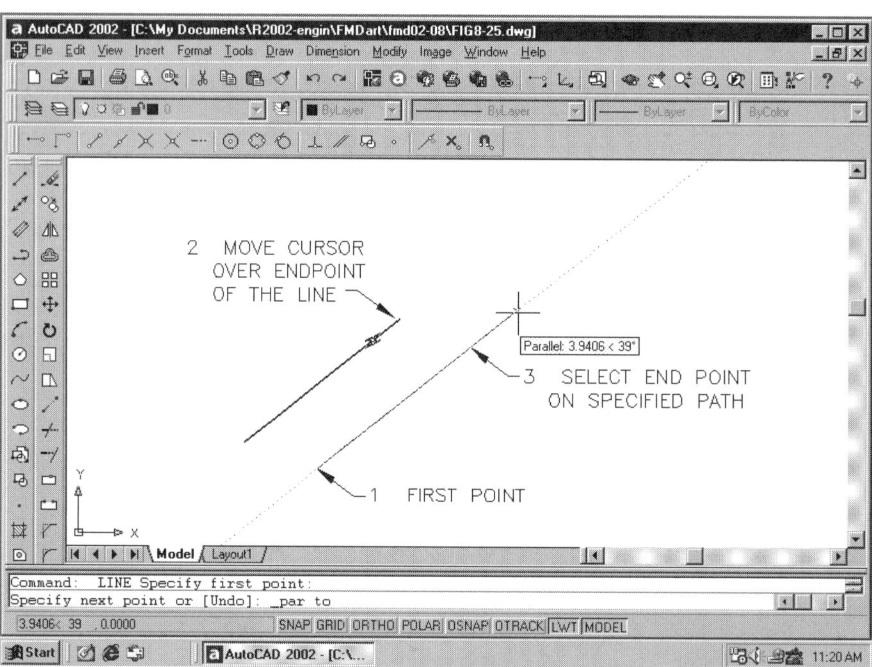

Figure 8.23 *The Parallel mode will draw a line parallel to an existing line. Unlike other modes, Parallel is specified after the "first" point is selected. Move the cursor over, but do not select the endpoint of the line that is to be the basis for the second line. Once the first line has been identified, specify the "next" point for the new parallel line.*

_qui,end of *(Select desired line.)*

Specify next point or [undo]: ENTER

Command:

FROM

The From mode can be used to establish a point for a line or other drawing objects based on a known base point. The base point can be located by selecting a point with the cursor, or by providing polar or relative coordinates for the point. Figure 8.24 shows the use of the From mode for locating a circle from the endpoint of an existing line. The command can be entered by selecting the **From** button on the **Object Snap** toolbar or by typing FRO (ENTER) at the Command prompt. The command sequence is as follows:

Command: **C** ENTER *(Or click the Circle button.)*

CIRCLE Specify center point or [3P/2P/Ttr (tan tan radius)]
FRO ENTER *(Or click the From button.)*

Base point <offset>: **END** ENTER *(Or click the Endpoint button.)*

of: *(Select desired line endpoint.)*

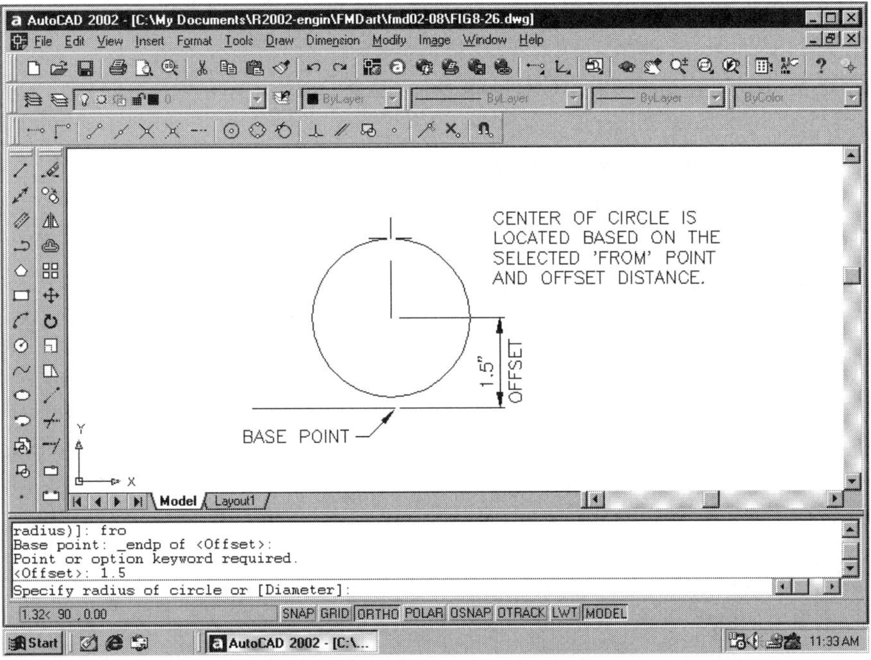

Figure 8.24 *The From mode of Object Snap allows a line, circle, or arc to be drawn from a base point by selecting the* **From** *button or by typing* **FRO** *(*ENTER*) for one of the selection points for the drawing object.*

of <offset>: **1.5** ENTER (Select location or provide coordinates from base point.)

Specify radius of circle or [Diameter]: *(Pick radius point or provide coordinates.)*

Command:

TRACKING

The Tracking mode can be used to place a first or next point that lies along an implied projection of a line. The option is also useful for finding the center of a rectangle. The option functions similarly to the Apparent Intersection mode. By selecting the midpoint on the bottom side of the rectangle and then selecting the midpoint of one of the sides, you can draw a line from the imaginary intersection of these two lines. Tracking can be started by selecting the **Tracking** button on the **Object Snap** toolbar or by typing TK (ENTER) at the Command prompt. The command sequence to draw a line from the center of a rectangle is as follows:

Command: **L** ENTER *(Or click the Line button.)*

LINE

Specify first point: **TK** ENTER *(Or click the Tracking button.)*

First tracking point: **MID** ENTER *(Or click the Midpoint button.)*

of *(Select the bottom edge of the rectangle.)*

Next point (Press ENTER to end tracking): **MID** ENTER *(Or click the Midpoint button.)*

of (Select the edge of the rectangle.)

Next point (Press ENTER to end tracking): ENTER *(Press ENTER to end tracking.)*

Specify next point or [undo]: *(Select the desired endpoint of the line.)*

Command:

As the tracking sequence is ended, the rubber-band line will now be centered in the rectangle and the prompt for a "To" point is given. The command sequence can be seen in Figure 8.25.

CONTINUOUS OSNAP

Up to this point, Object Snaps have been used in the override mode. The Object Snap mode was selected by button or keyboard for one specific point, and then the option was ended. It might be necessary to draw several lines to various quadrants of existing circles. Although you could continue to select the **Quadrant** button each time you draw a line, there is a faster way. Object Snap selection can be set to be in effect until the command is no longer needed.

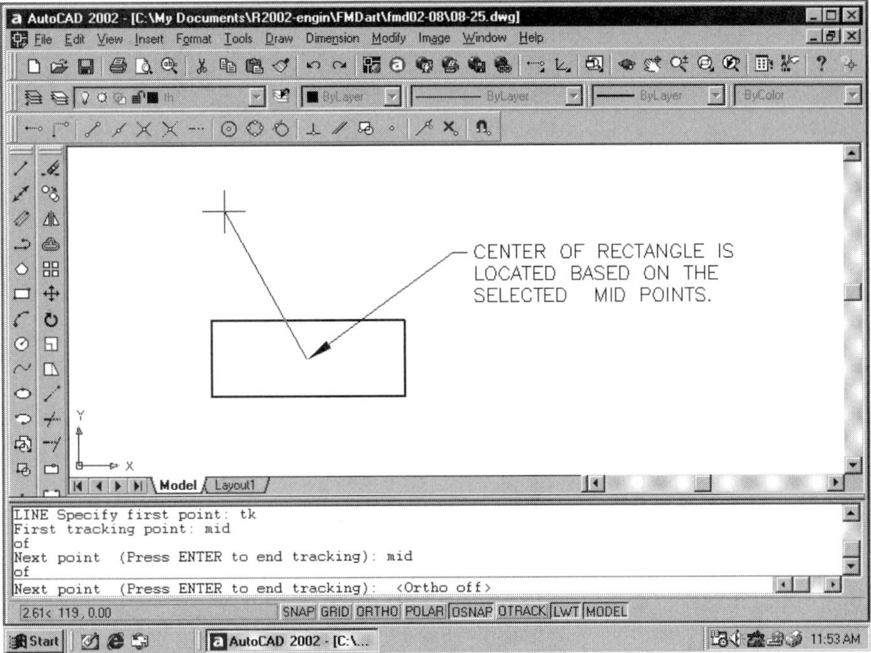

Figure 8.25 *Tracking can be used to determine the center point of a rectangle. After selecting the Tracking mode, select the midpoint of the bottom and then select the midpoint of one side of the rectangle. A line or arc can now be projected from the center of the rectangle.*

Continuous OSNAP can be set on the **Object Snap** tab of the **Drafting Settings** dialog box. To display the tab, choose the **Object Snap Settings** button on the **Object Snap** toolbar or type OS (ENTER) at the Command prompt. You can also display the dialog box by selecting **Drafting Settings** from the **Tools** menu or by typing DSETTINGS (ENTER) at the Command prompt. Each method will produce a dialog box similar to Figure 8.26.

Once the **Object Snap** tab is displayed, place a check in the mode to be activated, and then click the **OK** button. If **Quadrant** is selected, a quadrant point will automatically be selected if a circle is selected for a "first" or "next" point each time a line or arc is drawn. Figure 8.27 shows an example where the Center mode is used to place columns in the center of each pier and lines to represent beams that start at the center of each column. The selection can be ended or altered by returning to the dialog box and altering the active modes.

Continuous OSNAP can also be set to combine multiple options. This is useful, for instance, if several lines need to be constructed from the intersection of one object to the center of other objects, as seen in Figure 8.28. Multiple options are activated by selecting the desired options on the **Object Snap** tab of the **Drafting Settings** dia-

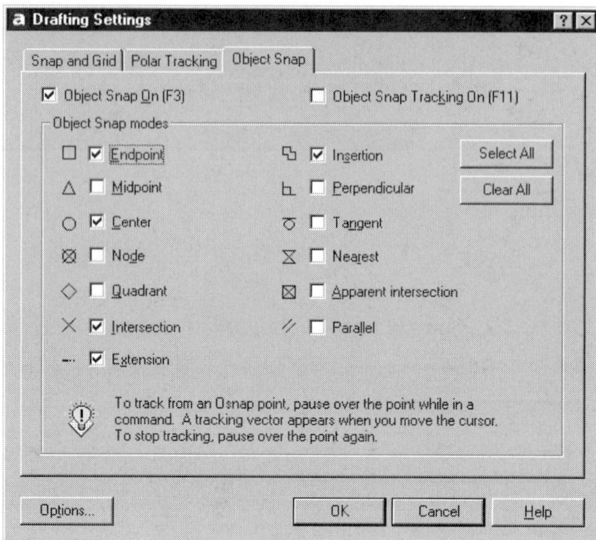

Figure 8.26 *Modes for continuous OSNAP can be set on the* **Object Snap** *tab of the* **Drafting Settings** *dialog box. Access the dialog box by selecting* **Drafting Settings** *from the* **Tools** *menu or by typing* **OS** *(ENTER) at the Command prompt.*

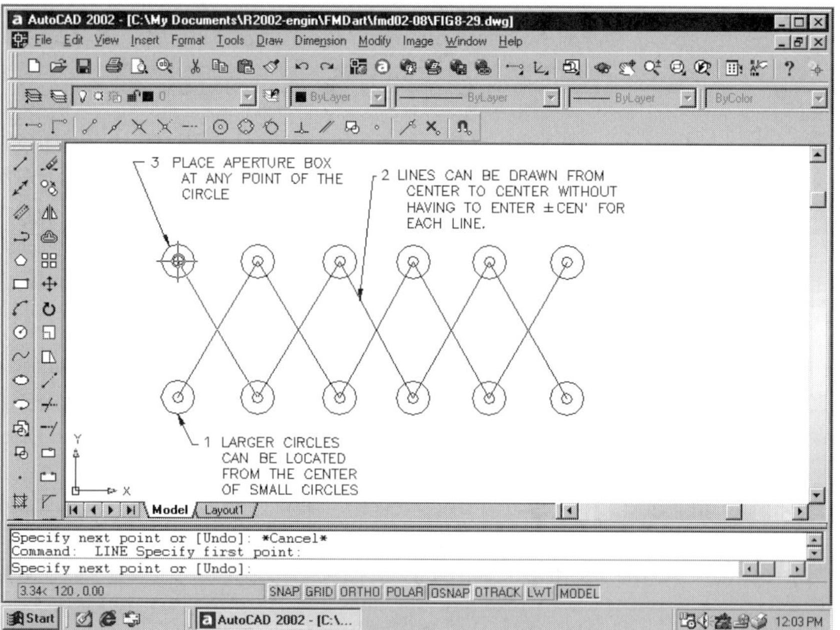

Figure 8.27 *An Object Snap mode can be used repeatedly if it is selected on the* **Object Snap** *tab of the dialog box. Active modes can be deactivated by clicking the OSNAP button on the status bar or by pressing F3.*

log box. There is no limit to the number of boxes that can be selected. A temporary override of continuous OSNAP can be achieved by selecting the **Snap to None** button on the **Object Snap** toolbar, or by typing NON (ENTER) at the prompt for an object snap. Deactivate continuous OSNAP by clicking the OSNAP button on the status bar or by pressing F3.

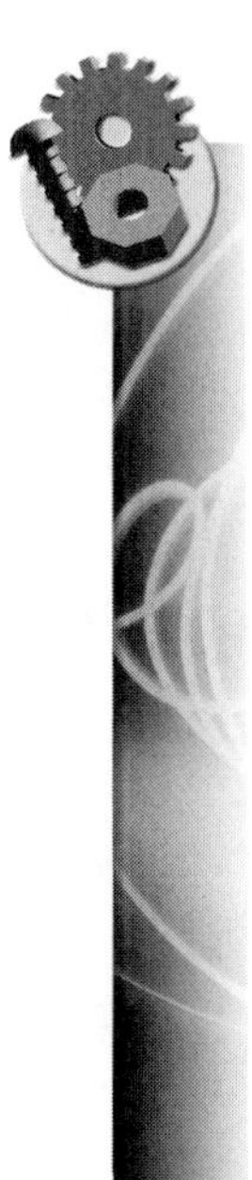

CHAPTER 8 EXERCISES

1. Draw an equilateral triangle inscribed in a 4" diameter circle. Use Object Snap Midpoint to form a second triangle inside the first. Draw a third triangle inside the second. Save the drawing as E-8-1.
2. Draw the base to a generator in plan view that is 60" × 32" with 8" wide walls. Draw an 8" wide support wall in the center of the base. Save the drawing as E-8-2.

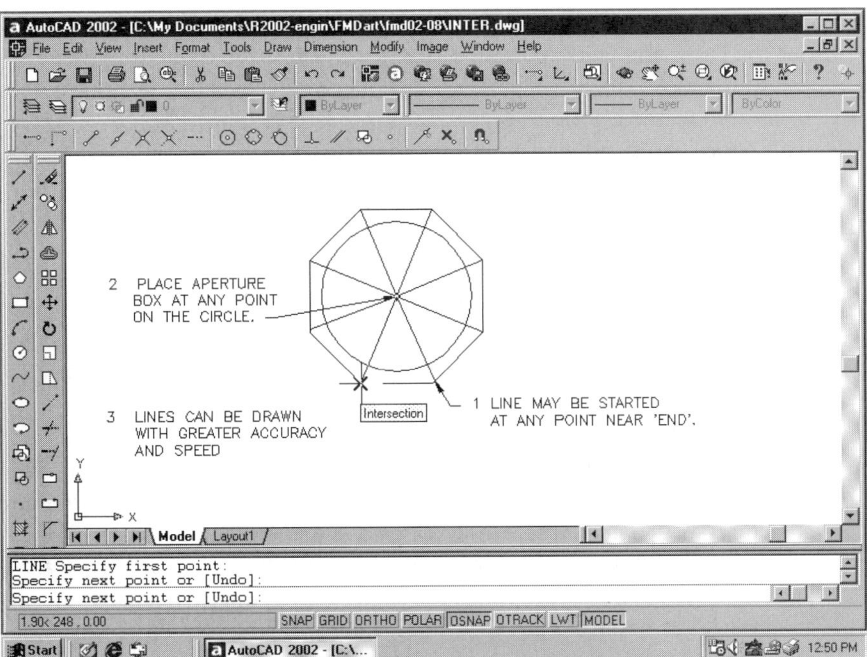

Figure 8.28 *Continuous OSNAP can be set to combine multiple options for "first" and "next" points. The drawing was completed using End, Center, and Intersections modes.*

3. Use the drawing below to draw a 1" diameter circle, a 2.25" diameter circle, a four-sided polygon, and a triangle. Complete the drawing using the CEN, MID, INT, and TAN modes. Save the drawing as E-8-3.

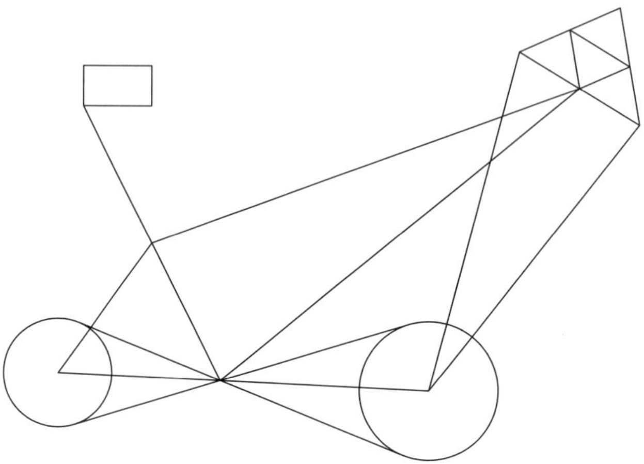

Problem 8-3

4. Set the limits and units to draw a 100' × 75' lot. Assume north to be at the top of the page. The north and south property lines will be 100' long. Five trees are located on the property. Each is located from the southwest corner with coordinates and diameters listed as follows:

	Coordinates	Trunk diameter	Branch diameter
a.	15',12'-9"	9" diameter	6' diameter
b.	30',17'	13" diameter	7.5' diameter
c.	42'-6", 19'-3"	19" diameter	9.5' diameter
d.	50', 28'-2"	15" diameter	8.5' diameter
e.	62'-0", 5'-3"	26" diameter	14' diameter

Locate each tree with a point. Use an "X" to locate each tree. Use the appropriate Object Snap modes to draw the diameter of the trunk and the branch structure around each tree center. Draw a line from the southwest property corner through the center of the trees to the southeast corner. Draw another line from the southeast property corner to the northern limits and tangent to the branch diameter of each tree and ending at the southwest corner. Save the drawing as E-8-4.

5. Open a suitable template and draw a 42" × 36" rectangle using continuous lines. Draw diagonal lines from opposite corners using hidden lines and show a

3" diameter circle at the intersection of the interior crossing lines. Save the drawing as E-8-5.

Exercises 6 through 10:

Use the attached isometric drawings to complete a minimum of three views for each problem. Use proper linetype and lineweight to represent each feature. Assume that each grid equals .25". Save each exercise using the corresponding exercise number.

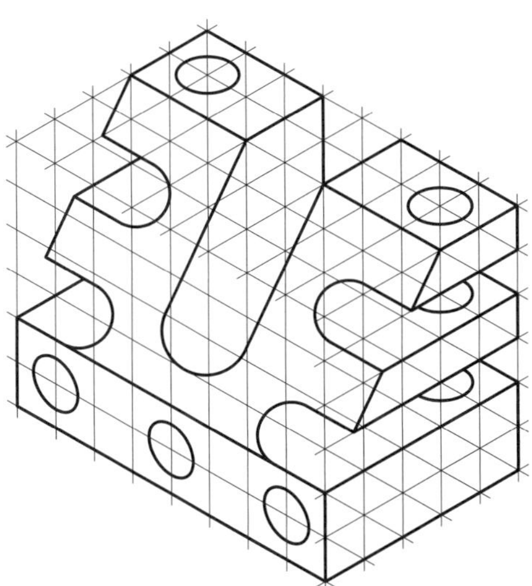

Problem 8-6

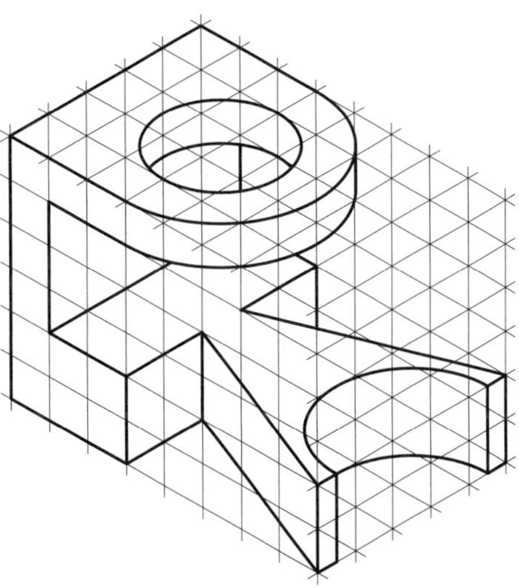

Problem 8-7

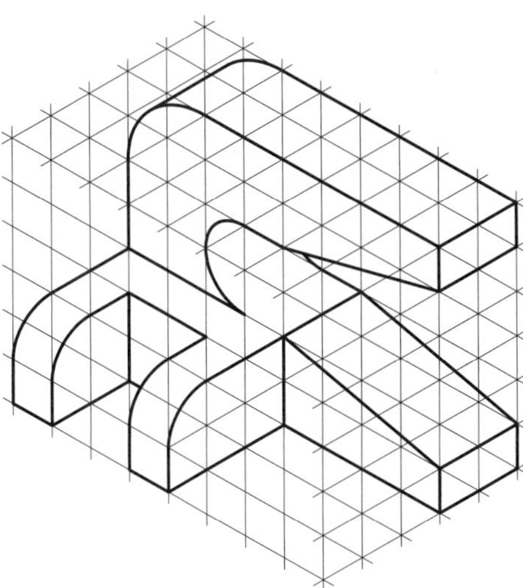

Problem 8-8

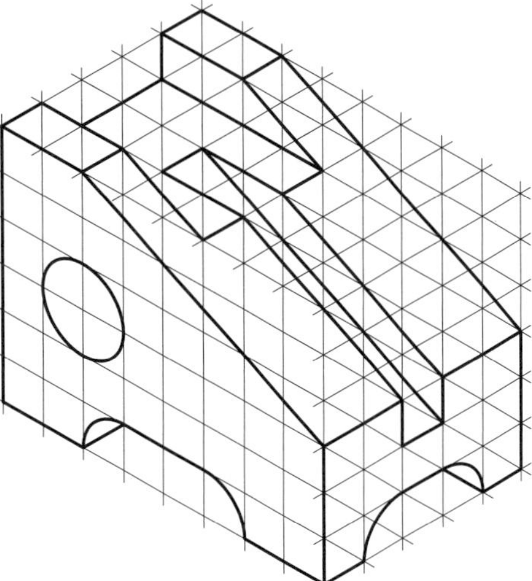

Problem 8-9

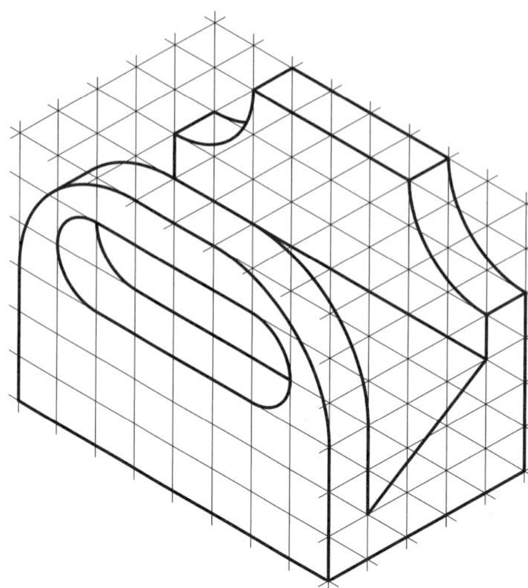

Problem 8-10

Exercises 11 through 15:

Use the attached drawings to complete a minimum of three views for each problem. Provide all missing lines. Provide an auxiliary view if necessary. Use proper linetype and lineweight to represent each feature. Assume that each grid equals .25". Save each exercise using the corresponding exercise number.

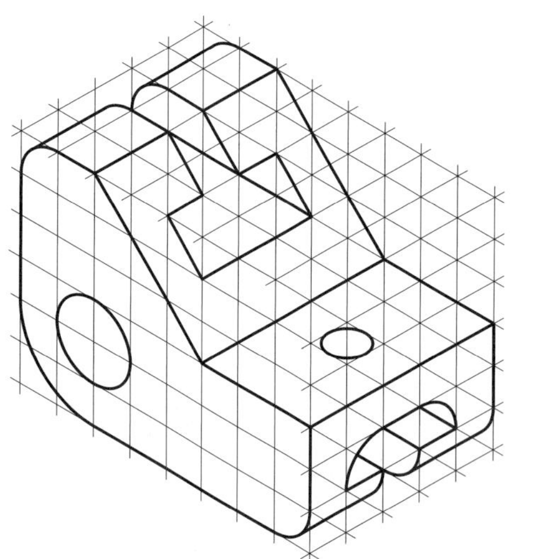

Problem 8-11

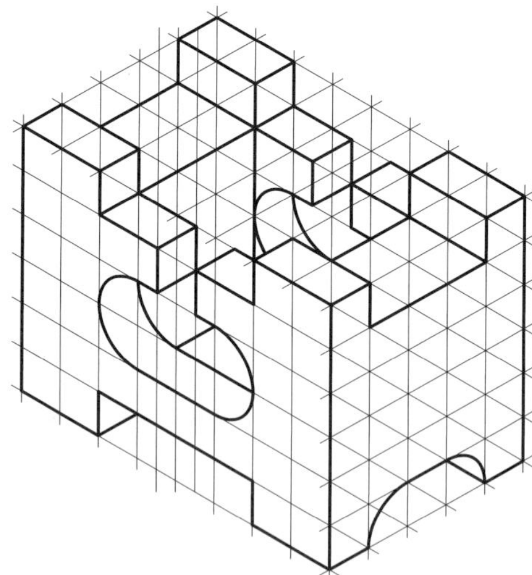

Problem 8-12

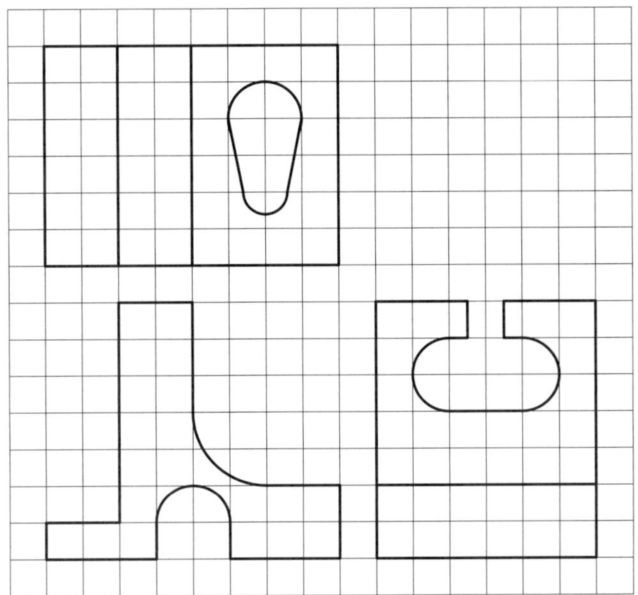

Problem 8-13

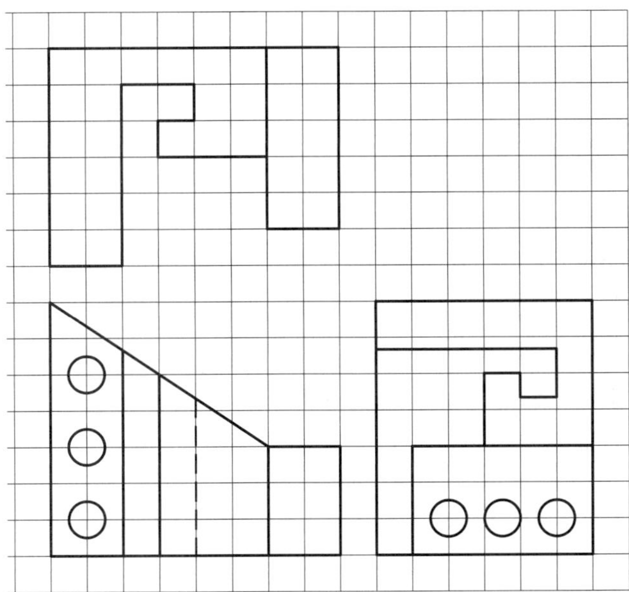

Problem 8-14

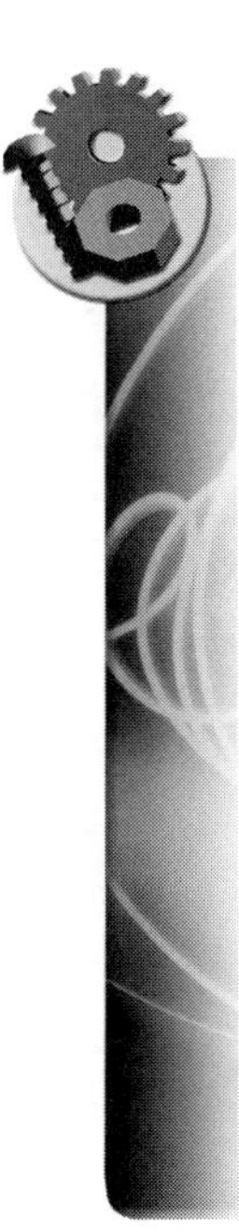

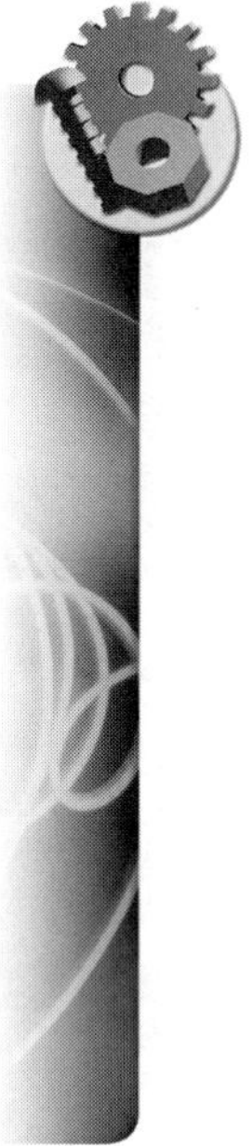

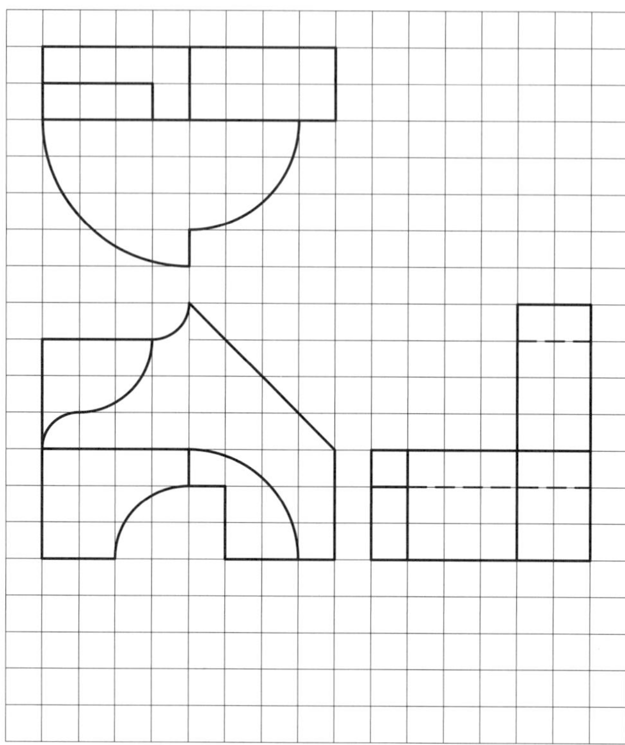

Problem 8-15

CHAPTER 8 QUIZ

1. What command is used to change the size of the aperture box?
2. What is the size range for the aperture box?
3. Type the command sequence to provide running snap to an endpoint.
4. List three Object Snap modes for connecting a line to a circle.
5. What can be done to save time when using Object Snap on a huge drawing?
6. List three methods to stop continuous OSNAP.
7. What steps would be needed to draw a perpendicular line to the midpoint of an existing line?
8. List the letters that are used to activate the Object Snap mode used to connect geometric shapes.
9. How can you get the dialog box for Object Snap to be displayed?
10. With the Quadrant mode of Object Snap, at what degree will lines be attached to a circle?

CHAPTER 9

Drawing Display Options

In the last chapter you were introduced to drawing accurately by snapping to specific points of an object. You can improve your drawing accuracy by changing the size of the view of the drawing and changing the drawing resolution. You were introduced to the **Zoom** and **Pan** commands in Chapter 2. This chapter will explore

- Options of the **Zoom** command
- Other commands that can be used to alter or move through the display

Commands to be explored in this chapter include

- **Zoom**
- **Pan**
- **View**
- **Aerial View**
- **Redraw**
- **Regen**
- **Regenall**
- **Viewres**

ZOOM

Similar to the zoom lens of a camera, AutoCAD's **Zoom** command has a potential zoom ratio of 10 trillion to 1. "Zooming in" on a drawing can magnify a portion of a drawing to allow for better visual quality. **Zoom** can be used in the middle of another command sequence, which can greatly improve accuracy on detailed drawings. As the apparent size of the object is increased, the area of the drawing that can be seen is reduced. This can be seen in Figures 9.1 and 9.2. The opposite holds true: when you reduce the apparent size of an object, more of the drawing can be seen. It

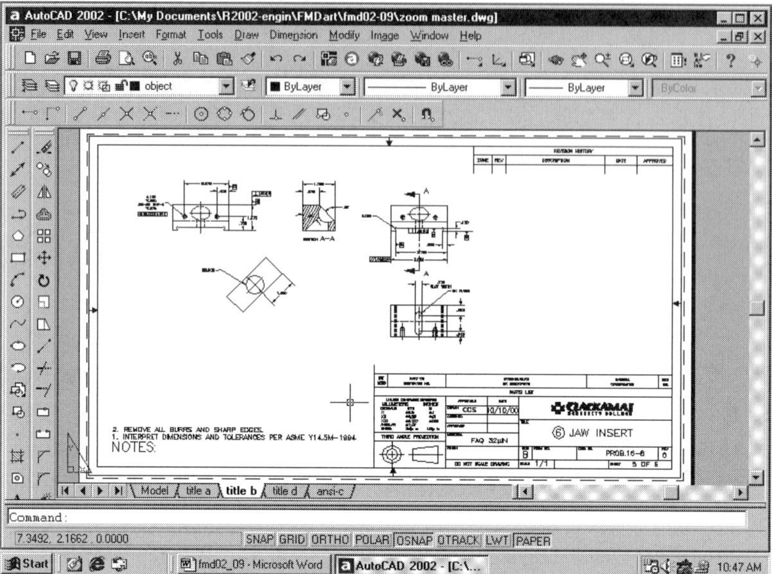

Figure 9.1 *As the drawing limits are enlarged, many features of the drawing might not be legible. Courtesy Cathy Schlesser.*

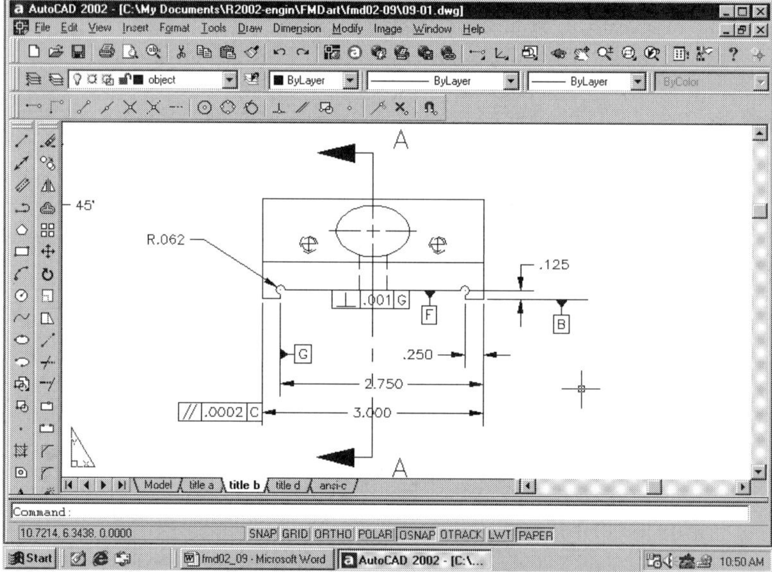

Figure 9.2 *The* **Zoom** *command alters the position of a drawing to be viewed, giving the sense that the portion of the drawing being viewed has been enlarged. Courtesy Cathy Schlesser.*

is important to keep in mind that the actual size of the object is not changing, only the magnification.

Zoom options can be accessed from the **Zoom** toolbar and by selecting one of three different buttons on the **Standard** toolbar: **Zoom Realtime**, **Zoom Flyout**, and **Zoom Previous**; these buttons will be discussed throughout the chapter. You can also access the **Zoom** command from the **View** menu, by the right-click shortcut menu, or by keyboard. Figure 9.3a shows the **Zoom** toolbar and the **Zoom Flyout** menu found on the **Standard** toolbar. Zoom is activated by selecting **Zoom** from the **View** menu; this will produce the selections seen in Figure 9.3b. Right-clicking will produce the **Realtime** option of **Zoom.** Typing Z (ENTER) at the Command prompt will also start the command sequence. Each will produce the prompt:

Command: **Z** ENTER *(Or click the Zoom button.)*

[All/Center/Dynamic/Extents/Previous/Scale/Window]<Real time>:

Each of the options listed at the Command prompt and in the menu can be selected from the **Zoom Flyout** menu on the **Standard** toolbar. If you're using a workstation equipped with a Microsoft IntelliMouse, the **Zoom** command can be executed using the wheel. Rolling the wheel forward will increase the magnification, and rolling the wheel backward will decrease the magnification.

REALTIME COMMAND

Realtime zooming means that as you move the cursor, you see the results. Start the **Zoom Realtime** command by the right-click shortcut menu, by selecting the **Zoom Realtime** button on the **Standard** toolbar, from the **Zoom** sub-

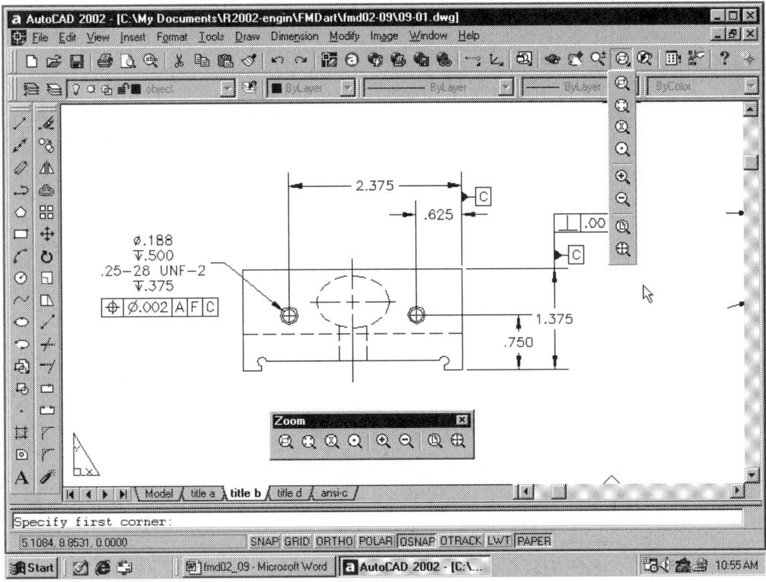

Figure 9.3a *The* **Zoom** *toolbar and the* **Zoom Flyout** *menu on the* **Standard** *toolbar can be used to select* **Zoom** *options. Courtesy Cathy Schlesser.*

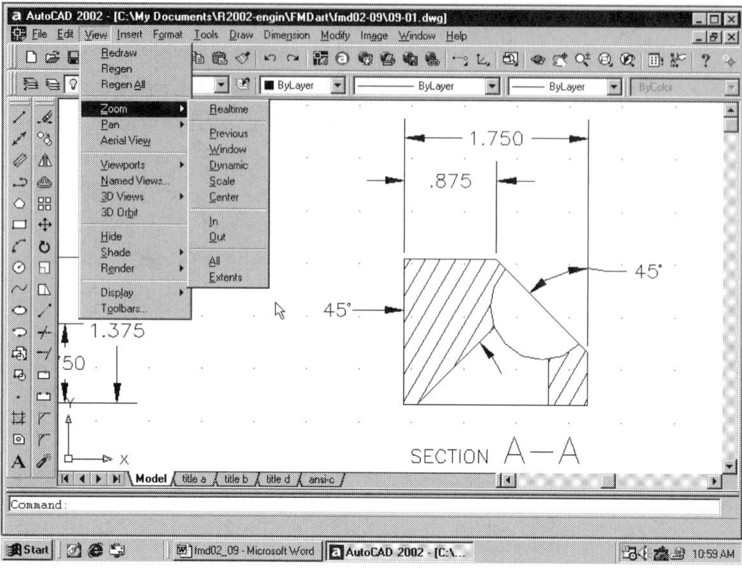

Figure 9.3b *The* **Zoom** *command can be selected by typing* **Z** *(ENTER) at the Command prompt or by selecting the command from the View menu. Courtesy Cathy Schlesser.*

menu of the **View** menu, or by typing **RTZOOM** (ENTER) at the Command prompt. As the command is activated, the cursor will change to a magnifying glass with a + and – symbol.

Move the cursor around on the screen and nothing happens. As you press and hold the select button, moving the cursor toward the top of the screen, the magnification of the drawings is enlarged. Placing the magnifying glass in the middle of the screen and moving to the top will enlarge the display 100%. Pressing and holding the select button and moving from the middle to the bottom of the screen will decrease the magnification 100%. Placing the magnifying glass at the bottom of the screen and moving to the top will enlarge the display 200%. Moving from the top to the bottom of the screen will decrease the magnification 200%. Each command can be repeated indefinitely, until the desired magnification is achieved.

Note: If you have trouble making **Realtime Zoom** alter the drawing size, remember that you must hold down the select button as the cursor is moved.

The command is ended by pressing ESC, SPACEBAR, or ENTER, or right-clicking and selecting **Exit** from the shortcut menu (see Figure 9.3c). Selecting **Exit** will end **Zoom** and return the Command prompt. The menu also allows other viewing options to be selected. By alternating between the **Pan** and the **Zoom** commands, you can enlarge a drawing while keeping the desired portion on the screen. Each of these options will be discussed throughout the balance of this chapter.

ZOOM WINDOW

The Window option of Zoom is a common option for zooming into a drawing. This option lets you select the area you wish to view by providing two opposite corner locations that will form the viewing window. Select the size of the view box by using the select button to select locations on the screen or by entering coordinates with the keyboard. Two methods are available for using this option of the **Zoom** command. By default, the command is waiting for you to enter two points that will form the corners of a rectangle. The command sequence is as follows:

> Command: **Z** ENTER *(Or click the Zoom button.)*
>
> ZOOM
>
> Specify corner of window, enter a scale factor (nX or nXP), or
>
> [All/Center/Dynamic/Extents/Previous/Scale/Window]<Real time>:
>
> *(Pick a corner to form the selection window.)*
>
> Specify other corner: *(Pick a corner.)*
>
> Command:

This process can be seen in Figure 9.4a. Objects within the view box will be redisplayed on the screen as the new display. You select the first corner of the zoom display by moving the cursor to the desired location. Once a corner is selected, the crosshairs will switch to a box that is enlarged or reduced as the mouse is moved across

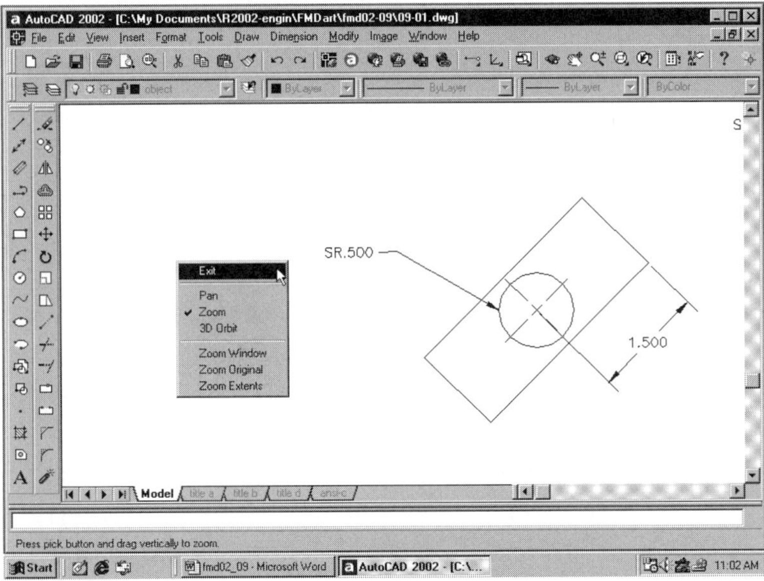

Figure 9.3c *Right-clicking while in the* **Zoom** *command will produce the Zoom shortcut menu. Courtesy Cathy Schlesser.*

the drawing area. Figure 9.4b shows the results of the selection made in Figure 9.4a. The process is similar when entering coordinates with a mouse.

A zoom window can also be created using the Window option from the **Zoom** command. The command process is as follows:

> Command: **Z** ENTER *(Or click the Zoom Window button.)*
>
> ZOOM
>
> Specify corner of window, enter a scale factor (nX or nXP), or
>
> [All/Center/Dynamic/Extents/Previous/Scale/Window]<Real time>:
> **W** ENTER
>
> Specify first corner: *(Pick a corner.)*
>
> Specify other corner: *(Pick a corner.)*
>
> Command:

Typing Z (ENTER), W (ENTER), or selecting the **Zoom Window** button are the most efficient methods of starting the command.

TRANSPARENT ZOOM

One of the really great features of Zoom is that it is transparent, meaning it can be used in the middle of another command. If you start to draw a line and realize that you're straining to see the line, the screen image can be magnified using **Zoom** in the

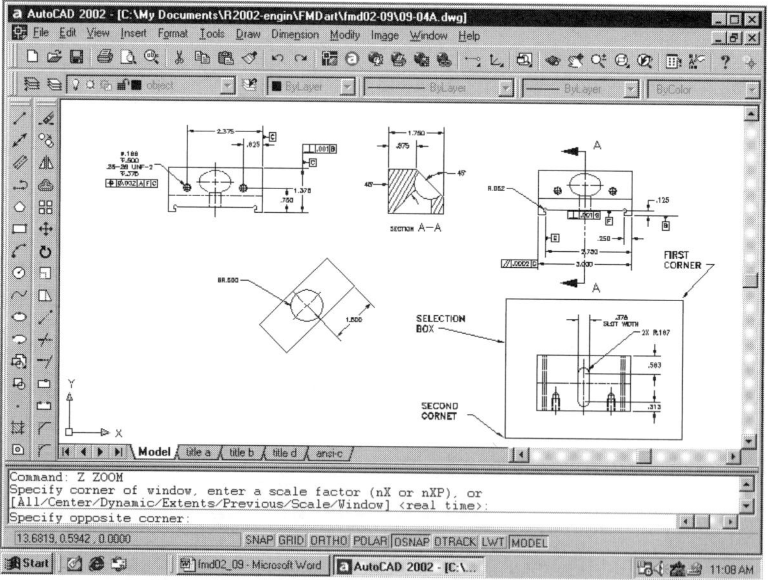

Figure 9.4a *The* **Zoom Window** *option allows the display area to be selected by specifying opposite corners of the "window." Courtesy Cathy Schlesser.*

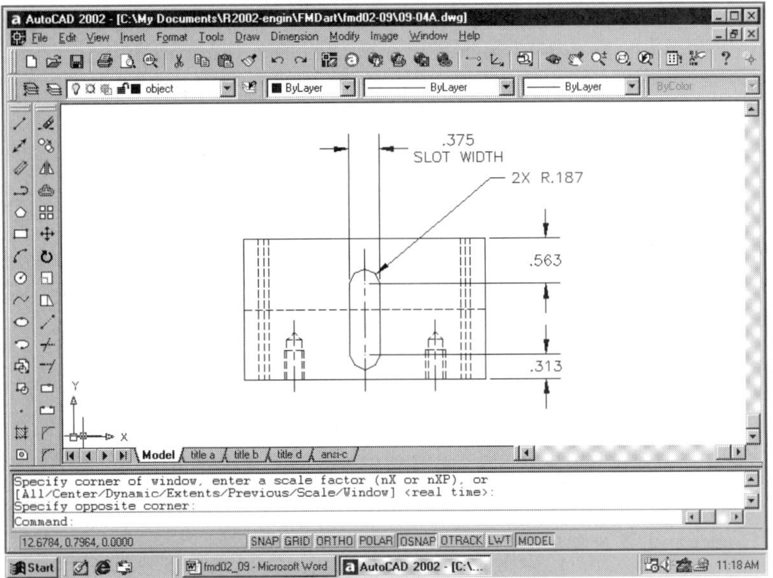

Figure 9.4b *The* **Zoom** *window created in Figure 9.4a now displays the selected area. Courtesy Cathy Schlesser.*

middle of a **Line** command sequence. Simply select the button of the desired Zoom option or enter the command by keyboard. To use a transparent command from the keyboard, type an ' (apostrophe) prior to the command. The command sequence for using the Window option of **Zoom** in the middle of the **Line** command would be as follows:

Command **L** ENTER *(Or click the Line button.)*

LINE

Specify first point: *(Select first line endpoint.)*

Specify next point or [Undo]: **'Z** ENTER

_zoom

>>Specify corner of window, enter a scale factor (nX or nXP), or

[All/Center/Dynamic/Extents/Previous/Scale/Window]<Real time>:

(Pick the desired start point for the selection window.)

>>Specify opposite corner: *(Select opposite corner to be enlarged.)*

Resuming LINE command.

Specify next point or [Undo]: *(Select desired next point.)*

Specify next point or [Undo]: ENTER

Command:

Zoom can be used at any point during the **Line** command or any other command when a prompt is given. In the **Line** command, the '**Zoom** command could have been started at the prompt for "first point" or at the "next point" prompt. Start the **Line**, **Circle**, or **Arc** command and then use '**Zoom** to become familiar with this option of zoom. Your eyes and back will love you as you zoom more and squint less.

ZOOM ALL

The **Zoom All** option will allow for an entire drawing to be viewed, based on the drawing limits. If the drawing extends beyond the drawing limits, the screen display will show the entire drawing. **Zoom All** should be used if you are zoomed in to a small area of a drawing, such as Figure 9.2. Using the All option returns all of the drawing to the screen so that the display resembles Figure 9.1. **Zoom All** can be selected from the **Zoom Flyout** menu on the **Standard** toolbar, from the right-click menu, from the **Zoom** submenu on the **View** menu, or by typing Z (ENTER), A (ENTER) at the Command prompt. The command sequence is as follows:

Command: **Z** ENTER *(Or click the Zoom All button.)*

ZOOM

Specify corner of window, enter a scale factor (nX or nXP), or

[All/Center/Dynamic/Extents/Previous/Scale/Window] <Realtime>: **A** ENTER

Command:

ZOOM CENTER

The **Zoom Center** option allows a zoom to be specified by its desired center point. Once a center point is selected, a prompt will be given for the height of the display. If the current default is maintained, the drawing is redisplayed based on the new center point, but the magnification is not changed. If a smaller value for the height is selected, the display magnification will be increased. If a greater value for the height is selected, the display magnification will be reduced. The command sequence is as follows:

Command: **Z** ENTER *(Or click the Zoom Center button.)*

ZOOM

Specify corner of window, enter a scale factor (nX or nXP), or

[All/Center/Dynamic/Extents/Previous/Scale/Window] <Realtime>: **C** ENTER

Specify center point: *(Select a point. or enter coordinates.)*

Enter magnification or Height < 0.50>: **5** ENTER

Command:

ZOOM EXTENTS

The **Zoom Extents** option will display the entire drawing based on the size of the drawing—not the size of the limits that were used for **Zoom All**. If a small object is being examined with large drawing limits, Extents will enlarge the object to fill the drawing screen. This can be quite helpful during the initial stages of a drawing. **Zoom Extents** is also helpful if the extent of the drawing exceeds the drawing limits, because the option allows the entire drawing to be examined. The command sequence is as follows:

Command: **Z** ENTER *(Or click the Zoom Extents button.)*

ZOOM

Specify corner of window, enter a scale factor (nX or nXP), or

[All/Center/Dynamic/Extents/Previous/Scale/Window] <Realtime>: **E** ENTER

Command:

ZOOM PREVIOUS

Each of the other **Zoom** options is used to enlarge an area of a drawing for easier viewing. The Previous option will allow you to go backward, so that the previous drawing area can be seen. When you use this option, the drawing displayed prior to the zoom will be returned to the screen. A maximum of 10 previous views can be restored using the Previous option. The command sequence for **Zoom Previous** is as follows:

Command: **Z** ENTER *(Or click the Zoom Previous button.)*

ZOOM

Specify corner of window, enter a scale factor (nX or nXP), or

[All/Center/Dynamic/Extents/Previous/Scale/Window] <Realtime>: **P** ENTER

Command:

ZOOM SCALE

The default selection for the **Zoom** command is Scale or magnification factor (X). The scale can be set to be relative to the full view or to a current view.

Scale Relative to Full View

When you enter a number for the scale factor, the entire drawing will be affected. A scale factor of 1 will display the current size. A scale factor of 3 will make objects appear three times as large. By entering a number smaller than 1, you will decrease the size of the object. If you enter .5, the object will appear half as big as the full display. The command sequence is as follows:

Command: **Z** ENTER *(Or click the Zoom Scale button.)*

ZOOM

Specify corner of window, enter a scale factor (nX or nXP), or

[All/Center/Dynamic/Extents/Previous/Scale/Window]<Realtime>:
S ENTER

Enter scale factor (nX or nXP): **.5** ENTER

Command:

Scale Relative to Current View

If you enter a numeric value followed by an "X", the scale of the zoom will be relative to the current view, rather than the entire drawing. The command sequence is as follows:

Command: **Z** ENTER *(Or click the Zoom Scale button.)*

ZOOM

Specify corner of window, enter a scale factor (nX or nXP), or

[All/Center/Dynamic/Extents/Previous/Scale/Window]<Realtime>:
S ENTER

Enter scale factor (nX or nXP): 1.5X ENTER

Command:

Scale Relative to Paper Space Units

AutoCAD thinks of space as model space and paper space. 2D drawings should be drawn using model space. Paper space is used to plot views of the model that have been created. **Zoom XP** can be used to scale each space relative to paper space units. This will be helpful, for instance, in printing drawings on the same sheet that are drawn at different scales. This topic will be covered in more detail in later chapters.

ZOOM IN

Zoom In automatically scales the current drawing screen display by a scale factor of 2. Selecting this Zoom option produces the same result as typing a scale factor of 2X (ENTER) from the command line. The **Zoom In** option is available on the **Zoom** toolbar, the **Zoom Flyout** menu on the **Standard** toolbar, and from the **Zoom** submenu of the **View** menu.

ZOOM OUT

Zoom Out automatically enlarges the drawing area that is displayed on the screen by a factor of 2. The drawing is, in effect, reduced by a scale factor of .5X. The **Zoom Out** option is available on the **Zoom** toolbar, the **Zoom Flyout** menu on the **Standard** toolbar, and from the **Zoom** submenu of the **View** menu.

ZOOM DYNAMIC

The **Zoom Dynamic** option will allow specific areas of the drawing to be selected for Zoom viewing. The command will seem complicated at first. Once you master it, you'll be able to select any portion of the drawing for the next view without requiring a **Regen**. The importance of this might seem lost on you as you compare a dynamic to a window zoom while working on the exercises. As you start to work on larger projects using hundreds of thousands of bytes of information, a **Zoom Dynamic** might be worth the effort. The original view box can be enlarged or reduced and moved about the drawing screen to select the exact portion of the drawing to be zoomed. The command sequence is as follows:

> Command: **Z** ENTER *(Or click the Zoom Dynamic button.)*
>
> ZOOM
>
> Specify corner of window, enter a scale factor (nX or nXP), or
>
> [All/Center/Dynamic/Extents/Previous/Scale/Window]<Realtime>:
> **D** ENTER

The screen will be cleared, and the original view will be returned, surrounded by three color-coded boxes. Each box can be seen in Figure 9.5 and represents a different viewing area.

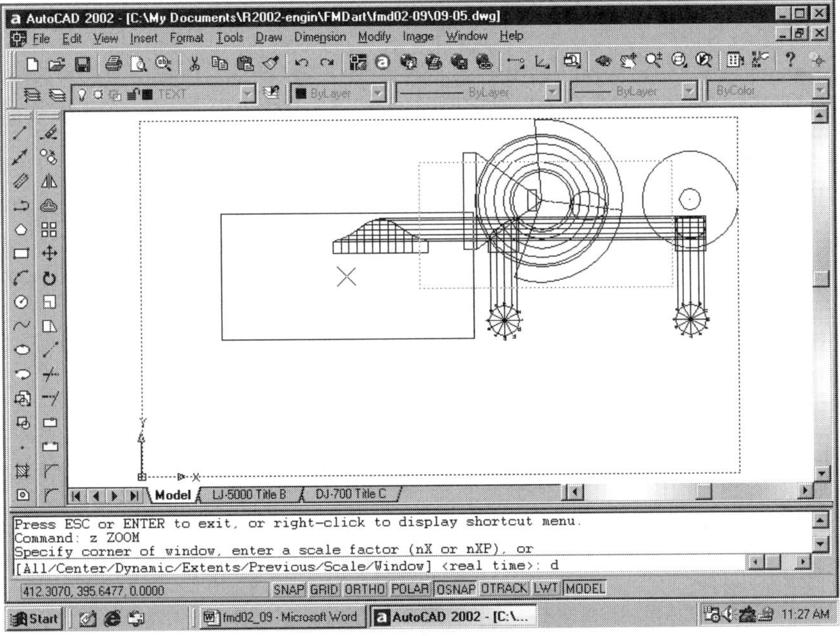

Figure 9.5 *Selecting the* **Zoom Dynamic** *option will display the original drawing area, as well as three view boxes. Courtesy Michael J. Easley.*

The first box that will be displayed is the Pan box. The Pan box allows you to drag the drawing across the screen to determine which portion of the drawing you would like to view. (See Figure 9.5.) Once the desired portion of the drawing is shown on the screen display, click the select button, and the zooming window, the view box, is displayed. The view box represents the current viewport (viewports will be discussed later). For now, you only have one viewport. The view box allows the size of the current viewport to be adjusted, either by shrinking or enlarging it, or by moving it at its current size around the screen. Adjust the size of the window and press ENTER. Click the select button if you need to readjust the portion of the drawing to be displayed. It sounds complicated, but it's not. However, it might not be worth the effort to learn it either. The **Dynamic Zoom** was a great method of moving through a drawing, until **Realtime Zoom** and **Pan** were added to AutoCAD. With these two options added to the **Standard** toolbar, **Zoom Dynamic** has been rendered almost obsolete.

PAN

You can best visualize the **Pan** command by thinking of your drawing as a large sheet of paper that is rolled up. As you pan across your drawing, it's just as if you were unrolling the drawing to view a different portion. The drawing magnification is not changed, just the portion of the drawing that is displayed. This can be helpful when you've zoomed in on a drawing, similar to the one in Figure 9.2, and you are looking at a specific detail. Rather than having to **Zoom All**, and then **Zoom** Window to a new detail, you can use **Pan** to bring in the desired detail. Start the **Pan** command by selecting the **Pan Realtime** button on the **Standard** toolbar, by selecting **Pan** from the **View** menu, by selecting **Pan** from the right-click shortcut menu or by typing P (ENTER) at the Command prompt. The submenu for **Pan** can be seen in Figure 9.6.

PAN REALTIME

The **Pan Realtime** command is similar to **Zoom Realtime**. The easiest method of starting the command is to select the **Pan Realtime** button on the **Standard** toolbar. If you're using a Microsoft IntelliMouse, the **Pan Realtime** command can be executed by pressing and holding the wheel down. No matter the access method, as the command is activated, the cursor will change to a hand cursor. The Command prompt displays the following:

Press Esc or ENTER to exit, or right-click to display shortcut menu.

Move the cursor (the hand) to the desired pick point and then press and hold the select button. This point now becomes the displacement point of the Pan Command. As the cursor is moved while the select button is being held, the screen display is altered. Panning can be repeated indefinitely, until the desired display is achieved.

End the command by pressing ESC or ENTER. Figures 9.7a and 9.7b compare the effects of using the **Pan** command.

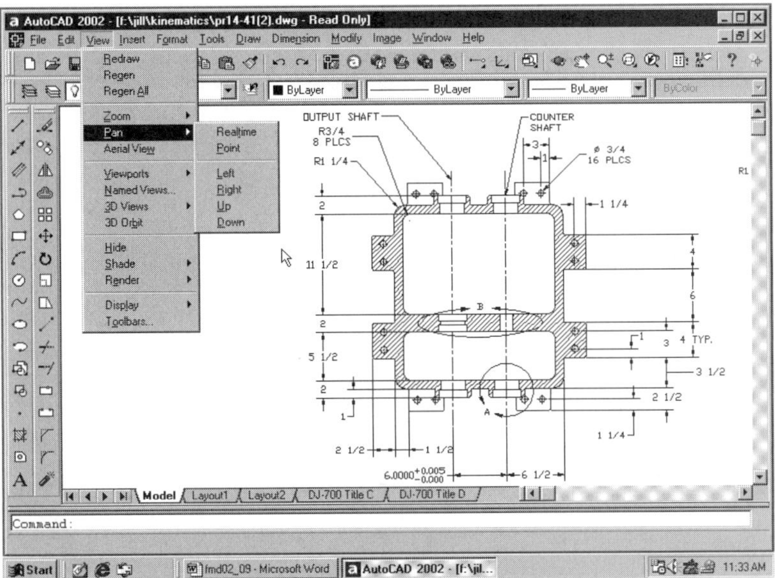

Figure 9.6 *Start the* **Pan** *command by selecting the* **Pan** *button on the* **Standard** *toolbar, by selecting* **Pan** *from the* **View** *menu, or by typing* **P** *(*ENTER*) at the Command prompt. Courtesy Jill Heilman.*

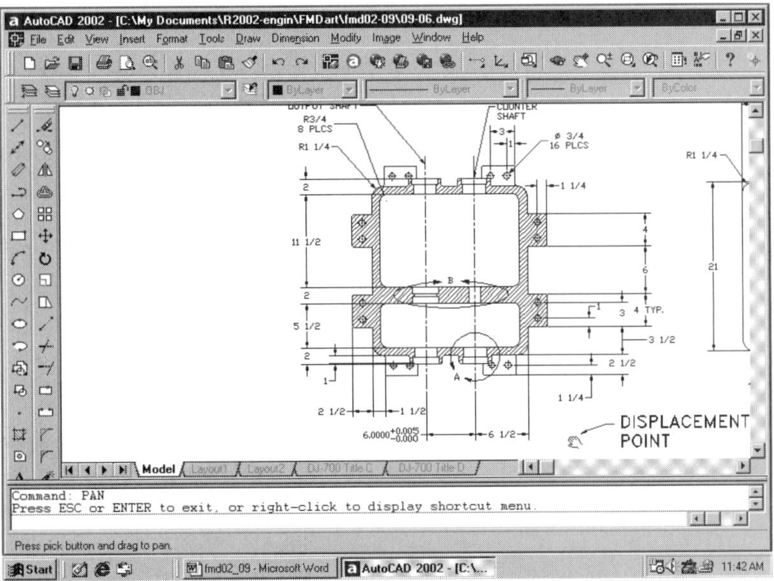

Figure 9.7a Pan *allows a drawing to be "scrolled" across the screen without the drawing magnification being changed. Courtesy Jill Heilman.*

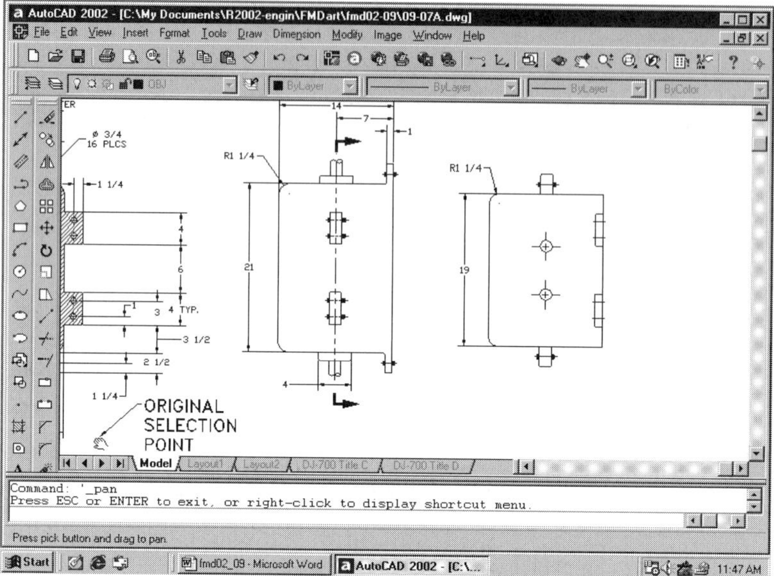

Figure 9.7b *The effects of the* **Pan** *that was started in Figure 9.7a. Courtesy Jill Heilman.*

Combining Realtime Zoom and Pan

By entering either **Pan Realtime** or **Zoom Realtime** using the appropriate button on the **Standard** toolbar or the shortcut menu, you can easily toggle between the two options and have a convenient method of moving between different portions of a large drawing. To experiment, try this:

1. Open a drawing and draw several lines and circles.
2. Click the **Zoom Realtime** button and use the select button to enlarge the drawing.
3. Now click the **Pan Realtime** button, and the cursor will change to the hand and allow the enlarged view to be panned.

Remember that when you right-click, a shortcut menu will be displayed that allows you to toggle between the **Zoom** and **Pan** commands.

Entering Realtime Pan from the Keyboard

You can begin **Realtime** Pan by typing -P (ENTER) at the Command prompt. Once started, the command will prompt you to select a point on the drawing screen and then ask, in effect, where would you like to move that point. You can enter a point by providing coordinates from the keyboard or by selecting a point on the screen. The drawing can be panned in greater increments by entering coordinates. Figure 9.7a shows the **Pan** process. The result can be seen in Figure 9.7b. The command sequence is as follows:

Command: **-P** ENTER *(Or click the Pan Realtime button.)*

Specify base point or displacement: *(Enter point or coordinates of point to be dragged.)*

Specify second point: *(Enter point or coordinates to bring first point to.)*

In most cases, the **Pan Realtime** button is much quicker to enter and easier to use.

VIEW

An alternative to **Zoom** and **Pan** is the use of predefined drawing views. The **View** command allows views to be named and saved so that they can be restored later. By default, AutoCAD begins each drawing as one view. As the limits are set, you are deciding how big the view is. The screen can be divided into smaller views for easy movement between drawings. On a drawing similar to Figure 9.1, each portion of the drawing could be named as a view so that each could be revised easily. By naming each detail as a view, you have in effect predefined the areas to be viewed. Instead of using **Zoom** or **Pan** to select a view, you could select the desired predefined View. The **View** command might seem like a waste of time on small exercises. However, the **View** command offers great flexibility for quickly viewing areas of a large, detailed drawing if you need to constantly move from one place to another.

Figure 9.8 shows the details that have been used throughout this chapter and the view names that will be used. Begin defining views by selecting the **Named Views** button on the **Standard** toolbar, or by selecting the **Named View** button on the **View** toolbar. Selecting **Named Views** from the **View** menu or typing V (ENTER) at the Command prompt can also be used to work with named views. Each method will produce the **View** dialog box shown in Figure 9.9. The **View** dialog box contains the **Named Views** and the **Orthographic & Isometric Views** tabs. This chapter will explore only the **Named Views** tab.

EXPLORING THE NAMED VIEWS TAB

The **Named Views** tab can be used to create, set, rename, and delete named views. The **Named Views** tab displays a listing of the current drawing views. In a new drawing, the only view listed is the Current view.

Creating and Saving a Named View

The following steps can be used to create a named view.

1. Select the **New** button from the **Named Views** tab of the **View** dialog box. This will display the **New View** dialog box shown in Figure 9.10.
2. Assign a view name in the **View name** edit box.
3. Choose the **Current display** radio button if the view on the screen is the view that you would like to have saved.

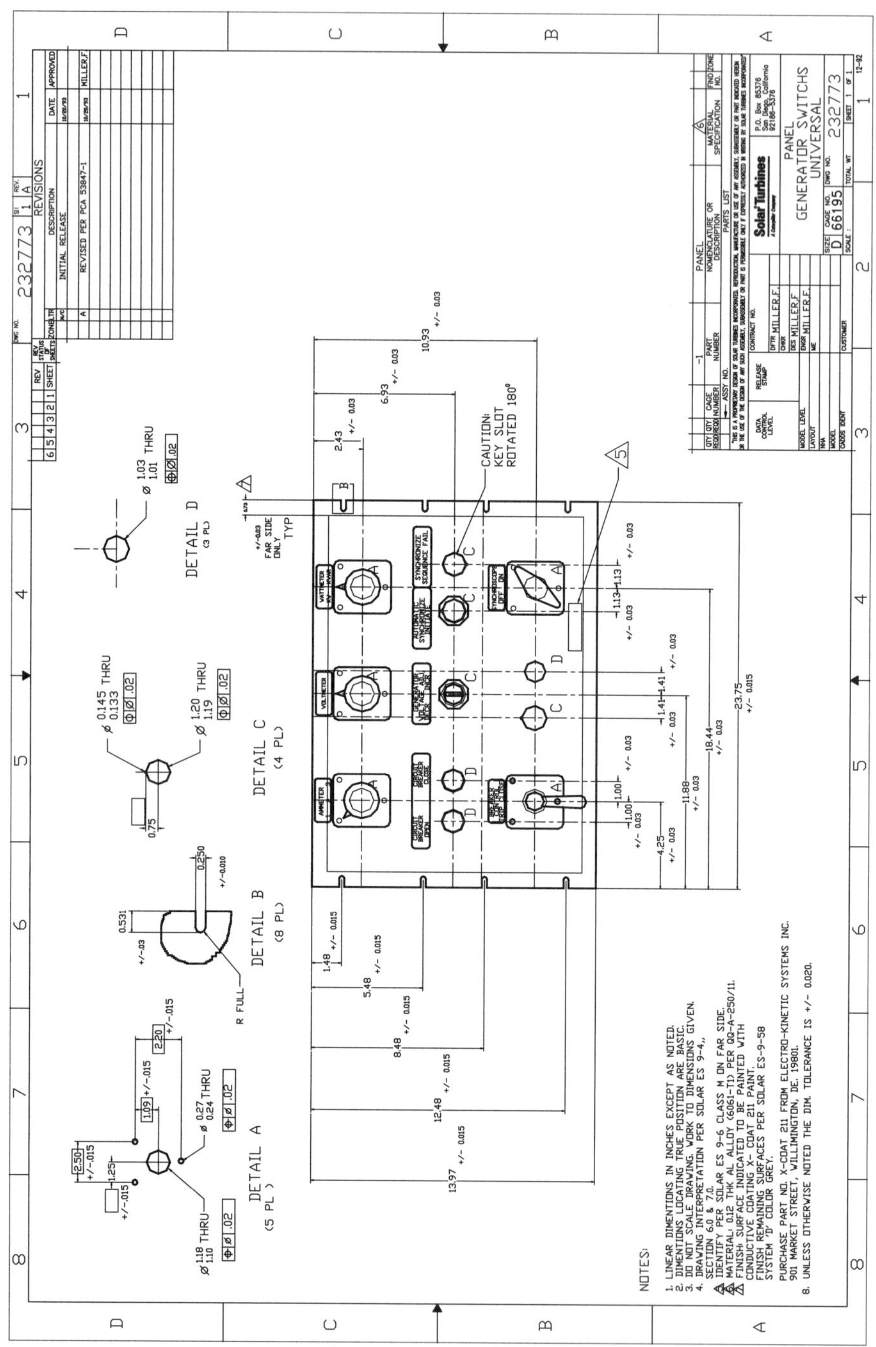

Figure 9.8 *The* **View** *command can make moving throughout a sheet with several drawings an easy task. Courtesy Floyd D. Miller.*

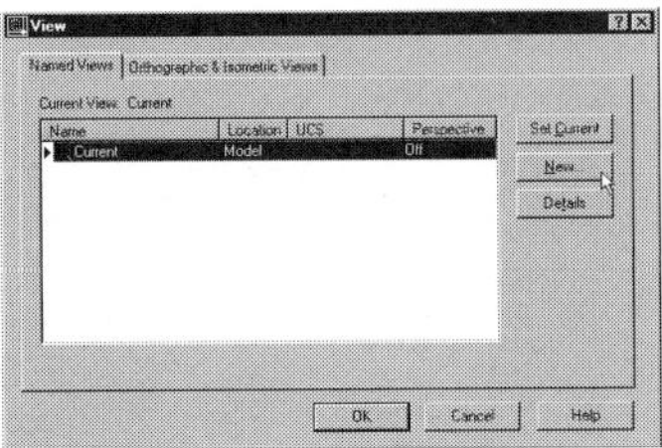

Figure 9.9 *The* **View** *command offers great flexibility for quickly viewing areas of a large, detailed drawing. Access the* **View** *dialog box by selecting the* **Named Views** *button on the* **Standard** *toolbar, by selecting the* **Named View** *button on the* **View** *toolbar, by selecting* **Named Views** *from the* **View** *menu or typing* **V** *(ENTER) at the Command prompt.*

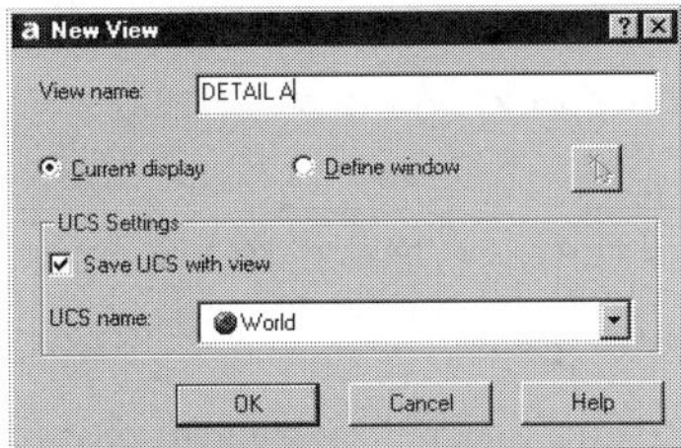

Figure 9.10 *The* **New View** *dialog box is used to assign drawing view names. Selecting the* **New** *button in the* **View** *dialog box displays this dialog box.*

Choose the **Define window** radio button if you need to select the objects to be included in the view.

4. If you selected the **Define window** button, use the **Define View Window** button to define the view. Selecting this button will remove the dialog box, returning you to the drawing area. Select the objects that will make up the view. Once the objects for the view are selected, the **New View** dialog box will be restored.
5. Click **OK**. The **View** dialog box will be restored, and the name of the selected view will be displayed.

Figure 9.11 shows the **View** dialog box with a partial listing of the twelve views described in Figure 9.8. The **New** option selects and saves the current screen display using a name that you supply. View titles can be up to 255 characters long, following the same guidelines used to name drawing files. Views are often named by their location in the job or by their specific function. If four details will be drawn, each could be given a name such as top, front, right, or auxiliary.

Altering the Current View

The screen display can easily be altered between named views. To make the RIGHT view of Figure 9.8 the current view, use the following steps:

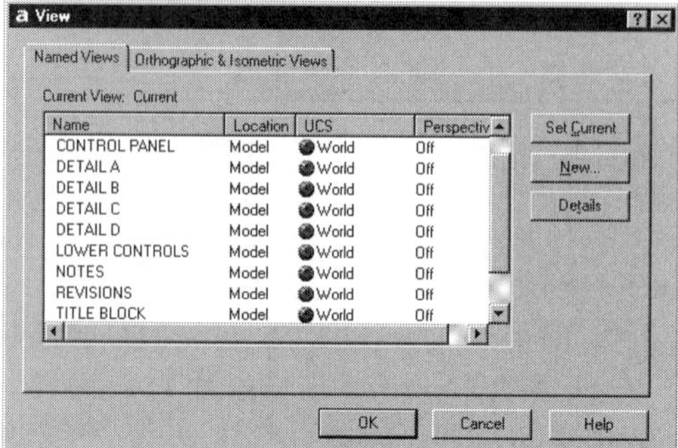

Figure 9.11 *The named views reflected in Figure 9.8 can be seen in the* **View** *dialog box.*

1. Display the **View** dialog box. Typing **V** (ENTER) at the Command prompt is the fastest method.
2. Highlight the name of the view that is to become the current view.
3. Click the **Set Current** button.
4. Click the **OK** button. The dialog box will be closed and the RIGHT view will now occupy the screen display area.

Renaming a Named View

The name of a view can be altered in the **View** dialog box. Use the following steps to change the existing name of a view:

1. Display the **View** dialog box.
2. Highlight the name of the view to be renamed.
3. Right-click to produce the shortcut menu and choose **Rename**.
4. Enter the desired name of the view.
5. Click the **OK** button to accept the new name, close the dialog box, and return to the drawing area.

An alternative to using the shortcut menu to rename a view is to use F2 once the name is highlighted. This will place the selected view name in an edit box and allow the new name to be entered.

Delete

This option will allow one or more views to be removed from the list of saved views. You're not removing the drawing, you're just removing the saved view from the list of named views. Delete views using a procedure similar to renaming them:

1. Display the **View** dialog box.

2. Highlight the name of the view to be deleted.
3. Right-click to produce the shortcut menu and choose **Delete**.
4. Click the **OK** button to close the dialog box and return to the drawing area.

View Details

Selecting the **Details** button in the **View** dialog box will display the **View Details** dialog box, similar to the display in Figure 9.12. This box provides information on the width and height of the selected view. Information provided includes the view name, the area of the view relative to the overall viewport, the coordinates of the view target location, and other information that is beyond the scope of entry-level discussion.

AERIAL VIEW

Aerial View is a useful method of viewing different portions of a large drawing file. An overall view of the entire drawing is displayed in a small window that is overlaid on the drawing screen. This is like the picture-in-picture feature of some TVs. Portions of the drawing can be selected for viewing from this small window and redisplayed while you can still see the entire drawing in the Aerial View window. This is especially helpful when you are working on a very large file. Many AutoCAD users often lose a sense of where they are in the drawing file when they have used the **Zoom** command several times in the drawing. The use of Aerial View provides that perspective by displaying the overall drawing in the smaller window. It also allows quick movement from one portion of the drawing to another, using **Realtime Zoom** and **Pan** options.

ACCESSING THE AERIAL VIEW WINDOW

Display the Aerial View window by typing AV (ENTER) at the Command prompt or by selecting **Aerial View** from the **View** menu. Each method will display the window

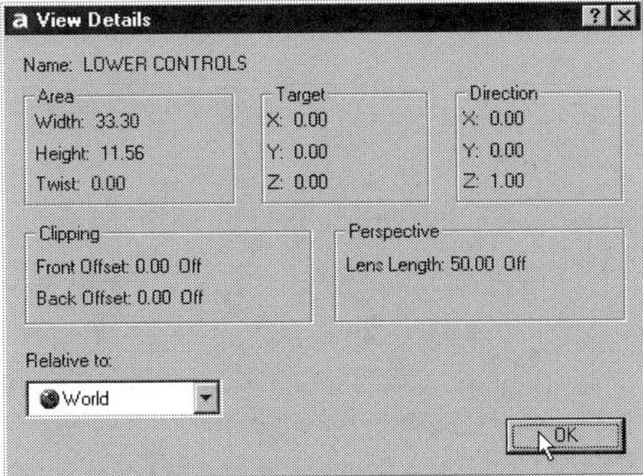

Figure 9.12 *Selecting the* **Details** *button of the* **View** *dialog box displays information including the view name, the area of the view relative to the overall viewport, and the coordinates of the view target location.*

seen in Figure 9.13. As the aerial view is displayed, it will consist of the current view surrounded by a thick black box. The size of the view can be altered using either **Zoom** or **Pan**.

AERIAL VIEW WINDOW MENUS

In addition to the standard **Help** menu, the Aerial View menus include **View** and **Options**.

View Menu

The options on the **View** menu include **Zoom In**, **Zoom Out**, and **Global**. If the entire drawing is displayed as **Aerial View** is started, the **Zoom In/Zoom Out** options will not be active. These options are used to change the magnification within the Aerial View window.

Zoom In

The **Zoom In** option of **Aerial View** is the same as the option found in the **Zoom** option of the **View** menu. The option increases the magnification of the image displayed in the Aerial View window.

Zoom Out

The **Zoom Out** option of **Aerial View** is the same as the option found in the **Zoom** option of the **View** menu. The option decreases the magnification of the image in the Aerial View window.

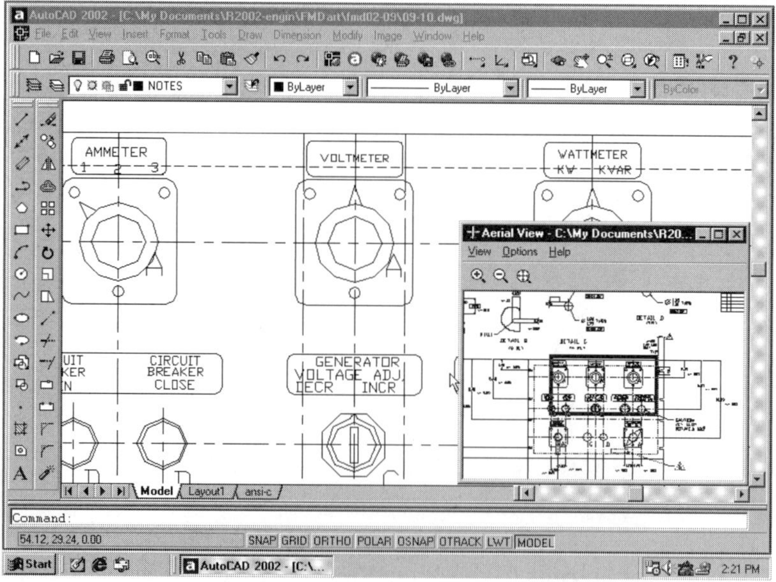

Figure 9.13 Aerial View *provides an option for moving between views of a drawing. The drawing extents are shown in the Aerial View window. Courtesy Floyd D. Miller.*

Global
Selecting this option will display the entire drawing and the current view in the Aerial View window.

Options Menu

The **Options** menu includes the options of **Auto Viewport**, **Dynamic Update**, and **Realtime Zoom**.

Auto Viewport
This option is a toggle switch between ON/OFF. When a check is displayed, the setting is ON. In the ON mode, switching to a different tiled viewport will automatically cause the new viewport to be displayed in the Aerial View window. In the OFF mode, a new viewport will not be displayed. Tiled viewports will be discussed in later chapters.

Dynamic Update
This option is a toggle switch between ON/OFF. When a check is displayed, the setting is ON. In the ON mode, the window will update the display after each change in the drawing.

Realtime Zoom
This option controls how the window will be updated. In the default setting, the window will be zoomed in real time. If you click this option, the view will be enlarged in **Zoom** rather than in **Realtime Zoom**.

USING THE AERIAL VIEW WINDOW

The **Aerial View** command allows the use of **Realtime Zoom** and **Pan** to move between areas of the drawings.

Zoom

The **Zoom** option of **Aerial View** changes the viewport center and view size by allowing points to be selected that will become the new corners of the current view. Use the following steps to alter the view with the **Zoom** option of the Aerial View window:

1. Display the Aerial View window.
2. Right-click and select **Realtime Zoom**. **Realtime Zoom** is active when the listing is preceded by a check. **Realtime Zoom** can also be activated from the **Options** menu of the Aerial View window.
3. In the Aerial View window, left-click to display the pan and zoom box. You can toggle between pan/zoom by clicking the select button. The Pan window will have an X in the middle of the box. The Zoom window will have an arrow on the right edge of the window.

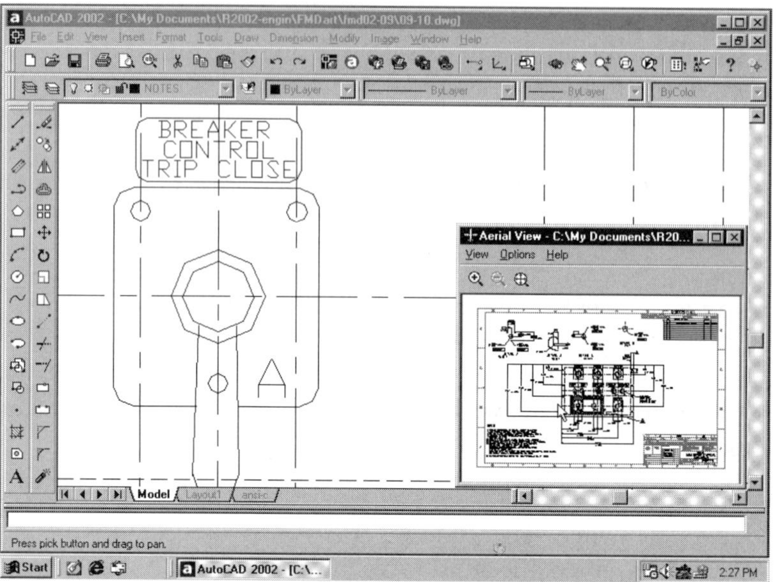

Figure 9.14 *With the* **Zoom** *option active, an Aerial View window can be created. As the cursor is moved, the window size is altered and then the area and size of the drawing are altered. Courtesy Floyd D. Miller.*

4. With the **Zoom** option active, move the cursor to control the size of the window (see Figure 9.14).
5. Press ESC to end **Realtime Zoom.**
6. Left-click to toggle to **Realtime Pan.**

Pan

This option allows the display to be altered from side to side, or top to bottom, without changing the drawing magnification. The Aerial View window provides a point of reference by showing the entire structure. If the **Pan** option is selected, a reference box will be displayed in the Aerial View window to pan the drawing. As the reference box is moved in the Aerial View window, the portion of the drawing displayed in the drawing area of the screen is altered. Move the reference box, and the drawing will snap to show the area contained in the reference box. Without pressing and holding any mouse buttons, the drawing will drag in real time to reflect the movement of the reference box. Figure 9.15 shows an example of the effect of the **Pan** option.

REFRESHING THE SCREEN

This chapter has introduced methods of controlling the drawing display using the **Zoom**, **Pan**, **View**, and **Aerial View** commands. Each command alters the view that is placed in the drawing area. The final aspect of controlling the drawing view is adjust-

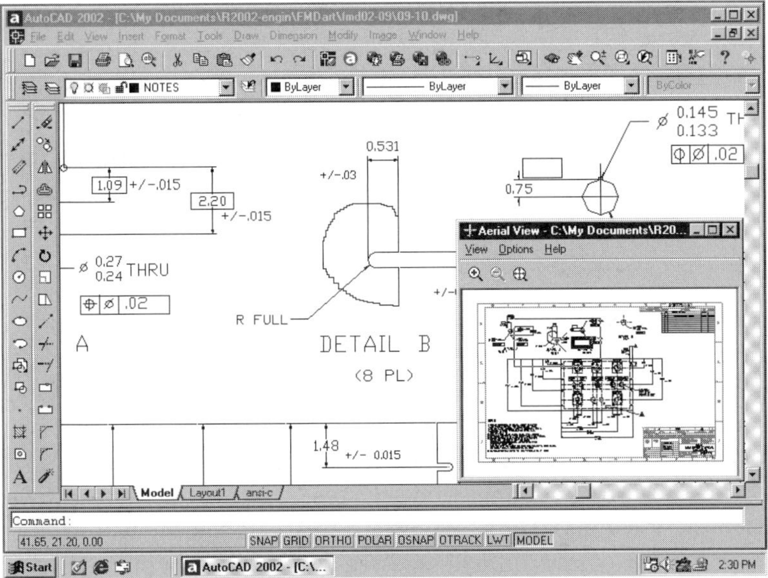

Figure 9.15 *Using the* **Pan** *option in the Aerial View window to pan the drawing. As the reference box is moved in the aerial view, the portion of the drawing displayed in the drawing area of the screen is altered. Courtesy Floyd D. Miller.*

ing the quality of the view that is displayed. The view quality can be controlled using redraw, regeneration, and view resolution commands.

REDRAW

The **Redraw** command removes the current screen display momentarily and then replaces the drawing. As the drawing is replaced, blips are removed from the drawing editor. Lines that might have been partially erased from view while other objects were edited will also be restored. Start the command by selecting the **Redraw All** button on the **Standard** toolbar or by typing R (ENTER) at the Command prompt. The same results can be achieved by selecting **Redraw** from the **View** menu. **Redraw** can be used while another command is in progress if it is entered at any non-text prompt. (Typing R (ENTER) in the middle of a command sequence where text is expected will confuse AutoCAD and frustrate you with a misspelled word.) On drawings similar in size to the exercises that you've been doing, **Redraw** will seem almost instantaneous. On larger drawings, **Redraw** can take slightly longer.

REGEN

Similar to **Redraw** is the slightly slower **Regen** command. AutoCAD clears the drawing screen and then regenerates the entire drawing. AutoCAD recalculates the current screen display with 14-place accuracy, keeping track of each pixel in the X and Y directions. If you have zoomed in on a specific part of a drawing, **Regen** is regenerating even the part of the drawing that cannot be seen. Start a screen regeneration by typing REGEN (ENTER) at the Command prompt or by selecting **Regen** from the **View** menu. To terminate the **Regen** operation, press ESC. **Regen** can also be used to improve the resolution of circular objects and polygons. Figures 9.16a and 9.16b compare the effects of **Redraw** and **Regen** on drawing objects. You'll learn how to create viewports in later chapters. Using viewports will speed up each **Regen** operation because only the current viewport is regenerated.

REGENALL

In later chapters you'll learn how to divide the drawing screen into different viewports so that specific areas of the drawing can be viewed quickly. When multiple viewports are created, the **Regen** command will only affect the current drawing viewport. All other viewports will remain unaffected. The **Regenall** command will regenerate all of the views, so that blips in each view will be removed, and reindex the drawing database to increase the accuracy of objection selection. Start the command by selecting **Regen All** from the **View** menu or by typing REA (ENTER) at the Command prompt. The significance of the command will be discussed in later chapters.

REGENAUTO

AutoCAD allows the regeneration mode to be set. With a small drawing, an automatic regeneration might not even be noticed. As the drawing size increases, using **Regen** might prove to be a frustration. AutoCAD allows a prompt to be given to remind you

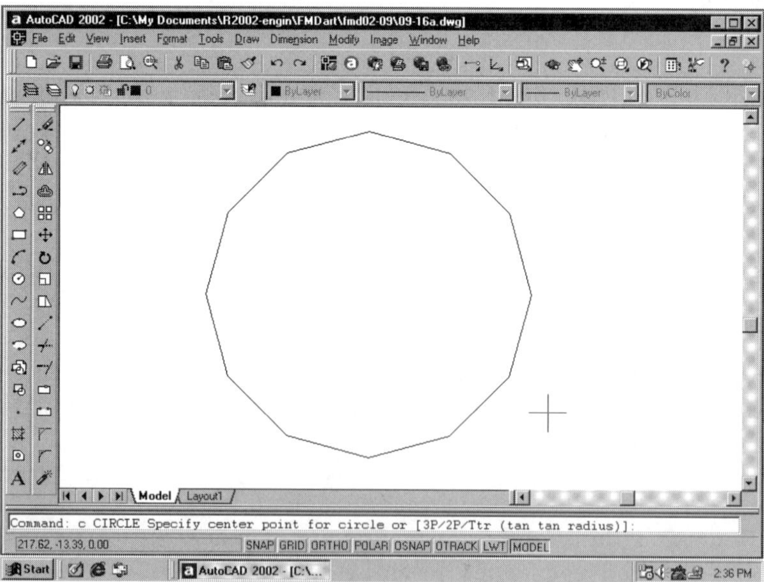

Figure 9.16a *Circular objects often appear as straight line segments when magnified or when a low view resolution is used.* **Redraw** *will not affect the smoothness of such features.*

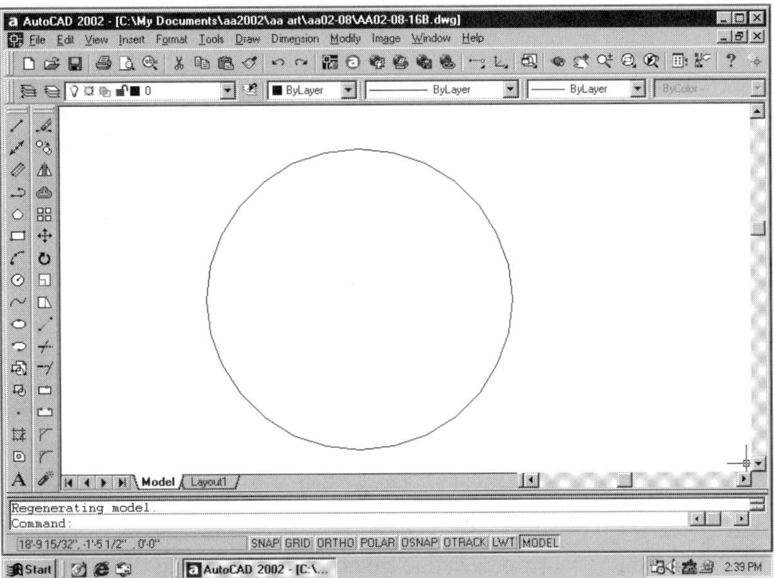

Figure 9.16b *The smoothness of circular objects will be improved with the* **Regen** *command.*

that the drawing is about to be regenerated. This aspect of **Regen** can be entered only by keyboard. The command sequence is as follows:

> Command: **REGENAUTO** ENTER
>
> Enter mode [ON/OFF] <ON>: *(Enter On or Off, or press* ENTER.*)*

If the ON option is retained, AutoCAD will perform an automatic regeneration when needed while altering the screen display in commands such as **Zoom**. When OFF is the current mode and a regeneration is needed as the screen is altered, AutoCAD will display the prompt:

> About to Regen—proceed <Y> Enter Y or N, or press ENTER

If you accept the default, the drawing regeneration will proceed, allowing the **Zoom** to take place. If the NO option is selected, the **Zoom** will not be completed. Because of AutoCAD's increased ability to handle huge amounts of information quickly and easily, you really can leave this option in the "that's nice" file of your mind and explore other viewing options.

VIEW RESOLUTION

The speed of each **Zoom** or **Pan** can be enhanced if a **Redraw** is used, rather than a **Regen**. The speed of all view changes also can be influenced by the view resolution. Resolution refers to the amount of detail that is represented when arcs and circles are drawn. The higher the resolution, the more lines that are used to represent an object and the smoother the arc or circle will appear. Higher resolution also means that more time is needed for zooms and pans.

The resolution of arcs and circles can be controlled from the **Options** dialog box or by keyboard. To control the resolution from the dialog box, select **Options** from the **Tools** menu and then select the **Display** tab. The **Arc and circle smoothness** option in the **Display Resolution** portion can be used to alter the smoothness of an arc. The **Segments in a polyline curve** option will be considered in later chapters. The other two options in this box affect 3D drawings and will not be considered in this text.

Arc and Circle Smoothness

The **Arc and circle smoothness** option controls the smoothness of circles. A value from 1 through 20,000 can be provided in the edit box, in place of the default value of 100. With a low number such as 25 or 50, circles will appear as a series of flat lines. As the value increases, the smoothness of the circle or arc is increased. Regeneration time on large drawings will slow slightly as the value is increased.

The resolution of a circular feature can also be adjusted using the **Viewres** command. The command is entered from the keyboard using the following sequence:

> Command: **VIEWRES** ENTER
>
> Do you want fast zooms? [yes/no] <Y> *(a Y or N is required.)* ENTER
>
> Enter circle zoom percent (1–20000) <100>: **150** ENTER
>
> Command:

In setting **Viewres**, the first decision to be made is for fast zooms.

Responding N (ENTER) will cause all **Zoom**, **Pan**, and **View** restorations to be done using regeneration.

Accepting the default will keep a large virtual screen and execute most **Zoom**, **Pan**, and **View** restorations at **Redraw** speed.

Zooming to the limits of a drawing might require regeneration, even though you have selected fast zooms. When a small circle is drawn with a low value, it will appear as a polygon. Using the **Regen** command will restore its circular shape.

CHAPTER 9 EXERCISES

1. Open drawing E-7-9. Use the **Zoom** command to make the view containing the triangles the current view. Save this view as TRIANGLES. Zoom into each of the remaining shapes, and make each shape a separate saved view. Save the entire drawing as E-9-1.

2. Using drawing E-9-1 as a base, make the view containing the squares the current view. Draw a square in the center of one of the existing squares using Midpoint Object Snap to form a diagonal square. Repeat this process four additional times, using **Zoom** as required. Inside the final square, draw a circle that is tangent to the smallest square. Save this view with the name TINY and save the entire exercise as E-9-2.

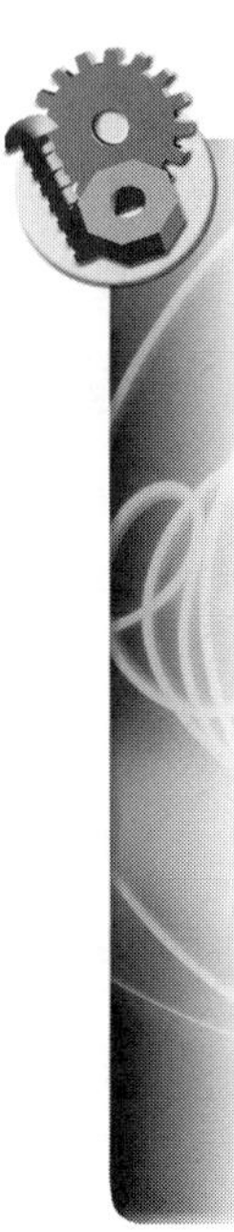

3. Load E-9-1 and enlarge the drawing view to twice its current size. Save this view as VIEW2X. Zoom out so that the edges of the largest triangle touch the edge of the screen. Name this view with the name of the command required to perform the zoom. Set the drawing so that screen markers will be produced and save the drawing as E-9-3. Write the name of the views created and the commands used in the space below.

VIEWS COMMANDS

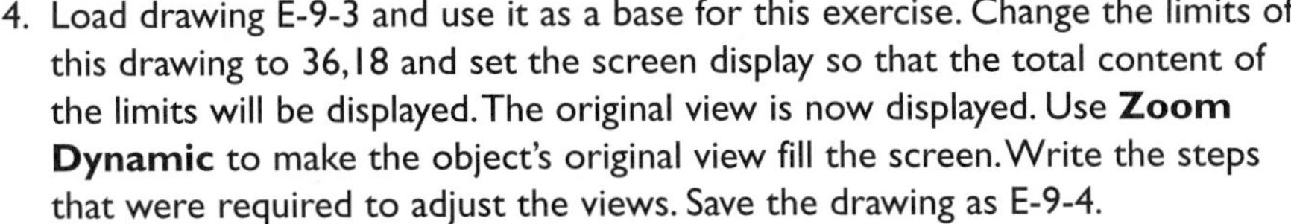

4. Load drawing E-9-3 and use it as a base for this exercise. Change the limits of this drawing to 36,18 and set the screen display so that the total content of the limits will be displayed. The original view is now displayed. Use **Zoom Dynamic** to make the object's original view fill the screen. Write the steps that were required to adjust the views. Save the drawing as E-9-4.

5. Start a new drawing using the ANSI D template with a grid of 1" and a snap of .25". Set units to measure two-place decimals. Set angles to be measured in decimal degrees with two-place accuracy. Set the view resolution to 150. Divide the drawing into the following seven views: All, Upper Left, Upper Cen, Upper Right, Low Right, Low Cen, and Low Left. Set All as the current view. Save the drawing as a drawing template.

6. Open a file of your choice containing multiview drawings and establish a minimum of three named views. Provide names that quickly describe the area captured in the viewport. Save the file as E-9-6.

7. Open a file of your choice containing multiview drawings and establish an Aerial View window. Use the window to zoom a portion of the drawing and then save the file as E-9-7.

CHAPTER 9 QUIZ

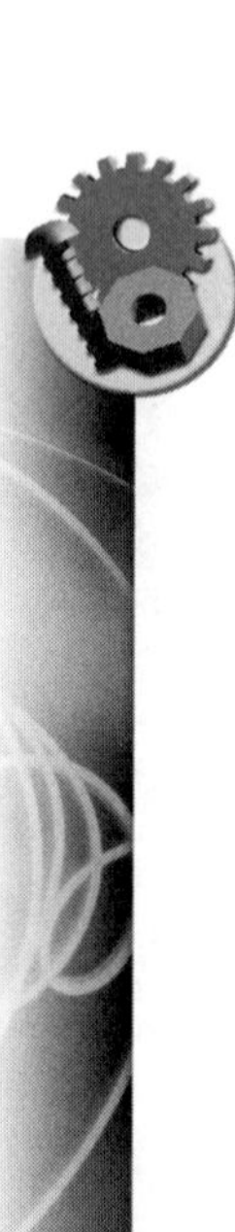

1. What command controls the use of screen markers, and how should it be set to provide markers?
2. Compare and explain the difference in **Viewres** settings between 50 and 500 and the default value.
3. Explain the difference between **Regen** and **Redraw**.
4. Describe the difference between **Zoom All** and **Zoom Extents**.
5. What commands will be needed to make a drawing exactly twice as large as the current screen display?
6. What option is used to scale space relative to paper space units?
7. Describe the meaning of "displacement point" and "second point."
8. List the steps to enlarge a portion of a drawing using the **Zoom** command.
9. List six options for accessing the **Zoom** command.
10. Describe the process that allows the **Zoom** command to be used in the middle of another command.
11. What term is used to describe a command that can be used while another command is being executed?
12. What command would you use if you wanted to create a picture-in-picture of an AutoCAD drawing?
13. What must be typed at the Command prompt to start **Aerial View**?
14. Describe how to reduce the size of the Aerial View window.
15. Other than putting you to sleep, what's the point of all of the options presented in this chapter?

SECTION 2
ENHANCING DRAWINGS

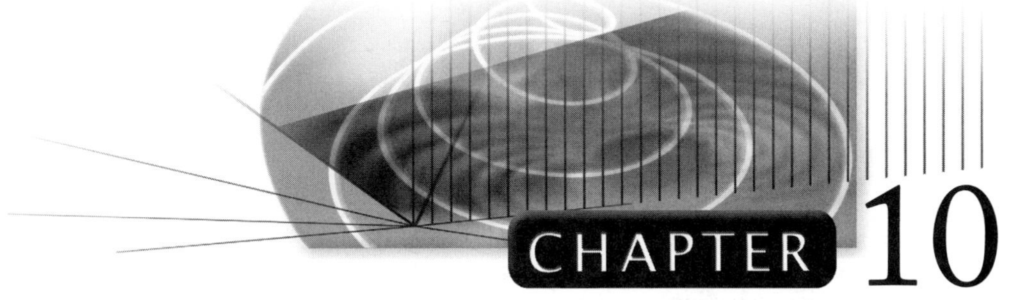

Basic Methods of Selecting and Modifying Drawing Objects

By now you've explored the basic components of a drawing. This chapter will explore methods for

- Selecting objects for editing, including the pick box, Window, Crossing window, automatic selection mode, Wpolygon, Cpolygon, Fence, All, and Last
- Modifying the selection process using Noun/verb, Shift to Add, Press and Drag, Implied Window, and Object grouping
- Modifying drawings by removing objects
- Modifying drawings by adding objects

Commands explored in this chapter include

- **Pickbox**
- **Erase**
- **Copy**
- **Mirror**
- **Mirrtext**
- **Offset**
- **Array**

Each command will reproduce all or parts of a drawing. A second group of editing commands for refining drawings and two new methods of sorting objects will be explored in the next chapter.

SELECTING OBJECTS FOR EDITING

AutoCAD provides many options for selecting objects. Individual objects can be selected using a pick box. Groups of objects can be selected for editing using one of the following options: Window, Crossing, Auto, WPolygon, CPolygon, Fence, All, Last, Add, Remove, or Multiple. As different methods of selecting objects for editing are discussed, each selection method will be introduced using the **Erase** command, although other commands can be used. These object selection options can be used any time an object selection needs to be made within an editing command sequence. (Selecting a button on a toolbar and entering the command at the keyboard are generally the fastest methods.) No matter which method is used to access the command, a prompt will be displayed asking you to

Select objects:

Each of the following selection methods can be entered at the Select objects prompt for any of the editing commands.

SELECTING OBJECTS USING A PICK BOX

As you've edited your drawings to this point, you selected objects to be edited one at a time. One or more objects have been selected using the pick box and clicking the select button. As the prompt for the edit command is entered, the crosshairs will change to an object selection target or pick box, and the object to be edited can be selected. Any object that is included in the pick box will be affected by the command. If the size of the pick box is altered, it will affect the accuracy of your selections. As the size of the box is enlarged, information that is not desired in the selection set might be added. With the box too small, you might have trouble selecting information.

The size of the box can be altered to enlarge or reduce the size of the selection area. Adjust the size of the pick box on the **Selection** tab of the **Options** dialog box (select **Options** from the **Tools** menu). As the slide bar is moved to the right, the size of the pick box is enlarged. Move the slide bar to the left to decrease the size of the pick box. The size can also be adjusted from the keyboard. Typing **PICKBOX** (ENTER) at the Command prompt will produce a prompt of

Command: **PICKBOX** ENTER

Enter new value for PICKBOX <3>: *(Enter desired value.)* ENTER

AutoCAD is now waiting for you to enter a numeric value from 0 to 50, to represent the size of the pickbox.

Selecting the Object

You can select objects to be edited by choosing any point on the object. Place the pick box on top of the object and click the select button. You can also select objects by entering coordinates for the location of the target. When you select objects with a thickness, it is important to place the pick box so that it touches an edge and not the center of the object. You also need to be careful not to select objects at their intersections with other objects.

SELECTING WITH A WINDOW

Rather than selecting objects to edit one at a time, you can select a group of objects by surrounding them completely in a window by responding W (ENTER) at the Select objects prompt. This procedure was used in Chapter 9 with the **Zoom** command, and is shown in Figure 10.1a. Any object that lies entirely inside the selection window will be edited. Any object that is partially inside the window will not be edited. The result of the selection set can be seen in Figure 10.1b. The command sequence to use the **Erase** command with a window is as follows:

Command: **E** ENTER *(Or click the Erase button.)*

ERASE

Select objects: **W** ENTER

Specify first corner: *(Select point for first corner.)*

Specify opposite corner: *(As the first point is selected, a window will drag across the screen until a second point is selected.)*

Select objects: *(Selected objects will be highlighted. Press* ENTER *when selection set complete.)*

When prompted to select objects, you now have an opportunity to alter the selection set by adding additional objects using another window, or by using the Add or Remove selection methods. If no changes need to be made, press ENTER. Making changes to the selection set will be introduced later in this chapter.

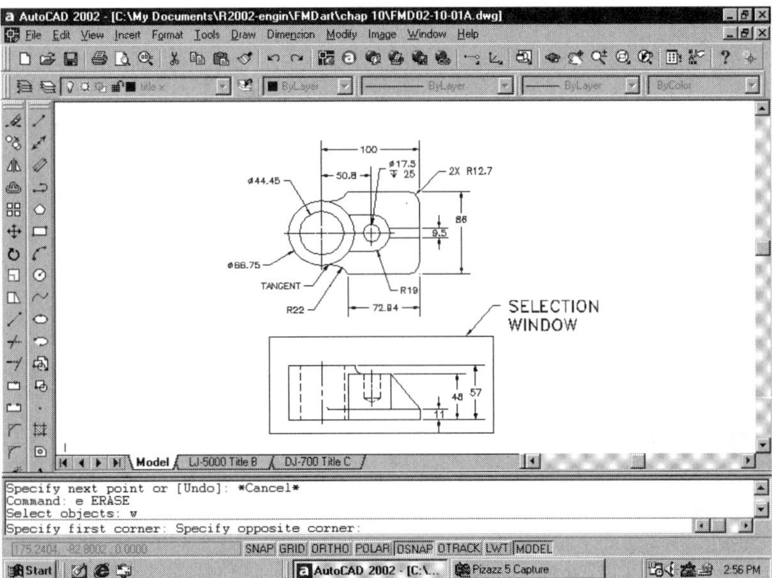

Figure 10.1a *If the Window option is used for selecting objects to be edited, all objects that lie entirely in the window will be edited. Courtesy Paul Bissell.*

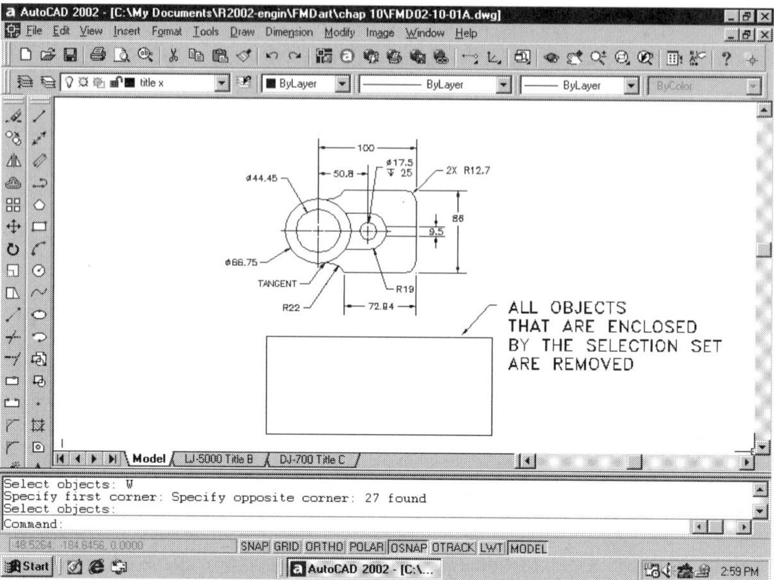

Figure 10.1b *The result of the* **Erase Window** *option started in Figure 10.1a. Courtesy Paul Bissell.*

AUTO SELECTION MODE

The auto selection mode is similar to editing using a window. The default mode for selecting objects is with the pick box. If the select button is clicked after the editing command is selected, the window method of selecting objects will be used. When prompted for the object to be erased, click the select button. This will turn the pick box into the corner of the selection box. As the cursor is moved, the window will be formed. Any items that the window encloses will be selected. When you're satisfied with the size of the selection box, click the select button to accept the objects. To remove the selected objects from the drawing, press ENTER. The command process is as follows:

Command: **E** ENTER *(Or click the Erase button.)*

ERASE

Select objects: *(Click the select button to select the first corner of the selection window.)*

Specify opposite corner: *(The window will drag across the screen until a second point is selected.)*

Select objects: ENTER

Command:

Implied Window Selection

The direction that is used to move the cursor as the selection window is defined will affect the selection process. If the cursor is moved from left to right when the selection set is picked, only objects completely enclosed in the window will be erased. When the selection window is created by moving from right to left, any object that the window crosses will be affected by the edit command. The results of implied window selection can be seen in Figure 10.2.

SELECTING WITH A CROSSING WINDOW

Choosing an object to edit with the Crossing option is similar to the Window option. You can select Groups of objects may be selected by surrounding them partially in a window by typing **C** (ENTER) at the Select objects prompt. The command process will be the same as the Window option. The crossing window will appear different from the Window option selection window. The crossing window is constructed using a dashed line, and a continuous line represents the Window option window. Any objects that lie entirely in the window will be edited. Any objects that are partially inside the window also will also be edited (see Figure 10.3).

SELECTING OBJECTS WITH THE WPOLYGON OPTION

The WPolygon option is especially useful for selecting an irregularly shaped group of objects to edit. The option is similar to the Window option, but you are allowed

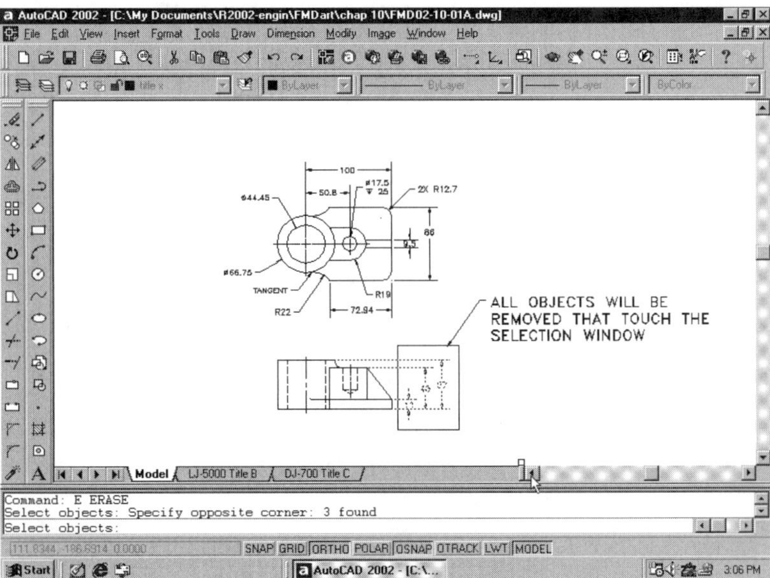

Figure 10.2 *If the cursor is moved from left to right, only objects completely enclosed will be in the selection set. When the selection window is created by moving from right to left, any object that the window crosses will be affected by the edit command. Courtesy Paul Bissell.*

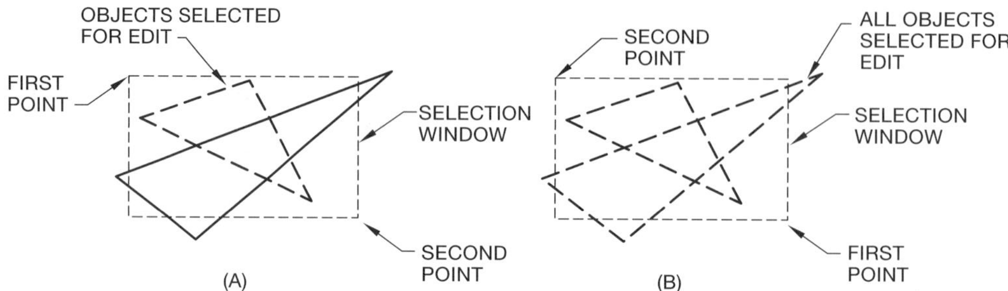

Figure 10.3 *Objects selected for editing with a window on the left and a crossing window on the right.*

to select points to form a polygon around objects to be edited. The polygon can be any shape except a shape that crosses itself, such as a figure eight. The command sequence is started once the editing option has been entered and you are prompted to select points to form the polygon. You can select points to define the polygon by using a pointing device or you can enter them as coordinates, as seen in Figure 10.4. Close the polygon by pressing ENTER. All objects lying totally inside the polygon will be edited. The process is completed using the following steps:

Command: **E** ENTER *(Or click the Erase button—any editing command can be used.)*

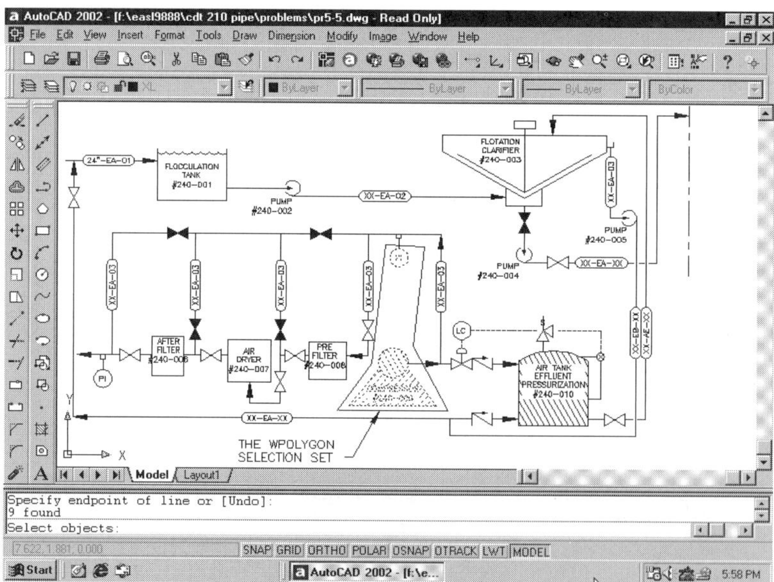

Figure 10.4 *The WPolygon option can be used as an alternative to the Closing option to select objects in an irregularly shaped group. Points for the polygon can be entered by selecting points or entering coordinates. Courtesy Michael J. Easley.*

ERASE

Select objects: **WP** ENTER

First polygon point: *(Enter a point.)*

Specify endpoint of line or [Undo]: *(Enter a point.)*

Specify endpoint of line or [Undo]: *(Enter a point.)*

Specify endpoint of line or [Undo]: *(Enter a point.)*

Specify endpoint of line or [Undo]: *(Continue until desired objects are selected.)* ENTER

100 found *(The number will vary as AutoCAD reports the number of objects in the selection set.)*

Select objects: ENTER

Command:

SELECTING OBJECTS WITH THE CPOLYGON OPTION

Objects can also be edited by typing **CP** (ENTER) at the Select objects prompt. This option will be similar to the effects of combining Crossing and WPolygon, in that it will edit all objects that are within or are crossed by the polygon. The command sequence is as follows:

Command: **E** ENTER *(Or click the Erase button—any editing command can be used.)*

ERASE

Select objects: **CP** ENTER

First polygon point: *(Enter a point.)*

Specify endpoint of line or [Undo]: *(Enter a point.)*

Specify endpoint of line or [Undo]: *(Enter a point.)*

Specify endpoint of line or [Undo]: *(Enter a point.)*

Specify endpoint of line or [Undo]: *(Continue until object set is complete.)* ENTER

100 found *(The number of objects in the selection set will vary.)*

Select objects: ENTER

Command:

SELECTING OBJECTS USING A FENCE

Objects can be selected for editing by Fence. This option is similar to WPolygon and CPolygon, except that it does not close the polygon. The fence might be a series of line segments or it could form a polygon that can cross itself. This allows long rows or columns of objects to be edited in one command sequence. As seen in Figure 10.5,

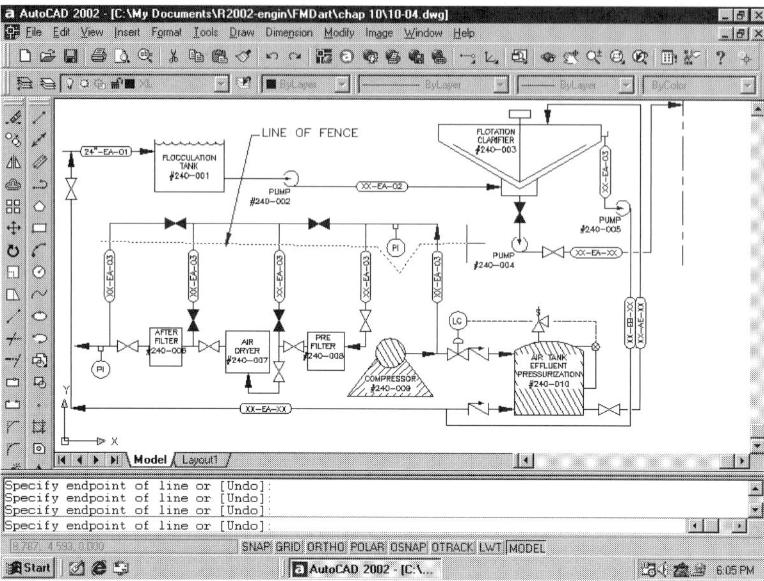

Figure 10.5 *The Fence option allows the use of straight lines to include objects in the selection set. Any objects that are touched by the fence will be included in the selection set. Courtesy Michael J. Easley.*

the selection fence edits only items that intersect or cross the fence. Fence points can be selected with a pointing device or by entering coordinates. The command sequence is as follows:

Command: **E** ENTER *(Or click the Erase button—any editing command can be used.)*

ERASE

Select options: **F** ENTER

First fence point: *(Enter a point.)*

Specify endpoint of line or [Undo]: *(Enter a point.)*

Specify endpoint of line or [Undo]: *(Enter a point.)*

Specify endpoint of line or [Undo]: *(Enter a point.)*

Specify endpoint of line or [Undo]: *(Continue until object set is complete.)* ENTER

5 found

Select objects: ENTER *(Pressing* ENTER *will execute the editing command.)*

Command:

SELECTING ALL DRAWING OBJECTS

When the All option is used, all objects that are current will be affected. You've learned how to freeze certain layers so that they cannot be seen. All objects except those on frozen or locked layers will be edited.

Command: **E** ENTER *(Or click the Erase button—any editing command can be used.)*

ERASE

Select objects: **ALL** ENTER *(All items will be highlighted.)*

137 found

Select objects: ENTER *(All items will be removed.)*

Command:

SELECTING THE LAST OBJECT

The Last option allows the most recently created object to be selected for editing.

Command: **E** ENTER *(Or click the Erase button—any editing command can be used.)*

ERASE

Select options: **L** ENTER

1 found

Select options: ENTER

Command:

ADDING OBJECTS TO THE SELECTION SET

When selection sets are defined, you might not have selected all objects intended for editing. Typing A (ENTER), for Add, at the Select objects prompt will allow for additional objects to be added to the selection set. Once Add is entered, objects can be added to the selection set individually or by any other selection method. When you are satisfied with the selection set, press ENTER to edit the drawing. Objects are added to a selection set using the following steps:

Command: **E** ENTER *(Or click the Erase button—any editing command can be used.)*

ERASE

Select objects: **W** ENTER *(Any selection method can be used.)*

Specify first corner: *(Select first corner.)*

Specify opposite corner: *(Specify second point.)*

5 found

Select objects: **A** ENTER

Select objects: **C** ENTER *(Any selection method can be used.)*

Specify first corner: *(Select first corner.)*

Specify opposite corner: *(Specify second point.)*

20 found, 25 total

Select objects: ENTER

Command:

REMOVING OBJECTS FROM THE SELECTION SET

This option will allow for objects to be removed from the selection set. Once objects are selected for editing, typing R (ENTER) will allow objects to be taken away from the selected objects to edit. Objects can now be removed from the selection set individually or with any other selection method. When you are satisfied with the selection set, click the select button to edit the drawing. The command sequence is as follows:

Command: **E** ENTER *(Or click the Erase button—any editing command can be used.)*

ERASE

Select objects: **C** ENTER *(Any selection method can be used.)*

Specify first corner: *(Select first corner.)*

Specify opposite corner: *(Specify second point.)*

5 found

Select objects: **R** ENTER

Select objects: *(Select objects to be removed using any selection method.)*

1 found, 1 removed 4 total

Remove objects: ENTER

Command:

MULTIPLE SELECTION SET

The Multiple selection option works well when drawing objects are stacked over each other. In Chapter 5 you learned to place parts of your drawing on different layers. The Multiple option will allow you to edit material on different layers at one time.

DEFINING THE SELECTION MODE

The **Selection** tab of the **Options** dialog box offers six modes that will affect the formation of the selection set. Choosing **Options** from the **Tools** menu opens the dialog box. An examination of this tab will reveal that three of the settings are active. Any combination of selection methods can be used.

NOUN/VERB SELECTION

Up to this point, as you have edited an object, you've used a process AutoCAD refers to as Verb/Noun. You've selected the verb (**Erase**) and then selected the object to be edited (the noun). With the **Noun/Verb selection** check box active, AutoCAD allows you to reverse the process. The object can be selected first, and then the editing command.

With **Noun/Verb selection** active, any object can be selected merely by choosing it with the pick box. Open a drawing and select a line. Notice the line is highlighted and a box appears in the middle and at each end of the line, but the Command prompt has not changed. Typing E (ENTER) will erase the selected line. The command sequence is as follows:

> Command: *(Select object to edit using any method.)*
>
> Command: **E** ENTER *(Or click the Erase button—any listed editing command can be used.)*

Noun/Verb selection can be used with the following commands to edit objects:

Array	**Ddchprop**	**Mirror**	**Wblock**
Block	**Dview**	**Move**	**Explode**
Change	**Erase**	**Rotate**	
Chprop	**Hatch**	**Scale**	
Copy	**List**	**Stretch**	

Some commands require that the edit command be executed before the objects can be selected. These commands include the following:

Break	**Divide**	**Fillet**	**Offset**
Chamfer	**Extend**	**Measure**	**Trim**

If you use these commands, they will ignore any selection set created prior to starting the command.

USING SHIFT TO ADD TO SELECTION

This selection method controls how objects are added to an existing selection set. Start a new drawing and draw a few lines. Toggle **Noun/Verb selection** to OFF (no X) and **Use Shift to add to selection** ON (X). Click OK and return to the drawing. Select an object to erase and it will be highlighted. Select a second object, and the first object no longer will be highlighted. To retain the first object, hold the SHIFT down while the second object is selected. Both objects will now be edited. If **Use Shift to add to selection** is not enabled, objects are added to the selection by selecting them individually or using a selection window. To remove objects from the selection set, press SHIFT while selecting the object.

PRESS AND DRAG SELECTION

This mode determines the method used for drawing a selection window. With the **Press and Drag** option active, the selection window can be made with one button selection rather than two. Pressing and holding the select button and moving the cursor diagonally creates the selection window. Releasing the select button completes the window when it is the size you want. When this option is not selected, objects can be selected for editing using the normal Window option.

IMPLIED WINDOWING

With the **Implied windowing** option in the default mode, a selection window is automatically created when the Select object prompt is displayed. Clicking the select button will automatically select the first corner point. A prompt will then be given requesting the other corner. If you draw the window from left to right, the window selects objects that lie entirely within the window. If the window is drawn from right to left, all objects in and touching the window will be selected. With the option off, the window must be set for each edit. In the OFF position, clicking the select button does nothing. A mode such as Window or Crossing needs to be entered to start the creation of a window.

OBJECT GROUPING

The **Object grouping** option determines if grouped objects will be recognized as individual objects or as a group. In the default setting of ON, grouped objects function as a group. When OFF, grouped objects can be edited individually. Grouping of objects will be discussed in later chapters.

ASSOCIATIVE HATCH

The **Associative hatch** option affects how the hatch patterns are edited and will be discussed in Chapter 19.

MODIFYING A DRAWING USING ERASE

The **Erase** command allows you to remove unwanted objects (mistakes) from a drawing. Because you're familiar with **Erase** by now, the command serves as a means of using some of the selection methods just covered. The selection prompts Point, Multiple, Window, Crossing, WPolygon, CPolygon, Fence, All, Last, Add, and Remove should start to seem more meaningful. Figure 10.6 shows an object that has been selected for editing by an automatic window. The selected object is highlighted, while AutoCAD waits for you to select other objects. When the selection process is complete, press ENTER to remove the object.

The Window option can be seen in Figure 10.7. Notice that at the Select object prompt, W was entered rather than WINDOW. Remember an object must be entirely inside the window to be edited. Figure 10.8 shows the objects to be erased as

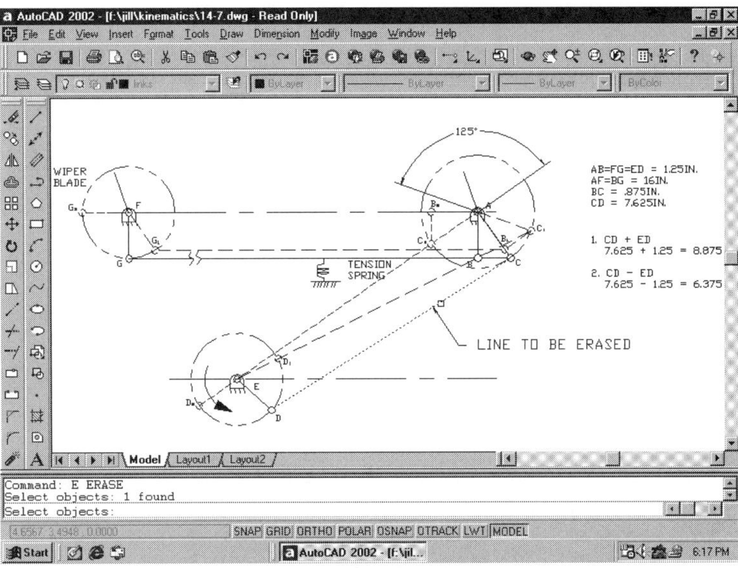

Figure 10.6 *Line C-D in the linkage drawing has been selected for* **Erase.** *Courtesy Jill Heilman.*

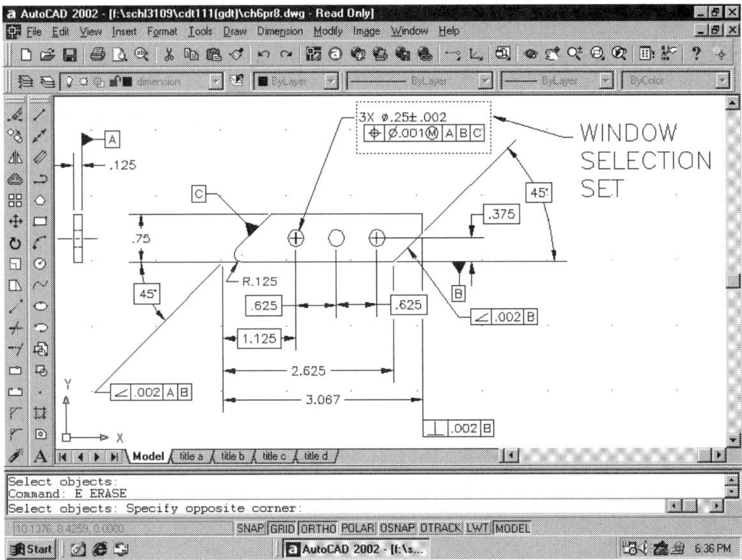

Figure 10.7 *Using automatic selection or the Window option to form a selection set for the* **Erase** *command. Courtesy Cathy Schlesser.*

highlighted. If you would like to Add or Remove objects from the selection set, enter the corresponding letter at the Select objects prompt. Figure 10.9 shows the effects of the Window **Erase** with no objects added or removed.

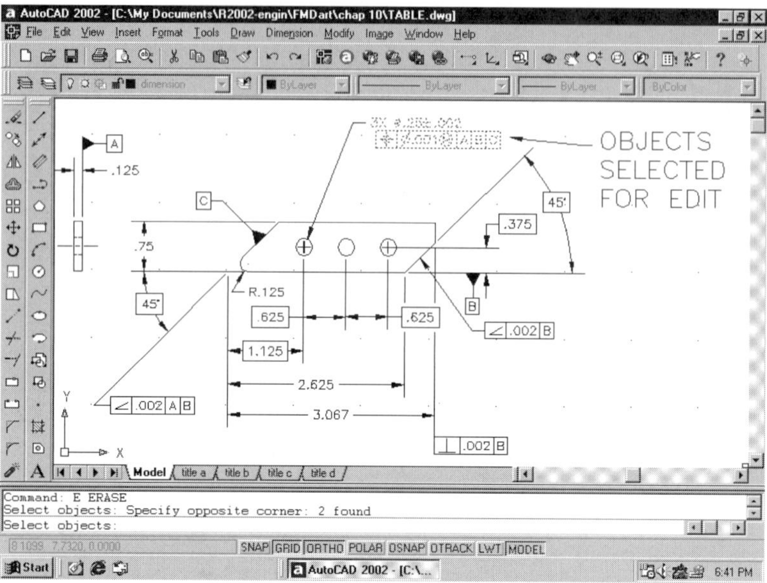

Figure 10.8 *Results of the Window selection set are now displayed to confirm the choice of objects to be included in the selection set. Notice that the number of objects included in the set is displayed above the Command prompt. Courtesy Cathy Schlesser.*

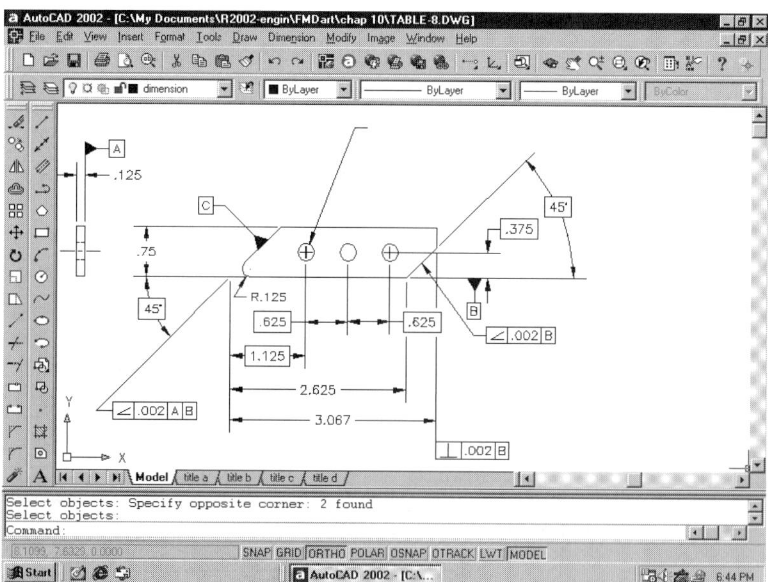

Figure 10.9 *The effect of the Window option used with the* **Erase** *command. Courtesy Cathy Schlesser.*

If you type C (ENTER) at the Select object prompt, objects that are within or are crossed by the window are selected to be erased, as seen in Figure 10.10. The result of a Crossing window selection can be seen in Figure 10.11. Because the engineer would like to keep the outer extension line and remove the dimensions, the Remove option was used to restore the extension line on the right. Figure 10.12 shows the final selection set.

MODIFYING A DRAWING BY ADDING OBJECTS

The balance of this chapter will introduce the skills needed to alter a drawing by adding objects to the drawing. Later chapters will help you edit a drawing by altering existing drawing components. The **Copy**, **Mirror**, **Offset**, and **Array** commands can each be used to modify a drawing by adding objects to the drawing base. Objects added using one of these commands can be selected with any of the selection methods introduced earlier in this chapter.

COPY

Whenever multiple objects need to be drawn, the **Copy** command is a convenient method of reproducing objects. **Array**, **Block**, and **Wblock** are also efficient means of reproducing objects (**Block** and **Wblock** will be covered in later chapters). Start the **Copy** command by selecting the **Copy** button on the **Modify** toolbar, by typing CP (ENTER) at the Command prompt, or by selecting **Copy** from

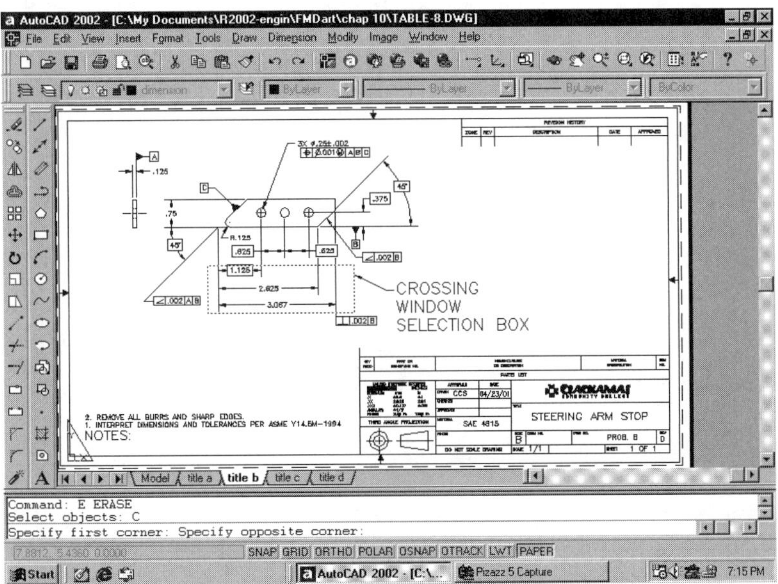

Figure 10.10 *Using the Crossing option to remove three dimensions. Any object touched by the window will be edited. Courtesy Cathy Schlesser.*

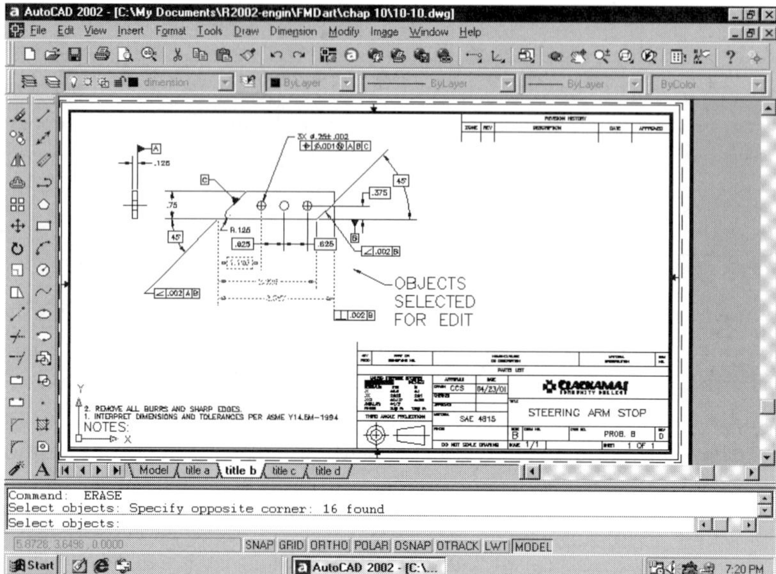

Figure 10.11 *Results of the Crossing window selection set are now displayed to confirm the choice of objects to be included in the selection set. Courtesy Cathy Schlesser.*

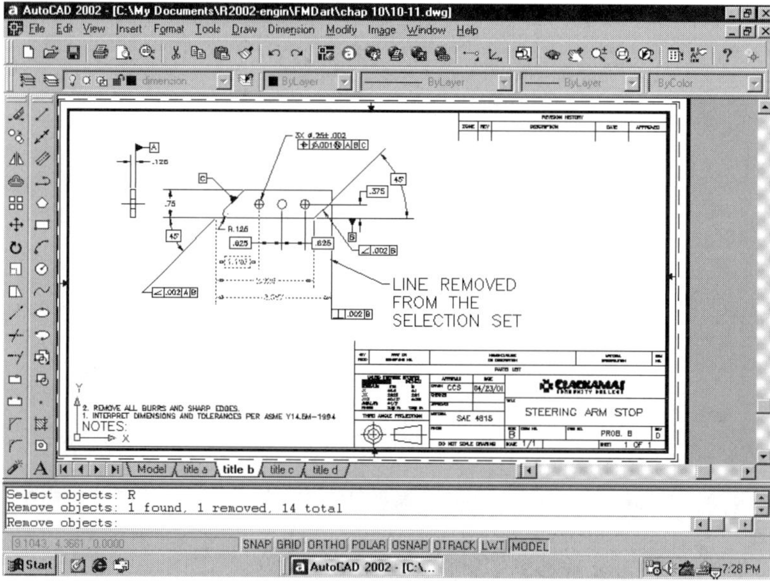

Figure 10.12 *If you would like to include more objects in the set, type* **A** *(*ENTER*) at the Select objects prompt. Typing* **R** *(*ENTER*) at the prompt will remove objects from the set. To accept the set, press* ENTER. *Courtesy Cathy Schlesser.*

the **Modify** menu. Figure 10.13 shows an example of a drawing created using the **Copy** command. The drawing on the left was drawn first and copied to produce the two components on the right. The command sequence is as follows:

Command: **CP** ENTER *(Or click the Copy button.)*

Select objects: *(Select objects by desired selection method.)*

Specify first corner: *(Select window corner.)*

Specify opposite corner: *(Select opposite window corner.)*

54 found

Specify base point or displacement or [Multiple]: *(Select base point or enter coordinates of object to be moved.)*

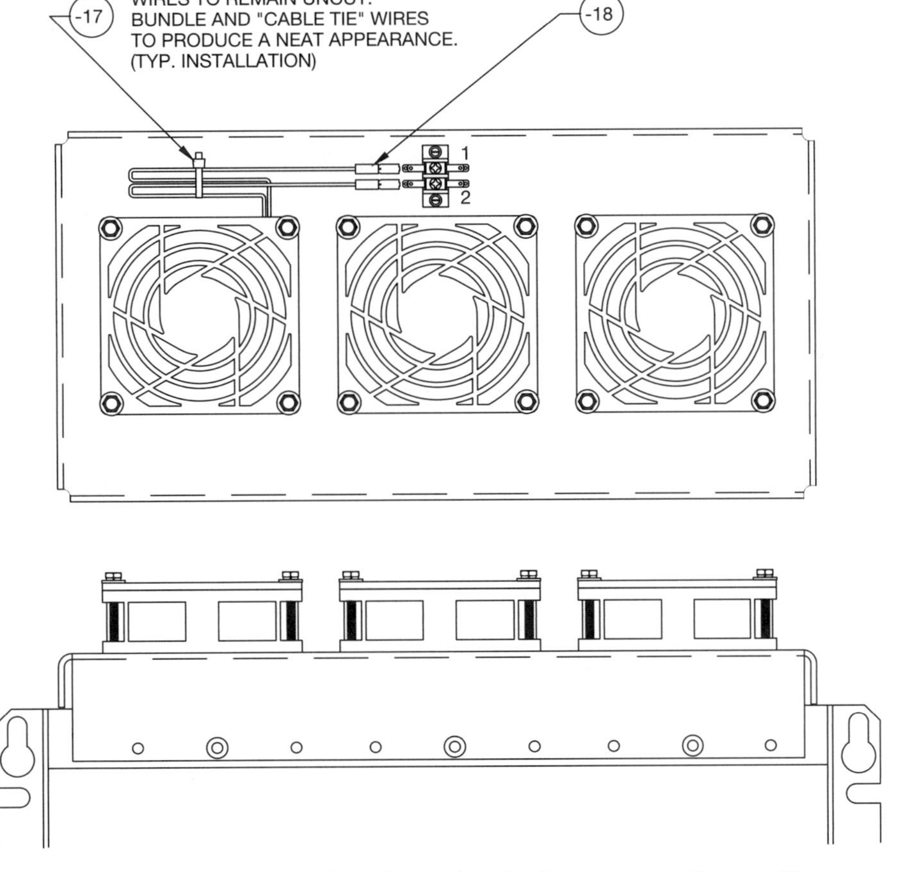

Figure 10.13 *The* **Copy** *command can be used to duplicate common features. The command can be typed, selected on the* **Modify** *toolbar, or selected from the* **Modify** *menu. Courtesy Floyd D. Miller.*

Specify second point of displacement or < use first point as displacement>: *(Select a point to relocate the base point to.)*

Command:

The Copy process can be seen in Figures 10.14a through 10.14c.

Multiple Copies

It's not uncommon to need to reproduce multiple copies of an object throughout a drawing. Multiple copies of an object or group of objects can be made once the selection set is selected. Notice in the **Copy** command sequence that the alternative to the first point of displacement is Multiple. When you type **M** (ENTER) at the prompt, the selection group will be placed at an unlimited number of locations. The command sequence to make four copies of a group of objects is as follows:

Command: **CP** ENTER *(Or click the Copy button.)*

COPY

Select objects: **C** ENTER *(Select objects by desired selection method.)*

Specify first corner: *(Select window corner.)*

Specify opposite corner: *(Select opposite window corner.)*

25 found

Select objects: ENTER

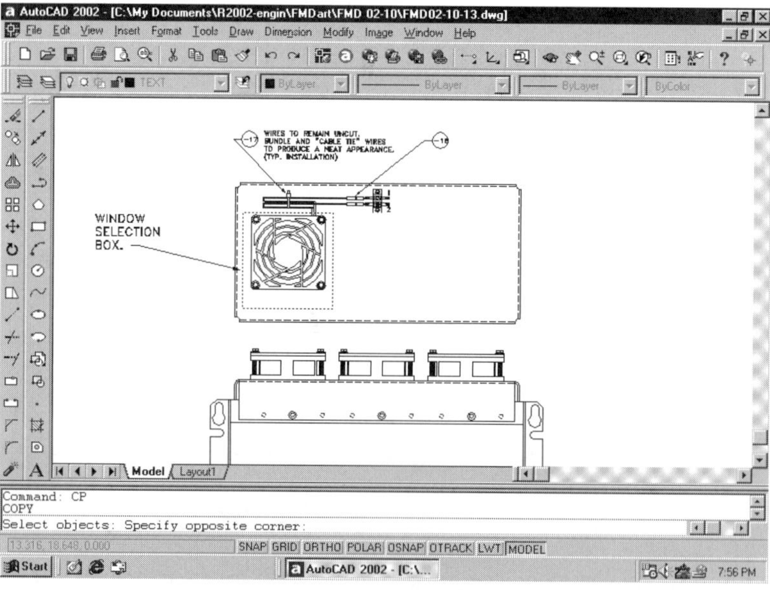

Figure 10.14a *The* **Copy** *command is started by defining the selection set. Courtesy Floyd D. Miller.*

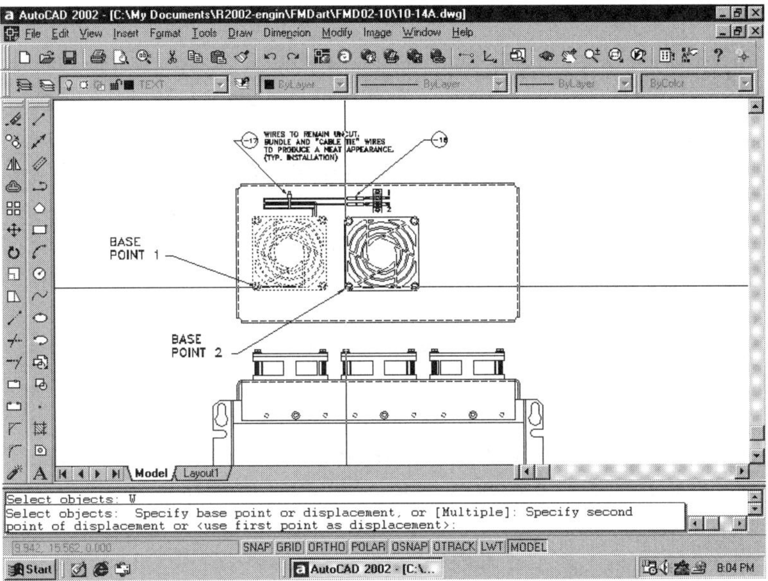

Figure 10.14b *The second step of the* **Copy** *command is to specify a base point and then specify a new location for the base point. Courtesy Floyd D. Miller.*

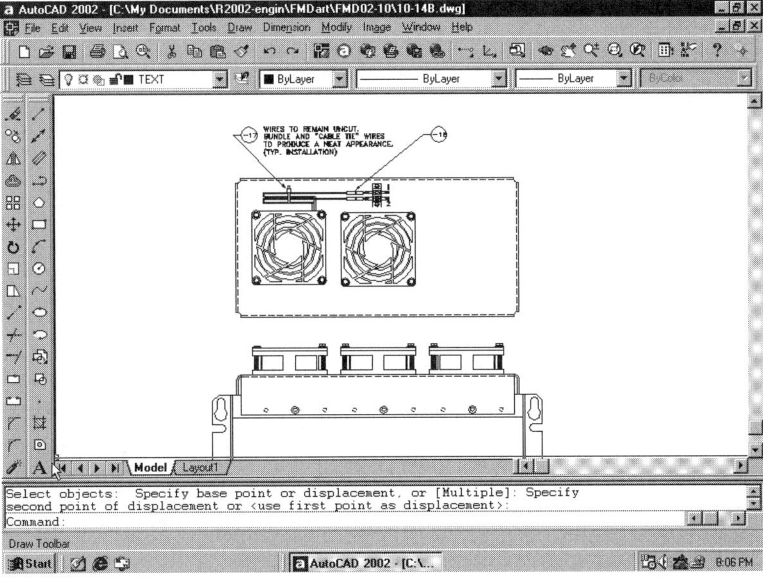

Figure 10.14c *Once* ENTER *is pressed for the new location, the objects in the selection set will be duplicated. Courtesy Floyd D. Miller.*

Specify base point or displacement or [Multiple]: **M** ENTER

Specify base point: *(Select base point or enter coordinates of object to be moved.)*

Specify second point of displacement or <use first point as displacement>: *(Select a point to relocate the base point to for the first copy.)*

Specify second point of displacement or <use first point as displacement>: *(Select a point to relocate the base point to for the second copy.)*

Specify second point of displacement or <use first point as displacement>: *(Select a point to relocate the base point to for the third copy.)*

Specify second point of displacement or <use first point as displacement>: *(Select a point to relocate the base point to for the final copy.)*

Specify second point of displacement or <use first point as displacement>: ENTER

Command:

The process can be continued until all of the desired copies are drawn. To end the sequence, press ENTER at the displacement prompt.

Practical Use for Multiple Copy

In a practical application, Figure 10.15 shows the effect of the Multiple **Copy** command. One fan housing was drawn and Multiple **Copy** was used to copy and place the two additional fans.

MIRROR

Objects in a drawing often need to be reversed. Something as simple as reversing to add a second hole to a part, or as complex as completing the opposite half of a symmetrical object (Figure 10.16) can be done using the **Mirror** command. The command can also be used any time objects are symmetrical. When you draw half the object, you can create the other half by using **Mirror**. Start the **Mirror** command by selecting the **Mirror** button on the **Modify** toolbar, by typing MI (ENTER) at the Command prompt, or by selecting **Mirror** from the **Modify** menu.

Once objects are selected to be mirrored (flipped), you will be asked to describe a "mirror line" by selecting each end point. The process can be seen in Figure 10.17. Think of the mirror line as a fold line between the old and the new objects to be mirrored. As you pick the first point of the mirror, drag the selection set into position and alter it as the cursor is moved. The use of ORTHO and Object Snaps can greatly aid in the placement of the mirrored image. The new placement can be seen in Figure 10.18.

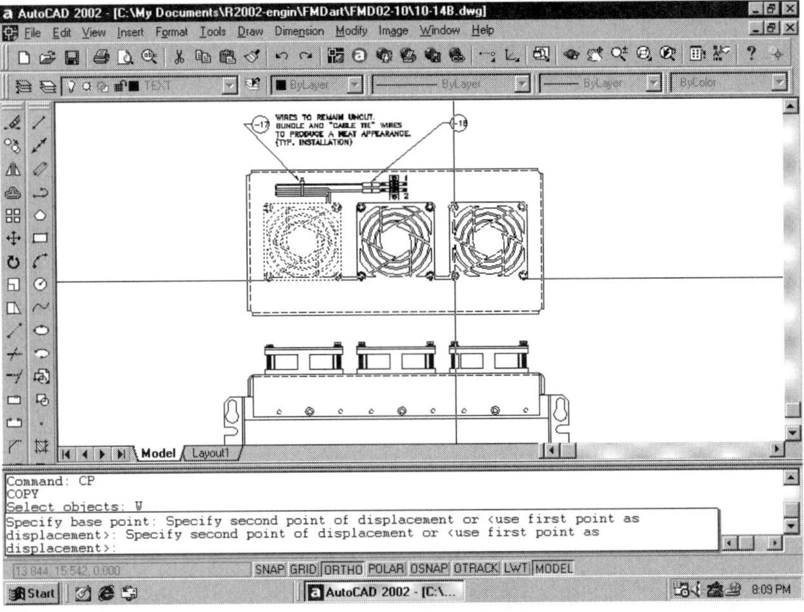

Figure 10.15 *The Multiple mode of* **Copy** *allows objects to be copied several times without the command having to be repeated. Courtesy Floyd D. Miller.*

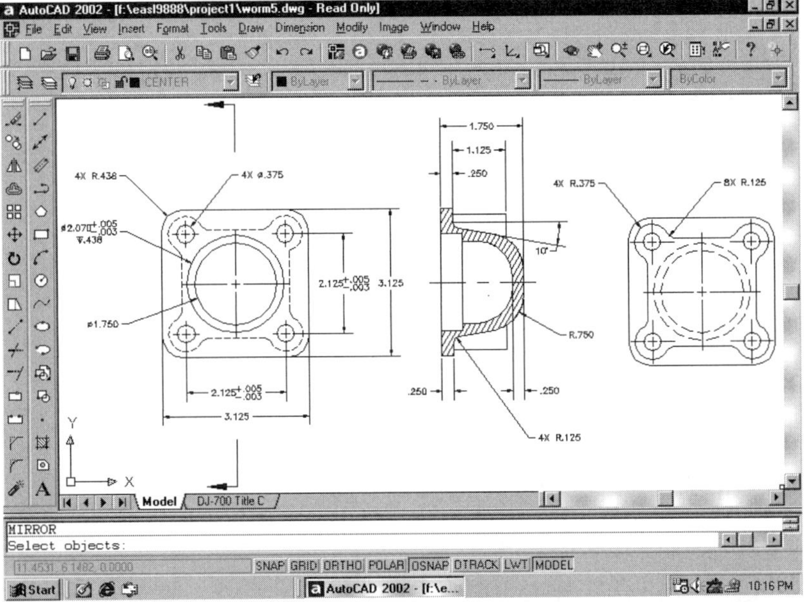

Figure 10.16 *The* **Mirror** *command can be used to flip existing objects. Each of the two views was created using the* **Mirror** *command. The command can be selected on the* **Modify** *toolbar, from the* **Modify** *menu, or entered by keyboard. Courtesy Michael J. Easley.*

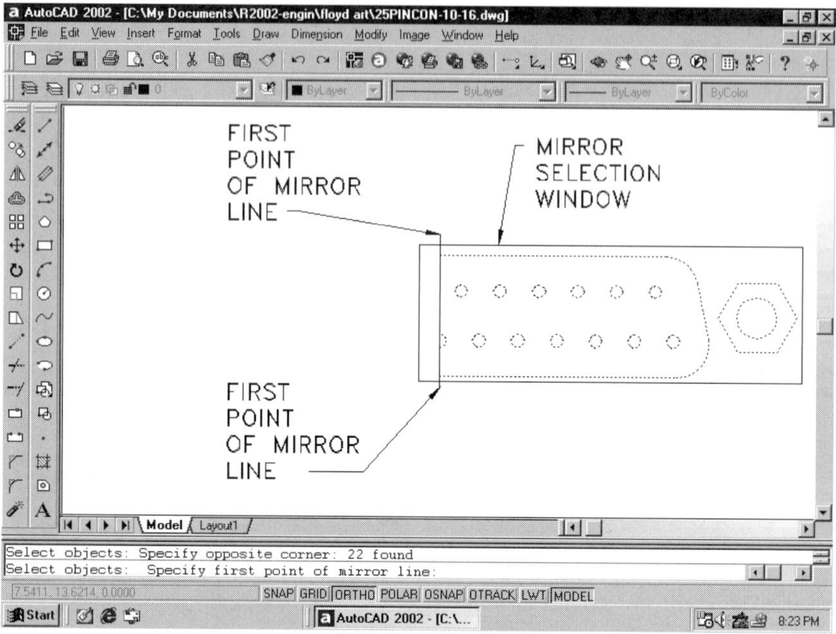

Figure 10.17 *Once the selection set has been defined, two points will be required to define the mirror line. Courtesy Floyd D. Miller.*

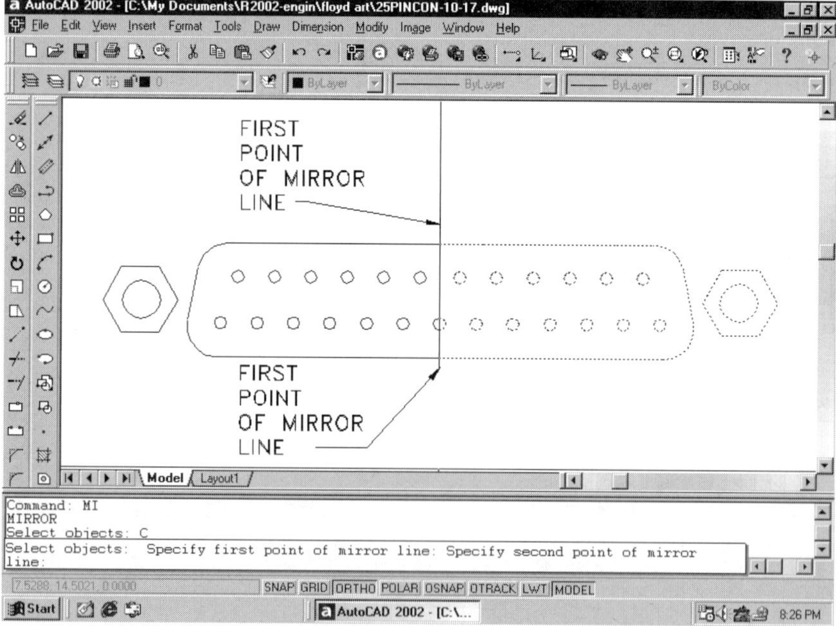

Figure 10.18 *As the second point is entered, the selection set will be mirrored into position. Courtesy Floyd D. Miller.*

You'll be given the option of keeping or discarding the original object prior to making the **Mirror** operation permanent. Your choice will depend on the use of the drawing. If the object is being completed by **Mirror**, don't delete the old object. The final result will resemble Figure 10.19. The command sequence is as follows:

Command: **MI** ENTER *(Or click the Mirror button.)*

Select objects: **C** ENTER *(Choose objects using the desired selection method or combination of methods.)*

Specify first corner: *(Select window corner.)*

Specify opposite corner: *(Select opposite window corner.)*

255 found

Select objects: ENTER

Specify first point of mirror line: *(Select a point or enter coordinates.)*

Specify second point of mirror line: *(Select a point or enter coordinates.)*

Delete source objects? [Yes/NO] <N>:ENTER

Command:

If the object is being flipped, you will want to remove the old object, as in Figure 10.20. The Command prompts would be similar to those just noted, until the last line is reached.

Delete source objects? [Yes/NO] <N>:Y ENTER

Typing Y (ENTER) will now delete the original object and display the mirrored copy of the original object.

One of the drawbacks of the **Mirror** command is how it affects text. Although you will not add text to a drawing for a few more chapters, an additional feature of **Mirror** will be helpful soon. As AutoCAD is currently set, if you were to mirror a drawing with text, the text would be reversed (see Figure 10.21). To mirror the drawing without flipping the text, adjust the **Mirrtext** value to zero. With **Mirrtext** set to zero, the drawing will be flipped, but the text will remain readable, as seen in Figure 10.22. This will be covered again as text is explored, but you might want to set the value in your drawing template now so that when you do add text, you won't need the **Undo** or **Oops** command. The command sequence is as follows:

Command: **MIRRTEXT** ENTER

New value for MIRRTEXT <1>: **0** ENTER

Command:

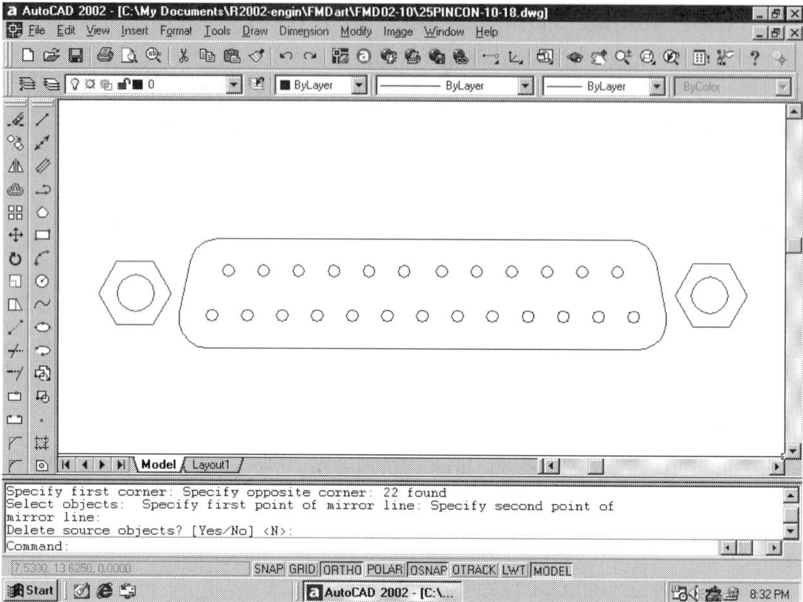

Figure 10.19 *Before the* **Mirror** *command is complete, an option will be given to "Delete source objects?" If you respond* **N** *(*ENTER*), the selection set will be mirrored and the original objects will be retained. Courtesy Floyd D. Miller.*

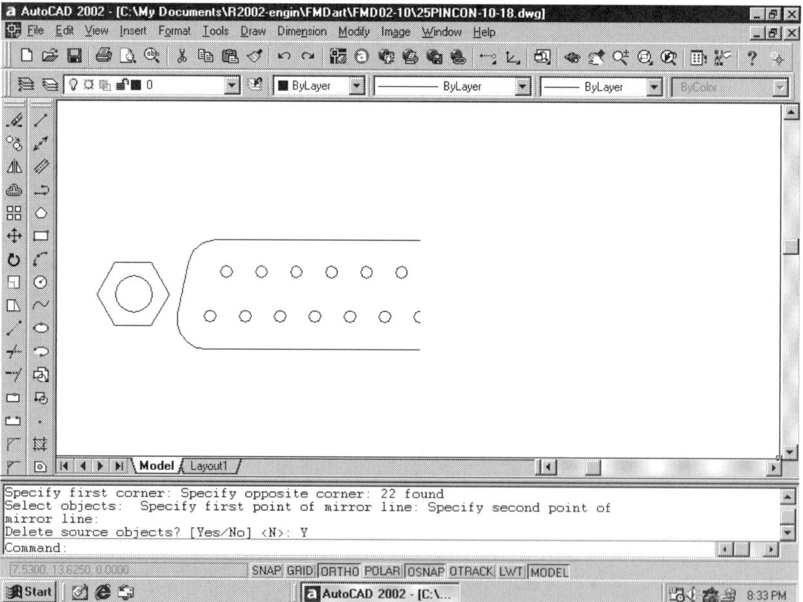

Figure 10.20 *If you respond* **Y** *(*ENTER*) to "Delete source objects?" the selection set will be mirrored and the original objects will be removed. Courtesy Floyd D. Miller.*

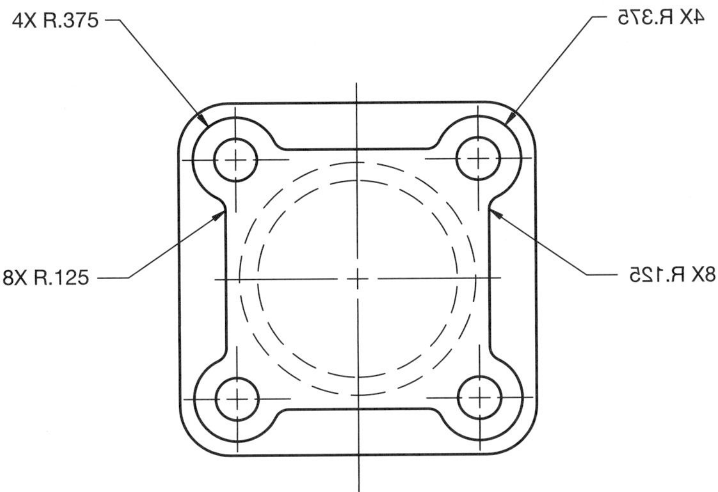

Figure 10.21 *The default setting of the* **Mirror** *command also will reverse text. Courtesy Michael J. Easley.*

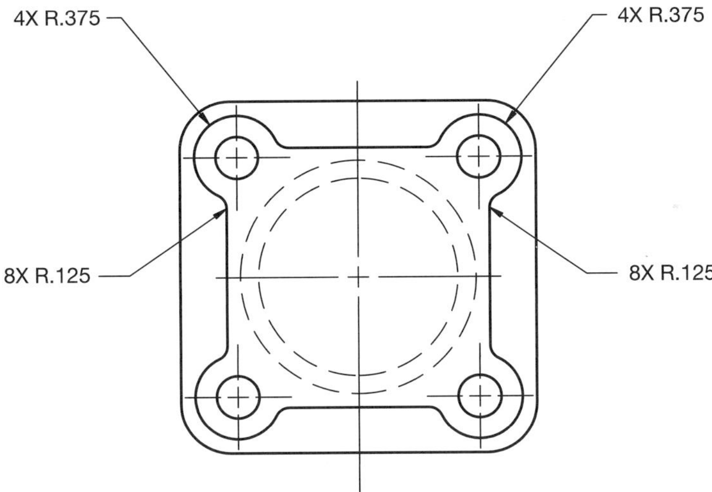

Figure 10.22 *If the* **Mirrtext** *variable is set to 0, text will not be mirrored. Courtesy Michael J. Easley.*

Practical Uses for Mirror

Mirror is useful whenever objects need to be grouped. Any drawing that contains objects that lie along a centerline can be created using the **Mirror** command. Drawing half of the symbol and then using **Mirror** to complete it will create each of these symbols. Later chapters will explain how to make and save each symbol so that it only needs to be created once. If you're really eager, explore Blocks in the **Help** menu.

OFFSET

The **Offset** command allows a line, circle, arc, or polygon to be copied and relocated parallel to and at a specific distance from the original object. The uses of this command are almost endless. A circle can be drawn and offset to represent threaded holes and machined features such as counterbores or countersinks. Starting with a horizontal and a vertical line, an entire rectangular object can be drawn by offsetting lines and then cleaned up using other editing commands, presented in later chapters.

Using **Offset**, lines can be drawn in the exact location without thought as to their exact length. Get the location right without worrying that the line might be too long, and then edit the length with **Fillet**, **Stretch**, or **Trim**. (Each command will be introduced in Chapter 11.) Figure 10.23 shows the use of **Offset** to lay out a simple object.

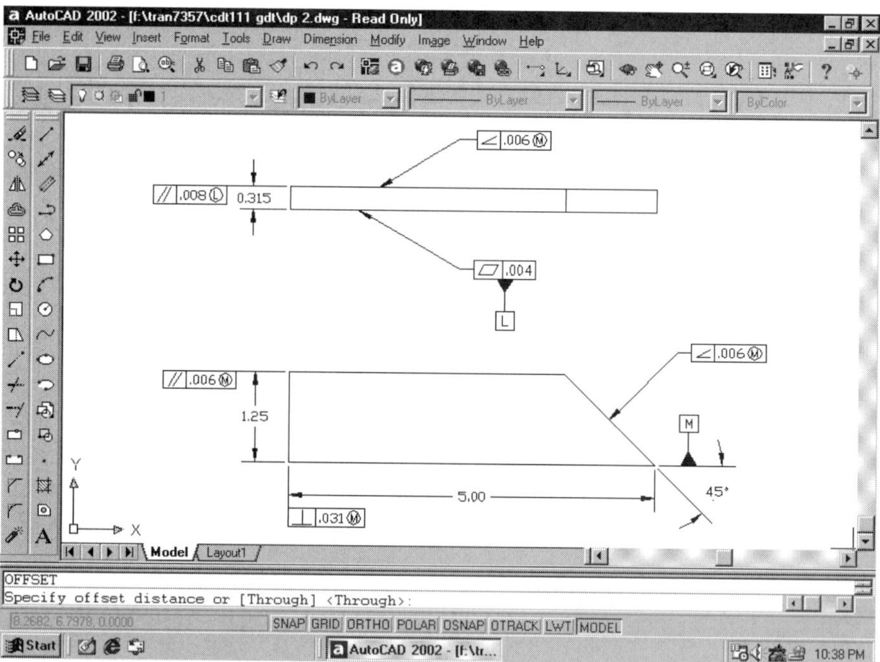

Figure 10.23 *With one horizontal and one vertical line, the* **Offset** *command can be used for drawing parallel lines at a specified distance apart to create an object. Once the basic lines have been put in the correct position using* **Offset**, *other editing commands can be used to finish the drawing. Courtesy Hanh Tran.*

Start the command process by selecting the **Offset** button on the **Modify** toolbar, by typing O (ENTER) at the Command prompt, or by selecting **Offset** from the **Modify** menu. The command sequence to offset a line 1" from the original line is as follows:

> Command: **O** ENTER *(Or click the Offset button.)*
>
> OFFSET
>
> Specify offset distance or [Through] <Through>: **1** ENTER

A distance can be entered by providing a numeric value or by typing T (ENTER). When the distance is entered, a new prompt will be displayed that prompts

> Select the object to offset or <exit>:

Lines, arcs, circles, ellipses, xlines, rays, and polylines can be offset (the last three objects will be introduced in later chapters). Once selected, the object will be highlighted and the prompt will read

> Specify point on side to offset:

Move the cursor slightly to the desired side and click the select button to offset the object and redisplay the prompt:

> Select the object to offset or <exit>:

The process can be continued indefinitely, with the offset distance that was entered originally remaining as the default. Terminate the **Offset** process by pressing ENTER. Once the process is terminated, a new default distance can be entered or a new command sequence can be started. The command sequence can be seen in Figure 10.24.

Typing T (ENTER) at the original prompt allows a point to be entered with a pointing device, and the object to be offset will pass through the indicated point. The command sequence is as follows:

> Command: **O** ENTER *(Or click the Offset button.)*
>
> OFFSET
>
> Specify offset distance or [Through] <1.00>: **T** ENTER
>
> Select the object to offset or <exit>: *(Select an object.)*
>
> Specify through point: *(Move cursor to the desired point and select.)*
>
> Select the object to offset or <exit>: ENTER
>
> Command:

The **Offset** process with the Through option can be seen in Figure 10.25.

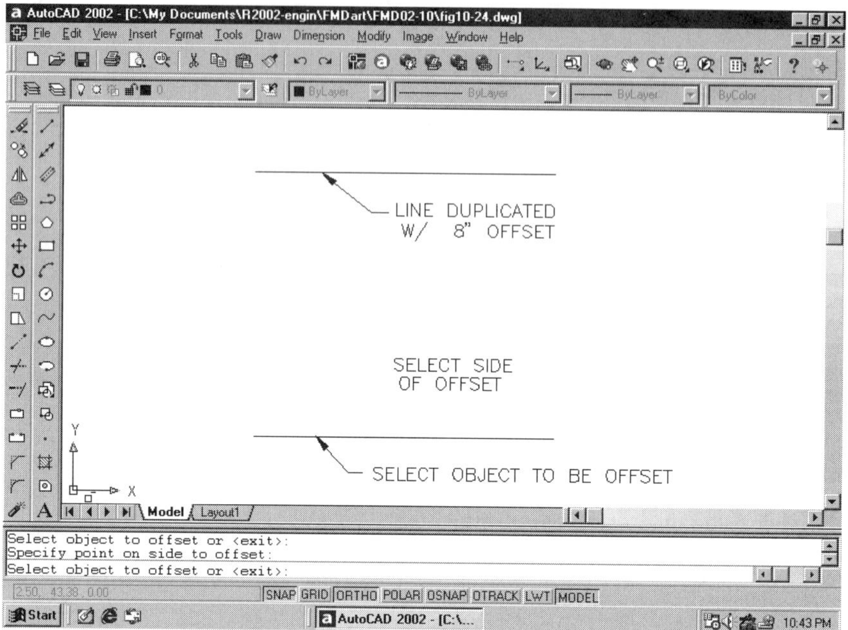

Figure 10.24 *Complete the* **Offset** *command by selecting the distance to be offset, the object to be offset, and the side of the original that the offset will occur on.*

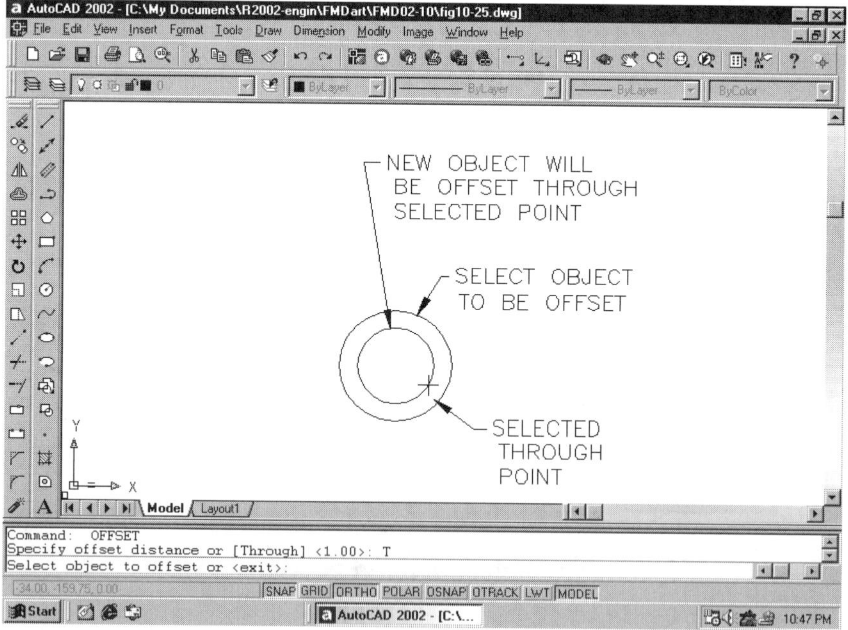

Figure 10.25 *The Through option of* **Offset** *allows the offset distance to be specified using the pointing device.*

Practical Uses for Offset

Offset is an efficient method of creating a drawing. By drawing two perpendicular lines, you can complete a rectangular view using **Offset**. The required width and height can be used for the offset distances. The space between views can be established using **Offset**, and then a top or side view can be completed using the same process. See Chapter 6 for a review of drawing layout using **Offset**.

ARRAY

The **Array** command will also allow for multiple copies of an object or group of objects to be reproduced. The command also provides several added features not available with **Copy**. **Array** reproduces objects in rectangular or circular (polar) patterns, allows the object to be rotated as it is reproduced, and allows for easy control of the spacing of the object during the **Array** operation. Figures 10.26 and 10.27 show examples of a rectangular and a polar array. Start the command by selecting the **Array** button on the **Modify** toolbar, by typing AR (ENTER) at the Command prompt, or by selecting **Array** from the **Modify** menu. Each method will produce the **Array** dialog box shown in Figure 10.28. Once the dialog box has been accessed, the type of array pattern should be selected.

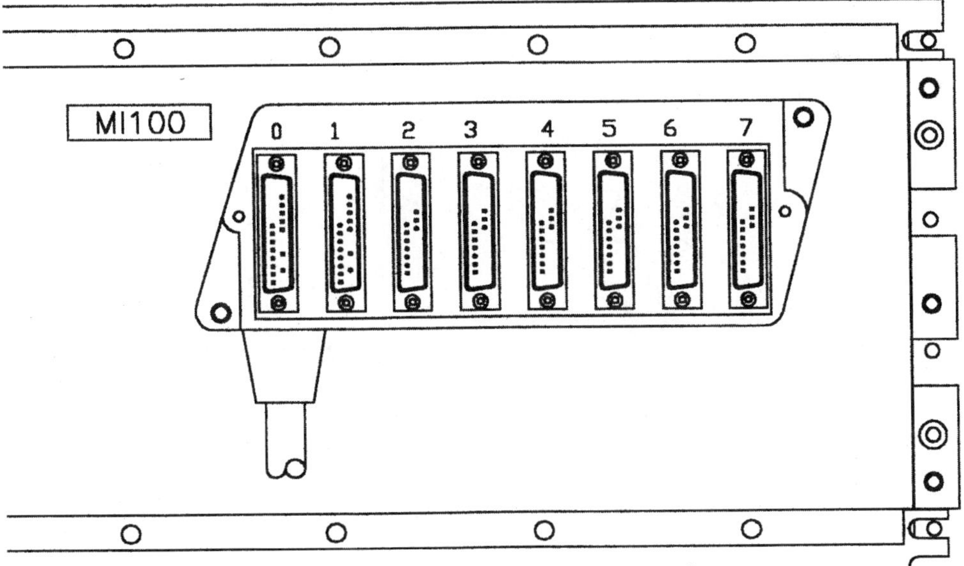

Figure 10.26 *The holes and connection plugs were placed using Rectangular option of the* **Array** *command. Courtesy Floyd D. Miller.*

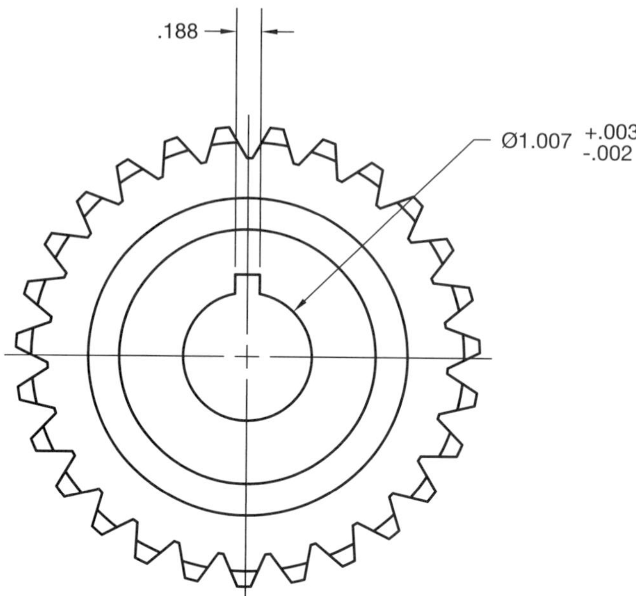

Figure 10.27 *The gear teeth were placed using the Polar option of the* **Array** *command. Courtesy Michael J. Easley.*

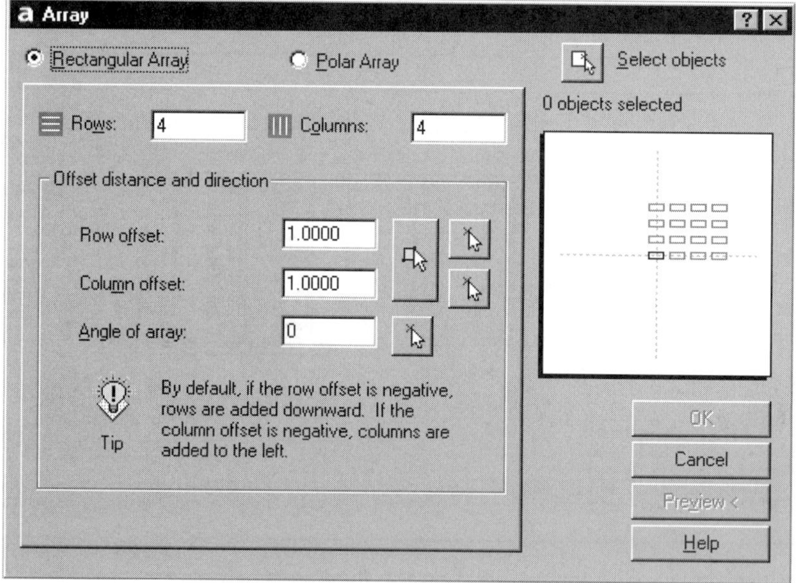

Figure 10.28 *The* **Array** *dialog box is used to recreate object in rectangular and circular patterns. Access the display dialog box by selecting the* **Array** *button or by typing* **AR** (ENTER) *at the Command prompt.*

Rectangular Array

A rectangular array will prove suitable for projects that are arranged in rows or columns. Selecting the **Rectangular Array** button starts the process. Six additional specifications must be provided to perform a rectangular array. These include selecting the object to be arrayed, the number of rows and columns, the row and column offsets, and the array angle. The elements can be seen in Figure 10.29.

Selecting the Object to Array

Selecting the **Select objects** button will remove the **Array** dialog box and allow objects to be selected for the array pattern. As the dialog box is removed, drawing objects can be selected using any of the drawing selection methods introduced earlier. When the selection set is complete, press ENTER or SPACEBAR.

Specifying the Number of Rows

In the **Rows** edit box, specify the quantity of horizontal rows to be created. Any whole number can be entered for the value. Figure 10.30 shows an example of an array with two, three, and four rows.

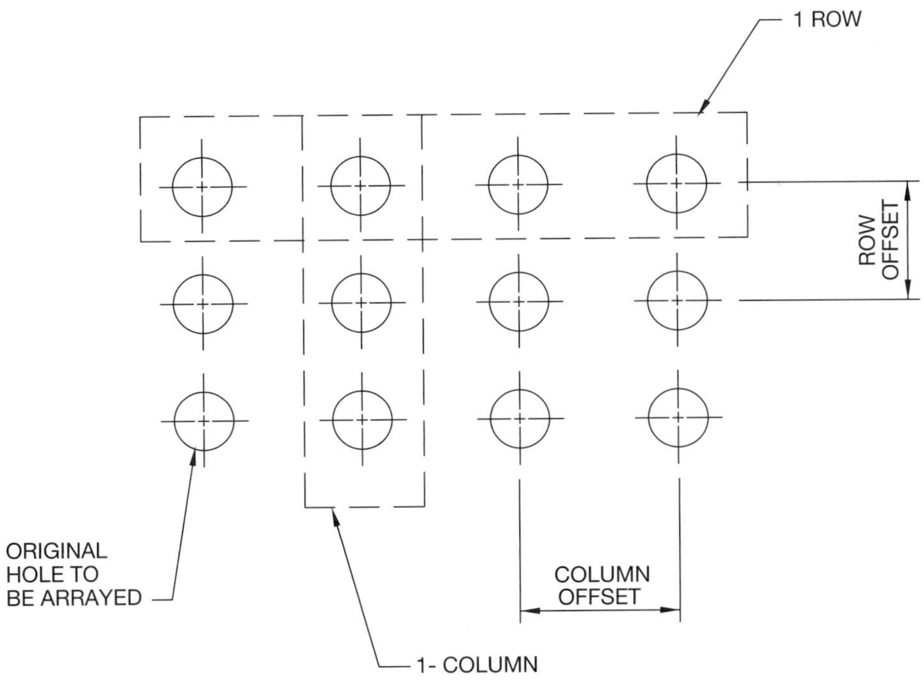

Figure 10-29 *The number of rows and columns and the spacing of each must be specified to use a rectangular array.*

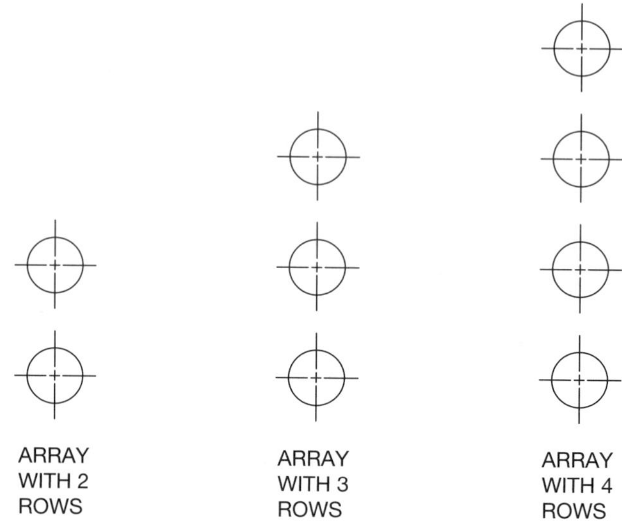

Figure 10.30 *Altering the number of rows of an array.*

Specifying the Number of Columns

In the **Columns** edit box, specify the number of vertical columns to be used. Any whole number can be entered for the value. If the value of 1 was used for the row value, a value other than 1 must be used for the column value. Otherwise, AutoCAD will respond with the prompt:

One-element array, nothing to do.

Figure 10.31 shows the effect of an array using two, three, and four columns.

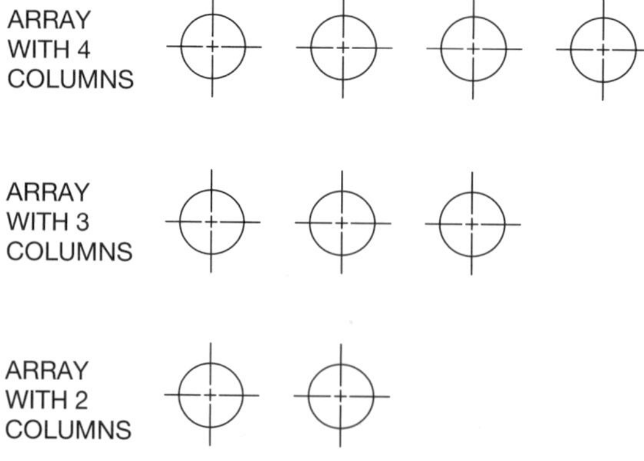

Figure 10.31 *Altering the number of columns in an array.*

Specifying the Row Offset

The **Row offset** edit box allows the offset distance, also called the unit cell distance, to be set for rows. The offset distance includes the size of the object to be arrayed. If you are arranging 16" wide chairs in rows, and would like 16" between rows, the spacing would require a 32" unit cell. You can set the distance between rows by selecting the **Pick Row Offset** button and then selecting a point with the mouse, or by selecting the **Pick Both Offset** button and then selecting two opposite points of a rectangle. This is shown in Figure 10.32. The rectangle will provide information for the row spacing. If more than one column is to be arrayed, the rectangle will provide information for both the row and the column distances.

Specifying the Column Offset

The **Column offset** edit box for a rectangular array provides the distance between columns. The same guidelines for spacing rows apply to column offsets. Figure 10.33 shows the results of an array pattern with three rows and eight columns.

Positive and Negative Offset Values

As values are provided for rows and columns, they can be positive, negative, or a combination of both. Using positive values for both will array the objects up and to the right of the original object, as seen in Figure 10.34. If a negative value is used for the row distance, new objects will be added below the original. If negative values are used for the column distance, new objects will be added to the left of the original object, as seen in Figure 10.35. Figures 10.36 and 10.37 show the effects of combining values.

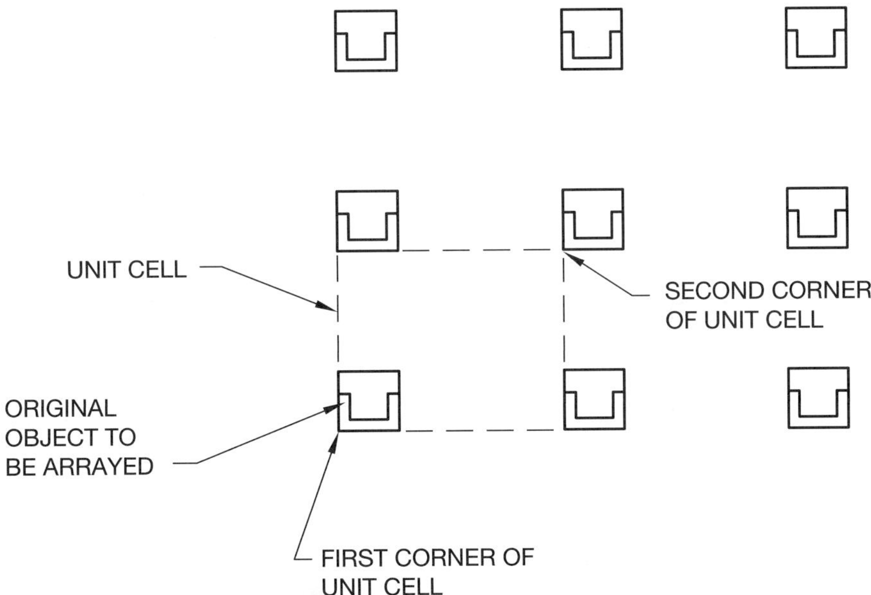

Figure 10.32 *The offsets for rows and columns (the unit cell distance) can be specified.*

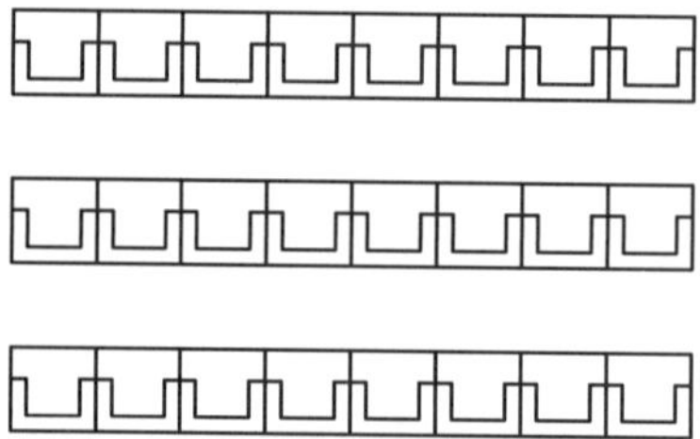

Figure 10.33 *The effect of row and column spacing in an array of three rows and eight columns with a row offset of 32" and a column offset of 16".*

Specifying an Angle for Rectangular Arrays

The array pattern is normally based on a horizontal or vertical baseline. Entering a numeric value in the **Angle of array** edit box will rotate the baseline angle. Selecting the **Pick Angle of Array** button will also allow the angle array to be altered. Selecting this button will prompt for two lines to be specified. The angle of the line drawn will become the angle for the array pattern. Figure 10.38 shows the effect of using an angled array.

Summary of Steps in the Rectangular Array Process

1. Display the **Array** dialog box by selecting the **Array** button on the **Modify** toolbar or by typing **AR** (ENTER) at the Command prompt.
2. Select the method of array to be used.
3. Select the objects to be arrayed using the **Select objects** button.

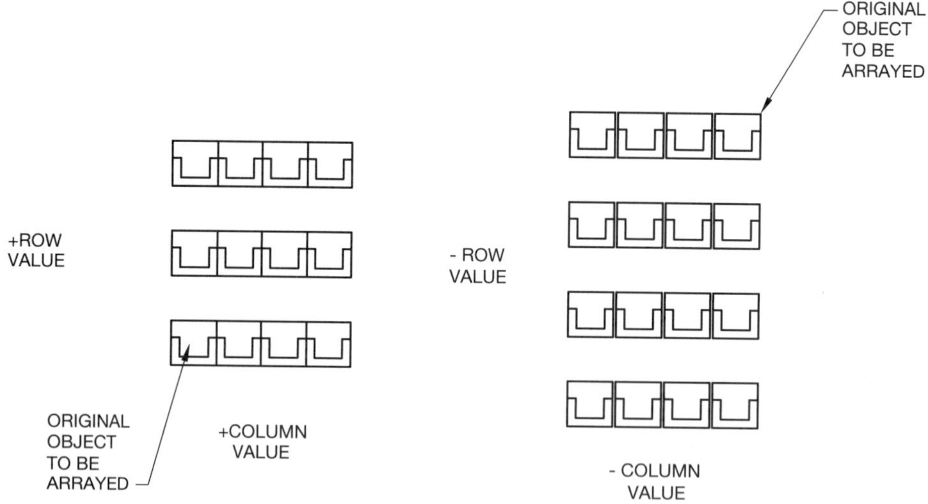

Figure 10.34 *An array based on entering two positive distances.*

Figure 10.35 *An array based on entering two negative values.*

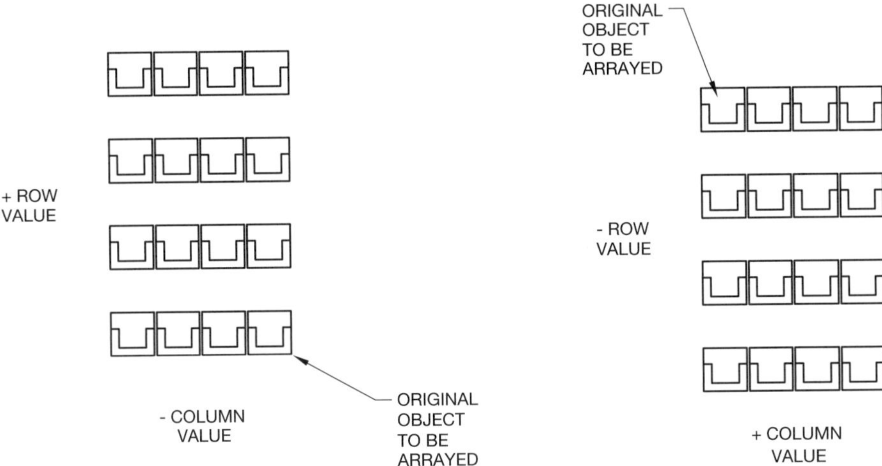

Figure 10.36 *An array based on a positive row value with a negative column value.*

Figure 10.37 *An array based on a negative row value with a positive column value.*

4. Specify the number of rows.
5. Specify the number of columns.

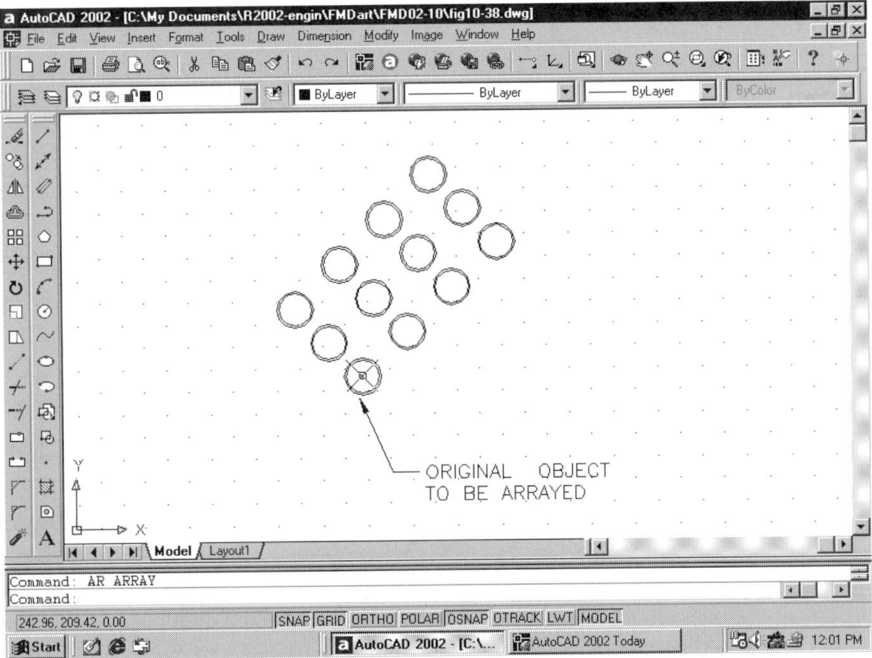

Figure 10.38 *A rotated array can be drawn by selecting the* **Pick Angle of Array** *button.*

6. Specify the row offset.
7. Specify the column offset.
8. Specify the array angle if required.
9. Preview the array by selecting the **Preview** button. If the pattern is okay, click the **Accept** button. If the pattern is not what you expected, select the **Modify** button, and alter the pattern as needed.
10. Click the **Accept** button to accept the array as specified. If you decide not to complete the pattern, select the **Cancel** button.

Polar Array

A polar array is a circular layout of objects based on a center point. Selecting the **Polar Array** button starts the process. Once **Polar Array** is selected, the dialog box will be altered as shown in Figure 10.39. Six additional specifications must be provided to perform a polar array. These include selecting the center point of the array pattern, the method of array, the rotation of objects and the object base point.

Selecting the Center Point

The center point is the point that the selected objects will be arrayed around. The center point can be selected by entering coordinates in the **Center point X** and **Y** edit boxes or by selecting the **Pick Center Point** button. **Selecting the Pick Center Point** button will temporarily remove the dialog box to allow a center point to be specified with the mouse.

Selecting the Method of Array

Three methods of completing a polar array pattern are available. Each method can be selected from the **Method** edit box. Options include the following:

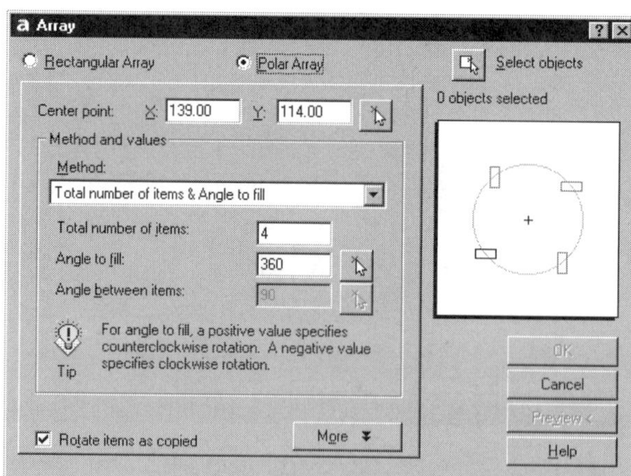

Figure 10-39 *Selecting the* **Polar Array** *button of the* **Array** *dialog box will alter the display of the dialog box, allowing objects to be created in a circular pattern.*

- Total number of items & Angle to fill
- Total number of items & Angle between items
- Angle to fill & Angle between items

The option selected will determine which of the three prompts in the **Method and values** area will be active.

Number of Items—The **Total number of items** edit box allows you to specify how many items, including the original, will be displayed around the center point.

Angle to Fill—The **Angle to fill** edit box controls the limits of the display around the center point. The default of 360° will locate the specified number of items around a full circle. A number less than 360° will array the indicated number of objects around a selected center point in a specified degree range. The pattern will start with the original object and spread the balance of the pattern from its location. A positive value will produce a counterclockwise rotation. A negative value will cause a clockwise rotation. Examples of each pattern can be seen in Figure 10.40.

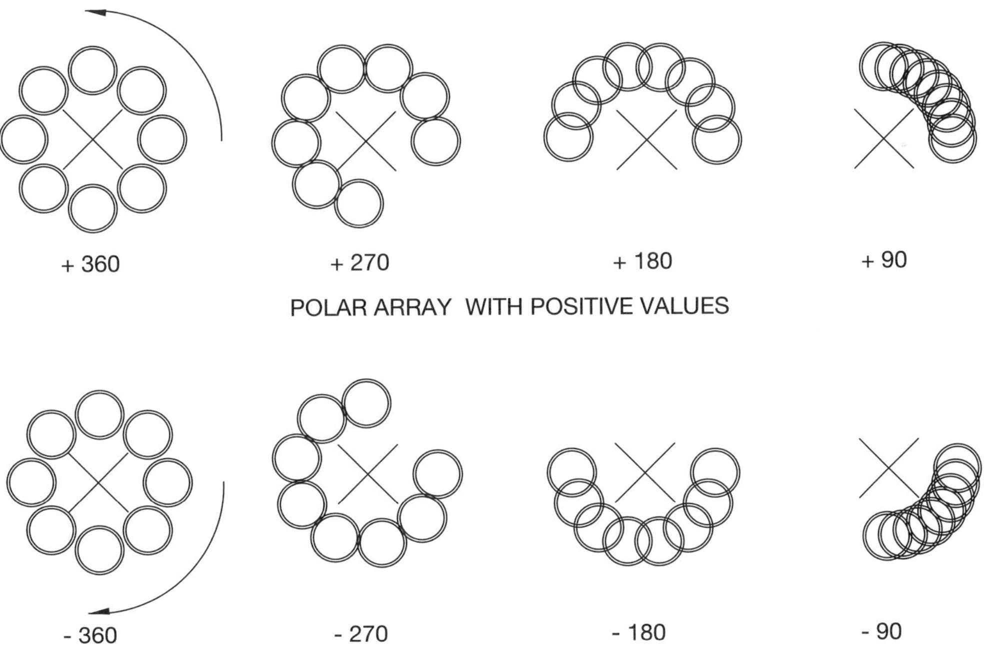

Figure 10.40 *The effect of adjusting the* **Angle to fill** *edit box for a polar array.*

Angle Between Items—The **Angle between items** edit box can be used to specify the angle between items. The number of items to be rotated, combined with the angle distance between each item, is often used to locate bolts or other similar features (see Figure 10.41). Materials also can be arrayed over a specified degree range and intervals, such as showing pipes that intersect a cooling unit over a 75° area at each 15°, as seen in Figure10.42.

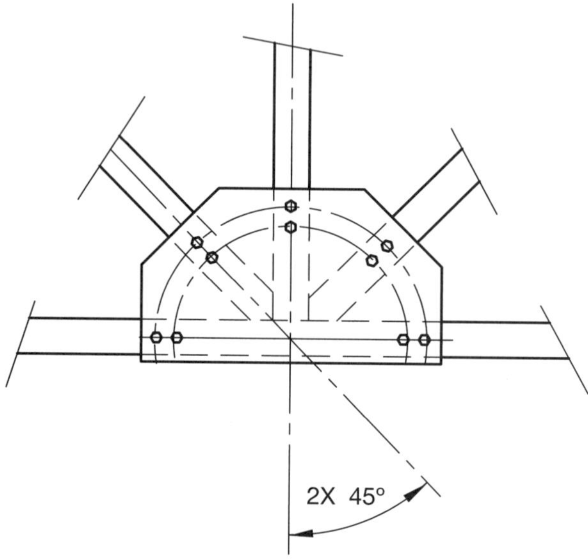

Figure 10.41 *Specifying the number of items to be rotated and the* **Angle between items** *is a common method for defining an array pattern.*

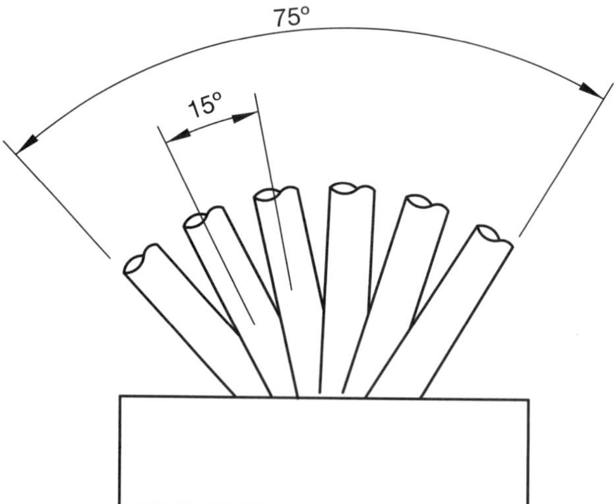

Figure 10.42 *Specifying the* **Angle to fill** *and the* **Angle between items** *is also used to define the array pattern.*

Rotating Arrayed Objects

When the **Rotate Items as Copied** check box is active, objects to be arrayed will be rotated as they are arrayed. The default is yes, which will rotate objects as they are arrayed around the center point. If the **Rotate Items as Copied** check box is inactive, the objects will not be rotated and will appear similar to the right side of Figure 10.43.

Selecting an Object Base Point

Selecting the **More** button displays the **Object Base Point** option. This option specifies a new base point relative to the selection set. The base point will remain a constant distance from the center point of the array as the objects to be arrayed. The point used depends on the type of the object to be arrayed. Default base points are listed as follows:

Object type	**Default base point**
Arc, circle, ellipse	Center point
Polygon, rectangle	First corner
Donut, line, polyline, ray	Start point
Block, paragraph text, single-line text	Insertion point
Construction lines	Midpoint
Region	Grip point

Although some of these terms are unfamiliar, they will be introduced in later chapters.

Summary of Steps in the Polar Array Process

The entire process for a polar array of 50 segments around a 250° arc is as follows:

1. Display the **Array** dialog box by selecting the **Array** button on the **Modify** toolbar or by typing **AR** (ENTER) at the Command prompt.

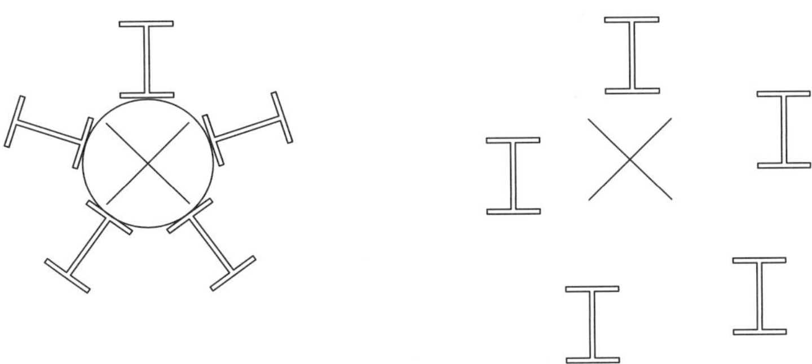

Figure 10.43 *Objects can be rotated as they are arrayed if the* **Rotate objects as copied** *check box is activated.*

2. Select the method of array to be used, **Polar Array**.
3. Select the objects to be arrayed using the **Select objects** button.
4. Select the center point of the array pattern.
5. Select the method. For this example, the Total number of items & Angle to fill option was used. Notice that the Angle between items edit box is inactive.
6. Enter 270 in the **Total number of items** edit box.
7. Enter 250 in the **Angle to fill** edit box.
8. Keep the **Rotate items as copied** check box active.
9. Preview the array by selecting the **Preview** button. If the pattern is okay, click the **Accept** button. If the pattern is not what you expected, select the **Modify** button, and alter the pattern as needed.
10. Click the **Accept** button to accept the array as specified. If you decide not to complete the pattern, select the **Cancel** button.

Practical Uses for Array

Array should be used whenever an object needs to be added to a drawing in a repetitive pattern. The symbols in Figure 10.5, and the holes in Figures 10.12 and 10.21 could easily have been placed using the **Array** pattern.

CHAPTER 10 EXERCISES

1. Start a new drawing and draw a horizontal line. Draw polygons with three, four, five, six, and eight sides with the centerline of each polygon on the horizontal line. Draw all polygons inscribed in a circle with a .5" radius. Make copies of the polygons with three, five, and eight sides and place them directly below their counterparts. Make copies of the four-sided and six-sided polygons and place them in a third row directly below their counterparts. Mirror the entire drawing with a horizontal mirror line, and delete the old objects. Save the drawing as E-10-1.
2. Open the existing file E-6-8. Draw a .50 diameter hole that goes through the base from front to back. Place the center of the hole .50 from the left end of the part. Mirror the hole so that a second hole is placed in the right side of the base. Add the necessary lines to the other views to reflect the holes. Save the drawing as E-10-2.
3. Draw half of a W 8" × 28" (I-shaped) beam. The beam has a total height of 8.28". The top and bottom flanges are .25" thick, with a total width of 5.25". The web is a total of .28" thick. Use the **Mirror** command to complete the drawing. Save the drawing as E-10-3.

4. Open exercise drawing E-6-13. Draw a .50" diameter hole that is centered on the inclined surface on the right side of the part. The centerline of the new hole is to be perpendicular to the inclined surface and extend through the part. Use the **Mirror** command to copy the hole to the inclined surface on the left. Add the necessary lines to the other views to reflect the holes. Save the drawing as E-10-4.

5. Using the following figure as a guide, draw a steel connector plate. The plate is 15.5" × 10"h. Draw the bolts as .75" diameter hexagons, 1.5" down and 1.5" in from the top and ends. Bolts are 3" O.C. Each end of the plate is symmetrical. Draw a 2.5" × 8" strap, centered in the bottom of the plate. Provide two bolts with similar spacing as the plate. Save the drawing as E-10-5.

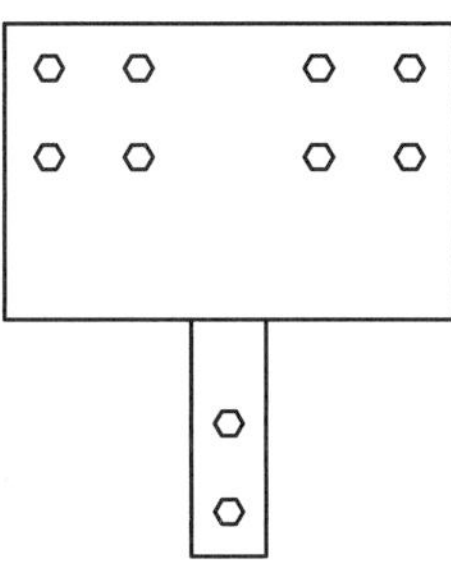

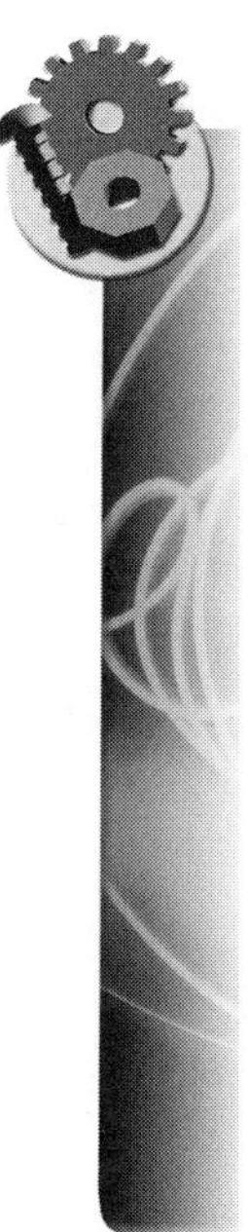

Problem 10-5

6. Open exercise E-7-18. Draw a 2.00" diameter circle tangent to the small end of the part. Use the **Array** command to place a total of 6 blades around and tangent to the circle. Rotate the blades as they are attached to the circle. Save the drawing as E-10-6.

7. Draw an equilateral triangle with legs that are 4.50 units long. Create 5 additional triangles inside the original triangle with each succeeding triangle .25 units smaller than the proceeding triangle. Save the drawing as E-10-7.

8. Use the attached drawing to create the drilling template. Assume that each grid is .50" units. The 12 holes are to be 1.25" diameter. Use proper line weights and line types to complete the object. Save the drawing as E-10-8.

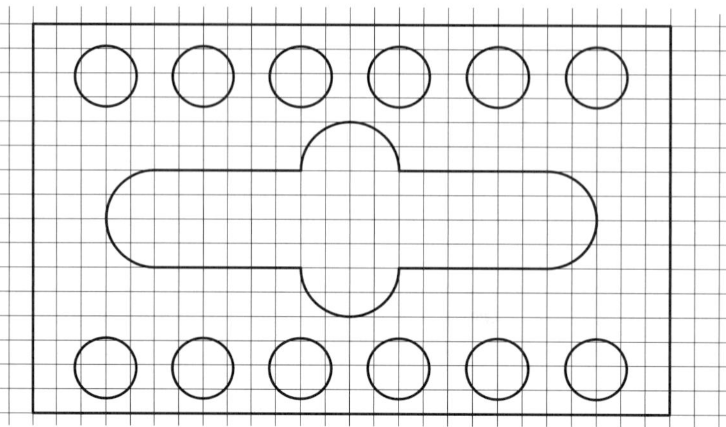

Problem 10-8

9. Use the attached drawing to create the drilling template. Assume that each grid is .50". The part has a diameter of 7.00" with a 4.00" diameter hole in the center. The 6 holes are to be 1.25" diameter and located on a bolt circle with a diameter of 7.25". The outer circles are 2.00" diameter with the center located on the 7.25" bolt circle. The 12 holes are to be .50" diameter located on a 5.00" diameter bolt circle. Use proper lineweights and linetypes to complete the object. Save the drawing as E-10-9.

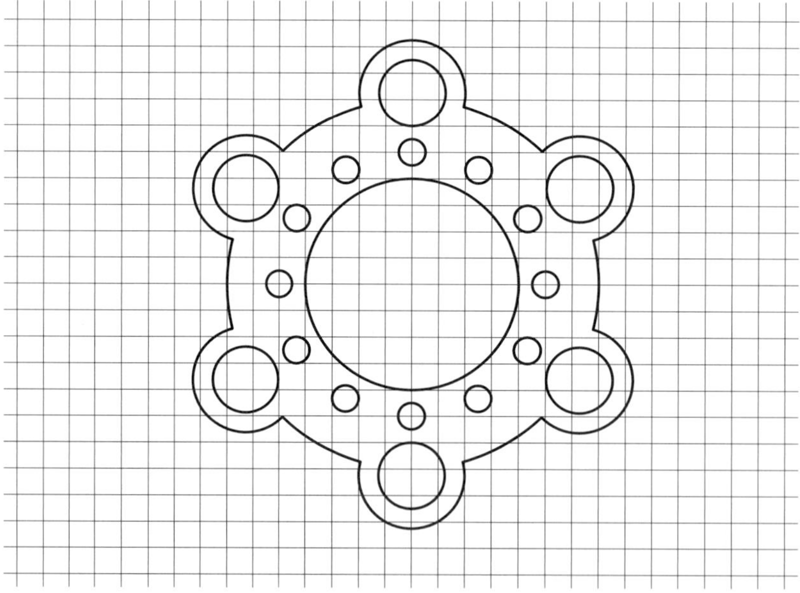

Problem 10-9

10. Open exercise E-6-14. Using the existing three views, draw a bottom view. Locate two holes with a diameter of .25" in each side of the bottom surface so that a total of four holes are added. Place the holes so that they are centered in both directions of each bottom surface and extend each hole through to the inclined surface. Add the necessary lines to the other views to reflect the holes. Save the drawing as E-10-10.
11. Open exercise E-8-7 and erase the .75" diameter hole that is shown in the top of the part. Erase all lines that represent this feature in other views. Draw a .50" diameter hole in the center of the top feature using the same center point that was used to draw the outer curve. Add eight holes with a diameter of .125" that are equally spaced around the center hole using a bolt circle of .75". Add the necessary lines to the other views to reflect the new features. Save the drawing as E-10-11.

12. Start a new drawing and set the limits as needed to draw a 4" square centered in a 15" diameter circle. Draw an 8" and a 12" diameter circle with the same center point as the 15" circle. Make three additional copies of this object and move each so that the four 15" circles are tangent. Save the drawing as E-10-12.
13. Open drawing E-8-8 and add two rows of five .0625" diameter holes to the top surface and extend through the part. The holes should be equally spaced from front to back and from left to right .035" from the edges of the part. Add a .25" diameter hole that passes through each of the two lower legs. Use the center point of the .50" diameter arch as the center point for the new holes. Add the necessary lines to the other views to reflect the new features. Save the drawing as E-10-13.

14. A circular storage tank is to be built. Use the drawing below as a guide and draw the support plan. Save the drawing as E-10-14.

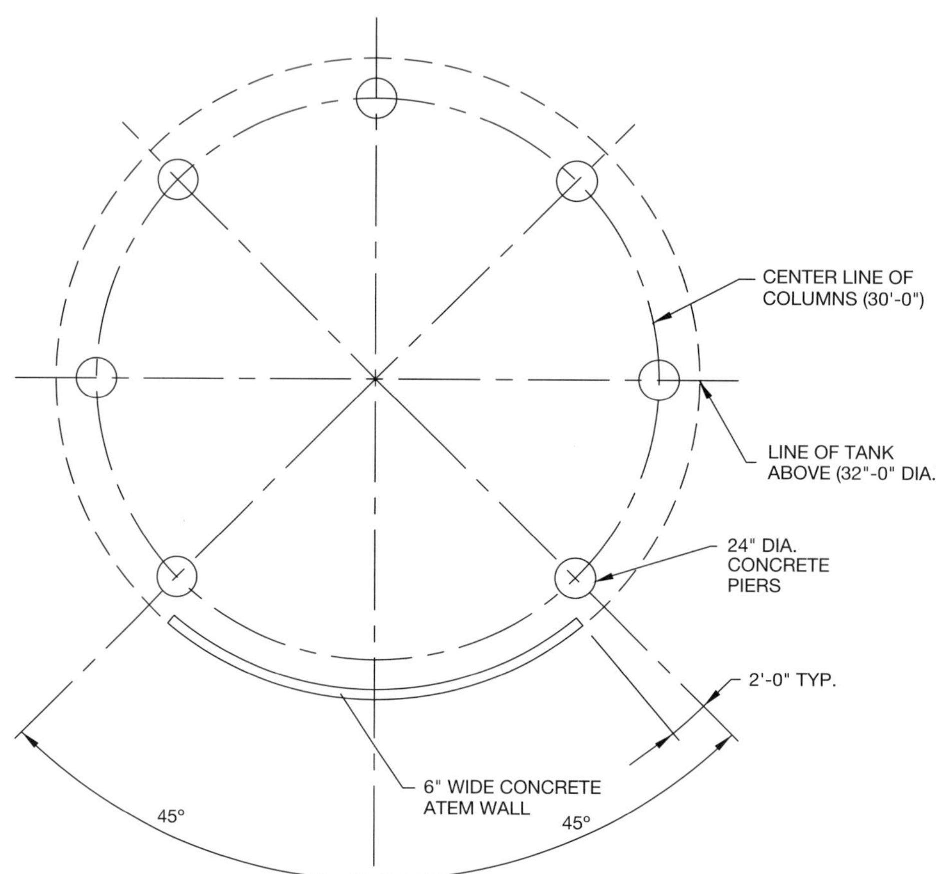

Problem 10-14

15. Draw a plan view of a 40' diameter water reservoir. The reservoir will be supported on (12) 6" diameter × .375" steel columns. The columns will be equally spaced, on a 36' diameter circle centered on an 18" diameter × .75" thick vertical inlet pipe in the center of the reservoir. Assume north to be at the top of the drawing, and locate a 12" diameter × .375" thick vertical outlet pipe at the outer edge of the tank, at 37° west of true north. Save the drawing as E-10-15.

CHAPTER 10 QUIZ

1. Describe the process to draw a rectangle that is 3" × 6" using only two lines.
2. Several copies of an object need to be made. What option is available so that the **Copy** command does not have to be repeated?
3. You have selected several objects for editing when you realize you have accidentally selected an object that you do not want changed. What can you do so that you do not have to start the edit sequence over again?
4. How will **Undo** affect objects that have just been moved?
5. List three methods suitable for selecting irregularly shaped objects for editing.
6. When you use the Fence option, what objects will it select for editing?
7. What editing mode will allow you to select objects and then decide how to edit them?
8. What process will display the dialog box that controls selection methods?
9. List the commands in sequence required to complete exercise E-10-9.
10. Explain when the Through option of **Offset** could be used.
11. Start the **Erase** command using each method and describe the advantages or disadvantages.
12. Experiment with the **Mirror** command. Describe the effects of ORTHO on the mirror line.
13. Mirror is being used to complete a symmetrical object. What options are available and how do they affect the **Mirror** sequence?
14. Where is **Erase** listed in a menu?
15. Describe how to change the size of the pick box.
16. What is the effect of base point on the **Copy** command, and how should it be selected?
17. List ways that the **Copy** command can be used to draw an object.
18. What is the benefit of having so many different methods of selecting objects for editing?
19. What are the drawbacks to enlarging the pick box to twice its normal size, or decreasing it to half its normal size?
20. Use the **Help** menu to explain the difference between the normal setting for an editing command and the Single option.
21. Explain the difference between Noun/Verb and Verb/Noun selection.
22. Explain the meaning and the effect of the following letters for selection groups: W, C, F, WP, CP, R, A, and P.

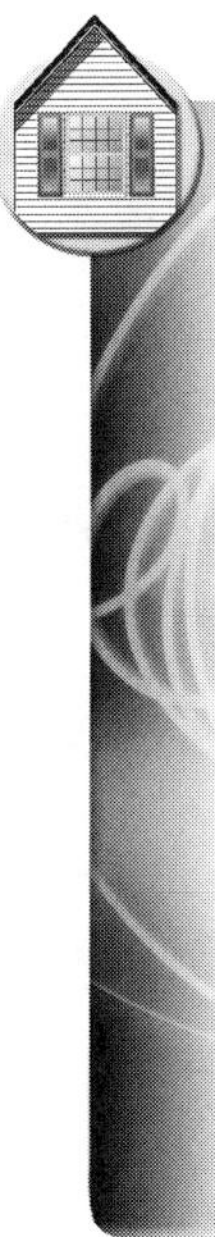

23. What effect does Press and drag have on selection?
24. Use the **Help** menu to list the effects of the **Pickadd**, **Pickauto**, and **Pickdrag** variables.
25. How can Object Snap be used with editing commands?
26. Explain the difference between the two types of array patterns.
27. Twelve bolts need to be drawn 2" on center in one horizontal line. What will the array sequence be?
28. Three horizontal lines of 10 holes need to be drawn 8" on center. The horizontal lines of columns will be 12" apart. What will the array sequence be?
29. Use the drawing for exercise E-10-14 and give the command sequence to array the supports.
30. Using an illustration as an example, define a unit cell.
31. How could an array pattern for objects aligned at 45° from horizontal be drawn?
32. Ten rows and three columns are to be arrayed below and to the left of the original object at 60" O.C. How can this be done?
33. Describe how to produce a clockwise rotation.
34. Give the combination of alternatives for describing a polar array.
35. If a positive column distance and a negative row distance are entered, where will the objects be arrayed in relation to the original object?

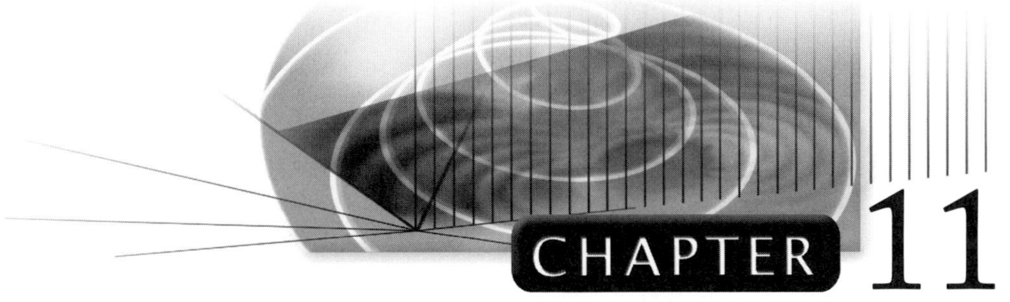

Modifying the Position and Size of Drawing Objects

In Chapter 10 you were introduced to editing commands that can affect an entire drawing. This chapter will explore

- Commands that can be used to modify part of an object by altering the position or size of the object
- Commands that can be used to edit objects by altering their length
- Methods for selecting objects for editing using grips
- Methods for selecting objects for multiple edits
- Methods for naming object groups

Commands explored in this chapter include

- **Move**
- **Align**
- **Rotate**
- **Scale**
- **Stretch**
- **Lengthen**
- **Trim**
- **Extend**
- **Break**
- **Chamfer**
- **Fillet**
- **Select**
- **Group**

Mastering these commands will greatly increase your drawing ability. All of the commands, with the exception of **Align**, can be selected on the **Modify** toolbar, or selected from the **Modify** menu. Each command can also selected from the **Modify** menu, entered by keyboard, or selected from the shortcut menu by selecting an object to edit and then right-clicking to select the editing command.

ALTERING THE POSITION OR SIZE OF OBJECTS

The **Modify** menus offer six options for modifying a drawing by changing the position of an object or by changing the size of the object. These commands include **Move**, **Align** (a **3D Operation** option), **Rotate**, **Scale**, **Stretch**, and **Lengthen**.

MOVE

The **Move** command allows one or more objects to be selected and moved from their current location to a new one. Access the command by selecting the **Move** button on the **Modify** toolbar, by typing M (ENTER) at the Command prompt, or by selecting **Move** from the Modify menu. Figures 11.1a and 11.1b compare the effects of the **Move** command.

The command sequence for **Move** is as follows:

> Command: **M** ENTER *(Or click the Move button.)*
>
> MOVE
>
> Select objects: *(Use any selection method.)*

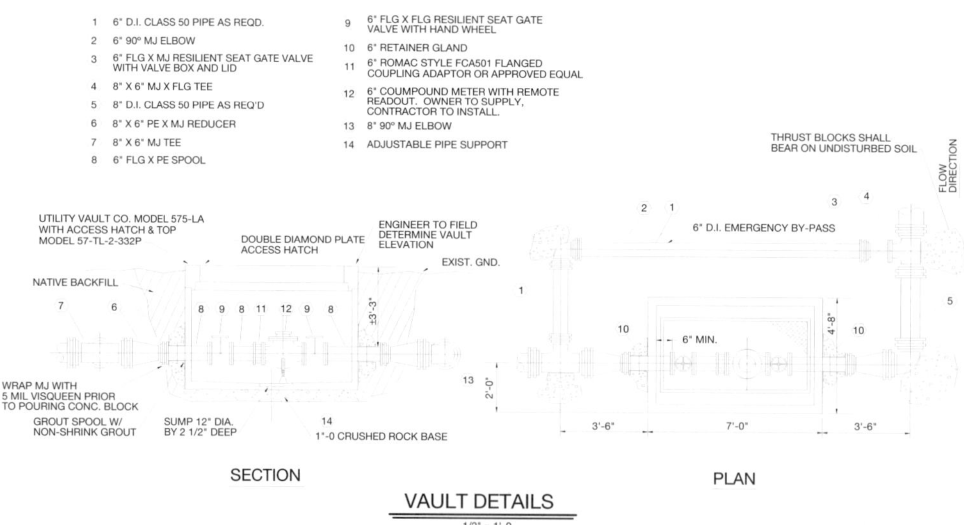

Figure 11.1a *The* **Move** *command can be used to alter the layout of a drawing. Courtesy Lee Engineering, Inc.*

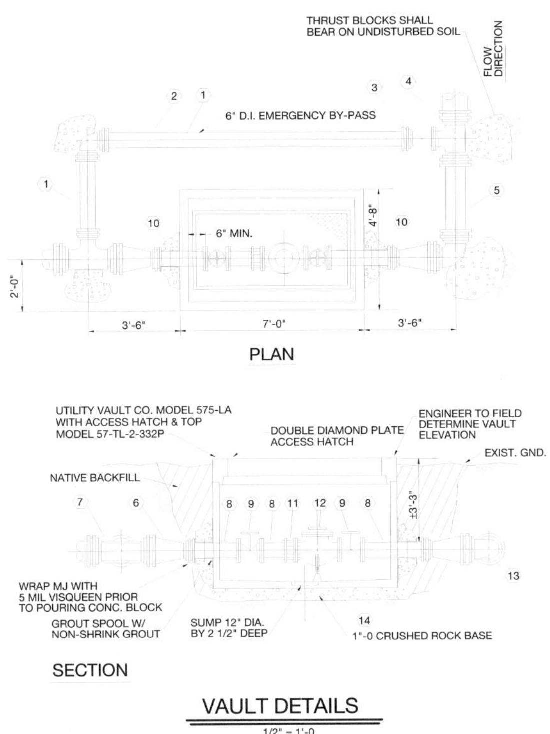

Figure 11.1b *The* **Move** *command was used to place the plan view of Figure 11.1a above the section drawing. Courtesy Lee Engineering, Inc.*

Select objects: ENTER

Specify base point or displacement: *(Enter point or coordinates.)*

Specify second point of displacement or <use first point as displacement>: *(Enter point or coordinates.)*

Command:

The process can be seen in Figures 11.2a through 11.2c.

- Use any selection method to select the objects to be moved.
- The "base point of displacement" prompt is asking for a reference point for moving the selection set. Any point may be selected, but it may help you to visualize the results of moving by selecting a point in the drawing, such as an intersection or endpoint. This is an excellent time to make use of OSNAP.
- The "second point of displacement" prompt is asking where you would like to move the first point.

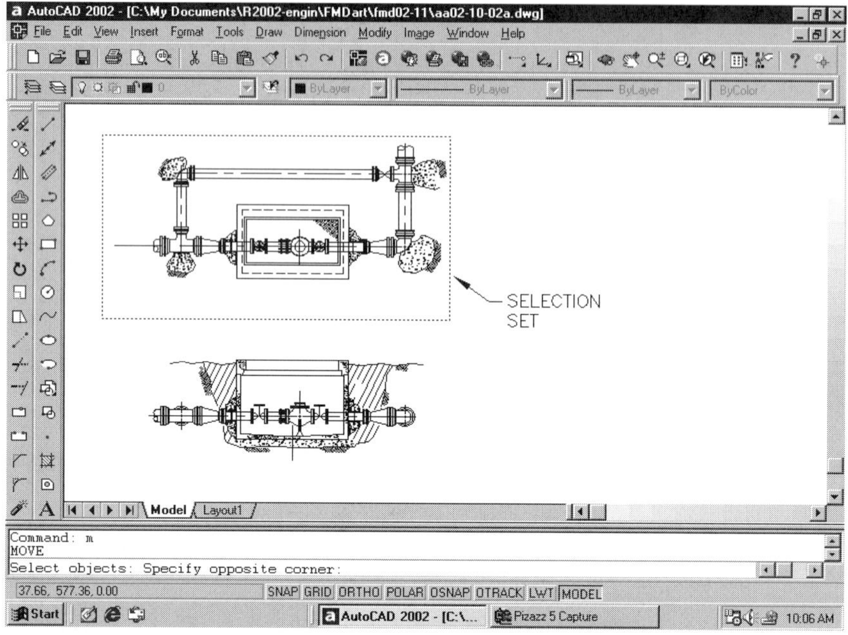

Figure 11.2a *The first step in the process for moving a drawing is to select the objects to be moved. Courtesy Lee Engineering, Inc.*

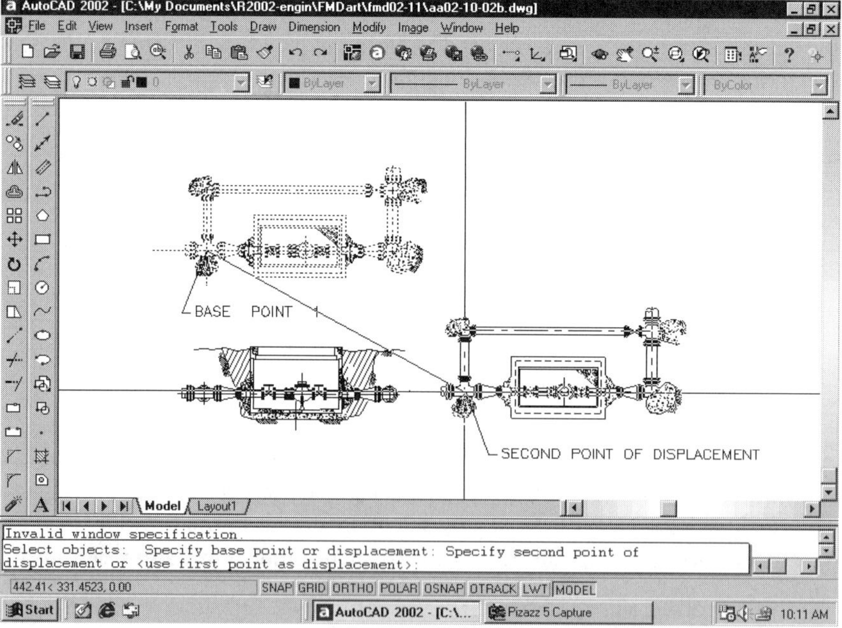

Figure 11.2b *The second step is to select the base point of displacement and the new location for this point. Courtesy Lee Engineering, Inc.*

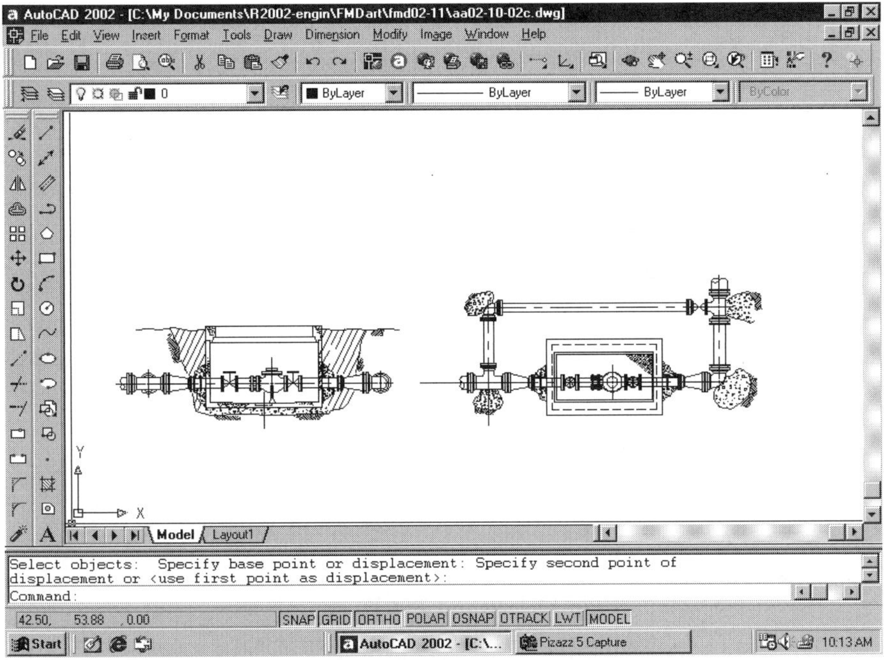

Figure 11.2c *Finally, the second point of displacement is selected, and the* **Move** *process is complete. Courtesy Lee Engineering, Inc.*

Each point can be entered by selecting a point with a pointing device or by entering coordinates. SNAP, ORTHO, OSNAP, and **Zoom** can be very helpful for aligning related objects as they are moved.

Note: Move should give you an immense sense of freedom, but not a sense of reckless abandonment. You still need to plan your drawings, but the **Move** command offers you great flexibility in drawing arrangement. Space between views can be altered as the needs of the drawing change. On the other hand, you have to keep in mind that each of your drawings will need to fit on a sheet of paper of a set size. If you're going to try and place six views on a sheet, there is only so much space where the views can be moved. AutoCAD still has not incorporated the CRAM command. Planning and wise use of the editing commands are the only hope for a good solution.

ROTATE

In Chapter 10 you learned how to array portions of a drawing. In a similar fashion, you can draw an object or group of objects and then rotate them around a base point to change their orientation. Start the command by selecting the **Rotate** button on the **Modify** toolbar, by typing RO (ENTER) at the Command prompt, or by selecting **Rotate** from the **Modify** menu. You can also start

Rotate from the shortcut menu by picking the object, right-clicking, and then selecting **Rotate.** Figure 11.3 shows an example of a feature that was rotated 90°.

To complete the command, you will need to select objects to be rotated, a base point to rotate the object around, and the desired degree of rotation.

- Entering a positive rotation angle rotates the selected object in a counter-clockwise direction.
- Entering a negative rotation angle produces a clockwise rotation.

The base point can be anywhere on the drawing screen, but selecting one corner of the object as the base point can ease visualization of the rotation. The command sequence is as follows:

Command: **RO** ENTER *(Or click the Rotate button.)*

ROTATE

Current positive angle in UCS: ANGDIR=counterclockwise
ANGBASE=0

Select objects: *(Use any selection method.)*

8 found

Select objects:ENTER

Specify base point: *(Select any point such as an endpoint or midpoint.)*

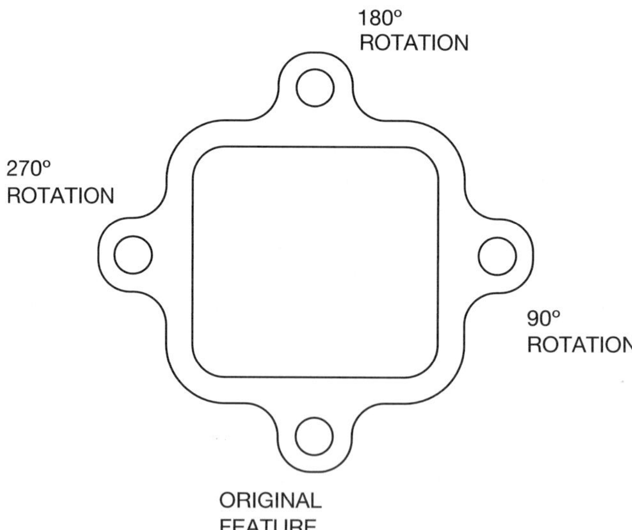

Figure 11.3 *The* **Rotate** *command allows objects to be drawn and rotated into a new position.*

Specify rotation angle or [Reference]: *(Enter angle or drag into position.)*

Command:

Note that instead of entering a precise angle, you can drag the selected objects into the new position as the cursor is moved. If ORTHO is toggled ON, the objects will be snapped at 90° intervals. With ORTHO off, the objects can be dragged into any desired angle. When you enter an angle, you can rotate objects to exact locations based on the current location. The process can be seen in Figure 11.4.

Objects also can be rotated based on absolute rotation angles. Suppose you've drawn an inclined plane at 22.5°. An engineering change notice (ECN) requires the incline to be altered to 27.5°. You could rotate the existing line 5° and hope you get it close enough, or use the Reference option and place the surface exactly where it should be. The correct option uses the following command sequence:

Command: **RO** ENTER *(Or click the Rotate button.)*

ROTATE

Current positive angle in UCS: ANGDIR=counterclockwise ANGBASE=0

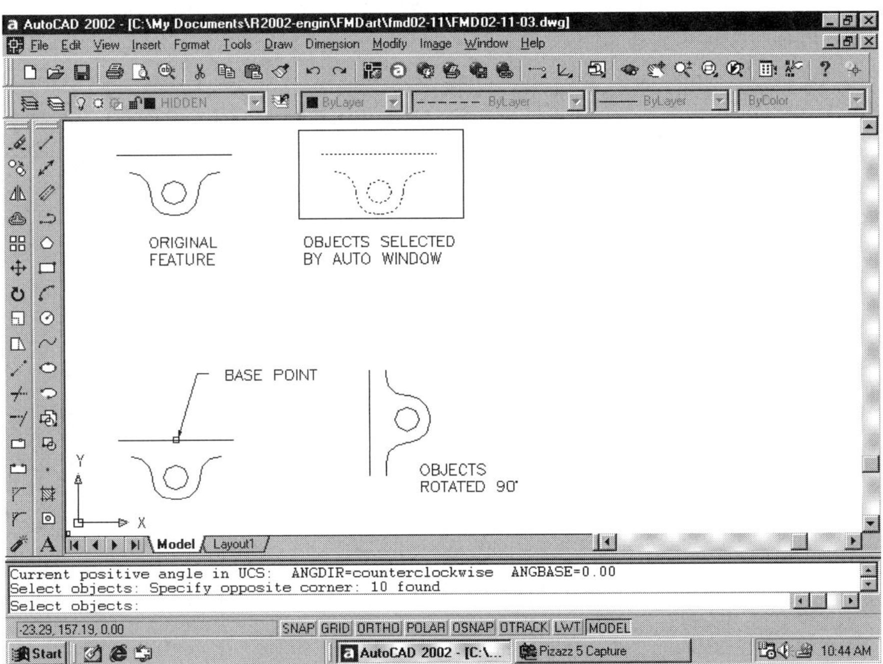

Figure 11.4 *Objects are rotated by defining the selection set, specifying a base point for the rotation, and providing the rotation angle.*

Select objects: *(Use any selection method.)*

5 found

Select objects:ENTER

Specify base point: *(Select any point.)*

Specify rotation angle or [Reference]: **R** ENTER

Specify the reference angle: **22.5** ENTER

Specify the new angle: **27.5** ENTER

Command:

The process can be seen in Figures 11.5a, 11.5b, and 11.5c.

ALIGN

Think of the **Align** command as a combination of the **Move** and **Rotate** commands. Although primarily a command for 3D drawing, it works well when an object or group of objects need to be moved and rotated. Start the command by typing AL (ENTER) at the Command prompt. The command will prompt you for two source points and two destination points. The source points will be used to define points on the original object in its original position. The destination points represent the locations where the object will be placed. The command can be used in two ways. If one pair of source and destination points is provided, the selected object will be moved in a fashion similar to when the **Move** command is used. The benefit of the **Align** command is in providing two pairs of points. If two pairs of source and destination points are provided, the selected object can be moved, rotated, and scaled to match other objects. This could be useful in moving, rotating, and align an object with another when redesigning a part. Use the following command sequence to move and rotate an object from one location to another:

Command: **AL** ENTER

ALIGN

Select objects: *(Select object using any selection method.)*

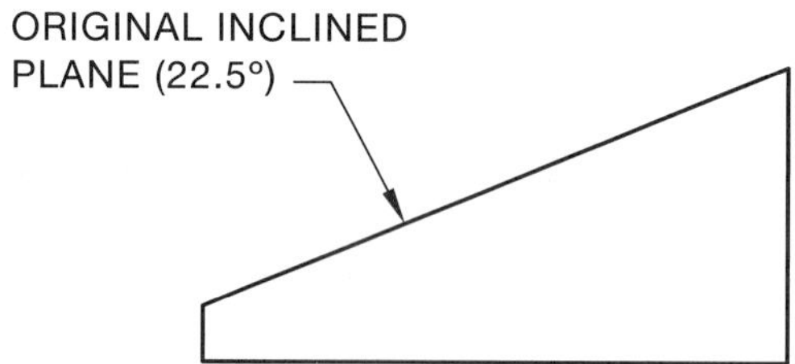

Figure 11.5a *The angle of an inclined surface can be altered by use of the Rotate command.*

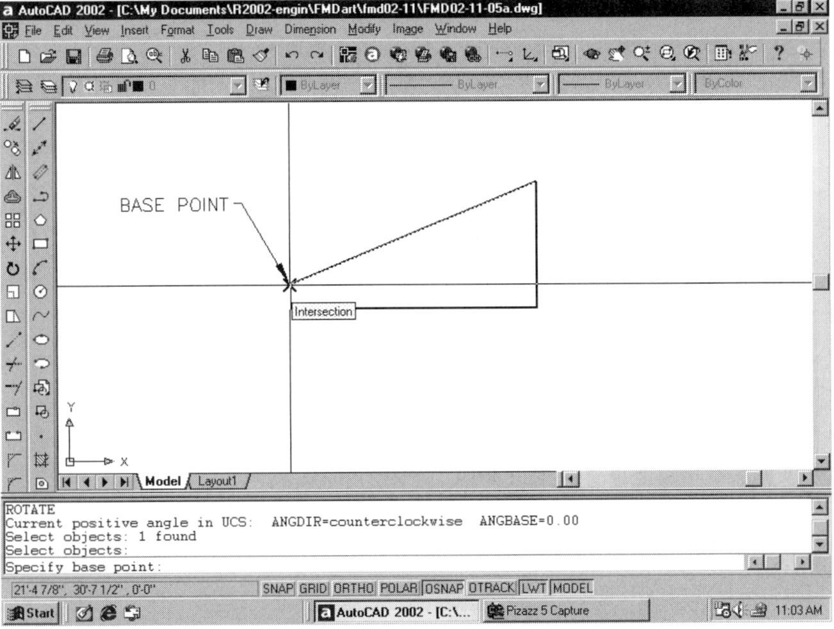

Figure 11.5b *The selection set is defined and the base point is specified.*

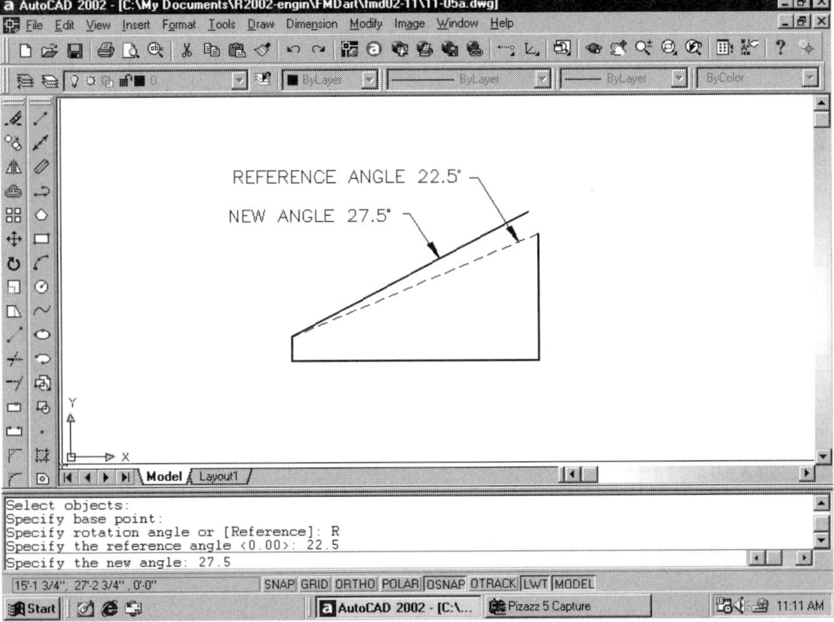

Figure 11.5c *With the reference angle specified (existing angle of 22.5°), the new angle of 27.5° is specified and the surface is rotated.*

Specify first source point: *(Select point with mouse or enter coordinates.)*

Specify first destination point: *(Select point with mouse or enter coordinates.)*

Specify second source point: *(Select point with mouse or enter coordinates.)*

Specify second destination point: *(Select point with mouse or enter coordinates.)*

Specify third source point or <continue>: ENTER

Scale objects based on alignment points? [Yes/No] <No>: ENTER

When ENTER is pressed, the object will be moved and rotated. The selection process can be seen in Figure 11.6. The result of the command can be seen in Figure 11.7. If Y (ENTER) is typed at the last prompt, the objects being rotated will be scaled to match the distance indicated by the first and second destination points. Figures 11.8a and 118b show the process of scaling the objects to meet the alignment points.

SCALE

Up to this point, your drawings have been drawn at full scale. If you needed to draw an airplane wing that is 50' long, you set the limits and units accordingly, zoomed out to the limits, and drew the wing. The whole process was done at real scale with the size of the zoom used to control the viewing screen. The

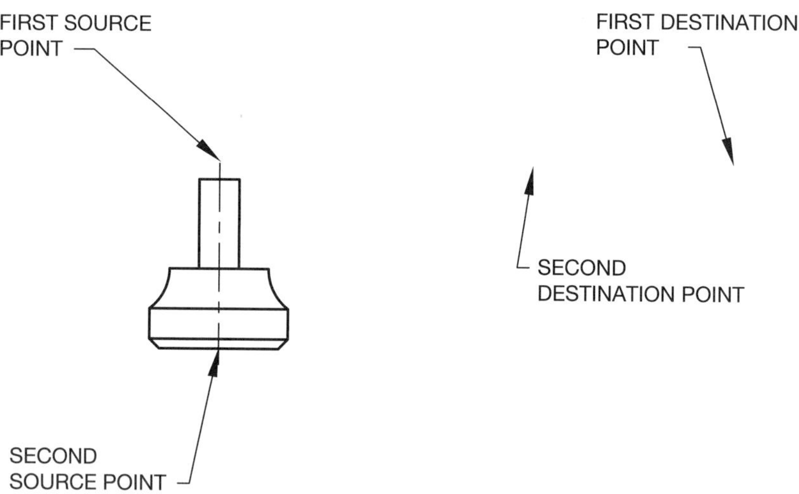

Figure 11.6 *The* **Align** *command allows portions of a drawing to be moved and rotated. The command will prompt you to provide source points, and then ask where you would like those points moved to. Courtesy Hanh Tran.*

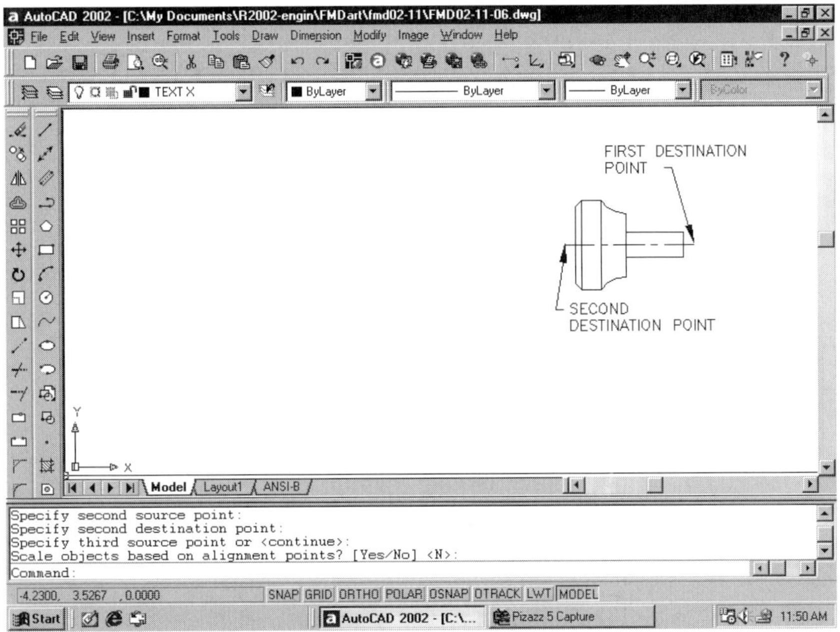

Figure 11.7 *By using the destination points shown in Figure 11.6, the object was moved and rotated to a new position with the* **Align** *command. Courtesy Hanh Tran.*

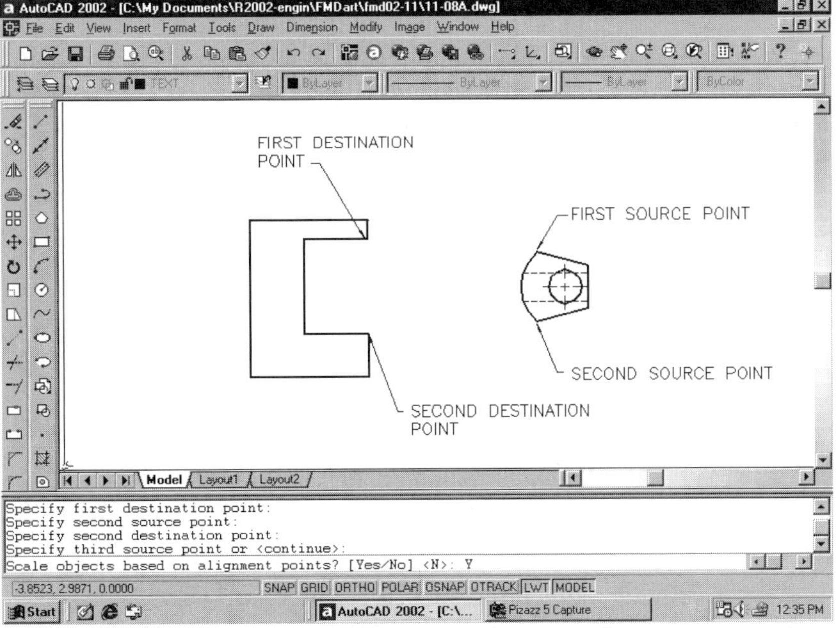

Figure 11.8a *If you choose to scale objects based on alignment points, an object or group of objects can be moved and scaled to match the selected destination points. Courtesy Jill Heilman.*

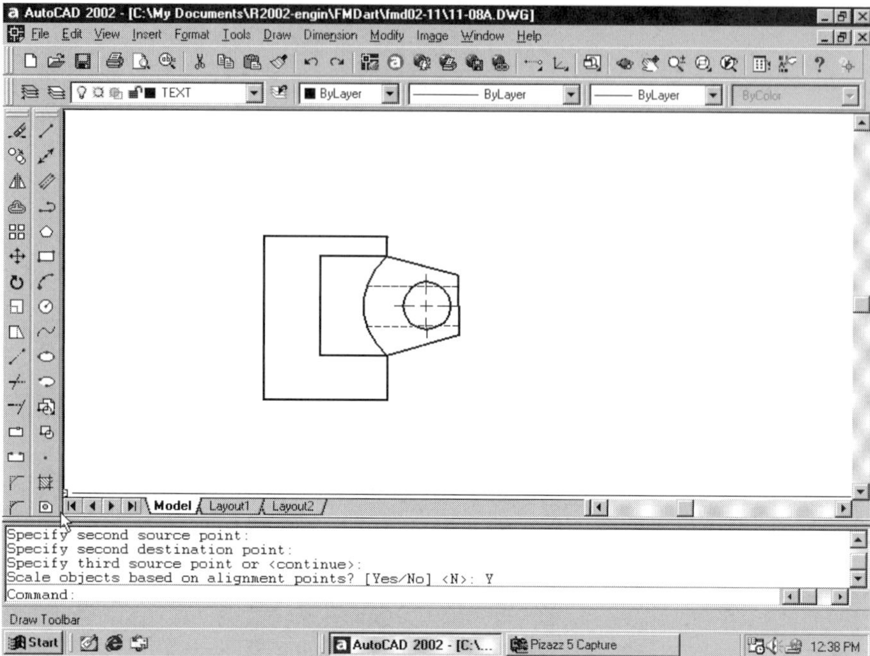

Figure 11.8b *If you choose to scale objects based on alignment points, the 5" long part is scaled to fit in a 6" space. Courtesy Jill Heilman.*

Scale command will allow you to change the size of an existing object or an entire drawing, as seen in Figure 11.9. Start the command by selecting the **Scale** button on the **Modify** toolbar, by typing SC (ENTER) at the Command prompt, or by selecting **Scale** from the **Modify** menu. The command can be completed using the following process:

Command: **SC** ENTER *(Or click the Scale button.)*

SCALE

Select objects: *(Select objects using any selection method.)*

1 found

Select objects:ENTER

Specify base point: *(Select base point.)*

Specify scale factor or [Reference]:

Scale Factor

The default for entering the effect of scaling is to enter a numeric value. To enlarge an object to twice its existing size, type **2** (ENTER) at the prompt. Using whole numbers will enlarge the selected objects by that factor. Type **3** (ENTER) at the prompt to

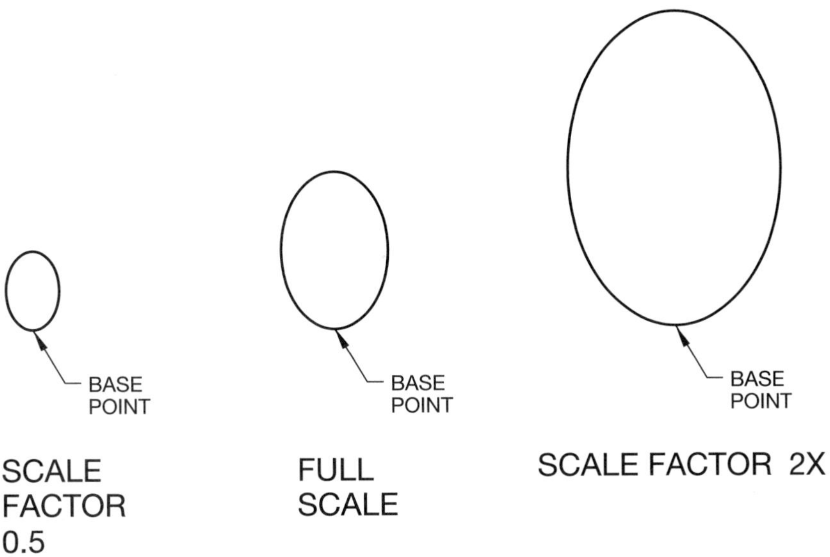

Figure 11.9 *The* **Scale** *command allows the size of objects to be altered. In this case, the ellipse on the left is half the size of the one in the middle.*

enlarge the objects to three times their existing size. Typing a fraction, such as .5 (ENTER), will reduce the selected object to half its original size.

Reference Factor

This option will allow an object to be enlarged to an absolute length rather than a relative scale. Rather than figuring the proportion of enlargement, type **R** (ENTER) at the Command prompt. The process can be seen in Figure 11.10. The command can be completed using the following process:

Command: **SC** ENTER *(Or click the Scale button.)*

SCALE

Select objects: *(Select objects using any selection method.)*

1 found

Select objects: ENTER

Specify base point: *(Select base point.)*

Specify scale factor or [Reference]: **R** ENTER

Specify reference length <1>: **5** ENTER

Specify new length: **6** ENTER

Command:

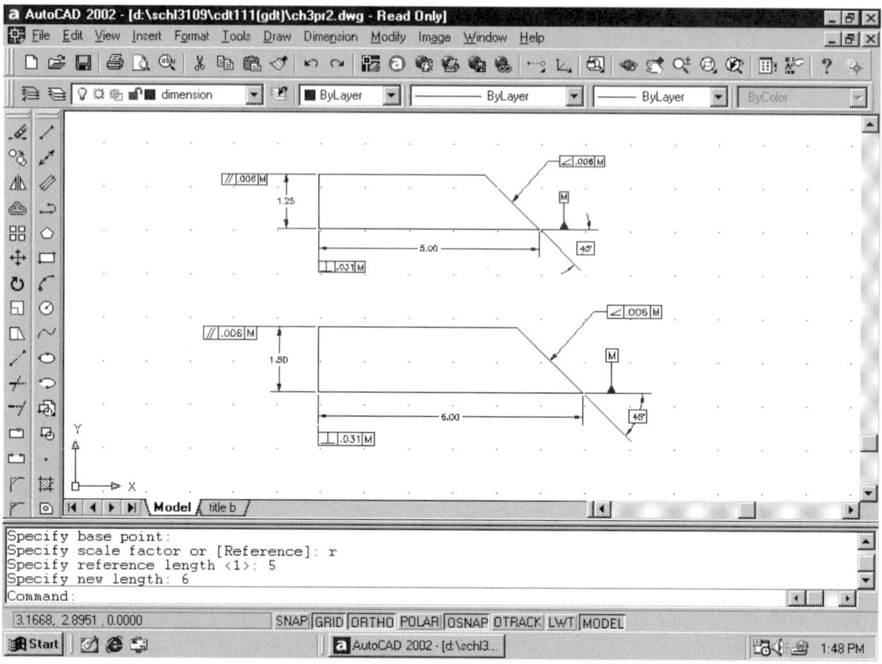

Figure 11.10 *Entering a reference factor allows an object to be scaled by an absolute length. Keep in mind that the total size will be altered, not just the length. Courtesy Cathy Schlesser.*

STRETCH

The **Stretch** command allows you to enlarge or shrink drawing objects. This can be especially helpful when editing the shape and size of an object to meet the changing needs of a client, as seen in Figure 11.11. Drawings made with lines, arcs, traces, solids (Chapter 20), and polylines (Web 1) can all be stretched. Start the command by selecting the **Stretch** button on the **Modify** toolbar, by typing S (ENTER) at the Command prompt, or by selecting **Stretch** from the **Modify** menu. The command sequence is as follows:

Command: **S** ENTER *(Or click the Stretch button.)*

STRETCH

Select objects to stretch by crossing-window or crossing-polygon...

Select objects: *(Select object to be stretched using automatic select window, crossing window, or a cpolygon options.)*

Specify first corner: *(Select a point for the corner of the window.)*

Specify opposite corner: *(Select the second point for the corner so that the object to be stretched is included in the window.)*

23 found

Select objects: ENTER

The automatic selection window only needs to cross a portion of the desired object because the default selection mode is Crossing. If the entire object is selected, **Stretch** will function like the **Move** command and move the entire object. Once the selection set is complete, the prompt will display the following:

> Specify base point or displacement: *(Select a point to serve as a reference. Be sure OSNAP is toggled ON.)*
>
> Specify second point of displacement: *(Select the location that you would like to stretch the base point to.)*
>
> Command:

As the cursor is moved, the effect is displayed as the object is stretched between the base and new points. The process can be seen in Figures 11.12a and 11.12b.

Not only can **Stretch** be used to lengthen objects, but it also can be used to stretch some objects and move others, as seen in Figures 11.13a, 11.13b, and 11.13c.

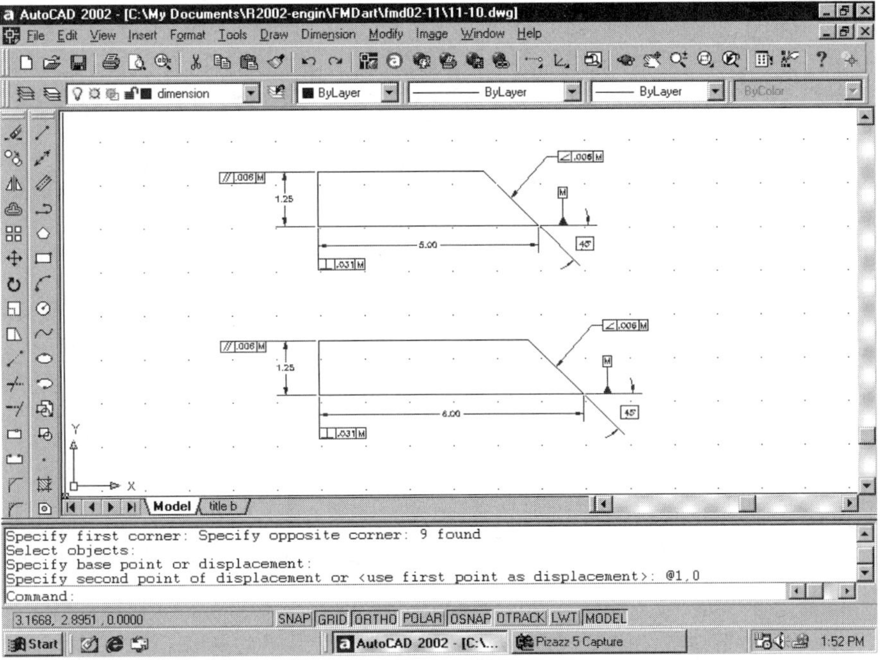

Figure 11.11 *The* **Stretch** *command can be used to elongate an object or group of objects. Courtesy Cathy Schlesser.*

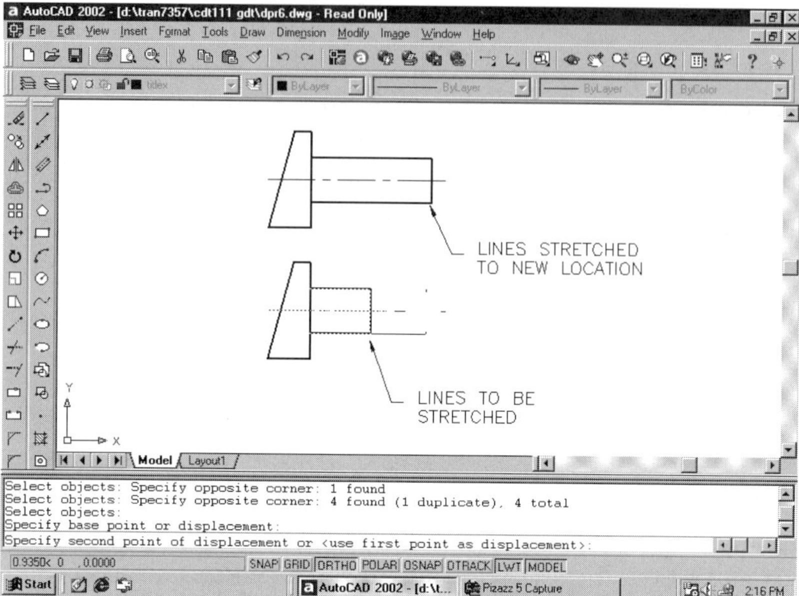

Figure 11.12a *The Window option is used to form the selection set of objects that are to be stretched. Once objects are selected, prompts will be displayed for the base point of the original objects and the new position they are to be stretched to. Courtesy Hanh Tran.*

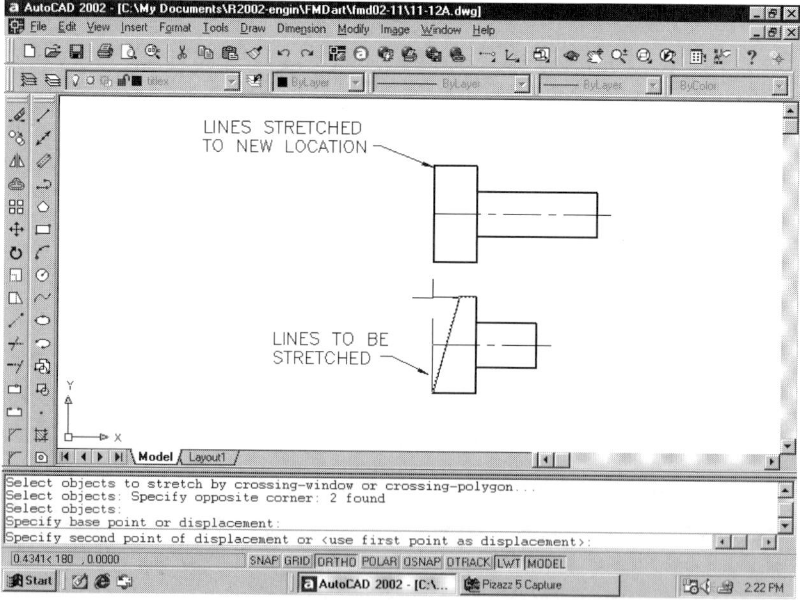

Figure 11.12b *In addition to adding length to line segments, the* **Stretch** *command can be used to alter the angle of the selected lines. Courtesy Hanh Tran.*

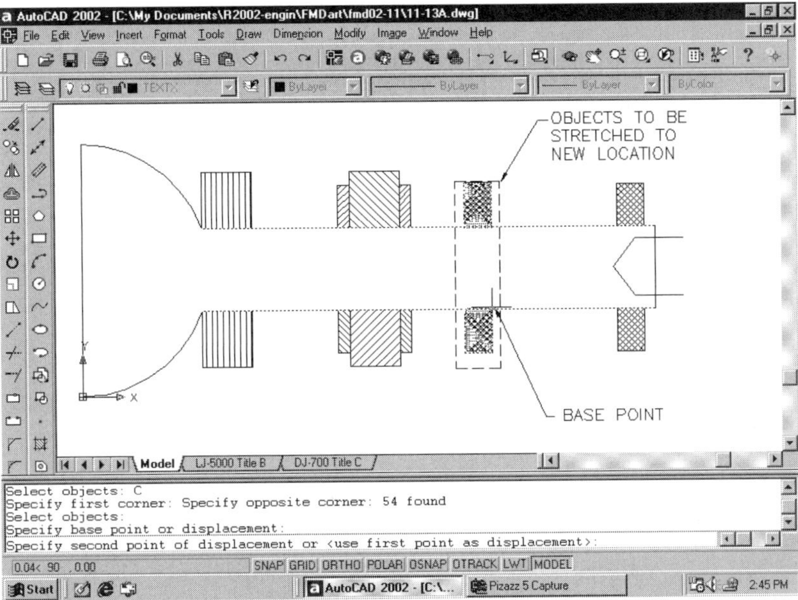

Figure 11.13a *The* **Stretch** *command can also be used to alter the location of objects. Courtesy Paul Bissell.*

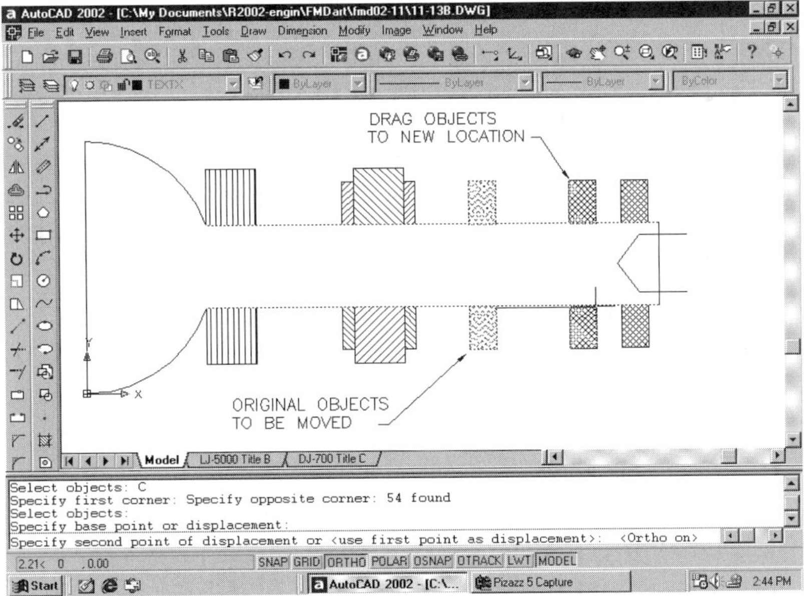

Figure 11.13b *The original objects shown in Figure 11.13a are stretched to alter their location. Lines on the left side of the object are elongated, while the lines on the right side are shortened. Courtesy Paul Bissell.*

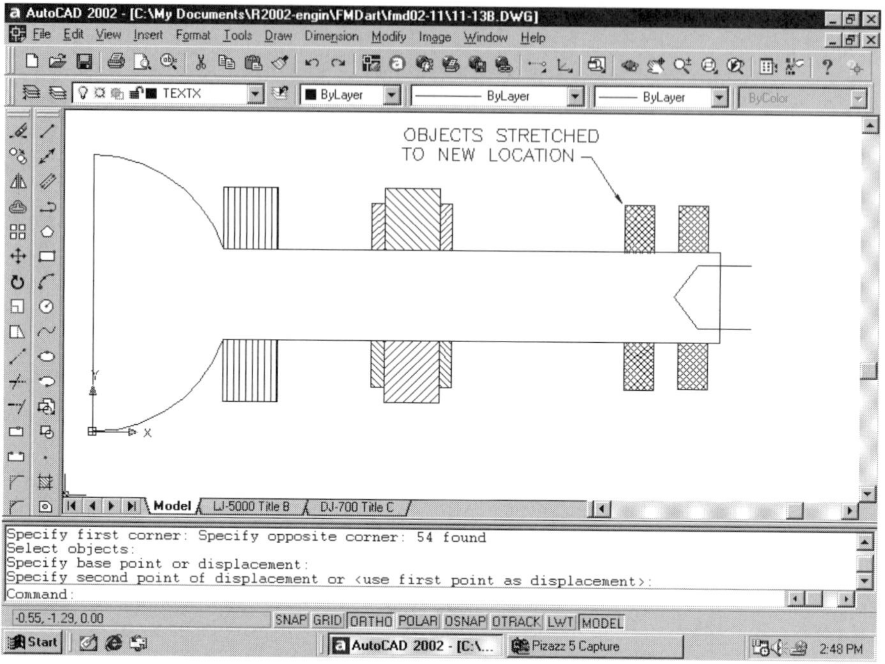

Figure 11.13c *The completed* **Stretch** *command. This is much more efficient than moving the objects, stretching the left lines, and trimming the right lines. Courtesy Paul Bissell.*

LENGTHEN

The **Lengthen** command is similar to the **Stretch** command. **Stretch** is used to edit the length of groups of objects and closed polygons. However, **Lengthen** is used to edit the length of objects, but not closed polygons. Start the **Lengthen** command by selecting the **Lengthen** button on the **Modify** toolbar, by typing LEN (ENTER) at the Command prompt, or by selecting **Lengthen** from the **Modify** menu. The command sequence is as follows:

Command: **LEN** ENTER *(Or click the Lengthen button.)*

LENGTHEN

Select an object or [DElta/Percent/Total/DYnamic]: *(Select an object to edit.)*

Current Length: 4.00 *(Length will vary.)*

Select an object or DElta/Percent/Total/DYnamic/<Select object>: **DE** ENTER

Using the DElta Option of Lengthen

Typing DE (ENTER) at the Select object: prompt will produce the prompt:

Enter delta length or [Angle] <0.00)>: **5** ENTER *(Enter the length to add.)*

Select an object to change or [Undo]: *(Select the object again.)*

Select an object to change or [Undo]:ENTER

Command:

The command sequence can be seen in Figure 11.14. Notice that when DElta was entered at the Select prompt, a choice could be made between length and angle. The default is to provide a length. Entering a positive number lengthens the line. Providing a negative response shortens the line. With each option of **Lengthen**, the end of the object that is selected will be the end that is edited.

Note: The length entered is the length that will be added to the selected line. The value will not become the length for the total line length.

The Angle option can be used to change the included angle of an arc. The effects of this command sequence can be seen in Figure 11.15. The command sequence is as follows:

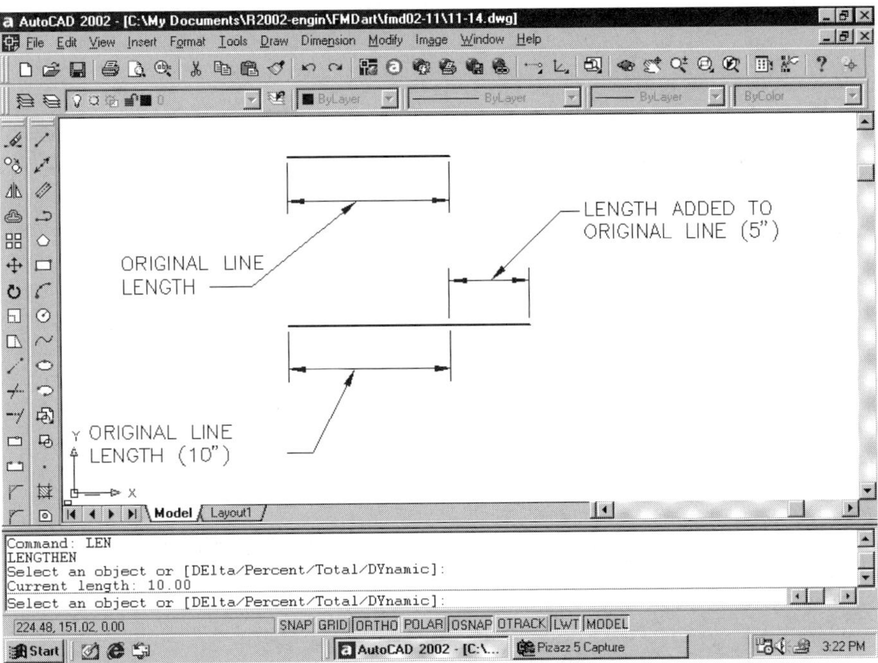

Figure 11.14 *The DElta option of the* **Lengthen** *command can be used to lengthen or shorten a line.*

Command: **LEN** ENTER *(Or click the Lengthen button.)*

LENGTHEN

Select an object or [DElta/Percent/Total/DYnamic]: *(Select an arc to edit.)*

Current Length: 4.100, included angle: 90 *(Length and angle will vary.)*

DElta/Percent/Total/DYnamic/<Select object>: **DE** ENTER

Enter delta length or [Angle] <0.00>: **A** ENTER

Enter delta angle<0.00>: **60** ENTER

Select an object to change or [Undo]: *(Select arc to be edited.)*

Select an object to change or [Undo]: ENTER

Command

Using the Percent Option of Lengthen

The Percent option allows the length of a line or arc to be altered by a percentage of the existing length. The existing line length represents 100%. To lengthen the line, provide a value greater than 100%. Providing a percentage of less than 100% will short-

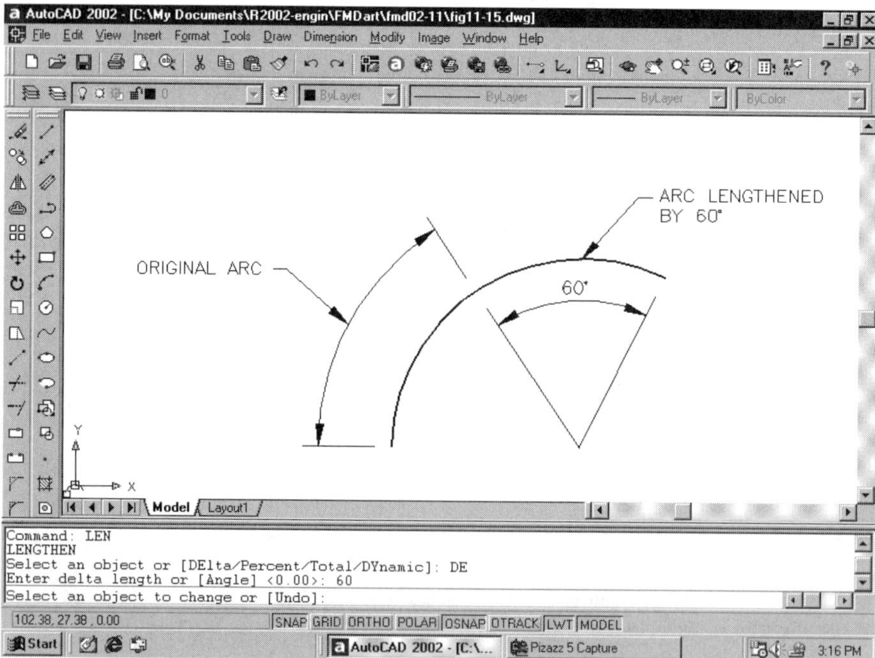

Figure 11.15 *The DElta and Angle options of the* **Lengthen** *command can be used to lengthen or shorten an angle.*

en the line. The command sequence to lengthen a 2" line to a 3" line using the Percent option is as follows:

Command: **LEN** ENTER *(Or click the Lengthen button.)*

LENGTHEN

Select an object or [DElta/Percent/Total/DYnamic]: *(Select an object to edit.)*

Current Length: 2.00 *(Length will vary.)*

Select an object or [DElta/Percent/Total/DYnamic]: **P** ENTER

Enter percentage length <100.0000>: **150** ENTER

Select an object to change or [Undo]: *(Select line to be edited.)*

Select an object to change or [Undo]: ENTER

Command:

The effects of the Percent option can be seen in Figure 11.16. When used with an arc, the angle of the arc will be edited based on a percentage of the total angle of the specified arc.

Using the Total Option of Lengthen

The Total option allows the length of a line or angle of an arc to be edited by providing the desired ending result. The command sequence to alter an existing line to a 5" line would be as follows:

Command: **LEN** ENTER *(Or click the Lengthen button.)*

LENGTHEN

Select an object or [DElta/Percent/Total/DYnamic]: *(Select an object to edit.)*

Current Length: 4.00 *(Length will vary.)*

Select an object or [DElta/Percent/Total/DYnamic]: **T** ENTER

Specify total length or [Angle]: **5** ENTER

Select an object to change or [Undo]:ENTER

Command:

The result of this sequence can be seen in Figure 11.17. When first prompted to select an object, you might prefer to select an object before choosing the Total option. When a line is selected first, the length of the existing line will be displayed.

Using the DYnamic Option of Lengthen

The DYnamic option allows the selected line or angle to be dragged to the selected point to achieve the required length. The results of the DYnamic option can be seen in Figure 11.18.

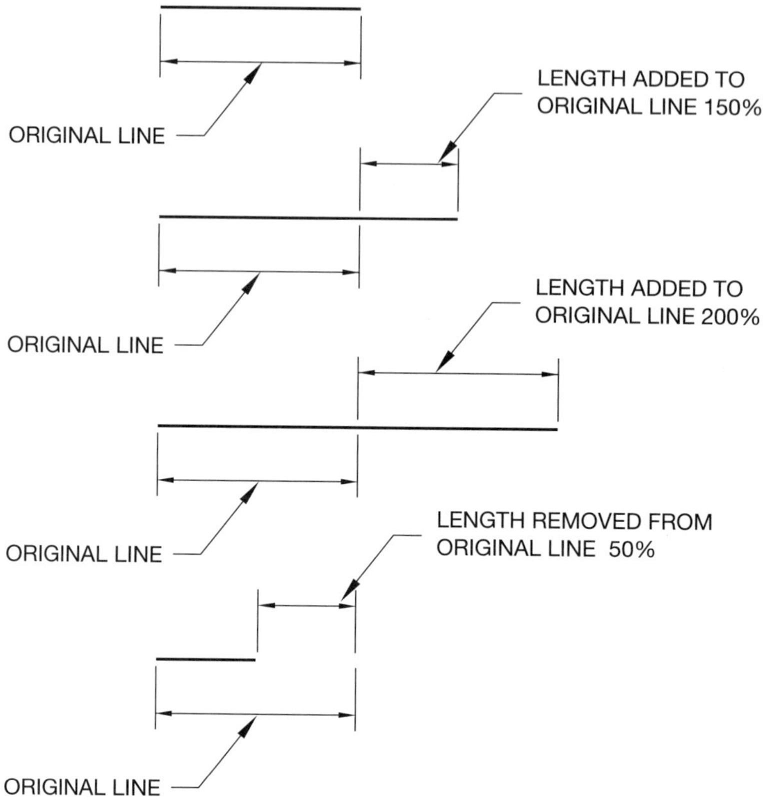

Figure 11.16 *The lengths of lines and arcs can be altered by entering a percentage of the original object. Percentages less than 100% will shorten the line or arc, and those greater than 100% will lengthen the line or arc.*

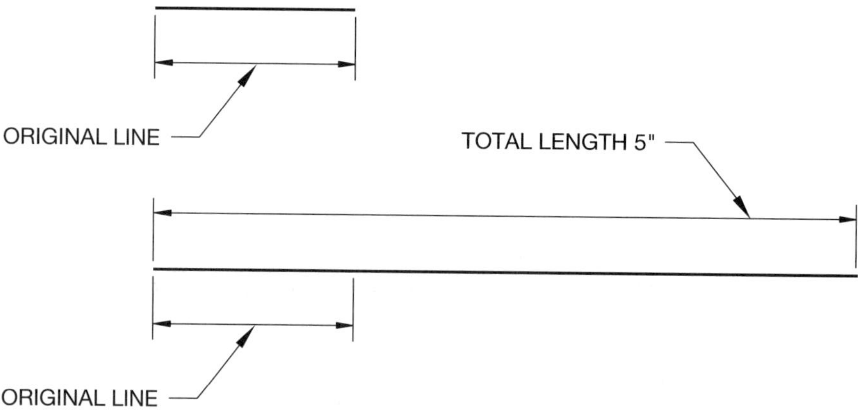

Figure 11.17 *The Total option of* **Lengthen** *allows a line or angle of an arc to be lengthened based on the desired total length or angle.*

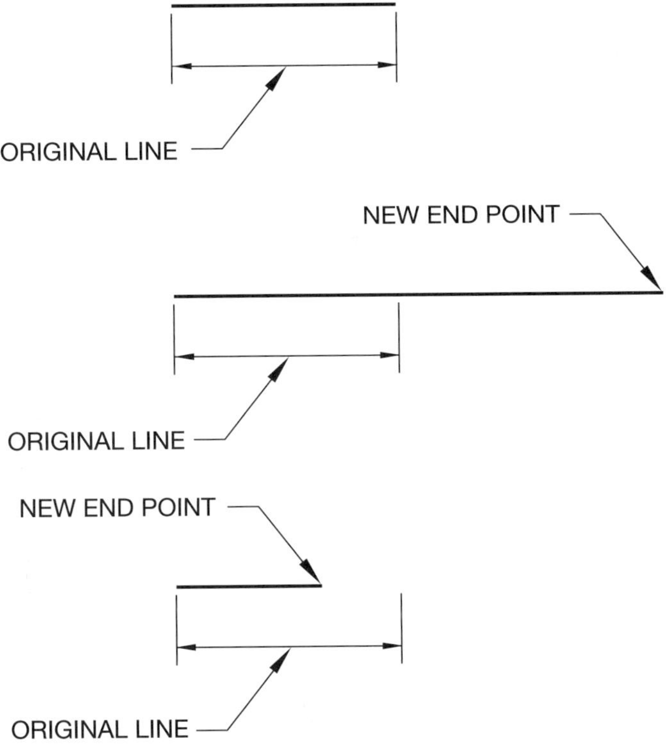

Figure 11.18 *The DYnamic option of* **Lengthen** *allows a line or angle of an arc to be dragged to the desired length by selecting a new ending point.*

Command: **LEN** ENTER *(Or click the Lengthen button.)*

LENGTHEN

Select an object or [DElta/Percent/Total/DYnamic]: *(Select an object to edit.)*

Current Length: 4.00 *(Length will vary.)*

Select an object or [DElta/Percent/Total/DYnamic]: **DY** ENTER

Select an object to change or [Undo]: *(Select line to be edited.)*

Specify new endpoint: *(Select new endpoint for line to be edited.)*

Select an object to change or [Undo]: ENTER

Command:

MODIFYING OBJECTS BY ALTERING LINES

Up to this point, you've learned to modify objects by making additional copies and by altering the position or size of an object. The commands contained in this portion of the **Modify** menu will allow you to modify a drawing by altering intersections of lines, circles or arcs, or by removing a portion of an object. The commands include **Trim**, **Extend**, **Break**, **Chamfer**, and **Fillet**.

TRIM

The **Trim** command was introduced in previous chapters as aids to drawing layout were explored. **Trim** was used for its ability to clean the corners of intersecting lines, as seen in Figure 11.19, or to modify the size of objects. This chapter will expand on the command, allowing you to make fuller use of the command's power. Start the **Trim** command by selecting **Trim** on the **Modify** toolbar, by typing TR (ENTER), or by selecting **Trim** from the **Modify** menu. The command allows you to edit lines, arcs, circles, or polylines using a cutting edge. (Polylines will be introduced in the next chapter.) Any portion of the selected object extending past the designated cutting edge will be removed. Lines, arcs, circles, and polylines can be used as cutting edges. Once the cutting edges are defined, the command will continue until ENTER is pressed. Any object that extends past the cutting edge can be trimmed. Notice that an object can be selected to be a cutting edge and it can also be an object to be trimmed.

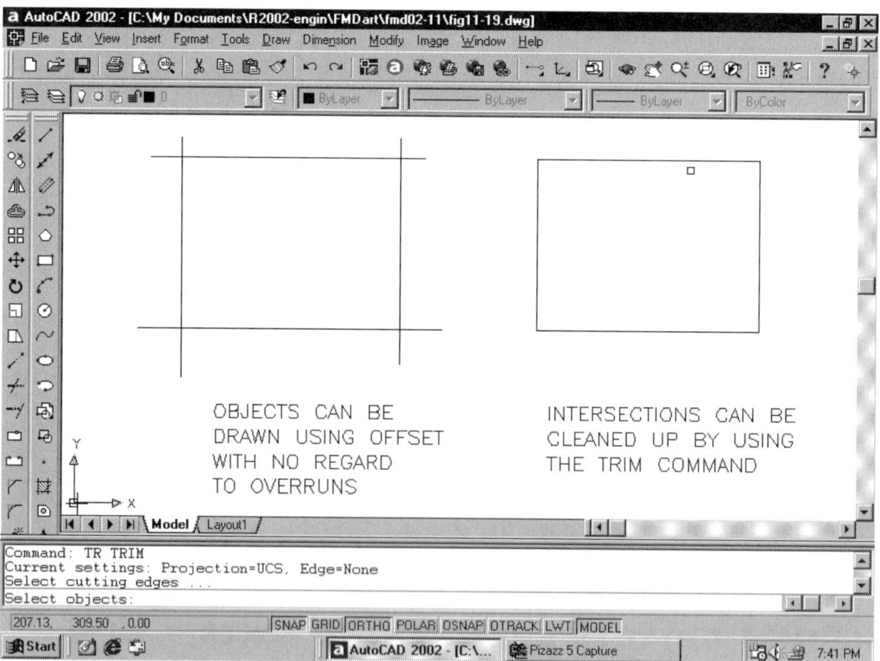

Figure 11.19 *The **Trim** command can be used to eliminate line overruns that were created using Offset.*

Multiple Trim Selections

To select more than one cutting edge, hold SHIFT while selecting the cutting edges (Remember that the **Use Shift to Add** button on the **Selection** tab of the **Options** dialog box must be active). The Window and Crossing window options work well for selecting multiple objects for cutting edges. Selecting the All option will use all drawing objects as trim edges. To use the All option, you must type **ALL** (ENTER) (not **A**).

Using the Trim Edge Option

The Edge option of **Trim** can be used to trim two objects that do not actually intersect, but would intersect if one of the objects were extended. AutoCAD refers to this as an implied extension Once this edge has been selected, type **E** (ENTER) when prompted to select another object. This will now display the option

```
Enter an implied edge extension mode [Extend/No extend]
    <Extend>:
```

When Extend is selected, AutoCAD checks the cutting edge object to verify if it will intersect when extended. If it would touch, the implied intersection is used to trim the second object. This option is shown in Figure 11.20. The command sequence is as follows:

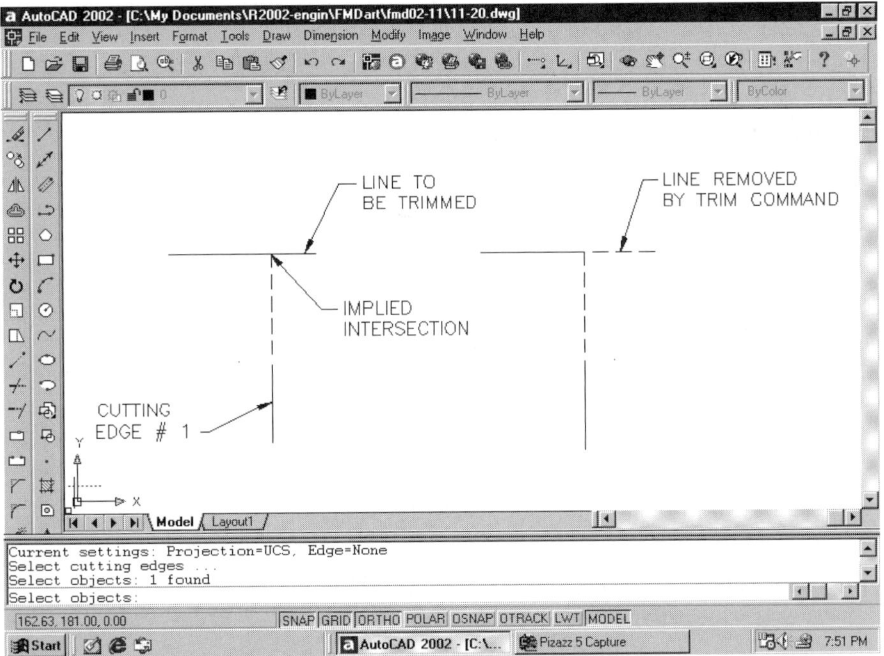

Figure 11.20 *Lines that would touch, if one were extended, can be trimmed by selecting the Extend mode of the* **Trim Edge** *option.*

Command: **TR** ENTER (Or click the Trim button.)

TRIM

Current settings: Projection=UCS Edge=None

Select cutting edges... *(Select the cutting edge using any object selection method.)*

1 found

Select objects: ENTER

Select object to trim or shift-select to extend or [Project/Edge/Undo]: **E** ENTER

Enter an implied edge extension mode [Extend/No Extend] <Extend>: ENTER

Select object to trim or shift-select to extend or [Project/Edge/Undo]: *(Select object to trim.)*

Select object to trim or shift-select to extend or [Project/Edge/Undo]: ENTER

Command:

Using the Trim Extend option

The Extend option of **Trim** can be used to extend objects that do not actually intersect, but would intersect if one of the objects were extended. AutoCAD refers to this as an implied intersection. To use the Extend option, select the line to be used as the destination. When prompted to select another object, press and hold SHIFT and select the line to be extended with the select button. The command sequence is as follows:

Command: **TR** ENTER *(Or click the Trim button.)*

TRIM

Current settings: Projection=UCS Edge=None

Select cutting edges... *(Select the cutting edge using any object selection method.)*

1 found

Select objects: ENTER

Select object to trim or shift-select to extend or [Project/Edge/Undo]: *(ShiftKey + select object to extend.)*

Select object to trim or shift-select to extend or [Project/Edge/Undo]: ENTER

Command:

Using the Trim Undo option

Occasionally you might trim an object and not be pleased with the results. The **Trim** Undo option will allow an unsatisfactory **Trim** to be undone while still in the command sequence. The command sequence is as follows:

Command: **TR** ENTER *(Or click the Trim button.)*

TRIM

Current settings: Projection=UCS Edge=Extend

Select cutting edges... *(Select the cutting edge using any object selection method.)*

I found

Select objects: ENTER

Select object to trim or shift-select to extend or [Project/Edge/Undo]: *(Select object to be trimmed.)*

Select object to trim or shift-select to extend or [Project/Edge/Undo]: **U** ENTER

Command has been completely undone.

Select object to trim or shift-select to extend or [Project/Edge/Undo]: *(Select object to be trimmed.)*

Select object to trim or shift-select to extend or [Project/Edge/Undo]:ENTER

Command:

Using the Trim Project Option

The **Trim Project** option is primarily a 3D option and that aspect of the command will be discussed only briefly in this text. Refer to the **Help** menu for further information. The option can be used just as the Edge option as a means of trimming objects that extend past a projection of the selected cutting edge. The command sequence is as follows:

Command: **TR** ENTER *(Or click the Trim button.)*

TRIM

Current settings: Projection=UCS Edge=None

Select cutting edges... *(Select the cutting edge using any object selection method.)*

I found

Select objects: ENTER

Select object to trim or shift-select to extend or [Project/Edge/Undo]: **P** ENTER

Enter a projection option [None/Ucs/View] <Ucs> **N** ENTER

Select object to trim or shift-select to extend or
[Project/Edge/Undo]: *(Select object to trim)*

Select object to trim or shift-select to extend or
[Project/Edge/Undo]: ENTER

Command:

Practical Use of Trim

As you discovered earlier, combining the **Line**, **Offset**, and **Trim** commands will allow almost any object to be drawn easily. Draw two perpendicular lines to represent the corner of an object without regard to the exact length. The exact size of the object can be determined with **Offset** or **Trim**.

EXTEND

Start the **Extend** command by selecting the **Extend** button on the **Modify** toolbar, by typing EX (ENTER), or by selecting **Extend** from the **Modify** menu. The command allows an object to be lengthened to an exact boundary point. The command will ask you to designate boundary edges, and then select objects to be extended to that edge. The boundary edge can be a line, arc, circle, or polyline. The process can be seen in Figure 11. 21. The command sequence is as follows:

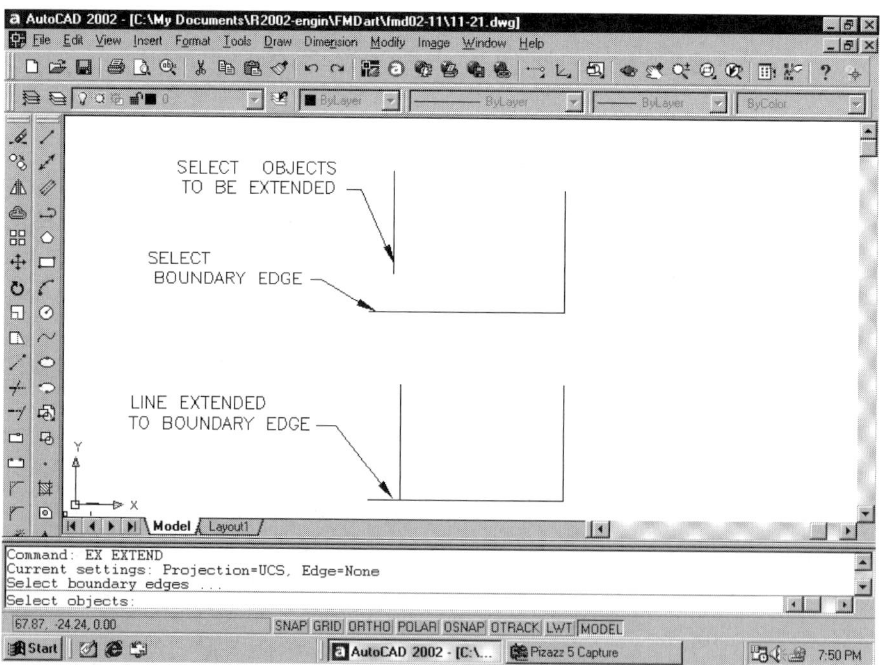

Figure 11.21 *The* **Extend** *command can be used to lengthen lines to a selected boundary edge.*

Command: **EX** ENTER *(Or click the Extend button.)*

EXTEND

Current settings: Projection=UCS Edge=Extend

Select boundary edges... *(Select the edge using any object selection method.)*

1 found

Select objects: ENTER

Select object to extend or shift-select to trim or [Project/Edge/Undo]: *(Select object to be extended using any selection method.)*

Select object to extend or shift-select to trim or [Project/Edge/Undo]: ENTER

Command:

As the object to be extended is selected, the line is extended to the other object that was selected as the boundary edge.

Using the Extend Edge Option

The **Extend Edge** option is similar to the Edge option presented with the **Trim** command. Edge will allow lines that do not presently touch but have an implied intersection to be extended. In the default setting, the command will automatically extend using an implied edge. If the Edge setting has been altered, the sequence to extend to an implied edge is as follows:

Command: **EX** ENTER *(Or click the Extend button.)*

EXTEND

Current settings: Projection=UCS Edge=Extend

Select boundary edges... *(Select the edge using any object selection method.)*

1 found

Select objects: ENTER

Select object to extend or shift-select to trim or [Project/Edge/Undo]: **E** ENTER

Enter an implied edge extension mode [Extend/No Extend] <Extend>: *(Select object to be extended using any selection method.)*

Select object to extend or shift-select to trim or [Project/Edge/Undo]:ENTER

Command:

The result of the command can be seen in Figure 11.22.

Using the Extend Trim Option

The Trim option of **Extend** allows an object to be trimmed while within the **Extend** command. The command sequence to use the Trim option is as follows:

Command: **EX** ENTER *(Or click the Extend button.)*

EXTEND

Current settings: Projection=UCS, Edge=Extend

Select boundary edges... *(Select the edge using any object selection method.)*

1 found

Select objects: ENTER

Select object to extend or shift-select to trim or [Project/Edge/Undo]: *(ShiftKey + Select object to be trimmed using any selection method.)*

Select object to extend or shift-select to trim or [Project/Edge/Undo]: ENTER

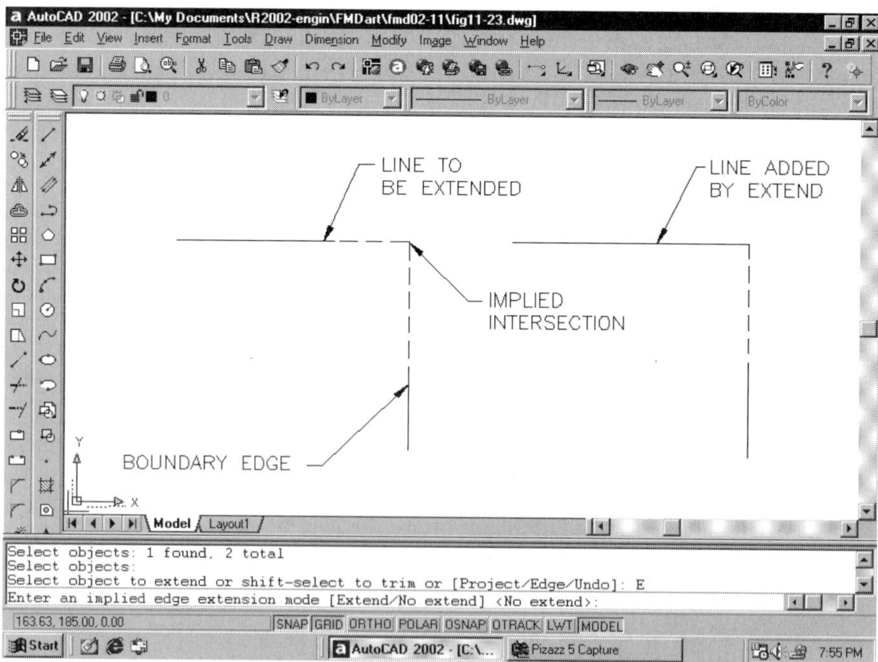

Figure 11.22 *Lines that would touch, if both were extended, can be extended by selecting the Edge option of* **Extend**.

Command:

Using the Extend Undo Option

The **Extend** Undo option allows a sequence that has not produced satisfactory results to be undone while still in the command sequence.

Using the Extend Project Option.

This mode is used for 3D drawings and will not be discussed in this text. Use the **Help** menu for further information on this subject.

BREAK

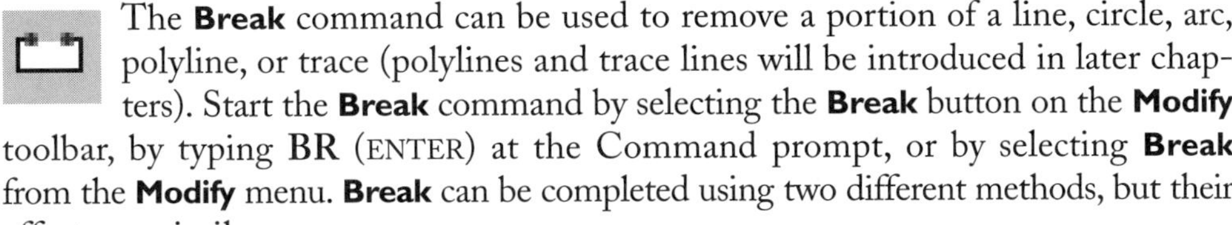

The **Break** command can be used to remove a portion of a line, circle, arc, polyline, or trace (polylines and trace lines will be introduced in later chapters). Start the **Break** command by selecting the **Break** button on the **Modify** toolbar, by typing BR (ENTER) at the Command prompt, or by selecting **Break** from the **Modify** menu. **Break** can be completed using two different methods, but their effects are similar.

Select Object, Second Break Point

The default option of **Break** allows you to remove a segment of an object by using two selection points. The first selection point selects the segment to be broken and specifies where the break will begin. The second point specifies where the Break will end. The command is as follows:

Command: **BR** ENTER *(Or click the Break button.)*

BREAK

Select object: *(Select an object to break.)*

Specify second break point or [First point]: *(Select a second point.)*

Command.

Once the second point is selected, the portion of the object from the first to the second point will be removed and the Command prompt will be returned. The sequence can be seen in Figure 11.23. If a circle is being edited, the break occurs in a counterclockwise direction from the first point to the second point, as seen in Figure 11.24. Type F (ENTER) when the second point is requested to allow the procedure to follow the Select Object, Two Break Points method.

Select Object, Two Break Points

This option allows you to remove a segment of an object by using three selection points. The first request selects the object to be broken. Once the segment has been defined for editing, the next selection point specifies where the break will begin, and the final selection point specifies where the break will end. The command sequence is as follows:

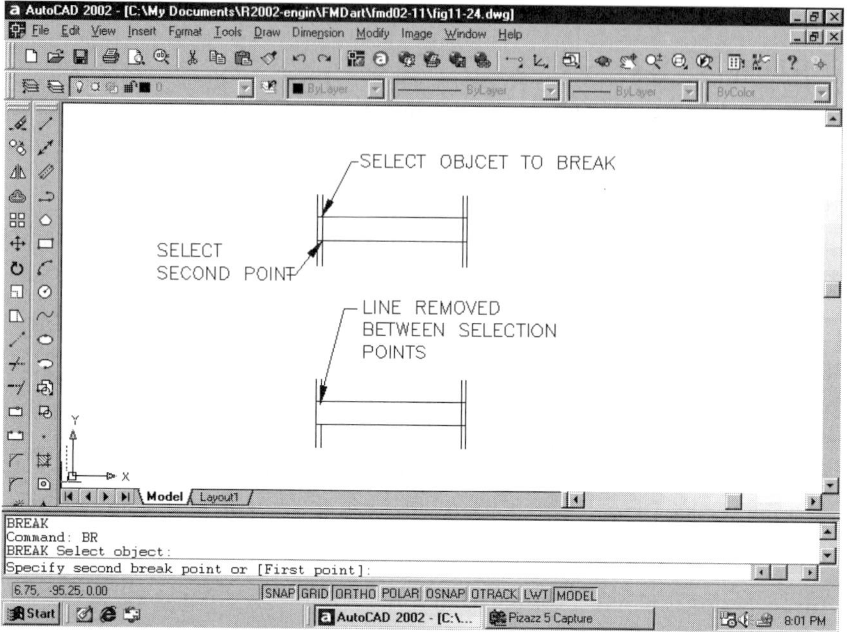

Figure 11.23 *The default option of* **Break** *allows for selecting the line to be broken and then specifying the segment to be removed.*

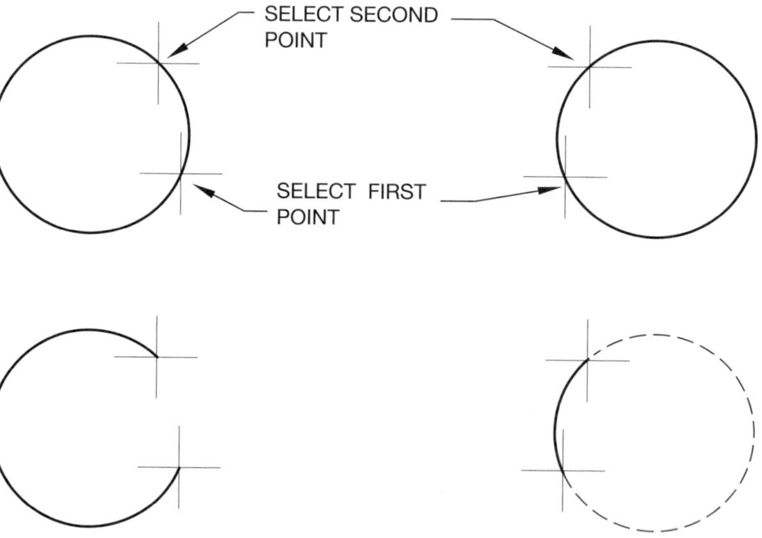

Figure 11.24 *When a break is specified for a circle, it will occur in a counterclockwise direction.*

Command: **BR** ENTER *(Or click the Break button.)*

BREAK

Select object: *(Select an object to break.)*

Specify second break point or [First point]: **F** ENTER

Specify first break point: *(Select a point to start the break.)*

Specify second break point: *(Select a point to end the break.)*

Command:

The command sequence can be seen in Figure 11.25.

Break @

This option will break a segment, but no apparent change will take place. **Break @** will divide an object into two portions, but no space will be left between the portions. The @ option is useful for erasing only a portion of an object by using **Break @** and then using the **Erase** command. The command sequence is as follows:

Command: **BR** ENTER *(Or click the Break at Point button on the Modify toolbar.)*

BREAK

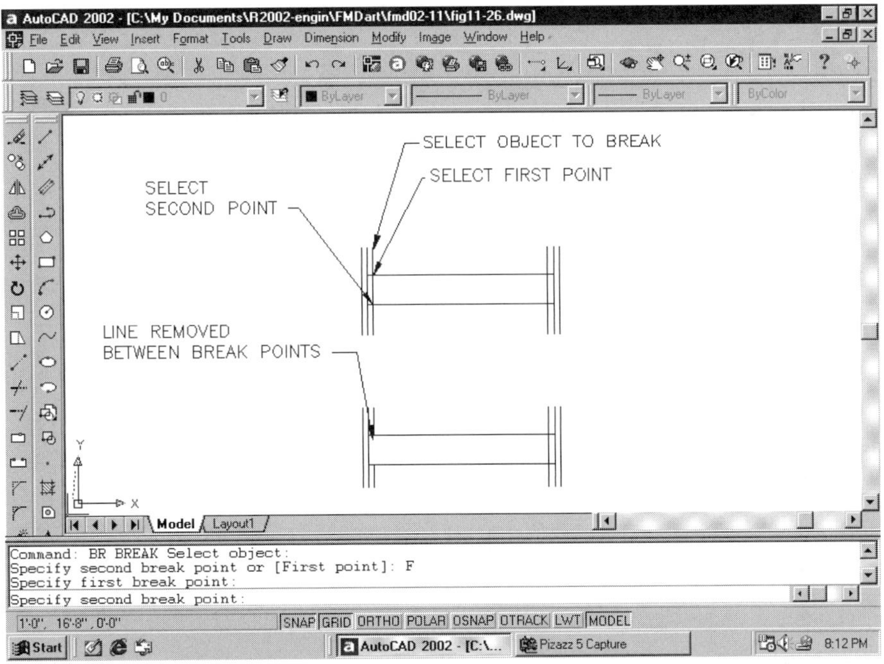

Figure 11.25 *The* **Break** *command's First option can be used to remove a segment from a line. Specified points are used to define which part of the line will be removed.*

Select object: *(Select an object to break.)*

Specify second break point or [First point]: @ ENTER

Command.

The sequence is shown in Figure 11.26.

EDITING WITH THE CHAMFER COMMAND

A chamfer is an angled edge formed between two intersecting surfaces, typical of what might be found on the edge of the cabinet counter shown in Figure 11.27. The **Chamfer** command trims two intersecting lines and provides a mitered corner. You'll be asked to select two lines and two distances. This will form a new line segment between the two existing line segments. Start the **Chamfer** command by selecting the **Chamfer** button on the **Modify** toolbar, by typing CHA (ENTER) at the Command prompt, or by selecting **Chamfer** from the **Modify** menu. The command sequence to set the default values to 0, which can be used to trim corners, is as follows:

Command: **CHA** ENTER *(Or click the Chamfer button.)*

CHAMFER

(TRIM mode) Current chamfer Dist1 = 0.50, Dist2 = 0.50

Select first line or [Polyline/Distance/Angle/Trim/Method]: **D** ENTER

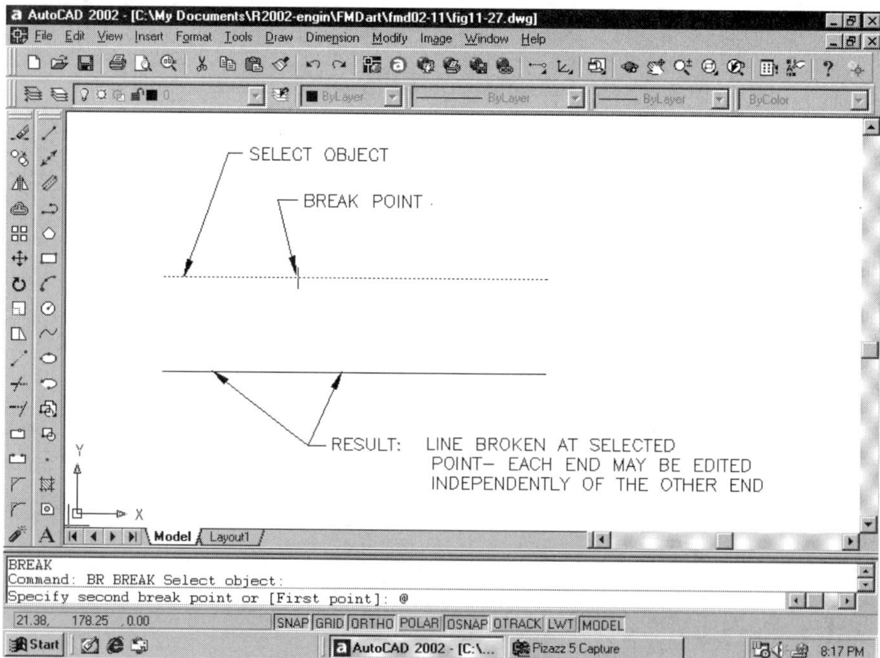

Figure 11.26 *The* **Break @** *command will break a line into two segments, but no visible change will occur.*

Specify first chamfer distance <0.50>: **0** ENTER

Specify second chamfer distance <0.00>: ENTER

Select first line or [Polyline/Distance/Angle/Trim/Method]:

If the command is continued, objects will be chamfered to form square corners. Because no chamfer distance has been entered, the process and the results would resemble Figure 11.28.

To provide a mitered corner, the sequence would be similar. The first value will become the default for the second value. When the distances are equal, the results will produce a 45° miter. For the following example, a distance of .375" will be used—although any numeric value can be used, as long as the value is no longer than that of the line being chamfered. If the value exceeds the length, the following message will be displayed and the command terminated:

Distance is too large

Invalid

The command sequence is as follows:

Command: **CHA** ENTER *(Or click the Chamfer button.)*

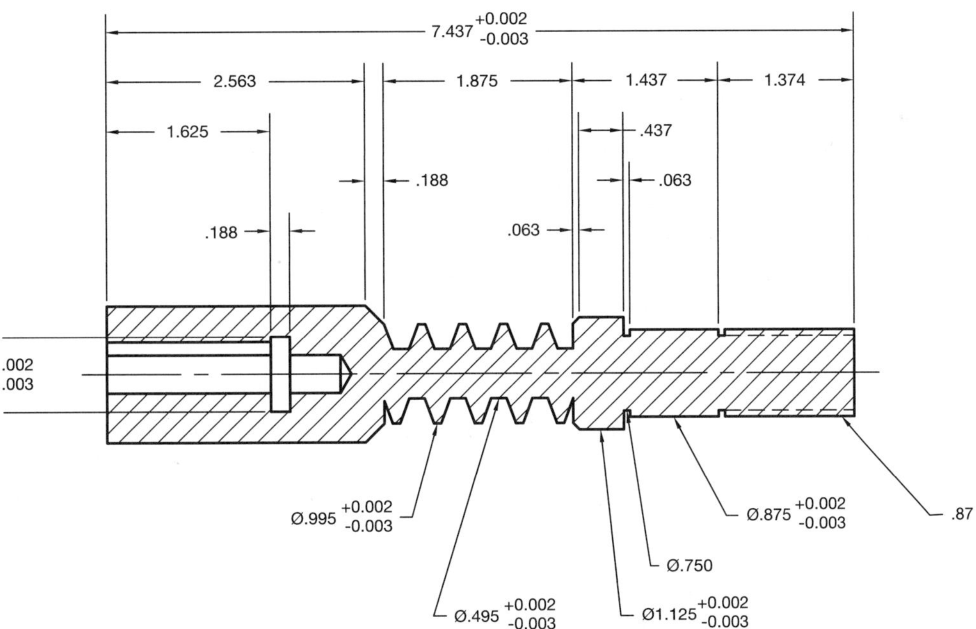

Figure 11.27 *The* **Chamfer** *command can be used to miter a corner. Courtesy Michael J. Easley.*

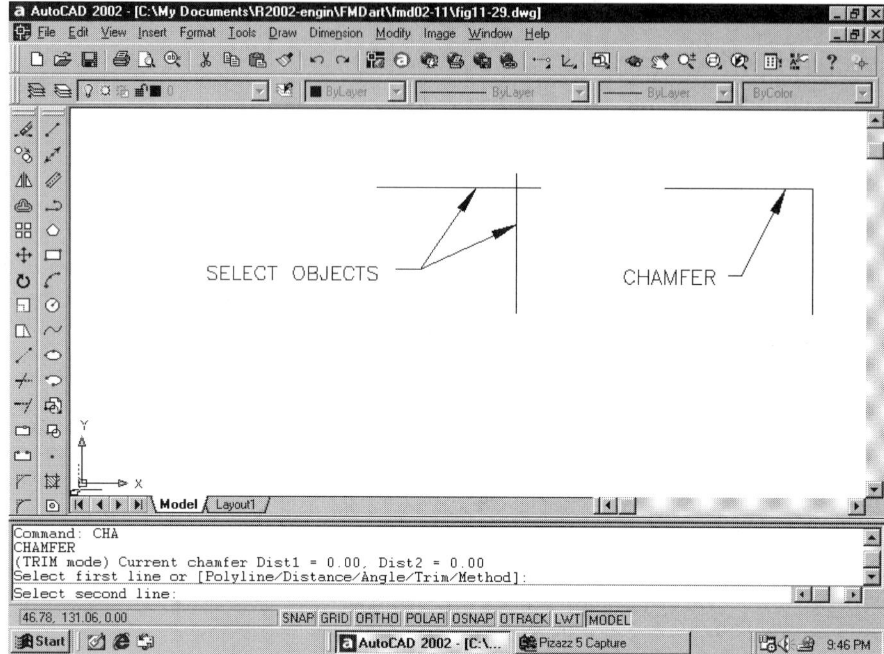

Figure 11.28 *With the value set at 0,* **Chamfer** *can be used to trim corners.*

CHAMFER

(TRIM mode) Current chamfer Dist1=0.00, Dist2=0.00

Select first line or [Polyline/Distance/Angle/Trim/Method]: **D** ENTER

Specify first chamfer distance <0.00>: **.375** ENTER

Specify second chamfer distance <0.375>: ENTER

Select first line or [Polyline/Distance/Angle/Trim/Method]: *(Select first line to be chamfered.)*

Select second line: *(Select second line to be chamfered.)*

As the second line is selected, the chamfer will be performed and the Command prompt returned. The effects of the command sequence can be seen in Figure 11.29. Figure 11.30 shows an example of two different distance values. **Chamfer** can be used to square or chamfer two intersecting lines, as well as to square or chamfer two lines that do not touch.

Selecting the Chamfer Angle

Chamfer also allows the chamfer to be created using one distance and an angle. Start this option by typing A (ENTER) when prompted for the first line. The command sequence to place a 30° angle 4" from a corner would be as follows:

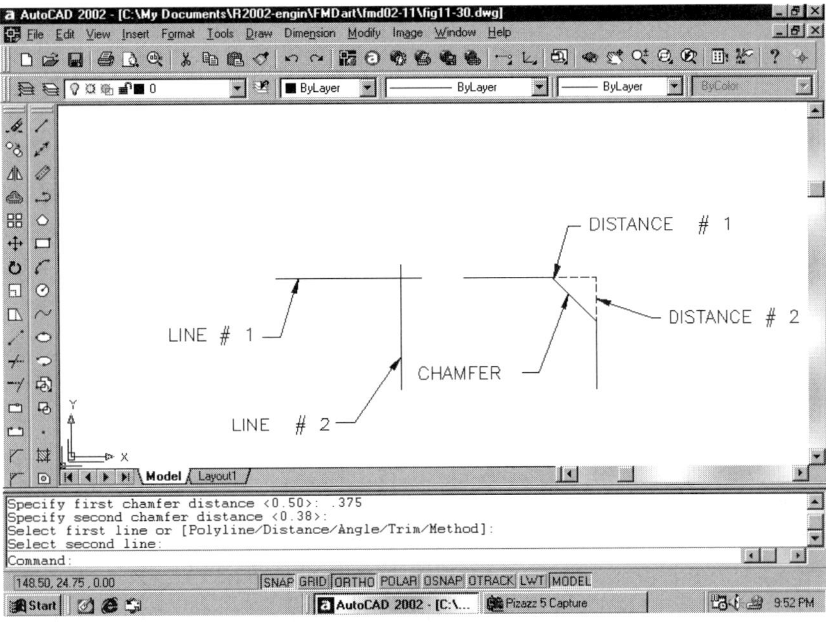

Figure 11.29 *In mitering a corner, prompts will be displayed for the "first distance," and the "second distance." Once the distances are provided, a new segment will be created between the two points.*

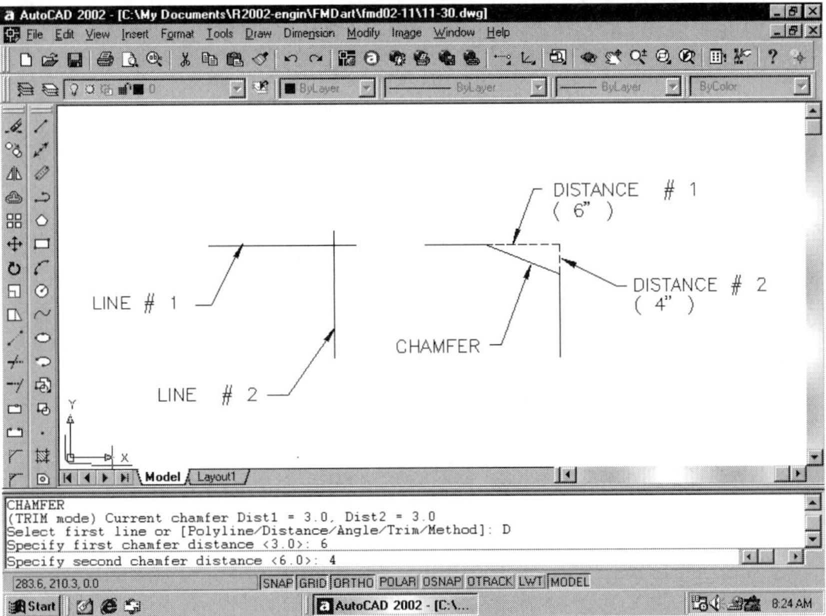

Figure 11.30 *Providing different values for the chamfer distances will affect the angle of the chamfer.*

Command: **CHA** ENTER *(Or click the Chamfer button.)*

CHAMFER

(TRIM mode) Current chamfer Dist1=0.375, Dist2=0.375

Select first line or [Polyline/Distance/Angle/Trim/Method]: **A** ENTER

Specify chamfer length on the first line <1.00>: **75** ENTER

Enter chamfer angle from the first line <0>: **30** ENTER

Select first line or [Polyline/Distance/Angle/Trim/Method]: *(Select first line.)*

Select second line: *(Select second line.)*

Command:

The process can be seen in Figure 11.31.

Selecting the Chamfer Trim Mode

The common method of using the **Chamfer** command is to create the chamfer between two lines and remove a portion of the lines where the chamfer is created. The No Trim option allows for the choice of not removing the lines by the chamfer (by default, the lines are trimmed). The command sequence is as follows:

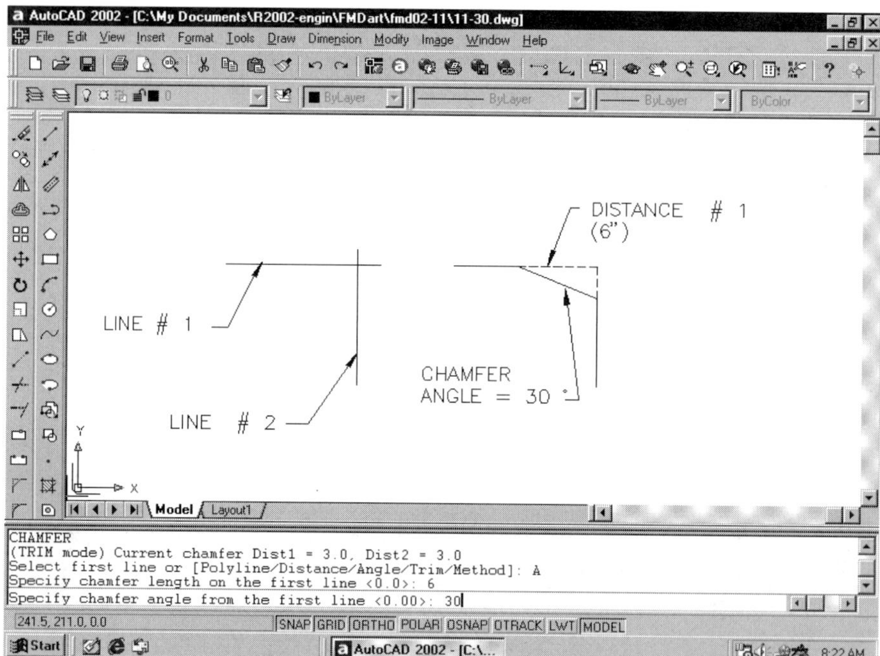

Figure 11.31 *Selecting the Angle option provides a prompt for one length and an angle. AutoCAD converts the angle to a length.*

Command: **CHA** ENTER *(Or click the Chamfer button.)*

CHAMFER

(TRIM mode) Current chamfer Dist1=0.75, Angle = 30

Select first line or [Polyline/Distance/Angle/Trim/Method]: **T** ENTER

Enter Trim mode option [Trim/No trim] <Trim>: **N** ENTER

Select first line or [Polyline/Distance/Angle/Trim/Method]: *(Select first line.)*

Specify second line: *(Select second line.)*

Command:

Note: Remember that if you alter the trim mode of **Chamfer**, the chamfer will be performed, but no lines will be removed. The **Chamfer Trim** mode will also affect the **Fillet** command. (**Fillet** will be discussed in the next section of this chapter.) Most users find the Trim option (the default) more useful than the No Trim option.

Selecting the Chamfer Method

When a chamfer is placed using the Distance or Angle option, the values are kept as the defaults until you change them. Using the Method option, the values selected for one option will not affect the other option, and you can switch between using the Distance or Angle method. The command sequence is as follows:

Command: **CHA** ENTER *(Or click the Chamfer button.)*

CHAMFER

(TRIM mode) Current chamfer Dist1=0.75, Angle 30

Select first line or [Polyline/Distance/Angle/Trim/Method]: **M** ENTER

Enter Trim method [Distance/Angle] <Trim>: **D** ENTER

Select first line or [Polyline/Distance/Angle/Trim/Method]: *(Select first line.)*

Specify second line: *(Select second line.)*

Command: ENTER

EDITING WITH THE FILLET COMMAND

You were introduced to the **Fillet** command as a tool for basic drawing construction. The command is used to connect two lines to form a corner, or allow two lines, arcs, or circles to be fitted together smoothly in rounded corners. Start the command by selecting the **Fillet** button on the **Modify** toolbar, by typing F (ENTER) at the Command prompt, or by selecting **Fillet** from the **Modify** menu. Once the command is started, set the desired radius using a process similar to setting the chamfer distance. Once the two objects are selected, the radius will be applied and the Command prompt will be returned. The **Fillet** process can be seen in Figure 11.32.

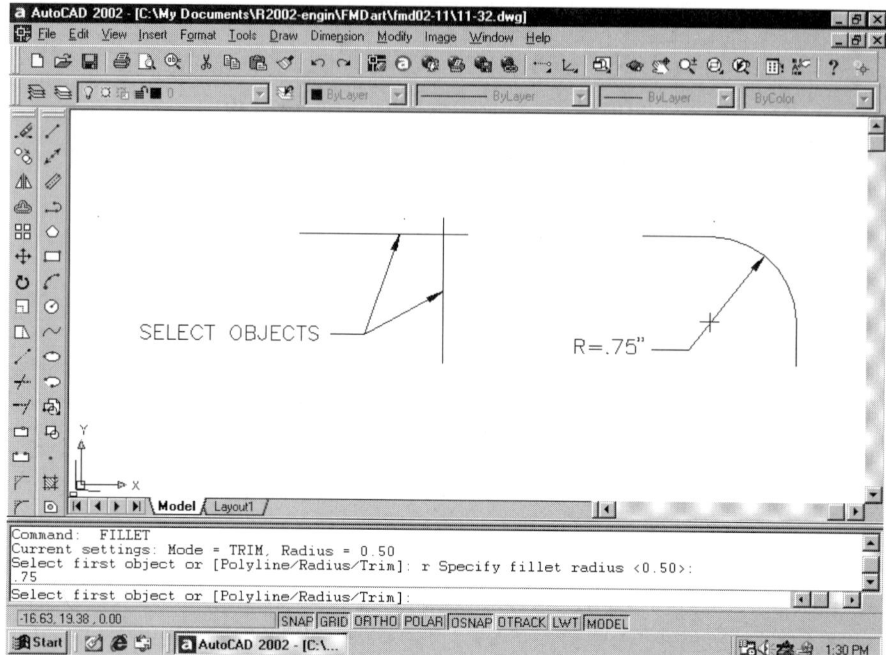

Figure 11.32 *By accepting the default or assigning a new value for the Radius, you can use the* **Fillet** *command to round square corners.*

If the Trim option is set to No Trim, the results of the **Fillet** will not be evident. Be sure the Trim option is set to Trim. The command can be used to form a radius between intersecting lines or two lines that do not touch.

For the command to function with crossing lines as you intend, you need to be careful which portion of the lines you select. Draw two more intersecting lines similar to those in Figure 11.32. Start the **Fillet** command and select the short ends of the lines. Once the **Fillet** command is completed, only the small portion of the lines will remain, as shown in Figure 11.33.

Altering the Fillet Radius

The **Fillet** radius can be altered using the Radius option. Once the first object is selected, use the Radius option to alter the radius of the fillet. Typing **R** (ENTER) at the first prompt allows a radius to be entered for rounded corners. To transform two existing perpendicular lines into a .75" radius corner, the command sequence is as follows:

Command: **F** ENTER *(Or click the Fillet button.)*

FILLET

Current settings: Mode=TRIM, Radius=0.00

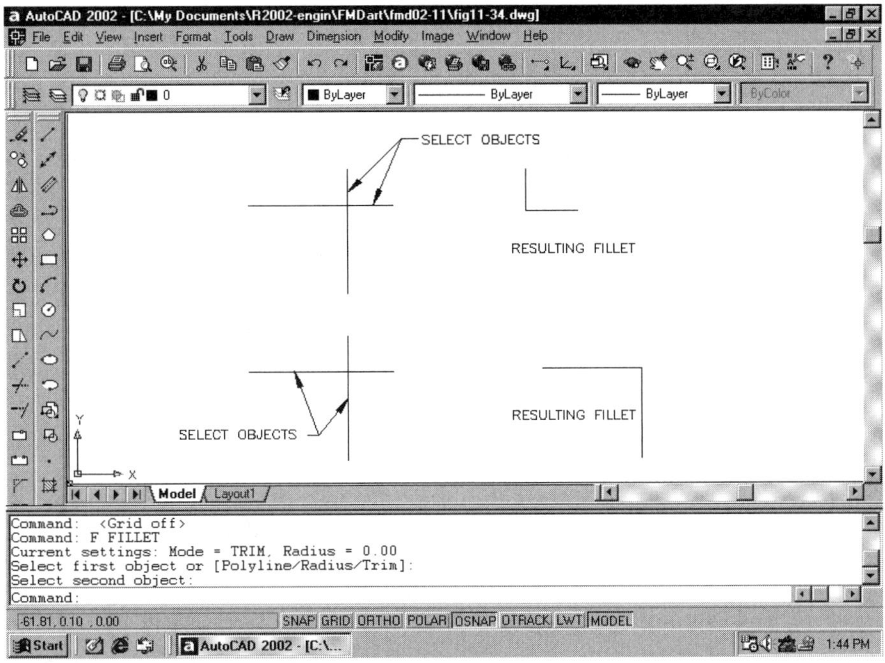

Figure 11.33 *Care must be taken when selecting objects to be edited by the* **Fillet** *command. Selecting the short ends of the lines removes the other portions of the lines.*

> Select first object or [Polyline/Radius/Trim]: **R** ENTER
>
> Specify fillet radius <0.50>: **.75** ENTER
>
> Select first object or [Polyline/Radius/Trim]: *(Select first line.)*
>
> Select second object: *(Select second line.)*
>
> Command:

As the second object is selected, the radius will be formed between the two selected lines, similar to Figure 11.32. The portion of the object that is selected will affect how the **Fillet** will be drawn. Examples of selections and their results can be seen in Figure 11.34. Don't speed past the Command prompt. If you've entered a radius value that exceeds the length of the lines, no change will be made to the drawing, and the following prompt will be displayed:

> Radius is too large
>
> Select second object:
>
> Command:

A smaller radius value will need to be selected to make the command work.

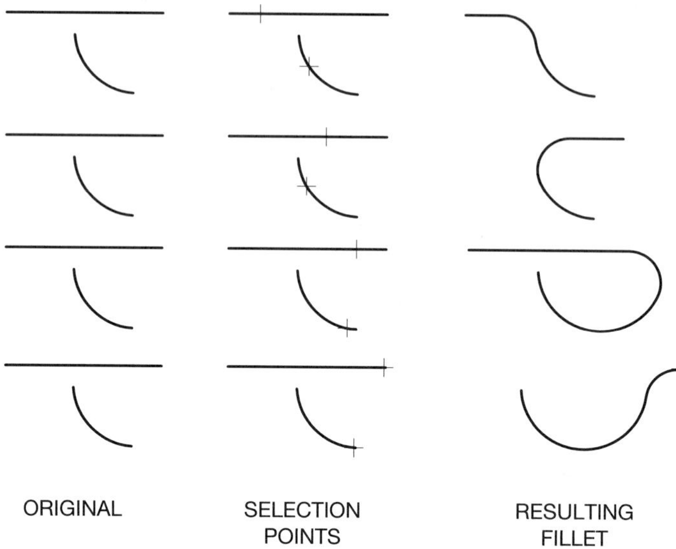

Figure 11.34 *Selecting different portions of the line segment will affect the results of the* **Fillet**.

SELECTING OBJECTS WITH GRIPS

In Chapter 10 you were introduced to several methods of selecting objects for editing. Grips provide an additional method of selecting objects for editing, as well as a method of combining several of the editing procedures introduced in this chapter.

Grips are similar to moving a heavy object by its handles. You can move the object without using the handles, but they make life so much easier. Grips are handles placed on objects for easy manipulation, similar to the OSNAP points, with **Stretch**, **Move**, **Rotate**, **Scale**, and **Mirror**. The default setting for grips is ON. Grips can be toggled ON/OFF on the **Selection** tab of the **Options** dialog box. Select **Options** from the **Tools** menu to display the dialog box (see Figure 11.35). With the **Enable grips** check box selected, grips will be displayed when an edit or inquiry command is given; grip boxes will be displayed at various object-specific positions, as seen in Figure 11.36. When the cursor is moved over a grip, the cursor will snap to the grip for editing. This is sometimes faster that using an OSNAP location.

USING GRIPS

Enter AutoCAD and draw several lines, arcs, and circles. Move the cursor so that it is over one of the lines and click the select button. Three boxes or grips are now displayed. Move the cursor to the center grip, and click the select button. The grip has changed from a hollow box to a solid square. This grip is now said to be *hot*. Notice that the Command prompt has also changed to display the following prompt:

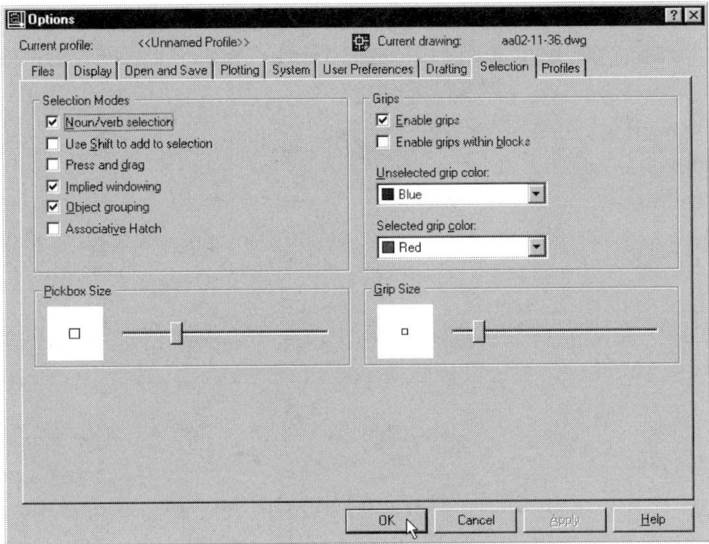

Figure 11.35 *Grips can be toggled ON/OFF on the* **Selection** *tab of the* **Options** *dialog box. Access the dialog box by selecting* **Options** *from the* **Tools** *menu or by typing* **DDGRIPS** (ENTER) *at the Command prompt.*

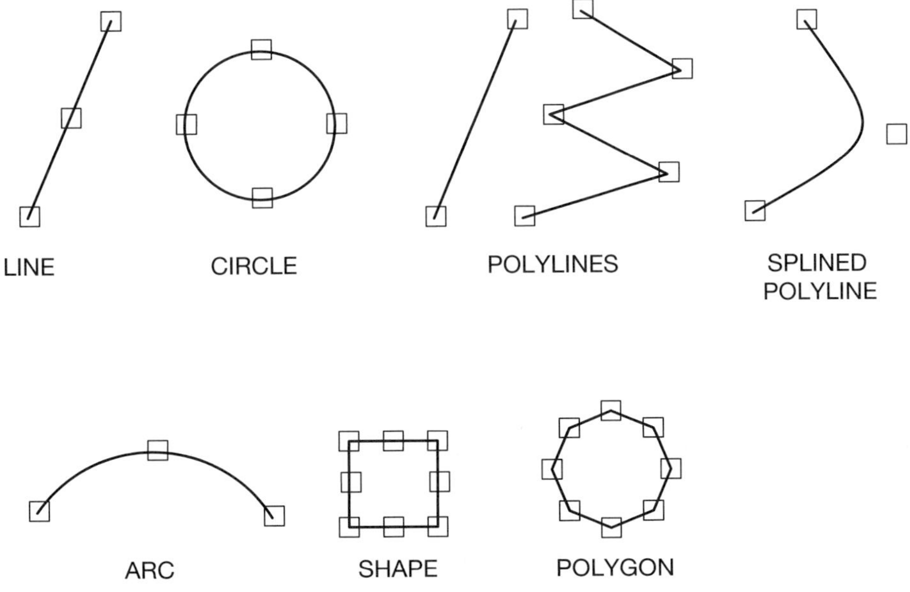

Figure 11.36 *With the default setting* **Enable grips** *check box toggled to ON, grips will be displayed for each geometric shape.*

** STRETCH **

Specify stretch point or [Base point/Copy/Undo/eXit]:

We'll come back to the prompt later. With the center grip hot, and the prompt reflecting **Stretch**, move the cursor slightly. The hot grip in the center of the line functions like the **Move** command. As the cursor moves, so does the line. Clicking the select button will move the line to this new location. Press ESC to cancel the command. Select the same line with the cursor and make one of the end grips hot. Now the grip at the other end remains a stationary point, and as the hot grip is moved, the selected line is stretched. The effect of **Stretch** on a line can be seen in Figure 11.37. Grips function in a similar manner with arcs and circles. Figures 11.38a and 11.38b show the effect of stretching each. If you decide you don't want to edit a selected line, press ESC to deactivate the grips.

GRIP OPTIONS

Earlier you saw that the command displayed in grips has several options. These grip options can be used once the grips for an object have been activated. These options include Base point, Copy, Undo, and eXit.

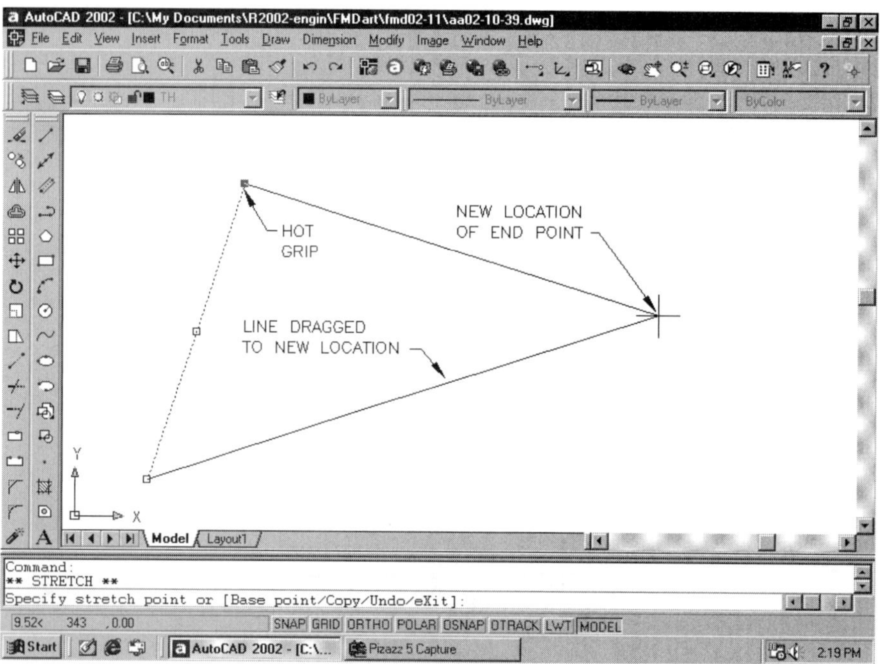

Figure 11.37 *Select a drawing object to display its grips. Moving to one of the grips and clicking the select button will make that grip hot. The object can now be edited by* **Stretch, Move, Rotate, Scale,** *and* **Mirror**.

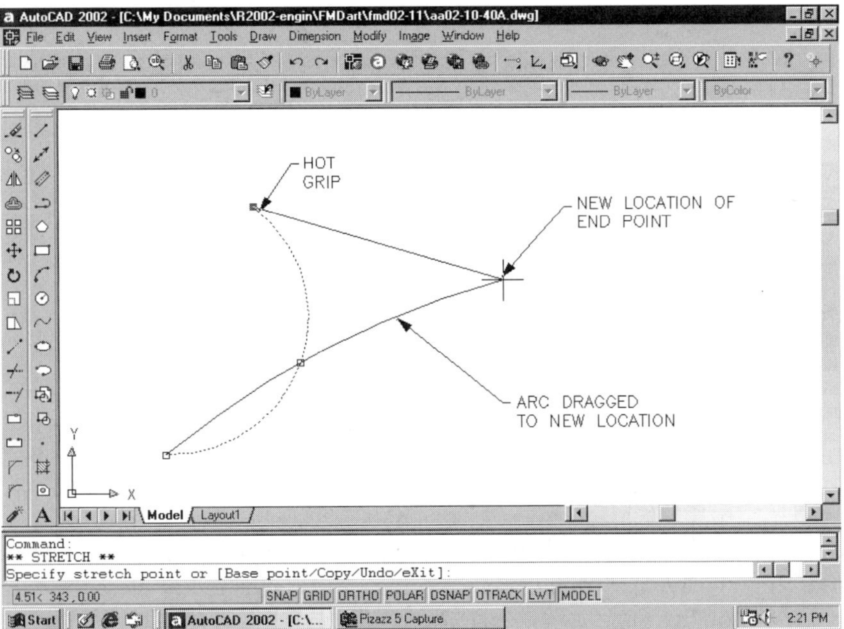

Figure 11.38a *The* **Stretch** *command applied to an arc using the hot grip.*

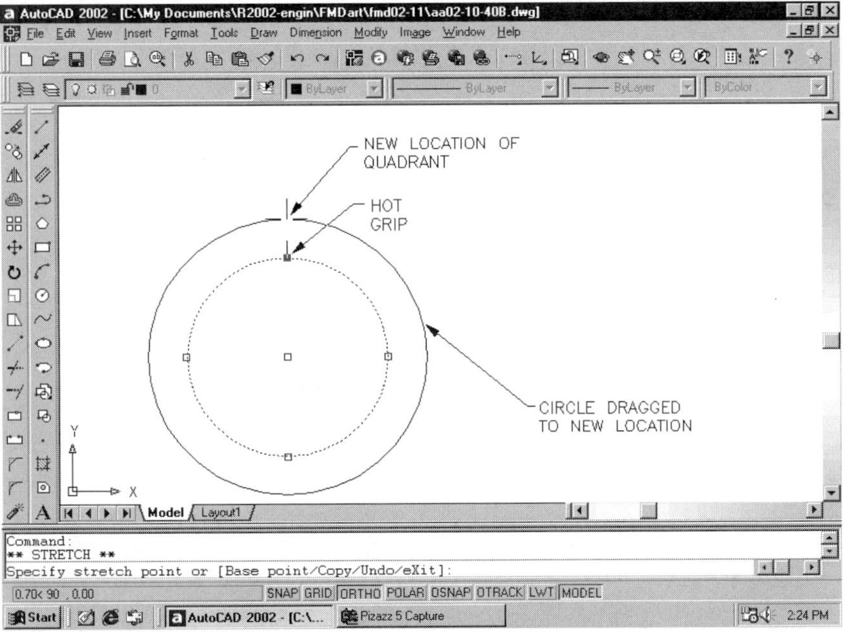

Figure 11.38b *The size of a circle can be altered using* **Stretch** *by selecting a hot grip and a new location.*

Base Point

Typing **B** (ENTER) at the prompt allows the hot box to be disregarded and a new base point to be selected.

Copy

Typing **C** (ENTER) at the prompt allows one or more copies of the selected object to be made.

Undo

Typing **U** (ENTER) at the prompt will undo the previous selection.

eXit

Typing **X** (ENTER) at the prompt terminates the grip edit and returns you to the Command prompt.

CHANGING GRIP OPTIONS

Once a grip box has been selected and made hot, right-clicking will produce the menu shown in Figure 11.39. Any one of the options can be selected to alter the function to be performed by the grip.

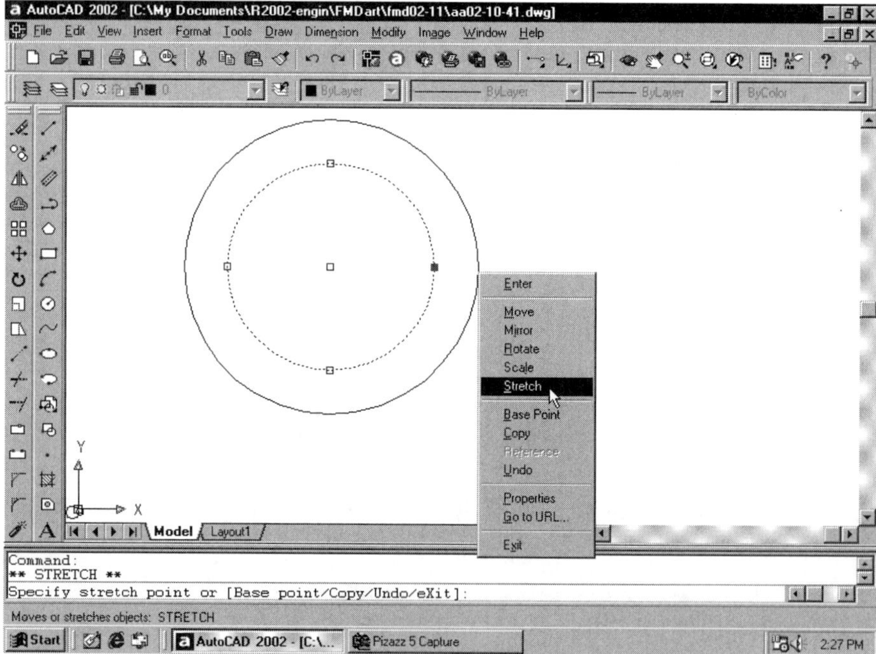

Figure 11.39 *Right-clicking produces this menu, which allows the grip edit option to be altered.*

SELECTING OBJECTS FOR MULTIPLE EDITS

Often an object or group of objects might need to be edited with several different editing commands. If you were arranging components in an assembly, a feature might need to be mirrored, moved, and rotated to form a new unit. Up to this point, you would have to select the objects to be mirrored, use the **Mirror** command, select the objects to be moved, use the **Move** command, select the objects to be rotated, and use the **Rotate** command. It works, but you'll never impress the boss as worthy of the big bucks. The **Select** command will allow an object or group of objects to be preselected and used for one or more editing functions. The command sequence to select is as follows:

Command: SELECT ENTER

Select objects: **W** ENTER *(Select objects using any of the selection methods.)*

First corner: Other corner: 100 found

Select objects: ENTER

Command:

The Command prompt is returned but no apparent changes have occurred. The objects that were selected have been grouped as a set and will be affected by any editing command. Start the **Mirror** command and when prompted to select objects, type P (ENTER) to automatically select the objects that were just placed in the set. The sequence to mirror, move, and rotate the objects is as follows:

Command: **MI** ENTER *(Or click the Mirror button.)*

MIRROR

Select objects: **P** ENTER

100 objects found

Select objects: ENTER

Specify first point of mirror line: *(Select first point.)*

Specify second point of mirror line: *(Select second point.)*

Delete source objects? [Yes/No]<N>: **Y** ENTER

Command: **M** ENTER *(Or click the Move button.)*

MOVE

Select objects: **P** ENTER

100 objects found

Select objects:ENTER

Specify base point or displacement or <use first point as displacement>: *(Select base point.)*

Specify second point of displacement or <use first point as displacement>: *(Select new location for base point.)*

Command: **RO** ENTER *(Or click the Rotate button.)*

ROTATE

Current positive angle in UCS: ANGDIR=counterclockwise ANGBASE=0

Select objects: **P** ENTER

100 objects found

Select objects: ENTER

Specify base point: *(Select base point.)*

Specify rotation angle or [Reference]: **45** ENTER

Command:

The **Group** command allows similar functions, but with more power.

Note: If you forget to use **Select** before the editing command, don't worry. The same results can be obtained by entering the commands in the following sequence: type MI (ENTER), select your objects, type RO (ENTER), and type P(ENTER) (for previous).

NAMING OBJECT GROUPS

While **Select** allows a group to be created, the **Group** command allows multiple groups to be created, named, and saved. For instance, the part that was just edited could be saved as a group called Base. It could be mirrored, and portions saved as a new group called Twinbase (a group that is part of another group is referred to as nested). The Twinbase group could also be mirrored and saved as a third group called Totalbase. Start the **Group** command by typing G (ENTER) at the Command prompt. This will produce the **Object Grouping** dialog box shown in Figure 11.40.

CREATING AND NAMING A GROUP

In Figure 11.40 you can see that this drawing has no named groups. To create a group, enter the desired name at the keyboard. As you type, the name will be entered in the **Group Name** box in the **Group Identification** area. Group names can be up to 255 characters long and include numbers, letters, or characters not used by Microsoft Windows or AutoCAD. Two separate words can't be used, but compound words can

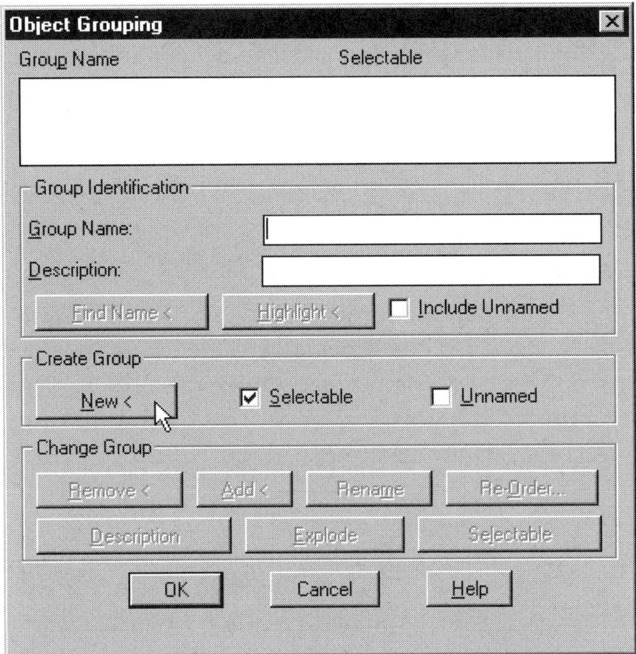

Figure 11.40 *Typing* **G** (ENTER) *at the command line will display the* **Object Grouping** *dialog box.* **Group** *can be used to select and save drawing objects in groups for editing.*

be joined by a dash or underline to create a name such as **base_unit.** As with the naming of drawing files, be sure to use names that will be easily recognized and that will clearly describe the contents of the group.

Description

A description of up to 64 characters can be created to help you remember the contents of a specific group. To create a description, click in the **Description** box and type a description.

New

Once the desired name and description have been entered, select the **New** button. This will remove the dialog box from the screen and prompt you to select objects. Select objects to be included in the group using any of the selection methods. When the selection process is complete, press ENTER to redisplay the dialog box. The group name will be listed in the **Group Name** box. As the group is created, the word *Yes* now displayed beside the group name indicates that it is selectable. If a group is selectable, selecting one object in the group selects the entire group. Once the group is created, the other areas of the dialog box can be used to alter or use the group.

IDENTIFYING A GROUP

Key elements of the **Group Identification** area can be seen in Figure 11.41. It is used to identify named groups.

Find Name

Selecting this button removes the dialog box from the screen and provides the prompt:

Pick a member of a group

When an object is selected, a **Group Member List** dialog box similar to the one shown in Figure 11.42 is displayed. This box shows all groups where the selected object is a member. Clicking the **OK** button returns the **Object Grouping** dialog box. (If you select an unnamed group, the **Add<** button in the **Change Group** area allows a name to be provided for an unnamed group.)

Highlight

The **Highlight** button is not active as the dialog box is displayed. Clicking one of the named groups activates the button. Clicking the **Highlight** button removes the dialog box from the screen and shows the named group in dashed lines as well as the

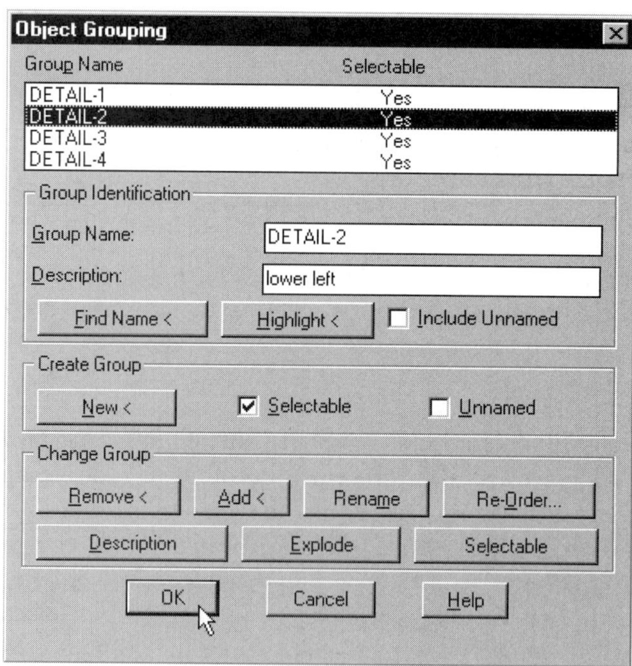

Figure 11.41 *Once groups are identified, selecting one of the groups will activate each portion of the dialog box.*

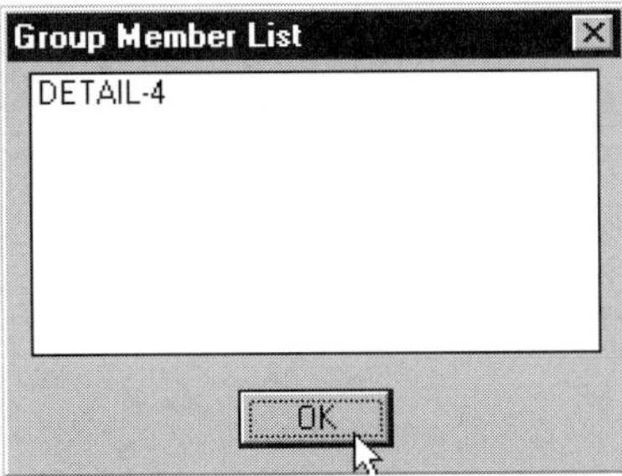

Figure 11.42 *Selecting* **Find Name** *will display the* **Group Member List** *box. Select an object on the drawing, and the group that it belongs to will be displayed.*

Continue button. This option is especially helpful on complicated drawings with multiple groups. Clicking the **Continue** button restores the **Object Grouping** dialog box.

Include Unnamed

Activating the **Include Unnamed** check box will display unnamed groups in the **Group Name** box. Unnamed groups can be renamed using the **Rename** button in the **Change Group** area.

CREATING A NEW GROUP

The key element of the **Create Group** area can be seen in Figure 11.41. The **New** box was covered previously in the section, "Creating and Naming a Group."

Selectable

Activating the **Selectable** check box determines the status of named groups as selectable. Remember that selecting one object in a named group selects the entire group.

Unnamed

If the **Unnamed** check box is active, AutoCAD will provide names for unnamed groups.

CHANGING A GROUP

Key elements of the **Change Group** area can be seen in Figure 11.41. All elements except **Re-Order** are inactive as the dialog box is displayed. Highlighting one of the named groups will activate all of the boxes in the **Change Group** area.

Add

The **Add** button allows drawing objects to be added to an existing group. Once this button is clicked, you will be prompted to select objects. If the selected object is part of an existing group, the group name and description boxes will display the current status of the object. Clicking the **OK** button completes the selection process.

Description

This **Description** button allows the description to be updated and saved. To alter a description, click in the **Description** box and type the desired material to be added. When you have finished, click the **Description** button to update the description and save it for future reference.

Explode

The **Explode** button is used to remove the selected group from the effects of **Group** selection. The objects that were in the group will not be altered, only the ability to select those objects as a group. Highlight the group to be removed in the **Group Name** list and then click the **Explode** button to explode the group.

Remove

Selection of the **Remove** button allows individual objects to be removed from the grouping. Highlight the group to be edited in the **Group Name** list and click **Remove** to display the Remove objects prompt. The dialog box will be removed and the objects in the group will be highlighted. Objects can be selected for removal using any of the selection methods. When finished removing objects from the set, press ENTER to return to the **Object Grouping** dialog box.

Rename

The **Rename** button can be used to rename an existing group. Highlight the name of an existing group so that the name and description are displayed in the **Group Identification** area. Edit the name as desired and click **Rename** to make the new name the current name.

Re-Order

Click the **Re-Order** button to produce the **Order Group** dialog box shown in Figure 11.43. Notice that the lower portion of the box includes three listings with numbered entries. Objects in a group are numbered in the same order that they were selected as the group was created. Re-ordering is typically used in CAD/CAM applications of manufacturing. Options include the following:

Group Name

Displays the name of the group to be re-ordered.

Description

This selection displays the description of the selected group.

Remove from position (0–x):

For each listing, x will vary depending on the size of the objects in the group. This selection specifies the current numbered position of the object to be reordered.

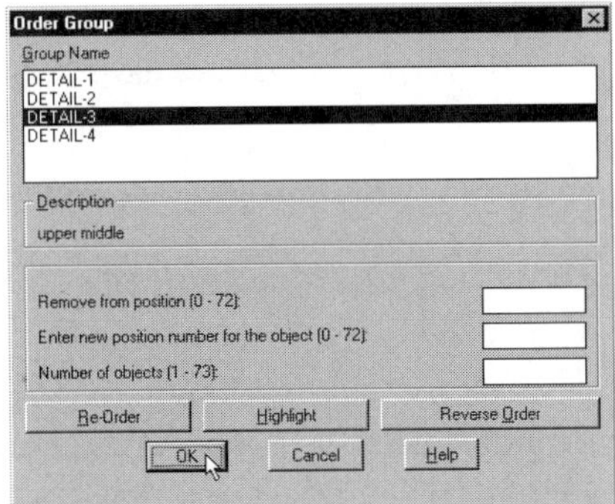

Figure 11.43 *The* **Order Group** *dialog box is displayed when the* **Re-Order** *button is selected. Re-ordering allows the order in which objects are selected for certain functions to be altered.*

Replace at position (0–x):

This selection specifies the new position the object will be moved to.

Number of objects (1–x):

This selection lists the range of numbers to be re-ordered.

CHAPTER 11 EXERCISES

1. Draw four sets of perpendicular crossing lines. On one set of lines, use the Fillet command to square the corners. On the second set of lines, provide a 2" fillet. On the third set of lines use a 1" chamfer. On the fourth set of lines use a 12" and a 24" chamfer. Save the drawing as E-11-1.
2. Open drawing E-10-5 and make a copy of the column cap. Adjust the base strap so that it is 3" longer and 1" wider. Add one additional bolt. Each end of the side plate must be 2" longer (4" total) and 1.5" higher. Save the drawing as E-11-2.
3. Open drawing E-8-1 and make three additional copies. In the original drawing, provide a 6" long line from each intersection and perpendicular to the opposite side of the triangle. Rotate the new triangles and place a point of each new triangle tangent to the existing triangle and perpendicular to the 6" lines. Draw a 10" diameter circle with a center point in the center of the original triangle. Trim all elements that extend beyond the circle. When complete, your drawing should resemble the attached drawing. Save the drawing as E-11-3.
4. Draw five concentric circles with the smallest diameter of 6" and the other sizes increasing in 3" increments. Remove a 1" wide strip going across each of the circles in both the vertical and horizontal directions. The center of the strip to be removed should be at the center of each quadrant. Rotate the

remaining portions of the second and fourth circles (starting at the inside) so that the remaining portions of these circles have been revolved 45° while remaining concentric. Save the drawing as E-11-4.

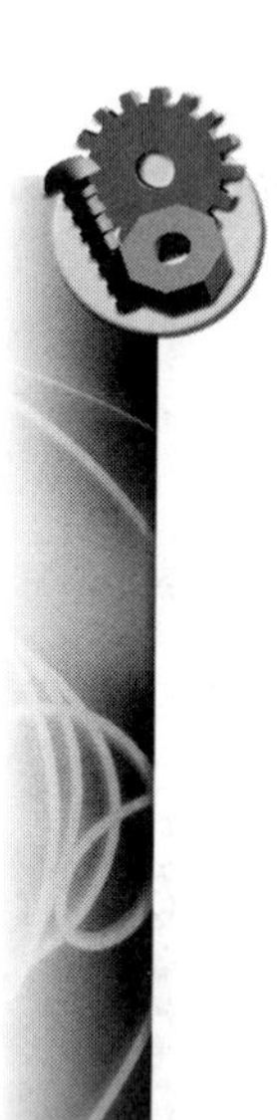

5. Draw concentric shapes of three, four, five, six, seven, and eight sides that are tangent to one circle. Make copies of these objects that are 1/4 size, 1/2 size, full size, 2× normal size, and 90% of the normal size. Save the drawing as E-11-5.
6. Draw a 3" diameter circle, a 6" line, and a polygon with eight sides that will fit in a 5.75" diameter circle. Save this portion of the drawing as E-11-6a. Make enough copies of each object and then use each option of each grips editing function to edit each object. Save the drawing as E-11-6b.
7. Use E-11-6a as a template drawing to complete this assignment. Make a second copy of these objects. Use **Select** to create a group consisting of the two circles and two polygons. Use **Mirror**, **Rotate**, and **Scale** to edit the group. Save the drawing as E-11-7.
8. Use drawing E-11-6a as a template drawing to complete this assignment. Make a second copy of these objects and then mirror both groups so that you have a total of four circles, lines and polygons. Use **Group** to name and save at least six different groups. Save the file as E-11-8.
9. Use the drawing below as a guide to draw three orthographic views. Do not place the dimensions on the drawing. Save the drawing as E-11-9.

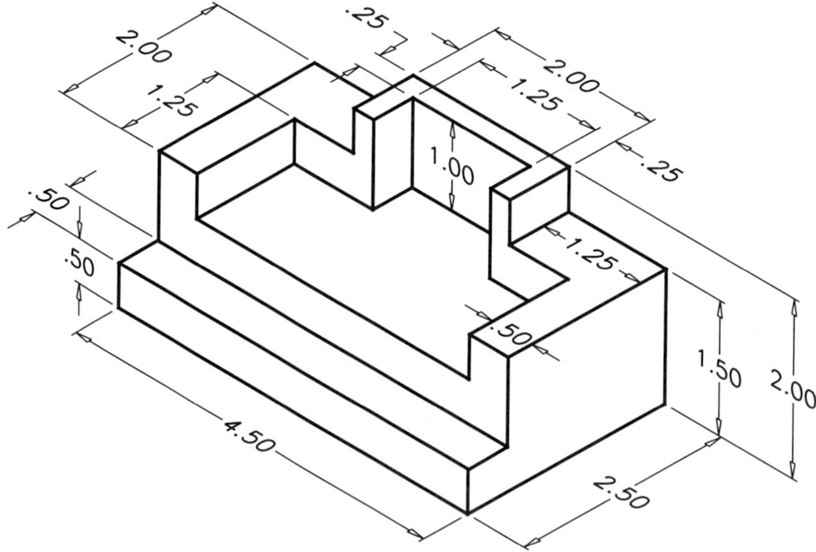

Problem 11-9

10. Open drawing E-8-8. Alter the length of the front view to be 2.25" long. Add the extra length to the right end of the part so that the 1.00" legs are altered. Reflect this change in the other views as necessary. Save the drawing as E-11-10.
11. Open drawing E-8-7. Alter the width of the rectangular portion of the upper flange from 1.25" to 1.75" without altering the arc. Reduce the size of the hole from .75" to .675". Move the outer edge of the lower flange from .50" to .375" from the front and rear edges. Reflect this change in the other views as necessary. Save the drawing as E-11-11.
12. Open drawing E-8-9. Alter the length of the front view to be 1.5 times larger than its current size. Add the length to the space between the two legs on the bottom, and the inclined edge of the top. Do not alter the inclined angle. Reflect this change in the other views as necessary. Save the drawing as E-11-12.
13. Open drawing E-7-18. Fillet the outer corner of the blade using a .375" radius. Alter the drawing to contain a total of four blades. Arrange four of the blades at 90° to each other so that the upper corner touches the adjacent blade. With the blades in position, alter the rectangle formed between each blade to be a .85" diameter circle. Rotate the entire fan blade around the center hole so that the blades are 45° off of horizontal or vertical. Save the drawing as E-11-13.
14. Open drawing E-7-25 and alter the height of the left side from .75" to 1.125". Add a .375" diameter hole in the center of the vertical plane on the right side of the part. Edit the circle to be a 1.00 long slot with .50" between the centerlines. Save the drawing as E-11-14.

Exercises 15 through 25: assume the grids are .50.

15. Use the drawing below as a guide to draw the template. Provide 14- .375" diameter circles, and 3-1.125" diameter holes. Save the drawing as E-11-15. Once the base drawing has been saved, alter the drawing so that it is 10.00" wide and 6.25" high. Provide a 1.00" × 60° chamfer at each lower corner. Save the drawing as E-11-15b.

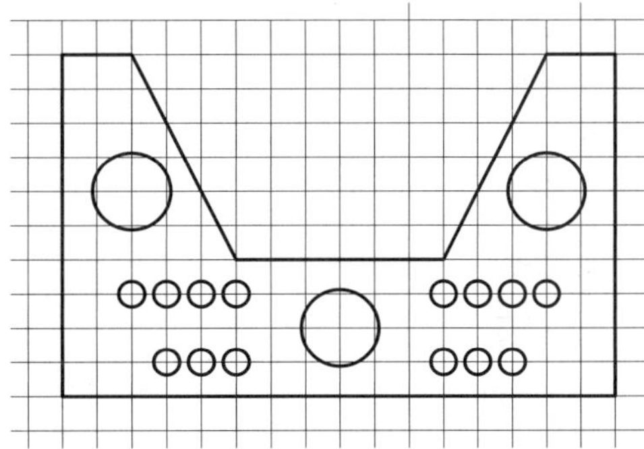

Problem 11-15

16. Use the drawing below as a guide to draw the template. Provide .875" diameter holes spaced at 15°. Save the drawing as E-11-16. Eliminate the rectangular edges and provide an arc that will be parallel to the interior arc with a 7.00" radius. Duplicate the finished part to form a circular part. Save the finished drawing as E-11-16b.

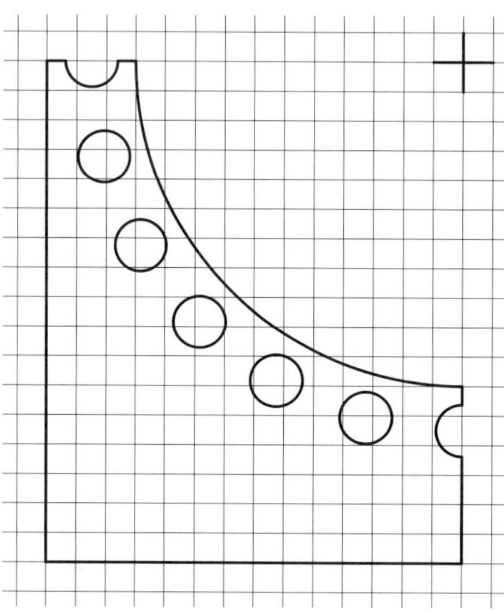

Problem 11-16

17. Use the drawing below as a guide to draw the template. All fillets are to be .375". The four corner lugs have a 1.50" diameter with a .75" diameter hole at the center. Provide 21- .50" diameter holes as indicated. Save the drawing as E-11-17. Edit the base drawing so that 1.00" is added to the length of each half of the drawing. Add the length so that the hole in the center of the part will be longer. Maintain the location of the .50" holes so that they continue to be at 1.00" on center. Edit the drawing so that .50" is removed from the height of each half of the drawing. Relocate the small vertical holes as needed to maintain there one inch spacing. Save the finished drawing as E-11-17b.

18. Use the drawing below as a guide to draw the template. Use a 2.00" diameter circle to draw the circle on the left. Provide a hexagon with 1.25" across the flats, centered in the circle. Use a circle with a radius of 1.25" to create the right end of the part. Provide a hexagon with 2.25" across the flats, centered in the circle. Use a radius of 6.00" to form the upper arc and a radius of 2.50" to form the lower arc. Save the drawing as E-11-18. Rotate the polygon on the left 36° from the horizontal centerline. Rotate the polygon on the right 30° from the vertical centerline. Using the center point of the right polygon, rotate the entire part –15°. Save the finished drawing as E-11-18b.

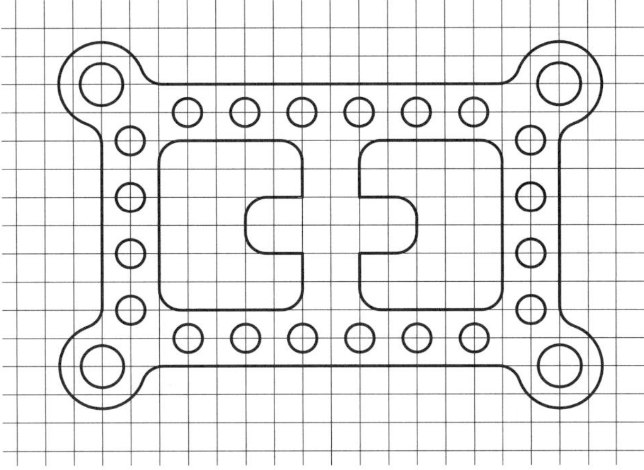

Problem 11-17

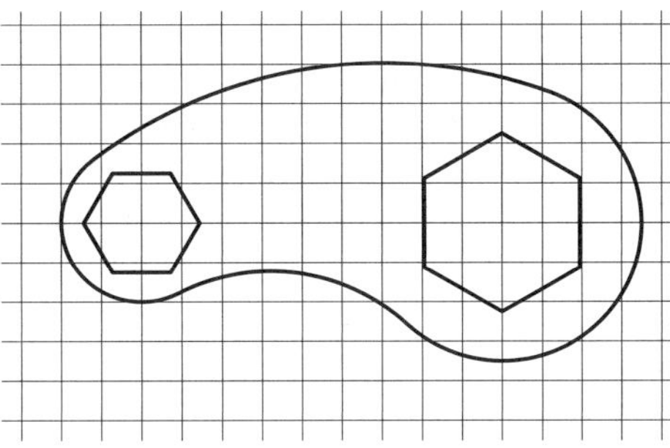

Problem 11-18

19. Use the drawing below as a guide to draw the template. Use an 8.50" diameter circle to establish the outside arc, and a 3.50 radius to establish the inside arc. Use a 1.125" radius to draw the circle around the octagon. The center of the circle is to be 1.75" down from the center point used to draw the two large diameters. The octagon is to be 1.50" across the points. Use a 1.00" diameter tangent circle to fillet the three circles. Draw a line that extends from the center point at 30° to provide an edge for the right side of the part. Draw a line that extends from the center point at 88° to provide an edge for the left side of the part. Save the drawing as E-11-19. Provide a second copy of the part so that it is 78% of the original size. Alter the left side angle from 88° to 30°. Save the finished drawing as E-11-19b.

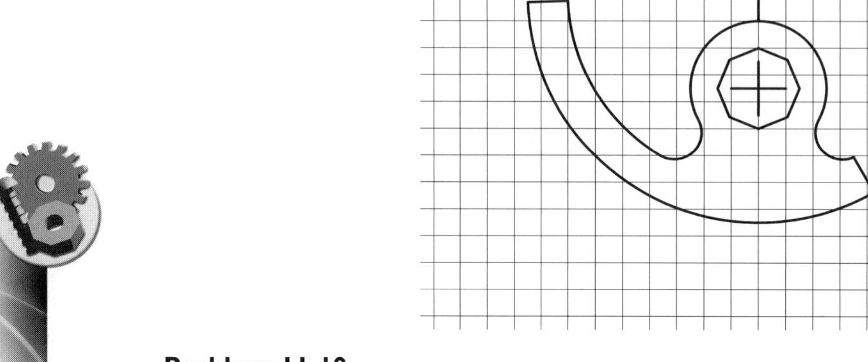

Problem 11-19

20. Use the drawing below as a guide to draw the template. Draw the ellipse with a major axis of 8.00" and a minor axis of 4.00". The exterior wall of the ellipse is to be .50" and intersections should have a .50" fillet radius. The sprocket teeth are to be .25" wide, with an outer diameter of 2.00" and an inner diameter of 1.50". Save the drawing as E-11-20. Use the center point of the sprocket and indicate the line of travel of the outer exterior edge of the part as it makes a complete revolution. Save the finished drawing as E-11-20b.

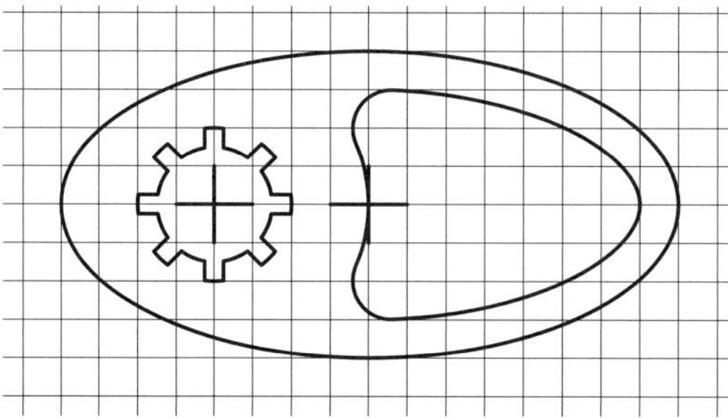

Problem 11-20

21. Use the drawing below as a guide to draw the template. Save the drawing as E-11-21. Alter the pattern so that all of the grids are changed from .50" wide to .375". Save the finished drawing as E-11-21b.

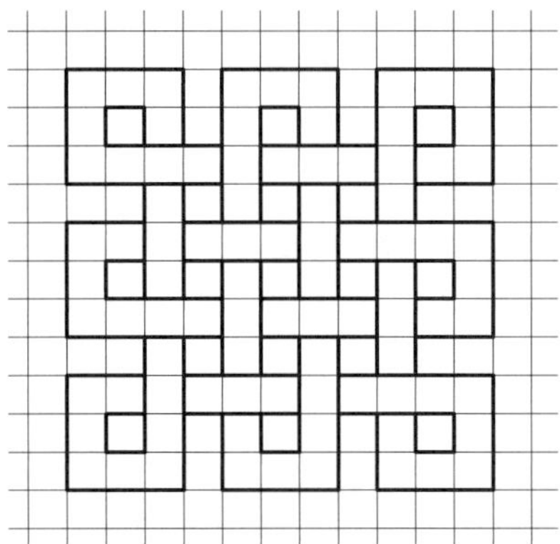

Problem 11-21

22. Use the drawing below as a guide to draw the gasket. The outer circle has a 4.00" diameter and the inner circle has a 1.50" radius. The bolt circle has a diameter of 3.75". The three lower arcs are based on .88" diameter circles. All of the smaller holes are .375" diameter. The two large holes have .75" radii. All of the straight edges are tangent to the adjacent circles. Save the drawing as E-11-22.

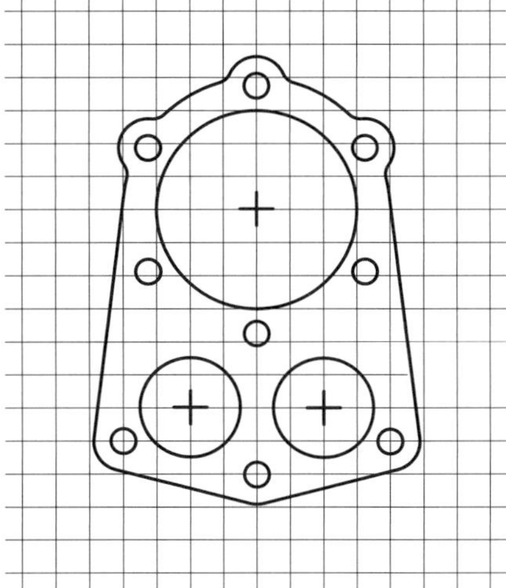

Problem 11-22

23. Use the drawing below as a guide to draw the template. Draw three equilateral triangles with a base of 2.50". Provide a space of .25" between each triangle. Use a width of .25" for each triangle wall. Save the drawing as E-11-23. Use the finished triangle to create a triangle comprised of three additional triangles with an 11.00" long base. Reduce the finished drawing so that the base is 9.50" long. Save the finished drawing as E-11-23b.

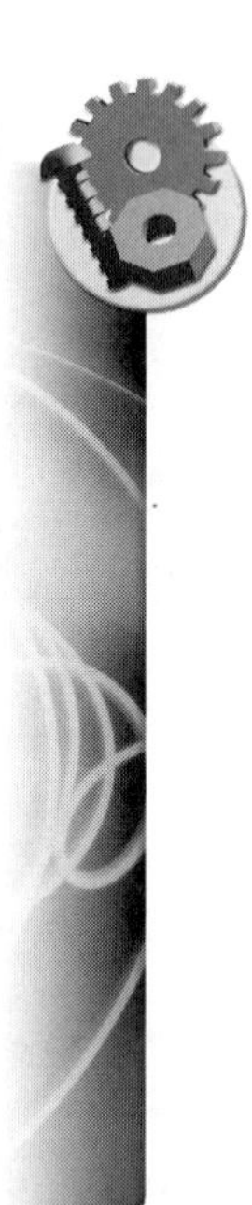

Problem 11-23

24. Open drawing E-10-8. Alter the ends to be radial, based on the current width. Shorten the inner slot by removing .50" from the left and right ends. Add 3-1.50" diameter equally spaced holes on a bolt circle that matches the edge distance of the existing holes. Rotate the entire part so that the right end is 28° above horizontal. Alter the left end of the part so that one hole is eliminated from each side, and remove 2.00" from the left side of the centerline. Alter the size of the part so that it is 80% of the original size. Save the drawing as E-11-24.
25. Open drawing E-10-9. Use the outer edge of the part to form the new outer diameter of the part and eliminate the six existing circular flanges. Divide the part along the horizontal centerline and add 1.30" to each side of the centerline so that the part is no longer circular. Add one large and one small hole centered on each side. Rotate the part 90° Use this top view to create front and side views. Save the drawing as E-11-25.

Exercises 11-26 through 11-35: use the attached isometric drawings as a guide to complete a minimum of three views for each project. Where required to clearly show all features, provide an auxiliary view. Do not place dimensions on the drawings. Save the drawing using the appropriate exercise number. Alter exercise 34 by disregarding the 63.5 mm width and using the 64 mm width for each leg. Alter exercise 35 by locating the top hole in the right leg to be .438" from the top of the inclined surface.

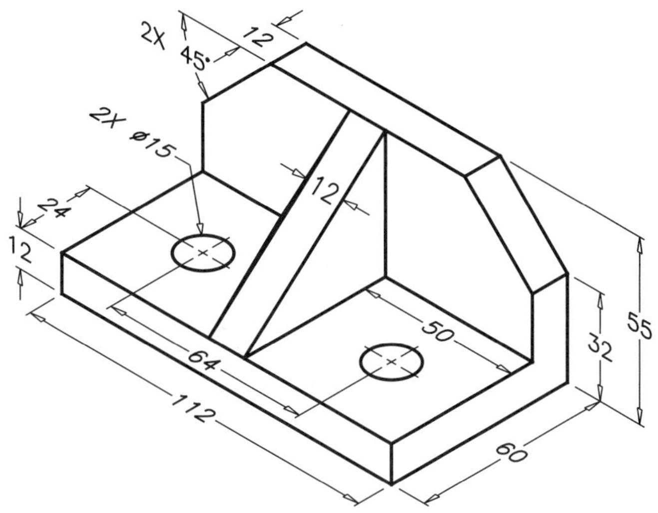

Problem 11-26

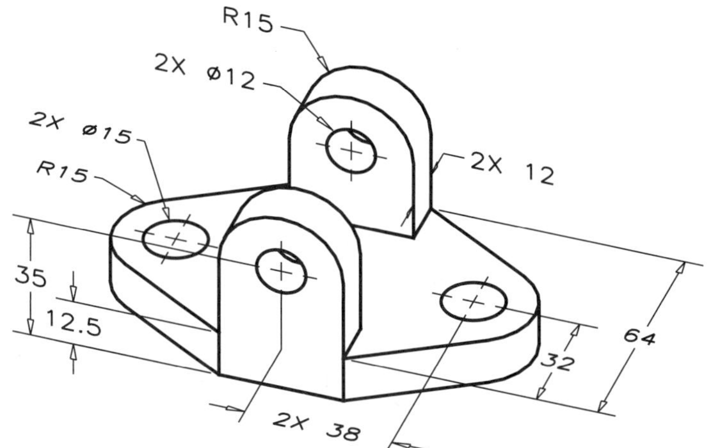

Problem 11-27

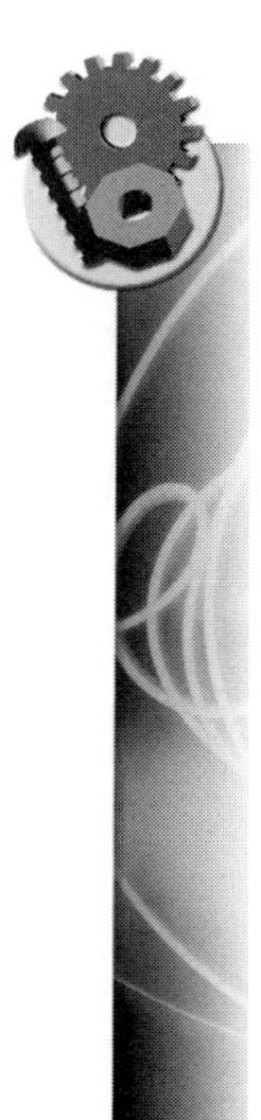

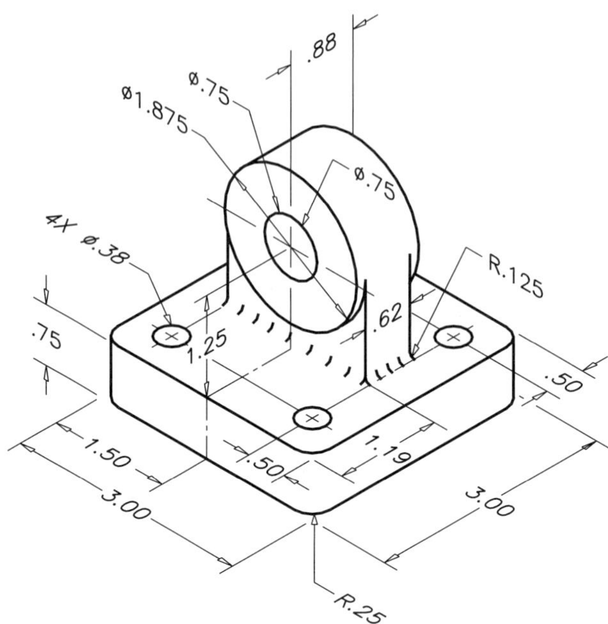

Problem 11-28

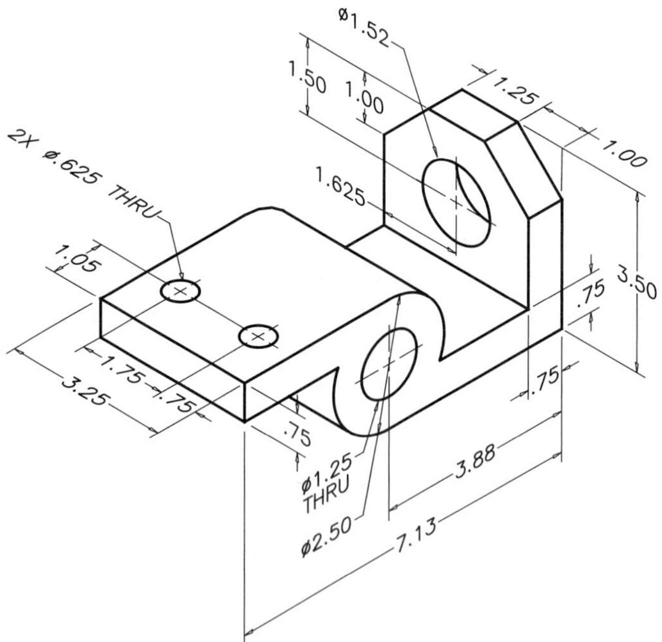

Problem 11-29

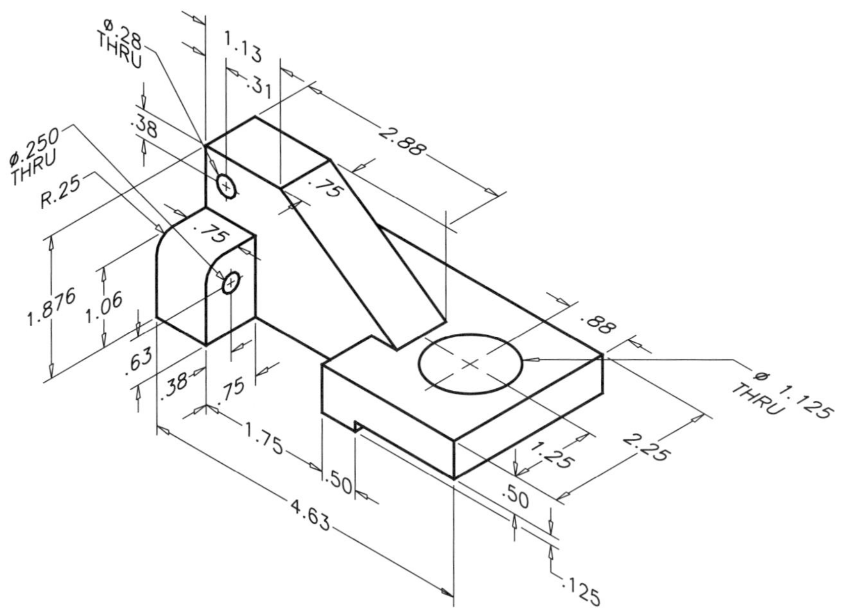

Problem 11-30

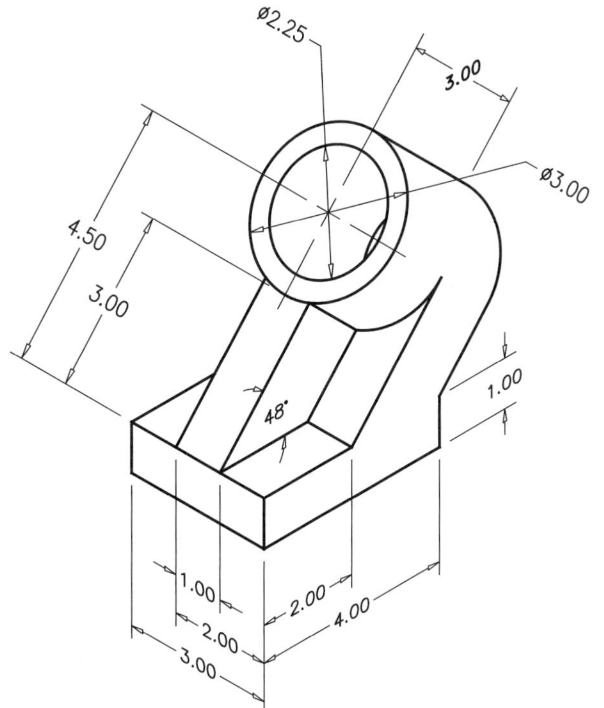

Problem 11-31

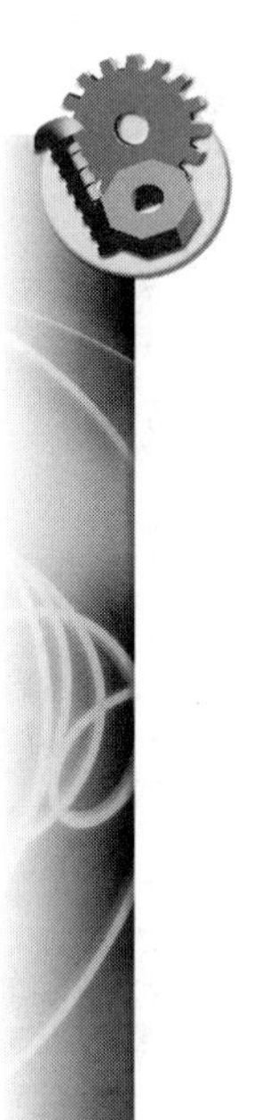

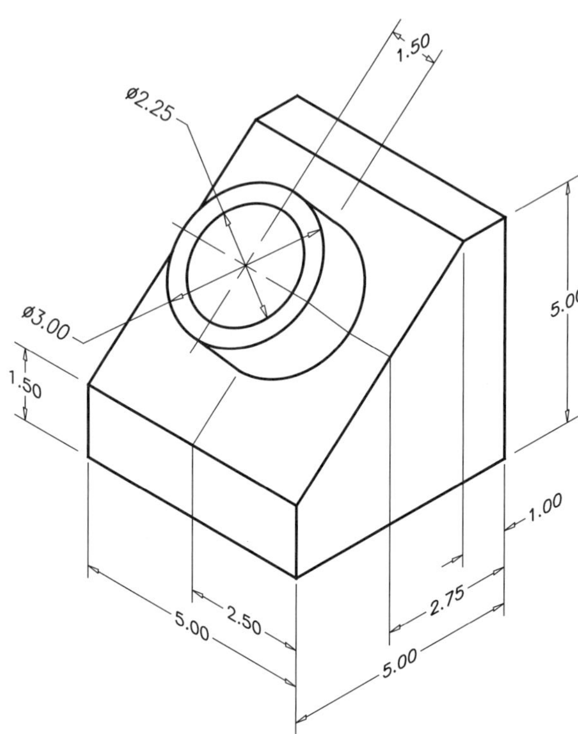

Problem 11-32

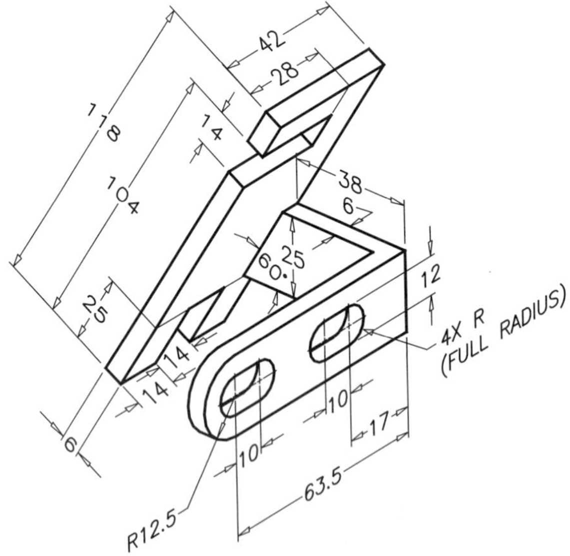

Problem 11-33

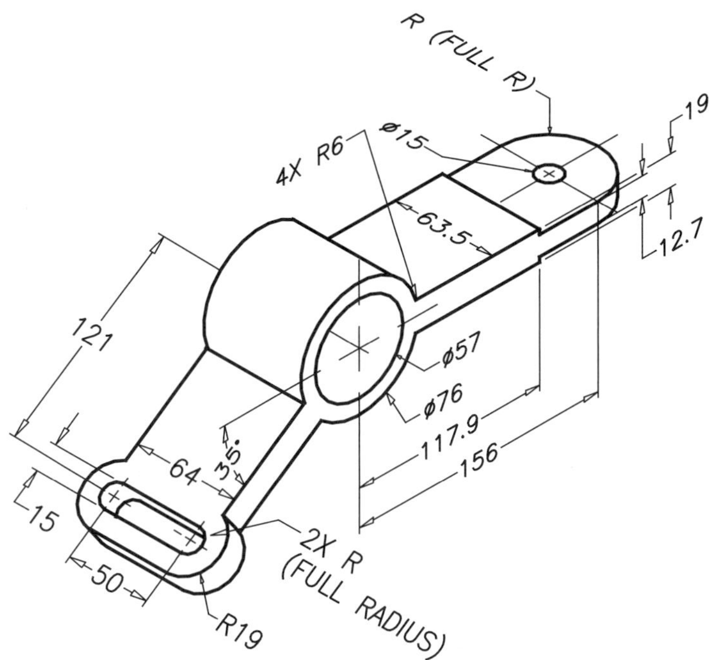

Problem 11-34

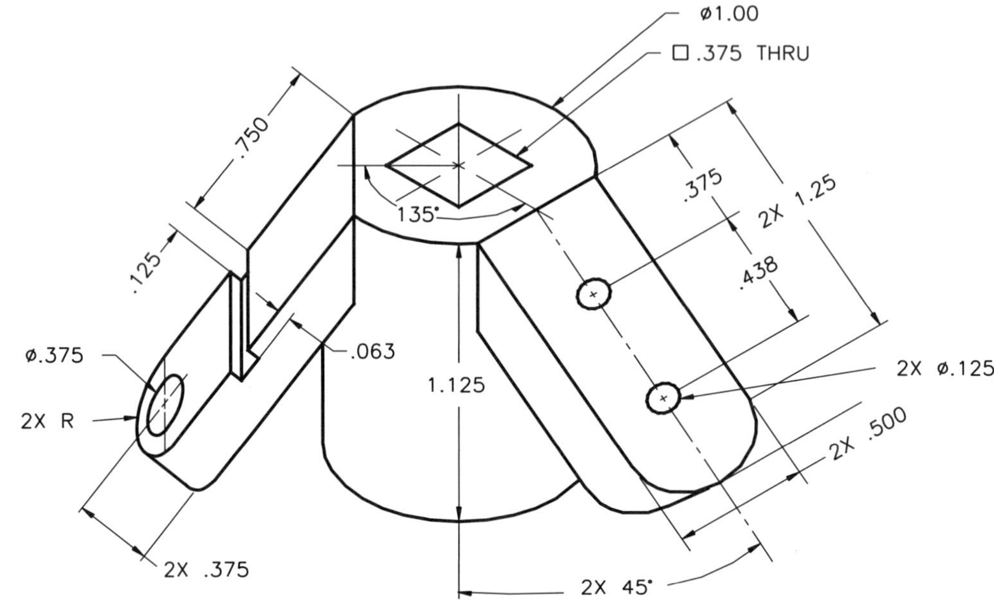

Problem 11-35

CHAPTER 11 QUIZ

1. Explain the uses of the **Break** command.
2. Describe two different uses of the **Fillet** command.
3. List the command sequence for **Rotate**.
4. Explain the difference between **Move** and **Copy**.
5. List the command sequence to **Extend** a line to meet another line and sketch and label the process.
6. Two perpendicular lines cross at a corner. List four methods to edit the lines to form a non-crossing corner.
7. A line has been drawn short of its required destination. How can the error be corrected?
8. What two options are available for offsetting an object?
9. In the **Chamfer** command, what is the relationship between the distances and the selection of lines?
10. You are preparing to **Move** some objects, so you type **P** at the Select objects prompt. What will be moved?
11. Use the **Help** menu to determine what **Setvar** controls a preset **Fillet** value.
12. How can an object be rotated in a clockwise direction?
13. An inclined plane is drawn at a 45° angle. It should be 49°. How can this error be corrected?
14. Six objects need to be moved. As you select the objects, an extra object is included in the selection set, and an intended object is not included. Describe the command process and prompts to move these objects.
15. List and describe the three **Break** options.
16. Describe how each of the editing commands presented in this chapter could be used in drawing a multiview drawing.
17. What will happen if you try to place a 60" radius fillet in two intersecting lines that are 48" long?
18. Explain how **Chamfer** will affect two lines that do not touch if they are selected as the objects of this command.
19. What will be the effect of the Trim option of **Chamfer** on two intersecting lines?
20. What will be the effect of the **Trim Edge** option on two lines that do not intersect?
21. How will the Edge option of **Extend** affect two lines?
22. How will Window and Crossing window affect objects selected for **Stretch**?

23. What are the options that are given if the DElta option of **Lengthen** is selected?
24. A 6" long line is selected to be edited using **Lengthen**. What will be the result of entering 50%? 150%? 200%?
25. How many selection points will be required when you break an arc using **Break @**?
26. Provide an example of how the **Align** command could save time.
27. List the commands used to create E-11-9.
28. What must be done to display grips on a line?
29. What five editing functions can be executed after a grip is hot?
30. What is a hot grip and how do you create one?
31. You're exploring options and select **Gripblock**. What will this option do and what can you do if you don't want this option?
32. What advantage does **Select** have over grips?
33. Describe the process of selecting and naming a group of objects using **Group**.
34. What is a nested group?
35. What is a selectable group?
36. You've named a group, but you come to realize the name does not describe the drawing well. How do you fix the problem?
37. Save a group of objects without providing a name. How does AutoCAD keep track of this group?
38. You're exploring again and you use the Explode option of **Group**. What effect will this have on your computer? What effect will it have on the current drawing? What effect will it have on the selected group?
39. How do you make the options in the **Change Group** area of the **Object Grouping** dialog box active?
40. Use the **Help** menu and determine what variables affect grips. What are the options and their effects?

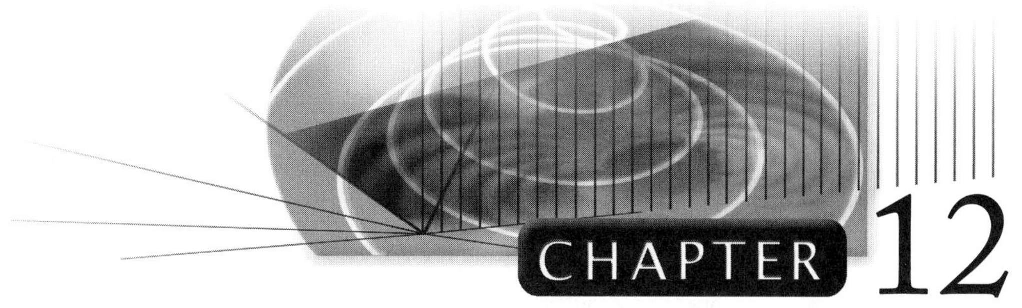

Supplemental Drawing Commands

This chapter will introduce methods for marking and dividing drawing space and explore five new methods for drawing lines.

Commands to be introduced in this chapter include

- **Point**
- **Divide**
- **Measure**
- **Ray**
- **Xline**
- **Sketch**
- **U**
- **Undo**
- **Redo**

MARKING AND DIVIDING SPACE

Three commands are available in AutoCAD for marking and dividing space. Each command can be used to mark space in a drawing and provide a location for placing lines.

PLACING POINTS

Often, a point needs to be marked on a drawing. AutoCAD will allow you to mark these load locations on a drawing by drawing a point. To place a point in a specific location, select the **Point** button on the **Draw** toolbar, type PO (ENTER) at the Command prompt, or select **Point** from the **Draw** menu. If the command is selected from the **Draw** menu, the command will allow for multiple points to be located. Each method will produce a prompt:

Command: **PO** ENTER *(Or click the Point button on the Draw toolbar.)*

POINT

Current point modes PDMODE=0 PDSIZE=–0.00

Specify a point: *(Select a point location or enter coordinates.)*

The prompt is now waiting for coordinates to be specified. Points are usually specified with the select button.

Controlling the Point Display

The shape that is used to identify a point can be altered through the **Point Style** dialog box, shown in Figure 12.1. Display the dialog box by selecting **Point Style** from the **Format** menu or by typing **DDPTYPE** (ENTER) at the Command prompt. The dialog box allows the style and size of the point to be controlled. The active setting is shown in a black box. In this case, a point is used for the marker. The marker will be displayed on the screen but will not be plotted. Notice the second box from the left in the top row contains no marker. If this option is selected, the point will be marked with an invisible marker. Although the point can't be seen, it will be selected when the Node mode of Object Snap is active.

The current point size is listed as 5% of the display screen size. Point size can be altered to meet the needs of the drawing. Just as important as the size is the display method. In the default setting, the size of the point will be displayed relative to the size of the display area. As you zoom in and out of a drawing, points will remain a set percent-

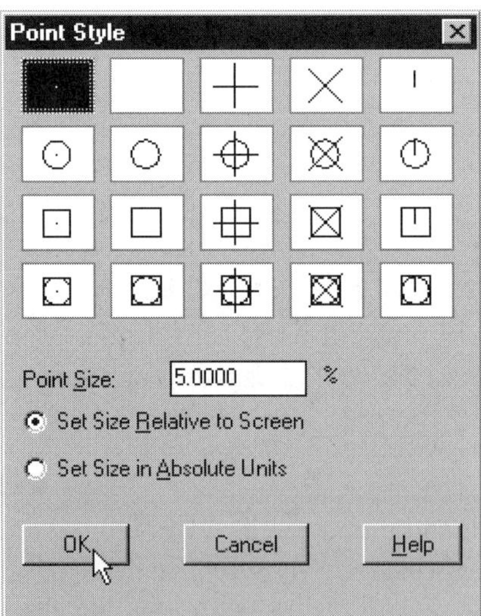

Figure 12.1 *Selecting* **Point Style** *from the* **Format** *menu or typing* **DDPTYPE** *(*ENTER*) at the Command prompt will produce the* **Point Style** *dialog box. The box can be used to alter the size and shape of the point marker.*

age of the drawing display size. If you select the display to be in absolute units, points will disappear as you zoom out and become huge as you zoom in. Entering a positive value for the point size will display the point relative to the drawing area. Use the following procedure to display a point that is marked by crosshairs surrounded by a circle and a square, and is displayed at 10% of the screen size:

1. Select **Point Style** from the **Format** menu.
2. Select a marker from the **Point Style** dialog box.
3. Change the display in the **Point Size** edit box to read 10%.
4. Click the **OK** button.

Note: You'll notice as you work through this chapter that several of the buttons noted by commands are not found on the standard toolbars. You can find these command buttons by selecting **Customize** from the **Toolbars** menu. You can also right-click when the cursor is in any open toolbar, and then select **Customize**. Select the **Categories** edit box on the **Commands** tab of the **Customize** dialog box to display a listing of command categories. To display the **Point** button, select **Draw** from the **Categories** edit box. Review Chapter 1 and the **Help** menu to place buttons on a toolbar.

EXPLORING THE DIVIDE COMMAND

The **Divide** command allows an object to be divided into any number of segments of equal length for easy editing. **Divide** will enable you to select points on a line, arc, circle, or polyline. Before using the **Divide** command to mark an object, you must decide how each mark will be indicated. The default marker for a point is a single point. The point style must be altered so that it can be seen on the line. The crosshairs, the X, and the crosshairs and circle, selected in the **Point Style** dialog box, make excellent markers (see Figure 12.1).

Once the marker is set, start the command sequence for **Divide** by typing DIV (ENTER) at the Command prompt, or by selecting **Divide** from the **Point** cascading menu on the **Draw** menu. The command sequence is as follows:

Command: **DIV** ENTER *(Or click the Divide button.)*

DIVIDE

Select object to divide: *(Select the objects.)*

This is your opportunity to select a single object to divide. If you select an object that cannot be divided, an error message will be displayed and the command will be exited. Once an object is selected, another prompt will be displayed.

Enter the number of segments or [Block]: *(Enter number.)* **14** ENTER

Typing a number between 2 and 32,767 and pressing ENTER will cause the object to be divided into the number of segments specified. The object is not physically divided into separate segments—the markers are just placed so that exact points can be selected. The process can be seen in Figure 12.2.

Another useful feature of the **Divide** command is its ability to insert a group of objects called a *block* at repeated distances. Blocks will be covered in a later chapter, but the idea of inserting them is not difficult to remember. You can insert blocks by dividing a line into the desired number of segments and then inserting the block at each marker. This can be done if the Block option is selected. Figure 12.3 shows the effect of aligning blocks with the object. The command sequence to divide a circle and place a block named SLOT is as follows:

> Enter the number of segments or [Block]: **B** ENTER
>
> Enter name of block insert: *(Type name of desired block.)* **SLOT** ENTER
>
> Align block with object? [Yes/No]<Y>: ENTER
>
> Number of segments: *(Enter number.)* **14** ENTER
>
> Command:

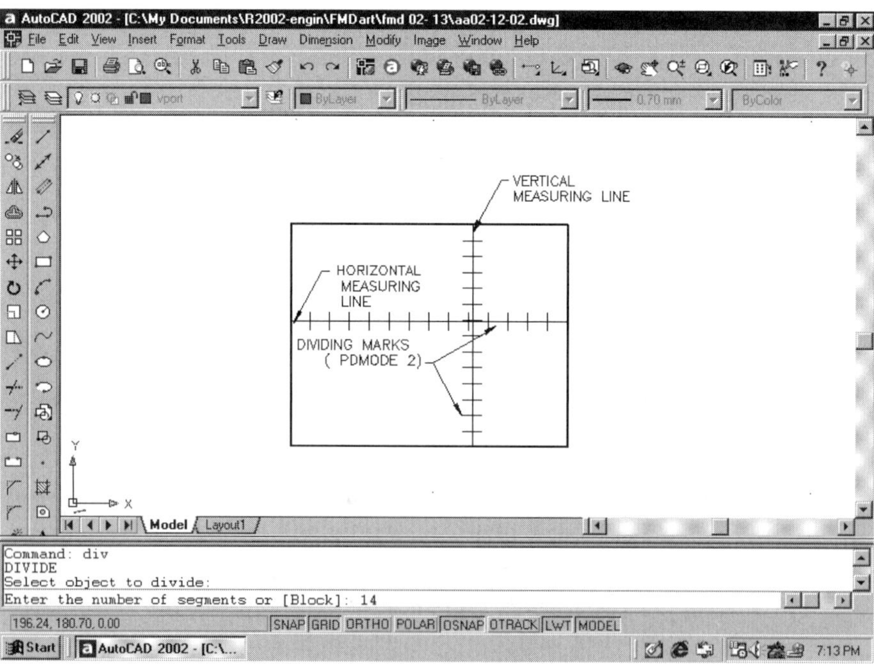

Figure 12.2 *Dividing a horizontal line into fourteen segments.*

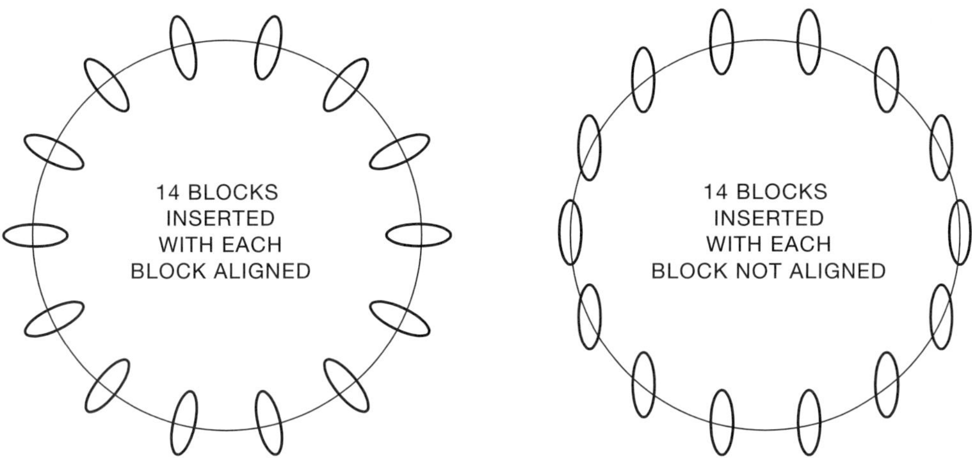

Figure 12.3 *Predetermined drawing components can be inserted in a drawing at each specified* **Divide** *marker using the Block option of* **Divide**.

EXPLORING THE MEASURE COMMAND

The **Measure** command is similar to the **Divide** command. **Divide** places a specified number of markers at equal distances along a segment. **Measure** places markers at specific distances along a line, arc, circle, or polyline. This command can be used whenever objects occur at a repetitive spacing, such as when you are placing objects at every 2.375" along a line. Start the command by typing ME (ENTER) at the Command prompt, or by selecting **Measure** from the **Point** cascading menu of the **Draw** menu. Be sure to change the point marker to something that will be visible when placed on a line. The command sequence to measure an object is as follows:

> Command: **ME** ENTER *(Or click the Measure button.)*
>
> MEASURE
>
> Select object to measure: *(Select object.)*

This is your opportunity to select a single object to measure. If you select an object that cannot be measured, an error message will be displayed and the command will be exited. Once an object is selected, another prompt will be displayed:

> Specify length of segment or [Block]: *(Enter distance.)* **2.375** ENTER
>
> Command:

You can enter any specific distance by keyboard or specify two points with the select button. The markers will be placed, starting at the selected end of the line of the object to be measured. This process can be seen in Figure 12.4.

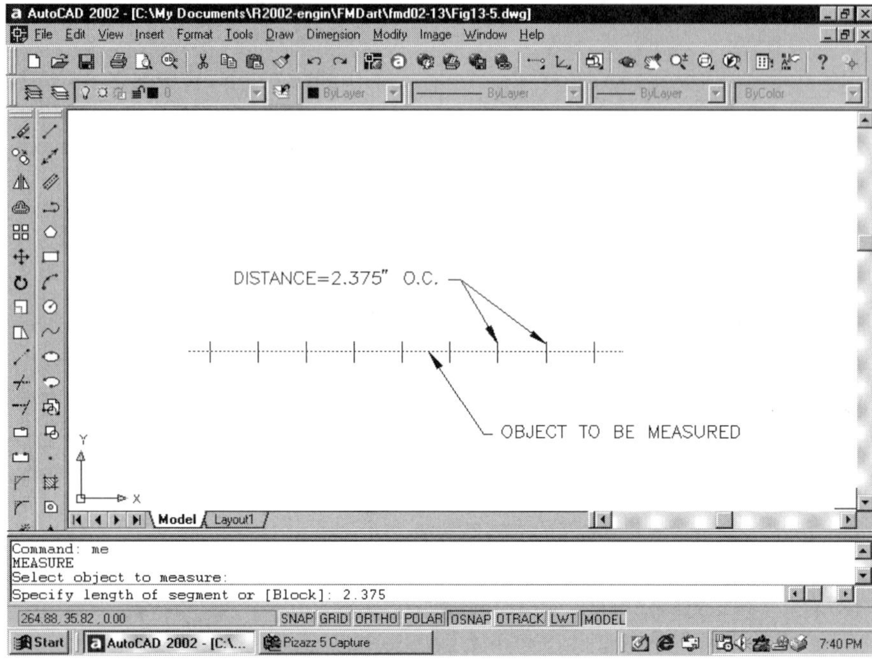

Figure 12.4 *Once a segment has been selected for measurement, the desired distance to be measured can be entered by keyboard or by selecting the distance with the select button.*

To use the select button to enter the distances, move the cursor to the desired location and click the select button. This will produce the following command sequence:

Command: **ME** ENTER *(Or click the Measure button.)*

MEASURE

Select object to measure: *(Select object.)*

Specify length of segment or [Block]: *(Select a location.)*

Second point: *(Enter a second point or coordinates.)* ENTER

Command:

DRAWING LINES OF VARIED WIDTH AND LENGTH

Each command introduced in this section can be used as an alternative to the **Line** command The **Ray** and **Xline** (**Construction Line**) commands allow semi-infinite and infinite lines to be added to a drawing. The **Sketch** command allows freehand sketching to be inserted into a drawing.

CREATING LINES USING THE RAY COMMAND

A ray is a semi-infinite line with a defined starting point that extends into infinity without changing the drawing area. Sounds technical, but they are really very helpful. Rays can be especially helpful in projecting features from one drawing view to another. See Figure 12.5. If you were to use a line to project from view to another it must have two endpoints. A zoom might be required to go from the "first" point to the "next" point. A transparent zoom ('Z) could be used, but this is time consuming when compared to using a ray. A ray will provide a guideline that extends from the selected point indefinitely, no matter how many **Zoom Previous** operations are required.

Start the **Ray** command by choosing **Ray** from the **Draw** menu, or by typing RAY (ENTER) at the Command prompt. The command sequence to draw a ray is as follows:

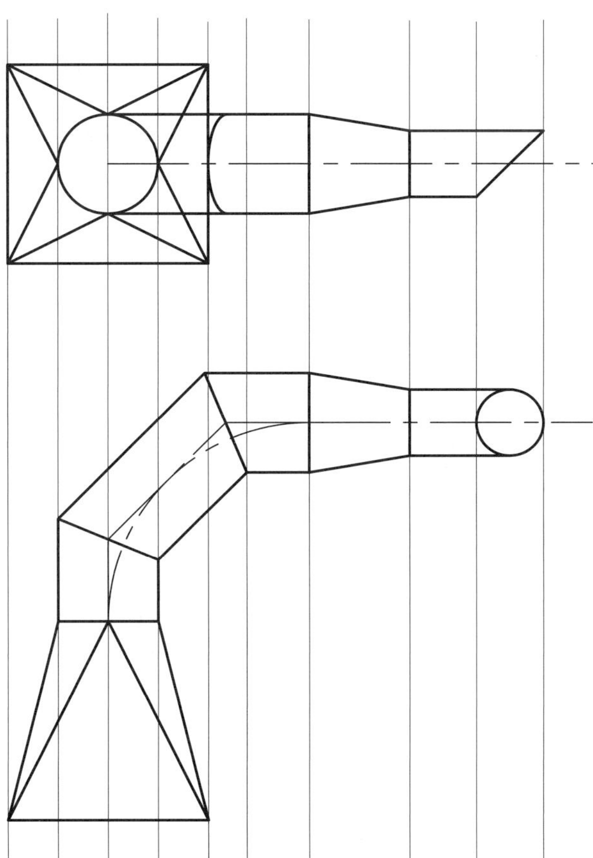

Figure 12.5 *Rays and xlines make excellent projection lines. Because of their length,* **Zoom** *and* **Stretch** *operations can be minimized. Courtesy Paul Bissell.*

Command: **RAY** ENTER *(Or click the Ray button.)*

RAY

Specify start point: *(Select desired starting point of ray.)*

Specify through point: *(Select desired direction of ray.)*

Specify through point: ENTER

Command:

A ray will show when the drawing is plotted. Rays should be kept on a layer separate from the drawing being created, so they can be easily removed from the drawing base by freezing the layer or by selecting the Do Not Plot option for that layer. They can also be erased easily using the Crossing window or Fence option of **Erase**.

CREATING LINES USING THE XLINE COMMAND

A line created using the **Xline** (**Construction** Line) command is similar to a ray, except that it is infinite in length. With **Ray**, you selected a starting point, and the line is extended from that point. With **Xline**, the line extends through the point an infinite distance in both directions and offers more options than with a ray. Access the command by selecting the **Construction Line** button on the **Draw** toolbar, by typing XL (ENTER) at the Command prompt, or by selecting **Construction Line** from the **Draw** menu. Unlike the **Ray** command, **Xline** offers six options to place the line. The default setting is to select a point for the line to pass through.

Through Point

By selecting a through point, you can display the line through a specified point. Once the point is selected, the line will be rotated to the desired angle through the selected point. Once a line location is selected, the command sequence will allow for additional lines to be drawn through the point. The command sequence is as follows:

Command: **XL** ENTER *(Or click the Construction Line button.)*

XLINE

Specify a point or [Hor/Ver/Ang/Bisect/Offset]: *(Select desired option.)*

Specify through point: *(Select a direction for the line to extend.)*

Specify through point: ENTER

Command:

Hor Option

Choosing the Hor option will allow a horizontal line to be drawn through a specified point. The line will remain horizontal, regardless of the ORTHO setting. The command sequence is as follows:

Command: **XL** ENTER *(Or click the Construction Line button.)*

XLINE

Specify a point or [Hor/Ver/Ang/Bisect/Offset]: **H** ENTER

Specify through point: *(Select the point for the line to pass through.)*

Specify through point: ENTERr

Command:

Ver Option

The Ver option will pass a vertical line through the selected point. The command functions in the same way as the horizontal option, once the Ver option is selected. The command sequence is as follows:

Command: **XL** ENTER *(Or click the Construction Line button.)*

XLINE

Specify a point or [Hor/Ver/Ang/Bisect/Offset]: **V** ENTER

Specify through point: *(Select the point for the line to pass through.)*

Specify through point: ENTER

Command:

Ang Option

The Ang option allows a specific angle for the line to be specified. The selected angle is relative to horizontal. The angle option can be useful in laying out roofs or other inclined surfaces. The command sequence is as follows:

Command: **XL** ENTER *(Or click the Construction Line button.)*

XLINE

Specify a point or [Hor/Ver/Ang/Bisect/Offset]: **A** ENTER

Enter angle of xline (0) or [Reference]: **30** ENTER

Specify through point: *(Select the point for the line to pass through.)*

Specify through point: ENTER

Command:

If the Reference option of the command is selected, a line can be drawn at an angle to any line. Once the reference angle is selected, a location for the line to pass through needs to be indicated. The command sequence is as follows:

Command: **XL** ENTER *(Or click the Construction Line button.)*

XLINE

Specify a point or [Hor/Ver/Ang/Bisect/Offset]: **A** ENTER

Specify angle of xline (0) or [Reference]: **R** ENTER

Select a line object: *(Select the line to be used as the reference.)*

Enter angle of xline <0>: **30** ENTER

Specify through point: *(Select the point for the line to pass through.)*

Specify through point: ENTER

Command:

Bisect Option

The Bisect option can be used to draw a construction line that bisects an existing angle. The line is created by selecting three points. The first point to be selected is the Angle vertex point. The second and third points are located on the lines forming the angle to be bisected. Figure 12.6 shows a construction line used to bisect an angle. The command sequence to bisect an angle is as follows:

Command: **XL** ENTER *(Or click the Construction Line button.)*

XLINE

Specify a point or [Hor/Ver/Ang/Bisect/Offset]: **B** ENTER

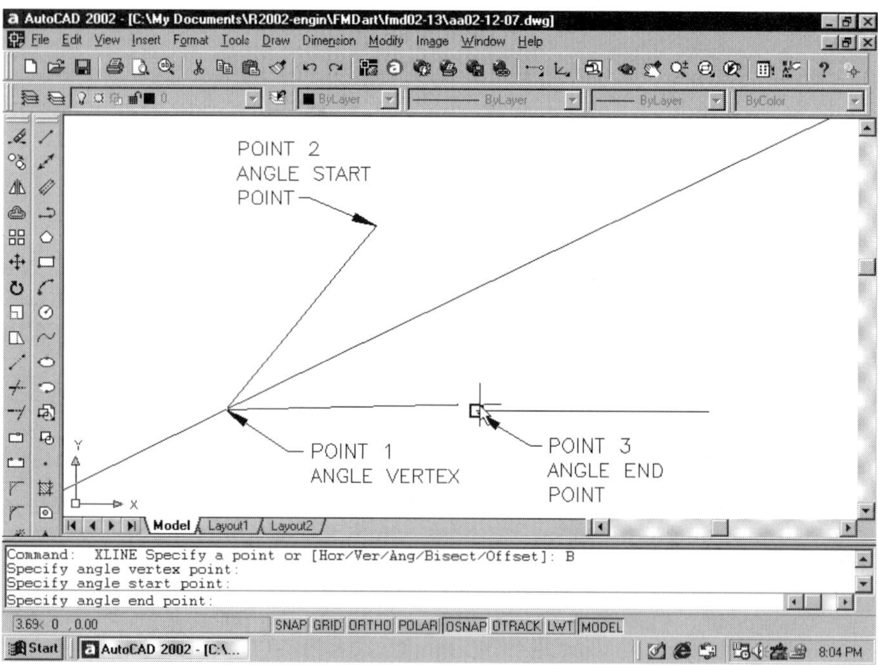

Figure 12.6 *The Bisect option of* **Xline** *can be used to divide an angle in half, by selecting three points.*

Specify angle vertex point: ENTER *(Select the point and press* ENTER.*)*

Specify angle start point: *(Select the end of one of the lines to be bisected.)*

Specify angle end point: *(Select the end of the other line to be bisected.)*

Specify angle end point: ENTER

Command:

Offset Option

The Offset option draws construction lines that are parallel to a selected line, based on a specified offset distance or point. This option functions like the **Offset** command. You'll be prompted for a line, a distance, and a side to offset. Select an offset distance, and then you will be prompted for a line to be offset and for a through point. The command sequence to offset a line 12" is as follows:

Command: **XL** ENTER *(Or click the Construction Line button.)*

XLINE

Specify a point or [Hor/Ver/Ang/Bisect/Offset]: **O** ENTER

Specify offset distance or [Through] <Through>: **12** ENTER

Select a line object: *(Select the line to be offset.)*

Specify side to offset: *(Select the side of the offset.)*

Select a line object: ENTER

Command:

CREATING LINES WITH THE SKETCH COMMAND

Using the **Sketch** command allows the equivalent of freehand drawings to be generated using a computer. Line segments are entered as the cursor is moved, rather than providing a "next" point. The command is useful for entering signatures or other irregular material on a drawing. Because each tiny segment of the sketch is recorded as a line segment, drawing file size can become extremely large in a very short time. The use of **Sketch** should be kept to a minimum to avoid filling up drawing space on a disk. The size of the drawing is limited only by your storage media.

Access the **Sketch** command by selecting the **Sketch** button or typing SKETCH (ENTER) at the Command prompt. Best results usually occur when ORTHO and SNAP are toggled to OFF, allowing greatest flexibility in line placement. With ORTHO mode ON, the **Sketch** command will draw only horizontal and vertical lines. With the SNAP mode on, the record increment (segment length) will equal the snap value.

Segment Length

As the command is entered, you will be prompted to provide information on the *record increment*, which is the segment length to be used while you are sketching. The increment should be set small enough to provide the needed resolution. Exceeding the resolution will only result in wasted disk space by storing lines that aren't necessary. The default line length is <0.10">. A new value can be entered by keyboard.

The increment that is chosen should be determined by the size of the reproduced drawing. If the drawing is to be reproduced at full scale on an 8 1/2" × 11" sheet of paper, .125" will not produce high-quality resolution. As the mouse is moved, a line segment will be recorded for each movement greater than .10". In a 10" wide space, 100 segments could be created. By changing the length to .5", a 10" long line could be generated with 20 segments in the same space, which would greatly decrease the resolution. However, if the drawing is going to be reproduced at a scale, a line segment of .5" would give extremely high resolution. A 1" long line segment might be more appropriate. Figure 12.7 compares four record increments. You can specify the width by providing a value by keyboard or by moving the cursor. To enter a value using the keyboard, type the value followed by ENTER. The command sequence is as follows:

> Command: **SKETCH** ENTER *(Or click the Sketch button.)*
>
> Record increment <0.10>: **.065** ENTER
>
> Sketch. Pen eXit Quit Record Erase Connect.

AutoCAD is now ready to sketch. To set the line increment using the cursor, click the select button when the prompt for the increment is provided. The sequence is as follows:

> Command: **SKETCH** ENTER *(Or click the Sketch button.)*
>
> Record increment <0.065>: *(Click the select button.)*
>
> Second point: *(Move the cursor the desired increment distance and click the select button.)*
>
> Sketch. Pen eXit Quit Record Erase Connect.

For a drawing requiring great accuracy, enter the line increment value using the keyboard. Once the width is entered, several options will be listed in the following prompt:

> Sketch. Pen eXit Quit Record Erase Connect.

Once the segment length is set, move the cursor into the drawing area. You'll notice that nothing happens as the cursor is moved throughout the drawing area.

Pen (P)

Think of the **Sketch** command as sketching on paper with a pencil. The **Sketch** command treats the cursor as a "pen" and allows for lifting the pen and moving it from one

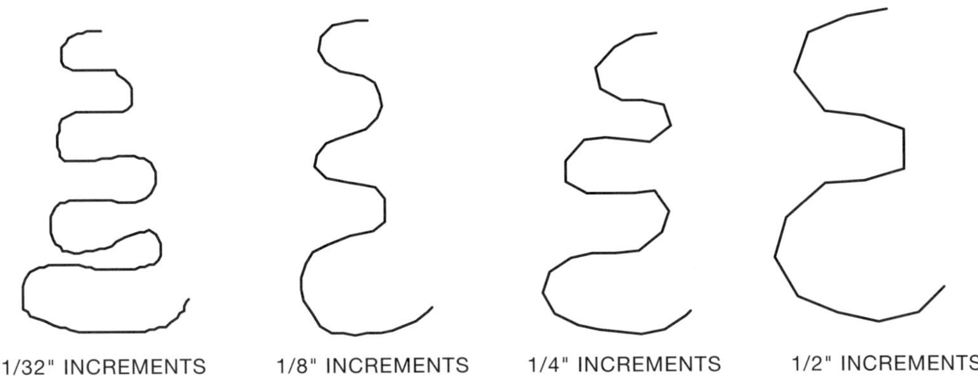

Figure 12.7 *The smaller the value for the segment length, the smoother the resulting curves will be. A scale should be selected that provides the desired accuracy without requiring unneeded lines.*

location to another without drawing a line. As **Sketch** is started, the drawing pen is in the "up" position, allowing for selection of a starting location. Once the start location is selected, the pen can be put down on the paper to allow for sketching. Click the select button or press P on the keyboard to raise and lower the pen. If you use the keyboard to control the pen, don't press ENTER after pressing P. Most users skip the keyboard once they've entered the command, and use the select button to control the pen. Each time the select button is clicked, the pen is raised or lowered to allow for movement and drawing. The process can be seen in Figure 12.8.

Period (.) Line to Point

The period (.) option can be used to draw a straight line from the endpoint of the last sketched line to the current location of the cursor. Sketch a few segments and raise the pen. Move the cursor to a new location and press the period (.) key. This will produce a straight line from the end of the sketch segment to the new cursor location. The effect can be seen in Figure 12.9.

Record (R)

When sketches are created, they are not actually part of the drawing until they are recorded. Enter the drawing area and sketch a few lines. You'll notice that the lines are green (color will vary depending on configuration of AutoCAD), even though you've done nothing to change the drawing color. The Record (R) option records all of the temporary sketch lines as a part of the drawing base and converts them to the present drawing layer and color. Once recorded, sketch lines cannot be edited using the sketch options, although they can be edited using any of the edit commands. Pressing R does not affect the position of the pen, and you can continue sketching after

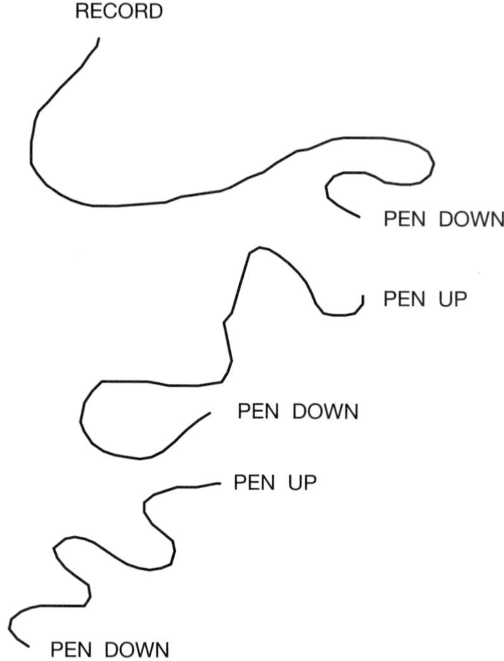

Figure 12.8 *Controlling the pen position of UP/DOWN allows breaks to be created between segments.*

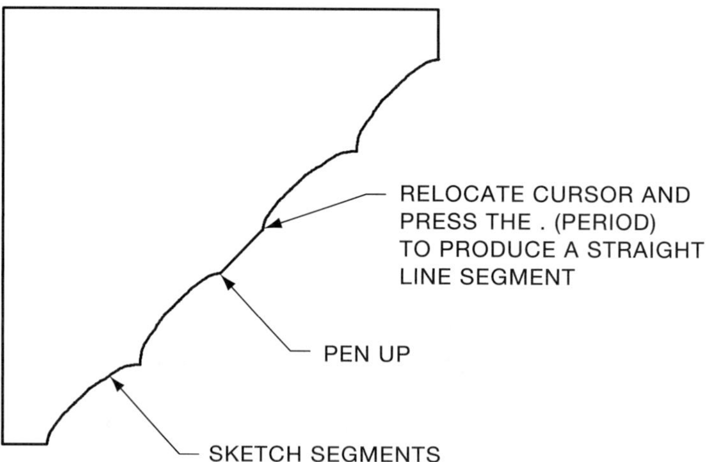

Figure 12.9 *The period (.) option of the* **Sketch** *command allows a straight line to be drawn between the end of the last sketch segment and the present cursor location.*

you've used it. Pressing ENTER will record the sketch segments and exit to the Command prompt.

Record and Exit (X)

Using the X will record the sketch segments to the drawing base, return the Command prompt to the screen, and re-establish the Command prompt. Using this option will display the number of lines recorded just above the command line.

Quit (Q)

Pressing Q on the keyboard will function similarly to the **Erase** command. Q will remove all temporary sketch segments entered since the last Record and return you to the Command prompt. This option can be used if you sketch an object and then decide that you don't need it.

Erase (E)

This option allows for part of the sketch segments to be removed prior to recording. Start this option by pressing E; it will remove a portion of the sketch from the selected point to the end of the sketch segments. Once the option is started, a message is displayed:

Erase: Select end of delete. < Pen up >

The pen will be placed in the up position to allow you to select the end of a segment. Move the crosshairs to the portion of the line to be removed. The line will be displayed with the indicated portion of the line removed. If the removed section is acceptable, press P to erase the segment. If the selection is unacceptable, press E. The message "Erase aborted" will be displayed and the **Sketch** prompt will be returned.

Connect (C)

Typing **C** (ENTER) is similar to the Endpoint mode of Object Snap when drawing lines, and will allow sketching to continue from the end of the last drawn sketch segment. When the option is selected, the following prompt will be displayed and the crosshairs will be displayed:

Connect: Move to endpoint of line

When the crosshairs are within one record increment of the end point, the pen will be placed in the down position at the end of the line. If the pen is already down when the Connect option is used, the following prompt will be displayed:

Connect Aborted

To terminate the Connect command, type **C** (ENTER) a second time.

THE U COMMAND

You've already used the Undo option to remove a drawing object. The U option has been used in the middle of a command routine to remove an object that might not be what you intended. The **U** command also can be used to backtrack through a command sequence once the sequence has been ended. Use the command carefully. Once it is used, objects are gone and will have to be redrawn if you want them back. Start a new drawing, draw a series of several lines, and press ENTER. Draw a circle, a polygon, and an ellipse in the order listed. Now if you type **U** (ENTER) at the Command prompt, you will see

Command: **U** ENTER

ELLIPSE GROUP

The last item drawn, the ellipse, will be removed from the screen. Notice that the most recent command (ellipse) will be listed and the Command prompt will reappear. If the process is repeated, the polygon will be removed.

Command: **U** ENTER

POLYGON GROUP

By continuing to enter **U**, you can remove the entire drawing.

Command: **U** ENTER

CIRCLE

Command: **U** ENTER

LINE

UNDO

The **Undo** command allows several command sequences to be undone at one time and permits several operations to be carried out as the objects are undone. The **Undo** command should give you the freedom to know you can try anything, and if the outcome is undesirable, type **UNDO** (ENTER) at the Command prompt to restore the drawing to its original state. The command sequence is as follows:

Command: **UNDO** ENTER

Enter the number of operations to undo or
[Auto/Control/BEgin/Mark/Back]<1>:

If a number is typed, such as **1** (ENTER), the last command will be undone. If you type **3** (ENTER), the last three command sequences will be removed.

AUTO

The Auto option issues a prompt to toggle between ON and OFF. When Auto is ON, the default value, any group of commands that is used to insert an item, or group of items, is treated as one item and removed by **U**. This will be helpful later when you work with blocks, wblocks, and macros. If Auto is OFF, each command in a group of commands is treated as an individual one.

CONTROL

The Control option is another method to limit the **Undo** command; it also can be used to completely disable **Undo**. The command sequence is as follows:

> Command: **UNDO** ENTER
>
> Enter the number of operations to undo or
> [Auto/Control/BEgin/Mark/Back]<1>: **C** ENTER
>
> Enter an UNDO control option [All/None/One] <All>:

If All is selected, **Undo** will not be limited, and it will erase back to the last Mark. The None option will disable **U** and **Undo** commands so that your machine will function as if these commands have never been placed in AutoCAD. You can reactivate **Undo** and **U** by entering **Undo** and typing A (ENTER). Selecting the One option limits **U** and **Undo** to a single operation before returning to the command line.

BEGIN AND END

The BEgin option of Undo groups a sequence of drawing operations. The group of drawing commands is defined by the use of the BEgin option of **Undo**. The end of the group is marked by the End option. Within a specified group, AutoCAD will treat grouped commands as a single operation.

BACK

The Back option of **Undo** will remove everything in the entire drawing. Not some, not most, we're talking *everything*. Do you get the feeling you need to use this option carefully? AutoCAD will even help you ponder the use of this option by displaying the prompt:

> This will undo everything. OK? <Y>

If ENTER is pressed, this will accept the default Yes value, and really clean up your drawing. Typing N (ENTER) will ignore the Back option. If you do use the Back option and then change your mind, use the **Redo** command to restore your drawing.

The Back option also can be used with the Mark option so that the entire drawing is not erased.

MARK

This option will limit the distance that Back will search through a drawing as it erases work. A Mark can be placed in a drawing prior to executing several commands. If the outcome of those commands is not what you expected, **Undo Back** will remove all of the commands until the Mark is reached. The command sequence is as follows:

Command: **UNDO** ENTER

Enter the number of operations to undo or
[Auto/Control/BEgin/Mark/Back]<1>: **M** ENTER

Command: **L** ENTER *(Now draw several features in an existing drawing.)*

Command: **UNDO** ENTER

Enter the number of operations to undo or
[Auto/Control/BEgin/Mark/Back]<1>: **B** ENTER

LINE

Mark encountered

Command:

NUMBER

The Number option of **Undo** will undo the specified number of drawing operations. The effect is the same as if the **U** command had be used the same number of times, but only a single regeneration is required.

REDO

Occasionally you will remove an object with **U** or **Undo** by mistake. The **Redo** command can be used to restore items that have been deleted. **Redo** will only restore objects, however, when used immediately after an **Undo** or **U**. If **U** or **Undo** is followed by a command such as **Line**, and then **Redo**, **Redo** will have no effect on the material that was deleted prior to the **Undo**. Start the command by selecting the **Redo** button on the **Standard** toolbar or by typing REDO (ENTER) at the Command prompt.

CHAPTER 12 EXERCISES

1. Start a new drawing and draw a line 27.50" long. Divide the line into five equal segments. Copy the original line and place markers at 5.375" o.c. Save the drawing as E-12-1.
2. Use the **Sketch** command and create a logo that can be used in the title block of a template drawing. Save the drawing as E-12-2.
3. Start a new drawing and draw three lines 10" long. Set the point marker to show a marker with a circle with an X in the center. Measure the line into seven equal segments. Divide the second line into 1.6" long segments using a square around the crosshairs marker. Show the third line with perpendicular lines at 1.92" o.c. Save the drawing as E-12-3.
4. Use the **Sketch** command and draw four curved lines using .065", .125", .25", and .5" long line segments. In at least two of the lines, include a straight-line segment between the sketched segments. Save the drawing as E-12-4.

5. Start a new drawing and draw a rectangle. Place a point using the crosshairs surrounded by a circle and a square in the center of the rectangle. Draw lines of infinite length that pass through the point in a horizontal, vertical, and at 30° to the horizontal and vertical planes. Trim the lines that extend beyond the rectangle. Save the drawing as E-12-5.
6. Start a new drawing and draw a circle with a semi-infinite line extending outward from each quadrant. Draw lines that extend from the center point to the edge of the circle and are placed at 15° intervals for a 360° pattern. Save the drawing as E-12-6.
7. Draw two lines that intersect. One of the lines is to be horizontal and the second line is to be 67° above the first line. Draw a line that bisects the intersection of the two lines. Save the drawing as E-12-7.
8. Assume a grid of .50 to complete the attached drawing.

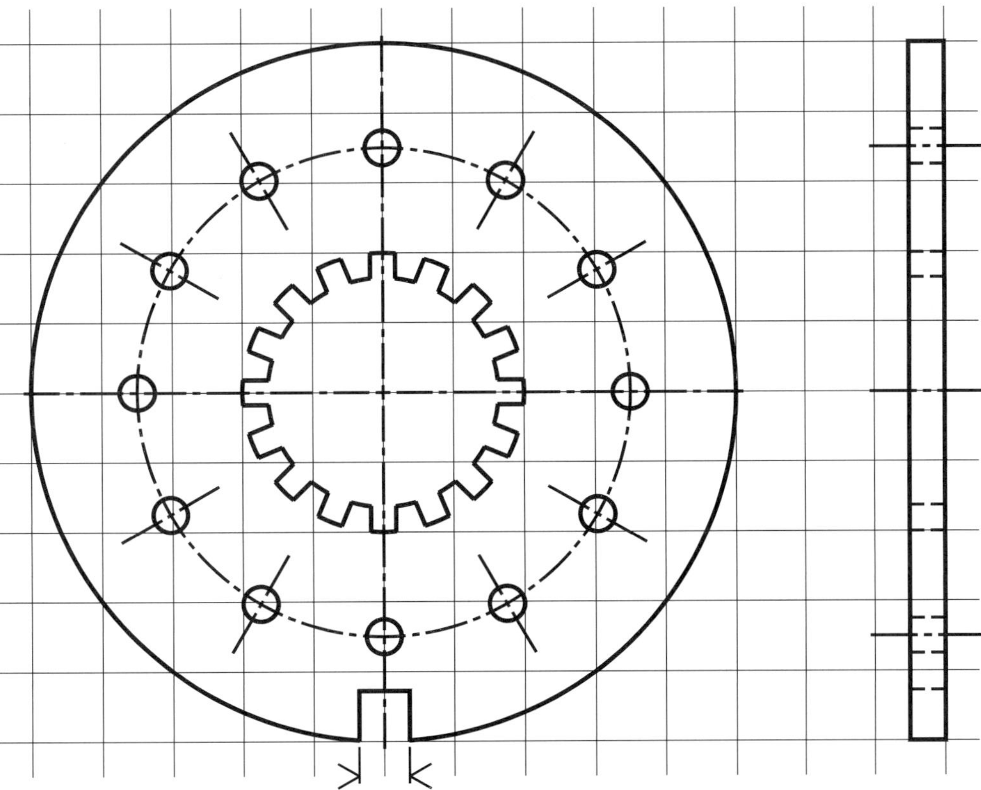

Problem 12-8

9. Use the attached isometric drawing to complete three orthographic projections.

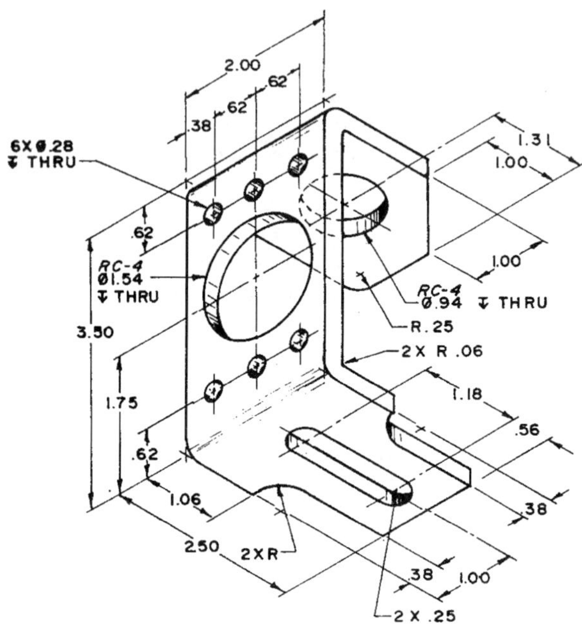

Thickness = 0.25

Problem 12-9

10. Use the attached isometric drawing to complete three orthographic projections.

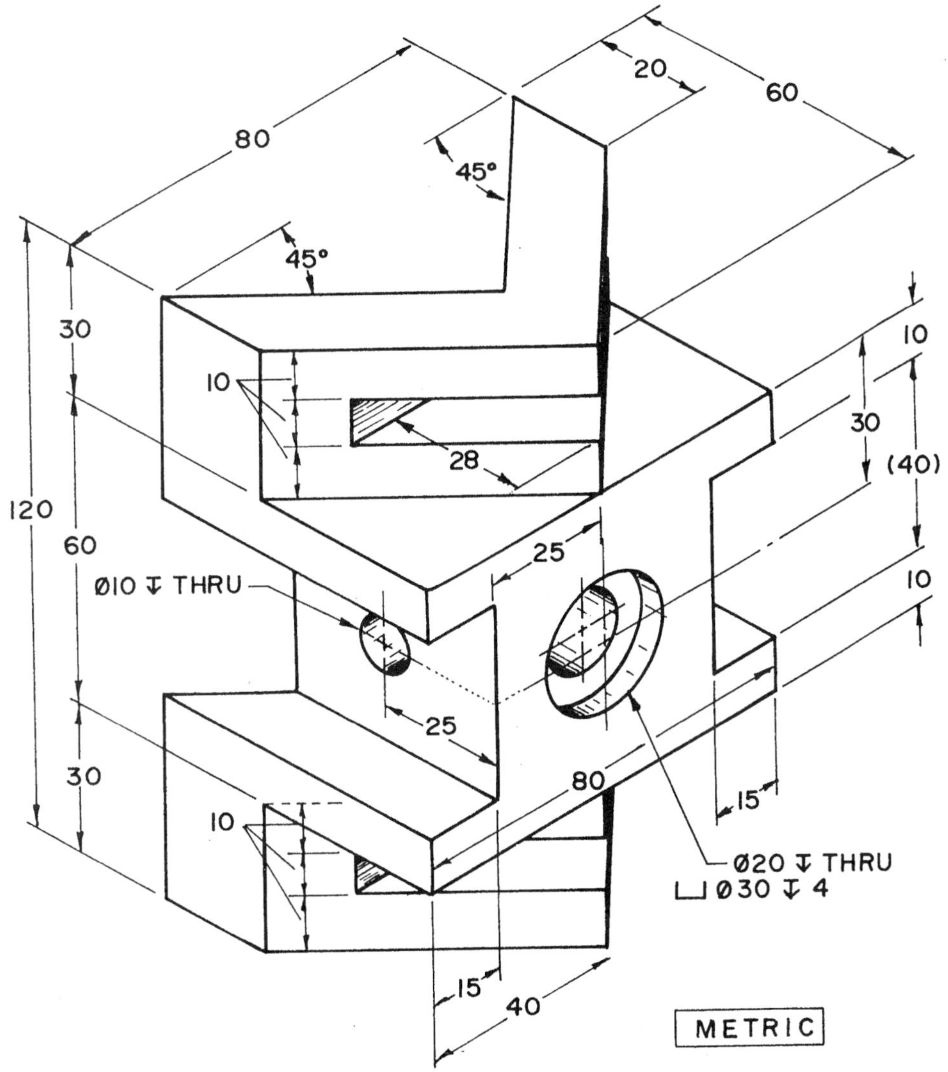

Problem 12-10

CHAPTER 12 QUIZ

Start a new drawing and draw the following objects in the order listed. Draw four pairs of line segments, a three-sided polygon, a circle, a polyline with three segments with a width of .125", and an ellipse. Copy the circle, move the ellipse, and enlarge the polygon to twice its size.

1. If the **U** command is used after the polygon is enlarged, what will be affected?
2. If **Undo 2** is used after the polygon was drawn, what will be affected?
3. If a mark is placed after the circle is drawn, what will the effect be if the **Undo Back** command is used after the circle is copied?
4. A whole drawing sequence is an experiment that you might want to delete. List the command and subcommands that would allow the entire sequence to be removed as one object. Explain when these subcommands are used.
5. If 5 is used with the **Undo Number** option, what will the effect be?
6. In using the **Undo Number 5** option, one too many objects were removed. How can this be corrected?
7. Explain the difference between **Divide** and **Measure** on a line.
8. Explain the difference between a Trace and a polyline line segment.
9. What is the default setting for a sketch line if the units are set to decimal units with a precision of 0.000?
10. What is the danger of having the sketch resolution too fine?
11. Explain how the correct option of **Sketch** will affect a drawing.
12. What is a ray and how could it be used?
13. What is the difference between a ray and an xline?
14. What is the effect of Hor on a ray if ORTHO is OFF?
15. Describe the difference between the two angle options of **Xline**.

Controlling Lines and Layers

This chapter will examine methods for

- Controlling lines in AutoCAD
- Organizing drawing information by the use of layers
- Altering linetypes, colors, and layers

Commands to be introduced include

- **Ltscale**
- **Celtscale**
- **Layer**
- **Layer Previous**
- **Match Properties**
- **Properties**
- **Qselect**

CONTROLLING LINETYPES

A linetype is a specific pattern of lines of varied length and spacing presented in a uniform pattern. Varied linetypes traditionally have been used throughout the manufacturing and engineering communities. Chapter 5 introduced methods for loading and drawing with varied linetypes. This chapter will introduce methods for altering factors that control linetypes.

CHANGING THE LINE SEGMENT SIZE

The dashes that comprise the line pattern are currently defined as drawing units that might be unsuitable for your current drawing needs. A centerline drawn in a 12" × 9" drawing area will appear as a centerline until you zoom in on the line. If the same centerline is used in a drawing area that is 50" wide, the centerline will be displayed as a continuous line. To have the centerline appear as a centerline requires that the scale of the line be altered. The **Global scale factor** setting of the **Linetype Manager** dialog box can be used to control the length of line segments in a drawing. Access the dialog box by typing LT (ENTER) or by choosing **Linetype** from the **Format** menu. Select the **Show Details** button to display the current line scale values.

You can also alter the segment length of each pattern by typing **LTS** (ENTER) at the Command prompt. (LTS represents linetype scale.) No matter how the value is entered, by default AutoCAD thinks of the selected length as one unit. This unit could be in inches, feet, millimeters, kilometers, or any other unit of measurement. Altering the value will allow you to alter the selected unit length in a way similar to the way the **Scale** command alters the size of objects. The distances specified in the linetype definition will be multiplied by the **Ltscale** value to produce a new length for all patterns. The command sequence is as follows:

Command: **LTS** ENTER

LTSCALE Enter new linetype scale factor<1.00>: **2** ENTER

If a value of 2 is entered, all line segments and spaces will be doubled in size. If the pattern appears too short, increase the value for the LTSCALE. If the line segments are too large decrease the value. The effect of **Ltscale** on line segments can be seen in Figure 13.1.

The line scale is determined by the size the drawing will be plotted at. Using an **Ltscale** of 2 will make the line pattern twice as big as the line pattern created by a scale of 1. For a drawing to be plotted at 1/2=1, a dashed line that should be .125" when plotted would need to be .25" long when drawn. This can be determined by multiplying .125 × 2 (the scale factor). You can also determine the **Ltscale** factor by using half of the scale factor. The drawing scale factor is always the reciprocal of the drawing scale. To use a drawing scale of 1/2=1, dividing 1 by .5 produces a scale factor of 2 with an **Ltscale** value of 1. Keep in mind that when the linetype scale is adjusted, it will affect ALL linetypes within the drawing. Usually this is not a problem, unless you are

ORIGINAL
LTSCALE

LTSCALE = 2

Figure 13.1 *The* **Ltscale** *command allows the length of line segments to be altered so that the pattern will still be seen.*

trying to create one unique line. Typically, the **Ltscale** factor can be preset as template drawings are set up. In most offices, template drawings are established based on the size of the paper or the scale the finished drawing will be plotted at. Common plotting scales and their respective scale factors include the following:

INCH VALUES		METRIC VALUES	
Drawing scale	**LTSCALE**	**Drawing scale**	**LTSCALE**
1/8 = 1	4	1 = 50	25
1/4 = 1	2	1 = 20	10
1/2 = 1	1	1 = 10	5
3/4 = 1	1.33	1 = 5	2
2 = 1	.50	1 = 2	1

CHANGING INDIVIDUAL LINE SEGMENT SIZES

Using **Ltscale** changes the values of all of the lines in a drawing. Using the **Celtscale** command allows the scale of existing lines to remain unaltered, but changes the scale of future lines. Typing CELTSCALE (ENTER) at the Command prompt will produce the prompt:

Enter new value for CELTSCALE <1.0000>:

The value that is assigned to celtscale is used in combination with the value assigned to the linetype scale. The celtscale acts as a multiplier of the linetype scale, producing a net scale effect on lines being drawn. With a linetype scale value of 2, and a celtscale of .25, the net scale factor would be 2 × .25=.50. Figure 13.2 shows the effects of using **Celtscale** to control the current line width.

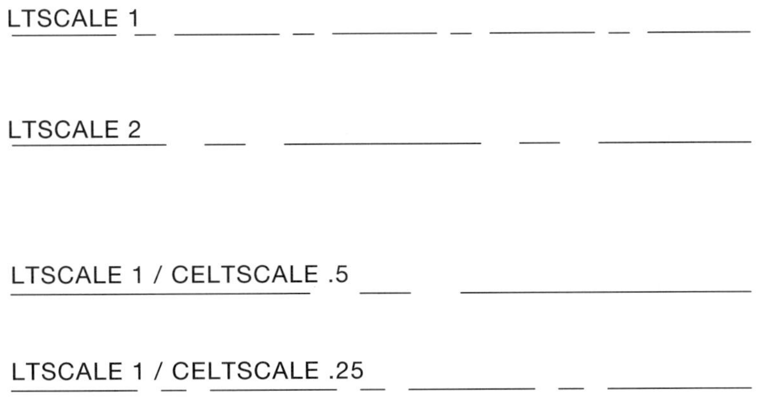

LTSCALE 1

LTSCALE 2

LTSCALE 1 / CELTSCALE .5

LTSCALE 1 / CELTSCALE .25

Figure 13.2 *The effects of* **Ltscale** *and* **Celtscale** *on line length.*

NAMING LAYERS

Layers were introduced in Chapter 5, along with methods to create, alter, and control a layer. As your drawing ability increases, it is crucial to fully understand the importance of using layers. The use of layers will be increasingly important as the techniques in the following chapters are introduced. Layers are used like separate sheets of paper to separate information within a drawing file. The base components of a drawing can be placed on one layer, and the text, dimensions, general notes, and revisions can be placed on separate layers. To help you control information, consideration must be given to layer names, controlling linetypes and colors by layer, as well as methods of displaying and hiding layer information.

LAYER NAMES

Names for layers have not become standardized throughout. Most offices have developed a standard method of assigning names, so that all members of the staff have easy access to information on a drawing. Even within a firm, names will vary depending on the type of project to be drawn. As you set up names for layers, use the same guidelines for naming layers that were used to name linetypes. Use a name that will still be meaningful in six months. This might include layer names such as OBJECT, HIDDEN, DIMENSION, TITLEBLOCK, HATCH, TEXT or VIEWPORT. Up to 255 characters (including spaces) can be used in naming a layer but it helps to keep layer names short and descriptive.

CONTROLLING LAYERS

Layers can be controlled by using the **Layer Properties Manager** dialog box or by using the **Layer Control** menu on the **Object Properties** toolbar. Access the **Layer Properties Manager** dialog box shown in Figure 13.3 by selecting the Layer button on the **Object Properties** toolbar, by typing LA (ENTER) at the Command prompt, or by selecting **Layer** from the **Format** menu. Previous chapters introduced each element of the layer and linetype properties. This chapter will introduce methods of setting and controlling each of the layer elements.

ASSIGNING COLOR BY LAYER

Colors and linetypes can be assigned by layers after a layer has been created. See Chapter 5 for a review of how to create a layer.

Figure 13.4 shows an example of the dialog box with the addition of the TITLE BLOCK, VIEWPORT, OBJECT, CENTER, and HIDDEN layers. Notice that the current layer is still 0, but new layers have been added. As new layers are created, they are initially placed at the bottom of the layer list. As the screen is regenerated, the names will be displayed in alphabetical order. Once a layer has been created, the qualities for that layer color and linetype can be assigned.

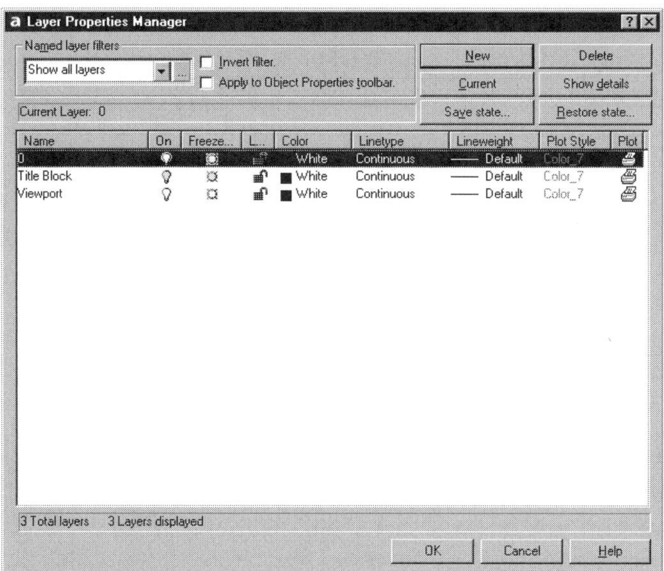

Figure 13.3 *The* **Layer Properties ManagerChange Properties** *dialog box can be used to control the characteristics of linetype and layers. It can be accessed by selecting the* **Layer** *button on the* **Object Properties** *toolbar or by typing* **LA** *(ENTER) at the Command prompt.*

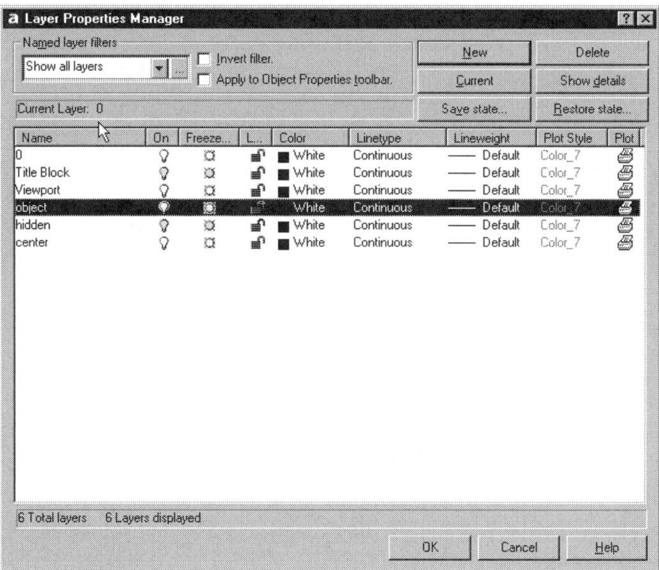

Figure 13.4 *To create new layers, click the* **New** *button. A layer named Layer 1 will be added. Highlight the name of Layer 1 and type in a more descriptive name. In this example, names of OBJECT, HIDDEN, and CENTER were added.*

The CENTER layer is currently listed as white. Any lines that are created on this layer will be drawn as white lines on a black background. Depending on the **Display** settings in the **Options** dialog box, you might be drawing with black lines on a white background, or any number of options for each color. To alter the color of a layer, select the layer so that the layer name is highlighted. You can alter the color of each layer by selecting the color button from the **Color** list or by choosing a color in the **Color** box in the **Details** area. (The **Details** area is not active until a layer is highlighted and the **Show Details** button is selected.) Selecting the color button for the selected layer will display the **Select Color** dialog box shown in Figure 13.5. If you're at a computer, select the red color button. As the desired color is selected from the palette, the name of the color will be displayed in the **Details** area, and the color will be displayed. Clicking **OK** will remove the color palette.

You can override the color for a layer or alter the color of individual objects on a layer by selecting the **Color** button on the **Object Properties** toolbar. Selecting this button will allow the color of the layer to remain unchanged but allow all items drawn after you select the button to have a color different from the specified layer color. You can select the **Color** button in the **Color** box on the **Object Properties** toolbar or the arrow beside the **Color** box. Either method will produce the **Color** menu shown in Figure 13.6. You can specify the color by selecting a color button or by entering the name or number of the color. Selecting the Other option of the menu will display the **Select Color** dialog box shown in Figure 13.5. If you want to select another color, simply select another color chip, and the display will be updated. When you're satisfied with the color, click the **OK** button, and the dialog box will be removed and the **Layer Control** box redisplayed.

Objects and layers can be assigned any color you desire. It may be helpful to assign like objects of different layers a different color. Assigning a different color to each layer might help you keep track visually of a drawing as different layers are set to ON/OFF for viewing or plotting. Color as it relates to plotting will be introduced in later chapters.

ASSIGNING LINETYPES TO LAYERS

The current linetype default for each layer is a continuous line. In previous chapters, methods of drawing different linetypes were introduced. The Linetype command allows for setting and drawing different linetypes without regard to the layer. The Linetype option of **Layer** will allow all lines on a specific layer to be drawn as one type of line. Linetypes can now be assigned to a specific layer and all lines on that layer will be drawn with the selected linetype.

The linetype can be altered using methods similar to altering layer color. Highlighting the name of the layer to be altered and then selecting the linetype name will display the **Select Linetype** dialog box shown in Figure 13.7. The dialog box can be used to

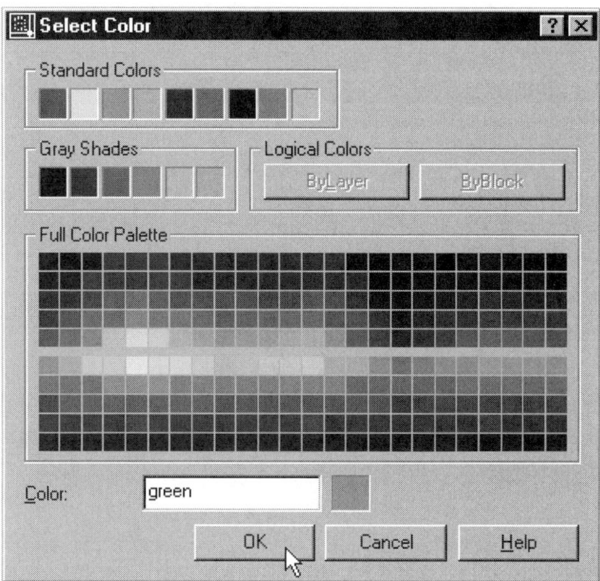

Figure 13.5 *Selecting the* **Color** *button in the* **Layer Properties Manager** *dialog box will display the* **Select Color** *dialog box, allowing the current layer color to be changed.*

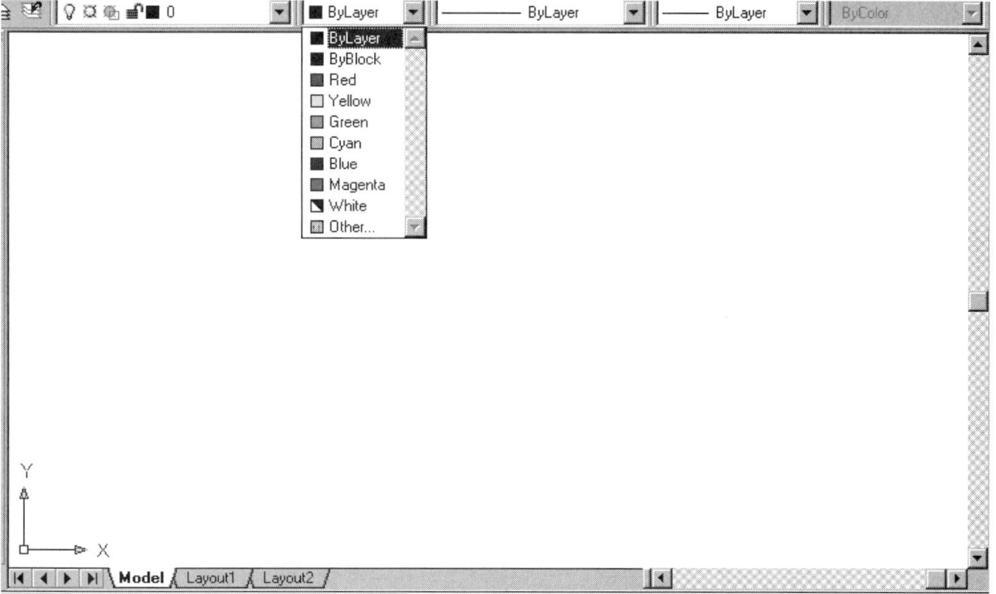

Figure 13.6 *You can override the color for a layer by selecting the* **Color** *button on the* **Object Properties** *toolbar or the down arrow beside the* **Color** *box.*

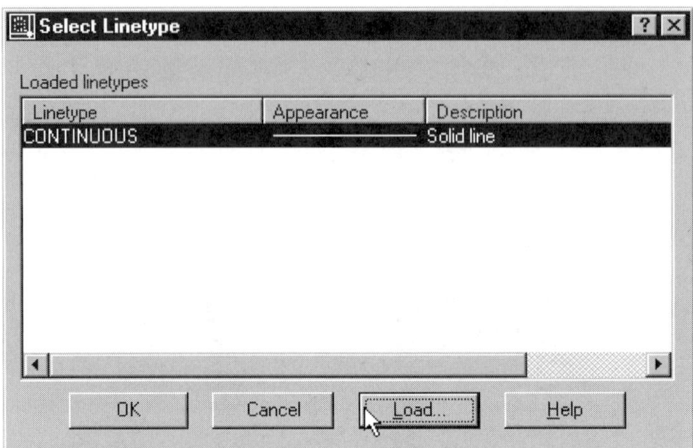

Figure 13.7 *Selecting the name of the current linetype in the* **Layer Properties Manager** *dialog box will display the* **Select Linetype** *dialog box, allowing for selection of one of the current linetypes.*

load and select linetypes. Selecting the **Load** button will display the **Load or Reload Linetypes** dialog box shown in Figure 13.8 and display a list of linetypes contained in AutoCAD. Highlight the names of the linetypes to load and click the **OK** button to remove the dialog box and restore the **Select Linetype** dialog box. As the box is restored, it will be updated to reflect the linetypes that were just loaded. One of these linetypes can now be selected from the **Select Linetype** dialog box and assigned to the layer that is highlighted in the **Layer Control** box. Assign a linetype of CENTER to the CENTER layer. Assign the linetype by selecting the linetype and clicking the

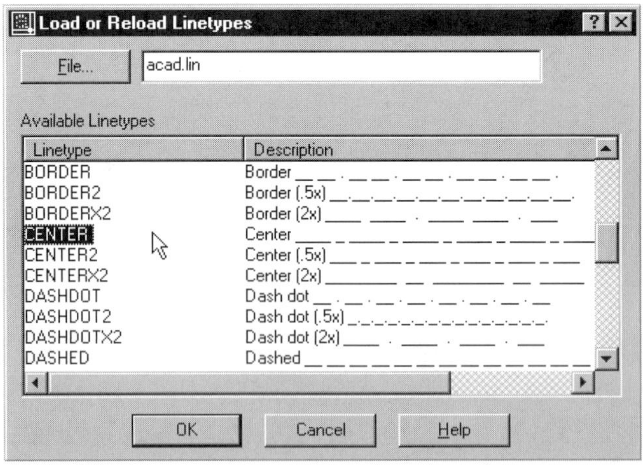

Figure 13.8 *You can add linetypes to the drawing base by highlighting the desired layer name and clicking the* **OK** *button.*

OK button. As the button is clicked, the **Layer Properties Manager** dialog box will be redisplayed, reflecting the selected linetype.

Note: To select multiple linetypes, hold down CTRL while you select the linetypes. To load a sequence of linetypes, hold down SHIFT and select the first and last listing. All linetypes that are listed between the first and last listings will be selected. These two selection methods can be used any time multiple objects are to be selected from a menu.

SETTING THE CURRENT LAYER

Up to this point, you've created layers with the name of OBJECT, CENTER, and HIDDEN, and assigned the color red and a linetype of CENTER. If you exit the dialog box and start to draw, nothing will appear to have changed. To draw red centerlines, you need to set the current layer to CENTER. To make the CENTER layer current, highlight the name of the layer in the **Layer Properties Manager** dialog box and click the **Current** button. Clicking the **OK** button will keep this change and restore the drawing screen. As you start the **Line** command, segments will be drawn using the selected color and linetype that were just added. A new layer can also be made current by selecting the down arrow beside the **Layer** box on the **Object Properties** toolbar and selecting the desired layer.

SETTING LAYERS ON/OFF

All of the new layers you have created up to this point are set to ON. If you were to switch to a layer such as HIDDEN and then draw, you would be able to see what is drawn. In a complex drawing it is sometimes helpful to remove construction and layout lines to view what will be the finished drawing. Construction lines can be removed from the display by the use of the ON/OFF button. If a layer is set to OFF, objects drawn on that layer will be invisible. The information is retained, but not displayed or plotted.

Layer Off

To turn layers off, display the **Layer Properties Manager** dialog box. The ON/OFF button is the lightbulb—with yellow representing ON and black representing OFF. Select the ON/OFF button for the desired layers to be turned off, and click the **OK** button. The dialog box will be removed and any information that is stored on a layer that is OFF will be removed from the screen. If you attempt to set the current layer to be turned OFF you will be given the following prompt:

The current layer is turned off.

If you click the **OK** button, the dialog box is removed, and the items on the current layer are removed from the drawing screen. To restore the current layer, redisplay the **Layer Properties Manager** dialog box and restore the current layer. If you accept the Current layer OFF, the current layer will be made invisible. Now if you draw objects, they will be drawn but not displayed. This situation occurs more by accident than by design. If you find yourself drawing a line but nothing is displayed, check the listing to see if the current layer has been accidentally switched OFF.

Layer On

Layers that have been made invisible by the OFF option can be restored by the ON option. Display the **Layer Properties Manager** dialog box and highlight the desired layers to be set to ON. Select the black lightbulb to turn it ON and click the **OK** button to restore the selected layers, remove the dialog box, and display the information on the selected layers.

THAWING AND FREEZING LAYERS

If you have set layers to OFF, you removed information from the display screen but not from the drawing base. The information merely became invisible. If you request a **Regen**, it would be performed with all of the information on the ON layers redisplayed. Information on layers that are set to OFF would also be processed, which will cause a slight slowdown on large drawings. Using the Freeze/Thaw options, you will be able to control the extent of information to be scanned during the **Regen** process. Layers that are in the Thaw mode are processed for a **Regen**; information that is frozen is not processed during a **Regen**, reducing the time required to **Regen**. This will be a factor only on huge drawing files. Figure 14.9a shows a drawing with the text and dimensions frozen. Figure 13.9b shows the same drawing with the text and dimensions displayed.

Freeze

Display the **Layer Properties Manager** dialog box, select the layer that you would like to freeze, and select the **Thaw** (sun) button, showing the Thaw mode. It will be changed to a snowflake, showing the Freeze mode. Clicking the **OK** button will activate the selection, restore the drawing screen, and remove the objects on the selected layer from the display. Unlike the OFF option, the current layer cannot be frozen.

Thaw

The Thaw option of the **Layer** command will allow for specified frozen layers to be thawed. To thaw a layer that is frozen, display the **Layer Properties Manager** dialog box and select the **Freeze** (snowflake) button of the layer to be thawed. As the snowflake is selected, it will change to the sun button, indicating that the layer has been thawed. Clicking the **OK** button will activate the selection, restore the drawing screen, and display the material on the layer that is now thawed.

Safeguarding Layers

The final pair of options of the **Layer** command is Lock and Unlock. As the name implies, a lock is used to protect something. AutoCAD allows you to lock layers so that they cannot be edited accidentally. Information on a locked layer is still visible;

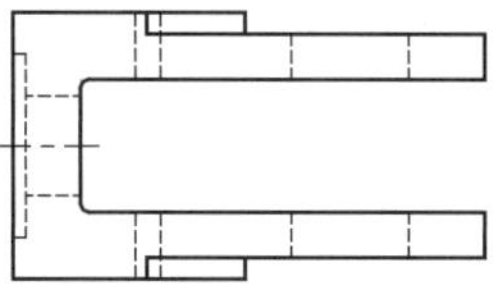

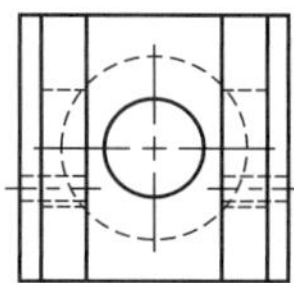

Figure 13.9a *A drawing with the text and dimensions frozen. Using the OFF setting for the DIMENSION layer would create a similar result. Courtesy Hanh Tran.*

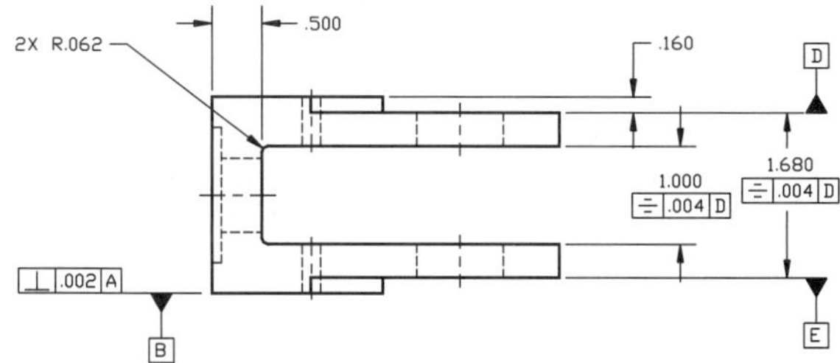

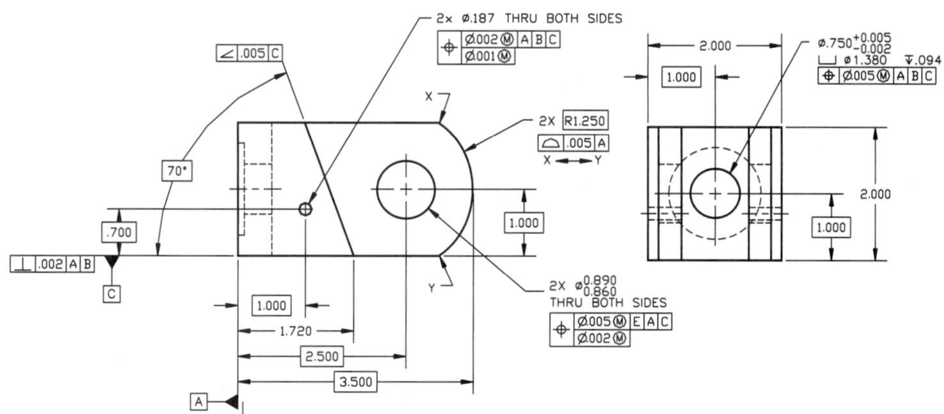

Figure 13.9b *The drawing shown in Figure 14.9a with the text and dimensions displayed. Using the ON setting for the DIMENSION layer would create a similar result. Courtesy Hanh Tran.*

it's just protected. To lock a layer, highlight the desired layer and select the **Lock** button followed by the **OK** button.

To unlock a layer, highlight the desired layer and select the **Unlock** button, and click the **OK** button. This process will reverse the consequences of the Lock. Remember that locking a layer is not a security device, but a method of protecting information on a layer from careless editing. Entities on a locked layer will not be included in a selection set of an edit command. Locking a layer will allow you to view the information on the layer, but it will not be affected as you edit a drawing. You can continue to draw on a locked layer and objects can still be selected for OSNAP locations, thawed or frozen, or altered using the **Linetype** or **Color** button. Objects on locked layers can't be edited.

Filters

Filters provide the ability to display certain layer names in the listing box, based on a specific attribute. Layers can be filtered based on layer name, color, lineweight, contents, locked or unlocked status, freeze or thaw status, and plot styles. There are several other filter options that will not be considered until viewports are explored in later chapters. Filters are controlled in the **Layer Properties Manager** dialog box using the **Named layer filters** edit box arrow. Choosing the **Filter** edit arrow will produce the list shown in Figure 13.10. Selecting the [...] button will produce the dialog box shown in Figure 13.11 and allow the parameters of the filter to be set. In the example, the filter is based on the color red. Clicking the **Add** button and then **Close** will shut the **Named Layer Filters** dialog box. Selecting the **Filter Edit** arrow will display the list shown in Figure 13.10, with the Red filter now shown in the list. Selecting Red will produce a layer listing of layers created using the specified criteria. (See Figure 13.12.) All of the existing features are still displayed on the screen—they just won't be listed in the **Layer** display until the filter is altered. Opening the **Named layer filters** list and selecting the **Show All Layers** option will restore the total layer listing.

RENAMING LAYERS

The current name of a layer can be altered to better define the use of the layer at any time throughout the life of a drawing. Only the 0 layer and XREF layers can't be renamed. XREF layers are layers that are created on a drawing that is attached to the current drawing, and they will be explained in later chapters. Rename layers inside the **Layer Properties Manager** dialog box by selecting the name of the layer to be renamed, and then selecting it again. As it is selected the second time, the name will be highlighted and a box will be placed around the original name with a blinking cursor. Pressing BACKSPACE while the name is highlighted will remove the existing name and allow a new name to be entered. Pressing ENTER once will accept the name. Clicking **OK** removes the dialog box and restores the drawing area.

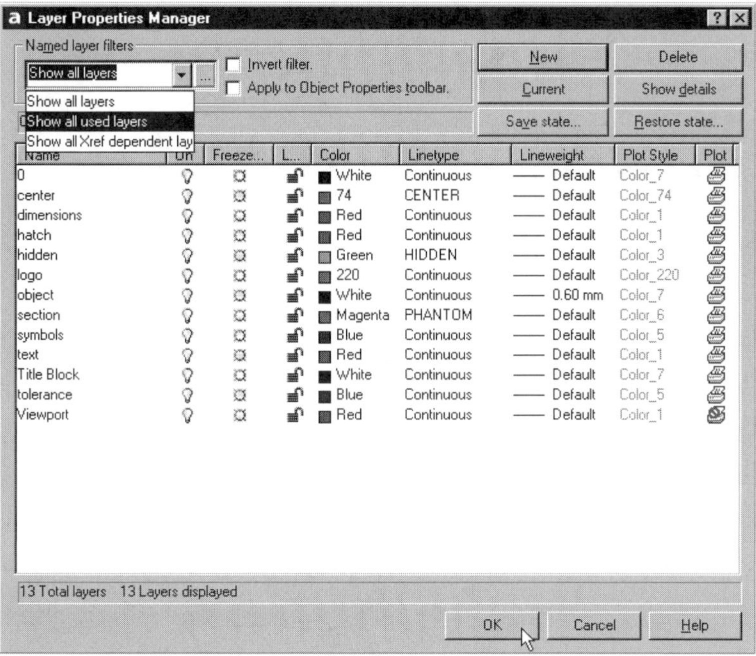

Figure 13.10 *Selecting the arrow beside the* **Named layer filters** *edit box will display this menu, which can be used to established filters to display.*

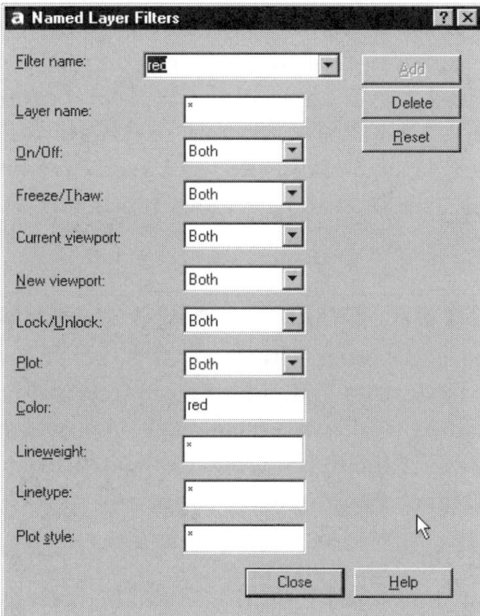

Figure 13.11 *Selecting the [...] button will produce the* **Named Layer Filters** *dialog box, allowing the scope of the filter to be defined.*

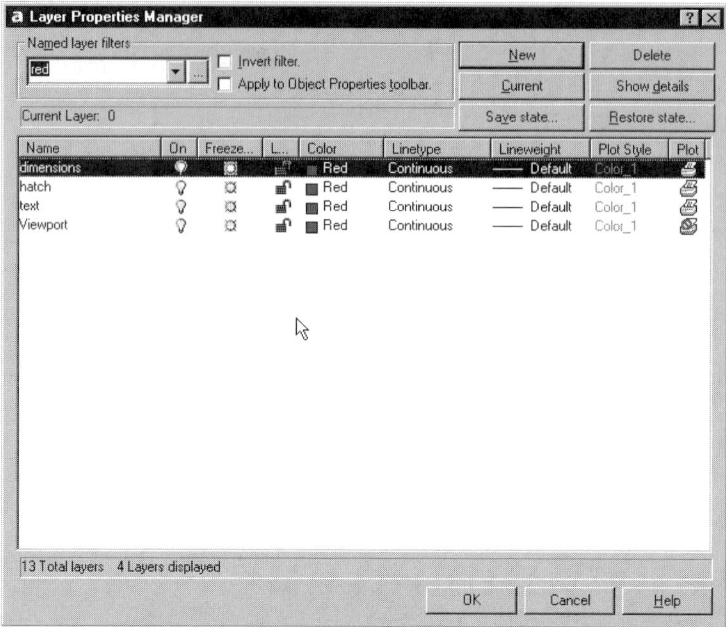

Figure 13.12 *Layers displayed based on the color red.*

DELETING LAYERS

An unused layer can be deleted from the drawing base at any time throughout the life of the drawing—with the exception of the 0 layer or an XREF layer. Layers that are referenced to a BLOCK and a layer named DEFPOINTS cannot be deleted, even if the layer is empty. XREF, DEFPOINTS, and BLOCK will be explored in later chapters. Delete an unused layer from inside the **Layer Properties Manager** dialog box by selecting the name of the layer to be removed and then clicking the **Delete** button—the layer will be removed from the list.

ALTERING LINETYPE, COLOR, AND LAYERS

AutoCAD allows individual properties within a layer to be altered, as well as the properties of the entire layer to be changed. The **Layer Control** button on the **Object Properties** toolbar can be used to alter the qualities of an entire layer. Individual objects on a layer can be changed using the **Make Object's Layer Current** button, or by using the **Match Properties**, **Object Properties**, **Properties**, and **Qselect** commands.

LAYER CONTROL MENU

The **Layer Control** menu shown in Figure 13.13 can be used to alter the visibility of an existing layer. Display the menu by selecting the **Layer Control** arrow to the right of the box on the **Object Properties** toolbar. This menu can be used to toggle

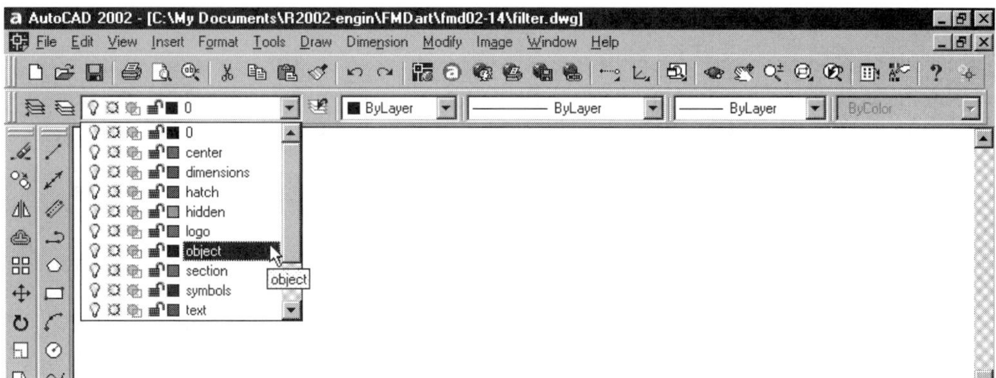

Figure 13.13 *Access the* **Layer Control** *menu by selecting the* **Layer Control** *arrow on the* **Object Properties** *toolbar. The menu can be used to toggle ON/OFF, Thaw/Freeze, Lock/Unlock, and Plottable/Non-plottable.*

ON/OFF, Thaw/Freeze, Lock/Unlock, Plottable/Non-Plottable, and to set the current layer.

- Adjust the visibility of layers by selecting the ON/OFF button or the Thaw/Freeze button. As the button for each layer is selected, the layer will be displayed or removed, depending on the setting of the button.
- Lock or unlock layers by selecting the button of the desired layer.
- The Plottable/Non–Plottable button allows a layer to remain visible on the screen, but to be removed from the plotting set. In the default setting, all layers will be plotted. Selecting a layer and choosing this button will set the layer to non-plottable.
- Alter the current layer by selecting the name of the layer that you want to make current. As soon as the layer name is selected, the menu will be closed, the layer will be changed, and the drawing area will be restored.

Note: The Plottable/Non-plottable control can be a great asset for displaying information that you need to see, but others might not. A non-plottable layer could be created for placing questions to ask the engineer or instructor. The questions can be seen on the display, but they will not be plotted.

MAKING AN OBJECT'S LAYER CURRENT

The **Make Object's Layer Current** button is located at the far left end of the **Object Properties** toolbar. The button can be used to alter the current layer. Selecting the button displays the following Command prompt:

```
Command: _ai_molc
```

Select object whose layer will become current:

The cursor will be changed to a pick box as the program waits for you to select a drawing object. Selecting a drawing object will make the layer that contains the selected object the current layer. The Command prompt will then display the name of the current layer, and drawing can continue.

ALTERING LAYERS USING LAYER PREVIOUS

The **Layerp** (**Layer Previous**) command can be used to undo changes made to layer settings. The command will affect settings such as On/OFF Freeze, Thaw, Lock, Color, Linetype, and Lineweight. **Layer Previous** will not undo the renamed layers, deleted layers, or added layers. Start the command by selecting the **Layer Previous** button on the **Object Properties** toolbar or by typing LAYERP (ENTER) at the Command prompt.

ALTERING THE DISPLAY WITH MATCH PROPERTIES

The properties of one layer object can be transferred to another object by the use of the **Match Properties** command. The **Match Properties** button is found on the **Standard** toolbar. Two other methods of executing the command include typing MA (ENTER) at the Command prompt or selecting **Match Properties** from the **Modify** menu. The command will copy the properties of the first object to the second object, including the layer, the linetype, and the color. Each method will produce the following prompt:

Command: **MA** ENTER *(Or click the Match Properties button.)*

MATCHPROP

Select Source Object: *(Select the object with the properties that you would like to copy.)*

Current active settings: Color Layer Ltype Ltscale Lineweight Thickness

PlotStyle Text Dim Hatch

Select destination object (s) or [Settings]: *(Select the object that you would like to change.)*

Select destination object (s) or [Settings]: ENTER

Command:

Instead of selecting the object that you would like to change, typing S (ENTER) for Settings at the second prompt will produce the **Property Settings** dialog box shown in Figure 13.14. The box can be used to specify the qualities that will be copied to selected objects. One or more of the properties can be assigned including the **Color**, **Layer**, **Linetype**, **Linetype Scale**, **Lineweight**, and **Thickness**—as well as several other qualities that will be discussed in later chapters. Once the qualities have been select-

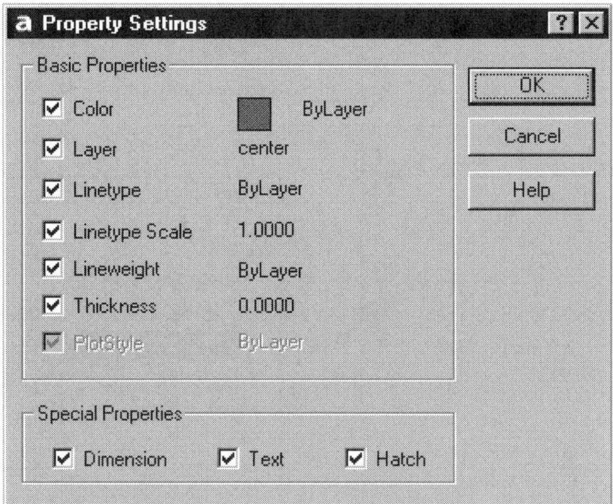

Figure 13.14 *Once a source object has been selected, typing* **S** *(ENTER) for Settings at the Command prompt will display the* **Property Settings** *dialog box. The dialog box allows specific properties of a layer to be altered without affecting other layer qualities.*

ed, click the **OK** button and resume the command sequence by selecting the object to receive the new properties.

CHANGING OBJECT PROPERTIES

The **Properties** command can be used to control the properties of existing objects. Start the command by selecting the **Properties** button on the **Standard** toolbar, by typing MO (ENTER) or CH (ENTER) at the Command prompt or by selecting **Properties** from the **Modify** menu. Each method will display the **Properties** dialog box similar to Figure 13.15, showing the qualities of the current layer. Object properties can be listed in alphabetical order or categorized by groups depending on which tab is selected. The dialog box allows drawing properties including the Color, Layer, Linetype, Thickness, and Linetype scale to be altered. To alter a property for the selected object, choose either the name or value of the desired property. Each property will display a specific flyout menu. Figure 14.16 shows the menu for altering lineweights. Once the desired properties are altered, select the **Close** button to remove the dialog box and return to the drawing.

CHANGING OBJECTS WITH QSELECT

The **Qselect** command can be used to create selection sets for altering the drawing objects. You can create selection sets to either include or exclude objects matching the specified object type and criteria. For instance, **Qselect** allows you to make a selection set made up of all objects that are green, or a selection set could be made to include everything except green objects. Start the command by

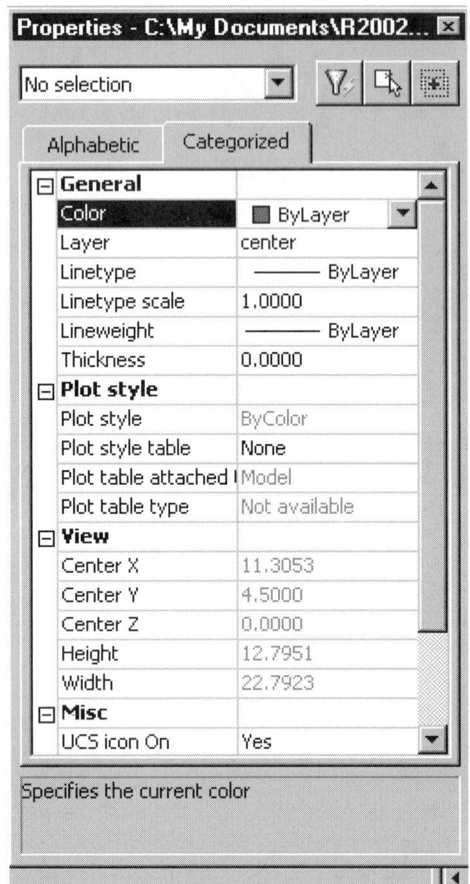

Figure 13.15 *The* **Properties** *dialog box can be used to control the properties of existing objects.*

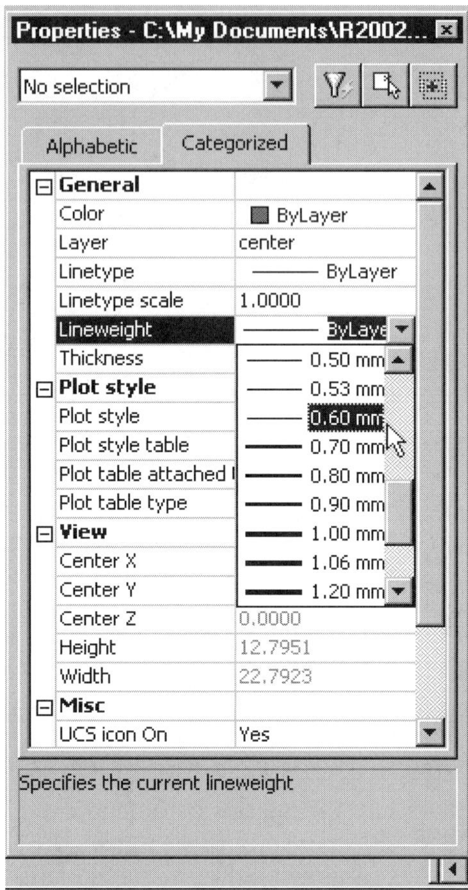

Figure 13.16 *To alter a property for the current object, choose either the name or value of the desired property. Each property will display a specific listing.*

right-clicking and choosing **Quick Select** from the shortcut menu, by selecting the **Quick Select** button in the **Properties** dialog box, or by selecting **Quick Select** from the **Tools** menu. You can also start the command by typing QSELECT (ENTER) at the Command prompt. Each method will produce the **Quick Select** dialog box shown in Figure 13.17. The dialog box contains seven key areas for adjusting the selection set:

Apply to—If a selection set has been defined and is current, it will be selected automatically in the **Apply to** box. If no selection set has been defined, by default the only option is to apply changes to the entire drawing. Choosing the **Select Objects** button creates selection options.

Select Objects—The **Select Objects** button is located beside the **Apply to** edit box. Selecting the button will remove the **Quick Select** dialog box temporarily so that items can be selected to apply the selection filters to.

Object type—This box specifies the object type for the filter. Until a selection set is created, the **Object type** box displays a list of each shape included in the drawing, similar to Figure 13.17. Once a selection set is created, the list will include only the object types contained in the selection set.

Properties—The **Properties** list shows all searchable properties for the selected object to be used with the filter. The **Properties** display will vary depending on the object selected in the **Object type** edit box. If a circle is selected in the **Object type** edit box, a listing similar to Figure 13.18 will be displayed.

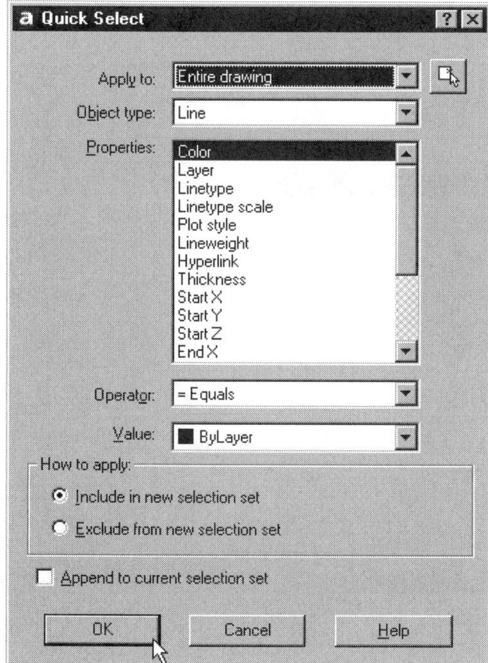

Figure 13.17 *Display the* **Quick Select** *dialog box by right-clicking and choosing* **Quick Select** *from the shortcut menu, by selecting the* **Quick Select** *button from the* **Properties** *dialog box, by selecting* **Quick Select** *from the* **Tools** *menu, or by typing* **QSELECT** *(ENTER) at the Command prompt.*

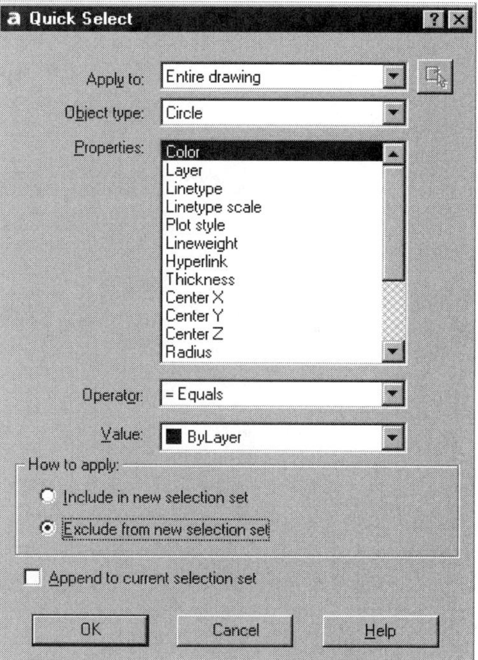

Figure 13.18 *The* **Object type** *box will only display a list of each shape included in the drawing. Once a selection set is created, the list will exclude object types contained in the selection set.*

Operator—The **Operator** box controls the filter range and is dependent on the selected property. Operator options include: =Equals, < >Not Equal, >Greater than, and <Less than (see Figure 13.19).

Value—The **Value** edit box allows a value for the filter to be entered. For instance, if the property is a circle, a value of 4" could be used to include all circles equal to, not equal to, greater than, or less than 4" to define the selection set.

How to Apply—The **How to apply** area contains radio buttons for including or excluding objects from the selection set.

> When **Include in new selection set** is active, a new selection set will be created composed of objects that match the filtering criteria.
>
> When **Exclude from new selection set** is active, a selection set will be created composed of objects that do not match the filtering criteria.

Append to current selection set—This check box determines if the selection set replaces or appends the current selection set.

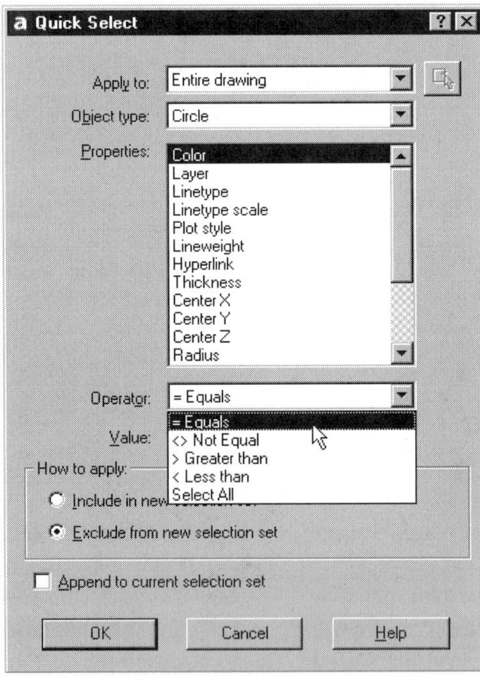

Figure 13.19 *The* **Properties** *display provides a list of all searchable properties for the selected object to be used with the filter. The display will vary depending on the object selected in the* **Object type** *edit box and the operator and value selected.*

Creating a Selection Set Using Qselect

A selection set for editing all green objects can be created using the following steps:

1. Display the **Quick Select** dialog box using one of the methods described earlier.
2. Using the **Apply to** edit box, select Entire drawing.
3. Select Multiple for the **Object type**.
4. Select Color in the **Properties** edit box.
5. Select =Equals in the **Operator** edit box.
6. Use the down arrow beside the **Value** edit box, and select Green.
7. Select **Include in new selection set** in the **How to apply** area.
8. Click the **OK** button to close the box.

AutoCAD will close the **Quick Select** edit box and provide grips on each green object in the entire drawing similar to Figure 13.20. When the process is complete, the Command prompt will show the number of items in the selection set.

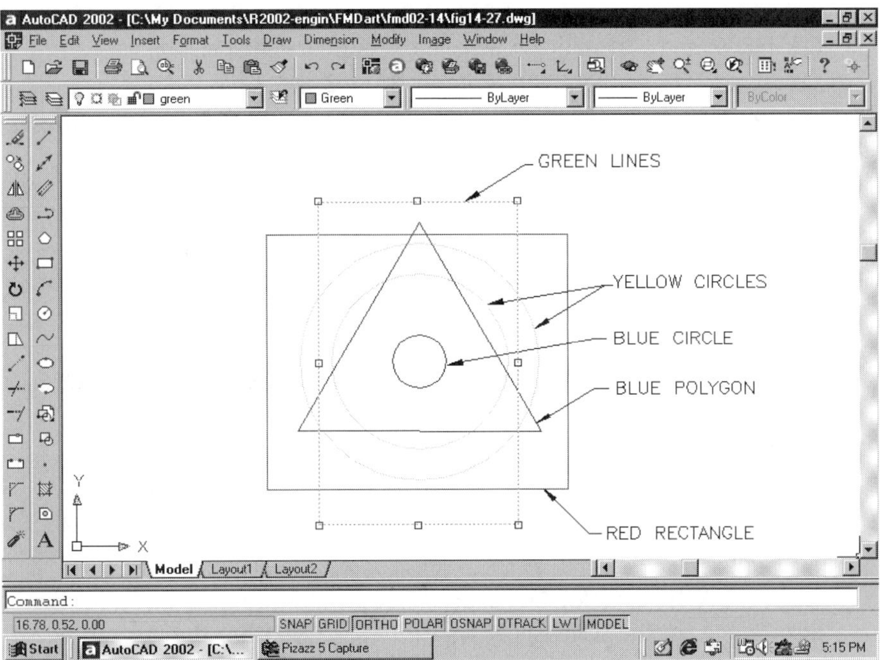

Figure 13.20 Qselect *can be used to create a selection set made up of all objects of a specific property description. In this example, all objects that are green have been selected for editing.*

Excluding Objects from the Selection Set

A selection set excluding objects of a specific property can be created. Using the objects in Figure 13.20, a selection set excluding all circles larger than 4" could be created using the following sequence:

1. Display the **Quick Select** dialog box using one of the methods described earlier.
2. Using the **Apply to** edit box, choose Current Selection.
3. Select Circle for the **Object type**.
4. Select Diameter in the **Properties** edit box.
5. Select >Greater Than in the **Operator** edit box.
6. Enter a value of 4.
7. Select **Exclude from new selection set** in the **How to apply** area.
8. Click the **OK** button to close the box.

AutoCAD will close the **Quick Select** edit box and remove from the selection set all circles with a diameter greater than 4".

CHAPTER 13 EXERCISES

1. Start a new drawing and load CENTER, CENTER2, CENTERX2, DASHED, DASHED2, and HIDDEN. Draw a line with each linetype, long enough so that the line pattern can be seen. Save the drawing as E-13-1.
2. Open E-11-4. Create different layers for each circle. Move the existing objects to the appropriate layer and save the drawing as E-13-2.
3. Open drawing E-8-4. Create layers for the specified layers and options:

PROPLINE	red	Centerx2	property line
TREES	Green	Dashed	tree shape
T-LAYOUT.	7	Dot2	point/baseline

   Set the **Ltscale** appropriate for future plotting at a scale of 1"=10'–0". Save the drawing as E-13-3.

4. Open drawing E-8-1. Assign six different layers, each with a different linetype and color. Change each of the existing shapes to a different layer. Draw a different geometric shape on the remaining layers. Use **Pline** and OSNAP. Save the drawing as E-13-4.

   Open the ANSI-C Color Dependent Plot Style drawing template, set the Ltscale, and set the scale values appropriate to .5=1. Set the units to two-place precision. Establish layers that would be suitable for object, center, hidden, and phantom lines with additional layers for text, notes, border, dimensions and symbols. Assign colors to each layer. Assign linetypes appropriate to the layer contents. Save the drawing as HALF. Create a similar template for drawings created at full scale and save with an appropriate name.

Exercises 6 through 10:

Complete the following orthographic projections by providing a minimum of three views. Create layers with separate colors and appropriate linetypes for each of the lines required to complete the drawing. Edit the drawing for exercise 10 so that the surface with the six holes is perpendicular to the base of the part.

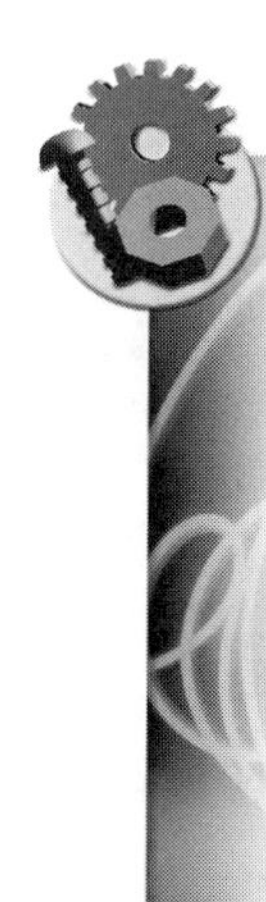

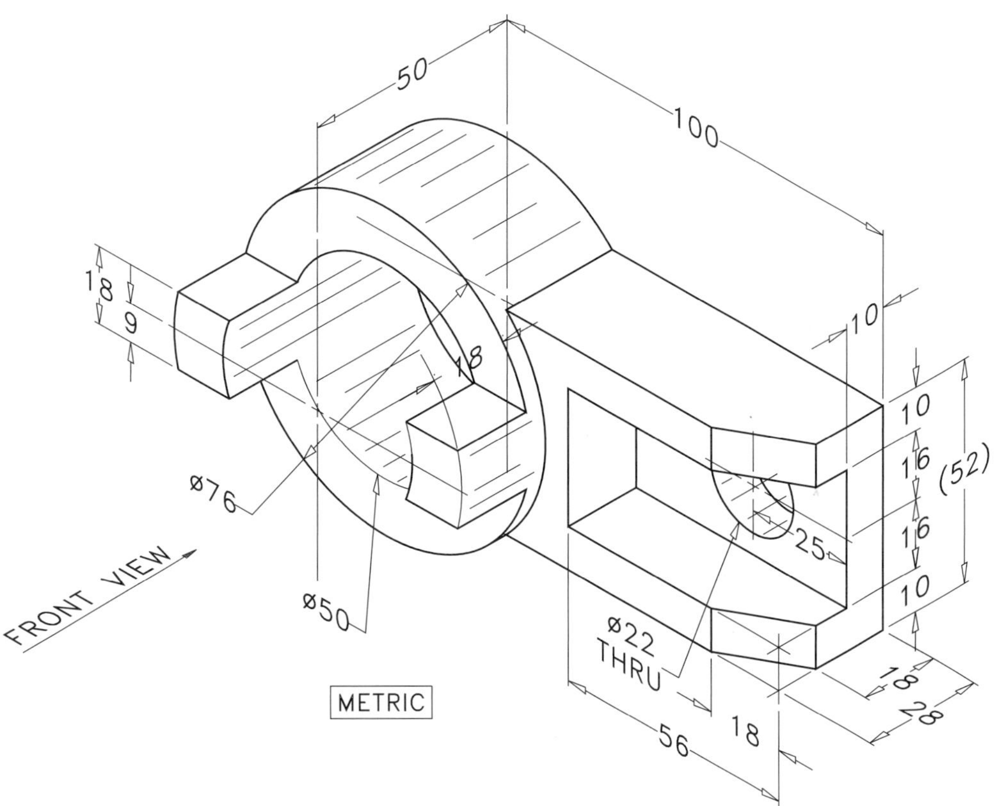

Problem 13-6

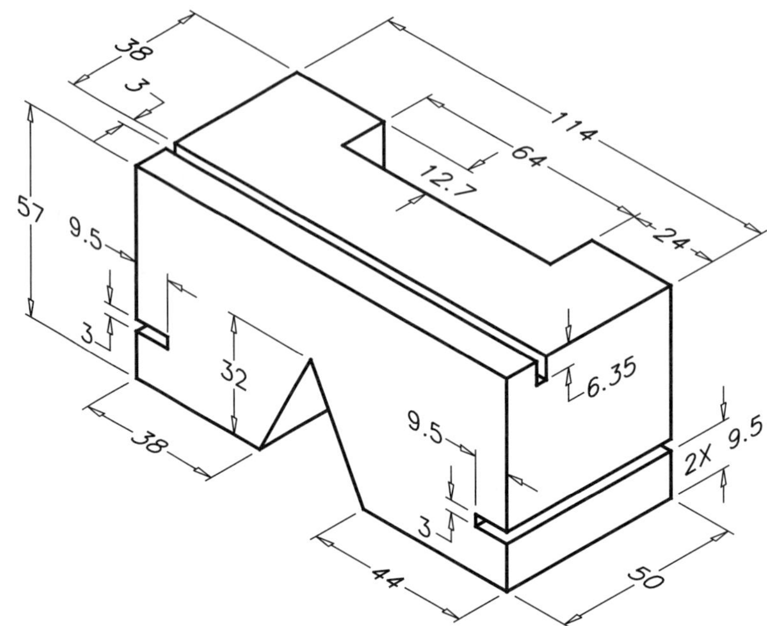

Problem 13-7

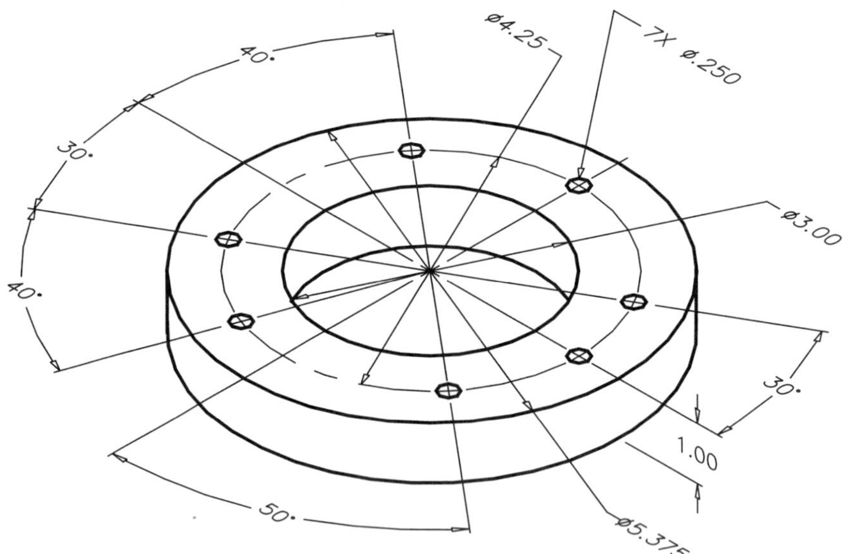

Problem 13-8

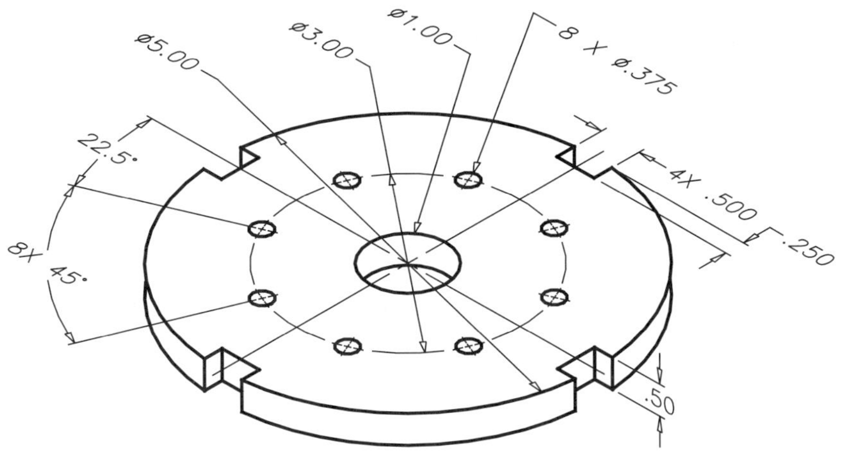

Problem 13-9

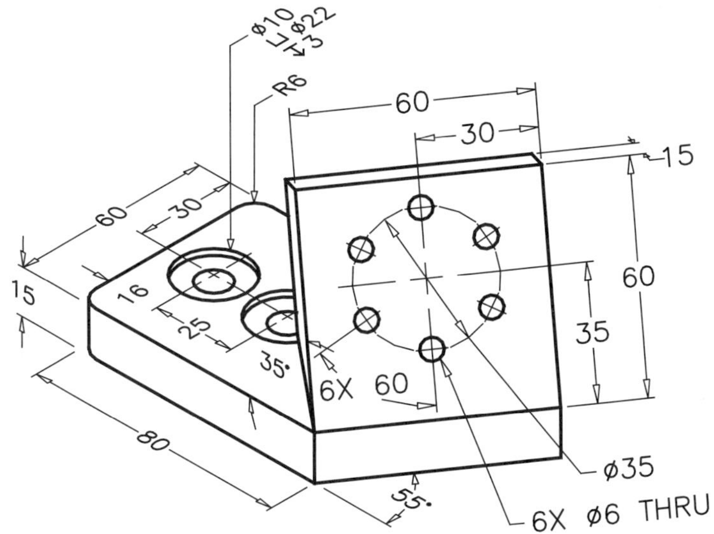

Problem 13-10

CHAPTER 13 QUIZ

1. How many linetypes are in the ACAD.LIN library?
2. List five commands that can be used to alter the appearance of a line.
3. List four objects that are affected by a change in linetype.
4. List the command sequence to draw with a centerline and a hidden line.
5. What would be the line scale value for a line that is to be plotted at a scale of 2 = 1?
6. When you describe the state of a layer, what qualities are considered?
7. What is the advantage of using **Qselect**?
8. What aspects of Change Properties apply to 2D drawing?
9. What are the uses for the include/exclude options of **Quick Select**?
10. You are working on a drawing containing several different linetypes and would like to find out their names. What command and option should be used?
11. How can the **Properties** dialog box be accessed and what can be done with it?
12. What will typing **DDLTYPE** at the Command prompt allow you to do?
13. Describe the difference between the effects of **Ltscale** and **Celtscale**.
14. You need to erase some information that is on a layer that is currently frozen. What is your plan?
15. Create a list of possible layer names and their contents, appropriate linetypes and colors for creating a drawing template suitable for orthographic projections.

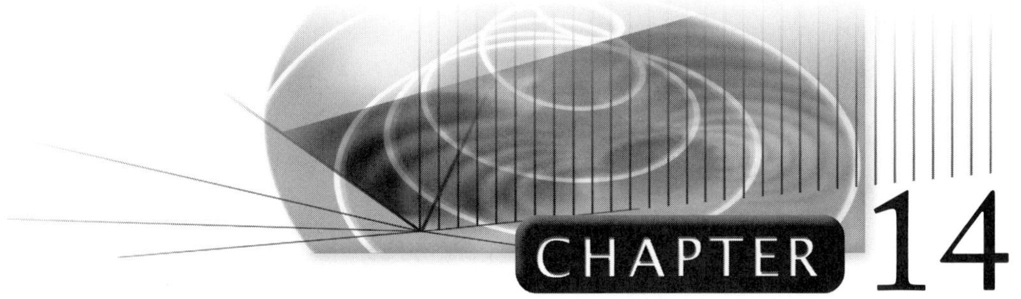

Inquiry Commands

As the drawings you're working with get more complex, you might find it hard to keep track of some of the drawing parameters you have established. This chapter will introduce methods for

- Determining information about drawing objects
- Determining information about drawing files
- Using the AutoCAD calculator

The **Inquiry** toolbar and the **Inquiry** menu on the **Tools** menu contain commands used to obtain information about drawing objects or an entire drawing. Commands to be introduced in this chapter include

- **Dist**
- **Area**
- **Massprop**
- **Region**
- **List**
- **Dblist (Data Base List)**
- **ID (ID Point)**
- **Time**
- **Status**
- **Purge**
- **Cal**

DETERMINING DISTANCE

The **Dist** command measures the distance and angle between two selected points. This command can be especially helpful when used with SNAP or OSNAP locations to measure exact locations. **Dist** can be useful in determining the size of a part during the design stage, or checking an angle of an object from a baseline. Access the command by selecting the **Distance** button on the **Inquiry** toolbar, by selecting **Distance** from **Inquiry** on the **Tools** menu, or by typing DI (ENTER) at the Command prompt. The command sequence is as follows:

Command: **DI** ENTER *(Or click the Distance button.)*

DIST: Specify first point: *(Select a point.)*

Specify second point: *(Select second point.)*

The resulting display for the points in Figure 14.1 would be

Distance=1.47 Angle in XY plane = 274. Angle from XY plane = 0

Delta X = 0.09, Delta Y = -1.46, Delta Z = 0.00

Command:

DETERMINING AREA

The **Area** command can be used to determine the area and perimeter of an object. If OSNAP is used, the **Area** command can be used to determine exact sizes of parts and, at the same time, determine lengths and locations of features. A running total can be kept and areas can be added to or deleted from the total. Access the command by selecting the **Area** button on the **Inquiry** toolbar, by selecting **Area** from **Inquiry** on the **Tools** menu, or by typing AA (ENTER) at the Command prompt. As the command is entered, four options are given:

Command: **AA** ENTER *(Or click the Area button.)*

AREA

Specify first corner point or [Object/Add/Subtract]: *(Specify a point or enter an option.)*

FIRST CORNER POINT

When you accept the **Area** "first corner point" option, you will be asked to supply a series of next corner points to specify the area to be defined. Points can be selected with the pointing device or by any of the other selection method. Using OSNAP will greatly increase the accuracy in computing the area. Figure 14.2 shows an example of computing the area of a part excluding the drilled features by using the point option. The command sequence is as follows:

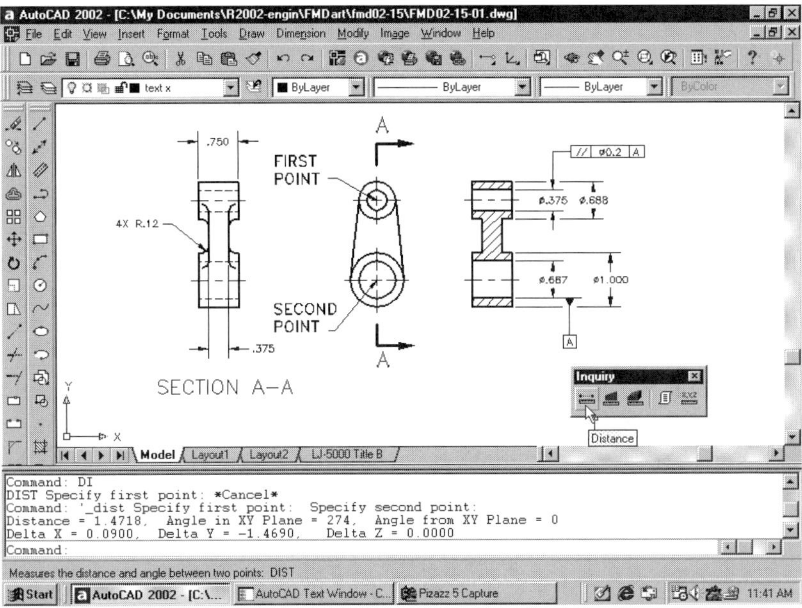

Figure 14.1 *The* **Dist** *command measures the distance and angle between two selected points. Courtesy Jill Heilman.*

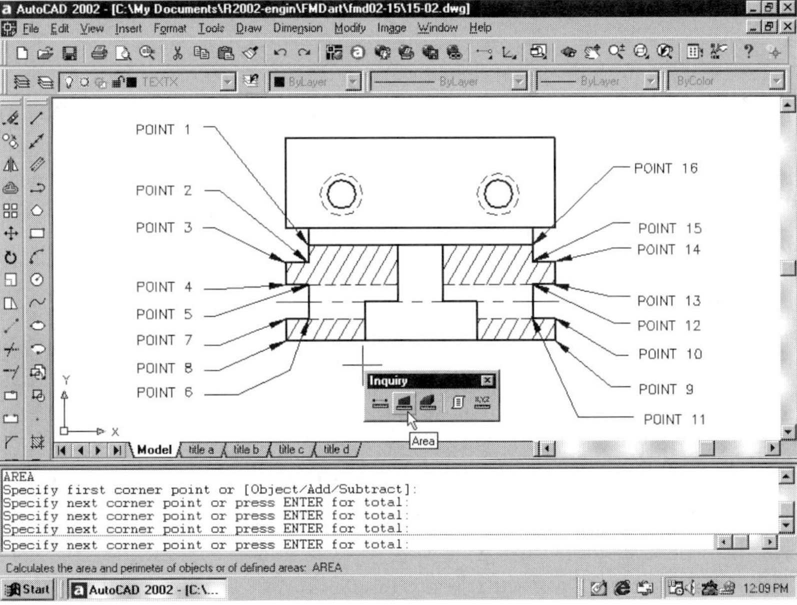

Figure 14.2 *The* **Area** *command can be used to compute the area of a specified shape. As the command is completed, the area and the perimeter are given at the Command prompt. Courtesy Cathy Schlesser.*

Command: **AA** ENTER *(Or click the Area button.)*

AREA

Specify first corner point or [Object/Add/Subtract]: *(Specify first point.)*

Specify next corner point or press ENTER for total:

Continue to select points as required. In the example in Figure 14.2, press ENTER at Point 16. This will produce a display listing the area and the perimeter. The same results will occur if Point 1 is entered as the ending point.

OBJECT

The Object option will compute the area of a circle, ellipse, spline, or an object composed of polylines. Selecting a circle will cause the area and the circumference to be displayed. When a closed polyline is selected, the area and perimeter will be displayed. If an open polyline is selected, the area is computed based on an imaginary line extending from each end of the polyline. The command sequence is as follows:

Command: **AA** ENTER *(Or click the Area button.)*

AREA

Specify first corner point or [Object/Add/Subtract]: **O** ENTER

Select object: *(Select object.)*

Area=27.63 Length 13.63

Command:

Once the object is selected, a display of the area and circumference will be displayed. If an open polyline is selected, AutoCAD will calculate the area as if a line were drawn between the endpoints. (See Figure 14.3.) The line that is projected by AutoCAD is not included in the calculation of the area or perimeter. The command will list the area confined by the polyline and the length of the line.

ADD AND SUBTRACT

Using the Add option will allow each area measured in this mode to be displayed and added to a running total. The running total will be displayed after each addition. To begin a running total, select the Add option before selecting the objects to be added. Using the Subtract option provides the opposite effect to that of Add. Each area measured in the Subtract mode will be displayed and then subtracted from the running total. With both the Add and Subtract options, the prompt will be altered to reflect the active mode. The selected option will stay in effect until the command is canceled. Figure 14.4 is an example of computing the area of a part with the Add and Subtract options. The area was determined using continuous OSNAP to select each corner of the lot, and then subtracting the area of the openings, which are both outlined by a polyline. The command sequence is as follows:

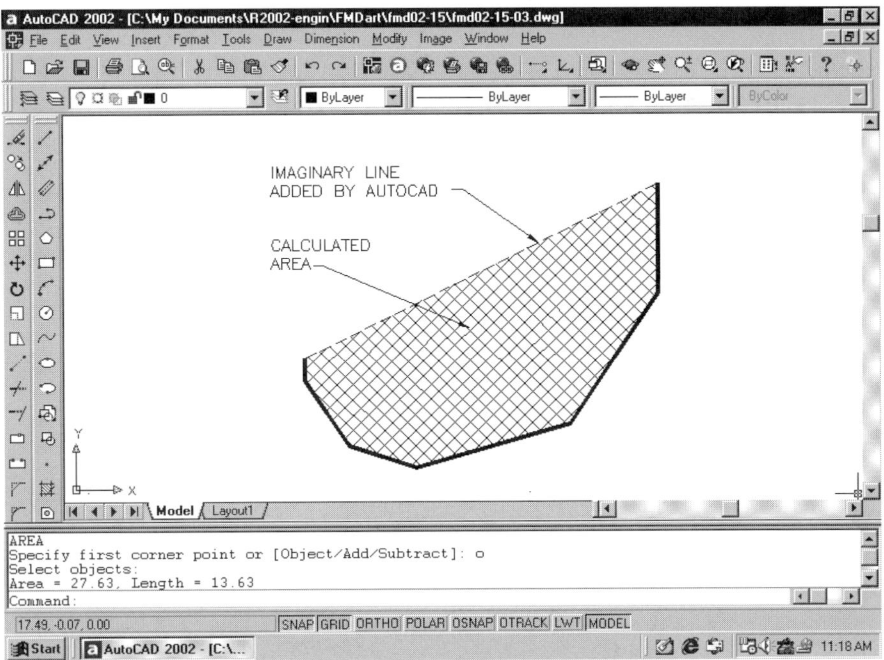

Figure 14.3 *The Object option of* **Area** *can be used to obtain information about an open or closed polyline. AutoCAD will provide the line required to close the polygon and calculate the area without displaying the line.*

Command: **AA** ENTER *(Or click the Area button.)*

AREA

Specify first corner point or [Object/Add/Subtract]: **A** ENTER

Specify first corner point or [Object/Subtract]: **O** ENTER

(ADD mode) Select objects: *(Select object.)*

Area = 1.9213, Perimeter = 6.9357

Total area = 1.9213

(ADD mode) Select objects: ENTER

Specify first corner point or [Object/Subtract]: **S** ENTER

Specify first corner point or [Object/Add]: **O** ENTER

(SUBTRACT mode) Select objects: *(Select object.)*

Area = 0.0491, Circumference = 0.7854

Total area = 1.8722

(SUBTRACT mode) Select objects: *(Select object.)*

Area = 0.0491, Circumference = 0.7854

Total area = 1.8231

(SUBTRACT mode) Select objects: *(Select object.)*

Area = 0.0491, Circumference = 0.7854

Total area = 1.7741

(SUBTRACT mode) Select objects: ENTER

Specify first corner point or [Object/Add]: **O** ENTER

Command:

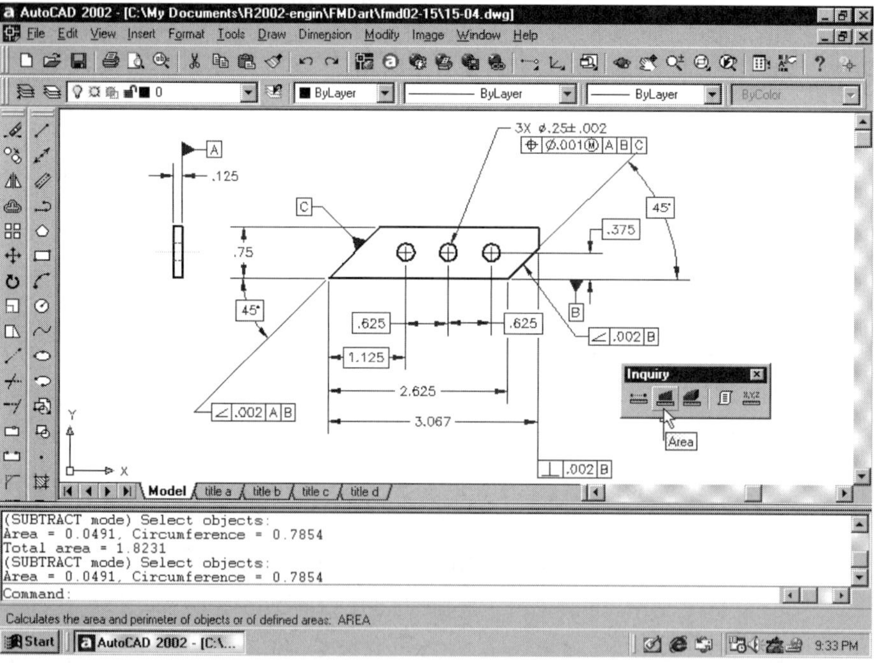

Figure 14.4 *The Add and Subtract options can be used with the* **Area** *command to add areas to or subtract areas from a running total. Courtesy Cathy Schlesser.*

MASS PROPERTIES

The **Massprop** command can be used to calculate the properties of 2D and 3D objects. Although primarily a 3D command, **Massprop** can be used to determine the area, perimeter and several other features of 2D shapes. Start the command by selecting the **Mass Properties** button on the **Inquiry** toolbar, by selecting **Mass Properties** from **Inquiry** on the **Tools** menu, or by typing MASSPROP (ENTER) at the Command prompt.

To use the **Massprop** command with a 2D object, first you need to convert the 2D object to a region. If an object that is not a region is selected, the following prompt is displayed:

Command: **MASSPROP** ENTER *(Or click the Mass Properties button.)*

Select Objects: *(Select desired objects using any selection method.)*

1 found

No solids or regions selected.

Command:

To change an object to a region, select the **Region** button on the **Draw** toolbar or type REG (ENTER) at the Command prompt and select the objects to be converted. **Massprop** can now be used to display the properties of the selected region. The command sequence is as follows:

Command: **REG** ENTER *(Or click the Region button.)*

Select Objects: *(Select desired objects using any selection method.)*

104 found

Select objects: ENTER

1 loop extracted

1 Region created

Command:

Once objects are selected, AutoCAD will display the mass properties in the text window. The mass properties for the shape in Figure 14.5 are seen below.

Regions

Area:	277.1929 sq in
Perimeter:	91.8271 in
Bounding box:	X: 23.5672 — 46.5362 in
	Y: -3.1626 — 18.0234 in
Centroid:	X: 39.2238 in

Y: 9.3193 in

Moment of inertia: X: 33838.1482 sq in sq in

Y: 433392.8323 sq in sq in

Product of inertia: XY: 97969.2639 sq in sq in

Radii of gyration: X: 11.0487 in

Y: 39.5412 in

Principal moments (sq in sq in) and X-Y directions about centroid:

I: 4704.5411 along [0.5527 -0.8334]

J: 11989.3890 along [0.8334 0.5527]

Write analysis to a file? [Yes/No] <N>: *(Press* ENTER *to continue)* ENTER

Although **Massprop** gives more information than is typically needed for most 2D situations, the command can be used to obtain the area and perimeter of an object. Although much of the information provided by this command is beyond the scope of this text, other information includes the following:

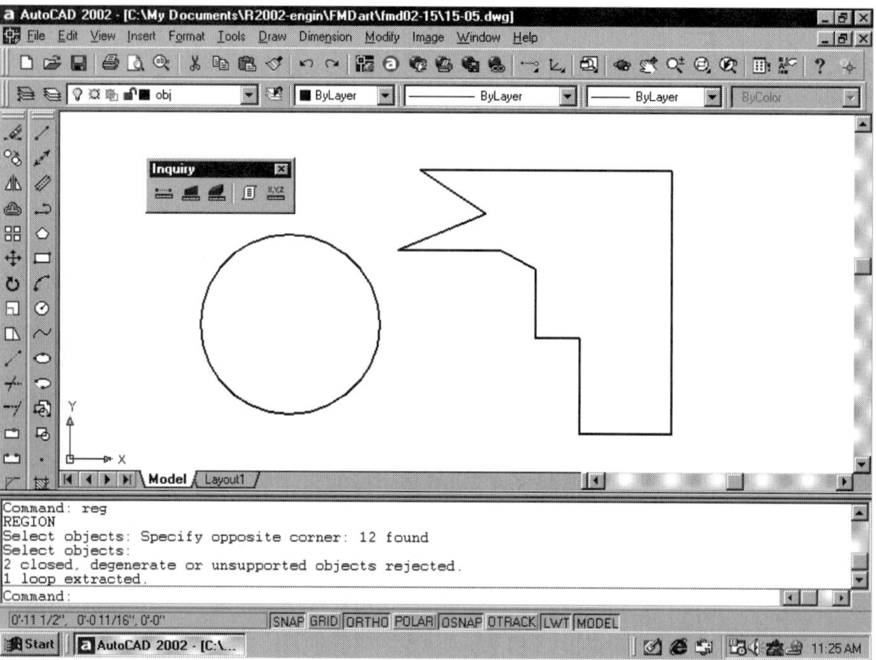

Figure 14.5 *The* **Massprop** *command can be used to calculate the properties of 2D objects, once they have been converted to regions. Once a region is selected, AutoCAD will display the mass properties in the text window.*

Bounding box—The two coordinates that define the bounding box are the diagonally opposite corners that enclose the rectangle.

Centroid—The center of the region.

Moment of inertia—A value used in engineering formulas to compute the distributed loads.

Product of inertia—A property used to determine forces causing motion in an object.

Radii of gyration—A value used to determine the moments of inertia of a solid.

Principal moments—The properties derived from the products of inertia.

LISTING INFORMATION ABOUT AN OBJECT

The **List** command displays a listing of information about an object in the drawing. Access the command by selecting the **List** button on the **Inquiry** toolbar, by selecting **List** from **Inquiry** on the **Tools** menu, or by typing LI (ENTER) at the Command prompt. The command sequence is as follows:

Command: **LI** ENTER *(Or click the List button.)*

Select objects: *(Select an object.)*

One or several objects can be selected for listing, although the information might not fit on the screen all at once. To view additional pages, press ENTER to resume the output display. Pressing F2 will terminate the text box and return the Command prompt. Selecting the **Close** button in the title bar will also close the window.

Figure 14.6 shows a shaft with a line selected for **List** and the resulting list for this line. The information that is displayed depends on the type of object that is selected—but the object's type, position relative to the current UCS, the layer location, and either model or paper space are always listed. The color and linetype are listed if either is not set by layer. Four values are always given to describe a line. The X and Y values specify the horizontal and vertical distances of the "first point" and the "next point," from the 0,0 point when the limits were established. The line in Figure 14.6 is 15.4837 to the right of the X axis. Notice that distances of the from point and the to point of the X axis is 12.2337. The difference between these two values is 3.2500, which is the specified length. Figure 14.7 shows an example of each of the listed values for a vertical and an inclined line.

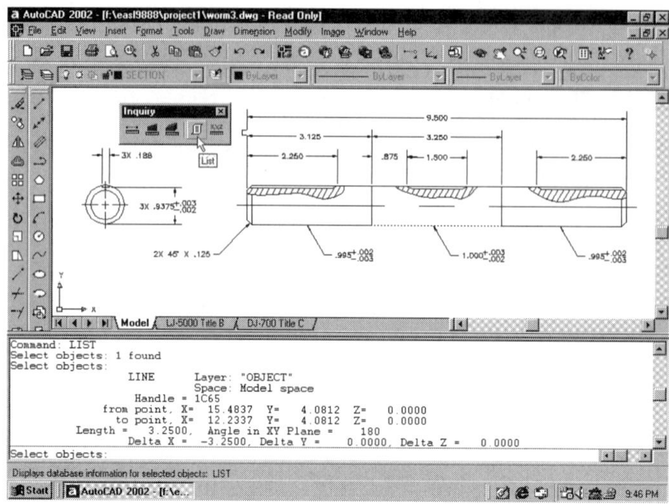

Figure 14.6 *The* **List** *command can be used to supply information about a drawing object. Courtesy Michael J. Easley.*

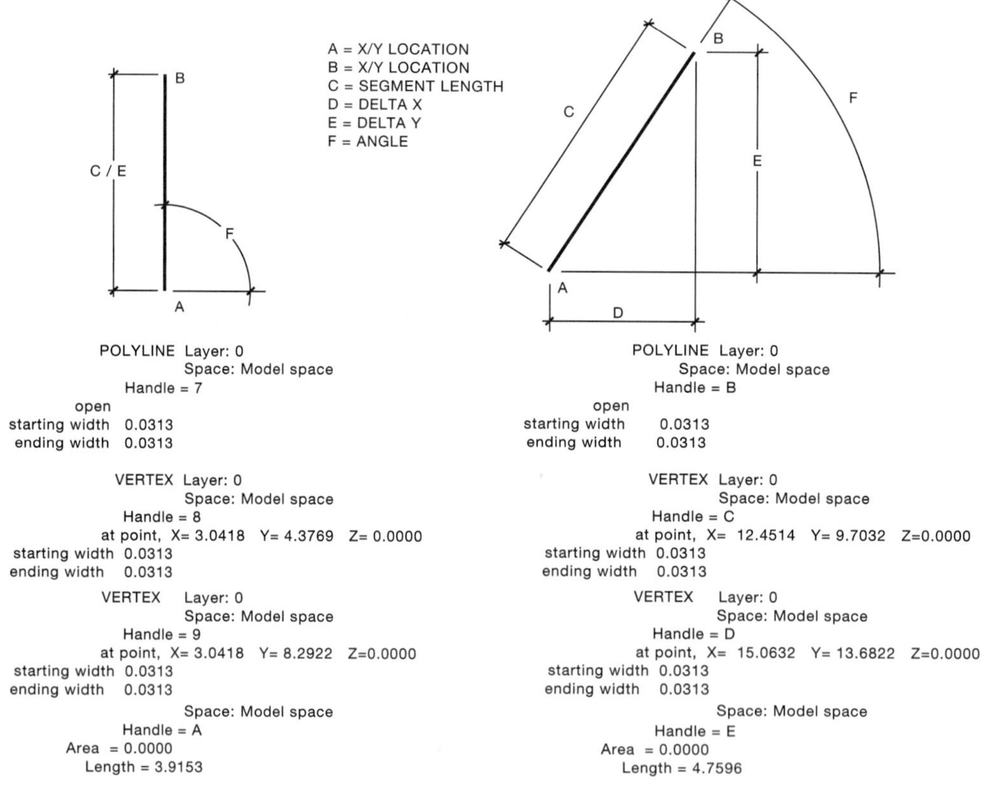

Figure 14.7 *List values for two lines.*

Listing circles will provide information based on the center point and radius, the area, and the circumference. Listing a circle is an easy way to determine the area of circular features. An example of a listing for the circle in Figure 14.8 is as follows:

CIRCLE	Layer: "OBJECT"
	Space: Model space
Handle=	1D28
Center point,	X= -23.72 Y= 0.00 Z=0.00
radius	55.00
circumference	345.58
area	9503.32

Press ENTER to Continue:

The **List** command can also be used to provide information about text and dimensions in the drawing base. (Each will be introduced in the next section.) The listing to describe text would list the following items:

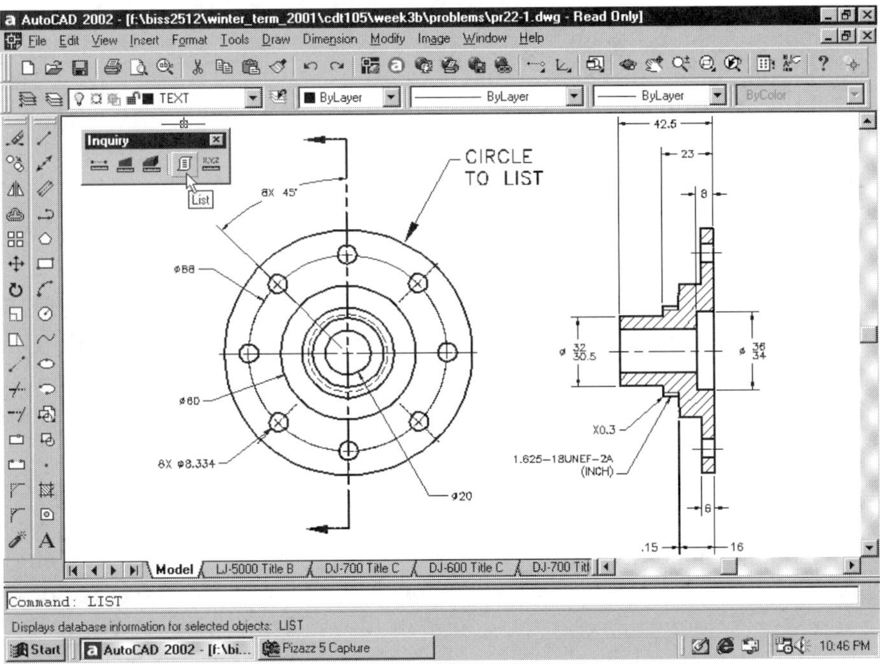

Figure 14.8 *The* **List** *command provides information based on the center point, the radius, and the diameter. Courtesy Paul Bissell.*

```
TEXT        Layer: "title block text"
            Space: Paper space
            Layout: title-c
    Handle=          73C
            Style=            "Standard"
            Font file =       gdt.shx
            start point, X=   0.71  Y = 1.34  Z = 0.00
    height  0.125
            text      1. INTERPRET DIMENSIONS AND TOLER-
                      ANCES PER ASME
            Y14.5M-1994.
    rotation angle            0
    width scale factor        1.00
    obliquing angle           0
generation normal
```

OBTAINING INFORMATION ABOUT A DRAWING

The **List** command is used to obtain information about selected objects. The **Dblist** command will produce information about the entire drawing. This command can be useful in comparing drawing sequences or information contained in two similar drawings. Information is presented in the same order as used for **List**, one page at a time. The command can be accessed only from the command line. The command sequence is as follows:

Command: **DBLIST** ENTER

Pressing ENTER will advance through a listing of information for the entire drawing. Pressing ESC will terminate the scrolling of the listing, and pressing F2 will remove the text window and restore the drawing screen. You can print listings from both the **Dblist** and **List** commands if your terminal is connected to a printer, by pressing PRINTSCREEN.

IDENTIFYING A POINT

Information about a specific point in a drawing can be obtained by using the **ID** command. This command will allow you to determine the exact location of a specific point. Access the command by selecting **Locate Point** on the **Inquiry** toolbar, by typing ID (ENTER) at the Command prompt, or by selecting **ID Point** from **Inquiry** on the **Tools** menu. The sequence is as follows:

Command: **ID** ENTER *(Or click the Locate Point button.)*

Point: *(Select a point.)*

Entering a point will display the corresponding X, Y, and Z coordinates. This can be helpful in determining the dimension of an object.

The **ID** command also can be used to identify a specific location in the drawing. If ID is entered at the Command prompt, coordinates can be supplied and that point will be identified in the drawing. This can be useful with relative coordinates entry methods to start an object at a specific location.

The **ID** listing for the location of intersection of the line in Figure 14.6 is as follows:

Command: **ID** ENTER *(Or click the Locate Point button.)*

Specify point: *(Select desired point. For this example the right end of the listed line in Figure 14.6 was used.)*

X = 15.48 Y = 4.08 Z = 0.00

TRACKING DRAWING TIME

The current time is displayed at the bottom of the drawing screen on the task bar. The **Time** command uses the date and time maintained by the computer to keep track of the time related to the drawing session. Access the command by selecting **Time** from **Inquiry** on the **Tools** menu, or by typing TIME (ENTER) at the Command prompt. Before the command can function properly, the date and time must be set. Using **Date/Time** on the **Control Panel**, accessed from **My Computer**, can set an accurate date or time. If you reenter a drawing, AutoCAD can now provide the current time, drawing creation time, last update time, and time elapsed while editing the drawing. Each might seem unimportant in an educational setting, but the information is highly useful in keeping track of hours spent on a project for billing purposes and minor bookkeeping functions, such as determining profit margins.

TIME DISPLAY VALUES

Access the time display by typing TIME (ENTER) on the command line. The display will resemble

Current time: 16 Aug, 2001 @ 11:15:25.510 AM

Times for this display:

Created:	Tuesday, Jul 21, 2001 at 3:3106.970 AM
Last updated:	Friday, Aug 10, 2001 at 2:2506.970 AM
Total editing time:	0 days 3:27:38.175
Elapsed timer (on):	0 days 1:59:51.510
Next automatic save in:	0 days 00:12:20.70

Enter option [Display/On/Off/Reset]:

Command:

Times are listed as day, month, year, hour, minute, seconds, and milliseconds.

- The Current time is the current date and time.
- Created time is marked when the drawing was initially started.
- The Last updated time lists the last editing session when either the **End** or **Save** command was used with the default file name.
- The Total editing time is the amount of time used editing the current drawing, since it was created until **Save** is used.

If **Quit** is used to end a drawing session, the time spent in that session is not added to the accumulated time. Time spent printing and plotting is excluded from the Editing time. This time is updated continuously and cannot be reset or stopped. This timer is most helpful to supervisors in tracking efficiency and logging total hours for billing.

The Elapsed time can be reset and turned On or Off. This timer is best suited for individual use each day to keep track of your time if you will be working on several different jobs. The Next automatic save time indicates when the next save is scheduled to occur (see Chapter 3).

PROMPT OPTIONS

After the time display, four options are listed on the prompt line.

- The Display option will redisplay the Time display and update the current values.
- The On option will activate the User Elapsed timer. The default value is set to On when a drawing is started.
- The Off option will stop the User Elapsed timer.
- The Reset option clears the User Elapsed timer to zero.

The command line can be returned by ENTER or F2.

STATUS

The **Status** command will list the current value settings of a drawing. Access the command by selecting **Status** from **Inquiry** on the **Tools** menu or by typing STATUS (ENTER) at the Command prompt. The listings of the drawing shown in Figure 14.8 are as follows:

Command: **STATUS** ENTER

```
1206 objects in C:\My Documents\R2002\Engineering\ pully.dwg
Model Space limits are:   X:  0.00       Y:  0.00 (off)
                          X:  12.00      Y:  9.00
Model space uses          X:  -107.90    Y:  -91.14 **Over
                          X:  162.63     Y:  105.41 **Over
Display shows             X:  -152.29    Y:  -92.60
                          X:  205.02     Y:  106.87
Insertion base is         X:  0.00       Y:  0.00 Z: 0.00
Snap resolution is        X:  0.25       Y:  0.25
Grid spacing is           X:  0.50       Y:  0.50
Current space:            Model space
Current layout            Model
Current layer:            OBJECT
Current color:            BYLAYER—7 (white)
Current linetype:         BYLAYER—"Continuous"
Current lineweight:       BYLAYER—
Current plot style:       ByLayer
Current elevation:        0.00 thickness: 0.00
Fill on   Grid off   Ortho on   Qtext off   Snap off   Tablet off
Object snap modes: Center, Endpoint, Intersection, Appoint
Free dwg disk (C:) space: 2047.7 Mbytes
Free temp disk (C:) space: 2047.7 Mbytes
Free physical memory: 2.6 Mbytes (out of 31.5M)
Free swap file space: 1996.6 Mbytes (out of 2016.5M).
Command:
```

Several elements in the list, such as color, linetype, Snap, Grid, and Ortho, should be very familiar by now. Other terms might seem a little foreign. An important value to keep track of is Free dwg disk. The value refers to the space that is available on the drive that contains the drawing. When this space is used up, AutoCAD will terminate the drawing session and save what has been done up to the point of termination. If you're working on the hard drive, this will rarely be a problem. Remember that much of the information provided by the **Status** command can be obtained using the **Properties** command.

PURGE

This command does not provide information about a drawing, but it can be a useful tool within a drawing file to eliminate unused items and minimize storage space used. Objects that have been named in a drawing session, such as Blocks, Dimstyles, Layer, Linetypes, Shapes, and Text styles, are examples of items that can be purged from a file. **Purge** can be used, for instance, if a template drawing is used as a base for a new drawing. The base drawing might contain layer listings for AUXILIARY, HATCH, SECTION, and TOLERANCE SYM. If information is not placed on these layers, each could be purged.

Items to be purged can be eliminated at any time after the drawing file is opened. Start the **Purge** command by typing PU (ENTER) at the Command prompt or by selecting **Purge** from **Drawing Utilities** on the **File** menu. Each method will produce a **Purge** dialog box similar to Figure 14.9.

Four choices can be made regarding information to be deleted from the drawing base:

> **View items you can purge**—In the default setting, this radio button, and the **Items not used in drawing** display window are displayed. The window lists categories of named items that have been defined in the drawing but are

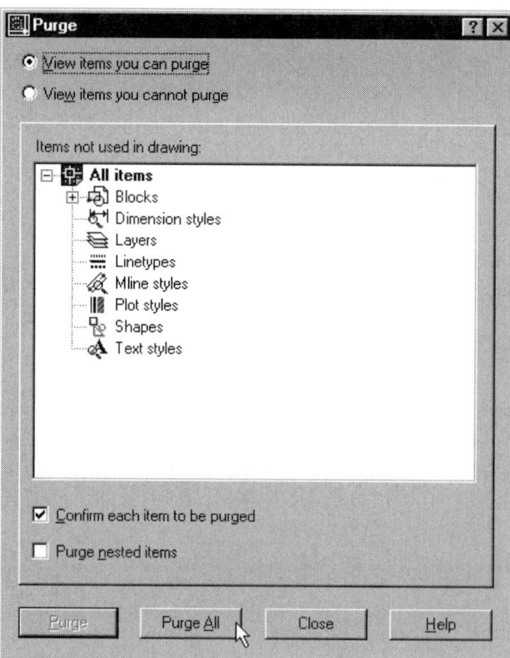

Figure 14.9 *The* **Purge** *dialog box can be used to delete unused objects and features from the drawings. Display the dialog box by typing* **PU** *(ENTER) at the Command prompt or by selecting* **Purge** *from* **Drawing Utilities** *on the File menu.*

not presently being used. Double-clicking any of the listed features will display a listing of individual named items for that category that have not been used. For instance, if you created a layer but placed nothing on that layer, the name of the layer would be listed as an item that is not in use. Notice in Figure 14.10 that layers Red and Hidden are listed. The **Purge** command will remove these and other unused items from the drawing. Execute the command by selecting the **Purge** button or the **Purge All** button.

View items you cannot purge—This radio button is OFF in the default setting. Notice in Figure 14.11 that the display window has been altered to list features that have been used in the drawing. The six layers are listed as layers that cannot be removed from the drawing. If one of these layers is selected for purge, a warning is displayed in the dialog box: "This Layer cannot be purged if it contains objects."

Confirm each item to be purged—Once an option has been entered, you can decide if you want verification of each object that is to be purged. With this check box active, a **Confirm Purge** warning box will be displayed, allowing a YES or NO answer to be given for each item to be purged.

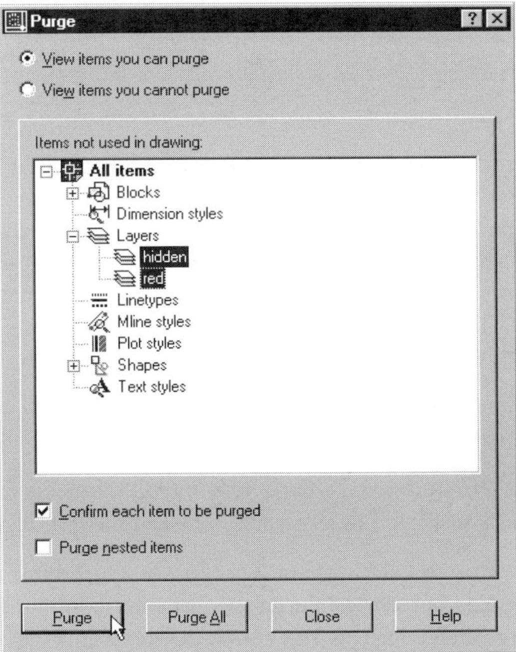

Figure 14.10 *With* **View items you can purge** *as the default setting, the* **Items not used in drawing** *list is displayed. The window lists categories of named items that have been defined in the drawing but are not presently being used. The layers Red and Hidden have been selected to be purged.*

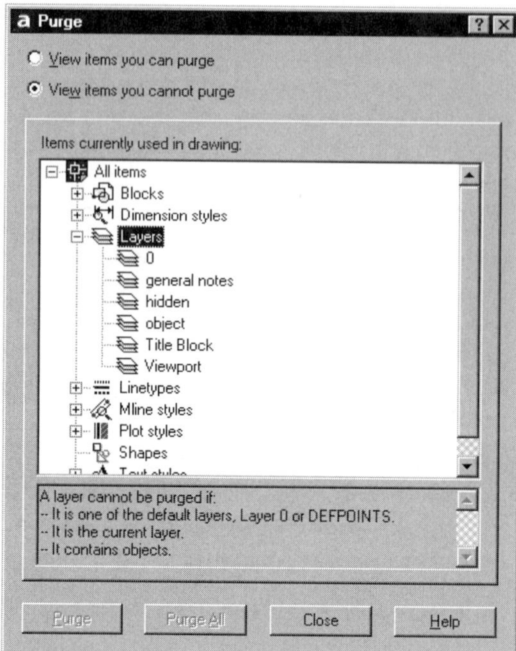

Figure14.11 *If the* **View items you cannot purge** *radio button is ON, the window lists features that have been used in the drawing. The six layers are listed as layers that cannot be removed from the drawing.*

Purge nested items—A nested object is an object that is part of another object. In later chapters you'll learn how to save components in groups called blocks. A repetitive feature such as a bolt, nut, or screw are examples of items that could be saved as a block. These blocks could then be saved in bigger blocks that represent typical arrangements. The bolt is said to be nested in the larger block. This option, if active, will purge drawing items even if they are contained within or referenced by other unused objects.

Once the desired settings have been entered, you can remove objects by selecting the **Purge** button or the **Purge All** button.

- Selecting the **Purge** button will remove multiple objects, one at a time. A warning will be displayed before each item is removed from the drawing base.
- Selecting the **Purge All** button still allows you to rethink the process. Selecting the **Yes** button will purge the item and continue prompting for items to purge. Selecting the **Yes to All** button will remove all items not used in the drawing.

If you decide that you've made a mistake, exit the command and select the **Undo** button to restore the purge.

USING THE AUTOCAD CALCULATOR

The calculator of AutoCAD can be used just as a hand-held calculator and also as a means of accurately locating points in a drawing. Typing CAL (ENTER) at the Command prompt starts the geometry calculator that will compute standard mathematical formulas. The prompt at the command line is as follows:

Command: **CAL** ENTER

>> Expression:

The double angle brackets (>>) preceding the Command prompt indicate that AutoCAD is waiting for input.

PERFORMING MATHEMATICAL CALCULATIONS

Mathematical calculations can be completed by AutoCAD, but specific symbols must be included as the values are entered. Symbols recognized by AutoCAD include the following:

+	Add	–	Subtract
*	Multiply	/	Divide
^	Exponents	[ ]	Grouped expressions

- To add 4 and 4, type **4+4** (ENTER) at the prompt.
- To subtract 50 from 100, type **100–50** (ENTER) at the prompt.
- To multiply 50 by 50 would require **50*50** (ENTER). Dividing 100 by 50 would require **100/ 50** (ENTER).

Of course, none of these problems requires a calculator, but the command is very useful for performing advanced calculations without having to clean up your work area to find a pocket calculator. By typing 'CAL (ENTER) at the Command prompt, you can use the calculator transparently. If the command line is currently set to display only one line of text, the answer will not be displayed. Adjust the display of text as needed to display the problem and the solution.

In addition to performing simple calculations, you can use the calculator to execute advanced formulas. Parentheses are used to group symbols and values of complex mathematical equations into sets. **Cal** evaluates expressions based on standard mathematical procedures using

- Left to right for operations of equal precedence.
- Expressions in parentheses first, starting with the innermost set.

Operations are performed by exponents first; followed by multiplication, division, addition, and subtraction.

Vector operators that can be used include the following:

+	Add vectors
–	Subtract vectors
*	Multiply a vector by a real number or scalar product of vector
/	Divide a vector by a real number
&	Vector product of vectors
[ ]	Signifies expression

Point/vector operators that can be used include the following:

[ ]	Encloses point or vector
,	Separates coordinate values
<	Precedes angle
@	Precedes distance
*	Denotes WCS (overrides UCS)

Entering Feet and Inches in the Calculator

Numbers are entered into the calculator just as they are at the Command prompt to enter coordinates. Although not common to mechanical drawings, ten feet three inches would be entered as **10'-3"** or **10'3"**. If you are curious about how many inches that would equal, press ENTER. Feet/inch values will be converted to inch values. The process is as follows:

Command: **CAL** ENTER

>> Expression: **10'3"** ENTER

123.0

Command:

Entering Angles

The default units of angle values are decimal degrees. Angles are entered using a format of 45d30'30". The minute (') and second (") symbols can be omitted if the values are zero. If the value is less than one degree, enter the value as **0d40'10"**. Angles can be entered in radians if the value is followed by the letter r (5.5r), or in grads when followed by the letter g (15g). No matter how the values are entered, they will be converted to decimal degrees.

Converting Units of Measurement

The calculator can be used to convert units of measurement from one system to another. Enter the value in the following format: **CVUNIT** (value, from unit, to unit). The process to convert 120 inches to millimeters is as follows:

Command: **CAL** ENTER

>> Expression: **CVUNIT(120,inch,mm)** ENTER

3048.0

Command:

You can find additional values accepted by the calculator in the **Help** menu by researching Standard Numeric Functions.

USING THE CALCULATOR TO CREATE OR EDIT POINTS

In addition to the normal functions of a calculator, the **Cal** command can be used to create or edit a drawing. Listed below are common functions.

Function	Calculation to be performed
CUR	Allows a point to be selected using a cursor.
DIST	(P1,P2) Distance between point P1 and P2.
DPL	(P,P1,P2) Distance between point P and line P1-P2.
ILL	(P1,P2,P3,P4) Gives the intersection between lines P1-P2 and P3-P4.
PLD	(P1,P2,Dist) Point on a line P1-P2, that is the specified distance from P1.
PLT	(P1,P2,T) Point on line P1-P2 defined by the parameter T.
RAD	Gives the radius of a selected circle.
ROT	(P,Origin, Angle) Rotates point P through an angle about the origin.
ROT	(P, P1ax, P2ax) Rotates point P through angle using line P1-P2 as the rotation axis.

In addition to the functions listed above, some commonly used functions have shortcuts that combine some of the previous functions with the Endpoint Object Snap mode. The shortcuts include the following:

Function	Shortcut for	Calculation to be performed
DEE	Dist(end,end)	Distance between two end points
ILLE	ill(end,end,end,end)	The intersection of two lines defined by four endpoints
MEE	(End+end)/2	Midpoint between two endpoints
NEE	Nor(end,end)	Unit vector in the XY plane and normal to two endpoints
VEE	Vec(end,end)	Vector from two endpoints
VEE1	Vec1(end,end)	Unit vector from two endpoints

To determine the length of a line, use the following sequence:

Command: **CAL** ENTER

>> Expression: **DEE** ENTER

>> Select one endpoint for DEE: *(Select desired endpoint.)*

>> Select another endpoint for DEE: *(Select the other endpoint.)*

10.2500 *(Length given for specified line.)*

Command:

To determine the intersection of two lines, use the following sequence:

Command: **CAL** ENTER

>> Expression: **ILLE** ENTER

>> Select one endpoint for ILLE: First line: *(Select first endpoint.)*

>> Select another endpoint for ILLE: First line: *(Select the other endpoint.)*

>> Select one endpoint for ILLE: Second line: *(Select first endpoint.)*

>>Select another endpoint for ILLE: Second line: *(Select the other endpoint.)*

(12.25 5.25 0.0) (Intersection given for specified lines in XYZ coordinates.)

Command:

To determine the radius of a circle, use the following sequence:

Command: **CAL** ENTER

>> Expression: **RAD** ENTER

Select circle, arc, or polyline segment for RAD function: *(Select circle.)*

1.13702 *(Radius for specified circle.)*

Command:

CHAPTER 14 EXERCISES

1. Open drawing E-10-14.

   Compute the following:

   The area the storage tank will cover _______________

   Area of each circular concrete support _______________

   Perimeter of the storage tank_______________

   Circumference of tank _______________

2. A steel fabricating company needs to order enough primer to cover (5,000) 21.5" × 8.5" steel plates. Each plate has 4- .75" diameter holes with the center points located 2" down, 2" from plate edge and 3" o.c. The plates will be made of 1/2" steel. Draw the top, front, and side views of the plate, and save the drawing as E-14-2. Use the drawing to determine the following information:

   Surface area 1 side ______________

   Surface area 2 of sides ______________

   Perimeter ______________

   Area of edges ______________

   Area of each hole ______________

   Total area of holes ______________

   Total area of 1 plate ______________

   Total area of 5,000 plates ______________

3. Open drawing E-14-2. Copy the listed times required to complete the project.

   Created ______________

   Last updated ______________

   Total editing time ______________

   Terminate the drawing session using End or Save.

4. Open drawing E-12-3 and provide the following information:

   Drawing name ______________

   Total objects ______________

   Limits ______________ ______________

   Limits ______________ ______________

   Current layer ______________

   Current linetype ______________

   Free disk space ______________

   Status of: Grid ______________ Snap ______________
   Ortho ______________

   Qtext ______________ Tablet ______________

   Terminate the drawing with **Save**.

5. Open drawing E-5-8 and determine the surface area of the

   AREA ______________ PERIMETER ______________

   List the total editing time ______________

   Current drawing session time ______________

   Terminate the drawing using **Save** or **End**.

Exercises 6 through 10:

Use the attached drawings to complete the required views. Determine the surface area of the object in exercises 6 through 8. Determine the perimeter of the holes to be placed in the object in exercise 9. Determine the surface area of the object in exercise 10.

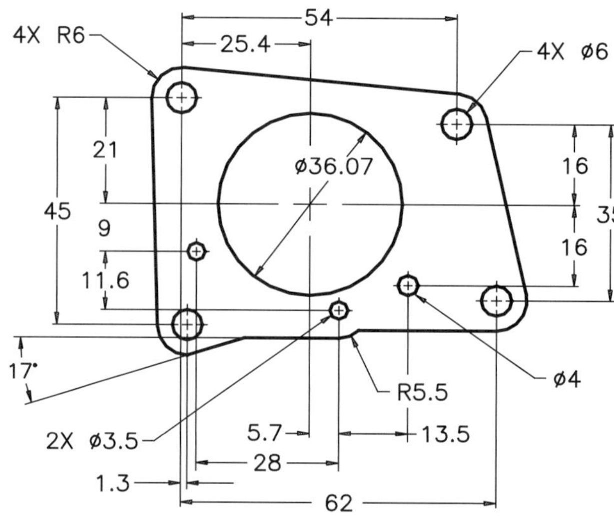

Problem 14-6

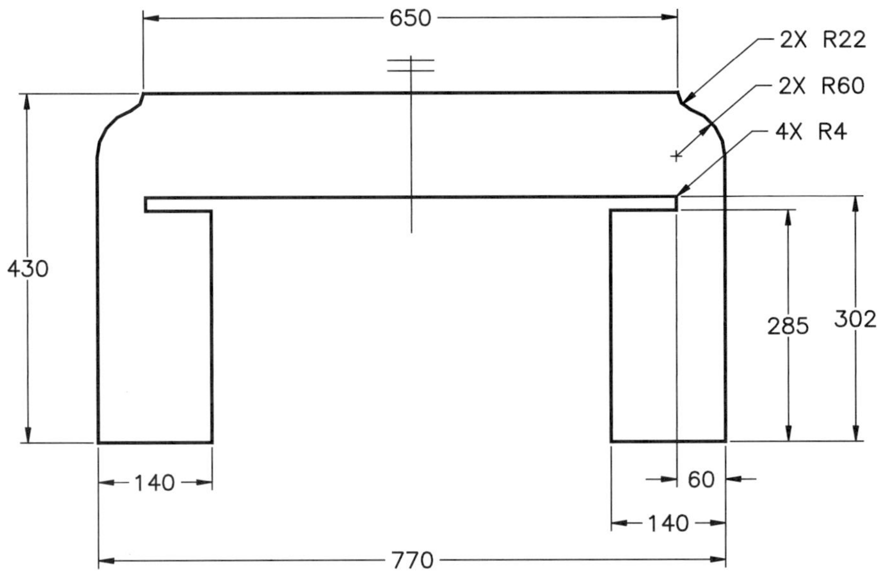

Problem 14-7

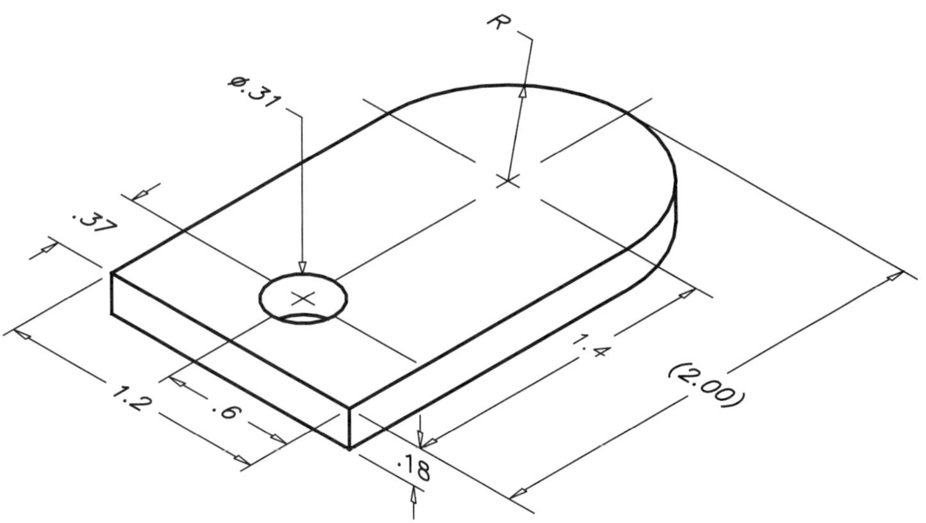

Problem 14-8

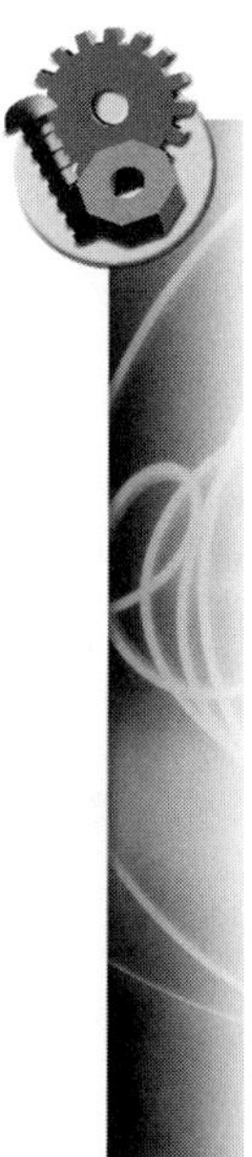

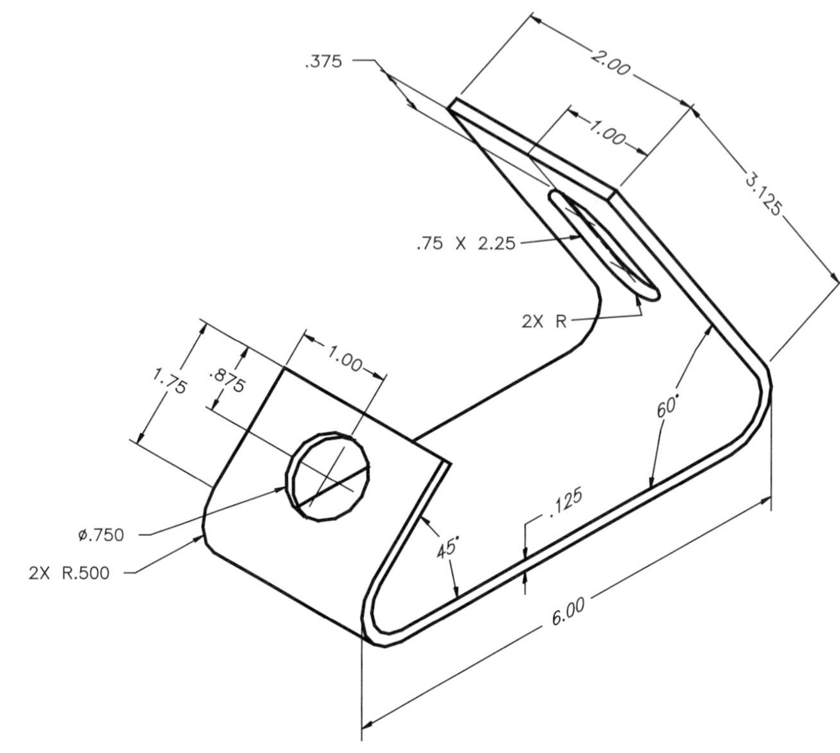
.375
2.00
1.00
3.125
.75 X 2.25
2X R
1.00
1.75
.875
60°
.125
Ø.750
2X R.500
45°
6.00

Problem 14-9

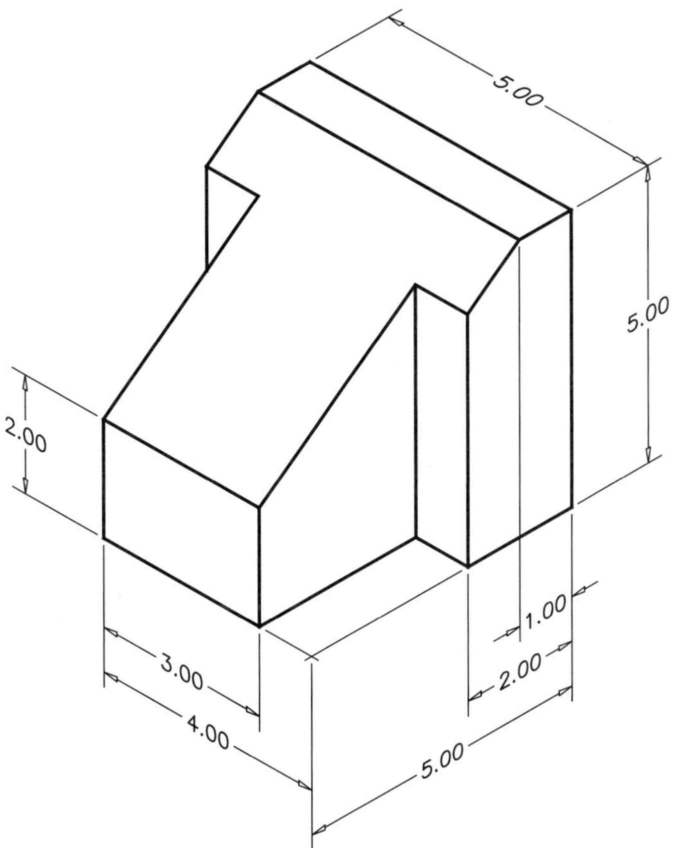

Problem 14-10

CHAPTER 14 QUIZ

1. Give four types of information that will be given if a circle is selected for a list.
2. Explain the difference between **List** and **Dblist**.
3. Write the name of the command and describe two different uses for identifying a point.
4. What are the two methods of selecting objects to determine their area?
5. What command is best suited to determine the length of a line between two points?
6. What is the process to set the internal clock of your computer?
7. What value of the status listing keeps track of available drawing space?
8. What is the meaning of DELTA X and DELTA Y?
9. You've opened a drawing, made a few changes, and then discarded the drawing. Will the time spent in this drawing be added to the accumulated time?
10. Explain each of the categories used to measure time.
11. Describe how the **Area** command can be used on an open polyline segment.
12. You need to determine the area of a rectangle. Explain the difference in selecting three and four points.
13. List the symbols used by calculator for add, subtract, multiply, divide, and to indicate a set.
14. Determine the following values for your workstation:

    Free drawing space

    Free temp disk space

    Free physical memory
15. Determine how many millimeters are in 500 inches, and how many feet are in 3000 mm. Explain the entries needed to determine the answer.
16. What is the circumference of a 4" diameter circle to the nearest 1/16"?
17. What function of the calculator should be used to determine the radius of an existing circle?
18. Research the operations required for the calculator to determine the following functions:

    The cosine of an angle

    The square root of a positive number
19. What will the ILLE calculator function do?
20. Research and describe the calculator function to rotate a point about an axis.

PLACING TEXT AND DIMENSIONING

Placing Text on a Drawing

This chapter will introduce methods for

- Placing small amounts of text using the **Text** command
- Placing large amounts of text using the **Mtext** command
- Creating varied styles and fonts of text
- Editing text on a drawing
- Finding and editing text
- Placing text in template drawings

New commands to be examined in this chapter include

- **Text**
- **Style**
- **Ddedit**
- **Mirrtext**
- **Qtext**
- **Mtext**
- **Spell**

GENERAL TEXT CONSIDERATIONS

Before text is placed on a drawing, several important factors should be considered. Each time a drawing is started, consider who will use the drawing, the information the text is to define, and the scale factor of the text. Once these factors have been considered, the text values can often be placed in a template drawing and saved for future use, similar to the stock notes seen in Figure 15.1. Most mechanical firms and offices that do civil, municipal, or government projects tend to use a simple block lettering shape.

Some offices use a forward or backward slant to make their office lettering more distinctive. You are unlikely to find all of these variations on one professional drawing. Figure 15.2 shows some of the variations that can be found on drawings.

TEXT SIMILARITIES

Although there are variations in styling, two major areas are uniform throughout mechanical drawings. Text height and capital letters are an industry standard. Text placed in the drawing area is required by ASME standards to be as follows:

Use	Height in inches	Height in mm
Drawing titles, Drawing size,	.25	6
Drawing number & revision number	.125	3
Section and View Letters	.25	6
Drawing block headings	.10	2.5
All other characters	.125	3

The other common feature is that text almost always comprises capital letters, although there are exceptions based on office practice. With CAPSLOCK activated, all lettering will be capitalized unless SHIFT is pressed. Pressing the SHIFT will make any letters typed while the key is depressed lowercase letters. If you have typing experience, it will take some time getting used to not having to press SHIFT at the start of each sentence.

8. UNLESS OTHERWISE NOTED THE DIM. TOLERANCE IS +/- 0.020.

7. FINISH: SURFACE INDICATED TO BE PAINTED WITH
CONDUCTIVE COATING X-COAT 211 PAINT.
FINISH REMAINING SURFACES PER STANDARD ES-9-58
SYSTEM 'D' COLOR GREY.
PURCHASE PART NO. X-COAT 211 FROM ELECTRO-KINETIC SYSTEMS INC.

6. MATERIAL: 0.12 THK AL ALLOY (6061-T1) PER QQ-A-250/11.

5. IDENTIFY PER STANDARD ES 9-6 CLASS M ON FAR SIDE.

4. DRAWING INTERPRETATION PER ES 9-4,
SECTION 6.0 & 7.0.

3. DO NOT SCALE DRAWING. WORK TO DIMENSIONS GIVEN.

2. DIMENSIONS LOCATING TRUE POSITION ARE BASIC.

1. LINEAR DIMENSIONS IN INCHES EXCEPT AS NOTED.

NOTES:

Figure 15.1 *General notes apply to the entire drawing and should be added to the drawing template to save drawing time. Courtesy Floyd D. Miller.*

MECHANICAL DRAFTERS USE MANY
STYLES OF LETTERING.
SOME TEXT IS ELONGATED.
SOME OFFICES USE A COMPRESSED STYLE.
NO MATTER THE STYLE THAT IS USED,
IT MUST BE EASY TO READ.

Figure 15.2 *Some common types of lettering found on drawings.*

Numbers

Numbers are the same size as the text they are placed with. Mechanical drawings are drawn with a minimum of two-place precision. Fractions are generally placed side by side (3/4") rather than one above the other ($\frac{3}{4}$) so that larger text size can be used.

If metric sizes are to be represented on drawings, the use of a comma as a number separator is somewhat different than with English units.

Metric sizes of four digits are often written with no comma. Three thousand one hundred and fifty millimeters is written as 3150 mm. When a number five digits or larger must be specified, a space is generally placed where the comma would normally be placed. Fifty-six thousand, three hundred and forty-five millimeters is written as 56 345 mm. Other common methods of placing numbers will be presented as dimension methods are introduced in later chapters.

TEXT PLACEMENT

Text placement refers to the location of text relative to the drawing and within the drawing file. Text on a mechanical drawing governed by ASME standards is unidirectional. All text is placed so that it is parallel to the bottom of the drawing. Text on a drawing is usually placed as a general note, a local note, as a revision, or in the title block. Figure 15.3 shows the specific areas of the drawing layout that are reserved for each type of text. The final consideration for text placement is to decide on layer placement. Text should be placed on a layer separate from other objects. Layer names might include LEGEND (legends and schedules), NOTE (general notes), NPLT (non-plotted text), REDLINE (required corrections), REVS (revisions), SYMB (symbols) TEXT (local notes), and TTLB (border and title block).

General Notes

Specifications that govern the entire drawing should be placed as a general note. General notes are placed in the title block or in a corner of the drawing. The exact location will depend on which standard is being followed. ASME standards place the gen-

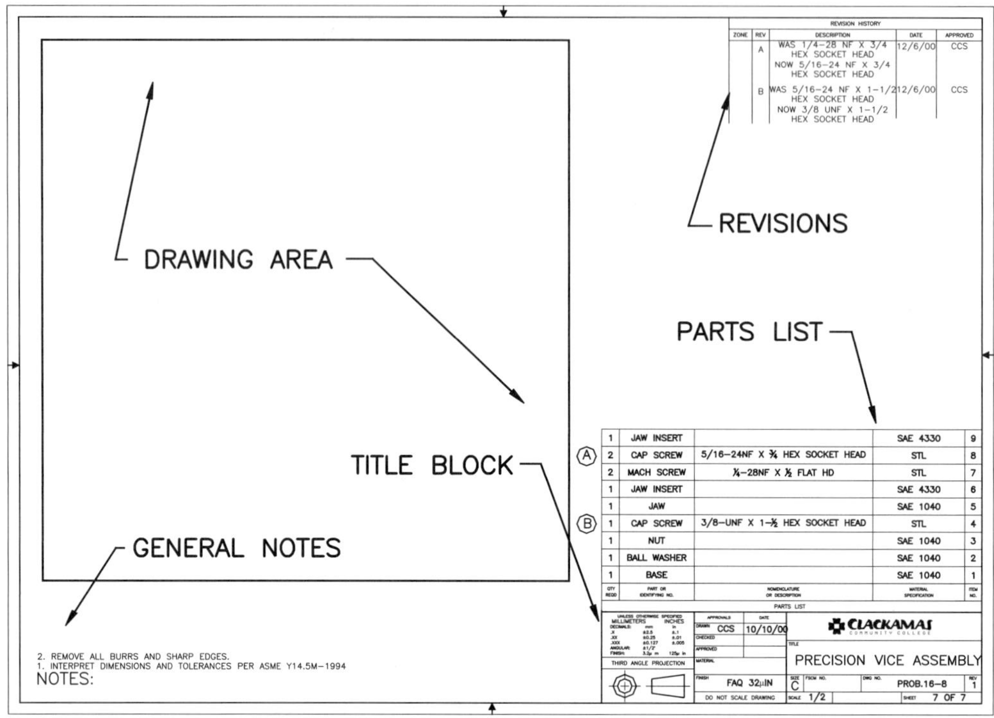

Figure 15.3 *Text locations for general notes, revisions parts list, and the title block are regulated by ASME standards. Courtesy Cathy Schlesser.*

eral notes in the lower left corner and are numbered from bottom to top. The general notes are identified by placing the word NOTES: .5" to the right and .5" above the lower border. Three common notes that will be on mechanical drawings include the following:

3. F.A.O.
2. REMOVE ALL BURRS AND SHARP EDGES.
1. INTERPRET ALL DIMENSIONS AND TOLERANCES PER ASME Y14.5M, 1994.

The letters F.A.O. represent 'FINISH ALL OVER' and will be explained in Chapters 16 and 18. Notes that describe the entire drawing can be placed as a general note. If all curved features on a part are a .25 radius, this could be listed as a general note similar to the following:

4. All radii to be .25.

If most features on a part are a similar size, a note could be

4. All radii to be .25 unless noted.

Radii that are a different size would now need to be specified with a local note referenced to the feature by a leader line. If the general notes exceed the area, the notes that will not fit should be placed on a separate sheet of A size material.

Local Notes

Local notes consist of specifications that are placed near the drawing and referenced to a specific feature using a leader line. The text should be placed so that it is totally contained within the drawing area indicated in Figure 15.3, but not so close to the object to hinder clarity. A neutral zone similar to Figure 15.4 should be established around the drawing where nothing but leader and extension lines are allowed. ASME requires a minimum of 6 mm of space to separate the object from text or dimensions.

Note: Keep in mind that this is a minimum size. A space of 10 mm or approximately .75" will make the drawing appear much cleaner.

The leader line can be either a straight line or an arc depending on office practice. Review Chapter 6 for other guidelines for placing leader lines. Text should be aligned to enhance clarity and can be either aligned left or right.

Revisions

Drawing revisions are represented in two locations on mechanical drawings. Revisions are listed in chronological order from top to bottom in the revision block. A brief description of the revision or an Engineering Change Notice (ECN) reference number can be used in the revision block to note the change. The revision is also represented in the drawing area near where the revision occurred. Revisions are represented by letters surrounded by a .375" diameter circle. A letter is assigned to the revisions based on each ECN, not the number of changes. If the first ECN requires five changes, all five changes will have an A placed beside the revised areas on the drawing. If five separate ECNs are issued, the letters A, B, C, D, and E will identify revisions. The letters I, O, S, X, and Z are not used to reference revisions. Revisions will also affect the REVISION section of the title block that will be discussed in the next section. Figure 15.5 shows a drawing with a revision.

Title Block

The text in the title block can generally be divided into fifteen areas. Each can be seen in Figure 15.6.

1. Drawn by, used by the drafter completing the drawing
2. Date, the date the drawing is completed, listed using the day, month, and year.

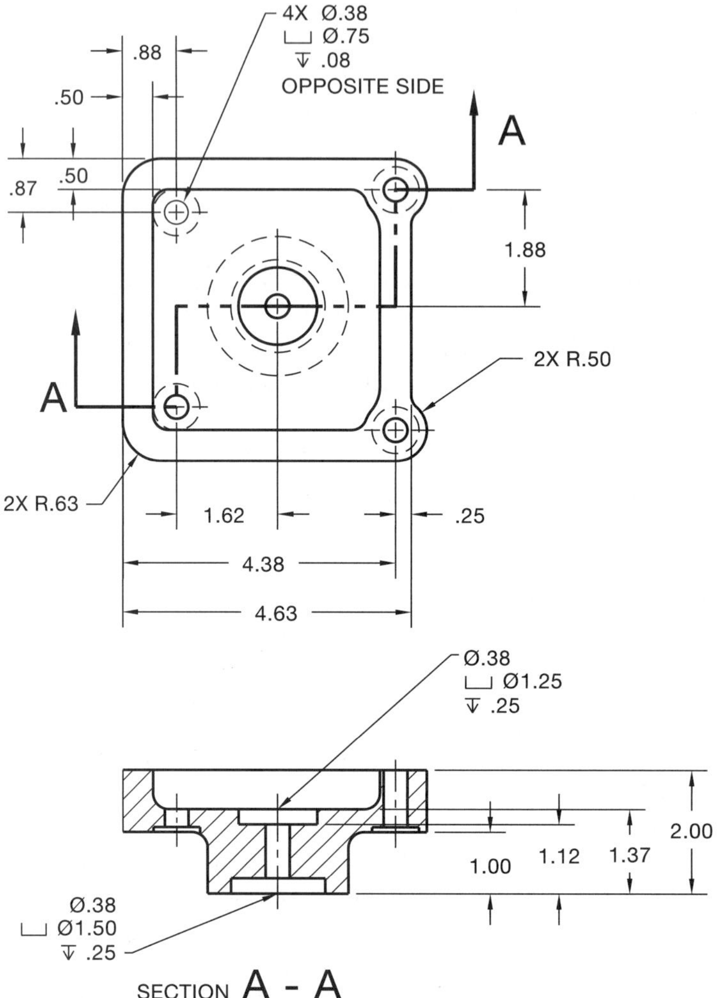

Figure 15.4 *A neutral zone should be established around the drawing where nothing but leader and extension lines are allowed. ASME requires a minimum of 6 mm of space to separate the object from text or dimensions. Notice the space between the object and all of the notes and dimensions. Courtesy Paul Bissell.*

3. Company or school name and logo.
4. Part name, the name used to describe the part.
5. Revision number, the number assigned to the drawing to represent the current revision. For drawings completed for a class project, the number should always be zero. The number is not altered unless the drawing has been revised after the part has gone into production.

REVISION HISTORY				
ZONE	REV	DESCRIPTION	DATE	APPROVED
	A	WAS 1/4-28 NF X 3/4 HEX SOCKET HEAD NOW 5/16-24 NF X 3/4 HEX SOCKET HEAD	12/6/00	CCS
	B	WAS 5/16-24 NF X 1-1/2 HEX SOCKET HEAD NOW 3/8 UNF X 1-1/2 HEX SOCKET HEAD	12/6/00	CCS

Figure 15.5 *Revisions are listed in chronological order from top to bottom in the revision block. A brief description of the revision or an ECN reference number can be used in the revision block to note the change. The revision is also represented in the drawing area near where the revision occurred. Courtesy Cathy Schlesser.*

6. Sheet number. For a detail drawing the page number will be 1 of 1. If the part is part of an assembly, the sheet number will be x of 0, where x denotes the page position in the set, and 0 represents the total number of pages in the set.
7. Drawing or Part number, the number used to describe the drawing for filing purposes. Numbering systems vary for each company, but are usually keyed to

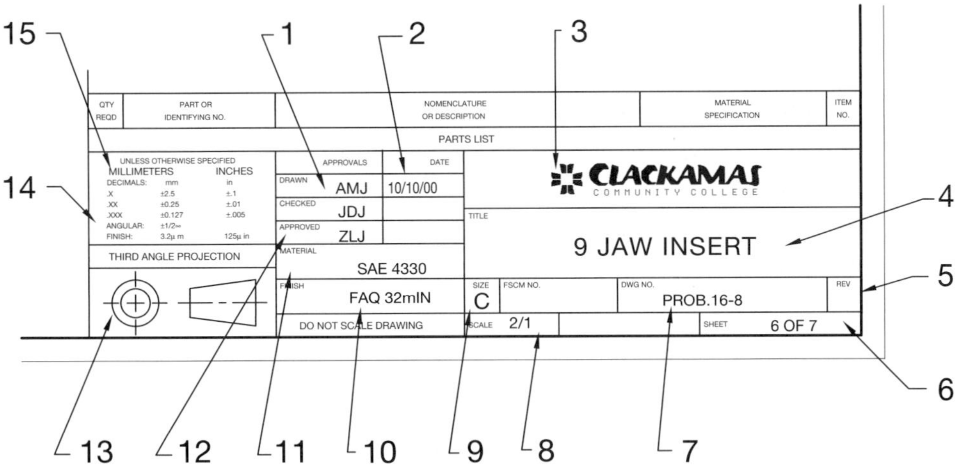

Figure 15.6 *The text in the title block can generally be divided into fifteen areas that are used to describe the part to be created.*

the manufacturing process used to make the part, to the originating department within the company, or to a numeral classification by the purchasing company.

8. Scale, the scale used to complete the drawing to initial the drawing.
9. Sheet size indicator, a letter used to represent the size of the drawing material.
10. Finish numbers represent the surface finish if a specific finish is not specified. See Chapter 18.
11. Material, describes the material to be used to make the part.
12. Checked /Approved, used by the supervisors who check and approve the drawing to initial the drawing.
13. Projection symbols representing third or first angle projection.
14. Unspecified tolerance. Tolerances to represent one, two or three place decimal precision should be specified in millimeters and inches. These tolerances apply to dimensions that do not have a specified tolerance. See Web Chapter 2.
15. Unless Otherwise Specified specifies the type of measurements used to complete the drawing. If the dimensions are in inches, the word metric will be blacked out or removed. If metric units are used, the word inches will be blacked out or removed.

Parts List

A parts list is required for assembly drawings to describe the components that comprise the assembly. The list is placed in an assembly drawing in the space normally occupied by the revision block of a detail drawing. An example of a parts list and ballooning can be seen in Figures 16.7a and 16.7b. A parts list generally includes the following:

Item number—listed in balloons referenced to each component.

Quantity—describes the number of a specific part needed to complete the assembly.

Part number—the number used in the title block of the individual detail drawing.

Description—the part name.

Material identification—the material the part is made of.

LETTERING HEIGHT AND SCALE FACTOR

One of the first prompts you'll be given as you place text will be for the height. The height of most text is .125" high. In previous chapters, the scale factor required to make line segments the desired size when plotted was considered. Similar factors must be applied to text height to produce the desired .125" high text when plotting is finished.

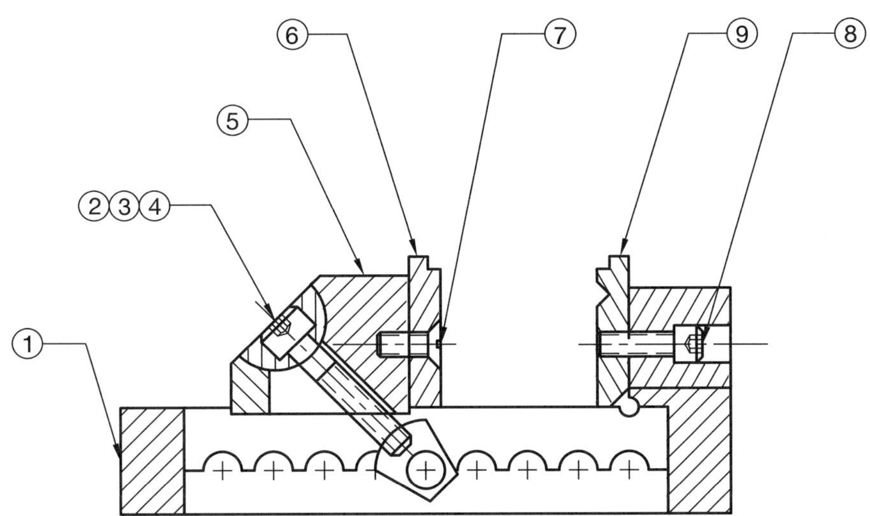

Figure 15.7a *An assembly drawing with the parts referenced to the parts list in Figure 15.7b. Courtesy Cathy Schlesser.*

Although text height can be adjusted before plotting, it is best to multiply the desired height (.125) by the scale factor. The text scale factor is the reciprocal of the drawing scale. For drawings at 1/2=1, this would be 2. By multiplying the desired height of .125" × 2 (the scale factor), you see that the text should be .25" tall.

QTY REQD	PART OR IDENTIFYING NO.	NOMENCLATURE OR DESCRIPTION	MATERIAL SPECIFICATION	ITEM NO.
1	JAW INSERT		SAE 4330	9
2	CAP SCREW	5/16-24NF X 3/4 HEX SOCKET HEAD	STL	8
2	MACH SCREW	1/4-28NF X 1/2 FLAT HD	STL	7
1	JAW INSERT		SAE 4330	6
1	JAW		SAE 1040	5
1	CAP SCREW	3/8-UNF X 1-1/2 HEX SOCKET HEAD	STL	4
1	NUT		SAE 1040	3
1	BALL WASHER		SAE 1040	2
1	BASE		SAE 1040	1

PARTS LIST

UNLESS OTHERWISE SPECIFIED	~~MILLIMETERS~~	INCHES	APPROVALS	DATE
DECIMALS:	mm	in	DRAWN CCS	10/10/00
.X	±2.5	±.1		

CLACKAMAS COMMUNITY COLLEGE

Figure 15.7b *The parts list used to describe the components of the assembly drawing in Figure 15.7a. Courtesy Cathy Schlesser.*

Quarter-inch high lettering should be .5" tall. Other common text scale heights include the following:

INCH VALUES		METRIC VALUES	
Drawing scale	**Text Scale**	**Drawing scale**	**Text Scale**
2 = 1	.5	1 = 2	50.8
1 =1	1	1 = 5	127
3/4 = 1	1.33	1 = 10	254
1/2 = 1	2	1 = 20	508
1/4 =1	4	1 = 50	1270
1/8 =1	8	1 = 100	2540

USING THE TEXT COMMAND

The **Text** command creates dynamic text that can be seen on the screen as it is entered at the keyboard. Multiple lines of text can be placed even though the command name on the **Draw** menu implies otherwise. **Text** allows some basic editing to be performed and multiple lines of text to be entered while in the same command sequence. Start the **Text** command by selecting the **Text** button on the **Text** toolbar, by typing TEXT (ENTER) at the Command prompt, or by selecting **Single-Line Text** from the **Draw** menu. The command sequence is as follows:

Command: **TEXT** ENTER *(Or click the Text button.)*

Current text style: "Standard" Text height: 0.20

Specify start point of text or [Justify/Style]: *(Select text starting point.)*

Specify height: <0.20>: ENTER

Specify rotation angle of text <0>: ENTER

Enter text: *(Enter desired text.)*

Enter text: ENTER

Command:

Once you know the other Justify options, the **Text** command can be continued with any one of the options. Notice that the first display is the text style. The current style is referred to as Standard. Methods for creating different styles will be explored after you've explored the **Text** command. Because the "start point" is the first setting requiring input, it is a logical place to start.

START POINT

The name says it all. This is your opportunity to dictate where the start of the text will be placed. Indicating a point with the cursor will make that point the lower left corner of the text location. This type of text placement is referred to as *left justified*. As the prompt for text is displayed, the crosshairs will display the starting point at the lower left corner of the text. The relation of the text to the starting point can be seen in Figure 15.8. If you don't like the location, move the cursor to a new location and select that location with the select button. Two other options are available instead of placing a start point. Each will be introduced after the basic sequence has been explored. Once the start point has been selected, a prompt for the height will be displayed.

HEIGHT

The second option specifies the text height, with the default of .20. If you'll be plotting this drawing at 1/2=1, type **.25** (ENTER) for the height. You can also adjust text height by dragging the pointing device until the distance between the cursor and the insertion point indicates the desired text height. If the height is set to zero, the text height will be controlled by the **Style** height value. (The **Style** command will be introduced later in this chapter.)

ROTATION ANGLE

The third option to be adjusted prior to entering text is the rotation angle of the text. The rotation angle can be visualized by thinking of a line extending from the start point at the specified angle, with the text above this line. The rotation angle can be entered at the Command prompt by entering X and Y coordinates, or by using the cursor. To use the cursor, drag the pointing device until the angle between the cursor and the insertion point represents the desired text rotation angle. Figure 15.9 shows an example of text rotated at various angles. Although text should be read looking from the bottom of the drawing, this option allows text to be aligned with specific drawing needs. You can accept the default of 0 rotation by pressing ENTER, or you can enter a numeric value.

PLACING TEXT

The desired text can now be entered by keyboard. Be sure to have CAPSLOCK activated. Notice that as you type, the text is displayed on the screen. Depending on the **Zoom** setting, you might be unable to read it. Either cancel the command and **Zoom** into

Figure 15.8 *The default placement with the* **Text** *command is to place text on the right side of the start point. This is called left justified.*

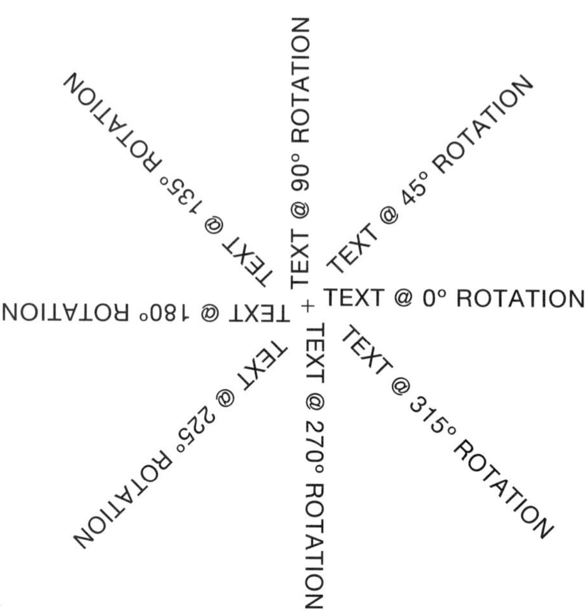

Figure 15.9 *Lines of text can be rotated so that they can be read from any angle.*

the area where you will be adding the text or read the text at the **Text** prompt line that replaces the command line. If any special characters are used, they will show up on the screen, but not at the text prompt. When the text entry is completed, press ENTER to add the text to the drawing. To add additional lines of text, press ENTER and the new start position will be adjusted to be directly below the first line of text. The prompts for start point, height, and rotation angle will be skipped and the values for the first line of text will be reused. The spacing between lines of text will be half the height of the text. The command sequence is as follows:

Command: **TEXT** ENTER *(Or click the Text button.)*

Current text style: "Standard" Text height: 0.20

Specify start point of text or [Justify/Style]: *(Select text starting point.)*

Specify height: <0.20: **.125** ENTER

Specify rotation angle of text <0>: ENTER

Enter text: **4. F.A.O.** ENTER

Enter text: **3. ALL FILLETS TO BE .25.** ENTER

Enter text: ENTER

Command:

Figure 15.10 shows the result of this command sequence. As you work with multiple lines of text, each succeeding line will have all of the attributes of the original text. Once the desired text has been entered, press ENTER twice to exit **Text**. Pressing ESC will destroy any text entered during the active command sequence. The **Mtext** command, which will be discussed later in this chapter, should be used to create large bodies of text.

JUSTIFYING TEXT

As the prompt for the start point was displayed, the options of Justify and Style were given. To justify text is to determine where and how the text will be located within a defined space. The default for placing text is for left justified text, which places the left edge of each new line of text directly in line with the line of text above. The lower left starting point is defined, and text will be placed to the right of the starting point. This is a common method of text placement on drawings. Fourteen other options are available for aligning text. Each option can be accessed by the following sequence:

Command: **TEXT** ENTER *(Or click the Text button.)*

Current text style: "Standard" Text height: 0.13

Specify start point of text or [Justify/Style]: **J** ENTER

Align/Fit/Center/Middle/Right/TL/TC/TR/ML/MC/MR/BL/BC/BR:
R ENTER

Once you feel comfortable with each option, the option can be accessed without displaying the Justify menu by typing the letter of the option at the Command prompt. The process to align text from the right can be started by the following sequence:

Command: **TEXT** ENTER *(Or click the Text button.)*

Current text style: "Standard" Text height: 0.13

Specify start point of text or [Justify/Style]: **R** ENTER

Specify right endpoint of text baseline:

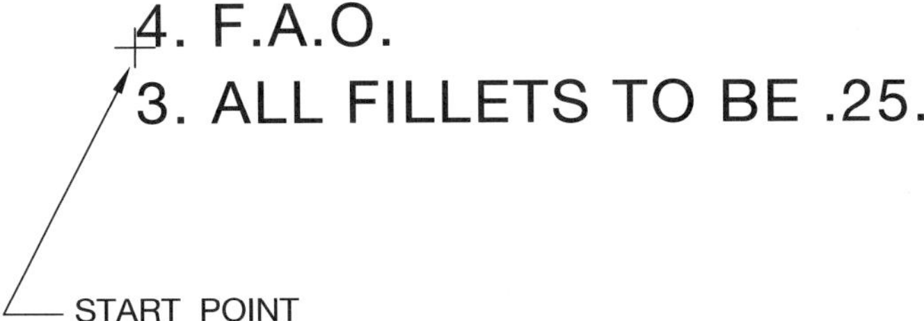

Figure 15.10 *The result of the* **Text** *command using a specified start point, a height of .125, and a rotation angle of 0.*

Even though Right is not listed as one of the current options, **R** will be accepted and will change the prompt to the right endpoint prompt for the text baseline. This prompt will then be followed by the height and rotation prompts. As text is entered using the Right option, it will be placed just as it was when Left justified was used. When ENTER is pressed to place the text, it will be added to the drawing as right justified text.

Align

The Align option will allow two useful options to be accessed. The starting and ending point can be selected so that text can be aligned with an inclined object similar to the way the rotation angle option is used. This method of selection is superior to using Rotation when the angle is not known. The second benefit of Align is the ability to select the start and end points of the line of text. The drawback to this option is that text height is assigned based on the amount of text to be placed between the start and end points. Figure 15.11 shows an example of aligned text. The command sequence is as follows:

Command: **TEXT** ENTER *(Or click the Text button.)*

Current text style: "Standard" Text height: 0.13

Specify start point of text or [Justify/Style]: **A** ENTER

Specify first endpoint of text baseline: *(Select starting point.)*

Specify second endpoint of text baseline: *(Select starting point.)*

Enter text: *(Enter desired text.)*

Fit

The Fit option is very similar to the Align option. With Fit, the text height will remain constant throughout several different lines of text, but the width will be varied. This

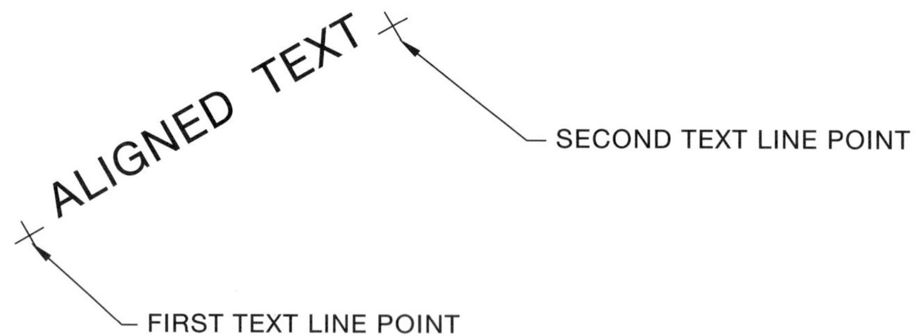

Figure 15.11 *Text can be aligned by specifying two points. The desired text will now be placed between these points. If more than one line of text is required, the height of each line may vary.*

option uses a command sequence similar to Align, but it allows for a height to be specified prior to placing the text. Figure 15.12 shows an example of text placed using the Fit option. The command sequence for Fit is as follows:

Command: **TEXT** ENTER *(Or click the Text button.)*

Current text style: "Standard" Text height: 0'-0 3/16"

Specify start point of text or [Justify/Style]: **F** ENTER

Specify first endpoint of text baseline: *(Select starting point.)*

Specify second endpoint of text baseline: *(Select starting point.)*

Specify height <0.13> *(Enter desired text height.)* ENTER

Enter text: *(Enter desired text.)*

Center

The Center text option will place a line of text centered on a specified point, as seen in Figure 15.13. The command sequence is as follows:

Command: **TEXT** ENTER *(Or click the Text button.)*

Current text style: "Standard" Text height 0.13 ENTER

Specify start point of text or [Justify/Style]: **C** ENTER

Specify center point of text: *(Select desired center point.)*

Specify height <0.13>: *(Enter desired text height.)*

Specify rotation angle of text <0>: *(Enter desired angle.)*

Enter text: *(Enter desired text.)*

To place several lines of text using the Center option, press ENTER to move the cursor to start the line below the first line. Type the second line of type and press ENTER when the line is complete. This process can be continued indefinitely. Remember that **Mtext** is better suited for entering large amounts of text.

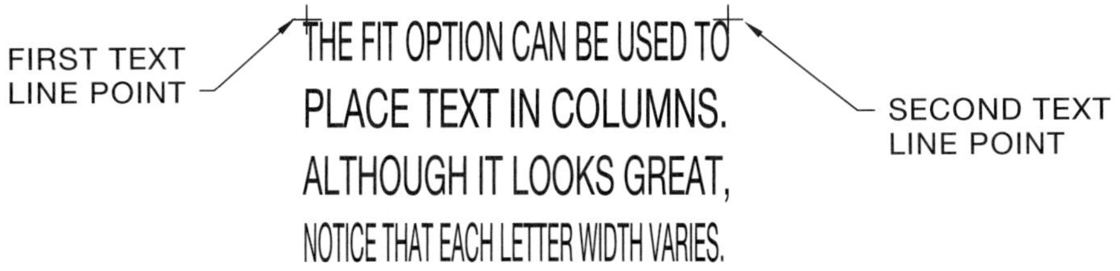

Figure 15.12 *The Fit option will place text between two specified points, but the height will remain constant if more than two lines of text are required.*

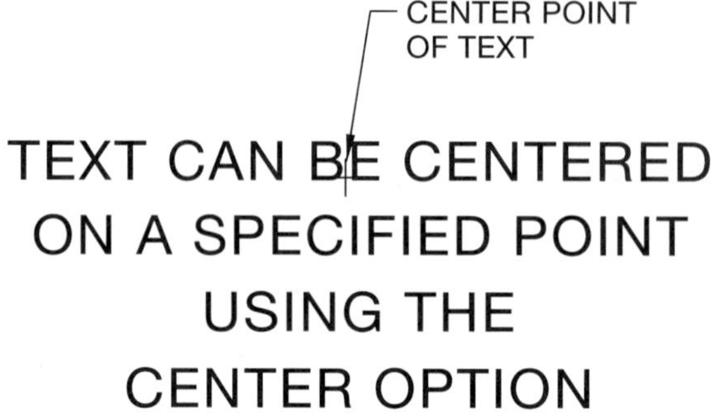

Figure 15.13. *The Center option allows text to be spaced equally around a specified center point.*

Middle

This option is similar to the effects of Center, but text is centered vertically and horizontally on the start point, as seen in Figure 15.14. The command sequence is similar to the sequence used for the Center option, once the Middle point is selected.

Right

This option is the reverse of the default left justified option. Figure 15.15 shows an example of right justified text.

Letter Options

The remaining nine options can be entered by typing the designated letters at the Justify prompt:

TL-top left	TC-top center	TR-top right
ML-middle left	MC-middle center	MR-middle right
BL-bottom left	BC-bottom center	BR-bottom right

The effects of each can be seen in Figure 15.16. The command sequence is similar to the sequence used for left justified text, although the start point is referred to by the new location. For instance, the command sequence for BR is as follows:

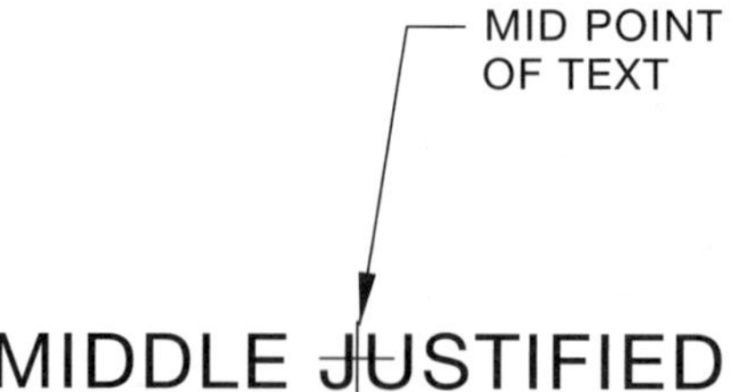

Figure 15.14 *The Middle justified option will center text along an imaginary line.*

Figure 15.15 *The Right justified option will align the right ends of lines of text.*

Command: **TEXT** ENTER *(Or click the Text button.)*

Current text style: "Standard" Text height: 0.13

Specify start point of text or [Justify/Style]: **BR** ENTER

Specify bottom–right point of text: *(Select desired start point.)*

Specify height <0.13>: *(Enter desired text height.)*

Specify rotation angle of text <0>: *(Enter desired angle.)*

Enter text: *(Enter desired text.)*

The result of this sequence can be seen in Figure 15.17.

STYLE OPTION

The last option of **Text** to be considered is the Style option that will determine the appearance of the text characters. The option is selected by typing S (ENTER) at the first text prompt. The sequence and options include the following:

Figure 15.16 *The start points for each of the nine letter options for placing text.*

Figure 15.17 *BR (bottom right) justification is often used when notes must be placed at the left side of an object.*

Command: **TEXT** ENTER *(Or click the Text button.)*

Current text style: "Standard" Text height: 0.13

Specify start point of text or [Justify/Style]: **S** ENTER

Enter style name or (?) <STANDARD>:

Standard will remain the default style until the **Style** command is used to create other styles. The **Style** command will be explored next in this chapter. If the question (?) option is selected, the following prompt will be displayed:

Enter text style(s) to list <*>: ENTER

Current text style: "STANDARD

Specify start point of text or [Justify/Style]:

WORKING WITH STYLE

The term *style* in CAD drafting describes the characteristics of a group of text such as the font, height, width, and the angle used to display the text. One style can be used for all drawing text, a second style used for the drawing titles, and a third style used for all information in the title block. The **Style** command can be used to add different fonts to your drawing as well as change the properties of the letter shapes. As styles are created, they should be stored in drawing templates for future use with all similar projects. Assigning a style name, selecting a font, and defining style properties are the three areas to be considered when setting a new style of text.

NAMING A STYLE

Start the command by selecting the **Text Style** button on the **Text** toolbar, by typing ST (ENTER) at the Command prompt, or by selecting **Text Style** from the **Format** menu. Each option will produce the dialog box shown in Figure 15.18, which can be used to name a style, assign a font, control the style options, and preview the results.

New styles can be created in the **Text Style** dialog box by selecting the **New** button. This will produce the **New Text Style** dialog box shown in Figure 15.19. The default new style is named Style1. Accept the name by clicking the **OK** button. The **New Text Style** box will be closed and Style1 is now displayed as the current name in the **Style Name** edit box. Style1 can also be altered before it is accepted as a style name by using the same procedure as in renaming a drawing or layer name. While the name is highlighted, type the desired name. Type the name of NOTES in the edit box. Clicking the **OK** button will remove the dialog box and change the name Style1 to the name provided. The name Style1 can also be renamed to something more meaningful by selecting the **Rename** button and using the **Rename Text Style** dialog box that can be seen in Figure 15.20. Selecting the **Cancel** button will remove the **Rename Text Style** dialog box and allow a different aspect of the **Style** command to be addressed, or allow the command to be terminated. The third option in the **Style Name** area is

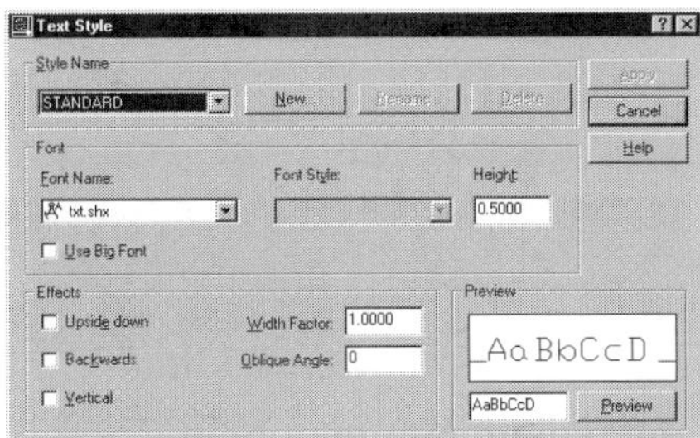

Figure 15.18 *Selecting the* **Text Style** *button on the* **Text** *toolbar, typing* **ST** *(*ENTER*) at the Command prompt, or selecting* **Text Style** *from the* **Format** *menu produces the* **Text Style** *dialog box, which can be used to name and control text styles.*

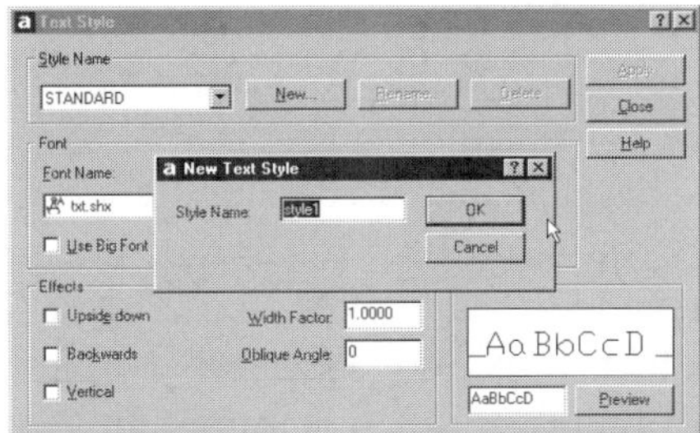

Figure 15.19 *Selecting the* **New** *button produces the* **New Text Style** *dialog box, allowing a new text style to be named.*

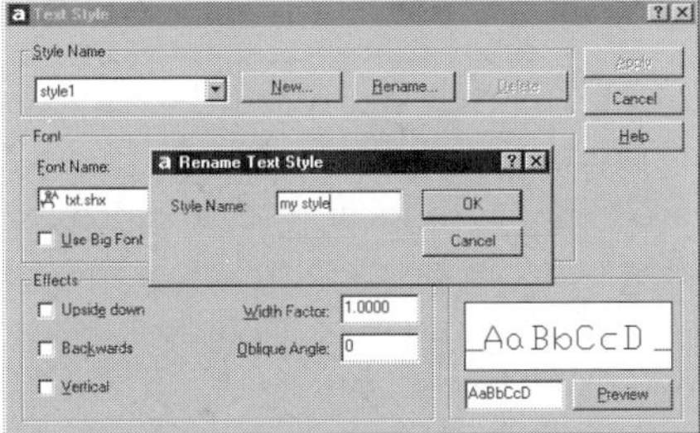

Figure 15.20 *Style1 can be renamed in the* **New Text Style** *dialog box before the name is assigned, or it can be altered by selecting the* **Rename** *button.*

the **Delete** button. Selecting this button allows a style to be deleted from the current list of text styles.

Choosing the Style Name

The name of the text style to be created should reflect the use of the text. The name can contain up to 255 letters, numbers, spaces, and special characters. (Names longer than 31 characters will spill out of the edit box.) The same guidelines that were used to name linetypes and layers should be used to name text styles. Some offices use the name of the text font for the style name. Names such as titleblock, gennotes (general notes), or notes (local notes), are examples of names that could be used to describe different styles of text. Names that include the text font and usage are helpful. Much of the lettering in this text was created using a font called GDT or ROMAND. You should create styles and save in the template drawing.

LETTERING FONTS

The word *font* is written in an italic font. The term font is used to describe a particular shape of text that includes all lower and upper case text and numerals. The shape of the letters is one of the defining differences between architectural and mechanical lettering. Mechanical drawings use gothic block lettering, although most offices use different fonts throughout a drawing to distinguish between titles and text or to highlight specific information.

AutoCAD allows all of the fonts loaded into your computer to be accessed by the **Style** command. Many of the fonts contained in AutoCAD can be seen in Figure 15.21. The illustrations in this chapter have been created using the default font of TXT, a simple font design, with letters consisting only of straight-line segments. Because of its simplicity, it is quick to regenerate and plot. Many engineering offices use fonts such as GDT, ROMANS, ROMANC, or ROMANT to letter drawings. Examples of these fonts can be seen in Figure 15.22. It is important to select fonts that are easy to read and quickly reproduced by a plotter. The complexity of a font also affects the amount of storage space required, the speed of regens, and the plotter speed.

Lettering Programs

One of the benefits of AutoCAD is that many third-party suppliers have developed material to supplement the standard program. Many offices add text libraries to the basic program. Consult advertisements in local trade magazines for other available text libraries. The **Text Style** dialog box can be used to select the font to be used with each lettering style. The box will be discussed in detail shortly, but it's important to remember that in addition to the fonts contained in AutoCAD, TrueType fonts contained in additional programs in your computer can be used in AutoCAD drawings. Methods of accessing these files will be discussed as ways of creating text styles are introduced.

Choosing a Font

The second aspect of defining a text style is choosing a font. Once the style name has been entered, selecting the down arrow beside the **Font Name** edit box will produce a list of all of the text fonts available to you. A sample of the list can be seen in Figure 15.23. The display will depend on the status of the **Use Big Font** check box. With the box inactive, all fonts in all programs on your computer will be displayed. With the box active, only the fonts contained in AutoCAD will be displayed in the list. The AutoCAD fonts are preceded by the Divider icon and followed by .SHX. The

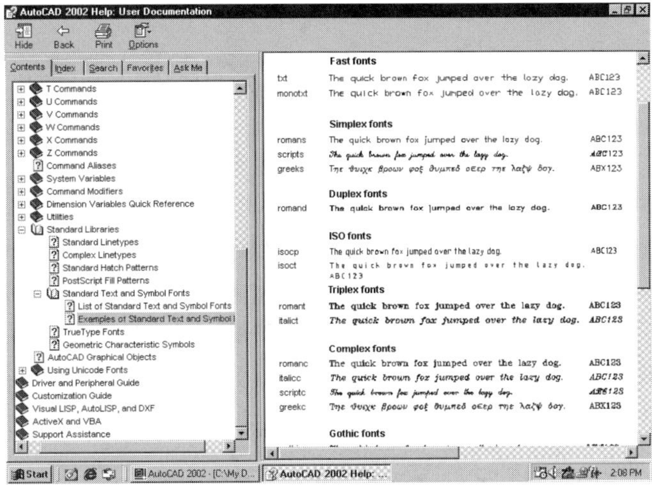

Figure 15.21 *Standard AutoCAD fonts.*

THE STYLE 'STANDARD' IS COMPOSED
OF THE TXT FONT.

THE STYLE 'NOTES' IS COMPOSED
OF THE ROMANS FONT.

THE 'TITLERESPONSE' STYLE IS COMPOSED
OF THE ROMANC FONT.

THE 'TBLOCK' STYLE IS COMPOSED
OF THE FONT ROMANT.

THE STYLE 'GENERALNOTES' IS
COMPOSED OF THE FONT ITALICC.

THE 'TITLEBLOCK' STYLE IS COMPOSED OF
THE FONT ROMANTIC

THE STYLE 'ENGINEERSTAMP'
IS COMPOSED OF THE SANSSERIF FONT.

Figure 15.22 *Common fonts used on mechanical drawings.*

TrueType icon precedes other fonts available to AutoCAD. Scroll through the list and select GDT font or a mechanical font of your choice. As the name of the font is selected in the list, the name will be displayed in the edit box, the list will be closed, and a sample of the font will be displayed in the **Preview** window.

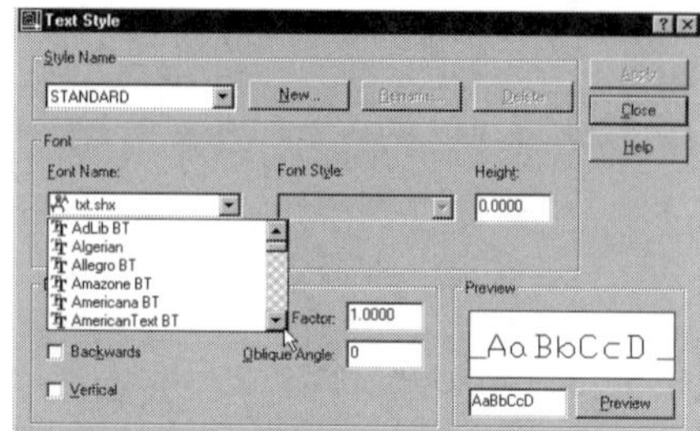

Figure 15.23 *Selecting the down arrow beside the* **Font Name** *edit box will produce a listing of the AutoCAD and the True Type font files. Scroll through the desired list and select the name of the font to be used with the style being created.*

Note: It's important to remember that even though you can add third-party fonts or fonts contained in other common Windows programs to your drawings, these fonts might cause problems if you send your drawings to firms that do not have these fonts. If a drawing is opened on a machine that does not contain the font, AutoCAD will substitute the TXT font, and some special symbols associated with the original font will be lost.

DEFINING STYLE PROPERTIES

Once the style has been named and the font has been assigned, seven options are available to further define the style being created. These options include **Font Style**, **Height**, **Upside-down**, **Backwards**, **Width Factor**, **Oblique Angle**, and **Vertical**.

Font Style

So far you've named a style (NOTES) and assigned the GDT font to the text style. The **Font Style** edit box is activated once a font is named. This edit box will allow character to be formatted with bold, italic, oblique or regular text. Options will vary depending on the selected font.

Height

If a height is entered at this prompt, all text created using this style of text will be the same height. If all local notes throughout the drawing are to be the same size, the height variable of the **Style** command can be set. When height is set with the **Style** command, the variable will not have to be set each time the font is used. The height of text is best controlled with the Height option of the command used to create the actual text. Entering a height now will delete the Height prompt in the **Text** and **Mtext** command sequences when this style is used. Responding with a 0 will allow the height to be determined each time this style is used.

Upside-Down

The default for this option will place text in the normal position. Activating this box will place text upside-down. Although this option is not typically used on drawings used for manufacturing, an example of upside-down text can be seen in Figure 15.24.

NORMAL ORIENTATION
UPSIDE-DOWN ORIENTATION

Figure 15.24 *Text can be printed in both normal and upside-down orientation.*

Backwards

The default value for this option, inactive, will produce text that is read from left to right. Activating this box will produce text that is read from right to left, or backwards. Although this option is not used often, it could be useful if a drawing is to be plotted on one side of a sheet of vellum or Mylar and text on the other side. Figure 15.25 shows an example of text printed backwards.

Width Factor

Font characters are displayed and plotted based on a ratio of their height and width. No matter what the actual height is, the width is assigned a value of 1. The default value for the width factor is also 1, making the width equal to the height of the letter. If a value of .5 is entered, text will be half as wide as it is tall. A value of 2 will produce an elongated style of lettering. Figure 15.26 shows a comparison of various width factors.

Oblique Angle

This option will allow text to be tilted from true vertical. The default angle of 0 will display text in the normal position, with the vertical leg of each letter in a vertical position. Entering a positive value will slant text to the right. Entering a negative value will produce a backslant. Figure 15.27 shows examples of altering the obliquing angle.

Vertical

This option is not active if a TrueType font has been selected. The option will display .SHX fonts in the vertical position, as shown in Figure 15.28.

Applying the Style

Once all of the parameters have been selected for the style, the options of **Apply** and **Close** remain to set the style. Choosing the **Apply** button will apply any style changes made in the dialog box to the text in that current style within this current drawing. Figure 15.29 shows an example of text created using the TXT.SHX font and the same

NORMAL ORIENTATION
BACKWARDS ORIENTATION

Figure 15.25 *Text can be printed in both the normal and backward orientation.*

WIDTH FACTOR = .5

NORMAL WIDTH FACTOR = 1.00

WIDTH FACTOR = 2.0

Figure 15.26 *Adjusting the* **Width Factor** *will affect the appearance of text.*

NORMAL OBLIQUING ANGLE = 0°

POSITIVE OBLIQUING ANGLE = 15°

NEGATIVE OBLIQUING ANGLE = -15°

Figure 15.27 *The* **Oblique Angle** *box of the* **Text Style** *dialog box affects the angle of the vertical stroke of each letter.*

VERTICAL TEXT ORIENTATION

Figure 15.28 *Placing text in a vertical position might be required on some drawings.*

TEXT STYLES CAN BE CHANGED ONCE THEY HAVE
BEEN CREATED. THIS IS TXT.SHX FONT.

TEXT STYLES CAN BE CHANGED ONCE THEY
HAVE BEEN CREATED. THIS IS ITALIC WITH
A WIDTH FACTOR OF 1.5.

Figure 15.29 *Text created using the THX.SHX font was changed to ITALIC.SHX with a width factor of 1.5 by changing the appropriate values and then selecting the* **Apply** *button.*

text after the font was changed to ITALIC.SHX with a width factor of 1.5. The text was altered by changing the appropriate values, selecting the **Apply** button, and then selecting the **Close** button. If you choose the **Close** option without selecting **Apply** first, it will accept the current style values and remove the dialog box. Existing styles will remain unchanged, and new text will be added using the current settings.

SPECIAL CHARACTERS

In addition to the symbols on a keyboard, it is often necessary to print special symbols, such as a degree or diameter symbol. This can be done by typing a code letter preceded by two % (percent) symbols. Common symbols of AutoCAD include the following:

%%C=diameter symbol

%%D=degree symbol

%%O=overscore text

%%P=plus/minus symbol

%%U=underscore text

The symbols for overscoring and underscoring text are toggles. Type %%U preceding the text to be underscored to activate underscoring. The symbol must also be typed at the end of the phrase to turn off underscoring. An example of the sequence would be

%%UALL CAPACITORS ARE LISTED AS MICROFARADS UNLESS NOTED.%%U

As the text line is displayed at the command line, the %%U will be displayed in addition to the desired text. When ENTER is pressed, the %%U will be removed from the screen and the text will be underscored. Although a single percent symbol can be created by using the percent key (SHIFT+5), when the percent symbol must precede another special character, %%% must be used. To write 50°+/–5% you must enter the following sequence:

50%%D%%P5%%%

Remember that these symbols apply to the fonts in AutoCAD and do not apply to text created by third-party vendors. It's also good to remember there is an easier method of applying special text. Use the **Mtext** command.

EDITING EXISTING TEXT

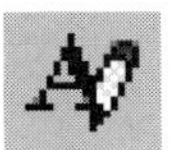

Text can be edited by using BACKSPACE and entering corrections before making the text part of the drawing base. Text can be edited once it is part of the drawing base using the **Ddedit** command. Start the command by selecting the **Edit Text** button on the **Text** toolbar, by typing ED (ENTER) at the Command prompt, or by selecting **Text** from the **Modify** menu. Typing ED (ENTER) will produce the following prompt:

> Command: **ED** ENTER *(Or click the Edit Text button.)*
>
> DDEDIT
>
> Select an annotation object or [Undo]:

Select the text to edit by moving the cursor to the line of text and select it with the select button. The error does not have to be selected, only the line of text containing the error. As the text is selected, the **Edit Text** dialog box is presented, containing the line of text. Figure 15.30a shows a line of text in a drawing and the line contained in the **Edit Text** dialog box. Because the line of text is too long to fit in the edit display, the left and right arrow keys can be used to alter the line of text that is displayed.

Once the proper portion of the text line is displayed, editing can be performed by moving the arrow to the desired location and using the select button. In the example, the following note is used:

> inTREPRET DIMNESIONS AND TOLERANCES PERAS ME
> Y14.5M 1994.

The "in" of "inTREPRET" should be in capital letters, INTREPRET and DMNESIONS are misspelled, a space should be placed between "PER" and "AS" and a space should be removed between "AS" and "ME" To fix the "in" move the arrow to the right

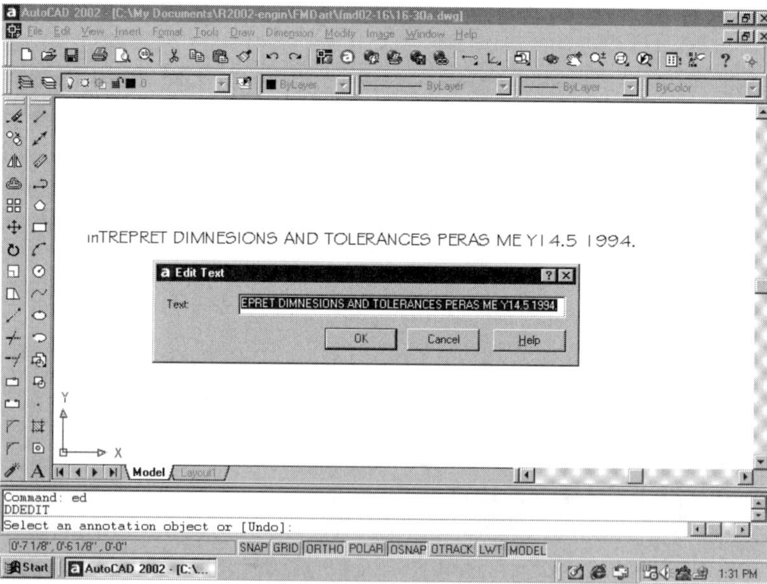

Figure 15.30a *The* **Ddedit** *command allows a string of text to be edited. Once selected, the text string will be placed in the* **Edit Text** *box.*

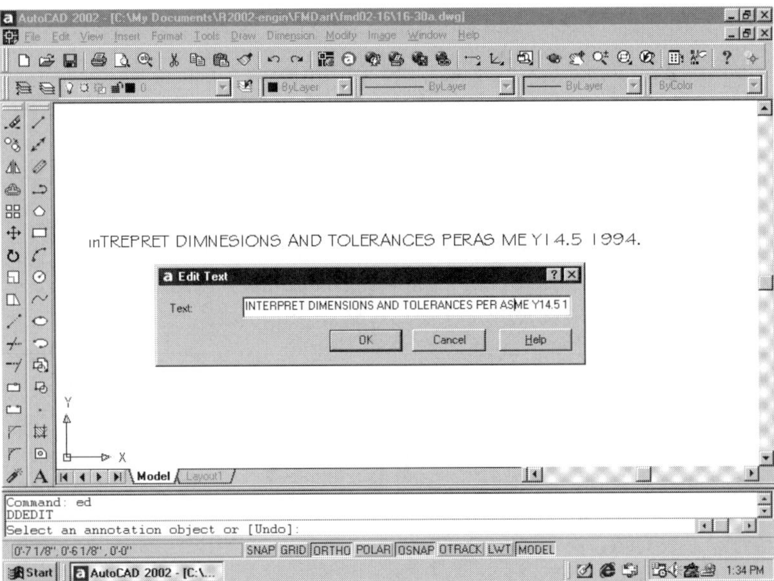

Figure 15.30b. *The note shown in Figure 15.30a is now as it was intended, but changes are not part of the drawings base until the* **OK** *button is clicked.*

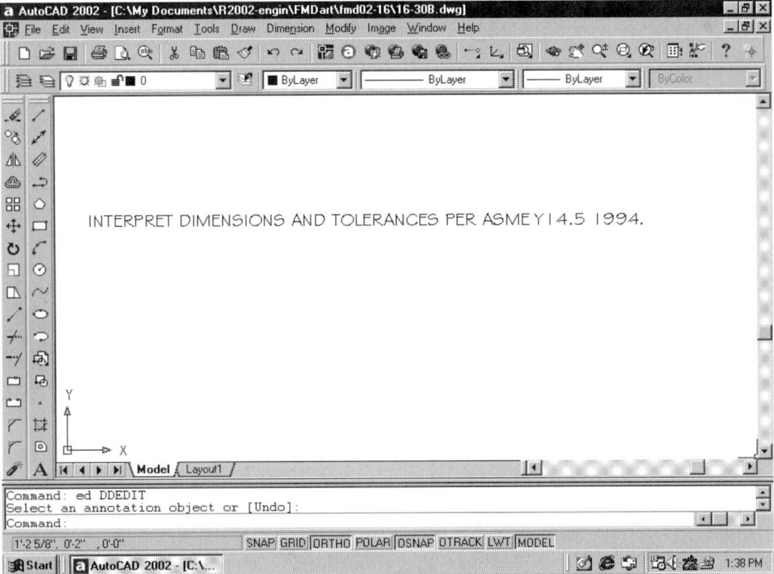

Figure 15.30c *The note shown in Figure 15.30a is now as it was intended.*

side of the "N" and click the select button. Press BACKSPACE to eliminate the "in". Without moving the cursor, the error can be fixed by typing an **IN**.

Use the arrow key to move the cursor to the right side of the first R in "INTREPRET" and click the select button. Press BACKSPACE, to eliminate the R. Now move the cursor to the right side of the first E, click the select button, and insert R in the proper location. Use the same technique to correct the spelling of DIMENSIONS. Move the cursor between the "PER" and "AS". You can insert a space by pressing SPACEBAR. "AS ME" can be corrected by placing the cursor in front of the M and pressing BACKSPACE. The text will now read as it was intended (See Figure 15.30b.) If the edit is not what you desired, the Undo option can be used to restore the line of text to its previous status. Once you are satisfied with the text, click the **OK** button to return to the drawing and display the corrected text. (See Figure 15.30c.) The **Ddedit** command is still active. If no other editing needs to be done, press ENTER to return the command line.

TEXT AND THE MIRROR COMMAND

The **Mirror** command was introduced in Chapter 10. If you mirror an object that includes text, the text will also be reversed. To avoid the problems this will create, use the **Mirrtext** system variable. Alter the variable by typing **MIRRTEXT** (ENTER) at the Command prompt. This will produce the prompt:

Command: **MIRRTEXT** ENTER

Enter new value for MIRRTEXT <1>:

Pressing ENTER at the prompt will accept the default value and the text will be reversed as the object is mirrored. Typing **0** (ENTER) at the prompt will mirror the object and the text, but keep the text readable.

QUICK TEXT

Quick text is not an additional way to create text, but a very useful way to display it. As you're editing drawings containing large amounts of text, you might find that **Regen** times start to slow. The **Qtext** command will display lines of text as rectangles, which are much quicker to regenerate. The height of the rectangle is the same as the text height, but the length of the rectangle is typically longer than the line of text. The command sequence is as follows:

Command: **QTEXT** ENTER

Enter mode [ON/OFF] <OFF>:

With **Qtext** set to ON, text will retain its current setting until a **Regen** is performed. Figures 16.31a and 16.31b show an example of **Qtext** in both the ON and OFF settings.

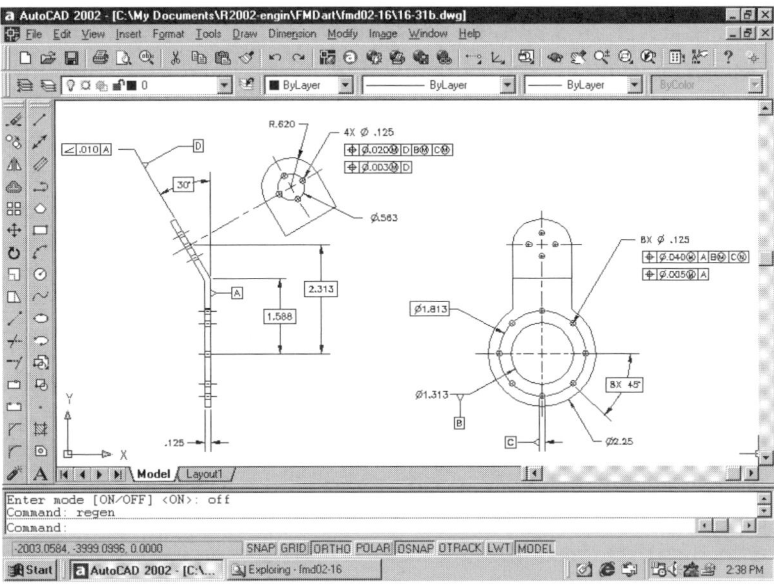

Figure 15.31a *An example of normal text. Figure 15.31b shows the same drawing with quick text. Courtesy David Dearth.*

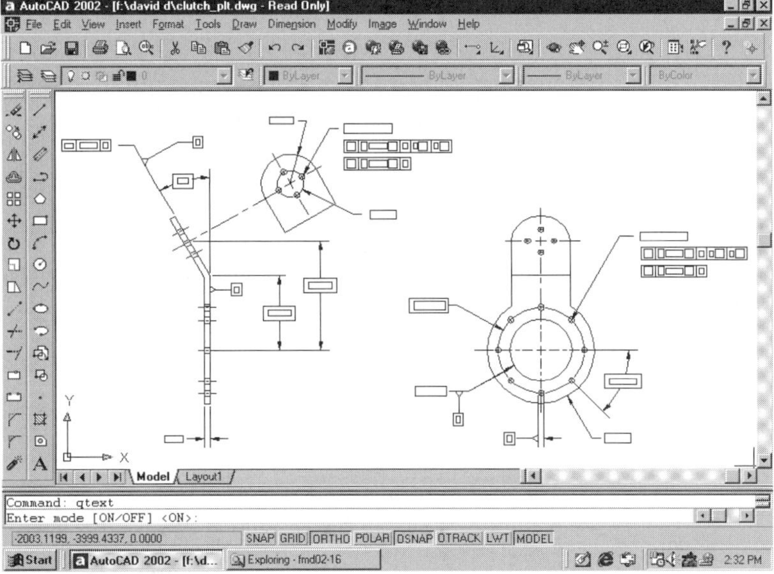

Figure 15.31b Qtext *can be used to speed the* **Regen** *of very large drawings. The effect of* **Qtext** *on the drawing in Figure 15.31a. Courtesy David Dearth.*

CREATING TEXT WITH MTEXT

A You might be feeling overwhelmed with text by now. You've worked with **Text**, **Style**, and **Qtext**, and you might be wondering why you need another way to enter text into a drawing. The answer is because this method offers more options than the previous method. **Text** is ideal for placing local notes in and around the drawing. However, paragraphs created in **Text** can be very frustrating to edit if they need to be moved and resized. **Qtext** is a tool for displaying text and offers no benefits for creating or editing text. **Mtext**, or multiline text, is ideal for placing large areas of text on a drawing. Multiline text can be created in the **Multiline Text Editor** dialog box, which offers greater flexibility in creation and editing of the text. Using **Mtext** is similar to using a word processing program within AutoCAD. **Mtext** also offers advantages when editing the drawing. Lines of text placed with **Text** act as individual lines when they are moved. Text placed with **Mtext** functions as one object. Multiline text also allows text to be easily underlined and varied fonts, colors, and heights to be assigned throughout the text. **Mtext** is ideal for entering and editing large bodies of text such as those seen in Figure 15.32. A paragraph can be moved by selecting one character rather than the entire paragraph. Access the command by selecting the **Multiline Text** button on the **Draw** or **Text** toolbar, by typing T (ENTER) at the Command prompt or by selecting **Multiline Text** from **Text** on the **Draw** menu. Each will produce the following prompt:

> Command: **T** ENTER
>
> _mtext Current text style: "titles" Text height: 13
>
> Specify first corner

By default, AutoCAD is waiting for you to select where you would like to place the text. Once you select the first corner, you will be prompted:

> Specify opposite corner or [Height/Justify/ Line spacing/ Rotation/Style/Width]:

Although you can respond to the second prompt by providing a value for one or more of the options, selecting the second corner to define the text window width will produce the **Multiline Text Editor** dialog box and allow each of these values and many

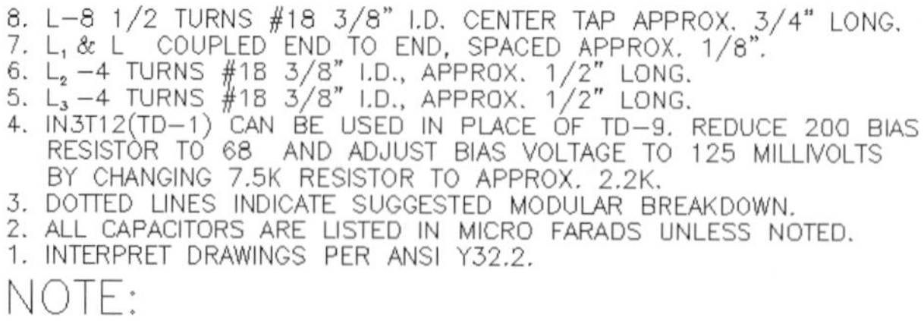

8. L–8 1/2 TURNS #18 3/8" I.D. CENTER TAP APPROX. 3/4" LONG.
7. L_1 & L COUPLED END TO END, SPACED APPROX. 1/8".
6. L_2 –4 TURNS #18 3/8" I.D., APPROX. 1/2" LONG.
5. L_3 –4 TURNS #18 3/8" I.D., APPROX. 1/2" LONG.
4. IN3T12(TD–1) CAN BE USED IN PLACE OF TD–9. REDUCE 200 BIAS RESISTOR TO 68 AND ADJUST BIAS VOLTAGE TO 125 MILLIVOLTS BY CHANGING 7.5K RESISTOR TO APPROX. 2.2K.
3. DOTTED LINES INDICATE SUGGESTED MODULAR BREAKDOWN.
2. ALL CAPACITORS ARE LISTED IN MICRO FARADS UNLESS NOTED.
1. INTERPRET DRAWINGS PER ANSI Y32.2.

NOTE:

Figure 15.32 Mtext *is ideal for creating large quantities of text. Courtesy Jill Heilman.*

more to be easily set. Notice in Figure 15.33, as the box is being dragged to indicate the second corner, an arrow is shown indicating the direction that excess text will flow out of the box. Methods of altering the direction will be introduced shortly. By prompting for a second point, AutoCAD is asking you to provide a window to place your text in. The width of the window you specify is very important. The width of the window will control the width of your paragraph. Extra text will spill out of the top or bottom of the window depending on the justification, so that the depth of the window is not critical. Once the other corner is selected, the **Multiline Text Editor** shown in Figure 15.34 will be displayed. Text can now be entered in the edit window using procedures similar to those used with other text editors. AutoCAD allows a third-party text editor to be used it you select the editor from the **File** tab in the **Options** dialog box (select **Options** on the **Tools** menu). See the AutoCAD User's Guide in the **Help** menu to select a third-party editor. The balance of this chapter will introduce the many capabilities of the AutoCAD program text editor.

As text is entered at the keyboard, it will be displayed in the edit box of the text editor. Text will wrap to the next line as the desired width is reached, just as with any word processing program. Notice in Figure 17.34 that the editor contains tabs for controlling text features. The default tab controls characters and there are also tabs for **Properties**, **Line Spacing**, and **Find/Replace**.

CONTROLLING TEXT CHARACTERS

As with **Text**, the qualities of Mtext can be easily altered. Qualities controlled on the **Character** tab include the font, height, bold, italic, underline, undo, stack, color, and

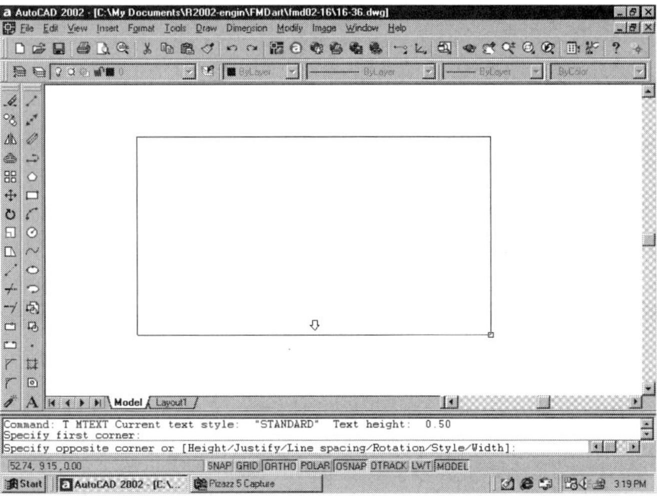

Figure 15.33 *Multiline text is started by selecting a box to define the width that the text will occupy. If the text does not fit into the box, it will overflow the bottom of the box in the current setting (the down arrow). The direction of overflow can be adjusted.*

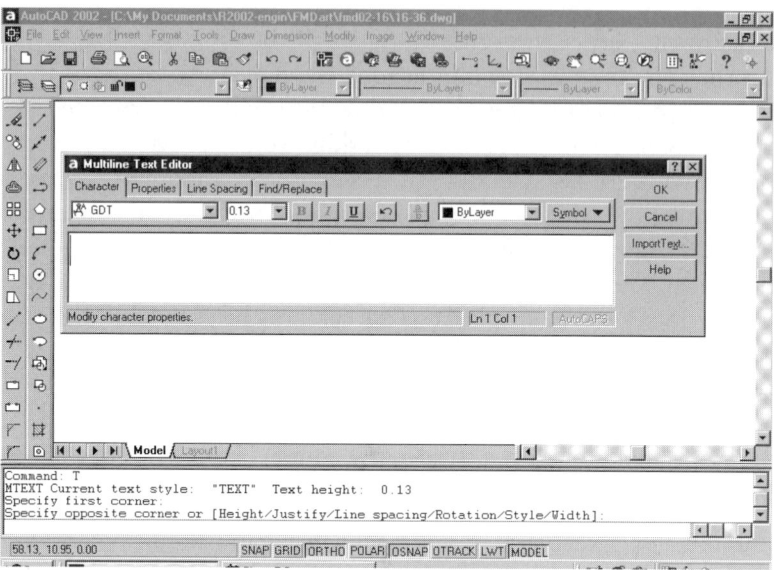

Figure 15.34 *The* **Multiline Text Editor** *is used to add large amounts of text to a drawing. By using the* **Character, Properties, Line Spacing,** *and the* **Find/Replace** *tabs, you can assign text various colors, and heights, underlining, and other alterations. In addition to using the AutoCAD text editor, you can use a third-part editor by selecting the desired editor using the* **File** *tab of the* **Options** *dialog box (select* **Options** *from the* **Tools** *menu).*

special symbols. This dialog box also includes the **?** (question mark) button in the title bar. Selecting the **?** button will produce a question mark button beside the cursor. Now if any of the buttons are selected, an on-screen explanation will be displayed. Figure 15.35 shows an example of selecting the **Stack** button.

Font

Clicking the down arrow beside the **Font** edit box displays a list similar to the font list displayed with the **Style** command. The list allows a font for new text to be selected or the font of selected text to be changed. The current font for multiline text is the same as the current font for the **Style** command. With **Mtext**, the style can be altered at any time throughout the use of the command, including mid sentence. Figure 15.36 shows an example of text that was altered to provide different fonts within the same command sequence.

Height

Just as with the **Text** command, this option controls the height of the text. The default value is the same as the current style. The value for text height can be changed at any point of the command, allowing words to be printed at varied heights. Selecting the down arrow will display the heights of all current styles. To provide a new height, select the **Height** edit box and click the cursor to highlight the current

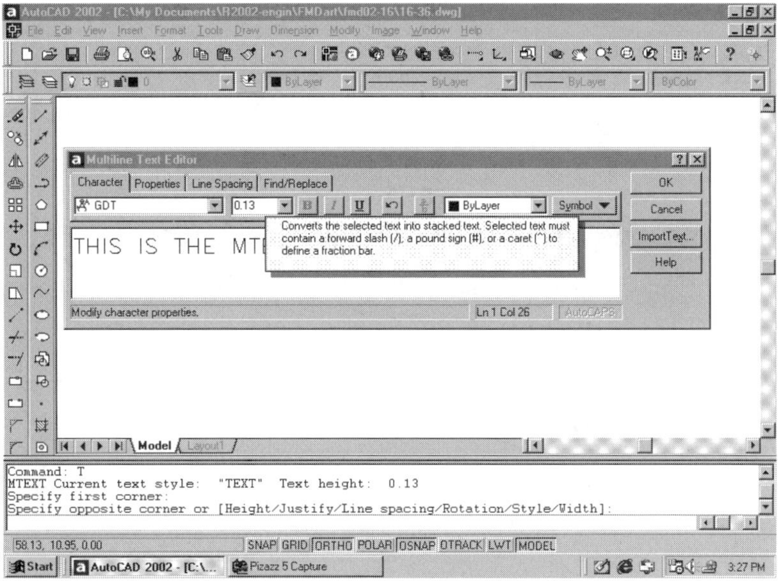

Figure 15.35 *Selecting the* **?** *button in the title bar and then selecting a box or button in the* **Multiline Text Editor** *will provide a brief explanation of the function.*

height. Remove the current height with the backspace arrow and then enter the desired height value. Text height can be altered before the text is added to the drawing by highlighting the text to be altered and then providing the required height in the edit box. Figure 15.37 shows an example of how text height can be altered.

Bold

Selecting the **Bold** button toggles between bold and normal text. The options are active only for TrueType fonts. Bold text can be created by selecting the **Bold** button before text is entered. Existing normal text can be changed to bold by highlighting the desired text to be changed, and then clicking the **Bold** button. Figure 15.38 shows an example of text that was altered to provide bold text.

Italic

Selecting the **Italic** key toggles italic formatting for text. As with other options, **Italic** can be applied to text being created as well as used to modify existing text. The option is available only for TrueType fonts.

Underline

Selecting this button allows the underline function to be toggled ON and OFF. This option of **Mtext** is so much easier than %%something that is required with the **Text** command.

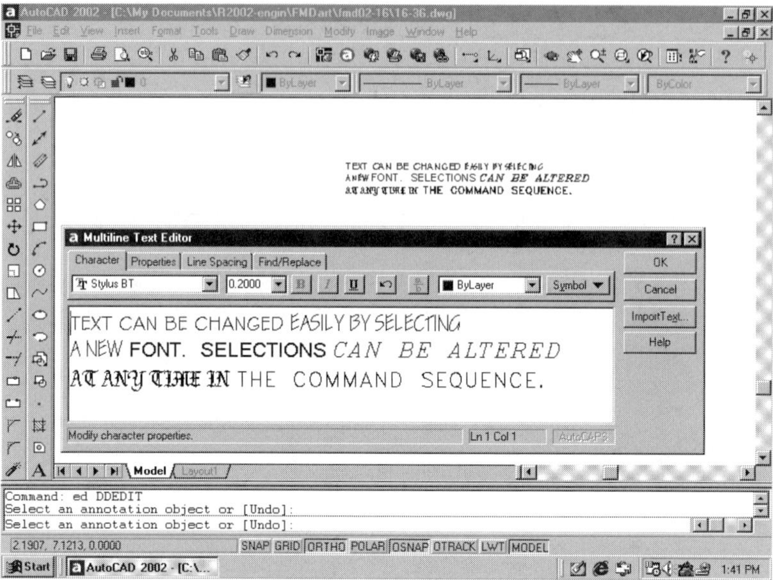

Figure 15.36 *The text font of one word or a string of words can be altered at any point of the drawing's life.*

Figure 15.37 *Multiline text height can be altered at any time throughout the life of the drawing. Selecting the down arrow will display a choice of heights. You can also enter a new height by selecting the* **Height** *box and replacing the current value with the desired value.*

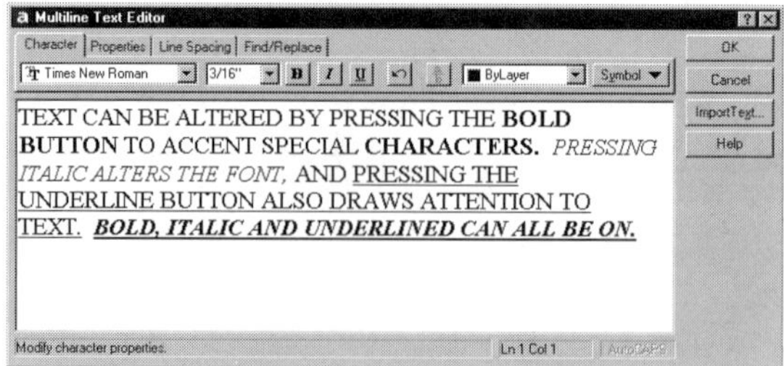

Figure 15.38 *Text can be highlighted by selecting the* **Bold, Italic,** *or* **Underline** *button. One or more of the features can be used in combination to affect the font.*

Undo

This option can be used to undo the last sequence of the text editor. Changes can be made to the text or to one of the formatting options.

Stack

The **Stack** option can be used to place a portion of text over other text. This is useful in writing stacked numbers such as $\frac{1}{4}$ rather than 1/4 or in writing equations.

With the text written as 1/4, highlight the fraction by pressing and holding the select button and moving the cursor over the text. Once the desired text is selected, selecting the **Stack** button will stack the indicated text. See Figure 15.39.

Color

Selecting the **Color** button allows the color of the text to be altered. By default, multiline text will be assigned the color of the current layer. To select a new text color, select the down arrow beside the **Color** edit box and select the desired color. Selecting the Other color option will produce the **Select Color** dialog box and allow for a full range of colors to be selected.

Symbol

Selecting the down arrow beside the **Symbol** button displays the list shown in Figure 15.40. The degree, plus/minus, and diameter are often used on construction documents. Each can be inserted into multiline text by selecting the down arrow to display the list, and then selecting the desired symbol. If this seems too easy, the symbols can also be inserted using the special characters that were discussed as **Text** was introduced. Selecting the Other option will display the **Character Map** dialog box shown in Figure 15.41 and allow one of the symbols from the **Character Map** to be inserted. Symbols will vary depending on the current font.

Control Buttons

As with other dialog boxes, control buttons are provided to implement the desired settings. This box features the **OK**, **Cancel**, **Import Text**, and **Help** buttons. The results of the **OK**, **Cancel**, and **Help** buttons should be familiar by now. The **Import Text** but-

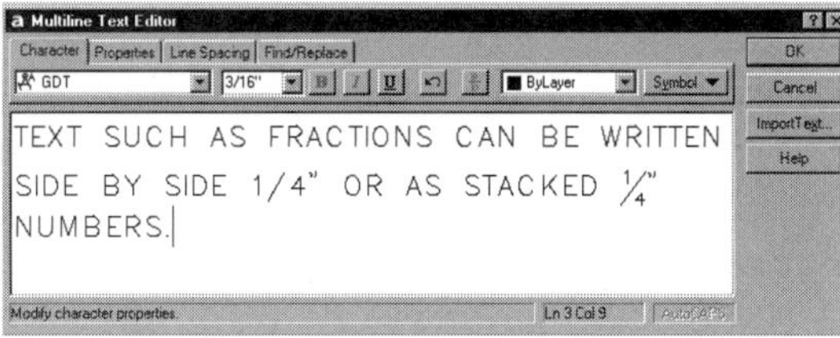

Figure 15.39 *Text such as fractions can be written side by side or stacked depending on office preference.*

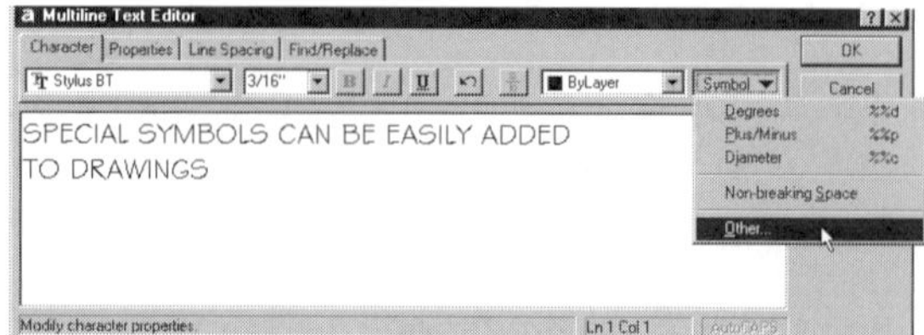

Figure 15.40 *Using* **Mtext***, special characters such as the degree, plus/minus, and diameter symbols can be added without using any %% special character.*

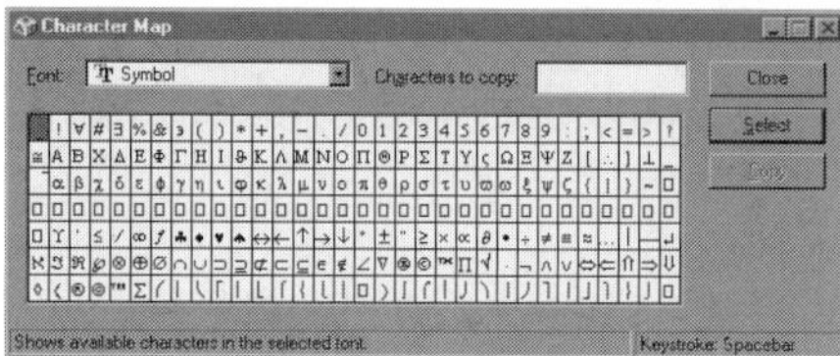

Figure 15.41 *Selecting the Other option from the* **Symbols** *list displays the character map.*

ton allows ASCII (American Standard Code for Information Interchange) and RTF (Rich Text Format) text files to be inserted in a drawing file. This might seem unimportant as you're learning AutoCAD now, but it means you can create schedules or large bodies of text up to 16 K in size using your second favorite software, and then import the file into AutoCAD once you feel more comfortable with multiline text. Imported text retains it original character formatting and style properties, but it can be edited as if it had been created within **Mtext**.

Note: Files larger than 16 K can be inserted into AutoCAD if they are not imported directly into AutoCAD. To insert a file larger than 16 K, minimize AutoCAD, and then open the file to be inserted using the word processor that was used to create the file. Now save the text file to the Clipboard. Maximize AutoCAD and paste the file into the drawing.

ADJUSTING TEXT PROPERTIES

Selecting the **Properties** tab of the **Multiline Text Editor** will provide options for controlling text properties by changing the style, justification, width, and rotation. Figure 15.42 shows the **Properties** tab. The tab can be selected at any point of the **Mtext** operation.

Style

The **Style** option allows an existing text style to be applied to a new or selected text body. Character formats for font, height, bold, and italic can be overridden by applying a new style to existing multiline objects. Information can be typed in one style, and altered before the text is placed in the drawing base. To alter the current text style, select the **Properties** tab, and then select the **Style** down arrow to display a listing of current text styles. Figure 15.43 shows an example of the **Style** list.

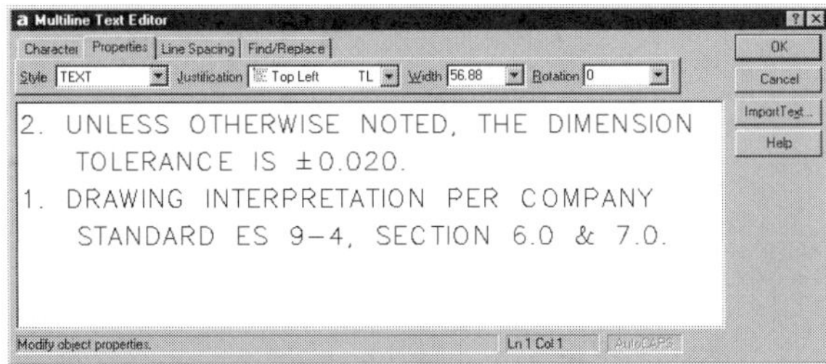

Figure 15.42 *Selecting the* **Properties** *tab displays the options for altering the style, justification, width, and rotation of multiline text.*

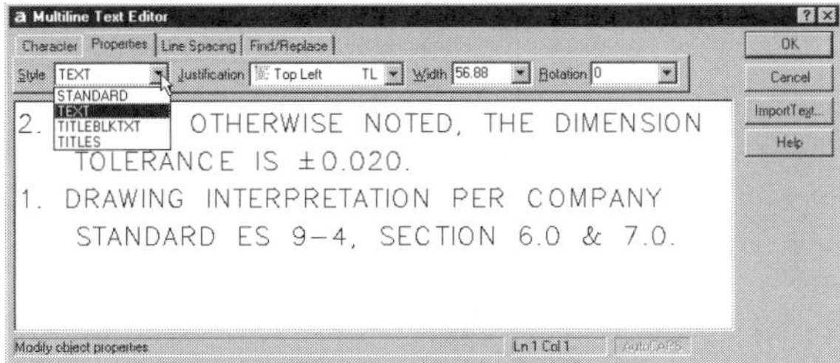

Figure 15.43 *Selecting the down arrow for* **Style** *displays a list of current text styles. Styles for multiline text are created using the same procedures as used to create a style for text.*

Justification

Justification with **Mtext** is similar to that of the **Text** command. Selecting the **Justification** option allows the alignment of text to be altered. The default setting places the cursor in the top left corner of the text box. Selecting the **Justification** down arrow will produce the list shown in Figure 15.44. Figure 15.45 shows examples of how text will be placed based on each setting.

Width

This option allows the width of the text to be redefined. Selecting the down arrow beside the **Width** edit box will display a list of current multiline text widths. One of the current widths can be selected or a new value can be added to the edit box.

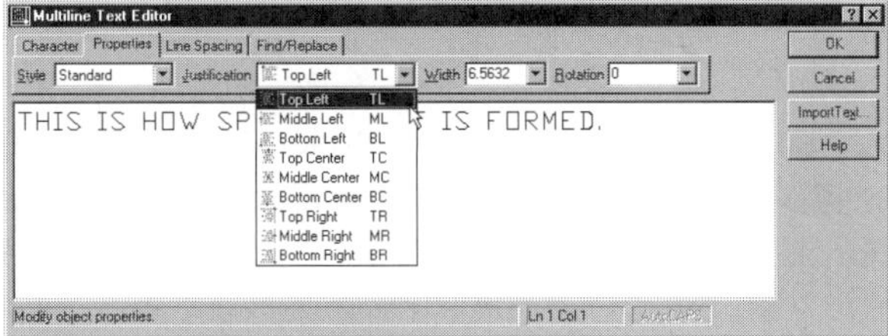

Figure 15.44 *Selecting the* **Justification** *down arrow will produce a list of options for arranging multiline text that are similar to the* **Text justification** *options.*

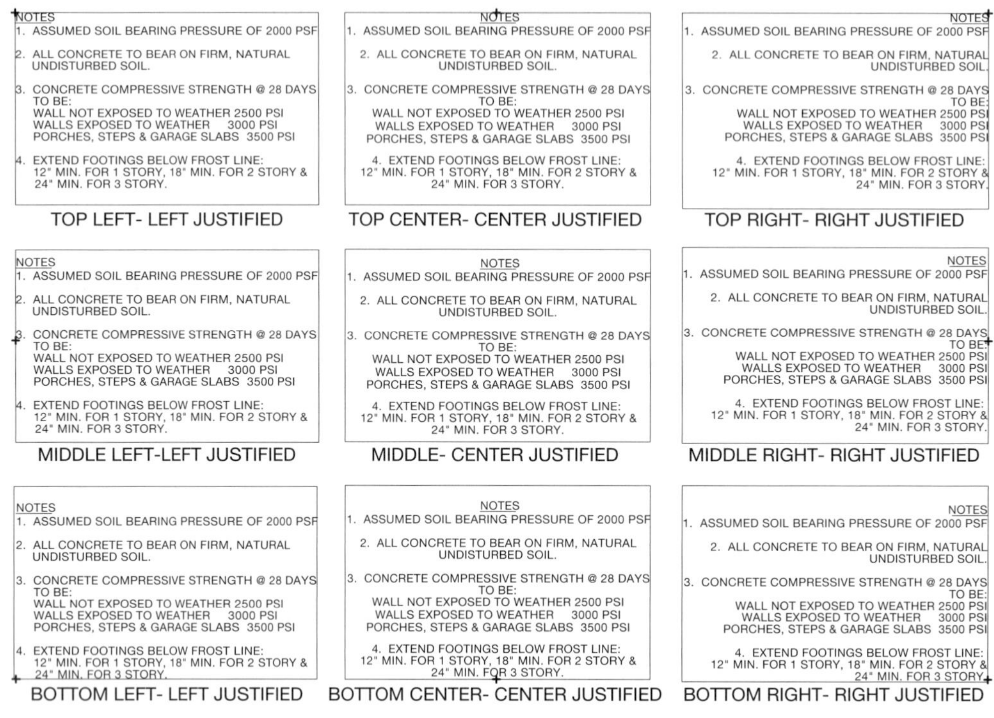

Figure 15.45 *The options for justification of multiline text.*

Rotation

This option functions just as the option did with the **Text** command, to set the text rotation angle. Selecting the **Rotation** down arrow displays a listing of angles in 15° increments. One of these options can be selected from the list, or any numerical value between 0 and 360 can be assigned. The option can be used with the current text before it is added to the drawing, or existing text can be selected and altered.

LINE SPACING

The **Line Spacing** tab includes two settings for controlling line spacing. These settings include the **At Least/Exactly** and the **Single (1.0x)** options.

At Least/Exactly

This setting specifies how the spacing between lines will be placed. The default setting of **At Least** is to allow AutoCAD to automatically add space between lines of text based on the height of the largest character in the line of text. The **Exact** option sets the spacing so that it will always be the same between lines of text. The **Exact** option maintains the exact spacing no matter the variation of individual letters in each line. The **Exact** option should be used when constructing tables such as a parts list.

Because the spacing between lines of text will always be the same, text placed with this option will have a very orderly appearance.

Single (1.0x)

This menu can be used to control the spacing increment between lines of text multiline text. This setting is the equivalent to choosing between single, double, or triple spaced text with a word processor. The spacing is the vertical distance measured from bottom to bottom of lines of text. The default distance is 1.66 times the height of the text. A setting of 1 represents single spacing, and 2x represents double spacing.

USING FIND/REPLACE

Selecting the **Find/Replace** tab of the **Multiline Text Editor** displays the options shown in Figure 15.46. The options help create search patterns for words or specified strings of text. Options include the **Find** box and button, the **Replace with** box and button, **Match Case**, and **Match Word**.

Find

The **Find** box can be used to enter a word or string of text that you wish to find. This can be especially helpful in finding a key word in a long list of specifications. To find a word, select the **Find** box by placing the cursor in the box and then type the desired word. Figure 15.47 shows an example of using **Find** to locate the phrase ES 9. Clicking the **Find** button to the right of the **Find** edit box starts the search for the text in the **Find** box. Press ENTER to find the next occurrence of the selected word.

Replace With

The **Replace with** option can be used in conjunction with **Find** to alter a word or word string. Once a word has been found, enter a word in the **Replace with** box and click the **Replace** button. Figure 15.48 shows the result of selecting ES 9 and replacing it with the phrase MD-5.

Match Case

With this box checked, only the exact word with the same case will be identified. With the box inactive, placing ES 9 in the Find box would identify es-9 in the body of text.

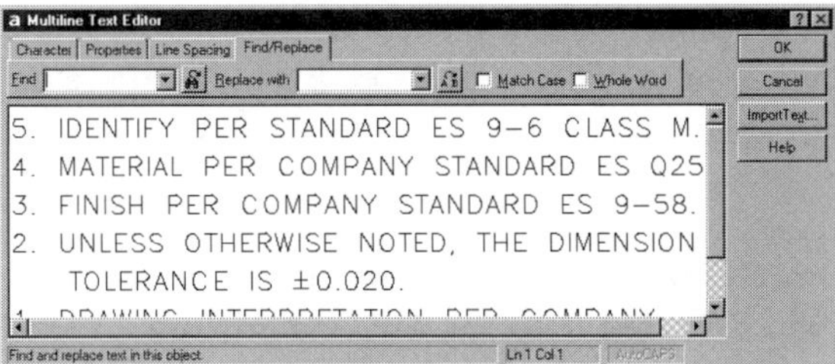

Figure 15.46 *Selecting the* **Find/Replace** *tab provide options for editing multiline text.*

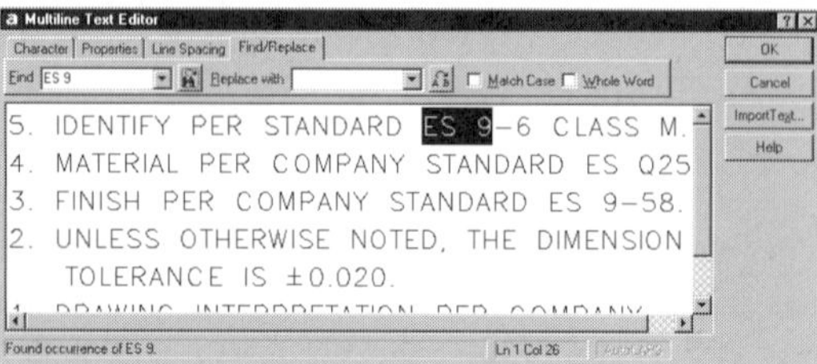

Figure 15.47 *Selecting the* **Find** *box allows a word or word string to be specified for a word search. The phrase ES 9 has been selected for the search. Clicking the* **Find** *button will highlight the first use of the specified word. Pressing* ENTER *will highlight the next occurence of the selected word or string.*

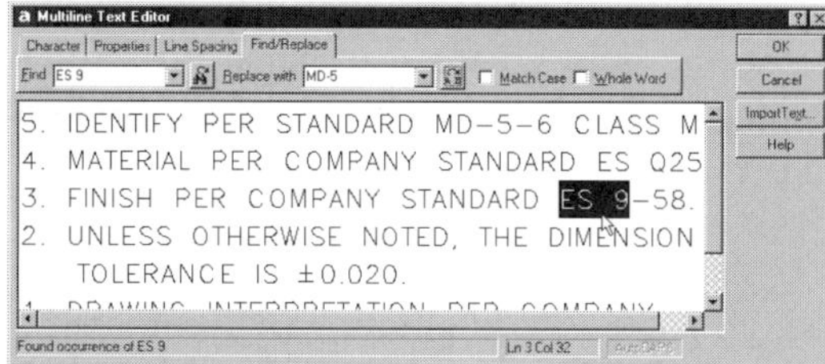

Figure 15.48 *With the phrase ES 9 as the active word for the search, the first occurrence of the word was altered to read MD-5 because MD-5 was typed in the* **Replace with** *box and the* **Replace** *with button was selected.*

With the **Match Case** box active, es-9 would not be identified. Only uses of the word or string that are identical will be located.

Whole Word

With this box inactive, any use of the selected word will be identified with the **Find** option. With ES 9 indicated in the **Find** box, ES 9-SEC:5.2 would be identified in the document. Selecting this option will identify the word only if it is a single word. When the word is part of another text string, it will be ignored.

EDITING MULTILINE TEXT

A multiline text object can be edited using the **Copy**, **Erase**, **Grips**, **Mirror**, **Move**, and **Rotate** commands. The text in multiline text can be edited using either the **Ddedit** or **Properties** command or by using the **Text Style**, **ScaleText**, and **Justify Text** buttons on the **Text** toolbar.

USING DDEDIT

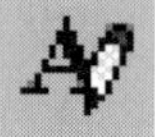

To use the **Ddedit** command with multiline text, you start the same way as when editing text created with the **Text** command. The quickest method of accessing the command is from the Command prompt, but you can also access the command by selecting **Edit Text** on the **Text** toolbar or by selecting **Text** from the **Modify** menu. The command sequence is as follows:

> Command: **ED** ENTER *(Or click the Edit Text button.)*
>
> Select an annotation object or [Undo]: *(Select the desired text to be edited.)*

When the paragraph to be edited is selected, the text will be displayed in the **Multiline Text Editor** dialog box. Each of the four tabs that were used to create the text can be used to alter the selected text. The mouse can also be used to move the cursor to the desired location. As the cursor is moved into the edit box, it changes to a blinking cursor. To edit text, move the cursor to highlight the text to be changed. With the text highlighted, any one of the features set using the text editor can be altered. When the changes are made, click the **OK** button to return to the drawing screen. As the screen is restored, the prompt for selecting additional annotation is still displayed. The program is waiting for you to select additional text to be edited. Press ENTER a second time to exit **Ddedit** and restore the Command prompt.

In addition to using the four tabs of the **Multiline Text Editor** dialog box to edit text, standard Windows control keys can be used to alter the text in the editor. The control keys include the following:

CTRL+A	Selects all text in the **Multiline Text Editor**.
CTRL+B	Applies or removes bold format for selected text.
CTRL+C	Copies selected text to the Clipboard.
CTRL+I	Applies or removes italic format for selected text.
CTRL+SHIFT+L	Converts selected text to lowercase.
CTRL+U	Applies or removes underline format for selected text.
CTRL+SHIFT+U	Converts selected text to uppercase.
CTRL+V	Pastes Clipboard contents to cursor location.
CTRL+X	Cuts selected text to the Clipboard.
CTRL+SPACEBAR	Removes character formatting in selected text.
ENTER	Ends the current paragraph and starts a new line.

EDITING WITH PROPERTIES

The **Properties** command can be used to alter the properties of multiline text. The command has been introduced earlier, but its power to alter text should not be forgotten. The properties of multiline text can be displayed using the following steps:

1. Select the body of text to be edited.
2. Right-click in the drawing area to display the shortcut menu.

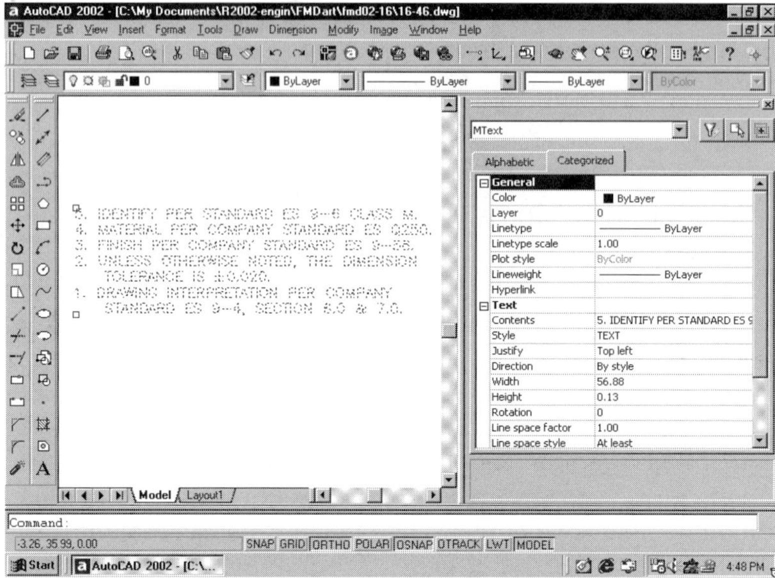

Figure 15.49 *The* **Properties** *dialog box can be used edit multiline text. Select the* **Properties** *button on the* **Standard** *toolbar or type* **MO** (ENTER) *at the Command prompt to produce the* **Properties** *dialog box.*

3. Select **Properties** from the shortcut menu.

You can also start the command by selecting the **Properties** button on the **Standard** toolbar or by typing MO (ENTER) at the Command prompt. Each method will display the **Properties** dialog box similar to Figure 15.49. When the properties are listed by category, seven general properties affecting the text and nine specific properties are listed for multiline text.

General Qualities

Each of the seven general properties are used to display or set the specified property. For instance, if you select a block of multiline text and then select **Color**, the color property will be highlighted, and the edit arrow will be displayed. Selecting the edit arrow will display the color palette, allowing the color of the text to be altered. Each of the other general properties of the text can be altered using the same procedure.

Specific Text Properties

Each of the multiline text features that are controlled in the **Multiline Text Editor** dialog box can be controlled using the **Properties** dialog box. To alter the current style, select the text to be altered. With the text highlighted, selecting **Style** will display a listing of the current styles. Selecting the name of the new style will automatically update the selected text to the indicated style. Each of the other properties can be altered using the same procedure. Other text specific properties include the following:

Contents—The edit box specifies the selected text string. Only a portion of the text will be displayed for long text strings.

Style—Specifies the style name of the selected text body and allows a new style to be specified. Selecting the **Text Style** button on the **Text** toolbar can also alter the style of existing text.

Justify—Specifies the attachment point of the text body using the nine common justification options for placement.

Direction—Allows the horizontal or vertical direction of the selected multiline text to be specified.

Width—Allows the width of multiline text to be edited. With the text highlighted, a grip will be displayed at each corner of the text box. To alter the width of the text, select a grip to make it hot. Move the hot grip to the desired location and the text will be redisplayed to the specified width.

Height—Specifies and alters the height of the selected text. Text can be enlarged or reduced by altering the current value. Selecting the **Scale Text** button on the **Text** toolbar can also alter the scale of existing text.

Rotation—This option specifies the rotation angle of the text body. Enter the desired rotation angle by keyboard and press ENTER. The text will be rotated around the current justification point.

Line space factor—Controls the line spacing factor of the text body. Enter the desired value by keyboard and press ENTER. The text will be altered to display the selected spacing value.

Line space style—Specifies the spacing based on text height using the Exact and At least options.

Note: Keep in mind that the **Properties** dialog box can be displayed beside the drawing area and used to alter more than just multiline text. With the box open, the **Properties** listing will be altered as each new object is selected. The name of the listed object will be displayed in the object edit box located at the top of the dialog box.

USING THE SPELL CHECKER

In addition to using **Ddedit** and **Properties** to edit text, you can use the **Spell** command to correct spelling errors in text and multiline text. Access **Spell** by selecting **Spelling** from the **Tools** menu or by typing SP (ENTER) at the Command prompt. Each option will prompt you to select objects to be checked for spelling. Once a word or body of text is selected, the **Check Spelling** dialog box similar to Figure 15.50 will be displayed. If no spelling errors are contained in the

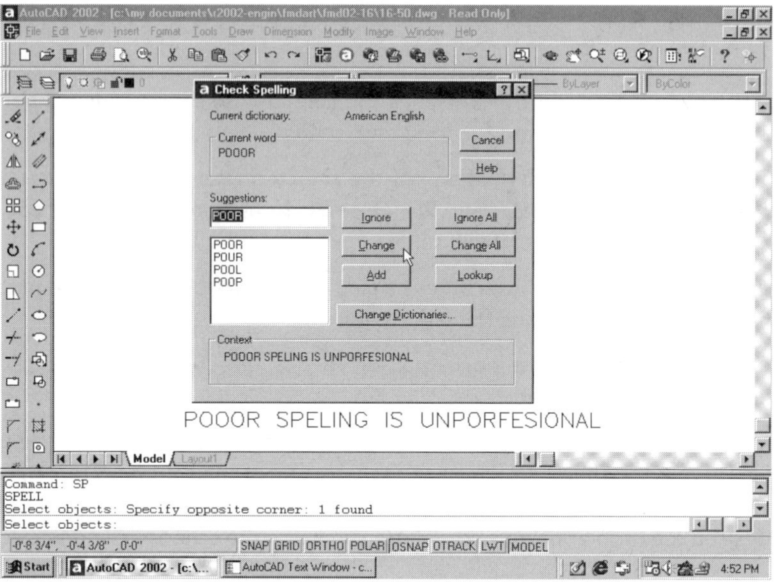

Figure 15.50 *The* **Spell** *command can be used to check text and multiline text. The current word "pooor," four suggestions, and the current context of the word are all displayed.*

selected text, you will see the display shown in Figure 15.51. Clicking **OK** or pressing ENTER will return you to the drawing. Pressing ENTER ENTER will remove the AutoCAD alert and reenter the **Spell** command so that additional text can be checked.

If errors are found in a selection, the line of text will be displayed in the **Context** box of the **Check Spelling** dialog box. The misspelled word will be displayed in the **Current word** box. A possible listing of alternatives can be found in the **Suggestions** box. If the current word is as you would like it, select the **Ignore** button, and the spell checker will move to the next word identified by the checker. If you would like to use one of the suggested words, move the cursor to highlight the word and then select **Change**. The editor will proceed through your text until the alert seen in Figure 15.51 is displayed.

CHANGING THE DICTIONARY

Selecting the **Change Dictionaries** button will produce the dialog box shown in Figure 15.52. The **Main dictionary** box will allow you to change the language that is stored in the dictionary. The **Custom dictionary** and **Custom dictionary words** boxes can be very useful for adding words, acronyms, client names or vendor names specific to your area of expertise to the dictionary. When creating a custom dictionary, use the same guidelines that were presented in Chapter 4 for naming drawing files. An extension of .CUS must be added after the name. To enter a custom name for a dic-

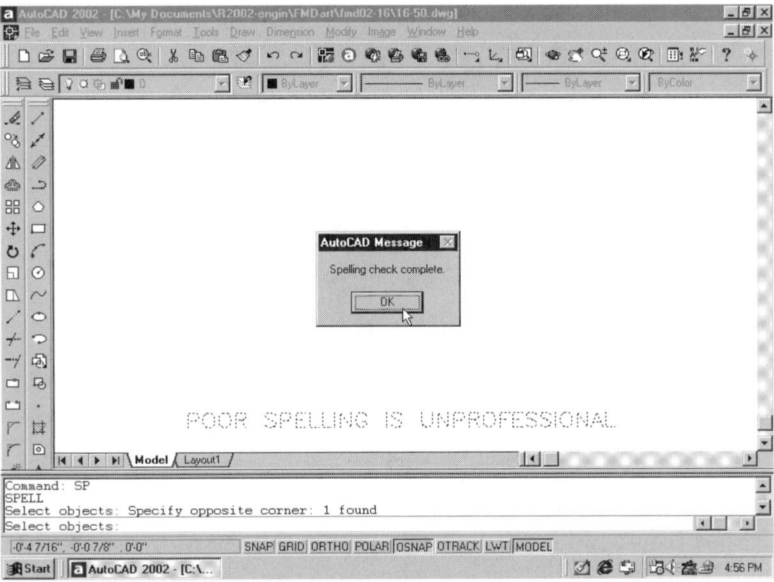

Figure 15.51 *If no errors are found, you will see this display.*

tionary, type a title such as MYWORDS.CUS in the **Custom dictionary** box. If several custom dictionaries have been created, the Browse button can be used to select a dictionary.

To add words to a specific dictionary, select the **Custom dictionary words** box. Names of up to 63 characters can be entered in the box. Multiple names can be entered at one time when separated by a comma. Selecting the **Add** button will add the names to the existing list of names contained in the dictionary. Figure 15.53 shows a custom dialog box for the American English dictionary with three words to be added to the existing list.

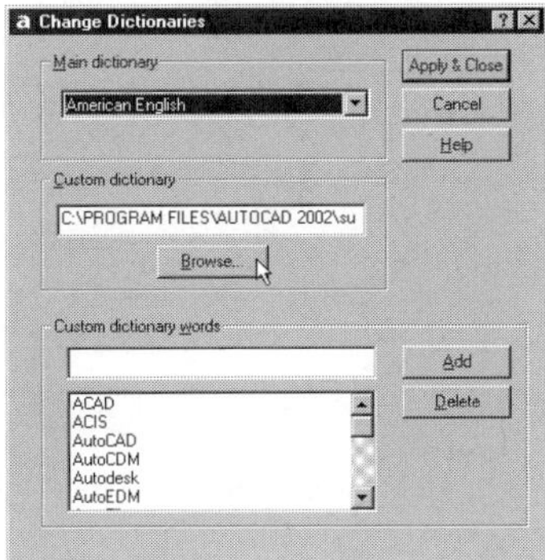

Figure 15.52 *A separate dictionary can be created to contain words specific to your field.*

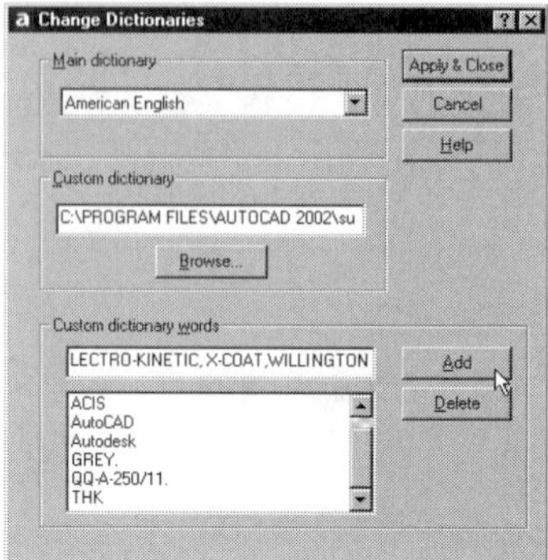

Figure 15.53 *Three words are to be added to the dictonary. A partial listing of the American English dictonary is shown in the lower left corner of the dialog box.*

FINDING AND REPLACING TEXT

The **Find** command is another tool that can be used to edit text. Start the command by right-clicking and choosing **Find** from the shortcut menu, by selecting the **Find and Replace** button on the **Standard** toolbar, by selecting the **Find and Replace** button on the **Text** toolbar, or by selecting **Find** from the **Edit** menu. Each will produce the **Find and Replace** dialog box similar to Figure 15.54. The **Find and Replace** dialog box can be used to select, find, replace, and zoom to text created by the **Text** and **Mtext** commands. The dialog box can also be used to edit dimension text, block attribute text, hyperlink descriptions, and hyperlinks. Each of these four types of text will be introduced in later chapters. The command can be used in a method similar to the **Find/Replace** tab of the **Multiline Text Editor** dialog box. To find text using the dialog box, enter the text or text string that you would like to find in the **Find text string** edit box. The search field can be the entire drawing or lim-

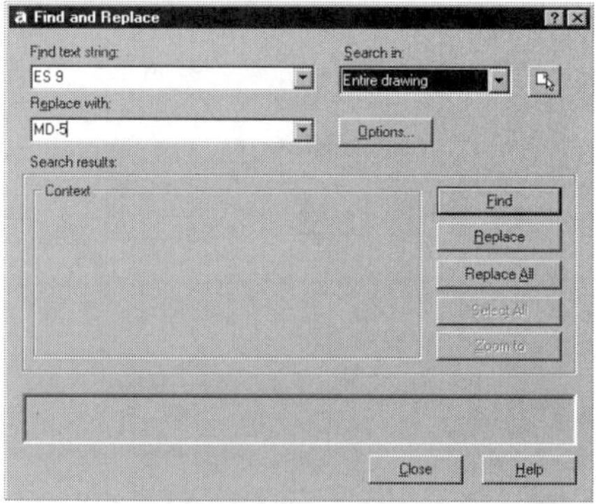

Figure 15.54 The **Find and Replace** *dialog box can be used to select, find, replace and zoom to text created by the* **Text** *and* **Mtext** *commands.*

ited to the current text selection. Once the text string and the search field are defined, select the **Find** button. If the text is found, it will be displayed in the **Search results** window. Choosing the **Zoom to** button will display the selected text on the screen.

After specific text is located in a drawing, the **Replace with** edit box can be used to replace a word or phrase. After entering the text to be found, enter a phrase to be used as the replacement text in the **Replace with** edit box. Choosing the **Replace** button will replace the specified text with the replacement text. Choosing the **Replace All** button will replace each usage of the selected text or phrase.

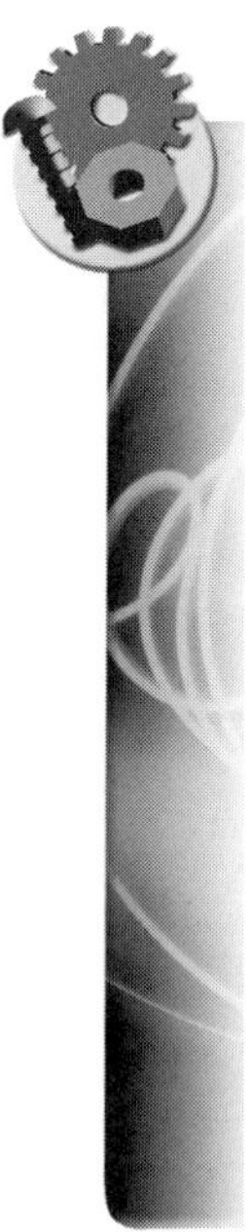

INCORPORATING TEXT INTO TEMPLATE DRAWINGS

Much of the text used to describe a drawing can be standardized and placed in a template drawing. Text can also be saved as a wblock and inserted into drawings as needed. (Wblocks will be discussed in later chapters.) Figure 15.55 shows an example of notes that a company includes as standard notes for all drawings. These notes can be typed once and saved in a template, or as a separate drawing so that they do not have to be retyped with each use. The wblock can then be either inserted into every schematic drawing or nested in the template drawing on the GENERAL NOTES layer. The layer can be frozen, thawed when needed, and then either the **Ddedit** or **Properties** command can be used to make minor changes based on specific requirements of the job.

CHAPTER 15 EXERCISES

1. Start a new drawing using A size material and type two sentences about your hobbies using left justified text with a height of .125"and a font of your choice. Continue typing and write two more sentences about yourself using ROMANS font with a height of .25". Finish this assignment by writing an additional two sentences about yourself using .375" high text and a font of GDT. Save the drawing as E-15-1.
2. Start a new drawing and make ROMANS the active font, with an obliquing angle of 10° and a height of .125". Type three sentences that describe the use of the Text command. Save the drawing as E-15-2.

4. TOLERANCE ON RADIAL DISPLACEMENT R = ±.0008
3. TOLERANCE ON ANGULAR DISPLACEMENT A° ±5°
2. REMOVE ALL BURRS AND SHARP EDGES.
1. INTERPRET DIMENSIONS AND TOLERANCES PER ASME Y14.5M–1994.

NOTES:

Figure 15.55 *Most companies have standard notes that are used for all common drawings. These notes can be typed once and saved as a separate drawing or stored in a template so that they do not have to be retyped with each use. Courtesy David Dearth.*

3. Open a new drawing and make GDT the active font and a height of .125". Use a rotation angle of 35°. Write a paragraph that describes the process for creating a style. Save the drawing as E-15-3.

4. Start a new drawing and make ROMANT the active font with a height of .125". Use a width factor of 1.25. Use the center option to type the following text:

   ATTENTION ALL CADD

   OPERATORS.

   DUE TO

   BUDGETARY PROBLEMS

   YOUR MOM DOES NOT WORK HERE.

   PLEASE CLEAN UP

   YOUR OWN MESS.

   Save the drawing as E-15-4.

5. Start a new drawing and create a style for text using ROMANS as the font with a height of .125". Use a width factor of 1.00. Create a second style named TITLES using ROMANT as the font, with a height of .25" and a width factor of 1.5. Assign each style a layer named for the style and provide a separate color for each. Use **Mtext** to create the text shown in Figure 15.1. Save the drawing as E-15-5.

6. Start a new drawing and create a line of text and make a copy of the text. Mirror one of the lines using a variable of 1 and one of the lines with a variable of 0. Save the drawing as E-15-6.

7. Start a new drawing and use the **Mtext** command to make a list of a minimum of three guidelines that describe the creation of multiline text. Use a second text style and justification point and type a second list of guidelines for describing text created with the Text command. Use **Spell** to check your spelling. Save the drawing as E-15-7.

8. Open drawing E-15-7 and make a copy of the text. Edit the new text to add additional text and change the color, style and width of some of the text. Save the drawing as E-15-8.

9. Use Figure 15.5 as a guide and create a copy of the revision block. Save the drawing as E-15-9.

10. Open your drawing template that has been created in previous chapters. Create separate layers, styles, and fonts for titles, text, and title text. Use the GDT font and add the word NOTES: in the appropriate location. Add the three general notes that were listed earlier in this chapter. Use the ASME text guidelines and alter the title block so that it includes each of the items that the text mentions as important to a title block. Save the template for future use.

CHAPTER 15 QUIZ

1. What text option will allow text to be placed between two selected points, at any angle?
2. What text option can be used to place multiple lines of text in a column?
3. If text is to be .125" high on a drawing that is to be plotted at a scale of 1=2, what should the text height be?
4. Explain the difference between rotation angle and oblique angle.
5. What is the difference between a style and a font?
6. What is the process for producing a degree symbol?
7. What commands are used to correct errors in text?
8. What effect will setting the style height option have on future text?
9. Explain how to correct the following problem: VERIFY EXATC hEIGHT AT JOBSITE.
10. When are lowercase letters appropriate for text on mechanical drawings?
11. Write the required entry to correctly specify three bolts that are three quarters of an inch in diameter at six inches on center. Include all special characters required if the style is set to be Romans.
12. A hole is located on a plane that is at 45° to horizontal. How should the text be oriented to describe the hole?
13. List and describe the four uses of text generally found on drawings.
14. What should be considered when choosing a font?
15. What are the effects of setting the text height in the **Style** command as opposed to setting the height in the **Text** command?
16. When would **Qtext** be used?
17. What is **Mtext** best suited for?
18. What should be the value for **Mirrtext** if you want to mirror the text with the object?
19. Your instructor wants you to create a body of text that starts in the upper right corner of a template drawing. The notes will be 4" wide and be .50" down and 4.50" from the border. What are the commands and options, and what is the proper sequence for placing these notes?
20. What is the process required to save a paragraph created using **Mtext**?
21. A paragraph of multiline text has been created using ROMANS but it should be in Italics. How can this be fixed?
22. Explain the major differences between **Ddedit** and **Properties** on multiline text.
23. What does the .CUS extension designate?

24. What is the advantage in placing stock notes on a template drawing?
25. A client has decided to change from brass to aluminum after all of the written specifications have been added to the drawing. How can this change be easily included?
26. What is the required height for the letters that represent a section according to ASME?
27. Where is the exact location that the word "NOTES:" should be placed? Why is this location important?
28. What is the difference between a local and a general note?
29. What is the difference in function and location between a parts list and a revision list?
30. Describe what different features you like and dislike in the use of **Text** and **Mtext**.

Introduction to Dimensions

In addition to the visual representation and the text used to describe a feature, dimensions are needed to describe the size and location of each feature of a part. Figure 16.1 shows a part and the dimensions used to describe the size and location of each feature. In this chapter you will be introduced to

- Dimensioning standards and goals
- Components of dimensions
- Methods of placing dimension text
- Methods of placing dimensions
- Methods of accessing dimensioning commands

Commands to be introduced in this chapter include

- **Dimlinear**
- **Dimaligned**
- **Dimordinate**
- **Dimradius**
- **Dimdiameter**
- **Dimcenter** (**Center Mark**)
- **Dimangular**
- **Dimbaseline**
- **Dimcontinue**
- **Quick Dimension**
- **Leader**

Several methods will be used to enter the dimension commands. Because so many options are available to place dimensions, this chapter will examine only the default values for dimensioning. Chapter 17 will introduce methods for creating different styles of dimensions and describe how to adjust the variables that control the size and placement of the dimensions. Web Chapter 2 will introduce methods of specifying tolerances, and Chapter 18 will introduce methods of dimensioning specific machine features. Because so many options are included in the software, some options that do not apply to mechanical drafting will be skipped entirely. Feel free to explore these options on your own, or consult the **Help** menu.

DIMENSIONING STANDARDS AND GOALS

Dimensions are placed on drawings to specify the size limitations of features on a part to be manufactured. Several different drawing standards were mentioned as paper sizes were discussed in Chapter 2. This chapter will introduce dimensions based on the ANSI/ASME standards. The standard for dimensions is Dimensioning and Tolerancing, ASME Y 14.5M-1994. This standard is adopted by the American National Standards Institute (ANSI) and is published by the American Society of

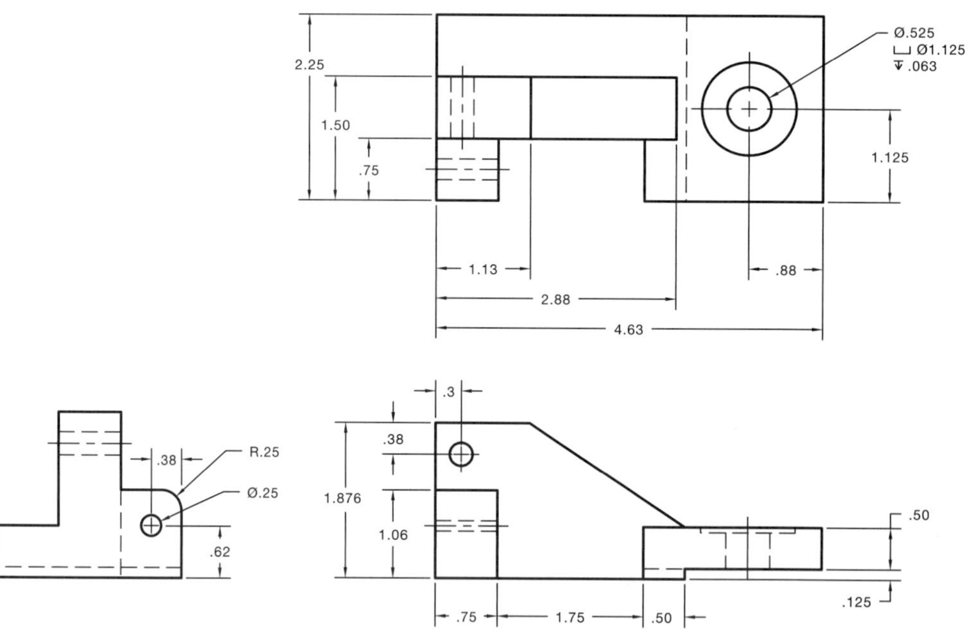

Figure 16.1 *Dimensions are placed to complement the drawings, notes, and title block. Courtesy Paul Bissell.*

Mechanical Engineers (ASME). The standard provides guidelines for the placement and style of dimensions to be used for most of the situations you are likely to encounter in the placement of dimensions. By following these standards, you will be able to place information on a drawing in a fashion that is common to most of the manufacturing world in North America.

DIMENSIONING GOALS

In Chapter 6 you were introduced to orthographic projection methods as a tool for completing multiview and auxiliary drawings. This chapter will introduce methods for placing dimensions on mechanical drawings. By adding notes based on the information in the previous chapter and dimensions, you are completing the process of specifying materials on a mechanical drawing. In Chapter 6, it was suggested that you draw three views of an object. Now as dimensions are considered, only the views needed to place dimensions to completely describe the part should be drawn. You have completely described the part if you have provided dimensions to locate the size and location of every feature of the part.

Size dimensions provide the information to describe the height, width, and depth of a part. Both notes and dimensions can be used to describe the size of a feature. Location dimensions are used to specify the placement of a feature as it relates to another feature or to an edge of a part. Figure 16.2 shows a feature with both size and location dimensions. Examining one of the holes shown in the top view, we can determine that there are four holes with a diameter of .40", a depth of .30", and the center of each hole is .30" from each edge and .50" from end of the part.

DIMENSIONING COMPONENTS

In considering the components of dimensioning, thought should be given to the elements that comprise a dimension, how the dimension text is placed relative to the object, and the text system to be used to dimension the part.

ELEMENTS OF A DIMENSION

Key elements in placing dimensioning include extension and dimension lines, text, and line terminators. Each component can be seen in Figure 16.3a.

Extension Lines

Extension lines are thin lines that extend out from the object being described and set the limits for the dimensions. Extension lines are offset .125" from the object, and they extend past the last dimension line .125". The offsets and extensions can be seen in Figure 16.3b.

Two different types of linetypes are typically used for extension lines. Solid lines are used to dimension to the edge of an object. A centerline is used to dimension to the center of a center circular objects. When dimensions need to be placed, the last long segment of the centerline is 'stretched' to become the extension line. Figure 16.3c shows examples of each.

Dimension Lines

Dimension lines are thin lines used to show the extent of the object being described. Dimension lines should be placed a minimum of .50" from the edge of a part. Three quarters of an inch will produce better clarity. A one-half inch offset should be allowed between consecutive dimensions line. Dimension lines should be placed so that

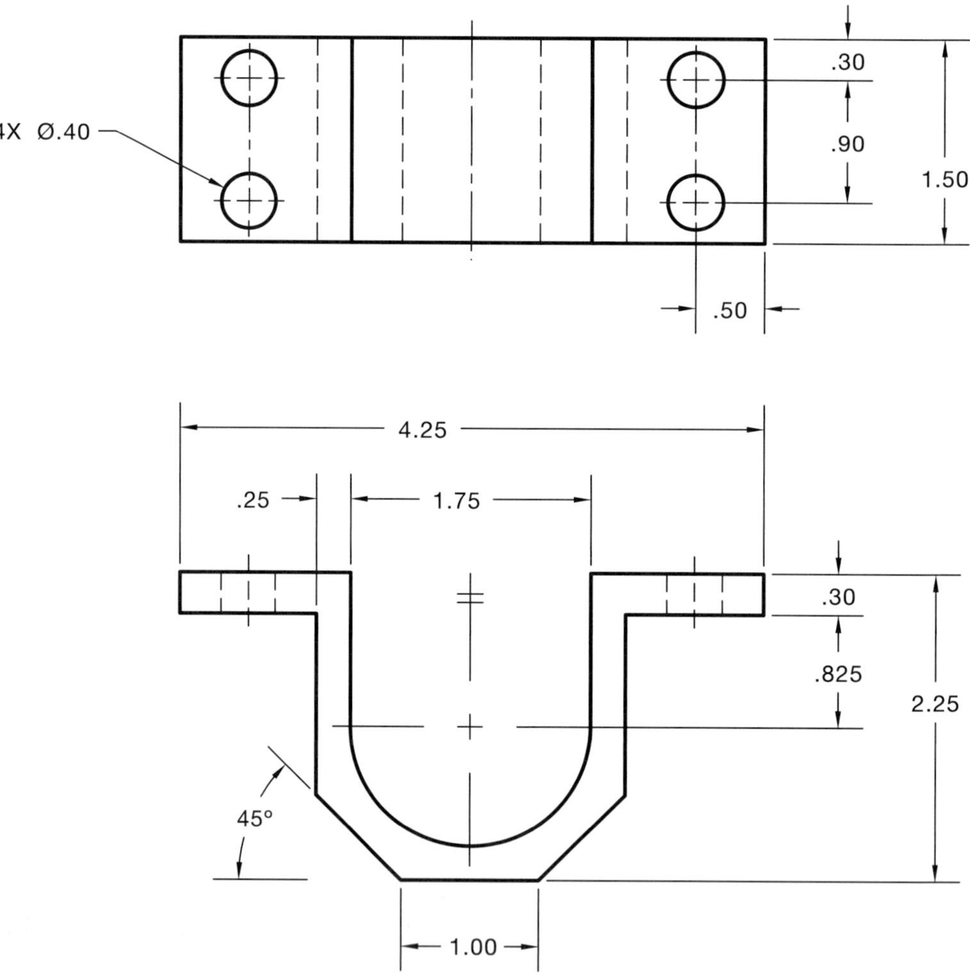

Figure 16.2 *Dimensions are used to show the size and location of each feature. When these two elements are provided for every feature, the dimensions are complete. Courtesy Paul Bissell.*

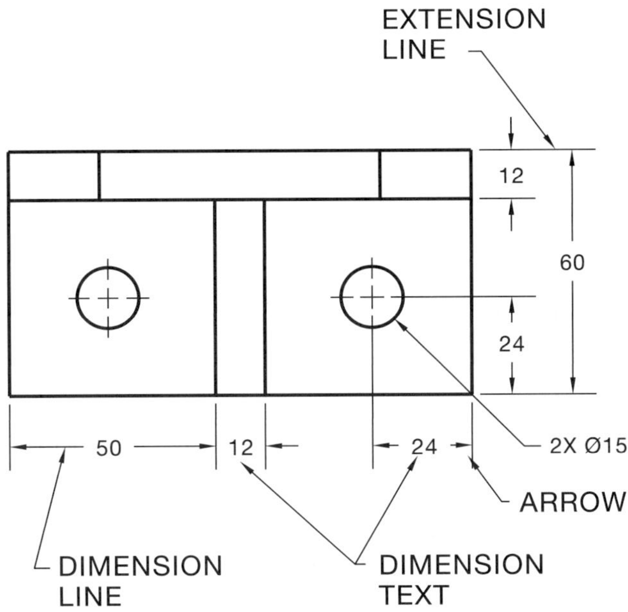

Figure 16.3a *Dimensions are composed of the dimension text, a dimension line, extension lines, and arrows (line terminators).*

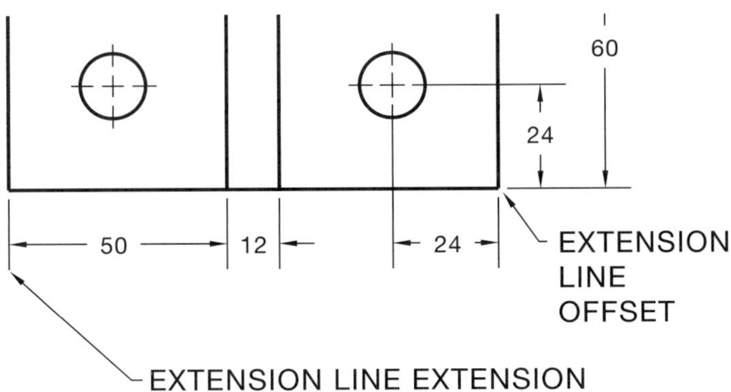

Figure 16.3b *Extension lines are offset .125" from the object, and they extend past the last dimension line .125".*

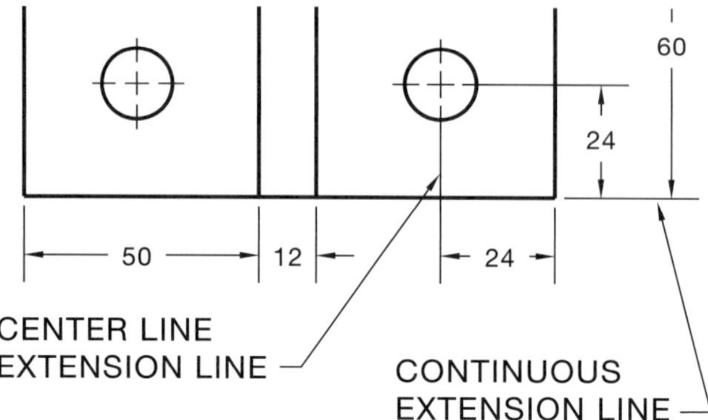

Figure 16.3c *Solid lines are used to dimension to the edge of an object. A centerline is used to dimension to the center of circular objects. When dimensions need to be placed, the last long segment of the centerline is 'stretched' to become the extension line.*

the shortest dimensions are placed closest to the part, with larger dimensions placed so that they do not cross other dimensions. Dimensions should also be placed between the views that they apply to similar to Figure 16.4. The dimension on the bottom of the front view is best because it is placed where the contour of the part it is describing is shown and it is between the top and front views, where that feature can be seen.

Guidelines for Placing Extension and Dimension Lines

Observe the following guidelines when placing extension and dimension lines:

- Relate dimensions to the view that shows the surface best. See Figure 16.4.
- Always place dimensions between the views where they apply. See Figure 16.4.
- Group dimensions for a feature in the same view when possible.
- An extension line can cross other extension lines.
- An extension line can never cross a dimension line. See Figure 16.5 at A.
- Place dimensions outside of a feature. Avoid placing dimensions inside a feature. See Figure 16.5 at B.
- Never dimension to a hidden feature. Draw a section so that hidden features are exposed. See Figure 16.5 at C.
- Keep extension lines as short as possible while maintaining .75" clearance from the object and .50" space between dimension lines.
- Always extend a dimension line to an extension line. Do not use the object as an extension line. See Figure 16.5 at B.

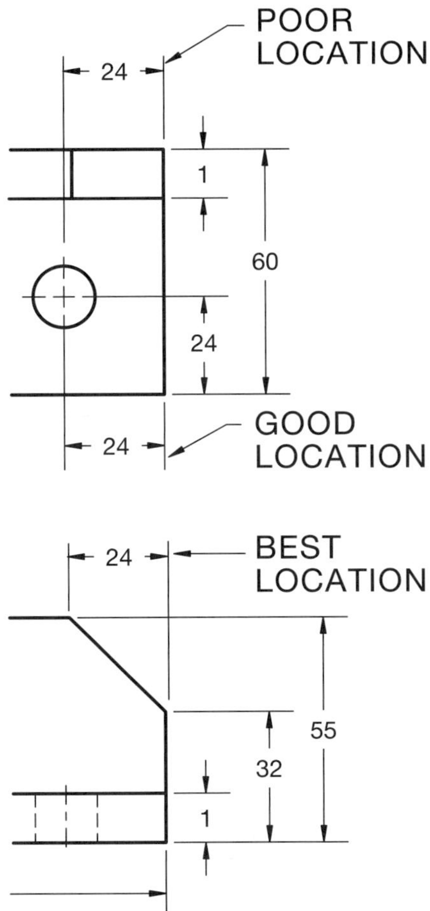

Figure 16.4 *Dimension lines should be placed on the view that shows the shape best and placed between the views that they apply to.*

Terminators

The default method for terminating dimension lines at an extension line is with an arrowhead. The length of the arrowhead should be three times the width. ASME standards recommend a length of .3". Most companies use arrowheads that are filled solid, but open arrows are an alternative. Figure 16.6 shows examples of alternatives that are used by some companies.

Text

The text used to represent the size or location of a feature is the dimension text. Later portions of this chapter will further define different types of text that can be placed with a dimension. The height for dimension text should be .125". Although this is not

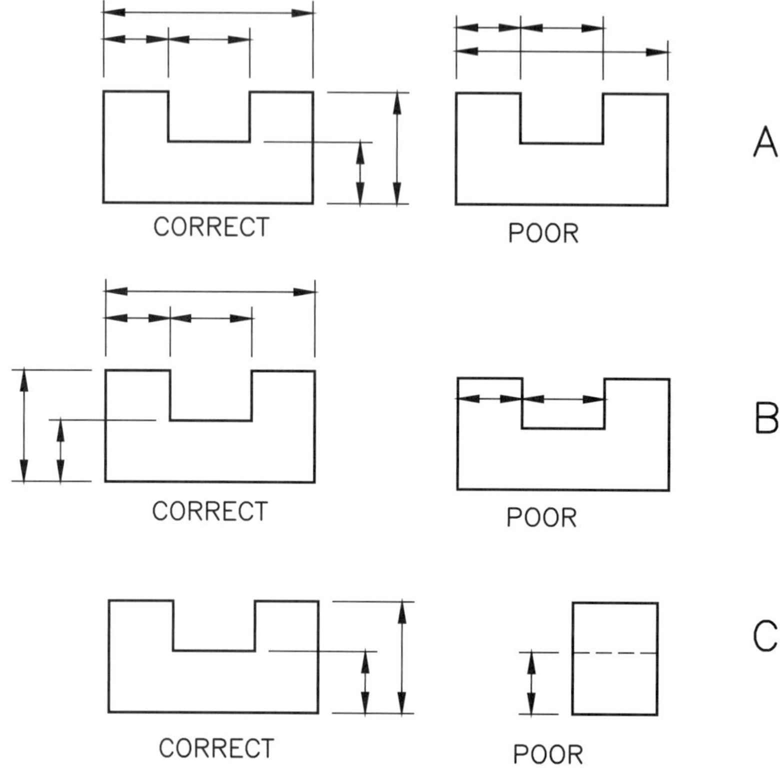

Figure 16.5 *The following guidelines should be followed when dimensioning: (A) extension lines should never cross a dimension line; (B) always place dimensions outside the feature; (C) always dimension to an object line, never to a hidden line.*

the default placement by AutoCAD, ASME standards require, when space allows, that the text for dimensioning be centered in a break between the dimension lines and centered between the two extension lines. See Figure 16.07. When a small space is to be dimensioned, the text is placed between the extension lines, and the dimension lines

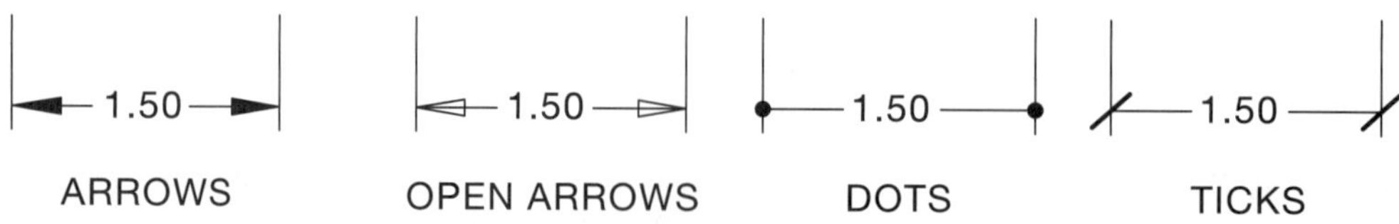

Figure 16.6 *Alternatives for terminating dimension lines.*

and arrows are placed outside the extension lines Figure 16.08 shows common dimension locations when space is limited. Notice in the example in the lower left corner that the arrow that is placed with the .065 dimension is also used for the .10 dimension.

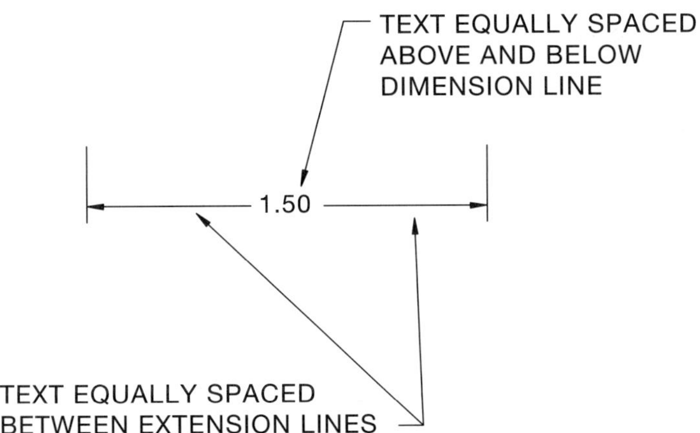

Figure 16.7 *When space allows, the text should be centered between the extension lines, and centered between the extension lines.*

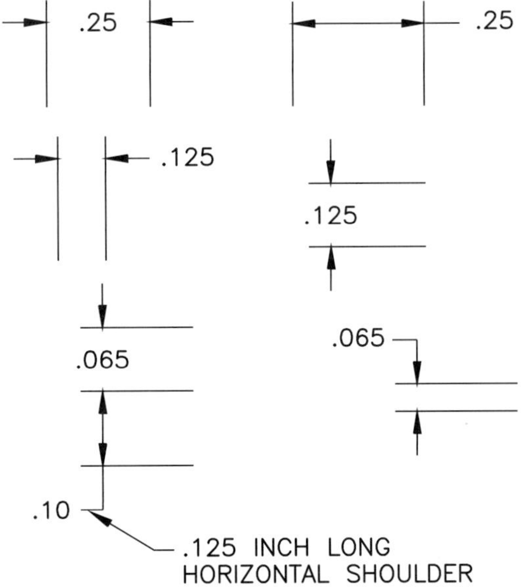

Figure 16.8 *Common professional alternatives for placing dimensions in small spaces.*

DIMENSIONING TEXT PLACEMENT

Two major systems are used to place text on drawing used for manufacturing: unidirectional and aligned text.

Unidirectional Dimensions

Most drawings used for manufacturing use unidirectional dimensions; these require that all dimension text and notes be placed in one direction. The direction of text is horizontal, and it is placed so that it is parallel to the long drawing border. Figure 16.9 shows examples of placing unidirectional dimensions.

Aligned Dimensions

Aligned dimensions are placed so that they are parallel to the surface that they describe. Aligned dimensions are inappropriate for use on drawings that are to be manufactured at a firm different from the firm that originated the drawings. Many small firms that both design and manufacture use aligned dimensions. Figure 16.10 shows a part with aligned metric dimensions.

METHODS OF PLACING DIMENSIONS

Unidirectional and aligned are methods of placing text as it relates to the part. This section will introduce methods of placing the text, dimensions, and extension lines as they relate to the part. These options include chain, datum, tabular, arrowless, and chart dimensions.

Chain Dimensions

Chain dimensions are also referred to as point-to-point dimensions because they locate features on a part from one feature to another. Figure 16.11 shows an example of chain dimensions. This is a common method of dimensioning when a high degree of precision is not required. Because the location of each feature is based on the preceding

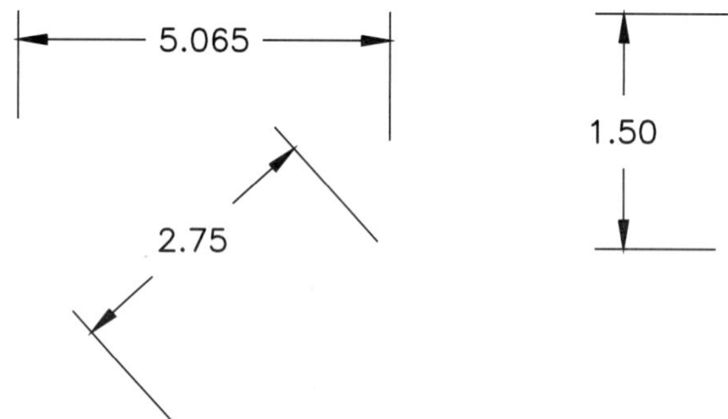

Figure 16.9 *Text on mechanical drawings should be unidirectional and parallel to the longest border.*

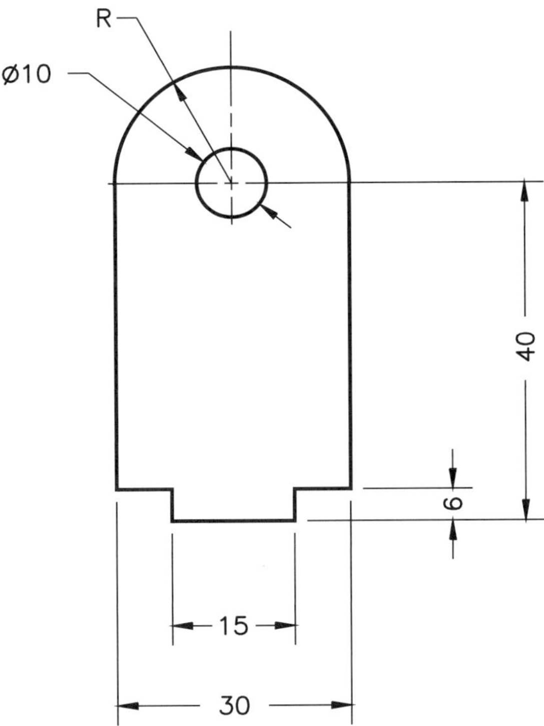

Figure 16.10 *Some fields of drafting not associated with manufacturing use aligned dimensions. With aligned dimensions, lines and text are placed parallel to the inclined surface.*

feature, if a mistake is made in locating the first feature, all subsequent feature locations will be affected. If the top surface that is 2.00" long is cut to 1.98", the feature that is 1.25" might still be the right size, but it will be in the wrong location by a distance of .02". Although a gate handle might still function if an error of .02" is made, a part in a jet engine can fail because of this 'tiny' error. This causes an effect referred to as tolerance buildup, which will be further discussed later in this chapter and in Web Chapter 2.

Datum Dimensions

A datum is a surface that is assumed to be perfect. Datum dimensions are dimensions that are referenced back to the surface. Rather than a feature being located from another feature, each feature of a part is located from the datum. The use of datum dimensioning increases the accuracy in producing the part by eliminating tolerance buildup. A part can have several datum surfaces, points, or assumed locations. Figure 16.12

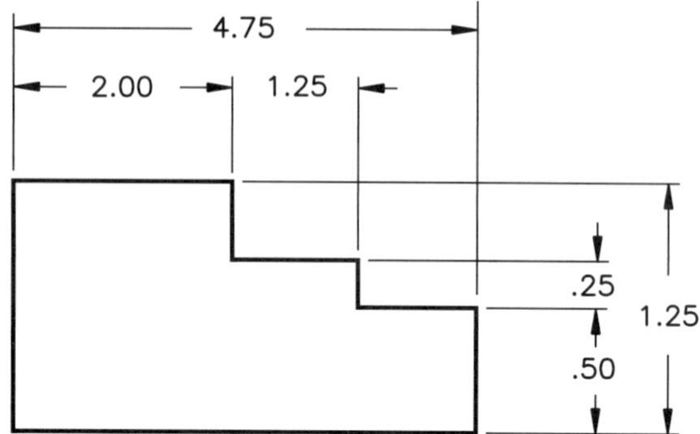

Figure 16.11 *Chain dimensions reference each feature from the preceding feature. If an error is made in locating the first feature, all successive features will be affected.*

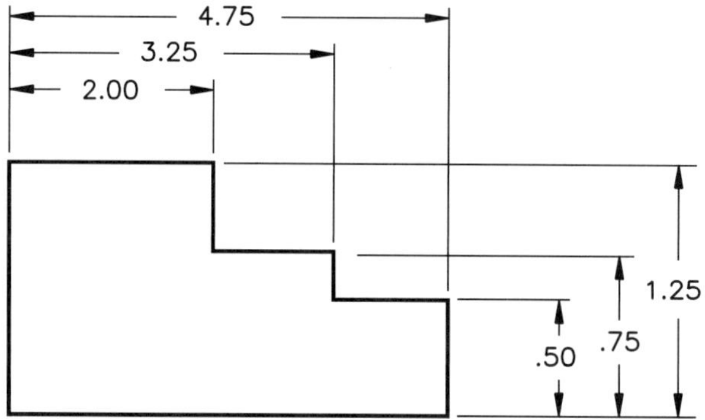

Figure 16.12 *Datum dimensions locate a feature from an assumed perfect surface referred to as a datum. Each successive feature is referenced back to this surface. With datum dimensioning, if an error is made, it only affects one feature.*

shows a part with the left side and the bottom used as assumed datum surfaces. Figure 16.13 shows a part dimensioned with an assumed datum center plane. Because the part is symmetrical, the width of each feature is assumed to be located based on an assumed centerline. The symmetrical object in Figure 16.14 shows datum dimensions from a centerline with the use of a symmetrical symbol. The lines used to make the symbol should be .195" long, with a space of .065" between them. The symmetrical symbol should be placed on the centerline so that it is .75" away from the object.

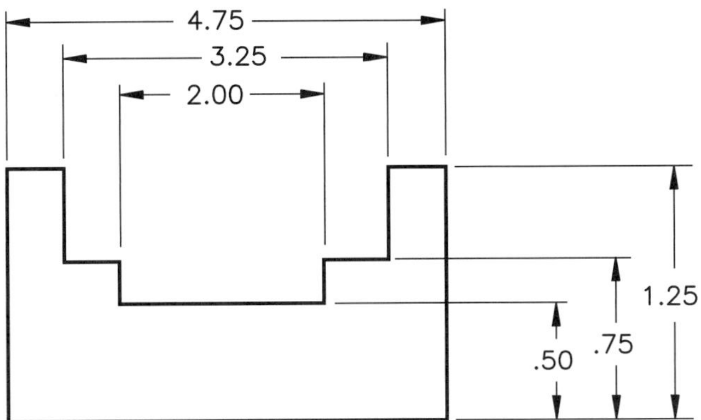

Figure 16.13 *Datum dimensions placed on an assumed centerline.*

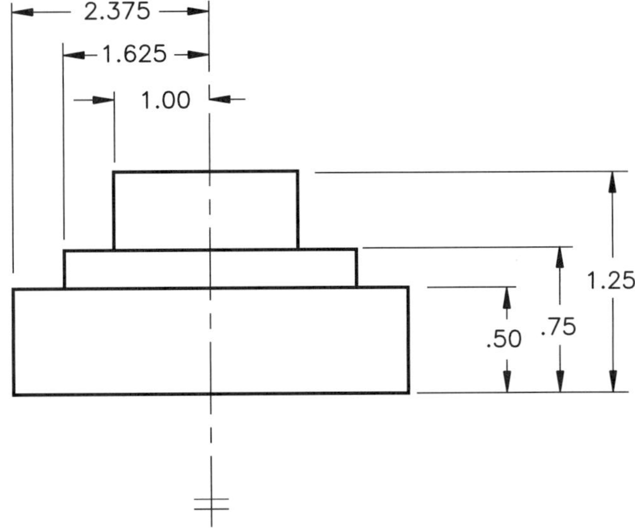

Figure 16.14 *Datum dimensions placed using a symmetrical symbol on the centerline to denote that each side is exactly the same.*

Tabular Dimensions

Tabular dimensioning locates dimensions from a datum for X, Y, and Z. The dimension text is placed in a table to maintain clarity. Figure 16.15 shows an example of tabular dimensions. The locations and size of each drilled feature can be found only in the table.

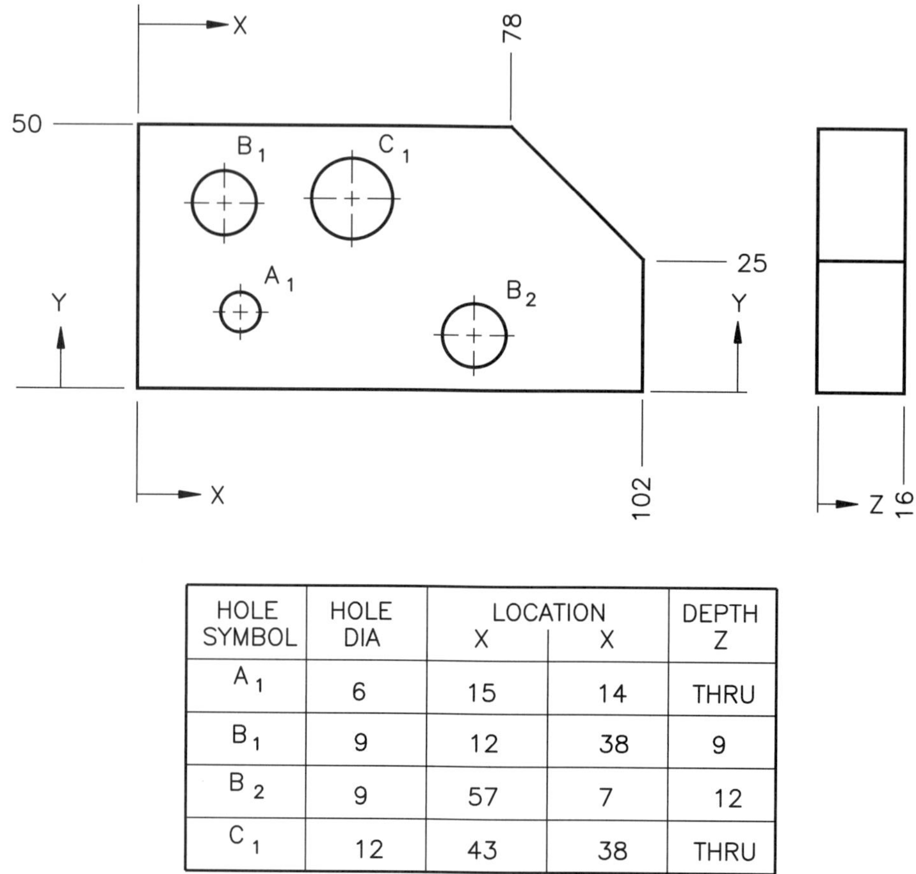

HOLE SYMBOL	HOLE DIA	LOCATION X	LOCATION X	DEPTH Z
A_1	6	15	14	THRU
B_1	9	12	38	9
B_2	9	57	7	12
C_1	12	43	38	THRU

Figure 16.15 *Tabular dimensions locate dimensions from a datum point but place the dimension in a table.*

Arrowless Dimensions

Arrowless dimensions are represented by a letter. One letter is placed by an object, and a corresponding letter is placed in a table to distinguish the size of the feature. Locations dimensions are located with extension lines as coordinates from a specified datum. Figure 16.16 shows an example of arrowless dimensions. Notice that aligned text is used on the part to describe the location.

Chart Dimensions

Chart dimensions are used when a part has one or more dimensions that change with different applications, such as the circular feature that is shown in Figure 16.17. The dimensions that vary are labeled on the drawing by a letter in place of dimension text. The dimension text is then referenced to the letter in a table. On assembly draw-

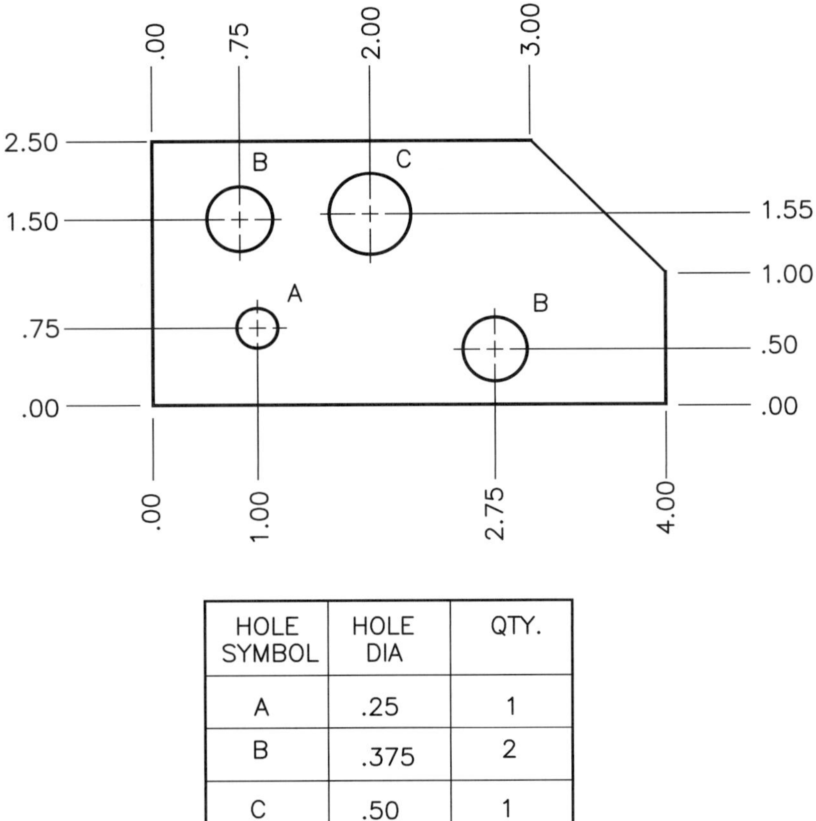

HOLE SYMBOL	HOLE DIA	QTY.
A	.25	1
B	.375	2
C	.50	1

Figure 16.16. *Arrowless dimensions are represented by a letter placed by an object that represents a dimension placed in a table. Location dimensions are located with extension lines as coordinates from a specified datum.*

ings, the purchase part number for each item is often listed. This method of dimensioning is also used in catalog drawings to represent alternative ordering options.

ACCESSING DIMENSIONING COMMANDS

Several methods are available for accessing the commands that control dimensions. Commands can be accessed by selecting buttons on the **Dimension** toolbar or by selecting an option from the **Dimension** menu. The menu and toolbar can be seen in Figure 16.18. Each of the options displayed on the menu and toolbar will be presented in this chapter. Dimensions can also be entered at the Command prompt. Options are also available in dialog boxes, which will be presented in Chapter 17.

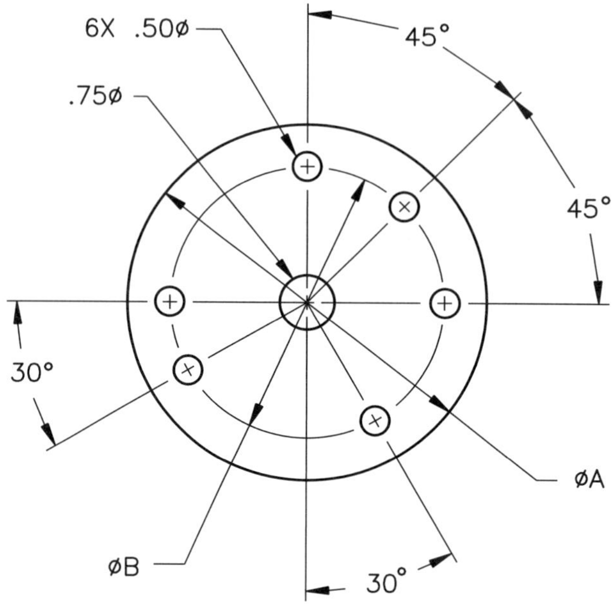

PART NO.	⌀A	⌀B
MD500	5.00	3.50
MD600	6.00	4.75
MD700	7.00	5.75
MD850	8.50	6.50

Figure 16.17 *Chart dimensions are used when a part has one or more dimensions that change with different applications.*

DIMENSIONING OPTIONS

Eleven methods are available for placing dimensions on a drawing. The **Dimlinear** and **Dimaligned** commands are used for providing linear dimensioning. **Dimdiameter, Dimradius, Dimangular**, and **Dimcenter** are used to describe circular features. Three other commands, **Dimbaseline, Dimcontinue**, and **Dimordinate,** are used to place dimensions and can be combined with other commands. The **Qleader** command can be used for referencing text and dimensions to a drawing. The **Tolerance** command can be used to access geometric dimensioning and tolerancing symbols. These

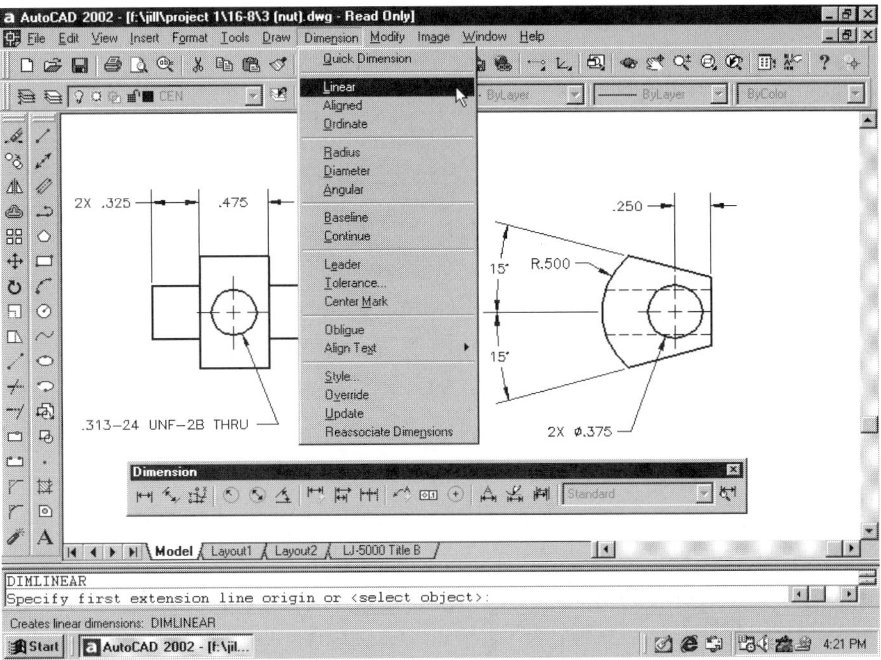

Figure 16.18 *The commands for placing dimensions can be selected on the Dimension toolbar or from the* **Dimension** *menu, or entered by keyboard. Courtesy Jill Heilman.*

symbols will be discussed in Web Chapter 2. **Style** can be used to access the **Dimension Style Manager** dialog boxes that are used to save dimensioning styles. Styles will be discussed in Chapter 17.

Note: In addition to the dimensioning commands listed above, AutoCAD allows dimensions to be placed using the **Quick Dimension** command. It can be used to create a series of dimensions using baseline or continued dimensions. The following portion of this chapter will introduce each dimensioning method. The final portion of this chapter will help you combine options using **Quick Dimension**. Chapter 17 will introduce methods of altering the features of each dimensioning method.

LINEAR DIMENSIONS

Linear dimensions are used to dimension straight surfaces. Selecting the **Linear Dimension** button on the **Dimension** toolbar or **Linear** from the **Dimension** menu will produce the **Dimlinear** Command prompt, which will be discussed shortly. You can also start the command by typing DLI (ENTER) at the Command prompt. This will produce the prompt:

Command: **DLI** ENTER *(Or click the Linear Dimension button.)*

DIMLINEAR

Specify first extension line origin or <select object>:

AutoCAD now offers two options to place a dimension. The default method is to specify two points to define the dimension. A second method allows selection of the object to be dimensioned. Once the object is selected, the end points are automatically defined.

Select Extension Line Origin

The start and end of a line or distance to be dimensioned can be specified. The command sequence for specifying each endpoint is as follows:

Specify first extension line origin or <select object>: *(Select point.)*

Specify second extension line origin: *(Select point.)*

Specify dimension line location or

[Mtext/Text/Angle/Horizontal/Vertical/Rotated]: *(Select dimension line location.)*

The default requires you to move the cursor and select the desired location to place the dimension line. Be sure to activate OSNAP settings before selecting dimension locations. As the second extension line location is selected, the dimension and extension lines are dragged into position to allow for visual inspection, but the coordinate display can be useful in spacing dimension lines. The dimension line location should be selected to allow .75" distance from the object. Place the dimension line in the desired location and click the select button. The process can be seen in Figure 16.19.

Mtext

Selecting the Mtext option allows the dimension value to be replaced by a body of text in a similar manner as the Text option. Typing **M** (ENTER) at the dimension prompt will produce a display similar to Figure 16.20 with the cursor in front of the brackets. The brackets represent the default dimension value. With the cursor preceding the brackets, text will be written followed by the dimension. An entry of **N.T.S.** <> would be displayed as **N.T.S. 4.00**. An entry of <>**N.T.S.** will be displayed as **4.00 N.T.S.** Using the cursor to remove the brackets will remove the default dimension value and allow only the text to be placed.

Text

One of the strengths of AutoCAD is also one of its weaknesses. AutoCAD is extremely accurate. Sometimes the dimension text that will be assigned is too accurate for the drawing purpose. Chapter 17 will introduce methods of controlling dimension accurately. If you haven't been accurate as you created the drawing, the text assigned to the dimension will not produce the desired results. Choosing the Text option before the dimension line location is selected allows you to override the

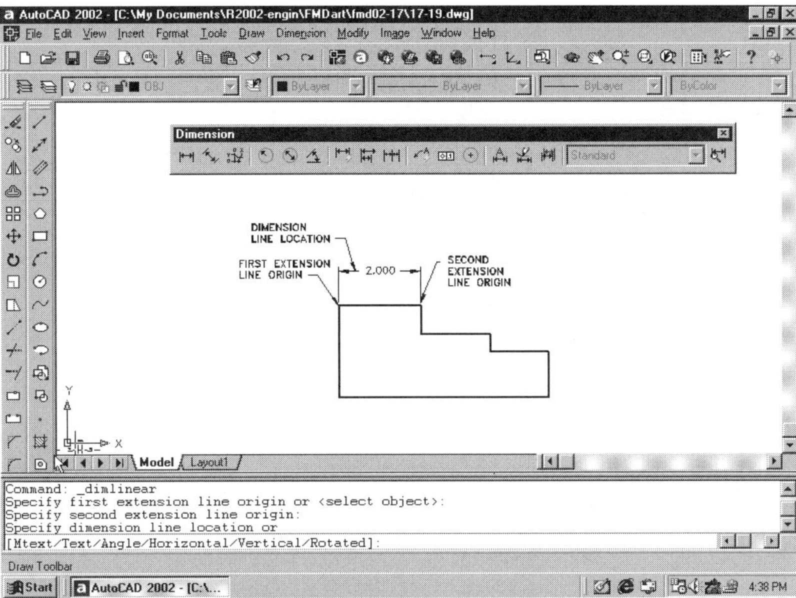

Figure 16.19 *Horizontal and vertical dimensions can be placed using the* **Dimlinear** *command. The dimension is placed by selecting the starting location for each extension line and then selecting the location for the dimension line. As the location for the second extension line is selected, the dimension is displayed and can be dragged into position.*

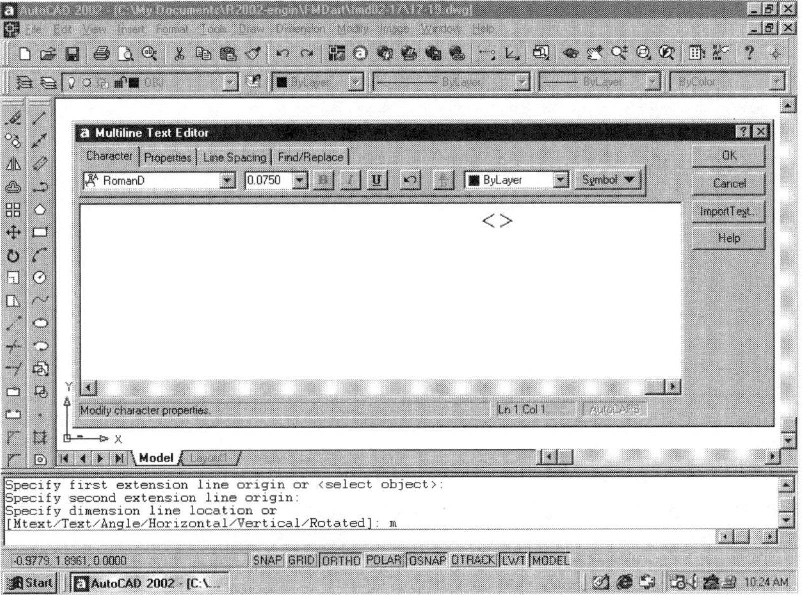

Figure 16.20 *Typing* **M** *(*ENTER*) at the Command prompt will produce the* **Multiline Text Editor**.

dimension text value by entering the value by keyboard. Use the following sequence to change the value from 3.50 to 4.00.

> Specify dimension line location or
>
> [Mtext/Text/Angle/Horizontal/Vertical/Rotated]: **T** ENTER
>
> Enter dimension text <3.50>: **4.00** ENTER
>
> Specify dimension line location or
>
> [Mtext/Text/Angle/Horizontal/Vertical/Rotated]: *(Select the dimension location.)*
>
> Command:

Pressing ENTER returns the dimension and allows it to be dragged into position. Pressing SPACEBAR and then ENTER rather than assigning a value at the prompt for dimension text displays the extension lines, dimension lines, and terminators but no text is assigned. This option can be useful for specifying the limits of a specific finish on a feature. See Figure 16.21.

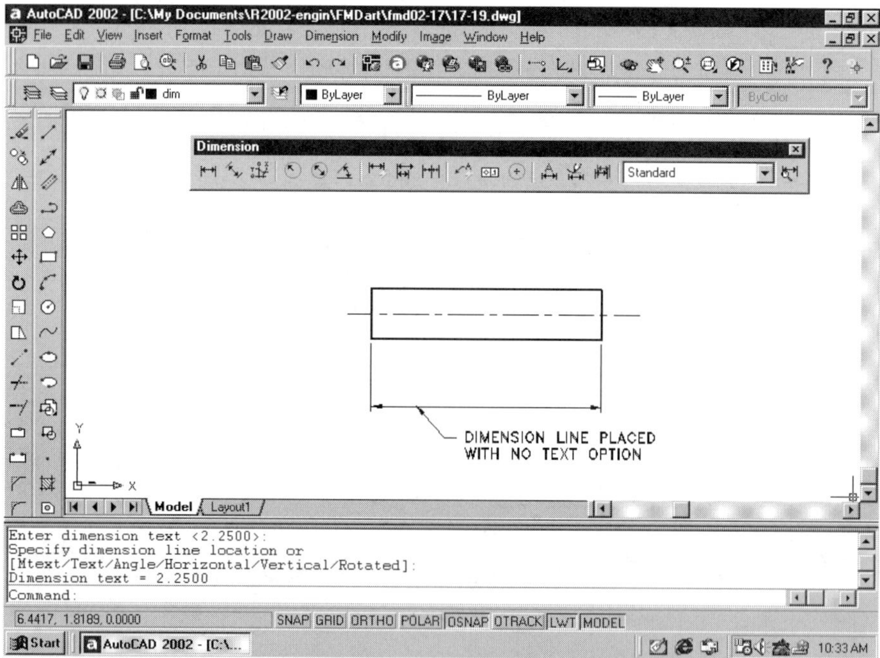

Figure 16.21 *Rather than entering a text value, pressing* SPACEBAR *followed by* ENTER *allows the dimension line to be placed with no dimension. The dimension line can be used for assigning specific information about finishing methods at a later date.*

Angle

Type A (ENTER) at the line location prompt to alter the angle used to display the text. Although this is not often done, it could be especially helpful to make a dimension for a small space stand out. Figure 16.22 shows an example of text rotated at 45° to the dimension line. The command sequence, once the two endpoints are selected, is as follows:

Specify dimension line location or

[Mtext/Text/Angle/Horizontal/Vertical/Rotated]: **A** ENTER

Specify angle of dimension text: **45** ENTER

Specify dimension line location or

[Mtext/Text/Angle/Horizontal/Vertical/Rotated]: *(Select the dimension location.)*

Dimension text = 3.375

Command:

Horizontal and Vertical

Typing H (ENTER) or V (ENTER) at the prompt allows the specified dimension direction to be entered. AutoCAD automatically places a vertical dimension if two vertical endpoints are selected or a horizontal dimension line if two horizontal endpoints are selected.

Rotated

Typing R (ENTER) at the prompt rotates the extension and dimension lines to a specified angle from the desired surface. This option places the dimension in a rotated position and can be used to draw attention to a dimension describing a small space. Use this option to place the extension and dimension lines for referencing a note to a specific area. The sequence to place the lines at 20° to a vertical surface is as follows:

Specify dimension line location or

[Mtext/Text/Angle/Horizontal/Vertical/Rotated]: **R** ENTER

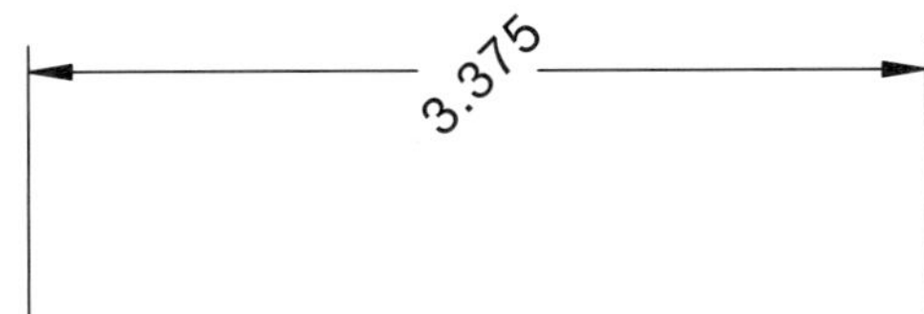

Figure 16.22 *The text angle can be rotated from the dimension line using the* **A** *option. Although not a common method of dimensioning, rotated text can be used to help draw attention to the dimension.*

Specify angle of dimension line <0>: **20** ENTER

Specify dimension line location or

[Mtext/Text/Angle/Horizontal/Vertical/Rotated]: *(Select the dimension location.)*

Dimension text = .50

Command:

Figure 16.23 shows an example of the rotated option.

The Select Object Option

Up to this point, each of the options discussed was made available once the two origins for the extension line endpoints were selected. By choosing the Select Object option, a specific line can be selected, and each endpoint will be determined automatically. The command sequence is as follows:

Command: **DLI** ENTER *(Or click the Linear Dimension button.)*

DIMLINEAR

Specify first extension line origin or <select object>: ENTER

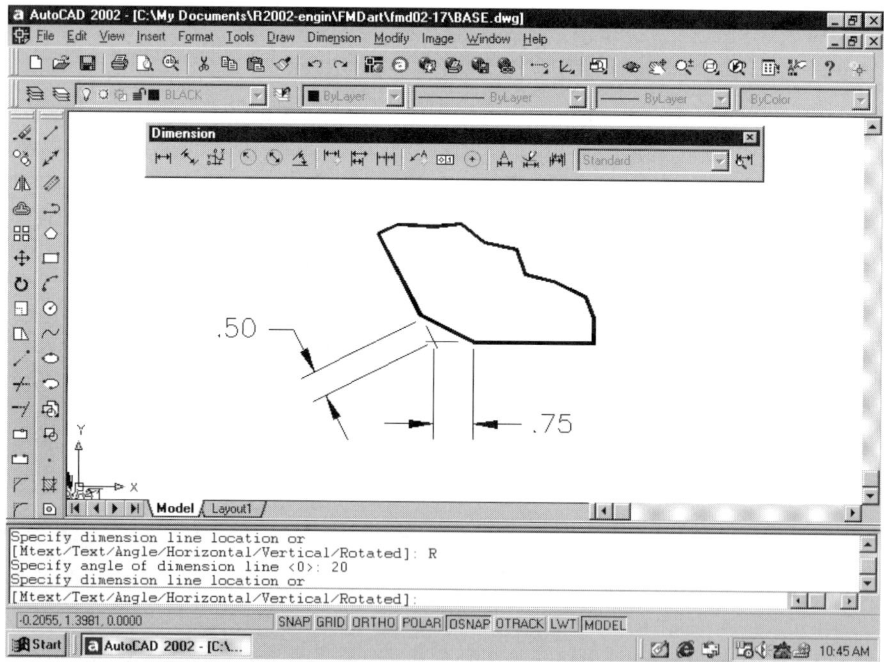

Figure 16.23 *The extension lines, dimension lines, and text can be rotated from the surface to be dimensioned using the* **R** *option.*

Once ENTER is pressed, the crosshairs are changed to a pick box and a specific object can be selected:

> Select object to dimension: *(Specify object to be dimensioned.)*
>
> Specify dimension line location or
>
> [Mtext/Text/Angle/Horizontal/Vertical/ Rotated]: *(Select the desired location.)*
>
> Dimension text = 2.25
>
> Command:

This process can be seen in Figure 16.24. The effect of the **Dimlinear** command on vertical dimensions is shown in Figure 16.25. If the area to be dimensioned is smaller than the length required to place the arrows and text, each will be placed outside the extension lines. Methods of altering placement will be discussed in Chapter 17.

ALIGNED DIMENSIONS

This method of placing dimensions is useful when the object such as an auxiliary view that is not parallel to the drawing borders must be described. An example of aligned dimensions can be seen in Figure 16.26. Notice that the

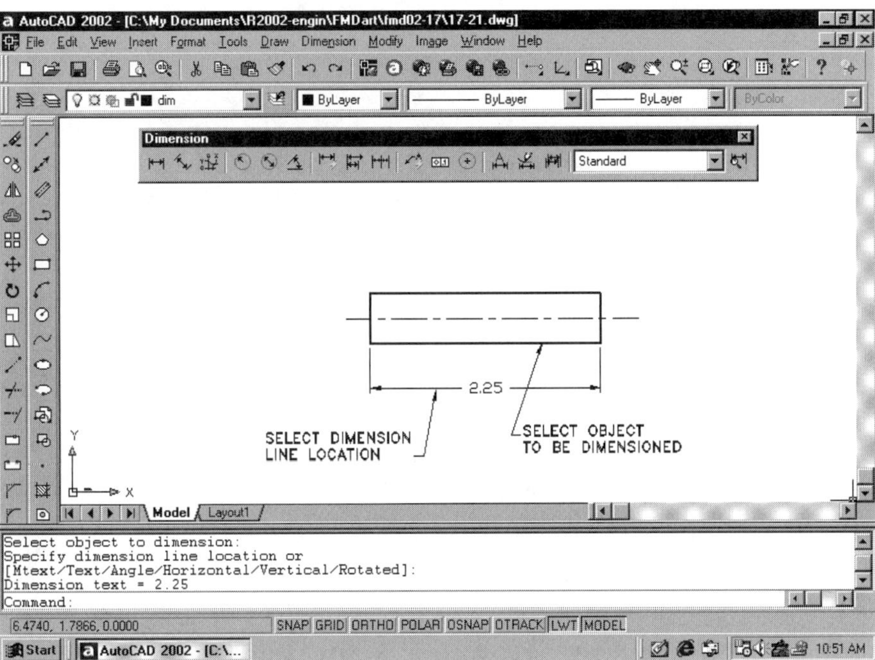

Figure 16.24 *Choosing the Select Object option rather than selecting an extension line location allows an object to be selected. Extension lines will automatically be placed at each end of the selected object.*

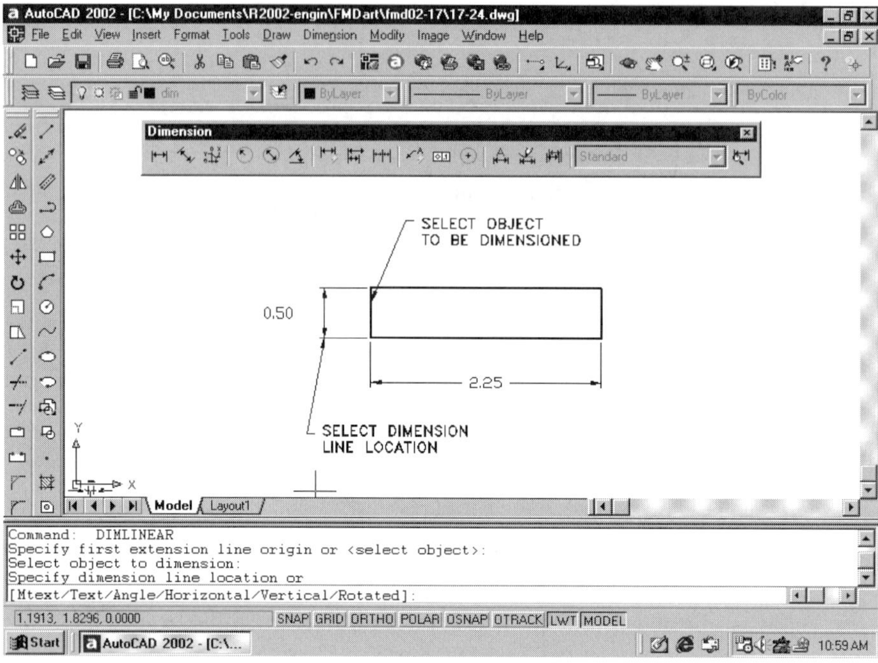

Figure 16.25 *Placement of vertical dimension lines using* **Dimlinear** *with the Select Object option.*

dimension lines are aligned, not the text. The command sequence is similar to the **Dimlinear** command. Start the **Dimaligned** command by selecting the **Aligned Dimension** button on the **Dimension** toolbar, selecting **Aligned** from the Dimension menu, or by typing DAL (ENTER) at the Command prompt:

> Command: **DAL** ENTER *(Or click the Aligned Dimension button.)*
>
> DIMALIGNED
>
> Specify first extension line origin or <Select object>: *(Select first point.)*
>
> Specify second extension line origin: *(Select second point.)*
>
> Specify dimension line location or
>
> [Mtext/Text/Angle]:
>
> Dimension text <1.750>:
>
> Command:

If a location is selected for the dimension line, the line will automatically be placed parallel to the surface defined by the endpoints. Selecting the Mtext or Text option allows the default text value to be altered. Selecting the Angle option allows the angle of the text to be altered. Each of the options is the same as the corresponding options for the **Dimlinear** command. Selecting the Select Object option automatically places the extension lines at the ends of the selected line and allows the dimension to be dragged into position.

The effects of the **Dimaligned** command can be seen in Figure 16.27.

ORDINATE DIMENSIONS

Ordinate dimensioning consists of placing dimensions without the use of dimension lines and arrows. An example of ordinate dimensions can be seen in Figure 16.28. Start the **Dimordinate** command by selecting the **Ordinate Dimension** button on the **Dimension** toolbar or **Ordinate** from the **Dimension** menu or by typing DOR (ENTER) at the Command prompt:

> Command: **DOR** ENTER *(Or click the Ordinate Dimension button.)*
>
> DIMORDINATE
>
> Select feature location: *(Select the surface to be dimensioned.)*

AutoCAD is prompting for the location that will serve as a base. It will be helpful to move the 0,0 location of the world coordinate system from the lower corner of the drawing to a corner of the object being dimensioned. The ORTHO mode should also

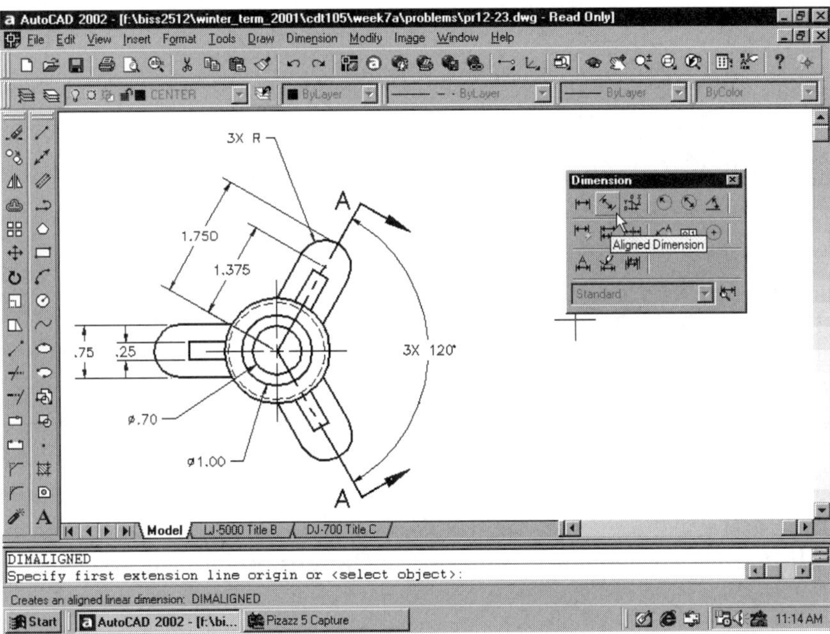

Figure 16.26 *When a surface is not horizontal or vertical, the* **Dimaligned** *command can be used to place dimensions. Courtesy Paul Bissell.*

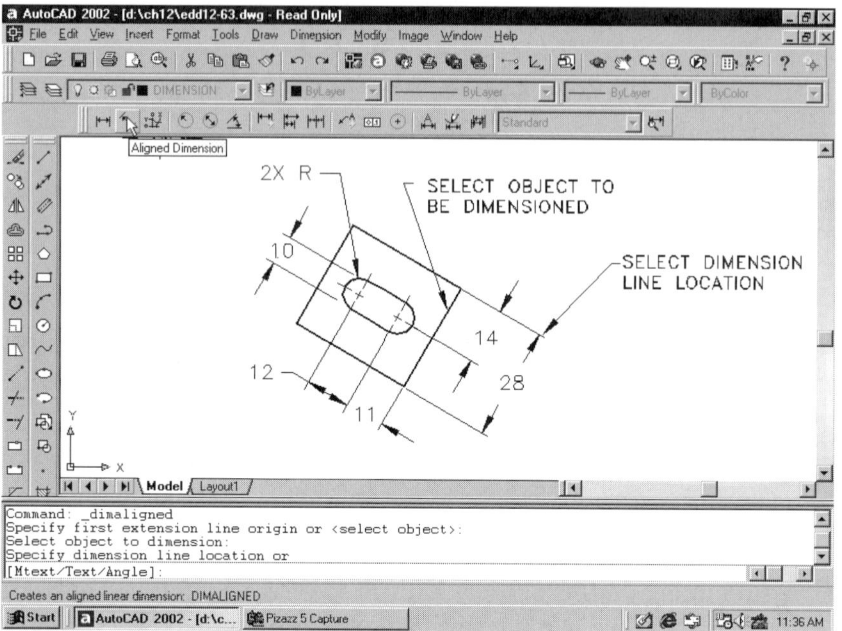

Figure 16.27 *Placing dimensions using the* **Dimaligned** *command is similar to using the Dimlinear command.*

be set to ON. Once the initial edge has been selected, the prompt will be altered to read

> Specify leader endpoint or [Xdatum/Ydatum/Mtext/Text/Angle]:
> *(Select the endpoint of the extension line.)*
>
> Dimension text: <.625> ENTER

The leader is a single horizontal line when the Ydatum ordinate is used and a single vertical line for the Xdatum.

RADIUS DIMENSIONS

Many curved features can be described by a radius using the **Dimradius** command. The way the information is placed will vary depending on the size of the circle or arc to be dimensioned. Figure 16.29 shows each option. The type of dimension also depends on the **Dimvar** settings. Start the command by selecting the **Radius Dimension** button on the **Dimension** toolbar, **Radius** from the **Dimension** menu, or by typing DRA(ENTER) at the Command prompt:

> Command: **DRA** ENTER *(Or click the Radius Dimension button.)*
>
> DIMRADIUS

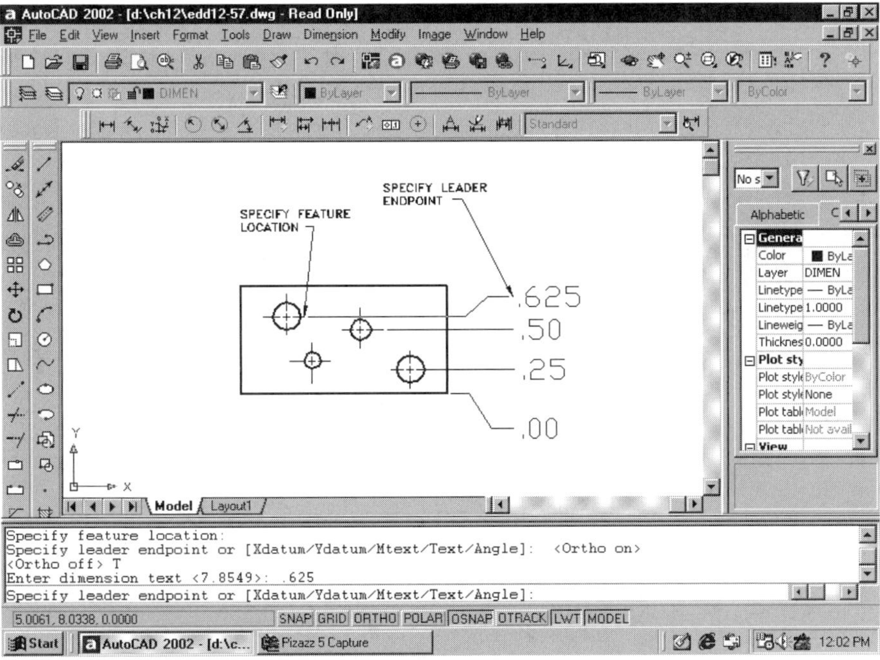

Figure 16.28 *Placement of dimensions using the* **Dimordinate** *command. All dimensions are placed using only a single extension line to represent each location.*

Select arc or circle: *(Select desired arc or circle.)*

Dimension text <10.75>

Specify dimension line location or [Mtext/Text/Angle]: *(Select location for dimension.)*

Command:

When you select the circle or arc to be dimensioned, the default dimension is dragged as you move the cursor. Entering **M**, **T** or **A** allows the text or angle value to be altered. The Mtext, Text, and Angle options function exactly as discussed with other commands. Once the desired Text and Angle prompts have been responded to, the prompt will be returned to locate the dimension line. Figure 16.30 shows an example of the use of the **Dimradius** command.

DIAMETER DIMENSIONS

A circular feature is most accurately described by providing a diameter. (Chapters 20 and Web Chapter 2 will present additional methods of describing circular features.) The **Dimdiameter** command will allow you to dimension the diameter of circular objects. Start the command by selecting the **Diameter Dimension** button on the **Dimension** toolbar or **Diameter** from the **Dimension** menu

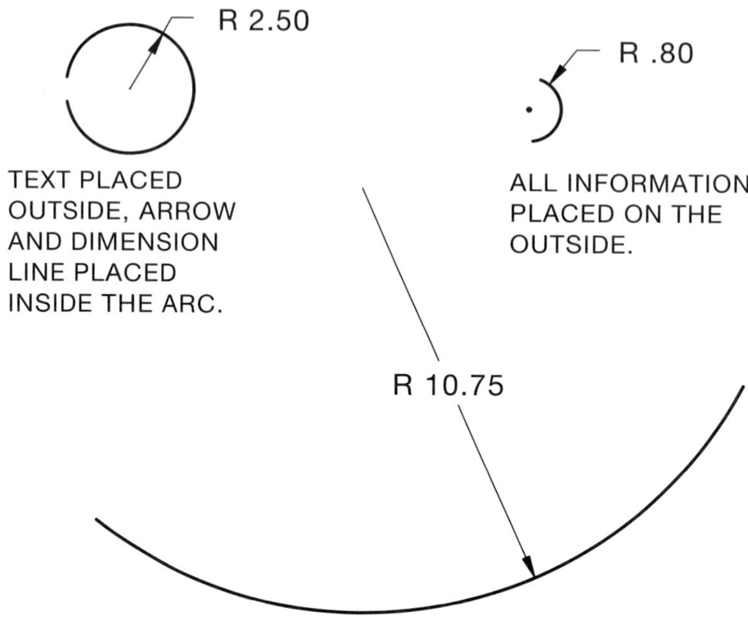

Figure 16.29 *The* **Dimradius** *command allows a radius to be placed using three different techniques.*

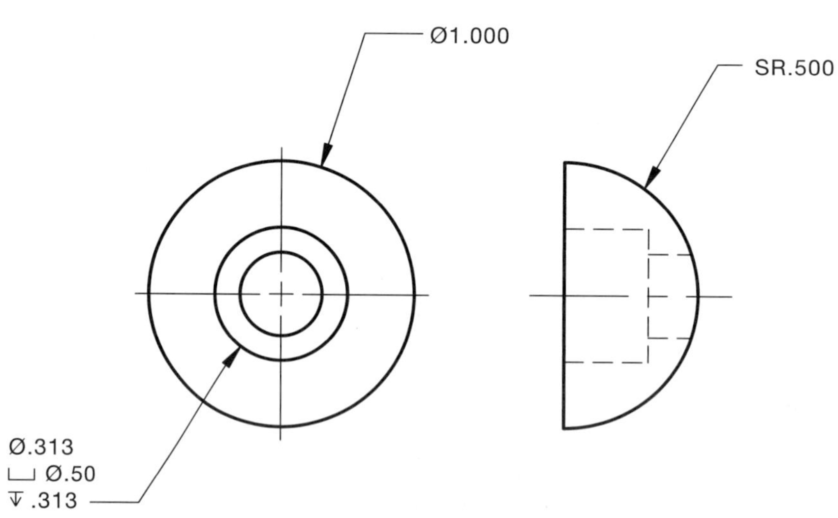

Figure 16.30 *Placing of dimensions using the* **Dimradius** *command. See Chapters 20 and Web Chapter 2 for an explanation of each of the symbols that are used in the notes. Courtesy Jill Heilman.*

or by typing **DDI** (ENTER) at the Command prompt. Two methods of placing the diameter specifications on a drawing can be seen in Figure 16.31. The default method is to use a leader line to display the dimension. If more than one feature is to be described, a note such as

3X .50 Ø

can be used to describe three holes with a diameter of 0.50. The command sequence to describe a diameter is as follows:

Command: **DDI** ENTER *(Or click the Diameter Dimension button.)*

DIMDIAMETER

Select arc or circle:

Dimension text <2.75>

Specify dimension line location or [Mtext/Text/Angle]: *(Specify a point or enter an option.)*

Command:

When you select the circle to be dimensioned, the default dimension is dragged as you move the cursor. Depending on the size of the circle, the dimension toggles between being placed inside or outside the circle. Placing the dimension inside or outside the circle depends entirely on the size of the circle and the room available. Chapter 17 will discuss methods for controlling where the text is placed. Typing **M** (ENTER), **T** (ENTER), or **A** (ENTER) allows the text or angle value to be altered. Mtext, Text, and Angle function exactly as discussed earlier for the **Dimlinear** command. Once the desired prompts have been responded to, the prompt will be returned for locating the dimension line.

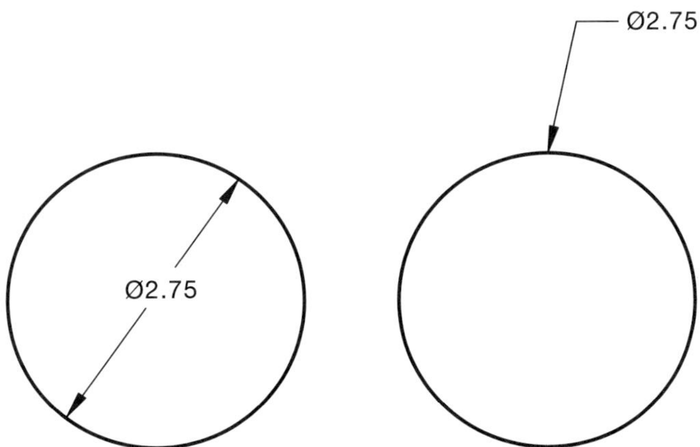

Figure 16.31 *The* **Dimdiameter** *command allows the diameter to be placed using two different methods.*

CREATING A CENTER MARK

The **Dimcenter** command will provide a center mark for a circle or an arc. This command, together with the **Dimcen** variable, can be used to control the center mark. The **Dimcen** variable allows the size of the mark to be altered. Enter a positive value, and a center mark will be located. Enter a negative value to place centerlines. The command sequence is as follows:

Command: **DIMCEN** ENTER

Enter new value for DIMCEN <0.065>: *(Provide desired value.)* **.125** ENTER

Command:

As the Command prompt is returned, nothing is changed. Selecting **Center Mark** on the **Dimension** toolbar or from the **Dimension** menu will now allow a circle or arc to be selected. The command sequence is as follows:

Command: **DCE** ENTER *(Or click the Center Mark button.)*

_DIMCENTER

Select arc or circle: *(Select desired circle to receive center mark.)*

Command:

ANGULAR DIMENSIONS

The **Dimangular** command can be used to dimension angles formed by selecting two straight nonparallel lines, an arc, a circle, and another point or three points. You can start the command by selecting the **Angular Dimension** button on the **Dimension** toolbar, or by selecting **Angular** from the **Dimension** menu. To enter the command at the keyboard, type DAN (ENTER) at the Command prompt. This will produce the following prompt:

Command: **DAN** ENTER *(Or click the Angular Dimension button.)*

DIMANGULAR

Select arc, circle, line, or <specify vertex>:

Choosing Two Lines

One of the most common situations for dimensioning angles is describing the angle formed between two intersecting lines. Respond to the prompt by selecting a line. The command sequence continues with

Select arc, circle, line, or <specify vertex>: *(Select first line.)*

Select second line: (Select second point.)

Specify dimension arc line location or [Mtext/Text/Angle] *(Select location.)*

Dimension text <4.50>

Command:

As the dimension arc line location (Mtext/Text/Angle) prompt is displayed, you will be able to choose the location of the dimension placement relative to the angle. Figure 16.32 shows alternatives for placing the angle dimensions.

Angles Based on an Arc

When you respond to the initial angle prompt by choosing an arc, dimensions can be placed to describe the tangent points of an arc. The command sequence is as follows:

Command: **DAN** ENTER *(Or click the Angular Dimension button.)*

DIMANGULAR

Select arc, circle, line, or <specify vertex>: *(Select first line.)*

Select second line: *(Select second point.)*

Specify dimension arc line location or [Mtext/Text/Angle] *(Select location.)*

Dimension text <156.00>

Command:

As you select the second line, the angle is displayed. Depending on the location of the cursor, the display can be located in one of four positions. The Text and Angle options are the same as with other dimensioning commands. Figure 16.33 shows an example of how text will be placed.

Angles Based on a Circle

When you respond to the initial angle prompt by choosing a circle, dimensions can be placed to describe angular patterns formed within a circle. The command sequence is as follows:

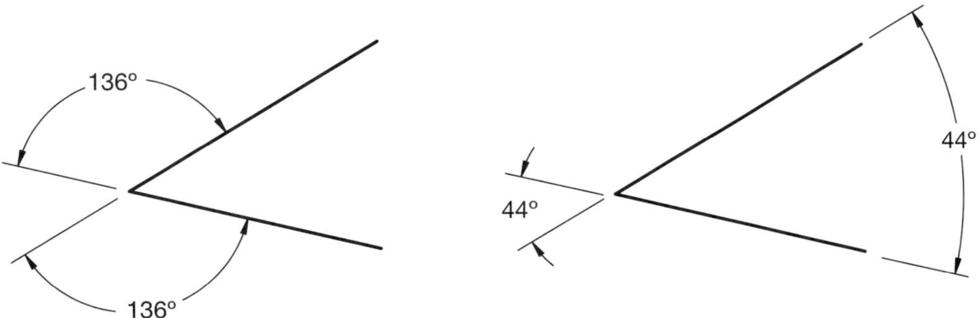

Figure 16.32 *Alternatives for placement of dimensions using the* **Dimangular** *command.*

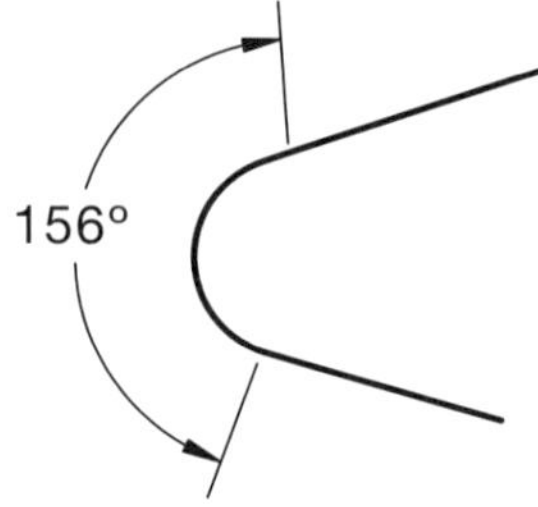

Figure 16.33 *Dimensioning of an arc using the* **Dimangular** *command.*

Command: **DAN** ENTER *(Or click the Angular Dimension button.)*

DIMANGULAR

Select arc, circle, line, or <specify vertex>: *(Select desired object.)*

Specify dimension arc line location or [Mtext/Text/Angle] *(Select location.)*

Dimension text <293>

Command:

Options for placing the dimension line can be seen in Figure 16.34.

Describing Three Points

When you respond to the original angular prompt by pressing ENTER, you will be allowed to select three different points that can be used to describe an angle. This option can be used to describe the center point of bolts around a center point. The two

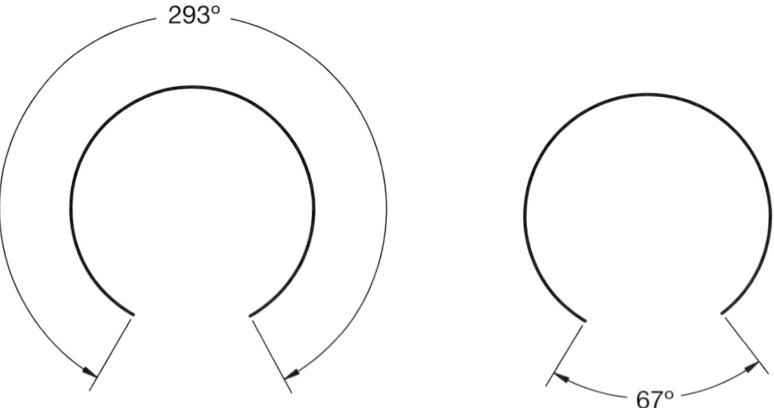

Figure 16.34 *Placement of angular dimensions using the circle option.*

options for placing the angle can be seen in Figure 16.35. The command sequence is as follows:

Command: **DAN** ENTER *(Or click the Angular Dimension button.)*

DIMANGULAR

Select arc, circle, line, or <specify vertex>: ENTER

Specify angle vertex: *(Select a circle center point.)*

Specify first angle endpoint: *(Select a point.)*

Specify second angle endpoint: *(Select a point.)*

Specify dimension arc line location or [Mtext/Text/Angle]: *(Select desired text location.)*

Dimension text <69>:

Command:

BASELINE

Occasionally, when the spacing of objects is extremely critical, the baseline system of placing dimensions can be used. Baseline dimensions (also called datum dimensions) were introduced earlier in this chapter and are used when a high degree of accuracy must be achieved. The **Dimbaseline** command sequence must be started by using a linear, angular or ordinate existing dimension. Start the command by selecting the **Baseline Dimension** button on the **Dimension** toolbar or **Baseline** from the **Dimension** menu. The command can also be started by typing DBA (ENTER) at the Command prompt. AutoCAD will use the first extension line from the existing dimension as the base point for other dimensions. The command sequence can be seen in Figure 16.36 and can be completed as follows:

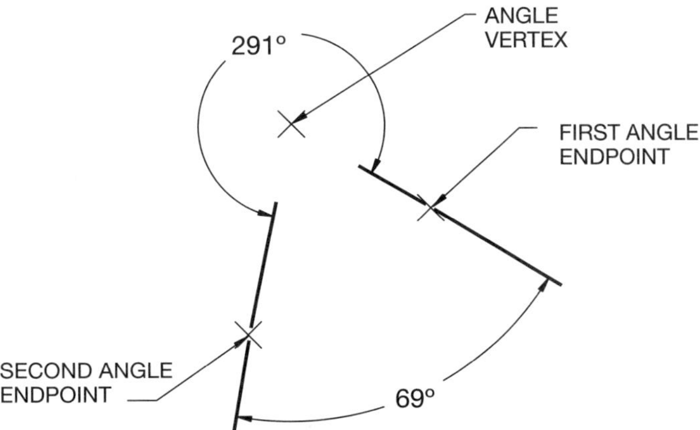

Figure 16.35 *Placement of dimensions using the three point option.*

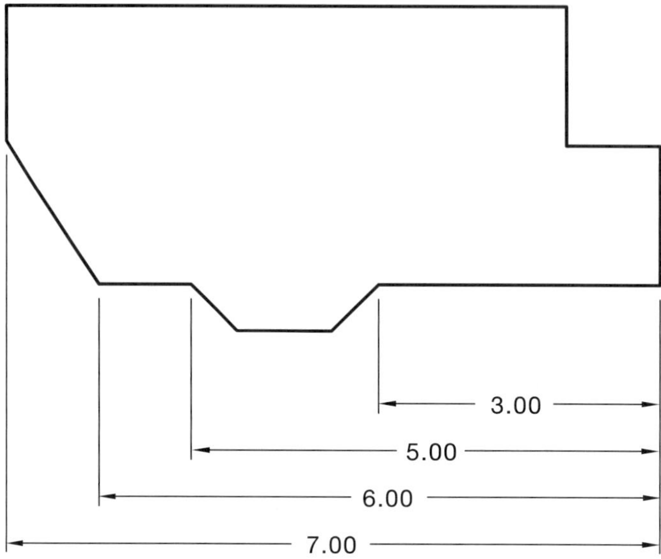

Figure 16.36 *Placement of dimensions using the* **Dimbaseline** *command.*

Command: **DLI** ENTER *(Or click the Linear Dimension button.)*

DIMLINEAR

Specify first extension line origin or <select object>: ENTER

Select object to dimension:

Specify dimension line location or

[Mtext/Text/Angle/Horizontal/Vertical/Rotated]: *(Select the desired dimension location.)*

Dimension text <3.00>:

Command:

Command: **DBA** ENTER *(Or click the Baseline Dimension button.)*

DIMBASELINE

Specify a second extension line origin or [Undo/Select] <Select>: *(Select base for second point.)*

Dimension text <5.00>

Specify a second extension line origin or [Undo/Select] <Select>: *(Select base point for the next dimension.)*

Dimension text <6.00>

Specify a second extension line origin or [Undo/Select] <Select>:
(Select base point for the next dimension.)

Dimension text <7.00>

Specify a second extension line origin or [Undo/Select] <Select>:
ENTER

Select base dimension: ENTER

Command:

CONTINUOUS DIMENSIONS

A series of dimensions is needed to dimension an object. In Figure 16.37, dimension lines extend from each edge across the part to describe all surfaces except one. (Web Chapter 2 will explain why one dimension is always omitted.) These dimensions can be placed with the **Dimcontinue** command. The command sequence starts by using the **Dimlinear** command to place the first dimension. The **Dimcontinue** command must be based on a linear, angular, or ordinate dimension. Start the **Dimcontinue** command by selecting the **Continue Dimension** button on the **Dimension** toolbar or by selecting **Continue** from the **Dimension** menu. You can also start the command by typing DCO (ENTER) at the Command prompt:

Command: **DCO** ENTER *(Or click the Continue Dimension button.)*

DIMCONTINUE

Select continued dimension: *(Select a dimension to continue from.)*

Specify a second extension line origin or [Undo/Select] <Select>:

After you select the starting point for the second extension line, AutoCAD uses the second extension line of the **Dimlinear** command sequence as the first extension line for the **Dimcontinue** sequence. Press ESC to end the **Dimcontinue** command sequence. Figure 16.37 shows an example of placing dimensions using the **Dimcontinue** command with the default settings. The command sequence to place the dimensions across the bottom is as follows:

Command: **DLI** ENTER *(Or click the Linear Dimension button.)*

DIMLINEAR

Specify first extension line origin or <select object>: ENTER

Select object to dimension:

Specify dimension line location or

[Mtext/Text/Angle/Horizontal/Vertical/Rotated]: *(Select the desired dimension location.)*

Dimension text 1.00

Command:

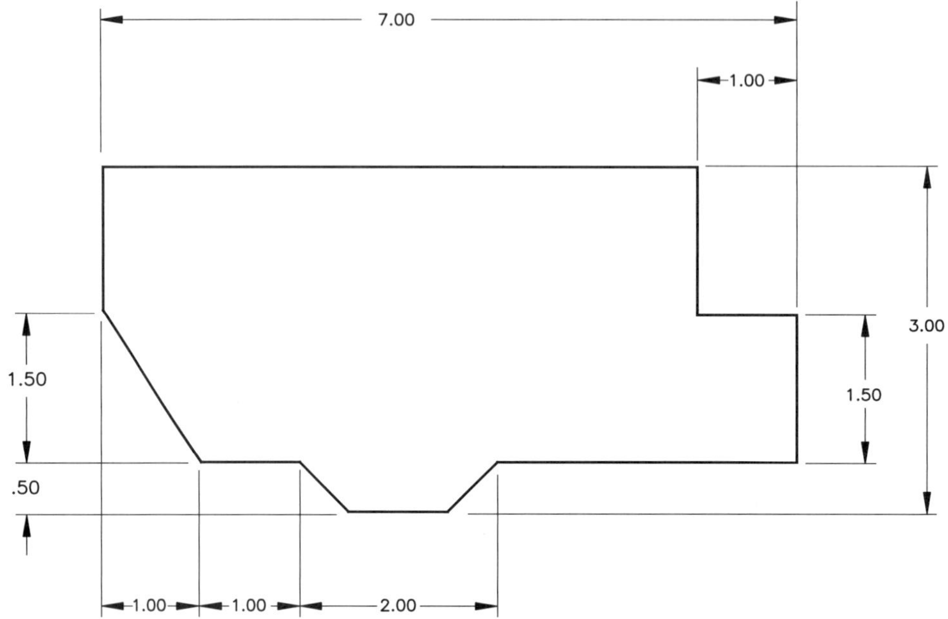

Figure 16.37 *Placement of dimensions using the* **Dimlinear** *and* **Dimcontinue** *commands. The 1.00 dimension on the bottom left was placed using the* **Dimlinear** *command. The next 1.00 dimension was placed using the* **Dimcontinue** *command. The* **Dimcontinue** *command uses the second extension line of the preceeding sequence as the first extension line of the* **Dimcontinue** *sequence.*

Command: **DCO** ENTER *(Or click the Continue Dimension button.)*

DIMCONTINUE

Specify a second extension line origin or [Undo/Select]: <Select>
(Select second point.)

Dimension text = 1.00

Specify a second extension line origin or [Undo/Select]: <Select>
(Select next point.)

Dimension text = 2.00

Specify a second extension line origin or [Undo/Select]: <Select>
(Select next point.)

Dimension text = 3.00

Specify a second extension line origin or [Undo/Select]:
<Select> ENTER

Select continued dimension: ENTER

Command:

QUICK DIMENSIONS

You've been exposed to several different methods for placing dimensions throughout this chapter. Each works well for a specific need. AutoCAD provides an additional method for placing dimensions that combines features of many of the previous options. As the name implies, the **Quick Dimension** command allows a single dimension or a series of dimensions to be placed quickly. **Quick Dimension** excels in placing strings of continuous or baseline dimensions with very few prompts required. Start the command by selecting the **Quick Dimension** button on the **Dimension** toolbar, by typing QDIM (ENTER), or by selecting **Quick Dimension** from the **Dimension** menu. Each will produce the following prompt at the command line:

Command: **QDIM** ENTER *(Or click the Quick Dimension button.)*

Select geometry to dimension: *(Select object or objects to be dimensioned.)*

AutoCAD is waiting for you to select an object to be dimensioned. Press ENTER to end the selection process. The Command prompt will now change to display the following:

Select Geometry to dimension: ENTER

Specify dimension line position, or

[Continuous/Staggered/Baseline/Ordinate/Radius/Diameter/datumPoint/Edit]

<Continuous>: *(Select the desired dimension location.)*

If only one object is selected, the command will function similarly to the Select Objects option of the **Dimlinear** command. An object to be dimensioned is selected, followed by a location for the dimension to be placed. As ENTER is pressed, the dimension will be displayed and can then be dragged into position as the cursor is moved. When the dimension is in the desired position, left-click to add the dimension to the drawing. The process can be seen in Figures 16.38a and 16.38b.

The best features of **Quick Dimension** can be seen if multiple objects are selected to be dimensioned. In Figure 16.39a, four objects have been selected to be dimensioned. Once the selection set is completed and ENTER is pressed, dimensions to describe each feature will be displayed. A dimension to describe each feature can be placed by identifying one location point. The process can be seen in Figures 16.39a and 16.39b. The command process to place the four dimensions is as follows:

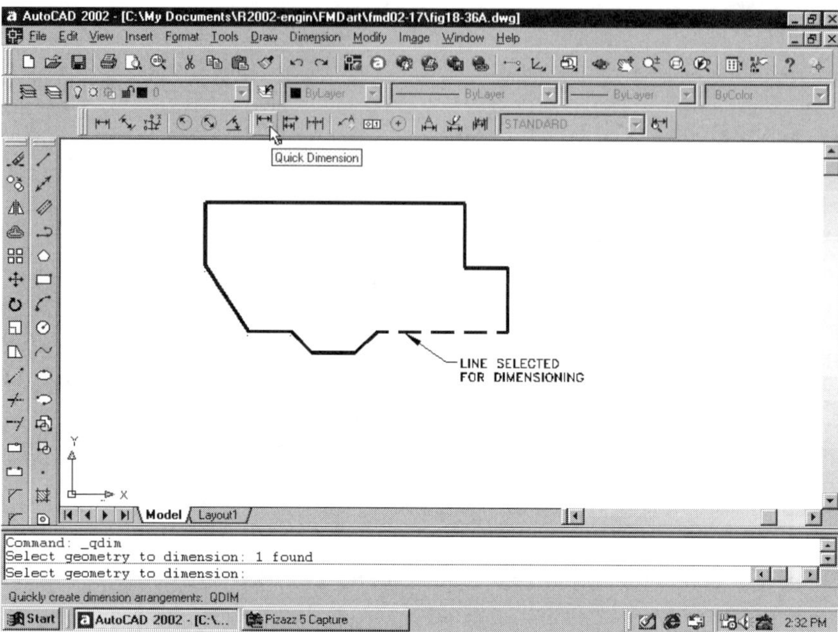

Figure 16.38a *The* **Quick Dimension** *command functions similarly to the Select Objects option of the* **Dimlinear** *command if only one object is selected to be dimensioned.*

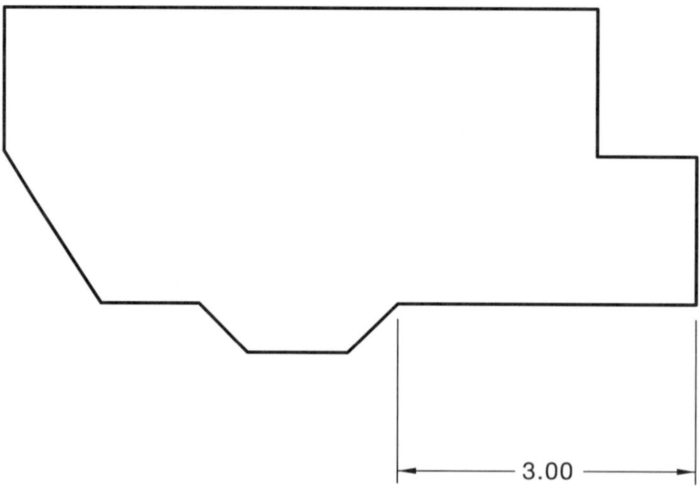

Figure 16.38b *With the* **Quick Dimension** *command, only an object and location are required to place a dimension.*

Command: **QDIM** ENTER *(Or click the Quick Dimension button.)*

Select Geometry to dimension: *(Select object or objects to be dimensioned.)*

Select Geometry to dimension: 1 found,

Select Geometry to dimension: 1 found, 2 total

Select Geometry to dimension: 1 found, 3 total

Select Geometry to dimension: 1 found, 4 total

Select Geometry to dimension: ENTER

Specify dimension line position, or [Continuous/Staggered/Baseline/Ordinate/Radius/Diameter/datumPoint/Edit]

<Continuous>: *(Select location to place the dimension string.)*

Command:

Note: Keep in mind that when defining a selection set with multiple objects, you need to hold down SHIFT to add selections. If you forget to hold SHIFT down, the objects to be dimensioned will be selected, but the previous selection will be deleted from the selection set.

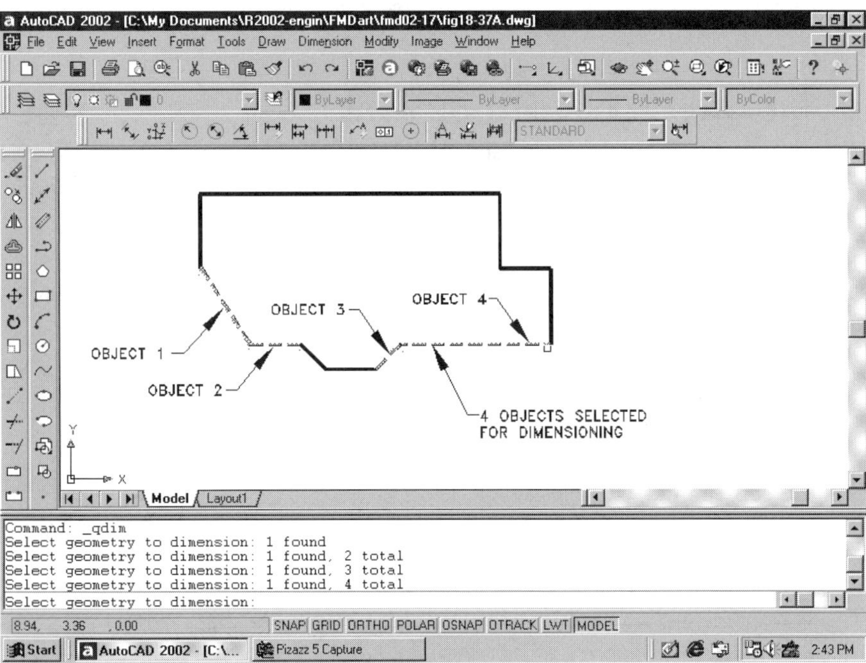

Figure 16.39a *The best features of* **Quick Dimension** *can be seen if multiple objects are selected to be dimensioned. In this example, four objects have been selected to be dimensioned.*

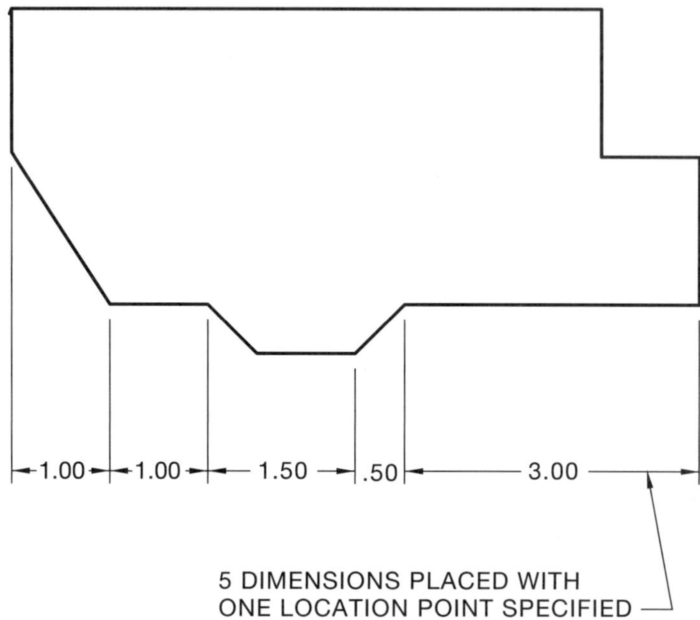

Figure 16.39b *A dimension to describe each selected feature is placed by identifying one location point.*

QUICK DIMENSION PLACEMENT OPTIONS

The Baseline option of **Quick Dimension** combines the features of the **Dimbaseline** dimension option and the Continuous process to place multiple dimensions. The command functions in a method similar to the Continuous option in that each object must be selected, followed by the location. Figure 16.40 shows an example of placing dimensions using the Baseline option of **Quick Dimension**. The command sequence is as follows:

Command: **QDIM** ENTER *(Or click the Quick Dimension button.)*

Select Geometry to dimension: *(Select object or objects to be dimensioned.)*

Select Geometry to dimension: 1 found,

Select Geometry to dimension: 1 found, 2 total

Select Geometry to dimension: 1 found, 3 total

Select Geometry to dimension: 1 found, 4 total

Select Geometry to dimension: ENTER

Specify dimension line position, or

[Continuous/Staggered/Baseline/Ordinate/Radius/Diameter/
datumPoint/Edit]

<Continuous>: **B** ENTER

Specify dimension line position, or

[Continuous/Staggered/Baseline/Ordinate/Radius/Diameter/
datumPoint/Edit]

<Baseline>: *(Select location to place the dimension string.)*

Command:

The Staggered option of **Quick Dimension** can be used to place staggered dimensions. Once objects to be dimensioned are selected, type S (ENTER) when prompted for an option. Figure 16.41 shows an example of staggered dimensions. Typing O (ENTER) at the prompt allows ordinate dimensions to be placed using methods for selection similar to other **Quick Dimension** options.

The Radius and Diameter options of **Quick Dimension** allow you to select and dimension multiple circles or arcs of various sizes in one command sequence rather that having to repeat a dimension sequence for each object.

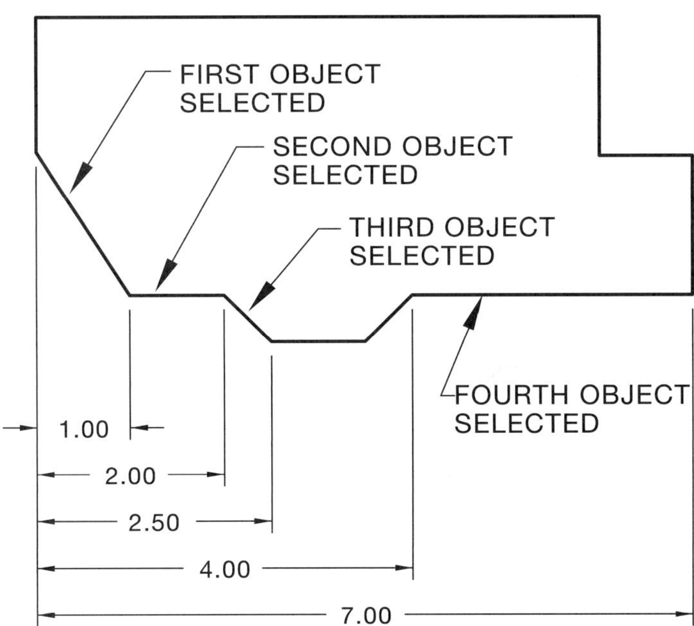

Figure 16.40 *The Baseline option of the* **Quick Dimension** *command functions similarly to the Continuous option in that each object must be selected, followed by the location.*

The datumPoint option will place dimensions in a pattern similar to the Ordinate option. The major difference between the two options is that datumPoint allows any point to be set as the base point. This option could be used to establish a datum, and allow dimensions to be placed above and below the datum. Figure 16.42 shows an example of dimensions placed using the datumPoint option. The option is started by typing **P** (ENTER) at the option prompt. The command sequence to place datumPoint dimensions is as follows:

Command: **QDIM** ENTER *(Or click the Quick Dimension button.)*

Select Geometry to dimension: *(Select object or objects to be dimensioned.)*

Select Geometry to dimension: 1 found,

Select Geometry to dimension: 1 found, 2 total

Select Geometry to dimension: 1 found, 3 total

Select Geometry to dimension: 1 found, 4 total

Select Geometry to dimension: ENTER

Specify dimension line position, or

[Continuous/Staggered/Baseline/Ordinate/Radius/Diameter/datumPoint/Edit]

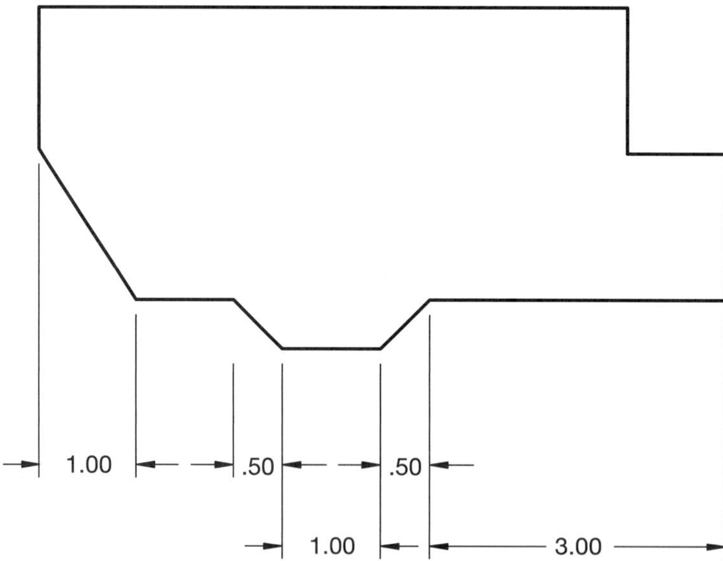

Figure 16.41 *Typing* **S** *(ENTER) when prompted for an option allows staggered dimensions to be placed.*

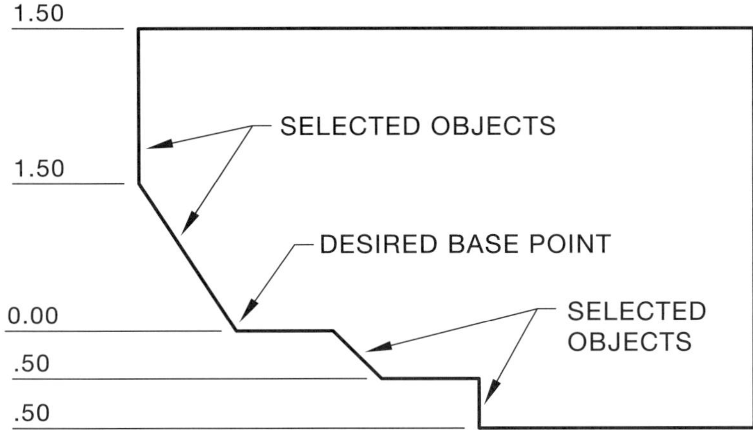

Figure 16.42 *The datumPoint option allows ordinate dimensions to be placed using any point to be set as the base point.*

<Continuous>: **P** ENTER

Select new datum point: *(Select the point to be the base of the dimensions.)*

Specify dimension line position, or

[Continuous/Staggered/Baseline/Ordinate/Radius/Diameter/ datumPoint/Edit]

<Ordinate>: *(Select location to place the dimension string.)*

Command:

LEADER LINES

The **Qleader** command can be used to place a leader line to connect notes or dimensions to a specific portion of the drawing. Figure 16.43 shows examples of the use of leader lines. Leaders are generally placed at any angle except horizontal or vertical. A horizontal leg approximately .125" is placed at the end of the leader line to tie the text to the leader. The horizontal leg should extend from the left side of the first line of text or from the right side of the last line of text. Figure 16.44 shows common methods of placing leader lines.

USING QUICK LEADERS

The **Qleader** command should be used when several local notes need to be referenced to the drawing, or when you want to control the width of the space where the notes will be placed. Start the command by selecting the **Quick Leader** button on the **Dimension** toolbar, by typing LE (ENTER) at the Command prompt, or by selecting **Leader** from the **Dimension** menu. Each will produce the prompt:

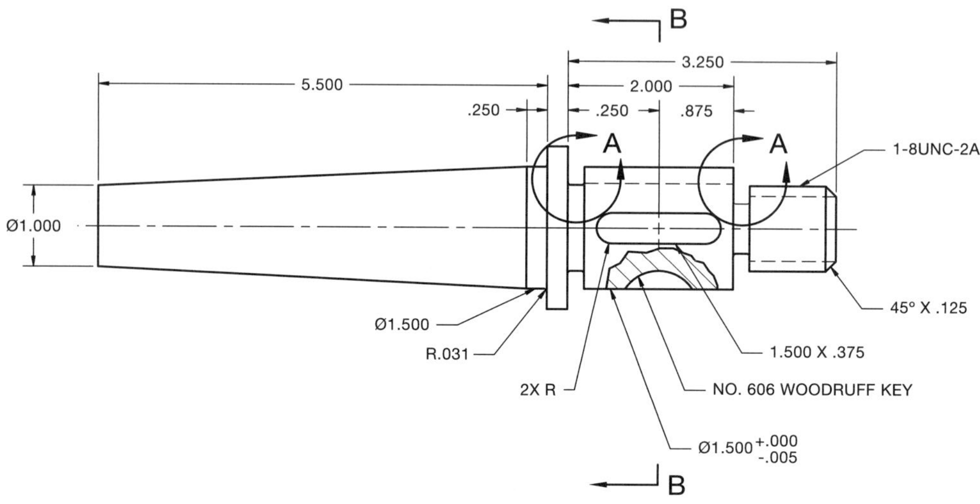

Figure 16.43 *Leader lines can be used to attach text, dimensions, and welding symbols to a specific portion of the drawing. Courtesy Paul Bissell.*

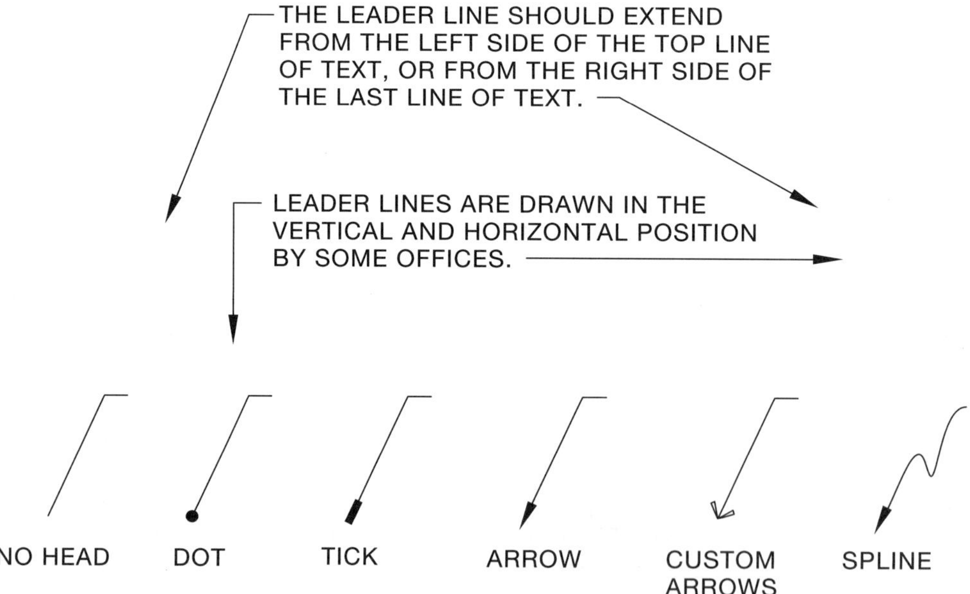

Figure 16.44 *Common methods of placing and terminating leader lines.*

Command: **LE** ENTER *(Or click the Quick Leader button.)*

QLEADER

Specify first leader point, or [Settings] <Settings>: *(Select desired start point for leader.)*

Specify next point: *(Select desired end point of leader.)*

Specify next point: *(Select desired end point for horizontal leader tail.)*

Specify text width <0.00>: ENTER

Enter first line of annotation text <Mtext>:

The first option to be considered is Settings; it allows several features of the leader to be altered. Settings will be explored after the default option of **Qleader** has been explored. For now, select a starting point for the leader. The "first" and "next" point prompts are like those of the Line command. AutoCAD will place a line with the tip of an arrow extending from the "first" point to the "next" point. The Settings option allows alternatives to the arrow to be used. A prompt for a second "next" point is also provided. As shown in Figure 16.45, this second "next" point produces a horizontal line.

Text Width

Once the location of the leader is specified, you will be prompted to provide the text width. The Text Width option allows the space to be specified where the annotation will be placed. The default setting for width is 0.00. A line of text will be of unspecified length with this setting. Text will be placed in one line until ENTER is pressed to

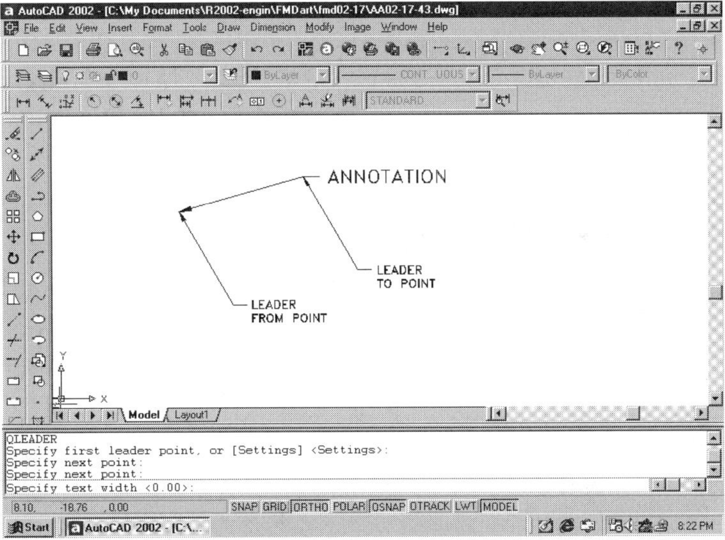

Figure 16.45 *When* **Qleader** *is used, the "first" and "next" point prompts are like those of the* **Line** *command.*

start a new line of text. If a value is provided for the text width, that value will specify the width of each line of text. Typing **4** (ENTER) will allow the text line to be 4" long. Once the specified width has been achieved, a new line of text will be started automatically.

Entering Annotation

The final prompt of the Qleader command is

Enter first line of annotation text <Mtext>:

Three options are available for entering text beside the leader. Pressing ESC will place the leader line and close the command without placing any text beside the leader. This option works well if stock notes have been stored in a drawing template, and they just need to be moved into the proper position. The notes can be placed, and then leader lines can be placed to reference the note to the drawing.

A second option as the text prompt is displayed is to begin entering the text at the Command prompt. As text is entered at the keyboard, it will not be displayed in the drawing area. Figure 16.46 shows how text will be displayed on the Command prompt. Pressing ENTER will display a prompt for additional lines of text. When the desired text has been entered, press ENTER to end the command and display the text. Figure 16.47 shows the leader and the placement of the text.

The third option of placing text at the leader line is to press ENTER at the text prompt. This option allows multiple lines of text, numbers or symbols to be placed at the end of the leader line using the **Multiline Text Editor**. Text entered by keyboard will be displayed in the edit box and will not be placed by the leader line until the **OK** button is clicked. Figure 16.48 shows an example of the edit box. The **Multiline Text Editor** was explored as multiline text was examined. See Chapter 15 for a review of the **Multiline Text Editor**.

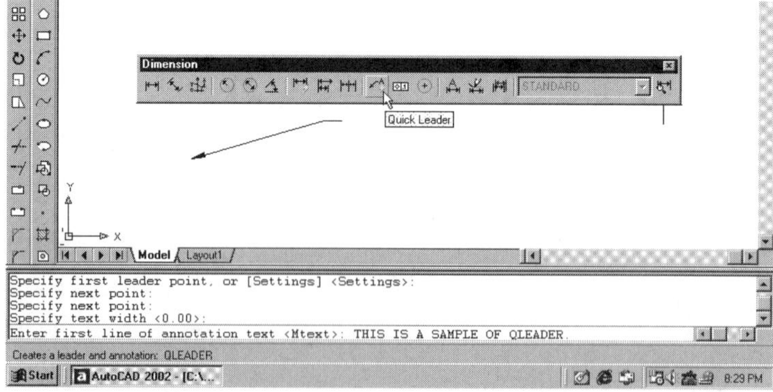

Figure 16.46 Qleader *allows the text to be entered at the Command prompt. As text is entered at the keyboard, it will be displayed at the Command prompt and not beside the leader line.*

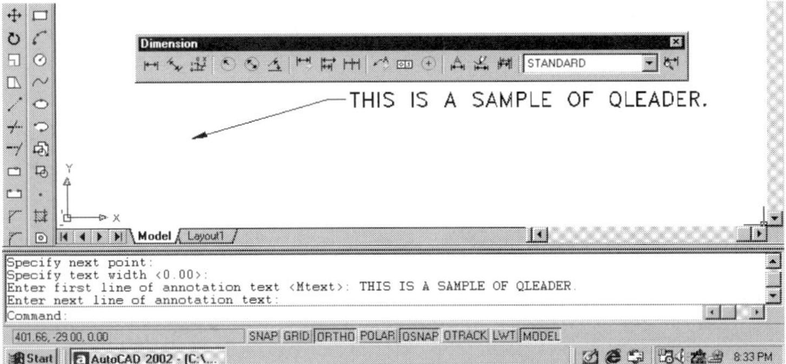

Figure 16.47 *When the text has been entered, pressing* ENTER *will close the command and display the text beside the leader line.*

ADJUSTING QLEADER SETTINGS

Earlier you were introduced to the Settings option of the **Qleader** command. Rather than adjust the settings, methods of placing text were explored. The Settings option will present controls for annotation, leader line, and arrow types, and methods of attaching the leader to the leader lines. **Qleader** settings can be altered by pressing ENTER at the first prompt.

> Command: **LE** ENTER *(Or click the Quick Leader button.)*
>
> QLEADER
>
> Specify first leader point, or [Settings] <Settings>: ENTER

Pressing ENTER will produce the **Leader Settings** dialog box shown in Figure 16.49.

Annotation Tab

The **Annotation** tab allows controls for **Annotation Type**, **MText options**, and **Annotation Reuse**.

Annotation Type

The default **Annotation Type** setting of this tab will place text using **Mtext**. Remember it's the default setting because it is meets a large variety of needs. Other options include the following:

- The **Copy an Object** option allows existing text to be copied to the new leader line. With this option active, once the leader line is placed, a prompt to select text to be copied is displayed. As the text body is selected, it will automatically be copied to the new location.
- The **Tolerance** option will display the **Tolerance** dialog box and allow the creation of a feature control frame. These are tolerance symbols that are used to control the accuracy of a dimension in relation to a specific surface or condition. Tolerances will be discussed in Web Chapter 2.

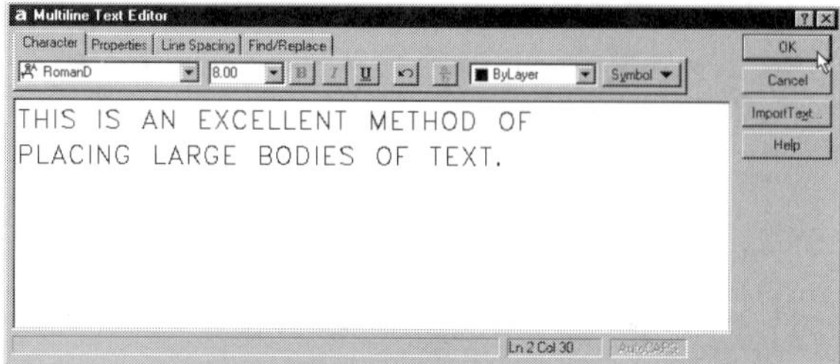

Figure 16.48 *Pressing* ENTER *at the text prompt allows text, numbers, or symbols to be placed at the end of the leader line using the* **Multiline Text Editor** *dialog box. Text entered by keyboard will be displayed in the edit box and not placed beside the leader line until the* **OK** *button is clicked.*

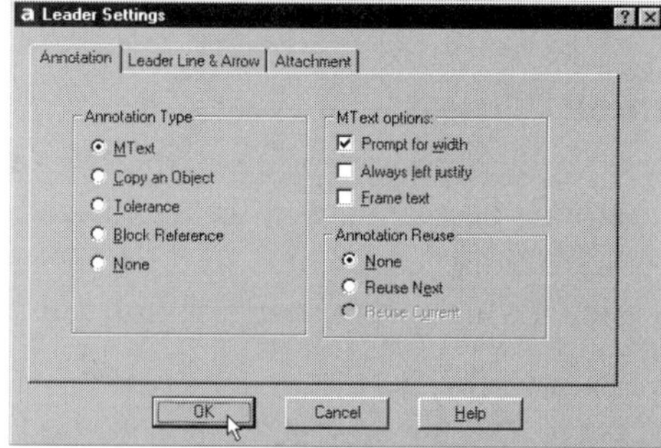

Figure 16.49 *Pressing* ENTER *at the first* **Qleader** *prompt will produce the Leader Settings dialog box.*

- The **Block Reference** option allows predefined groups of text to be inserted by the leader line. Blocks will be introduced in Chapter 20.
- The **None** option will place a leader with no text. Uses for this option have been introduced earlier in this chapter. The benefit of using this option is that it quickly produces the leader line with no prompt for text and no need to remember if you press ESC or ENTER.

MText Options

The **MText options** area is activated only if **MText** is the selected annotation type. The **Prompt for width** check box, if deactivated, can be used to suppress the width prompt. If you're placing text in a drawing and don't care how long the lines of text are, suppress the prompt. The **Always left justify** check box allows you to specify aligned text. Most professionals pride themselves on their neatness, and aligned text will add to a professional appearance. You've explored justification as multiline text was introduced. Further options will explore methods of altering text to leader line relationship. An option that you should explore is **Frame text**. With this option active, text will be displayed as seen in Figure 16.50.

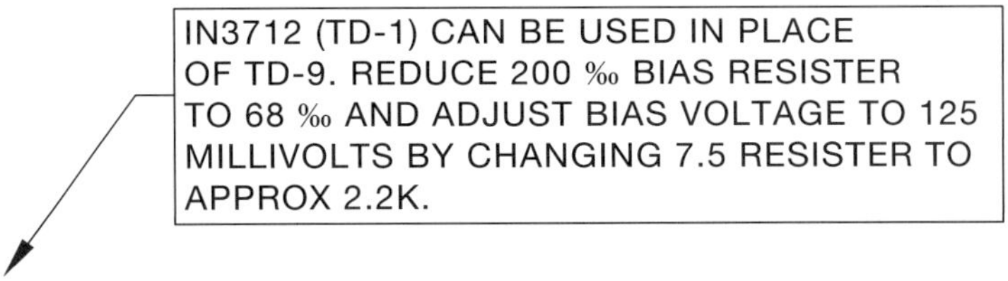

IN3712 (TD-1) CAN BE USED IN PLACE
OF TD-9. REDUCE 200 ‰ BIAS RESISTER
TO 68 ‰ AND ADJUST BIAS VOLTAGE TO 125
MILLIVOLTS BY CHANGING 7.5 RESISTER TO
APPROX 2.2K.

Figure 16.50 *The* **Frame text** *option will enclose annotation text in a box. The option provides an excellent method of drawing attention to important notes.*

Annotation Reuse

This portion of the **Annotation** tab sets options for reusing text. The default option is not to reuse text at a leader line. The **Reuse Next** option will reuse the text that is created with the current command sequence for all future leaders. The current text will be placed by each leader until the setting is altered. The **Reuse Current** option will reuse current annotation. With this option active, AutoCAD will automatically select this option when you reuse text after selecting the **Reuse Next** option.

Leader Line & Arrow Tab

The **Leader Line & Arrow** tab contains options that can be used to control the appearance of the leader line and arrow. The tab can be seen in Figure 16.51.

Leader Line

Two options are available for placing the leader line. Throughout this chapter, each example has been created using the **Straight** option. Figure 16.52 shows an example of a **Spline** leader line.

Number of Points

This option sets the number leader points that you will be prompted to provide. With the **No Limit** option selected, prompts for points will be provided until ENTER is pressed twice. Setting a number for the maximum number of prompts will limit the

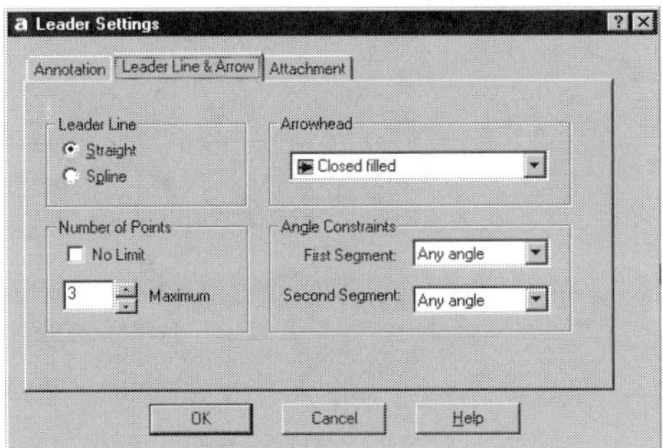

Figure 16.51 *The* **Leader Line & Arrow** *tab of the* **Leader Settings** *dialog box.*

number of prompts displayed as you proceed through the command. Set the value for one more than the number of leader lines you desire. In the default setting of 3, you will be prompted to place text after specifying the two lines formed by the three points.

Note: Adjusting the number of points will be a powerful tool when you create **Spline** leader lines.

Arrowhead

This portion of the tab allows the leader line terminator to be set. Figure 16.53 shows the options that are available. These options can be used to replace the filled arrowhead. To replace the filled arrow, select the desired terminator. As the selection is made, the list will be closed and the **Leader Line & Arrow** tab will be displayed. Notice that the last option in the list is User Arrow. This option will display the Select **Custom Arrow Block** dialog box that can be used to select predefined arrows that have been saved as a block. Chapter 20 will introduce the **Block** command.

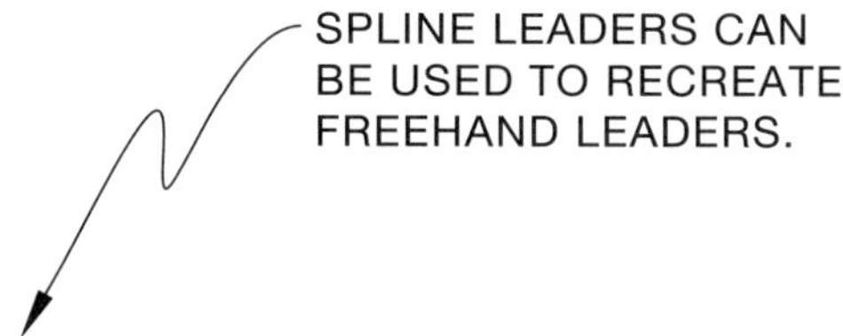

Figure 16.52 *A spline leader line.*

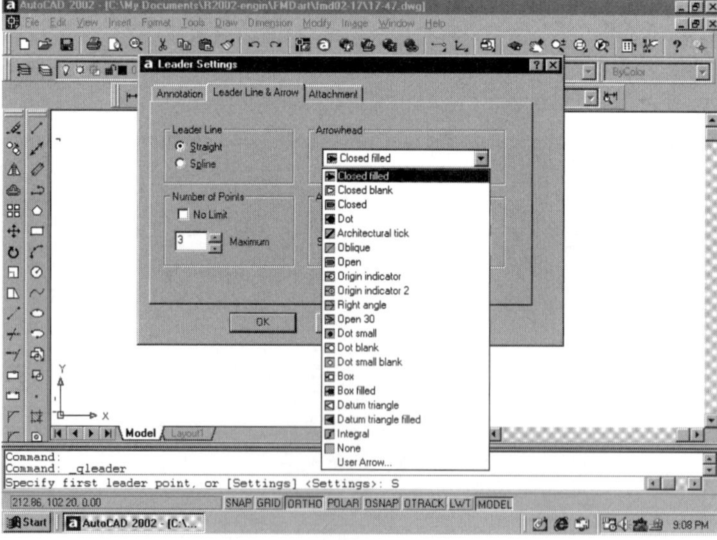

Figure 16.53 *Options that are available for leader line terminators.*

Angle Constraints

These options will allow the angle of each segment of the leader line to be controlled. Values for each option include Any angle, Horizontal, 90, 45, 30, and 15 degrees. Options are provided for the first and second lines of a leader line. Using the Any angle option for the **First Segment** provides freedom in placing text around a drawing. Selecting an angle provides uniformity in appearance, but it might make placing text more challenging. Specifying the Horizontal option for the **Second Segment** will provide a horizontal tail without requiring ORTHO to be active.

Attachment Tab

The **Attachment** tab shown in Figure 16.54 can be used to set the attachment of multiline text and the leader line. Settings are available to control the leader to text relationship depending on which side of the text the leader is placed. Options can be seen in Figure 16.55.

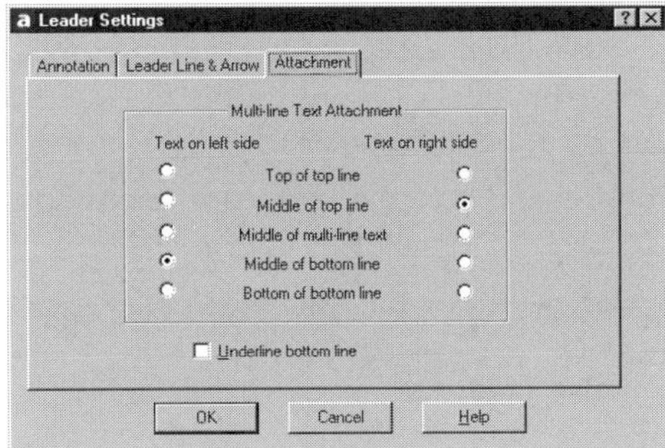

Figure 16.54 *The* **Attachment** *tab can be used to set the attachment of multiline text and the leader line.*

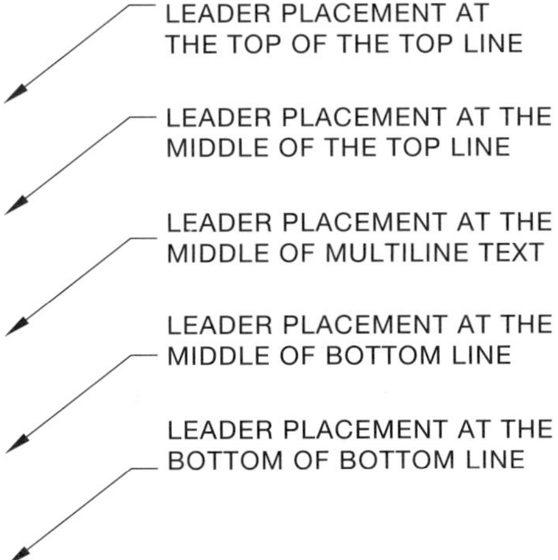

Figure 16.55 *Options for placing the leader line in relationship to the text.*

CHAPTER 16 EXERCISES

1. Start a new drawing and set the units to decimal with two-place accuracy. Draw a hexagon with a horizontal top and base inscribed in a 4" diameter circle. Use the Dimlinear command to provide overall dimensions. Save the drawing as E-16-1.
2. Open drawing E-16-1. Change the hexagon to a layer titled BASE. Move the dimensions to a layer titled DIMEN and turn the layer off. Create a layer titled ANGLE and make it current. Provide dimensions to indicate the angle of the sides. Save the drawing as E-16-2.
3. Open drawing E-8-4 and dimension objects a, c, e, and g using two-place decimal dimensions. Place the remaining objects on a layer titled EXTRA and freeze the layer. Save the drawing as E-16-3.
4. Open drawing E-10-3 and scale the drawing using a scale factor of .5. Create a layer for the dimensions. Completely dimension the beam based on the original sizes from your drawing. Save the drawing as E-16-4.
5. Open drawing E-10-5 and create a layer titled DIMEN. Rescale the drawing by a factor of .25 and completely dimension the drawing so that the metal hanger can be fabricated. Save the drawing as E-16-5.
6. Open drawing E-10-4 and create a layer for dimensions. Completely dimension the part and save the drawing as E-16-6.
7. Start a new drawing and draw the following object. Starting @ 2,2:

B. @1.5,0	K. @ –2,0
C. @ 0,1	L. @ 0,–.75
D. @ 1.25,0	M. @ –2,0
E. @ .75,.75	N. @ 0,–1.25
F. @ 1.5,0	P. @ –2.5,0
G. @ 0,.75	Q. @ .5,–1.5
H. @ 1.5,0	R. AND THEN CLOSE.
J. @ 0,1.75	

   Change the line segments to polylines. Completely dimension the drawing, using the **Dimcontinue** command where possible. Save the drawing as E-16-7.
8. Open drawing E-16-7 and create a new layer for the existing dimensions. Move the dimensions to the new layer that was just created, and set it to OFF. Create another layer for dimensions and completely dimension the object using baseline dimensioning. Assume the lower left corner as the starting point for horizontal and vertical dimensions. Save the drawing as E-16-8.
9. Open the ANSI-C template, and draw an object similar to Figure 16.37. Create a layer for placing dimensions. Dimension every surface using the Continuous option of **Quick Dimension**. Save the drawing as E-16-9.

10. Open drawing E-16-9 and create a new layer for dimensions. Dimension each surface using the Baseline option of **Quick Dimension**. Save the drawing as E-16-10.
11. Open drawing E-16-10 and create another layer for dimensions. Dimension each surface using the Staggered option of **Quick Dimension**. Save the drawing as E-16-11.
12. Open drawing E-6-1 and add layers appropriate for dimensions. Place dimensions to completely describe each feature using continuous dimensions. Erase any views that are not necessary. Save the drawing as E-16-12.
13. Open drawing E-6-3 and add layers appropriate for dimensions. Place dimensions to completely describe each feature using datum dimensions. Erase any views that are not necessary. Save the drawing as E-16-13.
14. Open drawing E-6-5 and add layers appropriate for dimensions. Place dimensions to completely describe each feature using arrowless dimensions. Create a table to represent the holes as 0.125, 0.25, 0.375, 0.50, and 0.75.Erase any views that are not necessary. Save the drawing as E-16-14.
15. Open drawing E-6-20 and add layers appropriate for dimensions. Place dimensions to completely describe each feature using tabular dimensions. Erase any views that are not necessary. Save the drawing as E-16-15.
16. Open drawing E-7-18 and add layers appropriate for dimensions. Place dimensions to completely describe each feature using assumed datum dimensions. Save the drawing as E-16-16.
17. Open drawing E-7-19 and add layers appropriate for dimensions. Place dimensions to completely describe each feature. Save the drawing as E-16-17.
18. Open drawing E-7-23 and add layers appropriate for dimensions. Place dimensions to completely describe each feature using point-to-point dimensions. Erase any views that are not necessary. Save the drawing as E-16-18.
19. Open drawing E-7-22 and add layers appropriate for dimensions. Place dimensions to completely describe each feature using point-to-point dimensions. Erase any views that are not necessary. Save the drawing as E-16-19.
20. Open drawing E-11-16 and add layers appropriate for dimensions. Place dimensions to completely describe each feature of E-11-16b. Save the drawing as E-16-20.
21. Open drawing E-11-17 and add layers appropriate for dimensions. Place dimensions to completely describe each feature of the original drawing. Use a system that will provide a high degree of accuracy. Save the drawing as E-16-21.
22. Open drawing E-11-19 and add layers appropriate for dimensions. Place dimensions to completely describe each feature of the original drawing. Use an appropriate system of dimensions to describe all features. Save the drawing as E-16-22.
23. Open drawing E-11-15b and add layers appropriate for dimensions. Place dimensions to completely describe each feature. Use an appropriate system of dimensions to describe all features. Save the drawing as E-16-23.

24. Open drawing E-11-31 and add layers appropriate for dimensions. Place dimensions to completely describe each feature. Erase any views that are not necessary. Use a system that will provide a high degree of accuracy. Save the drawing as E-16-24.
25. Open drawing E-11-33 and add layers appropriate for dimensions. Place dimensions to completely describe each feature. Erase any views that are not necessary. Use chain dimensions. Save the drawing as E-16-25.

CHAPTER 16 QUIZ

1. What common components comprise dimensioning?
2. How should text be placed on drawings to ease the job of the print reader?
3. List four options typically used for the intersections of dimension and extension lines.
4. Describe the two major methods of placing text on a drawing.
5. List four groups of commands that relate to placing dimensions on a drawing.
6. Describe two methods of locating extension lines for **Dimlinear** dimensions.
7. List four methods to describe angular dimensions.
8. Open one of the drawing exercises that were completed for this chapter, and provide the name of the layer that was added to the drawing base by AutoCAD.
9. What is the difference between tabular and chart dimensions?
10. Explain the difference between **Dimaligned** and **Dimrotated** dimensions.
11. What is the first prompt for the **Dimlinear** command asking for?
12. Should the default text value for a dimension always be accepted? Explain your answer.
13. What does the Angle option alter in the **Dimlinear** command?
14. What is the effect of pressing ENTER at the first **Dimlinear** command prompt rather than selecting a point?
15. Should a dimension be placed inside or outside a circular object?
16. What is the process for placing a .125 center mark in the center of a circular part?
17. List and explain the options for leader annotation.
18. Four lines of text need to be placed by a leader line. What option should be used?
19. Explain the options for ordinate dimensioning.
20. List five options of leader lines, besides the text, that can be altered.
21. Why does datum dimensioning offer better accuracy than chain dimensioning?
22. How are arrowless dimensions effective on a mechanical part?
23. How can you be sure a drawing is complete?
24. How many views of an object should be drawn?
25. What defines how dimensions will be placed on a drawing?

Placing Dimensions on Drawings

In Chapter 16 you were introduced to dimension requirements of different drawings and the methods of describing linear, angular, circular, continuous, and baseline dimensions. Each system was described using the default values. If all of your drawings could fit on a computer screen, everything would be fine. To allow for drawings of various sizes, the spacing of extension and dimension lines and the size of text and arrows must be altered. In this chapter you'll be introduced to

- Altering dimension variables
- Establishing dimension styles
- Editing existing dimensions

Commands to be explored include

- **Dimstyle**
- **Dimoverride**
- **Dimedit**
- **Dimtedit**
- **Dimension Update**

CONTROLLING DIMENSIONS VARIABLES WITH STYLES

As text was explained, you were introduced to the **Style** command. **Style** allows you to group variables such as the font, text height, width, oblique angle, and orientation so that several different styles of text can easily be created within one drawing. AutoCAD also allows dimension styles to be created within a drawing to meet the needs of various situations. To save time, individual styles as well as each dimensioning variable should be set on template drawings on the DIMENSIONS layer.

INTRODUCTION TO DIMENSION STYLES

A dimension style is a set of dimension variables that control various aspects of placing the extension lines, dimension lines, and annotation in relation to these lines and line terminators. The type of drawing you are working on will affect how the variables are to be set. Orthographic, isometric, and assembly drawings may require different styles of dimensions. Dimensions that are always inside extension lines, three-place precision, or the use of tolerances are all elements that can be stored in a dimension style and saved in a template drawing. You will be introduced to creating various styles later in this chapter.

Before a style can be created, you must have a thorough understanding of the dimension variables. Dimension variables are the qualities that control how dimensions will appear. You will be introduced to dimensioning variables through the default style of Standard. Later you'll be introduced to methods of creating specific styles to meet the needs of each drawing type.

CONTROLLING DIMENSION VARIABLES

Dimension variables are altered using the **Dimension Style Manager** dialog box, displayed in Figure 17.1. Access the dialog box by selecting **Dimension Style** on the **Dimension** toolbar, by selecting **Dimension Style** from the **Format** menu, by selecting **Style** from the Dimension menu, or by typing D (ENTER) at the Command prompt.

DIMENSION STYLE MANAGER

The main areas of the **Dimension Style Manager** dialog box include the **Styles** box, a **Preview** window, and the buttons **Set Current**, **New**, **Modify**, **Override**, and **Compare**.

Current Dimension Style

Notice that in Figure 17.1 the current style is named Standard, the name of the default settings for AutoCAD. As styles are created, the name of each style will be listed in the **Styles** box. The current style will be highlighted. Once multiple styles have been created, a different style can be made current by highlighting the name of the style and then selecting the **Set Current** button.

List

The **List** box displays the options that control the display of dimension styles. Options include All Styles and Styles in Use. With All styles selected, all of the dimension styles contained in the drawing will be displayed. With Styles in Use selected, the display will display only dimension styles that are referenced by dimensions in the drawing.

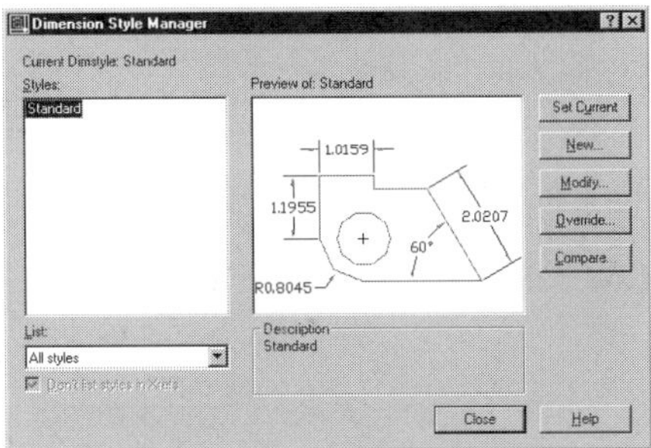

Figure 17.1 *Dimension variables are created and altered using the* **Dimension Style Manager** *dialog box. Access the dialog box by selecting* **Dimension Style** *on the* **Dimension** *toolbar, by selecting* **Dimension Style** *from the* **Format** *menu, by selecting* **Style** *from the* **Dimension** *menu, or by typing* **D** *(ENTER) at the Command prompt.*

Just below the **List** box is the check box for displaying dimension styles in externally referenced drawings. XREF drawings will be discussed in future chapters.

Control Buttons

Five buttons are available for creating and altering the dimension style. Each will be introduced now and explored in detail throughout the chapter.

Set Current

This option has already been briefly introduced. Once a style is created, you can set it as the current style by highlighting the style and selecting the **Set Current** button. Each of the settings of the drawing style will be applied to the new dimensions.

New

Selecting the **New** button will display the **Create New Dimension Style** dialog box shown in Figure 17.2. The box is used to name new dimension styles, to select the style that will be used as a base to build the new style, and to indicate the dimension types that the new style will be applied to. Components include the following:

New Style Name—By default, AutoCAD will assign Copy of Standard as the new style name. With the name highlighted, change the style name to provide a useful description.

Start With—This list box determines what will be used as the base to form the new style. Rather than starting from scratch, AutoCAD will use the Standard style as a base and allow these options to be adjusted to meet the drawing needs. As new styles are created, each style name will be listed, allowing it to be selected to be the base for a new style.

Use for—Selecting the **Use for** edit arrow will display the list shown in Figure 17.3, which allows a style to be created that will be applied only to certain types of dimensions. This option could be used if you would like to use

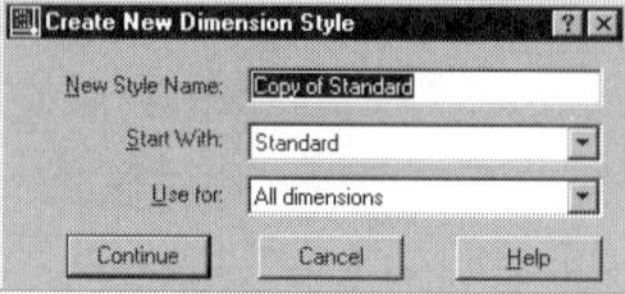

Figure 17.2 *Selecting the* **New** *button will display the* **Create New Dimension Style** *dialog box.*

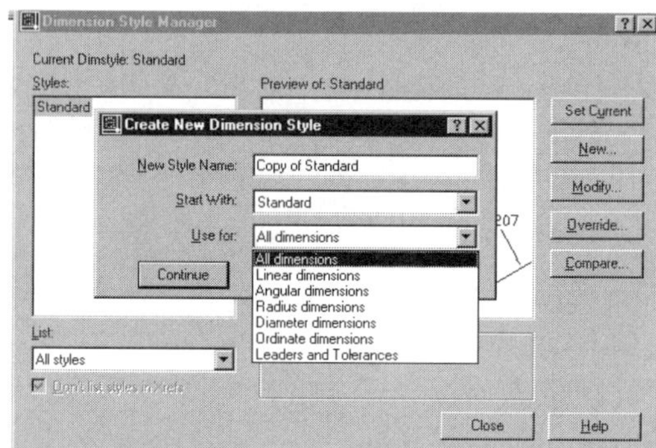

Figure 18.7 *Selecting the* **Use for** *edit arrow will display a list of dimension types that the style will be applied to.*

the Standard style with black text as a base but allow all radius dimensions to be red. Selecting Standard as the **Start With** option, and Radius Dimensions as the **Use for** selection will begin the process of creating a sub-style of the Standard style. With these options active, each time the Standard style is used, radius dimensions will be red, and all other dimensions will be black.

Continue—Selecting the **Continue** button will display the **New Dimension Style** dialog box. The display will be explored in the next section of this chapter.

Modify

Selecting the **Modify** button in the **Dimension Style Manager** dialog box allows an existing style to be modified. As the **Modify** button is selected, the **Dimension Style Manager** dialog box is removed and the **Modify Dimension Style** dialog box is displayed. The dialog box is identical to the **New Dimension Style** dialog box. Methods of altering an existing style will be explored later in this chapter.

Override

Selecting the **Override** button will produce the **Override Dimension Style** dialog box. The box is identical to the **New Dimension Style** dialog box. This option allows a temporary override to a dimension style. Methods of overriding an existing style will be explored later in this chapter.

Compare

Selecting this option displays the **Compare Dimension Styles** dialog box shown in Figure 17.4. (The contents of the display will depend on the styles being compared.) This dialog box can be very helpful in comparing the differences between two styles or in displaying all of the properties of one style.

CREATING A NEW STYLE

You've just explored the **Dimension Style Manager** and the **Create New Dimension Style** dialog boxes. Selecting the **New** button of the **Dimension Style Manager** dialog box and then the **Continue** button of the **Create New Dimension Style** dialog box will display the **New Dimension Style** dialog box shown in Figure 17.5. (The exact title will vary depending on the name of the style being created.) The dialog box contains six tabs for controlling the dimension style properties.

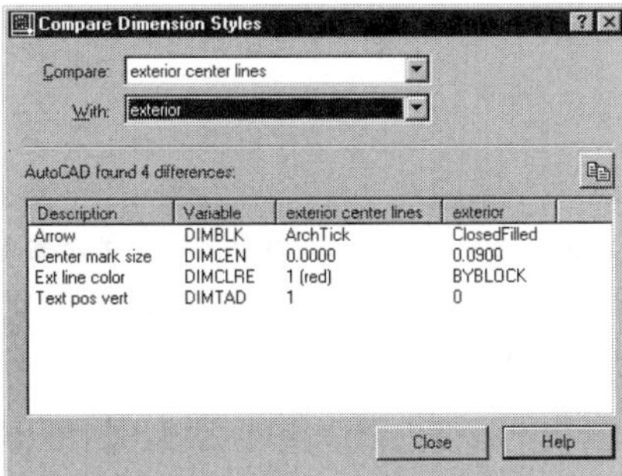

Figure 17.4 *The* **Compare Dimension Styles** *dialog box can be used to compare the differences between two styles or to display all of the properties of one style.*

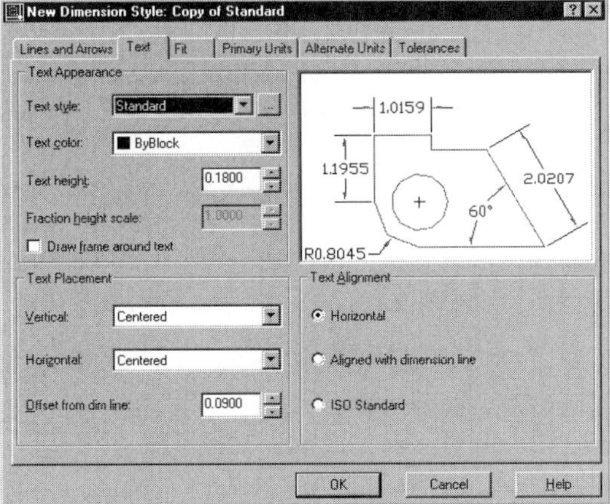

Figure 17.5 *Selecting the* **New** *button in the* **Dimension Style Manager** *dialog box and then the* **Continue** *button in the* **Create New Dimension Style** *dialog box will display a* **New Dimension Style** *dialog box and allow the properties of a dimension style to be set.*

LINES AND ARROWS TAB

The **Lines and Arrows** tab contains five major areas for adjusting the lines and arrowheads to be used with the new style. These areas are the display window, **Dimension Lines**, **Extension Lines**, **Arrowheads**, and **Center Marks**. A display tab is in each tab and will demonstrate the effect of the current tab settings.

Display Window

The display window presents a visual representation of how the dimension properties will be displayed in the drawing area using the current settings. Each time a property is modified on the **Lines and Arrows** tab, the display will be updated to reflect the new property setting.

Dimension Lines

The **Dimension Lines** area contains settings to control **Color**, **Lineweight**, **Extend beyond ticks**, **Baseline spacing**, and **Suppress**.

Color

The **Color** list displays and sets the color for dimension leader lines. Selecting the **Color** edit arrow displays the seven basic colors. Selecting the Other option displays the **Select Color** dialog box. Although a different color can be set for each aspect of a dimension, colors are best controlled by the layer that they are to be placed on.

Lineweight

Selecting the **Lineweight** edit arrow displays the lineweight list, allowing the lineweight of dimension lines to be set.

Extend Beyond Ticks

By default, this option is inactive. It will become active if the arrowhead is altered to be Oblique, Architectural Tick, Integral, or None. This option controls how far the dimension line will extend past the extension line. Figure 17.6 shows examples of adjusting the extension. For most mechanical drawings, this value should not be altered.

Baseline Spacing

Selecting the **Baseline spacing** edit box allows the distance between lines of dimensions to be set. AutoCAD refers to this increment as the baseline setting. The default setting for mechanical units is .5". Figure 17.7 shows an example of the baseline spacing.

Figure 17.6 *The default placement of dimension line extension on the left is often altered so that the dimension line extends past the extension line to simulate manual drafting methods.*

Suppress

These options works as toggles allowing parts of the dimension line to be omitted. The **Dim Line1** and **Dim Line2** check boxes allow the first or second portion of the dimension line to be omitted from the drawing. Figure 17.8 shows an example of the first or second dimension line suppressed. The first or second dimension line is relative to the selection of the first and second extension line. These options are generally not set active for mechanical drawings.

Extension Lines

This area has features similar to those in the **Dimension Lines** area, except that these settings control each aspect of the extension line in relation to the object being dimensioned.

Color

This option can be used to set the color of the extension lines. Selecting the **Color** edit arrow displays the color list. Although a different color can be set for extension lines, colors are best controlled by the layer that they are to be placed on.

Lineweight

Selecting the **Lineweight** edit arrow displays the lineweight list, allowing the lineweight of extension lines to be set. Lineweight is best described by the layer containing the dimension features.

Extend Beyond Dim Lines

The **Extend beyond dim lines** option allows the distance that the extension line extends past the dimension line to be controlled. The default value is .18". An extension offset of .125" is typical on most professional drawings. The effect of setting **Extend beyond dim lines** can be seen in Figure 17.9.

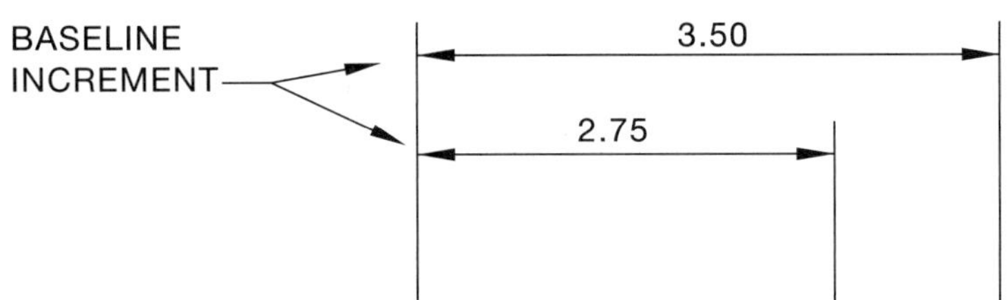

Figure 17.7 *The baseline increment controls the spacing between stacked lines of dimensions.*

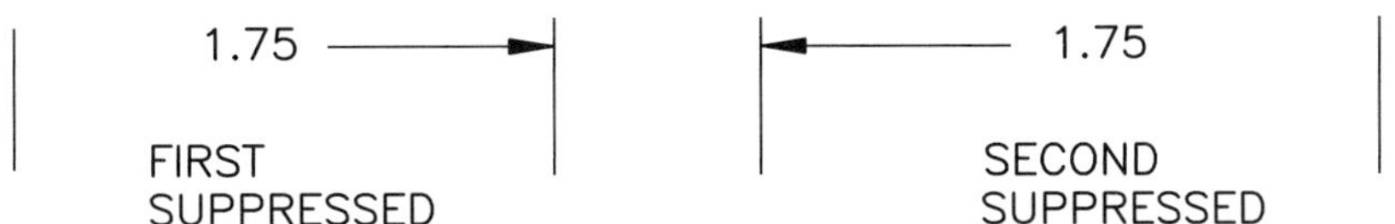

Figure 17.8 *Suppression of dimension lines is generally not required in mechanical drafting.*

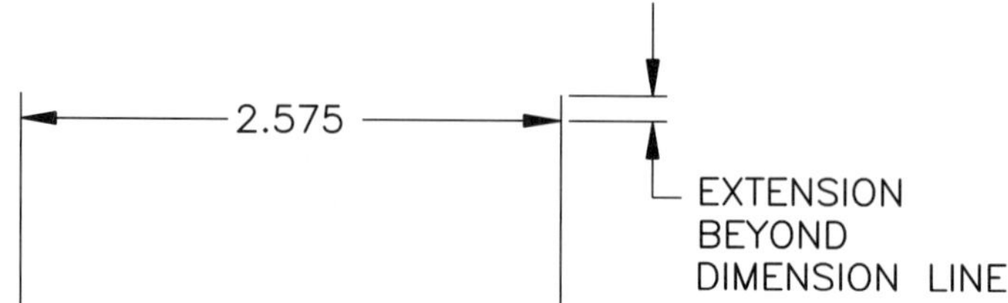

Figure 17.9 *The extension line is usually set to extend above the dimension line by a distance equal to the text height.*

Offset From Origin

The **Offset from origin** option is used to control the gap between the extension lines and the object being dimensioned. The default value is .065" but a space of .125" is common in many companies. The effect of setting the **Offset from origin** is shown in Figure 17.10.

Suppressing Extension Lines

This option allows for the suppression of the first and second extension lines, as shown in Figure 17.11. By default, extension lines will be displayed at each end of the dimension line. This option works well for establishing overall sizes. The first extension line can be omitted by selecting the **Ext Line1** box. With this option active, as the location for extension lines is selected, the first location will not receive an extension line. Selecting both boxes suppresses the extension lines at each end of the dimension line. This is useful if a dimension line is to be placed between two existing extension lines. Figure 17.12 shows examples of suppressing extension lines.

Arrowheads

This area controls the type of dimension line terminator to be used. The current selection for the first and second terminator, the closed filled arrow, is displayed in the edit box. By default, AutoCAD uses the terminator selected for the first dimension line used for the second line. The type of terminator can be selected by selecting the but-

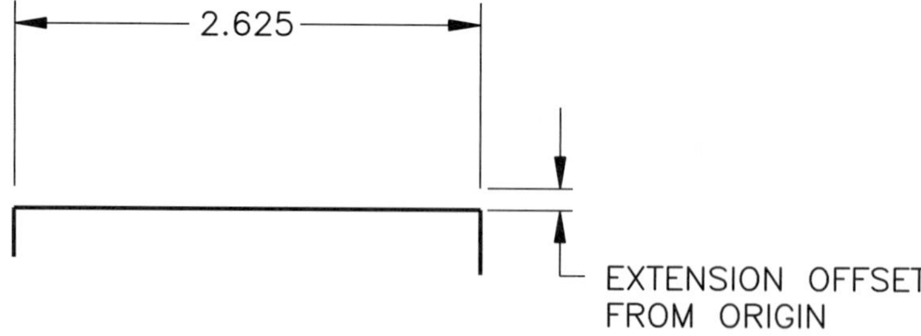

Figure 17.10 *The extension line should be offset from the object being described.*

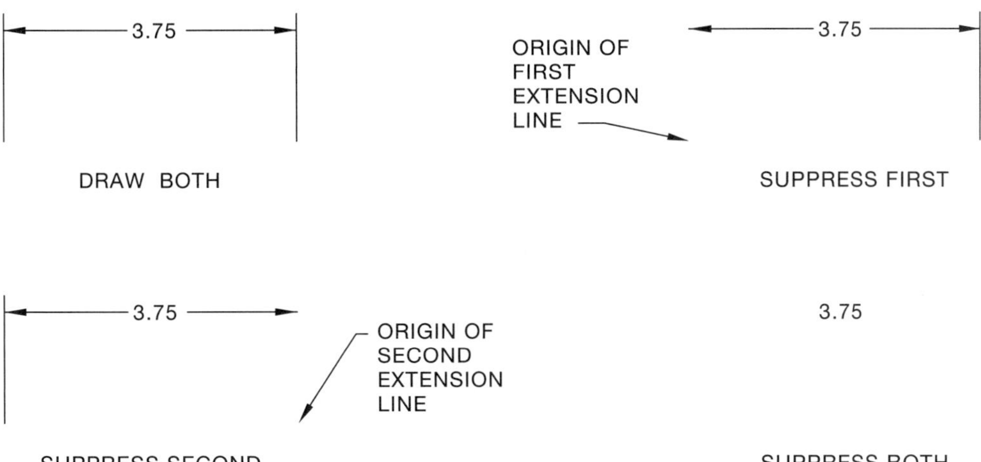

Figure 17.11 *Suppression of extension lines.*

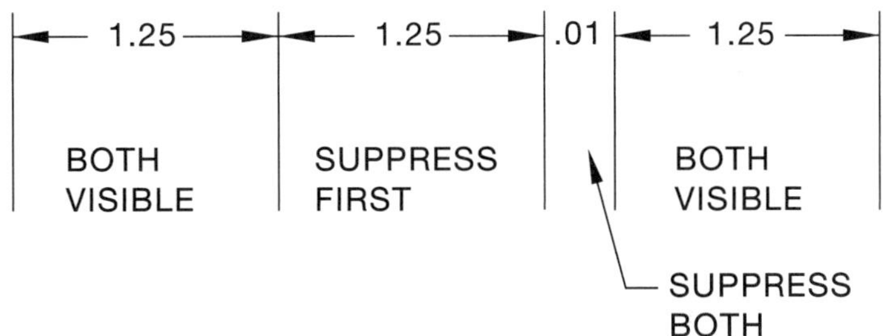

Figure 17.12 *Practical applications for suppressing extension lines.*

ton or the arrow on the right side of the **1st list** box. Selecting the button or the down arrow displays the list (Figure 17.13 shows a sample). Making the selection of OPEN 30 automatically alters the display of each arrow to the new selection. You can alter the second terminator to be different from the first by using the list for **2nd** option. Rarely is this done on professional drawings, and you should not decide to start a new trend! Examples of the variations of dimension line terminators can be seen in Figure 17.14.

User Arrow

Most companies use the filled, open 30, or closed blank arrows as a terminator. The user-defined tick (User Arrow) can be created using the **Block** command (discussed in Chapter 20). Once the desired block is drawn, it can be named and saved to the drawing base. Selecting User Arrow displays the dialog box shown in Figure 17.15. The name that was assigned to the block can be assigned in the **Select Custom Arrow Block** dialog box. The display window will now display the name of the block rather

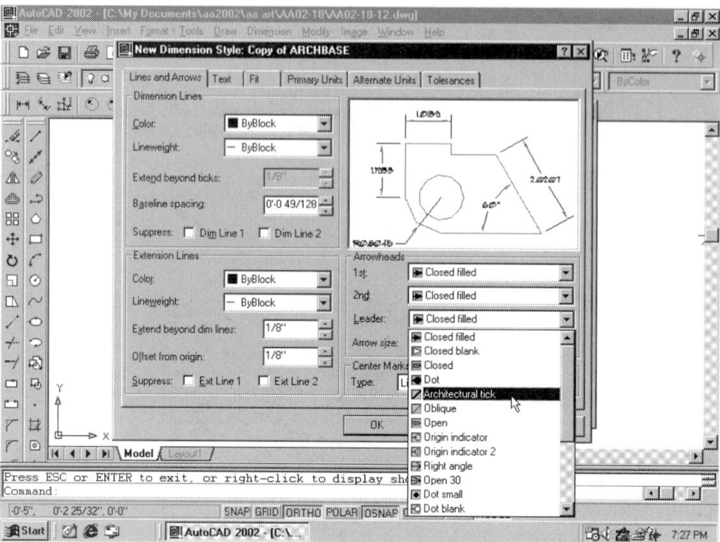

Figure 17.13 *Selecting the button or the down arrow displays the terminator list. The balance of the list can be seen by using the down arrow.*

than showing an icon.

Arrow Size

The **Arrow size** edit box can be used to determine the size of the terminator relative to other dimensioning features. Selecting a tick value that is the same height as the text is a safe guideline to follow. If .375" high text is to be used, change the value to .375.

Center Marks

The **Center Marks** box is used to control how and when the center mark for radial dimensions will be displayed. The center mark is drawn only if the dimensions for **Dimcenter**, **Dimradius**, or **Dimdiameter** are placed outside the circle or arc being described. Figure 17.16 shows the options for placement.

Displaying the Center Mark

The default setting of Mark displays the center mark using the center **Size** value. Selecting the Line option displays centerlines that project beyond the limits of the circle. Selecting the **None** radio button will display no markings for the centerpoint of a circle or arc. The **Size** edit box can be used to control the size of the center mark. Many professionals adjust the default setting of .065" to .125".

TEXT TAB

The **Text** tab is used to set the appearance, placement, and alignment of the text that is used with dimensions. Each of the four major areas of the tab can be seen in Figure 17.17.

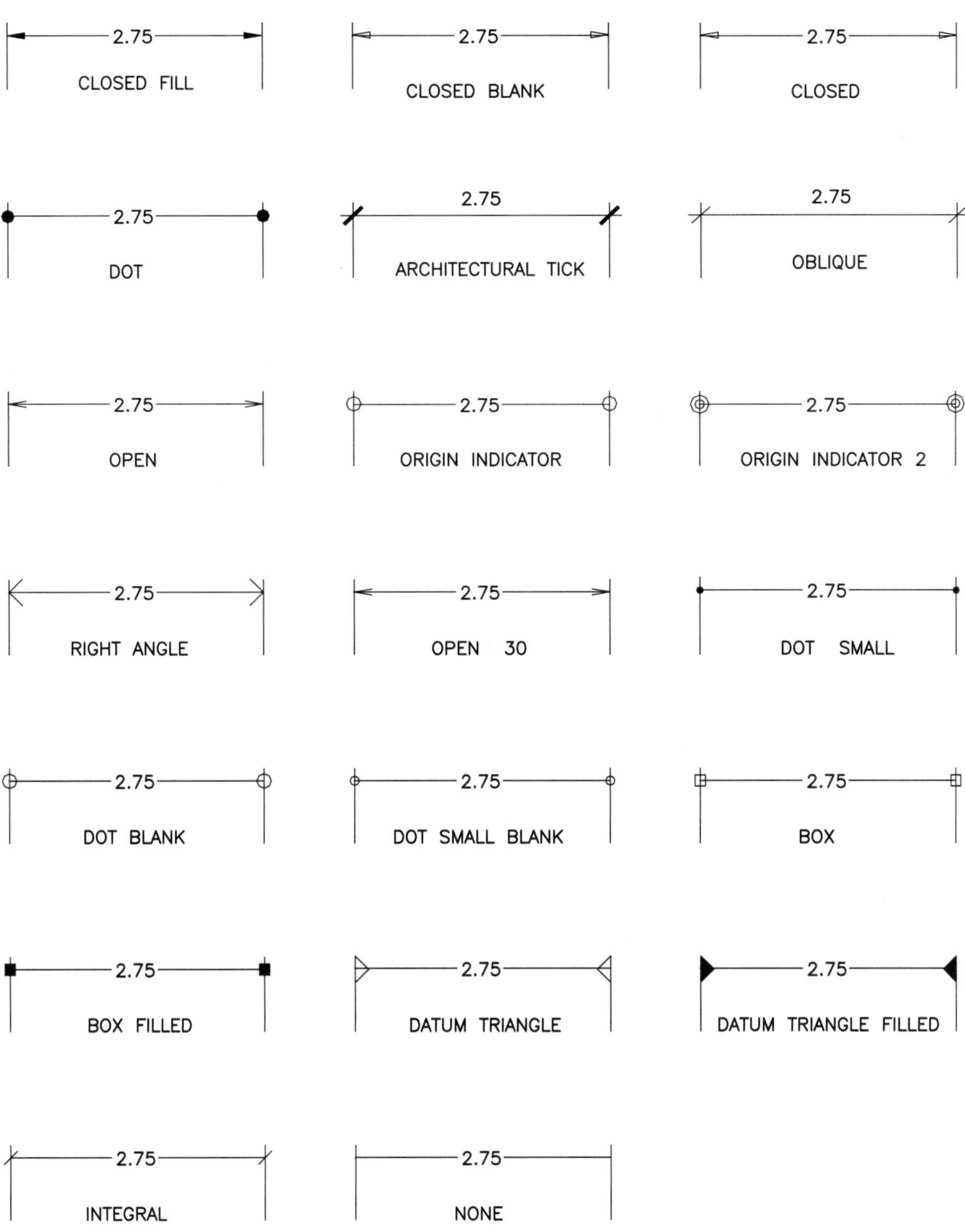

Figure 17.14 *Dimension line terminators available in AutoCAD.*

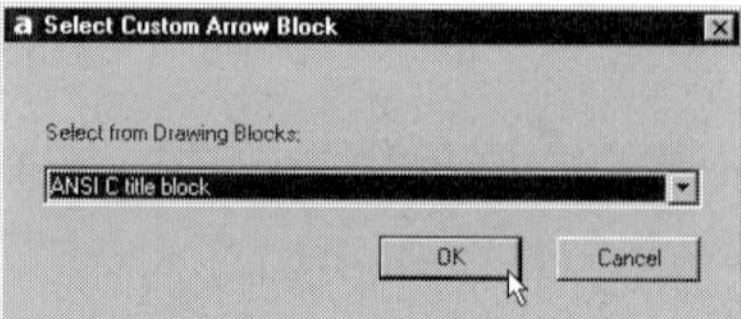

Figure 17.15 *A user-defined arrow can be created as a block and added to the drawing. Chapter 20 will introduce the* **Block** *command.*

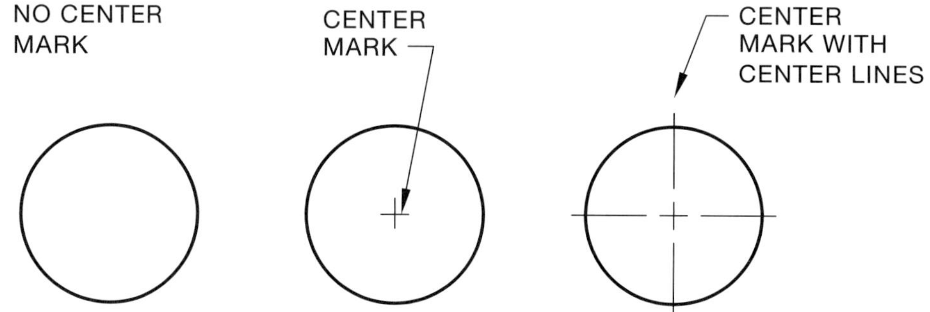

Figure 17.16 *Options for displaying center marks.*

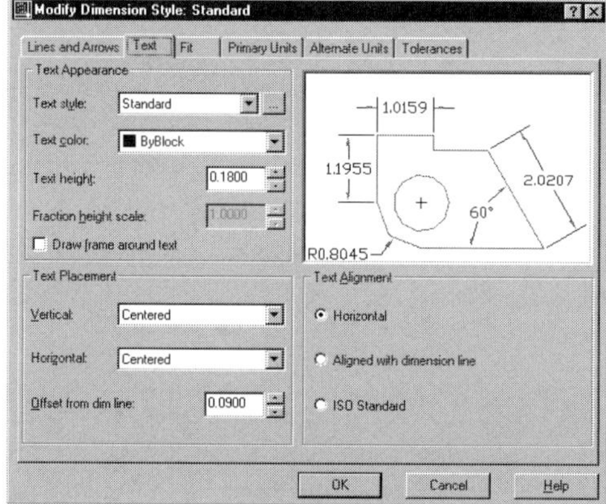

Figure 17.17 *The* **Text** *tab is used to set the appearance, placement, and alignment of the text that is used with dimensions.*

Text Appearance

The **Text Appearance** portion of the **Text** tab contains controls for the dimension text style, color, height, size, and frame.

Text Style

The **Text style** edit box displays the name of the current text style. It also allows one of the defined text styles to be used for dimension text. Selecting the [...] ellipsis button beside the edit box displays the **Text Style** dialog box, which allows a new text style to be created.

Text Color

The **Text color** box displays and sets the color for dimension text. Selecting the Color edit arrow displays the seven basic colors. Selecting the Other option displays the **Select Color** dialog box. Use a setting of ByLayer.

Text Height

The **Text height** box displays the current text height and allows the height of the dimension text to be altered. The height entered here must be coordinated with the height that is entered in the **Text Style** dialog box. If a text height is entered in the **Text Style** dialog box, that height will override the text entered in the **Text** tab. For the value of the **Text** tab to be active, a text value of 0 must be entered in the **Text Style** dialog box.

Fraction Height Scale

The **Fraction height scale** option sets the scale of fractions relative to the dimension text. In its current state, the option is inactive. For this option to be active, the **Precision** must be set on the **Primary Units** tab. Once a precision has been set, the scale factor can be altered. With the default value of 1.00, the height of fractions will be the same as the dimension text height.

Framing Text

The **Draw frame around text** check box is currently inactive. Selecting this option will place a box around the text. With the current settings, selecting the box will do nothing. Once values are altered in the **Text Placement** area, selecting this option box will display a box around a dimension similar to Figure 17.18.

Text Placement

The **Text Placement** area of the **Text** tab controls the vertical and horizontal placement of the dimension text, as well as the offset distance.

Vertical

This option controls the vertical justification of dimension text relative to the dimension line. Options include Centered, Above Outside, and JIS. Figure 17.19 shows the two options that relate to mechanical drawing. The Outside or the JIS (Japanese Industrial Standards) are not common on mechanical drawings.

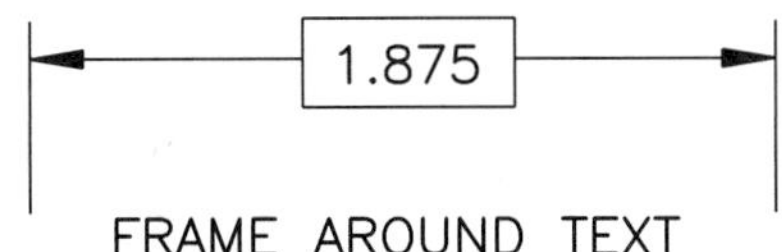

Figure 17.18 *Once values are altered in the* **Text Placement** *area, selecting the* **Draw frame around text** *check box will display a box around a dimensions.*

Figure 17.19 *Two common settings of the* **Vertical Text Placement** *list.*

Centered—This option breaks the dimension lines so that text is centered between the dimension lines.

Above—This option is often used for architectural and engineering drawings. The option places the dimension text above the dimension line. The distance from the dimension line to the bottom of text is controlled by the **Offset from dim line** option.

Outside—The Outside option places the dimension text on the side of the dimension farthest away from the defining point.

JIS—This option places dimension text to conform to Japanese Industrial Standards (JIS) representation.

Horizontal

The **Horizontal** box allows the location of text along the dimension line to be set. The box contains a button and an edit box that can be used to control where the text is placed. Selecting the down arrow allows the options to be altered. As the arrow is selected, the **Horizontal** list is displayed. Selecting an option makes it current and removes the list. Options include the following:

Centered—The default value places text centered along the dimension line.

At Ext Line 1—This option places the text close to the first extension line.

At Ext Line 2—This option places the text close to the second extension line.

Over Ext Line 1—This option places the text parallel and above the first extension line.

Over Ext Line 2—This option places the text parallel and above the second extension line.

Examples of each option can be seen in Figure 17.20.

Offset from the Dimension Line

The **Offset from dim line** edit box displays the current gap between the text and the dimension line. The setting also allows the gap between the dimension line and the text to be altered. The gap value should be set to equal half of the text height value.

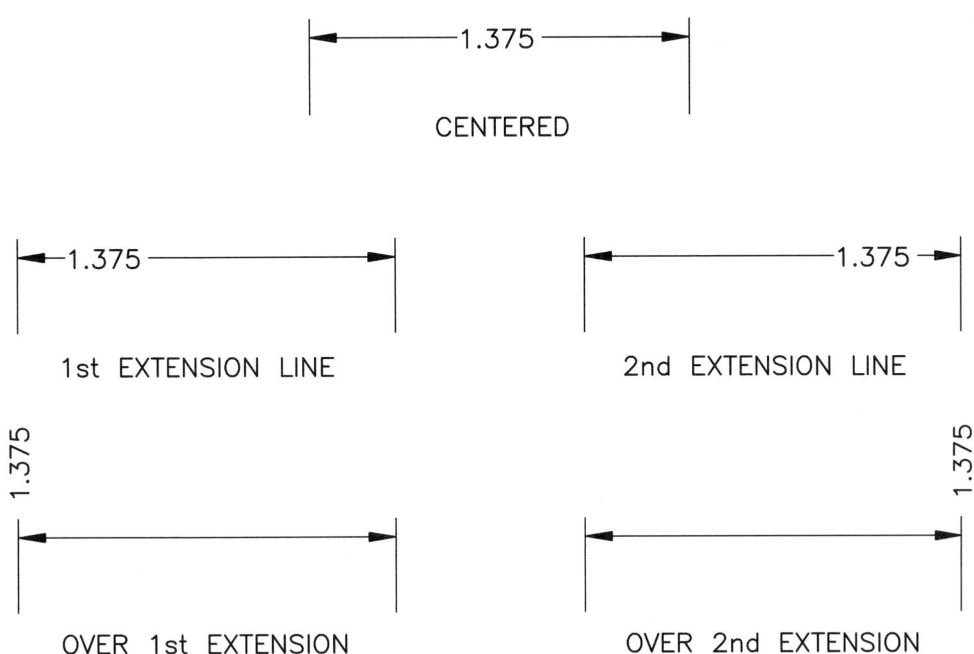

Figure 17.20 *Placement of the dimension text using the* **Horizontal Text Placement** *list.*

Text Alignment

The **Text Alignment** area of the **Text** tab can be used to control the orientation of dimension text to the dimension line. Choosing one of the inactive options alters the selection. Options include the default option of **Horizontal, Aligned with dimension line**, and **ISO Standard**. Figure 17.21 shows an example of each type of text placement.

Horizontal

With this box active, dimension text is placed horizontal, regardless of the angle of the dimension line. The option should be set to ON for most mechanical drawings, so that the text remains parallel to the long paper border.

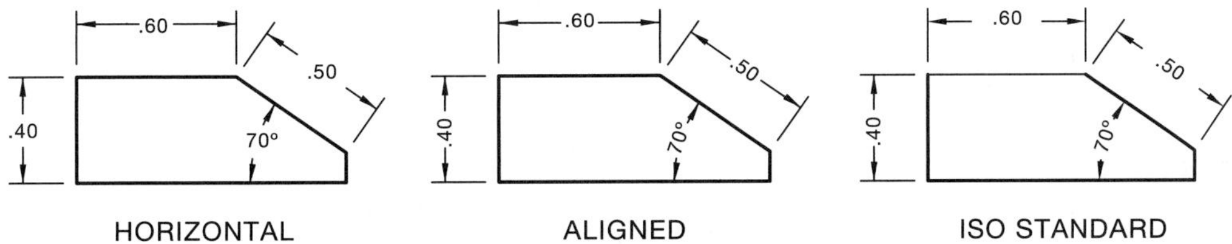

Figure 17.21 *The* **Text Alignment** *area of the* **Text** *tab can be used to control the orientation of dimension text to the dimension line. Most drawings use horizontal dimensions.*

Aligned with Dimension Line

This option is most typically associated with construction drawings. Selecting this option will align dimension text to be parallel with the dimension line. It should not be activated on mechanical drawings governed by ASME standards.

ISO Standard

This option aligns text with the dimension line if the space between the extension lines is wide enough to display the text. Text will be placed in a horizontal position outside the extension lines if the text will not fit between the extension lines. It should not be activated on drawings governed by ASME standards.

FIT TAB

The **Fit** tab provides controls for the placement of text in relationship to the dimension line. The **Fit** tab can be used to set the values that control the placement of text, arrows, and leaders for dimension placement. Figure 17.22 shows the **Fit** tab of the **Modify Dimension Style** dialog box. Figure 17.23 shows common text placements that might be seen.

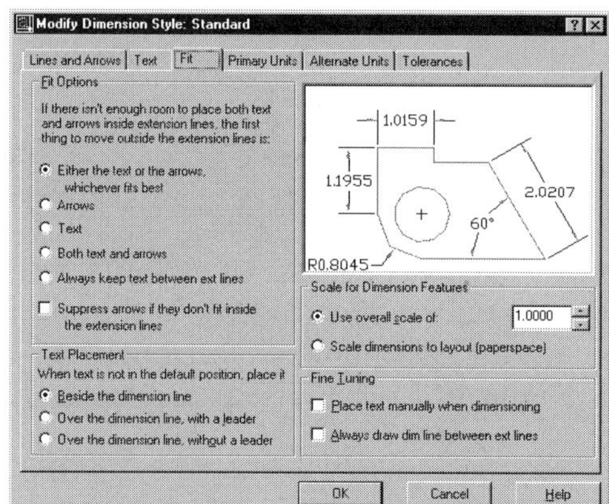

Figure 17.22 *The* **Fit** *tab of the* **Modify Dimension Style** *dialog box.*

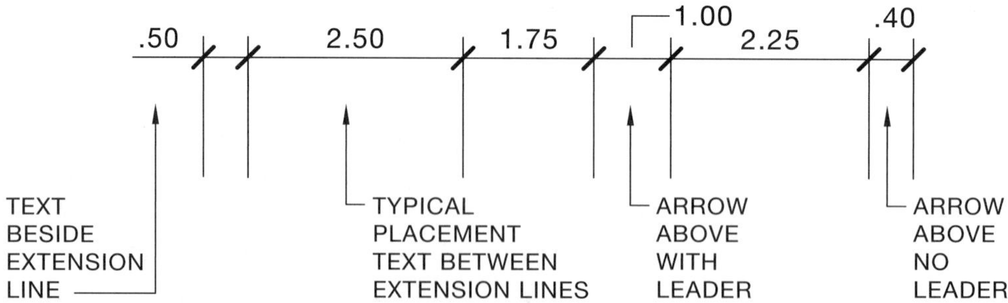

Figure 17.23 *Common dimension text placements.*

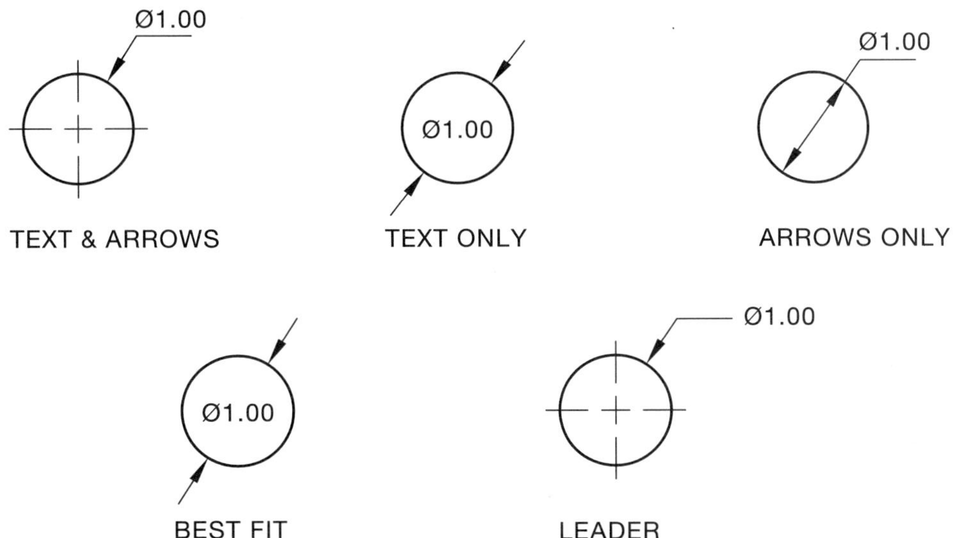

Figure 17.24 *Examples of* **Fit** *options. Notice that depending on the size of the space to be dimensioned, some options produce the same results.*

Fit Options

This portion of the **Fit** tab controls the placement of text and arrowheads if there is not enough space to place both inside the extension lines. Figure 17.24 shows examples of placement options.

Either the Text or the Arrows, Whichever Fits Best
Selecting the default option will display the arrows and the text inside the extension lines when space is available. If sufficient space is not available between the extension lines, the text is placed between the extension lines and the arrows are placed outside the extension lines. If no space is available for text, only the arrows are placed between the extension lines. If space is not available for either text or arrows, both will be placed outside the extension lines.

Arrows
This option displays the arrows and the text inside the extension lines when space is available. If both will not fit between the extension lines, the arrows will be placed inside and the text will be placed outside the extension lines. If sufficient space is not available between the extension lines, the arrows will be placed outside the extension lines.

Text
This option also displays the arrows and the text inside the extension lines when space is available. If sufficient space is not available between the extension lines, the text is

placed between the extension lines and the arrows are placed outside the extension lines. If no space is available for text, both will be placed outside the extension lines.

Both Text and Arrows
This option always displays the arrows and the text outside the extension lines. This option works well with small dimensions placed at the end of a dimension string.

Always Keep Text Between Ext Lines
This option always displays the text between the extension lines. This option works well with small dimensions placed in the middle of a dimension string.

Suppressing Arrows
The Suppress arrows if they don't fit inside the extension line option keeps the text between the extension lines and removes arrowheads when space is not available.

Text Placement

The **Text Placement** portion of the **Fit** tab allows for the control of the placement of text when it is not in the default position. Options include **Beside the dimension line**, **Over the dimension line, with a leader**, and **Over the dimension line, without a leader**. Remember that just because AutoCAD provides a setting, it might not meet ASME or your company standards.

Beside the Dimension Line
This option displays the text and the arrows outside the extension lines. This option works well in combination with using the **Either the text or the arrows, whichever fits best** option of the **Fit Options** area with small dimensions placed at the end of a dimension string.

Over the Dimension Line, with a Leader
This option displays the text above the normal position of text and to the side of the dimension line and uses an arrow to reference the text to the dimension line. This option works well in combination with using the **Either the text or the arrows, whichever fits best** option of the **Fit Options** area with small dimensions placed in the middle of a dimension string, when several small dimensions will be placed near each other.

Over the Dimension Line, without a Leader
This option displays the text above the normal position of text above the dimension line but does not use an arrow to reference the text to the dimension line. This option works well in combination with using the **Either the text or the arrows, whichever fits best** option of the **Fit Options** area with small dimensions placed in the middle of a dimension string.

Scale For Dimension Features

Up to this point of planning dimensions, each dimension will only be legible on a 12" x 9" screen. This area of the **Fit** tab sets the overall scale value or the paper space scaling for all dimensions. Options include **Use overall scale of** and **Scale dimensions to layout (paperspace)**.

Use Overall Scale of

The **Use overall scale of** edit box allows the scale of all dimensioning features to be increased or decreased based on the scale a drawing is to be plotted at. Typically, you will be drawing in model space with actual sizes. Text that is to be .125" high when plotted at a scale of 1/2=1 will need to be set to .25" in height. To have legible arrows, offsets, and text you'll have to convert every variable. If you actually have a life, you might want to just set the scale factor to 2. This will control all dimension variables and leave you a little free time. To find the desired scale factor, take the reciprocal of the drawing scale. A scale of 1/4"=1'–0" is 12/.25, a scale factor of 48. Other common scale factors include the following:

INCH VALUES		METRIC VALUES	
Plotting Scale	**Scale Factor**	**Plotting Scale**	**Scale Factor**
1/8=1	8	1=50	1270
1/4=1	4	1=20	508
1/2=1	2	1=10	254
3/4=1	1.3	1=5	127
2=1	.5	1=2	50.8

Scale Dimensions to Layout

Selecting the **Scale dimensions to layout (paperspace)** box will apply a scaling factor to dimensions plotted in paper space. The option sets a scale factor based on the scaling between the current model space viewport and paper space.

Fine Tuning

The final portion of the **Fit** tab allows minor text adjustments to be made. Options include **Place text manually when dimensioning** and **Always draw dim line between ext lines**.

Place Text Manually When Dimensioning

Activating this check box allows the location for dimension text to be specified as you place each dimension regardless of the justification. With this box active, AutoCAD ignores other justification settings and allows you to specify where the text will be placed. Adjusting the variable can be useful on radial dimensions, as shown in Figure 17.25.

Always Draw Dimension Line

When a small space is dimensioned, the text will be placed outside the extension lines. Selecting the **Always draw dim line between ext lines** option will place a dimension line between the extension lines even when the text and arrows are outside the extension lines. Figure 17.26 shows the effect of using this option on a radial dimension.

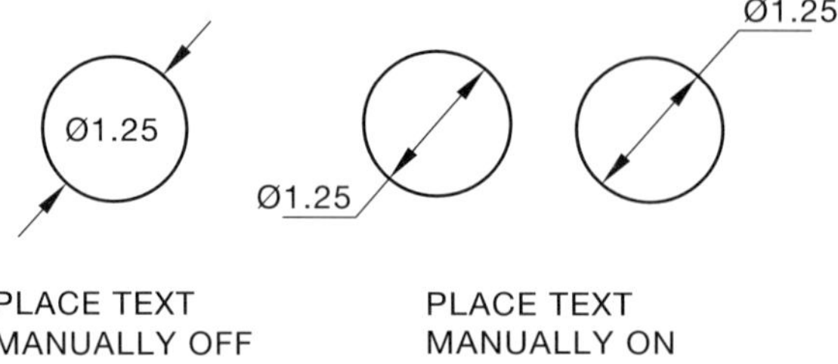

Figure 17.25 *With the* **Place text manually when dimensioning** *check box active, AutoCAD ignores other justification settings and allows you to specify where the text will be placed.*

Figure 17.26 *Selecting the* **Always draw dim line between ext lines** *option will place a dimension line between the extension lines even when the text and arrows are outside the extension lines.*

PRIMARY UNITS TAB

A dimension of 1.50" is considered a primary dimension unit. Some mechanical-related dimensions consist of primary units, with no other information required. Depending on the quality required in manufacturing, a tolerance might be added to the primary units. This portion of the **Modify Dimension Style** dialog box controls the display of the primary measurement units and any prefixes or suffixes for the dimension text. Selecting the **Primary Units** tab displays a dialog box similar to Figure 17.27. This tab is used to set the variables that control the appearance of text for dimension and leader lines. The tab includes the areas of **Linear Dimensions**, **Measurement Scale**, **Angular Dimensions**, and **Zero Suppression**.

Linear Dimensions

This portion of the **Primary Units** tab displays and sets the format and precision for linear dimensions. Key elements include the **Unit format**, **Precision**, **Fraction format**, **Decimal separator**, **Round off**, **Prefix**, and **Suffix**.

Unit Format

The default **Unit format** setting is decimal. Select the down arrow to display the list, which includes Scientific units, Engineering units, Architectural units, Fractional units, and Windows Desktop. The decimal units will meet the needs for most drawings.

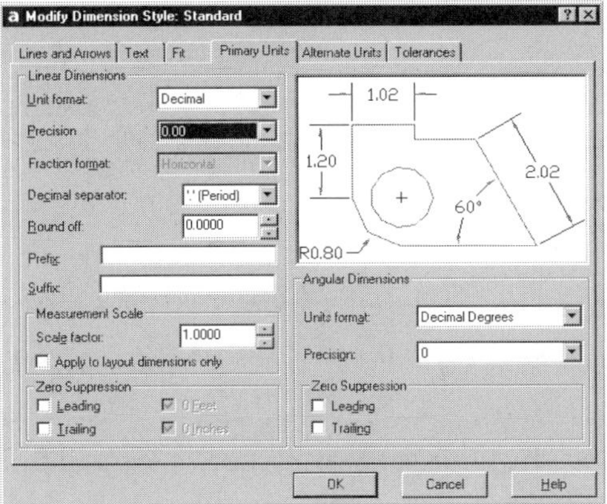

Figure 17.27 *The* **Primary Units** *tab.*

Precision

The default setting for decimal dimensions is four-place precision. Selecting the down arrow of the **Precision** edit box displays the precision of the dimension. Options for the Decimal format are shown in Figure 17.28. For most dimensions, a setting of 0.00 will provide the needed accuracy. Most engineers determine the precision based on calculations rather than relying on drawing accuracy. If the precision needs to be altered for a dimension, it can be added by editing the text once the dimension is displayed.

Fraction Format

The **Fraction format** edit box is inactive if decimal units are active. ASME standards require decimal units rather than fractions. To display fractions with the dimensions, set the unit format to Architectural or Fractional. With one of these settings

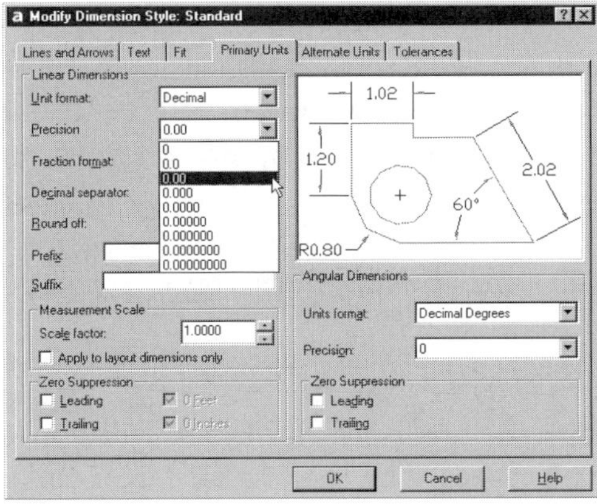

Figure 17.28 *The* **Precision** *list controls the accuracy assigned to a dimension.*

3'-3$\frac{3}{4}$"   3'-3¾"   3'-3 3/4"

HORIZONTAL   DIAGONAL   NOT STACKED

Figure 17.29 *The* **Fraction format** *options.*

active, Fractional format can be displayed. Options include Diagonal, Horizontal, and Not Stacked. Each option can be seen in Figure 17.29.

Decimal Separator
This option is active when the **Unit format** is set to Decimal units. Selecting the edit arrow displays the period, comma, and space options. Period is the default for decimal units, with comma and space options available for metric numbers.

Round Off
This option determines the value for rounding measurements for all dimensions types except for Angular. The default value is 0.00. With this setting, values will not be rounded off and will be placed exactly as measured.

Prefix and Suffix
A prefix or suffix can be added by typing in the appropriate box. Common suffixes might include TYP (typical), MM (millimeter), or MAX (maximum). This is one area where the control codes that were introduced in Chapter 16 might come in useful. The %%C code can be used to enter the diameter symbol as a prefix or suffix. Other control codes include %%D (degree symbol) and %%P (plus / minus symbol).

Measurement Scale

The **Measurement Scale** area of the **Primary Units** tab provides options for setting the scale factor and a method for limiting the scale values.

Scale Factor
The **Scale** box specifies the scale factor of a dimension without affecting the component, angles, or tolerance values. The option affects all linear distances except Angular. Alter the scale factor by using the up/down arrows or by highlighting the existing value and then typing a new value. A scale factor of 2 doubles the value of the dimension that is entered on the drawing. With a scale factor of 2, a dimension of 8.00" will be written when a 4.00" long object is dimensioned. For most drawings the default value of 1 does not need to be altered.

Apply to Layout Dimensions Only
The **Apply to layout dimensions only** check box will apply a scaling factor to dimensions plotted in paper space. The layout scale factor will be adjusted to reflect the zoom scale factor for objects created in model space viewports.

Zero Suppression

This area of the **Primary Units** tab controls the display of zeros for leading, trailing, feet and inches. When a check mark is in the box, it is activated.

Leading

With this box active, the zero is suppressed preceding a decimal point in units less than one. Seven tenths will be written .7 rather than 0.7. ASME standards require the zero to be suppressed for numbers smaller than one. When working with metric units, a zero should precede the decimal point for values less than one millimeter.

Trailing

This box controls the display of zeros after a decimal point. With the box active, no zeros will be displayed after the decimal. Seven tenths will be written .7 rather than .70.

0 Feet

This option will be available only when Architectural units are current. A dimension less than one foot can be displayed as 0'–8" or 8". Selecting the **0 Feet** box causes the zero representing feet to be suppressed when the dimension is less than twelve inches. For most drawings, this setting will not need consideration.

0 Inches

This option will be available only when Architectural units are current. With this box active, zeros will be suppressed. For most drawings, this setting will not need consideration.

Controlling the Display of Angles

The **Angular Dimensions** area controls the format of angular dimensions. Options include **Units format**, **Precision**, and **Zero Suppression**. Options are similar to their linear counterparts.

Units Format

The **Units format** for angular dimensions includes the options of Decimal Degrees, Degrees/Minutes/Seconds, Gradians, and Radians. The Decimal Degrees option will meet the needs for most drawings.

Precision

The **Precision** list will vary depending on the **Units format** setting. When units are measured in decimals, the precision will be measured in decimal places from zero to eight places.

Zero Suppression

This area suppresses the leading or trailing zero of angular dimensions. With the **Leading** option activated, 0.75 will be displayed as .75. The **Trailing** option, when activated, suppresses zeros that follow the decimal. .7500 will be displayed as .75.

ALTERNATE UNITS TAB

As you display the tab, each option is inactive. Selecting the **Display alternate units** check box activates the options. The **Alternate Units** tab is shown in Figure 17.30. Major components include the **Alternate Units**, **Zero Suppression**, and the **Placement** areas. Features of this tab function in a manner similar to their counterparts on the **Primary Units** tab. One option that should be considered on this tab is the ability to create and display metric units behind the decimal primary units or decimal units behind metric primary units. Selecting Decimal as the **Unit format**, a multiplier for all units of 25.4, and a suffix of mm will display dimensions similar to Figure 17.31. Current ASME standards prefer either metric or decimal units and no longer recommend dual dimensions.

TOLERANCES TAB

A tolerance is a range that is allowed to compensate for human error. If a note specifies that bolts in a steel plate are to be 3" o.c., and the holes are placed at 3.065" o.c., should the plate be scrapped? A tolerance specification will define what is an acceptable range of variation. The **Tolerances** tab of the **Modify Dimension Style** dialog box, shown in Figure 17.32, controls the display and precision of the tolerance. Major components include the **Tolerance Format**, **Zero Suppression**, and the **Alternate Unit Tolerance** areas. Features of this tab function in a manner similar to their counterparts on the **Primary Units** tab. With the default **Method** value of None, most of the box is inactive. Once a method of tolerance specification is selected, the balance of options on the tab will be activated. Figure 17.33 shows examples of the available tolerance methods. Web Chapter 2 will further explore the application of tolerances.

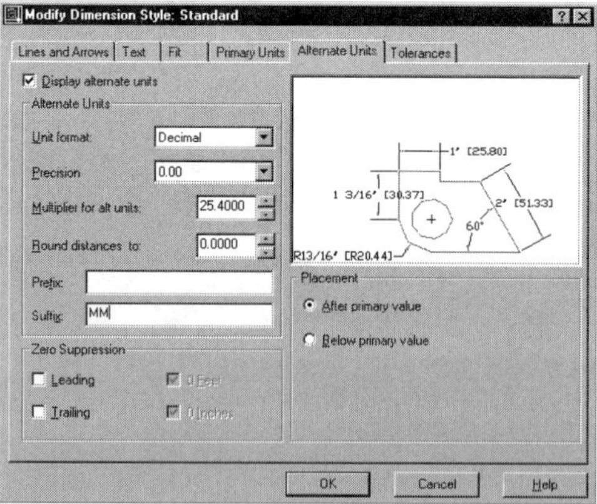

Figure 17.30 *The* **Alternate Units** *tab options are inactive as it is first displayed. Selecting the* **Display alternate units** *check box activates this tab, allowing alternate units to be displayed behind the primary units.*

COMBINING VARIABLES TO CREATE A STYLE

If you're feeling overwhelmed by styles, variables, and values, don't be discouraged. Remember, most of these values will be set on a template drawing and left alone. Throughout this chapter comparisons have been made using the **Modify Dimension Style** dialog box and its tabs. This section will help you combine these values to create a style. Dimensioning variables can be combined to create a particular style in much the same way that options are combined with **Text** to create different text styles. Styles can be created, saved, modified and quickly loaded by using the **Dimension Style Manager** dialog box shown in Figure 17.1.

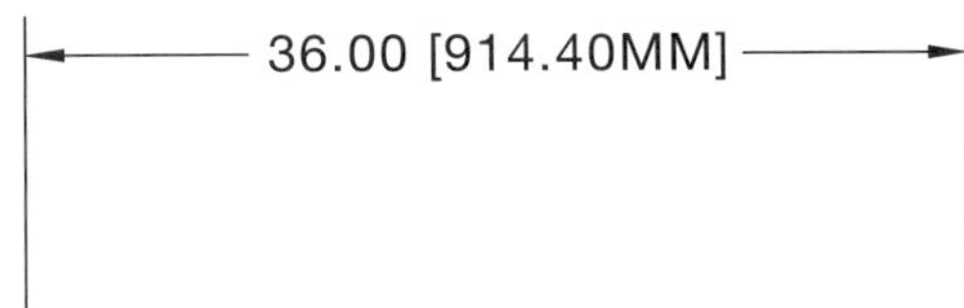

Figure 17.31 *Selecting Decimal as the alternate unit format, a multiplier for all units of 25.4, and a suffix of mm will display architectural units followed by metric dimensions.*

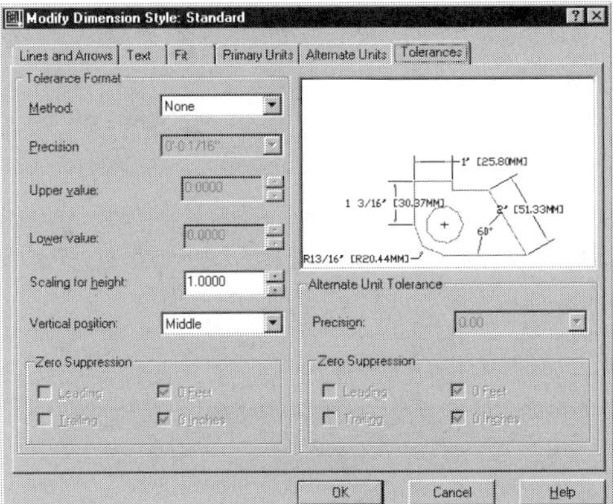

Figure 17.32 *The* **Tolerances** *tab of the* **Modify Dimension Style** *dialog box controls the display and precision of the tolerances.*

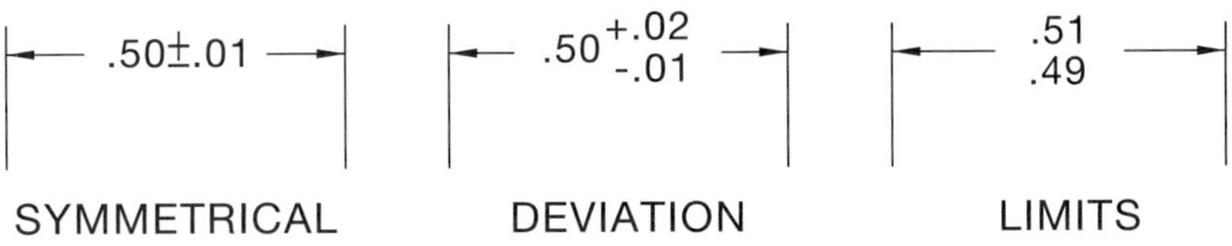

.50±.01 — SYMMETRICAL

$.50^{+.02}_{-.01}$ — DEVIATION

.51 / .49 — LIMITS

Figure 17.33 *Tolerance methods available in AutoCAD include symmetrical, deviation, and limits.*

CREATING DIMENSIONING STYLES

Dimensioning styles allow you to save groups of settings for dimension variables. If a company does both mechanical and civil engineering projects, two distinct dimensioning styles could be created and saved in template drawings to speed the dimensioning process. Even in a firm that does only mechanical drawings, varied styles can be useful to group option settings. Once created, these styles can be quickly loaded and interchanged.

Use the following steps to create a dimension style suitable for drawings to be plotted at a scale of 1=1:

1. Select the **Dimension Style** button on the **Dimension** toolbar. This will display a **Dimension Style Manager** dialog box similar to Figure 17.1.
2. Select the **New** button. This will display a **Create New Dimension Style** dialog box similar to Figure 17.2.
3. Enter the desired style name. In this case DRAWING BASE will be used.
4. Select an existing style to use as a base. In this case the Standard style will be used.
5. Select what dimensions that this style will be applied to in the **Use for** edit box. In this case All Dimensions will be used.
6. Select the **Continue** button. This will display a **New Dimension Style: DRAWING BASE** dialog box. Proceed through each tab and set the options to meet your specific situation. The next section will recommend values that you might want to consider using.
7. Once each option has been selected, click the **OK** button. This will return the **Dimension Style Manager** dialog box, similar to Figure 17.34.
8. Select DRAWING BASE from the Styles box.

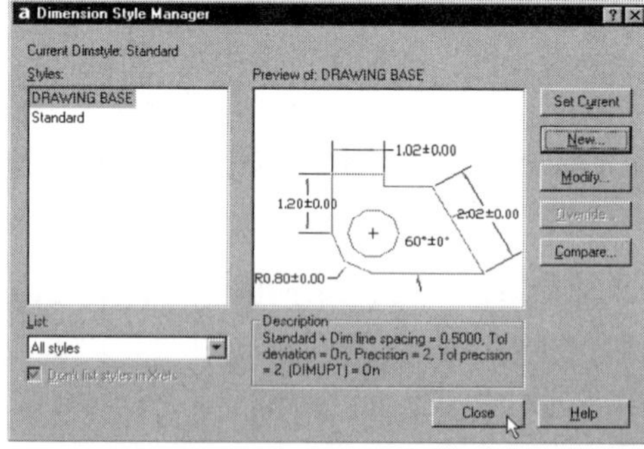

Figure 17.34 *In the* **Dimension Style Manager** *dialog box, select DRAWING BASE from the* **Styles** *box, select the* **Set Current** *button, and then select the* **Close** *button to return to the drawing area.*

9. Select the **Set Current** button.
10. Select the **Close** button.

This process will create the DRAWING BASE style and set it as the current dimension style.

Common Style Settings

Although offices often have individual dimensioning standards, several common options can be set to meet most needs. Use the Standard style as a base to create the desired settings to update your template:

Lines and Arrows Tab (Figure 17.13)

Color and Lineweight: Bylayer

Arrowheads: closed Filled

Extend beyond ticks: .125

Arrow size: .19

Baseline spacing: .50

Center Mark: Line

Center Mark Size: .125

Text Tab (Figure 17.17)

Text style: GDT

Text placement: Centered

Text color: ByLayer

Alignment: Horizontal

Text height: .125

Offset from dim line: .065

Fit Tab (Figure 17.22)

Fit Options: Text

Use overall scale of: 1

Text placement: Beside

Always draw dim line between ext lines: OFF

Primary Units Tab (Figure 17.27)

Unit format: Decimal

Precision: 0.00

Decimal Degrees

Precision: 0.0

Following the guidelines to create DRAWING BASE will produce dimensions similar to Figure 17.35a. The dimension style works well for the overall dimensions, but it is somewhat limited when an object has a jog in it. If the style is used to place dimension A, the desired results will be obtained, but two extension lines will be placed on the left side of space A. The style will give the desired results if you're careful to always use OSNAP when you select where the extension lines will be placed. Failure to use OSNAP might produce results similar to Figure 17.35b. Even if OSNAP or **Quick Dimension** is used to place the dimensions for space B and C, the desired results will not be achieved. The extension line for B will extend to be within 0.125 of surface B, leaving no offset for surface A as shown in Figure 17.35b. The same problem will occur at B as surface C is dimensioned (see Figure 17.35c).

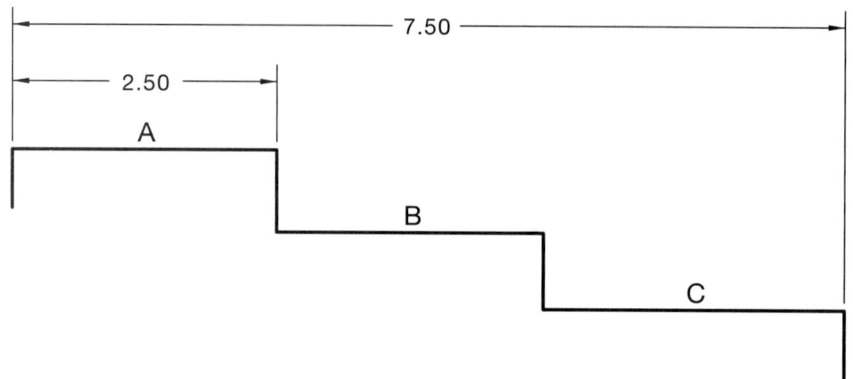

Figure 17.35a *Using the DRAWING BASE style will produce dimensions suitable for most applications.*

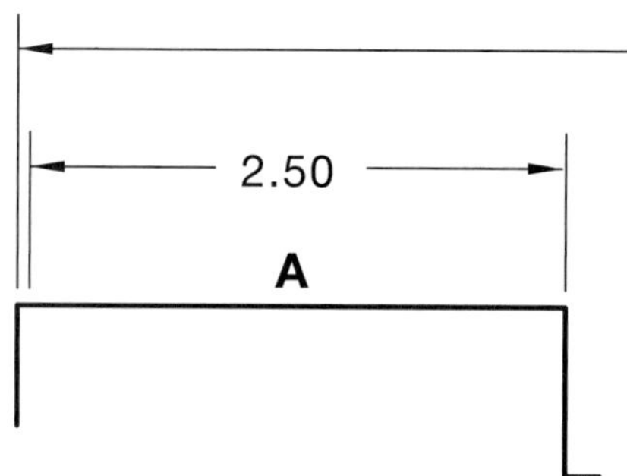

Figure 17.35b *Failure to use OSNAP might produce extension lines that do not align.*

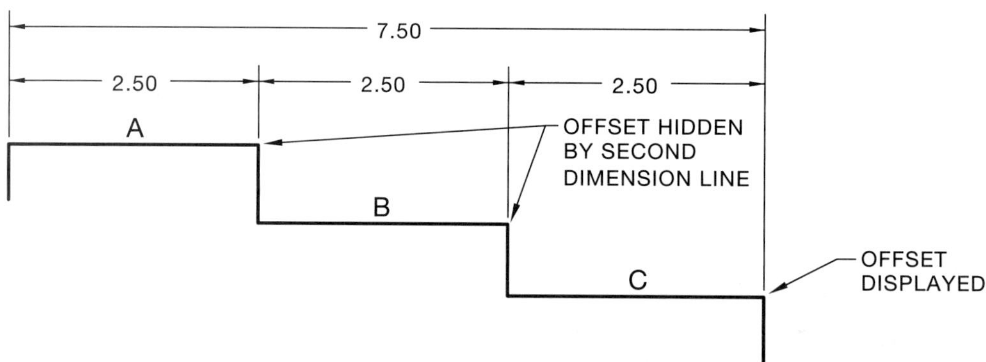

Figure 17.35c *Even if OSNAP or* **Quick Dimension** *is used to place the dimensions for space B and C, the desired results will not be achieved. The extension line for B will extend to be within .125" of surface B, leaving no offset for surface A. The same problem will occur at B as surface C is dimensioned.*

MODIFYING A STYLE

Once a style has been created, you might find that it is not all you hoped it would be. The **Modify** option of **Dimension Style Manager** dialog box allows an existing dimension style to be altered. The process for altering an existing style is similar to the process used to create a new style using an existing style as a base. An existing style can easily be modified using the following procedure:

1. Display the **Dimension Style Manager** dialog box and select the style to be modified from the **Styles** display box.
2. With the desired style highlighted, select the **Modify** button. This will display the **Modify Dimension Style** dialog box. The box is exactly the same as the **New Dimension Style** dialog box.
3. Use the necessary tabs to alter the existing values. For the example, the **Lines and Arrows** tab was used to alter the arrowheads.
4. Click the **OK** and the **Close** buttons to modify the existing style and return to the drawing area.

OVERRIDING EXISTING STYLES

Using the **Modify** option of the **Dimension Style Manager** dialog box changes an entire style to the specified parameters. Using the **Override** option allows an existing style to stay intact while altering new features that are added to the style. Use the following steps to override an existing style:

1. Display the **Dimension Style Manager** dialog box and select the style to override from the **Styles** display box.
2. With the desired style highlighted, select the **Override** button. This will display the **Override Current Style** dialog box. The box is exactly the same as the **New Dimension Style** dialog box.
3. Use the necessary tabs to alter the existing values. For the example, the **Lines and Arrows** tab was used to alter the arrowheads, and the **Text** tab was used to center the text between the dimension lines.
4. Click the **OK** and the **Close** buttons to override the existing style and return to the drawing area.

Styles can also be overridden at the Command prompt by typing DOV (ENTER). This will produce the prompt:

> Enter dimension variable name to override [or Clear overrides]:

Enter the name of the desired variable to be altered followed by the name of the new style.

COMPARING STYLES

The final option of the **Dimension Style Manager** dialog box is to compare styles. This option can be used to list the properties of a style or to compare the properties of two dimension styles. The function performed depends on the settings of the **Compare** and **With** edit boxes. To list the properties of a style, select the name of the desired style in the **Compare** edit box, and select <None> in the **With** list. Figure 17.36 displays a partial listing of the properties for the Standard style. Entering a style name in the **Compare** edit box and a different style name in the **With** edit box will display a listing of the property differences similar to Figure 17.37. The list of properties and the list of differences can be copied to the Windows Clipboard. Once copied to the Clipboard, the listings can be pasted into other Windows applications.

RENAMING OR DELETING STYLES

A style can be renamed or deleted using the **Style** portion of the **Dimension Style Manager** dialog box. Use the following procedure to rename a style:

1. Display the **Dimension Style Manager** dialog box.
2. Highlight a style in the **Style** portion of the dialog box.
3. With the style highlighted, right-click to produce a shortcut menu.
4. Select the **Rename** option and enter the desired new name.

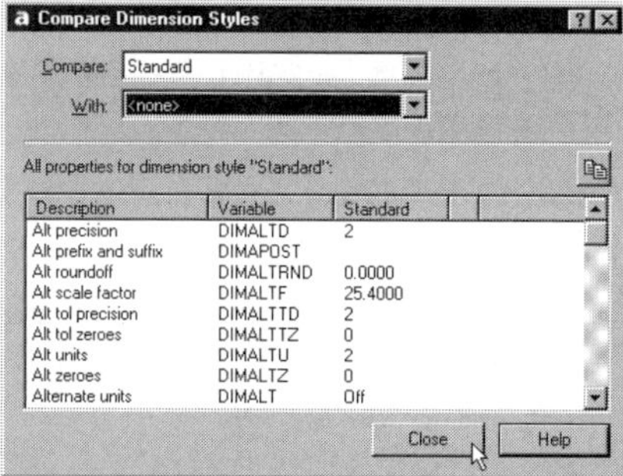

Figure 17.36 *A partial listing of the properties for the Standard style.*

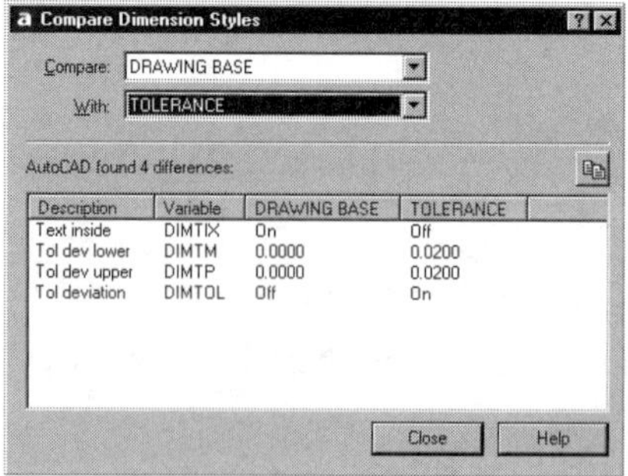

Figure 17.37 *Entering a style name in the* **Compare** *edit box and a different style name in the* **With** *edit box will display a listing of the property differences.*

5. Continue renaming other menu listings or select **Close** to remove the dialog box and return to the drawing area.

Similarly, deleting a dimension style is done in the same dialog box, using the following procedure:

1. Once the style to be deleted is highlighted, right-click and select the **Delete** option from the shortcut menu.
2. Continue deleting other styles or select **Close** to remove the dialog box and return to the drawing area.

APPLYING STYLES

Each dimension you create has a style. If you started to place dimensions prior to reading this chapter, the dimensions were created using the Standard style. Throughout this chapter, you've learned to create, modify, and override dimension styles. In addition to selecting the **Set Current** button in the **Dimension Style Manager** dialog box, two other methods are available to alter the dimension style. A new style can be applied by using the shortcut menu or by using the **Dim Style Control** box on the **Object Properties** toolbar.

Using the Shortcut Menu

Use the following steps to assign a new dimension style to an existing dimension with the shortcut menu:

1. Select the desired dimension and right-click.
2. Choose **Dim Style** from the shortcut menu.
3. Choose the name of the desired style to be applied to the selected dimension and press ENTER.

The process can be seen in Figure 17.38.

Using the Dim Style Control Box

Use the following steps to assign a new dimension style to an existing dimension with the **Dim Style** edit box:

1. Select the desired dimension.
2. Display the **Dim Style Control** menu on the **Object Properties** toolbar.
3. Choose the name of the desired style to be applied to the selected dimension and press ENTER.

EDITING DIMENSION PLACEMENT

The placement of dimensions can be edited using grips or the editing commands. The editing commands can be selected from the shortcut menu or by the methods presented in earlier chapters. To use the shortcut menu, select a dimension with the cursor. Each

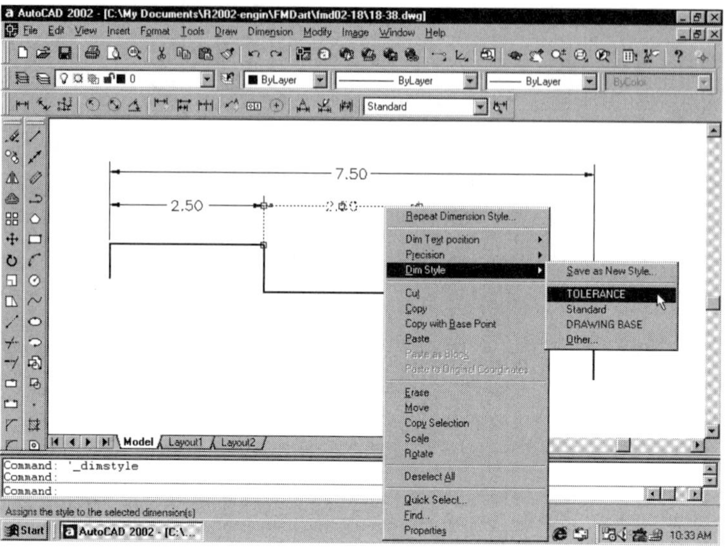

Figure 17.38 *A dimension style can be altered by activating the grips, right-clicking to produce the shortcut menu, and then selecting the name of the desired style.*

of the grips will be activated. Right-click to obtain the editing menu. Notice in Figure 17.39 that a grip is displayed for the dimension text, at each intersection of the dimension and extension lines, and at what appears to be the end of the extension line. Closer examination will show the grips that appear to be at the end of the extension lines actually lie beyond the lines. These grips mark the definition points, or Defpoints, for the dimensions. AutoCAD automatically assigns definition points for every dimension to define the dimension location. Use the **Layers** button to examine the layers for the current drawing and you'll find that the program created a layer titled Defpoints to display the definition points. The layer will not be plotted. You can edit dimensions using grips without knowing the location of the definition points. The Defpoints will automatically be displayed. If you alter dimensions using the edit commands, the Defpoints must be included in the selection set. Using the Node mode of Object Snap will aid in Defpoint selection.

STRETCHING DIMENSIONS

Dimensions can be stretched to alter their position using either grips or the **Stretch** command. The process of stretching a dimension can be seen in Figure 17.40. Use the following steps to alter the dimension position using grips:

1. Select the dimension to be stretched.
2. Make one of the grips on the dimension line the hot grip.
3. Drag the grip to the desired location. OSNAP will aid in accurate placement.

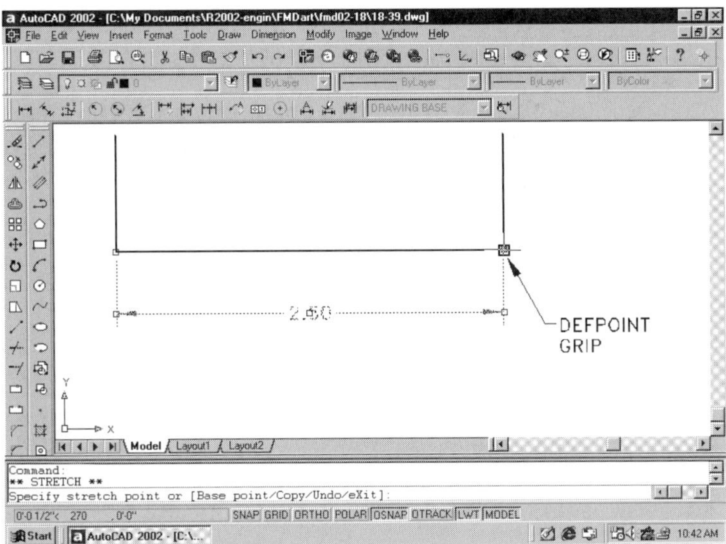

Figure 17.39 *A grip is displayed for the dimension text, at each intersection of the dimension and extension lines, and at the definition points (DefPoints) for the dimensions. AutoCAD automatically assigns definition points for every dimension to define the dimension location.*

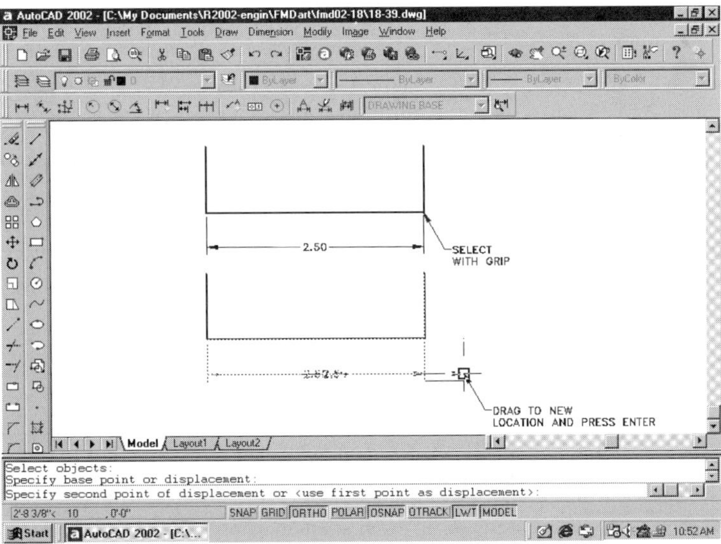

Figure 17.40 *Select the dimension to be stretched and then make one of the grips on the dimension line the hot grip. Drag the hot grip to the desired location. OSNAP will aid in accurate placement when you are aligning with other dimension lines.*

This process will stretch all features of the dimension. To move just the text, select the text grip. Make it the hot grip, and move the text to the desired location.

The **Stretch** command can be used to stretch dimensions using the same procedure used to move drawing objects. Use the Crossing option to select the dimension to be relocated. **Stretch** can't be used to relocate the text.

EDITING OBJECTS AND DIMENSIONS WITH THE MODIFY MENU

Objects and dimensions can be edited using the editing commands contained on the **Modify** menu as well as the grip editing functions. As an object is edited, the dimensions are also edited.

Stretch

Dimensions can be stretched in the same way that drawing objects are stretched. If the dimension is selected with the object, as seen in Figure 17.41, both will be stretched and the dimension text will be revised automatically.

Extend

The **Extend** command sequence also can be used to extend a dimension line and adjust the text accordingly. Figure 17.42 shows the effects of the **Extend** command on dimensions.

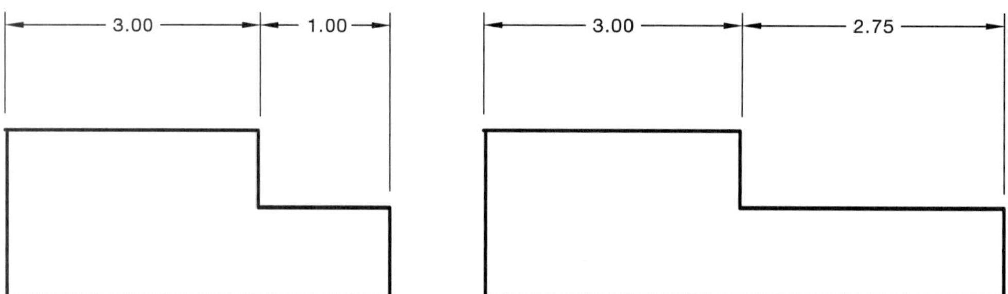

Figure 17.41 *Using the **Stretch** command to alter an object with dimensions.*

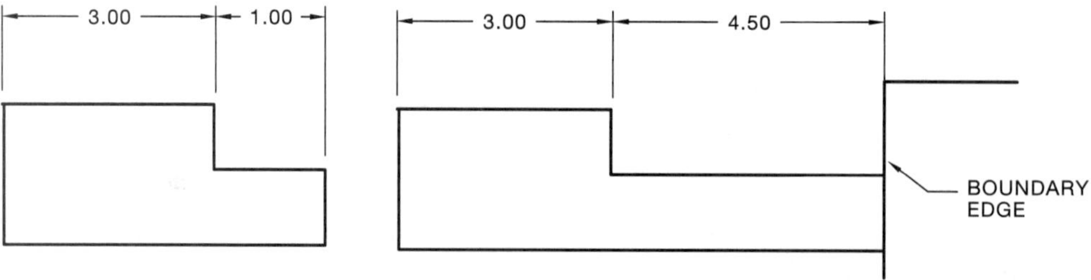

Figure 17.42 *The effect of the **Extend** command on an object with dimensions.*

Trim

The **Trim** command sequence can be used to shorten a dimension line and adjust the text accordingly. Figure 17.43 shows the effects of **Trim** on dimensions.

EDITING DIMENSION TEXT

Dimensions that have been placed on a drawing can be edited using the **Properties**, **Dimedit**, and **Dimtedit** commands.

EDITING DIMENSION TEXT USING PROPERTIES

The dimension text can easily be edited using the **Properties** command. Use the following steps to alter dimension text using the **Properties** command:

1. Select the dimension to be edited.
2. Select the **Properties** button on the **Standard** toolbar or use one of the other access methods.
3. From the **Alphabetic** tab, choose **Text Override**.
4. Enter the desired dimension in the **Text Override** edit box.
5. Continue to alter properties or close the **Properties** dialog box.

Text entered in the **Text Override** edit box will override the measured dimension of the feature. The true measurement will continue to be displayed in the **Measurement** box of the **Properties** dialog box. Dimension text properties can also be altered by selecting the text, right-clicking, and then selecting **Properties** from the shortcut menu. Other options for altering the location and text will be introduced later in this section.

Note: This feature can be especially helpful in dimensioning an object that must be a certain size, but you were unsure of the exact size when the part was drawn and dimensioned. If you know bolts are typically placed in steel plates at a 3.00" spacing, you can draw and dimension a detail prior to receiving the final calculation from the engineer. If the engineer determines that this application needs the bolts at 3.50" spacing, the dimension can be

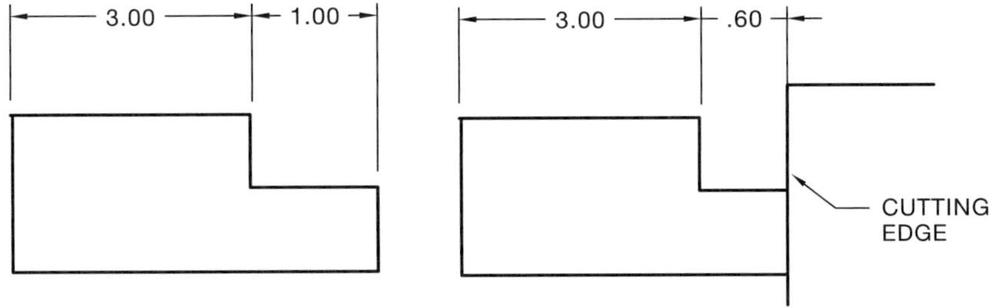

Figure 17.43 *The* **Trim** *command also affects an object, the dimension lines, and text.*

altered without editing the bolt locations on the drawing. The bolt locations will not be drawn to scale, but if the math works, so does the detail, and you've saved time.

EDITING DIMENSION TEXT USING DIMEDIT

The **Dimedit** command allows dimension text to be edited. The existing text can be edited, rotated, moved, or restored to its original position. The angle of the extension lines can also be altered from perpendicular to oblique. Start the command by selecting the **Dimension Edit** button on the **Dimension** toolbar, or by typing DED (ENTER) at the Command prompt:

> Command: **DED** ENTER *(Or click the Dimension Edit button.)*
>
> Enter type of dimension editing [Home/New/Rotate/Oblique]
> <Home>:

Changing the Dimension Text

To alter existing text, select the New option:

> Enter type of dimension editing [Home/New/Rotate/Oblique]
> <Home>: **N** ENTER

This will produce the **Multiline Text Editor** and allow the desired text to be entered. Remove the brackets to replace the existing text. If the new text is placed in front of the brackets, the new text will be placed in front of the existing dimension text. Enter the new text and select the **OK** button to remove the editor. The prompt will continue:

> Select objects: *(Select dimension text to be altered.)* 1 found
>
> Select objects: ENTER
>
> Command:

The result of the command can be seen in Figure 17.44.

Rotating Existing Text

Choosing the Rotate option allows existing text to be rotated. Selecting the text in Figure 17.44 to be rotated to a 45° angle would require the following command sequence:

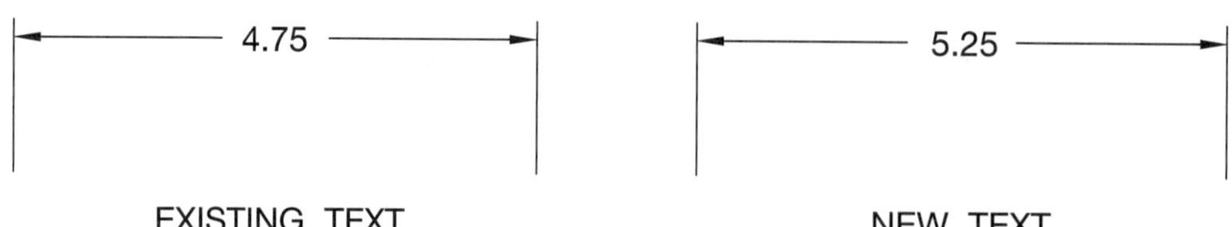

Figure 17.44 *The New option of the* **Dimedit** *command can be used to edit existing dimension text.*

Command: **DED** ENTER *(Or click the Dimension Edit button.)*

Enter type of dimension editing [Home/New/Rotate/Oblique]
<Home>: **R** ENTER

Specify angle for dimension text: **45** ENTER

Select object: *(Select text to be altered.)*

Select object: ENTER

Command:

This would produce text as shown in Figure 17.45.

Altering the Extension Line Angle

The angle of existing extension lines can be altered using the Oblique option of **Dimedit**. Although not recommended for drawings governed by ASME standards, the command sequence to alter the angle of the extension lines in Figure 17.46 is as follows:

Command: **DED** ENTER *(Or click the Dimension Edit button.)*

Enter type of dimension editing [Home/New/Rotate/Oblique]
<Home>: **O** ENTER

Select object: *(Select dimension to be altered.)* 1 found

Enter obliquing angle *(press ENTER for none)*: **-75** ENTER

Command:

The results of the sequence can be seen in Figure 17.46.

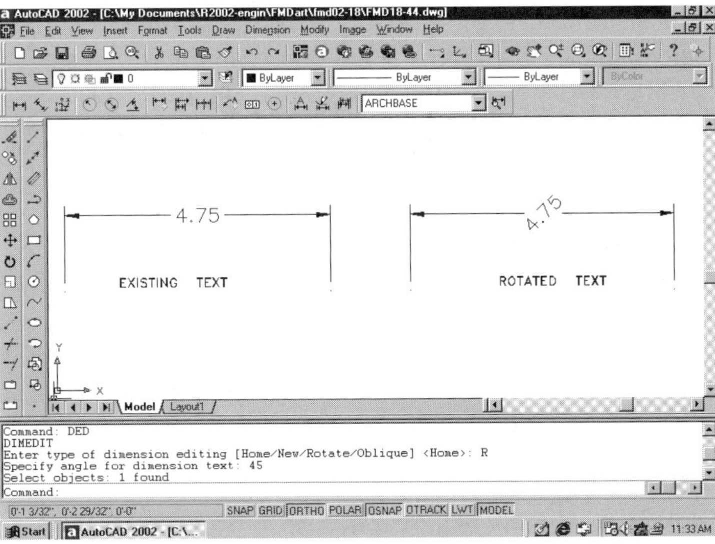

Figure 17.45 *The Rotate option of the* **Dimedit** *command can be used to rotate existing dimension text.*

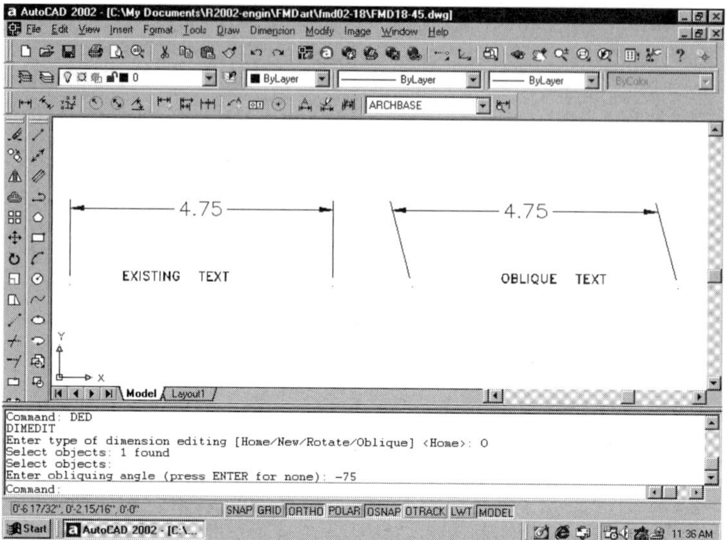

Figure 17.46 *The Oblique option of the* **Dimedit** *command can be used to alter the oblique angle of existing extension lines.*

Returning Home

Using the default option of Home will return the text to its default position.

EDITING DIMENSION TEXT USING DIMTEDIT

The **Dimtedit** command is used to position the text relative to the dimension line. Start the command by selecting the **Dimension Text Edit** button on the **Dimension** toolbar, by selecting **Align Text** from the **Dimension** menu, or by typing DIMTED (ENTER) at the Command prompt. Any access method will prompt you to select a dimension to be edited. Once a dimension is selected, the following prompt will be displayed.

Specify new location for dimension text or
[Left/Right/Center/Home/Angle):

Choosing the Left option will move the text toward the left end of the dimension line. Choosing Right will move the text toward the right end of the dimension line. Choosing Angle will allow the text to be rotated, choosing Center will center the text on the dimension line, and choosing Home will return the text to its default position. Each option can be seen in Figure 17.47.

UPDATING DIMENSIONS

The **Dimension Update** command allows existing dimensions to be selected and altered to a different style. The command is started by selecting the **Dimension Update** button on the **Dimension** toolbar or by selecting **Update**

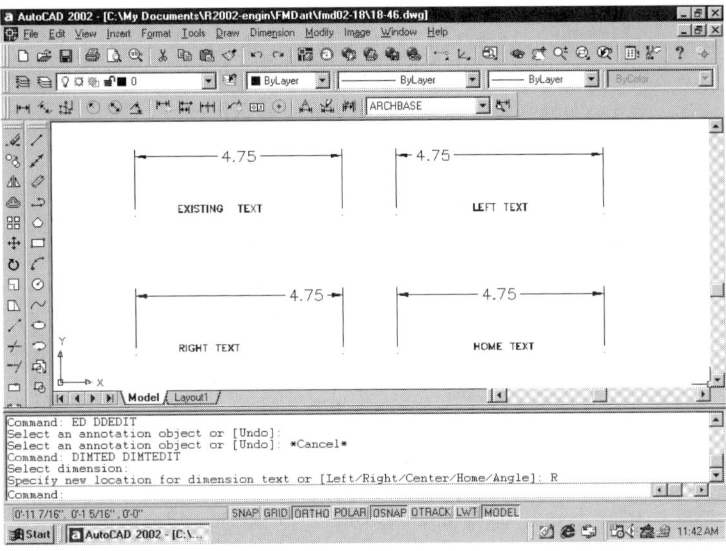

Figure 17.47 *The options of* **Dimtedit** *will alter the location of existing dimension text.*

from the **Dimension** menu. The command duplicates the action of the **Dimension Text Edit** command but automatically applies the Apply option. The command sequence is as follows:

Command: *(Select the Dimension Update button.)*

-dimstyle

Enter a dimension style option

Current dimension style: Standard

Enter dimension style option *(Enter the name of the desired style or press* ENTER *to update the dimension to the current style)*

[Save/Restore/Status/Variable/Apply/?] <Restore>:apply

Select objects: *(Select any dimension (s) whose settings are to be updated to conform to the current dimension style)* 1 found

Select objects: ENTER

Command:

WHAT IT ALL MEANS

More so than any area of AutoCAD, dimensioning seems to cause new users the most anxiety. You've read through what must seem like an unlimited number of commands and options, and you might be having a hard time getting all of this information organized. The good news is that once you set the variables using the dialog boxes and store them in template drawings, most of your work is done.

CHAPTER 17 EXERCISES

DIRECTIONS: Create suitable layers for the placement of dimensions for the following drawings where necessary. Remove drawing views not required for dimensioning, or add views if necessary.

1. Open your drawing template. Use the appropriate dialog box and a style suitable for dimensions with both extension lines, a style with one dimension line, and a style with no extension lines. Set the text offset from the dimension line for all styles to be .065", with a baseline increment of .50. Assign each style a different color. Mark circles with a .125 center mark and provide for centerlines. Use .125 for the feature offset and center, horizontal settings. Use .19 long arrows. Set the text to be two-place decimal with a height as .125, with the text forced inside the extension line, and centered on the dimension line. Save these styles and then save the drawing as BASE DIM.
2. Open drawing E-13-7. Stretch the length of the drawing from 114 mm to 124 mm by placing 5 mm on each side of the V. Do not alter the size of the existing V or the slot on the back side. Edit the dimensions to reflect the current sizes. Alter the dot line terminators to .19 arrows. Save the drawing as E-17-2.
3. Open drawing E-11-30. Create a dimension style suitable for use with this part. Completely dimension this drawing. Save the drawing as E-17-3.
4. Open drawing E-11-34. Create a dimension style suitable for use with this part. Completely dimension this drawing. Use the auxiliary drawing for as many dimensions as possible using unidirectional dimensions. Save the drawing as E-17-4.
5. Open drawing E-11-35. Provide a third hole centered between the two existing .25" Ø holes. Save the drawing as E-17-5.
6. Open drawing E-11-9. Completely dimension the part using an appropriate dimensioning system. Save the drawing as E-17-6a. Make a copy of the drawing. Stretch the right side of the drawing (the 1.25 feature) to be .50 longer. Do not alter the size or location of the highest feature. Edit the dimensions to reflect the required changes. Save this drawing as E-17.6b.
7. Open drawing E-11-26. Completely dimension the part using a dimensioning system that will ensure a high degree of accuracy. Save the drawing as E-17-7a. Make a copy of the drawing. Stretch the each end of the part to allow an additional 15 Ø hole to be placed in each end, with a distance of 20 mm from the existing center points. Do not alter the size or location of the inclined features relative to each edge. Edit the dimensions to reflect the required changes. Save this drawing as E-17.7b.
8. Open drawing E-11-27. Completely dimension the part using metric chain dimensions. Save the drawing as E-17-8.

9. Open drawing E-11-28. Completely dimension the part and save the drawing as E-17-9a. Make a copy of the drawing and alter the base to be 4.50 square. Provide 3 additional .38 Ø holes equally spaced between the existing holes. Place two of the eight holes to be added parallel to each edge. Save the drawing as E-17-9b.
10. Open drawing E-11-29 and completely dimension the part. Save the drawing as E17-10a. Using the existing drawing as a base, show the 3.50 feature on the right end of the part to be at a 40° angle from vertical. Draw the required views to reflect this change and completely dimension the part using chain dimensions. Save the drawing as E-17-10b.
11. Open drawing E-13-6 and dimension the drawing. Save the drawing as E-17-11.
12 Open drawing E-11-32 and dimension the drawing using chain dimensions. Save the drawing as E-17-12.
13. Open drawing E-13-9 and dimension the drawing. Save the drawing as E-17-13.
14. Open drawing E-16-24 and alter a minimum of five style features. Do not be concerned with maintaining ASME standards as you alter style features. Save the drawing as E-17-14.
15. Open drawing E-16-25 and completely dimension the drawing. Save the drawing as E-17-15a. Make a copy of the drawing and alter the text placement, use aligned text, and different line terminators. Do not be concerned with maintaining ASME standards as you alter style features. Save the drawing as E-17-15b.

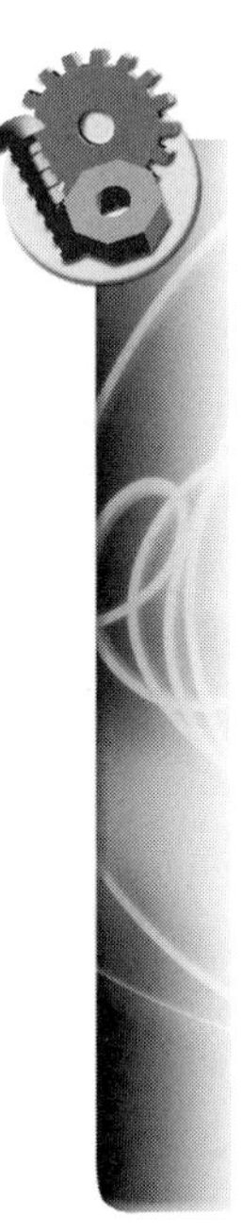

CHAPTER 17 QUIZ

1. A group of dimensioning controls that can be saved for future use is called a __________ .
2. A ______________________ is used to control individual aspects of dimensioning.
3. What command can be typed to access a listing of dimensioning features?
4. How can styles be used in an office that does both mechanical and electrical projects?
5. What value controls the size of each dimension feature and how should this value be set for a drawing that will be reproduced at a scale of 1/4=1?
6. What option will remove both extension lines from a dimension?
7. The spacing between two dimension lines is controlled by what option?
8. Use the **Help** menu and determine what the **Dimgap** controls.
9. List two features that can be displayed to describe round objects.
10. List options for terminating a dimension line that are suitable for a mechanical drawing.

11. List and describe two features of the extension line as it relates to the dimension line and the object being described, and give the value that controls each.
12. What are the options for controlling the height of text?
13. Name the dialog box that can be used to make changes to dimensioning text.
14. What style will be used as a base when dimensioning styles are created?
15. What option will allow for changing an attribute without changing the entire style?
16. What does the **Dimtedit** command affect?
17. Describe how associative dimensions will affect editing.
18. How can the **Leader** command be used to provide two arrows pointing to the same text?
19. What dimensioning command is used if the text is to be placed so that it is not parallel, perpendicular, or aligned with a surface?
20. What command can be used to change existing dimensions to the current style?
21. Sketch examples of right angle and closed terminators.
22. Explain the process for creating a user-defined terminator.
23. Explain the options to display a centerline in each circle to be dimensioned.
24. Give the proper scale factor for a drawing to be plotted at a scale of 1/8=1 and explain how the value is determined.
25. Explain the process to place text above the dimension line.
26. List and explain the options for controlling zeros in dimensions.
27. Explain the difference between associative and normal dimensions.
28. Explain the options of the **Dimedit** command.
29. Explain the effect of selecting an extension line to be erased.
30. Explain the effect of stretching a feature that is dimensioned 3.00" long to a length of 6.00".

GROUPING AND PRESENTING INFORMATION

Representing Machined Features and Finishes

Now that you've been exposed to the multitude of linetypes and views that can be used in a drawing, you need to understand the basic machine features that are associated with manufactured parts. This chapter will introduce

- Circular features such as holes created by drills, reams, bores, counterbores, spotface, countersinks, counterdrills, slotted holes, lugs, keys, key seats, keyways, neck, splines, and chamfers
- Rectangular features including dovetails, kerfs, pads, and knurls
- Surface texture and finishes

No new AutoCAD commands will be introduced in this chapter. Each of the features introduced in this chapter can be completed using commands from previous chapters. Figure 18.1 shows an example of a drawing with machined features.

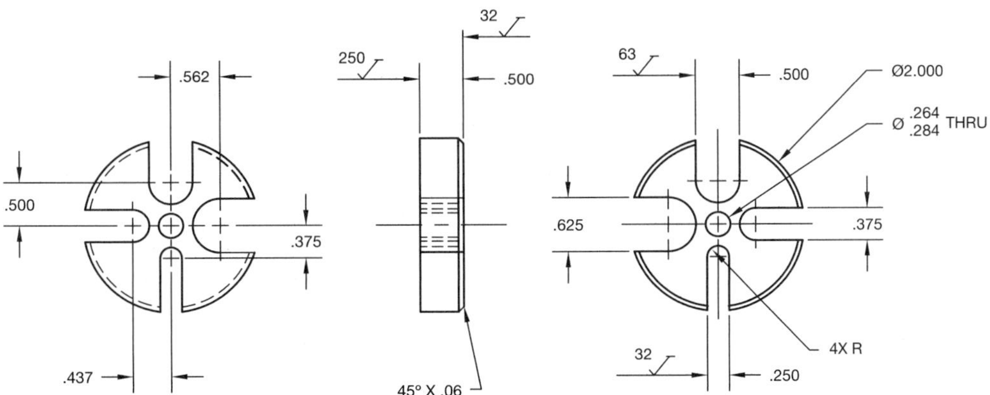

Figure 18.1 *Machined features that are drilled or cut are often required on mechanical drawings. Courtesy Cathy Schlesser.*

REPRESENTING MACHINED FEATURES

Circular features are typically machined into parts to aid connections, or to allow fluid or other parts to move through the circular feature. Circular features are machined into a component using drills, reamers, or a process known as boring. They are drawn using the **Circle** command.

DRILLS, REAMS, CORES, AND BORES

Drilling is the process of creating or enlarging a circular feature using a drill bit. *Reaming* is the process of finishing a drilled or cored hole using a reamer. Only a small amount of material is removed during the reaming process. Reaming removes a small amount of material from a hole to provide a smoother finish than can be achieved by drilling. A *core* is a hole created in a part that is cast in a mold. *Boring* is the process of enlarging an existing hole using a lathe.

REPRESENTING CIRCULAR FEATURES

Figure 18.2 at A shows how a drilled hole should be drawn and specified. Use the guidelines presented in Chapter 17 to place dimensions to locate the hole. Use a leader line that points to the center point of the circle to describe the diameter of the hole. Use the following ASME guidelines to represent drilled features.

- Holes are always referenced by centerlines with intersecting short lines at the center of the hole.
- Use hidden lines to represent the hole, and a centerline to represent the axis in the view where the axis of the hole is seen.
- Describe the diameter of the hole in the view where the hole appears as a circle. Use a diameter symbol preceding the size. Never describe the feature in the view where it is represented by hidden lines. The depth of a hole can often be placed in a section view where the lines show as continuous lines. See Chapter 15 for information about using leader lines to reference notes to drilled features.
- If the hole goes all the way through the part, and this is indicated in a view, no depth needs to be indicated. If the hole does not extend all the way through the part, as seen in Figure 18.2 at B, represent the bottom of the hole with a line that is perpendicular to the sides of the hole. Represent the bottom of the drilled area with lines that are drawn at 60° to the centerline. Represent the depth of the hole in a note using a depth symbol similar to Figure 18.2 at B. The depth symbol can be placed by typing a lowercase X using the GDT font.
- Describe the location of the hole with dimensions that extend to the center-line of the circular view. Never dimension circular features in views where the

feature is represented by a hidden line. See Chapter 15 for ANSI requirements for dimensioning holes.

- Dimension multiple holes by listing the number of holes, followed by an uppercase X, followed by a space, followed by the diameter. See Figure 18.3.

Counterbore

A *counterbore* is a hole drilled around the end of another hole to provide space to recess a bolt head. The bottom of the counterbore is perpendicular to the edge of the hole. The small hole is drilled first, and then used as a guide to drill the counterbore. Figure 18.4 shows proper representation for a counterbore. Figure 18.5 shows the process for drawing and specifying multiple counterbores. In addition to the guidelines for a circle, use the following guidelines for drawings showing a counterbore:

- Describe the diameter and depth of the counterbore in the view where it appears as a circle. Use a diameter symbol, a counterbore symbol, and a depth

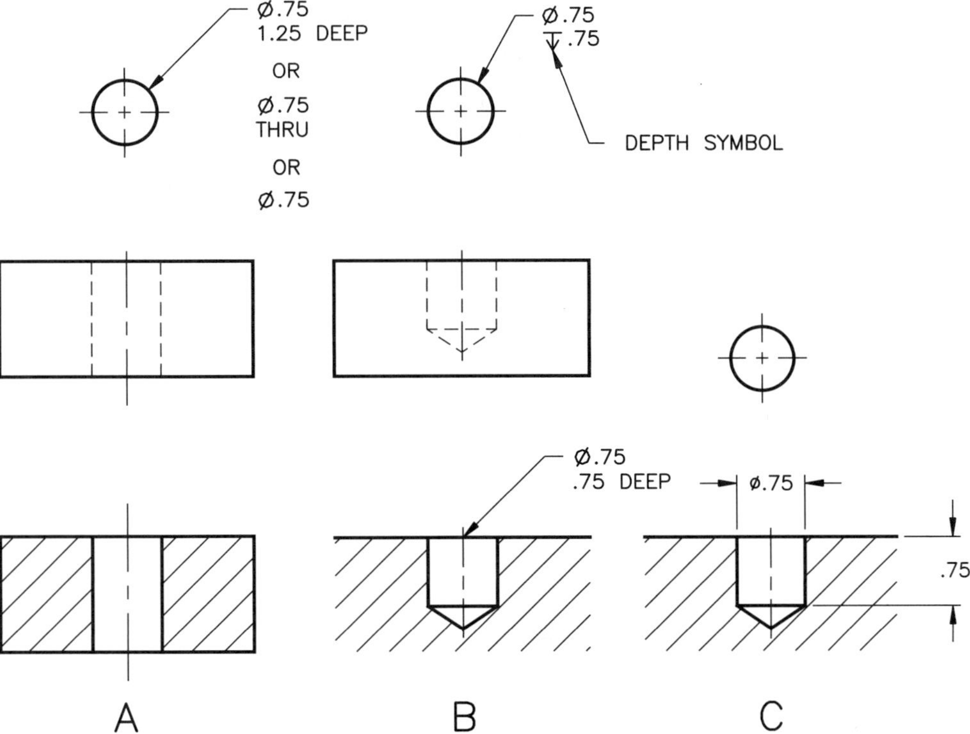

Figure 18.2 *A hole is represented by a circle in plan view, and with hidden lines when the axis is parallel to the viewing plane. If the hole does not go through the feature, the bottom is represented by hidden lines drawn at 60° to the centerline. When shown in section, the depth can be specified because no hidden lines are used.*

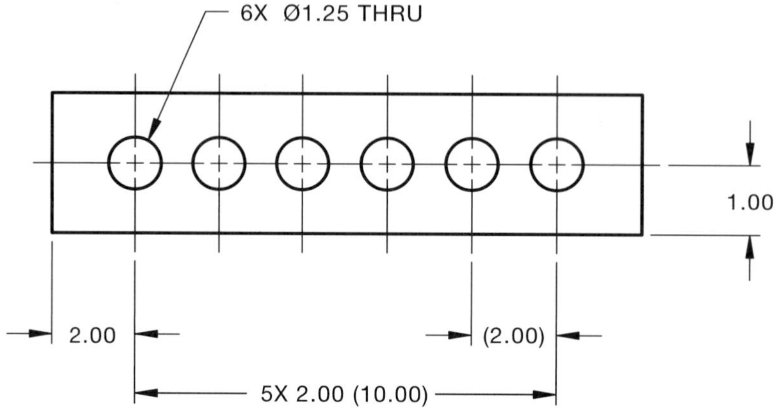

Figure 18.3 *Locating multiple drilled features.*

symbol for specifying the counterbore information. The counterbore symbol can be placed by typing a lowercase V using the GDT font.

- Use hidden lines to represent the hole and the counterbore, and a centerline to represent the axis in the view where the axis of each hole is seen.
- Describe the location of the hole and counterbore with dimensions that extend to the centerline of the circular view. Never dimension circular features in views where the feature is represented by a hidden line.

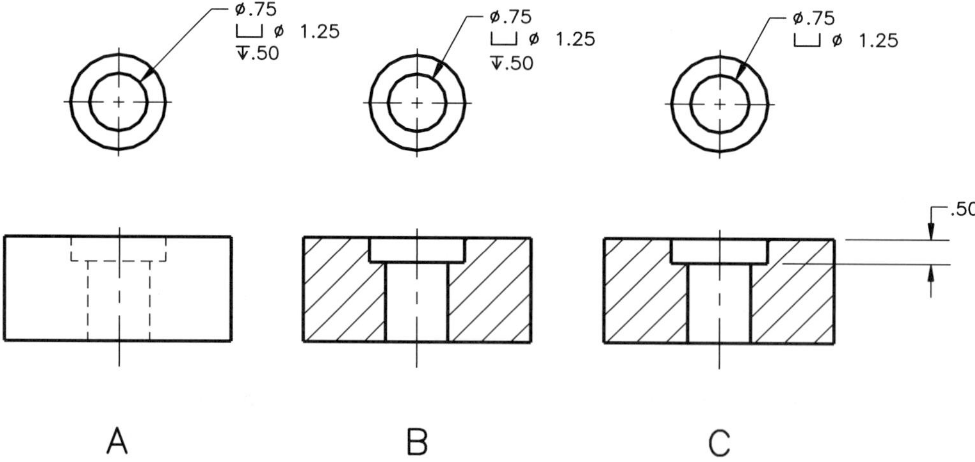

Figure 18.4 *Drawing and specifying a counterbore.*

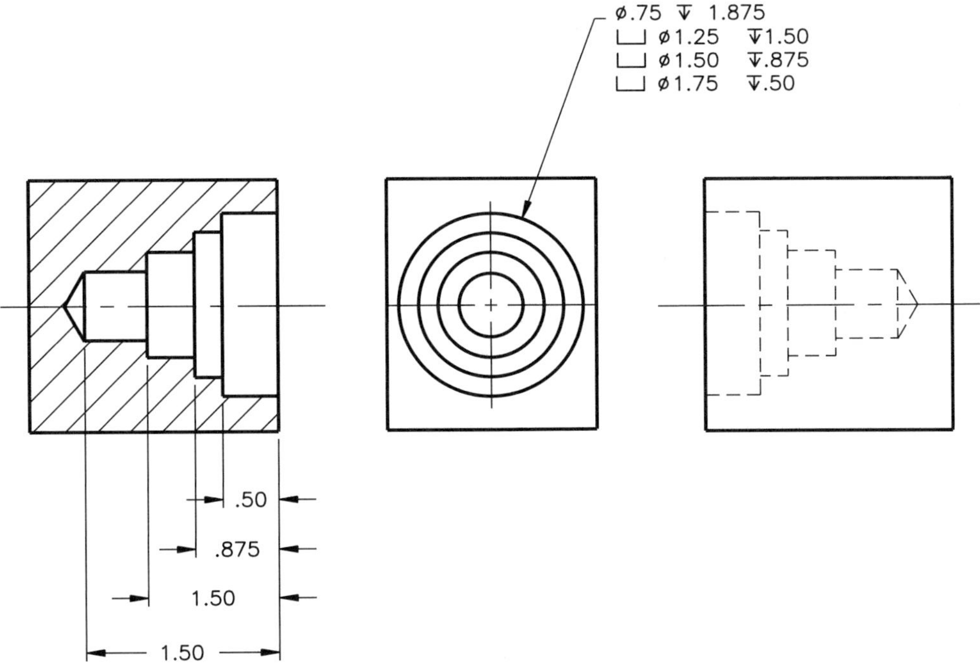

Figure 18.5 *Specifying multiple counterbores.*

Spotface

A *spotface* is a surface that has been machined to provide a flat surface for a bolt head or washer that will be applied to a curved surface. Although a spotface might appear similar to a counterbore, it is typically only deep enough to provide a flat surface, but not deep enough to fully recess the bolt head. Figure 18.6 shows how a spotface should be represented. Specify a spotface just as you would a counterbore.

Boss

A *boss* is a flat surface that has been raised above a curved surface to provide a bearing surface for a bolt. Figure 18.7 shows how a boss should be represented and how the height and diameter can be represented.

Countersink

A *countersink* is a conical feature placed around the end of another hole to provide space to recess a screw head. The sides of the countersink are typically at a 41° to the axis of the hole. The small hole is drilled first, and then used as a guide to locate the countersink. Figure 18.8 shows proper representation for a countersink. The countersink symbol can be placed by typing a lowercase W using the GDT font. In addition to

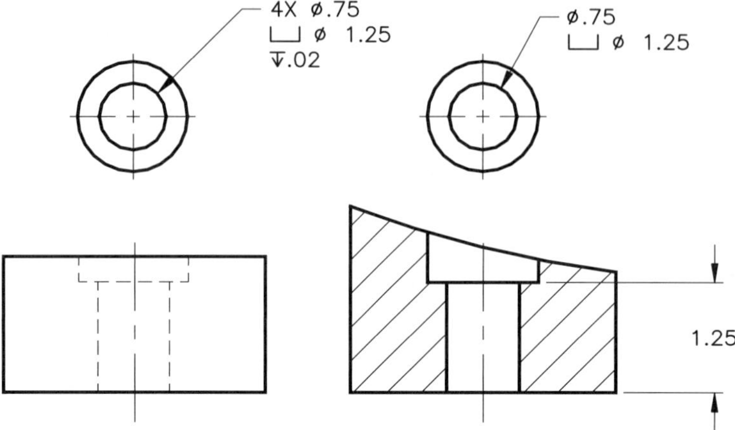

Figure 18.6 *A spotface provides a recessed flat surface for a bolt head.*

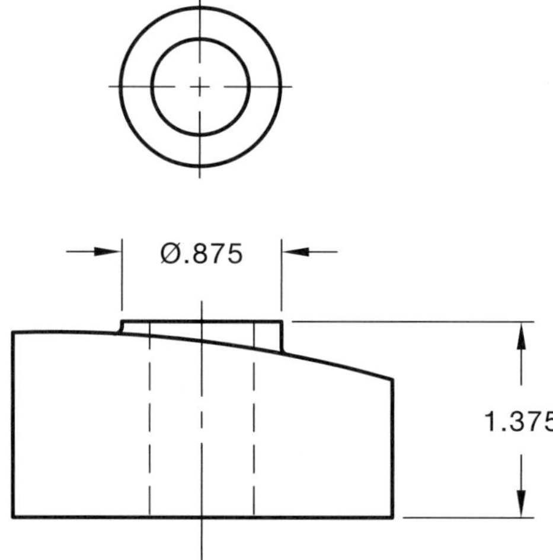

Figure 18.7 *A boss provides a raised flat surface for a bolt head.*

the guidelines for a circle, use the following guidelines for drawings showing countersinks:

- Use hidden lines to represent the hole and the countersink, and a centerline to represent the axis in the view where the axis of each hole is seen.
- Describe the hole diameter and the countersink diameter and angle in the view where it appears as a circle. Use a diameter symbol, a countersink sym-

bol, and an angle symbol for specifying the countersink information. Although the angle of a countersink is always 82°, it must be specified.

- Describe the location of the hole and countersink with dimensions that extend to the centerline of the circular view. Never dimension circular features in views where the feature is represented by a hidden line.

Counterdrill

A *counterdrill* is a drilled feature placed around the end of another hole. The sides of the counterdrill are placed at a 60° to the axis of the hole, but are specified by giving the total angle. Figure 18.9 shows proper representation for a counterdrill. In addition to the guidelines for a circle, use the following guidelines for drawings showing a counterdrill:

- Use hidden lines to represent the hole and the counterdrill, and a centerline to represent the axis in the view where the axis of each hole is seen.
- Describe the hole diameter and the counterdrill diameter and depth in the view where it appears as a circle. Use a diameter symbol, a counterdrill symbol for the large hole, and an angle symbol for specifying the inclined surface. Although the angle of the inclined surface is always 120°, it must be specified in a sectioned view.
- Describe the location of the hole and counterdrill with dimensions that extend to the centerline of the circular view.

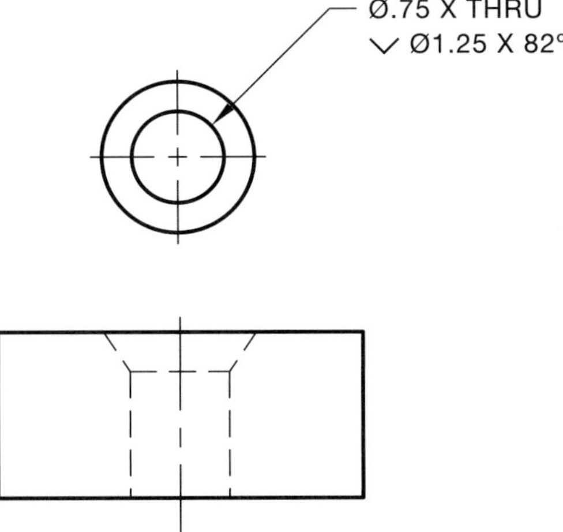

Figure 18.8 *A countersink provides a tapered recess around a hole.*

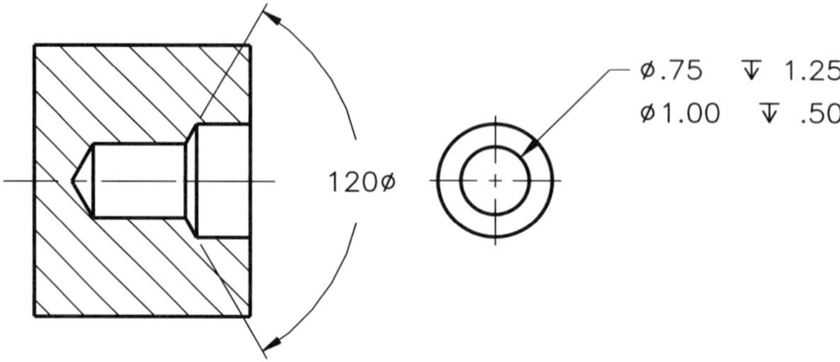

Figure 18.9 *Drawing and placing information for a counterdrill.*

Slotted Hole

A *slotted hole* is an elongated hole that allows movement in one direction. If a bolt were placed through the hole in Figure 18.10, the plate could move from left to right but not from top to bottom. Slotted holes can be drawn in plan view using a combination of the **Arc**, **Circle**, and **Line** commands. Use hidden lines to represent the diameter in views where the axis of the drilled feature is represented. Use centerlines to represent the axis of the drilled feature. Use one of the three methods to specify the size and location of the slotted hole, depending on what information the engineer has supplied. Figure 18.11 shows a slotted hole with an end radius larger than the width of the slot.

Lug

A *lug* is a projected portion of a part that is used as a mounting bracket. Figure 18.12 shows an example of a part with four lugs. Draw the drilled features in lugs as you would any other drilled feature, using either the **Arc** or **Circle** command and the **Line** command. The width of the lug can be specified as if it were a rectangular part, and the

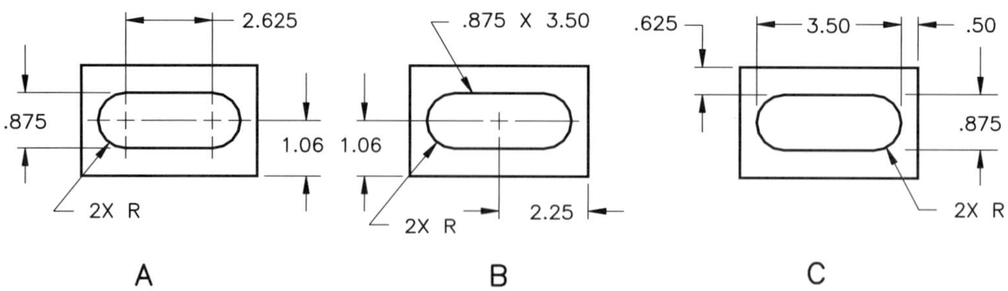

Figure 18.10 *Three options can be used to specify the size and location of a slotted hole.*

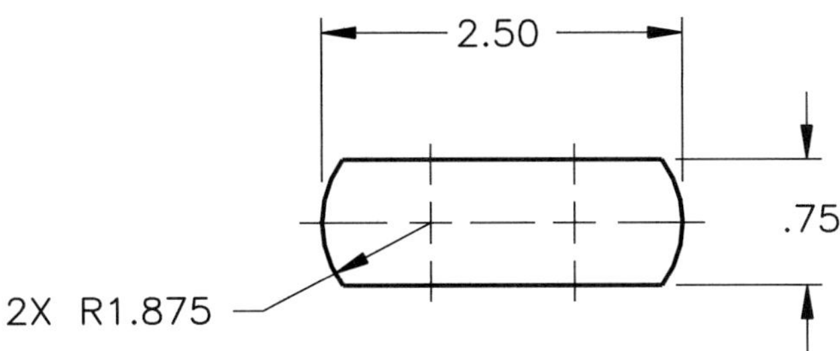

Figure 18.11 *A slotted hole with a radius that exceeds the width of the slot.*

radius defined as an R referenced to a leader line, or the radius of the lug can be given. The height of the part is described in the same way as any other rectangular feature.

Key, Keyseat, and Keyway

A *key* is a machined part that is used to provide a semi-permanent connection between two parts such as a gear and a shaft. The key will hold the gear in place, but if removed, it will allow the gear to be removed from the shaft. The key can be drawn using the **Line**, **Circle**, and **Trim** commands. The key is placed in a groove in

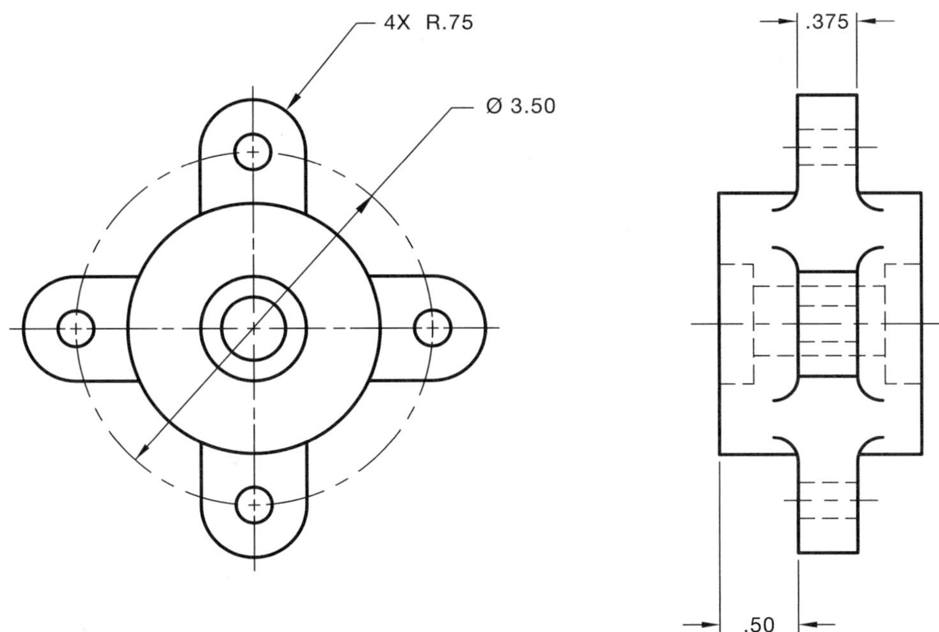

Figure 18.12 *A lug is a projection used as a mounting surface.*

the shaft called a *keyseat*. Occasionally a continuous groove or channel called a *keyway* is used to seat the key. As the gear or other connecting part is pulled over the key, the part is held in position and the torque of the shaft is transferred to the mating part. A keyseat and keyway can be drawn using the **Line** command. Figure 18.13 shows methods of representing keyseats and keyways.

Neck

A *neck* is a groove that is cut into a cylindrical part. Three different methods are shown in Figure 18.14 to represent a neck when the centerline of the cylinder is parallel to the viewing plane. When the neck has a flat or inclined surface, the **Line** command is used to represent the sides of the neck. The **Arc** command can be used to represent the curved surface of the neck. When the end of the cylinder is parallel to the viewing plane, the neck is represented using a circular hidden line.

Spline

The term *spline* is used in several different contexts in drafting. The term can be used to describe a gear-like cylinder, a curved line, or an AutoCAD command. As a machined feature, a *spline* is a gear-like pattern cut into a shaft to transfer torque and still allow movement along the shaft. It is used in place of a key. Figure 18.15 shows an example of a spline. A spline can be drawn using the **Circle**, **Line**, **Trim**, and **Array** commands.

Chamfer

A *chamfer* is an inclined surface placed on the edge of a part. When applied to an exterior straight edge, the chamfer will appear as in Figure 18.16. The angle can only be specified by leader line if the angle is 45°. Figure 18.16 shows alternative methods of

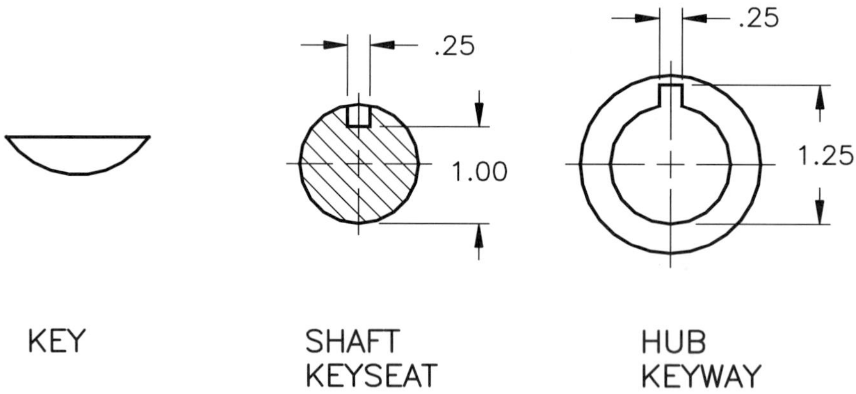

Figure 18.13 *Proper representation of a key, keyway, and keyseat. Keys are usually not drawn because they are considered stock parts.*

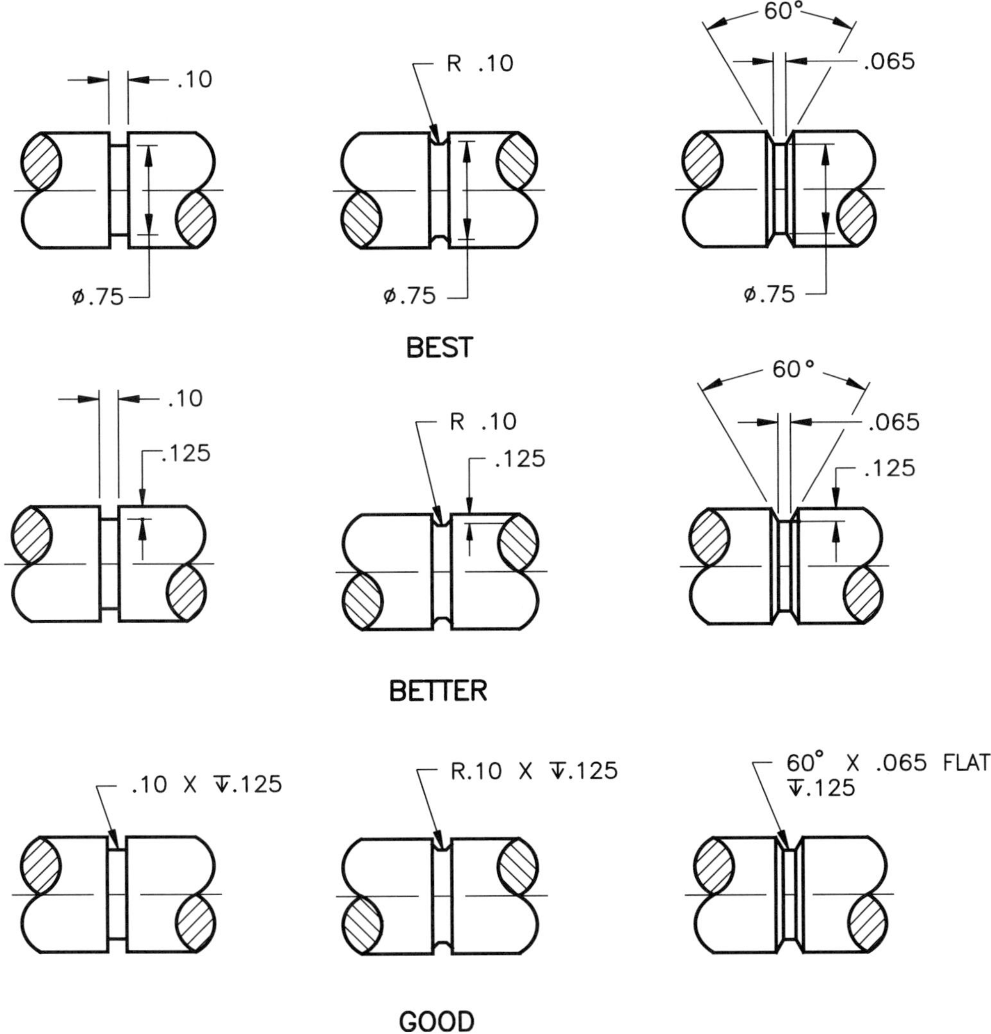

Figure 18.14 *Proper representation and specifications for a square, round, and inclined neck.*

dimensioning exterior chamfers and examples of interior chamfers. Although it appears as a countersink, the angle can vary from 82°.

REPRESENTING RECTANGULAR FEATURES

Several common machined features are rectangular or angular. Common features include kerf, rectangular slots, pad, and knurl.

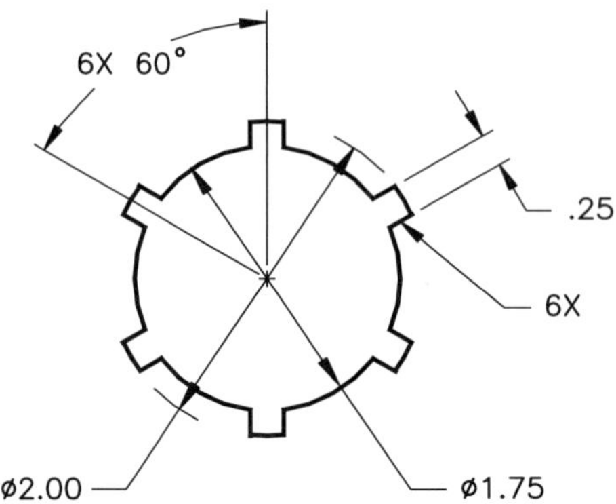

Figure 18.15 *A spline is a gear-like cylinder.*

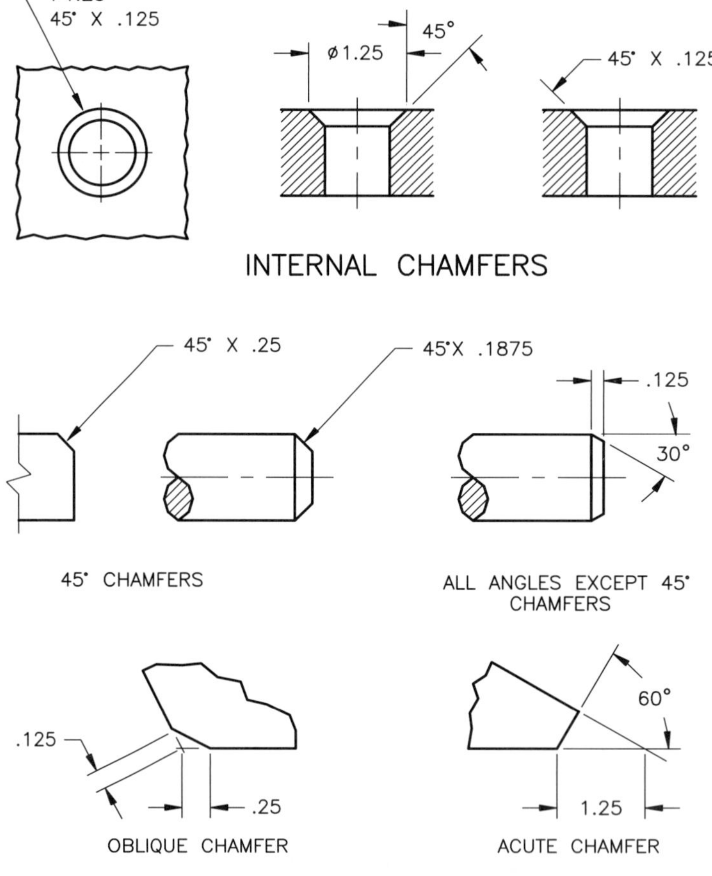

Figure 18.16 *Representation of interior and exterior chamfers.*

Kerf

A *kerf* is a rectangular slot that is cut into a feature to create a sliding connection between two mating parts. Figure 18.17 shows how it is represented. It can be drawn using the **Line** command with hidden and continuous linetypes.

Slots

Common rectangular slots include the t-slot and a dovetail, which are both machined into a feature to create a sliding connection between two mating parts. A *t-slot* is a slot with perpendicular edges, and a *dovetail* is an angled slot. Figure 18.18 shows how a

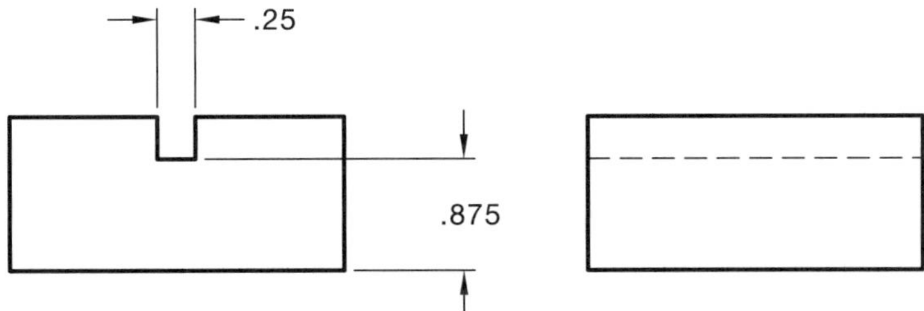

Figure 18.17 *A kerf is a rectangular slot that is cut into a feature.*

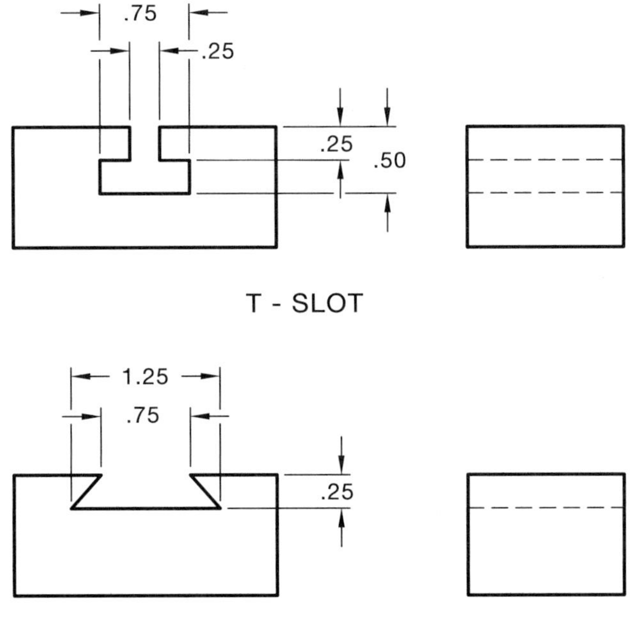

Figure 18.18 *T-slots and dovetail slots.*

t-slot and dovetail are represented. Each type of slot can be drawn using the **Line** command with hidden and continuous linetypes.

Pad

A *pad* is a raised surface provided to aid in mounting a mating part. Unlike a boss that is always circular, a pad can be rectangular, square, or round. Figure 18.19 shows an example of a pad. It can be drawn using the **Line** or **Rectangle** command, depending on the desired shape.

Knurl

Knurling is a pattern that is machined into a part to provide a roughed surface to aid in gripping the part. Handles or other gripping surfaces are often given a knurled surface to resist slipping. Figure 18.20 shows two of the most common knurling patterns. These patterns are not drawn on production drawings but are represented by note specifying the type, pitch, and diameter.

PROJECTION OF MACHINED FEATURES

In Chapter 6 you were introduced to projection of objects in multiview drawings. Circular features do not hold to all of the rules of true projection. Two additional considerations are incomplete views and aligned features.

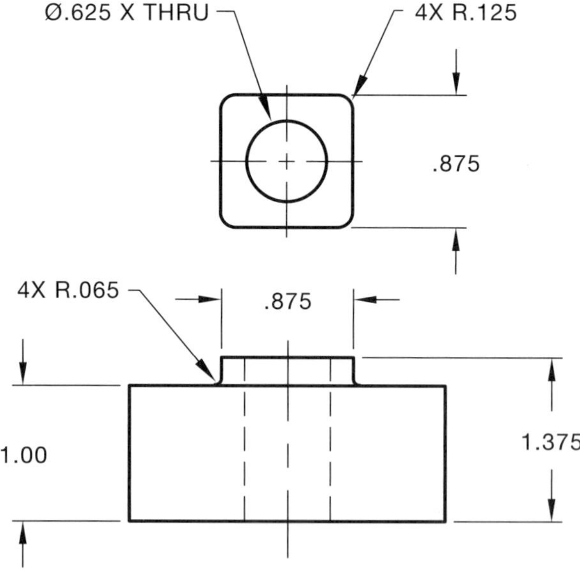

Figure 18.19 *A pad is a raised flat rectangular surface similar to a boss.*

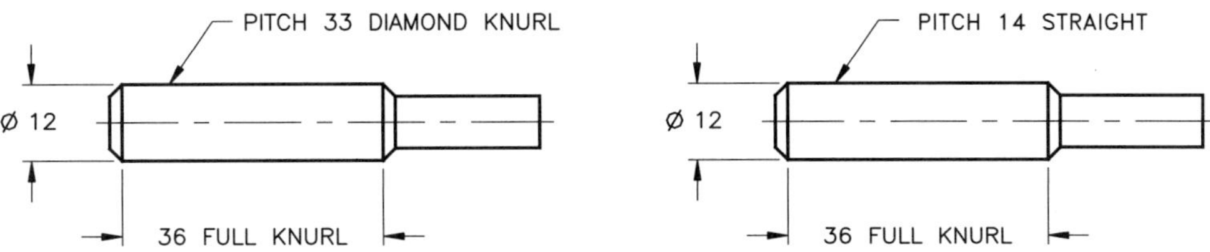

Figure 18.20 *Knurling is a pattern that is machined into a part to provide an aid for gripping the part.*

Incomplete Views

Your role as a CAD drafter is to clearly convey information from the engineering to the manufacturing department. Your goal is clarity. It is possible to show all of the features of a part, but lose clarity. Figure 18.21 shows an example of a technically correct drawing that lacks clarity. Figure 18.22 shows the same drawing with some features eliminated. Features can be eliminated to provide clarity using the following guidelines:

- The feature must be clearly specified in other views.

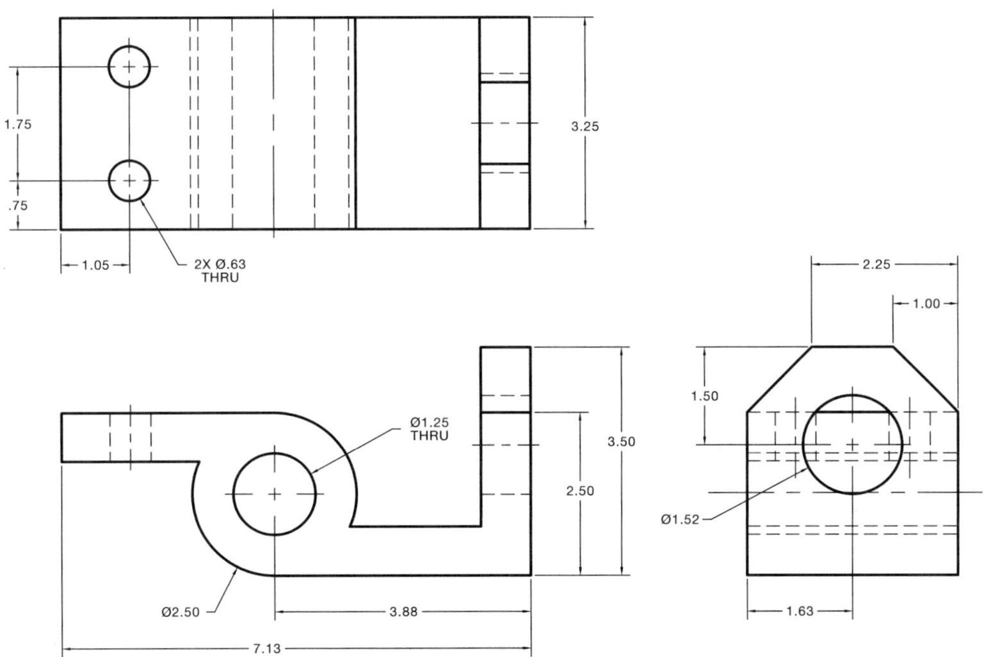

Figure 18.21 *Although technically accurate, showing all lines and features can hinder clarity.*

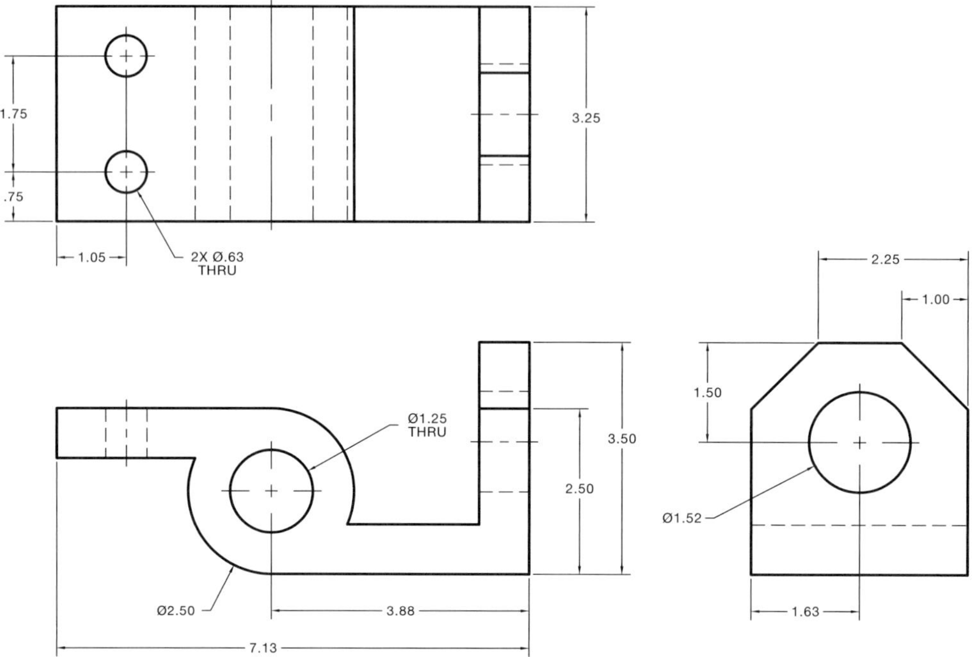

Figure 18.22 *Eliminating features that are clearly shown in another view can add to clarity. Courtesy Paul Bissell.*

- Omit hidden features that lie farthest from the cutting plane.

In Figure 18.22, the drilled features are specified in the top view and shown in the front view. Features that are farthest from the right side were omitted.

Aligned Features

Using true orthographic projection can be misleading when circular features are projected. Figure 18.23 shows how four holes would appear using true projection. The side view creates the illusion that the holes are at varied distances from the center of the part. The true position of circular features is not altered, but in theory, features are rotated to the nearest quadrant of the feature before projecting to provide a more accurate representation. Figure 18.23 shows aligned representation of the three holes.

SURFACE FINISHES

Surface finish refers to the roughness of a surface that results from a machining process. In its simplest form, surface finish is represented on a drawing with a number placed beside a check mark. The number represents the height in micro inches (millionths of a inch) or micrometers (millionths of a meter). Figure 18.24 shows simple callouts for metric and inch units of surface finish. The symbol specifies that a sur-

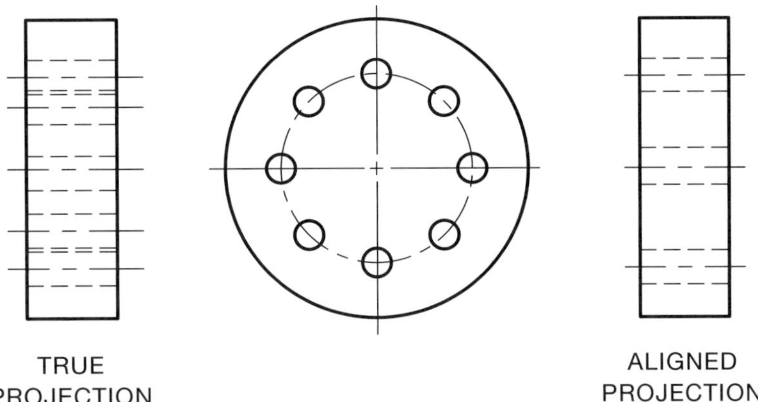

Figure 18.23 *Holes projected in true location will give a deceiving view of the part. Showing the feature using aligned projection provides a true description of how the features relates to the part.*

face must be machined to a finished specification. Depending on the requirements of the part, surface waviness, lay, and flaws of a surface can also be specified. Figure 18.24 shows how each of these items can be placed on the surface finish symbol.

SURFACE WAVINESS AND LAY

Surface roughness refers to the height of individual irregularities. *Surface waviness* refers to the height of a pattern of irregularities. Surface waviness is caused by machine vibration, wobble of a cutting blade, heat treatment, or work deflection. The direction of the pattern waviness is referred to as *lay*. If you've ever seen wood cut with an old circular saw blade, you've seen waviness and lay. Figure 18.25 shows the symbols used to describe the lay of the surface waviness.

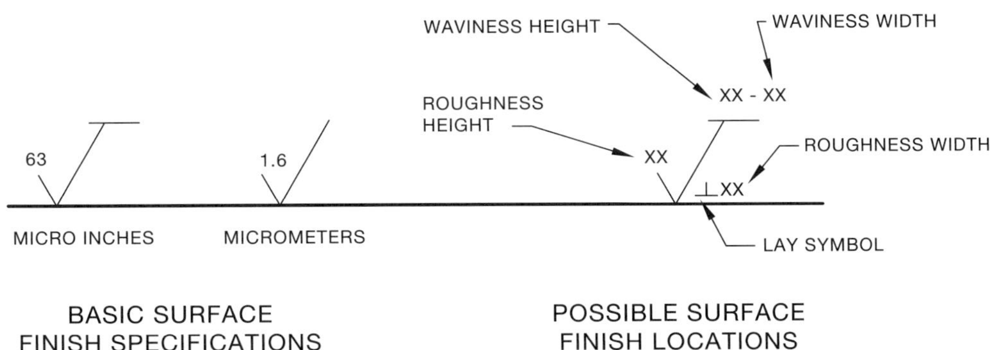

Figure 18.24 *A simple callout for an inch and metric surface finish and detailed surface finish symbols.*

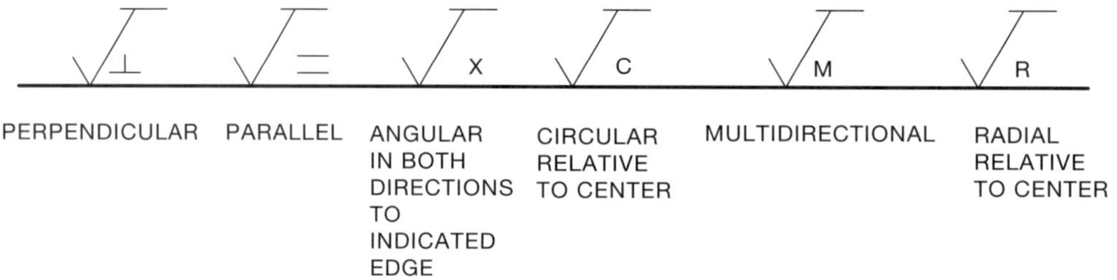

Figure 18.25 *Common symbols used to describe lay.*

SPECIFYING SURFACE FINISH

The placement of the surface finish symbol is very important to the finish a surface is to receive. To correctly display surface finish, the symbol must be placed where the surface to be machined shows as an edge. The finish symbol can be placed on the object line that represents the surface, or on an extension line referenced to the intended surface. Figure 18.26 shows common methods of placing the symbols. The number to represent the surface finish is placed using unidirectional text. Specifications for surface finish in inch and metric include the following:

Roughness Height Rating

Micro inches	Micrometers	Surface Description
1	0.025	Super Finish
2-4	0.050-0.100	Extremely smooth
8	0.20	Very fine
16	0.40	High grade
32	0.80	High grade
63	1.6	Good machine finish
125	3.2	Medium
250	6.3	Coarse
500	12.3	Rough machining
1000	25	Very rough

For most drawings, the surface finish is specified in the tolerance block located similar to Figure 18.27. A general note can also be used to specify a general tolerance. If either of these methods is used, a surface finish symbol with no height specification is placed on the all surfaces to be finished. If all surfaces of the part are to receive the specified finish, a general note is placed indicating F.A.O. (Finish All Over).

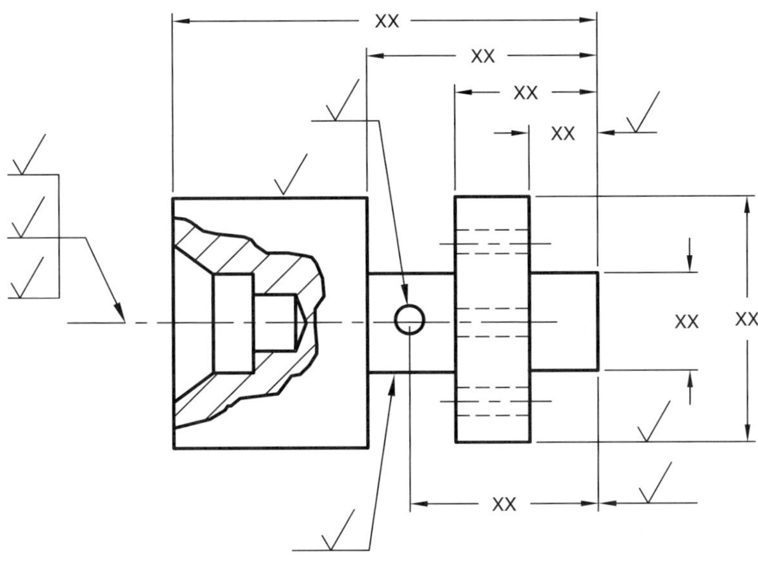

Figure 18.26 *Common methods of placing surface finish symbols.*

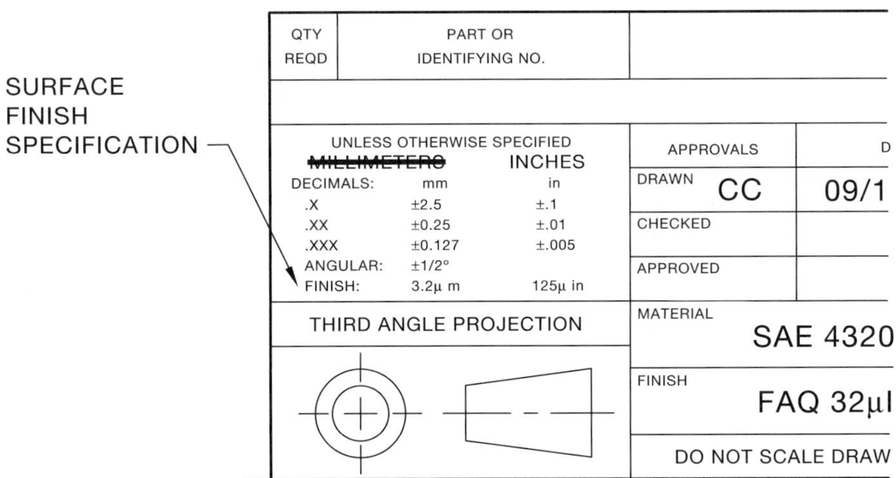

Figure 18.27 *Surface finish can be specified in the tolerance block or as a general note.*

CHAPTER 18 EXERCISES

1. Open drawing E-17-2-4 and add a counterbore to each end of the part. The counterbores will be placed on what is the back surface of the drawings. Create new views as needed to show a Ø5 mm hole through the feature with a Ø10 mm × 5mm counterbore. Locate the center of the hole down 20 mm and over 15 mm from each end. Specify the bottom of the feature to have a coarse machine finish, the front to be a high grade finish, and the balance of the part to be a medium finish. Save the drawing as E-18-1.
2. Open drawing Web E-2-5 and place five Ø 25" evenly spaced holes through the front feature from top to bottom. Place a Ø.75" hole in the center of the surface that is .625" from the front. Center the hole from left to right on the 2.00 wide feature at the top of the part. Extend the hole through the part with the axis running from top to bottom. Place a Ø1.00" countersink around the top edge of the hole. Update all views and provide new views as needed to completely dimension these changes. Save the drawing as E-18-2.
3. Open drawing Web E-2-6 and stretch the thickness of the base from 12 to 24 so that the part is a total of 67 mm high. Add a Ø20 × 5 deep counterdrill to each of the two Ø15 holes. Update all views and provide new views as needed to completely dimension these changes. Save the drawing as E-18-3.
4. Open drawing E-17-8 and stretch the thickness of the base from 12 mm to 24 mm so that the part is a total of 61 mm high. Do not alter any angled surfaces. Add a Ø20 mm × 3 mm deep spotface to the top of each of the two Ø15 mm holes. Place an Ø18 mm hole centered in both directions in the top of the upper flat surface. Place a Ø25 mm × 5 mm deep counterbore around the Ø18 mm holes. Place a Ø15 mm × 45° chamfer around the outer edge of each of the Ø12 mm holes. Show the outer face of each of the two protruding legs to have a very fine surface. Show the inner face of each of the two protruding legs to have a rough surface. Update all views and provide new views as needed to completely dimension these changes. Save the drawing as E-18-4.
5. Open drawing E-16-25. Add a radius to the end of the 14 mm slot in the bottom of the part. Add three Ø5 mm holes equally spaced and centered in both directions above the upper slot in the inclined surface. Add a Ø12 mm hole centered 75 mm from the bottom of the inclined edge, and centered from left to right. Provide a Ø20 mm × 1 mm spotface centered on this new hole. Update all views and provide new views as needed to completely dimension these changes. Save the drawing as E-18-5.
6. Open drawing E-17-4. Add a slotted hole centered on the existing slotted hole extending up the inclined surface. Show the distance from the existing centerline to the center point for the new radius as 50 mm, and show the radius at the end of the slot as equal to the width of the slot. Show a 5 mm × 45° chamfer around each of the Ø57 mm holes. Update all views and provide

new views as needed to completely dimension these changes. Save the drawing as E-18-6.

7. Open drawing E-14-10. Add a 1.25"Ø hole centered on the existing inclined surface 2.60" from the top. Show the axis for the hole perpendicular to the inclined surface and extending through the part. Provide a 1.50" × .50" counterbore centered on the new hole. Show the inclined surface to have a good finish of 63. Show the base with a coarse finish. Update all views and provide new views as needed to completely dimension these changes. Save the drawing as E-18-7.
8. Open drawing Web E-2-9. Stretch the part to be 45 mm long. Specify a straight knurl with a pitch of 14 starting 10 mm from the end of the part and extending up the shaft 25 mm. Update all views and provide new views as needed to completely dimension these changes. Save the drawing as E-18-8.

Use the attached drawings to complete exercises 9 through 15. Assume the grid to be .50". Complete the necessary views to completely dimension all features. Assign at least one threaded feature, three different surface finishes, and at least four geometric control features for each exercise. Save all drawings with the appropriate exercise number.

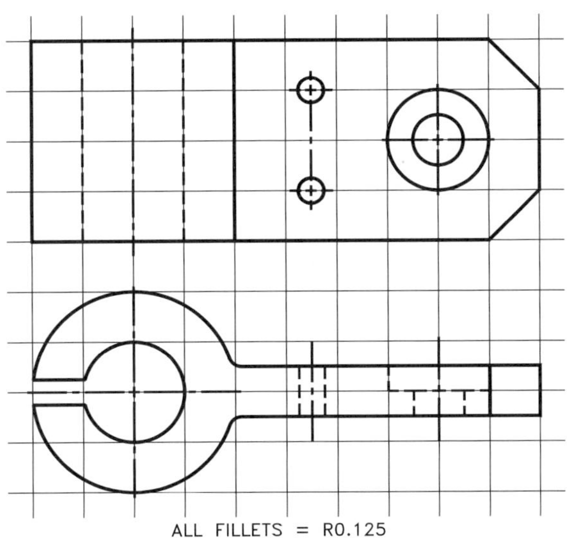

Problem 18.9

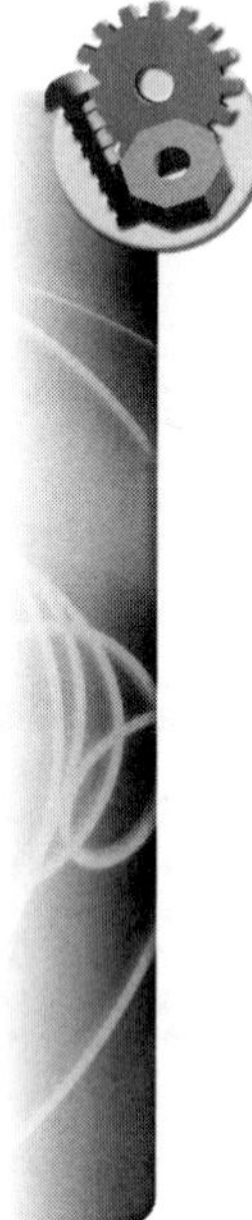

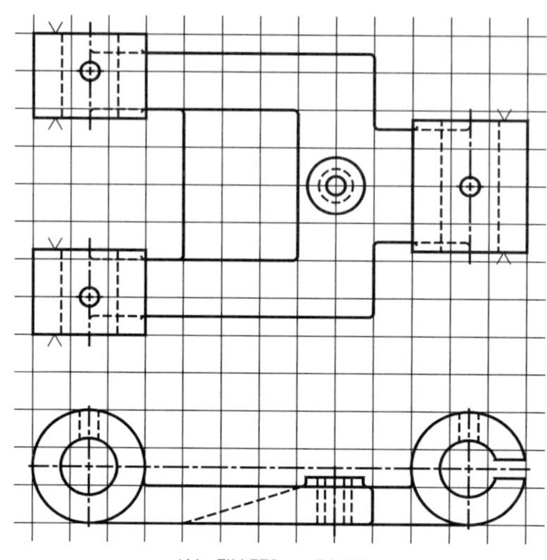

Problem 18.10

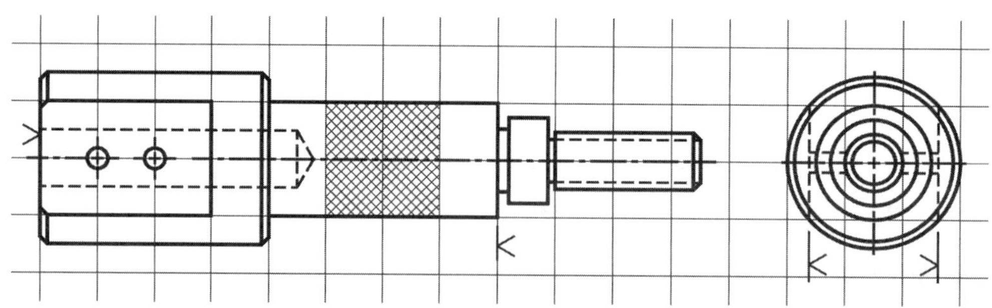

Problem 18.11

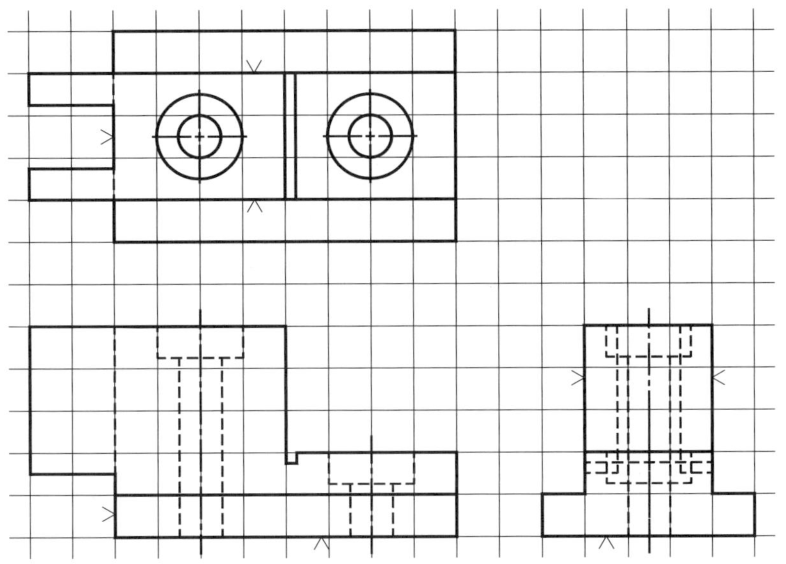

Problem 18.12

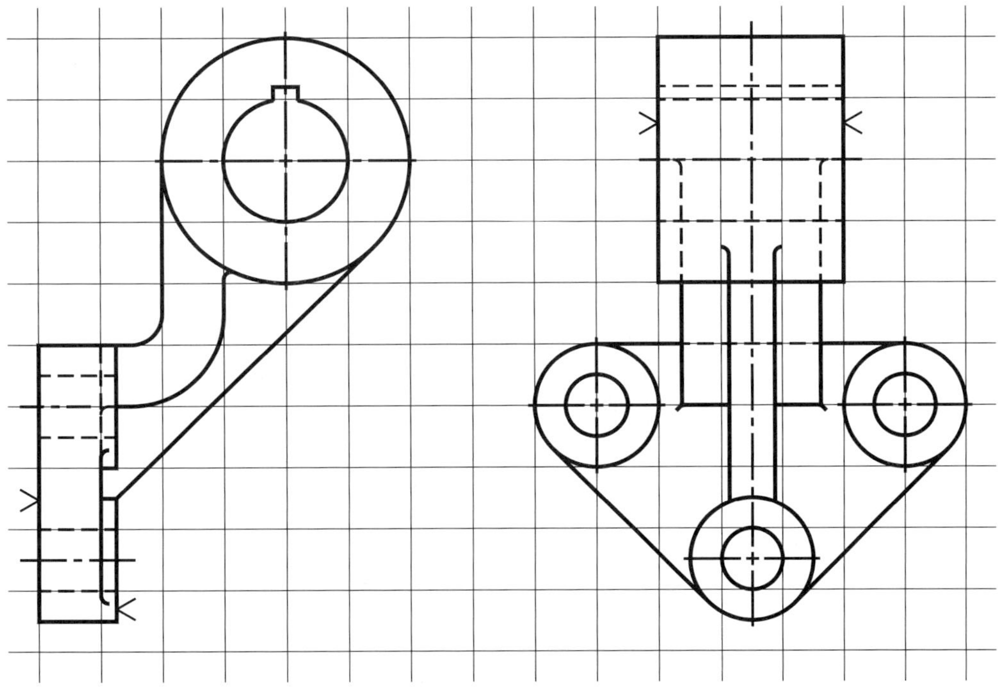

Problem 18.13

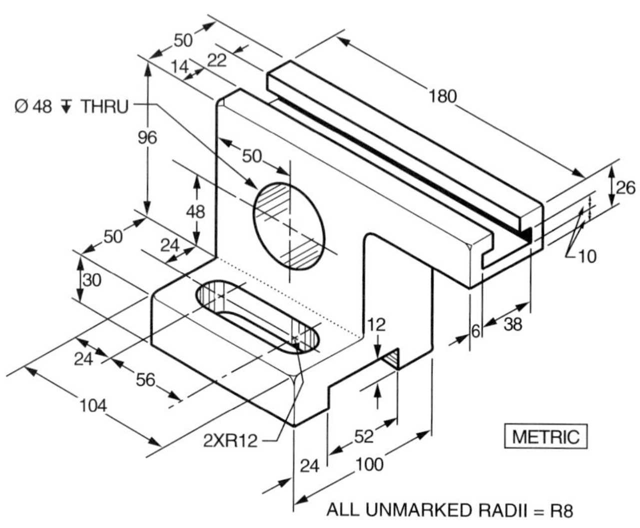

Problem 18.14

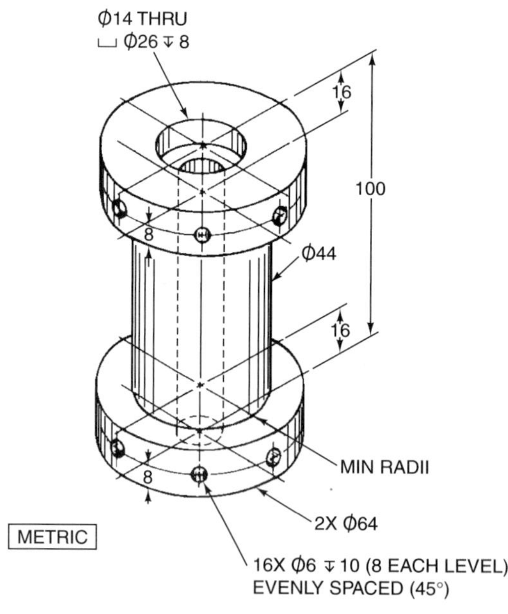

Problem 18.15

CHAPTER 18 QUIZ

1. Explain the difference between a drilled and a reamed hole.
2. Use two views to sketch a drilled featured that does not go through a feature using proper line quality.
3. What is the major difference between a counterbore and a countersink?
4. Sketch an example of a drilled feature in plan view and show the correct relationship of the centerline to the hole.
5. Draw a sketch to describe a 2" diameter hole that is .75" deep and centered in a 2" × 3" × 2" deep block. Add the required notation to completely locate and describe all features.
6. A bolt needs to be placed through a curved feature in such a manner that the bolt head will be on a flat surface. What are the options to provide even distribution of stress on the bolt head?
7. Describe and sketch the difference between a counterbore and a spotface.
8. Sketch and briefly describe the difference between a countersink and a counterdrill.
9. List and describe the types of lines required to describe a threaded hole.
10. What is the angle of incline of a counterdrill? For a Counterbore? For a countersink?
11. What is the purpose of a lug?
12. Sketch and describe a keyseat, a keyway, and a key.
13. Describe how the bottom of a drilled feature should be drawn.
14. A bolt will be used to bond two parts together. One of the parts will need its position altered. What would you recommend for easing the attachment and movement of the part?
15. What is the difference between a t-slot, a kerf, and a dovetail?
16. Sketch the symbols used to describe depth.
17. Describe a neck as it relates to mechanical drawing.
18. What can be done to make a part less slippery?
19. What is the purpose of a spline?
20. A drilled feature combining a counterbore and a countersink is a ______________ .
21. A hole extends completely through a feature. What is the best way to describe the depth of the hole?
22. Explain the difference between a boss, a pad, and a spotface.
23. What is the drawback to showing all features in each view and how can the problem be overcome?

24. What should be drawn to represent a knurled surface?
25. Explained aligned projection.
26. What is the difference between a chamfer and a countersink?
27. Where is a standard finish for a drawing produced using inch dimensions?
28. The top of a part is to have a surface finish of 32. Where should the symbol be placed?
29. Provide a sketch to explain the difference between surface roughness and surface waviness.
30. You're working on a metric drawing and need to specify a very fine finish. What symbol will be needed?

Sectioning and Hatching

Chapter 6 explored the process for making multiview drawings, and Chapters 16 and 17 introduced the dimensioning process for common machined features. The use of sections will expand on your multiview skills and make dimensioning placement easier. This chapter will explore

- The concepts involved in creating a section
- The ANSI standards that govern sections
- Planning for the placement of a cutting plane line
- Considerations in placing section lines
- Common types of sections found in mechanical drafting
- Methods of representing various materials using different line or shape patterns, referred to as hatch patterns

Commands to be introduced in this chapter include

- **Bhatch**
- **Region**
- **Hatchedit**

CREATING SECTION VIEWS

A section, or sectional view, is a view created to describe the internal features of a part that would otherwise be represented by hidden lines in one of the standard orthographic views. The sectional view theoretically removes a portion of the object to reveal internal features. Because part of the exterior has been removed, internal features can be represented using object lines rather than hidden lines. Because the internal features can now be seen and are represented with object lines, dimensions can be referenced to these features that otherwise could not have been placed when represented by hidden lines. The view on the left side of Figure 19.1 shows an example of a part that has interior features represented by hidden lines. The right side of Figure 19.1 shows an example of the same part that has been sectioned.

Drawing and placement of sections is regulated by the American National Standards Institute. *Lines Conventions and Lettering* (ASME Y14.2M) and *Multi and Sectional View Drawings* (ANSI Y14.3) are the specific documents that set the standard for sectional views. Both will be used as the basis for the information presented in this chapter.

PLACING THE CUTTING PLANE

Planning where the 'break' will occur is the first step in creating a section view. A cutting plane, introduced in Chapter 5, is used to represent where the imaginary break will occur. Figure 19.2 shows a top view with a cutting plane. This symbol represents a cutting plane passing through the part, removing the material below the cutting plane. The section that will result can be seen in Figure 19.3. Material that lies between the viewer and the cutting plane is removed, allowing interior features that lie beyond the cutting plane to be exposed. Figure 19.4 shows the section that would result from the cutting plane in Figure 19.2. The section is projected from the view containing the cutting plane, using the same projection methods that are used to create an orthographic projection. Other guidelines for placing the cutting plane include the following:

- On very simple parts similar to Figure 19.5, the cutting plane can be omitted if its location is obvious.

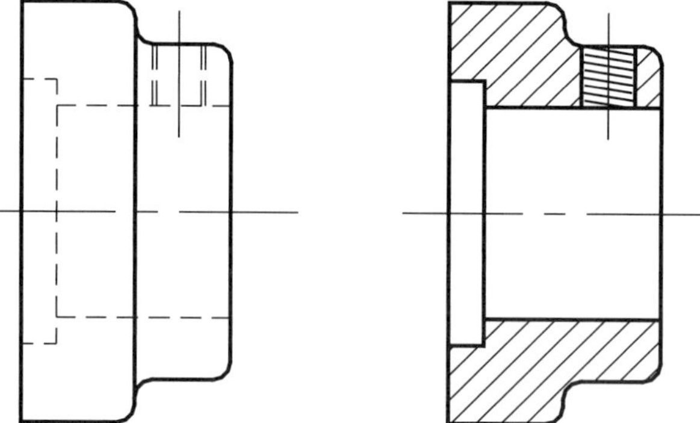

Figure 19.1 *A section can be used to replace one of the six orthographic views that may contain hidden features. With the use of a section, the hidden features are represented with dimensions, allowing dimensions to be referenced to the feature.*

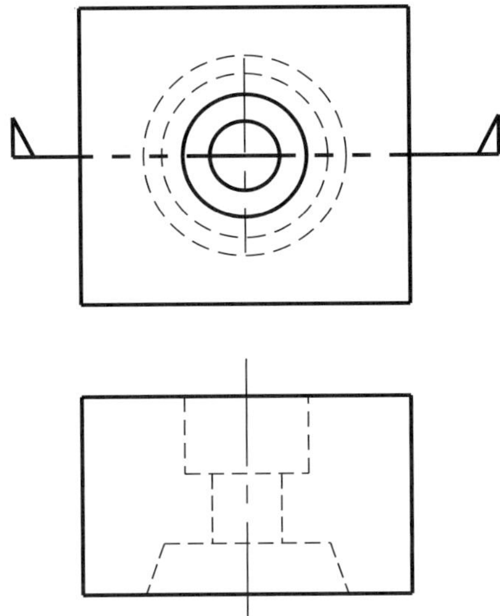

Figure 19.2 *A cutting plane is used to represent where the part will be sectioned. The arrows on the cutting plane indicate the direction of sight.*

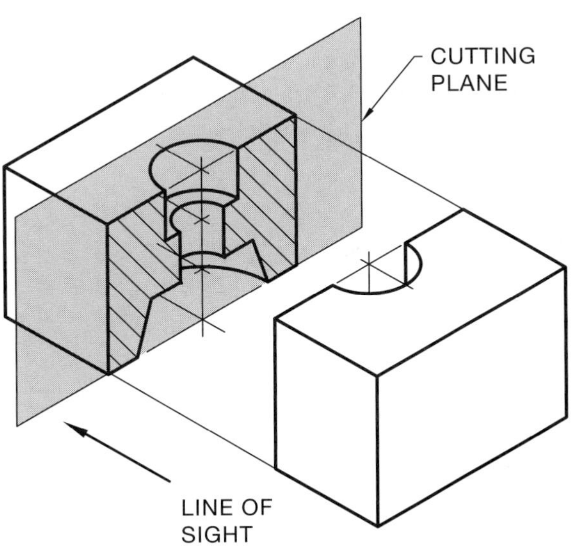

Figure 19.3 *An isometric representation of the results of the cutting plane in Figure 19.2.*

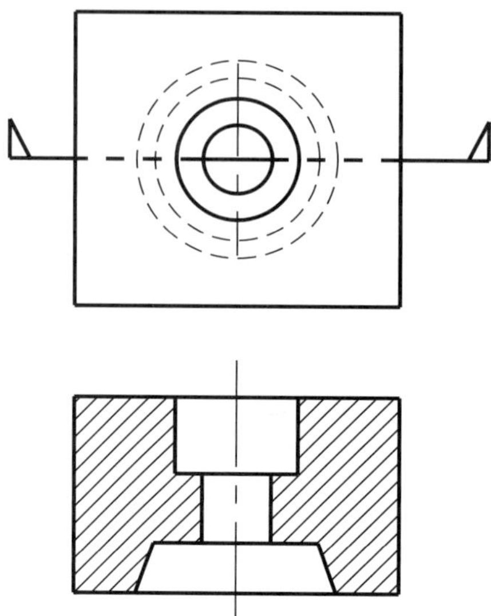

Figure 19.4 *The section that results from the cutting plane in Figure 19.2. Notice that the features that the cutting plane passes through are now represented by object lines instead of hidden lines.*

- On more complex parts, the cutting plane is represented, but it can be simplified similar to Figure 19.6.
- If more than one section is required, each cutting plane should be shown and labeled as shown in Figure 19.7, using the guidelines presented in Chapter 5.

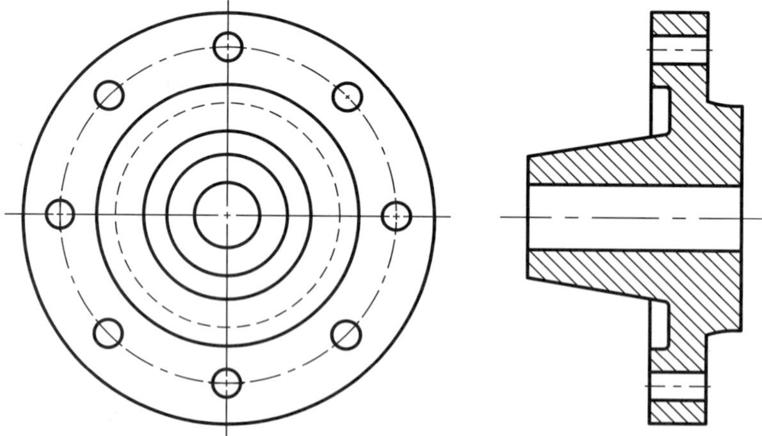

Figure 19.5 *The cutting plane can be omitted when the location is obvious.*

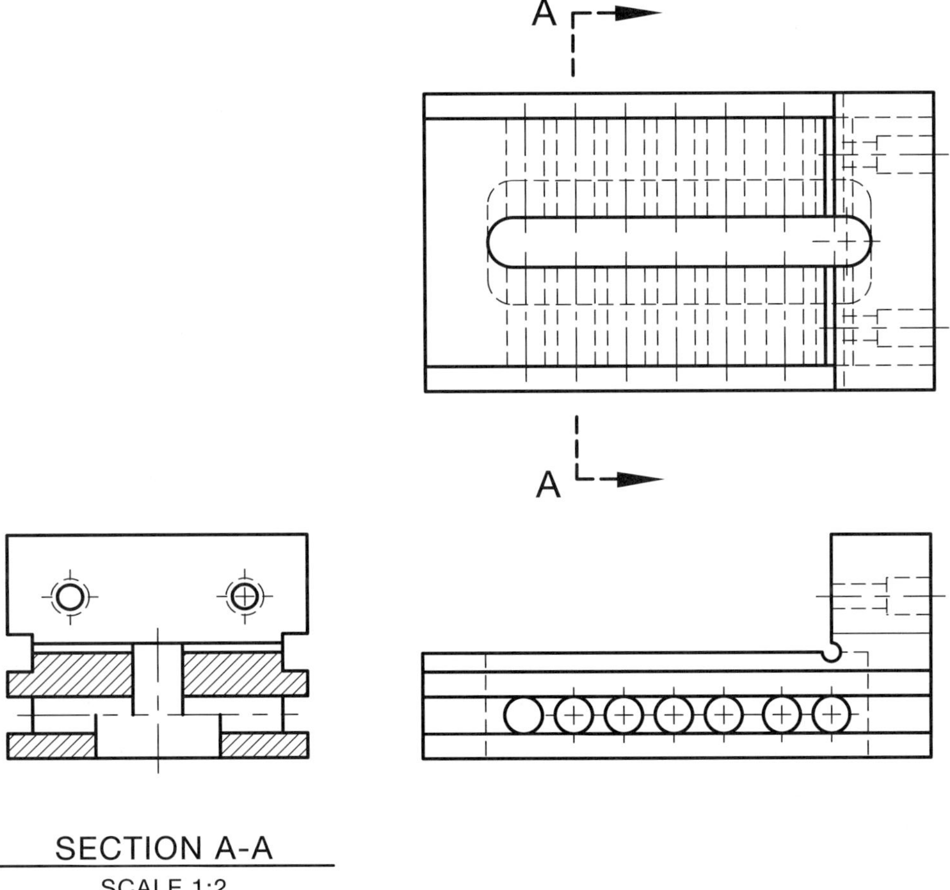

Figure 19.6 *A simplified cutting plane that does not extend through the part can be used to denote the location of the section. Courtesy Hanh Tran.*

- When shown in circular parts, the cutting plane will replace the centerline.
- If space is limited and the section must be moved from its projected position, the cutting plane must always be shown.

Non-Sectioned Features

Some features are left unsectioned even if the cutting plan passes through the object. Objects that are not sectioned include bolts, bearings, gear teeth, keys, pins, nuts, ribs, rivets, screws, spokes, and webs.

- If the cutting plane passes through and is parallel to one of these features, the feature is not sectioned. See Figure 19.7.

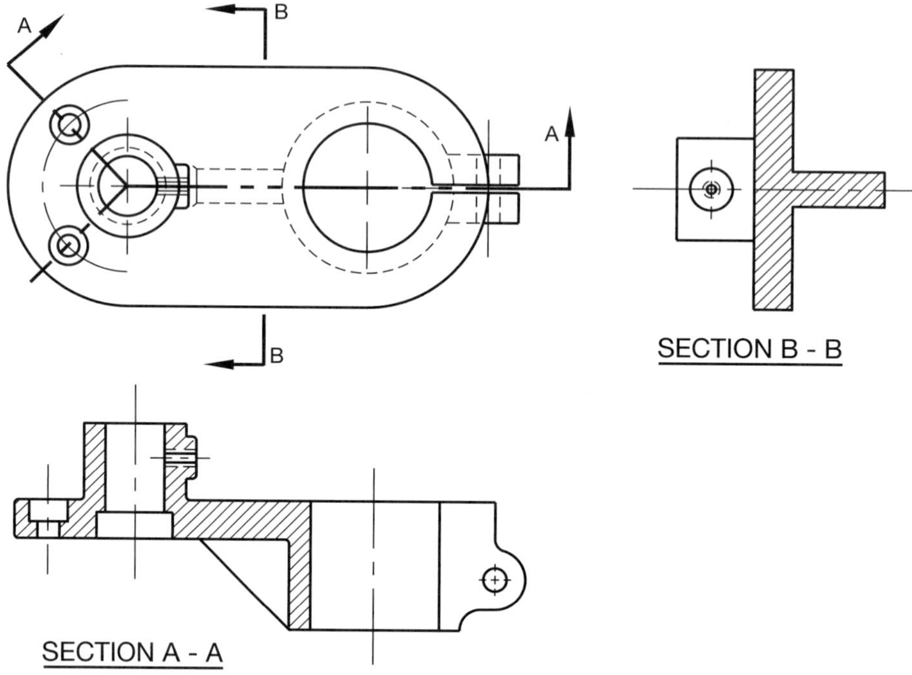

Figure 19.7 *Each cutting plane must be identified by a letter when more than one cutting plane is used.*

- If the cutting plane passes through and is perpendicular to one of these features, the feature is sectioned. See Figure 19.7.

Conventional Revolutions

When the true projection of a feature will result in the distortion of a feature, the feature should be revolved onto a plane perpendicular to the line of sight. Figure 19.8 shows the projection of a hole with true projection and revolved projection.

PLACING SECTION LINES

Section lines are thin lines used to indicate the portion of the part that the cutting plane has passed through. Common requirements for placing section lines include the following:

- Section lines should be drawn at a 45° angle but should not be parallel or perpendicular to any part of the object that has been sectioned.
- If an angle other than 45° is used, it should be between 15 and 75°.
- If two adjacent parts have been sectioned, section lines should be placed at opposing angles.

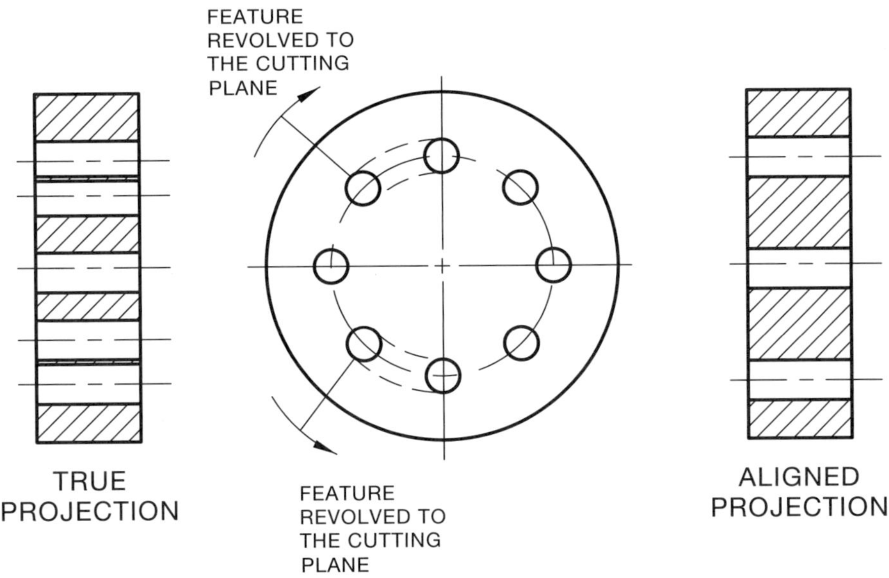

Figure 19.8 *Aligned projections rotate features to be parallel to the cutting plane, and then into the section view. Courtesy Michael J. Easley.*

- Although spacing between section lines can be altered depending on the size of the part that is sectioned, a spacing of .125" should be provided between section lines.
- An object line should never be placed through sectioned material.
- Hidden features that lie beyond the cutting plane should never be represented in the section view.

Figure 19.9 shows an example of placing section lines using a pattern named ANSI31. Methods of placing section lines using AutoCAD will be presented later in this chapter. As you explore the subject of sectioning, it is important to understand that the power of AutoCAD far exceeds what is usually needed for a mechanical drawing. Equally spaced lines placed at a 45° angle to the part are the most commonly used section lines, but specific patterns are available to represent various materials. When specific patterns are represented, equally spaced lines represent cast iron. Figure 19.10 shows an example of using coded section lines to represent aluminum magnesium. Common practice is to use an AutoCAD pattern ANSI31 to represent most materials and then specify the required material in the title block.

Note: Unless specifically told by your instructor or supervisor, use the simplified section lines and describe the material specified in the title block.

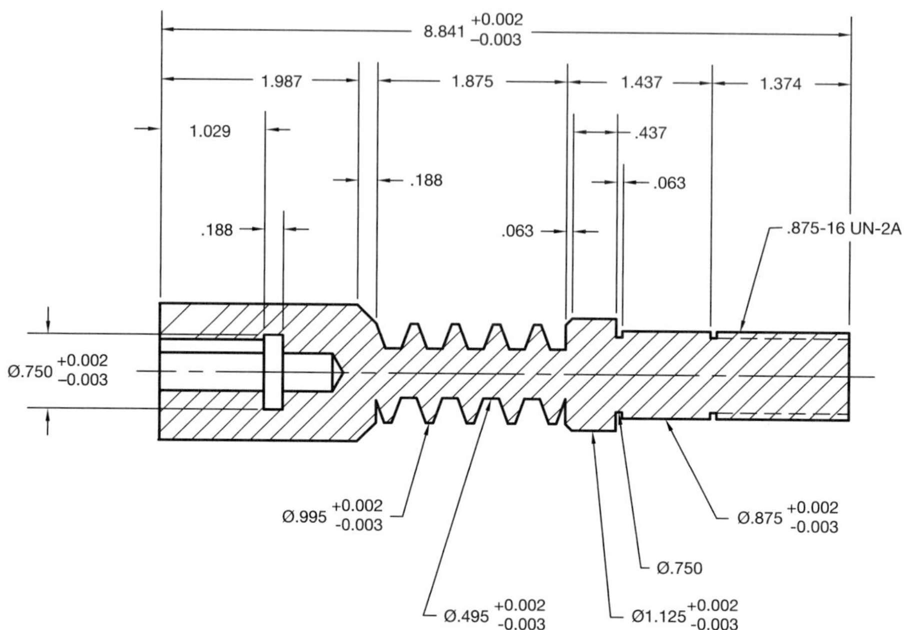

Figure 19.9 *Thin parallel lines are used to represent material cut by the cutting plane. Courtesy Michael J. Easley.*

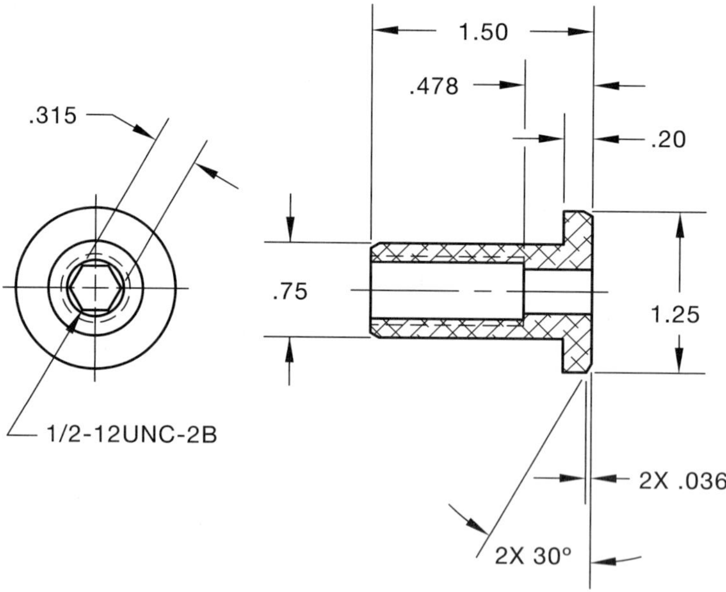

Figure 19.10 *Coded material patterns can be used to represent specific materials. This pattern represents aluminum magnesium.*

TYPES OF SECTIONS

The type of section to be drawn will depend on the amount of information that needs to be conveyed to the print reader. A key consideration in deciding the type of section to be drawn is to determine, by use of a sketch, which dimensions cannot be placed on the normal orthographic views. Once the missing dimensions have been determined, the type of section needed to provide the missing dimension can be determined. Common types of sections include full, half, offset, aligned, revolved, removed, and broken out.

Full

A full section results when the cutting plane passes through the entire object. Figure 19.11 shows an example of a full section. When a full section is drawn, it replaces one of the six orthographic views. In this example, the front view was not drawn, and the section was created instead.

Half

A half section is used with symmetrical parts. An L-shaped cutting plane is used to represent the material to be removed. Although approximately one quarter of the object is removed, it is called a half section because the cutting plane passes through half of

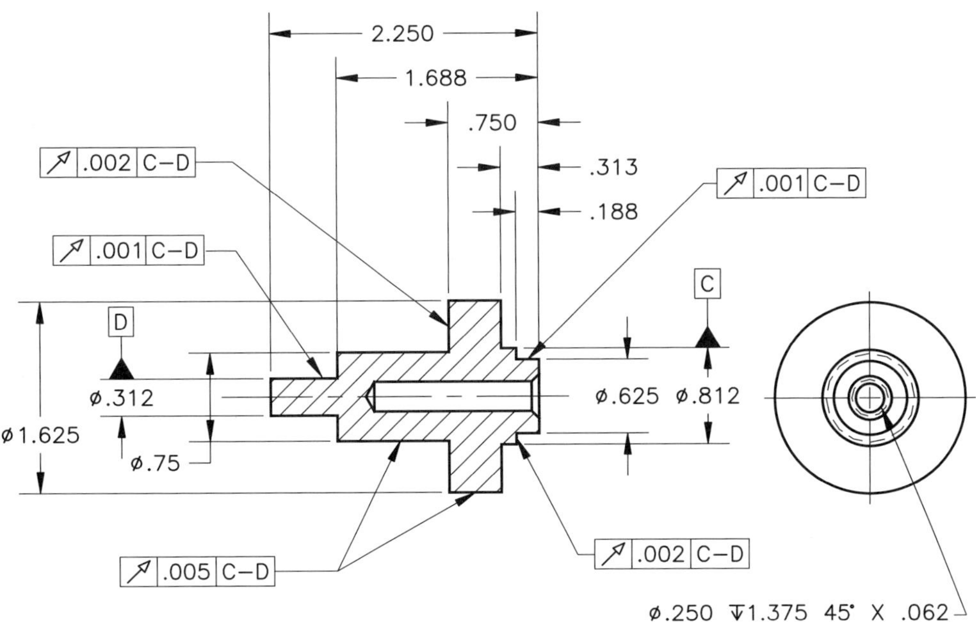

Figure 19.11 *A full section results when the cutting plane extends through the entire feature. Because this is a simple feature, the cutting plane is assumed to lie along the vertical centerline. Courtesy Cathy Schlesser.*

the object. Figure 19.12 shows the theory of a half section. The advantage of a half section is that it shows half of the view in section and half of the view as it would normally appear. Hidden lines are omitted from the normal portion of the view, because they can be seen clearly in the sectional view. To maintain clarity, hidden features that are beyond the cutting plane are never represented in the section view. If an object remains hidden, a second section will be required.

The object from Figure 19.11 could have been represented using a half section. Because the part is symmetrical, dimensions for the interior features can be placed from a centerline. If tolerance build up is critical, a full section would be better, because parts could be dimensioned from edge to edge, rather than from center to edge.

Offset

An offset section is similar to a full section in that it passes all the way through the object, but the cutting plane is staggered to allow viewing of objects that are at different distances from the outer edge of the object. Figure 19.13 shows an example of

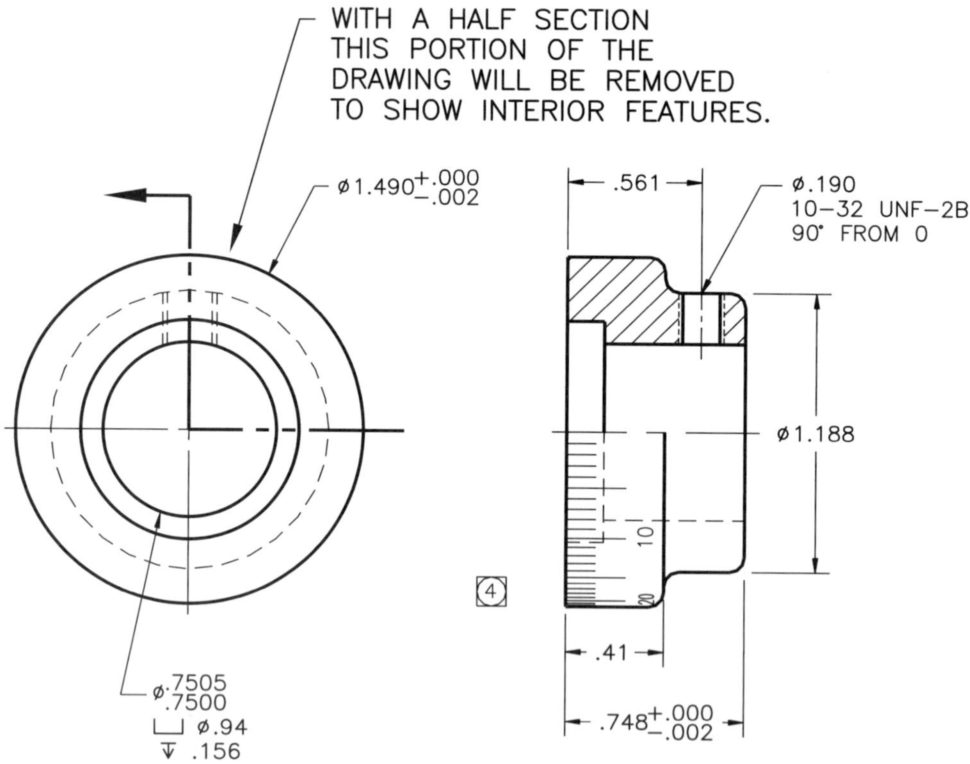

Figure 19.12 *A half section results from passing the cutting plane through half of the feature, and removing one quarter of the object. Courtesy Paul Bissell.*

an offset section. New drafters are often confused when they represent the results of the cutting plane in an offset section. Remember that the break is imaginary.

Note: Do not place an object line in the section view to represent the offsets in the cutting plane.

Aligned

An aligned section uses a staggered cutting plane to pass through staggered features on circular parts. It is similar to an offset section in that it has a cutting plane with an offset, but the cutting plane is placed at an angle of less than 90°. Figure 19.14 shows an example of an object with an aligned cutting plane. A key to drawing an

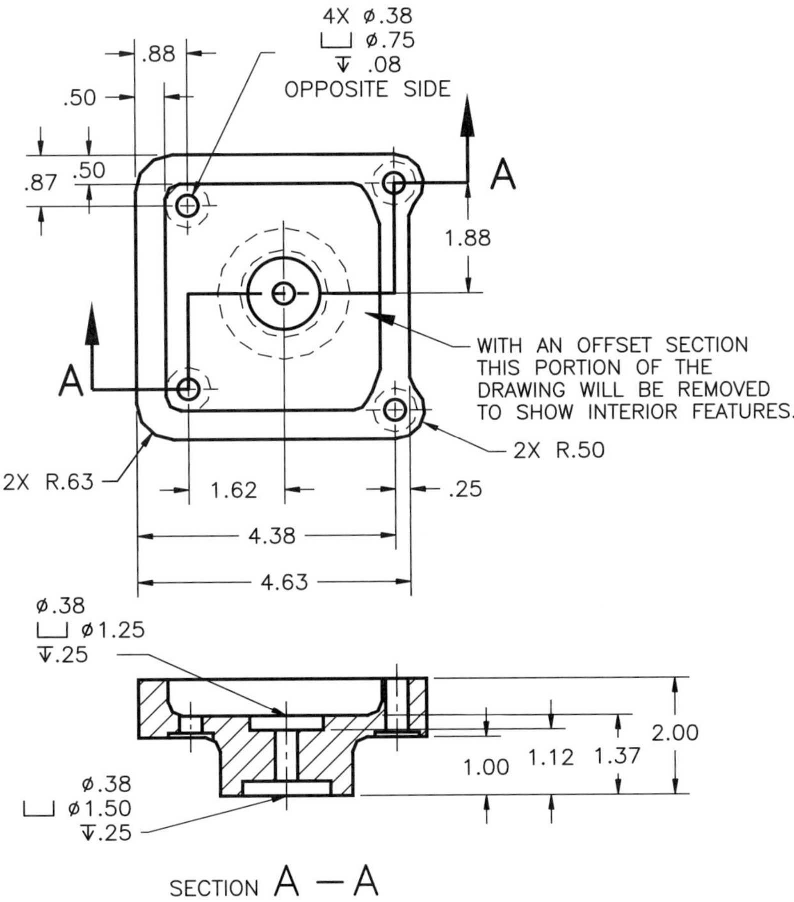

Figure 19.13 *An offset section is similar to a full section, but the cutting plane is staggered to pass through internal features. Courtesy Paul Bissell.*

aligned section correctly is that the plane is rotated to a plane that is perpendicular to the line of sight, and then projected.

Revolved

A revolved section is used to show an internal shape that remains constant, but that cannot be seen in one of the six orthographic views. The handle of a wrench is an example of when a revolved section would be useful. A revolved section is created within the part, and revolved 90° to reflect the interior profile. Figure 19.15 shows two methods used to display the section. The part can be broken to display the section or can be continuous, as long as clarity is maintained. Breaking the feature around the section usually provides more space to locate dimensions.

Removed

Removed sections are similar to both full and revolved sections. As the name implies, the section is removed from the interior of the feature and displayed as a separate view. Removed sections should be used instead of a revolved section when drawing clarity is hampered due to crowding, or if the section needs to be enlarged to show small details. Removed sections can be drawn at an enlarged scale and used as a detail. If a removed section is used, a cutting plane line is placed in the object to indicate where

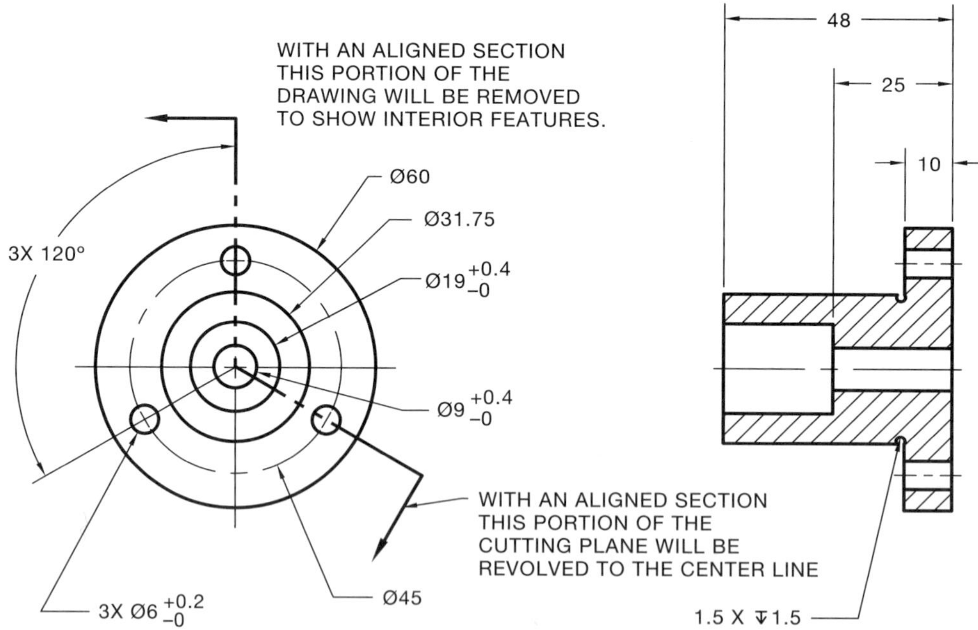

Figure 19.14 *An aligned section is similar to a full section, but is used with circular features to show internal features that do not lie on one of the four quadrants.*

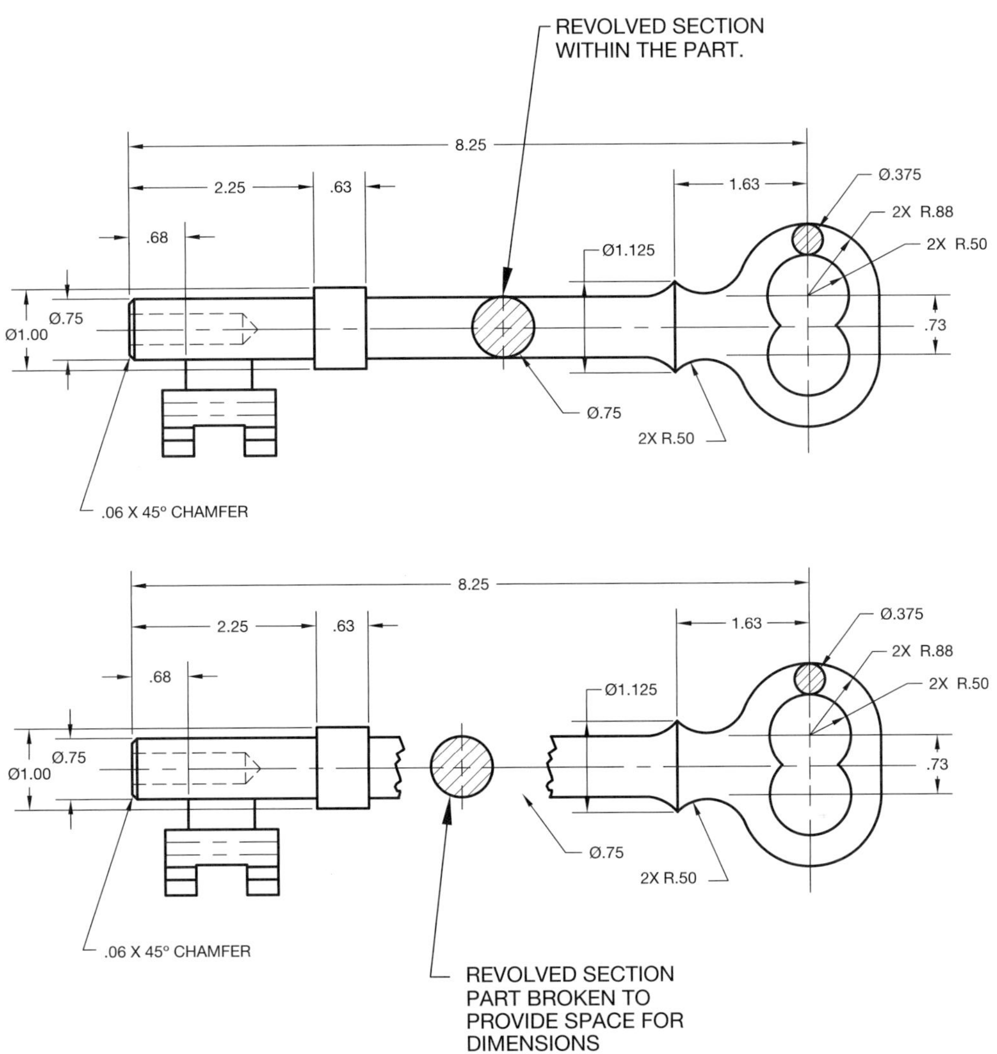

Figure 19.15 *A revolved section can be used to show the shape of a feature that cannot be seen in one of the six orthographic views.*

the part was sectioned. Removed sections should not be placed in one of the normal view locations. If multiple removed sections are drawn, they should be arranged in alphabetical order. Figure 19.16 shows examples of removed sections.

Broken Out

A broken out section is used when a hidden feature needs to be exposed for dimensioning. Rather than using one of the other sectioning methods and creating anoth-

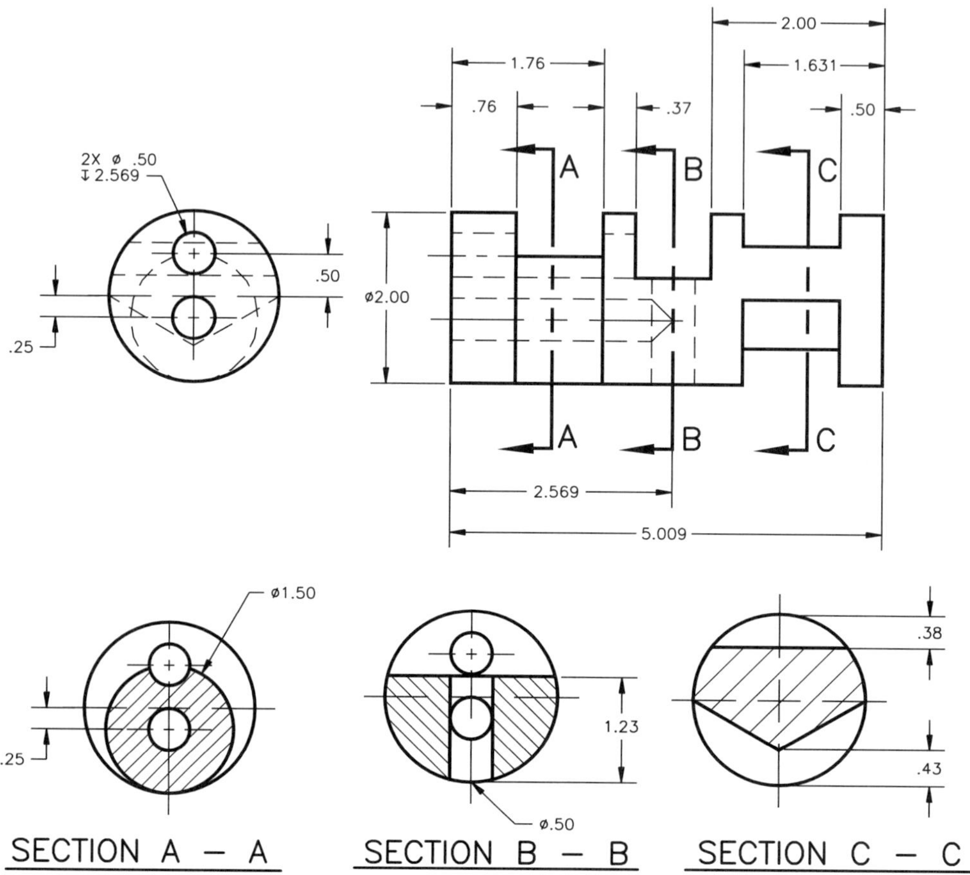

Figure 19.16 *Removed sections combine features of full and revolved sections. Courtesy Paul Bissell.*

er view, the material is removed from around the feature so that hidden lines become object lines and dimensions can be placed. Short break lines are used to indicate where the break occurs, similar to Figure 19.17.

HATCHING METHODS

Hatch patterns are repetitive patterns of lines, dots, and other symbols used to represent a surface or specific material in section. In addition to the many patterns used on manufacturing drawings, a solid pattern can be placed in an object. No matter what the pattern, it should be placed on a separate layer with a name such as HATCH or PATTERN. Hatch patterns are placed using either the **Bhatch** (**Boundary Hatch**) or the **Hatch** command. Although the patterns will look exact-

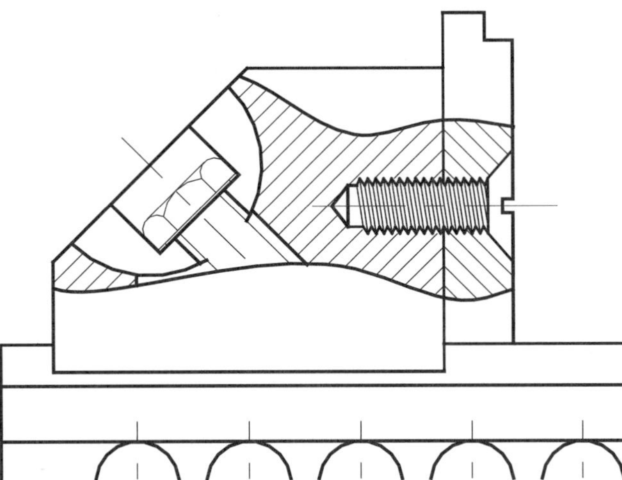

Figure 19.17 *A broken section removes just enough material to expose interior features. Courtesy Cathy Schlesser.*

ly alike, there is a major difference between the two commands. **Bhatch** creates patterns that can be associative or non-associative, while **Hatch** creates only non-associative patterns.

- Associative patterns are linked to their boundaries. If the boundary is altered, the pattern is altered to match the boundary.
- Non-associative patterns are not affected by alterations to their boundaries.

The other major difference between **Bhatch** and **Hatch** commands is in how they are added to the drawing. Boundary hatch patterns are applied using the **Boundary Hatch** dialog box. Display the **Boundary Hatch** dialog box by selecting **Hatch** on the **Draw** toolbar, by typing BH (ENTER) at the Command prompt, or by selecting **Hatch** from the **Draw** menu. Hatch patterns are added to a drawing from the command line. **Hatch** will not be discussed in this text. Consult the **Help** menu if you're interested in using the slower, harder method.

USING THE QUICK TAB

The **Boundary Hatch** dialog box consists of the **Quick** and **Advanced** tabs. The **Quick** tab of the **Boundary Hatch** dialog box is shown in Figure 19.18. The **Quick** tab can be used for placing patterns for many common shapes encountered on mechanical drawings. As you prepare to place section lines using **Bhatch**, you can take several steps to speed up the process. Zooming into an area rather than keeping the entire drawing in view will eliminate the number of lines that must be examined to determine the boundary set that will define the hatch area. Speed can also be gained if extra layers

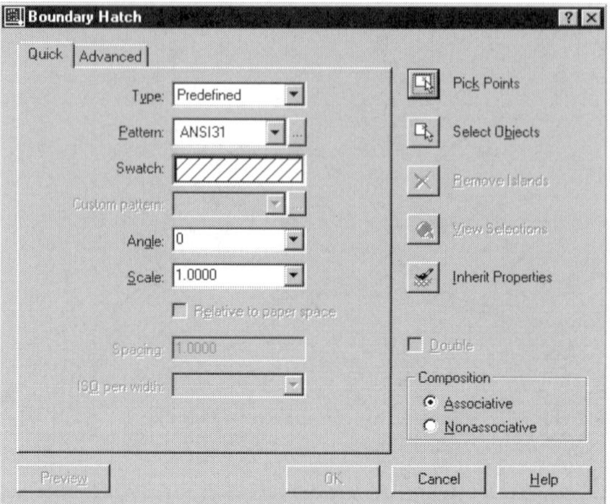

Figure 19.18 *Start The* **Bhatch** *command by selecting* **Hatch** *on the* **Draw** *toolbar, by typing* **BH** *(ENTER) at the Command prompt, or by selecting* **Hatch** *from the* **Draw** *menu. Each method will produce the* **Boundary Hatch** *dialog box.*

are turned OFF or Frozen so that objects on these layers are not examined as part of the selection set.

PATTERN TYPE

The **Type** edit box is used to set the pattern type. Options include Predefined, User defined, and Custom. The default value is to use a Predefined pattern. Selecting the down arrow displays the listings shown in Figure 19.19. Predefined patterns can be viewed in the **Hatch Pattern Palette** dialog box that will be introduced in the next section of this chapter. For most uses, the predefined patterns will be all that are ever needed. Selecting the User defined option creates a pattern based on the current linetype in the drawing. This option allows the angle and spacing of the pattern to be altered. The Custom option allows a user created pattern to be to be added to the drawing base. Custom patterns allow a drawing pattern to be saved in the acad.pat file or in its own PAT file. Methods of creating patterns will be discussed later.

SELECTING A PATTERN

Notice that the ANSI31 hatch pattern is the default setting shown in the **Pattern** edit box. It should be used to represent most materials. A different pattern can be selected for a **Bhatch**, and it will remain the default until a new pattern is selected. You can select a new pattern using the **Pattern** edit box using three different methods. Selecting the pattern name in the edit box or the edit box arrow will display a listing of the hatch pattern names similar to the list shown in Figure 19.20. The slide bar can be used to scroll through the list of predefined patterns. Clicking the name in the pattern box a second time will close the list without making any changes. Selecting a name

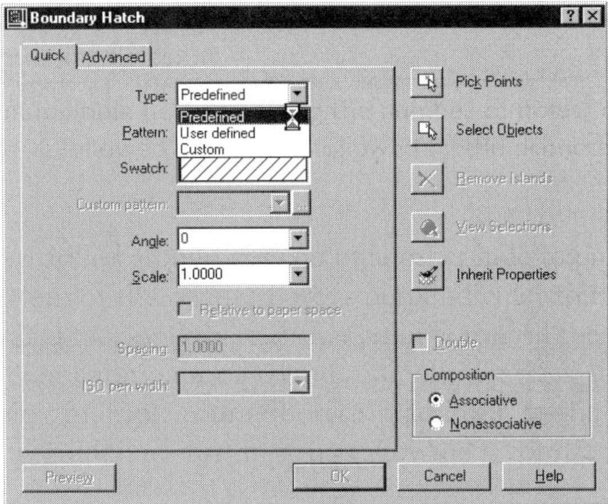

Figure 19.19 *Selecting the arrow beside the* **Pattern** *edit box will display options for selecting the pattern type.*

from the list will make that pattern the current pattern and close the list. This method of pattern selection works well if you are familiar with the names of patterns.

If you're a new user, names like ANSI31 or ANSI32 will be meaningless. Select the **Pattern [...]** button beside the edit box to see a display of each image in the **Hatch Pattern Palette** dialog box. See Figure 19.21a. The **ANSI** and **Other Predefined** (see Figure 19.21b) tabs provide patterns associated with a wide variety of drawings. The slide bar can be used to move through the listing to select the desired hatch pat-

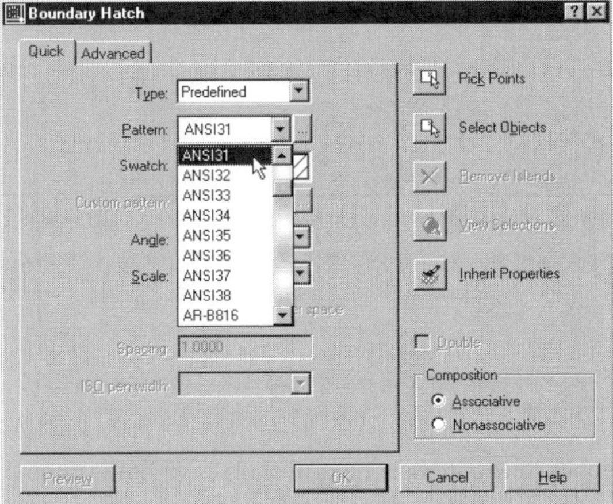

Figure 19.20 *The* **Pattern** *list can be used to select the pattern to be used for the* **Bhatch** *command.*

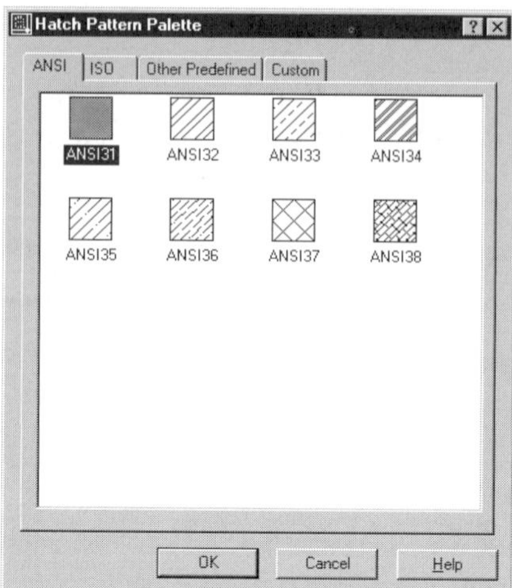

Figure 19.21a *Selecting the* **Pattern [...]** *button beside the* **Pattern** *edit box will display the* **Hatch Pattern Palette** *dialog box.*

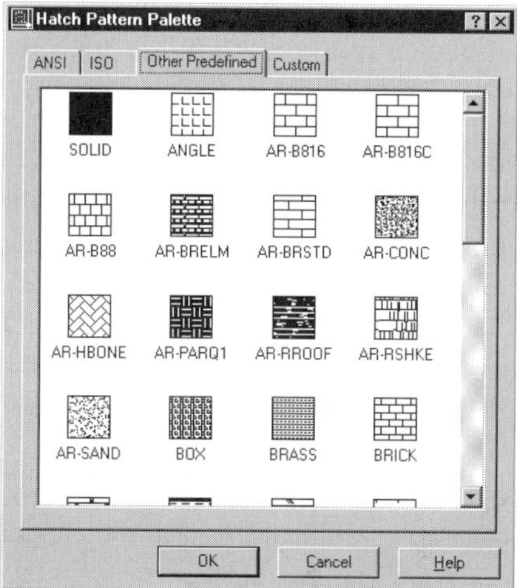

Figure 19.21b *The* **Other Predefined** *tab provides patterns associated with common types of drawings. The slide bar can be used to move through the palette to select the desired hatch pattern.*

tern. Once a pattern is selected for use in a drawing, click the **OK** button to return to the **Quick** tab.

A third method of selecting a hatch pattern is to use the **Swatch** display. Clicking the **Swatch** pattern will display the most recently used tab of the **Hatch Pattern Palette** dialog box.

PATTERN ANGLE

Once the pattern has been selected, the **Angle** edit box should be considered. By default, the pattern will be placed at an angle of 0. The pattern will be reproduced as shown in the **Swatch** display and in the **Hatch Pattern Palette** dialog box. You can enter a value by keyboard or select a value from the Angle list shown in Figure 19.22. Two important points to remember when placing the hatch pattern are as follows:

- Most section lines should be drawn at a 45° angle but should not be parallel or perpendicular to any part of the object that has been sectioned.
- If an angle other than 45° is used, it should be between 15 and 75°.

Entering a value other than zero allows the hatch pattern to be rotated. By entering an angle of 45°, you can use pattern ANSI31 to place vertical lines.

PATTERN SCALE

The **Scale** option sets the size of the pattern elements with the default of one drawing unit. The relationship of the pattern can be controlled by the drawing scale factor. Adjust the scale value using the same method as used to adjust the angle value, either by keyboard entry or by choosing from the **Scale** list. Figure 19.23 shows an example of three differently scaled patterns. Selecting the largest scale pattern that is still visually appealing will save **Regen** time. Drawing a line the length of the desired pattern to be inserted can help to ease the scale selection. Later in the command process, you'll have a chance to preview how the pattern will appear.

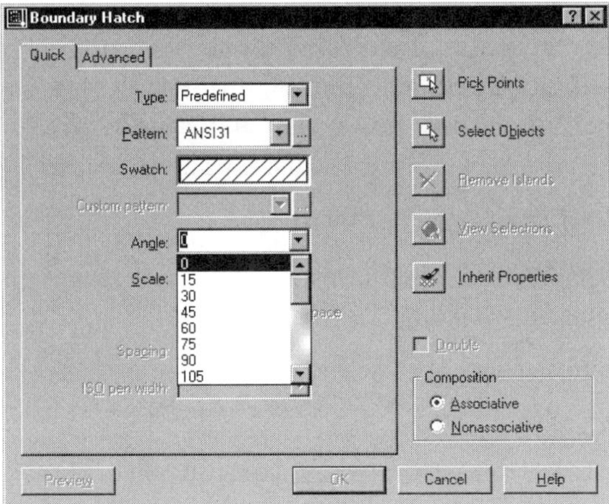

Figure 19.22 *A value for rotating the pattern can be entered by keyboard or by selecting a value from the Angle list.*

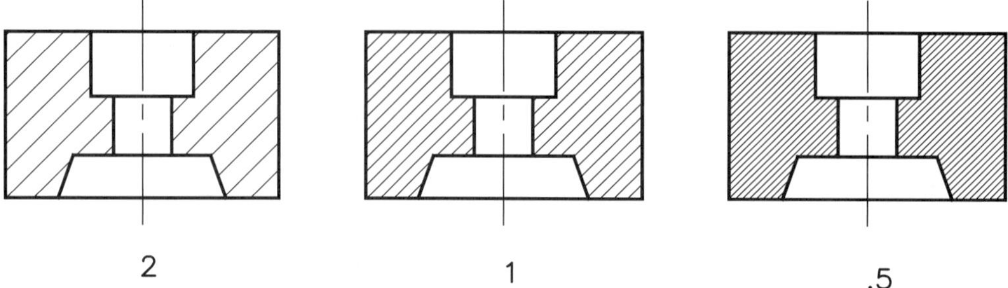

Figure 19.23 *The effect of altering the scale factor. Always use the largest size possible to save Regen time.*

SELECTING OBJECTS FOR HATCHING

Once the pattern, angle, and scale have been selected, use one of the methods from the right side of the dialog box to choose the area to be hatched. For now, the options are limited to **Pick Points** and **Select Objects**.

- If the **Pick Points button** is selected, only one point will need to be selected to place the pattern.
- If the **Select Objects** button is selected, any of the selection methods introduced in previous chapters can be used to define the area.

Other methods of selecting the boundary can be selected using the options on the **Advanced** tab, which will be discussed later in this chapter.

Using the Pick Points Option

The **Pick Points** option can be used to hatch drawing objects that form an enclosed boundary. Once you've selected the **Pick Points** button, the dialog box will be removed and the drawing will be restored. A prompt is now displayed requesting "Select internal point." Select any point that lies within the area to be hatched. Once it is selected, the command line will display the prompt:

Select internal point: Selecting everything ...

Selecting everything visible…

Analyzing the selected data...

Analyzing internal islands…

Select internal point: ENTER

Because the Pick Points button was selected, you can select the object to be hatched by selecting one point that lies within the object. As the point is selected, the boundary of the area to be hatched will be displayed as dashed lines. AutoCAD is waiting for you to select additional objects to be hatched. Press ENTER or right-click to redisplay the Boundary Hatch dialog box. Press ENTER a second time to apply the hatch pattern to the selected object. Figure 19.24 shows the resulting hatch pattern.

Using the Select Objects Option

Selecting the **Select Objects** button will remove the **Boundary Hatch** dialog box and prompt you to select objects. Figures 19.25a and 19.25b show the process for selecting an object to be hatched using the **Select Objects** option. The object was selected using the Window selection method. The right side shows the hatch pattern that would result. The object to be hatched does not need to form a closed boundary, or it can contain an internal area, referred to as an island that will not be hatched. Figure 19.26 shows the result of selecting an object with an island using the Window selection method.

If the object to be hatched is surrounded by a polyline, the object can be selected by picking the object at any point, as seen in Figure 19.27. Once the object to be hatched is selected, press ENTER to restore the **Boundary Hatch** dialog box. Pressing ENTER a second time will apply the pattern.

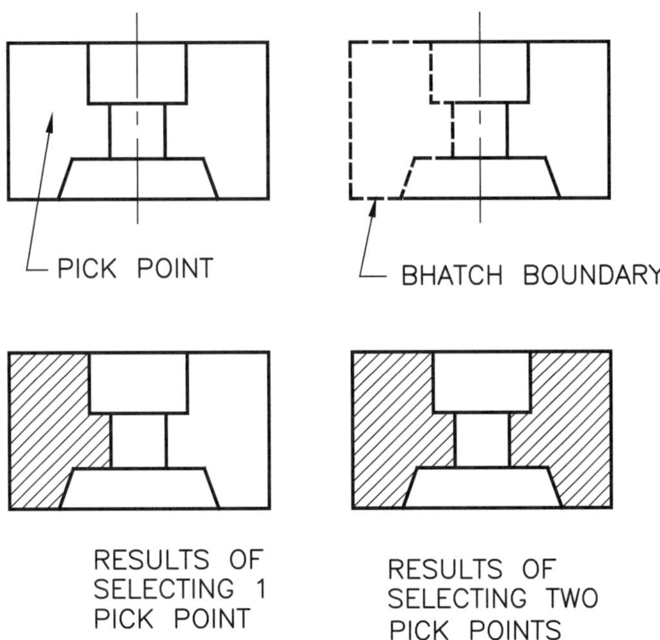

Figure 19.24 *Using* **Bhatch** *to add a pattern to an object. Once the area to receive the pattern is selected, the boundary will be highlighted. The ANSI31 pattern at a scale of .75 and an angle of 0 was selected.*

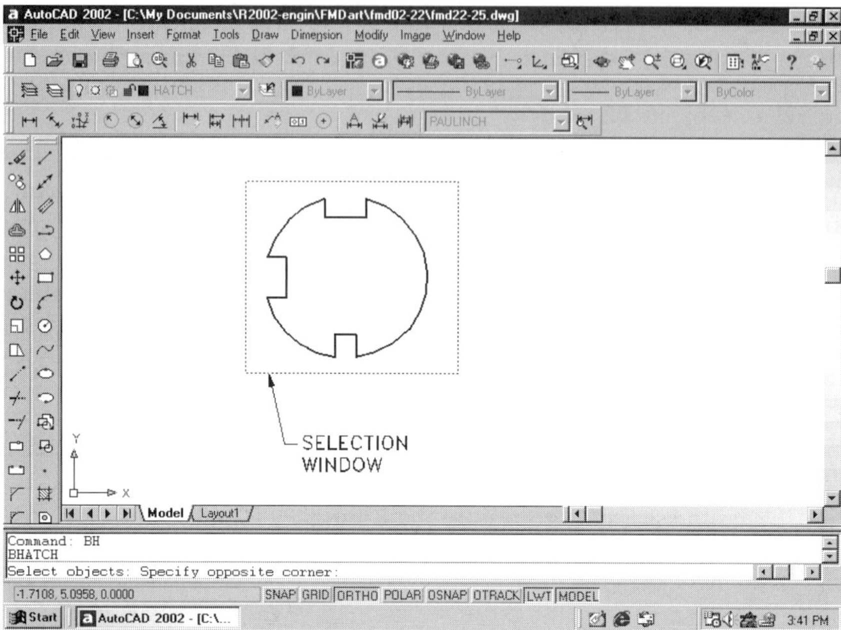

Figure 19.25a *With the* **Select Object** *option, an object can be selected using any selection method, such as Window or Crossing.*

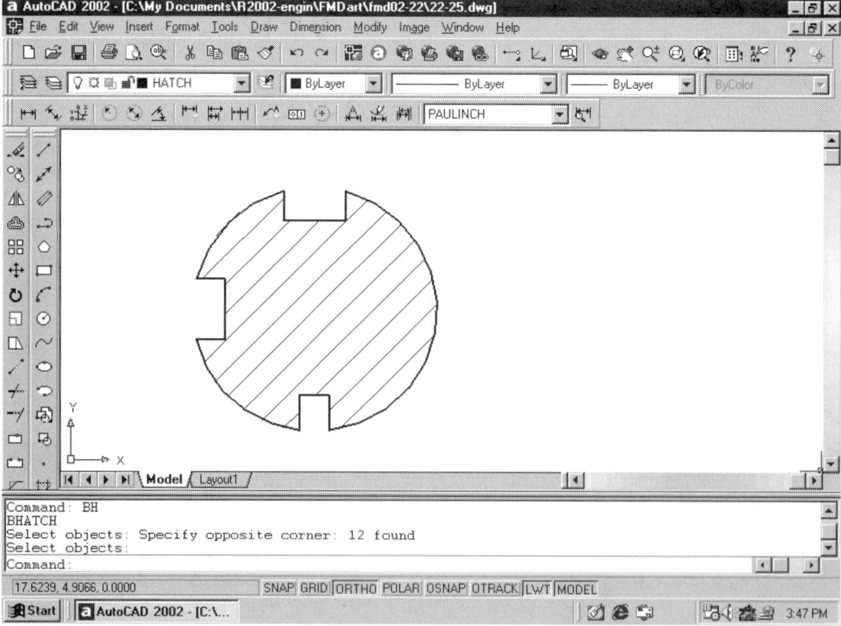

Figure 19.25b *The resulting hatch pattern for the selection set in Figure 19.25a.*

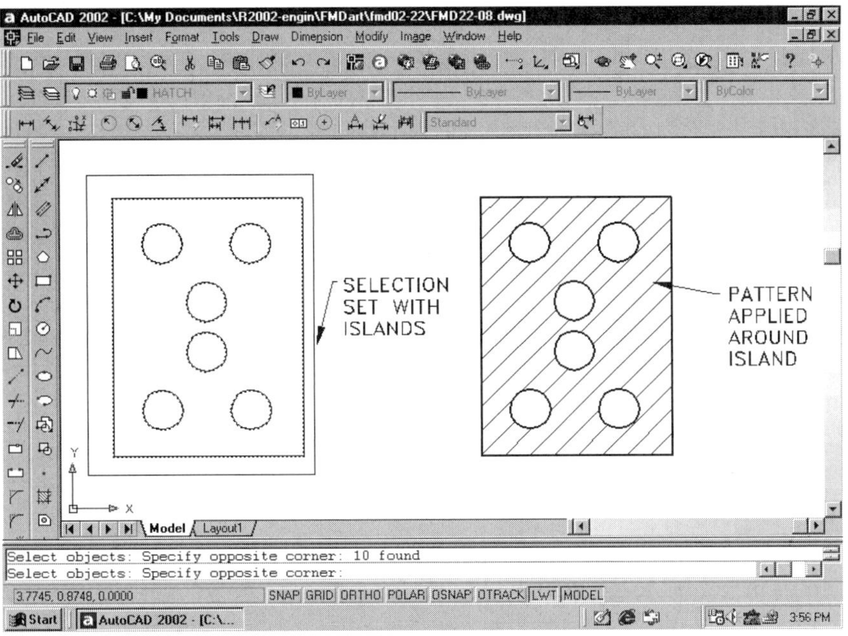

Figure 19.26 *The result of using the* **Bhatch Select Object** *option on an object with an island. The placement of the pattern can be altered with the* **Style** *option.*

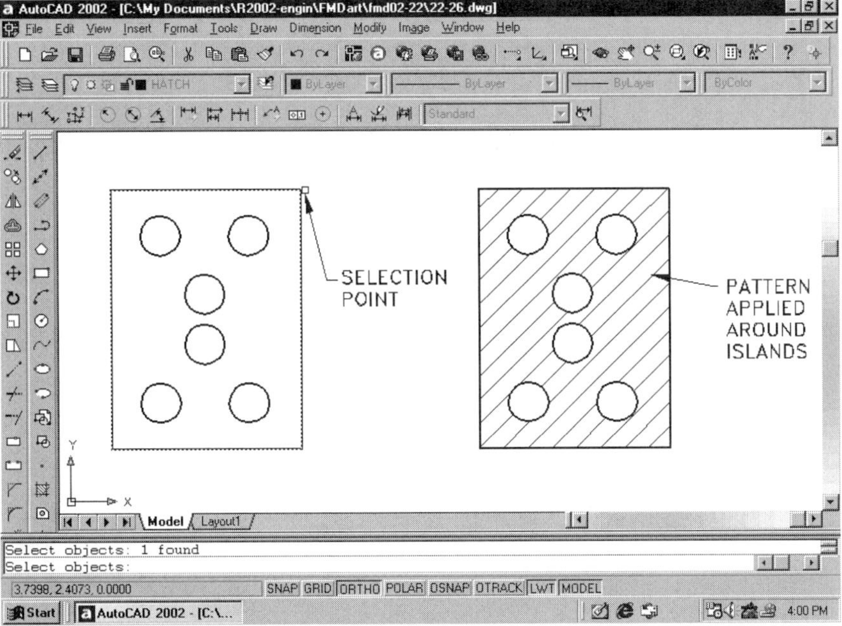

Figure 19.27 *When the boundary of an object is formed by a polyline, it can be selected by single point.*

USING THE PREVIEW OPTION

If you're unsure of any of the pattern parameters, select the **Preview** button prior to applying the hatch pattern. The **Preview** option provides a display of how the pattern will appear based on the current settings. **Preview** is especially helpful when a large, irregularly shaped boundary is to be hatched. When the **Preview** option is selected, the dialog box is removed and the object with the hatch pattern is displayed, as seen in Figure 19.28. Notice that the boundary is still dotted, which is a reminder that the hatch pattern has not yet been added to the drawing file. Press ENTER or right–click to restore the **Boundary Hatch** dialog box. If the pattern is acceptable, click the **OK** button or press ENTER. If the preview did not display the intended results, any of the pattern options can be altered. In Figure 19.29, a new hatch pattern was selected by returning to the **Boundary Hatch** dialog box. The **Preview** option was selected. Using **Preview** to test alternatives can go on indefinitely. The final change is to adjust the **Angle**. The result can be seen in Figure 19.30. Because this view was acceptable, ENTER was used to return to the **Boundary Hatch** dialog box, and the **OK** button was selected.

A Summary of Steps in Placing a Hatch Pattern

To place the hatch pattern in Figure 19.31 would require the following steps:

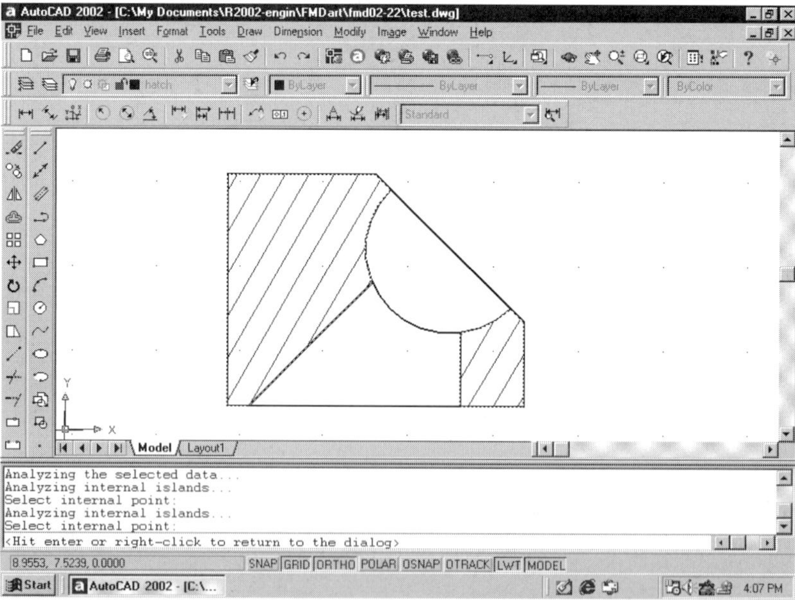

Figure 19.28 *You can display the proposed pattern by selecting the* **Preview** *button. If the pattern is not what you expected, the properties can be altered before you apply the pattern.*

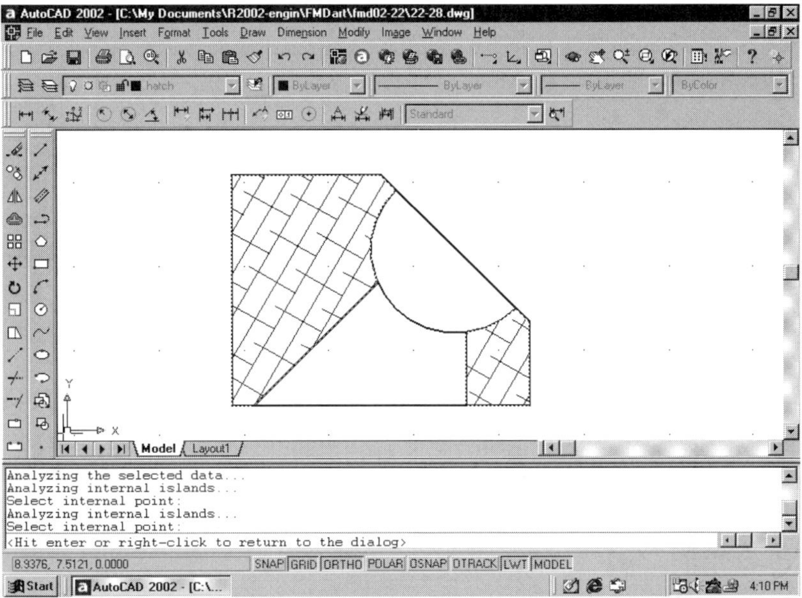

Figure 19.29 *If the* **Preview** *is acceptable, you can add it to the drawing base by clicking the* **OK** *button. You can select a new pattern by returning to the* **Boundary Hatch** *dialog box.*

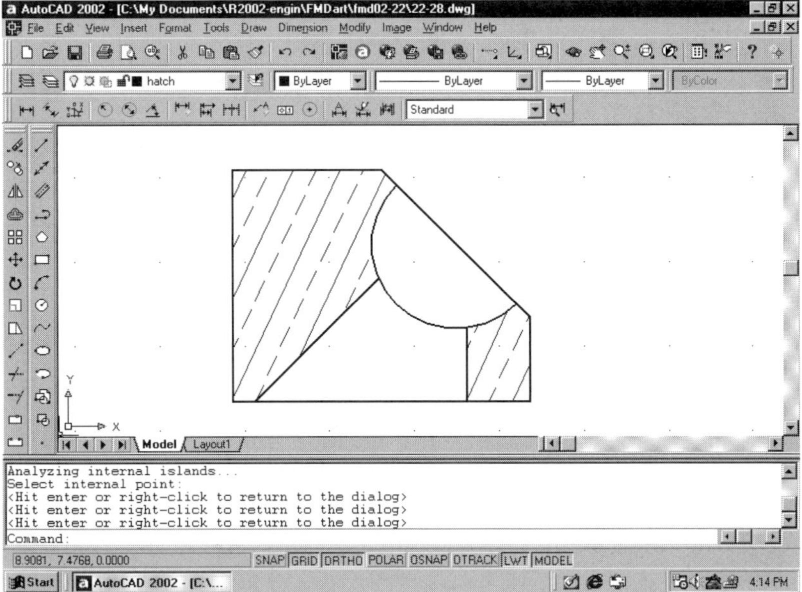

Figure 19.30 *The* **Preview** *option can be used to view the hatch pattern prior to adding it to the drawing base. If you are unhappy with the pattern, change the pattern, the scale, and the angle options and then reuse the* **Preview** *option. Courtesy Cathy Schesser.*

1. Display the **Boundary Hatch** dialog box by selecting **Hatch** on the **Draw** toolbar, by typing **BH** (ENTER) at the Command prompt, or by selecting **Hatch** from the **Draw** menu.
2. Choose the **Quick** tab.
3. Select the [...] button to display the **Hatch Pattern Palette** dialog box.
4. Choose the **ANSI** tab.
5. Select the ANSI31 display and click **OK**.
6. Select the **Pick Points** button in the **Boundary Hatch** dialog box.
7. Specify an internal point in the object to be hatched.
8. Select the **Preview** button to display the proposed hatch pattern.

   If the pattern is not acceptable, alter the hatch properties using the **Quick** tab, and then use **Preview** option again.
9. If the pattern is acceptable, click the **OK** button to apply the hatch pattern to the drawing.

REMOVING ISLANDS

Figure 19.32 shows an object with four islands (the holes). When the boundary is selected, each of the islands will automatically be excluded from the hatch. The hatch pattern was omitted from each of the cavities on the left side of Figure 19.32. Once the object to be hatched has been defined, selecting the **Remove Islands** button in the **Boundary Hatch** dialog box allows islands to be included and receive the hatch pattern. The example on the right side of Figure 19.32 shows the result of removing two of the islands from the boundary set.

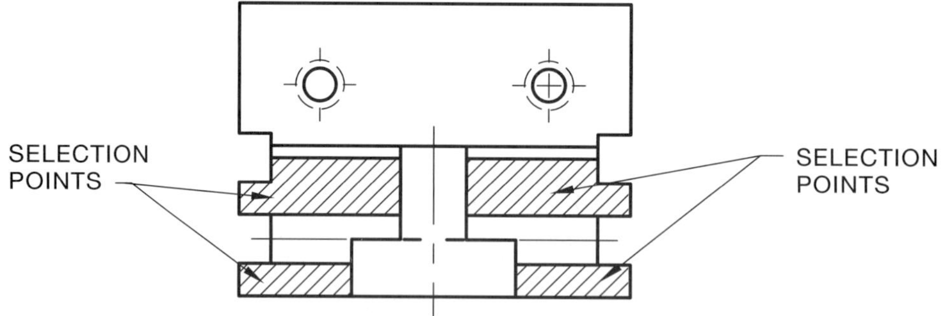

Figure 19.31 *Placing a hatch pattern is an important part of accurately drawing a section. Be sure to place the pattern so that lines are not parallel to any surface containing the pattern. Courtesy Hanh Tran.*

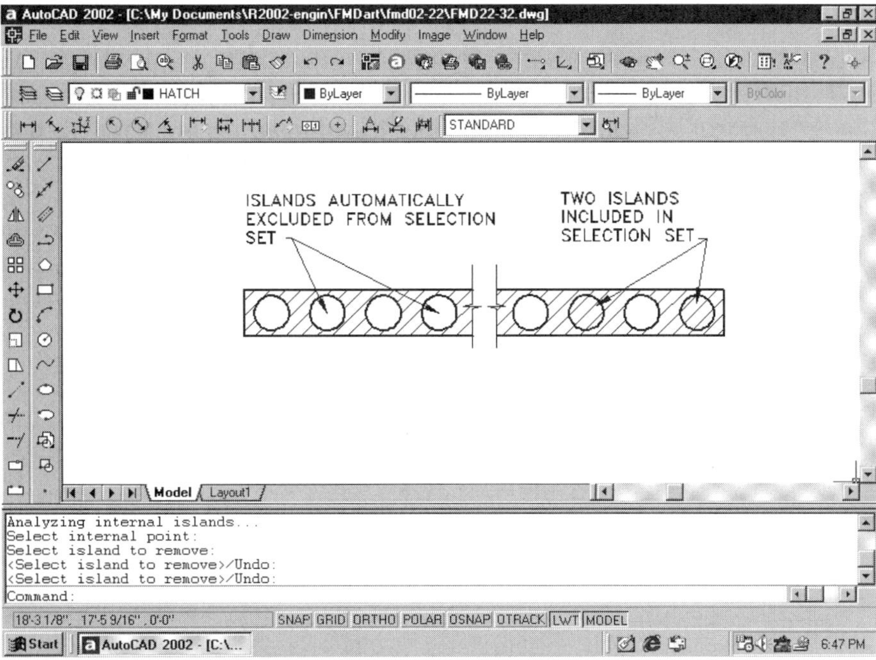

Figure 19.32 *The* **Remove Islands** *option can be used to control the placement of the pattern in islands within the boundary.*

VIEW SELECTION

With all of the options available for altering the boundary or patterns used in a boundary hatch, you might lose track of what options you've selected. The **View Selections** button will remove the dialog box and display the currently defined boundaries.

INHERIT PROPERTIES

This option is very useful when you are hatching an object using an existing pattern and you can't remember the qualities of the existing hatch pattern. Selecting the **Inherit Properties** button will close the **Boundary Hatch** dialog box and return the drawing area with the prompt:

Select associative hatch object: *(Select the hatched pattern that you would like to duplicate.)*

Inherited Properties: Name <ANSI31>, Scale <2.00>, Angle <0>

Select internal points: ENTER

Pressing ENTER will restore the **Boundary Hatch** dialog box.

COMPOSITION

The final portion of the **Boundary Hatch** dialog box to be examined is the **Composition** area, which contains toggles for **Associative** and **Nonassociative**. With the **Associative** button active, as an object containing a hatch pattern is edited, so is the hatch. With **Nonassociative** activated, if an object with a hatch pattern is altered, the object is altered, but the hatch pattern remains unaffected. Figure 19.33 shows the effect of stretching two steel beams and the effect of an associative pattern.

CONTROLLING PATTERN PLACEMENT

Although some hatch patterns such as the ANSI patterns can be placed without regard to the starting point of the boundary, other patterns such as brick or block should be placed relative to the hatch boundary. The **Snap** command can be used to control the relationship of the hatch pattern to the boundary. Use the following command sequence to make one of the hatch boundaries the starting point for the pattern.

> Command: **SNAP** ENTER
>
> Specify snap spacing or [ON/OFF/Aspect/Rotate/Style/Type] <0.50>: **R** ENTER
>
> Specify base point < 0.00,0.00>: *(Select a corner of the pattern boundary.)*
>
> Specify rotation angle <0>: ENTER
>
> Command: **BH** ENTER

The pattern can now be placed using the Bhatch command.

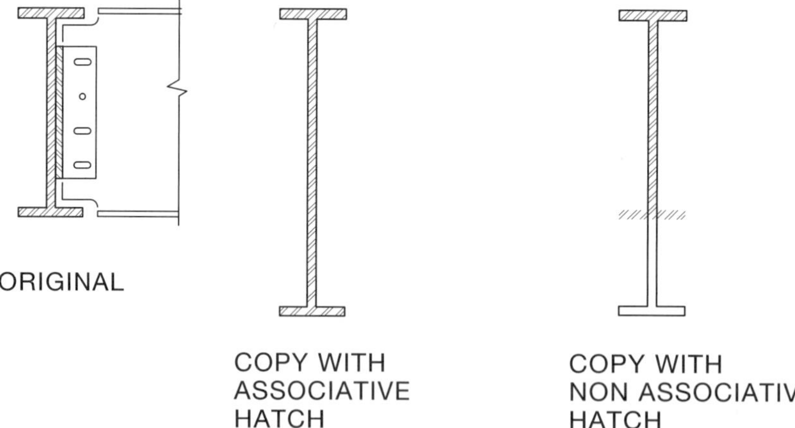

Figure 19.33 *With* **Associative** *active, stretching a feature such as this steel beam will alter the hatch pattern. With the* **Associative** *box inactive, the beam is stretched, but the hatch pattern is unaltered.*

DEFINING BOUNDARY HATCH USING THE ADVANCED TAB

Although the **Quick** tab can be used to hatch simple objects, the **Advanced** tab provides options to help increase the efficiency of selecting boundary hatch boundaries. Unless you're working on a very large drawing, you will probably not need this option. The **Bhatch** command evaluates all objects displayed on the screen when you select objects to be hatched. If you are working on a very large object, the options on the **Advanced** tab can speed the boundary selection process. Key elements of the **Advanced** tab include controls for island detection, object type, boundary set, and island detection method. The **Advanced** tab can be seen in Figure 19.34.

ISLAND DETECTION STYLE

The **Island Detection Style** area of the **Advanced** tab allows three styles of island detection for displaying the hatch pattern. The Normal setting is the default and the image tile reflects how the pattern will be displayed. You can change the setting by selecting one of the other radio buttons.

Normal

Figure 19.35 at A shows how hatching sets made up of multiple objects will be hatched when the default of **Normal** style is selected and each of the objects is selected as a boundary. **Normal** hatching style works inward, starting at the area boundary, and proceeds until another boundary is found. The pattern is turned off until another boundary is discovered.

Outer

Figure 19.35 at B shows how hatching sets made up of multiple objects will be hatched when the **Outer** option is selected and each of the objects is selected as a boundary. The **Outer** hatching style works inward, starting at the area boundary and

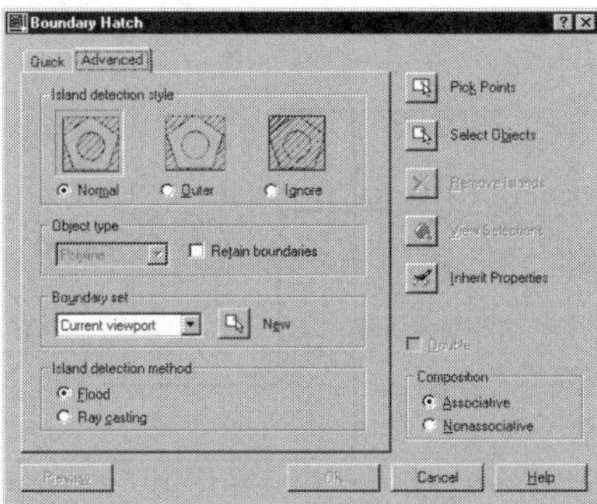

Figure 19.34 *The* **Advanced** *tab provides options to help place hatch patterns in complex shapes. Key elements of the tab include controls for island detection, object type, boundary set, and island detection method.*

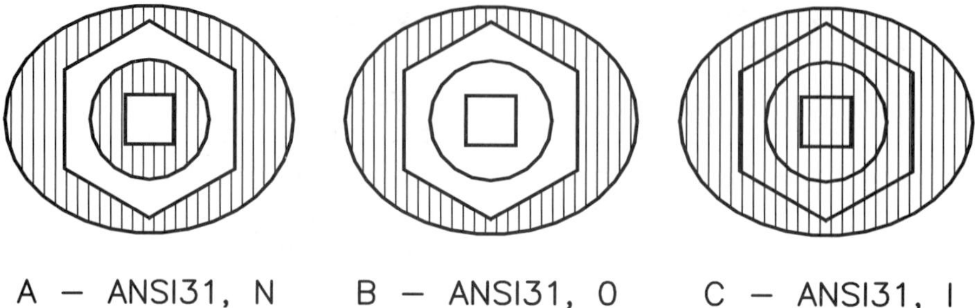

Figure 19.35 *The* **Island Detection Style** *area controls how the pattern will be displayed around multiple object sets. The* **Normal** *option works inward until another boundary is found. The pattern is not displayed until another boundary is found. The* **Outer** *option will display only the pattern in the outer boundary area. The* **Ignore** *option ignores all boundaries and hatches all areas within the outer boundary.*

proceeds until another boundary is found. The pattern is turned off at the second boundary.

Ignore

Figure 19.35 at C shows how multiple objects will be hatched when the **Ignore** option is selected. The **Ignore** hatching style works inward, starting at the area boundary, and passes through all objects within that boundary.

Note: With the **Normal** setting, if a hatch operation encounters another shape or text, the pattern will be placed around the text. Although text has not yet been added to the drawing, it will be important to remember that text inside a hatch boundary will still be legible (see Figure 19.36).

OBJECT TYPE

The **Object Type** area controls boundary options for applying the hatch pattern. The options for the boundary include the **Retain Boundaries** check box and the **Object Type** box. The **Retain Boundaries** option specifies whether or not a temporary boundary will be added to the drawing. Boundaries for hatch patterns can be specified by **Object Type**. These options are not active unless the **Retain Boundaries** option is active. **Object Type** controls the type of new boundary to be created. Options are Region and Polyline. You've explored polylines in a previous chapter. A closed shape comprised of a polyline can be used to form a boundary for hatch patterns.

Region

A region is an enclosed 2D area created from shapes called loops. Loops are a sequence of connected curves and lines that do not intersect and define an area on a plane. Sounds technical until you realize that loops are nothing more that the basic shapes that make up a drawing. Loops can be a combination of lines,

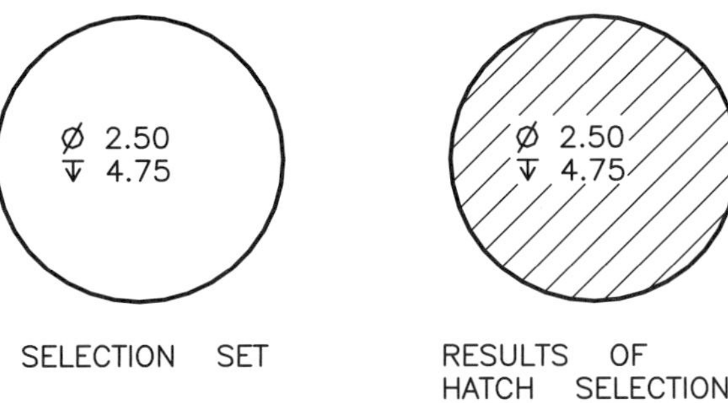

Figure 19.36 *Text is normally not placed inside a mechanical feature. If text is placed in an object, it will be treated like an island and automatically excluded from the area to be hatched.*

polylines, circles, arcs, elliptical arcs, splines, and solids. Loops or regions form the area where the hatch will be applied. Create a region by selecting the **Region** button on the **Draw** toolbar, by selecting **Region** from the **Draw** menu, or by typing REG (ENTER) at the Command prompt. Regions are created using the following command sequence:

Command: **REG** ENTER

Select objects: **W** ENTER *(Select objects to form the region using any selection method and press* ENTER.*)*

4 found

Select objects: ENTER

1 loop extracted.

1 Region created

Command:

The selected lines will now function as a polyline, forming a boundary for future hatch patterns.

BOUNDARY SET

This portion of the dialog box defines how the selection set for the hatch pattern will be selected when the **Pick Point** option is selected. By fine-tuning the selection set, you can have the **Bhatch** command function more quickly because AutoCAD has to examine fewer objects to define the selection set. The default selection to define boundaries is Current viewport.

Selecting **New** will remove the dialog box from the screen and allow a new boundary to be selected. The existing boundary set will be discarded and you will be allowed to construct a new boundary set to define the area to be hatched.

Once the new boundary set is selected, the dialog box will be returned and the default setting is now Existing set. An unlimited number of boundary sets can be selected. The last set selected will remain the default until a new boundary set is selected. When Current viewport is selected, the selection set will be created from everything that is visible in the current viewport. Selecting this option when there is a current boundary set will discard the current selection and place the pattern in everything in the current viewport. The normal setting is to place the pattern based on the Existing Boundary Set.

ISLAND DETECTION METHOD

The final option of the **Advanced** tab is the **Island Detection Method** area. In its current state of **Flood**, internal features (islands) are selected to be part of the boundary set. When the **Ray casting** option is current, AutoCAD sends an invisible object called a ray to the nearest object it can find and then traces the boundary in a counterclockwise direction. You might be tempted to store this with other boring facts about AutoCAD that you wish you never knew, but it really is important when you use **Bhatch**. If the ray hits text or an internal boundary of an object, the boundary definition error is displayed.

EDITING HATCH PATTERNS

Features of boundary hatch patterns can be edited using the **Hatchedit** command. Start the command by selecting a hatch pattern to edit and the right-clicking in the drawing area, by selecting **Edit Hatch** on the **MODIFY II** toolbar, by typing HE (ENTER) at the Command prompt, or by selecting **Hatch** from the **Modify** menu, or double-click the hatch pattern to be edited. Each method will produce the prompt:

Command: **HE** ENTER *(Or click the Edit Hatch button.)*

Select hatch object: *(Select hatch pattern to be edited.)*

As the pattern is selected, a **Hatch Edit** dialog box similar to the one seen in Figure 19.37 will be displayed. (The exact display will vary depending on which pattern is

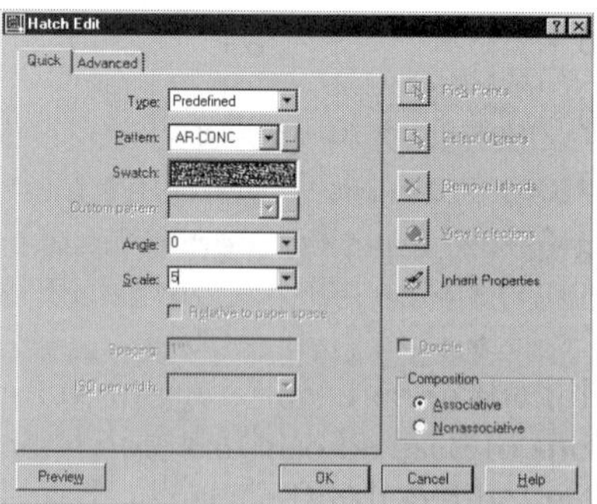

Figure 19.37 *The* **Hatch Edit** *dialog box can be used to edit the pattern once it has been added to the drawing base. Display the dialog box by selecting a hatch pattern to edit and then right-clicking.*

selected.) The box is very similar to the **Boundary Hatch** dialog box. Each of the features of the pattern can be altered, and the options function as they do in the **Boundary Hatch** box.

CHAPTER 19 EXERCISES

Create a new layer titled hatch for each of the following drawings.

1. Start a new drawing and use the following coordinates to draw the required object.

Start: 1,1	B. = @ 3,0	C. = @ –5,.25	D. = @ –.75,1	E.= @ –1,1
F. = @ .75,1	G. = @ .5,1	H. = @ 0,1	J. = @ .25,.25	
K. = –2.25,0	L. = @ .25,–.25	M. = @ 0,–1	N. = @ .75,–1	
P. = @ –1,–1	Q.= @ 0,–1	R. = @ 1,–.75	S. = C	

   Make a copy of the completed shape. Use the SOLID pattern to fill one of the objects. Turn Fill to OFF. Use the ANSI31 hatch pattern and adjust the scale factor to a suitable scale to hatch the object so that it appears solid. Save the drawing as E-19-1.

2. Start a new drawing and use the drawing below as a guide to draw a 15" diameter circle. Draw a 12" diameter circle with a center point 10" above the first center point. Array the 12" circle so that there are a total of six circles. Copy the pattern so that you have a total of three circular patterns. Create layers to separate the object from the hatching. Use the **Bhatch** command to form three completely different hatching styles. Use different hatch patterns for each area to be hatched. Save the drawing as E-19-2.

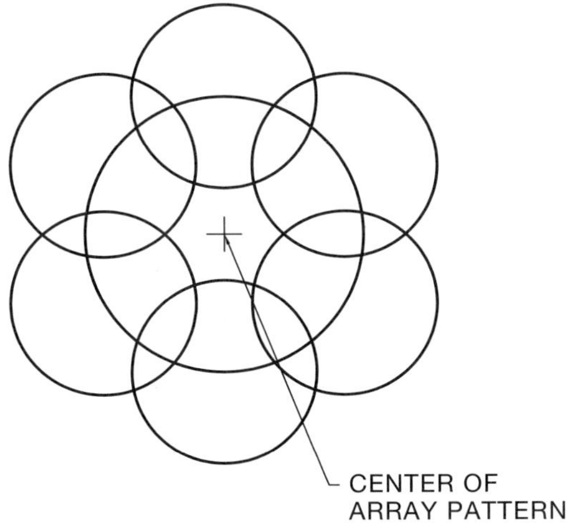

Problem 19.2

3. Start a new drawing and draw a red square, surrounded by a blue circle, surrounded by a green octagon, surrounded by a cyan triangle. Copy this pattern so that there are a total of three copies. Create a separate layer for hatching. Use ANSI34 and create a double hatching pattern. Provide examples of **Normal**, **Outer**, and **Ignore** styles of hatching. Save the drawing as E-19-3.
4. Open drawing E-13-3 and create a layer for oak (green), maple (red), and Dutch elm (yellow). Change the trees from their existing layer to these new layers. Tree E is an oak, A is a maple, and the other three are Dutch elm. Hatch each type of tree with a different pattern. Save the drawing as E-19-4.
5. Open Drawing E-18-10 and draw a full section. Save the drawing as E-19-5.
6. Open drawing E-18-12 and draw a full section. Save the drawings as E-19-6.
7. Open drawing E-18-15 and draw a full section. Provide appropriate drawings to describe the sixteen holes. Save the drawings as E-19-7.
8. Open drawing E-18-15 and draw a half section. Save the drawing as E-19-8.
9. Open drawing E-18-9 and add a Ø.75" × .125 deep counterbore to the bottom of the threaded feature. Use a broken section to complete this project. Save the drawing as E-19-9.
10. Open drawing E-18-11 and draw a broken out section. Save the drawing as E-19-10.

Used the attached drawings to complete exercises 11 through 25. Fully dimension the parts using as few views as possible. Unless noted, choose the best type of section to describe the object. Save all drawings with the appropriate exercise number.

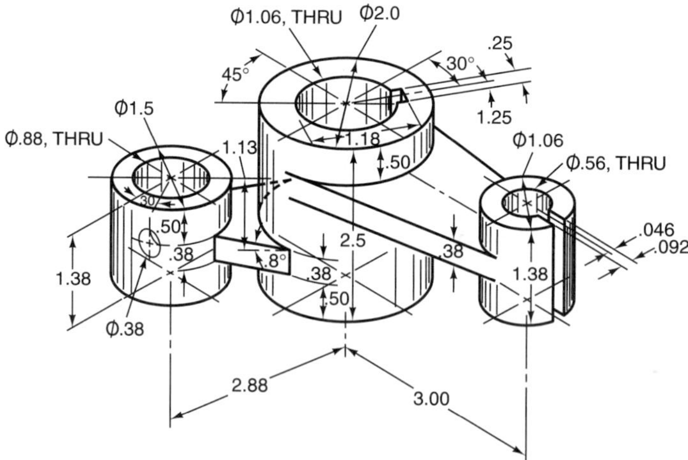

Problem 19.11

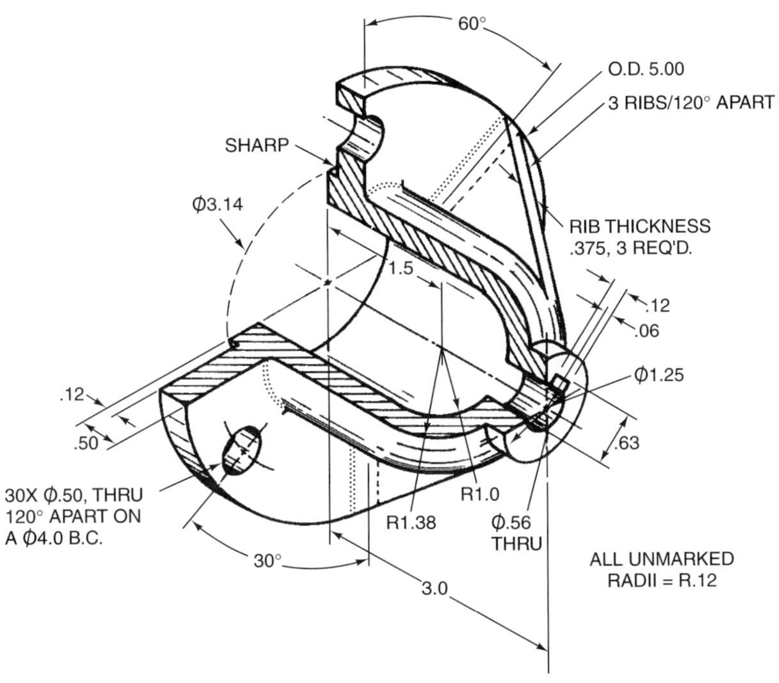

Problem 19.12

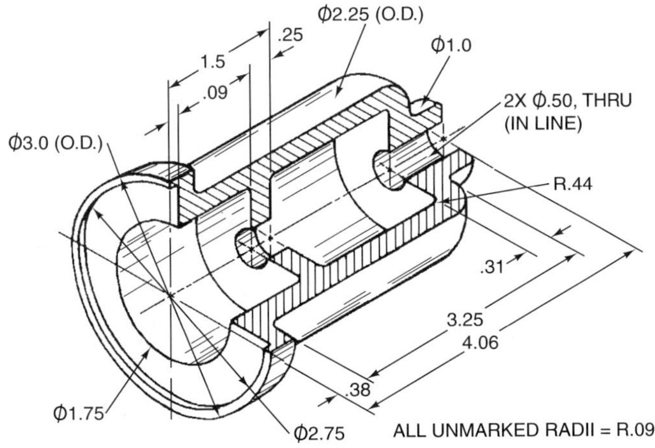

Problem 19.13

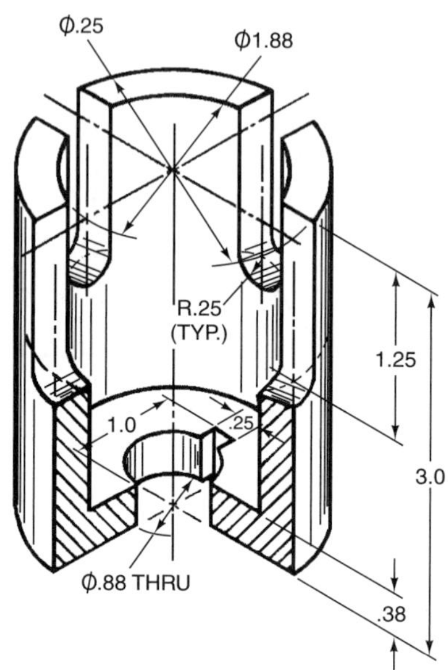

Problem 19.14

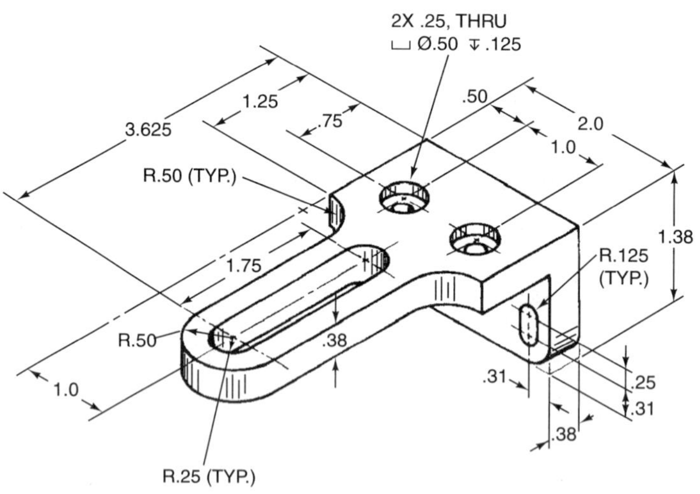

Problem 19.15

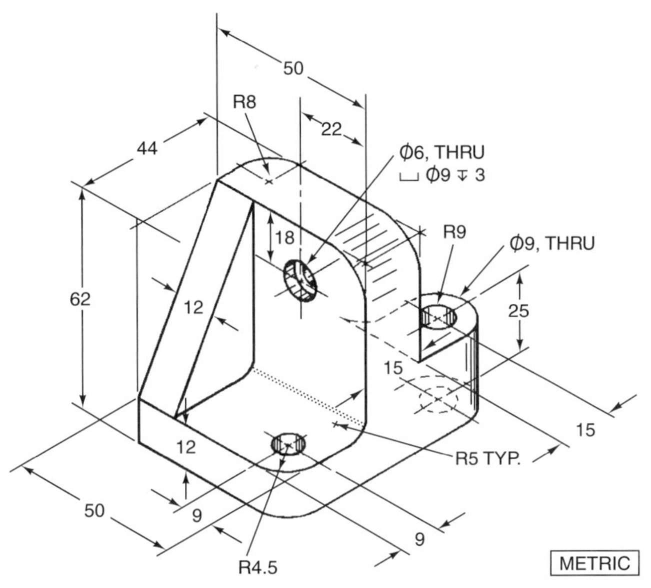

Problem 19.16

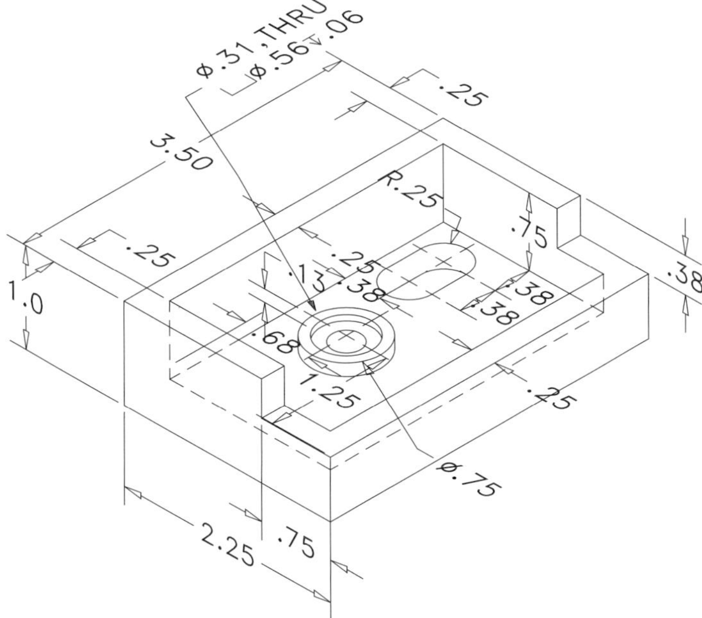

Problem 19.17

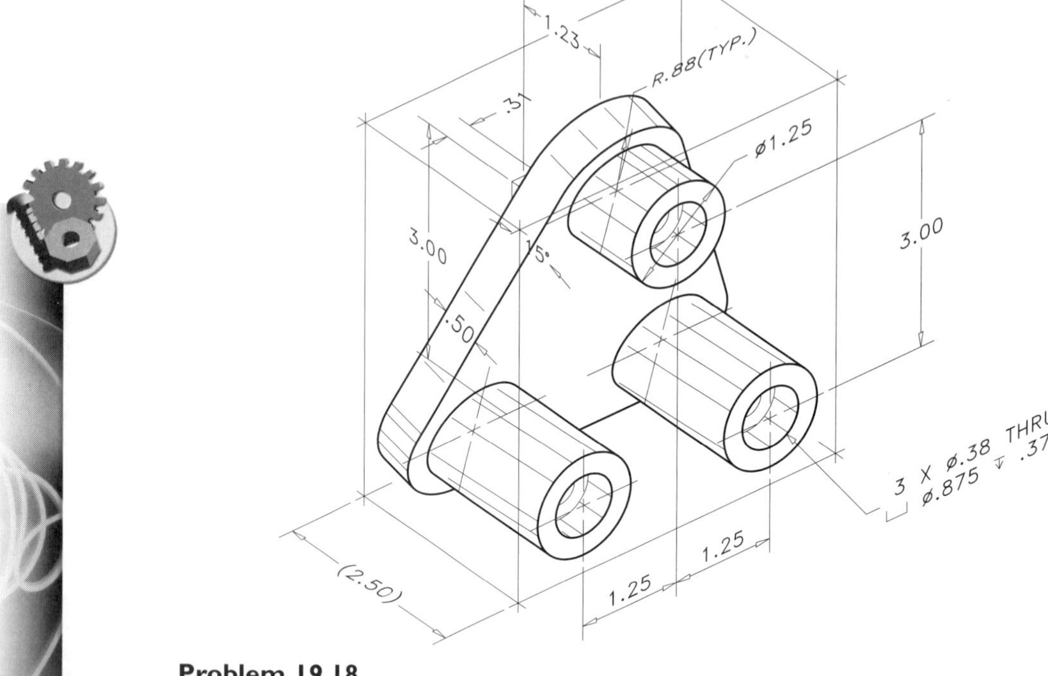

Problem 19.18

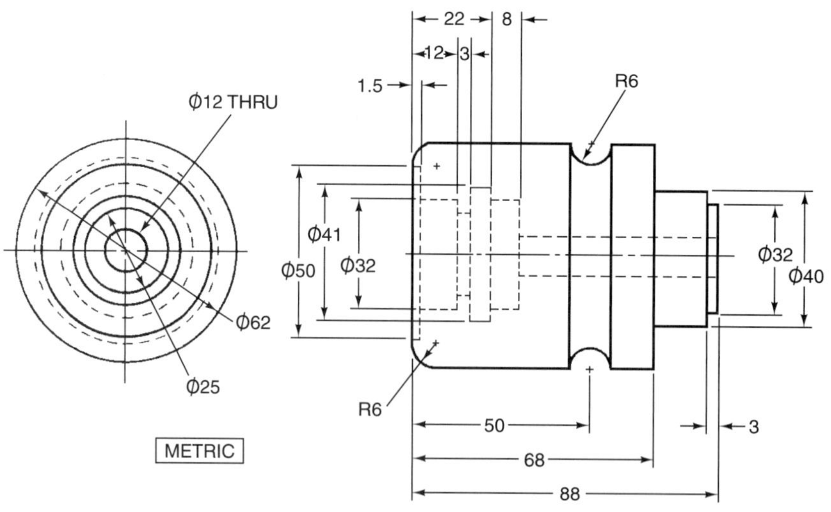

Problem 19.19

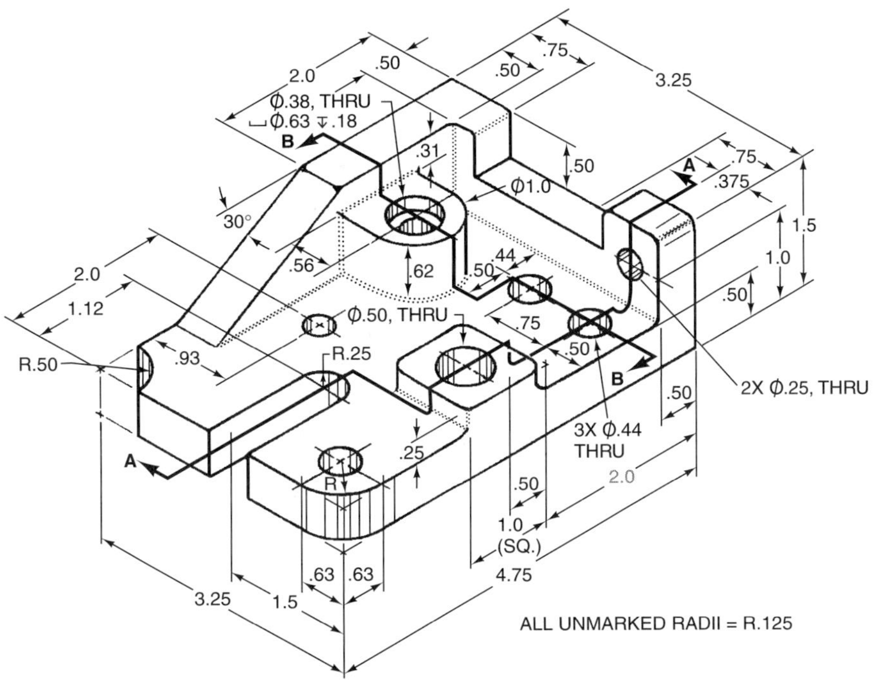

Problem 19.20

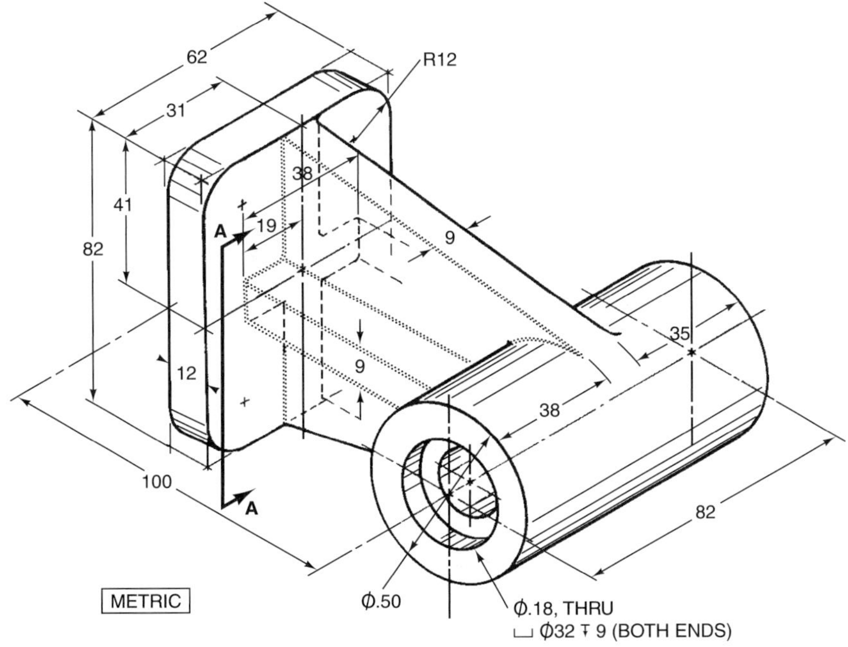

Problem 19.21

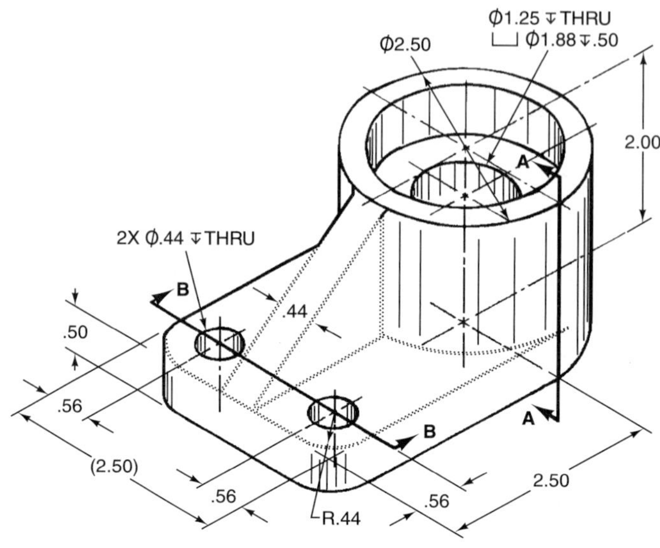

ALL UNMARKED RADII = R.06

Problem 19.22

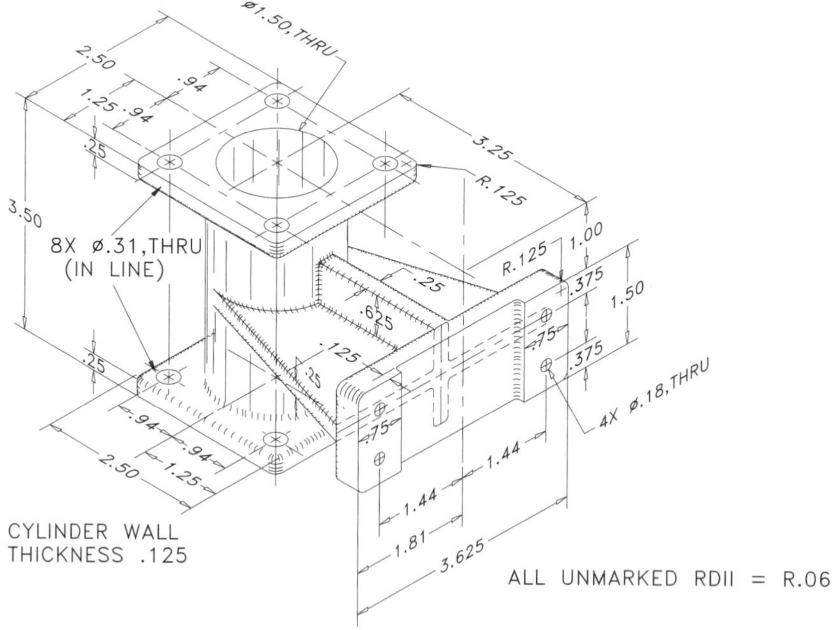

Problem 19.23

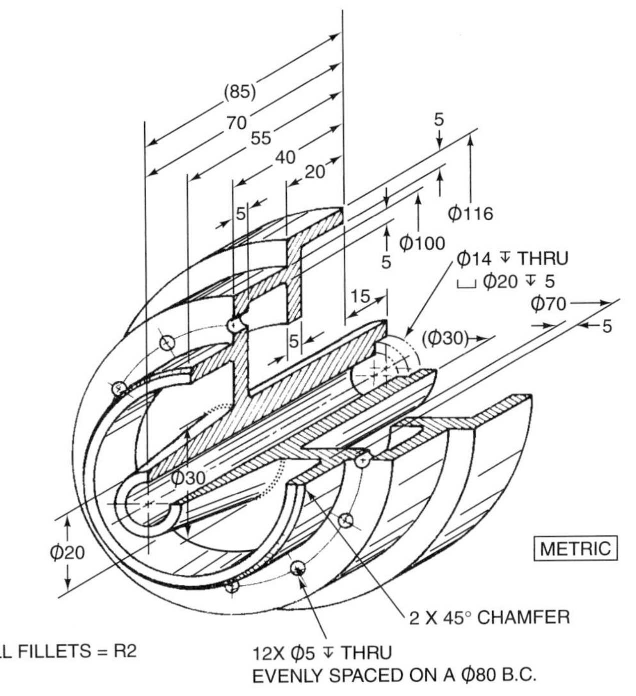

Problem 19.24

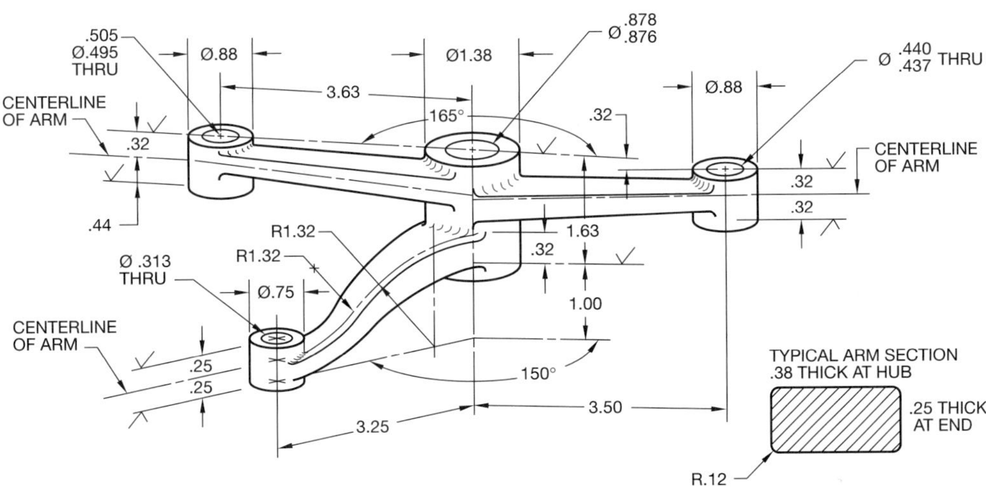

Problem 19.25

CHAPTER 19 QUIZ

1. A circle surrounds a square, which encloses a hexagon. What style is used to hatch the circle and hexagon?
2. In the object described in question 1, what effect will **Ignore** have on each shape?
3. What option should be used to fill an object with a hatch pattern?
4. What type of section is best to describe the interior of a symmetrical feature?
5. List the tabs that are associated with **Bhatch**.
6. Describe features that are never sectioned.
7. How can the **Hatch Pattern Palette** dialog box be accessed?
8. How can the scale factor of the **Bhatch** pattern be controlled by the plotted scale factor?
9. Describe how the location of the cutting plane affects how a web will be shown in section.
10. What process is used to keep the hatch pattern from crossing text?
11. Explain the differences in using the **Pick Point** option and the **Select Objects** option.
12. Explain a conventional revolution.
13. What is direct hatching?
14. Describe the process to create a hatch pattern of vertical lines at 1/16" spacing.
15. How is the location of the cutting plane determined?
16. What is the effect of using the ANSI32 pattern?
17. Use the **Help** menu to determine what effect the **Explode** command will have on a hatch pattern.
18. Does the spacing and angle of the section lines remain constant? Explain your answers.
19. What does associative have to do with **Bhatch**?
20. You've defined a hatch pattern, but you're not sure how it will appear. List two methods to view the pattern.
21. What is an island?
22. Explain the difference between a full and an offset section.
23. Explain the difference between removed and revolved sections.
24. A drawing shows a portion of a part that has been sectioned. An object line extends through the hatch pattern. Is this a mistake?
25. Describe when a removed section should be used.

Creating Blocks and Wblocks

Since Chapter 2 you've had the basic drawing skills to draw most projects you could imagine. Subsequent chapters have helped you to develop skills needed to ease the drawing and design process but still have not equipped you to master a complicated project easily. This chapter will

- Introduce you to creating blocks for repetitive use
- Explore methods of inserting blocks into a drawing
- Examine methods for editing blocks that have been inserted in a drawing
- Explore methods to redefine blocks that have been inserted in a drawing
- Examine the benefits of using wblocks in multiple windows

Commands to be introduced in this chapter include

- **Block**
- **Insert**
- **Minsert**
- **Base**
- **DesignCenter**
- **Explode**
- **Refedit**
- **Wblock**

BLOCKS

Drawings are composed of symbols. To show a threaded feature on a drawing, you don't draw a thread; instead you use a hidden line or circle to represent threads. A block is a group of objects that are treated as one object. The objects used to represent a bolt can be saved as a block and inserted throughout the drawing as one object. The **Block** command allows symbols that are used repeatedly to be created, saved, and inserted easily in the drawing in which they were created, which provides many benefits for the user.

BENEFITS OF BLOCKS

Symbols that are used frequently can be saved as a block, which will greatly increase drawing speed, uniformity, and efficiency.

Speed

If you're working on a drawing with repetitive features, these symbols will need to be represented many times. Rather than draw each symbol, you can draw one and save it as a block. As a block, repetitive features such as the electrical connection contact for a turbine control panel can be inserted quickly and the size altered or the position rotated, as seen in Figure 20.1. With a little more effort, these same operations can also be accomplished with the **Copy**, **Rotate**, **Mirror**, **Array**, and **Scale** commands, but why use several commands when you could use one?

Editing

Throughout the evolution of a drawing, many changes and revisions will take place. Blocks can greatly aid in this revision process. If a client decides that all of the bolt heads in a part are to be .375" wider, several hours might be required to change all of the features. If the threads are drawn as a block, the block can be redefined and all references to that block will be updated automatically.

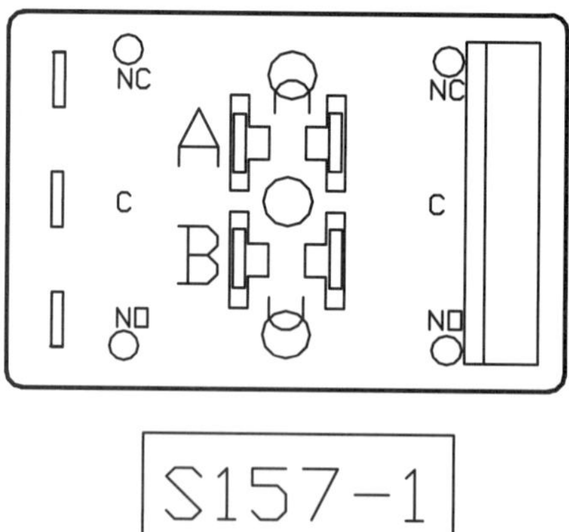

Figure 20.1 *Repetitive features such as the electrical connection contact for a turbine control panel can be inserted quickly and the size altered or the position rotated. Courtesy Floyd D. Miller.*

Attributes

An attribute is text that is associated with a block. Many blocks require text to explain variations. Size, material, quantity, and supplier are examples of information that can be saved as an attribute with a block. This text can vary with each application of the block. Attributes can be displayed on the screen as text or they can be stored with the drawing but made invisible. Attributes associated with blocks can also be extracted from a drawing and transferred to a database form, such as a parts list. Web Chapter 4 will explore adding attributes to blocks.

Efficiency

In addition to saving time and providing a quick method for revisions, the **Block** command combines the storage of groups of objects into one unit. In drawing an individual welding symbol, information must be stored about the start and end points of the original line and arc, the radius of the arc, the scale factor as the symbol is enlarged or reduced for various uses, and the location of each unit as it is displayed throughout the drawing. On something as simple as a welding symbol, this might not seem like a lot of savings, but if a whole component such as the connection bar in Figure 20.2 are saved as block and reused, savings in both time and efficiency can be achieved. When the symbol is installed as a block, only a block reference is stored, saving valuable storage space.

DRAWING OBJECTS FOR BLOCKS

Because a block is a group of objects that is saved as one object, there is no special method for drawing the block. The components that will comprise the block can be created using any of the drawing or editing commands. Some consideration should be given to the layer the block is created on and the size of the block while it is being constructed, to avoid some problems when the block is inserted in the drawing.

The Effects of Layer on Blocks

A block will take on the characteristics of the layer on which it is created. A block can also consist of objects that have been drawn on several layers with different linetypes and colors. When the block is inserted in a drawing, the block will retain its original

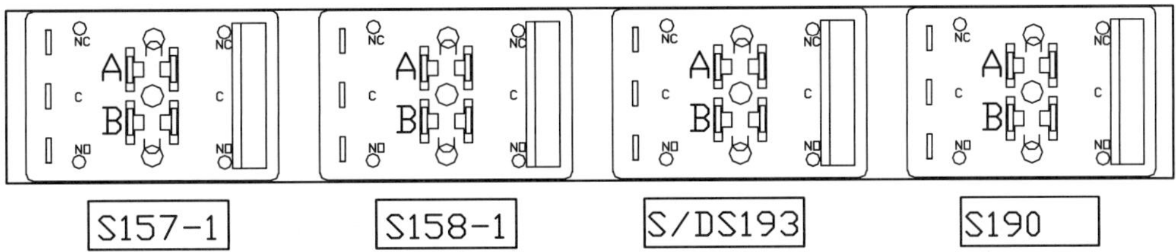

Figure 20.2 *A whole component such as the connection bar can be saved as a block and reused, gaining both time-savings and efficiency. Courtesy Floyd D. Miller.*

characteristics. If an electrical symbol is created on a layer titled SYMBOL and inserted in a drawing with the TEXT layer current, the symbol will be inserted in the drawing on the SYMBOL layer. If the block was created on a layer that does not exist in the current drawing, the layer containing the block will be created. There are, however, three exceptions to this guideline.

- Blocks that are created on the 0 layer will be placed on the current layer when placed in a drawing.
- Blocks drawn with the color created BYLAYER are drawn with the color of the current layer, but will reflect the color of the layer into which the block is inserted.
- Blocks drawn with the linetype created BYBLOCK are drawn with the linetype of the current layer, but will reflect the linetype of the layer in which the block is inserted.

Block Size

Once the block is saved, it can be inserted throughout a drawing. When the **Insert** command is introduced later in this chapter, you will have an opportunity to alter the size of the block in the X and Y directions. Altering the size of the block during the insertion process can be easier if the size of the block is considered during creation.

Some blocks such as electrical symbols are a standard size. These blocks can be drawn at their actual size and inserted. Other blocks, such as bolts, screws, bolt heads, nuts, pins, and keys vary in size with each application. Drawing these blocks in one-unit squares will allow the size to be adjusted each time it is inserted. Rather than having separate blocks for every possible bolt size, one bolt can be created, and the X value controlling its length, and the Y value controlling its width can be altered. See Figure 20.3.

CREATING A BLOCK

Figure 20.4 shows the drawing process for creating a threaded feature, which can then be saved as a block. Keep in mind that although the part has been drawn, it is not a block yet. Once the drawing is ready to be reproduced, start the **Block** command by selecting the **Make Block** button on the **Draw** toolbar, by typing B (ENTER) at the Command prompt or by selecting **Make** from the **Block** cascading menu on the **Draw** menu. Each method will produce the **Block Definition** dialog box shown in Figure 20.5.

Choosing a Name

Start the process by entering the name of the block to be created in the **Name** edit box. A name of up to 255 characters can be used to describe the block. The block name can contain letters, digits, and special characters such as $, -, and _. Any character not

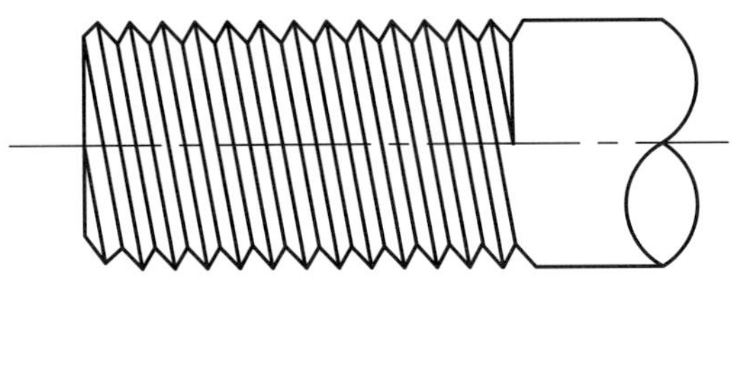

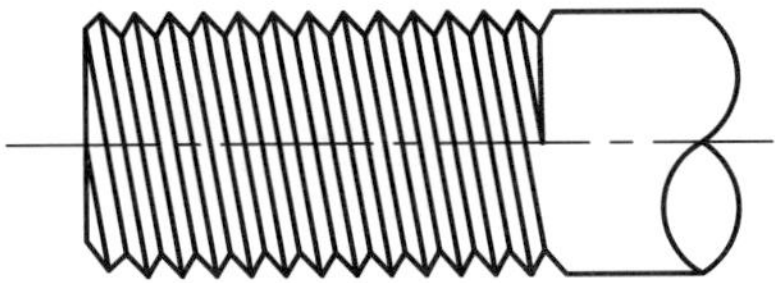

Figure 20.3 *Blocks that vary in size should be created using one unit, such as an inch, so the size of the block can be adjusted easily each time the block is inserted. The upper block was inserted at full scale, the lower block at .75.*

used by Microsoft Windows or AutoCAD, including blank spaces, can be used. Use a name that will describe the contents. For instance, RESIST would be an appropriate name to describe a resistor for an electrical drawing. If a name for an existing block is provided for the new block, the old block will be destroyed and the new block will take its place. Before it is vaporized, you will be given an alert message by AutoCAD warning that the block is about to be redefined.

Selecting the Objects

Once the name has been provided, click either the **Pick point** or **Select objects** button. Each option must be selected, but the order will not affect the

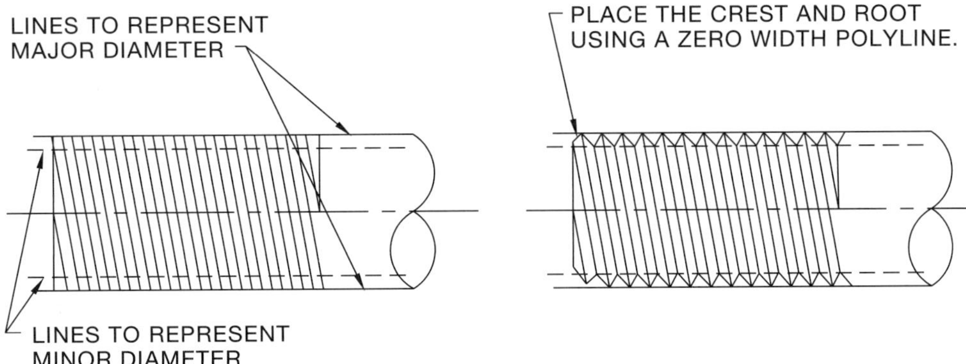

Figure 20.4 *A bolt can be created by drawing pairs of parallel lines and then adding the required inclined lines for the threads.*

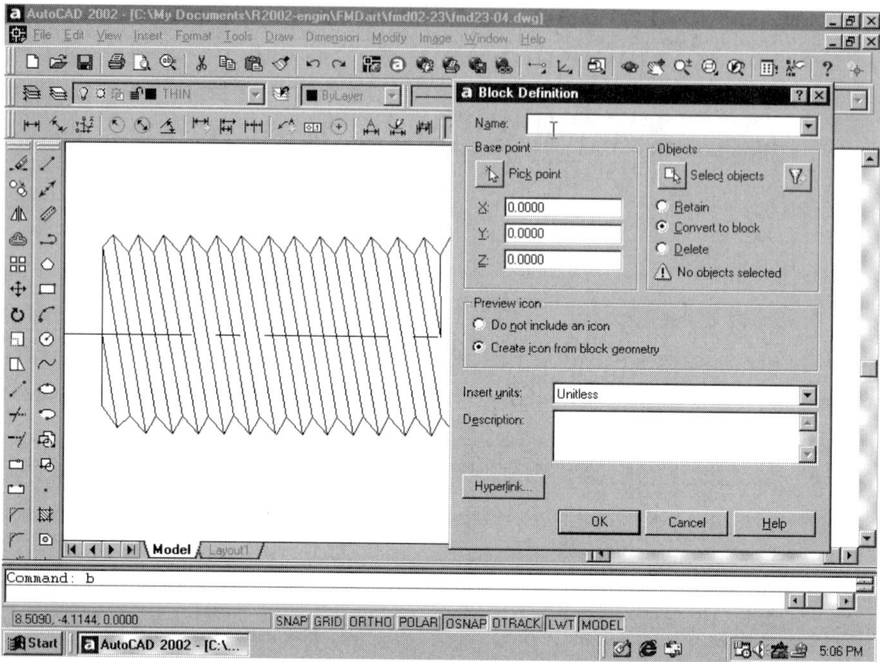

Figure 20.5 *Once the drawing is ready to be reproduced, start the* **Block** *command by selecting the* **Make Block** *button on the* **Draw** *toolbar, by typing* **B** *(ENTER) at the Command prompt, or by selecting* **Make** *from* **Block** *on the* **Draw** *menu.*

outcome of the command. For this discussion, the object will be selected first. Select the **Select objects** or **Quick Select** button to define the objects that will be in the block. (The use of the **Quick Select** dialog box will be explained shortly.) Selecting the **Select objects** button temporarily removes the dialog box from the screen and allows the objects that will make up the block to be selected. As the dialog box is removed, the command line will prompt to select objects. Objects to comprise the block can be selected using any object selection method. Once the desired objects have been selected, pressing ENTER will return the **Block Definition** dialog box. An image of the selected objects will be displayed in the **Preview icon** display similar to Figure 20.6. The option for no icon can be activated if you do not want to see the image you'll be using as a block.

Note: The **Blockicon** command can be used to display images of blocks stored in drawings created using AutoCAD 2000 or previous releases. The command can be entered at the Command prompt and will provide prompts for what blocks you would like to update. Pressing ENTER will update all blocks.

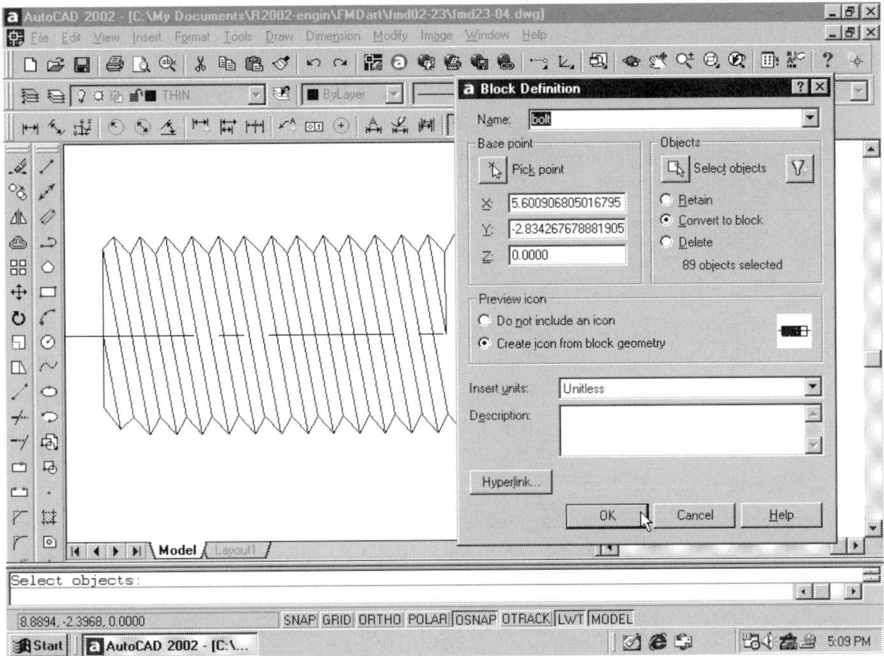

Figure 20.6 *Once the objects to comprise the block have been selected, pressing* ENTER *will return the* **Block Definition** *dialog box with an image of the selected objects displayed in the* **Preview icon** *display box.*

Using Quick Select

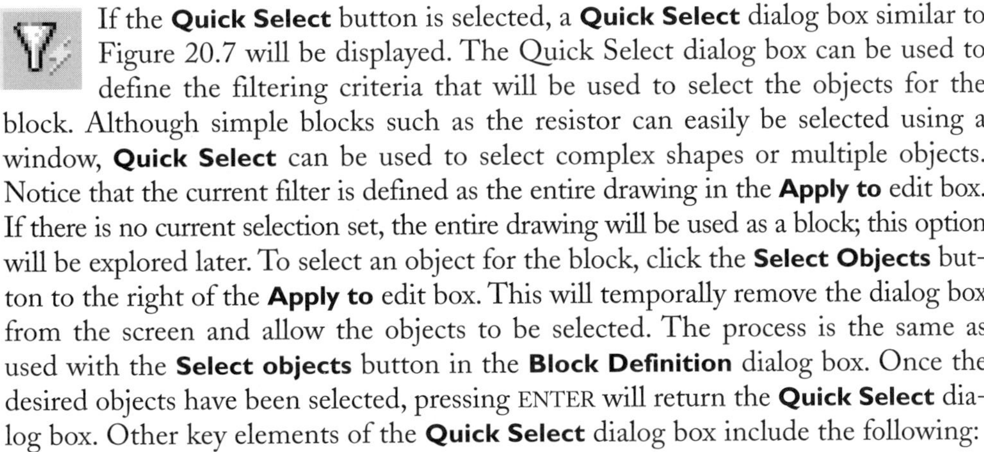

If the **Quick Select** button is selected, a **Quick Select** dialog box similar to Figure 20.7 will be displayed. The Quick Select dialog box can be used to define the filtering criteria that will be used to select the objects for the block. Although simple blocks such as the resistor can easily be selected using a window, **Quick Select** can be used to select complex shapes or multiple objects. Notice that the current filter is defined as the entire drawing in the **Apply to** edit box. If there is no current selection set, the entire drawing will be used as a block; this option will be explored later. To select an object for the block, click the **Select Objects** button to the right of the **Apply to** edit box. This will temporally remove the dialog box from the screen and allow the objects to be selected. The process is the same as used with the **Select objects** button in the **Block Definition** dialog box. Once the desired objects have been selected, pressing ENTER will return the **Quick Select** dialog box. Other key elements of the **Quick Select** dialog box include the following:

- **Object type** specifies the type of object to use to define the filter. The current choices are Multiple, and Block Reference. Since you're defining a block, choose Block Reference.

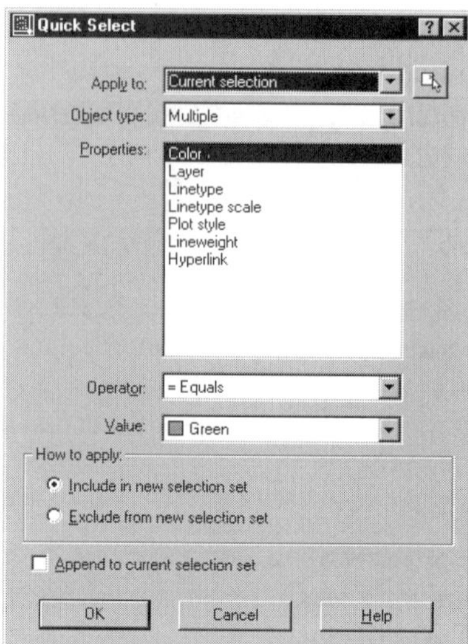

Figure 20.7 *The* **Quick Select** *dialog box can be used to define the filtering criteria that will be used to select the objects for the block.*

- The **Properties** box can be used to specify properties to be used as a filter. Objects to form the block can be defined by the color, lineweight, linetype, or any of the other listed properties.
- **Operator** controls the range of the selected property filter. Depending on the selected property, options can include Equals, Not Equal, Greater Than, Less Than, and Select All. Selecting the edit arrow displays a list to allow the operator value to be altered.
- **Value** defines the filter property value. If the selected property is a color, the value will specify a color. If the property is a layer, the 0 layer is the default value. Selecting the edit arrow displays a list to allow the value to be altered.
- The **How to apply** area of the box specifies if the selected objects will form or be excluded from the selected set to form the block. Choosing **Exclude from new selection set** will create a block of all objects that do not match the selection filter.

Object Selection Options

In addition to methods for selecting objects to define the block, the **Objects** area in the **Block Definition** dialog box contains options to retain, convert, and delete the selected objects.

Retain—Selecting **Retain** will keep the selected objects in their original state in the selection set. Once the object is selected to be a block, the original objects will remain in their current location.

Convert to block—Selecting this option will remove the selected objects from the selection set and replace them with a previously defined block.

Delete—Selecting this option will remove the objects that define the block from the drawing. This option works well if you create an object that is to become a block as part of another drawing. With **Delete** active, once the object is selected to be a block, the original object will be removed from the drawing. It will still exist as a block in the drawing base, but it can't be seen until the block is inserted.

SPECIFYING THE BASE POINT

The insertion point of the block will align with a point that will be designated as the block is inserted into other parts of a drawing. The insertion point can be selected by selecting any point of the object or by entering X and Y coordinates. The **Pick point** button temporally removes the dialog box from the screen and allows the insertion point to be selected. Figure 20.8 shows the effect of the insertion point of a block with the insertion point at the destination. The insertion point should be chosen using OSNAP based on the block's use. The insertion point also can be used as a rotation point for the block as the block is inserted in its new location. Figure 20.9 shows how a block can be rotated about its insertion point. Other options will be discussed as the Insert command is explored. Using a center point or the lower left corner of a block will help to keep track of where to insert the block.

DESCRIPTION

The **Description** edit box allows a description to be assigned to the block. A block can be created without providing a description, but on a complex block, the description might help you distinguish it from a similar block.

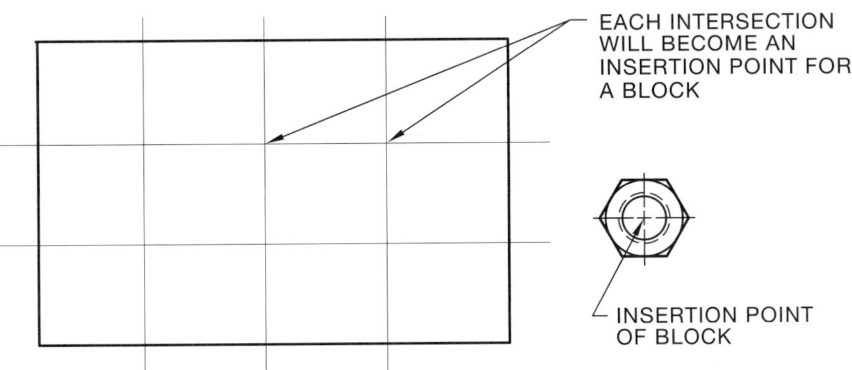

Figure 20.8 *The insertion point defines the location of the block to be used as a handle.*

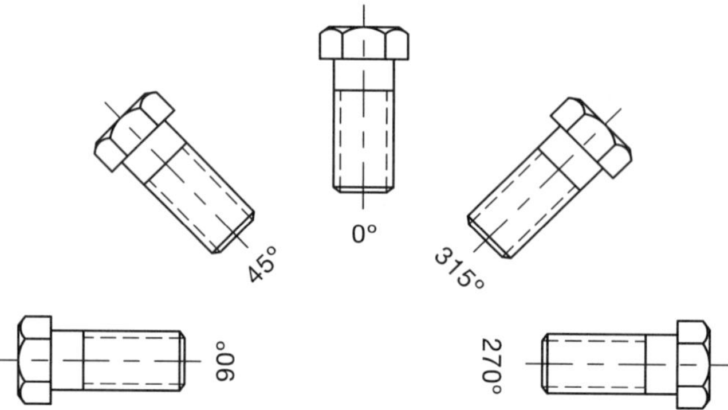

Figure 20.9 *A block can be rotated about its insertion point.*

Summary of Steps in Creating a Block

It's taken several pages to explain what a block is. Remember that a block is just a collection of objects that can be used over and over. Because they are so important to your drawing future, use the summary as you create blocks. Use the following steps to create blocks:

1. Display the **Block Definition** dialog box by selecting the **Make Block** button on the **Draw** toolbar, by typing **B** (ENTER) at the Command prompt, or by selecting **Make** from **Block** on the **Draw** menu.
2. Provide a name that will define the block.
3. Select the drawing objects to be reproduced as blocks.
4. Select a base point to be used as the block is inserted in the drawing.
5. Select modifiers in the **Objects** and **Preview** portions of the dialog box.
6. Click **OK** to save the block to the drawing base.

The object has now been saved as part of the drawing base and is ready to be reused throughout the drawing. If another bolt, or what ever object is used as a block, needs to be added to the drawing, you can add it to the drawing by inserting the block rather than redrawing the objects.

Note: You've just read through the process to create a block. As you create a block, it might leave you waiting for something exciting to happen. The creation of a block is nothing special. It's what you can do with it that will revolutionize your drawing process.

PREPARING BLOCKS FOR INSERTION

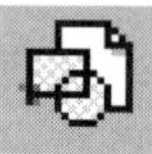

Once a block has been created and saved, it can be placed in a drawing using the **Insert** command. The command can be accessed by selecting the **Insert Block** button on the **Draw** toolbar, by typing **I** (ENTER) at the

Command prompt, or by selecting **Block** from the **Insert** menu. Each method will produce an **Insert** dialog box similar to Figure 20.10.

Choosing a Block

The **Name** edit box is used to select the block to be inserted. The name of the block to be inserted can be typed in the edit box or selected from a list of blocks contained in the drawing. Selecting the edit arrow beside the box will display a listing of all blocks contained in the drawing. Selecting the **Browse** button will open the **Select Drawing File** dialog box. The dialog box is the same as the file selection dialog boxes that you've used as drawing files are opened. The dialog allows a block or file to be selected for insertion. Inserting files will be introduced later in this chapter.

Specifying an Insertion Point

Activating the **Insertion Point Specify On-screen** check box allows an insertion point to be used to specify where the block will be located. With the **Specify On-screen** box inactive, the X and Y coordinate locations (Z is the vertical reference for 3D) can be used to locate the insertion point. As the block was created, an insertion point was determined. Once the desired insertion parameters are adjusted, a prompt will be displayed to provide a location to place the original insertion point. Figure 20.11 shows the relationship of the block insertion point to the destination insertion point. Once the insertion point is selected, the block will be dragged into the indicated position, and the scale can be altered. The parameters to control the insertion can each be adjusted before removing the dialog box to insert the block, or at the Command prompt before the block is inserted.

Scale

This area allows you to define scale factors for the block to be inserted. With the **Specify On-screen** check box active, a prompt will be displayed at the Command

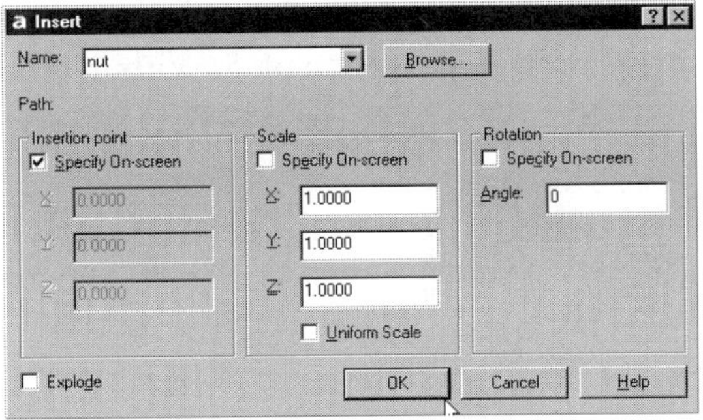

Figure 20.10 *Selecting the* **Insert Block** *button on the* **Draw** *toolbar, typing* **I** *(ENTER) at the Command prompt, or selecting* **Block** *on the* **Insert** *menu will display the* **Insert** *dialog box.*

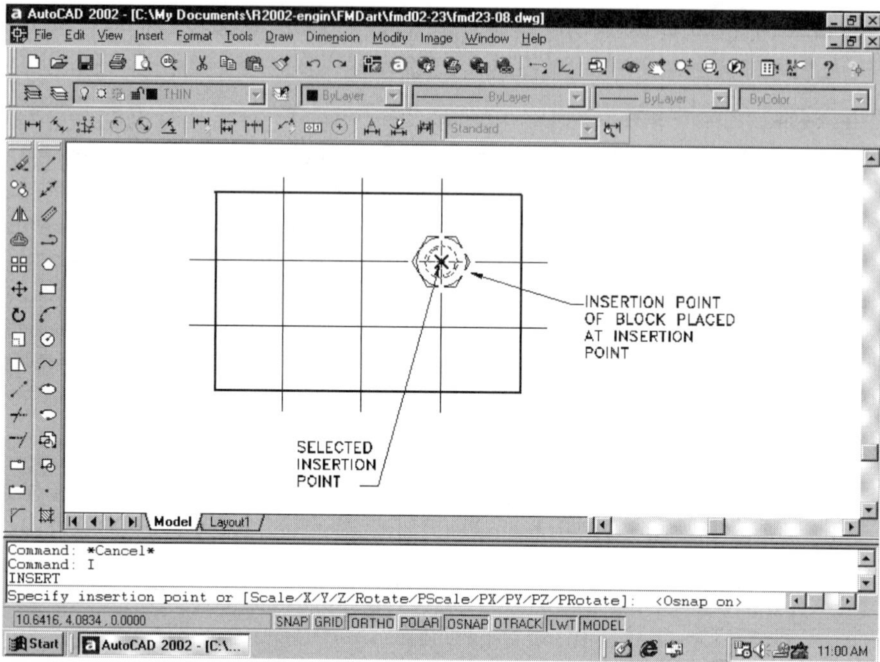

Figure 20.11 *The effect of the selected insertion point of a drawing and the insertion point of a block.*

prompt before the block is inserted. With the box inactive, the X, Y, and Z edit boxes are activated. Unless a different value is entered, the value entered for X will become the default value for Y. Any dimensions contained in the block will be multiplied by the scale factor as the block is inserted. If the nut in Figure 20.11 was drawn as a 1.00", the scale for inserting the hole would be 1. Inserting a nut created as a 1.00" block to be .50" would require an X scale factor of .5. With the **Uniform Scale** check box active, the Y value will equal the X value. The effects of the scale factor can be seen in Figure 20.12.

Negative Scale Factors

The location of a block can be altered by using negative X and Y values. Using negative values combines the **Insert** and **Mirror** commands. Figure 20.13 shows an example of negative value alternatives.

Rotation

The **Rotation** area allows you to supply a rotation angle. With the **Specify On-screen** box active, a prompt will be displayed at the Command prompt before the block

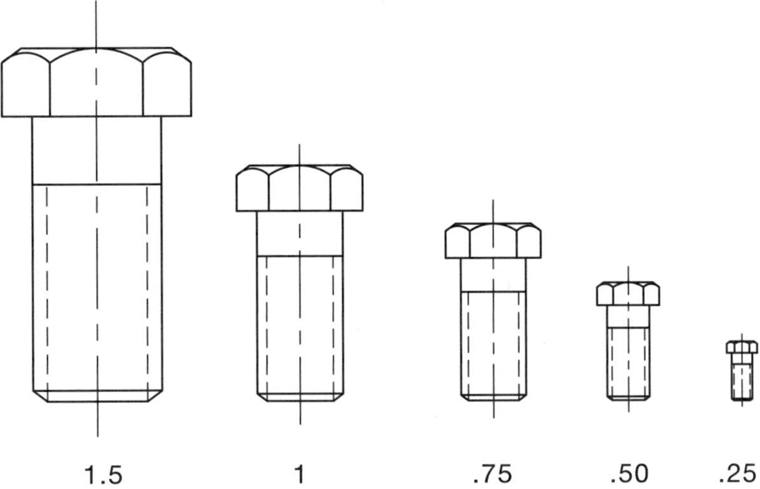

Figure 20.12 *The effect of scale factors on a block.*

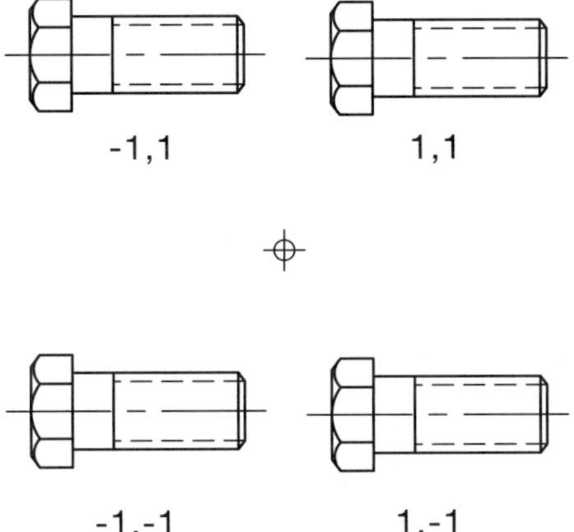

Figure 20.13 *Altering block location using negative scale factors.*

is inserted. With the box inactive, the **Angle** edit box is activated and an angle can be entered to control the rotation of the block at insertion. Figure 20.14 shows an object rotated around its insertion point. Any angle can be provided for the rotation angle.

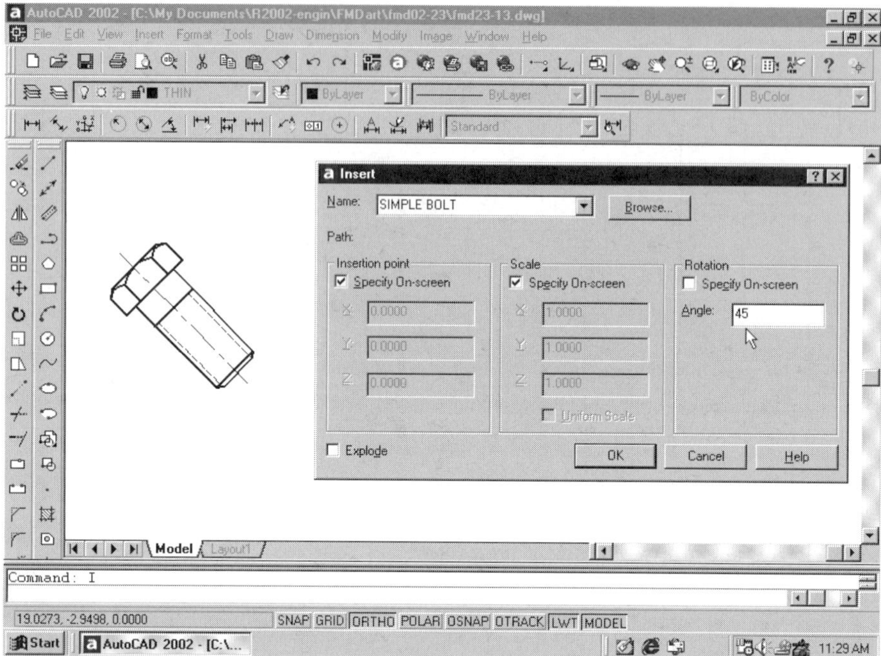

Figure 20.14 *Specifying an angle of rotation allows a block to be rotated relative to its insertion point.*

Explode

Blocks function as one object when they are inserted in a drawing. If the **Explode** check box is activated, as the block is inserted in the drawing, each object of the block will function as a separate object.

INSERTING BLOCKS

Once the parameters of the Insert dialog box have been set, selecting the OK button will close the dialog box and allow a selection point to be specified. With the **Insert** dialog box in the default settings, the following prompt will be displayed at the Command prompt:

> Command: **I** ENTER *(Or click the Insert Block button. Clicking the OK button will close the Insert dialog box and continue the Command prompt.)*
>
> Specify insertion point or [Scale/X/Y/Z/Rotate/PScale/PX/PY/PZ/PRotate]:

Selecting an insertion point will insert the block at the specified parameters. The scale, insertion coordinates, and rotation values that were set in the **Insert** dialog box can be altered at the Command prompt. Additional prompts are also shown at the Command prompt.

WORKING WITH BLOCKS

In addition to inserting blocks, several other options can be used to control the use of blocks. Useful tools include the Minsert command for multiple inserts, using an entire drawing as a block, building blocks by using other blocks in a process known as nesting, exploding blocks, and controlling blocks with the DesignCenter.

MINSERT—MAKING MULTIPLE COPIES

The **Minsert** command combines the features and the command sequences of **Insert** and **Array**. **Minsert** will allow blocks to be arrayed in a rectangular pattern. Start the **Minsert** command by typing MINSERT (ENTER) at the Command prompt. Figure 20.15 shows an example of a multiple insert. The drawing was created using the following command sequence:

Command: **MINSERT** ENTER

MINSERT

Enter block name or [?]:<bolt> **HOLE** ENTER

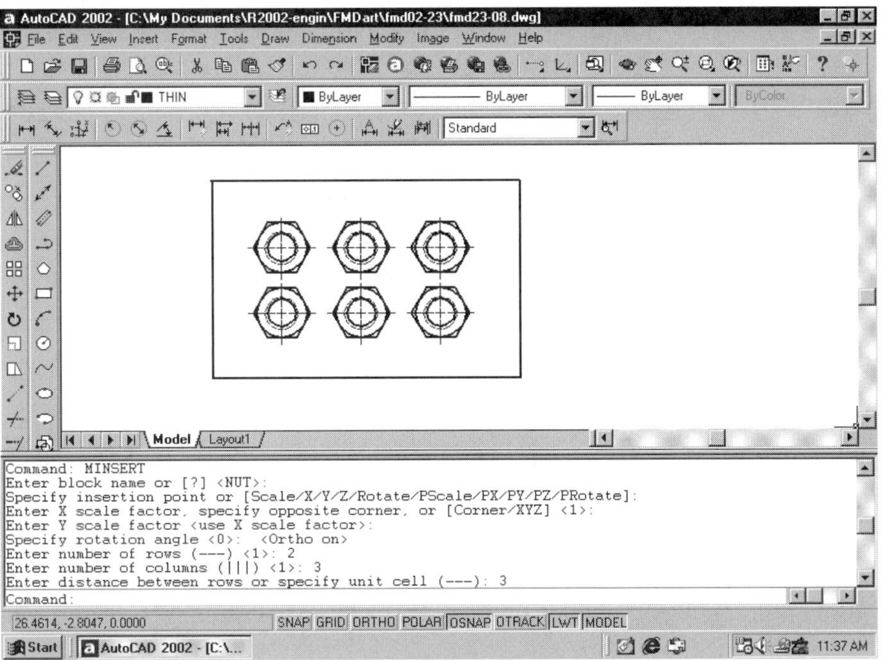

Figure 20.15 *The* **Minsert** *command allows blocks to be arrayed in a rectangular pattern.*

Specify insertion point or [Scale/X/Y/Z/Rotate/PScale/PX/PY/PZ/PRotate]: *(Select the insertion point.)*

Enter X scale factor, specify opposite corner, or [Corner/XYZ] <1>: ENTER

Enter Y scale factor, < use X scale factor>:ENTER

Specify rotation angle: <0>: ENTER

Enter number of rows (—-) <1>: **2** ENTER

Enter number of columns (|||) <1>: **3** ENTER

Enter distance between rows or specify unit cell (—-): **3** ENTER

Specify distance between columns (|||): **2** ENTER

Command:

The array pattern created with **Minsert** has the qualities of a block, except that it cannot be exploded.

MAKING AN ENTIRE DRAWING A BLOCK

An entire drawing can be used to form a block. When the entire drawing is to be a block, an insertion point other than 0,0 might be desired. The **Base** command can be used to define a new base point. The command can be accessed by typing BASE (ENTER) at the Command prompt or by selecting **Base** from **Block** on the **Draw** menu. Each will produce the following prompt:

Command: **BASE** ENTER

BASE

Enter base point: *(Select a point.)*

The base point will now become the insertion point.

A better alternative to this method of using an entire drawing as a block will be introduced later, as wblocks are introduced, and in Web Chapter 5, as externally referenced drawings are explored.

NESTING BLOCKS

So far you've been introduced to creating blocks to represent common symbols or any number of repetitive drawing elements. Blocks can also be combined with other objects or blocks to form a nested block. Figure 20.16 shows an example of nested blocks used to create the electrical components in a turbine control panel drawing. The electrical components in the cabinet drawing were created as a block started in Figure 20.1, nested in a component (Figure 20.2), and then inserted into the control panel. Joining blocks of individual components can create entire units of the control panel.

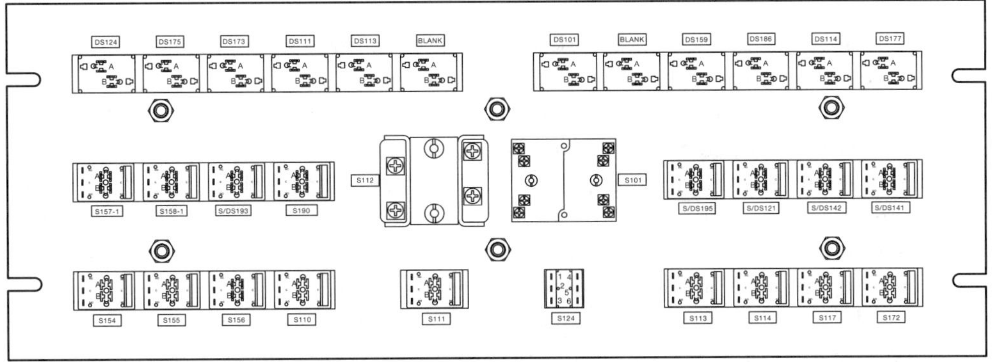

Figure 20.16 *The blocks created in Figure 20.1 and 20.2 were used to create the entire control panel. Courtesy Floyd D. Miller.*

Care must be taken in the use of nested blocks to avoid confusion in naming and layering. Use the following guidelines to avoid confusion:

- Assign properties individually rather than ByBlock if all instances of a block require the same color, layer, linetype and lineweight properties.
- To control the properties of each block using the properties of the layer the block is inserted on, draw the objects that form each block, including nested blocks, on the 0 layer. Set properties such as color, layer, linetype, and lineweight properties ByLayer.

USING BLOCKS IN MULTIPLE DRAWINGS

One of the benefits of working with multiple drawings is that you can move or copy blocks in one drawing to a different drawing. This can be done with opened or closed drawings. Use the following steps to copy blocks between two open drawings.

1. Select the block to be copied.
2. Right-click to display the shortcut menu.
3. Select **Copy**.
4. Move to the other drawing and make it the current drawing by selecting the title bar.
5. Select **Paste** from the shortcut menu or the **Edit** menu.
6. Insert the block in the desired location.

CONTROLLING BLOCKS WITH DESIGNCENTER

The DesignCenter allows blocks to be inserted from drawings that are not open. If you work with consulting firms, you can copy a block from a drawing in their office, if you have the proper access to their network. For now, we'll work on copying a block from one drawing to another. Later chapters will provide information on working with networked machines and consulting firms using the Internet. To copy a block from a drawing file on your computer, display the DesignCenter by selecting the **DesignCenter** button on the **Standard** toolbar. Chapter 21 will introduce additional methods for opening the DesignCenter. The DesignCenter will resemble Figure 20.17. (The exact appearance will depend on what drawings are opened.) Click the **Desktop** button in the upper left corner. This will produce a display exactly like that produced by Windows Explorer. Figure 20.18 shows a portion of this display. Double-clicking a drawing name will produce a display of the contents of the drawing similar to Figure 20.19. Double-clicking the Blocks listing will provide a listing of the blocks contained in the drawing similar to Figure 20.20. Highlighting the name of the desired block displays the block and the display window and allows that block to be dragged into the current drawing.

USING HYPERLINKS

Selecting the **Insert Hyperlink** button on the **Standard** toolbar opens the **Insert Hyperlink** dialog box, allowing a hyperlink to be associated with the block definition. Figure 20.21 shows the Insert Hyperlink dialog box. You can also display the dialog box by selecting **Hyperlink** from the **Insert** menu, or by typing HYPERLINK (ENTER) at the Command prompt. Hyperlinks are created in

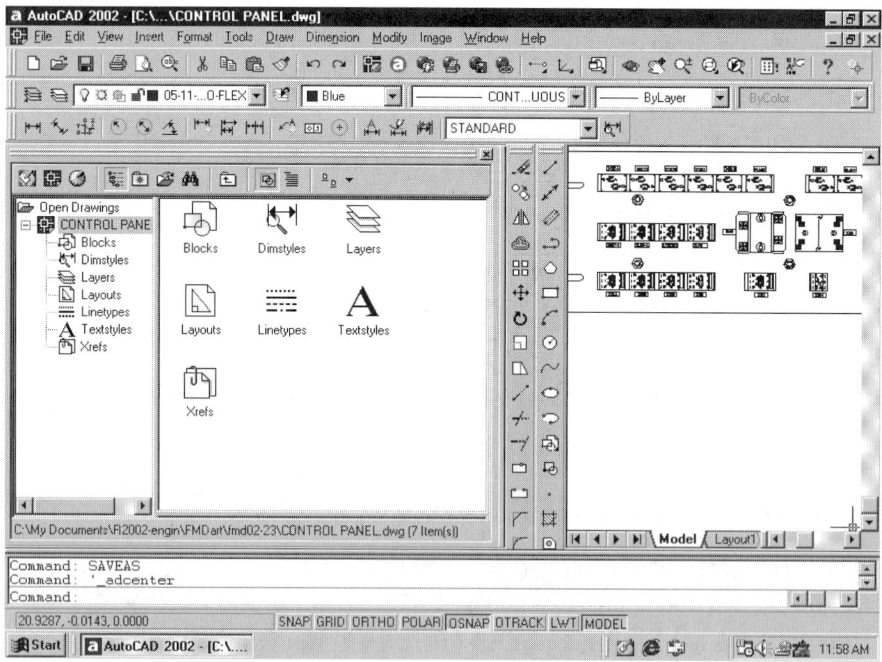

Figure 20.17 *The DesignCenter.*

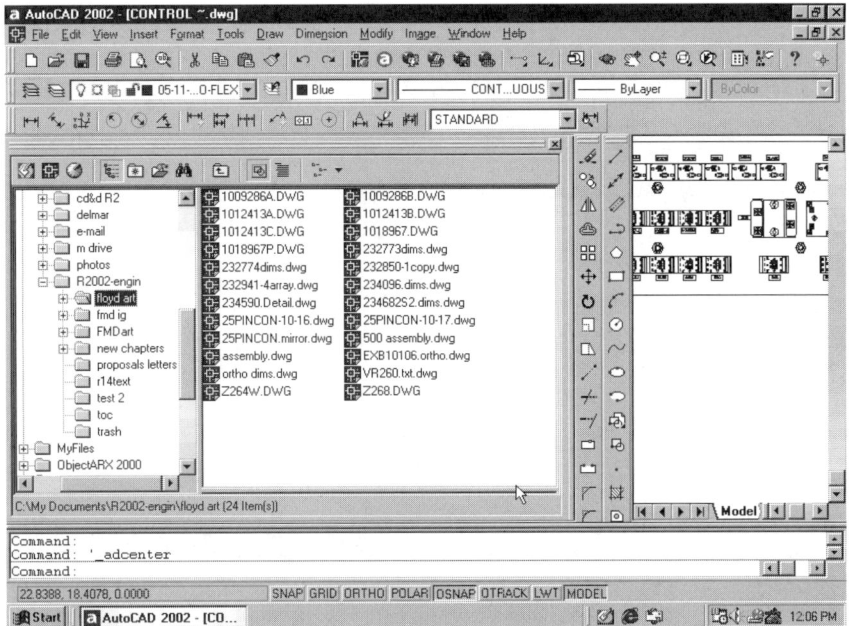

Figure 20.18 *Selecting the* **Desktop** *button in the DesignCenter will produce a display similar to Windows Explorer.*

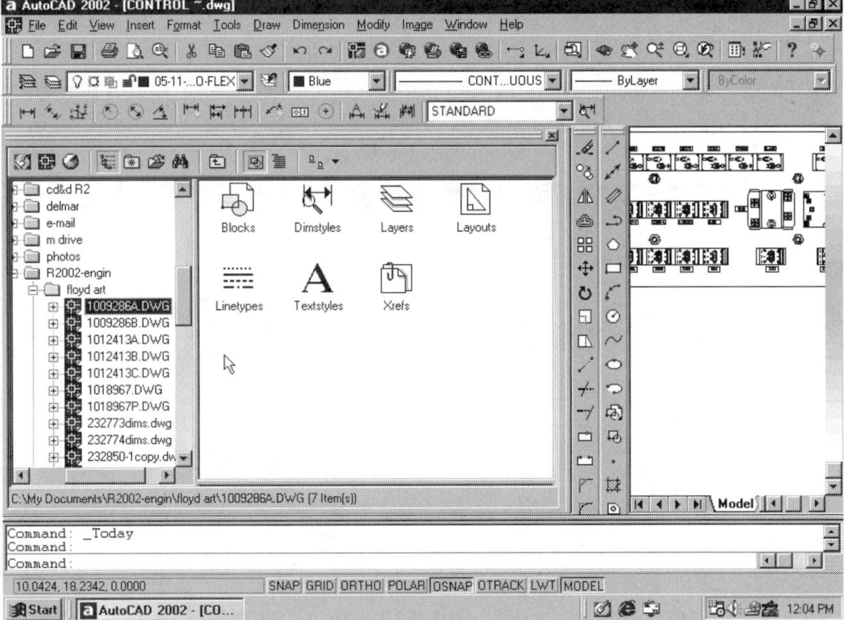

Figure 20.19 *Selecting a drawing from the list will display a list of the contents.*

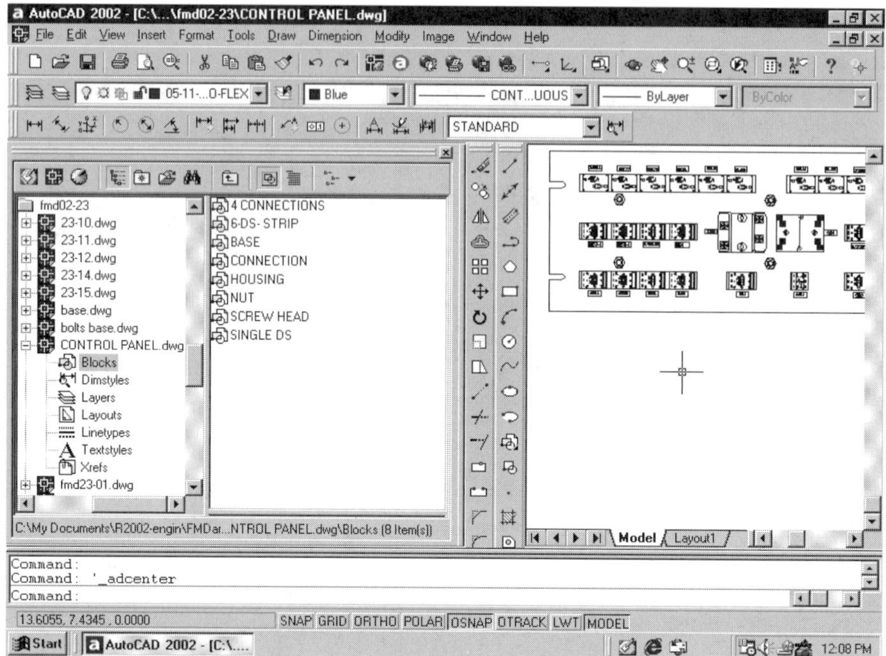

Figure 20.20 *Selecting blocks from the list in Figure 20.19 will display a list of blocks in the selected drawing. One or all of the listed blocks can be copied to a new drawing.*

AutoCAD drawings as pointers to associated files. Hyperlinks can be used for varied task such as launching a word processing program, opening a specific file, pointing to a named location in a file, or activating your Web browser and loading a specified HTML page. Hyperlinks can be used to point to locally stored files, files on a net-

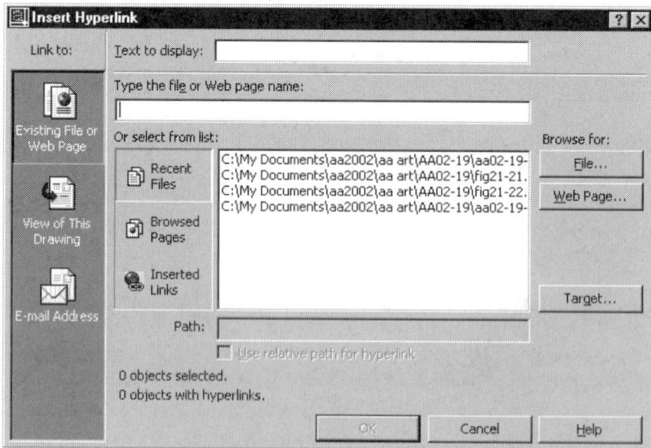

Figure 20.21 *Selecting the* **Insert Hyperlink** *button will display the* **Insert Hyperlink** *dialog box, allowing a hyperlink to be associated with a block definition.*

work drive, or files on the Internet. Cursor feedback is automatically provided to indicate when the crosshairs are over a graphical object that has an attached hyperlink. You can then select the object and use the **Hyperlink** shortcut menu to open the file associated with the hyperlink. Chapter 24 will also explore the **Hyperlink** command.

EDITING BLOCKS

Blocks can be edited by changing the description, by using the **Explode** command, or by using the Redefine option of the **Block** command.

EDITING BLOCK DESCRIPTIONS

A block description can be altered in the **Block Definition** dialog box. Display the dialog box by selecting the **Make Block** button on the **Draw** toolbar or using one of the methods mentioned earlier in this chapter. Select the edit arrow beside the **Name** box. This will display a listing of the current blocks. Select the name of the block that will be altered. Enter the changes that will be made. Typically, this might include adding objects to the block. Click **OK** when you're satisfied with the description.

REDEFINING A BLOCK

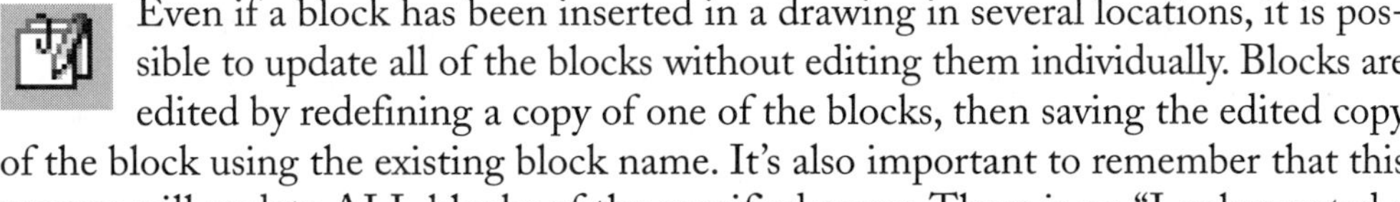

Even if a block has been inserted in a drawing in several locations, it is possible to update all of the blocks without editing them individually. Blocks are edited by redefining a copy of one of the blocks, then saving the edited copy of the block using the existing block name. It's also important to remember that this process will update ALL blocks of the specified name. There is no "I only wanted a few of them" option. Use the following sequence to redefine a block. For the following example, the NUT block of Figure 20.15 will be altered.

1. Select the **Edit Block or Xref** button on the **Refedit** toolbar, select **In-place Xref and Block Edit Reference** from the **Modify** menu or type **REFEDIT** (ENTER) at the command line.
2. Select the block to be edited. Selecting a block to be altered will produce a **Reference Edit** dialog box similar to Figure 20.22.
3. Select the name of the block from the **Reference name** edit box.
4. Select the **Enable unique layer and symbol names** check box.
5. Select the **Display attribute definitions for editing** check box.
6. Click the **OK** button. The command line will prompt to select nested objects.
7. Select the block to be edited. Other uses of the selected block might appear to be altered. Don't panic! The blocks will be restored shortly. As a block is selected, the **Refedit** toolbar will be displayed, if has not been previously opened.
8. Edit the block as desired.

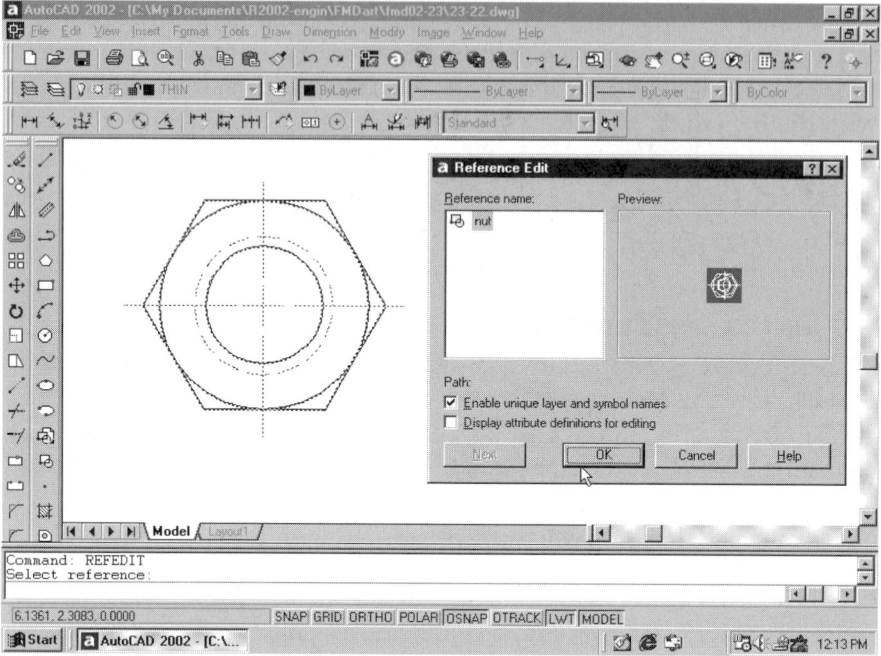

Figure 20.22 *The* **Reference Edit** *dialog box can be used to select a block to be edited.*

9. Select the **Save Back Changes to Reference** button on the **Refedit** toolbar. This will display a warning box indicating that all of the selected blocks are about to be updated.
10. Click the **OK** button in the warning box. The box will be closed and all of the selected blocks will be redisplayed to reflect the indicated changes.

Each nut used throughout the drawing will now be updated to reflect the current changes. Figure 20.23 shows an example of a redefined block.

EXPLODING BLOCKS

If the **Explode** option is activated in the **Insert** dialog box, each object in the block will function independently of other objects in the block. By default, as blocks are inserted in a drawing, all objects in the block will function as one object for editing purposes. Combining all of the objects of a block as one object is great if you want to **Move**, **Rotate**, or **Copy** a block. Only one object has to be selected and the entire block will respond to the command. Using other editing commands such as **Erase** to remove one object also will cause the entire block to be erased. To erase just one object in the block, you must use the **Explode** command (first introduced in Chapter 12). Access the command by selecting the **Explode** button on

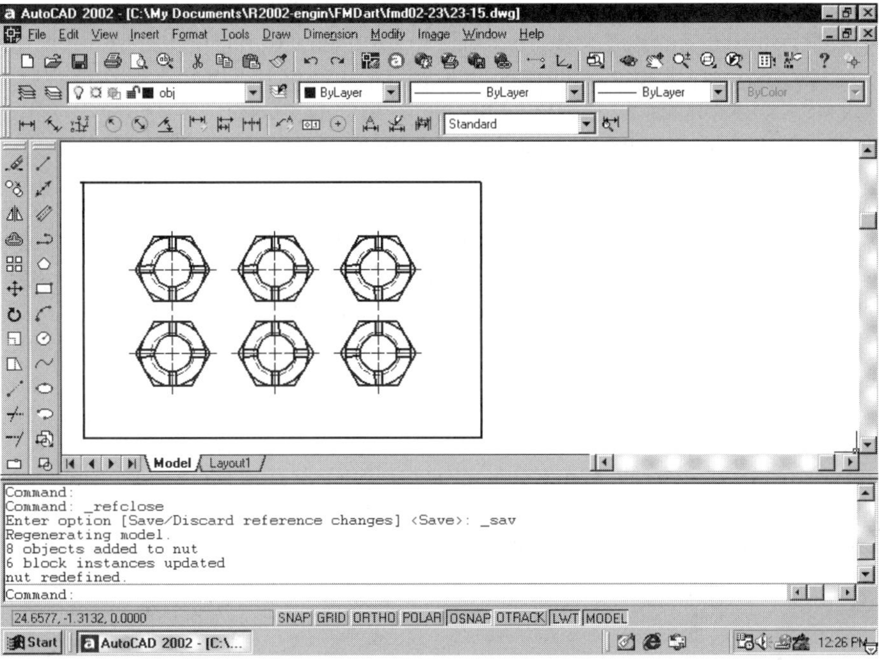

Figure 20.23 *When a block is redefined, all occurrences of the block in the drawing will be updated.*

the **Modify** toolbar, by typing X (ENTER) at the Command prompt, or by selecting **Explode** from the **Modify** menu. The command sequence is as follows:

> Command: **X** ENTER *(Or click the Explode button.)*
>
> Select objects: *(Select the desired blocks.)*
>
> Select object: *(This prompt will continue until all of the desired blocks are selected.)* ENTER
>
> Command:

The block will be displayed with no apparent change. Individual components of the block can now be edited. Blocks with equal X, Y, and Z values will keep their shape when exploded. Blocks with unequal X, Y, and Z values may change shapes when exploded. Blocks that are inserted with **Minsert** can't be exploded.

WBLOCKS

Blocks are a great way to make repeated copies of a symbol throughout a drawing. Blocks are saved in the drawing file they are created in, but with the aid of the AutoCAD DesignCenter, a block can be used on other drawings. If a block is saved on a template drawing, the block can be used each time the template drawing is used. The **Wblock** command provides another option for blocks to be used in multiple drawings no matter where the drawing was created, by saving the block as a .DWG file. A wblock will be displayed as a .DWG file each time a directory of drawings is made, but it will act as a block when inserted in a drawing.

CREATING WBLOCKS

Typing W (ENTER) at the Command prompt will start the command and display a **Write Block** dialog box similar to Figure 20.24. The dialog box is used to create and save blocks to a file for future use. The two major areas of the **Write Block** dialog box include the **Source** and **Destination** areas. The areas that are active will depend on the selections made in the **Source** portion of the dialog box.

Source

The **Source** area of the **Write Block** dialog box selects existing blocks or objects, writes them as a file, and specifies an insertion point for when they are inserted. Options are provided for selecting an existing block or an entire drawing, or specifying objects to form the wblock.

Block

Selecting the **Block** option allows an existing block to be selected to be stored as wblock file. With the **Block** setting active, the block edit box is activated and the **Base**

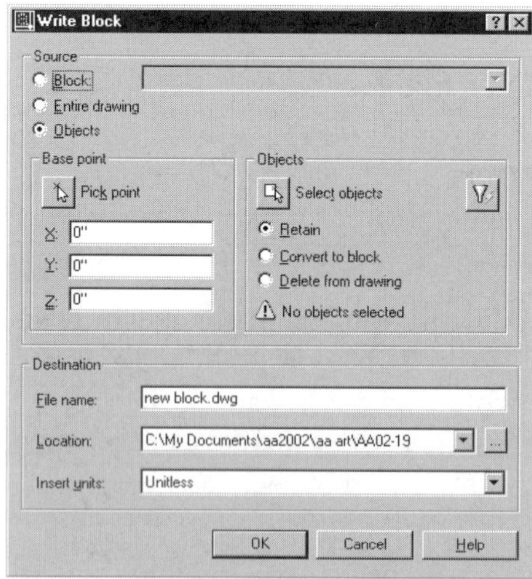

Figure 20.24 *Typing* **W** *(ENTER) at the Command prompt will display the* **Write Block** *dialog box, allowing blocks, drawings, or objects to be saved for future use.*

point and **Objects** portions of the dialog box are inactive. Selecting the edit arrow will display a listing of blocks contained in the current drawing. Selecting a name from the list will allow that block to be inserted. The **Destination** portion of the block displays where the wblock file will be stored. If the **OK** button is clicked, the selected block will be saved as a drawing file in the indicated destination. Methods of altering the destination will be examined in the next section of this chapter.

Entire Drawing

Selecting the **Entire drawing** option allows an entire drawing to be saved as a wblock. This option will have the same results as when **Saveas** is used. The benefit of using the option is that unreferenced objects such as blocks, layers, and linetypes that were not used will be removed from the drawing file. This is an excellent method for reducing the size of a drawing and for saving a drawing at the end of a work session.

Objects

Using the default option of **Objects** allows drawing objects to be selected to form the wblock. The process is very similar to the procedure used to create a block. With the **Object** button active, the **Base point** and **Objects** portions of the dialog box are activated so that objects and a base point can be selected. The order in which the base point or the objects are selected can be altered. Each of these functions is exactly the same as its counterpart in the **Block** command. As each button is selected, the dialog box is removed, allowing for a selection to be made. Once the selection is made, the dialog box is restored, allowing for other options to be altered.

Destination

The box allows the name, location, and units of the drawing file to be specified.

File Name

The **File name** display will vary depending on which of the three **Source** options is active. If the **Block** option is active, the name of the block entered in the **Block** edit box will be displayed in the **File name** edit box. If the **Entire Drawing** box is active, the current name of the drawing will be displayed in the **File name** edit box. With the **Objects** option active, the default name of New Block.dwg will be displayed. Best results will be achieved by providing a more descriptive name.

Location

Because blocks stored using the **Wblock** command are treated as a drawing file, a storage location must be given. The **Location** edit box displays the current drive and path that will be used to store the drawing. Selecting the **Browse for Folder** button allows the drive and folder to be altered. Selecting the button produces a display similar to Windows Explorer. Once a folder name has been selected, the **Create Drawing** dialog box will allow for selection of a file name and a location for the file.

Note: Great care should be given in the naming and storing of wblocks. The name should clearly describe the contents of the drawing. Wblocks should be stored in folders and subfolders that will define the use. For instance, a name of PROTOS could be used for the folder. Subfolder names would include names such as THREADS, ELECTRICAL, HVAC, PIPING or other common drawings to be used. Each subfolder would contain only wblocks for that specific type of drawing.

Summary of Steps in Creating a Wblock

Remember that a wblock is just a collection of objects that can be used over and over in a multitude of drawings. Use the following steps to create wblocks:

1. Display the **Write Block** dialog box by typing **W** (ENTER) at the Command prompt.
2. Choose the source of the block to be created.
3. If the wblock will be comprised of a pre-existing block, provide a name for the block.
4. Select the drawing objects to be reproduced as blocks.
5. Select a base point to be used as the block is inserted in the drawing.
6. Select a name to designate the file.
7. Specify the drive, folder, and subfolder to be used as the destination.
8. Click the **OK** button to save the block as a drawing file.

The object has now been saved as a drawing file and is ready to be reused throughout the existing drawing or in a new drawing. If the nut from Figure 20.8 needs to be added to the existing drawing or to a totally new drawing file, it can be added to the drawing using the **Insert** command. It can also be placed in other drawings using DesignCenter.

INSERTING WBLOCKS

Once a symbol or drawing has been saved as a wblock, it can be used just as any other drawing. It can be retrieved using the drawing **Startup** dialog box as a new drawing session is started, or it can be inserted in an existing drawing using the **Insert** command. The **Insert** command will bring an existing drawing into the current drawing from the Command prompt. Each aspect of inserting a wblock is the same as for inserting a block.

EDITING WBLOCKS

One of the strengths of using an entire drawing as a wblock is that when it is inserted, it will move as one object. A wblock can be altered using the same commands used to edit a block.

CHAPTER 20 EXERCISES

1. Open the tolerance block that was created in Web Chapter 2 and save this as a block.
2. Use an appropriate template to complete the following drawing exercise. Use the block Pipe Fittings library supplied with AutoCAD and the attached sketch as a guide to complete the attached non-scaled drawing. Save the drawing as E-20-2.

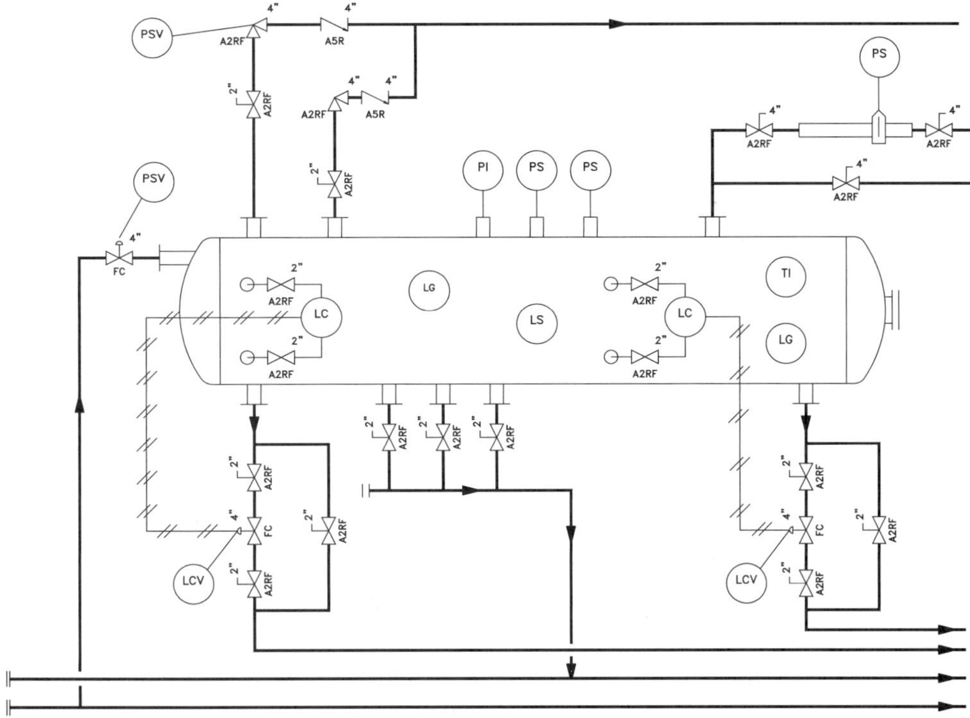

Problem 20.2

3. Open drawing Web E-3-1 and save this drawings as block. Save the drawing as E-20-3.

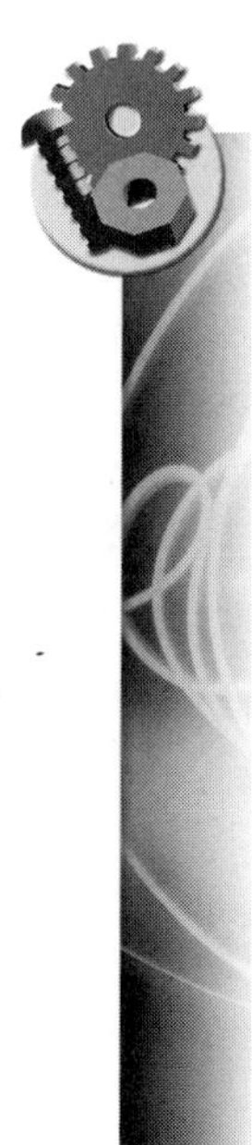

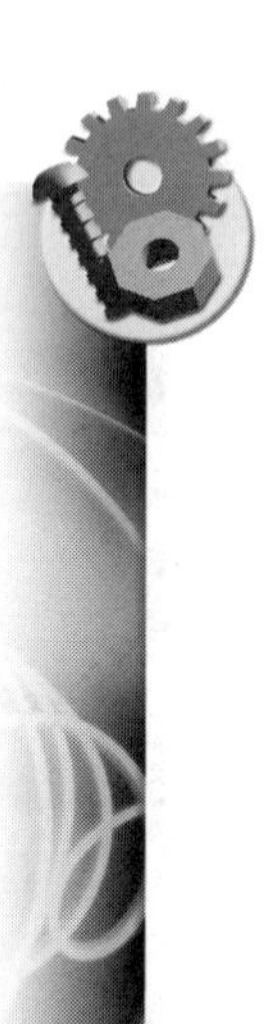

4. Use an appropriate template to complete the following drawing problem. Use the block Basic Electronics library supplied with AutoCAD and the attached sketch as a guide to complete the attached non-scaled drawing. Save the drawing as E-20-4.

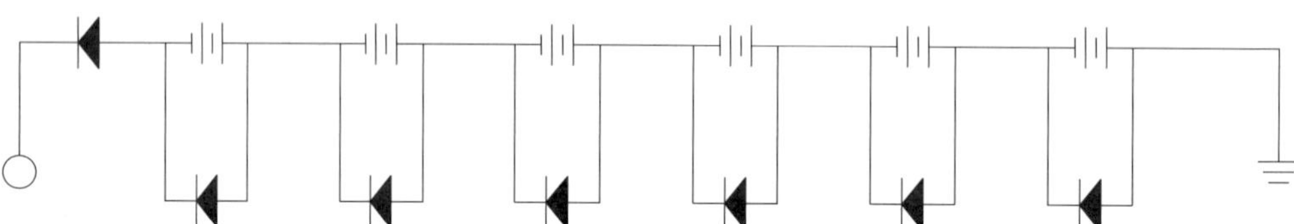

Problem 20.4

5. Open drawing Web E-3-3 and save this drawing as a block. Save the drawing as E-20-5.
6. Start a new drawing and insert four copies of the block created in E-20-3. Insert the block using four different values for X and Y. Save the drawing as E-20-6.
7. Start a new drawing and insert the block created in exercise 5 using five different rotation angles. Save the drawing as E-20-7.
8. Start a new drawing and insert the block created in exercise 5. Use the polar option of the **Array** command and array 15 copies of the bolt around a center point. Do not rotate the pattern around the center point. Save the drawing as E-20-8a. Make a second copy of the bolt and rotate the bolt around the center point. Save this copy as E-20-8b.
9. Use the drawing saved as E-20-4 and the attached sketch to create the following non-scaled drawing. Save the drawing as E-20-9.
10. Use the block Basic Electronics library supplied with AutoCAD and the attached sketch to complete the attached non-scaled drawing. Save the drawing as E-20-10.
11. Open the drawing saved as E-20-8b. Edit the length of the bolts and save the drawing as E-20.11.

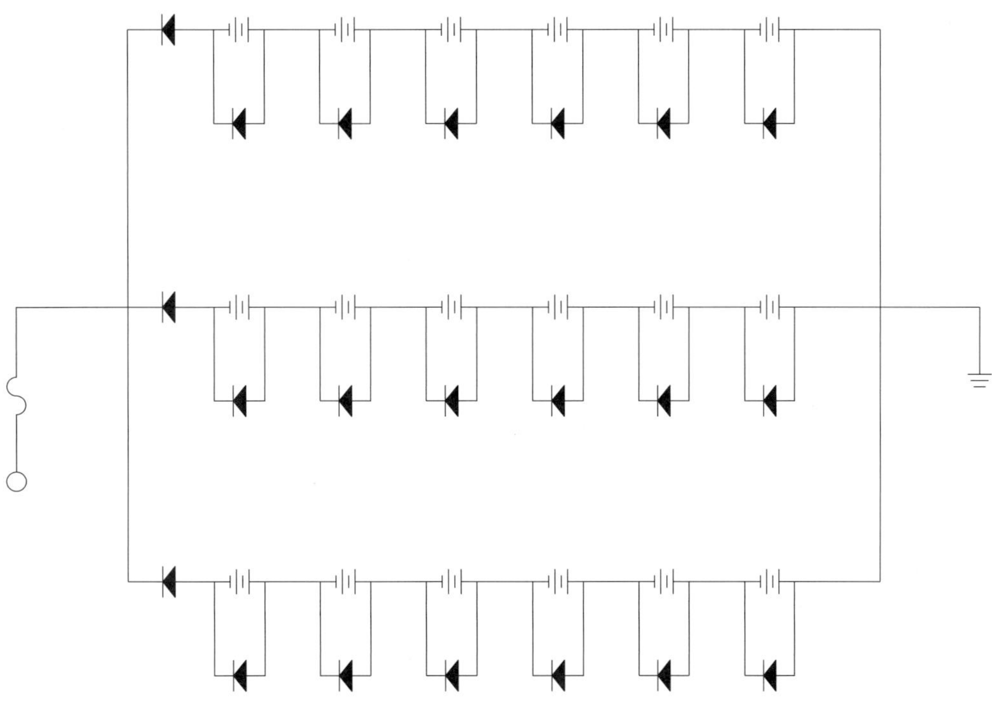

Problem 20.9

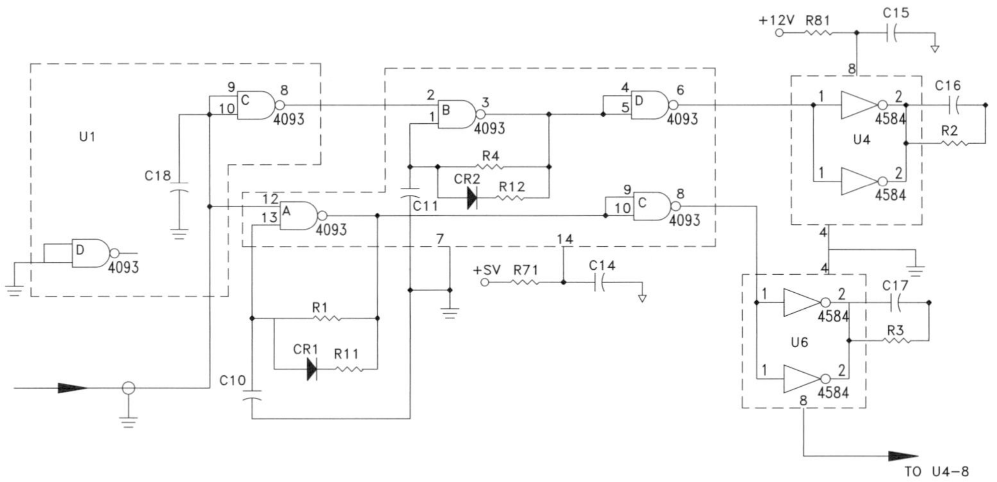

Problem 20.10

CHAPTER 20 QUIZ

1. How can a wblock increase drawing speed?
2. Why is using blocks more efficient than copying an object?
3. If a block is created on the 0 layer, where will the block be located when it is inserted?
4. When the color of a block is set "BY BLOCK," how will the block be affected?
5. What is the advantage of drawing blocks such as a bolt or nut in a 1" box rather than full size?
6. List the process to create a block.
7. How does the Insertion Point specified when the block is defined affect the insertion point requested when the block is inserted?
8. What command will allow for multiple blocks to be inserted?
9. What command will allow for editing blocks?
10. What is the quickest method for editing the same mistake in several blocks?
11. Explain the difference between a block and a wblock.
12. List the process for inserting a wblock in a drawing.
13. How can a block be mirrored before it is inserted?
14. What method can be used to insert a block by selecting two points?
15. What is the effect of activating the **Explode** check box in the **Insert** dialog box?

Working with Multiple Drawings

This chapter will introduce

- General considerations for multiple document environments
- Methods of moving information between two or more drawing files
- Features of DesignCenter

Commands to be introduced include

- **Cut**
- **Copy**
- **Paste**
- **Copybase**
- **Pasteblock**
- **Pasteorig**
- **Adcenter**

THE MULTIPLE DOCUMENT ENVIRONMENT

Template drawings have been used throughout this text to store frequently used styles, settings, and objects for use in similar drawings. This chapter will introduce methods for moving objects and information between two or more drawing files. Objects or drawing properties can be moved between drawings using cut/copy and paste, object dragging, the **Match Properties** command, and concurrent command execution.

GENERAL CONSIDERATIONS FOR MULTIPLE DOCUMENT ENVIRONMENTS

AutoCAD is by default set to work with multiple documents. The toggle for switching from multiple to single document interface is the **Single Drawing Compatibility Mode** button in the **General options** area of the **System** tab in the **Options** dialog box. (Access the dialog box by selecting **Options** from the **Tools** menu.) With the box inactive, multiple drawings can be opened in one AutoCAD session. Don't gloss over the concept of multiple drawings and one drawing session. Just as you can have several programs open on your desktop at once and rapidly switch from one program to another, AutoCAD allows you to have several drawings open at the same time. Chapter 1 introduced the techniques for displaying multiple drawings. Start AutoCAD and select **Open a Drawing**. Select an existing drawing to be opened. While holding down CTRL, select three other drawings to be opened, and then click the **Open** button. You have now used one of the most powerful tools of AutoCAD and have four drawings opened in one drawing session.

Now select **Cascade** from the **Window** menu. Your drawing display will resemble Figure 21.1. The title bar of each drawing will be displayed. The drawing with the blue title bar on top of the stack is the active drawing. Apart from being on top of the pile, it is the only drawing that can currently be used. Return to the **Window** menu and select **Tile Vertical**. Now select **Tile Horizontal** instead from the **Window** menu. The drawing display will now resemble Figure 21.2. With four drawings open, the horizontal and vertical displays will appear alike. With an odd number of drawings open, they will be displayed in vertical rows with the **Tile Vertical** option. Important considerations to keep in mind when multiple drawings are open are as follows:

- In any **Window** display selection, any one of the four drawings can be made active. Notice that each small drawing area has all of the components of a normal AutoCAD drawing.

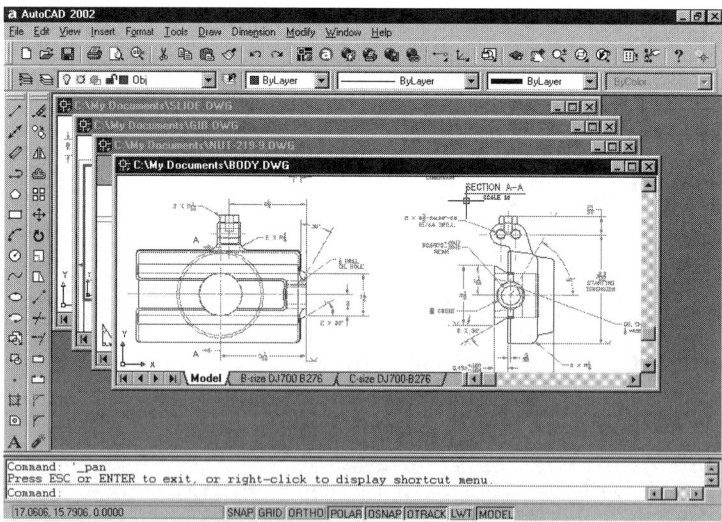

Figure 21.1 *Selecting* **Cascade** *from the* **Window** *menu will display the title bar of each drawing. The drawing with the blue title bar on top of the stack is the active drawing. Courtesy Hanh Tran.*

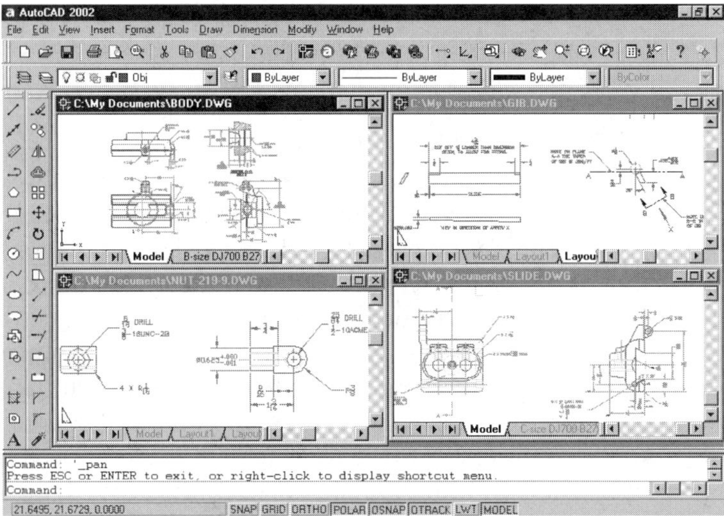

Figure 21.2 *Selecting* **Tile Horizontal** *from the* **Window** *menu will display each open drawing. In this orientation, any one of the four drawings can be made active. Notice that each small drawing area has all of the components of a normal AutoCAD drawing. Courtesy Hanh Tran.*

- Any one of the four drawings can be maximized or minimized to ease viewing, or all can be left open and you can switch from drawing to drawing as needed.
- You can start a command in one drawing, switch to another drawing and perform a different command, and then return to the original drawing and complete the command in progress. As you return to the original drawing, click anywhere in the drawing area, and the original command will be continued.
- Each drawing is saved as a separate drawing file, independent of the other open files. You can close a file without affecting the contents of the remaining open files.

Opening Multiple Drawings from the Windows Explorer

In addition to being opened from within AutoCAD, multiple drawings can also be opened from Windows Explorer. To start the process, open Windows Explorer and double-click the drawing to be opened. Once the drawing is opened, minimize the drawing and reduce the size of the AutoCAD drawing session by clicking the **Restore** button at the top right side of the toolbar so that the drawings in Windows Explorer can still be seen. With a drawing open, additional drawings can be dragged from Windows Explorer and dropped into the current drawing session. Drag Files by pressing and holding CTRL and the left mouse button while the desired drawing is selected. Once the Drawing icon is in the drawing area, release CTRL and select button to open the drawing. Drawings will be placed in a cascading display similar to Figure 21.1.

Note: If a drawing is dragged from Windows Explorer into a current drawing, the **Insert** command will be activated. Instead of opening multiple drawings, you'll be inserting a drawing in the current drawing.

USING THE CUT, COPY, AND PASTE COMMANDS

In addition to standard procedures that are allowed using the **Cut**, **Copy**, and **Paste** commands of any Windows software, AutoCAD allows drawing objects and properties to be transferred directly from one drawing to another using these same commands. Each command can be performed using the **Edit** menu, by keyboard, or by shortcut menu. **Cut** will remove objects from a drawing and move them to the Clipboard, from where they can be placed in a new drawing using **Paste**. **Copy** allows an object to be reproduced in another location. The object to be copied is reproduced in its new location using the **Paste** command. As the object is pasted to the new drawing, so are any new layers, linetypes, or other properties associated with the object. Figures 21.3a and 21.3b show examples of using **Copy** to place the bearing support detail into the second drawing. The bearing support detail was copied by using the following sequence:

From the Source Drawing:

Command: *(Select Copy from the shortcut menu.)*

_copyclip

Select objects: *(Select object to be copied using any selection method.)*

Select objects: ENTER

Command:

In the Second Drawing:

Command: *(Select Paste from the shortcut menu.)*

_pasteclip

Select insertion point: *(Select the location for the new object.)*

Command:

Copy with Base Point

The **Copy with Base Point** command on the shortcut menu is similar to the **Copy** command. You can also start the command by typing COPYBASE (ENTER) at the Command prompt. With this option, before you select objects to copy, you'll be prompted to provide a base point. This option works well when objects need to be inserted accurately. The piping diagram in Figure 21.4 was copied into the new drawing using the **Copy with Base Point** command. The piping diagram was copied using the following sequence:

From the Source Drawing:

Command: *(Select Copy with Base Point from the shortcut menu.)*

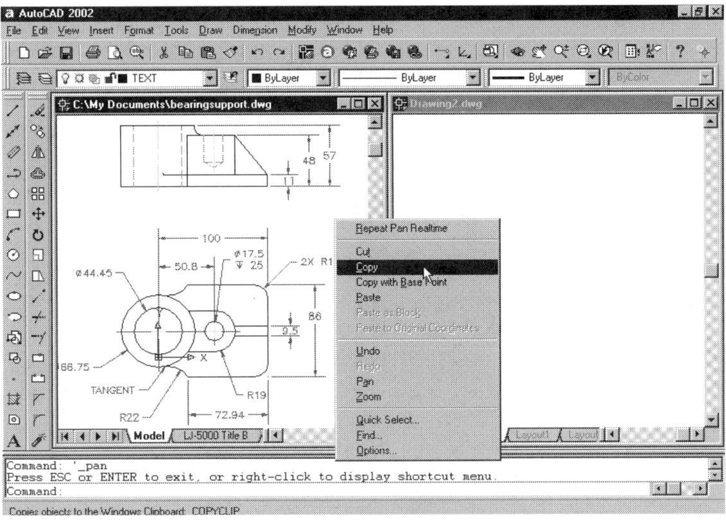

Figure 21.3a *The bearing support in the left drawing can be copied into the right drawing by selecting* **Copy** *from the shortcut menu. Courtesy Paul Bissell.*

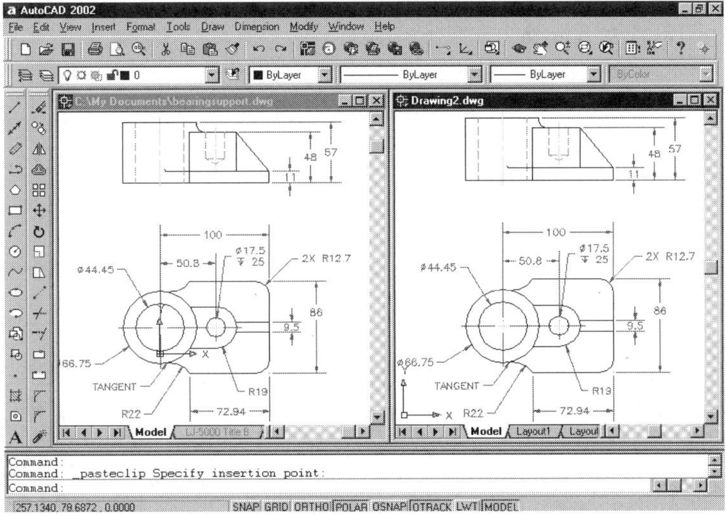

Figure 21.3b *Once the object is selected to be copied, move the cursor to the desired drawing, and single-click to make that drawing active. Right-click and choose* **Paste** *from the shortcut menu, and then select an insertion point. Courtesy Paul Bissell.*

_copybase

Specify base point: *(Select the base point for the objects to be copied.)*

Select objects: *(Select object to be copied using any selection method.)*

Select objects: ENTER

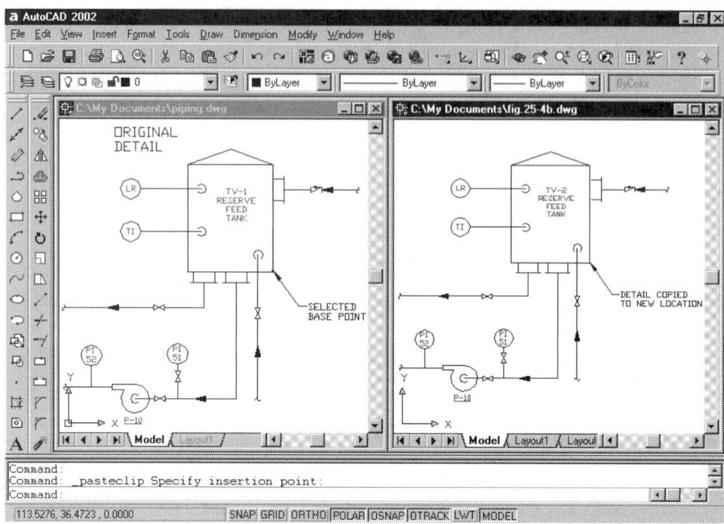

Figure 21.4 *The* **Copy with Base Point** *command, similar to the* **Copy** *command, works well when objects need to be inserted accurately. With this option, before you select objects to copy, you'll be prompted to provide a base point. Courtesy Manfull-Curtis Inc.*

Command:

In the Second Drawing:

Command: *(Select Paste from the shortcut menu.)*

_pasteclip

Select insertion point: *(Select the location for the new object.)*

Command:

Paste as Block

The **Paste as Block** command can be used to paste a block into a different drawing. It functions similarly to the **Paste** command. Select the objects to be copied in the original drawing and select the **Copy** command. Selected objects do not need to be a block, because the **Paste as Block** command will turn them into a block. Next activate the new drawing and right-click to select the **Paste as Block** command. The copied objects are now a block. To be edited, they must be exploded. You can also start the command by typing PASTEBLOCK (ENTER) at the Command prompt. The command sequence is the same as for the **Paste** command. You will notice that blocks that are copied will be renamed. A name such as CAP will be assigned a name such as A$C2AA9F67C4 by AutoCAD. The new block can be renamed using the **Rename** command.

Paste to Original Coordinates

This command can be used to copy an object in one drawing to the exact same location in another drawing. The command can be started by using the shortcut menu or by typing **PASTEORIG** (ENTER) at the Command prompt. The command is only active when the Clipboard contains AutoCAD data from a drawing other than the current drawing. Selecting an object to be copied starts the command sequence, and it is completed by making another drawing active and then selecting the **Paste to Original Coordinates** command from the shortcut menu.

DRAGGING OBJECTS

AutoCAD allows objects to be moved between drawings using a drag and drop sequence. Dragging allows the following operations:

- Move or copy objects to a new location in the same drawing.
- Move or copy objects to another drawing in the same drawing session.
- Copy objects to a drawing in another session.
- Copy objects to another application.

Dragging can be carried out using either the left or right mouse buttons. Slightly different results and procedures will be required using each method.

Selecting Objects to Drag with Left-Click

The procedure to drag using the left mouse button begins by selecting the objects to be moved or copied. Any object selection method can be used to select objects. Once the set is selected, move the cursor over one of the selected objects and press and hold down the left mouse button. Moving the cursor while the left button is held down will show the effect of moving the selected objects. Release the mouse button when the objects are in the desired location. If you're working with a single drawing, the selected objects will be moved. Objects can be copied in a single drawing by pressing CtrlKey prior to pressing the left mouse button to drag the object. (Objects will always be copied when dragged into a different drawing.) You are not allowed to drag objects into a drawing that has an active command. The unavailable cursor—a circle with a diagonal line through it—is displayed if you attempt to drag objects into a drawing with an active command. Figure 21.5 shows the steps to copy a diagram into another drawing using the left mouse button.

Selecting Objects to Drag with Right-Click

The procedure to drag using the right mouse button begins by selecting the objects to be moved or copied. Any object selection method can be used to select objects. Once the set is selected, move the cursor over one of the selected objects and press the right mouse button. Moving the cursor while the button is held down will show the effect of moving the selected objects. Release the mouse button when the objects are in the desired location. As the right button is released, a menu will be displayed. The menu will depend on what type of drawing you're working in. The menu options are as follows:

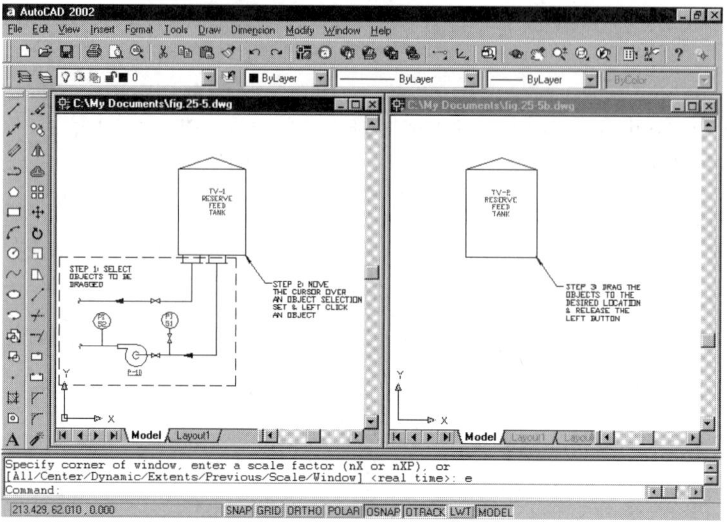

Figure 21.5 *Objects can be moved or copied easily using the drag and drop technique. Courtesy Manfull-Curtis Inc.*

Drag in the Same Drawing	Drag in a Different Drawing
Move Here	Copy Here
Copy Here	Paste as Block
Paste as Block	Paste to Orig Coords
Cancel	Cancel

Select a menu option to end the drag and drop sequence.

MATCHING PROPERTIES

The **Match Properties** command cannot be used to move or copy objects, but it can be used to copy properties from one object to another. The **Match Properties** command was introduced in previous chapters for transferring properties within a drawing. The command can also be used to transfer properties from an active drawing to other drawings. The process to change the layer and linetype of objects in a drawing to match another drawing can be completed using the following process:

1. Select the source object using any selection method.
2. Select **Match Properties** on the **Standard** toolbar.
3. Move to the target drawing and make it the current drawing. The **Match Properties** cursor will be displayed by the selection box.
4. Select the target objects. The command can be continued within the current drawing, or another drawing can be made current, and the properties of the original object can be transferred to additional objects.

EXPLORING THE AUTOCAD DESIGNCENTER

The DesignCenter can be used to locate, organize, and customize drawing information. It can be used to directly transfer data such as blocks, layers, and external references from one drawing to another without using the clipboard. Drawing content can be moved from other open files in AutoCAD or from files stored on your machine, on a network drive, or on an Internet site. Access the DesignCenter by selecting the **AutoCAD DesignCenter** button on the **Standard** toolbar, or by typing ADCENTER (ENTER) at the Command prompt. Either method will display a DesignCenter similar to Figure 21.6. The DesignCenter can also be displayed or closed using CTRL+2.

The default position of DesignCenter is docked on the left side of the drawing screen. You can resize the DesignCenter by clicking a border, the palette sidebar or by grabbing a corner and moving the cursor in a diagonal direction. If you are using the DesignCenter, the **Properties** dialog box, and a drawing, you might be tempted to shrink the size of the DesignCenter window. Resist shrinking the DesignCenter past the point where two rows of icons can't be seen in the palette display. The display can be reduced so no palette is displayed, but then you've limited the usefulness of the display. Clicking and dragging the title bar will float the DesignCenter, allowing it to be moved to any location in the drawing area. Figure 21.7 shows an example of a floating DesignCenter. You can dock the DesignCenter by double-clicking

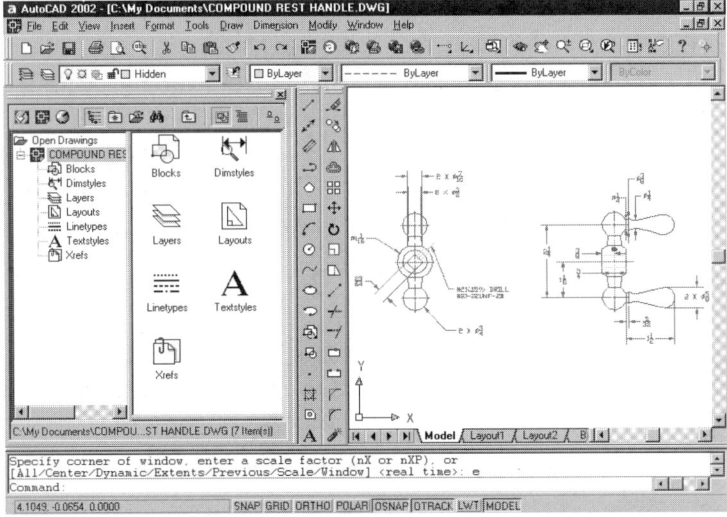

Figure 21.6 *Access the DesignCenter by selecting the* **AutoCAD DesignCenter** *button on the* **Standard** *toolbar or by typing* **ADCENTER** *(ENTER) at the Command prompt. The DesignCenter can be used to directly transfer data such as blocks, layers, and external references from one drawing to another. Courtesy Hanh Tran.*

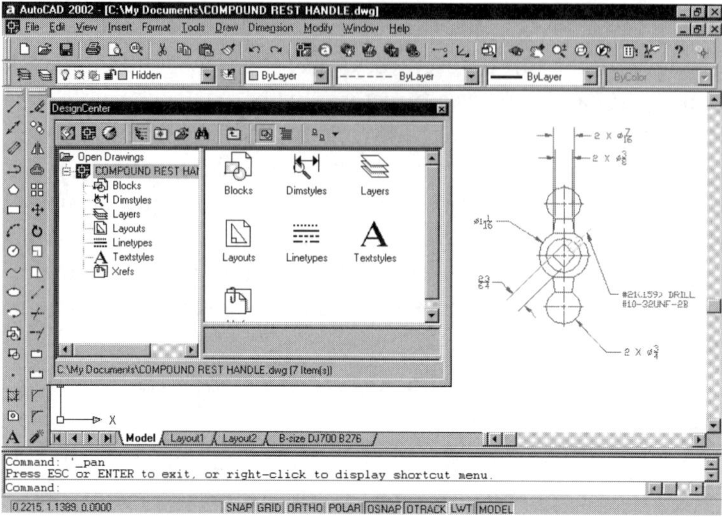

Figure 21.7 *Clicking and dragging the title bar will float the DesignCenter, allowing it to be moved to any location in the drawing area. You can dock the DesignCenter by double-clicking the title bar of the DesignCenter or by dragging it to the left or right side of the drawing area. Courtesy Hanh Tran.*

the title bar of the DesignCenter display or by dragging it to the left or right side of the drawing area.

FEATURES OF DESIGNCENTER

The DesignCenter provides direct access to any drawing file that you have on your machine, network, or the Internet. Files stored on diskette, the hard drive, network or Internet can be accessed and reused as a library for retrieving or coping blocks, referenced drawings, layouts, layers, linetypes, text styles or dimension styles. Other operations of the DesignCenter include the following:

- Create a Favorites folder to create a shortcut to frequently accessed drawings, folders, or Internet locations.
- View, attach, insert, or copy and paste layer and block definitions of any file you have access to into the current drawing. This feature could be used to update a drawing supplied by another user to match current standards.
- Open drawing files by dragging the file from the tree view into the current drawing area.
- Find drawing contents stored in any folder, drive, network, or from any Internet site that you have access to, based on search criteria for the source. Search tools such as a name or date can be used to locate materials to be dragged into the current drawing file.

- Use the DesignCenter palette to view aspects of the object, block, or drawing content.

EXAMINING THE DESIGNCENTER TOOLBAR

The **DesignCenter** toolbar is located across the top of the DesignCenter. Before examining the contents, you might need to adjust the size of the DesignCenter window to be sure that all of the contents are displayed. The toolbar consists of eleven buttons for controlling the contents of the DesignCenter. These buttons, from left to right, are **Desktop**, **Open Drawings**, **History**, **Tree View Toggle**, **Favorites**, **Load**, **Find**, **Up**, **Preview**, **Description**, and **Views**. Each can be seen in Figure 21.8.

Desktop and Open Drawings

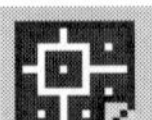

These buttons serve as toggles to control what is displayed in the DesignCenter. When **Desktop** is selected, the display will resemble Windows Explorer. All resources available to your workstation will be displayed, including internal drives, local network, and Internet files. The exact display will depend on the contents of the selected drive and the setting for the **Tree View Toggle** button. Selecting the **Open Drawings** button will display a list of drawings that are currently opened, similar to Figure 21.9. Notice that the drawings have a + sym-

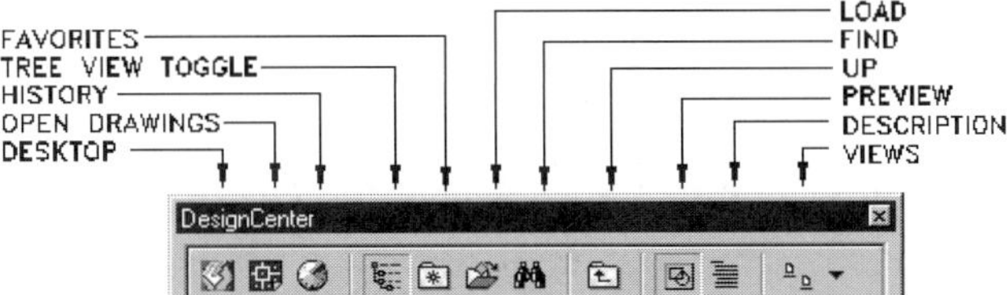

Figure 21.8 *The buttons on the* **DesignCenter** *toolbar.*

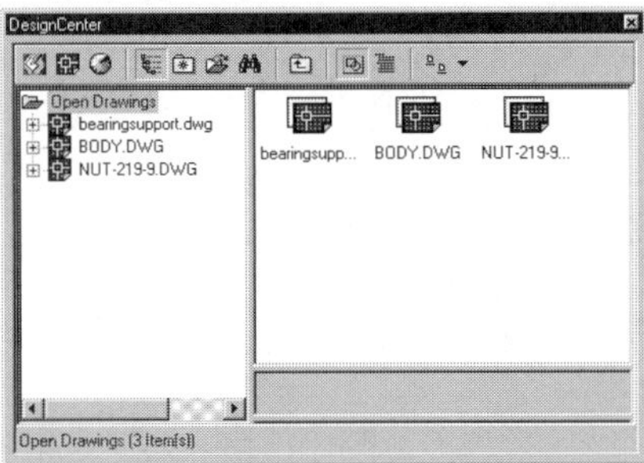

Figure 21.9 *Selecting the* **Open Drawings** *button will display a list of drawings that are currently opened. Notice that the drawings have a + symbol preceding the title. The + symbol indicates that additional information can be displayed about the drawing file or folder. Double-clicking the drawing name will display the drawing contents and double-clicking a drawing with a full display (–) will close the contents of that file.*

bol preceding the title. The + symbol indicates that additional information can be displayed about the drawing file or folder. Double-clicking the drawing name will display the drawing contents and double-clicking a drawing with a full display (–) will close the contents of that file.

History

Selecting this button will display a listing of the last twenty locations accessed through the DesignCenter. The history can also be displayed by right-clicking in the palette and then selecting **History** from the shortcut menu.

Tree View Toggle

This button displays or hides the tree view in much the same way that **Details** and **Large Icons** of the **View** menu of Windows Explorer alter that display. Tree view displays open drawings, history, and files or folders for all of the drives available to your computer. Figure 21.9 shows a display using the tree view. Figure 21.10 shows a display with the tree view hidden. You can hide the tree view by clicking the **Tree View Toggle** button or by right-clicking in the palette and selecting **Tree** from the shortcut menu. With the tree view hidden, you can then restore the tree by clicking the button or by right-clicking in the palette and then clicking **Tree** from the shortcut menu.

Favorites

This option of the DesignCenter displays shortcuts to files or folders that you need to access quickly on a regular basis. The tree, palette, and shortcut menus each include menus for accessing the Favorites menu.

Load

Selecting the **Load** button displays the **Load DesignCenter Palette** dialog box. The dialog box is similar to the **Select File** dialog box used to open drawings. You can also display it by right-clicking in the palette and then selecting **Load** from the shortcut menu. You can choose any available drive from the

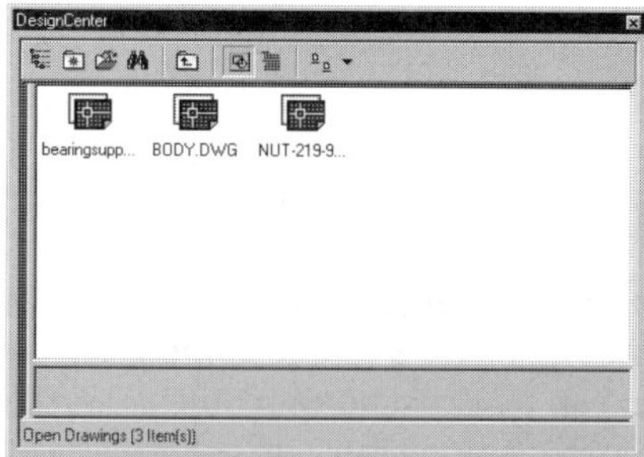

Figure 21.10 *The DesignCenter display with the tree view hidden.*

Look in edit box. The initial step in using the DesignCenter is to load the palette. Use the following steps to load the palette:

1. Select the **Load** button.
2. In the Load DesignCenter Palette dialog box, choose one of the following options:

   Look in—This source displays the files and folders on local and network drives.

   Favorites—The **Favorites** button displays the contents of the Favorites folder.
3. Select the listing that is to be loaded.

Find

In addition to selecting information from the desktop, you can search for drawings or parts of drawings using **Find**. Selecting the **Find** button displays a **Find** dialog box similar to Figure 21.11. The dialog box can be used to narrow the search to find drawings, blocks, layers, layouts, text styles, dim styles, and referenced drawings. Currently the search will be conducted in all drawings. Selecting the **Look for** edit box allows the search to be limited to Blocks, Dimstyles, Drawings, Drawing and Blocks, Hatch pattern files, Hatch patterns, Layers, Layouts, Linetypes, Text Styles, or Xrefs. Depending on the field, the tabs in the dialog box will be altered. The tabs for finding drawings include **Drawings**, **Date Modified**, and **Advanced**.

Drawings

This tab can be used to specify the name or text that will be the object of the search. In Figure 21.12, the name COVER PLATE is entered in the **Search for the word(s)** edit box. The **In the field(s)** edit box can be used to limit the search crite-

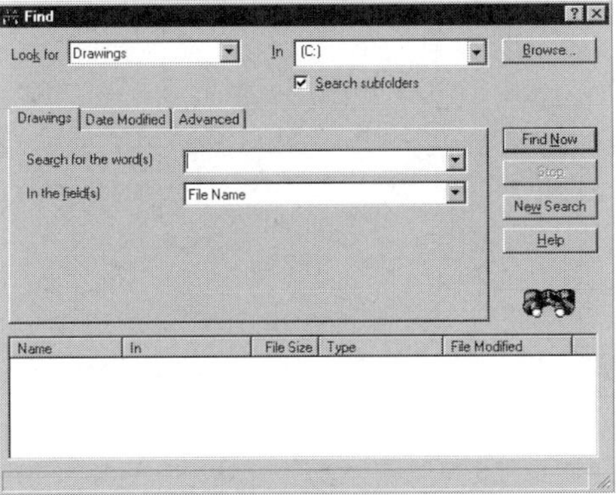

Figure 21.11 *Selecting the* **Find** *button displays a* **Find** *dialog box. The dialog box can be used to find drawings, blocks, dimstyles, layers, layouts, linetypes, text styles, and referenced drawings.*

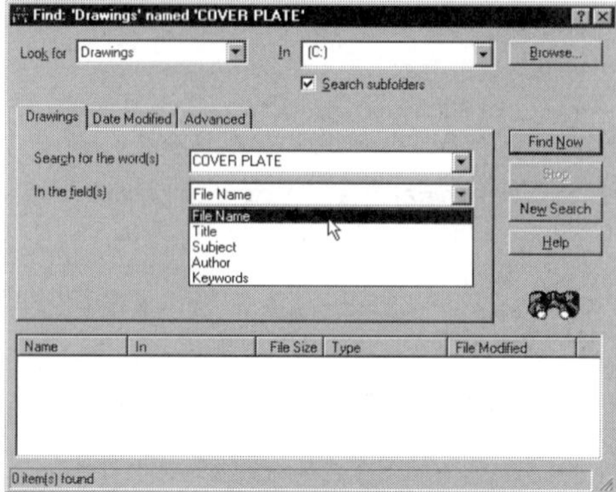

Figure 21.12 *Entering COVER PLATE in the* **Search for** *edit box and clicking the* **Find Now** *button will conduct a search for all file names containing the words COVER PLATE.*

ria. In the current setting, the search will be conducted in file names. Alternatives include looking in titles, subjects, author, and keywords.

Date Modified

The **Date Modified** tab allows a search to be conducted based on the date that the file was created or last modified. A date range specified in days or months can be used if you're not sure of the exact date the drawing was created.

Advanced

The **Advanced** tab allows additional search parameters to be set based on text contained in a portion of a file or by the file size. A search can be defined using the Containing edit box to search block name, block and drawing descriptions, attribute tag, or attribute value for text containing, but not limited to, a specified text. The search parameters can also be limited to looking in files of a minimum or maximum size.

Up

The **Up** button is used to move through the contents of the DesignCenter. The **Up** button performs the same function here as it does in the **Select File** dialog box. Each time the button is selected, the display is moved up one level from the current location.

Preview

The **Preview** window displays an image of a selected file or block.

Description

If a description was created for objects such as a block, this portion of the dialog box will display the description. The description will be given for the file or object that is displayed in the **Preview** window. A description can also be displayed by right-clicking in the palette, and then selecting **Description** from the shortcut menu.

Views

Selecting **Views** presents one of four options for displaying the format of the contents of the DesignCenter. Each time the button is selected, a different method of presentation will be displayed. Selecting the edit arrow beside the button will present a menu of the four options, so that the desired display method can be selected without having to scroll through each option. Display options include **Large icons**, **Small icons**, **List**, and **Details**.

Shortcut Menus

Right-clicking in the palette produces a shortcut menu similar to Figure 21.13. The shortcut menu contains the same options as found on the **DesignCenter** toolbar.

OPENING DRAWINGS USING THE DESIGNCENTER

The DesignCenter can be used to open drawings. Start AutoCAD using any of the options presented throughout this text. The DesignCenter can now be used to drag a drawing into an empty drawing area with the following steps:

1. Display the DesignCenter.
2. Select the drawing source from the Desktop.
3. Drag the icon of the drawing to be opened and drop it in the drawing area.
4. Provide the insertion point and responses to the scaling prompts.
5. Resize the drawing window as needed to display the drawing.

A drawing can also be placed in another drawing by using the shortcut menu. From the DesignCenter, select the file to be opened, right-click, and select the **Open to Drawing** option from the shortcut menu.

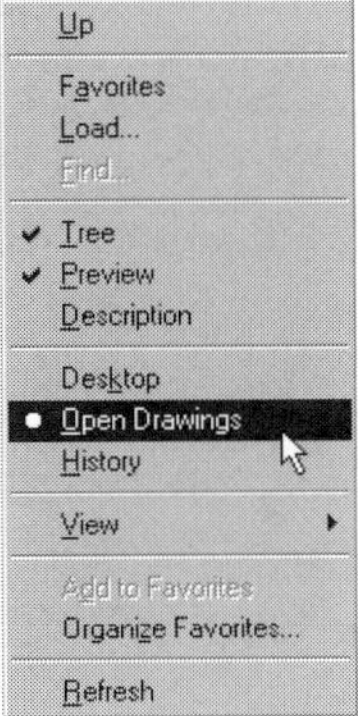

Figure 21.13 *Right-clicking in the palette produces a shortcut menu that contains the same options as found on the* **DesignCenter** *toolbar.*

ADDING CONTENT TO DRAWINGS USING THE DESIGNCENTER

One of the best uses for the DesignCenter is adding content from the palette to a new or existing drawing. Information can be selected from the palette or from **Find** and dragged directly into a drawing without opening the drawing containing the original. Previous chapters introduced methods of moving blocks using DesignCenter. DesignCenter can also be used to attach external referenced drawings and copy layers between drawings.

Inserting Blocks

The DesignCenter provides two methods for inserting blocks in a drawing: using the default scale and rotation of the block, or altering the block parameters as the block is dragged into the new drawing.

Providing Parameters for Block Insertion

For most purposes this will be the preferred method of inserting a block using the the DesignCenter. Use the following sequence to specify the insertion point, scale, and rotation angle as a block is inserted:

1. Display the DesignCenter.
2. Use the palette to select the block to be copied.
3. Use the right mouse button to drag the desired block to the new drawing.
4. Release the right button to drop the block into the drawing area. This will display a shortcut menu.
5. Choose **Insert Block** from the shortcut menu. This will display the **Insert** dialog box.
6. Enter the values for each parameter or choose the **Specify On-screen** option for each parameter.
7. Click the **OK** button to insert the block at the specified location.

Using the Default Scale

When blocks are inserted using the default scaling and rotation option, the block is inserted using Autoscaling. Autoscaling compares the units of the drawing with those of the block. As the block is inserted, AutoCAD scales the block as needed, based on the ratio of the block to the new drawing. Blocks inserted using this method might not retain their true size. Use the following sequence to insert a block using the default parameters:

1. Display the DesignCenter.
2. Use the palette to select the block to be copied.
3. Drag the desired block to the new drawing.
4. Drop the block at the desired location.

Attaching Referenced Drawings

The DesignCenter can be used to attach a referenced drawing using steps similar to those used to attach a block and provide the parameters. Use the following procedure to bring an xref into a new drawing.

1. Display the DesignCenter.
2. Use the palette to select the xref to be attached.
3. Use the right mouse button to drag the desired xref to the new drawing.
4. Release the right button to drop the xref into the drawing area. This will display a shortcut menu.
5. Choose **Attach Xref** from the shortcut menu. This will display the **External Reference** dialog box.
6. Select **Attach** or **Overlay** from the **Reference Type** box.
7. Enter the values for each parameter or choose the **Specify On-screen** option for each parameter.
8. Click the **OK** button to complete the command.

Working with Layers

The DesignCenter can be used to copy layers from one drawing to another. Typically, a template drawing containing stock layers can be used when you create new drawing files. The DesignCenter can be used to drag layers from a template, or any other drawing, into the new drawing and ensure drawing consistency. Drag layers into a new drawing using the following command sequence:

1. Display the DesignCenter.
2. Use the palette or **Find** to select the drawing that contains the layers to be copied.
3. Double-click the name of the drawing to display the options of DesignCenter.
4. Double-click Layers.
5. Highlight the names of the layers to be added to the new drawing (holding down CTRL allows you to select several layers at once).
6. Press the right mouse button and drag the layer names to the new drawing.
7. When the cursor is in the new drawing area, release the right button to display the shortcut menu.
8. Select **Add Layers(s)** from the shortcut menu.

ACCESSING FAVORITE CONTENTS

As with most programs, the AutoCAD DesignCenter allows frequently used sources to be stored for easy retrieval. The final section of this chapter will explore methods of adding shortcuts and displaying and organizing your favorite drawing tools.

Adding Shortcuts to Favorites

Folders, files, drawings, or drawing content can be added to the Favorites list. Selecting a drawing or other option to be placed in the Favorites list adds a shortcut to the AutoCAD Favorites folder. The selected item is not moved to the folder, only the directions for quickly fetching it. Use the following procedure for adding items to the AutoCAD favorites:

1. Right-click the item to be listed as a favorite.
2. Select **Add to Favorites.** The selected item will be listed the next time the Favorites listing is displayed.

Display the contents of the Favorites folder by selecting the **Favorites** button on the DesignCenter toolbar or by right-clicking in the palette background and selecting **Favorites.** The shortcut saved in the Favorites folder can be copied, moved, or deleted using the Autodesk Explorer. The explorer is accessed by right-clicking in the palette and choosing **Organize Favorites** from the shortcut menu. Figure 21.14 shows an example of the Autodesk Explorer.

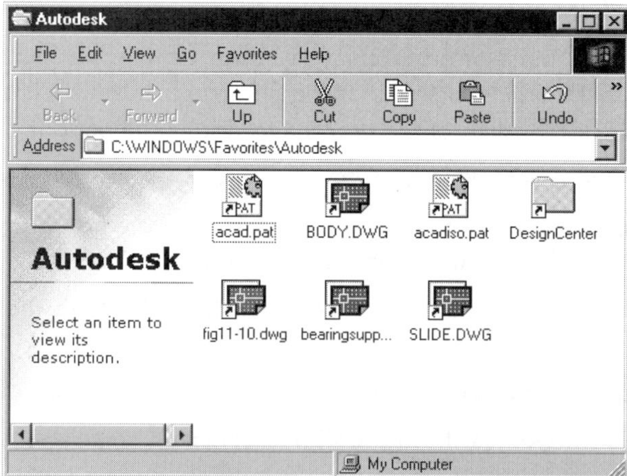

Figure 21.14 *Right-clicking in the palette and choosing* **Organize Favorites** *from the shortcut menu will display the Autodesk Explorer. Each of the four options is the same as their counterparts in Windows and AutoCAD.*

CHAPTER 21 EXERCISES

1. Open any three drawings and copy features from one drawing to the other two. Save the drawings as E-21-1a, E-21-1b, and E-21-1c.
2. Open the drawings created in exercise 1. Copy one object from one of the drawings using the base point option so that the new object aligns with an existing intersection. Save the drawing as E-21-2.
3. Open one of the drawings from exercise 2. Use the **Paste as Block** option to place two of the blocks created in Chapter 20 in the drawing. Save the drawing as E-21-3.
4. Open an ANSI-C template drawing. Use the DesignCenter and move layers, text styles, and dim styles to a new ANSI-D template. Save the drawing as E-21-4.
5. Open a drawing that you've created and move five tolerance symbols into the drawing. Save the drawing as TOLERANCE.

CHAPTER 21 QUIZ

1. Describe the options for inserting blocks using the DesignCenter.
2. Having multiple drawings opened in one drawing session is a powerful tool. Describe the different options used to display your drawings, and explain how to toggle from one drawing to another.
3. Explain the process to copy an object from one drawing to another in a multiple drawing session.
4. Explain the difference between **Paste with Original Coordinates** and **Paste as Block**.
5. How can the **Match Properties** command be used in a multiple drawing environment?
6. List two methods for opening the DesignCenter.
7. What determines whether an object will be moved or copied using drag and drop methods with the left mouse button?
8. Explain the effects of right-click on drag and drop methods.
9. How can the **History** button of the DesignCenter be used?
10. Describe the four displays that can be used in the DesignCenter.
11. You need to find a binder screw detail created in early September. Explain the process.
12. What do the + and – symbols mean when they are displayed by a folder or file name.
13. Explain the process to place a drawing in the Favorites folder.

14. A drawing template contains five of the best layer names ever created. How can these layers be transferred into a new drawing?
15. You've come to your senses and want to remove three of the layers that were just added to the drawing. How can this be done?

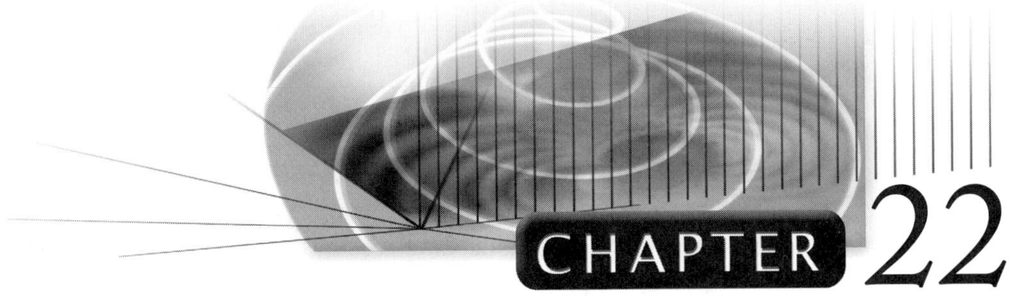

Working with Layouts and Viewports

This chapter will introduce

- Comparisons between model space and paper space
- Creating and using layouts
- Working with tiled viewports
- Working with floating viewports

Commands to be introduced include

- **Layout**
- **Vports**
- **Mspace**
- **Pspace**
- **Vpclip**

Chapter 2 introduced the layout tabs of the AutoCAD template drawings. A layout is a paper space drawing environment that represents how your drawing created in model space will appear on a sheet of paper when plotted. Throughout this text you've used the model space tab of the ANSI-C template. The template has been edited to include your favorite linetypes, lineweights, text fonts, and dimensioning styles. This chapter will introduce you to using the layout tabs to provide a predictable plotting setup. Methods of using existing layouts and creating new layouts will be explored. By completing the steps in this chapter, you'll be able to take drawings that have been created in previous chapters, place them in a viewport within the layout, and prepare them for plotting. Methods for creating multiple viewports within a drawing will also be explored. Multiple viewports allow objects that must be displayed at different scales within the same drawing to be plotted easily.

COMPARING MODEL AND PAPER SPACE

As you begin to prepare a drawing for plotting, it's important that you understand the use of the terms model space, paper space, and layout. In Chapter 2 you were introduced to drawing templates that were created in paper space. Throughout the text you've been drawing in model space. Model space is where you create the drawing; paper space is where you plot. A layout is a paper space tool that contains one or more viewports to aid in plotting. If you open the ANSI-C template and select the ANSI-C Title Block tab, your drawing is displayed in paper space. The paper space UCS icon is displayed in the lower left corner, and the ANSI-C title block surrounds your drawing. If you click the PAPER button on the status bar below the Command prompt, the UCS icon is changed to reflect model space and is placed inside the drawing template. Your drawing is now in a viewport or window that is in model space within a paper space border. Open a template or refer to Chapter 2, and you'll notice that the ANSI-C template contains a black line that overlays the border. This box is the viewport (see Figure 22.1). You can edit the drawing in the model space viewport without having to return to the model space tab. Later in this chapter you'll be introduced to methods for dividing model space into tiled viewports to represent different views of the drawing. Switch back to paper space by clicking the MODEL button, and the drawing is displayed in the paper space layout with viewports, ready for plotting.

Note: As a general rule, NEVER alter the location of the drawing in paper space. When the drawing location is altered in paper space, it can affect layouts by other drafters who might be working on the project. Alter the drawing in model space and then return to paper space to display the drawing with the title block and borders.

AN OVERVIEW OF PAPER SPACE

To understand the process of viewing the assembly detail in Figure 22.2, visualize a sheet of 11" × 17" vellum with a title block and border printed on the sheet. Imagine a hole, the viewport, cut in the vellum that allows you to look through the paper and see the detail. This is a simplified version of what is required to display a drawing for plotting. Figure 22.3 shows the theory of displaying a drawing in model space to a layout in a template drawing created in paper space. The detail shown in Figure 22.2 is 90" wide in model space. If you were to hold a sheet of B size vellum in front of the plan, the paper would seem minute. To make the assembly fit inside the viewport on the paper, you're going to have to hold the paper a great distance away from the assembly until the drawing is small enough to be seen through the hole. AutoCAD will automatically figure the distance. By entering the desired scale factor for plotting, you can reduce the drawings to fit inside the viewport and maintain a scale typically used in manufacturing. Multiple viewports can be placed in a single layout, and multiple layouts can be created within a single drawing file.

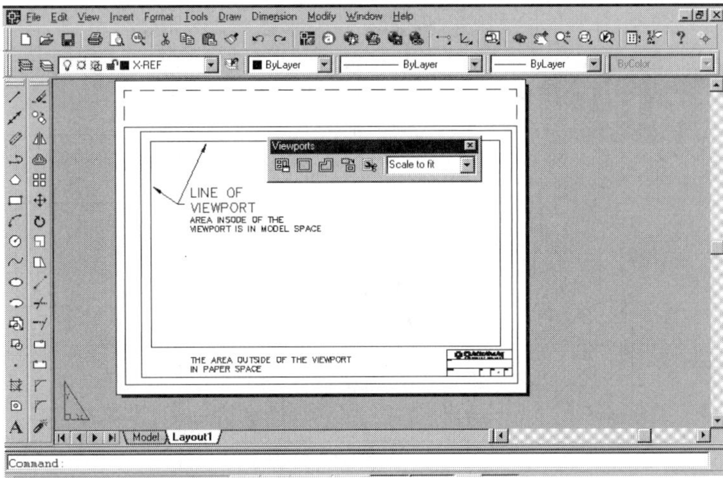

Figure 22.1 *The ANSI template contains a black line that overlays the border. This box is the viewport.*

WORKING WITH LAYOUTS

Throughout this text you've been using the ANSI-C template or a template that you've created to meet your needs. When your drawing is complete, you can select the layout tab to begin preparing the drawing for reproduction. If you're using the ANSI-C template, the layout tab is titled ANSI-C Title Block. As the tab is displayed for the first time, the **Page Setup** dialog box, introduced in an earlier chapter, is displayed. (Chapter 23 will deal with page setup in depth.) Layout settings for plotting can be

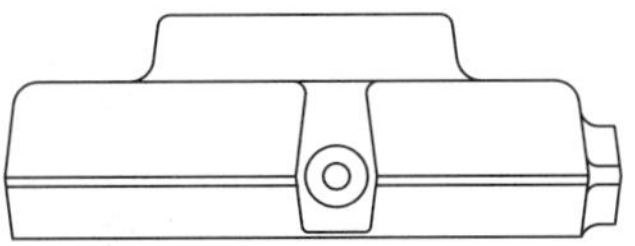

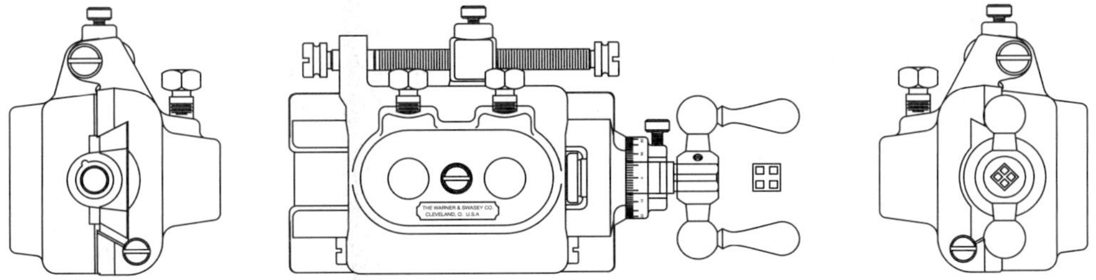

Figure 22.2 *An assembly drawing that is 90" long can be displayed in paper space using a paper space layout. Courtesy Hanh Tran.*

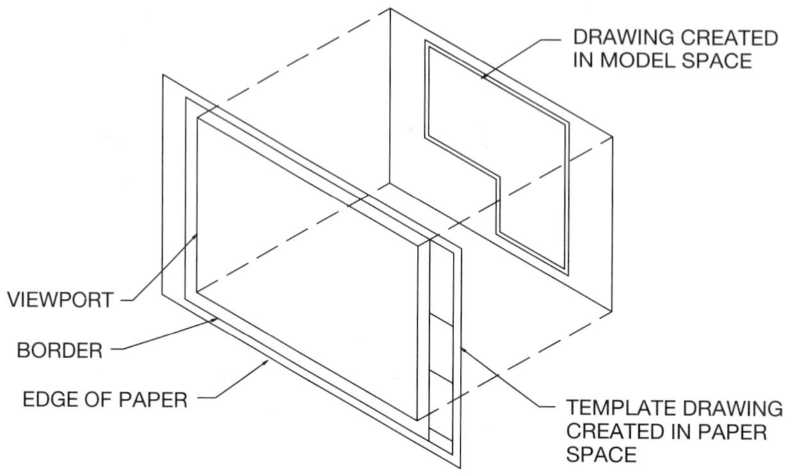

Figure 22.3 *A viewport is provided in each drawing layout created in paper space. Additional layouts can be created to allow for different plotting displays.*

set once and stored with the drawing. For now, the goal is to help you create drawing displays that are ready for plotting. When your drawing is complete, use the following steps to adjust a drawing in the viewport in paper space:

1. Select the ANSI C tab.
2. Click the PAPER button on the bottom of the screen to enter model space in the viewport.
3. Use the All or Extents option of **Zoom** to view the drawing.
4. Use **Pan** to center the drawing in the layout.
5. Click the MODEL button on the bottom of the screen to return to paper space.

The procedure to view the drawing in paper space is simple and should be familiar by now. In addition to using layouts for plotting, you can create several layouts for the same drawing so that different parameters can be emphasized in the plotting process. Different scales and paper sizes can be assigned to different layouts, or different plotting devices can be assigned to the layouts.

CREATING NEW LAYOUTS

The existing template and layout work well for displaying a single drawing for plotting. Additional paper space layouts can be created to plot various layers, or in some other way customize a drawing for a specific plotting need. Layouts can be assigned a unique name, and multiple paper space viewports can be assigned to one or more of the layouts. Independent settings can be assigned and stored in each layout for plotter and page setup. Once created, layouts can be copied, deleted, moved, or renamed. Selecting **Create Layout** from **Wizards** on the **Tools** menu allows a new layout to be created. The **Create Layout Wizard** contains a series of pages that walk you through the process of creating a new layout. Options are similar to the plotting options introduced in Chapter 5. Each setting will be further discussed in the next chapter.

Figure 22.4 shows an example of the beginning page of the **Create Layout Wizard**. This page is used to assign a name to the layout being created. Once the desired name is entered, selecting the **Next** button will produce a display for choosing a printer from a list of configured printing devices. Using multiple layouts with multiple plotter configurations will allow you to create check prints with a laser printer, or full size plots with a plotter without having to adjust plotter settings each time you make a plot. Once the plotter to be assigned to the layout is selected, choosing **Next** will display a **Create Layout - Paper Size** dialog box similar to Figure 22.5. Depending on the configured plotter, this dialog box allows the paper size and drawing units to be selected.

Selecting the **Next** button allows the orientation of the plot to the paper to be configured. Options include **Portrait** or **Landscape**. When you are finished, selecting the **Next** button allows a predefined template to be assigned to the layout. By default, the name of the template assigned to the previous layout will be highlighted and displayed in the **Preview** area. By using multiple layouts, you can display various title blocks with the drawing.

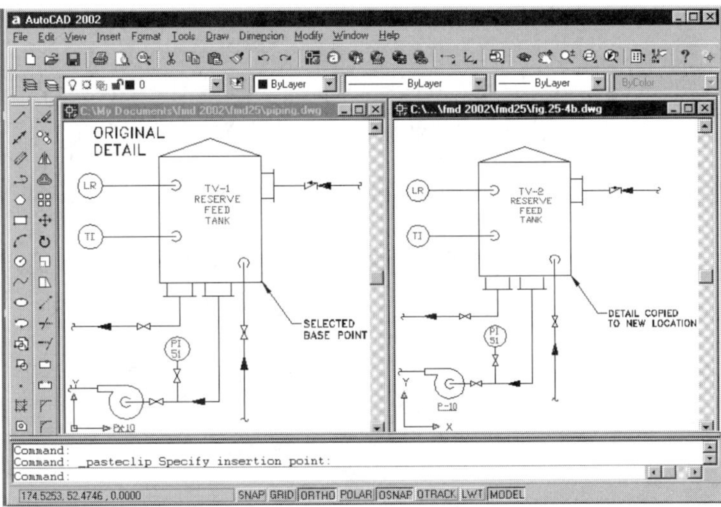

Figure 22.4 *The* **Create Layout Wizard** *contains a series of pages that walk you through the process of creating a new layout. Options are similar to the plotting options introduced in Chapter 5. The beginning page of the* **Create Layout Wizard** *is used to assign a name to the layout being created.*

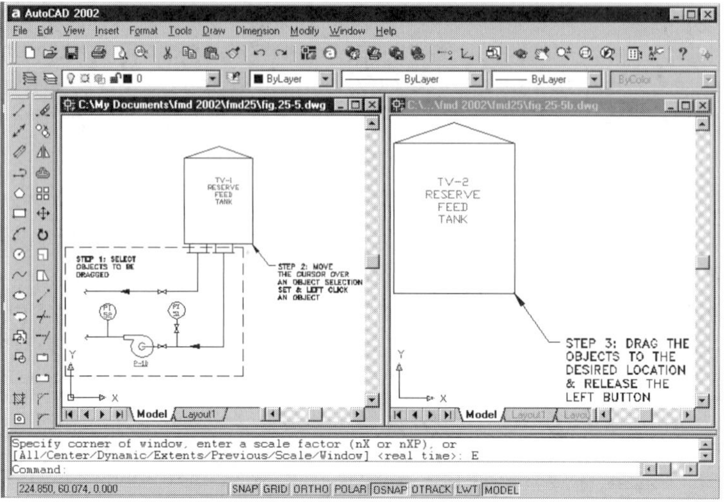

Figure 22.5 *Once the plotter to be assigned to the layout is selected, choosing* **Next** *will display a* **Create Layout - Paper Size** *dialog box that allows the paper size and drawing units to be selected.*

Once a title block is selected, clicking the **Next** button displays a **Create Layout - Define Viewports** dialog box similar to Figure 22.6. This dialog box can be used to define a viewport and select the scale for presenting the information in the viewport. By default, a single viewport is displayed. Creating multiple viewports will be discussed in the next section of this chapter. The default setting for the viewport scale factor is **Scaled to Fit**. Most drawings need to be plotted at a measurable scale. Selecting the **Next** button after selecting the desired scale will display the Create Layout - Pick Location dialog box, which prompts you to select the corner locations of the viewport. In the ANSI-C layout, the viewport edges align with the top, right, and bottom borders. This option allows a different viewport size to be selected for the new layout. The

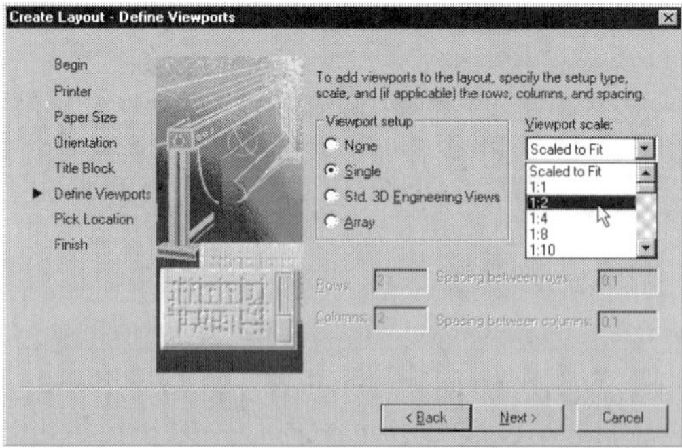

Figure 22.6 The **Define Viewports** *page allows the scale of the drawing in the viewport to be set and maintained in paper space.*

size of the viewport will depend on the size of the drawing to be displayed, the desired scale, and the number of viewports to be created in the layout. As the size is selected, the final page of the wizard will be displayed. This page will display the name of the layout you've created and methods for altering the layout. Selecting the **Finish** button will place the layout in the drawing base as the current layout.

Summary of Steps in Creating a New Layout

In summary, the steps to creating a new layout with a wizard are as follows:

1. Select Create Layout from Wizard on the Tools menu.
2. Provide a name for the layout being created.
3. Select the plotting device.
4. Select the paper size and the drawing units.
5. Select the orientation.
6. Select the title block to be used.
7. Select the number of viewports and the scale to be used in the viewport.
8. Select the corners of the viewport.
9. Click the **Finish** Button to return to the drawing area.

The new layout will now be displayed as the current layout, similar to Figure 22.7a. Notice that the viewport that was created is not large enough to display the detail. Selecting the viewport border will display the grips for the viewport. Make one of the

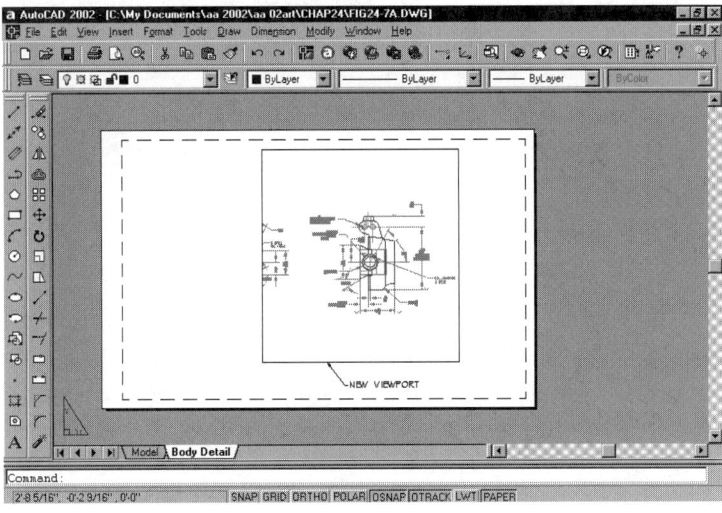

Figure 22.7a *The viewport that was created might not be large enough to display the entire drawing. Courtesy Hanh Tran.*

grips hot, and then stretch the viewport to the desired size, as shown in Figure 22.7b. Pressing ESC will end the **Stretch** command and restore the viewport. Figure 22.7c shows the results of stretching the viewport. Other modifications can be made to the layout by using the **Page Setup** dialog box. Access this dialog box by selecting the **Page Setup** button on the **Layouts** toolbar or by right-clicking the Model tab or a layout tab and choosing **Page Setup**.

Note: As several layouts are created, all of the layout names might not appear. To display more of the layout tabs, select and hold the edge of the slide bar, and then move the cursor to the right. Figures 22.8a and 22.8b show the process. You can also display the other layouts by using the arrow keys directly to the left of the Model tab. This allows you to scroll through the layout tabs and select the layout you need.

ALTERING LAYOUTS

You can create a new layout or alter an existing one by typing **LAYOUT** (ENTER) at the Command prompt. This method will present the following prompt at the command line:

Enter layout option
    [Copy/Delete/New/Template/Rename/SAveas/Set/?] <Set>:

These options can be used to create, define, and save a new drawing layout.

Copy

Selecting Copy will create a new layout by making a copy of an existing layout. A prompt for the name of the layout to copy will be displayed at the Command

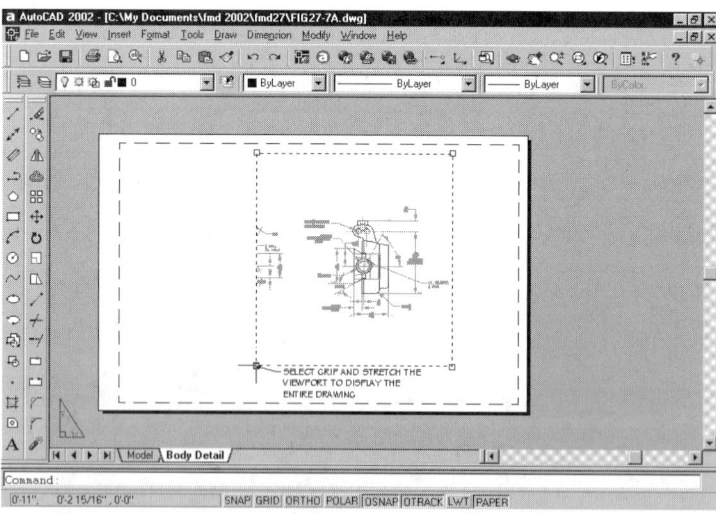

Figure 22.7b *Selecting the viewport border will display the grips for the viewport. Make one of the grips hot, and then stretch the viewport to the desired size. Courtesy Hanh Tran.*

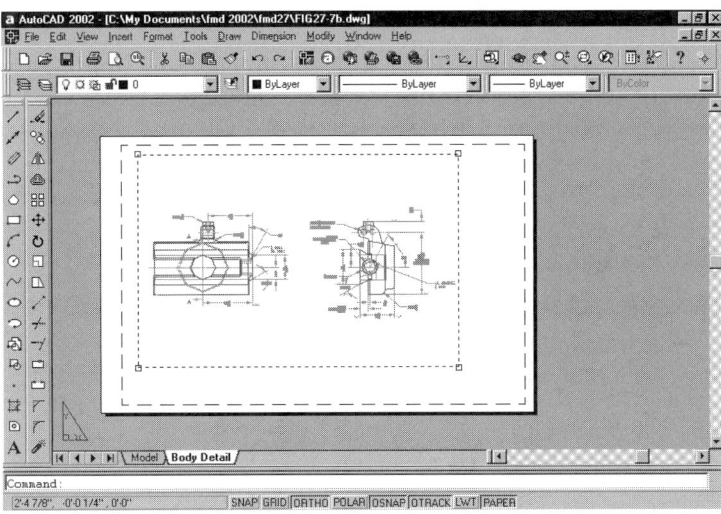

Figure 22.7c *A viewport can be stretched to display the full drawing. Pressing* ESC *will end the* **Stretch** *command and restore the viewport. Courtesy Hanh Tran.*

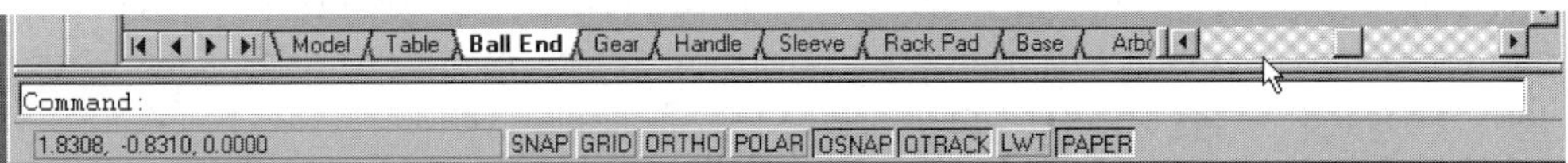

Figure 22.8a *Depending on the size of the slide bar, not all the layout tabs might be displayed.*

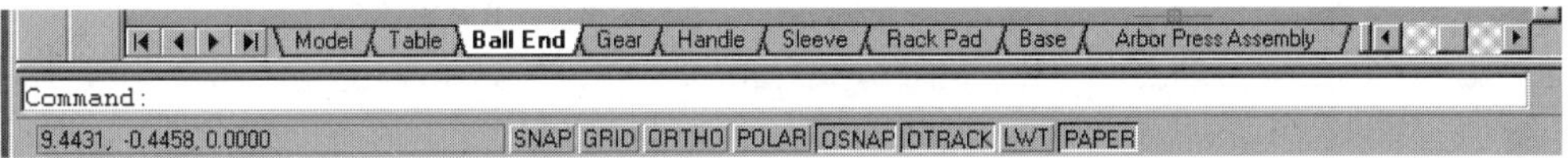

Figure 22.8b *Decrease the size of the slide bar to display more layout titles.*

prompt. The default name will be the name of the current layout—in this case, the ANSI-C Title Block is the default name. If a new layout name is not provided, the new layout assumes the name of the copied layout, with an incremental number added to the name. The command sequence is as follows:

Enter layout option
    [Copy/Delete/New/Template/Rename/SAveas/Set/?]
    <Set>: **C** ENTER

Enter layout to copy < ANSI-C Title Block>: ENTER

Enter layout name for copy < ANSI-C Title Block (2) >: ENTER

Layout "ANSI-C Title Block" copied to "ANSI-C Title Block (2)".

Delete

This option will delete a layout. By default, the current layout will be deleted. (The Model tab cannot be deleted.) One or more layouts can be deleted. If all layouts are removed, an empty layout titled Layout1 will be created and displayed.

New

This option creates a new layout tab without the use of the **Create Layout Wizard**. Selecting the New option will display the following prompts:

Enter layout option
[Copy/Delete/New/Template/Rename/SAveas/Set/?]
<Set>: **N** ENTER

Enter new Layout name <ANSI-C Title Block>: **COOL STUFF** ENTER

Command:

Template

Selecting the Template option creates a new layout based on an existing template (.DWT) or drawing file (.DWG). The layouts from an existing template or drawing file can be inserted in a new drawing. In addition to the layout, objects contained in the layout can also be made part of the layout being created. The command sequence is as follows:

Enter layout option
[Copy/Delete/New/Template/Rename/SAveas/Set/?]
<Set>: **T** ENTER

Pressing ENTER will display the **Select Template from File** dialog box, allowing a template to be selected. Selecting the **File type** edit box allows a drawing file to be selected. Once the template or file is selected, it will be displayed in the preview area. Selecting the **Open** button will close the **Select Template from File** dialog box and provide the **Insert Layout** dialog box that verifies the template to be added to the layout. Selecting **OK** will return the layout display with the template. You can also select a template by choosing the **Layout From Template** button on the **Layouts** toolbar.

Rename

This option can be used to rename a layout. The layout that was last used is shown as the default. Prompts will be displayed to select the layout to be renamed, and to select the new name. Prompts to rename the JUNK layout as COOL STUFF2 are as follows:

Enter layout option
[Copy/Delete/New/Template/Rename/SAveas/Set/?]
<Set>: **R** ENTER

Enter Layout to rename <ANSI-C Title Block>: **JUNK** ENTER

Enter new layout name: **COOL STUFF2** ENTER

Layout "JUNK" renamed to "COOL STUFF2".

Command:

Layout names are not case sensitive and can be up to 255 characters long. Only 32 characters are displayed on the layout tab.

Saveas

As you complete the **Create Layout Wizard**, the new layout is stored as part of the current drawing. Any existing drawing can be saved and used as a template drawing. You can save all of the objects, properties and settings as a .DWT file (template) by selecting the SAveas option from the **Layout** command. New drawings can be created from drawing templates or new layouts can be created in the current drawing based on a layout in the template file.

Set

The Set option makes an existing layout the current layout. Clicking the tab of a layout does the same thing.

The ? Option

Selecting this option will list all of the layouts defined in a drawing. This can be helpful when all layout names cannot be displayed below the drawing. Pressing F2 will also display a list of the current layouts.

WORKING WITH VIEWPORTS

AutoCAD allows two types of viewports to be created. Floating viewports are used to create plotting layouts in paper space. Tiled viewports can be created in model space to aid in the display of the drawing. Up to four tiled viewports can be created, but only one viewport can be active at a time. Tiled viewports allow you to divide the screen into multiple areas so that you have multiple views of different parts of the same object. Figure 22.9 shows an example of a dimensioned detail. Creating multiple viewports would allow you to zoom into a top view in one viewport, a side view in another viewport, and a section view in a third viewport, as shown in Figure 22.10. This multiple use of viewports allows for drawing and editing within the enlarged portions of the drawing without having to zoom in and out to move from one detail to another. It is important to remember that even though you are seeing multiple images on the screen, you're working with only one database. An object that is drawn or edited in one viewport will also affect the other viewports.

DISPLAYING MULTIPLE TILED VIEWPORTS

Multiple tiled viewports can be displayed in several configurations that are controlled by the **Viewports** dialog box. Display the **Viewports** dialog box by selecting the **Display Viewports Dialog** button on the **Viewports** toolbar, by

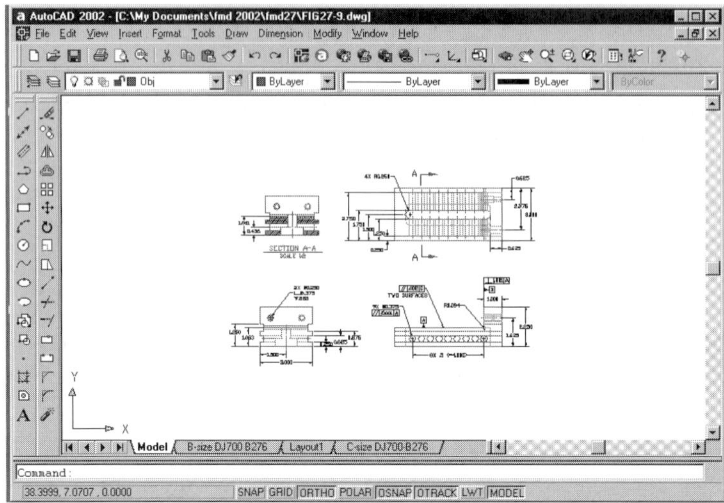

Figure 22.9 *When you work on different views of details, it is often hard to coordinate information from one view to another. Courtesy Hanh Tran.*

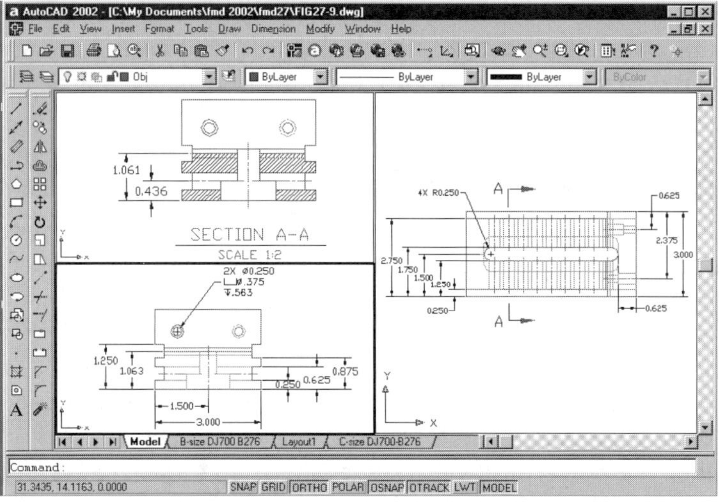

Figure 22.10 *Tiled viewports allow you to divide the screen into multiple images so that you have multiple zoom images of different parts of one drawing file. The details shown in Figure 22.9 can be easily viewed using multiple viewports. Courtesy Hanh Tran.*

selecting **New Viewports** from **Viewports** on the **View** menu, or by typing VPORTS (ENTER) at the Command prompt. Each method will produce the display shown in Figure 22.11. In the current setting, one viewport is displayed. Selecting a listing from the **Viewport** menu will alter the display in the preview to reflect the change. Figure 22.12 shows the display for each option. Use the following steps to display multiple tiled viewports in a new configuration:

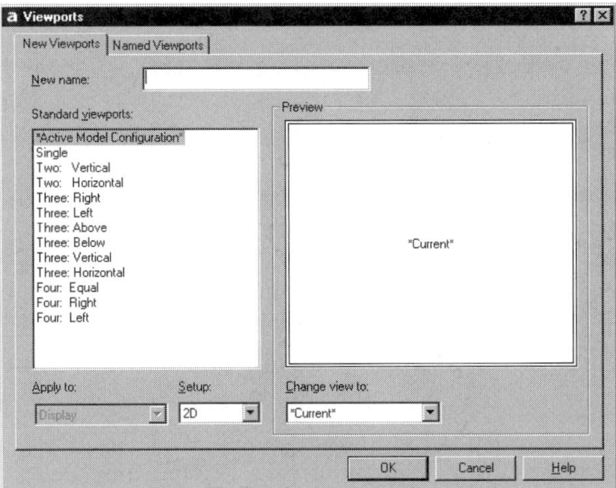

Figure 22.11 *Display the* **Viewports** *dialog box by selecting the* **Display Viewports Dialog** *button on the* **Viewports** *toolbar, by selecting* **New Viewports** *from* **Viewports** *on the* **View** *menu, or by typing* **VPORTS** (ENTER) *at the Command prompt.*

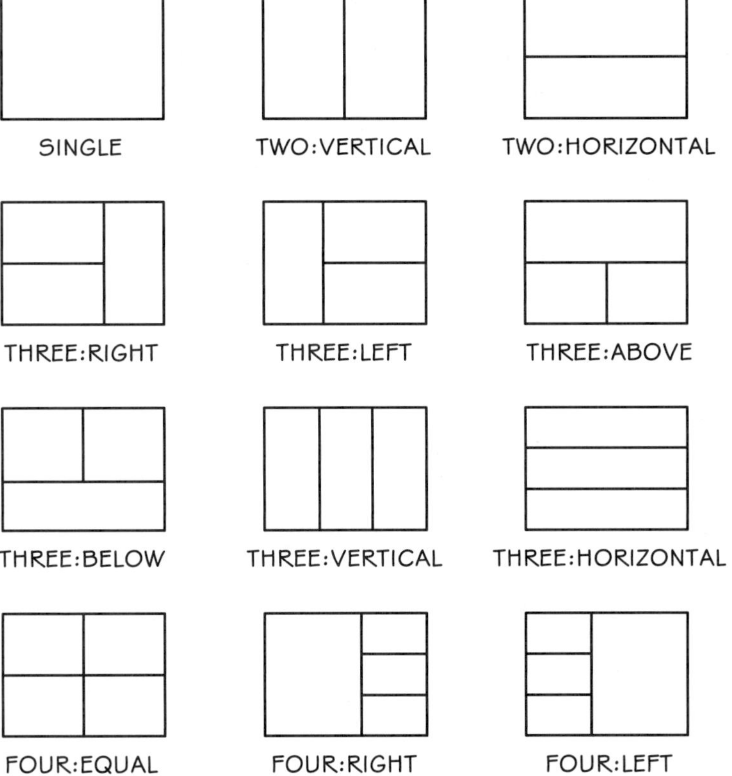

Figure 22.12 *Possible display options for tiled viewports.*

1. Display the **Viewports** dialog box.
2. Select the name of the desired viewport configuration from the **Standard viewports** list.
3. Select Display as the current setting for the **Apply to** list.
4. Select 2D as the current setting for the **Setup** list.
5. Click the **OK** button to apply the configuration and close the dialog box.

Notice, in Figure 22.10, that a bold black line surrounds the viewport on the right side. As you move the cursor around the screen, you'll notice that the crosshairs only work in the drawing on the right side. This view is called the active viewport. You can draw or edit in this viewport just as in any other drawing. As the cursor crosses the center of the drawing into the viewport on the left, the crosshairs change to an arrow. Notice, in comparing Figure 22.13 with Figure 22.14, that the title to the section detail was removed from the upper left viewport, which also removed the information from the database and the total view. Although the total detail is not legible, you can see that information is missing. The fact that the information is not legible is the point. Viewports make multiple enlarged views of one drawing that are legible for easy drawing or editing, while leaving a view of the overall drawing for perspective.

To make a different viewport active, move the cursor to the desired viewport and click. To convert a drawing like Figure 22.13 to one similar to Figure 22.14, change between active viewports and use the **Zoom** command. Remember, once active, all drawing and editing commands function normally. First make the upper left viewport active. Select the detail in the upper right quadrant for a **Zoom** window and enlarge

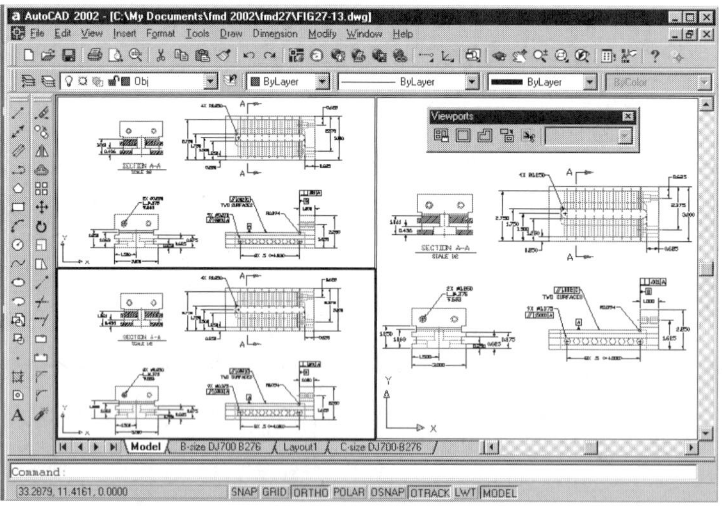

Figure 22.13 *Selecting Three: Right followed by the* **OK** *button will display the drawing shown in Figure W4.9 in three viewports. Courtesy Hanh Tran.*

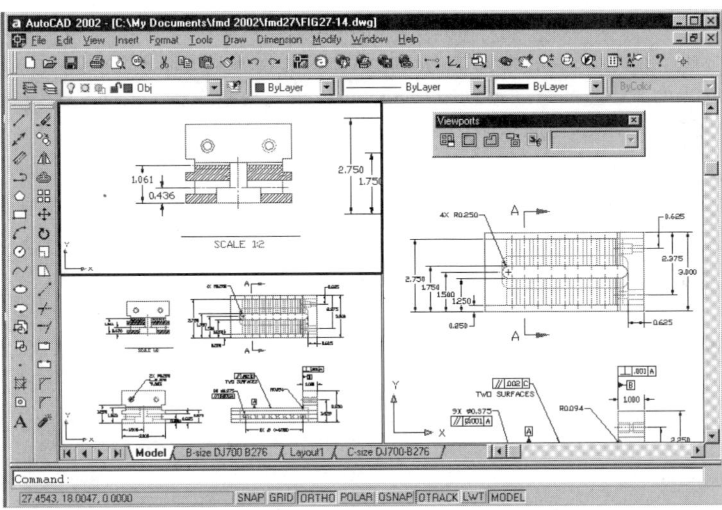

Figure 22.14 *The viewports in Figure 22.13 have each been edited to show this display. With the upper left viewport active, the detail was enlarged using the* **Zoom** *command to display the section detail. The same process was used in the lower left viewport to view the entire sheet. Courtesy Hanh Tran.*

to fill the viewport. Finally, make the viewport on the lower left active and select the detail in the lower right quadrant for a **Zoom**, using the Window option. With these details enlarged, you can make changes to each without having to zoom from one detail to another.

Working in Multiple Viewports

In addition to easing viewing, viewports can ease working. A command such as **Line** can be started in one viewport and completed in another viewport. This could be especially helpful on detailed drawings. Dimensioned details are typically created side by side so that heights can be projected from one view to another. If all views are to be displayed, each view will be tiny. Creating a viewport for each view allows for easy viewing. Being able to draw lines from one viewport to another will speed production. Use the following steps to draw a line between viewports:

1. Select the viewport that contains the start point, making it the active viewport.
2. Start the **Line** command sequence.
3. Select the viewport that will contain the ending point, making it the active viewport.
4. Select the ending point for the line.

A line will now be displayed between the two selected points, similar to Figure 22.15.

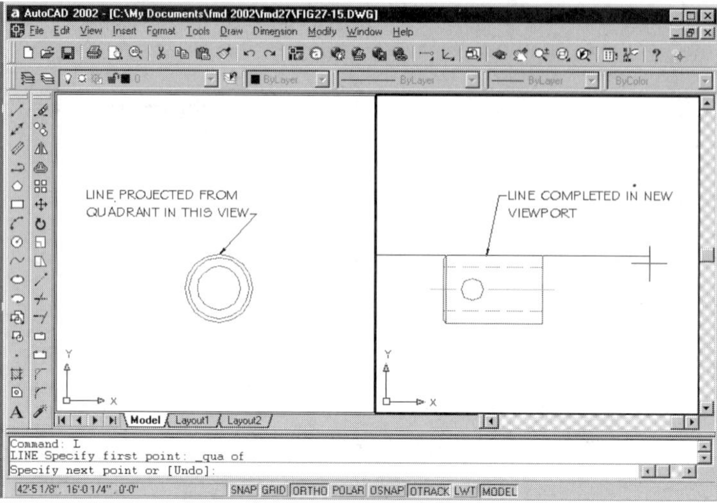

Figure 22.15 *A command can be started in one viewport and completed in another viewport. Select the viewport that contains the start point, making it the active viewport, and start the command sequence. Next select the viewport that will contain the endpoint, making it the active viewport, and select the endpoint for the line. Courtesy Hanh Tran.*

WORKING WITH FLOATING VIEWPORTS

Each time you display a layout, you're working with a floating viewport. The viewport is referred to as floating because it can be moved and the size can be altered, just like a floating dialog box. The viewport allows you to look through the paper and see the drawing created in model space. The floating viewport is used to display the drawing in the layout for plotting. An unlimited amount of viewports can be created in paper space, but only 64 viewports can be visible at one time. The balance of this chapter will explore floating viewports and how they can be used to prepare a drawing layout for plotting. A floating viewport can be considered as a drawing object that provides a view into model space. Several floating viewports can be created within a layout, and the viewports can be overlapped or separated from each other. A floating viewport can also be an irregular shape. Because floating viewports are considered objects, drawing objects displayed in the viewport cannot be edited. To edit objects within the viewport, model space must be restored. You can toggle between model space and paper space by choosing either the Model or layout tab, depending on which one is inactive. Other methods for toggling between model and paper space include the following:

Double-clicking in a floating viewport will switch the display to model space.

Note: Remember, while editing in model space in a layout, you risk changing the drawing setup. This should only be used by experienced AutoCAD users or when in the initial setup of the view inside the viewport.

Select the PAPER or MODEL button on the status bar. Remember, when you're in a layout, the page is in paper space, but the viewport can be toggled to model space or paper space. When PAPER button is displayed, the viewport is a paper space display of the drawing in a paper layout. Selecting PAPER will change the button to MODEL and display the drawing in a model space viewport within a paper space layout.

Type MS (ENTER) for **Mspace** or PS (ENTER) for **Pspace** at the Command prompt.

Most professionals work in model space and arrange the final drawing for output in paper space. While you work in a layout, with the viewport in paper space, any material added to the drawing will be added to the layout, but not shown in the model display. This will prove useful as the final plotting layout is constructed.

ALTERING VIEWPORT PROPERTIES

Because viewports are objects created by AutoCAD, they have properties such as color, layer, linetype, lineweight, and plot style. You can alter each of these properties using the **Properties** dialog box. As you create a viewport, a new layer can be created and displayed for the viewport. This layer should be frozen before the drawing is saved so that it is not plotted. Linetype and lineweight are ignored in rectangular viewports but are displayed in nonrectangular viewports.

VIEWPORT SCALING

Most users of AutoCAD plot their drawings displayed in a layout at a scale of 1:1. To do this, the drawing must be inserted or attached to the viewport at a predetermined scale factor. AutoCAD has created an aid to do this. Earlier it was pointed out that the scale can be set as a drawing is placed in a viewport. The scale can be locked using the **Properties** dialog box. Locking the viewport scaling will prevent the scale from being altered if the drawing is zoomed once it is displayed in the viewport. The scale can be locked by the following procedure:

1. Display the layout containing the viewport to be locked.
2. Select the edge of the viewport so that it is highlighted.
3. Display the **Properties** dialog box by selecting the **Properties** button on the **Standard** toolbar.
4. Select **Display locked** from the **Properties** dialog box.
5. Select Yes from the **Display locked** edit list.
6. Close the **Properties** dialog box and return to the drawing area.

The scale of the selected viewport is now locked. If you attempt to alter the scale while in model space, AutoCAD displays "warning: Viewport is view-locked". If you change the zoom factor in the viewport, only paper space objects are affected.

Note: It's been stressed not to edit object in viewports while in paper space. Once you've placed a lock on the viewport, it is safe to edit.

CREATING MULTIPLE FLOATING VIEWPORTS

The **Viewports** toolbar shown in Figure 22.16 can be used to create floating viewports. Buttons include **Display Viewports Dialog**, **Single Viewport**, **Polygonal Viewport**, **Convert Objects to a Viewport**, **Clip Existing Viewport**, and **Viewport Scale**. Up to this point, you've used layouts with viewports that take up the entire drawing area. Floating viewports can be created in a layout that fills the entire drawing area or only takes up a small portion of the drawing area. Floating viewports can be created using a process that is similar to the process for creating tiled viewports. Floating viewports are created using the **Viewports** dialog box shown in Figure 22.17. Compare the **Viewports** dialog box in Figure 22.17 with the Viewports dialog box in Figure 22.11. Although each dialog box is very similar, they create different types of viewports, depending on how they are accessed. The dialog box shown in Figure 22.11 was accessed while in model space. Use the following steps to create a floating viewport.

1. Display a layout by selecting an existing layout tab or by creating a new layout.
2. Display the **Viewports** dialog box by selecting the **Display Viewports Dialog** button from the **Viewports** toolbox, by selecting **New Viewports** from **Viewports** on the **View** menu, or by typing **VPORTS** (ENTER) at the Command prompt.
3. Select the desired viewport configuration from the **Standard viewports** list.
4. Enter the desired spacing to be maintained between viewports in the **Viewports Spacing** edit box.
5. Select 2D in the **Setup** edit box.
6. Click the **OK** button to apply the configuration and close the dialog box.

A process similar to placing tiled viewports in a layout can be used to place multiple floating viewports in an existing layout.

CREATING A SINGLE FLOATING VIEWPORT

Floating viewports provide an excellent means of assembling multi-scaled drawings for plotting. New floating viewports can be created without using the standard viewport configurations. Figure 22.18 shows a detail with multiple views and scale factors using floating viewports. Manufacturing projects typically comprise drawings that are drawn at a variety of scales. These details range in scale from 1:1 to 10:1. These details cannot be easily drawn or plotted without the use of multiple viewports and inserting or referencing each detail to a base drawing. As with drawings where a single object is attached, the use of scale factors to control text and

Figure 22.16 *The* **Viewports** *toolbar can be used to control floating viewports. Buttons include* **Display Viewports Dialog, Single Viewport, Polygonal Viewport, Convert Objects to a Viewport, Clip Existing Viewport,** *and* **Viewport Scale.**

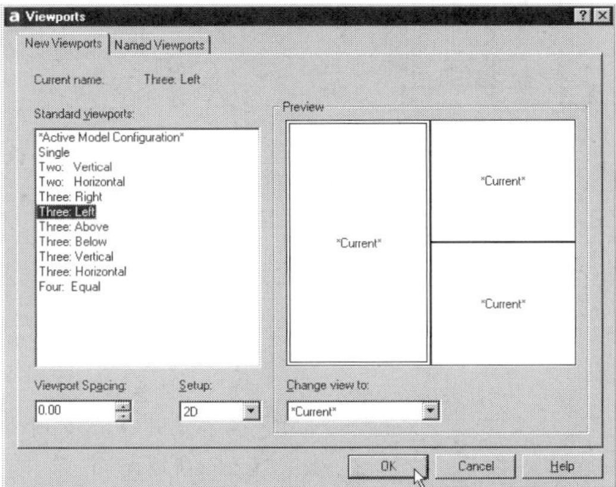

Figure 22.17 *The* **Viewports** *dialog box. Although this dialog box is very similar to the dialog box shown in Figure 22.11, they create different types of viewports, depending on how they are accessed.*

dimension variables will be critical to the success of the final outcome of the drawing. The process for creating multiple viewports will be similar to the creation of one viewport.

The following example will demonstrate the steps used to prepare the section and three details shown in Figure 22.19 for plotting. The HVAC plan will be displayed at a scale of 1/4"=1'–0", the exhaust hood detail will be displayed at a scale of 1"=1'–0". For most users, drawings such as a HVAC plan or details will be created in a drawing template using model space. The following guidelines will walk you through the process of creating multiple viewports in a template and inserting or attaching multiple drawings. For this discussion, the **Insert** command will be used. The drawings could also be attached using the **Xref** command, depending on the drawings needs.

Displaying Model Space Objects in the Viewport

The easiest method for displaying multiple drawings at multiple scales is by inserting each of the drawings to be plotted in the template in model space. With the details arranged in one drawing similar to Figure 22.19, switch to the ANSI-C Title Block layout. Use the following steps to display the drawings in multiple viewports:

1. In the ANSI-C Title Block layout, click the PAPER button in the status bar to toggle the viewport to model space. Use the **Zoom All** option to enlarge the

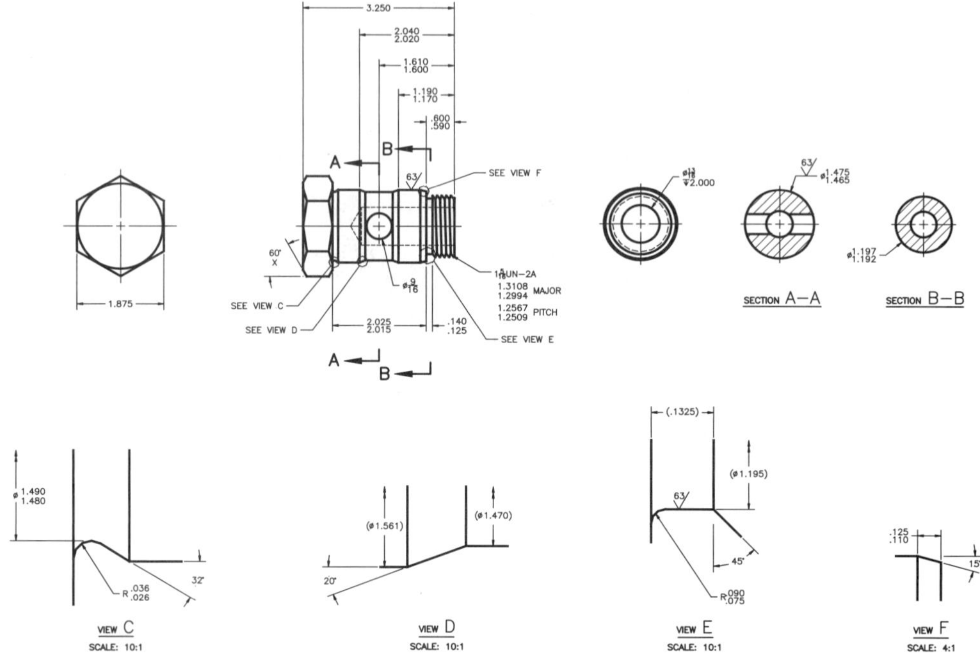

Figure 22.18 *Details drawn at varied scales can be assembled on one sheet for plotting using the* **Vports** *command. Courtesy Paul Bissell.*

area of model space to be displayed in the existing viewport. All the drawings will be displayed, but each will be at an unknown scale. Your drawing will resemble Figure 22.20.

2. Select what will be the largest drawing to work with first. In our example, the viewport for displaying the HVAC plan will be adjusted first.
3. Activate the **Viewports** toolbar.
4. Set the scale of the viewport to the appropriate value for displaying the section using the **Viewport Scale** menu on the **Viewports** toolbar. For this example a scale of 1/4"=1'–0" was used.
5. Click the MODEL button on the status bar to return to paper space in the ANSI-C Title Block layout.

Adjusting the Existing Viewport

Once the desired drawing is inserted or attached to the viewport, the size of the viewport can be altered if the entire drawing can't be seen. To alter the viewport size, move the cursor to touch the viewport and activate the viewport grips. Select one of the grips to make it hot and then use the hot grip to drag the window to the desired size. The

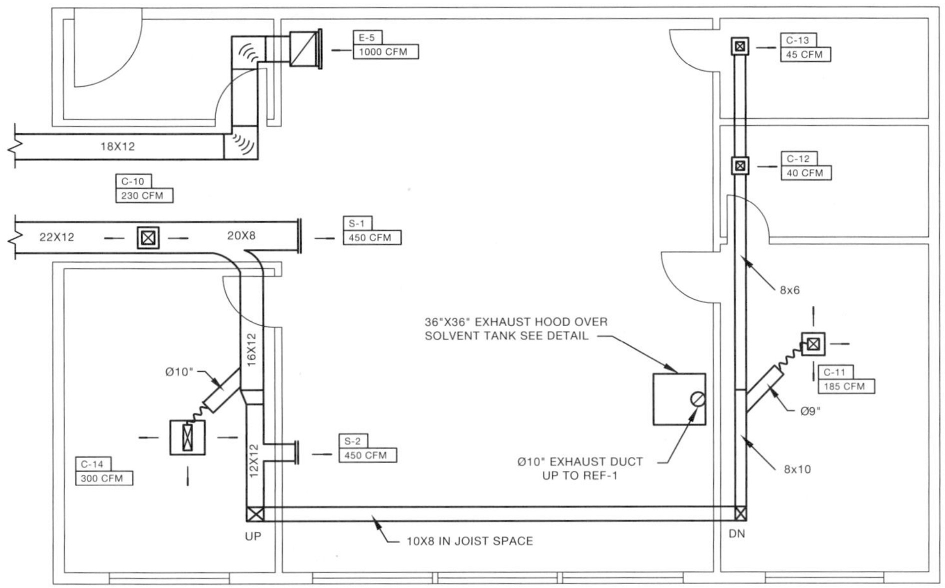

PROPOSED ADDITION HVAC PLAN

SCALE 1/4"=1'-0"

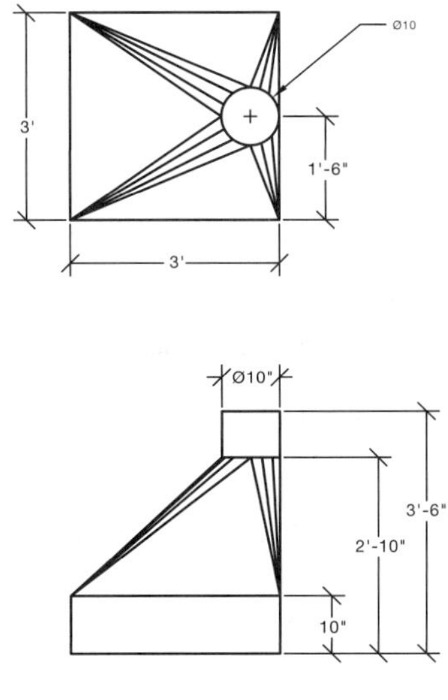

EXHAUST HOOD DETAIL

SCALE 1"=1'-0"

Figure 22.19 *Drawings to be attached to the base sheet. Courtesy Hanh Tran.*

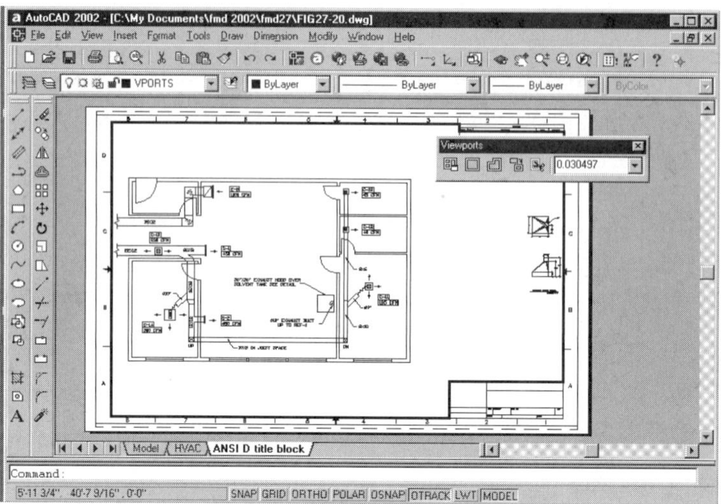

Figure 22.20 *Toggle to model space and use the* **Zoom All** *option to view all the drawings in your viewport, but each will be at an unknown scale. Courtesy Hanh Tran.*

HVAC plan has now been inserted in the template and is ready to be plotted at a scale of 1/4"=1'–0". The display would resemble Figure 22.21.

Note: All of the drawing commands can be used inside the viewport, but try to limit editing to changes that will affect only this plot. Changes to the drawing that will be needed on future drawings should be made to the original drawing. Remember, if the original is attached to this drawing by **Xref**, the changes made to the original drawing will be reflected in all uses of the referenced drawing. Enter model space prior to making changes to the drawing in the current viewport. This will allow you to add detail numbers and drawing titles that change with each use of the drawing.

Creating Additional Viewports

Use the following steps to prepare for additional viewports:

1. Set the current layer to Viewport. This layer is part of each template.
2. Select **Single Viewport** on the **Viewports** toolbar.
3. Select the corners for the new viewport. The coordinates for the viewport can be entered or specified by indicating the corners of the viewport—just as if it were a window used for determining a selection set.

The result of this command can be seen in Figure 22.22. Although you should try to size the viewport accurately, you can stretch it to enlarge or reduce its size once the scale has been set.

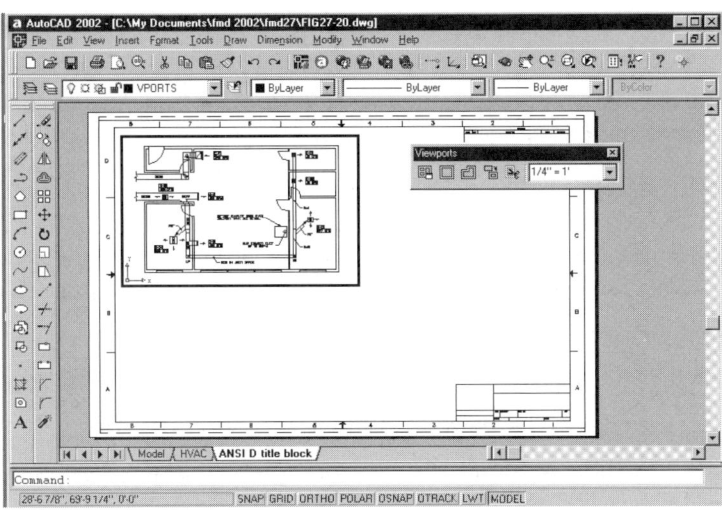

Figure 22.21 *Once the desired drawing is inserted or attached to the viewport, set the scale of the viewport to the appropriate value for displaying the drawing using the* **Viewport Scale** *menu on the* **Viewports** *toolbar. Next, alter the size of the viewport if the entire drawing can't be seen. The drawing is now ready to be plotted at a scale of 1/4"=1'–0". Courtesy Hanh Tran.*

Altering the Second Viewport

In its current state, if a new viewport is created, the display from the existing viewport will be shown in the new viewport as well. Use the following steps to alter the display.

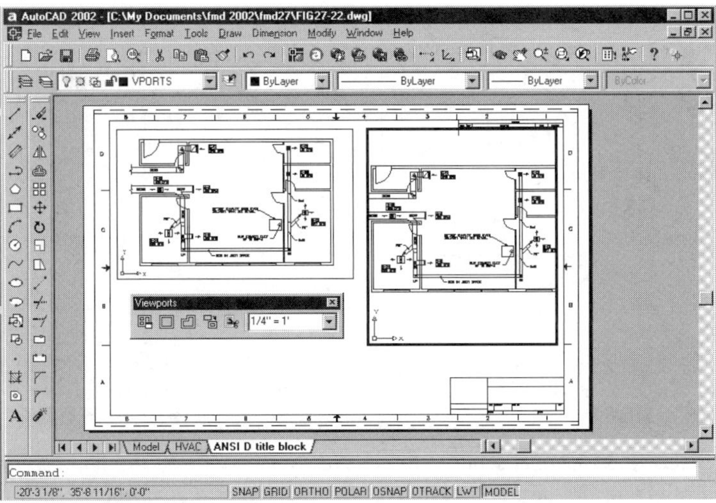

Figure 22.22 *If a new viewport is created, the display from the existing viewport will be shown in the new viewport as well. Courtesy Hanh Tran.*

1. Click the PAPER button on the status bar to toggle the viewport to model space.
2. Make the new viewport the current viewport by placing the cursor in the viewport and clicking.
3. Use the **Zoom All** option to display all of the model space contents in the second viewport.
4. Set the viewport scale to the appropriate value for displaying the detail. For this example, a scale of 1"=1'–0" will be used to display the exhaust hood detail.
5. Use the **Pan** command to center the detail in the viewport.
6. Click the MODEL button in the status bar to toggle the drawings to paper space.
7. Select the edge of the viewport to display the viewport grips.
8. Make a grip hot and shrink the viewport so that only the detail is shown.
9. Use the **Move** command to move the viewport to the desired position in the template.
10. Save the drawing for plotting.

The drawing should now resemble Figure 22.23.

Repeating the Process for Additional Viewports

Once the viewport for the exhaust hood detail has been adjusted, additional viewports can be created. Placing all of the viewports at once will eliminate having to keep

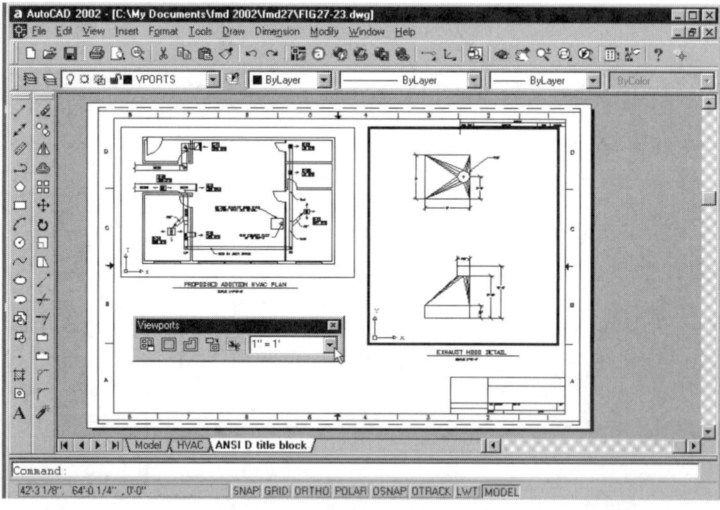

Figure 22.23 *Adjust the viewport scale to the desired value, and use the Pan command to move the drawings so that the desired information is centered in the viewport. Once it is centered, adjust the size of the viewport so only the desired material is displayed. Courtesy Hanh Tran.*

switching from paper space to model space. Once all of the desired details have been added to the layout, the viewport outlines can be frozen. Start the **Layer** command, select VIEWPORT as the layer to be edited, and then select the Freeze option. Freezing the viewports will eliminate the line of the viewport being produced as the drawing is reproduced. Any viewports that are still visible were created on a layer other than VIEWPORT. Change any existing viewports to the correct layer. The finished drawing, with the viewports frozen, will resemble Figure 22.24.

CONTROLLING THE DISPLAY IN FLOATING VIEWPORTS

As you placed drawings in viewports, the **Pan** and **Zoom** commands were used to alter what was displayed in each viewport. AutoCAD allows displayed objects to be altered by screening, turning viewports ON or OFF, and hiding lines. Hiding lines is typically associated with 3D drawings and will not be covered in this chapter. Consult the **Help** menu for additional information.

Screening Objects in Viewports

Screened objects use less ink when plotted, so they appear dimmer on screen and when plotted. Screening could be used to distinguish between objects, without plotting in color. To screen an object, you need to assign it a plot style and then assign the screening value to that plot style. (Plot styles will be discussed in the next chapter.) Values ranging from 0 to 100 can be assigned, with 100 being the default value. A value of 100 assigns no screening and will provide normal plotting values. A value of 0 would assign no ink to the object and have the same effect as freezing the viewport contents. Figure 22.25 shows an assembly drawing with multiple hatch patterns.

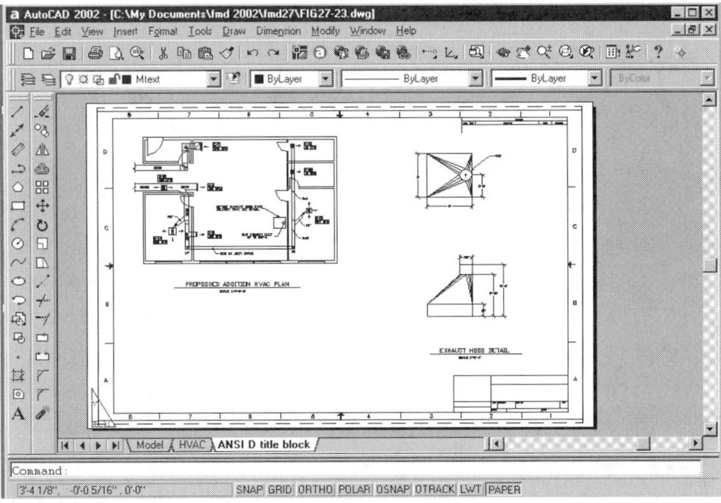

Figure 22.24 *The finished drawing, with the Viewports layer frozen, is ready for plotting. Courtesy Hanh Tran.*

Displaying a Floating Viewport

By default, a viewport is ON as it is created. The performance of your machine can be affected if a large number of complex viewports are created. Turning viewports OFF can save regeneration time for large drawings. Toggle viewports ON or OFF using the Properties dialog box and the following steps:

1. Select the viewport to alter.
2. Open the **Properties** dialog box.
3. Select **On** from the **Misc** listing and then select No from the edit list.
4. Close the **Properties** dialog box.

As the viewport is toggled to No, the contents of the viewport will be removed from the drawing display. The line representing the viewport can be removed from the display by using the **Layer Properties Manager** dialog box to freeze the layer. Figure 22.26 shows the drawing from Figure 22.24 with one of the viewports in the No setting.

Controlling Layer Display with the Layer Property Manager

The contents of specific layers in each viewport can be controlled using the **Layer Properties Manager**. Using the layer options works well when an object will be displayed in two or more views. Specific aspects of the drawings, such as layers containing text, can be frozen in one viewport and displayed in another. Options for freezing layers in floating viewports include **Freeze in All Viewports**, **Freeze in Active Viewport**, and **Freeze in New Viewport**. These options are inactive as the **Layer Properties**

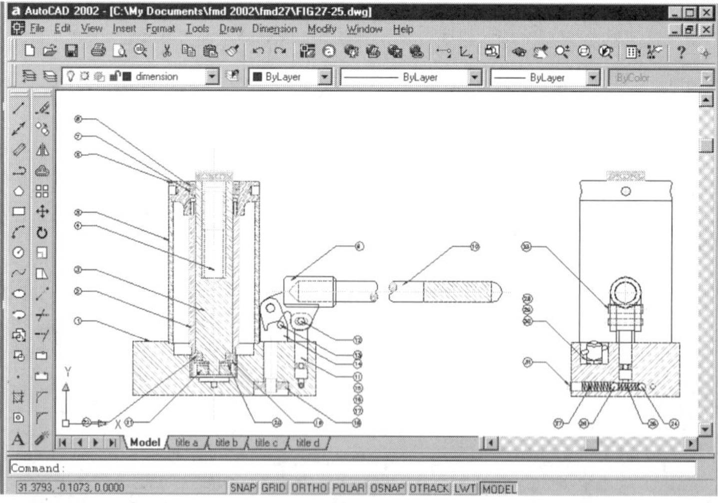

Figure 22.25 *Screened objects use less ink when plotted, so they appear dimmer when plotted. To screen an object, assign it a plot style, and then assign the screening value to that plot style. Courtesy Cathy Schlesser.*

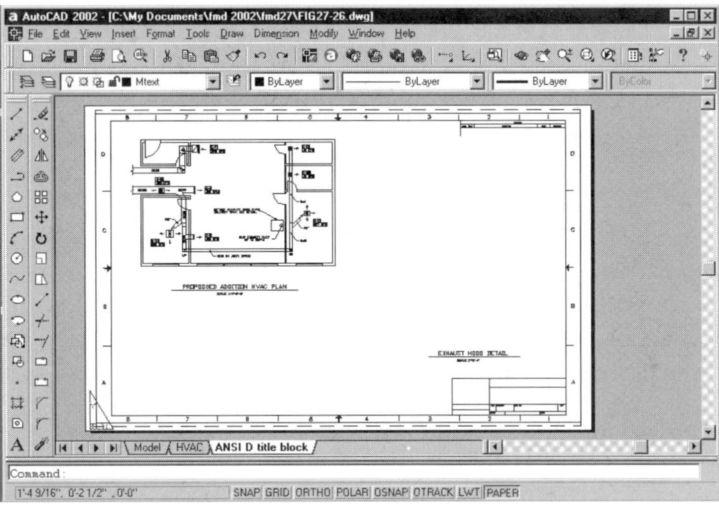

Figure 22.26 *One of the viewports from Figure W4.24 has been toggled to the NO setting. As the viewport is toggled to NO, the contents of the viewport will be removed from the drawing display. Courtesy Hanh Tran.*

Manager dialog box is displayed, but they are activated as a layer is highlighted. The **Layer Properties Manager** can also be used to turn plotting ON or OFF for visible layers. A layer created for interoffice notes can be displayed on the screen, but frozen for plotting. Layers are frozen for plotting by selecting the **Plot** icon of the desired layer.

Controlling Scaling in Viewports

Altering the size of a floating viewport does not change the scale of the drawing within the viewport. When the scale was adjusted for the viewport, you set a consistent scale for accurate display and plotting. Earlier in the text, selecting a scale for linetypes was discussed. If a hidden line is to be .125" long when plotted at 10:1, a scale factor must be applied to the line to create the desired result. The scale for linetypes can be based on the drawing units of the space where the object was created, or based on paper space units. The **Psltscale** system variable can be used to maintain a uniform linetype scaling for objects displayed at different zoom factors. The **Psltscale** variable can be set by the following steps:

1. Select **Linetype** from the **Format** menu.
2. In the **Linetype Manager** dialog box, select **Show Details**.
3. Enter the desired global scale factor to apply to the linetypes in the **Global Scale Factor** edit box.
4. Click **OK** to return to the drawing area.

CREATING VIEWPORTS WITH IRREGULAR SHAPES

AutoCAD allows viewports to be created in an irregular shape that best suits the object to be displayed. Create a nonrectangular viewport by selecting the **Polygonal Viewport** button, by clipping an existing viewport, or by converting an existing object to a viewport.

CREATING POLYGONAL VIEWPORTS

A polygonal viewport is created by selecting the **Polygonal Viewport** button on the **Viewports** toolbar, by selecting **Polygonal Viewport** from **Viewports** on the **View** menu, or by typing -VPORTS (ENTER) at the Command prompt. The command is completed by selecting a series of points using a command sequence similar to the sequence for creating a polyline. The command sequence is as follows:

Command: **-VPORTS** ENTER *(Click the Polygonal Viewports button.)*

_-vports

Specify corner of viewport or

[ON/OFF/Fit/Hideplot/Lock/Object/Polygonal/Restore 2/3/4/ <Fit>: _p

Specify start point: *(Specify viewport starting point.)*

Specify next point or [Arc/Length/Undo]: *(Select points to define polygon.)*

Specify next point or [Arc/Close/Length/Undo]: *(Select points until the polygon is defined. As the polygon is closed, the command will end with the following prompt.)*

Regenerating model

Command:

Figure 22.27 shows an example of an irregularly shaped viewport.

USING VPCLIP TO ADJUST THE VIEWPORT

The **Vpclip** command can be used to clip or redefine the shape of an existing viewport. Access the command by selecting the **Clip Existing Viewport** button on the **Viewports** toolbar or by typing VPCLIP (ENTER) at the Command prompt. Selecting the viewport to be accessed, right-clicking to display the shortcut menu, and then selecting **Viewport Clip** will also access the command. The command sequence is as follows:

Command: **–VPCLIP** ENTER *(Or click the Clip Existing Viewport button.)*

Select viewport to clip: *(Select desired viewport.)*

Select Clipping object or [Polygonal/Delete] <Polygonal>: ENTER

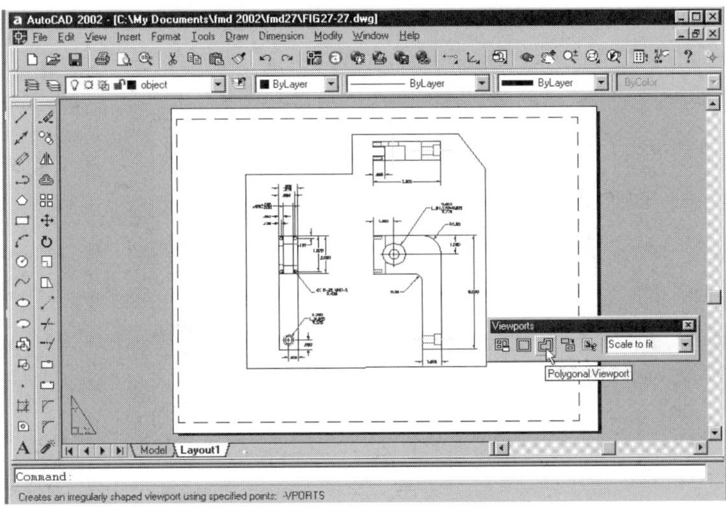

Figure 22.27 *A polygonal viewport can be created by selecting the* **Polygonal Viewport** *button on the* **Viewports** *toolbar. Courtesy Cathy Schlesser.*

Specify start point: *(Specify viewport starting point.)*

Specify next point or [Arc/Length/Undo]: *(Select a minimum of three points to define polygon.)*

Specify next point or [Arc/Close/Length/Undo]: *(Select points until the polygon is defined.)*

Command:

Objects that lie within the polygon will be displayed. Objects outside the polygon will be removed from the viewport and will not be displayed. The viewport in Figure 22.28 has been edited using **Vpclip**. Irregular viewports can be edited by selecting the edge of the viewport. With the viewport grips displayed, the viewport can be edited in the same way as any other object is altered with grips. (See Figure 22.29.)

CONVERTING EXISTING OBJECTS TO A VIEWPORT

AutoCAD allows existing shapes to be converted to viewports. The object can be a circle, ellipse, region, spline, or closed polyline comprised of lines and arc segments. Access the command by selecting the **Convert Object to a Viewport** button on the **Viewports** toolbar. The command is completed using the following command sequence:

Command: *(Click the Convert Object to a Viewport button on the Viewports toolbar.)*

Specify corner of viewport or

[ON/OFF/ Fit/Hideplot/ Lock/Object/Polygonal/Restore/2/3/4] <Fit>:

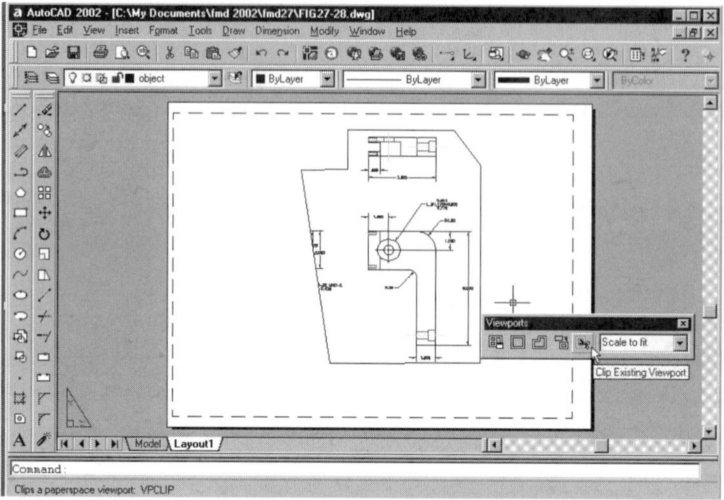

Figure 22.28 *The* **Vpclip** *command can be used to clip or redefine the shape of an existing viewport. You can also access the command by selecting the* **Clip Existing Viewport** *button on the* **Viewports** *toolbar. Courtesy Cathy Schlesser.*

Select object to clip viewport: *(Select desired object)* Regenerating model

Options include the following:

ON and OFF—These serve as toggle switches to display or hide the contents of the individual viewports. Selecting either option will provide a prompt

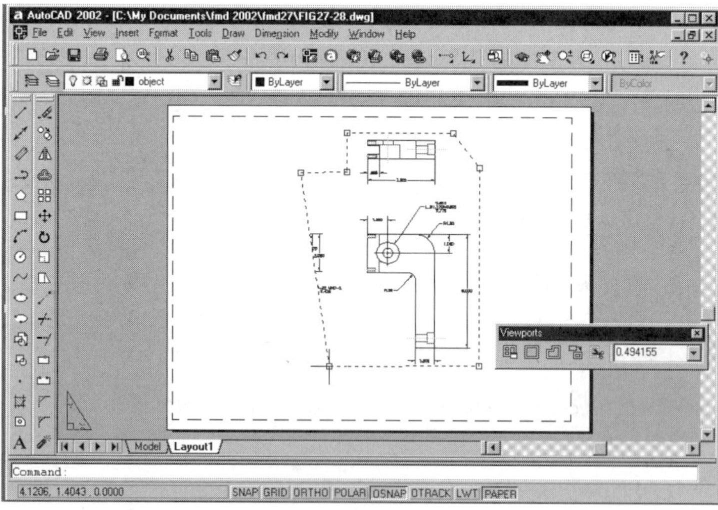

Figure 22.29 *A viewport can be edited by selecting a viewport grip. Courtesy Cathy Schlesser.*

to select objects so the desired viewport can be controlled. This option can be very useful when you plot stock sheets of details, if one or two attached details do not apply to the current use. Setting these viewports to OFF will eliminate the viewport from the plot. This option is also very useful for controlling **Regen** time. Inactive viewports can be set to OFF while the current viewport is edited. If all viewports are set to OFF, you will not be allowed to work in model space of a layout until a new viewport is created.

Fit—This option will create a viewport to fit the current screen display. The size of viewport will be determined by the dimensions of the paper space view. In the example used throughout this chapter, the objects were created first in base drawings and then drawings were inserted in a template. The Fit option will allow a viewport to be created around an existing drawing or a portion of a drawing.

Hideplot—This option will allow hidden lines to be removed from selected viewports when plotting in paper space. This option generally is not required in architectural and engineering drawings.

Lock—Locks the current viewport in a similar manner to locking a layer.

Object—Selects a circle, closed polyline, ellipse, region, or spline to be used as a viewport.

Polygonal—Creates an irregularly shaped viewport using the same process that is used when the **Polygonal Viewport** button is selected.

Restore—This option can be used to restore saved viewport layouts. The prompt is

?/Name of window configuration to insert <default name>:

The name of a saved view can be entered. Typing the question symbol (?) will provide a list of view names.

2, 3, 4—Selecting one of these options will create two, three, or four viewports without your having to reissue the command. The required number of viewports is entered, followed by specifications for how to arrange the viewport. Each option is similar to the arrangements used for tiled viewports.

CHAPTER 22 EXERCISES

1. Open the ANSI-C template and create a new layout titled COOL STUFF. Use the **Viewports** dialog box and create viewports using the Three: Above layout. Insert a drawing in model space. Adjust each viewport so that a different aspect of the drawing is centered in each viewport. Save the drawing as E-22-1.
2. Open the ANSI-C template and use the **Create Layout Wizard** to create a new layout titled VIEWPORT TEST. Set the printer as the default system printer, and if necessary select an appropriate paper size. Set the orientation to portrait. Assign a single viewport suitable for displaying a detail at 1:1. Select

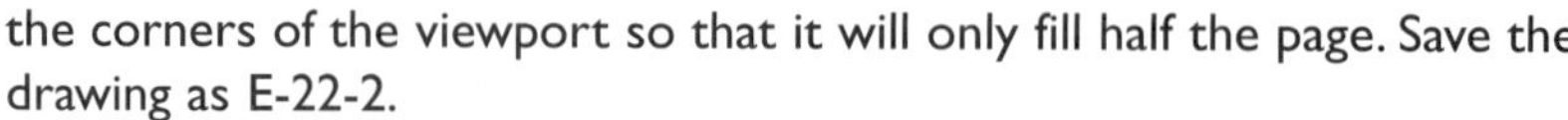

the corners of the viewport so that it will only fill half the page. Save the drawing as E-22-2.

3. Open any three drawings and insert them in a template drawing. Create a floating viewport to display the drawings at a suitable scale. Save the drawing as E-22-3

4. Open the ANSI-C drawing template. Attach any four drawings from your library. Each drawing is to be shown at a different scale when plotted. Provide a title and the scale for each detail. Save the drawing as E-22-4.

   Open drawing E-22-4 and clip two of the drawings so that they are not displayed. Save the drawing as E-22-5.

   Open the ANSI-C drawing template. Insert the tolerance block created in E-20-1. Save the drawing as E-22-6.

CHAPTER 22 QUIZ

1. Describe how a referenced drawing can be used in an office to complete a project.
2. Use the Help menu and determine how and what the **Tilemode** setting affects.
3. What layer is added to the drawing base when you work with viewports? How will it affect the display?
4. What command or process is used to create a viewport?
5. What are the benefits of using layouts?
6. How can the display in the second viewport be altered so that it does not display the same drawing as another viewport?
7. How does paper space differ from model space?
8. A drawing has been inserted in a template in model space but it can't be seen. Describe how to show it.
9. List the steps to create a layout using the **Create Layout Wizard**.
10. Explain the difference between floating and tiled viewports.
11. How is a scale applied to the display in a viewport?
12. List three options for controlling the display of the viewport.
13. A drawing is displayed at the proper scale in a viewport, but the entire drawing can't be seen. Describe how the problem can be fixed.
14. Three drawings are displayed in a viewport, but you would like to hide one from a meeting with a client. Describe how to hide the drawing.
15. Describe the process to create an irregularly shaped viewport.

Controlling Output

Chapter 22 walked you through the process of taking a drawing created in model space and placing it in a layout for plotting. As computers become more common in drafting offices, more of the projects will be transmitted between offices electronically. AutoCAD allows an electronic plot to be created that is accessible over the Internet. This allows the designers and drafters to have access to areas of the project that require contributions from them. In addition to providing access to the drawings, electronic communication can save many hours by eliminating the need to produce paper copies. Unfortunately, some offices have not made the transition to CAD and still require paper copies. This chapter will introduce methods for

- Controlling plotting
- Controlling the plotting device
- Assigning plot styles
- Configuring the drawing to be plotted

The commands to be introduced in this chapter include

- **Plot**
- **Stylemanager**

AN OVERVIEW OF PLOTTING

The commands that control plot devices and settings can be accessed from the **Plot** dialog box, which consists of the **Plot Device** and **Plot Settings** tabs. The **Plot Device** tab is used to assign plotters and plot style tables. The **Plot Settings** tab is used to set plotting parameters. Access the **Plot** dialog box by selecting the **Print** button on the **Standard** toolbar, by typing PLOT (ENTER) at the Command prompt, or by selecting **Print** from the **File** menu. Each method will produce a **Plot** dialog box similar to Figure 23.1. The **Plot** dialog box can be used to control either a printer or a plotter.

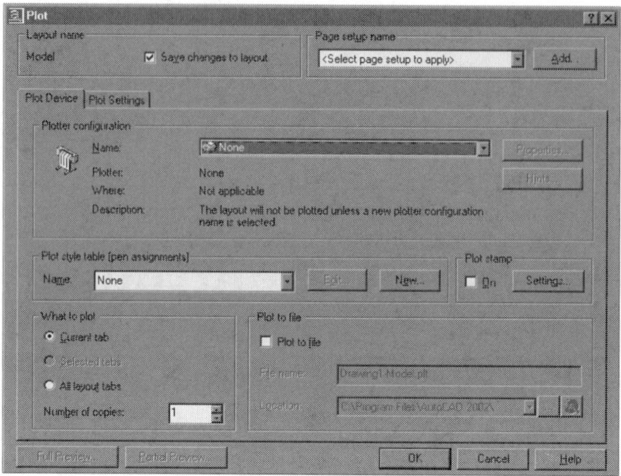

Figure 23.1 *The commands to control plotting can be accessed from the* **Plot** *dialog box. The dialog box consists of* **Plot Device** *and* **Plot Settings** *tabs. The* **Plot Device** *tab is used to select plotters and plot style tables. The* **Plot Settings** *tab is used to set plotting parameters. Access the* **Plot** *dialog box by selecting the* **Print** *button on the* **Standard** *toolbar, by typing* **PLOT** (ENTER) *at the Command prompt, or by selecting* **Print** *from the* **File** *menu.*

Each time a new layout was entered, the **Page Setup** dialog box was displayed. The **Page Setup** dialog box was first introduced when layout tabs were first examined. Although it was dealt with quickly when the initial drawing process was started, as you gain experience, it should be used to define plotting parameters. When the drawing is complete, it will be plotted based on the specified parameters during the page setup. As you gain experience, plotting parameters can be specified and saved within a layout. The **Plot** dialog box is similar to the **Page Setup** dialog box. Plotting parameters assigned in the **Plot** dialog box will override those set in the **Page Setup** dialog box.

Note: It's important to keep sight of the goal as you begin this chapter. Your goal is to produce a plot at a useable scale; so far you've been making scaled plots using the **Scaled to Fit** option. This produced scaled drawings, but the scale was most likely a scale known only to AutoCAD and not the rest of the known world. In the industry, most drawings are plotted at a scale such as 1/8"=1". AutoCAD produces a warning as the **Plot** dialog box is displayed that might seem a bit intimidating for new users. AutoCAD has added many aids and wizards to help make plotting easier, but at the same time has made the initial process seem more complex. Think of yourself in a new, top-of-the-line, bright red, convertible sports car. All you want to do is drive—you'll read the manual when the thrill wears off. The purpose of previous chapters was to get you driving. You've taken AutoCAD plotting on a test spin. This chapter will introduce the concepts needed to produce a drawing at a standard scale. Once the process to produce a plot has been explored, you'll be introduced to the many features that have been provided by AutoCAD to enhance plotting.

REVIEWING THE PLOTTING PROCESS

A scaled plot can be created easily using the **Plot** dialog box with the same procedure that was introduced in previous chapters. As the dialog box is displayed, you'll be prompted to specify the parameters that will define the plot. Once the **Plot** dialog box is displayed, the plotting process will include the following steps:

1. If a named page setup has been previously defined and saved, it can be selected from the **Page Setup Name** list.
2. Using the **Plot Device** tab, verify that the desired plotter is current.
3. Verify that the correct plot style table (pen assignments) is attached to the drawing.
4. Use the **Plot Settings** tab to define the plotting parameters. Settings include **Paper size**, **Plot area**, **Drawing orientation**, **Plot scale**, **Plot offset**, and **Plot options**.
5. Select a preview option.
6. Press ENTER. If the preview does not provide the expected results, redefine the plotting parameters on each tab as needed and repeat the preview process.
7. When the plot preview meets your expectation, select the **OK** button to produce the plot.

EXPLORING THE PLOT DIALOG BOX

The **Plot** dialog box consists of two tabs and three areas that are common to each tab. At the top of the dialog box are areas that describe the **Layout name** and **Page setup** name. At the bottom of the display are buttons to control preview options and buttons to continue or cancel the plotting process. The tabs are used to control the plotting devices and the plot settings.

Layout Name Display

The **Layout name** area of the **Plot** dialog box displays the name of the layout to be plotted. In Figure 23.1 the layout displayed is the Model. The selected layout names will be displayed in this area if multiple layouts are selected to be plotted. Model will be displayed if the Model layout is the current viewport. Model space will produce a plot of the drawing, without displaying the border and title block. For best results, select a layout that will display the drawing with a title block.

Saving Plotting Specifications

AutoCAD allows the plotting parameters that are defined in the **Plot** dialog box to be attached to the layout. The next time you need a plot of the selected drawing, instead of setting the parameters of the two tabs, you only need to click the **OK** button of the **Plot** dialog box. With the **Save changes to layout** check box active, changes made to each tab will be saved in the current layout. This option is not available if multiple layouts are selected for plotting in the **What to plot** area.

Page Setup Name
This portion of the **Plot** dialog box displays a listing of named and saved page setups. Attempting to plot a drawing displayed in the Mechanical Title Block would include the following options: Previous Plot, Mechanical Title Block no output device.

Add
Selecting the **Add** button will display the **User Defined Page Setups** dialog box, which is used to name the plot setup that is to be saved to the layout. In addition to naming the current setup, this option can be used to rename, delete, or import user-defined setups and apply them to the current drawing.

Preview Options
Once all of the plotting parameters have been set, the preview options can be used to view the outcome of the plot before the plot is produced. Options include the full and partial previews. Each will be explored in detail after all of the settings have been explored. A full preview will display the drawing just as it will appear when plotted. A partial preview will show a box that defines the limits of the objects to be plotted inside a box that defines the plotting area. The **Partial Plot Preview** dialog box displays the **Paper size**, **Printable area**, **Effective area**, and **Warnings**. The **Warnings** area displays a list of warnings that will affect the effective plotting area. The other three categories can each be set on the **Plot Settings** tab.

EXPLORING THE PLOT DEVICE TAB

As the **Plot** dialog box is displayed, the **Plot Settings** tab is displayed by default. Once a plotter is configured to your workstation, the plot device will not need to be accessed for each plot. The key elements of the **Plot Device** tab include settings for the current plotter and settings for using a plot style table. The plotting devices that are configured to your workstation or network can be used by AutoCAD to plot drawings.

As you've made check prints from model space, AutoCAD has used the plotter specified in the **Options** dialog box to determine the default plotter. As you plot using a layout, the plotting device specified on the **Layout** tab is used. The layout stores settings for plot devices, the plot style table, print area, rotation, paper size, and scale. These settings can be provided as the drawing is started, and stored as a named page setup. When the drawing is ready to be plotted, these values can used to control the plot production. The **Plotter Manager** controls existing plotters, modifies existing configurations, and adds new plotter configurations. To access this feature choose the **Plotter Manager** from the **File** menu. Users of previous releases of AutoCAD will appreciate the elimination of command line prompts and the addition of the **Add-A-Plotter Wizard** to configure plotters.

Defining Plotting Parameters

The **Plot Device** tab displays the settings and controls for the plotter to use, a plot style table, the layout to be plotted, and information for saving the drawing to a file. The major areas of the **Plot Device** tab include **Plotter configuration**, **Plot styles table (pen assignments)**, **What to plot**, **Plot to file**, and five buttons common to both tabs.

Plot Configuration

The **Plotter Configuration** area of the tab displays the current plotting device and its operating parameters. It also displays the port it is configured to. Selecting the **Name** edit box will display a list similar to Figure 23.2. The list shows each available plotter that is connected to either the workstation or the network. Notice that an icon is displayed by each listing to distinguish between system devices and those stored with a PC3 file name. A system printer or plotter is a device that is configured in the Printer folder of My Computer, or added by selecting **Printer** from the **Settings** menu. Selecting the **Add Printer** option from the Windows **Printer** dialog bcx will produce a prompt for the **Add Printer Wizard** and allow a printer to be added and accessed by other programs. You can select previously configured printers using the **Device and Default Information** box of the **Print/Plot Configuration** dialog box. Methods of configuring a printer or plotter and adding it to this list will be introduced later in this chapter. Selecting the **Properties** button will display a **Plotter Configuration (PC3)** editor. The editor is used to modify or view the current plotter parameters. Selecting

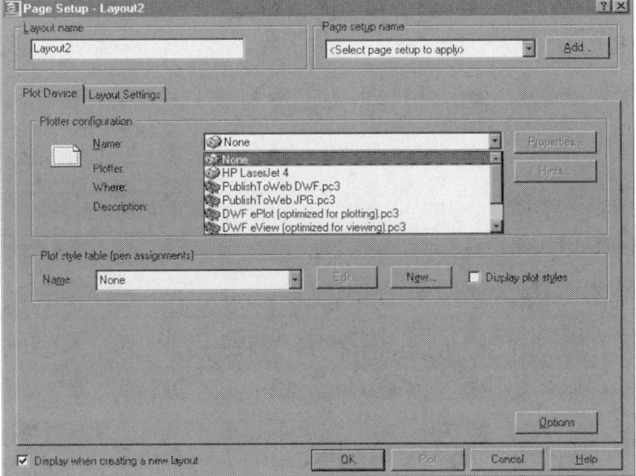

Figure 23.2 *Selecting the* **Name** *edit box will display a list of available plotting devices that are connected to either the workstation or the network.*

the **Hints** button will display information on improving the plot quality specific to the configured plotting device.

A PC3 file is a file created by AutoCAD that is drawing independent. PC3 files are just like any other file in that they are portable and can be shared between different workstations. A PC3 file contains all of the information on what to plot and how to plot, including the specified plotting device, paper size, and pen information. For users of previous releases of AutoCAD, PCP and PC2 files can be imported into AutoCAD 2002. These older formats will automatically be converted to PC3 format.

Plot Style Table (Pen Assignments)

If you've used an older release of AutoCAD, part of configuring the plotter was to configure pen assignments. AutoCAD 2002 replaces pen assignments with plot styles. Just as a style can be assigned to text or dimension, AutoCAD allows a style to be assigned to each layout and model space view. This portion of the **Plot Device** tab is used to create, edit, or set the plot style table assigned to the current model and layout tabs. Selecting the **Name** edit box displays a list of available styles. Later in this chapter you'll be introduced to the procedure to create and plot with plot styles. Selecting the **New** button will start the **Add Color-Dependent Plot Style Table Wizard**. The wizard can be used to create a new plot style table. Once a plot style table has been created, it will be listed in the **Name** box. Notice that in Figure 23.3, the listing shows several plot styles, even though none have been defined. These are plot styles defined by AutoCAD. Most mechanical drawings can be plotted by selecting the monochrome.ctb, Default R14 pen assignment.ctb, or the R14 cfg-1-Default System Printer.ctb.

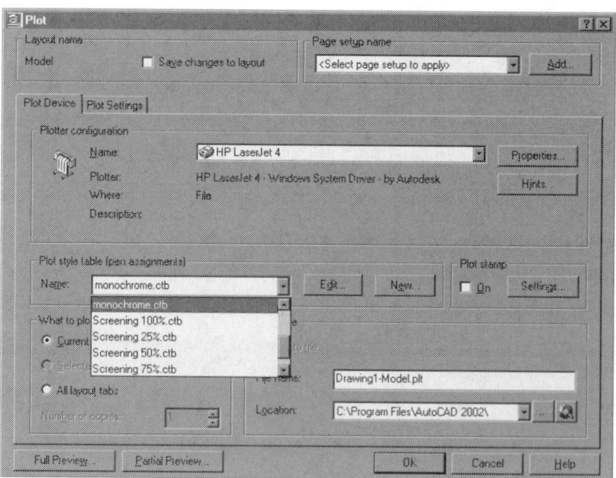

Figure 23.3 *The plot style* **Name** *listing shows several plot style tables defined by AutoCAD. As you create new styles, they will be added to the list.*

What to Plot

This portion of the **Plot Device** tab allows you decide if you will plot using model space or one of the layout tabs of the drawing. The default setting is the **Current tab**. If the model space view of the drawing is displayed on the screen, this is what will be plotted. Setting a layout as the current display will make it the current tab, selected as the drawing to be plotted. If **All layout tabs** is selected, all of the layouts contained in the current drawing will be plotted. The **Selected tabs** option is currently inactive when only one tab is selected for plotting. To make this option active, hold down CTRL and select the tabs to be plotted prior to displaying the **Plot** dialog box. The **Number of copies** setting determines the number of plots to be made of each layout.

Plot to File

When you complete all of the options on both tabs of the **Plot** dialog box, the information will be sent to the plotting device, and the information will be processed. The time required to produce the plot will depend on the size of the drawing file and the type of plotter. If the **Plot to file** check box is selected, the plot will be stored as a .PLT file instead of being sent to the plotting device. Saving a plot file allows the drawing to be plotted at a more convenient time. As the **Plot to file** check box is selected, the other options in this area will become active. A name such as Widget Layout1 will be displayed, indicating the drawing and layout names. The **Location** edit box shows the location of the current drawing. The PLT file will be stored in the same location as the drawing to be plotted. Selecting the [...] button will display the **Browse for Folder** dialog box, with a display similar to the Explorer display. The dialog box allows the storage location of the plot file to be specified using methods similar to that used by other dialog boxes for saving files. The file will be stored to disk with the indicated name followed by the .PLT extension. Any storage device in your machine or accessible to the network can be used to store the plot file. Selecting the **Browse the Web** button allows the drawing to be stored at a specified Web site.

EXPLORING THE PLOT SETTINGS TAB

The **Plot Settings** tab is used to control the paper size and paper units, plot area, drawing orientation, plot scale, plot offset, and plot options. Each area can be seen in Figure 23.4.

Paper Size and Paper Units

This portion of the **Plot Settings** tab displays the configured plotter, the current paper size, the printable area, and toggles from inch to metric. The plotter that was specified on the **Plot Device** tab is displayed as the **Plot device**. This display indicates the standard plotting sizes that can be accommodated by the current plotting device. Selecting the edit arrow will display a list similar to Figure 23.5. (The list will vary, based on the configured plotter.) The paper size to be used for the plot can be selected from the list. The sizes displayed are the actual paper sizes. Sizes are listed

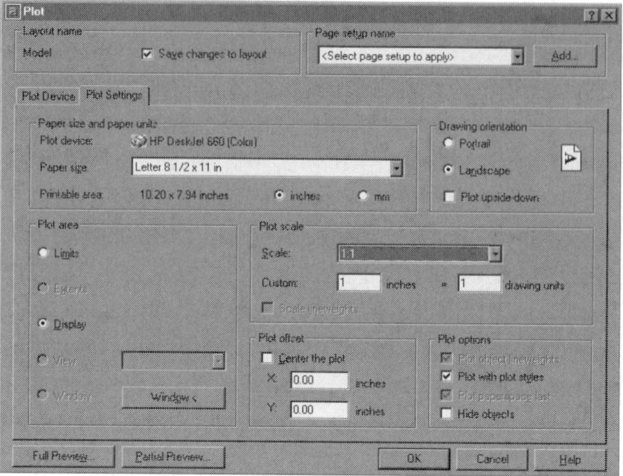

Figure 23.4 *The* **Plot Settings** *tab is used to control the paper size and paper units, plot area, drawing orientation, plot scale, plot offset, and plot options.*

as width (X) × height (Y). Once the desired value is selected, the **Printable area** will be updated. This is the maximum area available based on the current paper size. The final option of this area of the tab is to select the plotting units. Options include inches and mm.

The Plot Area

This area of the **Plot Settings** tab provides options for describing what portion of the current drawing will be plotted. Because the entire drawing often does not need to be plotted, five options are given to decide what portion of the drawing will be plotted

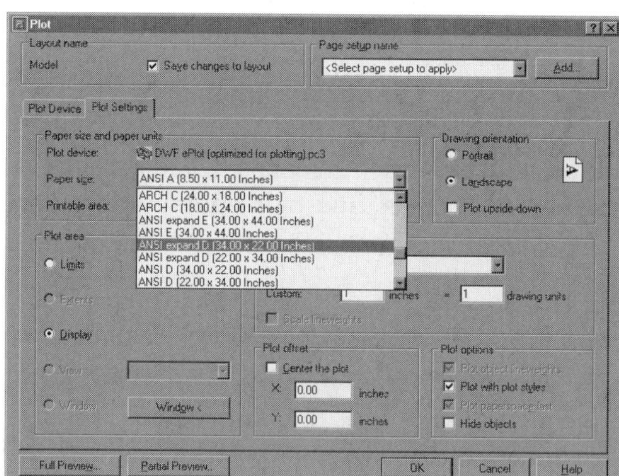

Figure 23.5 *The plotter that was specified in the* **Plot Device** *tab is displayed as the* **Plot device.** *The* **Paper size** *list indicates the standard plotting sizes that can be accommodated by the current plotting device. The list will vary, based on the configured plotter.*

Layout

This option will plot everything within the margins of the selected paper size. This will be the option of choice when the entire contents of the layout need to be plotted.

Extents

Choosing this option will plot the current drawing with the extent of the drawing objects as the maximum limits that will be displayed in the plot. The option is similar to the Extents option of the **Zoom** command. Using this option might eliminate objects at the perimeter of the drawing because the plotter cannot reach the edge of the plotting surface.

Display

This setting plots only the material that is currently displayed on the screen. If you have zoomed into a drawing, only that portion of the drawing will be plotted.

View

If specific views have been saved using the **View** command, the **View** option for plotting will be activated. This option will allow specific views to be plotted as they were saved. In the initial setting, the **View** radio button is inactive. When a drawing contains named views, the **View** button will be active and a listing showing each saved view name can be displayed. Selecting the desired view to be plotted will highlight that view, place the name in the edit box, close the list, and allow the selections to continue.

Window

This option will allow a specific area of a drawing to be defined for plotting. In the same way that a window is used to zoom or select objects for editing, a window can be specified here to define the selection plot. Initially the **Window** radio box is inactive. If the **Window<** button is selected, the **Plot** dialog box will be removed from the display area, and prompts for defining the window will be displayed. The prompts are as follows:

> Specify first corner: *(Select first window corner.)*
>
> Specify other corner: *(Select opposite window corner.)*

Once the window is defined, the **Plot** dialog box will be redisplayed.

Drawing Orientation

This area of the **Plot Settings** tab allows the drawing orientation to be specified. The box on the right side of the display is the orientation icon and not a selection box. The current orientation of the box is based on the configured plotter and indicates the drawing will be plotted in the normal orientation for construction drawing. This setting is referred to as *landscape*. When **Landscape** is used, the A will be placed on its side, indicating the print will be read from the long edge. When the plot is rotated so that it is read from the title block end, the plot is referred to as a portrait plot. The A will be in the vertical position indicating the print will be read looking from the title block end. A portrait plot is more likely to be used with laser printers. The orientation of the drawing to the paper can be adjusted by selecting the appropriate radio button. A check box is also available for plotting the drawing upside-down. Selecting this option will rotate the portrait or landscape layout 180°.

Plot Scale

The **Plot scale** area of the **Plot Settings** tab controls settings for the scale and lineweight scale. As drawings have been created throughout this text, they've been drawn at real size rather than at a reduced scale, as would be done in manual drawing. In earlier chapters you were introduced to methods of creating layouts for plotting. If you're plotting a layout, the scale was set as the layout was created. This will allow the scale factor to be at 1:1 here. If you want to make a plot of a drawing created in model space, you'll need to provide a scale factor for plotting. Select the **Scale** edit arrow and the desired scale from the list to provide a plotting scale. Figure 23.6 shows a portion of the **Scale** list showing common scale factors that can be used to produce a plot at a given size. Dimensions and text were both multiplied by a scale factor in anticipation of the scale at which the drawing would be plotted.

An alternative to selecting a precise scale is to activate the **Scaled to Fit** box on the **Plot Settings** tab of the **Options** dialog box. An alternative to selecting a precise scale is to activate the **Scaled to Fit** box on the **Plot Settings** tab of the **Plot** dialog box. This option will allow the desired plot to be sized to fit within the parameters of the paper. The resulting scale will be displayed in the **Custom** display box. Drawings that are scaled to fit are useful as check prints to determine completeness, but they are not

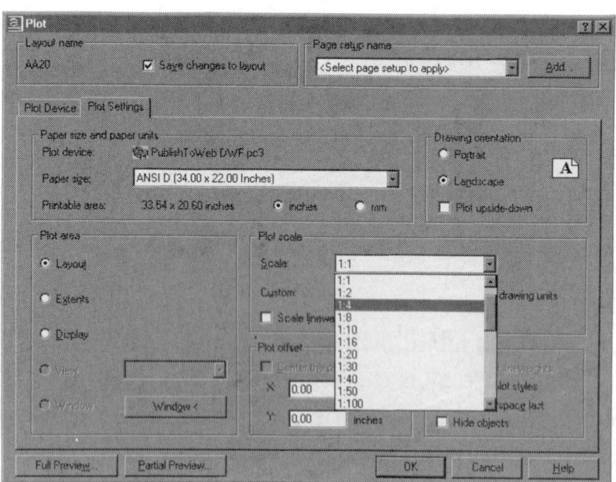

Figure 23.6 *Select the* **Scale** *edit arrow to display a list of common scale factors that can be used to produce a plot at a given size.*

suitable for most projects within the manufacturing industry. The **Custom** box is better suited for entering a scale that is not listed. The scale is created by entering the number of inches equal to the number of drawing units.

Use the **Scale lineweights** check box to scale lineweights in proportion to the plotted scale. Lineweights selected during drawing construction normally specify the line width of printed objects and are plotted with the specified width regardless of the plot scale.

Plot Offset

The **Plot offset** portion of the **Plot Settings** tab contains the **Center the plot** check box and a display for X and Y coordinates. Activating the **Center the plot** option will automatically center the plot in the center of the plotting paper. The plot offset values specify an offset distance of the drawing to be plotted from the lower left corner of the paper. If you're plotting a layout, the lower left corner of the plotting area is positioned at the lower left edge of the plotting material. You can offset the origin of the plot from the paper edge by entering a positive or negative value. The plotter unit values are in inches or millimeters on the paper. Entering a value in the **X edit** box specifies the plot origin in the X direction. Entering a value in the **Y edit** box specifies the plot origin in the Y direction. The default setting for plotting origin is X=0–Y=0. For most plotters, this will be the lower left corner of the paper. By selecting the X or Y origin box, new values can be entered. Entering an X value of 12 and a Y value of 10 will place the new origin 12 units to the right and 10 units above the original plotting origin.

Plot Options

This portion of the **Plot Settings** tab contains four check boxes for controlling the outcome of the plot. The options control lineweight, plot styles, and the current plot style table. Plot styles will be introduced in the next portion of this chapter.

> **Plot object lineweights**—With this box active, lineweights specified in the drawing will be plotted.
>
> **Plot with plot styles**—With this option active, plot styles will be applied to objects that are defined in the plot style table. All style definitions with different property characteristics are stored in the plot style tables.
>
> **Plot paperspace last**—With this option active, model space geometry will be plotted first.
>
> **Hide objects**—With this option active, hidden lines will be removed from layouts. Viewports can be used to remove hidden lines for model space drawings.

Previewing the Plot

You've now explored each tab of the **Plot** dialog box. Once values have been entered for each option, the drawing can be plotted. Selecting either preview button allows the outcome of the current plot settings to be viewed prior to the actual plot. Two options are available for previewing the plot: **Partial Preview** and **Full Preview**.

Partial Preview
Selecting the **Partial Preview** button will produce a display similar to the one seen in Figure 23.7. A white box represents the paper size, and a dashed line specifies the limits of the effective plotting area. A blue box represents the area that the drawing will fill. A warning also will be given of any problems that may be encountered during the plot. Common warnings displayed in the warning box include the following:

Effective area too small to display

Origin forced effective area off display

Plotting area exceeds plotter maximum

A small red triangular Rotation icon is displayed in the corner to represent the current rotation. The triangle will be moved, depending on the **Landscape**, **Portrait**, or **Upside-down** selection. Clicking **OK** will restore the **Plot** dialog box.

Full Preview
Choosing this option will produce a graphic display of the drawing as it will appear when plotted. Figure 23.8 shows an example of a full preview. A few seconds will elapse between the selection of the **Full Preview** button and the display of the proposed plot. An outline of the paper size will be drawn on the screen, which displays the portion of the drawing to be plotted. Once the drawing is displayed, the **Realtime Zoom** button is displayed. The **Pan** and **Zoom** options can be quite useful in viewing detailed areas of the drawing a final time prior to plotting. Once the indicated **Pan** or **Zoom** is performed, press ENTER to restore the **Plot** dialog box.

Producing the Plot

The two tabs of the **Plot** dialog box have been examined. If you change your mind on the need for a plot, click the **Cancel** button. The dialog box will be terminated, and

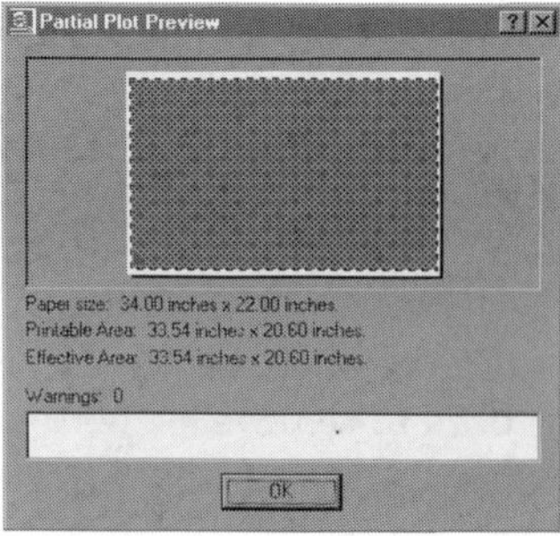

Figure 23.7 *Selecting the* **Partial Preview** *button will produce a display of the area to be plotted. The white box represents the paper size, and a dashed line specifies the limits of the effective plotting area. A blue box represents the area that the drawing will fill.*

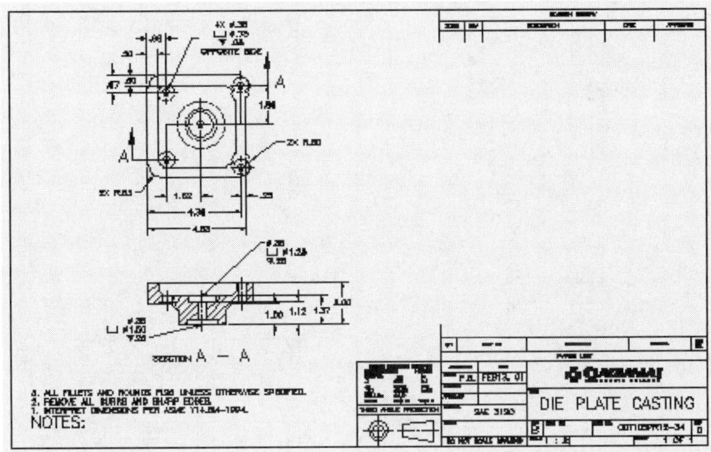

Figure 23.8 *A full preview shows each drawing object as it will appear in the finished plot. Courtesy Paul Bissell.*

the drawing will be redisplayed. If you are satisfied with the plotting parameters that have been established, click the **OK** button. The dialog box will be removed from the screen and the Command prompt will display the effective plotting area. With older pen plotters, you might be given a prompt to remind you to position paper in the plotter. Newer plotters use rolls of paper, but they also need to be reloaded occasionally. Don't laugh! After you labor through the **Plot** dialog box the first few times, simple things like turning the plotter on and installing the paper might tend to be forgotten. If you need to set up the plotter, you can do it before you click **OK**. Your plotter might have additional features, such as pen pressure and speed, to be adjusted. The completed plot can be seen in Figure 23.9.

WORKING WITH PLOT STYLES

Rather than setting pen assignments like past releases, AutoCAD 2002 allows a plot style table to be created and saved as a file. Plot styles are a new object property assigned to all drawing objects to control the appearance of the finished plot. Plot styles can be used as the drawing is reproduced to override the object properties assigned as the drawing was created. They can be used as overrides for color, dithering, gray scale, linetype, lineweight, pen assignments, screening, end styles, join styles, and fill styles. They can be assigned layer by layer or object by object. Each aspect of the style is defined in a plot style table that can be attached to the model or layout tab. By assigning different plot styles to the same layout, you can reproduce the drawing in different ways to meet the needs of different clients or subcontracting offices. The three steps required to use a plot style are as follows:

1. Define the plot style using a plot style table.

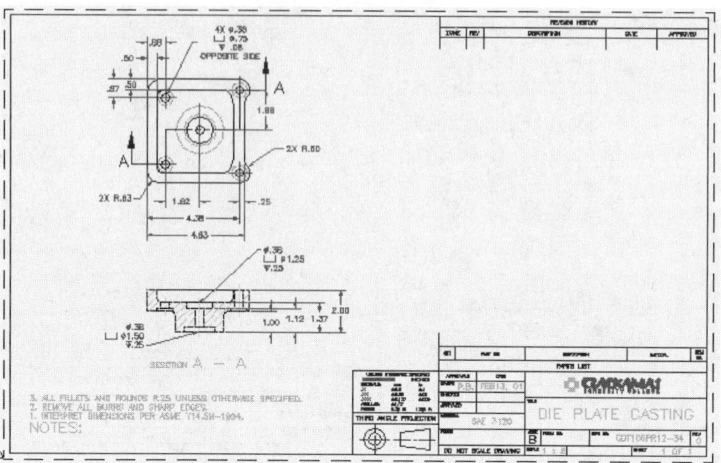

Figure 23.9 *The plot of the drawing previewed in Figure 23.8. Courtesy Paul Bissell.*

2. Attach the plot style table to the layout.
3. Assign the plot style to an object or layer in the drawing.

The effects of the attached plot style can be viewed on the **Plot Settings** tab of the **Plot** dialog box.

EXPLORING PLOT STYLE MODES

The two types of modes used for all drawings in AutoCAD 2002 are **Color Dependent** and **Named** plot styles. The mode is controlled using the **Plotting** tab of the **Options** dialog box. (Select **Options** from the **Tools** menu.) The mode is controlled in the **Default plot style behavior for new drawings** area of the **Plotting** tab. Users of previous releases of AutoCAD will be most familiar with the **Color-Dependent** mode.

Color-Dependent Plot Styles

A color-dependent plot style controls the plotting based on the color assigned to objects in the drawing. There are 255 color-dependent plot styles based on the 255 colors in the AutoCAD color index. Color-dependent plot styles are very similar to the pen assignments introduced with AutoCAD Release 14. Key characteristics of a color dependent plot style include the following:

- Plot characteristics are assigned to each color to determine how the drawing will be plotted.
- Color-dependent plot styles cannot be added, deleted, or renamed in a style table.

Using a color-dependent plot style, you can alter the appearance of all red drawing objects by changing the plot style that corresponds to RED. For most of the drawings that you've done in this text, a color-dependent plot style will work well. All red lines could be assigned to a specific pen width, all blue lines could be assigned to a different pen width. However, as your drawing become more complex, using the object color could limit the use of color in a drawing. By associating color with a specific pen, you lose the ability to work with a color independent of lineweight or linetype. Creating color-dependent plot styles will be explored as the use of the **Add Plot Style Table Wizard** is explored.

Named Plot Styles

A named plot style depends on an assigned name rather than the color of an object. Named styles are independent of object colors in the style. Key characteristics of a named plot style include the following:

- The Normal style assigned by AutoCAD cannot be modified.
- Each defined style in the plot style table has a name and specific plotting characteristics.
- A named plot style can be assigned to layers or objects.

CREATING A PLOT STYLE TABLE

One of the most common uses of a plot style table is to plot a layout containing a wide variety of colors on a plotter using black lines. The drawing can be plotted without using a plot style table, but each line might be displayed in a gray tone rather than black. By creating a plot style table, you can assign each color to use the color black for plotting, with the same color intensity. A plot style table can be created using the **Stylemanager** command. Access the command by selecting **Plot Style Manager** from **File** on the **Standard** toolbar or by typing STYLEMANAGER (ENTER) at the Command prompt. Each method will display the **Plot Styles** dialog box shown in Figure 23.10. You can also access the **Plot Styles** dialog box by selecting the **Add or Edit Plot Styles Tables** button on the **Plotting** tab of the **Options** Dialog box. Each of the available plot style tables is displayed in the **Plot Styles** dialog box by a name and icon. Selecting the **Add A Plot Style Table** option will start the wizard shown in Figure 23.11. After reading the introductory text, click the **Next** button.

Figure 23.12 shows the second dialog box prompt of the **Add Plot Style Table Wizard**. By default, you'll be starting the plot table from scratch. The dialog box also allows options for the following:

Use My R14 Plotter Configuration—this option allows a new plot style table to be created using pen assignments stored in an R14 drawing, but you don't have access to the PCP or PCP2 files associated to the drawing.

Use a PCP or PCP2 file—This option allows a new plot style table to be created using pen assignments stored in acadr14.cfg files.

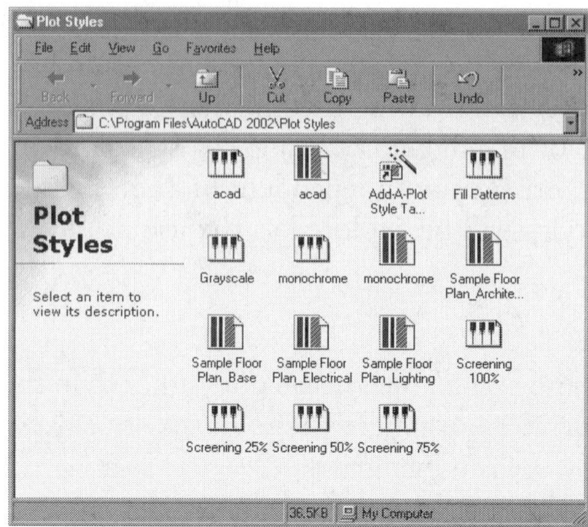

Figure 23.10 *A plot style table can be created using the* **Stylesmanager** *command. Access the command by selecting* **Plot Style Manager** *from* **File** *on the* **Standard** *toolbar or by typing* **STYLESMANAGER** *(ENTER) at the Command prompt. Each method will display the* **Plot Styles** *dialog box.*

As you proceed through the initial viewing of the wizard, accept the default option and select the **Next** button. The next dialog box of the **Add Plot Style Table Wizard** will display a prompt for the type of table to be used. Options include **Color-Dependent Plot Style Table** and **Named Plot Style Table**. Choose the **Named Plot Style Table** button. Selecting the **Next** button will display the **File Name** dialog box. Enter a name to describe the plot style in the **Name** edit box. For this example a name of MONO will be used. Once a name has been entered in the edit box, select the **Next** button to display the final display of the wizard.

Up to this point, you've started a plot style table named MONO. The final display of the setup wizard allows you to complete the naming process or to edit the plot style.

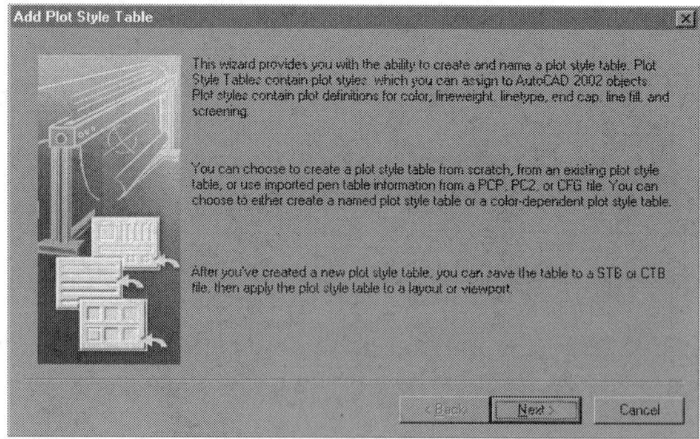

Figure 23.11 *Selecting the* **Add A Plot Style Table** *button will start the wizard. After reading the introductory text, click the* **Next** *button to proceed through the setup.*

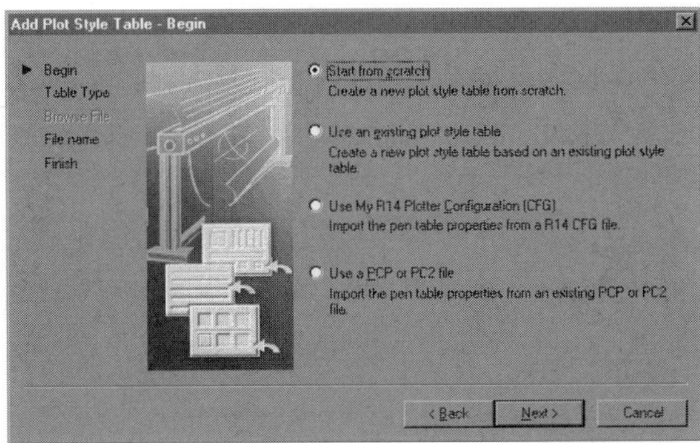

Figure 23.12 *The* **Add Plot Style Table - Begin** *dialog box shows the options for selecting a starting point for creating a plot style table. By default, you'll be starting the plot table from scratch.*

Selecting the **Plot Style Editor** button will display a **Plot Style Table Editor** dialog box similar to Figure 23.13. This table can be used to alter the properties of the drawing to be plotted. Later sections of this chapter will deal with altering drawing properties. Select the **Finish** button to create a named plot style table and exit the wizard. Selecting **Finish** will create a .STB file that is listed in the **Plot Styles** dialog box. Selecting the **Use this plot style table for layouts in new drawings and pre-AutoCAD 2002 drawings** will make the newly created plot style the default plot style for future drawings.

Creating a Color-Dependent Plot Style Table

The process to create a color-dependent plot style is similar to the process of creating a named style. The steps to create a color-dependent include the following:

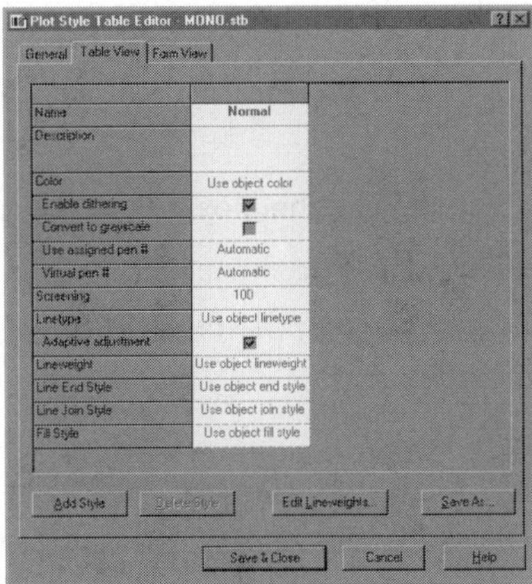

Figure 23.13 *Selecting the* **Plot Style Editor** *button will display a* **Plot Style Table Editor** *dialog box. This dialog box can be used to alter the properties of the drawing to be plotted.*

1. Select the **Add A Plot Style Table** button in the **Plot Styles** dialog box.
2. Select the **Next** button after reading the introductory material.
3. Select the option to be used to start the plotting style table and select the **Next** button.
4. Choose the type of plot style table to be created and select the **Next** button.
5. Enter the name of the plot style table and choose the **Next** button.
6. Set the options on the **Finish** page and click the **Finish** button.

The color-dependent plot style will be stored as a .CTB file and listed in the **Plot Styles** dialog box.

ATTACHING A PLOT STYLE TABLE TO A LAYOUT

A plot style created with the **Add Plot Style Table Wizard** can be attached to the model or layout tab using the **Page Setup** dialog box. Use the following steps to attach a plot style to a layout tab.

1. Select the desired layout tab to apply the plot style.
2. Select **Page Setup** from the **File** menu.
3. Using the **Page Setup** dialog box, select the **Plot Device** tab.
4. Select the desired plot style to be applied in the **Plot Styles** portion of the **Plot Device** tab.
5. Click the **OK** button.

The qualities assigned to the plot style can now be applied to the selected layout.

Viewing a Plot Style

Once a plot style has been created, it can be previewed before it is plotted. Changes made to object properties in a layout can be seen when a drawing is regenerated, or you can use the **Full Preview** option in the **Plot** dialog box. Plot styles can be displayed in a drawing using the following steps:

1. Select the layout tab containing the plot style to be used.
2. Select **Page Setup** from the **File** menu.
3. Select **Display Plot Styles** from the **Plot style table (pen assignments)** area on the **Plot Device** tab.
4. Click the **OK** button.

The properties of the plot style will now be displayed. If the changes are not displayed, type REGENALL (ENTER) at the Command prompt to display plot styles in a layout.

Adding Named Plot Styles

Once you've created your own plot style, it must be added to a plot style table before it can be applied to the layout to be plotted. Named plot styles are added to the plot style table using the **Plot Style Table Editor**. Plot styles are added using the following sequence:

1. Select **Plot Style Manager** from **File** on the Standard toolbar.
2. Use one of the following methods to display the **Plot Style Table Editor**:
   - Double-click an .STB file.
   - Right-click an .STB file and select **Open** from the shortcut menu.
3. In the **Active Plot Style Table** of the **Current Plot Style Table** dialog box or the **Select Plot Style Table** dialog box, select the desired plot style and choose **Editor**.
4. Select the **Table View** tab or **Form View** tab in the **Plot Style Table Editor** dialog box to view the style settings.
5. Select the **Add Style** button.

A style titled Style1 will be added to the table. While the new style is still highlighted, change the name to better describe the style contents. Figure 23.14 shows an example of a style that was added. Adjust the options as needed, and select the **Save & Close** button to return to the **Plot Styles** dialog box.

Deleting a Named Plot Style

As plotting requirements change throughout the life of a project, a plot style in a named plot style table can be deleted. The procedure to delete a style is similar to the procedure used to add the style. Use the following steps to delete a file:

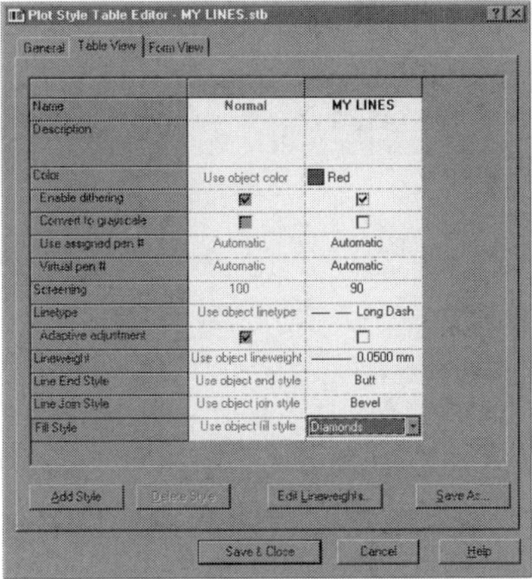

Figure 23.14 *The* **Table View** *tab of the* **Plot Style Table Editor** *dialog box can be used to add a plot style. Notice that the color, screening, linetype, lineweight, and line styles have been altered.*

1. Select **Plot Style Manager** from **File** on the Standard toolbar.
2. Double-click the .STB file to be edited.
3. Use the **Form View** tab of the **Plot Style Table Editor** dialog box to select the plot style to be deleted.
4. With the style highlighted, click the **Delete Style** button.
5. Select the **Save & Close** button or the **Cancel** button to return to the **Plot Styles** dialog box.

DEFINING A PLOT STYLE

Figure 23.14 shows an example of the MY LINES plot style. Properties of a style can be defined or edited, using similar methods used to alter a layer color in the **Layer Properties Manager**. These settings will provide overrides to be used while plotting for values defined in the drawing. Settings can be defined using the **Table View** or **Form View** tab of the **Plot Style Table Editor** dialog box. The **Table View** tab shown in Figure 23.14 works well if only a few styles are assigned to the table. Figure 23.15 shows an example of the **Form View** tab. The **Form View** tab works well if several styles have been assigned to the table. Notice that the style names are listed on the left side, and the values for the current style are listed on the right side.

To define the color for the plot style requires the following steps:

1. Using the **Table View** tab of the **Plot Style Table Editor** dialog box, select the current color setting for the plot style to be altered. As Use object color is selected, the standard color list will be displayed.
2. Select the desired color. As the color is selected, the list will be closed, allowing other options to be defined.

Other options can be defined using similar methods.

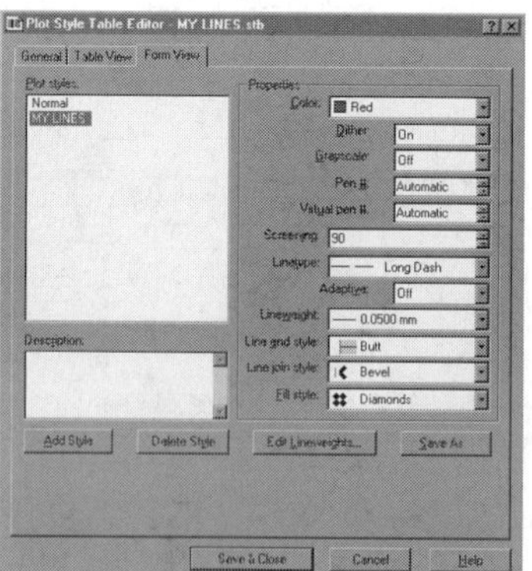

Figure 23.15 *The* **Form View** *tab works well if several styles have been assigned to the table. Notice that the style names are listed on the left side, and the values for the current style are listed on the right side.*

Enable Dithering

Dithering is used to approximate colors using a dot pattern. If the plotter does not support dithering, this option will be inactive. If the plotter supports dithering, dithering can be turned on to provide dim colors more visibility. With dithering off, AutoCAD maps colors to the nearest color.

Convert to Grayscale

With this option active and a plotter that supports grayscale, AutoCAD will convert colors to grayscale.

Use Assigned Pen Numbers

This option specifies which pen will be used with each plot style. The default setting is automatic. In the **Plotter Configuration Editor**, you can specify that a certain object will be plotted using pen #1 as black and .02", and pen #2 as blue and .031". With this option, pen #1 can be assigned to one plot style and pen #2 can be assigned a different plot style. This can be a very useful tool if the drafter has used pen assignments different from the desired standard. Using the Automatic or 0 setting, AutoCAD will select the pen closest in color to the color of the object you are plotting. The value can't be altered if a color-dependent plot table is being used or if the plot style color setting of Use object color is active. Values can be entered by keyboard or by using the range arrow. As the box is selected, arrows are displayed to adjust the range. With a setting of 0, only the up arrow is active. A pen value from 1 to 32 can be specified.

Virtual Pen Number

Most new plotters do not use pens but are capable of duplicating pen plotter output by using virtual pens. In its current setting, AutoCAD will specify a color from the AutoCAD Color Index closest to the color of the object to be plotted. A virtual pen number between 1 and 255 can be provided and used by plotters without pens. With virtual pens, all other style settings are ignored and only the virtual pen settings are used. Values can be entered by keyboard or by using the range arrow. As the box is selected, arrows are displayed to adjust the range.

Screening

This setting allows a color intensity to be specified by controlling the amount of ink to be placed during the plot. The default setting of 100 will plot at full intensity. A value between 0 and 100 can be entered. Entering 0 will have the same effect as freezing the assigned objects. Values can be entered by keyboard or by using the range arrow. As the box is selected, arrows are displayed to adjust the range. With a setting of 100, only the down arrow is active.

Linetype

Selecting the **Linetype** option will display a linetype list similar to Figure 23.16. The default setting for plot style linetype is Use object linetype. Selecting a linetype from the list will override the linetypes specified in the drawing.

Adaptive Adjustment

This setting adjusts the linetype scale to be used to display the linetype pattern. This setting should be OFF when the display of the linetype is important.

Lineweight

The **Lineweight** box displays a sample of the selected lineweight and the numeric value of the lines to be used with the style. An override value can be selected from the drop-down list, displayed as the box is selected.

Line End Style

In addition to using the line end style specified for drawing objects, AutoCAD allows four options to be assigned to override the line end style of drawing objects. Selecting this field will produce a list with four options in addition to the default Use object end style: Butt, Diamond, Round, and Square.

Line Join Style

AutoCAD allows four options to be assigned to override the line join style specified for drawing objects. Selecting this field will produce a menu with the four options in addition to the default Use object join style: Bevel, Diamond, Miter, and Round.

Assigning Fill Style

AutoCAD allows fill patterns to be assigned to override the fill style specified for drawing objects. The patterns will be applied to donuts, polylines, and solids. Selecting this field will produce a list with the options. In addition to the default of Use object fill style, options include Solid, Checkerboard, Crosshatch, Diamond, Horizontal Bars, Slant Left, Slant Right, Square Dots, and Vertical Bars.

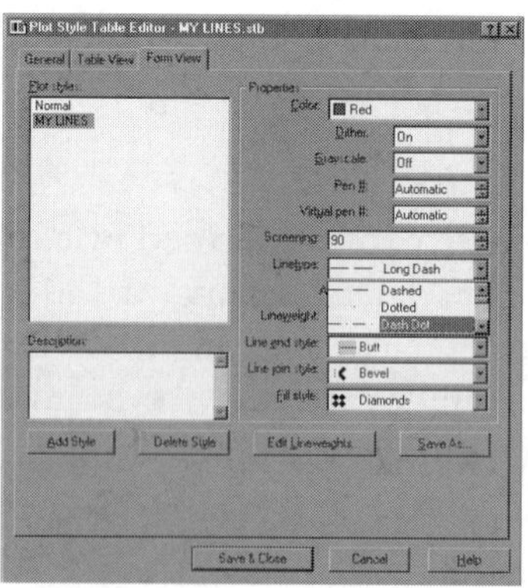

Figure 23.16 *Selecting the* **Linetype** *option will display a linetype list. The default setting for plot style linetype is Use object linetype.*

EDITING PLOT STYLES

A plot style can be edited using the same methods that were used to define it. Either the **Table View** or the **Form View** tab can be used to edit the plot style. Qualities in a plot style can be edited using the same methods that were used to define the style. In addition to altering the qualities, the **Plot Style Table Editor** dialog box can be used to add, delete, copy, paste, and modify plot styles.

Copying a Plot Style

A plot style can be copied using the **Table View** tab of the **Plot Style Table Editor** dialog box. A plot style can be moved from one table to another table by using the drag and drop technique. In addition, one or more settings from a plot style can be copied to another style using the following steps:

1. Open the **Plot Style Manager**.
2. Double-click the desired .CTB or .STB file to be opened.
3. Make the **Table View** tab current.
4. Select the plot style to be copied by selecting the gray area above the plot style name. (You can copy individual settings by selecting individual fields.)
5. Right-click and choose **Copy** from the shortcut menu.
6. Switch to the plot style table to be altered.
7. Select the plot style to be replaced by selecting the gray area above the plot style name.
8. Right-click and choose **Paste** from the shortcut menu.

Renaming Plot Styles

A named plot style can be renamed using the **Rename** command. Before a style can be renamed, it must exist in the plot style table associated with the layout. The new plot style name must be a name that is not currently used by any object or layer. Use the following steps to rename a named plot style:

1. Select **Rename** from the **Format** menu.
2. Select **Plot Style** from the **Named Objects** list.
3. Select the name of the plot style to be renamed from the Items list. The name will now be displayed in the **Old Name** box.
4. Provide a new name in the **Rename To** box.
5. Select the **Rename To** button. The new name will now be displayed in the list box.
6. Click the **OK** button to return to the drawing area.

CHAPTER 23 EXERCISE

1. Plot a drawing template containing a border and title block on a D size sheet of paper. Create a new base drawing that will fit within the limits of the borders. Attach any four drawings in your library. Each drawing is to be shown at a different scale when plotted. Provide a title and the scale used by the original drawing prior to plotting.

CHAPTER 23 QUIZ

1. How many colors can be configured for plotting?
2. Is it possible to have more than one plotting device connected to a single computer? Explain your answer.
3. What options are available to preview the information to be plotted prior to plotting, and how are they accessed?
4. Describe the meaning of the display used in a partial plot preview.
5. What tab contains a display of the current plotting scale?
6. What can be used to control how much of the current drawing will be displayed?
7. List five options for controlling what portion of the drawing will be plotted.
8. Describe the typical origin location for a plotter and a printer.
9. What is the advantage of saving the plot information with **Plot to file**?
10. Which takes precedence, the page setup or the plot style?
11. How does AutoCAD deal with pen assignments?
12. Explain the difference between portrait and landscape.
13. What plot style should be used to plot a drawing with 46 different colors using only black?
14. What command is used to create a plot style?
15. List two types of plot style tables and explain which will most typically meet your needs.
16. Explain the steps to attach a plot style to a layout.
17. What is the purpose of a plot style?
18. You've inserted five details in a drawing template. Each of the details was created for plotting at a different scale. Each was inserted in the template using the proper methods. What scale should be used to plot this drawing?
19. How are plotters configured using AutoCAD 2002?
20. What does the **What to plot** area control?

AutoCAD and the Internet

The Internet is an important way to exchange files and information. AutoCAD allows you to access and store drawings and related files and to interact through the Internet. The AutoCAD Internet features allow drawings to be opened and saved to an Internet location. The features in AutoCAD allow hyperlinks to be added to drawings, so that a variety of related documents can easily be accessed. AutoCAD allows drawings to be displayed in Web format (DWF) files that can be viewed with an Internet browser. DWF files can be viewed by anyone with an Internet browser using a free WHIP!® plug-in or Volo View Express provided by Autodesk. This chapter will introduce methods for

- Accessing and working with drawings on the Internet
- Launching a Web browser
- Using a hyperlink, or Uniform Resource Locator (URL)
- Using the Drawing Web Format (DWF) to display drawings on the Internet
- Publish drawings to the Web

Commands to be introduced include

- **Browser**
- **Inetlocation**
- **Hyperlink**
- **Pasteashyperlink**
- **Publishtoweb**
- **Etransmit**

INTERNET ACCESS

You are probably familiar with the best-known uses for the Internet: email (electronic mail) and the Web (short for the World Wide Web). Email lets users exchange messages and data at very low cost. The Web brings together text, graphics, audio, and movies in an easy-to-use format. To use the features of AutoCAD requires Internet access and an Internet browser to be installed on your workstation. To view and plot AutoCAD DWF files from an Internet browser requires the WHIP! 4.0 plug-in or the Volo Viewer Express from autodesk.com to be downloaded. An Internet browser like Microsoft Internet Explorer or Netscape Navigator must also be installed on your computer. With the required software loaded, AutoCAD can open, insert, and save drawings to and from the Internet. To save files to an Internet location requires access rights to the directory where the files are stored.

AutoCAD is able to launch a Web browser from within AutoCAD using the Browser command. The browser is used for viewing drawings in 2D format on Web pages. The **Hyperlink** command can be used to link a drawing with other documents on your computer or the Internet. The electronic plot option of the **Plot** command can create DWF files for viewing 2D drawings on Web pages. With the **Open**, **Insert**, and **Saveas** commands, AutoCAD can import or export drawings to and from the Internet.

WORKING WITH UNIFORM RESOURCE LOCATORS

A uniform resource locator (URL) is the file naming system of the Internet. This system allows files such as a text file, a Web page, a program file, an audio file, or movie clip to be located on the Internet. A typical URL would resemble *http://www.autodesk.com*. The prefix http:// is not typically required to be entered. Most Web browsers automatically add in the routing prefix, which saves you a few keystrokes. Common URLs you should be familiar with include the following:

URL	**Meaning**
http://www.autodesk.com	Autodesk Primary Web site.
http://www.autodeskpress.com	Autodesk Press Web site

URLs can access several different kinds of resources such as Web sites, email, or news groups. No matter the resource, the URL always takes on the same general format: "scheme://networkaddress."

The scheme accesses the specific resource on the Internet such as

file://	File located on your computer's hard drive or local network
ftp://	File Transfer Protocol (used for downloading files)
http://	HyperText Transfer Protocol (the basis of Web sites)

mailto://	Electronic mail (email)
news://	Usenet news (news groups)
telnet://	Telnet protocol
gopher://	Gopher protocol

The characters :// indicate a network address. Autodesk recommends these formats for specifying URL-style file names:

Web Site	*http://servername/pathname/filename*
FTP Site	*ftp://servername/pathname/filename*
Local File	*file:///drive:/pathname/filename*
or	*file:///drive/pathname\filename*
or	*file://\\localPC\pathname\filename*
or	*file://\\localPC\pathname\filename*
or	*file:////localPC/pathname/filename*
Network File	*file://localhost/drive:/pathname/filename*
or	*\\localhost\drive:\pathname\filename*
or	*file://localhost/drive|/pathname/filename*

The following definitions will help to clarify this confusing terminology.

Term	Meaning
Servername	A name or location of a computer on the Internet such as www.autodesk.com
Pathname	The same as a subdirectory or folder name
Drive	Your local drive
localPC	A file located on your computer
localhost	The name of the network host computer

If you are not sure of the name of the network host computer, use Windows Explorer to check the Network Neighborhood for the network names of computers.

OPENING AND SAVING INTERNET FILES

The **Browser** command lets you start a Web browser while inside AutoCAD. The two most common browsers are Microsoft Internet Explorer and Netscape Navigator. By default, the **Browser** command uses the Web browser program that is registered in your computer's Windows operating system. AutoCAD lists the name of the browser before prompting you for the uniform resource locator (URL). The URL is the Web site address, such as *http://www.autodesk.com*. The **Browser** command is used to launch a Web browser from within AutoCAD and requires that Netscape Navigator or Microsoft Internet Explorer be installed on your computer. The command can also be used in scripts, toolbar or menu macros, and AutoLISP routines to automatically access the Internet. The command can be accessed using the following command sequence:

Command: **BROWSER** ENTER

Enter Web location (URL) < http://www.autodesk.com >:
(Enter URL..)

Pressing ENTER will accept the default URL, an HTML file added to your computer during the installation of AutoCAD. This URL is Autodesk's own Web site. After you type the URL and press ENTER, AutoCAD launches the Web browser and contacts the Web site, such as the Autodesk Web site shown in Figure 24.1.

CHANGING THE DEFAULT WEB SITE

Altering the **Inetlocation** system variable setting can change the default Web page that your browser starts with from within AutoCAD. The variable stores the URL used by the **Browser** command and the **Browse the Web** dialog box. Make the change as follows:

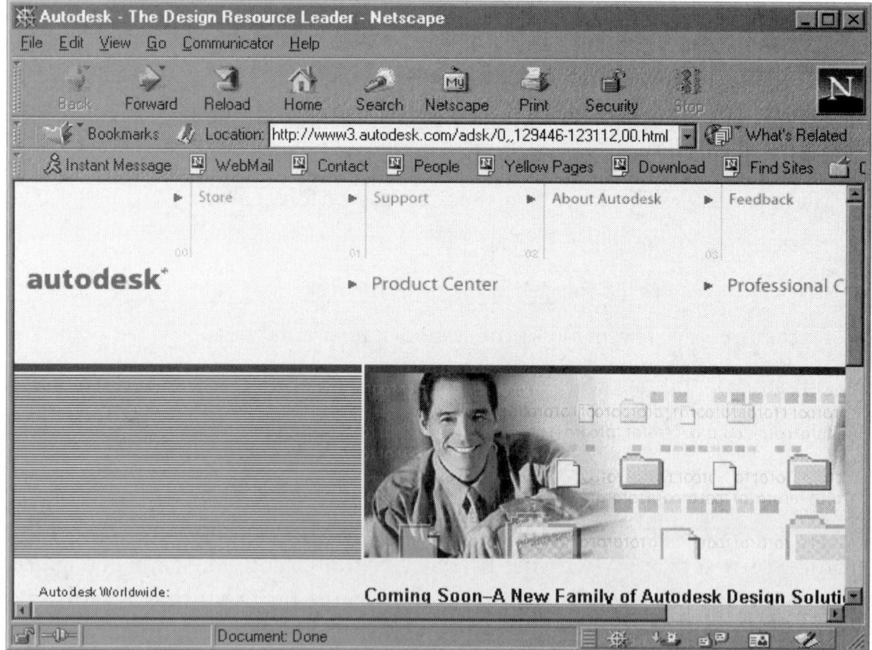

Figure 24.1 *Display the Autodesk Web site by typing* '**BROWSER** *(*ENTER*) at the command prompt and then entering the name of the URL.*

Command: **INETLOCATION** ENTER

Enter new value for INETLOCATION <" http://www.autodesk.com
">: *(Type URL..)*

DRAWINGS ON THE INTERNET

When a drawing is stored on the Internet, you access it from within AutoCAD using the standard **Open**, **Insert**, and **Save** commands. Instead of specifying the file's location with the usual drive-subdirectory-file name format, such as c:\acad\file-name.dwg, use the URL format. The URL is the universal file naming system used by the Internet to access any file located on any computer hooked up to the Internet.

OPENING DRAWINGS FROM THE INTERNET

Use the **Open** command from the **File** menu to open a drawing from the Internet. The options available are to search the Web, open from the saved favorites, browse my files at Point A, or browse the Buzzsaw or Redspark sites shown in figure 24.2.

Search the Web—Display the **Browse the Web** dialog box, a simplified Web browser. This is the button next to the look in box that is the image of the world.

Look in Favorites—Open the Favorites folder, which is the equivalent to bookmarks that store Web addresses.

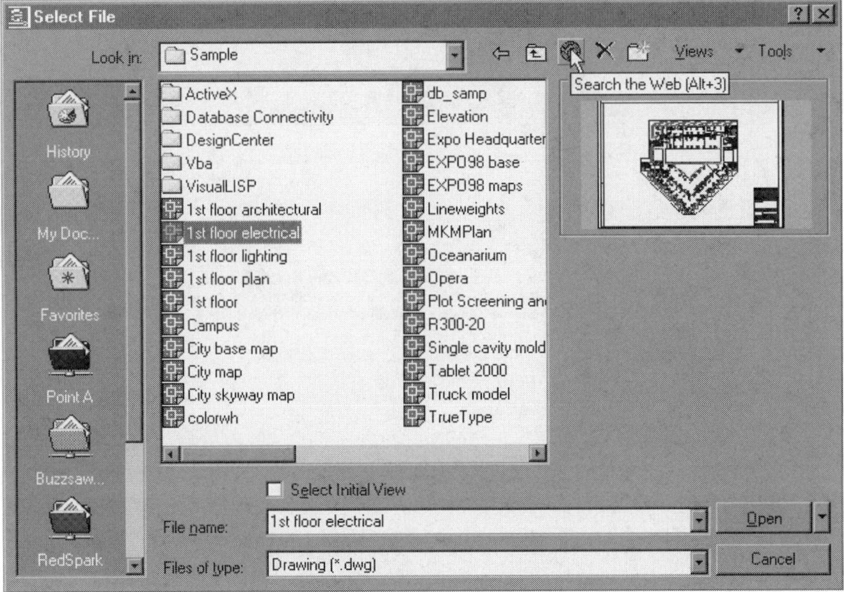

Figure 24.2 *The* **Select File** *dialog box contains the options to aid in working with the Internet.*

Point A—Browse for files at Point A Autodesk Web site.

Buzzsaw—Browse for files at the Buzzsaw Web site.

Redspark—Browse for files at the Redspark Web site.

Add to Favorites—Save the current URL (hyperlink) to the Favorites folder. Access this option from the **Tools** menu in the upper right corner of the dialog box.

When you click the **Search the Web** button, AutoCAD opens the **Browse the Web** dialog box. This dialog box is a simplified version of a Web browser. The purpose of this dialog box is to allow you to browse files at a Web site. By default, the **Browse the Web** dialog box displays the contents of the URL stored in the **Inetlocation** system variable. (See Figure 24.3a.) You can easily change this to another folder or Web site, as noted earlier. (See Figure 24.3b.)

The **Browse the Web** dialog box has six buttons across the top to aid in working with files. These include the following:

Back—Go back to the previous URL.

Forward—Go forward to the next URL.

Stop—Halt displaying the Web page. This option is useful if the connection is slow or the page is very large.

Refresh—Redisplay the current Web page.

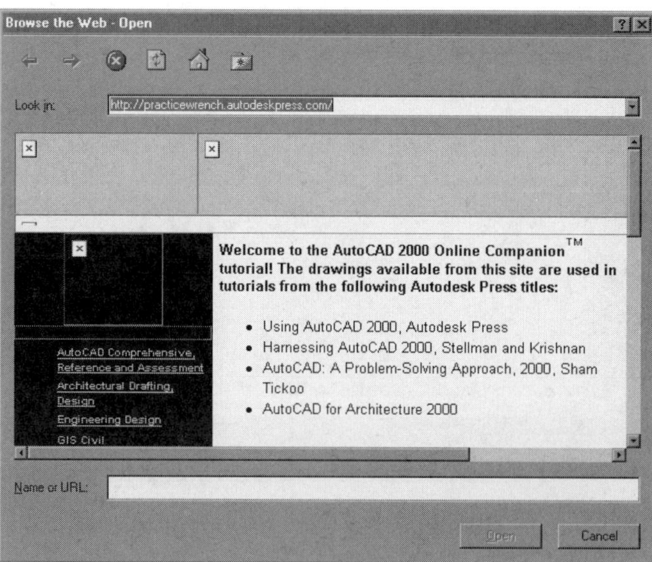

Figure 24.3a *By default, the* **Browse the Web - Open** *dialog box displays the contents of the URL stored in the* **Inetlocation** *system variable.*

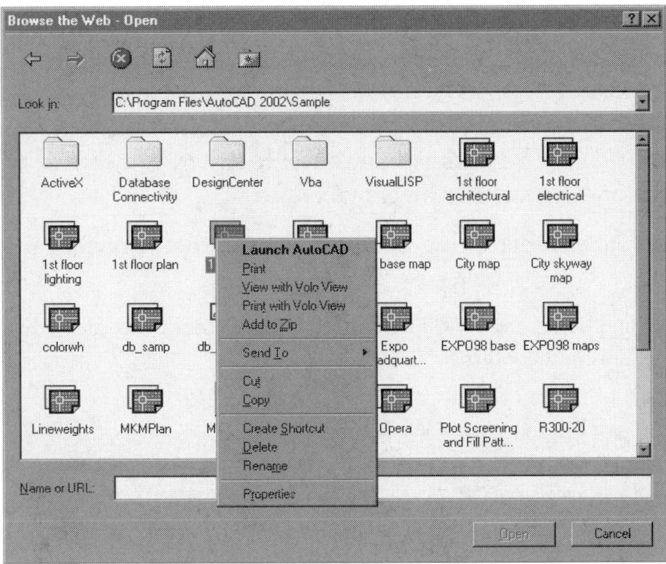

Figure 24.3b *The* **Browse the Web - Open** *dialog box and shortcut menu.*

Home—Return to the location specified by the **Inetlocation** system variable.

Favorites—List stored URLs (hyperlinks) or bookmarks. If you have previously used Internet Explorer, you will find all your favorites listed here. Favorites are stored in the \Windows\Favorites folder on your computer.

The **Look in** edit box allows you to type the URL. Alternatively, click the down arrow to select a previous destination. If you have stored Web site addresses in the Favorites folder, select a URL from that list. You can either double-click a file name in the window or type a URL in the edit box. The following table gives the templates for typing the URL to open a drawing file:

Drawing Location	**Template URL**
Web or HTTP Site	http://servername/pathname/filename.dwg
	http://practicewrench.autodeskpress.com/wrench.dwg
FTP Site	ftp://servername/pathname/filename.dwg
	ftp://ftp.autodesk.com
Local File	drive:\pathname\filename.dwg
	c:\acad 2002\sample\tablet2002.dwg
Network File	\\localhost\drive:\pathname\filename.dwg
	\\upstairs\e:\install\sample.dwg

When you open a drawing over the Internet, it will probably take much longer than opening a file found on your computer. During the file transfer, AutoCAD displays a dialog box to report the progress. If your computer uses a 28.8 Kbps modem, you should allow about five to ten minutes per megabyte of drawing file size. If your computer has access to a faster T1 connection to the Internet, you should expect a transfer speed of about one minute per megabyte.

The **Open** command will place the drawing in the current drawing file.

Note: The **Locate** and **Find File** buttons in the **Select File** dialog box do not work for locating files on the Internet.

Tutorial: Opening a Drawing at a Web Site

The Autodesk Press Web site has an area that allows you to practice using the Internet with AutoCAD. In this tutorial, you open a drawing file located at that Web site.

1. Start AutoCAD.
2. Ensure that you have a live connection to the Internet. If you normally access the Internet via a telephone (modem) connection, dial your Internet service provider now.
3. Select **Open** from the **File** menu or click the **Open** button on the **Standard** toolbar. The **Select File** dialog box is displayed.
4. Click the **Search the Web** button. The **Browse the Web** dialog box is displayed, as shown in Figure 24.4.
5. In the Look in edit box, type **PRACTICEWRENCH.AUTODESK-PRESS.COM** (ENTER). After a few seconds, the **Browse the Web** dialog box displays the Web site.
6. In the **Name or URL** edit box, type **WRENCH.DWG** (ENTER).

AutoCAD begins transferring the file. Depending on the speed of your Internet connection, this will take between a couple of seconds and half a minute. Notice the drawing of the wrench in AutoCAD that is shown in Figure 24.5.

INSERTING A BLOCK FROM THE INTERNET

When a block (symbol) is stored on the Internet, you can access it from within AutoCAD using the **Insert** command. When the **Insert** dialog box appears, click the **Browse** button to display the **Select Drawing File** dialog box. This is identical to the dialog box discussed above. After you select the file, AutoCAD downloads the file and continues with the **Insert** command's familiar prompts.

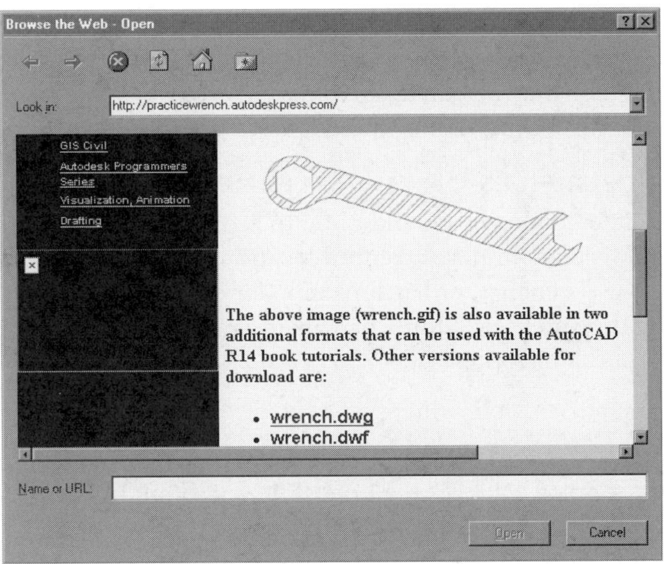

Figure 24.4 *The* **File Download** *display box.*

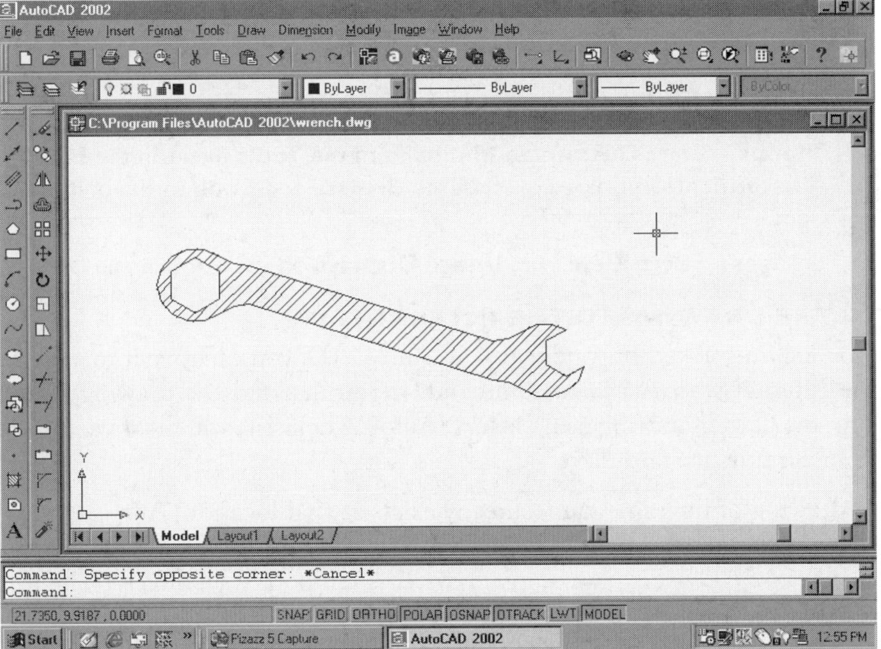

Figure 24.5 *The Practice Wrench is placed in a drawing.*

The process is identical for accessing external reference (xref) and raster image files. Other files that AutoCAD can access over the Internet include 3D Studio, SAT (ACIS solid modeling), DXB (drawing exchange binary), WMF (Windows metafile), and EPS (encapsulated PostScript). All of these options are found on the Insert menu.

ACCESSING OTHER FILES ON THE INTERNET

Most other file-related dialog boxes allow you to access files from the Internet. This allows your firm or agency to have a central location that stores drawing standards. When you need to use a linetype or hatch pattern, for example, you can access the LIN or PAT file over the Internet. More than likely, you would have the location of those files stored in the Favorites list.

Some examples include the following:

Linetypes—Select **Linetype** from the **Format** menu. In the **Linetype Manager** dialog box, click the **Load**, **File**, and **Look in Favorites** buttons.

Hatch Patterns—Use the Web browser to copy .PAT files from a remote location to your computer.

Layer Name—Select **Layer Properties Manager** from **Layers**. In the **Layer Manager** dialog box, click the **Import** and **Look in Favorites** buttons.

LISP and ARX Applications—Select **Load Applications** from the **Tools** menu.

Scripts—Select **Run Scripts** from the **Tools** menu.

Menus—Select **Customize Menus** from the **Tools** menu. In the **Menu Customization** dialog box, click the **Browse** and **Look in Favorites** buttons.

Images—Select **View** from **Image Displays** on the **Tools** menu.

SAVING THE DRAWING TO THE INTERNET

When you are finished editing a drawing in AutoCAD, you can save it to a file server on the Internet with the **Save** command. If you inserted the drawing from the Internet in the default drawing using **Insert**, AutoCAD insists you first save the drawing to your computer's hard drive.

When a drawing of the same name already exists at that location, AutoCAD warns you just as it does when you use the **Saveas** command. Recall from the **Open** command that AutoCAD uses your computer system's temporary subdirectory, hence the reference to it in the dialog box.

USING HYPERLINKS WITH AUTOCAD

AutoCAD allows you to employ URLs in two ways: directly within an AutoCAD drawing, and indirectly in DWF files displayed by a Web browser. (URLs are also known as hyperlinks, the term we will use from now on.) Hyperlinks are created, edited, and removed with the **Hyperlink** command, using a dialog box, and the **-Hyperlink** command, using prompts at the command line. The **Hyperlink** command (select **Hyperlink** from the **Insert** menu) prompts you to "Select objects" and then displays the **Insert Hyperlink** dialog box. (See Figure 24.6.) As a shortcut, you can press CTRL+K or click the **Insert Hyperlink** button on the **Standard** toolbar.

As an alternative, you can use the **-Hyperlink** command. This command displays its prompts at the command line, and is useful for scripts and AutoLISP routines. The **-Hyperlink** command has the following syntax:

Command: **-HYPERLINK** ENTER

Enter an option [Remove/Insert] <Insert>: ENTER

Enter hyperlink insert option [Area/Object] <Object>: ENTER

Select objects: *(Pick an object.)*

1 found Select objects: ENTER

Enter hyperlink <current drawing>: *(Enter the name of the document or Web site.)*

Enter named location <none>: *(Enter the name of a bookmark or AutoCAD view.)*

Enter description <none>: *(Enter a description of the hyperlink.)*

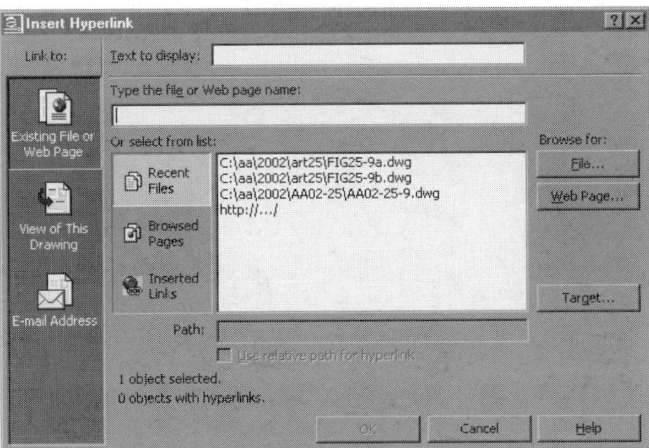

Figure 24.6 *The* **Insert Hyperlink** *dialog box.*

Notice that the command also allows you to remove a hyperlink. It does not, however, allow you to edit a hyperlink. To do this, use the Insert option and re-specify the hyperlink data.

In addition, this command allows you to create hyperlink areas, which are rectangular areas that can be thought of as a 2D hyperlink. The dialog box–based **Hyperlink** command does not create hyperlink areas. When you select the Area option, the rectangle is placed automatically on layer URLLAYER and colored red.

In the following sections, you'll learn how to apply and use hyperlinks in an AutoCAD drawing and in a Web browser through the dialog box **Hyperlink** command.

HYPERLINKS INSIDE AUTOCAD

AutoCAD allows you to add a hyperlink to any object in the drawing. An object is permitted just a single hyperlink. On the other hand, a single hyperlink can be applied to a selection set of objects.

You can tell an object has a hyperlink by passing the cursor over it. The cursor displays the "linked Earth" icon, as well as a tooltip describing the link, similar to Figure 24.7.

If, for some reason, you do not want to see the hyperlink cursor, you can turn it off. Select **Options** from the **Tools** menu and then click the **User Preferences** tab shown in Figure 24.8. The **Display hyperlink cursor and shortcut menu** check box toggles the display of the hyperlink cursor, and you can toggle the **Display hyperlink tooltip** option.

Tutorial: Attaching Hyperlinks

Let's see how that works by trying an example. In this example, we have the drawing of a floor plan. To this drawing, we will add hyperlinks to another AutoCAD drawing, a Word document, and a Web site. Hyperlinks must be attached to objects. For

Figure 24.7 *An example of a hyperlink.*

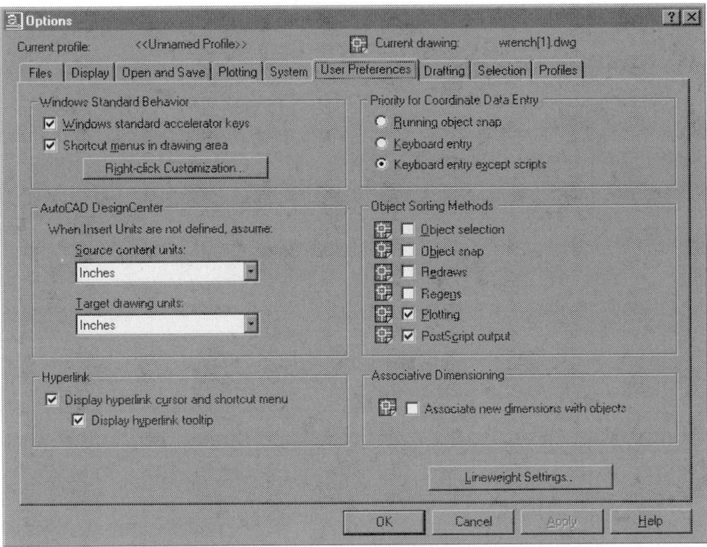

Figure 24.8 *The* **Options** *dialog box controls the display of hyperlinks.*

this reason, we will place some text in the drawing, and then attach the hyperlinks to the text.

1. Start AutoCAD.
2. Open the 1st floor plan.dwg file found in the AutoCAD Sample folder. (If necessary, click the Model tab to display the drawing in model space.)
3. Start the **Text** command and place the text in the drawing as shown in the following command sequence:

   Command: **TEXT** ENTER

   Current text style: "Standard" Text height: 0.20000

   Specify start point of text or [Justify/Style]: *(Pick a point in the drawing.)*

   Specify height <0.2000>: **2** ENTER

   Specify rotation angle of text <0>: ENTER

   Enter text: **Site Plan** ENTER

   Enter text: **Lighting Specs** ENTER

   Enter text: **Electrical** ENTER

   Enter text: ENTER

   The result up to this point can be seen in Figure 24.9

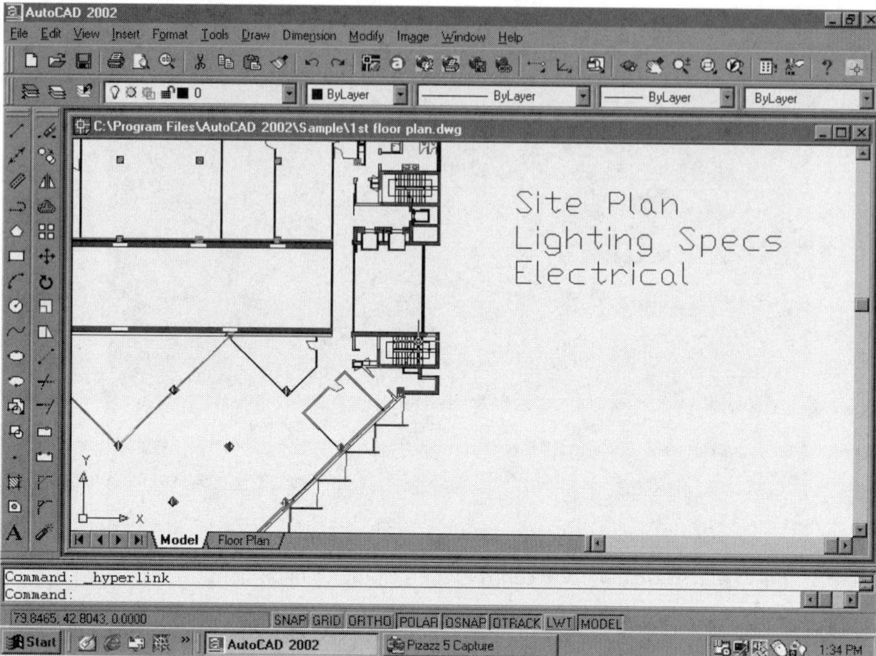

Figure 24.9 *Text is added to the drawing to be used for hyperlinks.*

4. Click the Site Plan text to select it.
5. Click the **Insert Hyperlink** button on the Standard toolbar. Notice the **Insert Hyperlink** dialog box.
6. Select the **File** button in the **Browse for** area.
7. Go to AutoCAD's Sample folder and select the City base map.dwg file. Click **Open**. AutoCAD does not open the drawing; rather, it copies the file's name to the **Insert Hyperlink** dialog box. Notice, in Figure 24.10, the name of the drawing you selected in the **Type the File or Web page name** edit box.
8. Click **OK** to close the dialog box. Move the cursor over the Site Plan text. Notice in Figure 24.11 that the display of the "linked Earth" icon appears; a moment later, the tooltip displays the URL.
9. To try out the link click "Site Plan" to select it. Right-click and select **Open "C:Program Files\AutoCAD 2002\Sample\City base map.dwg"** from **Hyperlink** on the shortcut menu as shown in Figure 24.12. Notice that AutoCAD opens "City base map.dwg". To see both drawings, select **Tile Vertically** from the **Window** menu. The screen will now appear as shown in Figure 24.13.

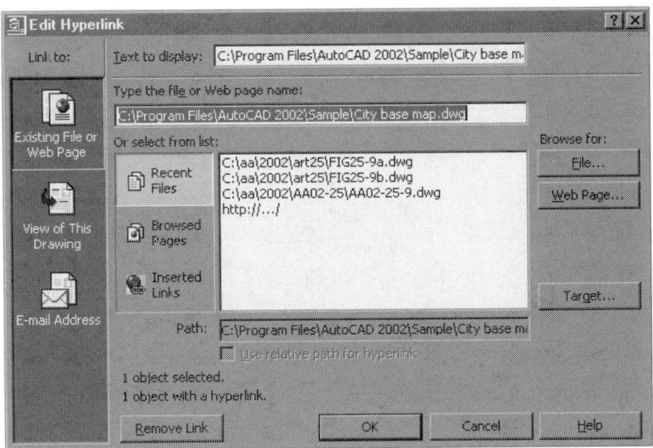

Figure 24.10 *The* **Edit Hyperlink** *dialog box.*

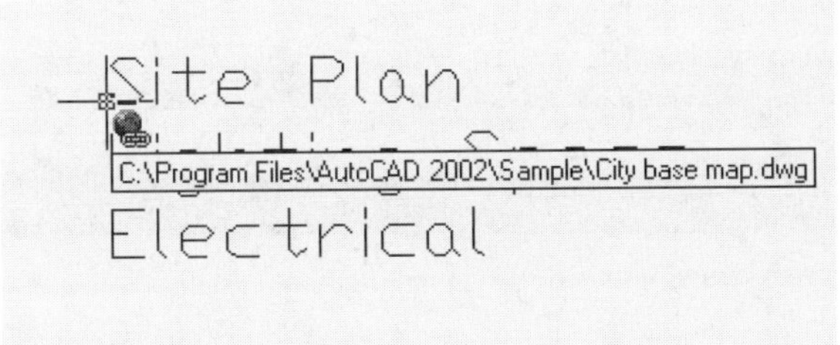

Figure 24.11 *The hyperlink cursor and tooltip.*

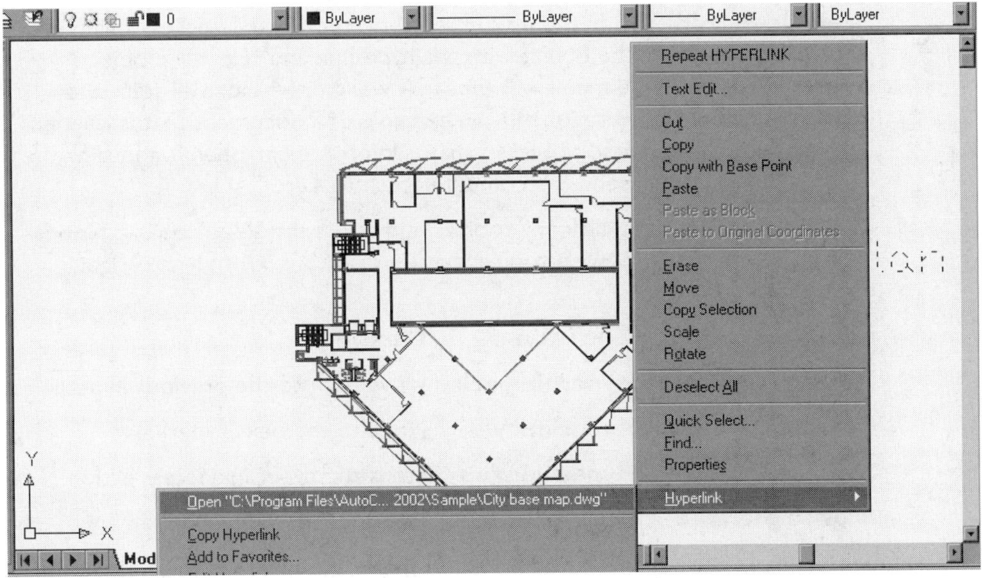

Figure 24.12 *The hyperlink shortcut menu.*

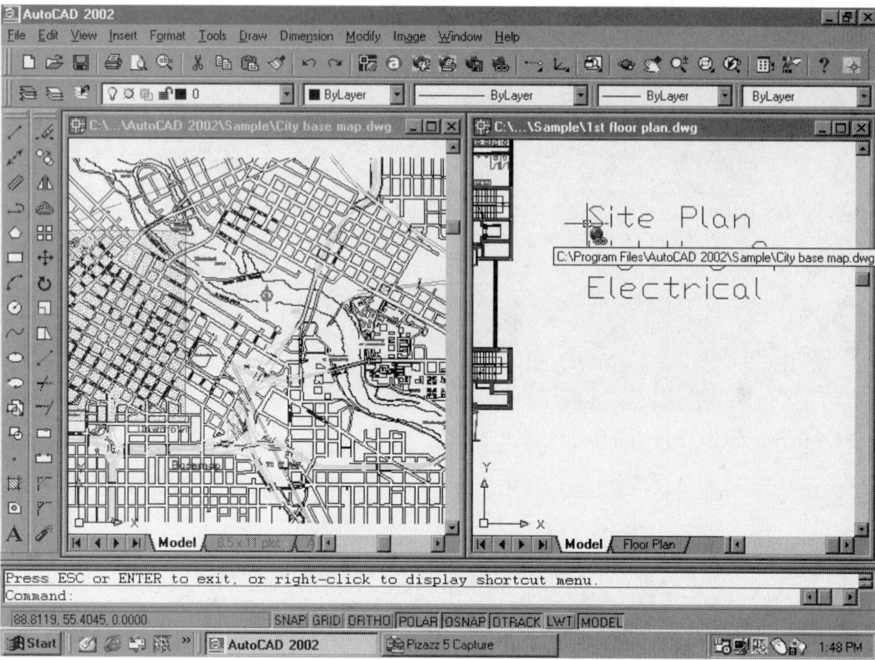

Figure 24.13 *Viewing two drawings.*

10. Hyperlinks can also be applied to a text document from a word processor. The process to create the hyperlink is the same process you have just seen. Select **Open from the hyperlink** in the shortcut menu. To access this menu place the cursor over the object with the hyperlink and select the object. Now right-click and the menu will appear. A word processor will start and open the document. Figure 24.14 is an example of a document that is opened by a hyperlink. If the Word processor has a **Web** toolbar, open it and click the **Back** button. This will send the computer back to AutoCAD.
11. Back in AutoCAD, right-click any toolbar and select the **Web** toolbar. Notice in Figure 24.15 that the **Web** toolbar has four buttons. From left to right, these are as follows:

Button	**Command**	**Meaning**
Go Back	**Hyperlinkback**	Move back to the previous hyperlink.
Go Forward	**Hyperlinkfwd**	Move to the next hyperlink.
Stop Navigation	**Hyperlinkstop**	Stops the hyperlink access action.
Browse the Web	**Browser**	Launches the Web browser.

When you work with hyperlinks in AutoCAD, you might come across these three limitations:

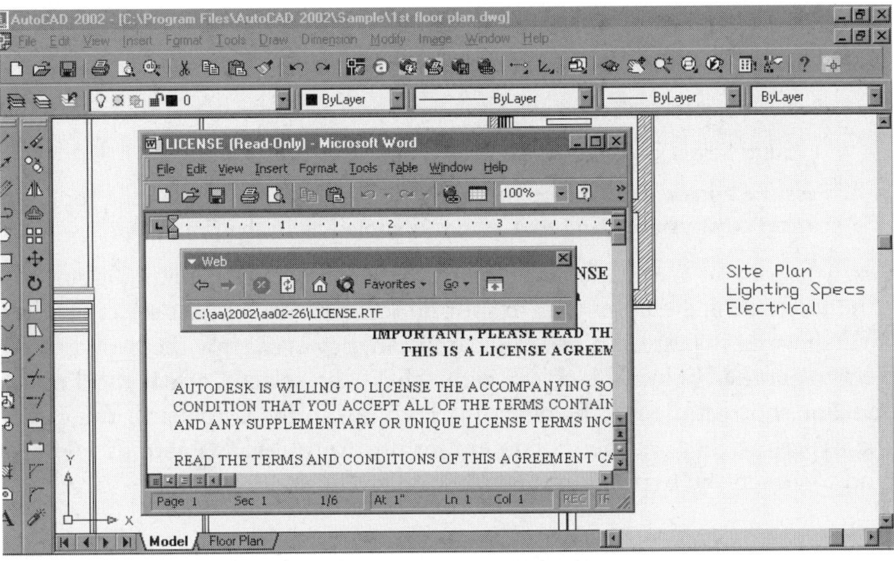

Figure 24.14 *Viewing a Word document.*

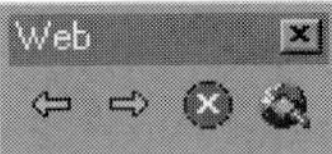

Figure 24.15 *Display AutoCAD's* **Web** *toolbar by moving the cursor to any open toolbar, right-clicking, and selecting* **Web**.

- AutoCAD does not check that the URL you type is valid.
- If you attach a hyperlink to a block, be aware that the hyperlink data is lost when you scale the block unevenly, stretch the block, or explode it.
- Wide polylines and rectangular hyperlink areas are only "sensitive" on their outline.

Pasting as Hyperlink

AutoCAD has a shortcut method for creating hyperlinks in the drawing. The **Pasteashyperlink** command pastes any text in the Windows Clipboard as a hyperlink to any object in the drawing. Here is how it works:

1. In a word processor, select some text and copy it to the Clipboard using either CTRL+C or the **Edit Copy** command. The text can be a URL such as *http://www.autodeskpress.com* or any other text.
2. Switch to AutoCAD and select **Paste as Hyperlink** from the **Edit** menu. Note that this command does not work (it is grayed out) if anything else is in the Clipboard, such as a picture.

3. Select one or more objects, as prompted:

   Command: _**PASTEASHYPERLINK** ENTER

   Select objects: *(Select an object.)*

   I found Select objects: ENTER

4. Pass the cursor over the object and note the hyperlink cursor and tooltip. The tooltip displays the same text that you copied from the document.

If the text you copy to the Clipboard is very long, AutoCAD displays only portions of it in the tooltip, using ellipses (…) to shorten the text. You cannot select text in the AutoCAD drawing to paste as a hyperlink. You can, however, copy the hyperlink from one object to another. Select the object, right-click, and select **Copy Hyperlink** from the **Hyperlink** shortcut menu. The hyperlink is copied to the Clipboard. You can now paste the hyperlink into another document, or use AutoCAD's **Paste as Hyperlink** command to attach the hyperlink to another object in the drawing.

Note: The **Match Properties** command does not work with hyperlinks.

Highlighting Objects with URLs

Although you can see the rectangle of area URLs, the hyperlinks themselves are invisible. To see them, you can use the **Qselect** command, which highlights all objects that match specifications. Select **Quick Select** from the **Tools** menu. AutoCAD displays a **Quick Select** dialog box similar to Figure 24.16

Enter these specifications: Select Entire Drawing in the **Apply to** edit box. Select Multiple for **Object type** and Hyperlink for **Properties**. Select * Wildcard as the **Operator** and enter '* in the **Value** edit box.

Click **OK**, and AutoCAD highlights all objects that have a hyperlink. Depending on your computer's display system, the highlighting shows up as dashed lines or as another color.

Editing Hyperlinks

Now that you know where the objects with hyperlinks are located, you can use the **Hyperlink** command to edit the hyperlinks and related data. Select the hyperlinked object and start the **Hyperlink** command. When the **Edit Hyperlink** dialog box appears (it looks identical to the **Insert Hyperlink** dialog box), make the changes and click **OK**.

Removing Hyperlinks from Objects

To remove a hyperlink from an object, use the **Hyperlink** command with the object selected. When the **Edit Hyperlink** dialog box appears, click the **Remove Hyperlink** button.

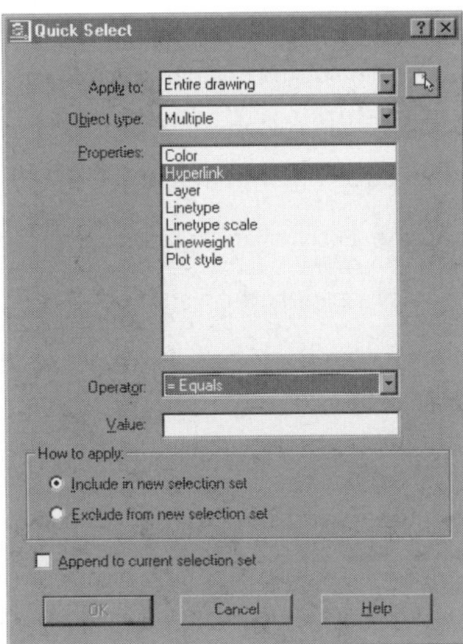

Figure 24.16 *Selecting* **Quick Select** *from the* **Tools** *menu displays the* **Quick Select** *dialog box.*

To remove a rectangular area hyperlink, you can simply use the **Erase** command. Start the command and select the rectangle, and AutoCAD erases the rectangle.

HYPERLINKS OUTSIDE AUTOCAD

The hyperlinks you place in the drawing are also available for use outside AutoCAD. The Web browser makes use of the hyperlinks when the drawing is exported in DWF format. To help make the process clearer, here are the steps that you need:

1. Open a drawing in AutoCAD.
2. Attach hyperlinks to objects in the drawing with the **Hyperlink** command. To attach hyperlinks to areas, use the **-Hyperlink** command's Area option.
3. Export the drawing in DWF format using the **Plot** command's DWF ePlot PC2 plotter configuration.
4. Copy the DWF file to your Web site.
5. Start your Web browser with the **Browser** command.
6. View the DWF file and click on a hyperlink spot.

THE DRAWING WEB FORMAT (DWF)

To display AutoCAD drawings on the Internet, Autodesk developed a new file format called drawing Web format (DWF). The DWF file has several benefits over DWG files and some drawbacks. The DWF file is compressed to make it as much as eight times smaller than the original DWG drawing file, so that it takes less time to transmit over the Internet, particularly with the relatively slow telephone modem connections. The DWF format is more secure, because the original drawing is not being displayed—another user cannot tamper with the original DWG file.

However, the DWF format has some drawbacks:

- You must go through the extra step of translating from DWG to DWF.
- DWF files cannot display rendered or shaded drawings.
- DWF is a flat 2D-file format; therefore, it does not preserve 3D data, although you can export a 3D view.
- AutoCAD itself cannot display DWF files.
- DWF files cannot be converted back to DWG format without using file translation software from a third-party vendor.
- Earlier versions of DWF did not handle paper space objects (version 2.x and earlier) or linewidths and non-rectangular viewports (version 3.x and earlier).

To view a DWF file on the Internet, your Web browser needs a *plug-in*, which is a software extension that lets a Web browser handle a variety of file formats. Autodesk makes the DWF plug-in freely available from its Web site at *http://www.autodesk.com/whip*. It's a good idea to regularly check for updates to the DWF plug-in, which is updated about twice a year.

OTHER DWF VIEWING OPTIONS

In addition to WHIP! with a Web browser, Autodesk provides two other options for viewing DWF files.

- CADViewer Light is designed to be a DWF viewer that works on all operating systems and computer hardware because it is written in Java. As long as your Windows, Mac, or Unix computer has access to Java (included with most Web browsers), you can view DWF files.
- Volo View Express is a stand-alone viewer that views and prints DWG, DWF, and DXF files.

Both products can be downloaded free from the Autodesk Web site.

VIEWING DWG FILES

To view AutoCAD DWG and DXF files on the Internet, your Web browser needs a DWG-DXF plug-in. At the time of publication, these plug-ins were available from the following vendors:

Autodesk: *http://www.autodesk.com/whip*

SoftSource: *http://www.softsource.com/*

California Software Labs: *http://www.cswl.com*

Cimmetry systems: *http://www.cimmetry.com*

CREATING A DWF FILE

Use the following steps to create a DWF file from AutoCAD.

1. Start the **Plot** command by typing **PLOT** (ENTER) at the Command prompt or select **Plot** from the **File** menu. The **Plot** dialog box is displayed.
2. In the **Name** list box of the **Plotter Configuration** area, select DWF ePlot pc3.
3. Click the **Properties** button. This will display the **Plotter Configuration Editor** dialog box similar to Figure 24.17.
4. Click Custom Properties in the tree view. Notice the **Custom Properties** button that appears in the lower half of the dialog box.
5. Click the **Custom Properties** button. The **DWF Properties** dialog box similar to Figure 24.18 is displayed.
6. Select the desired options, as described below, and click **OK** to exit the dialog boxes back to the **Plot** dialog box.
7. Accept the DWF file name listed in the **File name** text box, or type a new name. If necessary, change the location that the file will be stored in. Note the two buttons: one has an ellipsis [...] that displays the **Browse for Folder** dia-

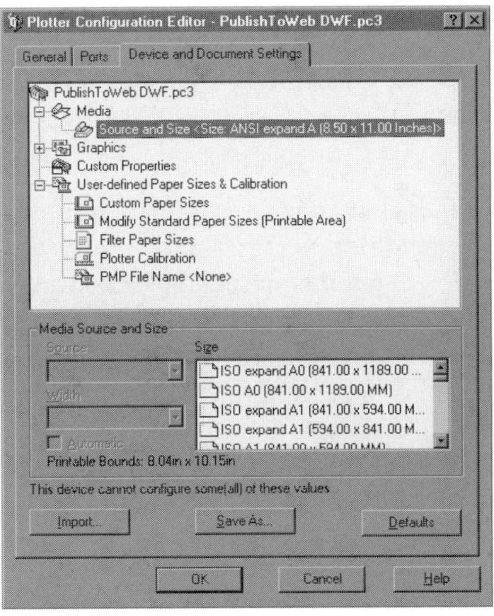

Figure 24.17 *Selecting the* **Properties** *button in the* **Plot** *dialog box displays the* **Plotter Configuration Editor** *dialog box.*

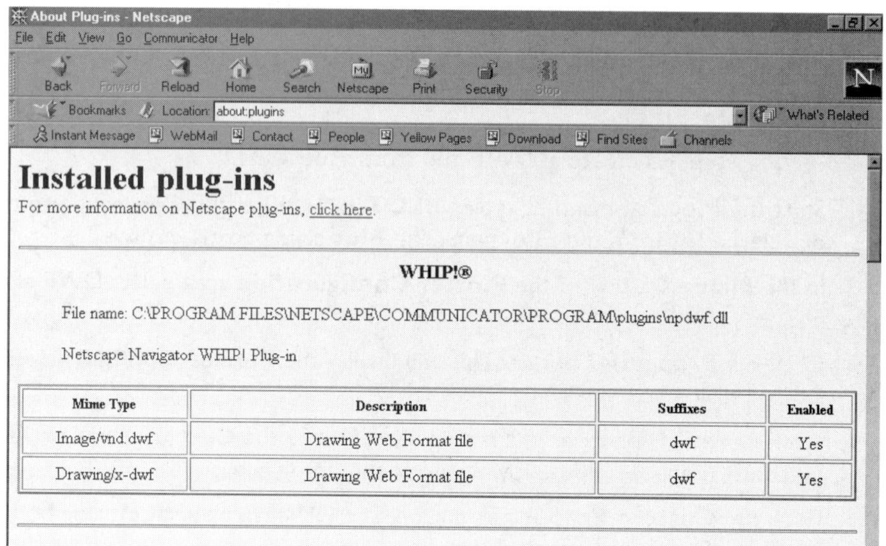

Figure 24.18 *Selecting the* **Custom Properties** *button displays the* **DWF Properties** *dialog box.*

log box. The second button brings up AutoCAD's internal Web browser, a simplified version of the Microsoft browser.

8. Click **OK** to save the drawing in DWF format.

DWF Properties

You can use the **DWF Properties** dialog box to set **Resolution** and **Format**, and you can determine other display options as well.

Resolution

Unlike AutoCAD DWG files that are based on real numbers, DWF files are saved using integer numbers. The **Medium** precision setting saves the drawing using 16-bit integers, which is adequate for all but the most complex drawings. **High** resolution saves the DWF file using 20-bit integers, while **Extreme** resolution saves using 32-bit integers. More significant is the difference in file size. When a DWF file from the "1st floor plan.dwg" file was created (342 KB), the medium resolution DWF file was 27 KB, while the extreme resolution file was 1,100 KB—some 40 times larger. The medium resolution DWF file would be transmitted over the Internet 40 times faster, a significant saving in time.

Format

Compression further reduces the size of the DWF file. You should always use compression, unless you know that another application cannot decompress the DWF file. Compressed binary format is seven times smaller than ASCII format (7,700 KB).

Therefore, a compressed binary DWF file transmits over the Internet seven times faster than an ASCII DWF file.

Other Options

In most cases, you would use all these options:

Background color shown in viewer—While white is probably the best background color, you can choose any of AutoCAD's 255 colors.

Include Layer information—This option includes layers, which allows you to toggle layers off and on when the drawing is viewed by the Web browser.

Include Scale and measurement information—This allows you to use the Location option in the Web browser's WHIP! plug-in to show scaled coordinate data.

Show Paper boundaries—Includes a rectangular boundary at the drawing's extents.

Convert .DWG hyperlink extensions to .DWF—Includes hyperlinks in the DWF file.

VIEWING DWF FILES

To view a DWF file, you need to use a Web browser with a special plug-in that allows the browser to correctly interpret the file. (Remember: you cannot view a DWF file with AutoCAD.) Autodesk named their DWF plug-in WHIP!, short for Windows HIgh Performance.

Autodesk updates the DWF plug-in approximately twice a year. Each update includes some new features. In summary, all versions of the DWF plug-in perform the following functions:

- DWF files created by AutoCAD are displayed within a browser.
- Right-clicking the DWF image displays a cursor menu with commands.
- Realtime pan and zoom let you change the view of the DWF file as quickly as a drawing file in AutoCAD.
- Embedded hyperlinks let you display other documents and files.
- File compression means that a DWF file appears in your Web browser much faster than the equivalent DWG drawing file would.
- Print the DWF file alone or along with the entire Web page.
- Works with Netscape Navigator or Microsoft Internet Explorer. A separate plug-in is required, depending on which of the two browsers you use.
- Allows you to drag a DWG file from a Web site into AutoCAD as a new drawing or as a block.

- Displays a named view stored in the DWF file.
- Allows you to specify a view using X,Y coordinates.
- Toggles layers off and on.

If you don't know whether the DWF plug-in is installed in your Web browser, select **About Plug-ins** from **Help** in the browser's menu bar. You might need to scroll through the list of plug-ins to find a display similar to Figure 24.19. If the plug-in is not installed or is an older version, then you need to download it from Autodesk's Web site at *http://www.autodesk.com*.

The WHIP! Viewer is in the product list of the Product area on this site. A better alternative is to use the Volo View Express from the same location.

For Netscape users, the file is quite large at 3.5 MB and it takes about half an hour to download it, using a typical 28.8 Kbaud modem. The file you download from the Autodesk Web site is either a self-installing file (with a .JAR extension) or a self-extracting installation file with a name such as Whip4.Exe. After the download is complete, follow the instructions on the screen. For Microsoft Internet Explorer users, the DWF plug-in is an ActiveX control.

DWF PLUG-IN COMMANDS

To display the DWF plug-in commands, move the cursor over the DWF image and right-click. This displays a shortcut menu with commands, such as **Pan**, **Zoom**, and **Named Views** similar to Figure 24.20. Options include the following:

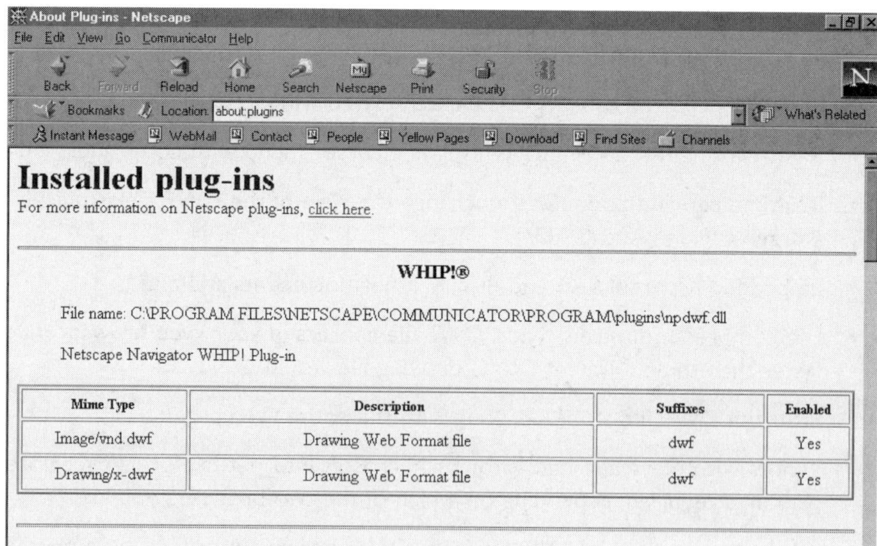

Figure 24.19 *Checking the status of browser plug-ins.*

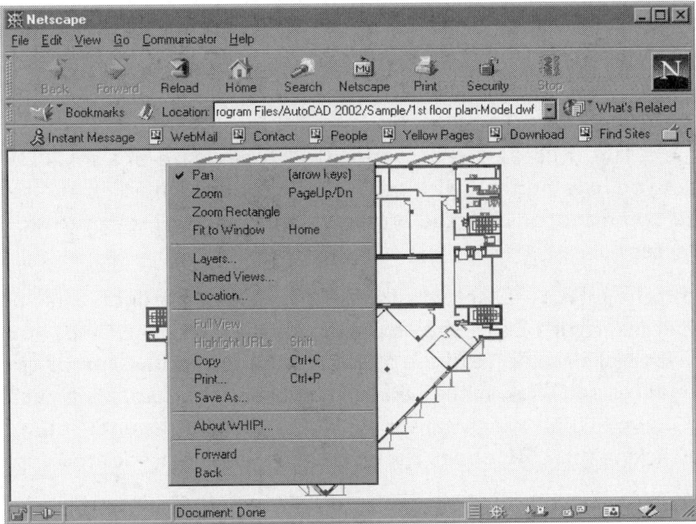

Figure 24.20 *Moving the cursor over the DWF image and right-clicking displays the WHIP!'s shortcut menu.*

Pan—The default command. Click the select button and move the mouse. The cursor changes to an open hand, signaling that you can pan the view around the drawing. This is exactly the same as realtime panning in AutoCAD. Panning works only when you are zoomed in; it does not work in **Full View** mode.

Zoom—Like the **Zoom** command in AutoCAD. The cursor changes to a magnifying glass. Hold down the select button and move the cursor up to zoom in and down to zoom out.

Zoom Rectangle—The same as **Zoom** Window in AutoCAD. The cursor changes to a plus sign. Click the select button at one corner, then drag the rectangle to specify the size of the new view.

Fit to Window—The same as AutoCAD's **Zoom Extents** command. You see the entire drawing.

Layers—Displays a non-modal dialog box similar to Figure 24.23 that lists all layers in the drawing. (Unlike AutoCAD's modal dialog boxes, a non-modal dialog box remains on the screen—you do not need to dismiss a non-modal dialog box to continue working.) Click a layer name to toggle its visibility between on (yellow lightbulb) and off (blue lightbulb).

Named Views—Works only when the original DWG drawing file contained named views created with the **View** command. Selecting **Named Views** displays a non-modal dialog box that allows you to select a named view. Click a named view to see it; click the small x in the upper-right corner to dismiss the dialog box.

Location—Displays a non-modal dialog box that shows the X,Y, Z coordinates of the cursor location. Because DWF files are only 2D, the Z coordinate is always 0.

Full View—Causes the Web browser to display the DWF image as large as possible. This is useful for increasing the physical size of a small DWF image. When you have finished viewing the large image, right-click and select the **Back** command, or click the browser's **Back** button to return to the previous screen.

Highlight URLs—Displays a flashing gray effect on all objects and areas with a hyperlink. Figure 24.24 shows the text highlighted. This helps you see where the hyperlinks are. To read the associated URL, pass the cursor over a highlighted area and look at the URL on the browser's status line. A shortcut is to hold down SHIFT, which causes the URLs to highlight until you release the key. Clicking the URL opens the document with its associated application.

Copy—Copies the DWF image to the Windows Clipboard in EMF (Enhanced Metafile or "Picture") format.

Print—Prints only the DWF image. To print the entire Web page, including the DWF image, use the browser's **Print** button.

Save As—saves the DWF file in three formats to your computer's hard drive:

- DWF
- BMP (Windows bitmap)
- DWG (AutoCAD drawing file)—Works only when a copy of the DWG file is available at the same subdirectory as the DWF file.

You cannot use the **Save As** command until the entire DWF file has been transmitted to your computer.

About WHIP—Displays information about the DWF file similar to Figure 24.21, including DWF file revision number, description, author, creator, source file name, creation time, modification time, source creation time, source modification time, current view left, current view right, current view bottom, and current view top.

Forward—Almost the same as clicking the browser's **Forward** button. When the DWF image is in a frame, only that frame goes forward.

Back—Almost the same as clicking the browser's **Back** button. It works differently when the DWF image is displayed in a frame.

Dragging from Browser to Drawing

The DWF plug-in allows you to perform several drag and drop functions. Hold down the (key to drag a DWF file from the browser into AutoCAD. Remember that AutoCAD cannot translate a DWF file to DWG drawing format. For this reason, this

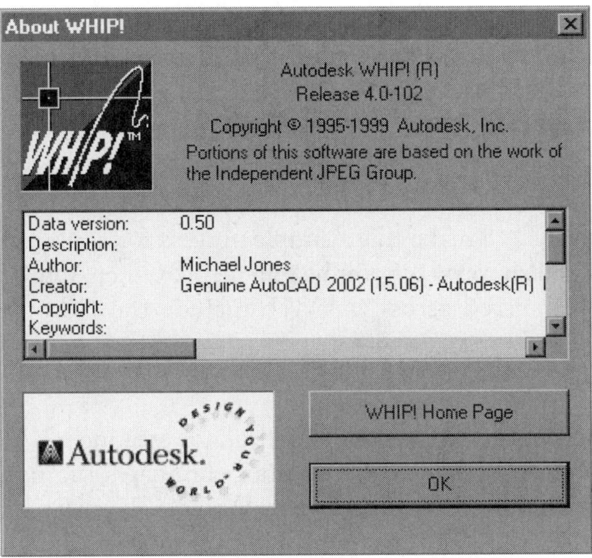

Figure 24.21 *Display of the WHIP! version number.*

form of drag and drop works only when the originating DWG file exists in the same subdirectory as the DWF file.

AutoCAD executes the Insert command using the following sequence:

> Command: _.-insert Enter block name or [?]: **C:\Program Files\AutoCAD 2002\Sample\1st floor plan.dwg** ENTER
>
> Specify insertion point or [Scale/X/Y/Z/Rotate/PScale/PX/PY/PZ/PRotate]: **0,0** ENTER
>
> Enter X scale factor, specify opposite corner, or [Corner/XYZ] <1>: ENTER
>
> Enter Y scale factor <use X scale factor>: ENTER
>
> Specify rotation angle <0>: ENTER

To see the inserted drawing, you might need to use the Extents option of the **Zoom** command.

Another drag and drop function is to drag a DWF file from the Windows Explorer into the Web browser. This causes the Web browser to load the DWF plug-in, and then display the DWF file. Once it is displayed, you can execute all of the commands listed in the previous section.

Finally, you can also drag a DWF file from Windows Explorer (Windows 95/98) or File Manager (Windows NT) into AutoCAD. This causes AutoCAD to launch another program that is able to view the DWF file. This does not work if you do not have other software on your computer system capable of viewing DWF files.

PUBLISHING DRAWINGS TO THE WEB

Publishing drawings to the Web allows the files to be accessed across the Internet. The drawings can be posted on a Web site in a number of different file formats. A wizard is available in AutoCAD to make this a simple process for any user. To begin to use the wizard, choose **Publish to Web** from the **File** menu. Other AutoCAD features that will allow files to be shared across the Web include the I-drop technology and **eTransmit**.

PUBLISH TO THE WEB

The **Publish to Web** command provides a method for creating a Web page that contains images of AutoCAD drawings. This tool eliminates the need to write programming code, which means that no previous Web development knowledge will be necessary with this method. To begin publishing your drawings to the Web, choose **Publish to Web** from the **File** menu. The following figures and instructions will walk you through the process.

The **Publish to Web - Begin** dialog box as seen in figure 24.22 has two options available. You can create a new Web page to contain the drawings or edit an existing Web page. You should be familiar with using wizards from previous examples, so simply choose the option needed and select the Next button.

In the **Publish to Web - Create Web Page** dialog box there are three areas to complete. Specify the name of your Web page, choose the storage location, and type in a description if needed.

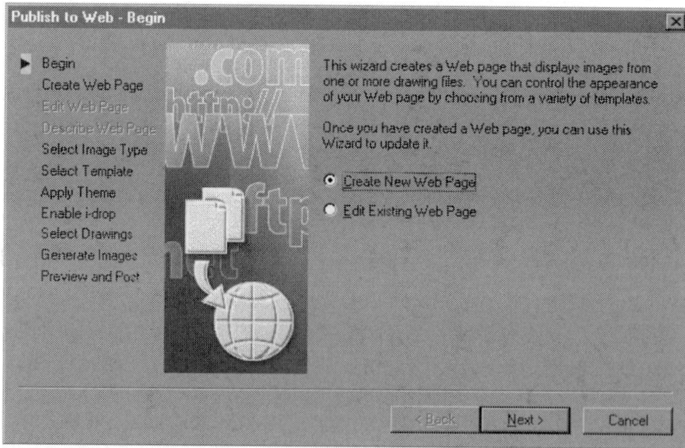

Figure 24.22 *The* **Publish to Web Wizard.**

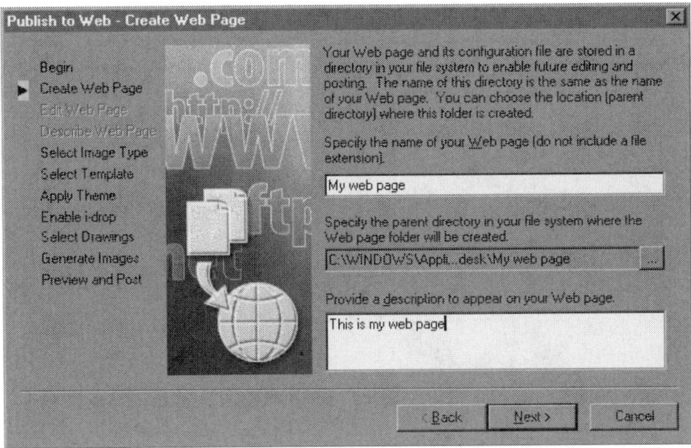

Figure 24.23 *Name and provide a description for your Web page.*

The next step is the option to choose the type of image to use. This is done using the **Publish to Web - Select Image Type** dialog box. There are three image types available in the list: a file that is stored as a .DWF, one that is JPEG, or a file format of PNG. The dialog box in Figure 24.24 gives a description of each file type. If the file is an image file type, the image size can be adjusted.

The **Publish to Web - Select Template** dialog box controls the display of the images in the Web page. A preview window shows how the images will appear in your Web page with the title on the top. See Figure 24.25.

The **Publish to Web - Apply a Theme** dialog box will control the color scheme to be used and improve the appearance of the Web page. The list shown in Figure 24.26 has several options, and the preview window shows how your Web page will look on the Web.

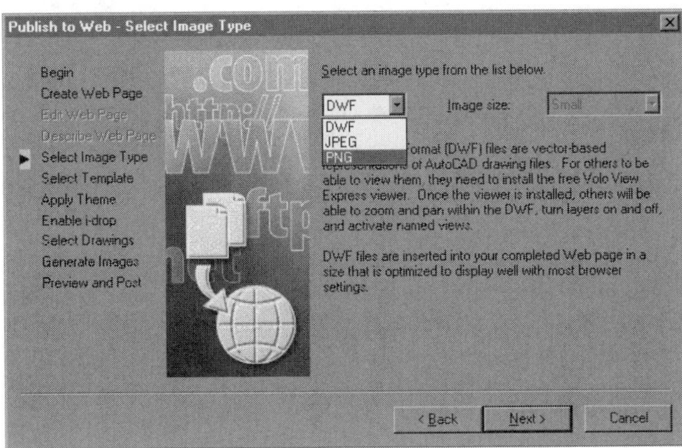

Figure 24.24 *Select the image type for the Web page.*

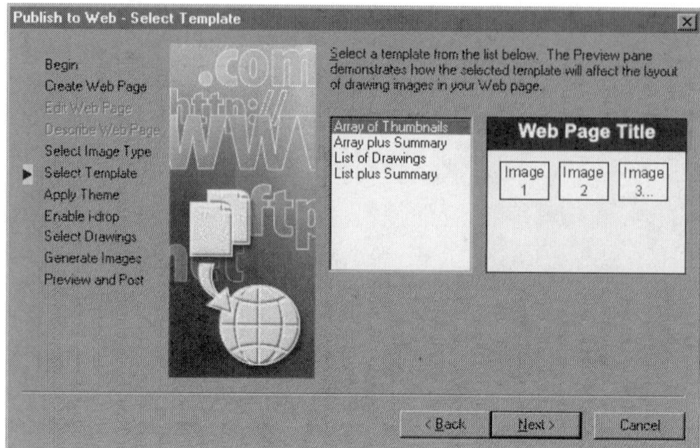

Figure 24.25 *Select a template for the Web page layout.*

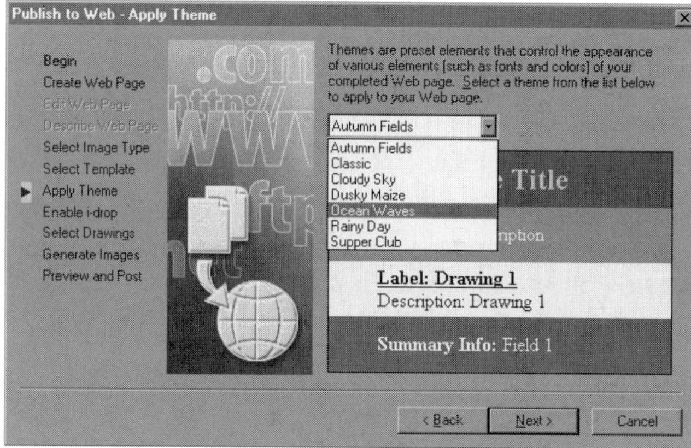

Figure 24.26 *Apply a theme for the color scheme of the Web page.*

With the **Publish to Web - Enable i-drop** dialog box in Figure 24.27 you can enable the feature that will allow the files in your Web page to be dragged from the Web into an AutoCAD drawing file. This will post .DWG files along with the images, to allow easy access to the drawings.

Now that the setup of the Web page is complete, it is time to select the drawings to be posted on the Web page. The **Publish to Web - Select Drawings** dialog box in Figure 24.28 is the way to make your drawing selections. The current drawing is displayed in the **Drawing** box and a browse [...] button is available to open a file dialog box to add additional files. When the file is selected, the layout list contains the name of the layout to be included from the drawing. To add additional layouts from the same

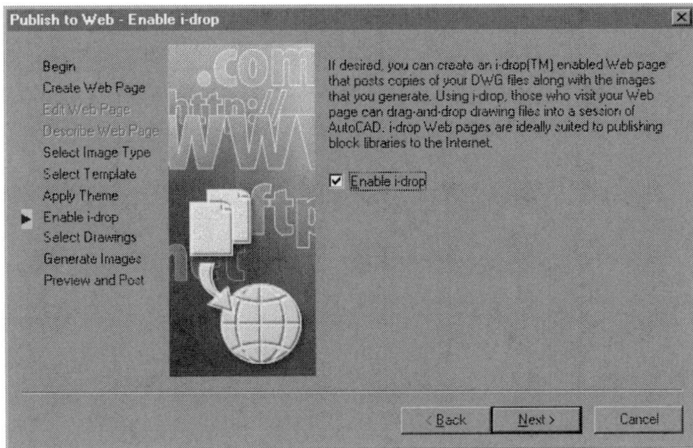

Figure 24.27 *Enable I-drop to allow dragging from the Web page.*

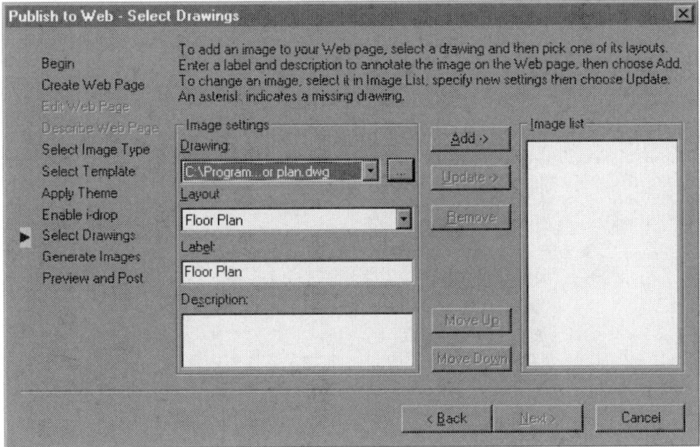

Figure 24.28 *Select the drawings to publish in the Web page.*

file, you will need to repeat the process and use the **Add** button to include each one. A description can be included for each drawing to make a comment about the drawing on the Web page. Figure 24.29 shows three layouts selected. Notice you can add drawings to the Web page, update the drawings if there have been modifications since the page was made, remove drawings from the Web page, or control the location of the drawing with the **Move Up** and **Move Down** buttons.

Now AutoCAD will finish the Web page drawings by creating all the images necessary for the Web page. The next step is to decide if all of the images need to be regenerated or only those that have changed since the page was created. Refer to Figure 24.30. The reason for the two settings is that this wizard is the tool used to edit existing Web

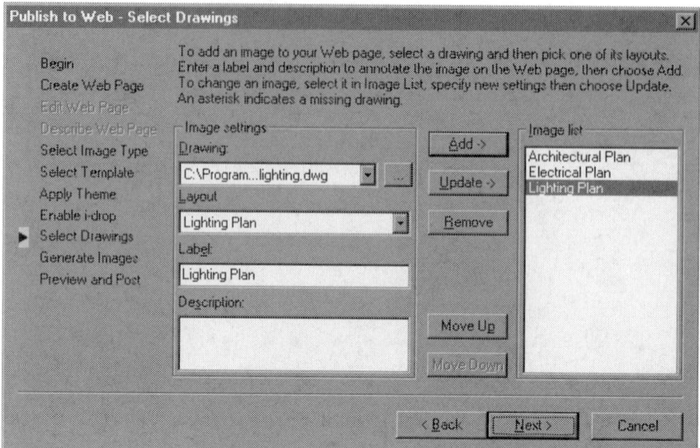

Figure 24.29 *The image list that is added to the Web page.*

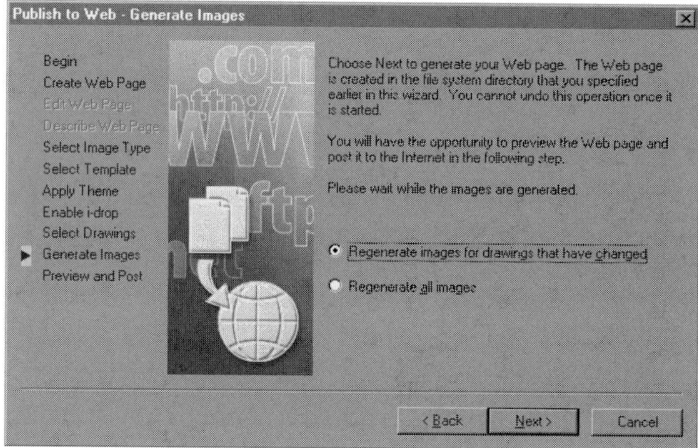

Figure 24.30 *Clicking* **Next** *in the* **Publish to Web - Generate Images** *dialog box will convert the files to Web file format.*

pages as well. When you select the **Next** button, AutoCAD will generate all the images requested. This might take a few minutes, depending on how much information is included.

The final display of the wizard is **Publish to Web - Preview and Post** dialog box, shown in Figure 24.31. It is used to notify you that your Web page is now available in the listed directory. You can either examine the new page using the **Preview** button or post the page by selecting the **Post Now** button. Once the page is posted, you can generate an email message that will include a hyperlinked URL to the page.

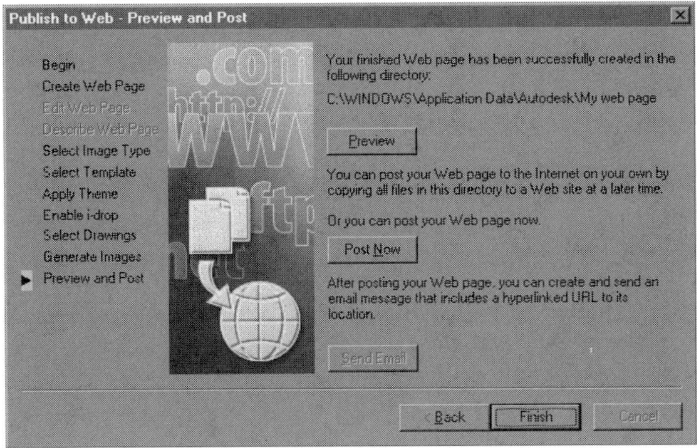

Figure 24.31 *Preview and post the images to the Web.*

ETRANSMIT

The **eTransmit** feature in AutoCAD enables you to pack currently open files with all of the associated files and xrefs into a transmittal set that can be sent to the desired location. This is very similar to sending email messages across the Web. This will allow files to be shared both internally in a company and externally to any location on the Web. To begin a transmittal, select **eTransmit** from the **File** menu.

Figure 24.32 is an example of the **eTransmit** dialog box. The following is a list of the features available:

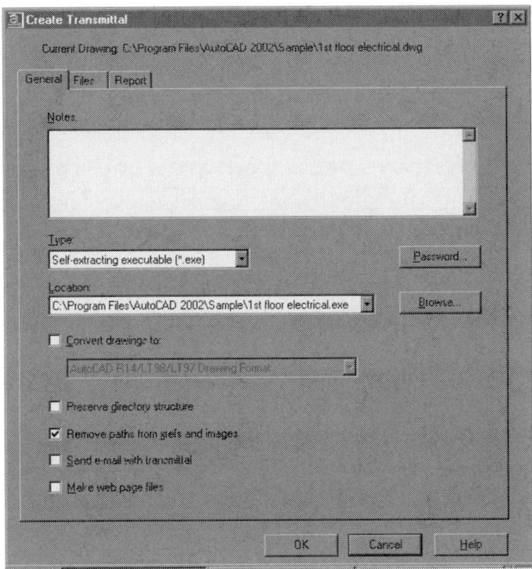

Figure 24.32 *Create a Transmittal to send a file to a client through the Web.*

Notes

If an explanation of the enclosed file is needed, type it in this box.

Compression

The file can be compressed to speed up the transfer rate in a self-extracting executable file (.exe) or a .zip file.

Security

You can protect the transmittal by applying a password. This will keep intruders on the Web from viewing the files. Select the **Password** button to protect the transmittal.

Email Notification

An automatic email message can be sent to let the recipient know that the transmittal is available for use. This is a good way to inform the receiving party about the drawings, the location of the information, or password needed to view the designs.

Web Posting

This allows the posting of the information to be handled across the Internet without tying up the email system. This would be useful if there are a lot of people involved with the project.

Drawing Conversion

Both Autocad 2000 and Release 14 drawing file formats are supported. Select the **Convert drawings to** check box and choose a format from the list.

Report Generation

This feature on the **Report** tab will generate a complete log about the transmittal set. This is a tracking system that is available to all the recipients to track transmittal activities, manage any problems, and determine which files have been included.

You should also activate the **Remove paths from xrefs and images** check box so there are no problems with sending the files that contain these objects.

MEET NOW

Meet now is a feature that allows hosting interactive meetings over the Internet using Microsoft's Net Meeting Technology. If you are familiar with chat sessions on the Internet, this feature is very similar. It allows voice, data, and video to be utilized in design team collaboration. Because it lies beyond the scope of what a new user would be expected to do in an office, it will not be further explored in this text. Refer to Microsoft and Autodesk documentation to learn more about this feature.

CHAPTER 24 QUIZ

1. Can you launch a Web browser from within AutoCAD?
2. What does DWF mean?
3. What is the purpose of DWF files?
4. What is URL short for?
5. Which of the following URLs are valid?

   www.autodesk.com

   http://www.autodesk.com

   Both of the above.

   Neither of the above.
6. What is the purpose of a URL?
7. What is FTP short for?
8. What is a "local host"?
9. Can URLs be used in an AutoCAD drawing?
10. The purpose of URLs is to let you create ______________________ between files.
11. When you attach a URL to a block, the URL data is ___________ when you scale the block unevenly, stretch the block, or explode it.
12. Can you attach a URL to rays and xlines?
13. The **Attachurl** command allows you to attach a URL to _________________ and _________________.
14. To see the location of URLs in a drawing, use the ___________________ command.
15. Rectangular URLs are stored on layer______________________________.
16. The **Listurl** command tells you ____________________________________.
17. The **Detachurl** command removes a _________________ from an object.
18. Compression in the DWF file causes it to take (less, more, the same) time to transmit over the Internet.
19. A DWF is created from a _______________ file using the ________________ or _______________ command.
20. What does a plug-in allow a Web browser do?
21. Can a Web browser view DWG drawing files over the Internet?

22. ________ is an HTML tag for embedding graphics in a Web page.
23. Are the WHIP! plug-in and the WHIP! display driver the same?
24. A file being transmitted over the Internet via a 28.8 Kbps modem takes about ____________ minutes per megabyte.
25. To open a drawing located on the Internet, use the __________________ command.

INDEX

? button, 41
? option, 843
>> button, 43
0 Feet option, 693
0 Inches option, 693
2, 3, 4 options, 863
3Points option, 27-28

AA. (*See* Area command)
About
  Plug-ins, 912
  WHIP option, 914
Above option, 684
Absolute option, 97
Accept button, 410, 414
Active
  Assistant, 8, 39-41, 49
  Plot Style Table, 883
Adaptive Adjustment, 886
Adcenter command, 813, 821
Add
  A Plot Style Table button, 880, 882
  button, 44, 471, 522, 611, 868, 919
  Color-Dependent Plot Style Table Wizard, 870
  Layer(s) option, 829
  option, 540-42
  or Edit Plot Styles Tables button, 879
  Plot Style Table
    - Begin dialog box, 881
    Wizard, 879-80, 882
  Printer
    option, 869
    Wizard, 869
  Profile dialog box, 124-25
  Style button, 883
  to
    Favorites, 830, 894
    List option, 124-25
Add< button, 470
Add-A-Plotter Wizard, 869
Additional angles, 96-97
Advanced
  Setup wizard, 11, 56, 58-62
  tab, 755, 760, 769-72, 825-26
Aerial View
  command, 345, 363-66
  option, 363
AL. (*See* Align command)
Alerts, 36
Alias for commands, 29
Align
  command, 421-22, 428-31
  Text option, 708
Aligned
  Dimension button, 640
  option, 640
  section, 751-52
  with dimension line option, 685-86
All
  commands option, 39
  layout tabs, 871
  option, 383, 445
Alphabetic tab, 705
Alternate
  Unit Tolerance area, 694
  Units tab, 694
Always
  draw dim line between ext lines option, 689-90
  Keep Text Between Ext Lines option, 688

left justify check box, 664
American National Standards Institute, 140-41
American Society of Mechanical Engineers, 56-57, 140
Ang option, 497-98
Angle
between items edit box, 412
controlling the display of, 693
direction, 61
display, 59-60
edit box, 77, 213-14, 759, 795
measurement, 60-61
of array edit box, 408
to fill edit box, 411, 414
Angular
Dimension button, 646
Dimensions area, 690, 693
option, 646
Annotation
entering, 662-63
Reuse, 663, 665
tab, 663, 665
type, 663-64
ANSI. (*See also* American National Standards Institute)
tab, 757, 766
Aperture
box, controlling the, 311-14
color, altering the, 314
Size
area, 313
option, 124
Apparent Intersection
button, 326-27
mode, 326-27
Append to current selection set check box, 530
Apply button, 589-91
Apply & Close button, 125, 161, 314
Apply to
box, 528-29, 531-32, 789, 906
layout dimensions only check box, 692
list, 846
AR. (*See* Array command)
Arc
button, 22, 264
center, start, angle, 275
center start, end, 273-75
center, start, length of chord, 275-76
command, 3, 16-17, 25, 27-28, 255, 266, 352, 722, 724
continue, 276-79
option, 264
start, center, angle, 269
start, center, end, 267-68
start, center, length, 269-70
start, end, angle, 270-71
start, end, direction, 271-72
start, end, radius, 272-73
three-point, 266
Arc and Circle Smoothness setting, 257, 370
Architectural - 1'-3 1/2" option, 57
Area
button, 538
command, 537-40
option, 538
specifying the drawing, 61
Array
button, 22, 403-4, 408, 413
command, 375, 385, 389, 403-4, 413-14, 724
dialog box, 403-5, 408, 410, 413
option, 403
Arrow
keys, 35. (*See also* Edit keys)
size edit box, 680
Arrowheads area, 676, 678-80
Arrows option, 687
Ask Me tab, 42, 44
ASME. (*See* American Society of Mechanical Engineers)
Assist command, 4, 40-41
Associative
button, 768
hatch option, 386
At
Ext Line 1 option, 684
Ext Line 2 option, 684
Least/Exactly option, 604-5

Attach
  option, 829
  Xref option, 829
Attachment tab, 667
Audit, 114, 136
Auditctl system variable, 136
Auto
  option, 504-5
  Selection mode, 378-79
  Viewport option, 365
AutoCAD
  Calculator, 555-58
  Learning Assistance tutorial, 44
  startup, 8-11
Autodesk Point A, 9
Automatic Save option, 122
Autosave, 130
AutoSnap
  Marker
    color edit box, 316
    Color option, 123, 316
    Size area, 315-16
    Size option, 123-24
  Settings area, 87, 314-18
AutoTrack Settings, 314, 317-18
Auxiliary views, 210-19
AV. (*See* Aerial View command)
Available Profiles list, 125

B. (*See* Block command)
Back
  button, 42, 894, 914
  command, 914
  option, 505, 914
Background color
  changing, 161
  shown in viewer option, 911
Backup files, 129
Backwards option, 588-89
Base
  command, 783, 798
  edit box, 78
  option, 798
  Point
    option, 466, 806-7
    specifying the, 791
Baseline
  Dimension button, 649
  option, 649
  spacing edit box, 676
BEgin option, 505
Beep on Error in User Input option, 123
Beside the Dimension Line option, 688
BH. (*See* Bhatch command)
Bhatch
  command, 741, 754-57, 761, 766, 769, 771-72
  Select Object option, 763
Bisect option, 498-99
Block
  cascading menu, 786
  command, 385, 389, 666, 679, 682, 783, 785-86, 788, 792, 807
  creating a, 786-91
  Definition dialog box, 786, 788-90, 792, 803
  descriptions, editing, 803
  edit box, 806-7
  inserting
    from the internet, 896-98
    using the DesignCenter, 828
  option, 786, 788, 792-93, 798, 806-7
  redefining a, 803-4
  Reference option, 664
  size, 786
Blockicon command, 788
Blocks, 783-97
  attributes, 785
  benefits of, 784-85
  drawing objects for, 785-86
  editing, 784, 803-5
  efficiency, 785
  exploding, 804-5
  in multiple drawings, 799
  inserting, 796-97
  nesting, 798-99
  speed, 784
  working with, 797-803

Bold button, 599-600
Boring, 716
Boss, 719
Both Text and Arrows option, 688
Boundary Hatch
  command. (*See* Bhatch command)
  dialog box, 28-29, 755-56, 761, 764-68, 773
Boundary set, 771-72
Bounding box, 545
BR. (*See* Break command)
Break
  button, 451
  command, 385, 421, 451-54
  lines, 147-48
  option, 451
  points, 451-53
Break@, 453-54
Broken Out section, 753-54
Browse
  button, 53-54, 63, 65, 793, 896
  For area, 902
  for Folder
    button, 807
    dialog box, 871, 909-10
  option, 53
  the Web
    button, 871, 904
    dialog box, 892-96
Browser command, 889, 891-92, 904, 907
Bulletin Board, 9
Buzzsaw option, 894
ByBlock button, 163
ByLayer button, 163-64

C. (*See* Circle command)
Cal command, 537, 555
Cancel
  button, 24, 28, 32-33, 37, 178, 410, 414, 601, 876, 884
  option, 83, 132
Cartesian coordinate system, 85-88
Cascade option, 68, 71, 814
Categories edit box, 491
Celtscale command, 511, 513
CEN. (*See* Center mode)
Center
  and diameter, 259
  and radius, 258-59
  button, 328-29
  Mark
    creating a, 646
    option, 646
  Marks
    area, 676
    box, 680
  mode, 327-28
  point
    x edit box, 410
    y edit box, 410
  the plot check box, 875
Centered option, 684
Centerlines, 143
Centroid, 545
CH. (*See* Properties command)
CHA. (*See* Chamfer command)
Chain lines, 148-49
Chamfer, 724-25
  angle, selecting the, 456-58
  button, 454
  command, 289-91, 385, 421, 454-59
  Method option, 459
  option, 454
  Trim mode, selecting the, 458-59
Change
  button, 610
  command, 385
  Dictionaries button, 610
  Group area, 470-71
  Profile dialog box, 125
Character
  Map dialog box, 601
  tab, 597-98
Check
  boxes, 34
  Spelling dialog box, 609-10
Chprop command, 385

Circle
button, 22, 91, 255-56
command, 25, 40, 123, 255-58, 352, 716, 722-24
option, 255-56
submenu, 257, 259, 262
three-point, 261
two-point, 259
Circle@ option, 263-64
Click, 7, 21
for more button, 43
to hide button, 43
Clip Existing Viewport button, 850-51, 860, 862
Clockwise button, 61
Close
box, 37
button, 11, 14-15, 23, 26, 43, 46, 132, 522, 527, 545, 589, 591, 696-97, 701
command, 113, 132
Closeall, 114, 132-33
Color
altering, 524-32
box, 516-17
button, 314-15, 516-17, 522, 601, 699
changing, 161
command, 139, 162
Control arrow, 163
Dependent
Plot Style Table option, 880
plot style, 878
edit
arrow, 676-77
box, 601
list, 516, 676
menu, 516
numbers and names, 162
Options dialog box, 161, 314-15
setting object, 162-64
Colors
button, 161-62
option, 122
dialog box, 122
Column offset edit box, 407
Columns edit box, 406
Command, 21
alias, 29
line, 16
Commands tab, 26, 491
Compare
button, 672
Dimension Style dialog box, 675
edit box, 700
Components to search edit box, 44
Composition area, 768
Compression, 922
Confirm
each item to be purged, 553
Purge warning box, 553
Connect (C) option, 503
Construction Line
button, 496
command. (*See* Xline command)
option, 84, 496
Contents
edit box, 609
tab, 42-44
Context box, 610
Continue
button, 471, 674-75
Dimension button, 651
option, 651
Control
buttons, 601-2
keys, 35, 38-39. (*See also* Edit keys)
option, 505
Panel, 549
Convert
drawings to check box, 922
.DWG hyperlink extensions to .DWF option, 911
Objects to a Viewport button, 850-51, 861
to block option, 791
Copies, multiple, 392-94, 797-98
Copy
an Object option, 663
button, 22, 389

command, 375, 385, 389-95, 804, 813, 816-18
Hyperlink, 906
option, 389, 391, 466, 799, 840, 887, 914
with Base Point command, 816, 818
Copybase command, 813, 816, 818
Coordinate display, 38
Core, 716
Counterbore, 717-18
Counterdrill, 721
Countersink, 719-21
CP. (*See* Cpolygon option and/or Copy command)
Cpolygon option, 381
Create
Backup Copy with Each Save, 123, 129
Drawing dialog box, 807
Drawings tab, 9, 50, 55-56, 62
Group area, 470-71
Layout
Define Viewports dialog box, 838
option, 837
Paper Size dialog box, 837-38
Wizard, 837, 842-43
New
Dimension Style dialog box, 673-75, 696
Drawing dialog box, 132
Crosshairs size
altering the, 19-20
option, 122
Crossing option, 379
Current
button, 155, 168-70, 519
display radio button, 359
Plot Style Table dialog box, 883
tab, 871
word box, 610
Cursor, 19
alternate, 20-21
Cursorsize command, 4, 20
Custom
Arrow Block dialog box, 666
dictionary box, 610-11
dictionary words box, 610-11
display box, 874-75
Properties button, 909-10
scale option, 175-76
Customize
dialog box, 23, 26, 31, 491
Menus, 898
option, 491
Cut
button, 22
command, 813, 816
Cutting plane
lines, 145-46
placing the, 742-46

D. (*See* Dimension Style Manager dialog box)
DAL. (*See* Dimaligned command)
DAN. (*See* Dimangular command)
Dashed lines. (*See* Hidden lines)
Date Modified tab, 825-26
Date/Time, 549
DBA. (*See* Dimbaseline command)
DCE. (*See* Dimcenter (Center Mark) command)
DCO. (*See* Dimcontinue command)
Dblist (Data Base List) command, 537, 548
Ddchprop command, 385
Ddedit command, 567, 591-94, 606-7, 613
DDGRIPS, 463
DDI. (*See* Dimdiameter command)
DDPTYPE, 490
Decimal
- 15.0000 option, 57
Degrees - 90.00 option, 60
separator option, 690, 692
DED. (*See* Dimedit command)
Default, 22
plot style behavior for new drawings area, 878
Define
window radio button, 361
View Window button, 361
Viewports, 838
Deg/Min/Sec - 90d00'00" option, 60

Delete
  button, 97, 125, 363, 524, 585
  option, 701, 791, 842
  Style button, 884
Description
  button, 472, 823, 826-27
  edit box, 125, 472, 791
DesignCenter, 821-30
  adding content to drawings using, 828-29
  button, 800, 821
  command, 783
  controlling blocks with, 799-800
  features, 822-23
  opening drawings using the, 827
  toolbar, 823-27
Desktop button, 800-1, 823
Destination area, 806-8
Details
  area, 516
  button, 363
  option, 824, 827
Device and Default Information box, 869
DI. (*See* Dist command)
Dialog boxes, 30-36
Dialogs only option, 40-41
Diameter
  Dimension button, 643
  option, 643
Dim
  Line1 check box, 677
  Line2 check box, 677
  Style
    Control box, 701
    Control menu, 701
    edit box, 701
    option, 701
Dimaligned command, 617, 632, 640-42
Dimangular command, 617, 632, 646-49
Dimbaseline command, 617, 632, 649-51
Dimcen variable, 646
Dimcenter (Center Mark) command, 617, 632, 646, 680
Dimcontinue command, 617, 632, 651-53
Dimdiameter command, 617, 632, 643, 645, 680
Dimedit command, 671, 706-8
Dimension
  aligned, 626, 639-41
  angular, 646-49
  arrowless, 630
  chain, 626-27
  chart, 630-31
  command, 150
    accessing, 631
  components, 619-31
  continuous, 651-53
  control buttons, 673
  datum, 627-28
  diameter, 643, 645
  Edit button, 706
  editing with the modify menu, 704-5
  elements of a, 619-25
  Lines area, 676-77
  lines, 151, 620-22
  menu, 631, 633, 640-43, 646, 649, 651, 653, 659, 672-73, 708-9
  methods of placing, 626-31
  options, 632-53
  ordinate, 641-42
  placement, editing, 701-5
  quick, 653-59
  radius, 642-44
  selection, 22-23
  standards and goals, 618-19
  stretching, 702-4
  Style
    button, 696
    Manager dialog box, 633, 672-75, 695-96, 699-701
    option, 672-73
  tabular, 629
  text, editing, 705-9
    using Dimedit, 706-8
    using Dimtedit, 708
  text, rotating existing, 706-7
  Test Edit button, 708-9
  toolbar, 22-23, 25, 151, 631, 633, 640-43, 646, 649, 651, 653, 659, 672-73, 696, 706, 708

unidirectional, 626
Update
button, 708
command, 671, 708
variables, controlling, 672-75
with styles, 671-72
Dimlinear command, 617, 632-39
Dimordinate command, 617, 632, 641-43
Dimoverride command, 671
Dimradius command, 617, 632, 642-44, 680
Dimstyle command, 671
DIMTED. (*See* Dimtedit command)
Dimtedit command, 671, 708-9
Dimvar settings, 642
Direction
button, 73-74
Control dialog box, 73
text specific property, 609
Diskette-handling procedures, 135
Display
alternate units check box, 694
attribute definitions for editing check box, 803
AutoSnap
aperture box check box, 313
tooltip check box, 316
AutoTrack tooltip check box, 318
button, 34, 43
full-screen tracking vector check box, 317
hyperlink
cursor and shortcut menu check box, 900
tooltip option, 900
Lineweight check box, 158-60
locked
edit list, 849
option, 849
polar tracking vector check box, 317
Resolution option, 122, 370
setting, 174, 516, 873
tab, 20, 30, 122, 161, 314-15, 370
Viewports Dialog button, 843, 845, 850-51
window, 676
Dist command, 537-39
Distance
button, 538
determining. (*See* Dist command)
option, 538
Dithering, 885
DIV. (*See* Divide command)
Divide
command, 385, 489, 491-93
option, 491
DLI. (*See* Dimlinear command)
DO. (*See* Donut command)
Docking toolbars, 26
Donut
button, 279
command, 255, 257, 279-81
drawing, 279-81
option, 279
DOR. (*See* Dimordinate command)
Dovetail slot, 727-28
DRA. (*See* Dimradius command)
Drafting
standards, 63-64
tab, 87, 123-24, 313, 315
Drafting Settings
dialog box, 17-18, 34, 76-78, 95-96, 214, 317-19, 337-39
option, 213, 317, 337-38
Dragmode command, 255
Draw
frame around text check box, 683
menu, 7, 16, 27-28, 36, 44, 81-82, 262, 264, 279-81, 286, 289-90, 294, 489, 491, 493, 495-96, 576, 755-56, 766, 771, 786, 788, 792, 798
toolbar, 22, 24-25, 27, 45, 81, 91, 255-57, 264, 281, 286, 289-90, 293-94, 489, 496, 543, 596, 755-56, 766, 771, 786, 788, 792-93, 803
Drawing
area, 8-9, 11-14
setting the, 58
box, 918
Conversion, 922
Limits option, 73

modifying
by adding objects, 389-414
using erase, 386-89
objects, selecting all, 383
opening multiple from Windows Explorer, 815-16
orientation, 174-75, 867, 873-74
placement of, 194-97
publishing to the web, 916-22
saving, 126-33
Settings dialog box, 17
types of, 189-90
Units dialog box, 73-74, 117
units, setting the, 56-57
Utilities, 136, 552
Web Format. (*See* DWF file format)
Drawings tab, 825-26
Drilling, 716
DSETTINGS command, 311, 317, 337
Dview command, 385
DWF
file format, 907-11
creating a, 909-10
viewing, 911-16
plug-in commands, 912-16
Properties dialog box, 909-10
DWG files, viewing, 908-9
Dynamic
Update option, 364
Zoom, 356

E. (*See* Erase command)
ED. (*See* Ddedit command)
Edit
Block button, 803
boxes, 35
Copy command, 905
Hatch option, 772
Hyperlink dialog box, 903, 906
keys, 35
menu, 101, 612, 799, 816, 905
Spline button, 298
Text
box, 592
button, 591
dialog box, 592
option, 607
Editor, 883
Effective area, 868
Either the text or the arrows, whichever fits best option, 687-88
EL command. (*See* Ellipse command)
Ellipse
button, 22, 281
command, 255, 281-85
Email Notification, 922
Enable
grips check box, 462-63
unique layer and symbol names check box, 803
End
command, 550
option, 505
Endpoint
button, 320
mode, 34, 320-21
Engineering - 1'-3.5000" option, 57
Enter, 22, 24
Entire drawing option, 807
Equipment, controlling your, 4
Erase
button, 22, 25, 99, 101
command, 4, 25, 98-101, 375-77, 385-88, 453, 804, 907
option, 99
Window option, 377-78
Erase (E) option, 503
Errors, correcting file, 136
Escape, 22
eTransmit
command, 889, 921-22
dialog box, 921
option, 921
EX. (*See* Extend command)
Exclude from new selection set, 530, 532, 790
eXit option, 466
Exit, 46, 113, 132, 178, 348

Explode
button, 472, 804
check box, 796
command, 385, 783, 804
option, 804-5
Export
button, 126
command, 127
Profile dialog box, 126
Ext Line 1 box, 678
Extend command, 385, 421
Extension
line angle, altering the, 707-8
lines, 150
mode, 324-26
Extend
beyond
dim lines option, 677-78
Ticks option, 676
button, 448
command, 448-51, 704
Edge option, 449-50
option, 448
Project option, 451
Trim option, 450-51
Undo option, 451
Extension
Lines area, 676-78
lines, 619-20
origin, selecting, 634-38
Extents
button, 34
option, 174, 873
External Reference dialog box, 829
Extreme resolution, 910

FATAL ERROR, 135
Favorites
area, 42-44
button, 823-25, 830, 895
Fence option, 381-82
File
button, 898, 902
Download display box, 897
menu, 35, 45-46, 55, 67, 69, 127, 129, 136, 172, 552, 865-66, 868, 879-80, 882-83-84, 893, 896, 909, 916
name, 127, 129, 807, 909
edit box, 35, 127, 130, 807
tab, 597-98
type edit box, 842
Filedia command, 4, 32
Files
backup, 129
managing, 133-37
tab, 122
Fill
command, 255, 280-81
style, assigning, 886
Fillet, 291-92
button, 207, 459
command, 189, 202, 205-8, 215-16, 385, 421, 459-62
option, 291-92, 459
radius, 460-61
Filter edit arrow, 522
Filters, 522
Find
and Replace dialog box, 612
box, 605
button, 605, 612-13, 823, 825-26
command, 612, 828
dialog box, 825
File button, 896
Name button, 470-71
Now button, 826
option, 612
text string edit box, 612
Find/Replace tab, 597-98, 605-6, 612
Finish
button, 58, 61, 839, 881-82
page, 882
First
angle projection, 197
Segment, 667
Fit
Data option, 299-300

Options area, 688
option, 687-88, 863
tab, 686-90
to Window option, 913
Flip Screen, 37
Floating toolbars, 25
Flood state, 772
Folders, managing, 133-37
Font
edit box, 598
Name edit box, 586, 588
Style option, 588
Fonts, lettering, 586-87
Form View tab, 883-84, 887
Format
menu, 16-17, 31-32, 73, 152-53, 158-59, 162, 166, 490-91, 512, 514, 584, 672-73, 887, 898
setting, 910-11
Forward
button, 894
option, 914
Fraction
format edit box, 690-92
height scale option, 683
Fractional - 15 1/2 option, 57
Frame text option, 664-65
Freehand sketching, 219-36
Freeze in
Active Viewport, 858
All Viewports, 858
New Viewport, 858
FRO. (*See* From mode)
From
button, 335
mode, 335-36
Front view, selecting the, 198-200
Full
Preview option, 177-78, 876, 882
section, 749
View option, 913
Function keys, 37-38
G. (*See* Group command)
General
notes, 569-71
Options, 9, 51, 123, 813
Glass cube, 191-94
Global
option, 364-65
scale factor, 157, 512, 859
Go Back button, 904
Go Forward button, 904
Grads - 100.00g option, 60
Grayscale, convert to, 885
Grid
button, 17, 96
command, 49, 75, 78
display, 38
setting, 75-78
Grip options, 464-66
Grips, selecting objects with, 462-66
Group
command, 421, 468-73
Identification area, 468-70, 472
Member List dialog box, 470-71
Name
box, 468-69, 471
list, 472

Habits, work, 5
Half section, 749-50
Hatch
command, 36, 754-55
Edit dialog box, 772
option, 755-56, 766, 772
Pattern Palette dialog box, 756-59, 766
Patterns, 898
Hatchedit command, 741, 772
Hatching methods, 754-55
HE. (*See* Hatchedit command)
Height
option, 577, 588, 598-600
text specific property, 609
Help
button, 39, 122, 601

command, 4, 41-43
menu, 70, 98, 364, 400, 491, 557, 597, 755, 857, 912
Hidden lines, 141
Hide
button, 42
objects check box, 875
Hideplot option, 863
High resolution, 910
Highlight
button, 470
URLs option, 914
Hints button, 870
History
(by Date) option, 52
(by Filename) option, 52
(by Location) option, 53
button, 823-24
Home button, 895
Hor option, 496-97
Horizontal
box, 684
list, 684
option, 685
Text Placement list, 685
Hot grip, 462
Hover Help option, 40-41
How to Apply area, 530-32, 790
Hyperlink, 800-3
command, 800, 803, 889-90, 899-900, 906-7
menu, 803
option, 800, 899
outside of AutoCAD, 907
shortcut menu, 906
-HYPERLINK, 899
Hyperlinkback command, 904
Hyperlinkfwd command, 904
Hyperlinkstop command, 904

I. (*See* Insert command)
Icon buttons, 5-6
ID (ID Point) command, 537, 548-49
Ignore
button, 610
hatching style, 770
Image
boxes, 35-36
Displays, 898
Images, 898
Implied windowing option, 386
Import
button, 126, 898
Profile dialog box, 126
Text button, 601-2
In the field(s) edit box, 825-26
Inches radio button, 159
Include
in new selection set, 530-31
Layer information option 911
Scale and measurement information, 911
Unnamed check box, 471
Increment angle edit box, 96, 317
Index tab, 42-44
Inetlocation command, 889, 892-95
Inherit Properties button, 767
Inplace Xref and Block Edit Reference, 803
Inquiry
menu, 537-38, 543, 545, 548, 549-50
toolbar, 537-38, 543, 545, 548
Insert
Block
button, 792-93
option, 828
command, 783, 786, 792, 794, 808, 851, 890, 893, 896, 898
dialog box, 793, 796-97, 804, 828, 896
Hyperlink
button, 800, 802, 899, 902
dialog box, 800, 802, 899, 902, 906
Layout dialog box, 842
menu, 793, 800, 899
mode, 330
Insertion Point Specify On-screen check box, 793
Internet
access, 890-91
files, opening and saving, 891-93

Intersection
  button, 323
  mode, 323-24
Island Detection
  Method area, 772
  Style area, 769-70
Islands, removing, 766
ISO Standard option, 685-86
Isometric
  drawings, 189-90
  sketches, 232-36
  Snap setting, 77
Isoplane crosshairs mode, 37
Italic key, 599-600
Items not used in drawing display window, 552-53

JIS option, 684
Justification option with Mtext, 603
Justify, 609

Kerf, 725, 727
Key, 723-24
Keyboard entry, 29
Keyseat, 723-24
Keyway, 723-24
Knurl, 725, 728

L command. (*See* Line command)
LA. (*See* Layer command)
Landscape, 873
  option, 837, 876
  position, 175
Large
  button option, 26
  icons display option, 827
  Icons, 824
Last object, selecting the, 383
Lay, 731
Layer, 164-72
  altering, 524-32
    qualities, 169-70
  assigning color by, 514, 516
  box, 519
  button, 515
  command, 32, 139, 166, 511, 857
  Control
    arrow, 524-25
    box, 172, 516, 518
    button, 524
    menu, 514, 524-25
  controlling the visibility, 170-72
  creating a new, 166-67
  deleting, 524
  freeze mode, 520
  linetype option, 516
  lock option, 520, 522
  making current, 168-69
  Name, 898
  naming, 514
  option, 31-32, 166, 514
  Previous
    button, 526
    command, 511. (*See also* Layerp command)
  Properties Manager
    Change Properties dialog box, 515
    dialog box, 31-32, 166, 168-70, 514, 517-20, 522-24, 858-59, 884, 898
  thaw option, 520
  thawing and freezing, 520-22
  unlock option, 520, 522
Layerp command, 526
Layers
  button, 31-32, 166, 168, 702
  option, 913
Layout
  command, 833, 843
  creating new, 836-40
  from Template button, 842
  name, 172, 867
  option, 873
  tab, 12-14, 868
  working with, 835-43
Layouts toolbar, 840, 842

LE. (*See* Qleader command)
Leader, 280
  button, 151
  command, 617
  Line & Arrow tab, 665-66
  lines, 151
  option, 659
  Settings dialog box, 663-65
Leading box, 693
LEN. (*See* Lengthen command)
Lengthen
  button, 438
  command, 421-22, 438-43
  DElta option, 438-440
  DYnamic option, 441, 443
  option, 438
  Percent option, 440-41
  Total option, 441
Lettering
  fonts, 586-87
  height, 574-76
LI. (*See* List command)
Limits
  button, 34
  command, 49, 73, 113, 117-18
  option, 173
  setting, 73-75
Line
  button, 16, 22-24, 27, 82
  closure, 91-94
  command, 3, 7, 16, 19, 23-24, 29, 39, 41-44, 67, 81-84, 116, 156, 160, 163, 207, 311, 315, 350-52, 519, 722-24, 727-28, 847
  continuation, 91
  End Style, 886
  Join Style, 886
  menu selection, 84
  precedence, 210
  segment size, changing the, 511-13
  space
    factor, 609
    style, 609
  Spacing tab, 597-98, 604-5
Lines, 82-84
  and Arrows tab, 676-80, 699
  controlling, 91-95
  editing, 98-102
  locating using UCS, 84-90
Linear
  Dimension button, 633
  Dimensions area, 690-92
  option, 633
Linetype, 139, 152
  altering, 524-32
  button, 522
  characteristics, 152-53
  command, 152-53
  controlling, 511-13
  Control menu, 152-545
  making current, 154-56
  Manager dialog box, 152-57, 512, 859, 898
  option, 512, 859, 885-86, 898
  scale, altering, 156-57
  standards, 140-51
Linetypes, 898
Lineweight, 139, 157-60
  Control
    arrow, 160
    box, 158
  dialog box, 170-71
  displaying, 160
  edit arrow, 676-77
  option, 158-59, 886
  Settings dialog box, 158-60
LISP and ARX Applications, 898
List
  box, 672-73
  button, 548
  command, 537, 545-48
  of components to search edit box, 44
  option, 548, 827
Load
  Applications, 898
  button, 153, 170, 518, 823-25, 898
  DesignCenter Palette dialog box, 824-25
  option, 824
  or Reload Linetypes dialog box, 153, 518

Local notes, 571
Locate button, 896
Location
edit box, 807, 871
option, 913
Lock option, 863
Look
for edit box, 825
in
edit box, 825, 895-96
Favorites option, 893, 898
list box, 53-54
LT command, 153, 512
LTS. (*See* Ltscale command)
Ltscale command, 141, 511-13
Lug, 722-23
LW. (*See* Lineweight)
LWEIGHT. (*See* Lineweight)
LWT button, 18

MA. (*See* Match Properties command)
Machined features
projection of, 728-30
representing, 716-30
Magnet check box, 316
Main dictionary box, 610
Make
Block button, 786, 788, 792, 803
Object's Layer Current button, 524-25
option, 786, 788, 792
Mark option, 505-6
Marker check box, 315
Markup tab, 72. (*See also* Volo View Express)
Mass Properties, 543-45
button, 543
option, 543
Massprop command, 537, 543-45
Match
Case, 605-6
Properties
button, 526
command, 155, 511, 524, 526-27, 813, 820, 906
option, 526, 820
similar words option, 43
WORD, 605-6
Maximize button, 12, 14-15
ME. (*See* Measure command)
Measure command, 385, 489, 493-94
Measurement Scale area, 690, 692
Medium resolution, 910
Meet Now, 922
Menu
bar, 15-16
Customization dialog box, 898
draw, 7, 16
Menus, 26-28, 898
Method
and values area, 411
edit box, 410
value, 695
Metric paper sizes, 65-66
MI. (*See* Mirror command)
Midpoint
button, 321-23
mode, 321-23
Minimize button, 12, 14-15, 67-69
Minsert command, 783, 797-98, 805
Minutes between Saves edit box, 131
Mirror
button, 22, 394
command, 375, 385, 394-95, 397-99, 467, 594, 794
option, 394-95
Mirrtext command, 375, 397, 399, 567, 594
Misc listing, 858
MO. (*See* Properties command)
Model
button, 856
space, 113, 115-16
tab, 12-14
Model Tab Pointer option, 314
Model/Paper button, 18
Modify
button, 410, 414, 672, 674, 699
Dimension Style dialog box, 674, 686, 690, 694-95, 699

menu, 391, 394-95, 401, 403, 421, 432, 434, 438, 444, 448, 451, 454, 459, 526-27, 592, 607, 704, 772, 803, 805
option, 699
toolbar, 22, 25, 99, 155-56, 202, 205, 207, 298, 391, 394-95, 401, 403, 408, 413, 421, 425, 432, 434, 438, 444, 448, 451, 454, 459, 805
Modify II toolbar, 298, 772
More button, 413
Most Recently Used option, 52
Mouse, 6-8
Move
button, 22, 422
command, 385, 421-25, 467, 804, 856
Down button, 919
option, 422
Up button, 919
Movement of inertia, 545
MS. (*See* Mspace command)
Mspace command, 833, 849
Mtext
command, 567, 579, 588, 591, 596-99, 603, 612, 663
option, 634, 663-65
Multiline Text
button, 596
Editor dialog box, 596-99, 602, 607-8, 612, 635, 662, 664, 706
option, 596
Multilines option, 84
Multiple
copies, 392-94
document environment, 813-20
editing, 606-9
edits, selecting objects for, 467-68
My Computer, 549

Name
box, 803, 909
edit box, 166, 786, 793, 869-70, 880
listing, 870
or URL edit box, 896
Named
layer filters
edit box, 522-23
list, 522
Layer Filters dialog box, 522-23
Objects list, 887
Plot Style Table option, 880
plot style, 878-79
View button, 359, 361
Views
button, 359, 361
option, 359, 361, 912-13
tab, 359-63
NEA. (*See* Nearest mode)
Neck, 724
Nearest
button, 332
mode, 332—33
New and enhanced commands radio button, 40
New
box, 471
button, 67, 69, 96, 166-67, 359, 361, 469, 515, 584-85, 672-75, 696, 771, 870
command, 49, 67, 69
Dimension Style
dialog box, 674-75, 699
DRAWING BASE dialog box, 696
option, 361, 842
Style Name edit box, 673-74
Text Style dialog box, 584-85
View dialog box, 359, 361
Viewports option, 844-45, 850
Next button, 837-38, 879-80, 882, 920
Next> button, 58-61
No Limit option, 665-66
Node
button, 329
mode ,328-29
Nonassociative button, 768
None
option, 664
radio button, 680
Normal hatching style, 769-70
North radio button, 61
Notes, 922

Noun/Verb selection check box, 385
Number
 of
  copies setting, 871
  objects (1-x): edit box, 473
 option, 506
Numbers, 569

O. (*See* Offset command)
Object
 Base Point option, 413
 dragging, 819-20
 edit using Noun/Verb selection, 385
 editing with the modify menu, 704-5
 Grouping dialog box, 468-72
 grouping option, 386
 lines, 141
 modifying by altering lines, 443-62
 option, 540-41, 863
 projection of, 207-9
 Properties
  command, 524
  toolbar, 22, 31-32, 152, 154-56, 158, 160, 162-63, 166, 168, 172, 319, 514-17, 524-26, 701
 selecting, 376-78
 Snap, 318-20
  activating, 98
  function key, 37
  modes, 320-36
  Settings button, 337
  tab, 34, 98, 319, 337-38
  toolbar, 18, 319-20, 323, 328-29, 331-32, 335-37, 339
  Tracking Settings option, 97
 Type
  area, 770-71
  box, 770, 789, 906
 type edit box, 529-30
Objects
 area, 790, 792, 806-7
 option, 807
 radio button, 807
Oblique Angle option, 588-90
Off option, 862
Office procedures for files and folders, 134
Offset
 button, 22, 202, 401
 command, 189, 201-4, 207, 211, 213, 215-16, 333, 375, 385, 400-3
 from
  dim line option, 684
  origin option, 678
 option, 202, 401, 499
 section, 750-51
OK button, 11, 20, 30, 32-33, 53, 63, 76, 97, 131, 154, 157, 159, 161, 163, 166, 168-70, 172, 179, 337, 361-63, 470-71, 491, 516, 518-20, 522, 527, 584, 593-94, 601, 607, 610, 662, 664, 696, 699, 706, 758, 764, 766, 792, 803-4, 807-8, 828-29, 842, 846, 850, 859, 867, 876-77, 882, 887, 902, 906, 909-10
Old Name box, 887
On
 Demand option, 40-41
 option, 858, 862
  layer, 520
Oops command, 98, 101
Open
 a Drawing, 814
  at a web site, 896
 an Existing Drawing option, 9
 and Save tab, 122-23, 129-31
 button, 22, 35, 54-55, 842, 883, 896, 902
 command, 49, 55, 890, 893, 896, 898
 Drawings
  button, 823-24
  tab, 50-55
 from the hyperlink option, 904
 option, 896
 to Drawing option, 827
Operator box, 530-32, 790, 906
Option dialog box, 30
Options, 22
 dialog box, 9, 19, 30-31, 51, 87, 98, 122, 125, 129-31, 161, 173, 313-15, 370, 376, 384, 445, 462-63, 516, 597-98, 813-14, 868, 874, 878-79
 menu, 364-65

option, 98, 113, 122, 124, 131, 161, 173, 313, 315, 370, 376, 384, 462-63, 597-98, 814, 878, 900
selection, 19, 30, 51
Order Group dialog box, 472-73
Ordinate
Dimension button, 641
option, 641
Organize Favorites shortcut, 830
Ortho, 38
button, 17
command, 96
Orthographic
& Isometric Views tab, 359
projections, 189-97
OS command, 317, 319, 337-38
Osnap
button, 17-18, 87
continuous, 336-39
selecting points with, 196-97
Other
button, 61, 162
Predefined tab, 757-58
Otrack button, 18
Outer hatching style, 769-70
Outside option, 684
Over Ext Line 1 option, 684
Over Ext Line 2 option, 684
Over the dimension line
with a leader option, 688
without a leader option, 688
Overlay, 829
Override
button, 672, 674, 699
Current Style dialog box, 699
Dimension Style dialog box, 674

Pad, 725, 728
Page Setup
button, 840
dialog box, 835, 840, 866, 882
option, 840, 867, 882
Page setup name, 172
list, 867
Pan
button, 118, 357
command, 49, 67, 113, 118-20, 345, 348, 356-58, 370, 836, 856-57, 912
option, 178, 356-57, 363, 366, 876, 913
Realtime
button, 66-67, 356, 358-59
command, 356, 358-59
submenu, 356-57
Paper
button, 834
size, 173, 867-68, 872
sizes, metric, 65-66
space, overview of, 834
units, 173
Paper Space, 113, 115
layout, 12-14
Parallel mode, 333-34
Partial
Plot Preview dialog box, 868
Preview button, 177, 876
Partiaload command, 49, 70
Partialopen command, 49, 70
Parts list, 574
Password button, 922
Paste
as
Block command, 818
Hyperlink option, 905-6
button, 22
command, 816-18
option, 799, 887
to Original Coordinates command, 819
Pasteashyperlink command, 889, 905-6
Pasteblock command, 813, 818
Pasteorig command, 813, 819
Pattern
angle, 759
edit box, 756-58
hatch, editing, 772-73
list, 757
placement, controlling, 768
Pattern [...] button, 757-58

Pen, 500-1
Pen numbers
  using assigned, 885
  virtual, 885
PER. (*See* Perpendicular mode)
Period (.) Line to Point, 501
Perpendicular
  button, 330
  mode, 330-31
Phantom lines, 148-49
Pick, 7, 22
  Angle of Array button, 408-9
  Both Offset button, 407
  Center Point button, 410
  Point
    button, 787, 791
    option, 771
  Points, 760-61, 766
  Row Offset button, 407
Pickbox command, 375-76
Place text manually when dimensioning option, 689
Placement area, 694
Pline command, 293
Plot
  area, 173-74, 867
  button, 33, 172-73
  command, 139, 172-73, 865-66, 890, 909
  Device, 871-72
    tab, 172, 865-62, 882
  dialog box, 34, 172-74, 176, 865-68, 871, 873-78, 882, 909
  icon, 859
  objects lineweights check box, 875
  offset, 176, 867, 875
  option, 172, 909
  options, 176, 867
  paperspace last check box, 875
  previewing the, 875-76
  scale, 175-77, 867, 874
  Settings tab, 172-73, 865-68, 871-75, 878
  Style
    Editor button, 881
    Manager option, 879-80, 883-84, 887
    Table Editor dialog box, 881, 883-84, 887
  style
    modes, 878-79
    table, creating a, 879-82
    table, adding to a layout, 882
  Styles
    dialog box, 879-84
    portion, 882
  styles
    adding named, 883
    editing, 887
    table (pen assignments), 869, 882
    working with, 877-87
  to file, 869, 871
  upside-down option, 174-75
  with lineweights option, 176-77
  with plot styles option, 177, 875
Plotter
  Configuration (PC3) editor, 869-70, 885, 909
  configuration, 869-70, 909
  Manager, 868
Plotting
  process, 867
  tab, 123, 173, 878-79
Plug-in, 908
PO. (*See* Point command)
Point
  A option, 894
  button, 22, 489, 491
  cascading menu, 491, 493
  command, 489-90
  option, 489
  Size edit box, 491
  Style
    dialog box, 490-91
    option, 490-91
  through, 496
POL command. (*See* Polygon command)
Polar
  Angle Measurements option, 97
  Array button, 410, 414
  button, 17, 96
  coordinates, 89-90

snaps, 80
Tracking
tab, 17, 96, 317-18
toggle, 38
Polygon
button, 22, 286
command, 255, 286
drawing, 285-89
option, 286
Viewport
button, 850-51, 860-61, 863
option, 860
Polygonal option, 863
Polyline, converting a spline-fitted, 298
Polylines option, 84
Portrait
option, 837, 876
rotation, 177
Post Now button, 920
Precision, 683, 690-91, 693
edit box, 58-59, 73
Preferences, saving user, 120-26
Prefix, 690, 692
Press and Drag option, 386
Preview
area, 837
box, 54-55
button, 410, 414, 764, 766, 823, 826, 920
icon display, 788, 792
option, 764-66
window, 587, 672, 827
Previewing, 177-78
Primary Units tab, 683, 690-93
Principal moments, 545
Print
button, 22, 865-66, 914
option, 865-66, 914
Print/Plot Configuration dialog box, 869
Printable area, 868, 872
Printer option, 869
Printing, 172-79
Product of inertia, 545
Profile
creating and saving, 124-26
name edit box, 124
Profiles tab, 124-25
Projecting auxiliary views, 215-16
Projection of objects, 207-9
Prompt
for width check box, 664
options, 550
Properties
button, 155-56, 527, 608, 705, 849, 869, 909
command, 139, 155, 511, 524, 527, 551, 607-9, 613, 705
dialog box, 155-56, 527-29, 608-9, 705, 821, 849, 858
display, 530
edit box, 531-32, 790, 906
list, 529, 609
option, 527, 608, 705
tab, 597-98, 602-3
Property Settings dialog box, 526-27
PS. (*See* Pspace command)
PSLTSCALE, 157, 859
Pspace command, 833, 849
PU. (*See* Purge command)
Publish to Web
- Apply a Theme dialog box, 917
- Begin dialog box, 916
command, 916
- Create Web Page dialog box, 916
- Enable I-drop dialog box, 918
- Generate Images dialog box, 920
option, 916
- Preview and Post dialog box, 920
- Select
Drawings dialog box, 918
Image Type dialog box, 917
Template dialog box, 917
Publishtoweb command, 889
Purge
All button, 553-54
button, 553-54
command, 136, 537, 552-54
dialog box, 552
nested items, 554
option, 552

QDIM. (*See* Quick Dimension command)
Qleader
  button, 659
  command, 632, 659-63
  prompt, 664
  settings, adjusting, 663-67
Qsave, 113, 127, 130-31
Qselect command, 511, 524, 527-32, 906
Qtext command, 567, 594-96
Quadrant
  button, 329, 336-37
  mode, 329-30
Quick
  Dimension
    Baseline option, 656
    button, 653
    command, 617, 633, 653, 697-98
    datumPoint option, 658
    Diameter option, 657
    option, 653
    placement options, 656-59
    Radius option, 657
    Staggered option, 657
    tab, 758, 766
  mode, 334-35
  Select
    button, 528-29, 788-89
    dialog box, 528-29, 531-32, 788-90, 906-7
    edit box, 531-32
    option, 528-29, 906-7
    selection, 528-29
  Setup wizard, 11, 55-56
  tab, 755
  Text, 594-95
Quit, 113, 132, 550
Quit (Q) option, 503

Radians - 1.57r option, 60
Radii of gyration, 545
Radio buttons, 34
Radius
  Dimension button, 642
  option, 642
Ray
  casting option, 772
  command, 489, 494-96
  option, 84, 495
REA. (*See* Regenall command)
Realtime
  option, 347
  Pan, 358, 366
  Zoom, 356, 363
    option, 365, 876
Reaming, 716
REC command. (*See* Rectangle command)
Record increment, 500. (*See also* Segment length)
Record (R) option, 501, 503
Record and Exit (X) option, 503
Recover, 114, 136-37
Rectangle
  button, 22, 289
  command, 25, 255, 289, 728
  drawing, 289-93
  option, 289
Rectangular
  Array button, 405
  slots, 725, 727-28
Redo
  button, 506
  command, 102, 489, 505-6
Redraw
  All button, 367
  command, 98, 345, 367, 370
  option, 367
Redspark option, 894
Refedit
  command, 783, 803
  toolbar, 803-4
Reference
  Edit dialog box, 803-4
  factor, 433
  name edit box, 803
  Type box, 829
Refresh
  button, 894
  screen, 366-70

REG. (*See* Region)
Regen
All option, 368
command, 255, 281, 345, 367-68, 370. 520, 594-95, 759
option, 368
Regenall command, 345, 368
Regenauto, 368-69
Region, 770-71
button, 543, 771
command, 537, 741
option, 771
Relative
coordinates, 88-89
to last segment option, 97
Remove
button, 44, 472
from position (0-x): edit box, 472-73
Hyperlink button, 906
Islands button, 766-67
paths from xref and images check box, 922
Removed section, 752-53
Rename
button, 125, 362, 471-72, 584-85
command, 818, 887
option, 700, 842-43, 887
Text Style dialog box, 584-85
To
box, 887
button, 887
Re-Order button, 471-73
Replace
All button, 613
at position (0-x): edit box, 473
button, 612-13
with
box, 605-6
button, 605-6
edit box, 613
Report
Generation, 922
tab, 922
Reset button, 126
Resolution, setting, 910
Restore
button, 14-15, 815
option, 863
Retain
Boundaries check box, 770
option, 790
Reuse
Current option, 665
Next option, 665
Revisions, 571
Revolved section, 752
RO. (*See* Rotate command)
Rotate
button, 22, 425
command, 385, 421-22, 425-28, 467-68, 804
Items as Copied check box, 413-14
option, 425-26
Rotation
angle option, 577
area, 794
option, 604
text specific property, 609
Round off option, 690, 692
Row offset edit box, 407
Rows edit box, 405
RTZOOM. (*See* Realtime option)
Run
menu, 8
Scripts, 898

S. (*See* Stretch command)
Save
as Type edit box, 127-28
Back Changes to Reference button, 804
button, 22, 45, 127, 130, 132
changes to layout check box, 867-68
command, 4, 44-45, 113, 126-31, 550, 893, 898
Drawing As dialog box, 45, 127-31
in edit box, 45, 127, 130
Save & Close button, 883-84
Saveas, 113, 122, 127-31, 807, 843, 890, 898, 914

Savetime, 113, 131
SC. (*See* Scale command)
Scale
  button, 432
  command, 385, 421-22, 430-34, 512
  dimension to layout (paperspace) option, 689
  edit box, 175, 874
  factor, 432-33, 574-76, 692
  lineweights check box, 875
  list, 759, 874
  option, 175, 759
  selecting a, 175-76
  Text button, 609
Scaled to Fit option, 175, 838, 866, 874
Scientific - 1.5500E+01 option, 57
Screen, refreshing the, 366-70
Screening, 885
Scripts, 898
Scroll bars, 18
Search
  for the word(s) edit box, 825-26
  previous results option, 43
  results window, 613
  tab, 42-43
  the Web option, 893-94, 896
  titles only option, 43
Second Segment, 667
Section
  lines, 146-47
    placing, 746-48
  types of, 749-54
  views, creating, 741-54
Security, 922
Segment length, 500
Segments in a polyline curve option, 370
Select, 22, 421
  a File list box, 53-54
  Color dialog box, 162-63, 171, 516-17, 601, 676, 683
  command, 467-68
  Custom Arrow Block dialog box, 679
  Drawing File dialog box, 896
  File dialog box, 35, 53, 63, 824, 826, 893, 896
  how to begin list box, 9, 52
  Linetype dialog box, 169-70, 516, 518
  Object option, 638
  Objects
    button, 405, 408, 414, 528-29, 761, 787-89
    option, 760-62, 788
  Plot Style Table dialog box, 883
  Template form File dialog box, 842
Selectable check box, 471
Selected tabs option, 871
Selection
  mode, defining the, 384-86
  set
    adding objects to the, 383-84
    multiple, 384
    removing objects from the, 384
  tab, 124, 376, 384, 445, 462-63
Set
  Current
    button, 125, 362, 672-73, 696-97, 701
    option, 672-73
  option, 843
Settings
  menu, 869
  option, 40, 76, 78
Setup
  edit box, 850
  list, 846
Sheet size, choosing a, 65-67
Shortcut menus, 36-37, 827
Shortcuts, adding to favorites, 830
Show
  All Layers option, 522
  Details button, 156-57, 512, 516, 859
  on Start check box, 40
  Paper boundaries option, 911
  selection, 42
  Startup Dialog
    button, 51
    setting, 123
Single (1.0x) option, 605
Single Drawing Compatibility Mode button, 813

option, 123
Single Viewport button, 850-51, 854
Single-Line Text option, 576
Size edit box, 680
Sketch
button, 499
command, 84, 219, 489, 494, 499-503
Sketches
arcs, 222-23
circles, 223-25
cylinders, 226
ellipses, 226
isometric, 232-36
keeping proportion in, 226-28
multiview, 228-31
techniques, 220-26
tools, 219-20
Slotted hole, 722
Small icons, 827
SN. (*See* Snap command)
Snap
and Grid tab, 17, 76-78, 213-14
Angle edit box, 78
button, 17, 95
command, 49, 78-80, 95, 189, 211, 213-14, 768
mode, 38
setting, 78-80
to None button, 339
Source area, 806-7
Space, marking and dividing, 489-94
SPE. (*See* Splinedit command)
Special characters, 591
Specify, 22
On-screen box, 793-94, 828-29
Spell command, 567, 609-10
Spelling option, 609
SPL. (*See* Spline command)
Spline, 724
button, 293
close option, 294-96
command, 293-302
editing, 298
fit tolerance, altering, 296-97
leader line, 665
option, 298
tangent points, altering, 297-98
Standard toolbar, 6, 22, 25, 33, 41, 55, 67, 69, 101-2, 118, 127, 130-31, 155-56, 172
Splinedit
command, 298-302
modes, 300-2
Move vertex option, 302
Spotface, 719
ST. (*See* Style command)
Stack
button, 601
option, 601
Standard
toolbar, 347, 352, 354, 356-59, 361, 367, 506, 527, 608, 612, 705, 800, 820-21, 849, 865-66, 879-80, 896
viewports list, 846, 850
Start
from Scratch option, 9, 50-51, 55, 67
point, 576-77
With list box, 673-74
Startup dialog box, 11, 31, 49-51, 56, 67, 114-15, 120, 123, 808
Status
bar, 17-18
command, 537, 550-51
option, 550
Stop
button, 894
Navigation button, 904
Storage problems, 134-35
Straight option, 665
Stretch
button, 434
command, 189, 202, 385, 421-22, 434-38, 464-65, 702, 704, 840
option, 434
Style
command, 567, 577, 584-91, 596
current dimension, 672
list, 602-3
modifying a, 699

Name
area, 584-85
choosing the, 585-86
edit box, 584
naming a, 584-86
option, 583-84, 602, 672-73, 763
overriding an existing, 699
portion, 700
properties, 609
defining, 588-91
settings, common, 697-98
Stylemanager command, 865, 879-80
Styles
box, 672, 696, 699
comparing, 699-700
Subtract option, 540-42
Suffix, 690, 692
Suggestions box, 610
Suppressing Arrows option, 688
Surface
finishes, 730-33
waviness, 731
Surveyor - N0d00'00"E option, 60
Swatch display, 758-59
Symbol button, 601
System tab, 9, 51, 123, 813

T. (*See* Mtext command)
Table View tab, 883-84, 887
Tablet mode, 37
Tan, Tan, Tan option, 262
TAN. (*See* Tangent mode)
Tangent
button, 331-32
mode, 331-332
Tangent, tangent, radius, 261-62
Tangent, tangent, tangent, 262—63
Template
creating a, 119-20
drawing, creating and saving, 114-20
incorporating text into, 613
option, 9, 50, 55, 62-67, 842
saving a, 131-32
selecting a, 63-67
tab, 114
Temporary Tracking flyout menu, 319
Terminators, 623
Text
Alignment area, 685-86
appearance, 682
button, 576
characters, controlling, 597-602
color box, 683
command, 567, 576-77, 588, 596, 901
considerations, 567-76
dimension, 623-25
editing existing, 591-94
finding, 612-13
height box, 683
justifying, 579-83
keys, 35
Lines in the Command Line Window option, 122
option, 585-85, 592, 596, 634, 687-88
Override, 705
placement, 569-74, 688-89
dimensioning, 626
Placement area, 683
placing, 577-84
properties, adjusting, 602-4
replacing, 612-13
similarities, 568-69
Style
button, 584-85, 609
dialog box, 584-86, 590, 683
edit box, 682-83
tab, 680, 682-86
toolbar, 576, 584-85, 591, 596, 609, 612
Width option, 661-62
Third angle projection, 197
Through point, 496
Tile
Horizontal option, 71, 814-15
options, 70, 72
Vertical option, 72, 814
Vertically, 902

Time
  command, 537, 549-50
  option, 549
Title
  Bar, 14-15
  Block, 571-74
TK. (*See* Tracking mode)
TODAY window, 8-9, 49-52
Tolerance
  command, 632
  dialog box, 663
  Format, 694
  option, 663
Tolerances tab, 694-95
Toolbar, 22-26
  altering, 25-26
  command, 22
  Object Properties, 22
  Standard, 6, 22
Toolbars
  menu, 491
  option, 22, 26, 319
Tools
  menu, 17, 19, 30, 34, 51, 76-77, 95-96, 98, 122, 124, 131, 161, 173, 213-14, 313-15, 317, 319, 337-38, 370, 376, 384, 462-63, 528-29, 537-38, 543, 545, 548-50, 597-98, 609, 814, 878, 898, 900, 906-7
  tab, 72. (*See also* Volo View Express)
Topics box, 44
Total number of items edit box, 411, 414
TR. (*See* Trim command)
Track
  orthogonally only option, 97
  using all polar angles option, 97
Tracking
  button, 336
  mode, 336
Trailing box, 693
Tree
  option, 824
  View Toggle button, 823-24
Trim
  command, 189, 202, 205-7, 215-16, 385, 421, 444-48, 705, 723-24
  Edge option, 445
  Extend option, 446
  option, 205, 444
  Project option, 447-48
  selections, multiple, 445
  Undo option, 447
T-slots, 727-28
Type
  list, 73
  the File or Web page name edit box, 902

U command, 489, 504-6
UCS. (*See* User Coordinate System icon)
UN. (*See* Units command)
Underline button, 599-600
Undo
  button, 554
  command, 94, 98, 101-2, 489, 504-6
  option, 466, 601
Uniform
  Resource Locator. (*See* URLs)
  Scale check box, 794
Unit format setting, 690, 692, 694
Units
  command, 49, 73, 113, 117
  control folder, 58-59
  drawing, setting the, 56-57
  format, 693
  option, 73
  precision, 58-59
Unnamed check box, 471
Up button, 823, 826
Update option, 708-9
Upside-down option, 588-89, 876
URLs, 890-91
Use
  a
    PCP or PCP2 file option, 879
    Template option, 10, 50, 65, 114
    Wizard option, 11, 50, 56
  Big Font check box, 586

for edit
arrow, 673-74
box, 696
My R14 Plotter Configuration option, 879
this plot style table for layouts in new drawings and pre-AutoCAD 2002 drawings, 881
overall scale of option, 689
Shift to add
button, 445
to selection, 385
User
Coordinate System icon, 18-19
Defined Page Setups dialog box, 868
Preferences
saving, 120-26
tab, 123, 900

V. (*See* View command)
Value edit box, 530, 790, 906
Variables, combining to create a style, 695-701
Ver option, 497
Vertical
option, 588-89
Text Placement list, 684
View
auxiliary, 210-19
button, 34, 873
command, 345, 359, 361, 370, 873, 913
creating, 201-10
Details dialog box, 363
determining number to be drawn, 200-1
dialog box, 359, 361-63
items you can purge, 552
items you cannot purge, 553-54
menu, 22, 26, 319, 347-48, 352, 354, 356-57, 359, 361, 363-64, 368, 824, 844-45, 850, 860
name edit box, 359
option, 174, 874, 898
resolution, 369-70
selecting the required, 197-201
Selections button, 767
toolbar, 359, 361
Viewing plane line, 143, 145
Viewport
Clip option, 860
menu, 844
properties, altering, 849
Scale
button, 850-51
menu, 852, 855
scaling, 849-50
Viewports, 113, 116-18
creating with irregular shapes, 860-63
dialog box, 843, 845-46, 850-51
floating, 848-59
working in multiple, 847-48
option, 844-45, 850, 860
Spacing edit box, 850
toolbar, 843, 845, 850-52, 854-55, 860-62
working with, 843-48
Viewres command, 345, 370
Views button, 823, 827
Volo View Express, 72
Vpclip command, 833, 860-62
-VPCLIP, 860
Vports command, 833, 844-45, 850
-VPORTS, 860

W. (*See* Wblock command)
Warning messages, 36
Warnings area, 868
Wblock
command, 385, 389, 783, 806-8
editing, 808
inserting, 808
WCS. (*See* User Coordinate System icon)
Web toolbar, 904-5
What to plot area, 868-69
Width
Factor option, 588-90
option, 292-93, 603
text specific property, 609
Window
button, 25, 34, 174, 873

  - Element display, 314
  - menu, 814-15
  - menu bar selection, 12, 68-70
  - option, 174-75, 873
  - radio box, 174
- Windows
  - Elements, 161
  - Printer dialog box, 869
- Windows, multiple, 11-12, 68-70
- Windows< button, 34, 174, 873
- With edit box, 700
- Wizards option, 9, 50, 55-61
- Work habits, 5
- WP. (*See* Wpolygon option)
- Wpolygon option, 379-81
- Write Block dialog box, 806, 808

- X. (*See* Explode command)
- X
  - edit box, 875
  - origin box, 176
- XL. (*See* Xline command)
- Xline command, 489, 494, 496-99
- Xref
  - button, 803
  - command, 851

- Y
  - edit box, 875
  - origin box, 176
- Yes button, 554
- Yes to All button, 554

- Z. (*See* Zoom command)
- Zero Suppression area, 690, 693-94
- Zoom
  - All option, 352-53, 356, 851-52, 854, 856
  - Center option, 352
  - command, 49, 67, 118, 120, 178, 311-12, 345-50, 352-53, 358, 369-70, 377, 425, 836, 846-47, 857, 873, 912, 915
  - Dynamic option, 355-56
  - Extents option, 353
  - Flyout
    - button, 347
    - menu, 347, 352, 354
  - In option, 354, 364
  - option, 347, 365-66, 876, 913
  - Out option, 354, 364
  - Previous
    - button, 347
    - option, 353
  - Realtime
    - button, 347-48, 358
    - command, 347-48, 356, 358
  - Rectangle option, 913
  - Scale, 353-54
  - submenu, 347-48, 352, 354
  - to button, 613
  - toolbar, 347, 354
  - transparent, 350-52
  - Window, 846
    - button, 350
    - option, 349-50, 356, 847
  - XP, 354